DNA

Born in the US, James Watson won the Nobel Prize for Physiology or Medicine in 1962 for elucidating the structure and function of DNA, along with the Englishmen Francis Crick and Maurice Wilkins. After studying and teaching at Harvard and Cambridge Universities, he has worked at the Cold Spring Harbor Laboratory in New York State for the last three decades, for most of that time as its director. He is the author of a number of books, including the international bestseller *The Double Helix*.

DNA

The Secret of Life

James D. Watson

with Andrew Berry

arrow books

Published in the United Kingdom in 2004 by Arrow Books

1 3 5 7 9 10 8 6 4 2

Copyright © 2003 by DNA Show LLC

James D. Watson has asserted his right under the Copyright, Designs
and Patents Act, 1988 to be identified as the author of this work

Published by arrangement with Alfred A. Knopf,
a division of Random House, Inc., New York
First published in the United Kingdom in 2003
by William Heinemann

Arrow Books
The Random House Group Limited
20 Vauxhall Bridge Road, London SW1V 2SA

Random House Australia (Pty) Limited
20 Alfred Street, Milsons Point, Sydney,
New South Wales 2061, Australia

Random House New Zealand Limited
18 Poland Road, Glenfield,
Auckland 10, New Zealand

Random House (Pty) Limited
Endulini, 5a Jubilee Road, Parktown 2193, South Africa

The Random House Group Limited Reg. No. 954009

www.randomhouse.co.uk

A CIP catalogue record for this book is available from the
British Library

Papers used by Random House are natural recyclable products
made from wood grown in sustainable forests.
The manufacturing processes conform to the
environmental regulations of the country of origin

ISBN 0 434 01116 9

Typeset by SX Composing DTP, Rayleigh, Essex
Printed and bound in Great Britain by
Bookmarque Ltd, Croydon, Surrey

For Francis Crick

CONTENTS

AUTHORS' NOTE

DNA: The Secret of Life was conceived over dinner in 1999. Under discussion was how best to mark the fiftieth anniversary of the discovery the double helix. Publisher Neil Patterson joined one of us, James D. Watson, in dreaming up a multifaceted venture including this book, a television series, and additional more avowedly educational projects. Neil's presence was no accident: he published JDW's first book, *The Molecular Biology of the Gene*, in 1965, and ever since has lurked genielike behind JDW's writing projects. Doron Weber at the Alfred P. Sloan Foundation then secured seed money to ensure that the idea would turn into something more concrete. Andrew Berry was recruited in 2000 to hammer out a detailed outline for the TV series and has since become a regular commuter between his base in Cambridge, Massachusetts, and JDW's at Cold Spring Harbor Laboratory on the north coast of Long Island, close to New York City.

From the start, our goal was to go beyond merely recounting the events of the past fifty years. DNA has moved from being an esoteric molecule only of interest to a handful of specialists to being the heart of a technology that is transforming many aspects of the way we all live. With that transformation has come a host of difficult questions about its impact – practical, social, and ethical. Taking the fiftieth anniversary as an opportunity to pause and take stock of where we are, we give an unabashedly personal view both of the history and of the issues. Moreover, it is JDW's personal view and is accordingly written in the first-person singular. The double helix was already ten years old when DNA was working its *in utero magic* on a fetal AB.

We have tried to write for a general audience, intending that someone with zero biological knowledge should be able to understand the book's every word. Every technical term is explained when first introduced. Should you need to refresh your memory about a term when you come across one of its later appearances, you can refer to the index, where such words are printed in bold to make locating them easy; a number also in bold will take you to the page on which the term is defined. We have inevitably skimped on many of the technical details and recommend that readers interested in learning more go to DNAi.org, the Web site of the multimedia companion project, DNA Interactive, aimed at high-schoolers and entry-level college students. Here you will find animations explaining basic processes and an extensive archive of interviews with the scientists involved. In addition, the Further Reading section lists books relevant to each chapter. Where possible we have avoided the technical literature, but the titles listed nevertheless provide a more in-depth exploration of particular topics than we supply.

We thank the many people who contributed generously to this project in one way or another in the acknowledgments at the back of the book. Four individuals, however, deserve special mention. George

Andreou, our preternaturally patient editor at Knopf, wrote much more of this book – the good bits – than either of us would ever let on. Kiryn Haslinger, our superbly efficient assistant at Cold Spring Harbor Lab, cajoled, bullied, edited, researched, nit-picked, mediated, wrote – all in approximately equal measure. The book simply would not have happened without her. Jan Witkowski, also of Cold Spring Harbor Lab, did a marvelous job of pulling together chapters 10, 11, and 12 in record time and provided indispensable guidance throughout the project. Maureen Berejka, JDW's assistant, rendered sterling service as usual in her capacity as the sole inhabitant of Planet Earth capable of interpreting JDW's handwriting.

James D. Watson
Cold Spring Harbor, New York

Andrew Berry
Cambridge, Massachusetts

PLATES

Plate 1. 1953: Francis Crick (right) and me with our model of the double helix.

Plate 2. The key to Mendel's triumph: genetic variation to pea plants.

Plate 3. Notoriously camera shy T. H. Morgan was photographed surreptitiously while at work in the fly room.

Plate 4. Eugenics as it was perceived during the first part of the twentieth century: an opportunity for humans to control their own evolutionary destiny.

Plate 5. "Large family" winner, Fitter Families Contest, Texas State Fair (1925).

Plate 6. Scientific racism: social inadequacy in the United States analyzed by Harry Laughlin as an umbrella term for a host of sins ranging from feeblemindedness to tuberculosis. Laughlin computed an institutional "quota" for each group on the basis of the proportion of that group in the U.S. population as a whole. Shown, as a percentage, is the number of institutionalized individuals from a particular group divided by the group's quota. Groups scoring over 100 per cent are overrepresented in institutions.

Plate 7. The physicist Erwin Schrödinger, whose book *What Is Life?* turned me on to the gene.

Plate 8. Erwin Schröder's *What Is Life?*, published in 1944.

Plate 9. A view through the microscope of blood cells treated with a chemical that stains DNA. In order to maximize their oxygen-transporting capacity, red blood cells have no nucleus and therefore no DNA. But white blood cells, which patrol the bloodstream in search of intruders, have a nucleus containing chromosomes.

Plate 10. Lawrence Bragg (left) with Linus Pauling, who is carrying a model of the α-helix.

Plate 11. Maurice Wilkins in his lab at King's College, London.

Plate 12. Rosalind Franklin on one of the mountain hiking vacations she loved.

Plate 13. X-ray photos of the A and B forms of DNA from, respectively, Maurice Wilkins and Rosalind Franklin. The differences in molecular structure are caused by differences in the amount of water associated with each DNA molecule.

Plate 14. The chemical backbone of DNA.

Plate 15. The insight that made it all come together: complementary pairing of the bases.

Plate 16. Bases and backbone in place: the double helix.

(A) is a schematic showing the system of base-pairing that binds the two strands together.

(B) is a "space-filling" model showing, to scale, the atomic detail of the molecule.

Plate 17. Short and sweet: our *Nature* paper announcing the discovery. The same issue also carried longer articles by Rosalind Franklin and Maurice Wilkins.

and red) transport amino acids to the ribosome for incorporation into the growing protein change.

THE SECRET OF LIFE

As was normal for a Saturday morning, I got to work at Cambridge University's Cavendish Laboratory earlier than Francis Crick on February 28, 1953. I had good reason for being up early. I knew that we were close – though I had no idea just how close – to figuring out the structure of a then little-known molecule called deoxyribonucleic acid: DNA. This was not any old molecule: DNA, as Crick and I appreciated, holds the very key to the nature of living things. It stores the hereditary information that is passed on from one generation to the next, and it orchestrates the incredibly complex world of the cell. Figuring out its 3-D structure – the molecule's architecture – would, we hoped, provide a glimpse of what Crick referred to only half-jokingly as "the secret of life."

We already knew that DNA molecules consist of multiple copies of a single basic unit, the nucleotide, which comes in four forms: adenine (A), thymine (T), guanine (G), and cytosine (C). I had spent the previous afternoon making cardboard cutouts of these various components, and now, undisturbed on a quiet Saturday morning, I could shuffle around the pieces of the 3-D jigsaw puzzle. How did they all fit together? Soon I realized that a simple pairing scheme worked exquisitely well: A fitted neatly with T, and G with C. Was this it? Did the molecule consist of two chains linked together by A-T and G-C pairs? It was so simple, so elegant, that it almost had to be right. But I had made mistakes in the past, and before I could get too excited, my pairing scheme would have to survive the scrutiny of Crick's critical eye. It was an anxious wait. But I need not have worried: Crick realized straightaway that my pairing idea implied a double-helix structure with the two molecular chains running in opposite directions (see Plate 1). Everything known about DNA and its properties – the facts we had been wrestling with as we tried to solve the problem – made sense in light of those gentle complementary twists. Most important, the way the molecule was organized immediately suggested solutions to two of biology's oldest mysteries: how hereditary information is stored, and how it is replicated. Despite this, Crick's brag in the Eagle, the pub where we habitually ate lunch, that we had indeed discovered that "secret of life," struck me as somewhat immodest, especially in England, where understatement is a way of life.

Crick, however, was right. Our discovery put an end to a debate as old as the human species: Does life have some magical, mystical essence, or is it, like any chemical reaction carried out in a science class, the product of normal physical and chemical processes? Is there something divine at the heart of a cell that brings it to life? The double helix answered that question with a definitive No.

Charles Darwin's theory of evolution, which showed how all of life is interrelated, was a major advance in our understanding of the world in materialistic – physicochemical – terms. The breakthroughs of biologists Theodor Schwann and Louis Pasteur during the second half of the nineteenth century were also an important step forward. Rotting meat did not spontaneously yield maggots; rather, familiar biological agents and processes were responsible – in this case egg-laying flies. The idea of spontaneous generation had been discredited.

Despite these advances, various forms of vitalism – the belief that physicochemical processes cannot explain life and its processes – lingered on. Many biologists, reluctant to accept natural selection as the sole determinant of the fate of evolutionary lineages, invoked a poorly defined overseeing spiritual force to account for adaptation. Physicists, accustomed to dealing with a simple, pared-down world – a few particles, a few forces – found the messy complexity of biology bewildering. Maybe, they suggested, the processes at the heart of the cell, the ones governing the basics of life, go beyond the familiar laws of physics and chemistry.

That is why the double helix was so important. It brought the Enlightenment's revolution in materialistic thinking into the cell. The intellectual journey that had begun with Copernicus displacing humans from the center of the universe and continued with Darwin's insistence that humans are merely modified monkeys had finally focused in on the very essence of life. And there was nothing special about it. The double helix is an elegant structure, but its message is downright prosaic: life is simply a matter of chemistry.

Crick and I were quick to grasp the intellectual significance of our discovery, but there was no way we could have foreseen the explosive impact of the double helix on science and society. Contained in the molecule's graceful curves was the key to molecular biology, a new science whose progress over the subsequent fifty years has been astounding. Not only has it yielded a stunning array of insights into fundamental biological processes, but it is now having an ever more profound impact on medicine, on agriculture, and on the law. DNA is no longer a matter of interest only to white-coated scientists in obscure university laboratories; it affects us all.

By the mid-sixties, we had worked out the basic mechanics of the cell, and we knew how, via the "genetic code," the four-letter alphabet of DNA sequence is translated into the twenty-letter alphabet of the proteins. The next explosive spurt in the new science's growth came in the 1970s with the introduction of techniques for manipulating DNA and reading its sequence of base pairs. We were no longer condemned to watch nature from the sidelines but could actually tinker with the DNA of living organisms, and we could actually read life's basic script. Extraordinary new scientific vistas opened up: we would at last come to grips with genetic diseases from cystic fibrosis to cancer; we would revolutionize criminal justice through genetic fingerprinting methods; we would profoundly revise ideas about human origins – about who we are and where we came from – by using DNA-based approaches to prehistory; and we would improve agriculturally important species with an effectiveness we had previously only dreamed of.

But the climax of the first fifty years of the DNA revolution came on Monday, June 26, 2000, with the announcement by U.S. president Bill Clinton of the completion of the rough draft sequence of the human genome: "Today, we are learning the language in which God created life. With this profound new knowledge, humankind is on the verge of gaining immense, new power to heal." The genome project was a coming-of-age for molecular biology: it had become "big science," with big money and big results. Not only was it an extraordinary technological achievement – the amount of information mined from the human complement of twenty-three pairs of chromosomes is staggering – but it was also a landmark in terms of our idea of what it is to be human. It is our DNA that distinguishes us from all other species, and that makes us the creative, conscious, dominant, destructive creatures that we are. And here, in its entirety, was that set of DNA – the human instruction book.

DNA has come a long way from that Saturday morning in Cambridge. However, it is also clear that the science of molecular biology – what DNA can do for us – still has a long way to go. Cancer still has to be cured; effective gene therapies for genetic diseases still have to be developed; genetic engineering still has to realize its phenomenal potential for improving our food. But all these things will come. The first fifty years of the DNA revolution witnessed a great deal of remarkable scientific progress as well as the initial application of that progress to human problems. The future will see many more scientific advances, but increasingly the focus will be on DNA's ever greater impact on the way we live.

DNA

CHAPTER ONE

BEGINNINGS OF GENETICS: FROM MENDEL TO HITLER

My mother, Bonnie Jean, believed in genes. She was proud of her father's Scottish origins, and saw in him the traditional Scottish virtues of honesty, hard work, and thriftiness. She, too, possessed these qualities and felt that they must have been passed down to her from him. His tragic early death meant that her only nongenetic legacy was a set of tiny little girl's kilts he had ordered for her from Glasgow. Perhaps therefore it is not surprising that she valued her father's biological legacy over his material one.

Growing up, I had endless arguments with Mother about the relative roles played by nature and nurture in shaping us. By choosing nurture over nature, I was effectively subscribing to the belief that I could make myself into whatever I wanted to be. I did not want to accept that my genes mattered that much, preferring to attribute my Watson grandmother's extreme fatness to her having overeaten. If her shape was the product of her genes, then I too might have a hefty future. However, even as a teenager, I would not have disputed the evident basics of inheritance, that like begets like. My arguments with my mother concerned complex characteristics like aspects of personality, not the simple attributes that, even as an obstinate adolescent, I could see were passed down over the generations, resulting in "family likeness." My nose is my mother's and now belongs to my son Duncan.

Sometimes characteristics come and go within a few generations, but sometimes they persist over many. One of the most famous examples of a long-lived trait is known as the "Hapsburg Lip." This distinctive elongation of the jaw and droopiness to the lower lip – which made the Hapsburg rulers of Europe such a nightmare assignment for generations of court portrait painters – was passed down intact over at least twenty-three generations.

The Hapsburgs added to their genetic woes by intermarrying. Arranging marriages between different branches of the Hapsburg clan and often among close relatives may have made political sense as a way of building alliances and ensuring dynastic succession, but it was anything but astute in genetic terms. Inbreeding of this kind can result in genetic disease, as the Hapsburgs found out to their cost. Charles II, the last of the Hapsburg monarchs in Spain, not only boasted a prize-worthy example of the family lip – he could not even chew his own food – but was also a complete invalid, and incapable, despite two marriages, of producing children.

Genetic disease has long stalked humanity. In some cases, such as Charles II's, it has had a direct impact on history. Retrospective diagnosis has suggested that George III, the English king whose principal claim to fame is to have lost the American colonies in the Revolutionary War, suffered from an inherited disease, porphyria, which causes periodic bouts of madness. Some historians – mainly

British ones – have argued that it was the distraction caused by George's illness that permitted the Americans' against-the-odds military success. While most hereditary diseases have no such geopolitical impact, they nevertheless have brutal and often tragic consequences for the afflicted families, sometimes for many generations. Understanding genetics is not just about understanding why we look like our parents. It is also about coming to grips with some of humankind's oldest enemies: the flaws in our genes that cause genetic disease.

Our ancestors must have wondered about the workings of heredity as soon as evolution endowed them with brains capable of formulating the right kind of question. And the readily observable principle that close relatives tend to be similar can carry you a long way if, like our ancestors, your concern with the application of genetics is limited to practical matters like improving domesticated animals (for, say, milk yield in cattle) and plants (for, say, the size of fruit). Generations of careful selection – breeding initially to domesticate appropriate species, and then breeding only from the most productive cows and from the trees with the largest fruit – resulted in animals and plants tailor-made for human purposes. Underlying this enormous unrecorded effort is that simple rule of thumb: that the most productive cows will produce highly productive offspring and from the seeds of trees with large fruit large-fruited trees will grow. Thus, despite the extraordinary advances of the past hundred years or so, the twentieth and twenty-first centuries by no means have a monopoly on genetic insight. Although it wasn't until 1909 that the British biologist William Bateson gave the science of inheritance a name, genetics, and although the DNA revolution has opened up new and extraordinary vistas of potential progress, in fact the single greatest application of genetics to human well-being was carried out eons ago by anonymous ancient farmers. Almost everything we eat – cereals, fruit, meat, dairy products – is the legacy of that earliest and most far-reaching application of genetic manipulations to human problems.

An understanding of the actual mechanics of genetics proved a tougher nut to crack. Gregor Mendel (1822–1884) published his famous paper on the subject in 1866 (and it was ignored by the scientific community for another thirty-four years). Why did it take so long? After all, heredity is a major aspect of the natural world, and, more important, it is readily, and universally, observable: a dog owner sees how a cross between a brown and black dog turns out, and all parents consciously or subconsciously track the appearance of their own characteristics in their children. One simple reason is that genetic mechanisms turn out to be complicated. Mendel's solution to the problem is not intuitively obvious: children are not, after all, simply a *blend* of their parents' characteristics. Perhaps most important was the failure by early biologists to distinguish between two fundamentally different processes, heredity and development. Today we understand that a fertilized egg contains the genetic information, contributed by both parents, that determines whether someone will be afflicted with, say, porphyria. That is heredity. The subsequent process, the *development* of a new individual from that humble starting point of a single cell, the fertilized egg, involves implementing that information. Broken down in terms of academic disciplines, genetics focuses on the information and developmental biology focuses on the use of that information. Lumping heredity and development together into a single phenomenon, early scientists never asked the questions that might have steered them toward the secret of heredity. Nevertheless, the effort had been under way in some form since the dawn of Western history.

The Greeks, including Hippocrates, pondered heredity. They devised a theory of "pangenesis," which claimed that sex involved the transfer of miniaturized body parts: "Hairs, nails, veins, arteries, tendons and their bones, albeit invisible as their particles are so small. While growing, they gradually separate from each other." This idea enjoyed a brief renaissance when Charles Darwin, desperate to support his theory of evolution by natural selection with a viable hypothesis of inheritance, put forward a modified version of pangenesis in the second half of the nineteenth century. In Darwin's scheme, each organ – eyes, kidneys, bones – contributed circulating "gemmules" that accumulated in the sex organs, and were ultimately exchanged in the course of sexual reproduction. Because these gemmules were produced throughout an organism's lifetime, Darwin argued any change that occurred in the individual after birth, like the stretch of a giraffe's neck imparted by craning for the highest foliage, could be passed on to the next generation. Ironically, then, to buttress his theory of natural selection Darwin came to champion aspects of Jean-Baptiste Lamarck's theory of inheritance of acquired characteristics – the very theory that his evolutionary ideas did so much to discredit. Darwin was invoking only Lamarck's theory of inheritance; he continued to believe that natural selection was the driving force behind evolution, but supposed that natural selection operated on the variation produced by pangenesis. Had Darwin known about Mendel's work (although Mendel published his results shortly after *The Origin of Species* appeared, Darwin was never aware of them), he might have been spared the embarrassment of this late-career endorsement of some of Lamarck's ideas.

Whereas pangenesis supposed that embryos were assembled from a set of minuscule components, another approach, "preformationism," avoided the assembly step altogether: either the egg or the sperm (exactly which was a contentious issue) contained a complete *preformed* individual called a homunculus. Development was therefore merely a matter of enlarging this into a fully formed being. In the days of preformationism, what we now recognize as genetic disease was variously interpreted: sometimes as a manifestation of the wrath of God or the mischief of demons and devils; sometimes as evidence of either an excess of or a deficit of the father's "seed"; sometimes as the result of "wicked thoughts" on the part of the mother during pregnancy. On the premise that fetal malformation can result when a pregnant mother's desires are thwarted, leaving her feeling stressed and frustrated, Napoleon passed a law permitting expectant mothers to shoplift. None of these notions, needless to say, did much to advance our understanding of genetic disease.

By the early nineteenth century, better microscopes had defeated preformationism. Look as hard as you like, you will never see a tiny homunculus curled up inside a sperm or egg cell. Pangenesis, though an earlier misconception, lasted rather longer – the argument would persist that the gemmules were simply too small to visualize – but was eventually laid to rest by August Weismann, who argued that inheritance depended on the continuity of germ plasm between generations and thus changes to the body over an individual's lifetime could *not* be transmitted to subsequent generations. His simple experiment involved cutting the tails off several generations of mice. According to Darwin's pangenesis, tailless mice would produce gemmules signifying "no tail" and so their offspring should develop a severely stunted hind appendage or none at all. When Weismann showed that the tail kept appearing after many generations of amputees, pangenesis bit the dust.

Gregor Mendel was the one who got it right. By any standards, however, he was an unlikely candidate for scientific superstardom. Born to a farming family in what is now the Czech Republic, he excelled at the village school and, at twenty-one, entered the Augustinian monastery at Brünn. After proving a disaster as a parish priest – his response to the ministry was a nervous breakdown – he tried his hand at teaching. By all accounts he was a good teacher, but in order to qualify to teach a full range of subjects, he had to take an exam. He failed it. Mendel's father superior, Abbot Napp, then dispatched him to the University of Vienna, where he was to bone up full-time for the retesting. Despite apparently doing well in physics at Vienna, Mendel again failed the exam, and so never rose above the rank of substitute teacher.

Around 1856, at Abbot Napp's suggestion, Mendel undertook some scientific experiments on heredity. He chose to study a number of characteristics of the pea plants he grew in his own patch of the monastery garden. In 1865 he presented his results to the local natural history society in two lectures, and, a year later, published them in the society's journal. The work was a tour de force: the experiments were brilliantly designed and painstakingly executed, and his analysis of the results was insightful and deft. It seems that his training in physics contributed to his breakthrough because, unlike other biologists of that time, he approached the problem quantitatively. Rather than simply noting that crossbreeding of red and white flowers resulted in some red and some white offspring, Mendel actually counted them, realizing that the ratios of red to white progeny might be significant – as indeed they are. Despite sending copies of his article to various prominent scientists, Mendel found himself completely ignored by the scientific community. His attempt to draw attention to his results merely backfired. He wrote to his one contact among the ranking scientists of the day, botanist Karl Nägeli in Munich, asking him to replicate the experiments, and he duly sent off 140 carefully labeled packets of seeds. He should not have bothered. Nägeli believed that the obscure monk should be of service to him, rather than the other way around, so he sent Mendel seeds of his own favorite plant, hawkweed, challenging the monk to re-create his results with a different species. Sad to say, for various reasons, hawkweed is not well-suited to breeding experiments such as those Mendel had performed on the peas. The entire exercise was a waste of his time.

Mendel's low-profile existence as monk-teacher-researcher ended abruptly in 1868 when, on Napp's death, he was elected abbot of the monastery. Although he continued his research – increasingly on bees and the weather – administrative duties were a burden, especially as the monastery became embroiled in a messy dispute over back taxes. Other factors, too, hampered him as a scientist. Portliness eventually curtailed his fieldwork: as he wrote, hill climbing had become "very difficult for me in a world where universal gravitation prevails." His doctors prescribed tobacco to keep his weight in check, and he obliged them by smoking twenty cigars a day, as many as Winston Churchill. It was not his lungs, however, that let him down: in 1884, at the age of sixty-one, Mendel succumbed to a combination of heart and kidney disease.

Not only were Mendel's results buried in an obscure journal, but they would have been unintelligible to most scientists of the era. He was far ahead of his time with his combination of careful experiment and sophisticated quantitative analysis. Little wonder, perhaps, that it was not until 1900 that the scientific community caught up with him. The rediscovery of Mendel's work, by three plant geneticists interested in similar problems, provoked a revolution in biology. At last the scientific world was ready for the monk's peas.

Mendel realized that there are specific factors – later to be called "genes" – that are passed from parent to offspring. He worked out that these factors come in pairs and that the offspring receives one from each parent.

Noticing that peas came in two distinct colors, green and yellow, he deduced that there were two versions of the pea-color gene. A pea has to have two copies of the G version if it is to become green, in which case we say that it is GG for the pea-color gene. It must therefore have received a G pea-color gene from both of its parents. However, yellow peas can result both from YY and YG combinations. Having only one copy of the Y version is sufficient to produce yellow peas. Y trumps G. Because in the YG case the Y signal dominates the G signal, we call Y "dominant." The subordinate G version of the pea-color gene is called "recessive."

Each parent pea plant has two copies of the pea-color gene, yet it contributes only one copy to each offspring; the other copy is furnished by the other parent. In plants, pollen grains contain sperm cells – the male contribution to the next generation – and each sperm cell contains just one copy of the pea-color gene. A parent pea plant with a YG combination will produce sperm that contain either a Y version or a G one. Mendel discovered that the process is random: 50 percent of the sperm produced by that plant will have a Y and 50 percent will have a G.

Suddenly many of the mysteries of heredity made sense. Characteristics, like the Hapsburg Lip, that are transmitted with a high probability (actually 50 percent) from generation to generation are dominant. Other characteristics that appear in family trees much more sporadically, often skipping generations, may be recessive. When a gene is recessive an individual has to have two copies of it for the corresponding trait to be expressed. Those with one copy of the gene are carriers: they don't themselves exhibit the characteristic, but they can pass the gene on. Albinism, in which the body fails to produce pigment so the skin and hair are strikingly white, is an example of a recessive characteristic that is transmitted in this way. Therefore, to be albino you have to have two copies of the gene, one from each parent. (This was the case with the Reverend Dr. William Archibald Spooner, who was also – perhaps only by coincidence – prone to a peculiar form of linguistic confusion whereby, for example, "a well-oiled bicycle" might become "a well-boiled icicle." Such reversals would come to be termed "spoonerisms" in his honor.) Your parents, meanwhile, may have shown no sign of the gene at all. If, as is often the case, each has only one copy, then they are both carriers. The trait has skipped at least one generation.

Mendel's results implied that *things* – material objects – were transmitted from generation to generation. But what was the nature of these things?

At about the time of Mendel's death in 1884, scientists using ever-improving optics to study the minute architecture of cells coined the term "chromosome" to describe the long stringy bodies in the cell nucleus. But it was not until 1902 that Mendel and chromosomes came together.

A medical student at Columbia University, Walter Sutton, realized that chromosomes had a lot in common with Mendel's mysterious factors. Studying grasshopper chromosomes, Sutton noticed that most of the time they are doubled up – just like Mendel's paired factors. But Sutton also identified one type of cell in which chromosomes were not paired: the sex cells. Grasshopper sperm have only a single set of chromosomes, not a double set. This was exactly what Mendel had described: his pea plant sperm cells also only carried a single copy of each of his factors. It was clear that Mendel's factors, now called genes, must be on the chromosomes.

In Germany Theodor Boveri independently came to the same conclusions as Sutton, and so the biological revolution their work had precipitated came to be called the Sutton-Boveri chromosome theory of inheritance. Suddenly genes were real. They were on chromosomes, and you could actually see chromosomes through the microscope (see Plate 2).

Not everyone bought the Sutton-Boveri theory. One skeptic was Thomas Hunt Morgan, also at Columbia. Looking down the microscope at those stringy chromosomes, he could not see how they could account for all the changes that occur from one generation to the next. If all the genes were arranged along chromosomes, and all chromosomes were transmitted intact from one generation to the next, then surely many characteristics would be inherited together. But since empirical evidence showed this not to be the case, the chromosomal theory seemed insufficient to explain the variation observed in nature. Being an astute experimentalist, however, Morgan had an idea how he might resolve such discrepancies. He turned to the fruit fly, *Drosophila melanogaster*, the drab little beast that, ever since Morgan, has been so beloved by geneticists.

In fact, Morgan was not the first to use the fruit fly in breeding experiments – that distinction belonged to a lab at Harvard that first put the critter to work in 1901 – but it was Morgan's work that put the fly on the scientific map. *Drosophila* is a good choice for genetic experiments. It is easy to find (as anyone who has left out a bunch of overripe bananas during the summer well knows); it is easy to raise (bananas will do as feed); and you can accommodate hundreds of flies in a single milk bottle (Morgan's students had no difficulty acquiring milk bottles, pinching them at dawn from doorsteps in their Manhattan neighborhood); and it breeds and breeds and breeds (a whole generation takes about ten days, and each female lays several hundred eggs). Starting in 1907 in a famously squalid, cockroach-infested, banana-stinking lab that came to be known affectionately as the "fly room," Morgan and his students ("Morgan's boys" as they were called) set to work on fruit flies (see Plate 3).

Unlike Mendel, who could rely on the variant strains isolated over the years by farmers and gardeners – yellow peas as opposed to green ones, wrinkled skin as opposed to smooth – Morgan had no menu of established genetic differences in the fruit fly to draw upon. And you cannot do genetics until you have isolated some distinct characteristics to track through the generations. Morgan's first goal therefore was to find "mutants," the fruit fly equivalents of yellow or wrinkled peas. He was looking for genetic novelties, random variations that somehow simply appeared in the population.

One of the first mutants Morgan observed turned out to be one of the most instructive. While normal fruit flies have red eyes, these had white ones. And he noticed that the white-eyed flies were typically male. It was known that the sex of a fruit fly – or, for that matter, the sex of a human – is determined chromosomally: females have two copies of the X chromosome, whereas males have one copy of the X and one copy of the much smaller Y. In light of this information, the white-eye result suddenly made sense: the eye-color gene is located on the X chromosome and the white-eye mutation, W, is recessive. Because males have only a single X chromosome, even recessive genes, in the absence of a dominant counterpart to suppress them, are automatically expressed. White-eyed females were relatively rare because they typically had only one copy of W so they expressed the dominant red eye color. By correlating a gene – the one for eye color – with a chromosome, the X, Morgan, despite his initial reservations, had effectively proved the Sutton-Boveri theory. He had also found an example of "sex-linkage," in which a particular characteristic is disproportionately represented in one sex.

Like Morgan's fruit flies, Queen Victoria provides a famous example of sex-linkage. On one of her X chromosomes, she had a mutated gene for hemophilia, the "bleeding disease" in whose victims proper blood clotting fails to occur. Because her other copy was normal, and the hemophilia gene is recessive, she herself did not have the disease. But she was a carrier. Her daughters did not have the disease either; evidently each possessed at least one copy of the normal version. But Victoria's sons were not all so lucky. Like all males (fruit fly males included), each had only one X chromosome; this was necessarily derived from Victoria (a Y chromosome could have come only from Prince Albert, Victoria's husband). Because Victoria had one mutated copy and one normal copy, each of her sons had a 50-50 chance of having the disease. Prince Leopold drew the short straw: he developed hemophilia, and died at thirty-one, bleeding to death after a minor fall. Two of Victoria's daughters, Princesses Alice and Beatrice, were carriers, having inherited the mutated gene from their mother. They each produced carrier daughters and sons with hemophilia. Alice's grandson Alexis, heir to the Russian throne, had hemophilia, and would doubtless have died young had the Bolsheviks not gotten to him first.

Morgan's fruit flies had other secrets to reveal. In the course of studying genes located on the same chromosome, Morgan and his students found that chromosomes actually break apart and re-form during the production of sperm and egg cells. This meant that Morgan's original objections to the Sutton-Boveri theory were unwarranted: the breaking and re-forming – "recombination," in modern genetic parlance – shuffles gene copies between members of a chromosome pair. This means that, say, the copy of chromosome 12 I got from my mother (the other, of course, comes from my father) is in fact a mix of my mother's two copies of chromosome 12, one of which came from her mother and one from her father. Her two 12s recombined – exchanged material – during the production of the egg cell that eventually turned into me. Thus my maternally derived chromosome 12 can be viewed as a mosaic of my grandparents' 12s. Of course, my mother's maternally derived 12 was itself a mosaic of her grandparents' 12s, and so on.

Recombination permitted Morgan and his students to map out the positions of particular genes along a given chromosome. Recombination involves breaking (and re-forming) chromosomes. Because genes are arranged like beads along a chromosome string, a break is statistically much more likely to occur between two genes that are far apart (with more potential break points intervening) on the chromosome than between two genes that are close together. If, therefore, we see a lot of reshuffling for any two genes on a single chromosome, we can conclude that they are a long way apart; the rarer the reshuffling, the closer the genes likely are. This basic and immensely powerful principle underlies all of genetic mapping. One of the primary tools of scientists involved in the Human Genome Project and of researchers at the forefront of the battle against genetic disease was thus developed all those years ago in the filthy, cluttered Columbia fly room. Each new headline in the science section of the newspaper these days along the lines of "Gene for Something Located" is a tribute to the pioneering work of Morgan and his boys.

The rediscovery of Mendel's work, and the breakthroughs that followed it, sparked a surge of interest in the social significance of genetics. While scientists had been grappling with the precise mechanisms of heredity through the eighteenth and nineteenth centuries, public concern had been mounting about the burden placed on society by what came to be called the "degenerate classes" – the inhabitants of poorhouses, workhouses, and insane asylums. What could be done with these people? It remained a matter of controversy whether they should be treated charitably – which, the less charitably inclined claimed, ensured such folk would never exert themselves and would therefore remain forever dependent on the largesse of the state or of private institutions – or whether they should be simply ignored, which, according to the charitably inclined, would result only in perpetuating the inability of the unfortunate to extricate themselves from their blighted circumstances.

The publication of Darwin's *Origin of Species* in 1859 brought these issues into sharp focus. Although Darwin carefully omitted to mention human evolution, fearing that to do so would only further inflame an already raging controversy, it required no great leap of imagination to apply his idea of natural selection to humans. Natural selection is the force that determines the fate of all genetic variations in nature – mutations like the one Morgan found in the fruit fly eye-color gene, but also perhaps differences in the abilities of human individuals to fend for themselves.

Natural populations have an enormous reproductive potential. Take fruit flies, with their generation time of just ten days, and females that produce some three hundred eggs apiece (half of which will be female): starting with a single fruit fly couple, after a month (i.e., three generations later), you will have $150 \times 150 \times 150$ fruit flies on your hands – that's more than 3 million flies, all of them derived from just one pair in just one month. Darwin made the point by choosing a species from the other end of the reproductive spectrum:

> The elephant is reckoned to be the slowest breeder of all known animals, and I have taken some pains to estimate its probable minimum rate of natural increase: it will be under the mark to assume that it breeds when thirty years old, and goes on breeding till ninety years old, bringing forth three pairs of young in this interval; if this be so, at the end of the fifth century there would be alive fifteen million elephants, descended from the first pair.

All these calculations assume that all the baby fruit flies and all the baby elephants make it successfully to adulthood. In theory, therefore, there must be an infinitely large supply of food and water to sustain this kind of reproductive overdrive. In reality, of course, those resources are limited, and not all baby fruit flies or baby elephants make it. There is competition among individuals within a species for those resources. What determines who wins the struggle for access to the resources? Darwin pointed out genetic variation means that some individuals have advantages in what he called "the struggle for existence." To take the famous example of Darwin's finches from the Galápagos Islands, those individuals with genetic advantages – like the right size of beak for eating the most abundant seeds – are more likely to survive and reproduce. So the advantageous genetic variant – having a bill the right size – tends to be passed on to the next generation. The result is that natural selection enriches the next generation with the beneficial mutation so that eventually, over enough generations, every member of the species ends up with that characteristic.

The Victorians applied the same logic to humans. They looked around and were alarmed by what they saw. The decent, moral, hardworking middle classes were being massively outreproduced by the dirty, immoral, lazy lower classes. The Victorians assumed that the virtues of decency, morality, and hard work ran in families just as the vices of filth, wantonness, and indolence did. Such characteristics must then be hereditary; thus, to the Victorians, morality and immorality were merely two of Darwin's genetic variants. And if the great unwashed were outreproducing the respectable classes, then the "bad" genes would be increasing in the human population. The species was doomed! Humans would gradually become more and more depraved as the "immorality" gene became more and more common.

Francis Galton had good reason to pay special attention to Darwin's book, as the author was his cousin and friend. Darwin, some thirteen years older, had provided guidance during Galton's rather rocky college experience. But it was *The Origin of Species* that would inspire Galton to start a social and genetic crusade that would ultimately have disastrous consequences. In 1883, a year after his cousin's death, Galton gave the movement a name: eugenics.

Eugenics was only one of Galton's many interests; Galton enthusiasts refer to him as a polymath, detractors as a dilettante. In fact, he made significant contributions to geography, anthropology, psychology, genetics, meteorology, statistics, and, by setting fingerprint analysis on a sound scientific footing, to criminology. Born in 1822 into a prosperous family, his education – partly in medicine and partly in mathematics – was mostly a chronicle of defeated expectations. The death of his father when he was twenty-one simultaneously freed him from paternal restraint and yielded a handsome inheritance; the young man duly took advantage of both. After a full six years of being what might be described today as a trust-fund dropout, however, Galton settled down to become a productive member of the Victorian establishment. He made his name leading an expedition to a then little known region of southwest Africa in 1850–52. In his account of his explorations, we encounter the first instance of the one strand that connects his many varied interests: he counted and measured everything. Galton was only happy when he could reduce a phenomenon to a set of numbers.

At a missionary station he encountered a striking specimen of steatopygia – a condition of particularly protuberant buttocks, common among the indigenous Nama women of the region – and realized that this woman was naturally endowed with the figure that was then fashionable in Europe. The only difference was that it required enormous (and costly) ingenuity on the part of European dressmakers to create the desired "look" for their clients.

> I profess to be a scientific man, and was exceedingly anxious to obtain accurate measurements of her shape; but there was a difficulty in doing this. I did not know a word of Hottentot [the Dutch name for the Nama], and could never therefore have explained to the lady what the object of my footrule could be; and I really dared not ask my worthy missionary host to interpret for me. I therefore felt in a dilemma as I gazed at her form, that gift of bounteous nature to this favoured race, which no mantua-maker, with all her crinoline and stuffing, can do otherwise than humbly imitate. The object of my admiration stood under a tree, and was turning herself about to all points of the compass, as ladies who wish to be admired usually do. Of a sudden my eye fell upon my sextant; the bright thought struck me, and I took a series of observations upon her figure in every direction, up and down, crossways, diagonally, and so forth, and I registered them carefully upon an outline drawing for fear of any mistake; this being done, I boldly pulled out my measuring tape, and measured the distance from where I was to the place she stood, and having thus obtained both base and angles, I worked out the results by trigonometry and logarithms.

Galton's passion for quantification resulted in his developing many of the fundamental principles of modern statistics. It also yielded some clever observations. For example, he tested the efficacy of prayer. He figured that if prayer worked, those most prayed for should be at an advantage; to test the hypothesis he studied the longevity of British monarchs. Every Sunday, congregations in the Church of England following the *Book of Common Prayer* beseeched God to "Endue the king/queen plenteously with heavenly gifts; Grant him/her in health and wealth long to live." Surely, Galton reasoned, the cumulative effect of all those prayers should be beneficial. In fact, prayer seemed ineffectual: he found that on average the monarchs died somewhat younger than other members of the British aristocracy.

Because of the Darwin connection – their common grandfather, Erasmus Darwin, too was one of the intellectual giants of his day – Galton was especially sensitive to the way in which certain lineages seemed to spawn disproportionately large numbers of prominent and successful people. In 1869 he published what would become the underpinning of all his ideas on eugenics, a treatise called *Hereditary Genius: An Inquiry into Its Laws and Consequences.* In it he purported to show that talent, like simple genetic traits such as the Hapsburg Lip, does indeed run in families; he recounted, for example, how some families had produced generation after generation of judges. His analysis largely ne-glected to take into account the effect of the environment: the son of a prominent judge is, after all, rather more likely to become a judge – by virtue of his father's connections, if nothing else – than the son of a peasant farmer. Galton did not, however, completely overlook the effect of the environment, and it was he who first referred to the "nature/nurture" dichotomy, possibly in reference to Shakespeare's irredeemable villain, Caliban, "a devil, a born devil, on whose nature/Nurture can never stick."

The results of his analysis, however, left no doubt in Galton's mind.

I have no patience with the hypothesis occasionally expressed, and often implied, especially in tales written to teach children to be good, that babies are born pretty much alike, and that the sole agencies in creating differences between boy and boy, and man and man, are steady application and moral effort. It is in the most unqualified manner that I object to pretensions of natural equality.

A corollary of his conviction that these traits are genetically determined, he argued, was that it would be possible to "improve" the human stock by preferentially breeding gifted individuals, and preventing the less gifted from reproducing.

> It is easy . . . to obtain by careful selection a permanent breed of dogs or horses gifted with peculiar powers of running, or of doing anything else, so it would be quite practicable to produce a highly-gifted race of men by judicious marriages during several consecutive generations.

Galton introduced the terms *eugenics* (literally "good in birth") to describe this application of the basic principle of agricultural breeding to humans. In time, eugenics came to refer to "self-directed human evolution": by making conscious choices about who should have children, eugenicists believed that they could head off the "eugenic crisis" precipitated in the Victorian imagination by the high rates of reproduction of inferior stock coupled with the typically small families of the superior middle classes.

Eugenics these days is a dirty word, associated with racists and Nazis – a dark, best-forgotten phase of the history of genetics. It is important to appreciate, however, that in the closing years of the nineteenth and early years of the twentieth centuries, eugenics was not tainted in this way, and was seen by many as offering genuine potential for improving not just society as a whole but the lot of individuals within society as well. Eugenics was embraced with particular enthusiasm by those who today would be termed the "liberal left." Fabian socialists – some the era's most progressive thinkers – flocked to the cause, including George Bernard Shaw, who wrote that "there is now no reasonable excuse for refusing to face the fact that nothing but a eugenic religion can save our civilisation." Eugenics seemed to offer a solution to one of society's most persistent woes: that segment of the population that is incapable of existing outside an institution (see Plate 4).

Whereas Galton had preached what came to be known as "positive eugenics," encouraging genetically superior people to have children, the American eugenics movement preferred to focus on "negative eugenics," preventing genetically inferior people from doing so. The goals of each program were basically the same – the improvement of the human genetic stock – but these two approaches were very different.

The American focus on getting rid of bad genes, as opposed to increasing frequencies of good ones, stemmed from a few influential family studies of "degeneration" and "feeblemindedness" – two peculiar terms characteristic of the American obsession with genetic decline. In 1875 Richard Dugdale published his account of the Juke clan of upstate New York. Here, according to Dugdale, were several generations of seriously bad apples – murderers, alcoholics, and rapists. Apparently in the area near their home in New York State the very name "Juke" was a term of reproach.

Another highly influential study was published in 1912 by Henry Goddard, the psychologist who gave us the word "moron," on what he called "The Kallikak Family." This is the story of two family lines originating from a single male ancestor who had a child out of wedlock (with a "feebleminded" wench he met in a tavern while serving in the military during the American Revolutionary War), as well as siring a legitimate family. The illegitimate side of the Kallikak line, according to Goddard, was bad news indeed, "a race of defective degenerates," while the legitimate side comprised respectable, upstanding members of the community. To Goddard, this "natural experiment in heredity" was an exemplary tale of good genes versus bad. This view was reflected in the fictitious name he chose for the family. "Kallikak" is a hybrid of two Greek words, *kalos* (beautiful, of good repute) and *kakos* (bad).

"Rigorous" new methods for testing mental performance – the first IQ tests, which were introduced to the United States from Europe by the same Henry Goddard – seemed to confirm the general impression that the human species was gaining downward momentum on a genetic slippery slope. In those early days of IQ testing, it was thought that high intelligence and an alert mind inevitably implied a capacity to absorb large quantities of information. Thus how much you knew was considered a sort of index of your IQ. Following this line of reasoning, early IQ tests included lots of general knowledge questions. Here are a few from a standard test administered to U.S. Army recruits during World War I:

Pick one of four:

The Wyandotte is a kind of:
1) *horse*   2) *fowl*   3) *cattle*   4) *granite*

The ampere is used in measuring:
1) *wind power*   2) *electricity*   3) *water power*   4) *rain fall*

The number of a Zulu's legs is:
1) *two*   2) *four*   3) *six*   4) *eight*

[Answers are 2, 2, 1]

Some half of the nation's army recruits flunked the test and were deemed "feebleminded." These results galvanized the eugenics movement in the United States: it seemed to concerned Americans that the gene pool really was becoming more and more awash in low-intelligence genes.

Scientists realized that eugenic policies required some understanding of the genetics underlying characteristics like feeblemindedness. With the rediscovery of Mendel's work, it seemed that this might actually be possible. The lead in this endeavor was taken on Long Island by one of my predecessors as director of Cold Spring Harbor Laboratory. His name was Charles Davenport.

In 1910, with funding from a railroad heiress, Davenport established the Eugenics Record Office at Cold Spring Harbor. Its mission was to collect basic information – pedigrees – on the genetics of traits ranging from epilepsy to criminality. It became the nerve center of the American eugenics movement. Cold Spring Harbor's mission was much the same then as it is now: today we strive to be at the forefront of genetic research, and Davenport had no less lofty aspirations – but in those days the forefront was eugenics. However, there is no doubt that the research program initiated by Davenport was deeply flawed from the outset and had horrendous, albeit unintended, consequences.

Eugenic thinking permeated everything Davenport did. He went out of his way, for instance, to hire women as field researchers because he believed them to have better observational and social skills than men. But, in keeping with the central goal of eugenics to reduce the number of bad genes, and increase the number of good ones, these women were hired for a maximum of three years. They were smart and educated, and therefore, by definition, the possessors of good genes. It would hardly be fitting for the Eugenics Record Office to hold them back too long from their rightful destiny of producing families and passing on their genetic treasure.

Davenport applied Mendelian analysis to pedigrees he constructed of human characteristics. Initially, he confined his attentions to a number of simple traits – like albinism (recessive) and Huntington disease (dominant) – whose mode of inheritance he identified correctly. After these early successes he plunged into a study of the genetics of human behavior. Everything was fair game: all he needed was a pedigree and some information about the family history (i.e., who in the line manifested the particular characteristic in question), and he would derive conclusions about the underlying genetics. The most cursory perusal of his 1911 book, *Heredity in Relation to Eugenics,* reveals just how wide-ranging Davenport's project was. He shows pedigrees of families with musical and literary ability, and of a "family with mechanical and inventive ability, particularly with respect to boat-building." (Apparently Davenport thought that he was tracking the transmission of the boat-building gene.) Davenport even claimed that he could identify distinct family types associated with different surnames. Thus people with the surname Twinings have these characteristics: "broad-shouldered, dark hair, prominent nose, nervous temperament, temper usually quick, not revengeful. Heavy eyebrows, humorous vein, and sense of ludicrous; lovers of music and horses."

The entire exercise was worthless. Today we know all the characteristics in question are readily affected by environmental factors. Davenport, like Galton, assumed unreasonably that nature unfailingly triumphed over nurture. In addition, whereas the traits he had studied earlier, albinism and Huntington disease, have a simple genetic basis – they are caused by a particular mutation in a particular gene – for most behavioral characteristics, the genetic basis, if any, is complex. They may be determined by a large number of different genes, each one contributing just a little to the final outcome. This situation makes the interpretation of pedigree data like Davenport's virtually impossible. Moreover, the genetic causes of poorly defined characteristics like "feeblemindedness" in one individual may be very different from those in another, so that any search for underlying genetic generalities is futile.

Regardless of the success or failure of Davenport's scientific program, the eugenics movement had already developed a momentum of its own. Local chapters of the Eugenics Society organized competitions at state fairs, giving awards to families apparently free from the taint of bad genes. Fairs that had previously displayed only prize cattle and sheep now added "Better Babies" and "Fitter Families" contests to their programs (see Plate 5). Effectively these were efforts to encourage positive eugenics – inducing the right kind of people to have children. Eugenics was even de rigueur in the nascent feminist movement. The feminist champions of birth control, Marie Stopes in Britain and, in the United States, Margaret Sanger, founder of Planned Parenthood, both viewed birth control as a form of eugenics. Sanger put it succinctly in 1919: "More children from the fit, less from the unfit – that is the chief issue of birth control."

Altogether more sinister was the growth of negative eugenics – preventing the wrong kind of people from having children. In this development, a watershed event occurred in 1899 when a young man called Clawson approached a prison doctor in Indiana called Harry Sharp (appropriately named in light of his enthusiasm for the surgeon's knife). Clawson's problem – or so it was diagnosed by the medical establishment of the day – was compulsive masturbation. He reported that he had been hard at it ever since the age of twelve. Masturbation was seen as part of the general syndrome of degeneracy, and Sharp accepted the conventional wisdom (however bizarre it may seem to us today) that Clawson's mental shortcomings – he had made no progress in school – were caused by his compulsion. The solution? Sharp performed a vasectomy, then a recently invented procedure, and subsequently claimed that he had "cured" Clawson. As a result, Sharp developed his own compulsion: to perform vasectomies.

Sharp promoted his success in treating Clawson (for which, incidentally, we have only Sharp's own report as confirmation) as evidence of the procedure's efficacy for treating all those identified as being of Clawson's kind – all "degenerates." Sterilization had two things going for it. First, it might prevent degenerate behavior, as Sharp claimed it had in Clawson. This, if nothing else, would save society a lot of money because those who had required incarceration, whether in prisons or insane asylums, would be rendered "safe" for release. Second, it would prevent the likes of Clawson from passing their inferior (degenerate) genes on to subsequent generations. Sterilization, Sharp believed, offered the perfect solution to the eugenic crisis.

Sharp was an effective lobbyist, and in 1907 Indiana passed the first compulsory sterilization law, authorizing the sterilization of confirmed "criminals, idiots, rapists, and imbeciles." Indiana's was the first of many: eventually thirty American states had enacted similar statutes, and by 1941 some sixty thousand individuals in the United States had duly been sterilized, half of them in California alone. The laws, which effectively resulted in state governments deciding who could and who could not have children, were challenged in court, but in 1927 the Supreme Court upheld the Virginia statute in the landmark case of Carrie Buck. Oliver Wendell Holmes wrote the decision:

> It is better for all the world if, instead of waiting to execute degenerate offspring for crime, or to let them starve for their imbecility, society can prevent those who are manifestly unfit from continuing their kind . . . Three generations of imbeciles is enough.

Sterilization caught on outside the United States as well – and not only in Nazi Germany. Switzerland and the Scandinavian countries enacted similar legislation.

Racism is not implicit to eugenics – good genes, the ones eugenics seeks to promote, can in principle belong to people of any race. Starting with Galton, however, whose account of his African expedition had confirmed prejudices about "inferior races," the prominent practitioners of eugenics tended to be racists who used eugenics to provide a "scientific" justification for racist views. Henry Goddard, of Kallikak family fame, conducted IQ tests on immigrants at Ellis Island in 1913 and found as many as 80 percent of potential new Americans to be certifiably feebleminded. The IQ tests he carried out during World War I for the U.S. Army reached a similar conclusion: 45 percent of foreign-born draftees had a mental age of less than eight (only 21 percent of native-born draftees fell into this category). That the tests were biased – they were, after all, carried out in English – was not taken to be relevant: racists had the ammunition they required, and eugenics would be pressed into the service of the cause.

Although the term "white supremacist" had yet to be coined, America had plenty of them early in the twentieth century. White Anglo-Saxon Protestants, Theodore Roosevelt prominent among them, were concerned that immigration was corrupting the WASP paradise that America, in their view, was supposed to be. In 1916 Madison Grant, a wealthy New Yorker and friend of both Davenport and Roosevelt, published *The Passing of the Great Race,* in which he argued that the Nordic peoples are superior to all others, including other Europeans. To preserve the United States' fine Nordic genetic heritage, Grant campaigned for immigration restrictions on all non-Nordics. He championed racist eugenic policies, too:

Under existing conditions the most practical and hopeful method of race improvement is through the elimination of the least desirable elements in the nation by depriving them of the power to contribute to future generations. It is well known to stock breeders that the color of a herd of cattle can be modified by continuous destruction of worthless shades and of course this is true of other characters. Black sheep, for instance, have been practically obliterated by cutting out generation after generation all animals that show this color phase.

Despite appearances, Grant's book was hardly a minor publication by a marginalized crackpot; it was an influential best-seller. Later translated into German, it appealed – not surprisingly – to the Nazis. Grant gleefully recalled having received a personal letter from Hitler, who wrote to say that the book was his Bible.

Although not as prominent as Grant, arguably the most influential of the era's exponents of "scientific" racism was Davenport's right-hand man, Harry Laughlin (see Plate 6). Son of an Iowa preacher, Laughlin's expertise was in racehorse pedigrees and chicken breeding. He oversaw the operations of the Eugenics Record Office, but was at his most effective as a lobbyist. In the name of eugenics, he fanatically promoted forced sterilization measures and restrictions on the influx of genetically dubious foreigners (i.e., non–northern Europeans). Particularly important historically was his role as an expert witness at congressional hearings on immigration: Laughlin gave full rein to his prejudices, all of them of course dressed up as "science." When the data were problematic, he fudged them. When he unexpectedly found, for instance, that immigrant Jewish children did better than the native-born in public schools, Laughlin changed the categories he presented, lumping Jews in with whatever nation they had come from, thereby diluting away their superior performance. The passage in 1924 of the Johnson-Reed Immigration Act, which severely restricted immigration from southern Europe and elsewhere, was greeted as a triumph by the likes of Madison Grant; it was Harry Laughlin's finest hour. As vice president some years earlier, Calvin Coolidge had chosen to overlook both Native Americans and the nation's immigration history when he declared that "America must remain American." Now, as president, he signed his wish into law.

Like Grant, Laughlin had his fans among the Nazis, who modeled some of their own legislation on the American laws he had developed. In 1936 he enthusiastically accepted an honorary degree from Heidelberg University, which chose to honor him as "the farseeing representative of racial policy in America." In time, however, a form of late-onset epilepsy ensured that Laughlin's later years were especially pathetic. All his professional life he had campaigned for the sterilization of epileptics on the grounds that they were genetically degenerate.

Hitler's *Mein Kampf* is saturated with pseudoscientific racist ranting derived from long-standing German claims of racial superiority and from some of the uglier aspects of the American eugenics movement. Hitler wrote that the state "must declare unfit for propagation all who are in any way visibly sick or who have inherited a disease and can therefore pass it on, and put this into actual practice," and elsewhere, "Those who are physically and mentally unhealthy and unworthy must not perpetuate their suffering in the body of their children." Shortly after coming to power in 1933, the Nazis had passed a comprehensive sterilization law – the law "for the prevention of progeny with hereditary defects" – that was explicitly based on the American model. (Laughlin proudly published a translation of the law.) Within three years, 225,000 people had been sterilized.

Positive eugenics, encouraging the "right" people to have children, also thrived in Nazi Germany, where "right" meant properly Aryan. Heinrich Himmler, head of the SS (the Nazi elite corps), saw his mission in eugenic terms: SS officers should ensure Germany's genetic future by having as many children as possible. In 1936, he established special maternity homes for SS wives to guarantee that they got the best possible care during pregnancy. The proclamations at the 1935 Nuremberg Rally included a law "for the protection of German blood and German honor," which prohibited marriage between Germans and Jews and even "extra-marital sexual intercourse between Jews and citizens of German or related blood." The Nazis were unfailingly thorough in closing up any reproductive loopholes.

Neither, tragically, were there any loopholes in the U.S. Johnson-Reed Immigration Act that Harry Laughlin had worked so hard to engineer. For many Jews fleeing Nazi persecution, the United States was the logical first choice of destination, but the country's restrictive – and racist – immigration policies resulted in many being turned away. Not only had Laughlin's sterilization law provided Hitler with the model for his ghastly program, but his impact on immigration legislation meant that the United States would in effect abandon German Jewry to its fate at the hands of the Nazis.

In 1939, with the war under way, the Nazis introduced euthanasia. Sterilization proved too much trouble. And why waste the food? The inmates of asylums were categorized as "useless eaters." Questionnaires were distributed among the mental hospitals where panels of experts were instructed to mark them with a cross in the cases of patients whose lives they deemed "not worth living." Seventy-five thousand came back so marked, and the technology of mass murder – the gas chamber – was duly developed. Subsequently, the Nazis expanded the definition of "not worth living" to include whole ethnic groups, among them the Gypsies and, in particular, the Jews. What came to be called the Holocaust was the culmination of Nazi eugenics.

Eugenics ultimately proved a tragedy for humankind. It also proved a disaster for the emerging science of genetics, which could not escape the taint. In fact, despite the prominence of eugenicists like Davenport, many scientists had criticized the movement and dissociated themselves from it. Alfred Russel Wallace, the co-discoverer with Darwin of natural selection, condemned eugenics in 1912 as "simply the meddlesome interference of an arrogant, scientific priestcraft." Thomas Hunt Morgan, of fruit fly fame, resigned on "scientific grounds" from the board of scientific directors of the Eugenics Record Office. Raymond Pearl, at Johns Hopkins, wrote in 1928 that "orthodox eugenicists are going contrary to the best established facts of genetical science."

Eugenics had lost its credibility in the scientific community long before the Nazis appropriated it for their own horrific purposes. The science underpinning it was bogus, and the social programs constructed upon it utterly reprehensible. Nevertheless, by midcentury the valid science of genetics, human genetics in particular, had a major public relations problem on its hands. When in 1948 I first came to Cold Spring Harbor, former home of the by-then-defunct Eugenics Record Office, nobody would even mention the "E word"; nobody was willing to talk about our science's past even though past issues of the German *Journal of Racial Hygiene* still lingered on the shelves of the library.

Realizing that such goals were not scientifically feasible, geneticists had long since forsaken the grand search for patterns of inheritance of human behavioral characteristics – whether Davenport's feeble-mindedness or Galton's genius – and were now focusing instead on the gene and how it functioned in the cell. With the development during the 1930s and 1940s of new and more effective technologies for studying biological molecules in ever greater detail, the time had finally arrived for an assault on the greatest biological mystery of all: what is the chemical nature of the gene?

CHAPTER TWO

THE DOUBLE HELIX:
THIS IS LIFE

I got hooked on the gene during my third year at the University of Chicago. Until then, I had planned to be a naturalist and looked forward to a career far removed from the urban bustle of Chicago's South Side, where I grew up. My change of heart was inspired not by an unforgettable teacher but a little book that appeared in 1944, *What Is Life?*, by the Austrian-born father of wave mechanics, Erwin Schrödinger (see Plate 7). It grew out of several lectures he had given the year before at the Institute for Advanced Study in Dublin. That a great physicist had taken the time to write about biology caught my fancy. In those days, like most people, I considered chemistry and physics to be the "real" sciences, and theoretical physicists were science's top dogs.

Schrödinger argued that life could be thought of in terms of storing and passing on biological information. Chromosomes were thus simply information bearers. Because so much information had to be packed into every cell, it must be compressed into what Schrödinger called a "hereditary code-script" embedded in the molecular fabric of chromosomes. To understand life, then, we would have to identify these molecules, and crack their code. He even speculated that understanding life – which would involve finding the gene – might take us beyond the laws of physics as we then understood them. Schrödinger's book was tremendously influential. Many of those who would become major players in Act 1 of molecular biology's great drama, including Francis Crick (a former physicist himself), had, like me, read *What Is Life?* and been impressed (see Plate 8).

In my own case, Schrödinger struck a chord because I too was intrigued by the essence of life. A small minority of scientists still thought life depended upon a vital force emanating from an all-powerful god. But like most of my teachers, I disdained the very idea of vitalism. If such a "vital" force were calling the shots in nature's game, there was little hope life would ever be understood through the methods of science. On the other hand, the notion that life might be perpetuated by means of an instruction book inscribed in a secret code appealed to me. What sort of molecular code could be so elaborate as to convey all the multitudinous wonder of the living world? And what sort of molecular trick could ensure that the code is exactly copied every time a chromosome duplicates?

At the time of Schrödinger's Dublin lectures, most biologists supposed that proteins would eventually be identified as the primary bearers of genetic instruction. Proteins are molecular chains built up from twenty different building blocks, the amino acids. Because permutations in the order of amino acids along the chain are virtually infinite, proteins could, in principle, readily encode the information underpinning life's extraordinary diversity. DNA then was not considered a serious candidate for the bearer of code-scripts, even though it was exclusively located on chromosomes and had been known about for some seventy-five years. In 1869, Friedrich Miescher, a Swiss biochemist working in Germany, had isolated from pus-soaked bandages supplied by a local hospital a substance he called "nuclein." Because pus consists largely of white blood cells, which, unlike red blood cells, have nuclei and therefore DNA-containing chromosomes, Miescher had stumbled on a good source of DNA (see Plate 9). When he later discovered that "nuclein" was to be found in chromosomes alone, Miescher understood that his discovery was indeed a big one. In 1893, he wrote: "Inheritance insures a continuity in form from generation to generation that lies even deeper than the chemical molecule. It lies in the structuring atomic groups. In this sense, I am a supporter of the chemical heredity theory."

Nevertheless, for decades afterward, chemistry would remain unequal to the task of analyzing the immense size and complexity of the DNA molecule. Only in the 1930s was DNA shown to be a long molecule containing four different chemical bases: adenine (A), guanine (G), thymine (T), and cytosine (C). But at the time of Schrödinger's lectures, it was still unclear just how the subunits (called deoxynucleotides) of the molecule were chemically linked. Nor was it known whether DNA molecules might vary in their sequences of the four different bases. If DNA were indeed Schrödinger's code-script, then the molecule would have to be capable of existing in an immense number of different forms. But back then it was still considered a possibility that one simple sequence like AGTC might be repeated over and over along the entire length of DNA chains.

DNA did not move into the genetic limelight until 1944, when Oswald Avery's lab at the Rockefeller Institute in New York City reported that the composition of the surface coats of pneumonia bacteria could be changed. This was not the result he and his junior colleagues, Colin MacLeod and Maclyn McCarty, expected.

For more than a decade Avery's group had been following up on another most unexpected observation made in 1928 by Fred Griffith, a scientist in the British Ministry of Health. Griffith was interested in pneumonia and studied its bacterial agent, *Pneumococcus*. It was known that there were two strains, designated "smooth" (S) and "rough" (R) according to their appearance under the microscope. These strains differed not only visually but also in their virulence. Inject S bacteria into a mouse, and within a few days the mouse dies; inject R bacteria and the mouse remains healthy. It turns out that S bacterial cells have a coating that prevents the mouse's immune system from recognizing the invader. The R cells have no such coating and are therefore readily attacked by the mouse's immune defenses.

Through his involvement with public health, Griffith knew that multiple strains had sometimes been isolated from a single patient, and so he was curious about how different strains might interact in his unfortunate mice. With one combination, he made a remarkable discovery: when he injected heat-killed S bacteria (harmless) *and* normal R bacteria (also harmless), the mouse died. How could two harmless forms of bacteria conspire to become lethal? The clue came when he isolated the *Pneumococcus* bacteria retrieved from the dead mice and discovered living S bacteria. It appeared the living innocuous R bacteria had acquired something from the dead S variant; whatever it was, that something had allowed the R in the presence of the heat-killed S bacteria to transform itself into a living killer S strain. Griffith confirmed that this change was for real by culturing the S bacteria from the dead mouse over several generations: the bacteria bred true for the S type, just as any regular S strain would. A *genetic* change had indeed occurred to the R bacteria injected into the mouse.

Though this transformation phenomenon seemed to defy all understanding, Griffith's observations at first created little stir in the scientific world. This was partly because Griffith was intensely private and so averse to large gatherings that he seldom attended scientific conferences. Once, he had to be virtually forced to give a lecture. Bundled into a taxi and escorted to the hall by colleagues, he discoursed in a mumbled monotone, emphasizing an obscure corner of his microbiological work but making no mention of bacterial transformation. Luckily, however, not everyone overlooked Griffith's breakthrough.

Oswald Avery was also interested in the sugarlike coats of the *Pneumococcus*. He set out to duplicate Griffith's experiment in order to isolate and characterize whatever it was that had caused those R cells to change to the S type. In 1944 Avery, MacLeod, and McCarty published their results: an exquisite set of experiments showing unequivocally that DNA was the transforming principle. Culturing the bacteria in the test tube rather than in mice made it much easier to search for the chemical identity of the transforming factor in the heat-killed S cells. Methodically destroying one by one the bio-chemical components of the heat-treated S cells, Avery and his group looked to see whether transformation was prevented. First they degraded the sugarlike coat of the S bacteria. Transformation still occurred: the coat was not the transforming principle. Next they used a mixture of two protein-destroying enzymes, trypsin and chymotrypsin, to degrade virtually all the proteins in the S cells. To their surprise, transformation was again unaffected. Next they tried an enzyme (RNase) that breaks down RNA (ribonucleic acid), a second class of nucleic acids similar to DNA and possibly involved in protein synthesis. Again transformation occurred. Finally, they came to DNA, exposing the S bacterial extracts to the DNA-destroying enzyme, DNase. This time they hit a home run. All S-inducing activity ceased completely. The transforming factor was DNA.

In part because of its bombshell implications, the resulting

February 1944 paper by Avery, MacLeod, and McCarty met with a mixed response. Many geneticists accepted their conclusions. After all, DNA was found on every chromosome; why shouldn't it be the genetic material? By contrast, however, most biochemists expressed doubt that DNA was a complex enough molecule to act as the repository of such a vast quantity of biological information. They continued to believe that proteins, the other component of chromosomes, would prove to be the hereditary substance. In principle, as the biochemists rightly noted, it would be much easier to encode a vast body of complex information using the twenty-letter amino-acid alphabet of proteins than the four-letter nucleotide alphabet of DNA. Particularly vitriolic in his rejection of DNA as the genetic substance was Avery's own colleague at the Rockefeller Institute, the protein chemist Alfred Mirsky. By then, however, Avery was no longer scientifically active. The Rockefeller Institute had mandatorily retired him at age sixty-five.

Avery missed out on more than the opportunity to defend his work against the attacks of his colleagues: He was never awarded the Nobel Prize, which was certainly his due, for identifying DNA as the transforming principle. Because the Nobel committee makes its records public fifty years following each award, we now know that Avery's candidacy was blocked by the Swedish physical chemist Einar Hammarsten. Though Hammarsten's reputation was based largely on his having produced DNA samples of unprecedented high quality, he still believed genes to be an undiscovered class of proteins. In fact, even after the double helix was found, Hammarsten continued to insist that Avery should not receive the prize until after the mechanism of DNA transformation had been completely worked out. Avery died in 1955; had he lived only a few more years, he would almost certainly have gotten the prize.

When I arrived at Indiana University in the fall of 1947 with plans to pursue the gene for my Ph.D. thesis, Avery's paper came up over and over in conversations. By then, no one doubted the reproducibility of his results, and more recent work coming out of the Rockefeller Institute made it all the less likely that proteins would prove to be the genetic actors in bacterial transformation. DNA had at last become an important objective for chemists setting their sights on the next breakthrough. In Cambridge, England, the canny Scottish chemist Alexander Todd rose to the challenge of identifying the chemical bonds that linked together nucleotides in DNA. By early 1951, his lab had proved that these links were always the same, such that the backbone of the DNA molecule was very regular. During the same period, the Austrian-born refugee Erwin Chargaff, at the College of Physicians and Surgeons of Columbia University, used the new technique of paper chromatography to measure the relative amounts of the four DNA bases in DNA samples extracted from a variety of vertebrates and bacteria. While some species had DNA in which adenine and thymine predominated, others had DNA with more guanine and cytosine. The possibility thus presented itself that no two DNA molecules had the same composition.

At Indiana I joined a small group of visionary scientists, mostly physicists and chemists, studying the reproductive process of the viruses that attack bacteria (bacteriophages – "phages" for short). The Phage Group was born when my Ph.D. supervisor, the Italian-trained medic Salvador Luria and his close friend, the German-born theoretical physicist Max Delbrück, teamed up with the American physical chemist Alfred Hershey. During World War II both Luria and Delbrück were considered enemy aliens, and thus ineligible to serve in the war effort of American science, even though Luria, a Jew, had been forced to leave France for New York City and Delbrück had fled Germany as an objector to Nazism. Thus excluded, they continued to work in their respective university labs – Luria at Indiana and Delbrück at Vanderbilt – and collaborated on phage experiments during successive summers at Cold Spring Harbor. In 1943, they joined forces with the brilliant but taciturn Hershey, then doing phage research of his own at Washington University in St. Louis.

The Phage Group's program was based on its belief that phages, like all viruses, were in effect naked genes. This concept had first been proposed in 1922 by the imaginative American geneticist Herman J. Muller, who three years later demonstrated that X rays cause mutations. His belated Nobel Prize came in 1946, just after he joined the faculty of Indiana University. It was his presence, in fact, that led me to Indiana. Having started his career under T. H. Morgan, Muller knew better than anyone else how genetics had evolved during the first half of the twentieth century, and I was enthralled by his lectures during my first term. His work on fruit flies (*Drosophila*), however, seemed to me to belong more to the past than to the future, and I only briefly considered doing thesis research under his supervision. I opted instead for Luria's phages, an even speedier experimental subject than *Drosophila:* genetic crosses of phages done one day could be analyzed the next.

For my Ph.D. thesis research, Luria had me follow in his footsteps by studying how X rays killed phage particles. Initially I had hoped to show that viral death was caused by damage to phage DNA. Reluctantly, however, I eventually had to concede that my experimental approach could never give unambiguous answers at the chemical level. I could draw only biological conclusions. Even though phages were indeed effectively naked genes, I realized that the deep answers the Phage Group was seeking could be arrived at only through advanced chemistry. DNA somehow had to transcend its status as an acronym; it had to be understood as a molecular structure in all its chemical detail.

Upon finishing my thesis, I saw no alternative but to move to a lab where I could study DNA chemistry. Unfortunately, however, knowing almost no pure chemistry, I would have been out of my depth in any lab attempting difficult experiments in organic or physical chemistry. I therefore took a postdoctoral fellowship in the Copenhagen lab of the biochemist Herman Kalckar in the fall of 1950. He was studying the synthesis of the small molecules that make up DNA, but I figured out quickly that his biochemical approach would never lead to an understanding of the essence of the gene. Every day spent in his lab would be one more day's delay in learning how DNA carried genetic information.

My Copenhagen year nonetheless ended productively. To escape the cold Danish spring, I went to the Zoological Station at Naples during April and May. During my last week there, I attended a small conference on X-ray diffraction methods for determining the 3-D structure of molecules. X-ray diffraction is a way of studying the atomic structure of any molecule that can be crystallized. The crystal is bombarded with X rays, which bounce off its atoms and are scattered. The scatter pattern gives information about the structure of the molecule but, taken alone, is not enough to solve the structure. The additional information needed is the "phase assignment," which deals with the wave properties of the molecule. Solving the phase problem was not easy, and at that time only the most audacious scientists were willing to take it on. Most of the successes of the diffraction method had been achieved with relatively simple molecules.

My expectations for the conference were low. I believed that a three-dimensional understanding of protein structure, or for that matter of DNA, was more than a decade away. Disappointing earlier X-ray photos suggested that DNA was particularly unlikely to yield up its secrets via the X-ray approach. These results were not surprising since the exact sequences of DNA were expected to differ from one individual molecule to another. The resulting irregularity of surface configurations would understandably prevent the long thin DNA chains from lying neatly side by side in the regular repeating patterns required for X-ray analysis to be successful.

It was therefore a surprise and a delight to hear the last-minute talk on DNA by a thirty-four-year-old Englishman named Maurice Wilkins from the Biophysics Lab of King's College, London. Wilkins was a physicist who during the war had worked on the Manhattan Project. For him, as for many of the other scientists involved, the actual deployment of the bomb on Hiroshima and Nagasaki, supposedly the culmination of all their work, was profoundly disillusioning. He considered forsaking science altogether to become a painter in Paris, but biology intervened. He too had read Schrödinger's book, and was now tackling DNA with X-ray diffraction.

He displayed a photograph of an X-ray diffraction pattern he had recently obtained, and its many precise reflections indicated a highly regular crystalline packing. DNA, one had to conclude, must have a regular structure, the elucidation of which might well reveal the nature of the gene. Instantly I saw myself moving to London to help Wilkins find the structure. My attempts to converse with him after his talk, however, went nowhere. All I got for my efforts was a declaration of his conviction that much hard work lay ahead.

While I was hitting consecutive dead ends, back in America the world's preeminent chemist, Caltech's Linus Pauling, announced a major triumph: he had found the exact arrangement in which chains of amino acids (called *polypeptides*) fold up in proteins, and called his structure the α-helix (alpha helix). That it was Pauling who made this breakthrough was no surprise: he was a scientific superstar. His book *The Nature of the Chemical Bond* essentially laid the foundation of modern chemistry, and, for chemists of the day, it was the Bible. Pauling had been a precocious child. When he was nine, his father, a druggist in Oregon, wrote to the *Oregonian* newspaper requesting suggestions of reading matter for his bookish son, adding that he had already read the Bible and Darwin's *Origin of Species*. But the early death of Pauling's father, which brought the family to financial ruin, makes it remarkable that the promising young man managed to get an education at all.

As soon as I returned to Copenhagen I read about Pauling's α-helix. To my surprise, his model was not based on a deductive leap from experimental X-ray diffraction data. Instead, it was Pauling's long experience as a structural chemist that had emboldened him to infer which type of helical fold would be most compatible with the underlying chemical features of the polypeptide chain. Pauling made scale models of the different parts of the protein molecule, working out plausible schemes in three dimensions. He had reduced the problem to a kind of three-dimensional jigsaw puzzle in a way that was simple yet brilliant.

Whether the α-helix was correct – in addition to being pretty – was now the question. Only a week later, I got the answer. Sir Lawrence Bragg, the English inventor of X-ray crystallography and 1915 Nobel laureate in Physics, came to Copenhagen and excitedly reported that his junior colleague, the Austrian-born chemist Max Perutz, had ingeniously used synthetic polypeptides to confirm the correctness of Pauling's α-helix (see Plate 10). It was a bittersweet triumph for Bragg's Cavendish Laboratory. The year before, they had completely missed the boat in their paper outlining possible helical folds for polypeptide chains.

By then Salvador Luria had tentatively arranged for me to take up a research position at the Cavendish. Located at Cambridge University, this was the most famous laboratory in all of science. Here Ernest Rutherford first described the structure of the atom. Now it was

Bragg's own domain, and I was to work as apprentice to the English chemist John Kendrew, who was interested in determining the 3-D structure of the protein myoglobin. Luria advised me to visit the Cavendish as soon as possible. With Kendrew in the States, Max Perutz would check me out. Together, Kendrew and Perutz had earlier established the Medical Research Council (MRC) Unit for the Study of the Structure of Biological Systems.

A month later in Cambridge, Perutz assured me that I could quickly master the necessary X-ray diffraction theory and should have no difficulty fitting in with the others in their tiny MRC Unit. To my relief, he was not put off by my biology background. Nor was Lawrence Bragg, who briefly came down from his office to look me over.

I was twenty-three when I arrived back at the MRC Unit in Cambridge in early October. I found myself sharing space in the biochemistry room with a thirty-five-year-old ex-physicist, Francis Crick, who had spent the war working on magnetic mines for the Admiralty. When the war ended, Crick had planned to stay on in military research, but, on reading Schrödinger's *What Is Life?*, he had moved toward biology. Now he was at the Cavendish to pursue the 3-D structure of proteins for his Ph.D.

Crick was always fascinated by the intricacies of important problems. His endless questions as a child compelled his weary parents to buy him a children's encyclopedia, hoping that it would satisfy his curiosity. But it only made him insecure: he confided to his mother his fear that everything would have been discovered by the time he grew up, leaving him nothing to do. His mother reassured him (correctly, as it happened) that there would still be a thing or two for him to figure out.

A great talker, Crick was invariably the center of attention in any gathering. His booming laugh was forever echoing down the hallways of the Cavendish. As the MRC Unit's resident theoretician, he used to come up with a novel insight at least once a month, and he would explain his latest idea at great length to anyone willing to listen. The morning we met he lit up when he learned that my objective in coming to Cambridge was to learn enough crystallography to have a go at the DNA structure. Soon I was asking Crick's opinion about using Pauling's model-building approach to go directly for the structure. Would we need many more years of diffraction experimentation before modeling would be practicable? To bring us up to speed on the status of DNA structural studies, Crick invited Maurice Wilkins, a friend since the end of the war, up from London for Sunday lunch. Then we could learn what progress Wilkins had made since his talk in Naples (see Plate 11).

Wilkins expressed his belief that DNA's structure was a helix, formed by several chains of linked nucleotides twisted around each other. All that remained to be settled was the number of chains. At the time, Wilkins favored three on the basis of his density measurements of DNA fibers. He was keen to start model-building, but he had run into a roadblock in the form of a new addition to the King's College Biophysics Unit, Rosalind Franklin.

A thirty-one-year-old Cambridge-trained physical chemist, Franklin was an obsessively professional scientist; for her twenty-ninth birthday all she requested was her own subscription to her field's technical journal, *Acta Crystallographica*. Logical and precise, she was impatient with those who acted otherwise. And she was given to strong opinions, once describing her Ph.D. thesis adviser, Ronald Norrish, a future Nobel Laureate, as "stupid, bigoted, deceitful, ill-mannered and tyrannical." Outside the laboratory, she was a determined and gutsy mountaineer, and, coming from the upper echelons of London society, she belonged to a more rarefied social world than most scientists (see Plate 12). At the end of a hard day at the bench, she would occasionally change out of her lab coat into an elegant evening gown and disappear into the night.

Just back from a four-year X-ray crystallographic investigation of graphite in Paris, Franklin had been assigned to the DNA project while Wilkins was away from King's. Unfortunately, the pair soon proved incompatible. Franklin, direct and data-focused, and Wilkins, retiring and speculative, were destined never to collaborate. Shortly before Wilkins accepted our lunch invitation, the two had had a big blowup in which Franklin had insisted that no model-building could commence before she collected much more extensive diffraction data. Now they effectively didn't communicate, and Wilkins would have no chance to learn of her progress until Franklin presented her lab seminar scheduled for the beginning of November. If we wanted to listen, Crick and I were welcome to go as Wilkins's guests.

Crick was unable to make the seminar, so I attended alone and briefed him later on what I believed to be its key take-home messages on crystalline DNA. In particular, I described from memory Franklin's measurements of the crystallographic repeats and the water content. This prompted Crick to begin sketching helical grids on a sheet of paper, explaining that the new helical X-ray theory he had devised with Bill Cochran and Vladimir Vand would permit even me, a former bird-watcher, to predict correctly the diffraction patterns expected from the molecular models we would soon be building at the Cavendish.

As soon as we got back to Cambridge, I arranged for the Cavendish machine shop to construct the phosphorous atom models needed for short sections of the sugar phosphate backbone found in DNA. Once these became available, we tested different ways the backbones might twist around each other in the center of the DNA molecule. Their regular repeating atomic structure should allow the atoms to come together in a consistent, repeated conformation. Following Wilkins's hunch, we focused on three-chain models. When one of these appeared to be almost plausible, Crick made a phone call to Wilkins to announce we had a model we thought might be DNA.

The next day both Wilkins and Franklin came up to see what we had done. The threat of unanticipated competition briefly united them in common purpose. Franklin wasted no time in faulting our basic concept. My memory was that she had reported almost no water present in crystalline DNA. In fact, the opposite was true. Being a crystallographic novice, I had confused the terms "unit cell" and "asymmetric unit." Crystalline DNA was in fact water-rich. Consequently, Franklin pointed out, the backbone had to be on the outside and not, as we had it, in the center, if only to accommodate all the water molecules she had observed in her crystals.

That unfortunate November day cast a very long shadow. Franklin's opposition to model-building was reinforced. Doing experiments, not playing with Tinkertoy representations of atoms, was the way she intended to proceed. Even worse, Sir Lawrence Bragg passed down the word that Crick and I should desist from all further attempts at building a DNA model. It was further decreed that DNA research should be left to the King's lab, with Cambridge continuing to focus solely on proteins. There was no sense in two MRC-funded labs competing against each other. With no more bright ideas up our sleeves, Crick and I were reluctantly forced to back off, at least for the time being.

It was not a good moment to be condemned to the DNA sidelines. Linus Pauling had written Wilkins to request a copy of the crystalline DNA diffraction pattern. Though Wilkins had declined, saying he wanted more time to interpret it himself, Pauling was hardly obliged to depend upon data from King's. If he wished, he could easily start serious X-ray diffraction studies at Caltech.

The following spring, I duly turned away from DNA and set about extending prewar studies on the pencil-shaped tobacco mosaic virus using the Cavendish's powerful new X-ray beam. This light experimental workload gave me plenty of time to wander through various Cambridge libraries. In the zoology building, I read Erwin Chargaff's paper describing his finding that the DNA bases adenine and thymine occurred in roughly equal amounts, as did the bases guanine and cytosine. Hearing of these one-to-one ratios Crick wondered whether, during DNA duplication, adenine residues might be attracted to thymine and vice versa, and whether a corresponding attraction might exist between guanine and cytosine. If so, base sequences on the "parental" chains (e.g., ATGC) would have to be complementary to those on "daughter" strands (yielding in this case TACG).

These remained idle thoughts until Erwin Chargaff came through Cambridge in the summer of 1952 on his way to the International Biochemical Congress in Paris. Chargaff expressed annoyance that neither Crick nor I saw the need to know the chemical structures of the four bases. He was even more upset when we told him that we could simply look up the structures in textbooks as the need arose. I was left hoping that Chargaff's data would prove irrelevant. Crick, however, was energized to do several experiments looking for molecular "sandwiches" that might form when adenine and thymine (or alternatively, guanine and cytosine) were mixed together in solution. But his experiments went nowhere.

Like Chargaff, Linus Pauling also attended the International Biochemical Congress, where the big news was the latest result from the Phage Group. Alfred Hershey and Martha Chase at Cold Spring Harbor had just confirmed Avery's transforming principle: DNA was the hereditary material! Hershey and Chase proved that only the DNA of the phage virus enters bacterial cells; its protein coat remains on the outside. It was more obvious than ever that DNA must be understood at the molecular level if we were to uncover the essence of the gene. With Hershey and Chase's result the talk of the town, I was sure that Pauling would now bring his formidable intellect and chemical wisdom to bear on the problem of DNA.

Early in 1953, Pauling did indeed publish a paper outlining the structure of DNA. Reading it anxiously I saw that he was proposing a three-chain model with sugar phosphate backbones forming a dense central core. Superficially it was similar to our botched model of fifteen months earlier. But instead of using positively charged atoms (e.g., Mg^{2+}) to stabilize the negatively charged backbones, Pauling made the unorthodox suggestion that the phosphates were held together by hydrogen bonds. But it seemed to me, the biologist, that such hydrogen bonds required extremely acidic conditions never found in cells. With a mad dash to Alexander Todd's nearby organic chemistry lab my belief was confirmed: The impossible had happened. The world's best-known, if not best, chemist had gotten his chemistry wrong. In effect, Pauling had knocked the A off of DNA. Our quarry was deoxyribonucleic acid, but the structure he was proposing was not even acidic.

Hurriedly I took the manuscript to London to inform Wilkins and Franklin they were still in the game. Convinced that DNA was not a helix, Franklin had no wish even to read the article and deal with the distraction of Pauling's helical ideas, even when I offered Crick's arguments for helices. Wilkins, however, was very interested indeed in the news I brought; he was now more certain than ever that DNA was helical. To prove the point, he showed me a photograph obtained more than six months earlier by Franklin's graduate student Raymond Gosling, who had X-rayed the so-called B form of DNA. Until that moment, I didn't know a B form even existed. Franklin had put this picture aside, preferring to concentrate on the A form, which she thought would more likely yield useful data. The X-ray pattern of this B form was a distinct cross (see Plate 13). Since Crick and others had already deduced that such a pattern of reflections would be created by a helix, this evidence made it clear that DNA had to be a helix! In fact, despite Franklin's reservations, this was no surprise. Geometry itself suggested that a helix was the most logical arrangement for a long string of repeating units such as the nucleotides of DNA. But we still did not know what that helix looked like, nor how many chains it contained.

The time had come to resume building helical models of DNA. Pauling was bound to realize soon enough that his brainchild was wrong. I urged Wilkins to waste no time. But he wanted to wait until Franklin had completed her scheduled departure for another lab later that spring. She had decided to move on to avoid the unpleasantness at King's. Before leaving, she had been ordered to stop further work with DNA and had already passed on many of her diffraction images to Wilkins.

When I returned to Cambridge and broke the news of the DNA B form, Bragg no longer saw any reason for Crick and me to avoid DNA. He very much wanted the DNA structure to be found on his side of the Atlantic. So we went back to model-building, looking for a way the known basic components of DNA – the backbone of the molecule and the four different bases, adenine, thymine, guanine, and cytosine – could fit together to make a helix. I commissioned the shop at the Cavendish to make us a set of tin bases, but they couldn't produce them fast enough for me: I ended up cutting out rough approximations from stiff cardboard.

By this time I realized the DNA density-measurement evidence actually slightly favored a two-chain, rather than three-chain, model. So I decided to search out plausible double helices. As a biologist, I preferred the idea of a genetic molecule made of two, rather than three, components. After all, chromosomes, like cells, increase in number by duplicating, not triplicating.

I knew that our previous model with the backbone on the inside and the bases hanging out was wrong. Chemical evidence from the University of Nottingham, which I had too long ignored, indicated that the bases must be hydrogen-bonded to each other. They could only form bonds like this in the regular manner implied by the X-ray diffraction data if they were in the center of the molecule (see Plate 14). But how could they come together in pairs? For two weeks I got nowhere, misled by an error in my nucleic acid chemistry textbook. Happily, on February 27, Jerry Donahue, a theoretical chemist visiting the Cavendish from Caltech, pointed out that the textbook was wrong. So I changed the locations of the hydrogen atoms on my cardboard cutouts of the molecules.

The next morning, February 28, 1953, the key features of the DNA model all fell into place. The two chains were held together by strong hydrogen bonds between adenine-thymine and guanine-cytosine base pairs (see Plate 15). The inferences Crick had drawn the year before based on Chargaff's research had indeed been correct. Adenine does bond to thymine and guanine does bond to cytosine, but not through flat surfaces to form molecular sandwiches. When Crick arrived, he took it all in rapidly, and gave my base-pairing scheme his blessing. He realized right away that it would result in the two strands of the double helix running in opposite directions.

It was quite a moment. We felt sure that this was it. Anything that simple, that elegant just had to be right. What got us most excited was the complementarity of the base sequences along the two chains. If you knew the sequence – the order of bases – along one chain, you automatically knew the sequence along the other. It was immediately apparent that this must be how the genetic messages of genes are copied so exactly when chromosomes duplicate prior to cell division. The molecule would "unzip" to form two separate strands. Each separate strand then could serve as the template for the synthesis of a new strand, one double helix becoming two.

In *What Is Life?* Schrödinger had suggested that the language of life might be like Morse code, a series of dots and dashes. He wasn't far off. The language of DNA is a linear series of As, Ts, Gs, and Cs. And just as transcribing a page out of a book can result in the odd typo, the rare mistake creeps in when all these As, Ts, Gs, and Cs are being copied along a chromosome. These errors are the mutations geneticists had talked about for almost fifty years. Change an "i" to an "a" and "Jim" becomes "Jam" in English; change a T to a C and "ATG" becomes "ACG" in DNA.

The double helix made sense chemically and it made sense biologically. Now there was no need to be concerned about Schrödinger's suggestion that new laws of physics might be necessary for an understanding of how the hereditary code-script is duplicated: genes in fact were no different from the rest of chemistry. Later that day, during lunch at the Eagle, the pub virtually adjacent to the Cavendish Lab, Crick, ever the talker, could not help but tell everyone we had just found the "secret of life." I myself, though no less electrified by the thought, would have waited until we had a pretty three-dimensional model to show off (see Plate 16).

Among the first to see our demonstration model was the chemist Alexander Todd. That the nature of the gene was so simple both surprised and pleased him. Later, however, he must have asked himself why his own lab, having established the general chemical structure of DNA chains, had not moved on to asking how the chains folded up in three dimensions. Instead the essence of the molecule was left to be discovered by a two-man team, a biologist and a physicist, neither of whom possessed a detailed command even of undergraduate chemistry. But paradoxically, this was, at least in part, the key to our success: Crick and I arrived at the double helix first precisely because most chemists at that time thought DNA too big a molecule to understand by chemical analysis.

At the same time, the only two chemists with the vision to seek DNA's 3-D structure made major tactical mistakes: Rosalind Franklin's was her resistance to model-building; Linus Pauling's was a matter of simply neglecting to read the existing literature on DNA, particularly the data on its base composition published by Chargaff. Ironically, Pauling and Chargaff sailed across the Atlantic on the same ship following the Paris Biochemical Congress in 1952, but failed to hit it off. Pauling was long accustomed to being right. And he believed there was no chemical problem he could not work out from first principles by himself. Usually this confidence was not misplaced. During the Cold War, as a prominent critic of the American nuclear weapons development program, he was questioned by the FBI after giving a talk. How did he know how much plutonium there is in an atomic bomb? Pauling's response was "Nobody told me. I figured it out."

Over the next several months Crick and (to a lesser extent) I relished showing off our model to an endless stream of curious scientists. However, the Cambridge biochemists did not invite us to give a formal talk in the biochemistry building. They started to refer to it as the "WC," punning our initials with those used in Britain for the toilet or water closet. That we had found the double helix without doing experiments irked them.

The manuscript that we submitted to *Nature* in early April was published just over three weeks later, on April 25, 1953 (see Plate 17). Accompanying it were two longer papers by Franklin and Wilkins, both supporting the general correctness of our model. In June, I gave the first presentation of our model at the Cold Spring Harbor symposium on viruses. Max Delbrück saw to it that I was offered, at the last minute, an invitation to speak. To this intellectually high-powered meeting I brought a three-dimensional model built in the Cavendish, the adenine-thymine base pairs in red and the guanine-cytosine base pairs in green.

In the audience was Seymour Benzer, yet another ex-physicist who had heeded the clarion call of Schrödinger's book. He immediately understood what our breakthrough meant for his studies of mutations in viruses. He realized that he could now do for a short stretch of bacteriophage DNA what Morgan's boys had done forty years earlier for fruit fly chromosomes: he would map mutations – determine their order – along a gene, just as the fruit fly pioneers had mapped genes along a chromosome. Like Morgan, Benzer would have to depend on recombination to generate new genetic combinations, but, whereas Morgan had the advantage of a ready mechanism of recombination – the production of sex cells in a fruit fly – Benzer had to induce recombination by simultaneously infecting a single bacterial host cell with two different strains of bacteriophage, which differed by one or more mutations in the region of interest. Within the bacterial cell, recombination – the exchange of segments of molecules – would occasionally occur between the different viral DNA molecules, producing new permutations of mutations – so-called "recombinants." Within a single astonishingly productive year in his Purdue University lab, Benzer produced a map of a single bacteriophage gene, *rII*, showing how a series of mutations – all errors in the genetic script – were laid out linearly along the virus DNA. The language was simple and linear, just like a line of text on the written page.

The response of the Hungarian physicist Leo Szilard to my Cold Spring Harbor talk on the double helix was less academic. His question was, "Can you patent it?" At one time Szilard's main source of income had been a patent that he held with Einstein, and he had later tried unsuccessfully to patent with Enrico Fermi the nuclear reactor they built at the University of Chicago in 1942. But then as now patents were given only for useful inventions and at the time no one could conceive of a practical use for DNA. Perhaps then, Szilard suggested, we should copyright it.

There remained, however, a single missing piece in the double helical jigsaw puzzle: our unzipping idea for DNA replication had yet to be experimentally verified. Max Delbrück, for example, was unconvinced. Though he liked the double helix as a model, he worried that unzipping it might generate horrible knots. Five years later, a former student of Pauling's, Matt Meselson, and the equally bright young phage worker Frank Stahl put to rest such fears when they published the results of a single elegant experiment.

They had met in the summer of 1954 at the Marine Biological Laboratory at Woods Hole, Massachusetts, where I was then lecturing, and agreed – over a good many gin martinis – that they should get together to do some science. The result of their collaboration has been described as "the most beautiful experiment in biology" (see Plate 18).

They used a centrifugation technique that allowed them to sort molecules according to slight differences in weight; following a centrifugal spin, heavier molecules end up nearer the bottom of the test tube than lighter ones (see Plate 19). Because nitrogen atoms (N) are a component of DNA, and because they exist in two distinct forms, one light and one heavy, Meselson and Stahl were able to tag segments of DNA and thereby track the process of its replication in bacteria. Initially all the bacteria were raised in a medium containing heavy N, which was thus incorporated in both strands of the DNA. From this culture they took a sample, transferring it to a medium containing only light N, ensuring that the next round of DNA replication would have to make use of light N. If, as Crick and I had predicted, DNA replication involves unzipping the double helix and copying each strand, the resultant two "daughter" DNA molecules in the experiment would be hybrids, each consisting of one heavy N strand (the template strand derived from the "parent" molecule) and one light N strand (the one newly fabricated from the new medium). Meselson and Stahl's centrifugation procedure bore out these expectations precisely. They found three discrete bands in their centrifuge tubes, with the heavy-then-light sample halfway between the heavy-heavy and light-light samples. DNA replication works just as our model supposed it would.

The biochemical nuts and bolts of DNA replication were being analyzed at around the same time in Arthur Kornberg's laboratory at Washington University in St. Louis. By developing a new, "cell-free" system for DNA synthesis, Kornberg discovered an enzyme (DNA polymerase) that links the DNA components and makes the chemical bonds of the DNA backbone. Kornberg's enzymatic synthesis of DNA was such an unanticipated and important event that he was awarded the 1959 Nobel Prize in Physiology or Medicine, less than two years after the key experiments (see Plate 20). After his prize was announced, Kornberg was photographed holding a copy of the double helix model I had taken to Cold Spring Harbor in 1953.

It was not until 1962 that Francis Crick, Maurice Wilkins, and I were to receive our own Nobel Prize in Physiology or Medicine. Four years earlier, Rosalind Franklin had died of ovarian cancer at the tragically young age of thirty-seven. Before then Crick had become a close colleague and a real friend of Franklin's. Following the two operations that would fail to stem the advance of her cancer, Franklin convalesced with Crick and his wife, Odile, in Cambridge.

It was and remains a long-standing rule of the Nobel Committee never to split a single prize more than three ways. Had Franklin lived, the problem would have arisen whether to bestow the award upon her or Maurice Wilkins. The Swedes might have resolved the dilemma by awarding them both the Nobel Prize in Chemistry that year. Instead, it went to Max Perutz and John Kendrew, who had elucidated the three-dimensional structures of hemoglobin and myoglobin respectively.

The discovery of the double helix sounded the death knell for vitalism. Serious scientists, even those religiously inclined, realized that a complete understanding of life would not require the revelation of new laws of nature. Life was just a matter of physics and chemistry, albeit exquisitely organized physics and chemistry. The immediate task ahead would be to figure out how the DNA-encoded script of life went about its work. How does the molecular machinery of cells read the messages of DNA molecules? As the next chapter will reveal, the unexpected complexity of the reading mechanism led to profound insights into how life first came about.

CHAPTER THREE

READING THE CODE: BRINGING DNA TO LIFE

Long before Oswald Avery's experiments put DNA in the spotlight as the "transforming principle," geneticists were trying to understand just how the hereditary material – whatever it might be – was able to influence the characteristics of a particular organism. How did Mendel's "factors" affect the form of peas, making them either wrinkled or round?

The first clue came around the turn of the century, just after the rediscovery of Mendel's work. Archibald Garrod, an English physician whose slow progress through medical school and singular lack of a bedside manner had ensured him a career in research rather than patient care at St. Bartholomew's Hospital in London, was interested in a group of rare diseases of which a common marked symptom was strangely colored urine. One of these diseases, alkaptonuria, has been dubbed "black diaper syndrome" because those afflicted with it pass urine that turns black on exposure to air. Despite this alarming symptom, the disease is usually not lethal, though it can lead in later life to an arthritis-like condition as the black-urine pigments accumulate in the joints and spine. Contemporary science attributed the blackening to a substance produced by bacteria living in the gut, but Garrod argued that the appearance of black urine in newborns, whose guts lack bacterial colonies, implied that the substance was produced by the body itself. He inferred that it was the product of a flaw in the body's chemical machinery, an "error in metabolism" in his words, suggesting there might be a critical glitch in some biochemical pathway.

Garrod further observed that alkaptonuria, though very rare in the population as a whole, occurred more frequently among children of marriages between blood relatives. In 1902, he was able to explain the phenomenon in terms of Mendel's newly rediscovered laws. Here was the pattern of inheritance to be expected of a rare recessive gene: two first cousins, say, have both received a copy of the "alkaptonuria" gene from the same grandparent, creating a one-in-four chance that their union will produce a child homozygous for the gene (i.e., a child with two copies of the recessive gene) who will therefore develop alkaptonuria. Combining his biochemical and genetic analyses, Garrod concluded that alkaptonuria is an "inborn error in metabolism." Though nobody really appreciated it at the time, Garrod was thus the first to make the causal connection between genes and their physiological effect. Genes in some way governed metabolic processes, and an error in a gene – a mutation – could result in a defective metabolic pathway.

The next significant step would not occur until 1941, when George Beadle and Ed Tatum published their study of induced mutations in a tropical bread mold. Beadle had grown up outside Wahoo, Nebraska, and would have taken over the family farm had a high-school science teacher not encouraged him to consider an alternative career. Through the thirties, first at Caltech in association with T. H. Morgan of fruit fly fame and then at the Institut de Biologie Physico-Chimique in Paris, Beadle had applied himself to discovering how genes work their magic in affecting, for example, eye color in fruit flies. Upon his arrival at Stanford University in 1937, he recruited Tatum, who joined the effort against the advice of his academic advisers. Ed Tatum had been both an undergraduate and graduate student at the University of Wisconsin, doing studies of bacteria that lived in milk (of which there was no shortage in the Cheese State). Though the job with Beadle might be intellectually challenging, Tatum's Wisconsin professors counseled in favor of the financial security to be found in a career with the dairy industry. Fortunately for science, Tatum chose Beadle over butter.

Beadle and Tatum came to realize that fruit flies were too complex for the kind of research at hand: finding the effect of a single mutation in an animal as complicated as *Drosophila* would be like looking for a needle in a haystack. They chose instead to work with an altogether simpler species, *Neurospora crassa,* the orange-red mold that grows on bread in tropical countries. The plan was simple: subject the mold to X rays to cause mutations – just as Muller had done with fruit flies – and then try to determine the impact of the resulting mutations on the fungi. They would track the effects of the mutations in this way: Normal (i.e., unmutated) *Neurospora,* it was known, could survive on a so-called minimal culture medium; on this basic "diet" they could evidently synthesize biochemically all the larger molecules they required to live, constructing them from the simpler ones in the nutrient medium. Beadle and Tatum theorized that a mutation that knocked out any of those synthetic pathways would result in the irradiated mold strain being unable to grow on minimal medium; that same strain should, however, still manage to thrive on a "complete" medium, one containing all the molecules necessary for life, like amino acids and vitamins. In other words, the mutation preventing the synthesis of a key nutrient would be rendered harmless if the nutrient were available directly from the culture medium.

Beadle and Tatum irradiated some five thousand specimens, then set about testing each one to see whether it could survive on minimal medium. The first survived fine; so did the second, and the third . . . It was not until they tested strain number 299 that they found one that could no longer exist on minimal medium, though as predicted it could survive on the complete version. Number 299 would be but the first of many mutant strains that they would analyze. The next step was to see what exact capacity the mutants had lost. Maybe 299 could not synthesize essential amino acids. Beadle and Tatum tried adding amino acids to the minimal medium, but still 299 failed to grow. What about vitamins? They added a slew of them to the minimal medium, and this time 299 thrived. Now it was time to narrow the field, adding each vitamin individually and then gauging the growth response of 299. Niacin didn't work, nor riboflavin, but when they added vitamin B_6, 299 was able to survive on minimal medium. 299's X-ray-induced mutation had somehow disrupted the synthetic pathway involved in the production of B_6. But how? Knowing that biochemical syntheses of this kind are governed by protein enzymes that promote the individual incremental chemical reactions along the pathway, Beadle and Tatum suggested that each mutation they discovered had knocked out a particular enzyme. And since mutations occur in genes, genes must produce enzymes. When it appeared in 1941, their study inspired a slogan that summarized what had become the understanding of how genes work: "One gene, one enzyme."

But since all enzymes were then thought to be proteins, the question soon arose whether genes also encoded the many cellular proteins that were *not* enzymes. The first suggestion that genes might provide the information for all proteins came from Linus Pauling's lab at Caltech. He and his student Harvey Itano studied hemoglobin, the protein in red blood cells that transports oxygen from the lung to metabolically active tissues, like muscle, where it is needed. In particular, they focused on the hemoglobin of people with sickle-cell disease, also known as sickle-cell anemia, a genetic disorder common in Africans, and therefore among African Americans as well. The red blood cells of sickle-cell victims tend to become deformed, assuming a distinctive "sickle" shape under the microscope, and the resulting blockages in capillaries can be horribly painful, even lethal. Later research would uncover an evolutionary rationale for the disease's prevalence among Africans: because part of the malaria parasite's life cycle is spent in red blood cells, people with sickle-cell hemoglobin suffer less severely from malaria. Human evolution seems to have struck a Faustian bargain on behalf of some inhabitants of tropical regions: the sickle-cell affliction confers some protection against the ravages of malaria.

Itano and Pauling compared the hemoglobin proteins of sickle-cell patients with those of non-sickle-cell individuals and found that the two molecules differed in their electrical charge. Around that time, the late forties, geneticists determined that sickle-cell disease is transmitted as a classical Mendelian recessive character. Sickle-cell disease, they therefore inferred, must be caused by a mutation in the hemoglobin gene, a mutation that affects the chemical composition of the resultant hemoglobin protein. And so it was that Pauling was able to refine Garrod's notion of "inborn errors of metabolism" by recognizing some to be what he called "molecular diseases." Sickle-cell was just that, a molecular disease.

In 1956, the sickle-cell hemoglobin story was taken a step further by Vernon Ingram, working in the Cavendish Laboratory where Francis

Crick and I had found the double helix. Using recently developed methods of identifying the specific amino acids in the chain that makes up a protein, Ingram was able to specify precisely the molecular difference that Itano and Pauling had noted as affecting the overall charge of the molecule. It amounted to a single amino acid: Ingram determined that glutamic acid, found at position 6 in the normal protein chain, is replaced, in sickle-cell hemoglobin, by valine (see Plate 21). Here, conclusively, was evidence that genetic mutations – differences in the sequence of As, Ts, Gs, and Cs in the DNA code of a gene – could be "mapped" directly to differences in the amino acid sequences of proteins. Proteins are life's active molecules: they form the enzymes that catalyze biochemical reactions, and they also provide the body's major structural components, like keratin, of which skin, hair, and nails are composed. And so the way DNA exerts its controlling magic over cells, over development, over life as a whole, is through proteins.

But how is the information encoded in DNA – a molecular string of nucleotides, As, Ts, Gs, and Cs – converted into a protein – a string of amino acids?

Shortly after Francis Crick and I published our account of the double helix, we began to hear from the well-known Russian-born theoretical physicist George Gamow. His letters – invariably handwritten and embellished with cartoons and other squiggles, some quite relevant, others less so – were always signed simply "Geo" (pronounced "Jo," as we would later discover). He'd become interested in DNA and, even before Ingram had conclusively demonstrated the connection between the DNA base sequence and the amino acid sequence of proteins, in the relationship between DNA and protein. Sensing that biology was at last becoming an exact science, Gamow foresaw a time when every organism could be described genetically by a very long number represented exclusively by the numerals 1, 2, 3, and 4, each one standing for one of the bases, A, T, G, and C. At first, we took him for a buffoon; we ignored his first letter. A few months later, however, when Crick met him in New York City, the magnitude of his gifts became clear and we promptly welcomed him aboard the DNA bandwagon as one of its earliest recruits.

Gamow had come to the United States in 1934 to escape the engulfing tyranny of Stalin's Soviet Union. In a 1948 paper, he explained the abundance of different chemical elements present throughout the universe in relation to thermonuclear processes that had taken place in the early phases of the Big Bang. The research, having been carried out by Gamow and his graduate student Ralph Alpher, would have been published with the byline of "Alpher and Gamow" had Gamow not decided to include as well the name of his friend Hans Bethe, an eminently talented physicist to be sure, but one who had contributed nothing to the study. It delighted the inveterate prankster Gamow that the paper appeared attributed to "Alpher, Bethe, and Gamow," no less than that its publication date was, fortuitously, April 1. To this day, cosmologists still refer to it as the αβγ (Alpha-Beta-Gamma) paper.

By the time I first met Gamow in 1954, he had already devised a formal scheme in which he proposed that overlapping triplets of DNA bases served to specify certain amino acids. Underlying his theory was a belief that there existed on the surface of each base pair a cavity that was complementary in shape to part of the surface of one of the amino acids. I told Gamow I was skeptical: DNA could not be the direct template along which amino acids arranged themselves before being connected into polypeptide chains, as lengths of linked amino acids are called. Being a physicist, Gamow had not, I supposed, read the scientific papers refuting the notion that protein synthesis occurs where DNA is located – in the nucleus. In fact, it had been observed that the removal of the nucleus from a cell has no immediate effect on the rate at which proteins are made. Today we know that amino acids are actually assembled into proteins in ribosomes, small cellular particles containing a second form of nucleic acid called RNA.

RNA's exact role in life's biochemical puzzle was unclear at that time. In some viruses, like tobacco mosaic virus, it seemed to play a role similar to DNA in other species, encoding the proteins specific to that organism. And in cells, RNA had to be involved somehow in protein synthesis, since cells that made lots of proteins were always RNA-rich. Even before we found the double helix, I thought it likely that the genetic information in chromosomal DNA was used to make RNA chains of complementary sequences. These RNA chains might in turn serve as the templates that specified the order of amino acids in their respective proteins. If so, RNA was thus an intermediate between DNA and protein. Francis Crick would later refer to this DNA → RNA → protein flow of information as the "central dogma." The view soon gained support with the discovery in 1959 of the enzyme RNA polymerase. In virtually all cells, it catalyzes the production of single-stranded RNA chains from double-stranded DNA templates.

It appeared the essential clues to the process by which proteins are made would come from further studies of RNA, not DNA. To advance the cause of "cracking the code" – deciphering that elusive relationship between DNA sequence and the amino acid sequence of proteins – Gamow and I formed the RNA Tie Club. Its members would be limited to twenty, one for each of the twenty different amino acids. Gamow designed a club necktie and commissioned the production of the amino-acid-specific tiepins. These were badges of office, each bearing the standardized three-letter abbreviation of an amino acid, the one the member wearing the pin was responsible for studying. I had PRO for proline and Gamow had ALA for alanine. In an era when tiepins with letters usually advertised one's initials, Gamow took pleasure in confusing people with his ALA pin. His joke backfired when a sharp-eyed hotel clerk refused to honor his check, noting that the name printed on the check bore no relation to the initials on the gentleman's jewelry.

The fact that most of the scientists interested in the coding problem at that time could be squeezed into the club's membership of twenty showed how small the DNA-RNA world was. Gamow easily found room for a nonbiologist friend, the physicist Edward Teller (LEU – leucine), while I inducted Richard Feynman (GLY – glycine), the extraordinarily imaginative Caltech physicist who, when momentarily frustrated in his exploration of inner atomic forces, often visited me in the biology building where I was then working.

One element of Gamow's 1954 scheme had the virtue of being testable: because it involved overlapping DNA triplets, it predicted that many pairs of amino acids would in fact never be found adjacent in proteins. So Gamow eagerly awaited the sequencing of additional proteins. To his disappointment, more and more amino acids began to be found next to each other, and his scheme became increasingly untenable. The coup de grâce for all Gamow-type codes came in 1956 when Sydney Brenner (VAL – valine) analyzed every amino acid sequence then available.

Brenner had been raised in a small town outside Johannesburg, South Africa, in two rooms at the back of his father's cobbler's shop. Though the elder Brenner, a Lithuanian immigrant, was illiterate, his precocious son discovered a love of reading at the age of four and, led by this passion, would be turned on to biology by a textbook called *The Science of Life*. Though he was one day to admit having stolen the book from the public library, neither larceny nor poverty could slow Brenner's progress: he entered the University of Witwatersrand's undergraduate medical program at fourteen, and was working on his Ph.D. at Oxford when he came to Cambridge a month after our discovery of the double helix. He recalls his reaction to our model: "That's when I saw that this was it. And in a flash you just knew that this was very fundamental."

Gamow was not the only one whose theories were biting the dust: I had my own share of disappointments. Having gone to Caltech in the immediate aftermath of the double helix, I wanted to find the structure of RNA. To my despair, Alexander Rich (ARG – arginine) and I soon discovered that X-ray diffraction of RNA yielded uninterpretable patterns: the molecule's structure was evidently not as beautifully regular as that of DNA. Equally depressing, in a note sent out early in 1955 to all Tie Club members, Francis Crick (TYR – tyrosine) predicted that the structure of RNA would not, as I supposed, hold the secret of the DNA → protein transformation. Rather, he suggested that amino acids were likely ferried to the actual site of protein synthesis by what he called "adaptor molecules," of which there existed one specific to every amino acid. He speculated that these adaptors themselves might be very small RNA molecules. For two years I resisted his reasoning. Then a most unexpected biochemical finding proved that his novel idea was right on the mark.

It came from work at the Massachusetts General Hospital in Boston, where Paul Zamecnik had for several years been developing cell-free systems for studying protein synthesis. Cells are highly compartmentalized bodies, and Zamecnik correctly saw the need to study what was going on inside them without the complications posed by their various membranes. Using material derived from rat liver tissue, he and his collaborators were able to re-create in a test tube a simplified version of the cell interior in which they could track radioactively tagged amino acids as they were assembled into proteins. In this way Zamecnik was able to identify the ribosome as the site of protein synthesis, a fact that George Gamow did not accept initially.

Soon, with his colleague Mahlon Hoagland, Zamecnik made the even more unexpected discovery that amino acids, prior to being incorporated into polypeptide chains, were bound to small RNA molecules. This result puzzled them until they heard from me of Crick's adaptor theory. They then quickly confirmed Crick's suggestion that a specific RNA adaptor (called transfer RNA) existed for each amino acid. And each of these transfer RNA molecules also had on its surface a specific sequence of bases that permitted it to bind to a corresponding segment of the RNA template, thereby lining up the amino acids for protein synthesis.

Until the discovery of transfer RNA, all cellular RNA was thought to have a template role. Now we realized RNA could come in several different forms, though the two major RNA chains that comprised the ribosomes predominated. Puzzling at the time was the observation that these two RNA chains were of constant sizes. If these chains were the actual templates for protein synthesis, we would have expected them to vary in length in relation to the different sizes of their protein products. Equally disturbing, these chains proved very stable metabolically: once synthesized they did not break down. Yet experiments at the Institut Pasteur in Paris suggested that many templates for bacterial protein synthesis were short-lived. Even stranger, the sequences of the bases in the two ribosomal RNA chains showed no correlation to sequences of bases along the respective chromosomal DNA molecules.

Resolution of these paradoxes came in 1960 with discovery of a third form of RNA, messenger RNA. This was to prove the true template for protein synthesis. Experiments done in my lab at Harvard and at both Caltech and Cambridge by Matt Meselson, François Jacob, and Sydney Brenner showed that ribosomes were, in effect, molecular factories. Messenger RNA passed between the two ribosomal subunits like ticker tape being fed into an old-fashioned computer. Transfer RNAs, each with its amino acid, attached to the messenger RNA in the ribosome so that the amino acids were appropriately ordered before being chemically linked to form polypeptide chains.

Still unclear was the genetic code, the rules for translating a nucleic acid sequence into an ordered polypeptide sequence. In a 1956 RNA Tie Club manuscript, Sydney Brenner laid out the theoretical issues. In essence they boiled down to this: how could the code specify which one of 20 amino acids was to be incorporated into a protein chain at a particular point when there are only four DNA letters, A, T, G, C? Obviously a single nucleotide, with only four possible identities, was insufficient, and even two – which would allow for 16 (4×4) possible permutations – wouldn't work. It would take at minimum three nucleotides, a triplet, to code for a single amino acid. But this also supposed a puzzling redundant capacity. With a triplet, there could exist 64 permutations ($4 \times 4 \times 4$); since the code needed only 20, was it the case that most amino acids could be encoded by more than one triplet? If that were so, in principle, a "quadruplet" code ($4 \times 4 \times 4 \times 4$) yielding 256 permutations was also perfectly feasible, though it implied even greater redundancy.

In 1961 at Cambridge University, Brenner and Crick did the definitive experiment that demonstrated that the code was triplet-based. By a clever use of chemical mutagens they were able to delete or insert DNA base pairs. They found that inserting or deleting a single base pair results in a harmful "frameshift" because the entire code beyond the site of the mutation is scrambled. Imagine a three-letter word code as follows: JIM ATE THE FAT CAT. Now imagine that the first "T" is deleted. If we are to preserve the three-letter word structure of the sentence, we have JIM AET HEF ATC AT – gibberish beyond the site of the deletion. The same thing happens when two base pairs are deleted or inserted: removing the first "T" and "E," we get JIM ATH EFA TCA T – more gibberish. Now what happens if we delete (or insert) *three* letters? Removing the first "A," "T," and "E," we get JIM THE FAT CAT; although we have lost one "word" – ATE – we have nevertheless retained the sense of the rest of the sentence. And even if our deletion straddles "words" – say we delete the first "T" and "E," and the second "T" – we still lose only those two words, and are again able to recover the intended sentence beyond them: JIM AHE FAT CAT. So it is with DNA sequence: a single insertion/deletion massively disrupts the protein because of the frameshift effect, which changes every single amino acid beyond the insertion/deletion point; so does a double insertion/deletion. But a triple insertion/deletion along a DNA molecule will not necessarily have a catastrophic effect; they will add/eliminate one amino acid but this does not necessarily disrupt all biological activity.

Crick came into the lab late one night with his colleague Leslie Barnett to check on the final result of the triple-deletion experiment, and realized at once the significance of the result, telling Barnett, "We're the only two who know it's a triplet code!" With me, Crick had been the first to glimpse the double helical secret of life; now he was the first to know for sure that the secret is written in three-letter words.

So the code came in threes, and the links from DNA to protein were RNA-mediated. But we still had to crack the code. What pair of amino acids was specified by a stretch of DNA with, say, sequence ATA TAT or GGT CAT? The first glimpse of the solution came in a talk given by Marshall Nirenberg at the International Congress of Biochemistry in Moscow in 1961.

After hearing about the discovery of messenger RNA, Nirenberg, working at the U.S. National Institutes of Health, wondered whether RNA synthesized in vitro would work as well as the naturally occurring messenger form when it came to protein synthesis in cell-free systems. To find out, he used RNA tailored according to procedures developed at New York University six years earlier by the French biochemist Marianne Grunberg-Manago. She had discovered an RNA-specific enzyme that could produce strings like AAAAAA or GGGGGG. And because one key chemical difference between RNA and DNA is RNA's substitution of uracil, "U," for thymine, "T," this enzyme would also produce strings of U, UUUUU . . . – poly-U, in the biochemical jargon. It was poly-U that Nirenberg and his German collaborator, Heinrich Matthaei, added to their cell-free system on May 22, 1961. The result was striking: the ribosomes started to pump out a simple protein, one consisting of a string of a single amino acid, phenylalanine. They had discovered that poly-U encodes polyphenylalanine. Therefore, one of the three-letter words by which the genetic code specified phenylalanine had to be UUU.

The International Congress that summer of 1961 brought together all the major players in molecular biology. Nirenberg, then a young scientist nobody had heard of, was slated to speak for just ten minutes, and hardly anyone, including myself, attended his talk. But when news of his bombshell began to spread, Crick promptly inserted him into a later session of the conference so that Nirenberg could make his announcement to a now-expectant capacity audience. It was an extraordinary moment. A quiet, self-effacing young no-name speaking before a who's who crowd of molecular biology had shown the way toward finding the complete genetic code.

Practically speaking, Nirenberg and Matthaei had solved but one sixty-fourth of the problem – all we now knew was that UUU codes for phenylalanine. There remained sixty-three other three-letter triplets (codons) to figure out, and the following years would see a frenzy of research as we labored to discover what amino acids these other codons represented. The tricky part was synthesizing the various permutations of RNA: poly-U was relatively straightforward to produce, but what about AGG? A lot of ingenious chemistry went into solving these problems, much of it done at the University of Wisconsin by Gobind Khorana. By 1966, what each of the sixty-four codons specifies (in other words, the genetic code itself) had been established (see Plate 22); Khorana and Nirenberg received the Nobel Prize for Physiology or Medicine in 1968.

Let's now put the whole story together and look at how a particular protein, hemoglobin, is produced.

Red blood cells are specialized as oxygen transporters: they use hemoglobin to transport oxygen from the lungs to the tissues where it is needed. Red blood cells are produced in the bone marrow by stem cells – at a rate of about two and a half million per second.

When the need arises to produce hemoglobin, the relevant segment of the bone-marrow DNA – the hemoglobin gene – unzips just as DNA unzips when it is replicating (see Plate 23). This time, instead of copying both strands, only one is copied or, to use the technical term, transcribed; and rather than a new strand of DNA, the product created with the help of the enzyme RNA polymerase is a new single strand of messenger RNA, which corresponds to the hemoglobin gene. The DNA from which the RNA has been derived now zips itself up again.

The messenger RNA is transported out of the nucleus and delivered to a ribosome, itself composed of RNA and proteins, where the information in the sequence of the messenger RNA will be used to generate a new protein molecule. This process is known as translation. Amino acids are delivered to the scene attached to transfer RNA. At one end of the transfer RNA is a particular triplet (in the case given in the diagram, CAA) that recognizes its opposite corresponding triplet in the messenger RNA, GUU. At its other end the transfer RNA is towing its matching amino acid, in this case valine. At the next triplet along the messenger RNA, because the DNA sequence is TTC (which specifies lysine), we have a lysine transfer RNA. All that remains now is to glue the two amino acids together biochemically. Do that 100 times, and you have a protein chain 100 amino acids long; the order of the amino acids has been specified by the order of As, Ts, Gs, and Cs in the DNA from which the messenger RNA was created. The two kinds of hemoglobin chains are 141 and 146 amino acids in length.

Proteins, however, are more than just linear chains of amino acids. Once the chain has been made, proteins fold into complex configurations, sometimes by themselves, sometimes assisted by "helper" molecules. It is only once they assume this configuration that they become biologically active. In the case of hemoglobin, it takes four chains, two of one kind and two of a slightly different kind, before the molecule is in business. And loaded into the center of each twisted chain is the key to oxygen transport, an iron atom.

It has been possible to use today's molecular biological tricks to go back and reconsider some of the classic examples of early genetics. For Mendel, the mechanism that caused some peas to be wrinkled and others round was mysterious; as far as he was concerned, these were merely characteristics that obeyed the laws of inheritance he had worked out. Now, however, we understand the difference in molecular detail.

In 1990, scientists in England found that wrinkled peas lack a certain enzyme involved in the processing of starch, the carbohydrate that is stored in seeds. It turns out that the gene for that enzyme in wrinkled-pea plants is nonfunctional owing to a mutation (in this case an intrusion of irrelevant DNA into the middle of the gene). Because wrinkled peas contain, as a result of this mutation, less starch and more sugar, they tend to lose more water as they are maturing. The outside seed coat of the pea, however, fails to shrink as the water escapes (and the volume of the pea decreases), and the result is the characteristic wrinkling – the contents being too little to fill out the coat.

Archibald Garrod's alkaptonuria has also entered the molecular era. In 1995, Spanish scientists working with fungi found a mutated gene that resulted in the accumulation of the same substance that Garrod had noted in the urine of alkaptonurics. The gene in question ordinarily produces an enzyme that turns out to be a basic feature of many living systems, and is present in humans. By comparing the sequence of the fungal gene to human sequences, it was possible to find the human gene, which encodes an enzyme called homogentisate dioxygenase. The next step was to compare the gene in normal individuals with the one in alkaptonurics. Lo and behold, the alkaptonurics' gene was nonfunctional, courtesy of single base pair mutations. Garrod's "inborn error in metabolism" is caused by a single difference in DNA sequence.

At the 1966 Cold Spring Harbor Symposium on the genetic code, there was a sense that we had done it all. The code was cracked, and we knew in outline how DNA exerted control of living processes through the proteins it specifies. Some of the old hands decided that it was time to move beyond the study of the gene per se. Francis Crick decided to move into neurobiology; never one to shy away from big problems, he was particularly interested in figuring out how the human brain works. Sydney Brenner turned to developmental biology, choosing to concentrate on a simple nematode worm in the belief that precisely so simple a creature would most readily permit scientists to unravel the connections between genes and development. Today, the worm, as it is known in the trade, is indeed the source of many of our insights into how organisms are put together. The worm's contribution was recognized by the Nobel Committee in 2002 when Brenner and two longstanding worm stalwarts, John Sulston at Cambridge and Bob Horvitz at MIT, were awarded the Nobel Prize in Physiology or Medicine.

Most of the early pioneers in the DNA game, however, chose to remain focused on the basic mechanisms of gene function. Why are some proteins much more abundant than others? Many genes are switched on only in specific cells or only at particular times in the life of a cell; how is that switching achieved? A muscle cell is hugely different from a liver cell, both in its function and in its appearance under the microscope. Changes in gene expression create this cellular diversity and differentiation: in essence, muscle cells and liver cells produce different sets of proteins. The simplest way to produce different proteins is to regulate which genes are transcribed in each cell. Thus some so-called housekeeping proteins – the ones essential for the functioning of the cell, such as those involved in the replication of DNA – are produced by all cells. Beyond that, particular genes are switched on at particular moments in particular cells to produce appropriate proteins. It is also possible to think of development – the process of growth from a single fertilized egg into a staggeringly complex adult human – as an enormous exercise in gene-switching: as tissues arise through development, so whole suites of genes must be switched on and off.

The first important advances in our understanding of how genes are switched on and off came from experiments in the 1960s by François Jacob and Jacques Monod at the Institut Pasteur in Paris. Monod had started slowly in science because, poor fellow, he was talented in so many fields that he had difficulty focusing. During the thirties, he spent time at Caltech's biology department under T. H. Morgan, father of fruit fly genetics, but not even daily exposure to Morgan's no-longer-so-boyish "boys" could turn Monod into a fruit fly convert. He preferred conducting Bach concerts at the university – which later offered him a job teaching undergraduate music appreciation – and in the lavish homes of local millionaires. Not until 1940 did he complete his Ph.D. at the Sorbonne in Paris, by which time he was already heavily involved in the French Resistance. In one of the few instances of biology's complicity in espionage, Monod was able to conceal vital secret papers in the hollow leg bones of a giraffe skeleton on display outside his lab. As the war progressed, so did his importance to the Resistance (and with it his vulnerability to the Nazis). By D-day he was playing a major role in facilitating the Allied advance and harrying the German retreat.

Jacob too was involved in the war effort, having escaped to Britain and joined General de Gaulle's Free French Army. He served in North Africa and participated in the D-day landings. Shortly thereafter, he was nearly killed by a bomb; twenty pieces of shrapnel were removed, but he retains to this day another eighty. Because his arm was damaged, his injuries ended his ambition to be a surgeon, and, inspired like so many of our generation by Schrödinger's *What Is Life?*, he drifted toward biology. His attempts to join Monod's research group were, however, repeatedly rebuffed. But after seven or eight tries, by Jacob's own count, Monod's boss, the microbiologist André Lwoff, caved in in June 1950 (see Plate 24):

> Without giving me a chance to explain anew my wishes, my ignorance, my eagerness, [Lwoff] announced, "You know, we have discovered the induction of the prophage!" [i.e., how to activate bacteriophage DNA that has been incorporated into the host bacterium's DNA].

I said, "Oh!" putting into it all the admiration I could and thinking to myself, "What the devil is a prophage?"

Then he asked, "Would it interest you to work on phage?" I stammered out that that was exactly what I had hoped. "Good; come along on the first of September."

Jacob apparently went straight from the interview to a bookshop to find a dictionary that might tell him what he had just committed himself to.

Despite its inauspicious beginnings, the Jacob-Monod collaboration produced science of the very highest caliber. They tackled the gene-switching problem in *E. coli,* the familiar intestinal bacterium, focusing on its ability to make use of lactose, a kind of sugar. In order to digest lactose, the bacterium produces an enzyme called beta-galactosidase, which breaks the nutrient into two subunits, simpler sugars called galactose and glucose. When lactose is absent in the bacterial medium, the cell produces no beta-galactosidase; when, however, lactose is introduced, the cell starts to produce the enzyme. Concluding that it is the presence of lactose that induces the production of beta-galactosidase, Jacob and Monod set about discovering how that induction occurs.

In a series of elegant experiments, they found evidence of a "repressor" molecule that, in the absence of lactose, prevents the transcription of the beta-galactosidase gene. When, however, lactose is present, it binds to the repressor, thereby keeping it from blocking the transcription; thus the presence of lactose enables the transcription of the gene. In fact, Jacob and Monod found that lactose metabolism is coordinately controlled: it is not simply a matter of one gene being switched on or off at a given time. Other genes participate in digesting lactose, and the single repressor system serves to regulate all of them. While *E. coli* is a relatively simple system in which to investigate gene-switching, subsequent work on more complicated organisms, including humans, has revealed that the same basic principles apply across the board.

Jacob and Monod obtained their results by studying mutant strains of *E. coli*. They had no direct evidence of a repressor molecule: its existence was merely a logical inference from their solution to the genetic puzzle. Their ideas were not validated in the molecular realm until the late sixties, when Walter (Wally) Gilbert and Benno Müller-Hill at Harvard set out to isolate and analyze the repressor molecule itself. Jacob and Monod had only predicted its existence; Gilbert and Müller-Hill actually found it. Because the repressor is normally present only in tiny amounts, just a few molecules per cell, gathering a sample large enough to analyze proved technically challenging. But they got it in the end. At the same time, Mark Ptashne, working down the hall in another lab, managed to isolate and characterize another repressor molecule, this one in a bacteriophage gene-switching system. Repressor molecules turn out to be proteins that can bind to DNA. In the absence of lactose, then, that is exactly what the beta-galactosidase repressor does: by binding to a site on the *E. coli* DNA close to the point at which transcription of the beta-galactosidase gene starts, the repressor prevents the enzyme that produces messenger RNA from the gene from doing its job. When, however, lactose is introduced, that sugar binds to the repressor, preventing it from occupying the site on the DNA molecule close to the beta-galactosidase gene; transcription is then free to proceed.

The characterization of the repressor molecule completed a loop in our understanding of the molecular processes underpinning life. We knew that DNA produces protein via RNA; now we also knew that protein could interact directly with DNA, in the form of DNA-binding proteins, to regulate a gene's activity.

The discovery of the central role of RNA in the cell raised an interesting (and long-unanswered) question: why does the information in DNA need to go through an RNA intermediate before it can be translated into a polypeptide sequence? Shortly after the genetic code was worked out, Francis Crick proposed a solution to this paradox, suggesting that RNA predated DNA. He imagined RNA to have been the first genetic molecule, at a time when life was RNA-based: there would have been an "RNA world" prior the familiar "DNA world" of today (and of the past few billion years). Crick imagined that the different chemistry of RNA (based on its possession of the sugar ribose in its backbone, rather than the deoxyribose of DNA) might endow it with enzymatic properties that would permit it to catalyze its own self-replication.

Crick argued that DNA had to be a later development, probably in response to the relative instability of RNA molecules, which degrade and mutate much more easily than DNA molecules. If you want a good stable, long-term storage molecule for genetic data, then DNA is a much better bet than RNA.

Crick's ideas about an RNA world preceding the DNA one went largely unnoticed until 1983. That's when Tom Cech at the University of Colorado and Sidney Altman at Yale independently showed that RNA molecules do indeed have catalytic properties, a discovery that earned them the Nobel Prize in Chemistry in 1989. Even more compelling evidence of a pre-DNA RNA world came a decade later, when Harry Noller at the University of California, Santa Cruz, showed that the formation of peptide bonds, which link amino acids together in proteins, is not catalyzed by any of the sixty different proteins found associated with the ribosome, the site of protein synthesis. Instead, peptide bond formation is catalyzed by RNA. He arrived at this conclusion by stripping away all the proteins from the ribosome and finding that it was still capable of forming peptide bonds.

Exquisitely detailed analysis of the 3-D structure of the ribosome by Noller and others shows why: the proteins are scattered over the surface, far from the scene of action at the heart of the ribosome (see Plate 25).

These discoveries inadvertently resolved the chicken-and-egg problem of the origin of life. The prevailing assumption that the original life-form consisted of a DNA molecule posed an inescapable contradiction: DNA cannot assemble itself; it requires proteins to do so. Which came first? Proteins, which have no known means of duplicating information, or DNA, which can duplicate information but only in the presence of proteins? The problem was insoluble: you cannot, we thought, have DNA without proteins, and you cannot have proteins without DNA.

RNA, however, being a DNA equivalent (it can store and replicate genetic information) as well as a protein equivalent (it can catalyze critical chemical reactions) offers an answer. In fact, in the "RNA world" the chicken-and-egg problem simply disappears. RNA is both the chicken and the egg.

RNA is an evolutionary heirloom. Once natural selection has solved a problem, it tends to stick with that solution, in effect following the maxim "If it ain't broke, don't fix it." In other words, in the absence of selective pressure to change, cellular systems do not innovate and so bear many imprints of the evolutionary past. A process may be carried out in a certain way simply because it first evolved that way, not because that is absolutely the best and most efficient way.

Molecular biology had come a long way in its first twenty years after the discovery of the double helix. We understood the basic machinery of life, and we even had a grasp on how genes are regulated. But all we had been doing so far was observing; we were molecular naturalists for whom the rain forest was the cell – all we could do was describe what was there. The time had come to become proactive. Enough observation: we were beckoned by the prospect of intervention, of manipulating living things. The advent of recombinant DNA technologies, and with them the ability to tailor DNA molecules, would make all this possible.

CHAPTER FOUR

PLAYING GOD:
CUSTOMIZED DNA MOLECULES

DNA molecules are immensely long. Only one continuous DNA double helix is present in any given chromosome. Popular commentators like to evoke the vastness of these molecules through comparisons to the number of entries in the New York City phone book or the length of the River Danube. Such comparisons don't help me – I have no sense of how many phone numbers there are in New York City, and mention of the Danube more readily suggests a Strauss waltz than any sense of linear distance.

Except for the sex chromosomes, X and Y, the human chromosomes are numbered according to size. Chromosome 1 is the largest and chromosomes 21 and 22 are the smallest. In chromosome 1 there resides 8 percent of each cell's total DNA, about a quarter of a billion base pairs. Chromosomes 21 and 22 contain some 40 and 45 million base pairs respectively. Even the smallest DNA molecules, those from small viruses, have no fewer than several thousand base pairs.

The great size of DNA molecules posed a big problem in the early days of molecular biology. To come to grips with a particular gene – a particular stretch of DNA – we would have to devise some way of isolating it from all the rest of the DNA that sprawled around it in either direction. But it was not only a matter of isolating the gene; we also needed some way of "amplifying" it: obtaining a large enough sample of it to work with. In essence we needed a molecular editing system: a pair of molecular scissors that could cut the DNA text into manageable sections; a kind of molecular glue pot that would allow us to manipulate those pieces; and finally a molecular duplicating machine to amplify the pieces that we had cut out and isolated. We wanted to do the equivalent of what a word processor can now achieve: to cut, paste, and copy DNA.

Developing the basic tools to perform these procedures seemed a tall order even after we cracked the genetic code. A number of discoveries made in the late sixties and early seventies, however, serendipitously came together in 1973 to give us so-called "recombinant DNA" technology – the capacity to edit DNA. This was no ordinary advance in lab techniques. Scientists were suddenly able to tailor DNA molecules, creating ones that had never before been seen in nature. We could "play God" with the molecular underpinning of all of life. This was an unsettling idea to many people. Jeremy Rifkin, an alarmist for whom every new genetic technology has about it the whiff of Dr. Frankenstein's monster, had it right when he remarked that recombinant DNA "rivaled the importance of the discovery of fire itself."

Arthur Kornberg was the first to "make life" in a test tube. In the 1950s, as we have seen, he discovered DNA polymerase, the enzyme that replicates DNA through the formation of a complementary copy from an unzipped "parent" strand. Later he would work with a form of viral DNA; he was ultimately able to induce the replication of all of the virus's 5,300 base pairs of DNA. But the product was not "alive"; though identical in DNA sequence to its parent, it was biologically inert. Something was missing. The missing ingredient would remain a mystery until 1967, when Martin Gellert at the National Institutes of Health and Bob Lehman at Stanford simultaneously identified it. This enzyme was named "ligase." Ligase made it possible to "glue" the ends of DNA molecules together.

Kornberg could replicate the viral DNA using DNA polymerase and, by adding ligase, join the two ends together so that the entire molecule formed a continuous loop, just as it did in the original virus. Now the "artificial" viral DNA behaved exactly as the natural one did: the virus normally multiplies in *E. coli,* and Kornberg's test-tube DNA molecule did just that. Using just a couple of enzymes, some basic chemical ingredients, and viral DNA from which to make the copy, Kornberg had made a biologically active molecule. The media reported that he had created life in a test tube, inspiring President Lyndon Johnson to hail the breakthrough as an "awesome achievement."

The contributions of Werner Arber in the 1960s to the development of recombinant DNA technology were less expected. Arber, a Swiss biochemist, was interested not in grand questions about the molecular basis of life but in a puzzling aspect of the natural history of viruses. He studied the process whereby some viral DNAs are broken down after insertion into bacterial host cells. Some, but not all (otherwise viruses could not reproduce), host cells recognized certain viral DNAs as foreign, and selectively attacked them. But how – and why? All DNA throughout the natural world is the same basic molecule, whether found in bacteria, viruses, plants, or animals. What kept the bacteria from attacking their own DNA even as they went after the virus's?

The first answer came from Arber's discovery of a new group of DNA-degrading enzymes, restriction enzymes. Their presence in bacterial cells *restricts* viral growth by cutting foreign DNA. This DNA-cutting is a sequence-specific reaction: a given enzyme will cut DNA only when it recognizes a particular sequence. *Eco*R1, one of the first restriction enzymes to be discovered, recognizes and cuts the specific sequence of bases GAATTC.

But why is it that bacteria do not end up cutting up their own DNA in every place where the sequence GAATTC appears? Here Arber made a second big discovery. While making the restriction enzyme that targets specific sequences, the bacterium also produces a second enzyme that chemically modifies those very same sequences in its own DNA wherever they may occur.* Modified GAATTC sequences present in the bacterial DNA will pass unrecognized by *Eco*R1, even as the enzyme goes its marauding way, snipping the sequence wherever it occurs in the viral DNA.

The next ingredient of the recombinant DNA revolution emerged from studies of antibiotic resistance in bacteria. During the sixties, it was discovered that many bacteria developed resistance to an antibiotic not in the standard way (through a mutation in the bacterial genome) but by the import of an otherwise extraneous piece of DNA, called a "plasmid." Plasmids are small loops of DNA that live within bacteria (see Plate 26) and are replicated and passed on, along with the rest of the bacterial genome, during cell division. Under certain circumstances plasmids may also be passed from bacterium to bacterium, allowing the recipient instantly to acquire a whole cassette of genetic information it did not receive "at birth." That information

*The enzyme achieves this chemical modification by adding methyl groups, CH_3, to the bases.

often encompasses the genes conferring antibiotic resistance. Natural selection imposed by antibiotics favors those bacterial cells that have the resistance factor (the plasmid) on board.

Stanley Cohen, at Stanford University, was a plasmid pioneer. Thanks to the encouragement of his high-school biology teacher, Cohen opted for a medical career. Upon graduation from medical school, his plans to practice internal medicine were shelved when the prospect of being drafted as an army doctor inspired him to accept a research position at the National Institutes of Health. He soon found that he preferred research over practicing medicine. His big breakthrough came in 1971, when he devised a method to induce *E. coli* bacterial cells to import plasmids from outside the cell. Cohen was, in effect, "transforming" the *E. coli* as Fred Griffith, forty years before, had converted strains of nonlethal pneumonia bacteria into lethal ones through the uptake of DNA. In Cohen's case, however, it was the plasmid, with its antibiotic resistance genes, that was taken up by a strain that had previously been susceptible to the antibiotic. The strain would remain resistant to the antibiotic over subsequent generations, with copies of the plasmid DNA passed along intact during every cell division.

By the early seventies, all the ingredients to make recombinant DNA were in place. First we could cut DNA molecules using restriction enzymes and isolate the sequences (genes) we were interested in; then, using ligase, we could "glue" that sequence into a plasmid (which would thus serve as a kind of floppy disk containing our desired sequence); finally, we could copy our piece of DNA by inserting that same plasmid floppy into a bacterial cell. Ordinary bacterial cell division would take care of replicating the plasmid with our piece of DNA just as it would the cell's own inherited genetic materials. Thus, starting with a single plasmid transplanted into a single bacterial cell, bacterial reproduction could produce enormous quantities of our selected DNA sequence. As we let that cell reproduce and reproduce, ultimately to grow into a vast bacterial colony consisting of billions of bacteria, we would be simultaneously creating billions of copies of our piece of DNA. The colony was thus our DNA factory.

The three components – cutting, pasting, and copying – came together in November 1972, in Honolulu. The occasion was a conference on plasmids. Herb Boyer, a newly tenured young professor at the University of California, San Francisco, was there, and, not surprisingly, so was Stanley Cohen, first among plasmid pioneers. Boyer, like Cohen, was an East Coast boy. A former high-school varsity lineman from western Pennsylvania, Boyer was perhaps fortunate that his football coach was also his science teacher. Like Cohen, he would be part of a new generation of scientists who were reared on the double helix. His enthusiasm for DNA even inspired him to name his Siamese cats Watson and Crick. No one, certainly not the coach, was surprised when after college he took up graduate work in bacterial genetics (see Plate 27).

Though Boyer and Cohen both now worked in the San Francisco Bay Area, they had not met before the Hawaii conference. Boyer was already an expert in restriction enzymes in an era when hardly anyone had even heard of them: it was he and his colleagues who had recently figured out the sequence of the cut site of the *Eco*R1 enzyme. Boyer and Cohen soon realized that between them they had the skills to push molecular biology to a whole new level, the world of cut, paste, and copy. In a deli near Waikiki, they set about late one evening dreaming up the birth of recombinant DNA technology, jotting their ideas down on napkins. That visionary mapping of the future has been described as "from corned beef to cloning."

Within a few months, Boyer's lab in San Francisco and Cohen's forty miles to the south in Palo Alto were collaborating. Naturally Boyer's carried out the restriction enzyme work and Cohen's the plasmid procedures. Fortuitously a technician in Cohen's lab, Annie Chang, lived in San Francisco and was able to ferry the precious cargo of experiments in progress between the two sites. The first experiment intended to make a hybrid, "a recombinant," of two different plasmids, each of which was known to confer resistance to a particular antibiotic. On one plasmid there was a gene, a stretch of DNA, for resistance to tetracycline, and on the other a gene for resistance to kanamycin. (Initially, as we might expect, bacteria carrying the first type of plasmid were killed by kanamycin while those with the second were killed by tetracycline.) The goal was to make a single "super-plasmid" that would confer resistance to both.

First, the two types of unaltered plasmid were snipped with restriction enzymes (see Plate 28). Next the plasmids were mixed in the same test tube and ligase added to prompt the snipped ends to glue themselves together. For some molecules in the mix, the ligase would merely cause a snipped plasmid to make itself whole again – the two ends of the same plasmid would have been glued together. Sometimes, however, the ligase would cause a snipped plasmid to incorporate pieces of DNA from the other type of plasmid, thus yielding the desired hybrid. With this accomplished, the next step was to transplant all the plasmids into bacteria by using Cohen's plasmid-importing tricks. Colonies thus generated were then cultured on plates coated with both tetracycline and kanamycin. Plasmids that had simply re-formed would still confer resistance to only one of the antibiotics; bacteria carrying such plasmids would therefore not survive on the double-antibiotic medium. The only bacteria to survive were those with recombinant plasmids – those that had reassembled themselves from the two kinds of DNA present, the one coding for tetracycline resistance *and* the one coding for resistance to kanamycin.

The next challenge lay in creating a hybrid plasmid using DNA from a completely different sort of organism – a human being, for example. An early successful experiment involved putting a gene from the African clawed toad into an *E. coli* plasmid and transplanting that into bacteria. Every time cells in the bacterial colony divided, they duplicated the inserted segment of toad DNA. We had, in the rather confusing terminology of molecular biology, "cloned" the toad DNA.* Mammal DNA, too, proved eminently clonable. This is not terribly surprising, in retrospect: a piece of DNA after all is finally still DNA, its chemical properties the same irrespective of its source. It was soon clear that Cohen and Boyer's protocols for cloning fragments of plasmid DNA would work just fine with DNA from any and every creature.

Phase 2 of the molecular biology revolution was thus under way. In phase 1 we aimed to describe how DNA works in the cell; now, with recombinant DNA,† we had the tools to intervene, to manipulate DNA. The stage was set for rapid progress, as we spied the chance to "play God." It was intoxicating: the extraordinary potential for delving deep into the mysteries of life and the opportunities for making real progress in the fight against diseases like cancer. But while Cohen and Boyer may indeed have opened our eyes to extraordinary

---

*"Cloning" is the term applied to producing multiple identical pieces of a piece of DNA inserted into a bacterial cell. The term is confusingly also applied to the cloning of whole animals, most notably Dolly the sheep. In the first type we are copying just a piece of DNA; in the other, we are copying an entire genome.

†The term "recombinant DNA" may present a little confusion in light of our encounter with "recombination" in the context of classical genetics. In Mendelian genetics, recombination involved the breaking and re-forming of chromosomes, with the result of a "mixing and matching" of chromosomal segments. In the molecular version, "mixing and matching" occurs on a much smaller scale, *recombining* two stretches of DNA into a single composite molecule.

scientific vistas, had they also opened a Pandora's box? Were there undiscovered perils in molecular cloning? Should we go on cheerfully inserting pieces of human DNA into *E. coli*, a species predominant in the microbial jungle in our guts? What if the altered forms should find their way into our bodies? In short, could we in good conscience simply turn a deaf ear to the cry of the alarmists, that we were creating bacterial Frankensteins?

In 1961 a monkey virus called SV40 ("SV" stands for "simian virus") was isolated from rhesus monkey kidneys being used for the preparation of polio vaccine. Although the virus was believed to have no effect on the monkeys in which it naturally occurs, experiments soon showed that it could cause cancer in rodents and, under certain laboratory conditions, even in human cells. Because the polio vaccination program had, since its inception in 1955, infected millions of American children with the virus, this discovery was alarming indeed. Had the polio prevention program inadvertently condemned a generation to cancer? The answer, fortunately, seems to be "no"; no epidemic of cancer has resulted, and SV40 seems to be no more pernicious in living humans than it is in monkeys. Nevertheless, even as SV40 was becoming a fixture in molecular biology laboratories, there remained doubts about its safety. I was particularly concerned since I was by this time head of the Cold Spring Harbor Laboratory, where growing ranks of young scientists were working with SV40 to probe the genetic basis of cancer.

Meanwhile, at Stanford University Medical School, Paul Berg was more excited by the promise than by the dangers of SV40; he foresaw the possibility of using the virus to introduce pieces of DNA – foreign genes – into mammalian cells. The virus would work as a molecular delivery system in mammals, just as plasmids had been put to work in bacteria by Stanley Cohen. But whereas Cohen used bacteria essentially as copy machines, which could amplify up a particular piece of DNA, Berg saw in SV40 a means to introduce corrective genes into the victims of genetic disease. Berg was ahead of his time. He aspired to carry out what today is called gene therapy: introducing new genetic material into a living person to compensate for inherited genetic flaws.

Berg had come to Stanford as a junior professor in 1959 as part of the package deal that also brought the more eminent Arthur Kornberg there from Washington University in St. Louis. In fact, Berg's connections to Kornberg can be traced all the way back to their common birthplace of Brooklyn, New York, where each in his time was to pass through the same high-school science club run by a Miss Sophie Wolfe. Berg recalled: "She made science fun, she made us share ideas." It was an understatement really: Miss Wolfe's science club at Abraham Lincoln High School would produce three Nobel laureates – Kornberg (1959), Berg (1980), and the crystallographer Jerome Karle (1985) – all of whom have paid tribute to her influence.

While Cohen and Boyer, and by now others, were ironing out the details of how to cut and paste DNA molecules, Berg planned a truly bold experiment: he would see whether SV40, implanted with a piece of DNA not its own, could be made to transport that foreign gene into an animal cell. For convenience he would use as the source of his non-SV40 DNA a readily available bacterial virus, a bacteriophage. The aim was to see whether a composite molecule consisting of SV40 DNA and the bacteriophage DNA could successfully invade an animal cell. If it could, as Berg hoped, then the possibility existed that he could ultimately use this system to insert useful genes into human cells.

At Cold Spring Harbor Laboratory in the summer of 1971, a graduate student of Berg's gave a presentation explaining the planned experiment. One scientist in the audience was alarmed enough to phone Berg straightaway. What if, he asked, things happened to work in reverse? In other words, what if the SV40 virus, rather than taking up the viral DNA and then inserting it into the animal cell, was itself manipulated by the bacteriophage DNA, which might cause the SV40 DNA to be inserted into, say, an *E. coli* bacterial cell? It was not an unrealistic scenario: after all, that is precisely what many bacteriophages are programmed to do – to insert their DNA into bacterial cells. Since *E. coli* is both ubiquitous and intimately associated with humans, as the major component of our gut flora, Berg's well-meaning experiment might result in dangerous colonies of *E. coli* carrying SV40 monkey virus, a potential cancer agent. Berg heeded his colleague's misgivings, though he did not share them: he decided to postpone the experiments until more could be learned about SV40's potential to cause human cancer (see Plate 29).

Biohazard anxieties followed hard on the heels of the news of Boyer and Cohen's success with their recombinant DNA procedures. At a scientific conference on nucleic acids in New Hampshire in the summer of 1973, a majority voted to petition the National Academy of Sciences to investigate without delay the dangers of the new technology. A year later a committee appointed by the National Academy and chaired by Paul Berg published its conclusions in a letter to the journal *Science*. I myself signed the letter, as did many of the others – including Cohen and Boyer – who were most active in the relevant research. In what has since come to be known as the "Moratorium Letter" we called upon "scientists throughout the world" to suspend voluntarily all recombinant studies "until the potential hazards of such recombinant DNA molecules have been better evaluated or until adequate methods are developed for preventing their spread." An important element of this statement was the admission that "our concern is based on judgements of potential rather than demonstrated risk since there are few experimental data on the hazards of such DNA molecules."

All too soon, however, I found myself feeling deeply frustrated and regretful of my involvement in the Moratorium Letter. Molecular cloning had the obvious potential to do a fantastic amount of good in the world, but now, having worked so hard and arrived at the brink of a biological revolution, here we were conspiring to draw back. It was a confusing moment. As Michael Rogers wrote in his 1975 report on the subject for *Rolling Stone*, "The molecular biologists had clearly reached the edge of an experimental precipice that may ultimately prove equal to that faced by nuclear physicists in the years prior to the atom bomb." Were we being prudent or chickenhearted? I couldn't quite tell yet, but I was beginning to feel it was the latter.

The "Pandora's Box Congress": that's how Rogers described the February 1975 meeting of 140 scientists from around the world at the Asilomar conference center in Pacific Grove, California. The agenda was to determine once and for all whether recombinant DNA really held more peril than promise. Should the moratorium be permanent? Should we press ahead regardless of potential risk, or wait for the development of certain safeguards? As chair of the organizing committee, Paul Berg was also nominal head of the conference, and so had the almost impossible task of drafting a consensus statement by the end of the meeting.

The press was there, scratching its collective head as scientists bandied about the latest jargon. The lawyers were there, too, just to remind us that there were also legal issues to be addressed: for example, would I, as head of a lab doing recombinant research, be liable if a technician of mine developed cancer? As to the scientists, they were by nature and training averse to hazarding predictions in the absence of knowledge; they rightly suspected that it would be impossible to reach a unanimous decision. Perhaps Berg was equally doubtful; in any case, he opted for freedom of expression over firm leadership from the chair. The resulting debate was therefore something of a free-for-all, with the proceedings not infrequently derailed by some speaker intent only on rambling irrelevantly and at length about the important work going on in his or her lab. Opinions ranged wildly, from the timid – "prolong the moratorium" – to the gung ho – "the moratorium be damned, let's get on with the science." I was definitely on the latter end of the spectrum. I now felt that it was more irresponsible to defer research on the basis of unknown and unquantified dangers. There were desperately sick people out there, people with cancer or cystic fibrosis – what gave us the right to deny them perhaps their only hope?

Sydney Brenner, then based in the United Kingdom, at Cambridge, offered one of the very few pieces of relevant data. He had collected colonies of the *E. coli* strain known as K-12, the favorite bacterial workhorse for this kind of molecular cloning research. Particular rare strains of *E. coli* occasionally cause outbreaks of food poisoning, but in fact the vast majority of *E. coli* strains are harmless, and Brenner assumed that K-12 was no exception. What interested him was not his own health but K-12's: could it survive outside the laboratory? He stirred the microbes into a glass of milk (they were rather unpalatable served up straight), and went on to quaff the vile mixture. He monitored what came out the other end to see whether any K-12 cells had managed to colonize his intestine. His finding was negative, suggesting that K-12, despite thriving in a petri dish, was not viable in the "natural" world. Still, others questioned the inference: even if the K-12 bacteria were themselves unable to survive, this was no proof they could not exchange plasmids – or other genetic information – with strains that could live perfectly well in our guts. Thus "genetically engineered" genes could still enter the population of intestine-dwelling bacteria. Brenner then championed the idea that we should develop a K-12 strain that was without question incapable of living outside the laboratory. We could do this by a genetic alteration that would ensure the strain could grow only when supplied with specialized nutrients. And of course we would specify a set of nutrients that would never be available in the natural world; the full complement of nutrients would occur together only in the lab. A K-12 thus modified would be a "safe" bacterium, viable in our controlled research setting, but doomed in the real world.

With Brenner's urging, this middle-ground proposal carried the day. There was plenty of grumbling from both extremes, of course, but the conference ended with coherent recommendations allowing research to continue on disabled, non-disease-causing bacteria and mandating expensive containment facilities for work involving the DNA of mammals. These recommendations would form the basis for a set of guidelines issued a year later by the National Institutes of Health.

I departed feeling despondent, isolated from most of my peers. Stanley Cohen and Herb Boyer found the occasion disheartening as well; they believed, as I did, that many of our colleagues had compromised their better judgment as scientists just to be seen by the assembled press as "good guys" (and not as potential Dr. Frankensteins). In fact, the vast majority had never worked with disease-causing organisms and little understood the implications of the research restrictions they wanted to impose on those of us who did. I was irked by the arbitrariness of much of what had been agreed: DNA from cold-blooded vertebrates was, for instance, deemed acceptable, while mammalian DNA was ruled off-limits for most scientists. Apparently it was safe to work with DNA from a toad but not with DNA from a mouse. Dumbstruck by such nonsense, I offered up a bit of my own: didn't everyone know that toads cause warts? But my facetious objections were in vain.

The guidelines led many participants in the Asilomar conference to expect clear sailing for research based on cloning in "safe bacteria." But anyone who set off under such an impression very soon hit choppy seas. According to the logic peddled by the popular press, if scientists themselves saw cause for concern, then the public at large should *really* be alarmed. These were, after all, still the days, though waning, of the American counterculture. Both the Vietnam War and Richard Nixon's political career had only recently petered out; a suspicious public, ill-equipped to understand complexities that science itself was only beginning to fathom, was only too eager to swallow theories of evil conspiracies perpetrated by the Establishment. For our part, we scientists were quite surprised to see ourselves counted among this elite, to which we had never before imagined we belonged. Even Herb Boyer, the veritable model of a hippie scientist, would find himself named in the special Halloween issue of the *Berkeley Barb,* the Bay Area's underground paper, as one of the region's "ten biggest bogeymen," a distinction otherwise reserved for corrupt pols and union-busting capitalists.

My greatest fear was that this blooming public paranoia about molecular biology would result in draconian legislation. Having experimental dos and don'ts laid down for us in some cumbersome legalese could only be bad for science. Plans for experiments would have to be submitted to politically minded review panels, and the whole hopeless bureaucracy that comes with this kind of territory would take hold like the moths in Grandmother's closet. Meanwhile, our best attempts to assess the real risk potential of our work continued to be dogged by a complete lack of data and by the logical difficulty of proving a negative. No recombinant DNA catastrophe had ever occurred, but the press continued to outdo itself imagining "worst case scenarios." In his account of a meeting in Washington, D.C., in 1977, the biochemist Leon Heppel aptly summed up the absurdities scientists perceived in the controversy.

> I felt the way I would feel if I had been selected for an *ad hoc* committee convened by the Spanish Government to try to evaluate the risks assumed by Christopher Columbus and his sailors, a committee that was supposed to set up guidelines for what to do in case the earth was flat, how far the crew might safely venture to the earth's edge, etc.

Even withering irony, however, could little hinder those hell-bent on countering what they saw as science's Promethean hubris. One such crusader was Alfred Vellucci, the mayor of Cambridge, Massachusetts. Vellucci had earned his political chops championing the common man at the expense of his town's elite institutions of learning, namely, MIT and Harvard. The recombinant DNA tempest provided him with a political bonanza. A contemporary account captures nicely what was going on.

> In his cranberry doubleknit jacket and black pants, with his yellow-striped blue shirt struggling to contain a beer belly, right down to his crooked teeth and overstuffed pockets, Al Vellucci is the incarnation of middle-American frustration at these scientists, these technocrats, these smartass Harvard eggheads who think they've got the world by a string and wind up dropping it in a puddle of mud. And who winds up in the puddle? Not the eggheads. No, it's always Al Vellucci and the ordinary working people who are left alone to wipe themselves off.

Whence this heat? Scientists at Harvard had voiced a desire to build an on-campus containment facility for doing recombinant work in strict accordance with the new NIH guidelines. But, seeing his chance and backed by a left-wing Harvard-MIT cabal with its own anti-DNA agenda, Vellucci managed to push through a several months' ban on all recombinant DNA research in Cambridge. The result was a brief but pronounced local brain drain, as Harvard and MIT biologists headed off to less politically charged climes. Vellucci, meanwhile, began to enjoy his newfound prominence as society's scientific watchdog. In 1977 he would write to the president of the National Academy of Sciences:

> In today's edition of the *Boston Herald American*, a Hearst Publication, there are two reports which concern me greatly. In Dover, MA, a "strange, orange-eyed creature" was sighted and in Hollis, New Hampshire, a man and his two sons were confronted by a "hairy, nine foot creature."
>
> I would respectfully ask that your prestigious institution investigate these findings. I would hope as well that you might check to see whether or not these "strange creatures" (should they in fact exist), are in any way connected to recombinant DNA experiments taking place in the New England area.

Though much debated, attempts to enact national legislation regulating recombinant DNA experiments fortunately never came to fruition. Senator Ted Kennedy of Massachusetts entered the fray early on, holding a Senate hearing just a month after Asilomar. In 1976, he wrote President Ford to advise that the federal government should control industrial as well as academic DNA research. In March of '77, I testified before a hearing of the California state legislature. Governor Jerry Brown was in attendance, and so I had the occasion to advise him in person that it would be a mistake to consider any legislative action except in the event of unexplained illnesses among the scientists at Stanford. If those actually handling recombinant DNA remained perfectly healthy, the public would be better served if lawmakers focused on more evident dangers to public health, like bike riding.

As more and more experiments were performed, whether under NIH guidelines or under those imposed by regulators in other countries, it became more and more apparent that recombinant DNA procedures were not creating Frankenbugs (much less – *pace* Mr. Vellucci – "strange orange-eyed creatures"). By 1978 I could write, "Compared to almost any other object that starts with the letter D, DNA is very safe indeed. Far better to worry about daggers, dynamite, dogs, dieldrin, dioxin, or drunken drivers than to draw up Rube Goldberg schemes on how our laboratory-made DNA will lead to the extinction of the human race."

Later that year, in Washington, D.C., the Recombinant DNA Advisory Committee (RAC) of the NIH proposed much less restrictive guidelines that would permit most recombinant work – including tumor virus DNA research – to go forward. And in 1979, Joseph Califano, Secretary of Health, Education, and Welfare, approved the changes, thus ending a period of pointless stagnation for mammalian cancer research.

In practical terms, the outcome of the Asilomar consensus was ultimately nothing more than five sad years of delay in important research, and five frustrating years of disruption in the careers of many young scientists.

As the 1970s ended, the issues raised by Cohen and Boyer's original experiments turned gradually into non-issues. We had been forced to take an unprofitable detour, but at least it showed that molecular scientists wanted to be socially responsible.

Molecular biology during the second half of the 1970s, however, was not completely derailed by politics; these years did in fact see a number of important advances, most of them building upon the still controversial Boyer-Cohen molecular cloning technology.

The most significant breakthrough was the invention of methods for reading the sequence of DNA. Sequencing depends on having a large quantity of the particular stretch of DNA that you are interested in, so it was not feasible – except in the case of small viral DNA – until cloning technologies had been developed. As we have seen, cloning, in essence, involves inserting the desired piece of DNA into a plasmid, which is then itself inserted into a bacterium. The bacteria, allowed to divide and grow, will then produce a vast number of copies of the DNA fragment. Once harvested from the bacteria, this large quantity of the DNA fragment is then ripe for sequencing.

Two sequencing techniques were developed simultaneously, one by Wally Gilbert in Cambridge, Massachusetts (Harvard), and the other by Fred Sanger in Cambridge, England (see Plate 34). Gilbert's interest in sequencing DNA stemmed from his having isolated the repressor protein in the *E. coli* beta-galactosidase gene regulation system. As we have seen, he had shown that the repressor binds to the DNA close to the gene, preventing its transcription into RNA chains. Now he wanted to know the sequence of that DNA region. A fortuitous meeting with the brilliant Soviet chemist Andrei Mirzabekov suggested to Gilbert a way – using certain potent combinations of chemicals – to break DNA chains at just the desired, base-specific sites.

As a high-school senior in Washington, D.C., Gilbert used to cut class to read up on physics at the Library of Congress. He was then pursuing the Holy Grail of all high-school science prodigies: a prize in the Westinghouse Talent Search.* He duly won his prize in 1949. (Years later, in 1980, he would receive a call from the Swedish Academy in Stockholm, adding to the statistical evidence that winning the Westinghouse is one of the best predictors of a future Nobel.)

*In 1998, as the Old Economy gave way to the New, the honor was renamed the Intel Prize.

Gilbert stuck with physics as an undergraduate and graduate student, and a year after I arrived at Harvard in 1956 he joined the physics faculty. But once I got him interested in my lab's work on RNA, he abandoned his field for mine. Thoughtful and unrelenting, Gilbert has ever since been at the forefront of molecular biology.

Of the two sequencing methods, however, it is Sanger's that has better withstood the test of time. Some of the DNA-breaking chemicals required by Gilbert's are difficult to work with; given half a chance, they will start breaking up the researcher's own DNA. Sanger's method, on the other hand, uses the same enzyme that copies DNA naturally in cells, DNA polymerase. His trick involves making the copy out of base pairs that have been slightly altered. Instead of using only the normal "deoxy" bases (As, Ts, Gs, and Cs) found naturally in DNA (deoxyribonucleic acid), Sanger also added some so-called "dideoxy bases." Dideoxy bases have a peculiar property: DNA polymerase will happily incorporate them into the growing DNA chain (i.e., the copy being assembled as the complement of the template strand), but it cannot then add any further bases to the chain. In other words, the duplicate chain cannot be extended beyond a dideoxy base.

Imagine a template strand whose sequence is GGCCTAGTA. There are many, many copies of that strand in the experiment. Now imagine that the strand is being copied using DNA polymerase, in the presence of a mixture of normal A, T, G, and C plus some dideoxy A. The enzyme will copy along, adding first a C (to correspond to the initial G), then another C, then a G, and another G. But when the enzyme reaches the first T, there are two possibilities: either it can add a normal A to the growing chain, or it can add a dideoxy A. If it picks up a dideoxy A, then the strand can grow no further, and the result is a short chain that ends in a dideoxy A (ddA): CCGGddA. If it happens to add a normal A, however, then DNA polymerase can continue adding bases: T, C, etc. The next chance for a dideoxy "stop" of this kind will not come until the enzyme reaches the next T. Here again it may add either a normal A or a ddA. If it adds a ddA, the result is another truncated chain, though a slightly longer one: this chain has a sequence of CCGGATCddA. And so it goes every time the enzyme encounters a T (i.e., has occasion to add an A to the chain); if by chance it selects a normal A, the chain continues, but in the case of a ddA the chain terminates there.

Where does this leave us? At the end of this experiment, we have a whole slew of chains of varying lengths copied from the template DNA; what do they all have in common? They all end with a ddA.

Now, imagine the same process carried out for each of the other three bases: in the case of T, for instance, we use a mix of normal A, T, G, and C plus ddT; the resultant molecules will be either CCGGAddT or CCGGATCAddT.

Having staged the reaction all four ways – once with ddA, once with ddT, once with ddG, and once with ddC – we have four sets of DNA chains: one consists of chains ending in ddA, one with chains ending with ddT, and so on. Now if we could only sort all these mini-chains according to their respective, slightly varying lengths, we could infer the sequence. How? A moment, please. First, let's see how we could do the sorting. We can place all the DNA fragments on a plate full of a special gel, and place the plate of gel in an electric field. In the pull of the electric field the DNA molecules will be forced to migrate through the gel, and the speed with which a particular mini-chain will travel is a function of its size: short chains travel faster than long ones. Within a fixed interval of time, the smallest mini-chain, in our case a simple ddC, will travel furthest; the next smallest, CddC, will travel a slightly shorter distance; and the next one, CCddG, a slightly shorter one still. Now Sanger's trick should be clear: by reading off the relative positions of all these mini-chains after a timed race through our gel, we can infer the sequence of our piece of DNA: first is a C, then another C, then a G, and so on.

In 1980, Sanger shared the Nobel Prize in Chemistry with Gilbert and with Paul Berg, who was recognized for his contribution to the development of the recombinant DNA technologies. (Inexplicably neither Stanley Cohen nor Herb Boyer has been so honored.)

For Sanger, this was his second Nobel.* He had received the chemistry prize in 1958 for inventing the method by which proteins are sequenced – that is, by which their amino acid sequence is determined – and applying it to human insulin. But there is absolutely no relation between Sanger's method for protein sequencing and the one he devised for sequencing DNA; neither technically nor imaginatively did the one give rise to the other. He invented both from scratch, and should perhaps be regarded as the presiding technical genius of the early history of molecular biology.

Sanger is not what you might expect of a double Nobel laureate. Born to a Quaker family, he became a socialist and was a conscientious objector during the Second World War. More improbably, he does not advertise his achievements, preferring to keep the evidence of his Nobel honors in storage: "You get a nice gold medal, which is in the bank. And you get a certificate, which is in the loft." He has even turned down a knighthood: "A knighthood makes you different, doesn't it? And I don't want to be different." Having retired, Sanger is content these days to tend his garden outside Cambridge, though he still makes the occasional self-effacing and cheerful appearance at the Sanger Centre, the genome-sequencing facility near Cambridge that opened in 1993 (see Plates 31 & 32).

*As a double Nobelist, Sanger is in exalted company. Marie Curie received the prize in physics (1903) and then in chemistry (1911); John Bardeen received the physics prize twice, for the discovery of transistors (1956) and for superconductivity (1972); and Linus Pauling received the chemistry prize (1954) and the peace prize (1962).

Sequencing would confirm one of the most remarkable findings of the 1970s. We already knew that genes were linear chains of As, Ts, Gs, and Cs, and that these bases were translated three at a time, in accordance with the genetic code, to create the linear chains of amino acids we call proteins. But remarkable research by Richard Roberts, Phil Sharp, and others revealed that, in many organisms, genes actually exist in pieces, with the vital coding DNA broken up by chunks of irrelevant DNA. Only once the messenger RNA has been transcribed is the mess sorted out by an "editing" process that eliminates the irrelevant parts. It would be as though this book contained occasional extraneous paragraphs, apparently tossed in at random, about baseball or the history of the Roman Empire. Wally Gilbert dubbed the intrusive sequences "introns" and the ones responsible for actual protein-coding (i.e., functionally part of the gene) he named "exons." It turns out that introns are principally a feature of sophisticated organisms; they do not appear in bacteria.

Some genes are extraordinarily intron-rich. For example, in humans, the gene for blood clotting factor VIII (which may be mutated in people with hemophilia) has twenty-five introns. Factor VIII is a large protein, some two thousand amino acids long, but the exons that code for it constitute a mere 4 percent of the total length of the gene. The remaining 96 percent of the gene is made up of introns.

Why, then, do introns exist? Obviously their presence vastly complicates cellular processes, since they always have to be edited out to form the messenger RNA; and that editing seems a tricky business, especially when you consider that a single error in excising an intron from the messenger RNA for, say, clotting factor VIII would likely result in a frameshift mutation that would render the resulting protein useless. One theory holds that these molecular intruders are merely vestigial, an evolutionary heirloom, left over from the early days of life on earth. Still it remains a much-debated issue how introns came to be and what if any use they may have in life's great code.

Once we became aware of the general nature of genes in eukaryotes (organisms whose cells contain a compartment, the nucleus, specialized for storing the genetic material; prokaryotes, such as bacteria, lack nuclei), a scientific gold rush was launched. Teams of eager scientists armed with the latest technology raced to be the first to isolate (clone) and characterize key genes. Among the earliest treasures to be found were genes in which mutations give rise to cancers in mammals. Once scientists had completed the DNA sequencing of several well-studied tumor viruses, SV40 for one, they could then pinpoint the exact cancer-causing genes. These genes were capable of transforming normal cells into cells with cancerlike properties, with for instance a propensity for the kind of uncontrolled growth and cell division that results in tumors. It was not long until molecular biologists began to isolate genes from human cancer cells, finally confirming that human cancer arises because of changes at the DNA level and not from simple nongenetic accidents of growth, as had been supposed. We found genes that accelerate or promote cancer growth and we found genes that slow or inhibit it. Like an automobile, a cell, it seems, needs both an accelerator and a brake to function properly.

The treasure hunt for genes took over molecular biology. In 1981, Cold Spring Harbor Laboratory started an advanced summer course that taught gene-cloning techniques. *Molecular Cloning*, the lab manual that was developed out of this course, sold more than eighty thousand copies over the following three years. The first phase of the DNA revolution (1953–72) – the early excitement that grew out of the discovery of the double helix and led to the genetic code – eventually involved some three thousand scientists. But the second phase, inaugurated by recombinant DNA and DNA sequencing technologies, would see those ranks swell a hundredfold in little more than a decade.

Part of this expansion reflected the birth of a brand new industry: biotechnology. After 1975, DNA was no longer solely the concern of biologists trying to understand the molecular underpinnings of life. The molecule moved beyond the academic cloisters inhabited by white-coated scientists into a very different world populated largely by men in silk ties and sharp suits. The name Francis Crick had given his home in Cambridge, the Golden Helix, now had a whole new meaning.

CHAPTER FIVE

DNA, DOLLARS, AND DRUGS: BIOTECHNOLOGY

Herb Boyer has a way with meetings. We have seen how his 1972 chat with Stanley Cohen in a Waikiki deli led to the experiments that made recombinant DNA a reality. In 1976, lightning struck a second time: the scene was San Francisco, the meeting was with a venture capitalist named Bob Swanson, and the result was a whole new industry that would come to be called biotechnology.

Only twenty-seven when he took the initiative and contacted Boyer, Swanson was already making a name for himself in high-stakes finance. He was looking for a new business opportunity, and with his background in science he sensed one in the newly minted technology of recombinant DNA. Trouble was, everyone Swanson spoke to told him that he was jumping the gun. Even Stanley Cohen suggested that commercial applications were at least several years away. As for Boyer himself, he disliked distractions, especially when they involved men in suits, who always look out of place in the jeans-and-T-shirt world of academic science. Somehow, though, Swanson cajoled him into sparing ten minutes of his time one Friday afternoon.

Ten minutes turned into several hours, and then several beers when the meeting was adjourned to nearby Churchill's Bar, where Swanson discovered he had succeeded in rousing a latent entrepreneur. It was in Derry Borough High School's 1954 yearbook that class president Boyer had first declared his ambition "to become a successful businessman."

The basic proposition was extraordinarily simple: find a way to use the Cohen-Boyer technology to produce proteins that are marketable. A gene for a "useful" protein – say, one with therapeutic value, such as human insulin – could be inserted into a bacterium, which in turn would start manufacturing the protein. Then it would just be a matter of scaling up production, from petri dishes in the laboratory to vast industrial-size vats, and harvesting the protein as it was produced. Simple in principle, but not so simple in practice. Nevertheless, Boyer and Swanson were optimistic: each plunked down $500 to form a partnership dedicated to exploiting the new technology. In April 1976 they formed the world's first biotech company. Swanson's suggestion that they call the firm "Her-Bob," a combination of their first names, was mercifully rejected by Boyer, who offered instead "Genentech," short for "genetic engineering technology."

Insulin was an obvious commercial first target for Genentech. Diabetics require regular injections of this protein since their bodies naturally produce either too little of it (Type II diabetes) or none at all (Type I). Before the discovery in 1921 of insulin's role in regulating blood-sugar levels, Type I diabetes was lethal. Since then, the production of insulin for use by diabetics has become a major industry. Because blood-sugar levels are regulated much the same way in all mammals, it is possible to use insulin from domestic animals, mainly pigs and cows. Pig and cow insulins differ slightly from the human version: pig insulin by 1 amino acid in the 51-amino-acid protein chain, and cow insulin by 3. These differences can occasionally cause adverse effects in patients; diabetics sometimes develop allergies to the "foreign" protein. The biotech way around these allergy problems would be to provide diabetics with the real McCoy, human insulin.

With an estimated 8 million diabetics in the United States, insulin promised a biotech gold mine. Boyer and Swanson, however, were not alone in recognizing its potential. A group of Boyer's colleagues at the University of California, San Francisco (UCSF), as well as Wally Gilbert at Harvard, had also realized that cloning human insulin would prove both scientifically and commercially valuable. In May 1978, the stakes were raised when Gilbert and several others from the United States and Europe formed their own company, Biogen. The contrasting origins of Biogen and Genentech show just how fast things were moving: Genentech was envisioned by a twenty-seven-year-old willing to work the phones; Biogen was put together by a consortium of seasoned venture capitalists who head-hunted top scientists. Genentech was born in a San Francisco bar, Biogen in a fancy European hotel. Both companies, however, shared the same vision, and insulin was part of it. The race was on.

Inducing a bacterium to produce a human protein is tricky. Particularly awkward is the presence of introns, those noncoding segments of DNA found in human genes. Since bacteria have no introns, they have no means for dealing with them. While the human cell carefully "edits" the messenger RNA to remove these noncoding segments, bacteria, with no such capacity, cannot produce a protein from a human gene. And so, if E. coli were really going to be harnessed to produce human proteins from human genes, the intron obstacle needed to be overcome first.

The rival start-ups approached the problem in different ways. Genentech's strategy was to chemically synthesize the intron-free portions of the gene, which could then be inserted into a plasmid. They would in effect be cloning an artificial copy of the original gene. Nowadays, this cumbersome method is seldom used, but at the time Genentech's was a smart strategy. The Asilomar biohazard meeting had occurred only a short time earlier, and genetic cloning, particularly when it involved human genes, was still viewed with great suspicion and fell under heavy regulation. However, by using an artificial copy of the gene, rather than one actually extracted from a human being, Genentech had found a loophole. The company's insulin hunt could proceed unimpeded by the new rules.

Genentech's competitors followed an alternative approach – the one generally used today – but, working with DNA taken from actual human cells, they would soon find themselves stumbling into a regulatory nightmare. Their method employed one of molecular biology's most surprising discoveries to date: that the central dogma governing the flow of genetic information – the rule that DNA begets RNA, which in turn begets protein – could occasionally be violated. In the 1950s scientists had discovered a group of viruses that contain RNA but lack DNA. HIV, the virus that causes AIDS, is a member of this group. Subsequent research showed that these viruses could nevertheless convert their RNA into DNA after inserting it into a host cell. These viruses thus defy the central dogma with their backward RNA → DNA path. The critical trick is performed by an enzyme, reverse transcriptase, that converts RNA to DNA. Its discovery in 1970 earned Howard Temin and David Baltimore the 1975 Nobel Prize in Physiology or Medicine.

Reverse transcriptase suggested to Biogen and others an elegant way to create their own intron-free human insulin gene for insertion in bacteria. The first step was to isolate the messenger RNA produced by the insulin gene. Because of the editing process, the messenger RNA lacks the introns in the DNA from which it is copied. The RNA itself is not especially useful because RNA, unlike DNA, is a delicate molecule liable to degrade rapidly; also the Cohen-Boyer system calls for inserting DNA – not RNA – into bacterial cells. The goal, therefore, was to make DNA from the edited messenger RNA molecule using reverse transcriptase. The result would be a piece of DNA without the introns but with all the information that bacteria would require to make the human insulin protein – a cleaned-up insulin gene.

In the end Genentech would win the race, but just barely. Using the reverse transcriptase method, Gilbert's team had succeeded in cloning the rat gene for insulin and then coaxing a bacterium into producing the rat protein. All that remained was to repeat the process with the human gene. Here, however, is where Biogen met its regulatory Waterloo. To clone human DNA, Gilbert's team had to find a P4 containment facility – one with the highest level of containment, the sort required for work on such unpleasant beasts as the Ebola virus. They managed to persuade the British military to grant them access to Porton Down, a biological warfare laboratory in the south of England.

In his book about the race to clone insulin, Stephen Hall records the almost surreal indignities suffered by Gilbert and his colleagues.

Merely *entering* the P4 lab was an ordeal. After removing all clothing, each researcher donned government-issue white boxer shorts, black rubber boots, blue pajama-like garments, a tan hospital-style gown open in the back, two pairs of gloves, and a blue plastic hat resembling a shower cap. Everything then passed through a quick formaldehyde wash. Everything. All the gear, all the bottles, all the glassware, all the equipment. All the scientific recipes, written down on paper, had to pass through the wash; so the researchers slipped the instructions, one sheet at a time, inside plastic Ziploc bags, hoping that formaldehyde would not leak in and turn the paper into a brown, crinkly, parchment-like mess. Any document exposed to lab air would ultimately have to be destroyed, so the Harvard group could not even bring in their lab notebooks to make entries. After stepping through a basin of formaldehyde, the workers descended a short flight of steps into the P4 lab itself. The same hygienic rigmarole, including a shower, had to be repeated whenever anyone left the lab.

All this for the simple privilege of cloning a piece of human DNA. Today, in our less paranoid and better informed times, the same procedure is often performed in rudimentary labs by undergraduates taking introductory molecular biology. The whole episode was a bust for Gilbert and his team as they failed to clone the insulin gene. Not surprisingly they blamed their P4 nightmare (see Plate 30).

The Genentech team faced no such regulatory hurdles, but their technical challenges in inducing *E. coli* to produce insulin from their chemically synthesized gene were considerable all the same. For Swanson the businessman, the problems were not merely scientific. Since 1923, the U.S. insulin market had been dominated by a single producer, Eli Lilly, which by the late seventies was a $3 billion company with an 85 percent share of the insulin market. Swanson knew Genentech was in no position to compete with the 800-pound gorilla, even with a genetically engineered *human* insulin, a product patently superior to Lilly's farm-animal version. He decided to cut a deal and approached Lilly, offering an exclusive license to Genentech's insulin. And so as his scientist partners beavered away in the lab, Swanson hustled away in the boardroom. Lilly, he was sure, would agree; even such a giant could ill afford to miss out on what recombinant DNA technology represented, namely the very future of pharmaceutical production.

But Swanson wasn't the only one with a proposal, and Lilly was actually funding one of the competing efforts. A Lilly official had even been dispatched to Strasbourg, France, to oversee a promising attempt to clone the insulin gene using methods similar to Gilbert's. However, when the news came through that Genentech had gotten there first, Lilly's attention was instantly diverted to California. Genentech and Lilly signed an agreement on August 25, 1978, one day after the final experimental confirmation. The biotech business was no longer just a dream. Genentech would go public in September 1980. Within minutes its shares rose from a starting price of $35 to $89. At the time, this was the most rapid escalation in value in the history of Wall Street. Boyer and Swanson suddenly found themselves worth some $66 million apiece.

Traditionally in academic biology, all that mattered was precedence: who made the discovery first. One was rewarded in kudos, not cash. There were exceptions – the Nobel Prize, for instance, does come with a hefty financial award – but in general we did biology because we loved it. Our meager academic salaries certainly did not offer much of an inducement.

With the advent of biotechnology, all that changed. The 1980s would see changes in the relationship of science and commerce that were unimaginable a decade before. Biology was now a big-money game, and with the money came a whole new mind-set, and new complications (see Plate 33).

For one thing, the founders of biotech companies were typically university professors, and not surprisingly the research underpinning their companies' commercial prospects typically originated in their university labs. It was in his Zurich University lab, for instance, that Charles Weissmann, one of Biogen's founders, cloned human interferon, which, as a treatment for multiple sclerosis, has since become the company's biggest moneymaker. And Harvard University hosted Wally Gilbert's ultimately unsuccessful attempt to add recombinant insulin to Biogen's roster of products. Certain questions were soon bound to be asked: Should professors be permitted to enrich themselves on the basis of work done in their university's facilities? Would the commercialization of academic science create irreconcilable conflicts of interest? And the prospect of a new era of industrial-scale molecular biology fanned the still-glowing embers of the safety debate: with big money at stake, just how far would the captains of this new industry push the safety envelope?

Harvard's initial response was to form a biotech company of its own. With plenty of venture capital and the intellectual capital of two of the university's star molecular biologists, Mark Ptashne and Tom Maniatis, the business plan seemed a sure thing; a major player was about to enter the biotech game. In the fall of 1980, however, the plan fell apart. When the measure was put to a vote, the faculty refused to allow Fair Harvard to dip its lily-white academic toes into the murky waters of commerce. There were concerns that the enterprise would create conflicts of interest within the biology department: with a profit center in place, would faculty continue to be hired strictly on the basis of academic merit or would their potential to contribute to the firm now come into consideration? Ultimately, Harvard was forced to withdraw, giving up its 20 percent stake in the company. Sixteen years later, the cost of that call would become apparent when the firm was sold to the pharmaceutical giant Wyeth for $1.25 billion. And to this day, Harvard's Department of Molecular and Cellular Biology lacks a designated endowment to support research above the cost of salaries.

The decision of Ptashne and Maniatis to press on regardless precipitated a fresh set of obstacles. Mayor Vellucci's moratorium on recombinant DNA research in Cambridge was a thing of the past, but anti-DNA sentiment lingered on. Carefully avoiding a flashy high-tech name like Genentech or Biogen, Ptashne and Maniatis named their company Genetics Institute, hoping to evoke the less threatening fruit fly era of biology, rather than the brave new world of DNA. In the same spirit, the fledgling company decided to hang its shingle not in Cambridge but in the neighboring city of Somerville. A stormy hearing in Somerville City Hall, however, demonstrated that the Vellucci effect extended beyond the Cambridge city limits: Genetics Institute was denied a license to operate. Fortunately the city of Boston, just across the Charles River from Cambridge, proved more receptive, and the new firm set up shop in an empty hospital building in Boston's Mission Hill district. As it became more and more apparent that recombinant methods posed no health or environmental risk, the Vellucci brand of antibiotech fanaticism could not endure. Within a few years, Genetics Institute would move to North Cambridge, just down the road from the university parent that had abandoned it at birth.

Over the past twenty years, the suspicion and sanctimoniousness attending the early days of the relationship between academic and commercial molecular biology has given way to something approaching a productive symbiosis. For their part universities now actively encourage their faculty to cultivate commercial interests. Learning from Harvard's mistake with Genetics Institute, they have developed ways to cash in on the lucrative applications of technology invented on campus. New codes of practice aim to prevent conflicts of interest for professors straddling both worlds. In the early days of biotech, academic scientists were all too often accused of "selling out" when they became involved with a company. Now involvement in commercial biotech is a standard part of a hotshot DNA career. The money is handy, and there are intellectual rewards as well because, for good business reasons, biotech is invariably on the scientific cutting edge.

Stanley Cohen proved himself a forerunner not only in technology but also in the evolution from a purely academic mind-set to one adapted to the age of big-bucks biology. He had known from the beginning that recombinant DNA had potential for commercial applications, but it had never occurred to him that the Cohen-Boyer cloning method should be patented. It was Niels Reimers in Stanford's technology licensing office who suggested that a patent might be in order when he read on the front page of the *New York Times* about the home team's big win. At first Cohen was dubious; the breakthrough in question, he argued, was dependent on generations of earlier research that had been freely shared, and so it seemed inappropriate to patent what was merely the latest development. But every invention builds on ones that have come before (the steam locomotive could only come after the steam engine); and patents rightly belong to those innovators who extend the achievements of the past in decisive and influential ways. In 1980, six years after Stanford first submitted the application, the Cohen-Boyer process was granted its patent.

In principle the patenting of methods could stifle innovation by restricting the application of important technologies, but Stanford handled the matter wisely, and there were no such negative consequences. Cohen and Boyer (and their institutions) were rewarded for their commercially significant contribution, but not at the expense of academic progress. In the first place, the patent ensured that only corporate entities would be charged for use of the technology; academic researchers could use it free of charge. Second, Stanford resisted the temptation to impose a very high licensing fee, which would have prevented all but the wealthiest companies and institutions from using recombinant DNA. For a relatively modest $10,000 a year with a maximum 3 percent royalty on the sales of products based on the technology, the Cohen-Boyer method was available to anyone who wanted to use it. This strategy, good for science, proved to be good for business as well: the patent has contributed some quarter of a billion dollars to the coffers of UCSF and Stanford. And both Boyer and Cohen generously donated part of their shares of the proceeds to their universities.

It was only a matter of time before organisms genetically altered by technology would themselves be patented. The test case had in fact originated in 1972; it involved a bacterium that had been modified using not recombinant DNA technology but traditional genetic methods. The implications for the biotech business were clear nevertheless: if bacteria modified with conventional techniques were patentable, then those modified by the new recombinant methods would be too.

In 1972, Ananda Chakrabarty, a research scientist at General Electric, applied for a patent on a *Pseudomonas* bacteria strain he had developed as an all-in-one oil-slick degrader. Before this, the most efficient way to break down an oil spill was to use a number of different bacteria, each of which degraded a different component of the oil. By combining different plasmids, each coding for a different degradation pathway, he managed to produce a superdegrader strain of *Pseudomonas*. Chakrabarty's initial patent application was turned down, but after wending its way through the legal system for eight years it was finally granted in 1980, when the Supreme Court ruled five to four in his favor, concluding that "a live, human-made micro-organism is patentable subject matter" if, as in this case, it "is the result of human ingenuity and research."

Despite the clarification supplied by the Chakrabarty case, the early encounters between biotechnology and the law were inevitably messy. The stakes were high and – as we shall see in the case of DNA fingerprinting in chapter 10 – lawyers, juries, and scientists too often speak different languages. By 1983, both Genentech and Genetics Institute had successfully cloned the gene for tissue plasminogen activator (t-PA), which is an important weapon against the blood clots that cause strokes and heart attacks. Genetics Institute did not, however, apply for a patent, deeming the science underlying the cloning of t-PA "obvious" – in other words, unpatentable. Genentech, however, applied for and was granted a patent, on which, by definition, Genetics Institute had infringed.

The case first came to court in England. The presiding judge, Mr. Justice Whitford, sat behind a large stack of books for much of the trial, appearing to be asleep. The basic question was whether the first party to clone a gene should be granted all subsequent rights over the production and use of the protein. In finding for Genetics Institute and its backers, the drug company Wellcome, Justice Whitford concluded that Genentech could justify a narrow claim for the limited process used by them to clone t-PA but could not justify broad claims for the protein product. Genentech appealed. In England when such esoteric technical cases are appealed they are heard by three specialist judges, who are led through the issues by an independent expert – in this instance, Sydney Brenner. The judges turned down Genentech's appeal, agreeing with Genetics Institute that the "discovery" was indeed obvious, and therefore the Genentech patent was invalid.

In the United States, such cases are argued in front of a jury. Genentech's lawyers ensured that no member of the jury had a college education. Thus what might be obvious to a scientist or to legal experts trained in science was not obvious to members of that jury. The jury found against Genetics Institute, deeming the broad-based Genentech patent valid. Not, perhaps, American justice's finest hour, but the case did nevertheless establish a precedent: from then on, people applied for patents on their products regardless of whether or not the science was "obvious." In future disputes, all that would matter was who cloned the gene first.

Good patents, I would suggest, strike a balance: they recognize and reward innovative work and protect it from being ripped off, but they also make new technology available to do the most good. Unfortunately, Stanford's wise example has not been followed in every case of important new DNA methodology. The polymerase chain reaction (PCR), for instance, is an invaluable technique for amplifying small quantities of DNA. Invented in 1983 at the Cetus Corporation, PCR – about which we shall hear more in chapter 7, in connection with the Human Genome Project – quickly became one of the workhorses of academic molecular biology. Its commercial applications, however, have been much more limited. After granting one commercial license to Kodak, Cetus sold PCR for $300 million to the Swiss giant Hoffmann-LaRoche, makers of chemical, pharmaceutical, and medical diagnostic products. Hoffmann-LaRoche in turn decided that, rather than granting further licenses, the way to maximize the return on their investment was to establish a monopoly on PCR-based diagnostic testing. As part of this strategy, it cornered the AIDS testing business. And only as the patent expiration date drew near did the firm grant any licenses for the technology; those granted have generally been to other major diagnostic companies that can afford the commensurably large fees. To create a subsidiary revenue stream from the same patent, Hoffmann-LaRoche has also levied hefty charges on producers of machines that carry out PCR. And so, to market a simple device for schoolchildren to use, the Cold Spring Harbor Dolan DNA Learning Center must pay the company a 15 percent royalty.

An even more pernicious effect on the productive availability of new technologies has been exerted by lawyers moving aggressively to patent not only new inventions but also the general ideas underpinning them. The patent on a genetically altered mouse created by Phil Leder is a case in point. In the course of their cancer research, Leder's group at Harvard produced a strain of mouse that was particularly prone to developing breast cancer. They did this using established techniques for inserting a genetically engineered cancer gene into a fertilized mouse egg cell. Because the factors inducing cancer in mice may be similar to those at work in humans, this "onco-mouse" was expected to help us understand human cancer. But instead of applying for a patent limited to the specific mouse Leder's team had produced, Harvard's lawyers sought one that covered all cancer-prone transgenic animals – they didn't even draw the line at mice. This umbrella patent was granted in 1988, and so was born the cancerous little rodent dubbed the "Harvard mouse." In fact, because the work in Leder's laboratory was underwritten by Du Pont, the commercial rights resided not with the university but with the chemical giant. The "Harvard mouse" might have been more aptly called the "Du Pont mouse." But whatever its name, the impact of the patent on cancer research has been profound and counterproductive (see Plate 35).

Companies interested in developing new forms of cancer-prone mice have been put off by the fees demanded by Du Pont, and those keen to use existing cancer mouse strains to screen experimental drugs have likewise curtailed their programs. Du Pont has begun demanding that academic institutions disclose what experiments are being performed using the company's patented onco-mice. This represents an unprecedented, and unacceptable, intrusion of big business into academic laboratories. UCSF, MIT's Whitehead Institute, and Cold Spring Harbor Laboratory, among other research institutions, have refused to cooperate.

When patents involve "enabling technologies" that are funda-mental to carrying out the necessary molecular manipulations, the patent holders can literally hold an entire area of research for ransom. And while every patent application should be treated on its particular merits, there are nevertheless some general rules that should be observed. Patents on methods clearly vital to scientific progress should follow the precedent set by the Cohen-Boyer case: the technology should be generally available (not controlled by a single licensee) and should be reasonably priced. These limitations by no means go against the ethic of free enterprise. If a new method is a genuine step forward, then it will be extensively used and even a modest royalty will result in substantial revenue. Patents on *products*, however – drugs, transgenic organisms – should be limited to the specific product created, not the entire range of additional products the new one might suggest.

Genentech's insulin triumph put biotechnology on the map. A quarter of a century later, genetic engineering with recombinant DNA technology is a routine part of the drug-discovery industry. These procedures permit the production in large quantities of human proteins, which are otherwise difficult to acquire. In many cases, the genetically engineered proteins are safer for therapeutic and diagnostic uses than their predecessors. Extreme short stature, dwarfism, often stems from a lack of human growth hormone (HGH). In 1959, doctors first started treating dwarfism with HGH, which then could be obtained only from the brains of cadavers. The treatment worked fine, but it was later recognized to carry the risk of a terrible infection: patients sometimes developed Creutzfeldt-Jakob disease, a ghastly brain-wasting affliction, similar to so-called mad cow disease. In 1985, the FDA banned the use of HGH derived from cadavers. By happy coincidence, Genentech's recombinant HGH – which carries no risk of infection – was approved for use that same year.

During the biotech industry's first phase, most companies focused on proteins of known function. Cloned human insulin was bound to succeed; after all, people had already been injecting themselves with some form of insulin for more than fifty years when Genentech introduced its product. Another example was epoetin alpha (EPO), a protein that stimulates the body to produce red blood cells. The target population for EPO is patients undergoing kidney dialysis who suffer from anemia caused by loss of red blood cells. To meet the need for this product, Amgen, based in Southern California, and Genetics Institute both developed a recombinant form of EPO. That EPO was a useful and commercially viable product was a given; the only unknown was which company would come to dominate the market. Despite being trained in the arcane subtleties of physical chemistry, Amgen CEO George Rathmann has adapted well to the rough and tumble of the business world. Competition brings out a decidedly unsubtle side in him: negotiating with him is like wrestling with a large bear whose twinkling eye assures you that it is only mauling you because it is obliged to. Amgen and its backer, Johnson & Johnson, duly won the court battle with Genetics Institute, and EPO is now worth $2 billion a year to Amgen alone. Amgen is accordingly today the biggest player in the biotech stakes, worth some $64 billion.

After biotech's pioneers had rounded up the "obvious" products, proteins with known physiological function like insulin, t-PA, HGH, and EPO, a second, more speculative phase in the industry got under way. Having run out of surefire winners, companies hungry for further bonanzas began to back possible contenders, even long shots. From knowing that something worked, they went to merely *hoping* that a potential product would work. Unfortunately, the combination of longer odds, technical challenges, and regulatory hurdles to be cleared before a drug is approved by the FDA has taken its toll on many a bright-eyed biotech start-up.

The discovery of growth factors – proteins that promote cell proliferation and survival – provoked a proliferation of new biotech companies. Among them, both New York–based Regeneron and Synergen, located in Colorado, hoped to find a treatment for ALS (amyotrophic lateral sclerosis or Lou Gehrig's disease), the awful degenerative affliction of nerve cells. Their idea was fine in principle, but in practice there was simply too little known at the time about how nerve growth factors act for these efforts to be anything more than shots in the dark. Trials on two groups of ALS patients failed, and the disease remains untreatable today. The experiments did, however, reveal an interesting side effect: those taking the drugs lost weight. In a twist that illustrates just how serendipitous the biotech business can be, Regeneron is today developing a modified version of its drug as a weight-loss therapy.

Another initially speculative enterprise that has seen more than its fair share of dashed commercial hopes is monoclonal antibody (MAb) technology. When they were invented in the mid-1970s at the MRC Laboratory of Molecular Biology at Cambridge University by César Milstein and Georges Köhler, MAbs were hailed as the silver bullets that would quickly change the face of medicine. Nevertheless, in an oversight that would today be unthinkable, the MRC failed to patent them. Silver bullets they proved not to be, but, after decades of disappointment, they are just now coming into their own.

Antibodies are molecules produced by the immune system to bind to and identify invading organisms. Derived from a single line of antibody-producing cells, MAbs are antibodies programmed to bind to a unique target. They can be readily produced in mice by injecting animals with the target material, inducing an immune response, and culturing the blood cells from the mouse that produced the MAb. Because MAbs can recognize and bind to specific molecules, it was hoped that they could be used with pinpoint accuracy against any number of pernicious intruders – tumor cells, for instance. Such optimism prompted the founding of a slew of MAb-based companies, but they quickly ran into obstacles. Ironically, the most significant of these was the human body's own immune system, which identified the mouse MAbs as foreign and duly destroyed them before they could act on their targets. A variety of methods have since been devised to "humanize" MAbs – to replace as much as possible of the mouse antibodies with human components. And the latest generation of MAbs represents the biggest growth area in biotech today.

Centocor, based near Philadelphia, now owned by Johnson & Johnson, has developed ReoPro, an MAb specific to a protein on the surface of platelets, which promote the formation of blood clots. By preventing platelets from sticking together, ReoPro reduces the chance of lethal clot formation in patients undergoing angioplasty, for instance. Genentech, never one to lag in the biotech stakes, now markets Herceptin, an MAb that targets certain forms of breast cancer. Immunex in Seattle produces an MAb-based drug called Enbrel, which fights rheumatoid arthritis, a condition associated with the presence of excessive amounts of a particular protein, tumor necrosis factor (TNF), involved in regulating the immune system. Enbrel works by capturing the excess TNF molecules, preventing them from provoking an immune reaction against the tissue in our joints.

Still other biotech companies are interested in cloning genes whose protein products are potential targets for new pharmaceuticals. Among the most eagerly sought are the genes for proteins usually found on cell surfaces that serve as receptors for neurotransmitters, hormones, and growth factors. It is through such chemical messengers that the human body coordinates the actions of any individual cell with the actions of trillions of others. Drugs developed blindly in the past through trial and error have recently been found to operate by affecting these receptors. And that same new molecular understanding has also explained why so many of these drugs have side effects. Receptors often belong to large families of similar proteins. A drug may indeed effectively target a receptor relevant to the disease in question, but it also may wind up inadvertently targeting similar receptors, thus producing side effects. Intelligent drug design should permit more specific targeting of the receptors so that *only* the relevant one is blocked. However, as with MAbs, what seems a great idea on paper is too often hard to apply in practice, and even harder to make big bucks from.

This depressing lesson was learned by SIBIA, a San Diego start-up associated with the Salk Institute. The discovery of membrane receptors for the neurotransmitter nicotinic acid promised a breakthrough treatment for Parkinson disease, but as so often in biotech a good idea was only the beginning of a long scientific process. Ultimately, after giving promising results in monkeys, SIBIA's drug candidate failed in humans.

Like the unexpected weight loss associated with Regeneron's nerve growth factor, breakthroughs in this area too are often born of pure luck rather than the scientific calculus of rational drug design. In 1991, for instance, a Seattle-based company, ICOS, led by George Rathmann of Amgen fame, was working with a class of enzymes called "phosphodiesterases," which degrade cell-signaling molecules. Their quarry was new drugs to lower blood pressure, but one of their test drugs had a surprising side effect. They had stumbled onto a Viagra-like therapy for erectile dysfunction, which may well yield a bigger jackpot than any they previously dreamed of.*

The market for easier erections notwithstanding, the search for cancer therapies has, not surprisingly, become the single greatest driving force for the biotech industry. The classic "cell-killing" approach to attacking cancer, using radiation or chemotherapy,

*Viagra itself has a similar history. Also originally developed to combat high blood pressure, trials on male medical students convinced researchers that it had other properties.

invariably also kills healthy normal cells, typically with dreadful side effects. With developing DNA methodologies researchers are finally closing in on drugs that can target only those key proteins – many of them growth factors and their receptors on the cell surface – that promote cancer cell growth and division. Developing a drug that inhibits a desired target without disabling other vital proteins is a formidable challenge even for the best of medicinal chemists. And the uncertain journey from a successfully cloned drug target gene to the widespread availability of an FDA-approved pharmaceutical is a veritable odyssey that seldom takes less than ten years.

Success stories are hard to come by, but will, I am sure, become more common. Discovered by chemists at the Swiss company Novartis, Gleevec works against a blood cancer called chronic myeloid leukemia (CML) by specifically blocking the growth-stimulating activity of membrane receptor proteins that are overproduced by cancerous cells of this type. If given early in the course of CML, Gleevec generally leads to long disease-free remissions, and hopefully in many cases to true cures. For some unlucky individuals, though, the disease reappears when new mutations in the gene encoding the membrane-receptor proteins render Gleevec ineffective.

One of the most important anticancer-drug target proteins may be the receptor for epidermal growth factor (EGFR). This receptor frequently shows up in much higher quantities in cancer cells (particularly in breast and lung cancers) than in normal ones, which suggests that it may well be a winner as a drug target. Several potent drugs that specifically block EGFR action are now in late-stage clinical testing. But while the arrival of target-specific drugs will certainly introduce big new guns in the war against cancer, the likelihood is that, after initial remission, many patients will suffer a relapse as resistance to the new drugs evolves among the cancer cells colonizing the body.

For this reason, many have come to believe that a better long-term way of fighting cancer cells may involve targeting their nutritional lifelines. They, like all cells in the body, need nutrients to grow, and they receive these nutrients from blood vessels that grow near them. If you block the growth of blood vessels into tumors, you can eventually starve to death the cancer cells they serve. The idea that small tumors become dangerous only once they are infiltrated by newly formed blood vessels (a process called "angiogenesis") first occurred to Judah Folkman in the early 1960s while he was doing his military service in the Naval Medical Research Institute outside Washington, D.C. The precocious son of an Ohio rabbi, Folkman was the first graduate of Ohio State University to enter Harvard Medical School. By the time he went to high school he had already assisted in surgery on a dog, and in college he invented a surgical device to cool the liver when its blood supply was temporarily cut off. At thirty-four, he became the youngest professor of surgery in the history of Harvard University. Folkman's anti-angiogenesis ideas could not, however, be explored therapeutically until the recent discovery of three specific growth factors that play vital roles in the growth of "endothelial" cells, those that line blood vessels. Inhibitors developed against these growth factors – anti-angiogenesis drugs – might very well prove effective against many forms of cancer. Some forty years after Folkman's original insight, we may at last be able in the foreseeable future to cure most cancers, including those that have become resistant to the best conventional anticancer drugs.

Already Sugen, a firm outside San Francisco, has developed two highly specific small-molecule drugs that work against distinct angiogenesis growth factors and inhibit tumors in model animal systems. Neither drug given separately has yet proved effective against advanced human cancers. However, preliminary data from experiments with cancer-prone mice done by Doug Hanahan at UCSF suggest the Sugen drugs might have worked had they been administered in tandem. Unfortunately the future of onco-mouse experiments at UCSF and elsewhere is jeopardized by the ongoing dispute provoked by Du Pont's aggressive onco-mouse licensing policies.

Blood vessel infiltration into mouse tumors has also been prevented by a newly discovered group of proteins that are likely naturally occurring inhibitors of blood vessel formation. Two such proteins, angiostatin and endostatin, isolated by Michael O'Reilly in Judah Folkman's lab, are currently in clinical trials. While neither is present in blood in amounts large enough to be extracted for human testing, recombinant DNA procedures permit both proteins to be made in yeast cells in quantities sufficient for clinical use. And while neither angiostatin nor endostatin alone has yet demonstrated miracle-like anticancer effects in humans, mouse experiments suggest that, as with Sugen's drugs, an efficacious combination of the two may soon be discovered. Over the next decade, a virtual armada of small-molecule and protein inhibitors will probably be ready to sail through the systems of cancer sufferers, thwarting blood vessel formation before tumors have a chance to become lethal. And if tumor growth can indeed be curtailed in this way, we may come to regard cancer as we do diabetes, as a disease that can be controlled rather than completely cured outright.

Since recombinant technologies allow us to harness cells to produce virtually any protein, the question has logically arisen: Why limit ourselves to pharmaceuticals? Consider the example of spider silk. So-called dragline silk, which forms the radiating spokes of a spider web, is an extraordinarily tough fiber. By weight, it is five times as strong as steel. Though there are ways spiders can be coaxed to spin more than their immediate needs require, unfortunately, attempts to create spider farms have foundered because the creatures are too territorial to be reared en masse. Now, however, the silk-protein-producing genes have been isolated and can be inserted into other organisms, which can thus serve as spider-silk factories. This very line of research is being funded by the Pentagon, which sees Spiderman in the U.S. Army's future: soldiers may one day be clad in protective suits of spider-silk body armor.

Another exciting new frontier in biotechnology involves improving on natural proteins. Why be content with nature's design, arrived at by sometimes arbitrary and now irrelevant evolutionary pressures, when a little manipulation might yield something more useful? Starting with an existing protein, we now have the ability to make slight alterations in its amino acid sequence. The limitation, unfortunately, is in our knowledge of what effect altering even a single amino acid in the chain is likely to have on the protein's properties.

Here we can return to nature's example for a solution: a procedure known as "directed molecular evolution" effectively mimics natural selection. In natural selection new variants are generated at random by mutation and then winnowed by competition among individuals; successful – better adapted – variants are more likely to live and contribute to the next generation. Directed molecular evolution stages this process in the test tube. After using biochemical tricks to introduce random mutations into the gene for a protein, we can then mimic genetic recombination to shuffle the mutations to create new sequences. From among the resulting new proteins our system selects the ones that perform best under the conditions specified. The whole cycle is repeated several times, each time with the "successful" molecules from the previous cycle competing in the next.

For a nice example of how directed molecular evolution can work, we need look no farther than the laundry room. Here disasters occur when a single colored item finds its way accidentally into a load of whites: some of the dye inevitably leaches out of that red T-shirt and before you know it every sheet in the house is a pale pink. It so happens that a peroxidase enzyme naturally produced by a toadstool – the inkcap mushroom, to be specific – has the property of decolorizing the dyes that have leached out of clothing. The problem, however, is that the enzyme cannot function in the hot soapy environment of a washing machine. By using directed molecular evolution, however, it been possible to improve the enzyme's capacity for coping with these conditions: one specially "evolved" enzyme, for instance, demonstrated an ability to withstand high temperatures 174 times greater than that of the toadstool's own enzyme. And such useful "evolutions" do not take long. Natural selection takes eons, but directed molecular evolution in the test tube does the job in just hours or days.

Genetic engineers realized early that their technologies could also have a positive impact on agriculture. As the biotech world now knows all too well, the resulting genetically modified (GM) plants are now at the center of a firestorm of controversy. So it's interesting to note that an earlier contribution to agriculture – one that increased milk production – also led to an outcry.

Bovine growth hormone (BGH) is similar in many ways to human growth hormone, but it has an agriculturally valuable side effect: it increases milk production in cows. Monsanto, the St. Louis-based agricultural chemical company, cloned the BGH gene and produced recombinant BGH. Cows naturally produce the hormone, but, with injections of Monsanto's BGH, their milk yields increased by about 10 percent. In late 1993 the FDA approved the use of BGH, and by 1997 some 20 percent of the nation's 10 million cows were receiving BGH supplements. The milk produced is indistinguishable from that produced by nonsupplemented cows: they both contain the same small amounts of BGH. In fact, a major argument against labeling milk as "non-BGH-supplemented" versus "BGH-supplemented" is that it is impossible to distinguish between milk from supplemented and nonsupplemented cows, so there is no way to determine whether or not such advertising is fraudulent. Because BGH permits farmers to reach their milk production targets with fewer cattle, it is in principle beneficial to the environment because it could result in a reduction in the size of dairy herds. Because methane gas produced by cattle contributes significantly to the greenhouse effect, herd reduction may actually have a long-term effect on global warming. Methane is twenty-five times more effective at retaining heat than carbon dioxide, and on average a grazing cow produces six hundred flatulent liters of the stuff a day – enough to inflate forty party balloons.

At the time I was surprised that BGH provoked such an outburst from the anti-DNA lobby. Now, as the GM food controversy drags on, I have learned that professional polemicists can make an issue out of anything. Jeremy Rifkin, biotechnology's most obsessive foe, was launched on his career in naysaying by the U.S. Bicentennial in 1976. He objected. After that he moved on to objecting to DNA. His response in the mid-1980s to the suggestion that BGH would not likely inflame the public was, "I'll *make* it an issue! I'll find something! It's the first product of biotechnology out the door, and I'm going to fight it." Fight it he did. "It's unnatural" (but it's indistinguishable from "natural" milk). "It contains proteins that cause cancer" (it doesn't, and in any case proteins are broken down during digestion). "It'll drive the small farmer out of business" (but, unlike with many new technologies, there are no up-front capital costs, so the small farmer is not being discriminated against). "It'll hurt the cows" (nearly nine years of commercial experience on millions of cows has proved this not to be the case). In the end, rather like the Asilomarera objections to recombinant techniques, the issue petered out when it became clear that none of Rifkin's gloom-and-doom scenarios were realistic.

The spat over BGH was a taste of what was to come. For Rifkin and like-minded DNA-phobes, BGH was merely the appetizer: genetically modified foods would be the protesters' main course.

CHAPTER SIX

TEMPEST IN A CEREAL BOX: GENETICALLY MODIFIED AGRICULTURE

In June 1962, Rachel Carson's book *Silent Spring* created a sensation when it was serialized in *The New Yorker*. Her terrifying claim was that pesticides were poisoning the environment, contaminating even our food. At that time I was a consultant to John Kennedy's President's Scientific Advisory Committee (PSAC). My main brief was to look over the military's biological warfare program, so I was only too glad to be diverted by an invitation to serve on a subcommittee that would formulate the administration's response to Carson's concerns. Carson herself gave evidence, and I was impressed by her careful exposition and circumspect approach to the issues. In person, too, she was nothing like the hysterical ecofreak she was portrayed as by the pesticide industry's vested interests. An executive of the American Cyanamid Company, for instance, insisted that "if man were to faithfully follow the teachings of Miss Carson, we would return to the Dark Ages, and the insects and diseases and vermin would once again inherit the earth." Monsanto, another giant pesticide producer, published a rebuttal of *Silent Spring*, called *The Desolate Year*, and distributed five thousand copies free to the media.

My most direct experience of the world Carson described, however, came a year later when I headed a PSAC panel looking into the threat posed to the nation's cotton crop by herbivorous insects, especially the boll weevil. Touring the cotton fields of the Mississippi Delta, West Texas, and the Central Valley of California, one could hardly fail to notice the utter dependence of cotton growers on chemical pesticides. En route to an insect research laboratory near Brownsville, Texas, our car was inadvertently doused from above by a crop duster. Here billboards featured not the familiar Burma-Shave ads but pitches for the latest and greatest insect-killing compounds. Poisonous chemicals seemed to be a major part of life in cotton country.

Whether Carson had gauged the threat accurately or not, there had to be a better way to deal with the cotton crop's six-legged enemies than drenching huge tracts of country with chemicals. One possibility promoted by the U.S. Department of Agriculture scientists in Brownsville was to mobilize the insects' own enemies – the polyhedral virus, for instance, which attacks the bollworm (soon to become a greater threat to cotton than the boll weevil) – but such strategies proved impracticable. Back then, I could not have conceived of a solution that would involve creating plants with built-in resistance to pest insects: such an idea would simply have seemed too good to be true. But these days that is exactly how farmers are beating the pests while at the same time reducing dependence on noxious chemicals.

Genetic engineering has produced crop plants with onboard pest resistance. The environment is the big winner because pesticide use is decreased, and yet paradoxically organizations dedicated to protecting the environment have been the most vociferous in opposing the introduction of these so-called genetically modified (GM) plants.

As with genetic engineering in animals, the tricky first step in plant biotechnology is to get your desired piece of DNA (the helpful gene) into the plant cell, and afterwards into the plant's genome. As molecular biologists frequently discover, nature had devised a mechanism for doing this eons before biologists even thought about it.

Crown gall disease results in the formation of an unattractive lumpy "tumor," known as a gall, on the plant stem. It is caused by a common soil bacterium called *Agrobacterium tumefaciens,* which opportunistically infects plants where they are damaged by, say, the nibbling of a herbivorous insect. How the bacterial parasite carries out the attack is remarkable. It constructs a tunnel through which it delivers a parcel of its own genetic material into the plant cell. The parcel consists of a stretch of DNA that is carefully excised from a special plasmid and then wrapped in a protective protein coat before being shipped off through the tunnel. Once the DNA parcel is delivered, it becomes integrated, as a virus's DNA would be, into the host cell's DNA. Unlike a virus, however, this stretch of DNA, once lodged, does not crank out more copies of itself. Instead, it produces both plant growth hormones and specialized proteins, which serve as nutrients for the bacterium. These promote simultaneous plant cell division and bacterial growth by creating a positive feedback loop: the growth hormones cause the plant cells to multiply more rapidly, with the invasive bacterial DNA being copied at each cell division along with the host cell's, so that more and more bacterial nutrients *and* plant growth hormones are produced.

For the plant the result of this frenzy of uncontrolled growth is a lumpy cell mass, the gall, which for the bacterium serves as a kind of factory in which the plant is coerced into producing precisely what the bacterium needs, and in ever greater quantities. As parasitic strategies go, *Agrobacterium*'s is brilliant: it has raised the exploitation of plants to an art form.

The details of *Agrobacterium's* parasitism were worked out during the 1970s by Mary-Dell Chilton at the University of Washington in Seattle and by Marc van Montagu and Jeff Schell at the Free University of Ghent, Belgium. At the time the recombinant DNA debate was raging at Asilomar and elsewhere. Chilton and her Seattle colleagues later noted ironically that, in transferring DNA from one species to another without the protection of a P4 containment facility, *Agrobacterium* was "operating outside the National Institutes of Health guidelines."

Chilton, van Montagu, and Schell soon were not alone in their fascination with *Agrobacterium*. In the early eighties Monsanto, the same company that had condemned Rachel Carson's attack on pesticides, realized that *Agrobacterium* was more than just a biological oddity. Its bizarre parasitic lifestyle might hold the key to getting genes into plants. When Chilton moved from Seattle to Washington University, St. Louis, Monsanto's hometown, she found that her new neighbors took a more than passing interest in her work. Monsanto may have made its entry late in the *Agrobacterium* stakes, but it had the money and other resources to catch up fast. Before long both the Chilton and the van Montagu/Schell laboratories were being funded by the chemical giant in return for a promise to share their findings with their benefactor.

Monsanto's success was built on the scientific acumen of three men, Rob Horsch, Steve Rogers, and Robb Fraley, all of whom joined the company in the early eighties. Over the next two decades they would engineer an agricultural revolution. Horsch always "loved the smell of [the soil], the heat of it" and, even as a boy, wanted "always to grow things better than what I could find at the grocery store." He instantly saw a job at Monsanto as an opportunity to follow that dream on an enormous scale. By contrast, Rogers, a molecular biologist at Indiana University, initially discarded the company's letter of invitation, viewing the prospect of such work as "selling out" to industry. Upon visiting, however, he discovered not only a vigorous research environment but also an abundance of one key element that was always in short supply in academic research: money. He was converted. Fraley was possessed early on by a vision for agricultural biotechnology. He came to the company after approaching Ernie Jaworski, the executive whose bold vision had started Monsanto's biotechnology program. Jaworski proved not only a visionary but also an affable employer. He was unfazed by his first encounter with the new man when they were both passing through Boston's Logan Airport: Fraley announced that one of his goals was to take over Jaworski's job.

All three *Agrobacterium* groups – Chilton's, van Montagu and Schell's, and Monsanto's – saw the bacterium's strategy as an invitation to manipulate the genetics of plants. By then it wasn't hard to imagine using the standard cut-and-paste tools of molecular biology to perform the relatively simple act of inserting into *Agrobacterium*'s plasmid a gene of one's choice to be transferred to the plant cell. Thereafter, when the genetically modified bacterium infected a host, it would insert the chosen gene into the plant cell's chromosome. *Agrobacterium* is a ready-made delivery system for getting foreign DNA into plants; it is a natural genetic engineer. In January 1983, at a watershed conference in Miami, Chilton, Horsch (for Monsanto), and Schell all presented independent results confirming that *Agrobacterium* was up to the task. And by this time, each of the three groups had also applied for patents on

Agrobacterium-based methods of genetic alteration. Schell's was recognized in Europe, but in the United States, a falling-out between Chilton and Monsanto would rumble through the courts until 2000, when a patent was finally awarded to Chilton and her new employer, Syngenta. But having now seen a bit of the Wild West show that is intellectual property patents, one shouldn't be surprised to hear that the story does not end so neatly there: as I write, Syngenta is in court suing Monsanto for patent infringement.

At first *Agrobacterium* was thought to work its devious magic only on certain plants. Among these, we could not, alas, count the agriculturally important group that includes cereals such as corn, wheat, and rice. However, in the years since it gave birth to plant genetic engineering, *Agrobacterium* has itself been the focus of genetic engineers, and technical advances have extended its empire to even the most recalcitrant crop species. Before these innovations, we had to rely upon a rather more haphazard, but no less effective, way of getting our DNA selection into a corn, wheat, or rice cell. The desired gene is affixed to tiny gold or tungsten pellets, which are literally fired like bullets into the cell. The trick is to fire the pellets with enough force to enter the cell, but not so much that they will exit the other side! The method lacks *Agrobacterium*'s finesse, but it does get the job done.

This "gene gun" was developed during the early 1980s by John Sanford at Cornell's Agricultural Research Station. Sanford chose to experiment with onions because of their conveniently large cells; he recalls that the combination of blasted onions and gunpowder made his lab smell like a McDonald's franchise on a firing range. Initial reactions to his concept were incredulous, but in 1987 Sanford unveiled his botanical firearm in the pages of *Nature*. By 1990, scientists had succeeded in using the gun to shoot new genes into corn, America's most important food crop, worth $19 billion in 2001 alone.

Corn is not only a valuable food crop; unique among major American crops, it also has long been a valuable seed crop. The seed business has traditionally been something of a financial dead-end: a farmer buys your seed, but then for subsequent plantings he can take seed from the crop he has just grown, so he never needs to buy your seed again. American corn seed companies solved the problem of nonrepeat business in the twenties by marketing hybrid corn, each hybrid the product of a cross between two particular genetic lines of corn. The hybrid's characteristic high yield makes it attractive to farmers. Because of the Mendelian mechanics of breeding, the strategy of using seed from the crop itself (i.e., the product of a hybrid × hybrid cross) fails because most of the seed will lack those high-yield characteristics of the original hybrid. Farmers therefore must return to the seed company every year for a new batch of high-yield hybrid seed.

America's biggest hybrid corn seed company, Pioneer Hi-Bred International (now owned by Du Pont), has long been a midwestern institution. Today it controls about 40 percent of the U.S. corn seed market, with $1 billion in annual sales. Founded in 1926 by Henry Wallace, who went on to become Franklin D. Roosevelt's vice president, the company used to hire as many as forty thousand high-schoolers every summer to ensure the hybridity of its hybrid corn. The two parental strains were grown in neighboring stands, and then these "detasselers" removed by hand the male pollen-producing flowers (tassels) before they became mature from one of the two strains. Therefore, only the other strain could serve as a possible source of pollen, so all the seed produced by the detasseled strain was sure to be hybrid. Even today, detasseling provides summer work for thousands: in July 2002, Pioneer hired thirty-five thousand temps for the job.

One of Pioneer's earliest customers was Roswell Garst, an Iowa farmer who, impressed by Wallace's hybrids, bought a license to sell Pioneer seed corn. On September 23, 1959, in one of the less frigid moments of the Cold War, the Soviet leader Nikita Khrushchev visited Garst's farm to learn more about the American agricultural miracle and the hybrid corn behind it. The nation Khrushchev had inherited from Stalin had neglected agriculture in the drive toward industrialization, and the new premier was keen to make amends. In 1961, the incoming Kennedy administration approved the sale to the Soviets of corn seed, agricultural equipment, and fertilizer, all of which contributed to the doubling of Soviet corn production in just two years.

As the GM food debate swirls around us, it is important to appreciate that our custom of eating food that has been genetically modified is actually thousands of years old. In fact, both our domesticated animals, the source of our meat, and the crop plants that furnish our grains, fruits, and vegetables, are very far removed genetically from their wild forebears (see Plate 36).

Agriculture did not suddenly arise, fully fledged, ten thousand years ago. Many of the wild ancestors of crop plants, for example, offered relatively little to the early farmers: they were low-yield and hard to grow. Modification was necessary if agriculture was to succeed. Early farmers understood that modification must be bred in ("genetic," we would say) if desirable characteristics were to be maintained from generation to generation. Thus began our agrarian ancestors' enormous program of genetic modification. And in the absence of gene guns and the like, this activity depended on some form of artificial selection, whereby farmers bred only those individuals exhibiting the desired traits – the cows with the highest milk yield, for example. In effect, the farmers were doing what nature does in the course of natural selection: picking and choosing from among the range of available genetic variants to ensure that the next generation would be enriched with those best adapted for consumption, in the case of farmers; for survival, in the case of nature. Biotechnology has given us a way to generate the desired variants, so that we do not have to wait for them to arise naturally; as such, it is but the latest in a long line of methods that have been used to *genetically modify* our food.

Weeds are difficult to eliminate. Like the crop whose growth they inhibit, they are plants too. How do you kill weeds without killing your crop? Ideally, there would be some kind of pass-over system whereby every plant lacking a "protective mark" – the weeds, in this case – would be killed, while those possessing the mark – the crop – would be spared. Genetic engineering has furnished farmers and gardeners just such a system in the form of Monsanto's "Roundup Ready" technology. "Roundup" is a broad-spectrum herbicide that can kill almost any plant. But through genetic alteration Monsanto scientists have also produced "Roundup Ready" crops that possess built-in resistance to the herbicide, and do just fine as all the weeds around them are biting the dust. Of course, it suits the company's commercial interests that farmers who buy Monsanto's adapted seed will buy Monsanto's herbicide as well. But such an approach is also actually beneficial to the environment. Normally a farmer must use a range of different weed killers, each one toxic to a particular group of weeds but safe for the crop. There are many potential weed groups to guard against. Using a single herbicide for all the weeds in creation actually reduces the environmental levels of such chemicals, and Roundup itself is rapidly degraded in the soil.

Unfortunately, the rise of agriculture was a boon not only to our ancestors but to herbivorous insects as well. Imagine being an insect that eats wheat and related wild grasses. Once upon a time, thousands of years ago, you had to forage far and wide for your dinner. Then along came agriculture, and humans conveniently started laying out dinner in enormous stands. It is not surprising that crops have to be defended against insect attack. From the elimination point of view at least, insects pose less of a problem than weeds because it is possible to devise poisons that target animals, not plants. The trouble is that humans and other creatures we value are animals as well.

The full extent of the risks involved with the use of pesticides was not widely apparent until Rachel Carson first documented them. The impact on the environment of long-lived chlorine-containing pesticides like DDT (banned in Europe and North America since 1972) has been devastating. In addition, there is a danger that residues from these pesticides will wind up in our food. While these chemicals at low dosage may not be lethal – they were, after all, designed to kill animals at a considerable evolutionary remove from us – there remain concerns about possible mutagenic effects, resulting in human cancers and birth defects. An alternative to DDT came in the form of a group of organophosphate pesticides, like parathion. In their favor, they decompose rapidly once applied and do not linger in the environment. On the other hand, they are even more acutely toxic than DDT; the sarin nerve gas used in the terrorist attack on the Tokyo subway system in 1995, for instance, is a member of the organophosphate group.

Even solutions using nature's own chemicals have produced a backlash. In the mid-1960s, chemical companies began developing synthetic versions of a natural insecticide, pyrethrin, derived from a small daisylike chrysanthemum. These helped keep farm pests in check for more than a decade until, not surprisingly, their widespread use led to the emergence of resistant insect populations. Even more troubling, however, pyrethrin, though natural, is not necessarily good for humans; in fact, like many plant-derived substances it can be quite toxic. Pyrethrin experiments with rats have produced Parkinson-like symptoms, and epidemiologists have noted that this disease has a higher incidence in rural environments than in urban ones. Overall – and there is a dearth of reliable data – the Environmental Protection Agency estimates that there may be as many as 300,000 pesticide-related illnesses among U.S. farmworkers every year.

Organic farmers have always had their tricks for avoiding pesticides. One ingenious organic method relies on a toxin derived from a bacterium – or, often, the bacterium itself – to protect plants from insect attack. *Bacillus thuringiensis* (Bt) naturally assaults the cells of insect intestines, feasting upon the nutrients released by the damaged cells. The guts of the insects exposed to the bacterium are paralyzed, causing the creatures to die from the combined effects of starvation and tissue damage. Originally identified in 1901, when it decimated Japan's silkworm population, *Bacillus thuringiensis* was not so named until 1911, during an outbreak among flour moths in the German province of Thuringia. First used as a pesticide in France in 1938, the bacterium was originally thought to work only against lepidopteran (moth/butterfly) caterpillars, but different strains have subsequently proved effective against the larvae of beetles and flies. Best of all, the bacterium is insect-specific: most animal intestines are acidic – that is, low pH – but the insect larval gut is highly alkaline – high pH – just the environment in which the pernicious Bt toxin is activated.

In the age of recombinant DNA technology the success of *Bacillus thuringiensis* as a pesticide has inspired genetic engineers. What if, instead of applying the bacterium scattershot to crops, the gene for the Bt toxin were engineered into the genome of crop plants? The farmer would never again need to dust his crops because every mouthful of the plant would be lethal to the insect ingesting it (and harmless to us). The method has at least two clear advantages over the traditional dumping of pesticides on crops. First, only insects that actually eat the crop will be exposed to the pesticide; non-pests are not harmed, as they would be with external application. Second, implanting the Bt toxin gene into the plant genome causes it to be produced by every cell of the plant; traditional pesticides are typically applied only to the leaf and stem. And so bugs that feed on the roots or that bore inside plant tissues, formerly immune to externally applied pesticides, are now also condemned to a Bt death.

Today we have a whole range of Bt designer crops, including "Bt corn," "Bt potato," "Bt cotton," and "Bt soybean," and the net effect has been a massive reduction in the use of pesticides. In 1995 cotton farmers in the Mississippi Delta sprayed their fields an average of 4.5 times per season. Just one year later, as Bt cotton caught on, that average – for all farms, including those planting non-Bt cotton varieties – dropped to 2.5 times. It is estimated that since 1996 the use of Bt crops has resulted in an annual reduction of 2 million gallons of pesticides in the United States. I have not visited cotton country lately but I would wager that billboards there are no longer hawking chemical insect-killers; in fact, I suspect that Burma-Shave ads are more likely to make a comeback than ones for pesticides. And other countries are starting to benefit as well: in China in 1999 the planting of Bt cotton reduced pesticide use by an estimated 1,300 *tons* (see Plate 37).

Biotechnology has also fortified plants against other traditional enemies in a surprising form of disease prevention superficially similar to vaccination. We inject our children with mild forms of various pathogens to induce an immune response that will protect them against infection when they are subsequently exposed to the disease. Remarkably when a plant, which has no immune system properly speaking, has been exposed to a particular virus, it often becomes resistant to other strains of the same virus. Roger Beachy at Washington University, in St. Louis, realized that this phenomenon of "cross-protection" might allow genetic engineers to "immunize" plants against threatening diseases. He tried inserting the gene for the virus's protein coat into the plants to see whether this might induce cross-protection without exposure to the virus itself. It did indeed. Somehow the presence in the cell of the viral coat protein prevents the cell from being taken over by invading viruses.

Beachy's method saved the Hawaiian papaya business. Between 1993 and 1997, production declined by 40 percent thanks to an invasion of the papaya ringspot virus; one of the islands' major industries was thus threatened with extinction. By inserting a gene for just part of the virus's coat protein into the papaya's genome, scientists were able to create plants resistant to attacks by the virus. Hawaii's papayas lived to fight another day.

Scientists at Monsanto later applied the same harmless method to combat a common disease caused by potato virus X. (Potato viruses are unimaginatively named. There is also a potato virus Y.) Unfortunately, McDonald's and other major players in the burger business feared the use of such modified spuds would lead to boycotts organized by the anti-GM food partisans. Consequently, the fries they now serve cost more than they should.

Nature conceived onboard defense systems hundreds of millions of years before human genetic engineers started inserting Bt genes into crop plants. Biochemists recognize a whole class of plant substances, so-called secondary products, that are not involved in the general metabolism of the plant. Rather, they are produced to protect against herbivores and other would-be attackers. The average plant is, in fact, stuffed full of chemical toxins developed by evolution. Over the ages, natural selection has understandably favored those plants containing the nastiest range of secondary products because they are less vulnerable to damage by herbivores. In fact, many of the substances that humans have learned to extract from plants for use as medicine (digitalis from the foxglove plant, used in precise doses, can treat heart patients), stimulants (cocaine from the coca plant), or pesticides (pyrethrin from chrysanthemums) belong to this class of secondary products. Poisonous to the plant's natural enemies, these substances constitute the plant's meticulously evolved defensive response.

Bruce Ames, who devised the Ames test, a procedure widely relied upon for determining whether or not a particular substance is carcinogenic, has noted that the natural chemicals in our food are every bit as lethal as the noxious chemicals we worry about. Referring to tests on rats, he takes coffee as an example:

> There are more rodent carcinogens in one cup of coffee than pesticide residues you get in a year. And there's still a thousand chemicals left to test in a cup of coffee. So it just shows our double standard: If it's synthetic we really freak out, and if it's natural we forget about it.

One ingenious set of chemical defenses in plants involves furanocoumarins, a group of chemicals that become toxic only when directly exposed to ultraviolet light. By this natural adaptation, the toxins are activated only when a herbivore starts munching on the plants, breaking open the cells and exposing their contents to sunlight. Furanocoumarins present in the peel of limes were responsible for a bizarre plague that struck a Club Med resort in the

Caribbean. The guests who found themselves afflicted with ugly rashes on their thighs had all participated in a game that involved passing a lime from one person to the next without using hands, feet, arms, or head. In the bright Caribbean sunlight the activated furanocoumarins in the humiliated lime had wreaked a terrible revenge on numerous thighs.

Plants and herbivores are involved in an evolutionary arms race: nature selects plants to be ever more toxic, and herbivores to be ever more efficient at detoxifying the plant's defensive substances while metabolizing the nutritious ones. In the face of furanocoumarins, some herbivores have evolved clever countermeasures. Some caterpillars, for example, roll up a leaf before starting to munch. Sunlight does not penetrate the shady confines of their leaf roll, and thus the furanocoumarins are not activated.

Adding a particular Bt gene to crop plants is merely one way the human species as an interested party can give plants a leg up in this evolutionary arms race. We should not be surprised, however, to see pest insects eventually evolve resistance to that particular toxin. Such a response, after all, is the next stage in the ancient conflict. When it happens, farmers will likely find that the multiplicity of available Bt toxin strains can furnish them yet another exit from the vicious evolutionary cycle: as resistance to one type becomes common, they can simply plant crops with an alternative strain of Bt toxin onboard.

In addition to defending a plant against its enemies, bio-technology can also help bring a more desirable product to market. Unfortunately, however, sometimes the cleverest biotechnologists can fail to see the forest for the trees (or the crop for the fruits). So it was with Calgene, an innovative California-based company. In 1994 Calgene earned the distinction of producing the very first GM product to reach supermarket shelves. Calgene had solved a major problem of tomato growing: how to bring ripe fruit to market instead of picking them when green, as is customary. But in their technical triumph they forgot fundamentals: their rather unfortunately named "Flavr-Savr" tomato was neither tasty nor cheap enough to succeed. And so it was that the tomato had the added distinction of being one of the first GM products to disappear from supermarket shelves.

Still, the technology was ingenious. Tomato ripening is naturally accompanied by softening, thanks to the gene encoding an enzyme called polygalacturonase (PG), which softens the fruit by breaking down the cell walls. Because soft tomatoes do not travel well, the fruit are typically picked when they are still green (and firm) and then reddened using ethene gas, a ripening agent. Calgene researchers figured that knocking out the PG gene would result in fruit that stayed firm longer, even after ripening on the vine. They inserted an inverted copy of the PG gene, which, owing to the affinities between complementary base pairs, had the effect of causing the RNA produced by the PG gene proper to become "bound up" with the RNA produced by the inverted gene, thus neutralizing the former's capacity to create the softening enzyme. The lack of PG function meant that the tomato stayed firmer, and so it was now possible in principle to deliver fresher, riper tomatoes to supermarket shelves. But Calgene, triumphant in its molecular wizardry, underestimated the trickiness of basic tomato farming. (As one grower hired by the company commented, "Put a molecular biologist out on a farm, and he'd starve to death.") The strain of tomato Calgene had chosen to enhance was a particularly bland and tasteless one: there simply was not much "flavr" to save, let alone savor. The tomato was a technological triumph but a commercial failure.

Overall, plant technology's most potentially important contribution to human well-being may involve enhancing the nutrient profile of crop plants, compensating for their natural shortcomings as sources of nourishment. Because plants are typically low in amino acids essential for human life, those who eat a purely vegetarian diet, among whom we may count most of the developing world, may suffer from amino acid deficiencies. Genetic engineering can ensure that crops contain a fuller array of nutrients, including amino acids, than the unmodified versions that would otherwise be grown and eaten in these parts of the world.

To take an example, in 1992 UNICEF estimated that some 124 million children around the world were dangerously deficient in vitamin A. The annual result is some half million cases of childhood blindness; many of these children will even die for want of the vitamin. Since rice does not contain vitamin A or its biochemical precursors, these deficient populations are concentrated in parts of the world where rice is the staple diet.

An international effort, funded largely by the Rockefeller Foundation (a nonprofit organization and therefore protected from the charges of commercialism or exploitation often leveled at producers of GM foods), has developed what has come to be called "golden rice." Though this rice doesn't contain vitamin A per se, it yields a critical precursor, beta-carotene (which gives carrots their bright orange color and golden rice the fainter orange tint that inspired its name). As those involved in humanitarian relief have learned, however, malnutrition can be more complex than a single deficiency: the absorption of vitamin A precursors in the gut works best in the presence of fat, but the malnourished whom the golden rice was designed to help often have little or no fat in their diet. Nevertheless golden rice represents at least one step in the right direction. It is here that we see the broader promise of GM agriculture to diminish human suffering.

We are merely at the beginning of a great GM plant revolution, only starting to see the astonishing range of potential applications. Apart from delivering nutrients where they are wanting, plants may also one day hold the key to distributing orally administered vaccine proteins. By simply engineering a banana that produces, say, the polio vaccine protein – which would remain intact in the fruit, which travels well and is most often eaten uncooked – we could one day distribute the vaccine to parts of the world that lack public health infrastructure. Plants may also serve less vital but still immensely helpful purposes. One company, for example, has succeeded in inducing cotton plants to produce a form of polyester, thereby creating a natural cotton-polyester blend. With such potential to reduce our dependence on chemical manufacturing processes (of which polyester fabrication is but one) and their polluting by-products, plant engineering will provide ways as yet unimagined to preserve the environment.

Monsanto was definitely the leader of the GM food pack, but naturally its primacy was challenged. The German pharmaceutical company Hoechst developed its own Roundup equivalent, an herbicide called Basta (or Liberty in the United States), with which they marketed "LibertyLink" crops genetically engineered for resistance. Another European pharmaceutical giant, Aventis, produced a version of Bt corn called "Starlink."

But Monsanto, aiming to capitalize on being biggest and first, aggressively lobbied the big seed companies, notably Pioneer, to license Monsanto's products. But Pioneer was still wed to its long-established hybrid corn methods so its response to the heated courtship was frustratingly lukewarm and, in deals made in 1992 and 1993, Monsanto looked inept when it was able to exact from the seed giant only a paltry $500,000 for rights to Roundup Ready soybeans and $38 million for Bt corn. When he became CEO of Monsanto in

1995, Robert Shapiro aimed to redress this defeat by positioning the company for all-out domination of the seed market. For a start, he broadened the attack on the old seed-business problem of farmers who replant using seed from last year's crop rather than paying the seed company a second time. The hybrid solution that worked so well for corn was unworkable for other crops. Shapiro, therefore, proposed that farmers using Bt seed sign a "technology agreement" with Monsanto, obliging them both to pay for use of the gene and to refrain from replanting with seed generated by their own crops. What Shapiro had engineered was a hugely effective way to make Monsanto anathema in the farming community.

Shapiro was an unlikely CEO for a midwestern agrichemical company. Working as a lawyer at the pharmaceutical outfit Searle, he had the marketing equivalent of science's "Eureka!" moment. By compelling Pepsi and Coca-Cola to put the name of Searle's brand of chemical sweetener on their diet soft drink containers, Shapiro made NutraSweet synonymous with a low-calorie lifestyle. In 1985, Monsanto acquired Searle and Shapiro started to make his way up the parent company's corporate ladder. Naturally, once he was appointed CEO, Mr. NutraSweet had to prove he was no one-trick pony.

In an $8 billion spending spree in 1997–98, Monsanto bought a number of major seed companies, including Pioneer's biggest rival, Dekalb, as Shapiro schemed to make Monsanto into the Microsoft of seeds. One of his intended purchases, the Delta and Pine Land Company, controlled 70 percent of the U.S. cottonseed market. Delta and Pine also owned the rights to an interesting biotech innovation invented in a U.S. Department of Agriculture research lab in Lubbock, Texas: a technique for preventing a crop from producing any fertile seeds. The ingenious molecular trick involves flipping a set of genetic switches in the seed before it is sold to the farmer. The crop develops normally but produces seeds incapable of germinating. Here was the real key to making money in the seed business! Farmers would *have* to come back every year to the seed company.

Though it might seem in principle counterproductive and something of an oxymoron, nongerminating seed is actually of general benefit to agriculture in the long run. If farmers buy seed every year (as they do anyway, in the case of hybrid corn), then the improved economics of seed production promote the development of new (and better) varieties. Ordinary (germinating) forms would always be available for those who wished them. Farmers would buy the nongerminating kind only if it were superior in yield and other characteristics farmers care about. In short, nongerminating technology, while closing off one option, provides farmers with more and ever improved seed choices.

For Monsanto, however, this technology precipitated a public relations disaster. Activists dubbed it the "terminator gene." They evoked visions of the downtrodden third world farmer, accustomed by tradition to relying on his last crop to provide seeds to sow for the new one. Suddenly finding his own seeds useless, he would have no choice but to return to the greedy multinational and, like Oliver Twist, beg pathetically for more. Monsanto backed off, a humiliated Shapiro publicly disavowed the technology, and the terminator gene remains out of commission to this day. Through the public relations fallout, its only real impact to date has been the termination of Monsanto's grandiose ambitions of the late 1990s.

Much of the hostility to GM foods, as we saw in the last chapter with bovine growth hormone, has been orchestrated by professional alarmists like Jeremy Rifkin. His counterpart in the United Kingdom, Lord Peter Melchett, was equally effective until he lost credibility in the environmental movement by quitting Greenpeace to join a public relations firm that has in the past worked for Monsanto. Rifkin, the son of a self-made plastic-bag manufacturer from Chicago, may differ in style from Melchett, a former Eton boy from a grand family, but they share a vision of corporate America as conspiratorial juggernaut pitted against the helpless common man.

Nor has the reception of GM foods been aided by the knee-jerk, politically craven attitudes and even scientific incompetence typical of governmental regulatory agencies – in the US the Food and Drug Administration (FDA) and the Environmental Protection Agency (EPA) – when they have been confronted with these new technologies. Roger Beachy, who first identified the "cross-protection" phenomenon that saved Hawaii's papaya farmers from ruin, remembers how the EPA responded to his breakthrough:

> I naively thought that developing virus-resistant plants in order to reduce the use of insecticides would be viewed as a positive advance. However the EPA basically said, "If you use a gene that protects the plant from a virus, which is a pest, that gene must be considered a pesticide." Thus the EPA considered the genetically transformed plants to be pesticidal. The point of the story is that as genetic sciences and biotech developed, the federal agencies were taken somewhat by surprise. The agencies did not have the background or expertise to regulate the new varieties of crop plants that were developed, and they did not have the background to regulate the environmental impacts of transgenic crops in agriculture.

An even more glaring instance of the government regulators' ineptitude came in the so-called Starlink episode. Starlink, a Bt corn variety produced by the European multinational Aventis, had run afoul of the EPA when its Bt protein was found not to degrade as readily as other Bt proteins in an acidic environment, one like that of the human stomach. In principle, therefore, eating Starlink corn *might* cause an allergic reaction, though there was never any evidence that it actually would. The EPA dithered. Eventually it decided to approve Starlink for use in cattle feed, but not for human consumption. And so under EPA "zero-tolerance" regulations, the presence of a single molecule of Starlink in a food product constituted illegal contamination. Farmers were growing Starlink and non-Starlink corn side by side, and non-Starlink crops inevitably became contaminated: even a single Starlink plant that had inadvertently found its way into the harvest from whole fields of non-Starlink was enough. Not surprisingly, Starlink began to show up in food products. The absolute quantities were tiny, but genetic testing to detect the presence of Starlink is supersensitive. In late September 2000, Kraft Foods launched a recall of taco shells deemed to be tainted with Starlink, and a week later Aventis began a buy-back program to recover Starlink seed from the farmers who had bought it. The estimated cost of this "cleanup" program: $100 million.

Blame for this debacle can only be laid at the door of an overzealous and irrational EPA. Permitting the use of corn for one purpose (animal feed) and not another (human consumption), and then mandating absolute purity in food is, as is now amply apparent, absurd. Let us be clear that if "contamination" is defined as the presence of a single molecule of a foreign substance, then every morsel of our food is contaminated! With lead, with DDT, with bacterial toxins, and a host of other scary things. What matters, from the point of view of public health, is the concentration levels of these substances, which can range from the negligible to the lethal. It should also be considered a reasonable requirement in labeling something a contaminant that there be at least minimal evidence of demonstrable detriment to health. Starlink has never been shown to harm anyone, not even a laboratory rat. The only positive outcome of this whole sorry episode has been a change in EPA policy abolishing "split" permits: an agricultural product will hereafter be approved for all food-related uses or not.

That the anti-GM food lobby is most powerful in Europe is no accident. Europeans, the British in particular, have good reason both to be suspicious about what is in their food and to distrust what they are told about it. In 1984, a farmer in the south of England first noticed that one of his cows was behaving strangely; by 1993, 100,000 British cattle had died from a new brain disease, bovine spongiform encephalopathy (BSE), commonly known as mad cow disease. Government ministers scrambled to assure the public that the disease, probably transmitted in cow fodder derived from remnants of slaughtered animals, was not transmissible to humans. By February 2002, 106 Britons had died from the human form of BSE. They had been infected by eating BSE-contaminated meat.

The insecurity and distrust generated by BSE has spilled over into the discussion of GM foods, dubbed by the British press "Frankenfoods." As Friends of the Earth announced in a press release in April 1997, "After BSE, you'd think the food industry would know better than to slip 'hidden' ingredients down people's throats." But that, more or less, is exactly what Monsanto was planning to do in Europe. Certain the anti-GM food campaign was merely a passing distraction, management pressed ahead with its plans to bring GM products to European supermarket shelves. It was to prove a major miscalculation: through 1998, the consumer backlash gained momentum. Headline writers at the British tabloids had a field day: "GM Foods Are Playing Games with Nature: If Cancer Is the Only Side-Effect We Will Be Lucky"; "Astonishing Deceit of GM Food Giant"; "Mutant Crops." Prime Minister Tony Blair's halfhearted defense merely provoked tabloid scorn: "The Prime Monster; Fury As Blair Says: I Eat Frankenstein Food and It's Safe." In March 1999, the British supermarket chain Marks and Spencer announced that it would not carry GM food products, and soon Monsanto's European biotech dreams were in jeopardy. Not surprisingly other food retailers took similar actions: it made good sense to show supersensitivity to consumer concerns, and no sense at all to stick one's neck out in support of an unpopular American multinational.

It was around this time of the Frankenfood maelstrom in Europe that news of the terminator gene and Monsanto's plans to dominate the global seed market began to circulate on the home front. With much of the opposition orchestrated by environmental groups, the company's attempts to defend itself were hamstrung by its own past. Having started out as a producer of pesticides, Monsanto was loath to incur the liability of explicitly renouncing these chemicals as environmental hazards. Yet one of the greatest virtues of both Roundup Ready and Bt technologies is the extent to which they reduce the need for herbicides and insecticides. The official industry line since the 1950s had been that proper use of the right pesticides harmed neither the environment nor the farmer applying them: Monsanto still could not now admit that Rachel Carson had been right all along. Unable to simultaneously condemn pesticides and sell them, the company could not make use of one of the most compelling of arguments in defense of the use of biotechnology on the farm.

Monsanto was never able to reverse this unfortunate momentum. In April 2000, the company effected a merger but its partner, the pharmaceutical giant Pharmacia & Upjohn, was primarily interested in acquiring Monsanto's drug division, Searle. The agricultural business, later spun off as an independent entity, still exists today under the name Monsanto. Gone, however, are the company's pioneering bravado and aura of invincibility.

The GM foods debate has conflated two distinct sets of issues. First, there have been the purely scientific questions of whether GM foods pose a threat to our health or to the environment. Second, there are economic and political questions centered on the practices of aggressive multinational companies and the effects of globalization. Much of the rhetoric has focused on agribusiness, Monsanto in particular. Having seemed throughout the 1990s to view the technology as little more than a means of dominating the world food supply, the company may indeed have harbored unwholesome dreams of becoming the Microsoft of the food industry, but since its stunning reversal of fortunes, this aspect of the controversy has been rendered largely baseless. It is not likely that another company with as much to lose will stumble into the same minefield. A meaningful evaluation of GM food should be based on scientific considerations, not political or economic ones. Let us therefore review some of the common claims.

It ain't natural. Virtually no human being, save the very few remaining genuine hunter-gatherers, eats a strictly "natural" diet. *Pace* Prince Charles, who famously declared in 1998 that "this kind of genetic modification takes mankind into realms that belong to God," our ancestors have in fact been fiddling in these realms for eons.

Early plant breeders often crossed different species, bringing into existence entirely new ones with no direct counterparts in nature. Wheat, for example, is the product of a whole series of crosses. Einkorn wheat, a naturally occurring progenitor, crossed with a species of goat grass, produced emmer wheat. And the bread wheat we know was produced by a subsequent crossing of emmer with yet another goat grass. Our wheat is thus a combination – perhaps one nature would have never devised – of the characteristics of all these ancestors.

Furthermore, crossing plants in this way results in the wholesale generation of genetic novelty: every gene is affected, often with unforeseeable effects. Biotechnology, by contrast, allows us to be much more precise in introducing new genetic material into a plant species, one gene at a time. It is the difference between traditional agriculture's genetic sledgehammer and biotech's genetic tweezers.

It will result in allergens and toxins in our food. Again, the great advantage of today's transgenic technologies is the precision they allow us in determining how we change the plant. Aware that certain substances tend to provoke allergic reactions, we can accordingly avoid them. But this concern persists, stemming to some degree from an oft-told tale about the addition of a Brazil nut protein to soybeans. It was a well-intentioned undertaking: the West African diet is often deficient in methionine, an amino acid abundant in a protein produced by Brazil nuts. It seemed a sensible solution to insert the gene for the protein into West Africa's soybean, but then someone remembered that there is a common allergic reaction to Brazil nut proteins that can have serious consequences, and so the project was shelved. Obviously the scientists involved had no intention of unleashing a new food that would promptly send thousands of people into anaphylactic shock; they halted the project once the serious drawbacks were appreciated. But for most commentators it was an instance of molecular engineers playing with fire, heedless of the consequences. In principle, genetic engineering can actually *reduce* the instance of allergens in food: perhaps the Brazil nut itself will one day be available free of the protein that was deemed unsafe to import into the soybean.

It is indiscriminate, and will result in harm to nontarget species. In 1999 a now-famous study showed that monarch butterfly caterpillars feeding on leaves heavily dusted with pollen from Bt corn were prone to perish. This was scarcely surprising: Bt pollen contains the Bt gene, and therefore the Bt toxin, and the toxin is intentionally lethal to insects. But everyone loves butterflies, and so environmentalists opposed to GM foods had found an icon. Would the monarch, they wondered, be but the first of many inadvertent victims of GM technology? Upon examination, the experimental conditions under which the caterpillars were tested were found to be so extreme – the levels of the Bt pollen so high – as to tell us virtually nothing of practical value about the likely mortality of caterpillar populations in nature. Indeed, further study has suggested that the impact of Bt plants on the monarch (and other nontarget insects) is trivial. But even if it were not, we should ask how it might compare with the effects of the traditional non-GM alternative: pesticides. As we have seen, in the absence of GM methods, these substances must be applied liberally if we are to have agriculture that is as productive as modern society requires. Whereas the toxin built into Bt plants affects only those insects that actually feed off the plant tissue (and to some lesser degree, insects exposed to Bt pollen), pesticides unambiguously affect all insects exposed, pest and non-pest alike. The monarch butterfly, were it capable of weighing in on the debate, would assuredly cast its vote in favor of Bt corn.

It will lead to an environmental meltdown with the rise of "superweeds." The worry here is that genes for herbicide resistance (like those in Roundup Ready plants) will migrate out of the crop genome into that of the weed population through interspecies hybridization. This is not inconceivable, but it is unlikely to occur on a wide scale for the following reason: interspecies hybrids tend to be feeble creations, not well equipped for survival. This is especially true when one of the species is a domesticated variety bred to thrive only when mollycoddled by a farmer. But let us suppose, for argument's sake, that the resistance gene does enter the weed population and is sustained there. It would not actually be the end of the world, or even of agriculture, but rather an instance of something that has occurred frequently in the history of farming: resistance arising in pest species in response to attempts to eradicate them. The most famous example is the evolution of resistance to DDT in pest insects. In applying a pesticide, a farmer is exerting strong natural selection in favor of resistance, and evolution, we know, is a subtle and able foe: resistance arises readily. The result is that the scientists have to go back to the drawing board and come up with a new pesticide or herbicide, one to which the target species is not resistant; the whole evolutionary cycle will then run its course before culminating once more in the evolution of resistance in the target species. The acquisition of resistance, therefore, is the potential undoing of virtually all attempts to control pests; it is by no means peculiar to GM strategies. It's simply the bell that signals the next round, and summons human ingenuity to invent anew.

Despite her concern about the impact of multinational corporations on farmers in countries like India, Suman Sahai of the New Delhi-based Gene Campaign has pointed out that the GM food controversy is a feature of societies for which food is not a life-and-death issue. In India, where people literally starve to death, as Sahai points out, up to 60 percent of fruit grown in hill regions rots before it reaches market. Just imagine the potential good of a technology that delays ripening, like the one used to create the Flavr-Savr tomato. The most important role of GM foods may lie in the salvation they offer developing regions, where surging birthrates and the pressure to produce on the limited available arable land lead to an overuse of pesticides and herbicides with devastating effects upon both the environment and the farmers applying them; where nutritional deficiencies are a way of life and, too often, of death; and where the destruction of one crop by a pest can be a literal death sentence for farmers and their families.

As we have seen, the invention of recombinant DNA methods in the early 1970s resulted in a round of controversy and soul-searching centered on the Asilomar conference. Now it is happening all over again. At the time of Asilomar, it may at least be said, we were facing several major unknowns: we could not then say for certain that manipulating the genetic makeup of the human gut bacterium, *E. coli*, would not result in new strains of disease-causing bacteria. But our quest to understand and our pursuit of potential for good proceeded, however haltingly. In the case of the present controversy, anxieties persist despite our much greater understanding of what we are actually doing. While a considerable proportion of Asilomar's participants urged caution, today one would be hard-pressed to find a scientist opposed in principle to GM foods. Recognizing the power of GM technologies to benefit both our species and the natural world, even the renowned environmentalist E. O. Wilson has endorsed them: "Where genetically engineered crop strains prove nutritionally and environmentally safe upon careful research and regulation . . . they should be employed."

The opposition to GM foods is largely a sociopolitical movement whose arguments, though couched in the language of science, are typically unscientific. Indeed, some of the anti-GM pseudoscience propagated by the media – whether in the interests of sensationalism or out of misguided but well-intentioned concern – would be actually amusing were it not evident that such gibberish is in fact an effective weapon in the propaganda war. Monsanto's Rob Horsch has had his fair share of run-ins with protesters:

> I was once accused of bribing farmers by an activist at a press conference in Washington, D.C. I asked what they meant. The activist answered that by giving farmers a better performing product at a cheaper price those farmers profited from using our products. I just looked at them with my mouth hanging open.

Let me be utterly plain in stating my belief that it is nothing less than an absurdity to deprive ourselves of the benefits of GM foods by demonizing them; and, with the need for them so great in the developing world, it is nothing less than a crime to be governed by the irrational suppositions of Prince Charles and others.

In fact, a few years from now, when the West inevitably regains its senses and throws off the shackles of Luddite paranoia, it may find itself seriously lagging in agricultural technology. Food production in Europe and the United States will come to be more expensive and less efficient than elsewhere in the world. Meanwhile, countries like China, which can ill afford to entertain illogical misgivings, will forge ahead. The Chinese attitude is entirely pragmatic: With 23 percent of the world's population but only 7 percent of its arable land, China *needs* the increased yields and added nutritional value of GM crops if it is to feed its population.

On reflection, we erred too much on the side of caution at Asilomar, quailing before unquantified (indeed, unquantifiable) concerns about unknown and unforeseeable perils. But after a needless and costly delay, we resumed our pursuit of science's highest moral obligation: to apply what is known for the greatest possible benefit of humankind. In the current controversy, as our society delays in sanctimonious ignorance, we would do well to remember how much is at stake: the health of hungry people and the preservation of our most precious legacy, the environment.

In July 2000 anti-GM-food protesters vandalized a field of experimental corn at Cold Spring Harbor Lab. In fact there were no GM plants in the field; all the vandals managed to destroy was two years' hard work on the part of two young scientists at the lab. But the story is instructive all the same. At a time in which the destruction of GM crops has become positively fashionable in parts of Europe, when even the pursuit of knowledge on that continent and this one can come under attack, those in the vanguard of the cause might do well to ask themselves: what are we fighting *for*?

CHAPTER SEVEN

THE HUMAN GENOME:
LIFE'S SCREENPLAY

The human body is bewilderingly complex. Traditionally biologists have focused on one small part and tried to understand it in detail. This basic approach did not change with the advent of molecular biology. Scientists for the most part still specialize on one gene or on the genes involved in one biochemical pathway. But the parts of any machine do not operate independently. If I were to study the carburetor of my car engine, even in exquisite detail, I would still have no idea about the overall function of the engine, much less the entire car. To understand what an engine is for, and how it works, I'd need to study the whole thing – I'd need to place the carburetor in context, as one functioning part among many. The same is true of genes. To understand the genetic processes underpinning life, we need more than a detailed knowledge of particular genes or pathways; we need to place that knowledge in the context of the entire system – the genome.

The genome is the entire set of genetic instructions in the nucleus of every cell. (In fact, each cell contains *two* genomes, one derived from each parent: the two copies of each chromosome we inherit furnish us with two copies of each gene, and therefore two copies of the genome.) Genome sizes vary from species to species. From measurements of the amount of DNA in a single cell, we have been able to estimate that the human genome – half the DNA contents of a single nucleus – contains some 3.1 billion base pairs: 3,100,000,000 As, Ts, Gs, and Cs (see Plate 38).

Genes figure in our every success and woe, even the ultimate one: they are implicated to some extent in all causes of mortality except accidents. In the most obvious cases, diseases like cystic fibrosis and Tay-Sachs are caused directly by mutations. But there are many other genes whose work is just as deadly, if more oblique, influencing our susceptibility to common killers like cancer and heart disease, both of which may run in families. Even our response to infectious diseases like measles and the common cold has a genetic component since the immune system is governed by our DNA. And aging is largely a genetic phenomenon as well: the effects we associate with getting older are to some extent a reflection of the lifelong accumulation of mutations in our genes. Thus, if we are to understand fully, and ultimately come to grips with, these life-or-death genetic factors, we must have a complete inventory of all the genetic players in the human body.

Above all, the human genome contains the key to our humanity. The freshly fertilized egg of a human and that of a chimpanzee are, superficially at least, indistinguishable, but one contains the human genome and the other the chimp genome. In each, it is the DNA that oversees the extraordinary transformation from a relatively simple single cell to the stunningly complex adult of the species, comprised, in the human instance, of 100 trillion cells. But only the chimp genome can make a chimp, and only the human genome a human. The human genome is the great set of assembly instructions that governs the development of every one of us. Human nature itself is inscribed in that book.

Understanding what is at stake, one might imagine that to champion a project seeking to sequence all the human genome's DNA would be no more controversial than sticking up for Mom and apple pie. Who in his right mind would object? In the mid-1980s, however, when the possibility of sequencing the genome was first discussed, this was viewed by some as a decidedly dubious idea. To others it simply seemed too preposterously ambitious. It was like suggesting to a Victorian balloonist that we attempt to put a man on the moon.

It was a telescope, of all things, that inadvertently helped inaugurate the Human Genome Project (HGP). In the early 1980s, astronomers at the University of California proposed to build the biggest, most powerful telescope in the world, with a projected cost of some $75 million. When the Max Hoffman Foundation pledged $36 million, a grateful UC agreed to name the project for its generous benefactor. Unfortunately, this way of saying thank-you complicated the business of raising the remaining money. Other potential donors were reluctant to put up funds for a telescope already named for someone else, so the project stalled. Eventually, a second, much wealthier California philanthropy, the W. M. Keck Foundation, stepped in with a pledge to underwrite the entire project. UC was happy to accept, Hoffman or no. (The new Keck telescope, on the summit of Mauna Kea in Hawaii, would be fully operational by May 1993.) Unprepared to play second fiddle to Keck, the Hoffman Foundation withdrew its pledge, and UC administrators sensed a $36 million opportunity. In particular, Robert Sinsheimer, chancellor of

UC Santa Cruz, realized that the Hoffman money could bankroll a major project that would "put Santa Cruz on the map."

Sinsheimer, a biologist by training, was keen to see his field enter the major leagues of big-money sciences. Physicists had their pricey supercolliders, astronomers their $75 million telescopes and satellites; why shouldn't biologists have their own high-profile, big-money project? So he suggested that Santa Cruz build an institute dedicated to sequencing the human genome; in May 1985, a conference convened at Santa Cruz to discuss Sinsheimer's idea. Overall it was deemed too ambitious and the participants agreed that the initial emphasis should instead be on exploring particular regions of the genome that were of medical importance. In the end, the discussion was moot because the Hoffman money did not actually make its way into the University of California's coffers. However, the Santa Cruz meeting had sown the seed.

The next step toward the Human Genome Project also came from deep in left field: the U.S. Department of Energy (DOE). Though its brief naturally concentrated on the nation's energy needs, the DOE did have at least one biological mandate: to assess the health risks of nuclear energy. In this connection, it had funded monitoring of long-term genetic damage in survivors of the atomic blasts at Nagasaki and Hiroshima and their descendants. What could be more useful in identifying mutations caused by radiation than a full reference sequence of the human genome? In the fall of 1985, the DOE's Charles DeLisi called a meeting to discuss his agency's genome initiative. The biological establishment was skeptical at best: Stanford geneticist David Botstein condemned the project as "DOE's program for unemployed bomb-makers," and James Wyngaarden, then head of the National Institutes of Health (NIH), likened the idea to "the National Bureau of Standards proposing to build the B-2 bomber." Not surprisingly, the NIH itself was eventually to become the most prominent member of the Human Genome Project coalition; nevertheless, the DOE played a significant role throughout the project, and, in the final reckoning, would be responsible for some 11 percent of the sequencing.

By 1986 the genome buzz was getting stronger. That June, I organized a special session to discuss the project during a major meeting on human genetics at Cold Spring Harbor Laboratory. Wally Gilbert, who had attended Sinsheimer's meeting the year before in California, took the lead by making a daunting cost projection: 3 billion base pairs, 3 billion dollars. This was big-money science for sure. It was an inconceivable sum to imagine without public funding, and some at the meeting were naturally concerned that the megaproject, whose success was hardly assured, would inevitably suck funds away from other critical research. The Human Genome Project, it was feared, would become scientific research's ultimate money pit. And at the level of the individual scientific ego, there was, even in the best case, relatively little career bang for the buck. While the HGP promised technical challenges aplenty, it failed to offer much in the way of intellectual thrill or fame to those who actually met them. Even an important breakthrough would be dwarfed by the size of the undertaking as a whole and who was going to dedicate his life to the endless tedium of sequencing, sequencing, sequencing? Stanford's David Botstein, in particular, demanded extreme caution: "It means changing the structure of science in such a way as to indenture us all, especially the young people, to this enormous thing like the Space Shuttle."

Despite the less than overwhelming endorsement, that meeting at Cold Spring Harbor Laboratory convinced me that sequencing the human genome was destined soon to become an international scientific priority, and that, when it did, the NIH should be a major player. I persuaded the James S. McDonnell Foundation to fund an in-depth study of the relevant issues under the aegis of the National Academy of Sciences (NAS). With Bruce Alberts of UC San Francisco chairing the committee, I felt assured that all ideas would be subject to the fiercest scrutiny. Not long before, Alberts had published an article warning that the rise of "big science" threatened to swamp traditional research's vast archipelago of innovative contributions from individual labs the world over. Without knowing for sure what our group would find, I took my place, along with Wally Gilbert, Sydney Brenner, and David Botstein, on the fifteen-member committee that during 1987 would hammer out the details of a potential genome project.

In those early days, Gilbert was the Human Genome Project's most forceful proponent. He rightly called it "an incomparable tool for the investigation of every aspect of human function." But having discovered the allure of the heady biotech mix of science and business at Biogen, the company he had helped found, Gilbert saw in the genome an extraordinary new business opportunity. And so, after serving briefly, he ceded his spot on the committee to Washington University's Maynard Olson to avoid any possible conflict of interest. Molecular biology had already proved its potential as big business, and Gilbert saw no need to go begging at the public trough. He reasoned that a private company with its own enormous sequencing laboratory could do the job and then sell genome information to pharmaceutical manufacturers and other interested parties. In spring 1987, Gilbert announced his plan to form Genome Corporation. Deaf to the howls of complaint at the prospect of genome data coming under private ownership (thus possibly limiting its application for the general good), Gilbert set about trying to raise venture capital. Unfortunately, he was handicapped at the outset by his own less-than-golden track record as a CEO. Following his resignation in 1982 from the Harvard faculty to take the reins of Biogen, the company promptly lost $11.6 million in 1983 and $13 million in 1984. Understandably, Gilbert took refuge behind ivy-covered walls, returning to Harvard in December 1984, but Biogen continued to lose money after his departure. It was hardly the stuff of a mouth-watering investment prospectus, but ultimately Gilbert's grand plan foundered owing more to circumstances beyond his control than to any managerial shortcoming: the stock market crash of October 1987 abruptly terminated Genome Corp.'s gestation.

In fact, Gilbert was guilty of nothing as much as being ahead of his time. His plan was not so different from the one Celera Genomics would implement so successfully a full ten years after Genome Corp. was stillborn. And the concerns his venture provoked about the private ownership of DNA sequence data would come into ever sharper focus as the HGP progressed.

The plan our Gilbert-less NAS committee devised under Alberts made sense at the time – and indeed the Human Genome Project has been carried out more or less according to its prescriptions. Our cost and timing projections have also proved respectably close to the mark. Knowing, as any PC owner has learned, that over time technology gets both better and cheaper, we recommended that the lion's share of actual sequencing work be put off until the techniques reached a sensibly cost-effective level. In the meanwhile, the improvement of sequencing technologies should have high priority. In part toward this end, we recommended that the (smaller) genomes of simpler organisms be sequenced as well. The knowledge gained thereby would be valuable both intrinsically (as a basis for enlightening comparisons with the eventual human sequence) and as a means for honing our methods before attacking the big enchilada. (Of course the obvious nonhuman candidates were the geneticists' old flames: *E. coli*, baker's yeast, *C. elegans* [the nematode worm popularized for research by Sydney Brenner], and the fruit fly.)

1. 1953: Francis Crick (right) and me with our model of the double helix

2. *The key to Mendel's triumph: genetic variation in pea plants*

3. *Notoriously camera shy T. H. Morgan was photographed surreptitiously while at work in the fly room.*

LIKE A TREE
EUGENICS DRAWS ITS MATERIALS FROM MANY SOURCES AND ORGANIZES
THEM INTO AN HARMONIOUS ENTITY.

4. *Eugenics as it was perceived during the first part of the twentieth century: an opportunity for humans to control their own evolutionary destiny*

5. *"Large family" winner, Fitter Families Contest, Texas State Fair (1925)*

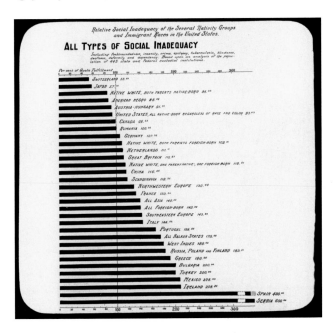

6. *Scientific racism: social inadequacy in the United States analyzed by national group (1922). "Social inadequacy" is used here by Harry Laughlin as an umbrella term for a host of sins ranging from feeblemindedness to tuberculosis. Laughlin computed an institutional "quota" for each group on the basis of the proportion of that group in the U.S. population as a whole. Shown, as a percentage, is the number of institutionalized individuals from a particular group divided by the group's quota. Groups scoring over 100 percent are overrepresented in institutions.*

7. *The physicist Erwin Schrödinger, whose book* What Is Life? *turned me on to the gene*

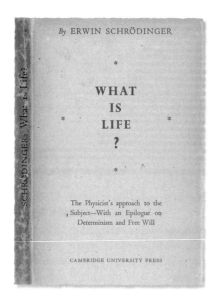

8. *Erwin Schrödinger's* What Is Life? *published in 1944.*

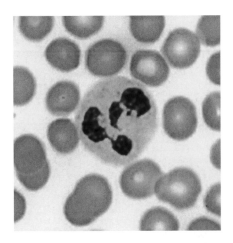

9. *A view through the microscope of blood cells treated with a chemical that stains DNA. In order to maximize their oxygen-transporting capacity, red blood cells have no nucleus and therefore no DNA. But white blood cells, which patrol the bloodstream in search of intruders, have a nucleus containing chromosomes.*

10. *Lawrence Bragg (left) with Linus Pauling, who is carrying a model of the α-helix*

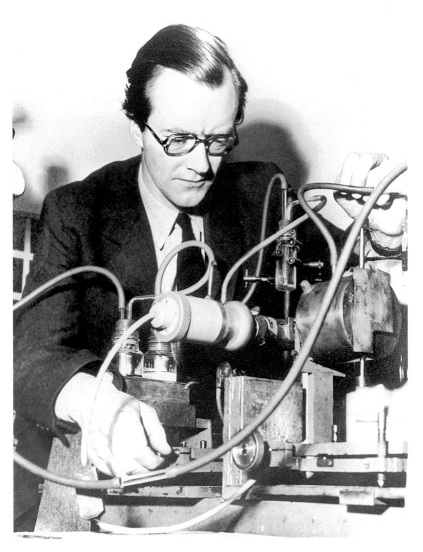

11. *Maurice Wilkins in his lab at King's College, London*

12. Rosalind Franklin on one of the mountain hiking vacations she loved

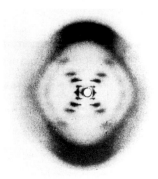

13. X-ray photos of the A and B forms of DNA from, respectively, Maurice Wilkins and Rosalind Franklin. The differences in molecular structure are caused by differences in the amount of water associated with each DNA molecule.

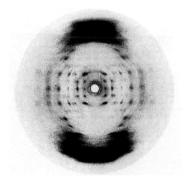

14. The chemical backbone of DNA

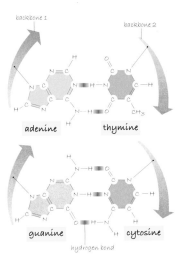

15. The insight that made it all come together: complementary pairing of the bases

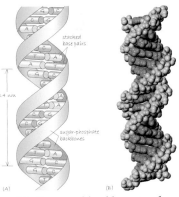

16. Bases and backbone in place: the double helix. (A) is a schematic showing the system of base-pairing that binds the two strands together. (B) is a "space-filling" model showing, to scale, the atomic detail of the molecule.

MOLECULAR STRUCTURE OF NUCLEIC ACIDS

A Structure for Deoxyribose Nucleic Acid

WE wish to suggest a structure for the salt of deoxyribose nucleic acid (D.N.A.). This structure has novel features which are of considerable biological interest.

A structure for nucleic acid has already been proposed by Pauling and Corey[1]. They kindly made their manuscript available to us in advance of publication. Their model consists of three intertwined chains, with the phosphates near the fibre axis, and the bases on the outside. In our opinion, this structure is unsatisfactory for two reasons : (1) We believe that the material which gives the X-ray diagrams is the salt, not the free acid. Without the acidic hydrogen atoms it is not clear what forces would hold the structure together, especially as the negatively charged phosphates near the axis will repel each other. (2) Some of the van der Waals distances appear to be too small.

Another three-chain structure has also been suggested by Fraser (in the press). In his model the phosphates are on the outside and the bases on the inside, linked together by hydrogen bonds. This structure as described is rather ill-defined, and for this reason we shall not comment on it.

We wish to put forward a radically different structure for the salt of deoxyribose nucleic acid. This structure has two helical chains each coiled round the same axis (see diagram). We have made the usual chemical assumptions, namely, that each chain consists of phosphate diester groups joining β-D-deoxyribofuranose residues with 3′,5′ linkages. The two chains (but not their bases) are related by a dyad perpendicular to the fibre axis. Both chains follow right-handed helices, but owing to the dyad the sequences of the atoms in the two chains run in opposite directions. Each chain loosely resembles Furberg's[2] model No. 1 ; that is, the bases are on the inside of the helix and the phosphates on the outside. The configuration of the sugar and the atoms near it is close to Furberg's 'standard configuration', the sugar being roughly perpendicular to the attached base. There

This figure is purely diagrammatic. The two ribbons symbolize the two phosphate—sugar chains, and the horizontal rods the pairs of bases holding the chains together. The vertical line marks the fibre axis

is a residue on each chain every 3·4 A. in the z-direction. We have assumed an angle of 36° between adjacent residues in the same chain, so that the structure repeats after 10 residues on each chain, that is, after 34 A. The distance of a phosphorus atom from the fibre axis is 10 A. As the phosphates are on the outside, cations have easy access to them.

The structure is an open one, and its water content is rather high. At lower water contents we would expect the bases to tilt so that the structure could become more compact.

The novel feature of the structure is the manner in which the two chains are held together by the purine and pyrimidine bases. The planes of the bases are perpendicular to the fibre axis. They are joined together in pairs, a single base from one chain being hydrogen-bonded to a single base from the other chain, so that the two lie side by side with identical z-co-ordinates. One of the pair must be a purine and the other a pyrimidine for bonding to occur. The hydrogen bonds are made as follows : purine position 1 to pyrimidine position 1 ; purine position 6 to pyrimidine position 6.

If it is assumed that the bases only occur in the structure in the most plausible tautomeric forms (that is, with the keto rather than the enol configurations) it is found that only specific pairs of bases can bond together. These pairs are : adenine (purine) with thymine (pyrimidine), and guanine (purine) with cytosine (pyrimidine).

In other words, if an adenine forms one member of a pair, on either chain, then on these assumptions the other member must be thymine ; similarly for guanine and cytosine. The sequence of bases on a single chain does not appear to be restricted in any way. However, if only specific pairs of bases can be formed, it follows that if the sequence of bases on one chain is given, then the sequence on the other chain is automatically determined.

It has been found experimentally[3,4] that the ratio of the amounts of adenine to thymine, and the ratio of guanine to cytosine, are always very close to unity for deoxyribose nucleic acid.

It is probably impossible to build this structure with a ribose sugar in place of the deoxyribose, as the extra oxygen atom would make too close a van der Waals contact.

The previously published X-ray data[5,6] on deoxyribose nucleic acid are insufficient for a rigorous test of our structure. So far as we can tell, it is roughly compatible with the experimental data, but it must be regarded as unproved until it has been checked against more exact results. Some of these are given in the following communications. We were not aware of the details of the results presented there when we devised our structure, which rests mainly though not entirely on published experimental data and stereochemical arguments.

It has not escaped our notice that the specific pairing we have postulated immediately suggests a possible copying mechanism for the genetic material.

Full details of the structure, including the conditions assumed in building it, together with a set of co-ordinates for the atoms, will be published elsewhere.

We are much indebted to Dr. Jerry Donohue for constant advice and criticism, especially on interatomic distances. We have also been stimulated by a knowledge of the general nature of the unpublished experimental results and ideas of Dr. M. H. F. Wilkins, Dr. R. E. Franklin and their co-workers at King's College, London. One of us (J. D. W.) has been aided by a fellowship from the National Foundation for Infantile Paralysis.

J. D. WATSON
F. H. C. CRICK

Medical Research Council Unit for the
  Study of the Molecular Structure of
    Biological Systems,
Cavendish Laboratory, Cambridge.
April 2.

[1] Pauling, L., and Corey, R. B., Nature, 171, 346 (1953) ; Proc. U.S. Nat. Acad. Sci., 39, 84 (1953).
[2] Furberg, S., Acta Chem. Scand., 6, 634 (1952).
[3] Chargaff, E., for references see Zamenhof, S., Brawerman, G., and Chargaff, E., Biochim. et Biophys. Acta, 9, 402 (1952).
[4] Wyatt, G. R., J. Gen. Physiol., 36, 201 (1952).
[5] Astbury, W. T., Symp. Soc. Exp. Biol. 1, Nucleic Acid, 66 (Camb. Univ. Press, 1947).
[6] Wilkins, M. H. F., and Randall, J. T., Biochim. et Biophys. Acta, 10, 192 (1953).

17. Short and sweet: our Nature *paper announcing the discovery. The same issue also carried longer articles by Rosalind Franklin and Maurice Wilkins.*

18. *Matt Meselson beside an ultra-centrifuge, the hardware at the heart of "the most beautiful experiment in biology"*

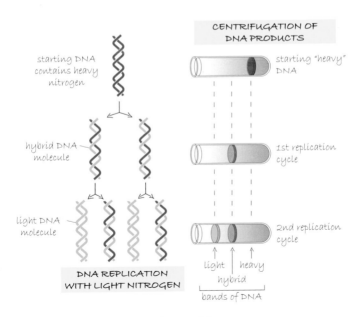

CENTRIFUGATION OF DNA PRODUCTS

starting DNA contains heavy nitrogen

hybrid DNA molecule

light DNA molecule

DNA REPLICATION WITH LIGHT NITROGEN

starting "heavy" DNA

1st replication cycle

2nd replication cycle

light
hybrid
heavy

bands of DNA

19. *The Meselson-Stahl experiment*

20. Arthur Kornberg at the time of winning his Nobel Prize

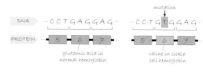

21. The impact of mutation. A single base change in the DNA sequence of the human beta hemoglobin gene results in the incorporation of the amino acid valine rather than glutamic acid into the protein. This single difference causes sickle-cell disease, in which the red blood cells become distorted into a characteristic sickle shape.

22. The genetic code, showing the triplet sequences for messenger RNA. An important difference between DNA and RNA is that DNA uses thymine and RNA uracil. Both bases are complementary to adenine. Stop codons do what their name suggests: they mark the end of the coding part of a gene.

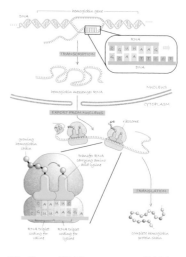

23. From DNA to protein. DNA is transcribed in the nucleus into messenger RNA, which is then exported to the cytoplasm for translation into protein. Translation occurs in ribosomes: transfer RNAs complementary to each base pair triplet codon in the messenger RNA deliver amino acids, which are bonded together to form a protein chain.

24. François Jacob, Jacques Monod, and André Lwoff

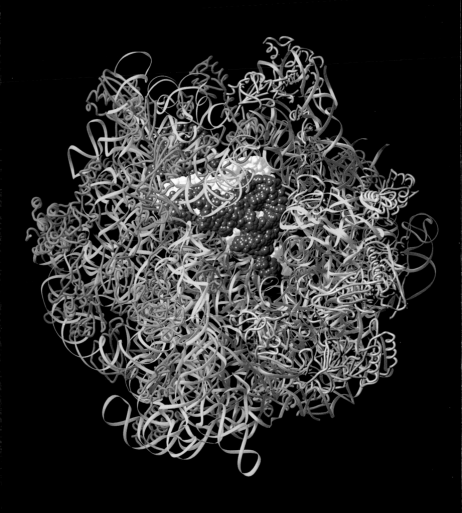

25. The cell's protein factory, the ribosome, in all its 3-D glory as revealed by X-ray analysis. (For simplicity, this computer-generated image does not show individual atoms.) There are.millioms of ribosomes in every cell. It is here that the information encoded in DNA is used to produce proteins, the actors in life's molecular drama. The ribosome consists of two subunits (orange and yellow), each composed of RNA, plus some sixty proteins (blue and green) plastered over the outside. Here the ribosome is caught in the act of producing a protein. Specialized small RNA molecules (purple, white, and red) transport amino acids to the ribosome for incorporation into the growing protein chain.

Meanwhile, we should concentrate on *mapping* the genome as accurately as possible. Mapping would be both genetic and physical. Genetic mapping entails determining relative positions, the order of genetic landmarks along the chromosomes, just as Morgan's boys had originally done for the chromosomes of fruit flies. Physical mapping entails actually identifying the absolute positions of those genetic landmarks on the chromosome. (Genetic mapping tells you that gene 2, say, lies between genes 1 and 3; physical mapping tells you that gene 2 is 1 million base pairs from gene 1, and gene 3 is located 2 million base pairs further along the chromosome.) Genetic mapping would lay out the basic structure of the genome; physical mapping would provide the sequencers, when eventually they were let loose on the genome, with fixed positional anchors along the chromosomes. The location on a chromosome of each separate chunk of sequence could then be determined by reference to those anchors.

We estimated that the entire project would take about fifteen years and cost about $200 million per year. We did a lot more fancy arithmetic, but there was no getting away from Gilbert's $1 per base pair estimate. Each space shuttle mission costs some $470 million. The Human Genome Project would cost six space shuttle launches.

Our report was published in February 1988. The rough draft of the genome was published in 2001. The gaps continue to be filled in by sequencing labs around the world as I write, and in 2003 – the fiftieth anniversary of the discovery of the double helix and the fifteenth of the committee's report – we will have seen the completion of the sequence.

While the NAS committee was still deliberating, I went to see key members of the House and Senate subcommittees on health that oversee the NIH's budget. James Wyngaarden, head of NIH, was in favor of the genome project "from the very start," as he put it, but less farsighted individuals at NIH were opposed. In my pitch for $30 million to get NIH on the genome track, I emphasized the medical implications of knowing the genome sequence. Lawmakers, like the rest of us, have all too often lost loved ones to diseases like cancer that have genetic roots, and could appreciate how knowing the sequence of the human genome would facilitate our fight against such diseases. In the end we got $18 million.

Meanwhile the DOE was able to secure $12 million for its own effort, mainly by playing up the project as a technological feat. This, one must remember, was the era of Japanese dominance in manufacturing technology; Detroit was in peril of being run over by Japan's automobile industry, and many feared the American edge in high-tech would be the next domino to fall. Rumor had it that three giant Japanese conglomerates (Matsui, Fuji, and Seiko) had combined forces to produce a machine capable of sequencing 1 million base pairs a day. It turned out to be a false alarm, but such anxieties ensured that the U.S. genome initiative would be pursued with the sort of fervor that put Americans on the moon before the Soviets.

In May 1988 Wyngaarden asked me to run NIH's side of the project. When I expressed reluctance to forsake the directorship of the Cold Spring Harbor Laboratory, he was able to arrange for me to do the NIH job on a part-time basis. I couldn't say no. Eighteen months later, with the HGP fast becoming an irresistible force, NIH's genome office was upgraded to the National Center for Human Genome Research; I was appointed its first director.

It was my job both to pry the cash away from Congress and to ensure that it was wisely spent. A major concern of mine was that the HGP's budget be separate from that of the rest of NIH. I thought it vitally important that the Human Genome Project not jeopardize the livelihood of non-HGP science; we had no right to succeed if by our success other scientists could legitimately charge that their research was being sacrificed on the altar of the megaproject. At the same time, I felt that we, the scientists embarking on this unprecedented enterprise, ought to signal somehow our awareness of its profundity. The Human Genome Project is much more than a vast roll call of As, Ts, Gs, and Cs: it is as precious a body of knowledge as humankind will ever acquire, with a potential to speak to our most basic philosophical questions about human nature, for purposes of good and mischief alike. I decided that 3 percent of our total budget (a small proportion, but a large sum nevertheless) should be dedicated to exploring the ethical, legal, and social implications of the Human Genome Project. Later at Senator Al Gore's urging, this was increased to 5 percent.

It was during these early days of the project that a pattern of international collaboration was established. The United States was directing the effort and carrying out more than half the work; the rest would be done mainly in the United Kingdom, France, Germany, and Japan. Despite a long tradition in genetics and molecular biology, the U.K.'s Medical Research Council was only a minor contributor. Like the whole of British science, it was suffering from Mrs. Thatcher's myopically stingy funding policies. Fortunately, the Wellcome Trust, a private biomedical charity, came to the rescue: in 1992 it established a purpose-built sequencing facility outside Cambridge – the Sanger Centre named, as we have seen, for Fred Sanger. In managing the international effort, I decided to assign distinct parts of the genome to different nations. In this way, I figured, a participating nation would feel that it was invested in something concrete, say, a particular chromosome arm, rather than laboring on a nameless collection of anonymous clones. The Japanese effort, for example, focused largely on chromosome 21. Sad to say, in the rush to finish, this tidy order broke down, and it proved to be not so easy after all to superimpose the genome map on a map of the world.

From the start I was certain that the Human Genome Project could not be accomplished through a large number of small efforts – a combination of many, many contributing labs. The logistics would be hopelessly messy, and the benefits of scale and automation would be lost. Early on, therefore, genome mapping centers were established at Washington University in St. Louis, Stanford and UCSF in California, the University of Michigan at Ann Arbor, MIT in Cambridge, and Baylor College of Medicine in Houston. The DOE's operations, first centered at their Los Alamos and Livermore National Laboratories, in time came to be centralized in Walnut Creek, California.

The next order of business was to investigate and develop alternative sequencing technologies with a view to reducing overall cost to about 50 cents a base pair. Several pilot projects were launched. Ironically, the method that eventually paid off, fluorescent dye-based automated sequencing, did not fare especially well during this phase. In retrospect, the pilot automated machine effort should have been carried out by Craig Venter, an NIH staff researcher who had already proved adept at getting the most out of the procedure. He had applied to do it, but Lee Hood, as the technology's original developer, was preferred. This early rebuff of Venter was to have repercussions later.

In the end, the HGP did not involve the wholesale invention of new methods of analyzing DNA; rather, it was the improvement and automation of familiar methods that ultimately enabled a progressive scaling up from hundreds to thousands and then to millions of base pairs of sequence. Critical to the project, however, was a revolutionary technique for generating large quantities of particular DNA segments (you need large quantities of a given segment, or gene, if you are going to sequence it). Until the mid-eighties, amplifying a particular DNA region depended on the Cohen-Boyer method of molecular cloning: you would cut out your piece of DNA, insert it into a plasmid, and then insert the modified plasmid into a bacterial cell. The cell would then replicate, duplicating each time your inserted DNA segment. Once sufficient bacterial growth had occurred, you would purify your DNA segment out from the total mass of DNA in the bacterial population. This procedure, though refined since Boyer and Cohen's original experiments, was still cumbersome and time-consuming. The development of the polymerase chain reaction (PCR) was therefore a great leap forward: it achieves the same goal, selective amplification of your piece of DNA, within a couple of hours, and without any need to mess around with bacteria.

PCR was invented by Kary Mullis, then an employee of Cetus Corporation. By his own account, "The revelation came to me one Friday night in April, 1983, as I gripped the steering wheel of my car and snaked along a moonlit mountain road into northern California's redwood country." It is remarkable that he should have been inspired in the face of such peril. Not that the roads in Northern California are particularly treacherous, but as a friend – who once saw the daredevil Mullis in Aspen skiing down the center of an icy road through speeding two-way traffic – explained to the *New York Times:* "Mullis had a vision that he would die by crashing his head against a redwood tree. Hence he is fearless wherever there are no redwoods." Mullis received the Nobel Prize in Chemistry for his invention in 1993 and has since become ever more eccentric. His advocacy of the revisionist theory that AIDS is not caused by HIV has damaged both his credibility and public health efforts (see Plate 39).

PCR is an exquisitely simple process. By chemical methods, we synthesize two primers – short stretches of single-stranded DNA, usually about twenty base pairs in length – that correspond in sequence to regions flanking the piece of DNA we are interested in. These primers bracket our gene. We add the primers to our template DNA, which has been extracted from a sample of tissue. The template effectively consists of the entire genome, and the goal is to massively enrich our sample for the target region. When DNA is heated up to 95°C, the two strands come apart (see Plate 40). This allows each primer to bond to the twenty-base-pair stretches of template whose sequences are complementary to the primer's. We have thus formed two small twenty-base-pair islands of double-stranded DNA along the single strands of the template DNA. DNA polymerase – the enzyme that copies DNA by incorporating new base pairs in complementary positions along a DNA strand – will only start at a site where the DNA is already double-stranded. DNA polymerase therefore starts its work at the double-stranded island made by the union of the primer and the complementary template region. The polymerase makes a complementary copy of the template DNA starting from each primer, and therefore copying the target region. At the end of this process, the total amount of target DNA will have doubled. Now we repeat the heating step, and the whole process occurs again; once more, the number of copies of the DNA bracketed by the two primers doubles. Each cycle of this process results in a doubling of the target region. After twenty-five cycles of PCR – which means in less than two hours – we have a 2^{25} (about a 34 million–fold) increase in the amount of our target DNA. In effect, the resulting solution, which started off as a mixture of template DNA, primers, DNA polymerase enzyme, and free As, Ts, Gs, and Cs, is a concentrated solution of the target DNA region.

A major early problem with PCR is that DNA polymerase, the enzyme that does the work, is destroyed at 95°C. It was therefore necessary to add it afresh in each of the process's twenty-five cycles. Polymerase is expensive, and so it was soon apparent that PCR, for all its potential, would not be an economically practical tool if it involved literally burning huge quantities of the stuff. Happily Mother Nature came to the rescue. Plenty of organisms live at temperatures much higher than the 37°C that is optimal for *E. coli*, the original source of the enzyme; and these creatures' proteins – including enzymes like DNA polymerase – have adapted over eons of natural selection to cope with serious heat. Today PCR is typically performed using a form of DNA polymerase derived from *Thermus aquaticus*, a bacterium that lives in the hot springs of Yellowstone National Park.

PCR quickly became a major workhorse of the Human Genome Project. The process is basically the same as that developed by Mullis, but it has been automated. No longer dependent on legions of bleary-eyed graduate students to effect the painstaking transfer of tiny quantities of fluid into plastic tubes, a state-of-the-art genome lab features robot-controlled production lines. PCR robots engaged in a project on the scale of sequencing the human genome inevitably churn through vast quantities of the heat-resistant polymerase enzyme. HGP scientists therefore especially resented the unnecessarily hefty royalties added to the cost of the enzyme by the owner of the PCR patent, the European industrial-pharmaceutical giant Hoffmann-LaRoche.

The other workhorse was the DNA sequencing method itself. Again, the underlying chemistry was not new: the HGP used the same method worked out by Fred Sanger in the mid-seventies. Innovation came as a matter of scale, through the mechanization of sequencing.

Sequencing automation was initially developed in Lee Hood's Caltech lab. As a high-school quarterback in Montana, Hood led his team to successive state championships; he would carry the lesson of teamwork into his academic career. Peopled by an eclectic mixture of chemists, biologists, and engineers, Hood's lab became a leader in technological innovation.

Automated sequencing was actually the brainchild of Lloyd Smith and Mike Hunkapiller. Then in Hood's lab, Hunkapiller approached Smith about a sequencing method using a different colored dye for each base type. In principle the idea promised to make the Sanger process four times more efficient: instead of four separate sets of sequencing reactions, each run in a separate gel lane, color-coding would make it possible to do everything with a single set of reactions, and run the result in a single gel lane. Smith was initially pessimistic, fearing the quantities of dye implied by the method would be too small to detect. But being an expert in laser applications, he soon conceived a solution using special dyes that fluoresce under a laser.

Following the standard Sanger method, a procession of DNA fragments would be created and sorted by the gel according to size. Each fragment would be tagged with a fluorescent dye corresponding to its chain-terminating dideoxy nucleotide; the color emitted by that fragment would thereby indicate the identity of that base. A laser would then scan across the bottom of the gel, activating the fluorescence, and an electric eye would be in place to detect the color being emitted by each piece of DNA. This information would be fed straight into a computer, obviating the excruciating data-entry process that dogged manual sequencing.

Hunkapiller left Hood's lab in 1983 to join a recently formed instrument manufacturer, Applied Biosystems, Inc. (ABI). It was ABI that produced the first commercial Smith-Hunkapiller sequencing machine. Since then, the efficiency of the process has been enormously improved: gels – unwieldy and slow – have been discarded and replaced with high-throughput capillary systems – thin tubes in which the DNA fragments are size-sorted very rapidly. Today, the latest generation of ABI's sequencing machines is phenomenally fast, some thousand times speedier than the prototype. With minimal human intervention (about fifteen minutes every twenty-four hours), these machines can produce as much as half a million base pairs of sequence per day. It was ultimately this technology that made the genome project doable.

While DNA sequencing strategies were being optimized during the first part of the Human Genome Project, the mapping phase forged ahead. The immediate goal was a rough outline of the entire genome that would guide us in determining where each block of eventual sequence was located. The genome had to be broken up into manageable chunks, and it would be those chunks that would be mapped. Initially we pursued this objective using yeast artificial chromosomes (YACs), a means devised by Maynard Olson of importing large pieces of human DNA into yeast cells. Once implanted, YACs are replicated together with the normal yeast chromosomes. But attempts to load up to a million base pairs of human DNA into a single YAC exposed methodological problems. Segments, it was discovered, were getting shuffled, and since mapping is all about the order of genes along the chromosome, this shuffling of sequences was just about the worst thing that could happen. BACs (bacterial artificial chromosomes), developed by Pieter de Jong in Buffalo, came to the rescue. These are smaller, just 100,000 to 200,000 base pairs long, and much less prone to shuffling.

For those attacking the human genome map head on – groups in Boston, Iowa, Utah, and France – the critical first steps involved finding genetic markers – locations where the same stretch of DNA drawn from two different individuals differed by one or more base pairs. These sites of variation would serve as landmarks for orienting our efforts throughout the genome. In short order the French effort, under Daniel Cohen and Jean Weissenbach, produced excellent maps at Généthon, a factorylike genomic research institute funded by the French Muscular Dystrophy Association. Like the Wellcome Trust across the English Channel, the French charity took up some of the slack created by insufficient government support. When, in the final push, detailed physical mapping of BACs became necessary, John McPherson's program at the genome center at Washington University was the major contributor.

As the HGP lurched into high gear, the debate persisted about the best way to proceed. Some pointed out that a large portion of the human genome is what we in the trade call "junk," stretches of DNA that apparently don't code for anything. Indeed those stretches that encode proteins – genes – constitute only a small fraction of the total. Why therefore, these critics asked, should we sequence the entire genome – why bother with the junk? There is actually an extremely quick-and-dirty way to secure a snapshot of all the coding genes in the genome, using the reverse transcriptase technology described in chapter 5. Purify a sample of messenger RNA from any type of tissue; if your source is the brain, you will have a sample of RNA for all the genes expressed in the brain. Using reverse transcriptase, you can then create DNA copies (known as cDNAs) of these genes and the cDNAs can then be sequenced.

This quick and dirty approach, however, was no substitute for doing the whole thing. As we now know, many of the most interesting parts of the genome lie outside genes, constituting the control mechanisms that switch the genes on and off. And so, in the case of the cDNA analysis of brain tissue just described, you will have an overview of the genes switched on in the brain but no idea how they are switched on: the hugely important regulatory regions of DNA are not transcribed into RNA by the RNA polymerase enzyme that copies the DNA strand into messenger RNA.

Working at the relatively cash-strapped Medical Research Council (MRC) in Britain, Sydney Brenner pioneered this cDNA-based approach to large-scale gene discovery. With limited research funds, he figured that sequencing cDNAs was the most cost-effective way of using what little money he had. Keen to reap the commercial benefits of the sequences, the MRC prevented Brenner from publishing them until British pharmaceutical firms had a chance to position themselves to profit from them.

On a visit to Sydney Brenner's lab, Craig Venter was impressed by this cDNA strategy. He could hardly wait to return to his NIH lab outside Washington, D.C., where he would apply the technique himself to produce a treasure trove of new genes. By sequencing even a small part of each one, Venter could determine whether or not it was new to science. In June 1991 an NIH official urged him to apply for patents on 337 of these new genes, although he had, in many instances, no clue about their function. A year later, having applied the technique more broadly, Venter added 2,421 sequences to the list submitted to the patent office. In my judgment, the very notion of blindly patenting sequences without knowledge of what they do was outrageous: what precisely was one protecting? This conduct could only be seen as a preemptive financial claim on a truly meaningful discovery someone else might yet make. I expounded my objections to the higher-ups at NIH, but to no avail. And the agency's persistence in endorsing the practice – a policy that was later reversed – spelled the beginning of the end of my career as a government bureaucrat. I had mixed feelings when Bernadine Healy, head of NIH, forced me to resign in 1992. Four years in the Washington pressure cooker had been enough. But what really mattered to me was that by the time of my departure, the Human Genome Project was undeflectably on course.

Venter's taste of the commercial possibilities of patenting chunks of the genome whetted his appetite for more. But he wanted it both ways: to remain a part of the academic community, in which information was freely shared and salaries were small; and also to enter the business arena, in which his discoveries could be kept under wraps until the patent cleared and he could cash in. With the help of a fairy godfather, venture capitalist Wallace Steinberg (the inventor of the Reach toothbrush), Venter got his wish in 1992. Steinberg supplied $70 million to set up not one but two organizations: a nonprofit, The Institute for Genomic Research (TIGR, pronounced "tiger"), to be headed by Venter, and a sister company, Human Genome Sciences (HGS), to be headed by commercially inclined molecular biologist William Haseltine. It would work this way: TIGR, the research engine, would crank out cDNA sequences, and HGS, the business arm, would market the discoveries. HGS would always have six months to review TIGR's data prior to publication, except when the findings indicated potential to develop a drug, in which case HGS would have a year.

Having grown up in California, Venter initially chose surfing over higher education. But a traumatic yearlong tour as a medical assistant in Vietnam during the war seemed to focus his mind, and on his return to the United States he acquired in short order an undergraduate degree and a Ph.D. in physiology and pharmacology from the University of California, San Diego. His migration from academia into the commercial sector made sense viewed in relation to his personal finances: by his own reckoning, he had $2,000 in the bank when he founded TIGR. But he was quick to turn his fortunes around: early in 1993 the British pharmaceutical company SmithKline Beecham, anxious for a stake in the genome gold rush, paid $125 million for the exclusive commercial rights to Venter's growing list of new genes. And a year later, the *New York Times* revealed that Venter's 10 percent share of HGS was itself worth $13.4 million. Not afraid to spend it, he dropped $4 million on an eighty-two-foot racing yacht, whose spinnaker he adorned with a twenty-foot image of himself.

In the 1970s William Haseltine had been at Harvard as a graduate student under the joint direction of Wally Gilbert and myself. Afterward, he would run an innovative HIV research center at the medical school's Dana Farber Cancer Center. But it was his marriage to the multimillionaire socialite Gale Hayman (creator of the 1980s must-have perfume Giorgio Beverly Hills) that gave him the most visibility and ensured Haseltine had rather more than $2,000 in the bank when he set up HGS. Even before he went corporate, his jet-setting had provoked comment from members of his Harvard Medical School laboratory. "What's the difference between Bill Haseltine and God?" Answer: "God is everywhere; Haseltine is everywhere but Boston, where he's supposed to be."

Precious little skill or ingenuity was involved in Venter and Haseltine's scramble to patent every human gene they could find on the basis of cDNA sequencing. TIGR and HGS were simply the biotech equivalent of the kids who round up all the toys at the playground just so no other kid can play with them (see Plate 41).

In 1995, HGS filed a patent for a gene called CCR5. HGS's preliminary sequence analysis suggested that the gene encoded a cell-surface protein in the immune system, and was therefore worth "owning" since such proteins may potentially serve as targets for drugs affecting the immune system. CCR5 was one of a batch of 140 patents for similar genes that HGS applied for. But in 1996 researchers discovered the role of CCR5 in the pathway by which HIV, the virus that causes AIDS, invades the immune system's T cells. They also found that mutations in CCR5 were responsible for AIDS resistance: it had been observed that some gay men – who turned out to have mutated CCR5 genes – never contracted the disease despite repeated exposure to HIV. Thus CCR5 was and remains clearly destined to play an important part in our assault on HIV. But although it made no contribution whatsoever to the hard work and solid science that determined CCR5's central role in AIDS infection, HGS stands to profit enormously from simply having got its hands on the gene first; and by exacting a fee for every attempted application of the knowledge, its CCR5 patent will sorely tax an area of medical research that desperately needs every penny it has. Haseltine's response is by turns unapologetic – "If somebody uses this gene in a drug discovery program after the patent has been issued . . . and does it for commercial purposes, they have infringed the patent" – and indignant: "We'd be entitled not just to damages, but to double and triple damages."

This kind of speculative gene patenting creates a terrible drag on medical research and development, leading in the long run to fewer and poorer treatment options. The trouble is that the speculators are in effect patenting potential drug targets – the proteins upon which any drug or treatment yet to be invented might act. For most big pharmaceutical firms gene patents on drug targets, filed by biotech companies with little or no biological information on function, become a poison pill. The large royalties demanded by gene-finding monopolies tip the economic balance against drug development; cloning a drug target is at most 1 percent of the way to an approved drug. Furthermore, if a company produces a drug with a particular target for which it also holds the patent on the underlying gene, that company has no immediate incentive to develop better drugs for that target. Why invest in R&D when your patent makes it prohibitively costly – if not simply illegal – for other companies to get in on the act?

The prospect of the TIGR/HGS/SmithKline Beecham triumvirate having a commercial stranglehold on human gene sequences alarmed both the academic and commercial molecular biology communities. In 1994 Merck, one of SmithKline Beecham's traditional rivals in the pharmaceutical business, provided the genome center at Washington University with $10 million to sequence human cDNAs and publish them openly, thus delivering an open-access riposte to HGS.

At about the time that TIGR and HGS were taking their first steps to commercialize the genome, Francis Collins was appointed to succeed me as the director of the NIH's genome effort. Collins was an excellent choice. He had proven himself a top-notch gene mapper, with several major disease genes under his belt, including the ones for cystic fibrosis, neurofibromatosis (the so-called Elephant Man disease), and, as part of a multipronged effort, Huntington disease. Had prizes been awarded in the early matches of the HGP tournament – those contests for the mapping and characterization of important genes – the palm would surely have gone to Collins. He did himself keep score after his own fashion: a Honda Nighthawk motorcycle being his preferred mode of transport, his colleagues added a decal to his helmet every time a new gene was mapped in his lab.

Collins was raised in Virginia's Shenandoah Valley on a ninety-five-acre farm without plumbing. Initially home-schooled by his parents, a drama professor and a playwright, he wrote and directed his own stage production of *The Wizard of Oz* at age seven. The wicked witch of science, however, dragged Collins away from a career in the theater; after completing a Ph.D. in physical chemistry at Yale, he went to medical school and from there into a research career in medical genetics. Collins is a member of a rare species, the devoutly religious scientist. In college, he recalls, "I was a pretty obnoxious atheist," but that changed in medical school, when "I watched people in terrible medical circumstances who were engaged in battles for survival, which many of them lost. I watched how some people leaned on their faith and saw what strength it gave them." To the Human Genome Project Collins brought scientific excellence as well as a spiritual dimension singularly lacking in his predecessor.

By the mid-nineties, with the initial mapping of the human genome accomplished and sequencing technologies fast developing, it was time to get down to the nitty-gritty of As, Ts, Gs, and Cs – time to start sequencing. Sticking to the game plan outlined at the outset by our NAS committee, we would first attack an array of model organisms: bacteria to start with, and then on to more complicated creatures (with more complicated genomes). The lowly nematode worm, *C. elegans*, was the first big nonbacterial challenge, and as the joint achievement of John Sulston at Britain's Sanger Centre and Bob Waterston at Washington University it provided an excellent model of international collaboration. The worm's sequence was published in December 1998, all 97 million base pairs of it. No bigger than a comma on this page, and comprising a fixed number of cells, just 959, the worm nevertheless has some 20,000 genes.

At first sight, Sulston appeared ill-suited to a leadership role in Big Science. He had spent most of his professional life staring down a microscope in order to produce in astonishing detail a complete description, cell by cell, of the development of the worm. Bearded and avuncular, he is the son of a Church of England vicar and also a lifelong socialist who believes passionately that business and the human genome should have nothing in common. Like Francis Collins, he is a motorcycle enthusiast; he used to commute on his 550cc machine from his home outside Cambridge to the Sanger Centre until, just as the HGP was gathering speed, an accident left him severely injured and his bike, as he put it, "little more than nuts and bolts." The Wellcome Trust, which was funding the Sanger Centre, was horrified to learn that the project's scientific leader was taking his life in his hands every time he came to work: "After we'd invested all that money in this bloke!" complained Bridget Ogilvie, then the trust's director.

Sulston's U.S. partner, Waterston, was an engineering major at Princeton and imported plenty of engineering savvy to the big sequencing center he ran at Washington University. Waterston has the capacity to extrapolate – to start small and finish big. Accompanying his daughter on a jog, he found he liked running, and is now an accomplished marathon runner. During its first year of operation, his sequencing group produced just forty thousand base pairs of worm sequence, but within a few years it was cranking out enormous amounts, and Waterston was one of the earliest to urge an all-out human sequencing effort (see Plate 42).

But even as those in the international HGP collaboration began to sequence model organisms, gearing up for the big one, the molecular biological equivalent of an earthquake shook the whole enterprise.

Craig Venter and TIGR had been doing well. Having milked his cDNA gene discovery strategy for several years, Venter became interested in sequencing whole genomes. In this, too, he was persuaded of the superiority of his own approach. The HGP had been carefully mapping the location of the different chunks of DNA on the chromosomes before actually sequencing them. That way, you already knew that chunk A was adjacent to chunk B and could look for overlaps between them when it came to knitting together the final sequence. Venter preferred a "whole genome shotgun" (WGS) approach, in which there was no initial mapping: you simply broke the genome up into random chunks, sequenced them all, fed all the sequences into a computer, and relied upon the computer to put them all in the right order on the basis of overlaps, without benefit of any prior positional information. Venter and his team at TIGR showed that this brute force method could indeed work, at least for simple genomes: in 1995 they published the genome sequence of a bacterium, *Haemophilus influenzae,* using this method.

It remained problematic, however, whether WGS would work for a large and complex genome like the human one. The problem is repeats – segments of the same sequence occurring in different places in the genome – which could in principle scupper a WGS sequencing attempt. These repeats might well mislead even the most sophisticated computer algorithm. If, for instance, a repeat occurs in chunks A and P, the computer could mistakenly situate A next to Q rather than in its proper position, next to B. For its part, the HGP itself had discussed this scenario when it considered using a WGS approach, and, based on careful calculations by Phil Green in Seattle, the consortium concluded that such an effort would likely be confounded by the human genome's massive amount of long-repeating sequences of junk DNA.

In January 1998, Mike Hunkapiller of ABI, maker of automated sequencing machines, invited Venter to check out his newest model, the PRISM 3700. Venter was impressed, but nothing could have prepared him for what was to follow. Hunkapiller suggested that Venter form a new company, funded by ABI's parent company, PerkinElmer, to sequence the human genome. Venter had no misgivings about forsaking TIGR – relations had long since soured with Haseltine at HGS. And so he wasted no time in founding the firm that was later to be called Celera Genomics. The company motto: "Speed matters. Discovery can't wait." The plan: to sequence the entire human genome by WGS using three hundred of Hunkapiller's machines and the single greatest concentration of computing power outside the Pentagon. The project would take two years and cost between $200 million and $500 million.

The news broke just before the leaders of what would come to be called the public (as opposed to private) Human Genome Project were meeting at Cold Spring Harbor Laboratory. To put it mildly, the news was not well received. The worldwide public project had already spent some $1.9 billion (of public money), and now, as the *New York Times* was spinning it, we might have nothing to show for the money except for the sequence of the mouse genome, while Venter waltzed off with the holy grail, the human genome. What was especially galling was Venter's flouting of what had come to be known as the Bermuda principles. In 1996, at an HGP conference in Bermuda – a meeting Venter attended – the HGP had agreed that sequence data should be released as soon as it was generated. The genome sequence, we all concurred, should be public property. Now a renegade, Venter had different ideas: he claimed he would defer releasing new sequence data for three months, selling licenses to pharmaceutical companies and any other parties seriously interested in buying a preview.

Fortuitously, the Wellcome Trust's Michael Morgan was able to give the public project a welcome boost just days after Venter's announcement by declaring that it would be doubling its support for the Sanger Centre, bringing the total up to around $350 million. Though the timing of the announcement made this look like a direct response to Venter's challenge, the increase in funding had in fact been in the works for quite some time. Shortly afterwards the U.S. Congress beefed up its own contribution to the public HGP's coffers. The race was on. In fact, from the outset there were always going to be at least two winners. Science only stood to benefit from two human genome sequences, one against which to check the other. (With over 3 billion base pairs involved, there was bound to be a typo or two.) Another winner would surely be ABI: they stood to sell a lot more PRISM sequencing machines, which most labs in the public project would now have to buy to keep up with Venter!

The acrimonious exchanges between the leaders of the private and public projects would become a fixture of newspaper science pages for the next couple of years. The back and forth got to a pitch that moved President Clinton to direct his science adviser, "Fix it . . . make these guys work together." But through it all, the sequencing moved ahead, and Venter did demonstrate that a WGS approach could work on a respectably sized genome when, in collaboration with the fruit fly wing of the public consortium, he announced the completion of an advanced draft of the *Drosophila* genome early in 2000. It, however, contains relatively little repetitive junk DNA, and Celera's success in assembling it in no way guaranteed WGS would work on the human genome.

No individual was more vital to meeting the Celera challenge than Eric Lander. It was he who envisioned an almost entirely automated sequencing process in which robots would take the place of technicians, and it was he who had the drive to make this vision a reality. Lander's résumé indeed shows he knew a thing or two about drive. A Brooklyn boy, he was a curve-busting math whiz at Stuyvesant High in Manhattan who went on to win first prize in the Westinghouse Talent Search; he then became valedictorian of his class at Princeton ('78) before earning his Ph.D. at Oxford on a Rhodes fellowship. A MacArthur "genius" award in 1987 seemed almost redundant. His mother, incidentally, has no idea how it all happened: "I'd love to say I'm responsible, but it's not true. . . . I'd have to say it was dumb luck."

Ultimately finding pure mathematics "an isolated, monastic kind of field," Lander, notably gregarious by the standards of his discipline, joined the jollier faculty of the Harvard Business School, but he soon found himself distracted and intrigued by the labors of his younger brother, a neuroscientist. Inspired, Lander taught himself biology by moonlighting in Harvard's and MIT's biology departments, all the while scarcely missing a beat on his day job at the B-school: "I pretty much picked up molecular biology on street corners," he says. "But around here, there are a lot of very good street corners." In 1989 he became a professor of biology on one of those street corners, MIT's Whitehead Institute.

Even among the so-called G5 – the public effort's five major centers, which also included the Sanger Centre, Washington

University's Genome Sequencing Center, Baylor College of Medicine, and the DOE in Walnut Creek – Lander's lab would be the largest single contributor of DNA sequences. His team at MIT would also be responsible for much of the enormous acceleration of productivity in the home stretch leading up to the release of the draft genome (see Plate 43). On November 17, 1999, the public project celebrated its billionth base pair, with the sequencing of a G. Just four months later, on March 9, 2000, a T was base number 2 billion. The G5 was cranking. Because Celera was using the public project's data, which were posted immediately on the Internet and were now pouring in thick and fast, Venter, perhaps finally breaking a sweat, halved the amount of sequencing he had originally projected Celera would do.

As the public/private race reached a climax in the media, behind the barricades the focus was increasingly shifting to the effort's mathematical brain trust, scientists hidden in back rooms among banks of computers. They were the ones who had to make sense of all those As, Ts, Gs, and Cs of crude sequence. They had two major tasks. First: To assemble a whole final sequence from the many, many discrete chunks on hand. Most parts had been sequenced numerous times, so there were several genomes' worth of sequence to sort out, all of which had to be distilled down into a single canonical genome sequence. Computationally, this was an enormous undertaking. Second: To figure out what was what in the final sequence, and above all where the genes were. Identifying the genome's components – distinguishing between one stretch of As, Ts, Gs, and Cs that encoded nothing but junk and another that encoded a protein – depended on extremely computer-intensive approaches.

At the heart of Celera's computer operations was Gene Myers, the computer scientist who had been the WGS approach's first and most forceful advocate. With James Weber of the Marshfield Medical Research Foundation in Wisconsin, he had proposed that the public effort adopt WGS long before Celera even came into existence. And so for Myers, the success of Celera's bid was a point of pride and vindication.

Anchored as it was by previously mapped genetic landmarks, the public project's job in assembling the sequence, though immense, seemed less daunting than the one confronting Myers in the landmark-free world of WGS. (In its final analysis, Celera used the public project's freely available map information.) In fact, in counting on these very landmarks, the public project had rather under-estimated its own computational challenge, so, as Celera added computer muscle, the public project stayed focused on gearing up its sequencing operation. Only very late in the day did the leaders of the public project realize that, despite the map, they, too, like the proverbial father facing the parts of a new bike on Christmas Eve, had a major assembly problem on their hands. A date for completion (and assembly) of the "rough draft" had been fixed for the end of June. But at the beginning of May, the public project still had no working means of assembling all their sequences. Their deus ex machina took a strange form: a graduate student from UC Santa Cruz.

His name was Jim Kent, and he looked like a member of the Grateful Dead. He had been programming computers since the beginning of the PC era, writing code for graphics and animations, but then decided on graduate school so he could be a part of bioinformatics, the new field dedicated to analyzing DNA and protein sequences. He realized that he was through with commercial programming when he received Microsoft's bulky twelve-CD-ROM package for developers of programs for Windows 95: "I was thinking to myself that the whole human genome could fit on one CD-ROM and didn't change every three months." Confident in May that he had a good way to crack the much-talked-about assembly problem, he induced his university to let him "borrow" 100 PCs recently bought for teaching purposes. He then embarked on a four-week pro-gramming marathon, icing his wrists at night to prevent them from seizing up as he churned out computer code by day. His deadline was June 26, when the completion of the rough draft was to be announced. The program finished, he set his 100 PCs to work, and, on June 22, his gang of PCs solved the public project's assembly problem. Myers at Celera cut it even closer, completing his assembly on the night of June 25.

Then came June 26, 2000. Bill Clinton at the White House and Tony Blair at 10 Downing Street simultaneously proclaimed the first draft of the HGP complete. The race was called as a tie, the honors to be shared equally. Happily the opposing parties managed to put the bad feelings behind them, for the morning at least. Clinton declared, "Today, we are learning the language in which God created life. With this profound new knowledge, humankind is on the verge of gaining immense, new power to heal." Grand words for a grand occasion. It was impossible not to feel some pride in an accomplishment that the press promptly compared to the first Apollo moon landing, even if the "official" date of the triumph was somewhat arbitrary. The sequencing was by no means over, and it would be more than six months before the scientific papers summarizing the genome were published. It has been suggested that the timing was dictated not by the HGP's timetable but by Clinton's and Blair's schedules.

Overlooked in the blaze of White House publicity was the fact that the object of celebration was but a *rough draft* of the human genome. Much work remained to be done. In fact, the sequences of only two of the smallest chromosomes, 21 and 22, were reasonably complete and had been published. And even these could not boast unbroken chromosome-tip-to-chromosome-tip sequences. As to the other chromosomes, some of their sequences were riddled with gaps. Since that big announcement, sights were set on a new deadline of April 2003 for filling in the gaps and securing a full, accurate sequence. Some small regions, however, have proved literally unsequenceable, and in practice the goal has become to obtain an "essentially complete" sequence: at least 95 percent of the sequence finished with an error rate of less than 1 in 10,000 bases.

One of those responsible for coaxing the international herd of sequencing centers over the final hurdles was Rick Wilson, the bluff midwesterner who succeeded Bob Waterston as head of the Washington University center. Quality control is the name of the game, so each chromosome has been assigned a coordinator to oversee progress and ensure that his or her charge meets the project's overall specifications. Occasional glitches occur – for example, an errant piece of rice sequence crept mysteriously into one submission to the database – but screening procedures have proved effective in removing such contaminants. When I wrote this, the Human Genome Project was well on course to being "essentially complete" by the April 2003 deadline, which is also the fiftieth anniversary of the publication of the double helix.

The Human Genome Project is an extraordinary technological achievement. Had anyone suggested in 1953 that the entire human genome would be sequenced within fifty years, Crick and I would have laughed and bought them another drink. And such skepticism would have still seemed valid more than twenty years later when the first methods for sequencing DNA were finally devised. Those methods were, to be sure, a technical breakthrough but sequencing was a painfully slow business all the same – in those days it was a major undertaking to generate the sequence of even one small gene a few hundred base pairs in length. And now there we were, just another twenty-five years further on, celebrating the completion of some 3.1 billion base pairs of sequence. But we must also bear in mind that the genome is much more than a monument to our technological wizardry, astonishing though that may be: whatever its immediate political motivation, that White House celebration was perfectly justified in hailing the possibilities of a marvelous new weapon in our fight against disease and, even more, a whole new era in our understanding of how organisms are put together and how they operate, and of what it is that sets us apart biologically from other species – what, in other words, makes us human.

CHAPTER EIGHT

READING GENOMES: EVOLUTION IN ACTION

I used to wish that when the human genome was finally completely sequenced it would turn out to contain 72,415 genes. My enthusiasm for this obscure number stemmed from the Human Genome Project's first big surprise. In December 1999, sandwiched between two major sequencing landmarks – billions one and two – came the first completed chromosome, number 22. Although a small one, constituting only 1.1 percent of the total genome, chromosome 22 was still 33.4 million base pairs long. This was our first glimpse of what the genome as a whole might look like; as one commentator for *Nature* wrote, it was like "seeing the surface or the landscape of a new planet for the first time." Most interesting was the density of genes along the chromosome. We had no reason to believe that chromosome 22 would not be representative of the entire genome, so we expected to find about 1.1 percent of all human genes in its sequence. That is to say, given the standard textbook estimate of about 100,000 human genes in total, we should have expected to see about 1,100 of them on chromosome 22. Almost exactly half that number were found: 545. Here was the first big hint that the human genome was not as gene-rich as we had supposed.

Suddenly the human gene count was a hot topic. At the Cold Spring Harbor Laboratory conference on the genome in May 2000, Ewan Birney, who was spearheading the Sanger Centre's computer analysis of the sequence, organized a contest he called Genesweep. It was a lottery based on estimating the correct gene count, which would finally be determined with the completion of the sequence in 2003; the winner would be the one who had come closest to the right answer. (That Birney should have become the HGP's unofficial bookie wasn't entirely surprising: numbers are his thing. After Eton, he took a year to tackle quantitative problems in biology while living in my house on Long Island – a far cry from trekking in the Himalayas or tending bar in Rio, just two of the more likely ways a young Briton might spend the "gap year" before university. Birney's CSHL work yielded two important research papers before he even set foot in Oxford.)

Originally Birney charged $1 per entry, but the price of admission to the pool increased with every published estimate that brought us closer to a final count. I was able to get in on the ground floor, putting $1 down on 72,415. My bet was a calculated attempt to reconcile the textbook figure, 100,000, and the new best guess of around 50,000, based on the chromosome 22 result. Birney announced the result in May 2003: 21,000 genes – many fewer than anyone had guessed. Lee Rowen, a Seattle-based expert in biological computing, won Genesweep with her bet of 25,947, still some 4000 off. Of course I was wildly over the mark so I'm now a dollar down on the genome.

Perhaps the only question to generate as much idle speculation as the gene count was that of whose genes we were sequencing. The information was in principle confidential, so money was not going to change hands on this one, but many wondered all the same. In the case of the public project, the DNA sample we sequenced had come from a number of randomly selected individuals from around Buffalo, New York, the same area where the processing work – isolating the DNA and inserting it into bacterial artificial chromosomes for mapping and sequencing – was taking place. Initially Celera claimed that its material too had been derived from six anonymous donors, a multicultural group, but in 2002 Craig Venter could not resist letting the world know that the main genome sequenced was actually his own. Today that sequence is Venter's last remaining connection to the company. Concerned that sequencing genomes, though glamorous and newsworthy, was not proving viable from a business perspective, Celera reinvented itself as a drug company and bid farewell to its founder in 2002. As for Venter, he has established two new institutes, one to study the ethical issues raised by modern genetics, and the other to use the genomes of bacteria to find fresh sources of renewable energy.

With the whole rough draft in hand, it is now confirmed that there is nothing atypical about the gene density on chromosome 22. If anything, in fact, chromosome 22 with its 545 genes was for its size gene-*rich* rather than gene-poor. Only 236 genes have been definitively located on chromosome 21, which is about the same size. As we've seen, there are only an estimated 21,000 genes in total from the entire complement of 24 human chromosomes. And while we should stress that the final number can only rise as we make more discoveries, we are all but certain to end well below the 30,000 mark, never mind the 100,000 one.

As to how far, only time will tell. Finding genes is not actually such a straightforward task: protein-coding regions are but strings of As, Ts, Gs, and Cs embedded among all the other As, Ts, Gs, and Cs of the genome – they do not stand out in any obvious way. And remember, only about 2 percent of the human genome actually codes for proteins; the rest, unflatteringly referred to as "junk," is made up of apparently functionless stretches of varying length, many of which occur repeatedly. And junk can even be found strewn within the genes themselves; studded with noncoding segments (introns), genes can sometimes straddle enormous expanses of DNA, the coding parts like so many towns isolated between barren stretches of molecular highway. The longest human gene found so far, dystrophin (in which mutations cause muscular dystrophy), sprawls over some 2.4 million base pairs. Of these, a mere 11,055 (0.5 percent of the gene) encode the actual protein; the rest consists of the gene's seventy-nine introns (a typical human gene has eight). It is this awkward architecture of the genome that makes gene identification so difficult.

But human gene spotting has become less tricky now that the genome of the mouse is better known. The credit goes to evolution: in their functional parts the human and mouse genomes, like the genomes of all mammals, are remarkably similar, having diverged relatively little over the eons intervening since the two species' common ancestor. The junk DNA regions, by contrast, have been evolution's wild frontier; without natural selection to keep mutation in check, as it does in coding segments, mutations aplenty have accumulated so that there is substantial genetic divergence between the two species in these regions. Looking for similarity in sequence between the human and mouse data is therefore an effective way of identifying functional areas, like genes.

Identifying human genes has also been facilitated by the completion of a rough draft of the puffer fish genome. Fugu, as it is better known to aficionados of Japanese cuisine, contains a potent neurotoxin; a competent chef removes the poison-containing organs, so dinner should produce only a little numbness in the mouth. But some eighty people die each year from poorly prepared fugu, and the Japanese imperial family is forbidden by law from enjoying this delicacy. More than a decade ago, Sydney Brenner developed a taste for the puffer, at least as an object of genetic inquiry. Its genome, just one-ninth the size of the human one, contains much less junk than ours: approximately one-third of it encodes proteins. Under Brenner's leadership, the fugu genome rough draft was completed for some $12 million, a genuine bargain by genome-sequencing standards. The gene count at present seems to fall between 32,000 and 40,000, in the same ballpark as humans. Interestingly, though, while fugu genes have roughly the same number of introns as human and mouse genes, the fugu introns are typically much shorter.

Even if we assume that plenty of genes remain to be discovered and generously increase our estimate of the human gene count from 21,000 to 25,000, we still get a somewhat exaggerated impression of our essential genetic complexity. Over evolution, certain genes have spun off sets of related ones, resulting in groups of similar genes of like, but subtly different, function. These so-called gene families originate by accident when, in the course of producing egg or sperm cells, a chunk within a chromosome is inadvertently duplicated, so that there are now two copies of a particular gene on that chromosome. As long as one copy continues to function, the other is unchecked by natural selection, free to diverge in whatever direction evolution may choose as mutations accumulate. Occasionally the mutations will result in the gene acquiring a new function, usually one closely related to that of the original gene. In fact, many of our human genes consist of slight variations upon a relatively few genetic themes. Consider, for example, that 575 of our genes (nearly 2 percent of our total complement) are responsible for encoding different forms of protein kinase enzymes, chemical messengers that pass signals around the cell. Then, there are the 900 human genes underlying your nose's capacity to smell: the proteins encoded are odor receptors, each one recognizing a different smell molecule or class of molecules. Roughly these same 900 genes are present in the mouse as well. But here is the difference: the mouse, having adapted to a mainly nocturnal existence, has greater need of its sense of smell – natural selection has favored the keener sniffer and kept most of the 900 odor-detecting genes in service. In the human case, however, some 60 percent of these genes have been allowed to deteriorate over evolution. Presumably, as we became more dependent on sight, we have needed fewer smell receptors, so natural selection did not intervene when mutations caused many of our smell genes to be incapable of producing functioning proteins, making us relatively inept smellers compared with other warm-blooded creatures.

How does our gene number compare with that of other organisms?

COMMON NAME	SPECIES NAME	NUMBER OF GENES
Human	*Homo sapiens*	25,000
Mustard plant	*Arabidopsis thaliana*	27,000
Nematode worm	*Caenorhabditis elegans*	20,000
Fruit fly	*Drosophila melanogaster*	14,000
Baker's yeast	*Saccharomyces cerevisiae*	6,000
Gut bacterium	*Escherichia coli*	4,000

In terms of gene complement, then, we are only fractionally more complex than a weedy little plant. Even more sobering is the comparison with the nematode, a creature composed of only 959 cells (against our own estimated 100 trillion), of which some 302 are nerve cells that form the worm's decidedly simple brain (ours consists of 100 billion nerve cells) – orders of magnitude difference in structural complexity and yet we have not even double the worm's gene complement. How can we account for this embarrassing discrepancy? It's no cause for embarrassment at all: humans, it would appear, are simply able to do more with their genetic hardware.

In fact, I would propose there is a correlation between intelligence and low gene count. My guess is that being smart – having a decent nerve center like ours or even the fruit fly's – permits complex functioning with relatively few genes (if indeed "few" has any real meaning in relation to the number 25,000). Our brain gives us sensory and neuromotor capabilities far beyond those of the eyeless, inching nematode, and thus a greater range of behavioral response options. And the plant, being rooted, has fewer options still: it requires a full onboard set of genetic resources for dealing with every environmental contingency. A brainy species by contrast can respond to, say, a cold snap by using its nerve cells to seek out more favorable conditions (a warm cave will do).

Vertebrate complexity may also be enhanced by sophisticated genetic switches that are typically located near genes. With the genome sequencing accomplished, we can now analyze in detail these regions flanking genes. It is here that regulation occurs, with regulatory proteins binding to the DNA to turn the adjacent gene on or off. Vertebrate genes seem to be governed by a much more elaborate set of switching mechanisms than those of simpler organisms. It is this nimble and complicated coordination of genes that permits the complexities of vertebrate life. Moreover, a given gene may in addition yield many different proteins, either because different exons are coupled together to create slightly different proteins (a process known as "alternative splicing") or because biochemical changes are made to the proteins after they have been produced.

The unexpectedly low human gene count provoked several op-ed page ruminations on its significance. These tended toward a common theme. Stephen Jay Gould (whose recent premature death tragically silenced an impassioned voice), writing in the *New York Times,* hailed the low count as the death knell of reductionism, the reigning doctrine of virtually all biological inquiry. This doctrine holds that complex systems are built from the bottom up. Put another way: To understand events at complex levels of organization, we must first understand them at simpler levels and piece together these simpler dynamics. And so it follows that by understanding the workings of the genome, we will ultimately understand how organisms are assembled. Gould and others took the surprisingly small human gene count as evidence that such a bottom-up approach is not only unworkable but also invalid. In light of its unexpected genetic simplicity, the human organism, argued the antireductionists, was living proof that we cannot begin to understand ourselves in relation to a sum of smaller processes. To them, our low gene number implied that nurture, not nature, must be the primary determinant of who each one of us is. It was, in short, a declaration of independence from the tyranny supposedly exercised by our genes.

Like Gould, I well appreciate that nurture plays an important part in shaping each of us. His evaluation of nature's role, however, is utterly wrong: our low gene count by no means invalidates a reductionist approach to biological systems; nor does it justify any logical inference that we are *not* determined by our genes. A fertilized egg containing a chimp genome still inevitably produces a chimp, while a fertilized egg containing a human genome produces a human. No amount of exposure to classical music or violence on TV could make it otherwise. Yes, we have a long way to go in developing our understanding of just how the information in those two remarkably similar genomes is applied to the task of producing two apparently very different organisms, but the fact remains that the greatest part of what each individual organism will be is programmed ineluctably into its every cell, in the genome. In fact, I see our discovery of a low human gene count as good news for standard reductionist approaches to biology: it's much easier to sort through the effects of 25,000 genes than 100,000.

While humans may not have an enormous number of genes, we do have, as the sprawling dystrophin gene illustrates, a large, messy genome. Returning again to the worm comparison: while we have not even twice as many genes, our genome is thirty-three times larger. Why the discrepancy? Gene mappers describe the human genome as a desert spotted with occasional genetic oases – genes. Fifty percent of the genome is constituted of repetitive junklike sequences of no apparent function; a full 10 percent of our DNA consists of a million scattered copies of a single sequence, called *Alu:*

    GGCCGGGCGCGGTGGCTCACGCCTGTAATCCCAGCACTTTGG
    GAGGCCGAGGCGGGCGGATCACCTGAGGTCAGGAGTTCGAGA
    CCAGCCTGGCCAACATGGTGAAACCCCGTCTCTACTAAAAATA
    CAAAAATTAGCCGGGCGTGGTGGCGCGCGCCTGTAATCCCAG
    CTACTCGGGAGGCTGAGGCAGGAGAATCGCTTGAACCCGGGA
    GGCGGAGGTTGCAGTGAGCCGAGATCGCGCCACTGCACTCCA
    GCCTGGGCGACAGAGCGAGACTCCGTCTCAAAAAA

Writing it out a million times would give a sense of the scale of the *Alu* presence in our DNA. In fact, levels of repetitive sequence are even higher than they would appear: sequences that would once have been instantly identifiable as repeats have, over many generations of mutation, diverged beyond recognition as members of a particular class of repetitive DNA. Imagine a set of three short repeats: ATTG ATTG ATTG. Over time mutation will change them, but if the period is short, we can still see where they came from: ACTG ATGG GTTG. Over a longer period, their original identity is completely lost in the welter of mutation: ACCT CGGG GTCG. Proportions of repetitive DNA are much lower in many other species: 11 percent of the mustard weed genome is repetitive, 7 percent of the nematode worm's, and just 3 percent of the fruit fly's. The large size of our genome is mostly due to its containing more junk than that of many other species (see Plate 44).

These differences in the amounts of junk DNA explain a long-standing evolutionary conundrum. The basic expectation is that more complex organisms should have bigger genomes – they need to encode more information – than simple ones. There is indeed a correlation between genome size and an organism's level of complexity: the yeast genome is bigger than that of *E. coli* but smaller than ours. It is, however, only a weak correlation.

COMMON NAME	SPECIES NAME	APPROX. GENOME SIZE (MILLIONS OF BASE PAIRS)
Fruit Fly	*Drosophila melanogaster*	180
Fugu (puffer)	*Fugu rubripes*	400
Snake	*Boa constrictor*	2,100
Human	*Homo sapiens*	3,100
Locust	*Schistocerca gregaria*	9,300
Onion	*Allium cepa*	18,000
Newt	*Amphiuma means*	84,000
Lungfish	*Protopterus aethiopicus*	140,000
Fern	*Ophioglossum petiolatum*	160,000
Amoeba	*Amoeba dubia*	670,000

It is reasonable to suppose that natural selection operates to keep genome size as low as possible. After all, every time a cell divides, it must replicate all its DNA; the more it has to copy, the greater the room for error, and the more energy and time the process requires. It is quite an undertaking for the amoeba (or newt, or lungfish). So what could have caused the amount of DNA in these species to get so out of hand? In cases of unusually large genomes, we can only infer that some other selective forces must have negated the selection-driven impulse to keep the genome slim. It could be, for instance, that large genomes are advantageous to species likely to be exposed to environmental extremes. Lungfish live at the interface of land and water, and they can survive protracted periods of drought by burying themselves in mud; it could be they need more genetic hardware than a species adapted to a single medium.

Two major evolutionary mechanisms account for this DNA excess: genome doubling, and the proliferation of particular sequences within a genome. Many species, particularly in the plant kingdom, are actually the product of a cross between two preexisting ones. The new species often simply combines the DNA complement from each of its parent species, yielding a double genome. Alternatively, through some kind of genetic accident, a genome may get doubled without input from another species. For example, one of the standbys of molecular biology, baker's yeast, has about 6,000 genes. But close inspection reveals that a large proportion of those genes are duplicates – baker's yeast often has two divergent copies of many of its genes. At some early stage in its evolutionary history, the yeast genome apparently got doubled. Initially the gene copies would have been identical, but, over time, they have diverged.

An even richer source of excess DNA has arisen from the multi-plication of genetic sequences capable of replicating and inserting themselves at more than one site in a given genome. These so-called mobile elements have been found to come in many varieties. But when their discovery was first announced by Barbara McClintock in 1950, the very idea of "jumping" genes was too far-fetched for most scientists accustomed to the simple logic of Mendel. McClintock, a superb corn geneticist, had already endured something of a bumpy career ride. When it became clear in 1941 that she would not be granted tenure at the University of Missouri, she came to Cold Spring Harbor Laboratory, where she would remain an active member of the staff until her death in 1992, at the age of ninety. McClintock once told a colleague, "Really trust what you see." This was exactly how she did her science: her revolutionary idea that some genetic elements could move around genomes followed simply from observable facts. She had been studying the genetics underlying the development of different-colored kernels in corn, and noticed that sometimes, part way through the development of an individual kernel, the color would switch. A single kernel might then turn out variegated, with both patches of the expected yellow cells and patches of purple ones. How to account for this sudden switch? McClintock inferred that a genetic element – a mobile element – had hopped into or out of the pigment gene.

Only with the advent of recombinant DNA technologies have we come to appreciate just how common mobile elements are; we now recognize them as major components of many, if not most, genomes, including our own (see Plate 44). And some of the most common mobile elements, those that appear again and again in different sites in the same genome, have earned names reflecting their itinerant lifestyles: two fruit fly mobile elements, for example, are called "gypsy" and "hobo." And among those who study a simple plant called *Volvox* one mobile element is honored for its extraordinary capacity to jump around the genome: it is known as the "(Michael) Jordan element."

Mobile elements contain DNA sequences that code for enzymes that, through their capacity to cut and paste chromosomal DNA, work to ensure that copies of their particular element are inserted into new chromosomal sites. If a jump carries a mobile element into a junk sequence, the functioning of the organism is unaffected, and the only result is more junk DNA. But when the jump lands the mobile element in a vital gene, thereby disabling its function, then selection intervenes: the organism may die or otherwise be prevented from passing on the new jumped-in gene. Very rarely the movements of mobile elements may either create new genes or alter old ones in a way that benefits the host organism. Over the course of evolution, therefore, the effect of mobile elements seems mainly to have been the generation of novelty. And curiously, in recent human history, there is little evidence of active jumping: most of our junk DNA, it appears, was generated long ago. In contrast, the mouse genome contains many actively reinserting mobile elements, making for a much more dynamic genome. But this seems not to trouble the mouse species unduly; the intrinsically high reproductive potential of mice likely helps the species as a whole tolerate the genetic disasters attending frequent jumps into vitally functioning genetic regions (see Plate 45).

Having been used to establish many of the basic facts about how DNA functions, E. coli's track record as a model organism was unparalleled. Not surprisingly, its genome therefore ranked high on the Human Genome Project's early "to do" list. It was Fred Blattner of the University of Wisconsin who was most eager to start sequencing E. coli. But his grant proposals went nowhere until the HGP got funded and he was awarded one of the first substantial sequencing grants. Were it not for his initial reluctance to adopt automated sequencing, his lab would have been the first to sequence a complete bacterial genome. But in 1991 his strategy for scaling up the operation was an old-fashioned one: employ more under-graduates. Another latecomer to automation was Wally Gilbert, whom I had urged two years before to have a go at the smallest known bacterial genomes, those of the parasitic *Mycoplasma* – tiny bacteria that live within cells. Sadly, when a clever new manual sequencing strategy of his came to naught, his *Mycoplasma* project died with it. Blattner did, however, accept automation in time to establish in 1997 that the *E. coli* genome contains some 4,100 genes.

But the broader race to complete the first bacterial genome had been won two years before at The Institute for Genomic Research (TIGR) by a large team led by Hamilton Smith, Craig Venter, and his wife, Claire Fraser. And the bacterium they sequenced was *Haemophilus influenzae,* from which twenty years earlier Smith – a towering six-foot-six one-time math major who had gone on to medical school – had isolated the first useful DNA cutting (restriction) enzymes, a feat that won him the Nobel Prize in Physiology or Medicine in 1978. With *Haemophilus* DNA prepared by Smith, Venter and Fraser used a whole genome shotgun approach to sequence its 1.8 million base pairs. Just documenting the first "small" genome was enough to suggest the awesome size of the awaiting larger ones: if all the As, Ts, Gs, and Cs of the *Haemophilus* genome were printed on paper of this size, the resulting book would run some four thousand pages. Two pages on average would be needed for each of its 1,727 genes. Of these, only 55 percent have readily identifiable func-tions: for example, energy production involves at least 112 genes, and DNA replication, repair, and recombination requires a minimum of 87. We can tell from their sequences that the remaining 45 percent are functioning genes, but we simply can't at this stage be sure what it is they do.

By bacterial standards, the *Haemophilus* genome is pretty small.

The size of a bacterial genome is related to the diversity of environments a particular species is likely to encounter. A species that leads a dull life in a single uniform setting – say, the gut of another creature – can well get by with a relatively small genome. One that hopes to see the world, however, and is apt to encounter more varied conditions, must be equipped to respond, and flexibility of response usually depends on having alternative sets of genes, each tailored to particular conditions, and ready at all times to be switched on.

Pseudomonas aeruginosa, a bacterium that can cause infections in humans (and poses a particular danger for cystic fibrosis [CF] patients), lives in many different environments. We saw in chapter 5 how a genetically doctored form of a related species became the first living organism to be patented; in that case, it was adapted to life in an oil slick, an environment notably different from the human lung. The *Pseudomonas aeruginosa* genome contains 6.4 million base pairs and 5,570 genes. About 7 percent of those genes encode transcription factors, proteins that switch genes on or off; a respectable proportion of its entire genetic complement is thus devoted to regulation. The *E. coli* "repressor" whose existence was predicted by Jacques Monod and François Jacob in the early sixties (see chapter 3) is just such a transcription factor. A rule of thumb then would go as follows: The greater the range of environments potentially encountered by a bacterial species, the larger its genome, and the greater the proportion of that genome dedicated to gene-switching.

TIGR did not stop at *Haemophilus.* In 1995, collaborating with Clyde Hutchison at the University of North Carolina, the institute sequenced the genome of *Mycoplasma genitalium* as part of what has been dubbed the "minimal genome project." *M. genitalium* (which, despite its ominous name, is a benign inhabitant of human plumbing) has the smallest known nonviral genome, some 580,000 base pairs. (Viruses have smaller genomes but, by co-opting the genomes of their hosts, can get away with not having the genetic wherewithal for many fundamental processes.) And that relatively short sequence was found to comprise 517 genes. So a question naturally arose: Is that the minimal gene complement necessary to sustain life? Subsequent research has set about knocking out *M. genitalium*'s genes to see which are absolutely vital and which are not. Currently it appears that the minimal genome contains no more than 350 genes and possibly as few as 260. Admittedly, this is a somewhat artificially defined "minimum" since the enfeebled bugs are supplied through their growth medium with every substance they could conceivably need. It's a bit like claiming kidneys are not necessary for life because patients can survive on dialysis machines.

Will we ever be able to construct a functioning minimal cell from scratch, by artificially combining its separate purified components? Considering there are more than a hundred *Mycoplasma genitalium* proteins whose functions remain a mystery, the achievement of such a goal seems for now a long way off. Even the five hundred proteins of *Mycoplasma*, some represented in the cell by a huge number of molecules, some by just a handful, constitute an enormously complex living system. I, for one, have enough difficulty following a movie like *Gosford Park* in which there are more than four or five major characters; the thought of blocking out the complexity of interactions among the vital players inside a living cell is nothing short of mind-blowing. For the living cell is no neat miniature machine; it is rather, as Sydney Brenner put it, "a snake pit of writhing molecules." Still, Craig Venter is confident that the era of the artificial cell is just around the corner, and he has wasted no time in assembling a panel of bioethicists to counsel him on whether to venture forth. They, like me, see no moral dilemma in trying to "create life" in this way. If such a feat were ever achieved, it would merely reaffirm what most of us in molecular biology have long known to be the truth: the essence of life is complicated chemistry and nothing more. Such a finding would have made headlines a century ago; today it's no big deal. Only the opposite conclusion – that there is more to the life of the cell than the sum of its basic components and processes – could generate deep excitement in today's scientific world.

DNA analysis has already changed the face of microbiology. Before DNA techniques were broadly applied, methods of identifying bacterial species were extremely limited in their powers of resolution: you could note the form of colonies growing in a petri dish, view the shape of individual cells through the microscope, or use such relatively crude biochemical assays as the Gram test, by which species can be sorted as either "negative" or "positive" depending on features of their cell wall. With DNA sequencing, microbiologists suddenly had an identification factor that was discernibly, definitively different in every species. Even species, like those inhabiting the ocean depths, that cannot be cultured in a laboratory because of the difficulty of mimicking their natural growing conditions are amenable to DNA analysis, providing a sample can be collected from the deep.

Now led by Claire Fraser, TIGR remains the leader of the bacterial genomics pack. In short order they have polished off the genomes of more than twenty different bacteria, including that of an ulcer-causing *Helicobacter,* a cholera-causing *Vibrio,* a meningitis-causing *Neisseria,* and a respiratory-disease-inducing *Chlamydia.* Their biggest competitor is a group at the Sanger Centre. The British contingent is led by Bart Barrell, who had the luck not to be in the United States, where his limited academic credentials would have barred him from top-gun status: he has no Ph.D., having come into science straight out of high school to work as Fred Sanger's assistant long before DNA sequencing became a reality. Before moving on to bacteria, Barrell made his name as an automation pioneer, having used several ABI sequencing machines to crank out some 40 percent of the baker's yeast genome of 14 million bases while the largely European yeast-sequencing consortium remained wedded to manual methods. Barrell's group later had the satisfaction of being the first to complete the sequence of *Mycobacterium tuberculosis*, the agent of the fearsome affliction once known as consumption.

In high school, Claire Fraser "had felt like an outcast because it wasn't cool to be a woman taking so many science courses." After studying at Rensselaer Polytechnic Institute, where she first became interested in microbes, she applied to medical school. Rather than accepting a place at prestigious Yale, she opted for SUNY Buffalo because her boyfriend was moving to Toronto. The director of admissions at Yale was nonplussed: "Well, young lady, I hope you know what you're doing." The Toronto connection would, alas, prove ephemeral; in 1981 Fraser married Venter, then a young assistant professor at SUNY Buffalo. "We went to a [scientific] meeting for our honeymoon," she recalled, "and wrote a grant proposal there."

The power of DNA analysis of microbes has been harnessed with great success in medical diagnostics: to treat an infection effectively, doctors must first identify the microbe causing it. Traditionally the identification has required culturing the bacteria from infected tissue – a process that is maddeningly slow, particularly in cases when time is of the essence. Using a fast, simple, and more accurate DNA test to recognize the microbe, doctors can start appropriate treatment that much sooner. And recently the same technology was pressed into service to deal with a national emergency: the hunt for the perpetrator of the anthrax outrage in the United States in the fall of 2001. By sequencing the anthrax bacteria from the first victim, TIGR investigators obtained a genetic fingerprint of the precise strain used.

The hope is that this precise information on the source of the anthrax will lead eventually to the culprit.

As we learn more about microbial genomes, a striking pattern is emerging. As we have seen, vertebrate evolution is a story of progressive genetic economy: through a widening array of mechanisms for gene regulation, it has become possible to do more and more with the same genes. And even when new genes do appear, they tend to be merely variations on an existing genetic theme. Bacterial evolution, by contrast, is proving itself a saga of far more radical transformation, a dizzying process that favors the importation or generation of whole new genes, as opposed to merely tinkering with what already exists.

Indeed, recombinant technology owes its very existence to the extraordinary ability of bacteria to incorporate new pieces of DNA (usually plasmids). Not surprisingly, then, microbial evolution too bears the footprint of dramatic gene-importing events of the past. *E. coli*, normally a benign inhabitant of our intestines (and of petri dishes), has morphed through gene importation into a killer variant. The toxins produced by one strain that occasionally causes outbreaks of food poisoning (killing twenty-one people in Scotland in 1996–97) and headlines about "Killer Burgers" are attributable to massive genetic "borrowing" from other species.

Genetic material normally moves *vertically* down a lineage – from ancestor to descendant – so this importation of DNA from outside is known as "horizontal transfer." Comparison of the genome sequence of normal *E. coli* to that of the pathogenic strain has revealed a shared genetic "backbone," identifying both strains as members of a common species, but there are many "islands" of divergent DNA unique to the pathogen. Overall, the pathogen lacks 528 of the normal strain's genes and has instead a staggering 1,387 genes not present in the normal strain. In that 528-for-1,387 exchange lies the key to the transformation of one of nature's most innocuous products into a killer.

Other bacterial nasties also show similar evidence of wholesale horizontal transfer. *Vibrio cholerae,* the agent of cholera, is unusual for a bacterium in that it has two separate chromosomes. The larger one (about 3 million base pairs in length) appears to be the microbe's original equipment, containing most of the genes essential to the functioning of the cell. The smaller one (about 1 million base pairs in length) seems to be a mosaic, made up of bits and pieces of DNA imported from other species.

Complex organisms, especially large ones like humans, are by design fairly inviolable gatekeepers of their own internal bio-chemistry: in most cases, if we don't ingest or inhale a substance, it cannot alter us profoundly. And so the biochemical processes of all vertebrates have tended over time to remain very similar. Bacteria, on the other hand, are much more exposed to the chemical vagaries of the environment; a colony may find itself suddenly awash in a noxious chemical – say, a disinfectant like household bleach. Little wonder these highly vulnerable organisms have evolved a stunning variety of chemistries. Indeed bacterial evolution has been driven by chemical innovation, the invention of enzymes (or the retrofitting of old ones) to do new chemical tricks. One of the most fascinating and instructive instances of this evolutionary pattern occurs among bacteria whose secrets we have only recently begun to learn about, a group known collectively as the "extremophiles" because of its members' predilection for the most inhospitable environments.

Bacteria have been found in Yellowstone hot springs (*Pyrococcus furiosus* thrives in boiling water and freezes to death at temperatures below 70°C [158°F]) and in the superheated water of deep-sea vents (where the high pressure at depth prevents the water from boiling). They have been found living in environments as acidic as concentrated sulfuric acid and in acutely alkaline environments as well. *Thermophila acidophilum* is an all-around extremophile, withstanding, as its name suggests, both high temperatures and low pH. Some species have been discovered in rocks associated with oil deposits, converting oil and other organic material into sources of cellular energy, rather like so many tiny sophisticated automobiles. One of these species inhabits rocks a mile or more down and dies in the presence of oxygen; appropriately, it is named *Bacillus infernus*.

Perhaps the most remarkable microbes discovered in recent years are the ones that subvert what was once considered a key dogma of biological science – that all energy for living processes comes ultimately from the sun. Whereas even *Bacillus infernus* and oil-consuming bacteria found in sedimentary rocks are connected to the organic past – the sun shone eons ago on the plants and animals whose remains are today's fossil fuels – so-called lithoautotrophs are capable of extracting the nutrients they need from rocks created de novo by volcanoes. These rocks – granite is an example – bear no traces of organic material; they contain no vestige of the energy of sunny prehistoric days. Lithoautotrophs have to construct their own organic molecules out of these inorganic materials. They live, literally, on a diet of rock.

There has been no more persuasive indicator of our general ignorance of the microbial universe than our belated discovery of the bacterial genus *Prochlorococcus*, whose planktonic cells photo-synthesize as they float in the open ocean. As many as 200,000 may inhabit a single milliliter of seawater, making this arguably the most abundantly represented species on the planet. It is certainly responsible for a huge proportion of the ocean's contribution to the global food chain. And yet *Prochlorococcus* was unknown to us until 1988.

The extraordinary microbial universe around us reflects the phenomenal power of eons of natural selection. Indeed the history of life on our planet can be told mostly as a tale of bacteria; more complicated organisms, ourselves included, are embarrassingly late arrivals – a virtual afterthought. Life appears to have originated as bacteria some 3.5 billion years ago. The first eukaryotes – cells whose genes are enclosed within nuclei – arose around 800 million years later, but they remained single-celled for about a billion years after that. Only about half a billion years ago did the breakthroughs occur that would ultimately give rise to the likes of the earthworm, the fruit fly, and *Homo sapiens*. The predominance of bacteria is reflected in the DNA-based reconstruction of the tree of life first carried out by Carl Woese at the University of Illinois: the tree of life is a bacterial tree, with a few multicellular beings forming a late-growth twig. Now generally accepted, Woese's ideas were at first strenuously opposed within the biological establishment. Still some of the implications of the DNA-based approach to the tree of life have been difficult to take: they have shown, for instance, that animals are not, as was once supposed, closely related to plants; rather, the closest relatives of animals are fungi. Humans and mushrooms stem from the same evolutionary root.

The Human Genome Project has proved Darwin more right than Darwin himself would ever have dared dream. Molecular similarities stem ultimately from the way in which all organisms are related through common descent. A successful evolutionary "invention" (a mutation or set of mutations that is favored by natural selection) is passed down from one generation to the next. As the tree of life diversifies – existing lineages splitting to produce new ones (reptiles persist as such, but also bud off into both bird and mammal lineages) – that invention may eventually appear in a huge range of descendant species. Some 46 percent of the proteins we see in yeast, for example, also appear in humans. The yeast (fungal) lineage and the one that ultimately gave rise to humans probably split about 1 billion years ago. Since each has subsequently developed independently, free to follow its own evolutionary trajectory, there have been in effect 1 billion years of evolutionary activity since that yeast/human common ancestor; and yet, *through all that time,* that set of proteins that existed in the common ancestor has changed only minimally. Once evolution solves a particular problem – for example, designing an enzyme to catalyze a particular biochemical reaction – it tends to stick with that solution. We have seen how this kind of evolutionary inertia is responsible for the centrality of RNA in cellular processes: life started in an "RNA world," and the legacy remains with us to this day. And the inertia extends to the biochemical details: 43 percent of worm proteins, 61 percent of fruit fly proteins, and 75 percent of fugu proteins have marked sequence similarities to human proteins.

Comparing genomes has also revealed how proteins evolve. Protein molecules can typically be envisioned as collections of distinct *domains* – stretches of amino acid chains that have a particular function, or form a particular three-dimensional structure – and evolution seems to operate by shuffling domains, creating new permutations. Presumably most new permutations are as useless as they are random, doomed to be eliminated by natural selection; but in the rare instance that a new permutation proves beneficial, a new protein is born. Some 90 percent of the domains that have been identified in human proteins are also present in fruit fly and worm proteins. In effect, therefore, even a protein unique to humans is likely nothing more than a reshuffled version of one found in *Drosophila*.

There is no better demonstration of this fundamental biochemical similarity among organisms than so-called rescue experiments, the aim of which is to eliminate a particular protein in one species and then use the corresponding protein from another species to "rescue" the missing function. We have already seen this strategy implemented in the case of insulin. Because human and cow insulins are so similar, diabetics who fail to produce their own can be given the cow version as a substitute.

In an example evocative of B-movie science fiction, researchers have been able to induce fruit flies to grow eyes on their legs by manipulating a particular gene that specifies where an eye should go. That gene then induces the many genes involved in producing a complete eye to go to work in that designated location. The mouse's corresponding gene is so similar to the fruit fly's that it will perform the same function when situated – by the genetic engineer's sleight of hand – in a fruit fly whose gene has been eliminated. That this can be done is nothing less than remarkable. Fruit flies and mice have been separated by evolution for at least half a billion years, so – following the logic applied above to humans and yeast evolving simultaneously along independent lines – the gene has in fact been conserved over a billion years of evolution. This is all the more astonishing when we consider that fruit fly and mouse eyes have fundamentally different structures and optics. Presumably each lineage perfected an eye appropriate for its respective purposes, but the basic machinery for determining the location of that eye, needing no improvement, stayed the same.

The most humbling aspect of the Human Genome Project so far has been the realization that we know remarkably little about what the vast majority of human genes do. To use the hard-won information properly requires us to devise methods for studying the function of genes on a genomewide scale.

In the wake of the HGP, two new postgenomic fields have duly emerged, both of them burdened with unimaginative names incorporating the "-omic" of their ancestor: proteomics and transcriptomics. Proteomics is the study of the proteins encoded by genes. Transcriptomics is devoted to determining where and when genes are expressed – that is, which genes are transcriptionally active in a given cell. If the genome is ultimately to be understood in its more dynamic reality, not as a mere set of instructions for life's assembly but as the screenplay for life's movie – all the drama described in the precise order it is meant to occur – then proteomics and transcriptomics provide the keys to glimpsing the live action. The more we learn, the more we see of *Life, the Movie.*

We have long appreciated that a protein is a great deal more in biological terms than the linear string of amino acids that compose it. How the string folds up to produce a distinctive three-dimensional configuration is really the key to its function – what proteomics seeks to know. Structural analysis is still done using X-ray diffraction: the molecule is bombarded with X rays that bounce off its atoms and scatter in a pattern from which the three-dimensional shape may be inferred. In 1962, my one-time colleagues at the Cavendish Lab at Cambridge University, John Kendrew and Max Perutz, received the Nobel Prize in Chemistry for their elucidation of the structures of, respectively, myoglobin (which stores oxygen in muscle) and hemoglobin (which transports oxygen in the bloodstream). Theirs was a monumental effort. The complexity of the X-ray diffraction images they had to interpret made me appreciate the relative simplicity of DNA!

Knowledge of a protein's three-dimensional structure greatly assists the work of medical chemists in their hunt for new drugs that work, as many do, by inhibiting protein functioning. In the ever more specialized and automated world of pharmaceutical research, several companies now offer to determine the structure of proteins as if they were production-line commodities. And the work is now immeasurably easier than it was in the day of Perutz and Kendrew: with more powerful X-ray sources, automated data recording, and faster computers driven by increasingly clever software, the time needed for solving a structure can be reduced from many years to a matter of weeks.

All too often, however, the three-dimensional structure itself provides no particular indication of that protein's function. Important clues may come instead from studying how the mystery protein interacts with other known ones. A simple way to identify such interactions involves spotting out samples of a set of known proteins on a microscope slide and then dousing them with the mystery protein, which has been previously treated so it will fluoresce under UV light. Where our test protein "sticks" to a particular spot on the slide's protein grid, it has become bound to the protein in that spot, causing it too to become fluorescent. Presumably, then, these two proteins are engineered to interact within the cell.

Ideally, to know life's screenplay, to "see" life's movie, we need to discover all the precise changes in protein composition that occur over the individual's development, from the moment of fertilization all the way through to adulthood. Though many proteins will be found to be active throughout the process, some will prove specific to a particular developmental stage, so in each growth phase we should expect to see different sets of proteins. Adult and fetal hemoglobins, for example, are subtly different. Similarly, each variety of tissue produces its own profile of proteins.

The most reliable way to sort out the various proteins from a given tissue sample is still the long-established method that uses two-dimensional gels to separate protein molecules on the basis of differences in their electrical charge and molecular weight. The several thousand protein spots thus differentiated can then be analyzed with a mass spectrometer, an instrument that can determine each one's amino acid sequence. Unfortunately, to apply proteomics like this to the vast number of proteins coded by an entire genome requires more funding than academic scientists typically have. For the most part, such expensive enterprises are left to the better-endowed researchers of large pharmaceutical companies. But because of the method's limitations, even their labs can't routinely find proteins that are present in very small amounts (see Plate 47).

This type of high-throughput proteomics, with all its expensive hardware and industrial-scale automation of complicated procedures, is therefore not the way most scientists nowadays study gene function at the genome level. Instead, methods of transcriptomics have been adopted, because they are cheaper and easier to apply: the functioning of all genes in a genome can be tracked by measuring the relative amounts of their respective messenger RNA (mRNA) products. If you are interested in the genes being expressed in, say, a human liver cell, you isolate a sample of mRNAs from liver tissue. This represents a snapshot of the mRNA population in the liver cell: very active genes, those most heavily transcribed and that produce many mRNA molecules, will be more abundantly represented, whereas genes that are rarely transcribed will contribute only a few copies to the mRNA sample.

The key to transcriptomics is a surprisingly simple invention known as a DNA microarray. Imagine a microscope slide with a grid of 25,000 tiny dot-shaped wells etched onto it. Using precise micropipetting techniques, DNA sequences from just one gene are deposited in each well so that the grid contains every gene in the human genome. Critically, the location on the microscope slide of each gene's DNA is known. Affymetrix, a company near Stanford, has managed to miniaturize these arrays even further by etching them onto a sliver of silicon the size of a small computer chip, yielding a "DNA chip."

Using standard biochemical techniques, you can tag your liver mRNAs with a chemical marker so, like the proteins mentioned above, they will fluoresce obligingly under UV light. Then comes the step where the power and simplicity of the technique becomes wonderfully apparent: you simply dump your sample of mRNAs onto the microarray with its minuscule chessboard of 25,000 gene-filled wells. The very same base-pairing bonds that hold together the two strands of the double helix will compel each mRNA molecule to pair off with the gene from which it was derived. The complementarity is precise and foolproof: the mRNA from gene X will bond only to the very spot occupied by gene X on the microarray. The next step is merely to observe which spots have picked up the fluorescent mRNAs. One spot on the microarray may show no fluorescence, implying that there was no complementary mRNA in the sample – and thus, we may infer, no active transcription of that gene in the liver cell. On the other hand, a number of spots do fluoresce, some with particular intensity; this indicates that many mRNA molecules have bound to it. Conclusion: a very active gene. Thus, with a single simple experimental assay, you have identified every one of the genes active in the liver. And such molecular panoramas have been made possible thanks to the success of the Human Genome Project and the new mind-set it has ushered into biology: we no longer need be content to study bits and pieces – we can now see the whole picture in all its spectacular glory.

It is hardly surprising that Stanford's Pat Brown, one of the method's leading practitioners, sees DNA microarrays as "a new kind of microscope." Marveling at the technology's potential to reveal a whole new genetic universe, he has declared: "We're toddlers now just starting to discover our world."

Transcriptomics is more than just another brilliant technical innovation. It promises to take us to a new level in the hunt for the genes that cause illness: using microarray technology we can discover the chemical basis for particular afflictions by studying the differences between healthy and diseased tissue as a function of gene expression. The logic is simple. We carry out microarray gene expression analysis on both normal and cancerous tissue, and spot the difference between the two, the genes being expressed in one and not the other. Once we can identify which genes are malfunctioning – either over- or under-expressing themselves in the cancerous tissue, for instance – we may be able to establish a target that can be attacked with pinpoint molecular therapies as opposed to broadly toxic radio- and chemotherapies that destroy healthy as well as diseased cells.

And we can apply the same technologies to distinguish among different forms of the same disease. Standard microscopy has offered limited assistance in this task: cancers that look alike to the pathologist peering through the eyepiece can in fact be critically different at the molecular level. Lymphoma cells, for instance, come in varieties that are hard to tell apart visually, even with the highest powers of resolution, but the differences in their gene expression profiles are clear, and vitally important in devising the most effective treatment. Referring to the earlier tendency to assume all cancers of a particular tissue have the same root, Brown said, "It was like thinking a stomachache has only one cause. Recognizing the distinctions makes it possible for us to do a better job of treating these cancers."

At Cold Spring Harbor Laboratory, Michael Wigler is using the method in yet another way: rather than adding RNA to a microarray and looking for gene expression, he is adding DNA from cancer cells to create a profile of the genetic diversity present in tumors. Many cancers are caused by chromosomal rearrangements – such as might occur when segments of a chromosome are inadvertently duplicated, leading to an excess in the number of genes that code for growth-promoting proteins. Other cancers arise due to the loss of genes coding for proteins that repress cell growth. Applying Wigler's technique, clinicians biopsy cancerous and healthy tissues from the same person. DNA from the cancerous tissue is chemically tagged with a red dye while the DNA from the normal tissue is tagged green. DNA microarrays, containing all 35,000 of the known human genes, are exposed to a mixture of the two samples. Like mRNA in a standard microarray experiment, the labeled DNA molecules bind base pair to base pair to their complementary sequences in the array. Genes amplified in cancer cells are marked by red spots (because there are many more red-tagged molecules binding to that spot than green-tagged ones) while genes deleted in cancer cells show up as green spots on the microarray (because there is no red-tagged molecule to bind there). Such experiments have already greatly expanded the list of genes known to contribute to breast cancer.

Whenever we tackle a specific human disease, we realize the extent to which we are probing in the dark. We could move so much more quickly to the heart of the problem – know the exact nature of what is wrong and how we might fix it – if only we had a more detailed knowledge of how our genes express themselves when all is well. With a fully formed dynamic understanding of when and where each of our 25,000 genes functions during normal development from fertilized egg to functioning adult, we would have a basis of comparison by which to understand every affliction: what we need is the complete human "transcriptome." This is the next holy grail of genetics, the next big quest in need of superfunding. In the short term, a likelier, even more important objective will be to obtain the complete transcriptome for the mouse, whose advantage over humans is that we can both observe and intervene experimentally during the course of prenatal development. Even collecting all such relevant data from the mouse will require major investments of money and time. And, as proved by the experience of DNA sequencing, we will be well served to take the time to gain what expertise we can by completing transcriptomes for simpler model organisms before taking on the mouse, much less the human.

Microarray studies of gene expression during the yeast cell cycle have already revealed the staggering complexity inherent in the molecular dynamics of cell division alone. More than eight hundred genes are involved, each called into action at its precisely specified time in the cell cycle. Here too we may depend on evolution's reluctance to fix what ain't broke: a biological process, once successfully evolved, will likely continue to employ the same basic molecular actors for as long as life persists on earth. As far as we can tell, those same proteins that direct development through the course of the yeast cell cycle carry out similar roles in the human cell.

Ultimately the goal of all three "-omics" (gen-, prote-, and transcript-) is to create a full picture, detailed right down to the level of the individual molecule, of how living things are assembled and operate. As we have seen, in even the simplest cases, the complexity is bewildering, and, despite the spectacular progress of the last decade, there remain many daunting challenges. As they relate to complex organisms, the molecular underpinnings of development – that extraordinary egg-to-adult journey that is governed by a linear code strand composed of just four letters – are for now best understood in the case of the fruit fly.

The fly has, of course, been the focus of intensive genetic investigation ever since its adoption by T. H. Morgan, and through the ensuing years of continual innovation *Drosophila melanogaster* has remained a genetic gold mine. In the late seventies at the European Molecular Biology Laboratory in Heidelberg, Germany, Christiane "Janni" Nüsslein-Volhard and Eric Wieschaus undertook a spectacularly ambitious fruit fly project. They used chemicals to induce mutations and then looked for disruptions in the very early embryonic stages of the flies' progeny. Classically, the quarry of the fruit fly geneticist was mutations affecting adults, like the one Morgan found to produce white (rather than red) eyes. In focusing on embryos, Nüsslein-Volhard and Wieschaus were not only condemning themselves to years of eyestrain as they stared down microscopes in pursuit of those elusive mutants, they were also venturing into utterly uncharted territory. The payoff, however, was spectacular. Their analysis uncovered several suites of genes that lay out the fundamental body plan of the developing fly larva.

The more universal message of their work is that genetic information is hierarchically organized. Nüsslein-Volhard and Wieschaus noticed that some of their mutants showed very broad effects while others evinced more restricted ones; from this they inferred correctly that the broad-effect genes operate early in development – at the top of a switching hierarchy – while the restricted-effect genes operated later. What they had found was a cascade of transcription factors: genes switching on other genes that in turn switch on others still, and so on. Indeed, hierarchical gene-switching of this kind is the key to the construction of complex bodies. A gene producing the biological equivalent of a brick will, left to its own devices, produce a pile of bricks; with proper coordination, however, it can produce a wall, and ultimately a building.

Normal development depends on cells "knowing" where they are in a body. A cell in the tip of a fly's wing, after all, should develop along very different lines than one located in the region that will give rise to the fly's brain. The first piece of essential positional information is the simplest: How does the developing fruit fly embryo know which end is which? Where should the head go? Bicoid, a protein produced by a gene in the mother, is distributed in varying concentrations through the embryo. The effect is called a "concentration gradient": the protein levels are highest at the head end and fall off as you travel toward the rear. Thus the bicoid concentration gradient instructs all cells within the embryo as to where they fall on the head-to-tail axis. Fruit fly development is segmental, meaning that the body is organized into compartments, all of which have much in common but each of which has some features unique to it. In many respects, a head segment is organized just like one in the thorax (the middle part of the insect body), but the former has head-specific organs, like eyes, and the latter thorax-specific ones, like legs. Nüsslein-Volhard and Wieschaus found groups of genes that specify the identities of different segments. For instance, "pair-rule" genes encode transcription factors – genetic switches – expressed in alternating segments. Pair-rule gene mutants result in an embryo with developmental problems in every second segment.

In 1995, Nüsslein-Volhard and Wieschaus received the Nobel Prize in Physiology or Medicine for their pioneering work. Unlike most laureates, both have remained active lab scientists – not for them the retreat into a big diploma-festooned office. For Wieschaus, science is still irresistible: "Because embryos are beautiful and because cells do remarkable things, I still go into the lab every day with great enthusiasm." As a child in Birmingham, Alabama, he dreamed of becoming an artist. Short of money as a sophomore at the University of Notre Dame, however, he took on one of the smelliest and most menial jobs in all of science: making the "fly food" (a noxious gelatinous concoction consisting largely of molasses) for a research lab's experimental population of fruit flies. Most people who serve as chef to a few hundred thousand messy and unappreciative insects would likely develop a lifelong aversion to the critters. For Wieschaus the result was the opposite: a lifelong commitment to the fruit fly and the mysteries of its development.

Born into an artistic German family, Nüsslein-Volhard was one of those students who excels at everything that interests them but puts absolutely no effort into anything else. Her hard work in illuminating the fruit fly's developmental genetics would have been achievement enough to justify two careers, but in the wake of her Nobel she has redirected her formidable attention to the development of another species altogether, the zebra fish: new work that promises to unlock many of the secrets of vertebrate development. At the 2001 event marking the centenary of the Nobel Prize it struck me that she was the only woman scientist present among the throngs of gray-haired males. Indeed, she is one of only ten women ever to win a Nobel in science.

One of those no-longer-youthful men was Caltech's Ed Lewis, an old fruit fly hand who shared the prize with Nüsslein-Volhard and Wieschaus. Actually Lewis doesn't much fit the gray-hair stereotype: though in his eighties at that Stockholm event, when he wasn't obliged to wear tails he was often seen in running gear! He too had long been concerned with the genetic control of fruit fly development, but his special interest was "homeotic mutations." These produce a most bizarre result: one developing segment mistakenly acquires the identity of a neighboring segment. His long and painstaking dedication to the Hox genes, in which these mutations occur, exemplifies values vanishing in an era when fads too often set science's agenda.

Homeotic mutations – which we now know disrupt transcription factor-encoding genes (the genetic switches) – can have drastic effects. The "antennapedia" mutation results in the fly's growing legs where its antennae belong: a fully formed pair of legs protruding from its forehead (see Plate 46). The "bithorax" mutation is almost as weird. Normally one of the segments making up the thorax produces the fly's pair of wings while the next thoracic segment toward the rear generates a pair of small stabilizing structures called "halteres." In a bithorax fly, the haltere segment mistakenly produces wings, so a fly that should have two wings in fact has four, the second pair just as perfectly formed as the first.

When they function properly, the genes regulating segment identity ensure that each body section acquires organs appropriate to its position: a head segment acquires antennae, and a thoracic segment acquires wings and legs. In the event of homeotic mutations, however, there is a confusion of segment identity. Thus, in the case of antennapedia, a head segment imagines itself a thoracic one and duly produces a leg rather than an antenna. Note, though, that while the leg is in the wrong place, it's still a perfectly good leg. Implication: The antennapedia positional gene switches on a whole suite of genes, typically those that produce an antenna, or, aberrantly, those that produce a leg; but the coordination within the suite is unhindered even when these genes are activated in the wrong place at the wrong time. Here again we see how genes high up in the developmental hierarchy control the fate of many, many genes farther down the line. As any librarian knows, hierarchical organization is an efficient way in which to store and retrieve information. With such a cascade arrangement, a surprisingly few genes can take you a long way.

Now that we are in the new era of comprehensiveness in biology ushered in by the once-unimaginable feat of the Human Genome Project, it may seem curious that we should find ourselves following the cutting edge of one of the next frontiers – that of developmental genetics – back into the realm of the fruit fly. But there is nowhere for us to go but back to the future, for even with the entire human genome in hand, the program and cues according to which its instructions are carried out remain a colossal mystery. Eventually we shall know the screenplay of human life as well as we know that of the fly. A comprehensive description of the patterns of human gene expression (the transcriptome) will be developed. A full inventory of the actions of all our proteins (the proteome) will be produced. And we will have a full and spectacularly complex picture of how each one of us is put together, and how each one of the multitudinous molecules we are made of figures in the functioning of you and me.

CHAPTER NINE

OUT OF AFRICA:
DNA AND THE HUMAN PAST

In August 1856 German quarry workers discovered part of a skeleton as they blasted their way into a limestone cave in the Neander Valley outside Dusseldorf. At first the remains appeared to be those of an extinct bear species whose bones often showed up in caves, but a local schoolteacher realized that the creature in fact belonged to a species much closer to our own. The exact identity of the owner of the bones, however, would prove a point of controversy. Particularly puzzling was the skull's thick brow ridge. One bizarre suggestion was that the bones belonged to an injured Cossack cavalryman who had crawled into the cave to die during the Napoleonic wars. Chronic pain from a preexisting condition, so the crackpot theory went, had produced a permanent furrow in the poor fellow's brow, deforming the bones of the skull to create the distinctive ridge. In 1863, in the midst of the debate about human origins provoked by the publication of Darwin's *Origin of Species* four years earlier, the original owner of the bones was given a name: *Homo neanderthalensis*. The bones belonged to a species distinct from, but similar to, *Homo sapiens*.

Though the German bones were the first to be officially designated Neanderthal, others found earlier in Belgium and Gibraltar were now recognized as being from members of the same species. More than a century later, many more specimens of *H. neanderthalensis* have been unearthed, and we now believe that Neanderthals settled throughout Europe, the Middle East, and parts of North Africa until about 30,000 years ago. French paleontologist Marcellin Boule is largely responsible for the popular image of Neanderthals as dim-witted and hulking. But his reconstruction, which used material from a French site at La Chapelle-aux-Saints, was based on a single individual who turns out to have been elderly and arthritic. In fact, Neanderthal brains were slightly larger than ours (and of a different shape due to a flatter cranium) and evidence from burial sites suggests that Neanderthals were culturally sophisticated enough to engage in funeral rituals; they may then have even believed in an afterlife.

The biggest debate triggered by the discovery of Neanderthals, however, centered not on how smart they were but on how they might be related to us. Are we descended from them? Paleontology suggests that modern humans arrived in Europe at roughly the same time as the last of the Neanderthals disappeared. Did the two groups interbreed or were the Neanderthals simply eliminated? Because the events in question happened in the ancient past and the surviving evidence is fragmentary – little beyond the odd bone – debates like this can drag on and on, keeping academic paleontologists and anthropologists endlessly entertained. Is a particular bone specimen perhaps intermediate between the thick bones typical of Neanderthals and the lighter bones of modern humans? Such specimens may have belonged to a hybrid individual produced by interbreeding between the two groups – a missing link. But then again they might just as well have come from a full Neanderthal with atypically light bones, or, for that matter, a fully modern human with unusually thick ones.

To everyone's surprise, the debate has been resolved by DNA: 30,000-year-old DNA extracted in 1997 from the very bones that started it all in 1856. Having evolved in order to store information securely and transmit it from one generation to the next, DNA, no surprise, shows great chemical stability. It doesn't degrade spontaneously or react readily with other molecules. But it is not impervious to chemical damage. At the moment of death, the body's genetic materials, like all its other constituents, become susceptible to a horde of would-be degraders: reactive chemicals, and enzymes that break down the molecular fabric. These chemical reactions require the presence of water, so DNA may be preserved if a corpse dehydrates fast enough. But even under ideal preservation conditions, the molecule is likely to survive perhaps 50,000 years at the absolute maximum. To obtain a legible DNA sequence from 30,000-year-old Neanderthal remains, preserved imperfectly, was therefore a tall order at best.

But Svante Pääbo, a tall, laconic Swede at the University of Munich, decided to have a crack at the problem. If anyone could do it, he was the one. Pääbo had pioneered work on the retrieval of so-called ancient DNA; he had scored sequences from Egyptian mummies, frozen mammoths, and the 5,000-year-old "Ice Man" who melted out of an Alpine glacier in 1991. Despite this impressive résumé, though, the prospect of drilling into a precious Neanderthal relic to look for intact DNA, if indeed any was to be found inside, was daunting. As his archaeologist colleague Ralf Schmitz recalls, "It was like getting permission to cut into the Mona Lisa."

Matthias Krings, Pääbo's graduate student, took on the project. He was pessimistic at first, but favorable early analyses to assess the bones' state of preservation emboldened Krings to press ahead. His search for viable DNA was focused not in the cells' nuclei, as one might expect, but in the little bodies called mitochondria, which are scattered throughout the cell outside the nucleus and produce the cell's energy. Each mitochondrion contains a small loop of DNA, some 16,600 base pairs in length. And because there are from 500 to 1,000 mitochondria in every cell, but only two copies of the genome proper (in the nucleus), Krings knew that those decaying Neanderthal bones were much more likely to yield intact mitochondrial sequences than intact nuclear ones. Furthermore, since mitochondrial DNA (mtDNA) had long been a staple of studies of human evolution, Krings would have plenty of modern human sequences against which to make comparisons.

A major worry for Krings and Pääbo was contamination. In the past a number of claimed successes at sequencing ancient DNA had proved to be erroneous when the sequence turned out to be from a modern source that had contaminated the sample. Every day each of us sloughs off a vast number of dead skin cells, showering our DNA into the environment to wind up we know not where. The polymerase chain reaction (PCR), with which Krings expected to amplify the stretch of mtDNA he hoped to find, is so sensitive that it can act upon a single molecule, amplifying any DNA it might encounter regardless of whether the source is ancient or still kicking. What if the Neanderthal DNA was too degraded for PCR to work, but the reaction proceeded nevertheless, amplifying a DNA sequence from an invisible contaminating particle that had flaked off Krings himself? Krings might then have to explain how he and the Neanderthal happened to have the same mtDNA sequence – a result unlikely to please the young man's boss, and even less his parents. To insure against this possibility, Krings and Pääbo arranged for a separate laboratory, Mark Stoneking's at Pennsylvania State University, to replicate the study. Contamination might occur there, too, but probably not with DNA from Krings, a continent away. And if both labs obtained the same result from the sample, it would be reasonable to suppose they had found a bona fide Neanderthal sequence.

"I can't describe how exciting it was," says Krings of the moment he first glimpsed the sequencing results. "Something started to crawl up my spine." Although, as feared, some sequences showed evidence of contamination, in others he could see something wondrous: a collection of intriguing similarities to, and differences from, the modern human sequence. Piecing together segments, he was able to reconstruct a Neanderthal mtDNA stretch running 379 base pairs. But the results weren't yet in from Penn State. Those sequences, however, proved to be the same: the identical 379 base pairs. "That's when we opened the champagne," Krings recalls.

The Neanderthal sequence had more in common with modern human mtDNA sequences than with those of chimpanzees, telling us that Neanderthals were unquestionably part of the human evolutionary lineage. At the same time, however, there were dramatic differences between the Neanderthal sequences and all 986 available sequences of modern human mtDNA to which Krings compared his sample. And even the most similar of those 986 sequences still differed from the Neanderthal one by at least 20 base pairs (or 5 percent). Subsequently, mtDNA has been sequenced from two other Neanderthals, one found in southwest Russia, the other in Croatia. The sequences, as expected, are not identical to the original one – we would expect to see variation among Neanderthal individuals just as we would among modern humans – but they are similar. The sum of the genetic evidence leads us to conclude that while Neanderthals do have their place on the evolutionary tree of humans and their relatives, the Neanderthal branch is a long way from the modern human limb. If, when they encountered each other in Europe 30,000 years ago, Neanderthals and moderns had indeed interbred, Neanderthal mtDNA sequences would have entered the modern human gene pool. That we see no evidence of such Neanderthal input implies that modern humans eliminated the Neanderthals rather than interbreeding with them. But whether they achieved the lethal result by direct confrontation or by more subtle means is something the DNA can't tell us.

Studies of Neanderthal DNA have shown that we are genetically distinct from Neanderthals. But the overall lesson of molecular studies of human evolution has tended to run in the opposite direction, revealing just how astonishingly close we are genetically to the rest of the natural world. In fact, molecular data have often challenged (and overthrown) long-held assumptions about human origins.

The great chemist Linus Pauling was the father of modern molecular approaches to evolution. During the early 1960s, he and Emile Zuckerkandl compared the amino acid sequences of corresponding proteins from several species. These were the early days of protein sequencing, and their data were inevitably limited. Nevertheless, the pair noticed a striking pattern: the more closely related two species are in evolutionary terms, the more similar are the sequences of their corresponding proteins. For example, comparing one of the protein chains of hemoglobin molecules, Pauling and Zuckerkandl noted that over its total length of 141 amino acids, there is only one difference between the human version and the chimpanzee, but the difference between humans and horses is 18 amino acids. The molecular sequence data reflect the fact that horses have been evolutionarily separated from humans longer than chimpanzees. Unearthing evolutionary history buried in biological molecules is now common practice; at the time, however, the idea was novel and controversial.

Molecular approaches to studying evolution depend on the correlation of two variables: the length of time two species (or populations) have been separated and the extent of molecular divergence between them. The logic of this "molecular clock" is simple. To illustrate it, let us imagine some matchmaking between two pairs of identical twins, one of genetically identical females and one of identical males. Each female is wed to one of the males, and each couple is then placed on its own otherwise uninhabited island. From a genetic perspective, the populations of the two islands are at the outset indistinguishable. Now leave each couple and its descendants alone for a few million years. At the end of this period, mutations will have occurred in the population on one island that will not have occurred in the population on the other. And vice versa. Because mutations occur at a low rate and because individual genomes, being large, offer huge numbers of possible sites where mutations might occur, it is inconceivable that both populations will have acquired the same set of mutations. So when we sequence DNA from the descendants of each couple, we will find that many differences between the once-identical genomes have accumulated. We say that the populations have "diverged" genetically. The longer they have been separated, the more divergent they will be.

But how do we tell time, so to speak, by looking at this "molecular clock"? Put another way, how can we measure the genetic divergence between ourselves, say, and the rest of the natural world? In the late sixties, long before the advent of DNA sequencing, Allan Wilson, a whimsical New Zealander at UC Berkeley, together with his colleague Vince Sarich, set about applying the Pauling-Zuckerkandl logic to humans and their closest relatives. But at a time when protein sequencing was still a dauntingly cumbersome and laborious affair, Wilson and Sarich found an ingenious shortcut.

The strength of an immune reaction to a foreign protein reflects *how* foreign the protein is: if it is relatively similar to the body's own protein, then the immune reaction is relatively weak, but if it is very different the reaction is proportionately stronger. Wilson and Sarich compared reaction strengths by taking a protein from one species and measuring the immune responses it triggered in others. This gave them an index of the molecular divergence between two species, but to introduce a time dimension to this "molecular clock" they needed to calibrate it. Fossil evidence implied that New and Old World monkeys (the two major groups of monkeys) separated from their common ancestor around 30 million years ago – and so Wilson and Sarich set the immunological "distance" between New and Old World monkeys as equivalent to 30 million years' separation. Where did this put humans in relation to their closest evolutionary kin, chimpanzees and gorillas? In 1967 Wilson and Sarich published their estimate that the human lineage had separated from that of the great apes about 5 million years ago. Their claim provoked an uproar: in paleoanthropological circles conventional wisdom held that the divergence had occurred around 25 million years ago. Between humans and apes, the establishment insisted, there is clearly much more than 5 million years' worth of difference. It was, for many, cause enough to dismiss the Berkeley team's newfangled genetic method as untrustworthy, and, to declare that, anyway, geneticists should stick to their fruit flies, and leave humans to the anthropologists! Wilson and Sarich, however, weathered the storm. And subsequent research has shown that their dating of the human/great ape split was remarkably accurate.

When the time came to extend his analysis of the human/ape divide from proteins to DNA, Wilson entrusted the effort to his graduate student Mary-Claire King (see Plate 48). The product, in 1975, was one of the outstanding scientific papers of the twentieth century. For a long time, though, such a triumphant outcome seemed unlikely, especially from King's perspective. Her work had not been going well, owing in part to the enormous distraction created at Berkeley by the anti–Vietnam War movement in the early 1970s. King considered going off to Washington, D.C., to work for Ralph Nader, but fortunately she sought Wilson's advice. "If everyone whose experiments failed stopped doing science," he wisely counseled, "there wouldn't be any science." King stuck with it.

King and Wilson's comparison of the chimpanzee and human genomes combined a number of methods, including a clever technique called "DNA hybridization." When two complementary strands of DNA come together to form a double helix, they can be separated by heating the sample to 95°C – a phenomenon called "melting" in the molecular geneticist's jargon. What happens when the two strands are not perfectly complementary – when there are mutations in one of them? It turns out that two such strands will melt apart at a temperature lower than 95°C. How much lower will depend upon the degree of difference between the two strands: the greater the difference, the less the heat required to pry them apart. King and Wilson used this principle to compare human and chimpanzee DNAs. The closer the two were in sequence, the closer the double helix's melting point would be to the perfect-match standard of 95°C. The closeness observed was surprising indeed: King was able to infer that human and chimpanzee DNA differ in sequence by a mere 1 percent. In fact humans have more in common with chimpanzees than chimpanzees do with gorillas, the genomes of the latter two differing by about 3 percent.

So striking was the result that King and Wilson felt obliged to put forward an explanation for the apparent discrepancy between the rates of genetic evolution – slow – and of anatomical and behavioral evolution – fast. How could so little genetic change account for the substantial difference we see between the chimpanzee at the zoo and the species on the other side of the glass? They suggested that most of the important evolutionary changes had occurred in the pieces of DNA that control the switching on and off of genes. This way, a small genetic change could have a major effect by changing, say, the timing of the expression of a gene. In other words, nature can create two very different-looking creatures by orchestrating the same genes to work in different ways.

The next, and biggest, bombshell from Wilson's Berkeley lab came in 1987. Using patterns of DNA sequence variation, he and his colleague Rebecca Cann figured out the family tree for our entire species (see Plate 49). It was one of the very few pieces of science ever to make the cover of *Newsweek*.

As Krings would in his analysis of Neanderthals a decade later, Cann and Wilson relied on mitochondrial DNA. There were several reasons for using mtDNA, but as usual the practical ones were most important. In the days before PCR technology had entered the research mainstream, getting enough DNA to probe a particular gene or region could be quite a headache. And Cann and Wilson's study called for analyzing not one but 147 samples. They therefore needed as much DNA as they could get their hands on. A human tissue sample is massively rich in mtDNA compared with the chromosomal DNA found in cell nuclei. Still, Cann and Wilson would need plenty of tissue if they were to have any hope of extracting even mtDNA in sufficient quantities. Their solution: placentas. Usually discarded by hospitals after babies are delivered, these are a rich source of mtDNA. All Cann and Wilson had to do was persuade 147 pregnant women to donate their babies' placentas to science – 146, actually, because

Mary-Claire King was more than willing to contribute her daughter's placenta. And they knew that to reconstruct the human family as completely as possible they would need tissue from the most genetically diverse range of donors they could assemble. Here America's melting-pot population offered a distinct advantage: they would not have to travel to Africa to get hold of African DNA – the slave trade had brought African genes to our shores. But Cann and Wilson would have to depend on collaborators in New Guinea and Australia to find Aboriginal women (not much represented in the U.S. gene pool) who were willing to participate.

Your mtDNA is inherited from your mother. Your father's genetic contribution, contained in the head of a single sperm, did not include mitochondrial material. The sperm's DNA is injected into an egg cell that already contains mitochondria derived from the mother. Cann and Wilson would therefore be tracing the history of the human female line. Inherited from just one parent, mtDNA never gets an opportunity to undergo recombination, the process by which segments of chromosome arms are exchanged so that mutations are shuffled from one chromosome to another. The absence of recombination in mtDNA is a major advantage when we come to reconstruct the family tree based on similarity of DNA sequences. If two sequences have the same mutation, we know that they must be descended from a common ancestor (in whom that mutation originally arose). Were recombination occurring, however, one of the lineages could have acquired the mutation just recently through a recombinational shuffling event, so having a mutation in common would not necessarily indicate common ancestry. Now the logic for using mtDNA to make the family tree is simple. Similar sequences – those with plenty of mutations in common – indicate close relationship; sequences with many differences indicate a more distant relationship. In visual terms, close relatives – those that derive from a relatively recent common ancestor – will cluster close together on the family tree; distant relatives are more spread out, because their common ancestor is relatively far back.

Cann and Wilson found that the human family tree has two major branches, one comprising only various groups within Africa and the other consisting of some African groups plus everyone else. This implies that modern humans arose in Africa – that is where the ancestors common to all of us lived. This idea was hardly new. Noting that both our closest relatives, chimpanzees and gorillas, are native to Africa, Charles Darwin himself inferred that humans had evolved there too. The most striking, and controversial, aspect of Cann and Wilson's family tree is how far back it goes in time. By making a number of simple assumptions about the rate at which mutations accumulate through evolution, it is possible to calculate the age of the family tree – the time back to the great-great-great-great- . . . -grandmother of us all. Cann and Wilson came up with an estimate of about 150,000 years. Even the most distantly related currently living humans shared a common ancestor as recently as 150,000 years ago.

Like Sarich and Wilson's result two decades earlier, Cann and Wilson's was greeted by many in the anthropological community with outraged disbelief. One widely accepted view of human evolution held that our species was descended from individuals who left Africa about 2 million years ago before settling throughout the Old World. Such a model implied that the family tree should be about thirteen times deeper. Cann and Wilson's alternative, dubbed by the media "The Eve Hypothesis" or, less misleadingly, "Out of Africa," did not deny the more ancient migration, but rather implied that when modern humans arrived in Europe they displaced those populations of early hominids derived from the original exodus nearly 2 million years before. *Homo erectus,* the species that spread out from Africa 2 million years ago, migrated through the Old World and gave rise, about

700,000 years ago, to Neanderthals, who were thus in effect their European descendants. Then, no more than about 150,000 years ago, another group, *Homo sapiens* or modern humans – also descendants of *Homo erectus* but a group that had evolved without ever having left the mother continent – now chose to repeat the odyssey out of Africa made eons before by their *H. erectus* ancestors. We have seen how the Neanderthals failed to interbreed with the new arrivals in Europe, and the same seems to have been true whenever *H. sapiens* encountered *H. erectus*. Wherever they met, the former displaced the latter. And the disappearance of the last Neanderthal, around 29,000 years ago, represents the extinction of the last of the nonmodern descendants of *H. erectus*.

Cann, Wilson, and their colleagues had changed fundamentally the way we understand our human past.

Subsequent research has confirmed Cann and Wilson's conclusion. Much of the newer work has come out of the Stanford laboratory of Luigi Luca Cavalli-Sforza, who pioneered the application of genetic approaches to anthropological problems. Raised in a distinguished Milanese family, Cavalli-Sforza was fascinated with microscopes. And in 1938, he enrolled as a precocious sixteen-year-old in medical school at the University of Pavia. "It turned out to be a very lucky choice," he notes: the alternative would have been service in Mussolini's army. When I first met him in 1951, Cavalli-Sforza was still an up-and-coming bacterial geneticist. But a chance remark made by a graduate student would inspire a turn away from the genetics of bacteria toward the genetics of humans. The graduate student, who had trained to be a priest, mentioned that the Catholic Church had kept detailed records of marriages over the past three centuries. Realizing that in these records there lurked a wealth of research possibilities, Cavalli-Sforza began to apply himself more and more to human genetics, and he probably remains one of a very few human geneticists who can legitimately claim to have found their vocation via the Church.

Cavalli-Sforza understood that the most convincing confirmation of Cann and Wilson's assertions about human evolution would ideally come from genes only transmitted from father to son, i.e., some component of the human genome passed down through the male line. If one could arrive at their conclusions tracking the male lineage – taking a patrilineal route as opposed to the matrilineal path Cann and Wilson found through mtDNA analysis – one could be assured of a truly independent corroboration. The male-specific component of the genome is, of course, the Y chromosome. By definition, the possessor of a Y is male (the Y chromosome, remember, is inherited by men from their fathers, whose sperm cells can contain either an X or a Y; upon fusing with the egg cell, which always contains an X, the sperm thus determines our sex, XX combinations producing females and XY males). The Y chromosome, then, holds the key to the genetic history of men. In addition, because recombination occurs only between paired chromosomes, the use of the Y allows us to avoid that dreaded pitfall of evolutionary analysis, recombination: a Y is unique whenever it is present, and so there is never a matching Y with which it might trade material.

In a blockbuster paper published in 2000, Cavalli-Sforza's colleague Peter Underhill did for the Y chromosome what Cann and Wilson had done for mtDNA. The findings were strikingly similar. Again the family tree was found to be rooted in Africa, and again it was shown to be remarkably shallow: not the ancient mighty oak imagined by anthropologists, but the shrub of Cann and Wilson's analysis, around 150,000 years old.

The existence of two independent data-sets yielding a similar picture of the human past is extremely compelling. When only one region, say mtDNA, is studied, the results, while suggestive, are still inconclusive; the pattern may simply reflect the peculiarities of the history of that particular region of DNA, rather than the impact of some major historical event on our species as a whole. Critically, the point at which a family tree converges – the most recent common ancestor of all the sequences in the study, that great-great- . . . -grandfather/mother of us all – is *not* necessarily associated with any particular event in human history. Though it *may* connote the origin of our species or some other historically significant demographic episode, it may just as likely signify something much more trivial from the point of view of human history – perhaps nothing more, say, than the effect of past natural selection on mtDNA. If, however, the same pattern of change can be observed in more than one region of the genome, the chances are that one has indeed found the genetic footprint of an important past event.

To better understand how natural selection can affect patterns of genetic variation (and the overall age of a family tree), imagine the following scenario: 150,000 years ago, the tribe of protohumans boasted a plethora of mtDNA sequences, just as our species does today, but then a beneficial mutation – one favored by natural selection – arose on one of those sequences. The mutation would increase in frequency until, after many generations, every member of the species would have it. Because there is no recombination in mitochondria, no exchange between mtDNAs, the selective process would affect the entire sequence in which a favored mutation first appeared, so every member of the species would end up with the same mtDNA sequence. So by the time that natural selection has finished its job and every individual possesses the favored mutation, there would be no mtDNA genetic variation in the species. Gradually over subsequent years, though, mutations would occur and variation would build up again, but all these new mtDNA sequences would ultimately be descended from that single sequence: the family tree's convergence point, the most recent common ancestor of all the sequences. The pattern would be exactly what Cann and Wilson found, but in this case the convergence point represents nothing more than an episode of evolution's fine-tuning of mtDNA.

This was the ambiguity that dogged Cann and Wilson's result: Was it produced by evolutionary tinkering, or by something much more significant in the overall scheme of human prehistory? But when Underhill observed a similar pattern for the Y chromosome, that ambiguity vanished. The coincidence suggested forcefully that at the moment in question (150,000 years ago), human populations did indeed undergo a radical genetic alteration, one capable of affecting mtDNAs and Y chromosomes simultaneously. The phenomenon involved, to which we shall turn in a moment, is called a "genetic bottleneck."

How can demographic factors affect a family tree? Any genealogy is the outcome of the waxing and waning of the lineages composing it: over time, some will thrive and others become extinct. Think of surnames. Assume that a thousand years ago on some remote island everyone had one of three surnames: Smith, Brown, and Watson. Assume, too, that small errors of transcription – "mutations" – occasionally occurred when the names of newborns were inscribed in the birth registers. The errors are infrequent and slight, so we can still tell which of the original names the altered forms derived from: "Browne" is clearly a mutation of "Brown." Now let us imagine that in the population today, a thousand years later, we find that everyone is called Brown, Browne, Bowne, Frown, or Broun. Smith and Watson have gone extinct while the Brown line has thrived (and diversified through mutation). What has happened? Pure chance has led to the loss of the Smith and Watson lines. Perhaps, for instance, several

Mr. & Mrs. Smiths of one generation managed to produce mainly daughters. Assume (in accordance with tradition, though not the modern alternative convention) that surnames are transmitted along the male line; the bumper crop of daughters would thus have the effect of reducing the representation of Smiths in the next generation. Now say that the new generation of Smiths also overproduced daughters, and the demographic effect was heightened once again – well, you get the picture: eventually, the Smith name disappeared altogether. So did Watson.

This kind of random extinction is, in fact, statistically inevitable. Usually, however, it happens so slowly that its impact can be felt only over huge periods of time. Sometimes, though, a bottleneck – a period of very much reduced population size – will massively accelerate the process. With only three couples (six individuals) on the island at the beginning of its population history, it was reasonably likely that we would lose Smith and Watson within a single generation, the chances being fairly good that both the Smiths and the Watsons would have only daughters, or fail to procreate at all. In a large population, such abrupt disappearances of lines cannot occur; it is statistically inconceivable, given a population with many Smith couples, that they could all wind up producing only girls or simply fail to have children. Only over the course of many generations would the effects of the dwindling ranks gradually mount up. Indeed a real-life example of this hypothetical name-extinction process actually occurred in the South Pacific, when the six *Bounty* mutineers colonized Pitcairn Island with their thirteen Tahitian brides. Within seven generations, the number of surnames had dwindled to three.

When we look today at the surnames in our theoretical population, Brown, Browne, Bowne, Frown, or Broun, we can infer that they are all descended from just one of the three starting lineages, Brown. And so the implication of the human mtDNA and Y chromosome data should hardly surprise us: 150,000 years ago there were many different mtDNA sequences and many different Y chromosome sequences, but today's sequences are all descended from just one of each. All the others went extinct, most probably disappearing during some ancient bottleneck event – a population crash caused by plague, a change in climate, what have you. But whatever this cataclysmic event in our early history, one thing is clear: some time afterward, groups of our ancestors started to head out of Africa, beginning the epic saga of the human colonization of the planet (see Plate 50).

Another interesting finding confirmed by both the mtDNA and Y chromosome data is the position on the human family tree of the San of southern Africa.* Theirs is the longest, and therefore the oldest, branch on the tree. This by no means implies that they are more "primitive" than the rest of us: every human is at the same evolutionary and molecular remove from our closest relatives among the great apes. If we trace lineages back to the last common ancestor of both chimpanzees and humans, my lineage is about 5 million years old, and so is a San's. In fact, our two lineages are the same for most of those eons; only 150,000 years ago did the San lineage separate from other human lines.

It appears, from the genetic evidence, that after an initial migration into southern and eastern Africa, the San remained relatively isolated throughout history. This pattern is borne out by sociolinguistics when we consider the distribution of the San's unusual (at least to my ears) "click" languages. Their current distribution is extremely limited owing to the expansion of Bantu-speaking people from west central Africa starting about 1,500 years

*The San are also known as Bushmen (Sanqua, in Dutch), a derogatory term given to them by Dutch settlers in the late seventeenth century.

ago. The Bantu expansion displaced the San to marginal environments like the Kalahari Desert.

Given their relatively stable history, do the San provide a snapshot of what the ancestors of all modern humans were like? Possibly, but not necessarily – substantial change may well have occurred along the San lineage over the past 150,000 years. Even inferences from the San about our early ancestors' ways of living are questionable: the San's present lifestyle is an adaptation to the harsh desert environment to which they have been confined since the relatively recent arrival of the Bantu speakers. In 2000 I experienced the unique thrill of living for several days in a San community in the Kalahari. I was struck by their remarkable pragmatism, their efficient no-nonsense way of taking on all tasks before them, even those outside their normal experience, like fixing a flat tire. I found myself wishing that more of my colleagues were likewise adaptable. And if, in genetic terms, these people are as genetically "different" from me as any on the planet, I could not fail to be impressed by just how like-minded we were.

The genetic and cultural uniqueness of the San will disappear shortly. Young people in the Kalahari show little desire to continue the simple hunter-gatherer lifestyle of their nomadic parents. When, for instance, the group I visited staged a traditional "trance dance," the younger members were visibly embarrassed by their elders' antics. They will move away from their communities and marry into other groups.

In fact, history has already recorded a trend toward mixture between the San and other groups. Nelson Mandela's Xhosa tribe, for one, represents a biological mix of Bantu and San peoples, as the Xhosa language, though Bantu-based, reveals in its many typically San clicks. In our technologically accelerated day and age, it is unlikely that the genetic and cultural integrity of the San will survive much longer. It is, therefore, fortunate indeed that considerable efforts have been made over the past few decades to understand and document this unique people and their way of life. Philip Tobias of the University of Witwatersrand in Johannesburg both initiated these studies and, for many years, championed the San as an unofficial spokesman during the dark days of apartheid. And Trefor Jenkins, a voluble Welshman who arrived in South Africa after working as a doctor in Zambian copper-mining towns, has long spearheaded genetic studies of the San and other indigenous groups.

Sadly it currently remains beyond the reach of even the most sophisticated genetic methods to elucidate the origins of human culture. Archaeological evidence shows that our ancestors were up to much the same activities as other hominids, Neanderthals included, during the first phase of their evolution. Indeed, a cave site at Skhul in Israel offers proof that about 100,000 years ago populations of *Homo sapiens* and *Homo neanderthalensis* coexisted, neither apparently endangering the other. But, as we have seen, modern humans subsequently wiped out their heavy-browed cousins around 30,000 years ago. It therefore seems likely that in the intervening 70,000-year period modern humans, through technological and/or cultural advances, somehow acquired the edge.

Independent archaeological information supports this hypothesis. It would appear that, around 50,000 years ago, modern humans suddenly became *culturally* modern: we see in the remains from this time the first indisputable ornaments, the first routine use of bone, ivory, and shell to produce familiar useful artifacts, and the first of many improvements in hunting and gathering technology. What happened? We shall probably never know. But one is tempted to speculate that it was the invention of language that made all of this – and all we have accomplished since – possible.

Prehistory by definition refers to the period prior to written records, and yet we find written in every individual's DNA sequences a record of our ancestors' respective journeys. The new science of molecular anthropology uses patterns of genetic variation among different groups to reconstruct this history of human colonization. Human "prehistory" has thus become accessible.

Studies of the distribution of genetic variation across the continents combined with archaeological information have revealed some details of our ancestors' global expansion. The journey along the fringes of Asia and through the archipelagoes of modern Indonesia to New Guinea and Australia was accomplished by about 60,000 years ago. Getting to Australia required crossing several substantial bodies of water, suggesting that our ancestors were already using boats at that early stage. Modern humans arrived in Europe around 40,000 years ago, and penetrated northern Asia, including Japan, some 10,000 years later.

Like so many other leaders in this field (including Rebecca Cann and Svante Pääbo), Michael Hammer, at the University of Arizona, received his training in Allan Wilson's Berkeley lab. And though Hammer's initial interest was mice, the publication of Cann and Wilson's mtDNA study diverted him from rodents to the human past. He was among the first to realize that information from the Y chromosome would provide the crucial test of Cann and Wilson's overall hypothesis. But the Y proved reluctant at first to yield its secrets. One study (done in Wally Gilbert's lab) sequenced the same chunk of DNA drawn from multiple individuals, only to find the sequence identical in every instance – a laborious effort that yielded zero information about genetic interrelations. Hammer persisted, however, and eventually he and others turned the Y chromosome into an anthropological gold mine, whose payoff culminated in Underhill's landmark paper.

A major vein in the Y chromosome mine has enriched our attempts to reconstruct the human colonization of the New World, a relatively late development. The identity of the oldest human settlement in the Americas remains contentious: a site in Clovis, New Mexico, is the traditional titleholder, dating back some 11,200 years; but fans of a site in Monte Verde, Chile, claim it to be at least 12,500 years old. It is also debated whether the first Amerindians crossed a land bridge across the Bering Strait during the last Ice Age or took a more southerly route in boats. What the genetic data make clear, however, is that the founding group was small: with only two major classes of Y chromosome sequences detected, there appear to have been just two distinct arrivals, each perhaps involving no more than a single family. Among Amerindians mtDNA variation is much more extensive than Y chromosome variation, suggesting that there were more women than men in each founding group. Probably the more common of the two Y chromosome sequences represents the first arrival; the descendant population would then already have been established before the arrival of the second group, which included the ancestors of today's Navajo and Apache. The more common sequence also boasts another distinction: the presence (first noted in 2002) of a mutation that is rarely found elsewhere on the planet. Giving further evidence of its bearers' precedence as pioneers, this mutation is calculated to be about 15,000 years old, not much older than the earliest known archaeological sites.

Genetic analyses have permitted the reconstruction of more recent phases of prehistory as well. Hammer, for example, has shown that modern Japanese are a mix of the Jomon ancient hunter-gatherers, currently represented by Japan's aboriginal Ainu population, and relatively recent immigrants, the Yayoi, who arrived about 2,500 years ago from the Korean peninsula, bringing with them weaving, metalworking, and rice-based agriculture. In Europe, too, we see evidence of waves of migration, often associated with advances in agricultural technology. Groups like the Basques (who live in the mountainous Pyrenees on the French-Spanish border) and the Celts (who arrived later and are found throughout the northwest margin of Europe, from Brittany in France through Ireland and western Britain) are genetically distinct from the rest of Europe. One explanation is that each of these groups was displaced to relatively far-flung regions by more recent arrivals.

Bryan Sykes at Oxford has done much to reveal the complexity of the genetic map of modern Europe. Conventional wisdom had held that modern Europeans were largely derived from the Middle Eastern populations that invented agriculture in the Fertile Crescent, between the Mediterranean and the Persian Gulf. Sykes, however, has found that most European ancestry can be traced not to the Fertile Crescent but to older indigenous lines predating the incursions of Middle Easterners and to migrant groups from Central Eurasia. Such groups include the Celts and the Huns, who swept into Europe from the East around 500 B.C. and A.D. 400 respectively. And taking his analysis of mtDNA a step further, Sykes has argued as well that virtually all Europeans are descended from one of seven "daughters of Eve," his term for the surprisingly few major ancestral nodes in the European mtDNA family tree. A company he founded, called Oxford Ancestors, will, for a fee, sequence part of your mtDNA to determine from which of the seven "daughters" you are descended.

Another key to understanding the human past may rest with an observation fruitfully exploited by Cavalli-Sforza and others: patterns of genetic evolution often correlate with those of linguistic evolution. There are, of course, the obvious parallels between genes and words. Both are transmitted from one generation to the next; both undergo change, which in the case of language can be particularly fast, as any parent of a teenager knows. Likewise, American English is similar to but distinct from British English even though the two have been evolving separately for only a few hundred years. On the basis of the similarities and differences, then, the family tree of languages can be reconstructed in much the same way the genetic family tree can. But even more important, in many cases, as Darwin himself first predicted,* we can identify instructive correspondences between the two trees, such that what we learn about the one can deepen our understanding of the other. Both the Celts and the Basques offer dramatic cases in point: each people is genetically isolated from the rest of Europe, and each one's languages are correspondingly distinct from those of the rest of the continent. As for the New World, a controversial linguistic theory proposes that there are but three major language groups native to the Americas, and two of these correlate with the two early immigration events discerned in the Amerindian Y chromosome data. The third, by far the smallest, involves the isolated Inuit.

The availability of sex-specific genetic data – mtDNA for women, Y chromosomes for men – invites comparisons between male and female history. Mark Seielstad, a graduate student of Cavalli-Sforza's, chose to compare patterns of migration between the sexes. The logic is simple. Imagine a mutation that arises on a Y chromosome in Cape Town, South Africa. The speed with which it reaches, say, Cairo, is an index of rates of male migration. Similarly, the speed with which a Cape Town mutation in mtDNA reaches Cairo can be said to measure the rates of female migration.

For good or ill, history has been much more the chronicle of men,

---

*In *The Origin of Species,* Darwin notes: "If we possessed a perfect pedigree of mankind, a genealogical arrangement of the races of man would afford the best classification of the various languages now spoken throughout the world."

rather than women, on the move. Typically they were in search of plunder or empire: think of Alexander the Great's march from Macedonia into the northern reaches of India; of the Vikings and their sea-borne rampages from Scandinavia to Iceland and America beyond; or of Genghis Khan and his horsemen pouring across the steppe of Central Asia. But even without warfare as an excuse for travel, we still think of men as the more mobile members of human society. Men traditionally do the hunting, an activity that can often take them a long way from the hearth, whereas women in traditional hunter-gatherer societies stay close to home, gleaning food locally and raising the children. Therefore, Seielstad had reason to expect that men would be our species's genetic prime movers. The data proved him startlingly wrong. Women, on average, are *eight times more* mobile than men.

In fact, counterintuitive though it may be, the pattern can be simply explained. Almost universally, across all traditional societies, we humans engage in something anthropologists call "patrilocality": when individuals from two different villages get married, the woman moves to the man's village, and not vice versa. Imagine that a woman from village A has married a man from village B, and she moves to B. They have a daughter and a son. The daughter marries a man from village C and moves to C; the son marries a woman from village D and she moves to join him in B. Thus the male line stays put in B whereas the female line has moved, in two generations, from A to C via B. This process is carried out generation after generation, and as a result female migration proves extensive, but male migration does not. Men do indeed occasionally rush off to conquer distant lands, but these events are unimportant in the grand scheme of human migratory patterns: it's actually that step-by-step village-to-village migration of women that has shaped human history, at least on the genetic level.

Detailed regional studies of Y chromosome and mtDNA variation may also reveal something of the patterns of sexual relations and mating customs promoted in the course of colonization. In Iceland, for instance, which was uninhabited before the arrival of the Vikings, we find a marked asymmetry when we compare mtDNA and Y chromosomes. Most Ys are predictably Norse, but a large proportion of the mtDNA types are derived from Ireland. Apparently, the Norsemen colonizing Iceland took Irish women with them. Unfortunately, how the Irish women felt about this cannot be extracted from the mtDNA data.

A recent study of Y chromosome and mtDNA variation in Colombia shows a similar effect. In most segments of society, Colombian Y chromosomes are Spanish Y chromosomes, a direct biological legacy of the European conquest of the Spanish Main. In fact, approximately 94 percent of the Y chromosomes studied have a European origin. Interestingly, however, the mitochondrial pattern is quite different: modern Colombians have a range of Amerindian mtDNA types. The implication is clear: the invading Spaniards, who were men, took local women for their wives. The virtual absence of Amerindian Y chromosome types reveals the tragic story of colonial genocide: indigenous men were eliminated, while local women were sexually "assimilated" by the conquistadors.

Sometimes, however, enduring asymmetries are more a matter of cultural continuity than violent clash of cultures. The Parsees, a minority group in India, believe themselves to be descended from the Zoroastrians, an Indo-European Aryan people who fled religious persecution in Iran in the seventh century. Genetic analysis of modern Parsees indeed reveals that they have retained "Iranian" Y chromosomes, but their mtDNA tends to be of the "Indian" type. In this case the asymmetry is maintained by tradition. To be accepted as a true Zoroastrian Parsee, one has to have a Zoroastrian Parsee father.

Thus membership in the Parsee community is paternally transmitted together with a Y chromosome. Here genetics confirms the hold of tradition.

Tradition has informed patterns of genetic variation among Jews as well. A recent study has shown that members of the priestly caste, the *kohanim* (and their descendants, usually identifiable today by the surname Cohen), have a Y chromosome distinctive enough to set them apart from all other groups. Even among the most obscure populations, those flung farthest by the Jewish Diaspora, such as South Africa's Lemba, the Cohen Y has been preserved – almost like a sacred religious text. Its source is thought to be Aaron, according to Scripture the founder of the *kohanim* caste and the brother of Moses. It is certainly not impossible that the *kohanim* Y chromosome sequence was indeed his and that it has been passed down intact, father to son, in every generation since. Such have been the rigors of tradition over the course of Jewish history.

Hammer and others have been able to use Y chromosomes to track the entire Diaspora with interesting results. The Ashkenazim, for example, who have lived in Europe for the past twelve hundred years (and now the United States and elsewhere), have nevertheless maintained the genetic indications of their Middle Eastern origins. In fact, molecular studies have made plain that the Jews, genetically at least, are virtually indistinguishable from all other Middle Eastern groups, including the Palestinians. So, too, is it written. Abraham, the great patriarch, is said to have had two sons by different women: Isaac, from whom the Jews are descended, and Ishmael, forefather of the Arabs. That such a deadly enmity should have arisen between the descendants of one man is an irony that grows only more bitter when genes seem to verify tradition's narrative.

A simple stroll down a Manhattan street would suggest that ours is the most genetically variable species on the planet. In fact, though, the human genome is markedly less variable than those of most species for which we have genetic information. Only about 1 in every 1,000 human base pairs varies among individuals. Genetically, then, we are 99.9 percent alike, a minute degree of difference by the standards of other species. Fruit flies – even if they all look the same to us – have levels of variation some 10 times higher. Even Adélie penguins, those icons of sameness in their vast Antarctic colonies of indistinguishable individuals, are more than twice as variable as we are. Nor is this lack of variability found in our nearer relatives: chimpanzees are about 3 times as variable as we are, gorillas 2 times, and orangutans 3.5.

With the mtDNA and Y chromosome family results at hand, it is readily apparent why we humans are so alike. It's because our common ancestor was so recent; 150,000 years is a blink of an eye by evolutionary standards – insufficient time for substantial variation to arise through mutation.

Another counterintuitive finding about human variation, what little there may be, is that it does not correlate, for the most part, with race. Prior to Cann and Wilson's demonstration of humankind's surprisingly recent flight out of Africa, it was assumed that different groups had been isolated from one another on different continents for ages and ages, up to two million years. This would have permitted the accumulation of substantial genetic difference, in accordance with the Pauling-Zuckerkandl model, whereby the extent of genetic divergence between isolated populations is a function of the time over which they have been isolated. In light of Cann and Wilson's conclusion that we all share a much more recent common ancestor, it is clear that there has simply not been time enough for geographically separate populations to diverge significantly. Thus, though genetic differences, like skin color, are manifest across groups, race-specific genetic differences tend to be very limited. Most of our scant variation is actually spread rather uniformly across populations: one is as likely to find a particular genetic variant in an African population as in a European one. One is left to surmise that much of the genetic variation in our species arose in Africa *before* the out-of-Africa event, and so was already present in the groups that went forth to colonize the rest of the world.

As a final blow to any pride we may take in our own genetic variety: the Human Genome Project's conclusion that only about 2 percent of our DNA encodes genes would suggest that at least 98 percent of our variation falls in regions of the genome where it has no effect. And because natural selection very efficiently eliminates mutations that affect functionally important parts of the genome (such as genes), variation accumulates preferentially in noncoding (junk) regions. The difference between us is small; the difference it makes is even smaller.

Because of the short evolutionary timescales involved, most of the consistent differences we do see among groups are probably products of natural selection: skin color, for one.

Under their dense matted hair, the skin of our closest relatives, the chimpanzee, is largely unpigmented. (Chimpanzees, you might say, are white.) And presumably the common ancestor of chimpanzees and humans from which the human lineage spun off five million years ago was similar. And so we infer that the heavy skin pigmentation characteristic of Africans (and of the earliest modern humans, in Africa born) arose in the course of subsequent human evolution. With the loss of body hair, pigment became necessary to protect skin cells from the sun's damaging ultraviolet (UV) radiation. We now know at a molecular level how UV rays can cause skin cancer: they make the thymine bases of the double helix stick to one another, creating a kink, so to speak, in the DNA molecule. When that DNA replicates itself, this kink often promotes the insertion of a wrong base, producing a mutation. If, by chance, that mutation is in a gene that regulates patterns of cell growth, cancer may result. Melanin, the pigment produced by skin cells, reduces UV damage. As anyone with as hopelessly fair a complexion as mine knows too well, sunburn, though typically not lethal, can be a much more immediate health threat than skin cancer. Thus it is easy to imagine natural selection favoring the acquisition of dark skin in order to prevent not only cancer, but also the infections that can easily result from a severe sunburn.

Why did people living in higher latitudes lose melanin? The best explanation involves vitamin D_3 synthesis, a process carried out in the skin and requiring UV light. D_3 is essential for calcium uptake, which in turn is a critical ingredient of strong bones. (A deficiency of D_3 can result in rickets and osteoporosis.) It is possible that, as our ancestors moved out of Africa into highly seasonal environments, with less year-round UV radiation, natural selection favored pale-skinned variants because they, with less sun-blocking pigment in their skin, synthesized D_3 more efficiently with the limited UV available. The same logic may apply to the movements of our ancestors *within* Africa. The San, for instance, in South Africa, where UV intensities are similar to those of the Mediterranean, have a strikingly pale skin. But what about the Inuit peoples, who live in or close to the hardly sunny Arctic but are surprisingly dark? Their opportunities for producing the vitamin would appear to be further limited by the need to be fully clothed all the time in their climate. In fact, the selective pressure favoring lightness seems not to have asserted itself among them, and the reason appears to be that they have solved the D_3 problem in their own way: a diet with plenty of fish, a rich source of the essential nutrient.

Given what a powerful determinant, mostly for ill, skin color has been in human history and individual experience, it is surprising indeed how little we know about its underlying genetics. This deficit, however, may have less to do with the limitations of our science and more with the intrusion of politics into science; in an academic world tyrannized by political correctness, even to study the molecular basis of such a characteristic has been something of a taboo. What little we understand about it depends on old studies of mixed-race children, which established that several genes contribute to pigmentation. But our knowledge of other species and the similarity of basic biochemical processes among all mammals suggest a more complicated picture. We know, for instance, that many genes affect coat color in mice, and it is likely that these have direct human equivalents. So far, though, we have managed to identify only two genes involved in human pigmentation: the one that, when mutated, causes albinism, and the other, the "melanocortin receptor," associated with red hair and a pale (often freckled) complexion. The melanocortin receptor gene is variable among Europeans and Asians, but invariant among Africans, suggesting that there has been strong natural selection in Africa against mutations in the gene, i.e., against red-haired, fair-skinned individuals. Albinos, who lack pigment altogether, occasionally appear today in African populations (probably through de novo mutation) but their acute sensitivity to sunlight puts them at a severe disadvantage.

Another morphological trait likely determined by natural selection is body shape. In hot climates, where dissipating body heat is a priority, two basic types have evolved. The "Nilotic form," represented by the East African Masai, is tall and slender, maximizing the surface-area-to-volume ratio and thus facilitating heat loss. The Pygmy form, on the other hand, though still lightly built, is very short. In this case, a physically strenuous hunter-gather lifestyle has selected for small size to minimize the energy expended in movement – why lug a big body around to look for food? In high latitudes, by contrast, selection has favored body forms that promote heat retention: those with the lower ratios of surface area to volume. Neanderthals from Northern Europe were therefore heavily built, and so too on average are today's inhabitants of the same boreal climes. Some of the variation in athletic performance we see among groups is presumably attributable to these body-form differences. It should come as no surprise that in the high jump, for instance, a tall Nilotic body is better adapted than a short robust one.

If there is a trait whose distribution among human populations is hard to fathom, it is lactose intolerance. Mammalian milk, including the human variety, is rich in a sugar called lactose, and newborn mammals typically produce a special enzyme, lactase, to break it down in the intestine. Upon weaning, however, most mammals, including humans – at least, most Africans, Native Americans, and Asians – stop making lactase and so as adults cannot digest lactose. "Lactose intolerance" means that drinking a glass of milk can have unpleasant consequences, including diarrhea, gas, and abdominal bloating. Most Caucasians and the members of a few other groups, on the other hand, continue to produce lactase throughout their lives, and can therefore handle a lifelong dairy diet. The explanation has been advanced that lactose tolerance evolved in those groups historically most dependent upon dairy products, but the pattern of the trait is by no means fully convincing; there are, for example, groups of Central Asian animal herders – cheese for everyone – who are lactose intolerant. And despite belonging to an ethnic group that is typically lactose tolerant, I am intolerant. If natural selection had favored tolerance in a particular group, why would it leave its job undone? The most compelling evidence yet in support of the standard explanation is the presence of lactose tolerance in African groups traditionally associated with livestock. We may never fully understand the adaptive dimension of this trait, but molecular biologists working on a Finnish population have recently identified the mutation responsible for it. And so while we are by no means fighting a killer here, it is now possible, with a simple genetic test, to determine whether a newborn will grow up to face a choice between ice-cream deprivation and chronic gastric cramps.

More interesting than the relatively few differences we see among the races is what we all have in common – what it is that makes us so different from our closest relatives. As we have seen, our lineal split from the chimpanzee about 5 million years ago has barely given us enough time apart to evolve a 1 percent genetic difference. But in that 1 percent lie the critical mutations that make us the remarkable thinking, speaking creatures we are. It may be debated whether other species possess some limited form of consciousness, but clearly none of them has produced a Leonardo da Vinci or a Francis Crick.

The chromosomes of humans and chimpanzees are very similar. Chimpanzees, however, have 24 pairs whereas we have 23. It turns out that our chromosome 2 was produced by the fusion of two chimpanzee chromosomes. There are also differences in the human and chimpanzee versions of chromosomes 9 (bigger in humans) and 12 (bigger in chimpanzees) and several examples of inversions (or flips) within chromosomes that differ in humans and chimpanzees. Whether these chromosomal differences will prove significant is hard to say.

The relative merits are not much clearer at the biochemical level, where so far we know of only two differences between humans and chimpanzees. Difference 1: In both species a sugar molecule called sialic acid appears on the outside of every cell. But while the molecule is subtly modified in chimpanzees through the action of an enzyme, in humans, the gene encoding that enzyme is always mutated: no enzyme is produced, and human cell-surface sialic acid is unmodified. We have no clue at all as to whether this is significant. Difference 2: This one, discovered in 2002 by Svante Pääbo's group, is more suggestive: a difference in FOXP2, a gene known to be involved somehow in human language. (Because mutations in the human version have been found to cause linguistic impairment, FOXP2 has been misleadingly dubbed by the media as "the grammar gene.") Out of a chain running 715 amino acids, just two changes distinguish humans from chimpanzees and gorillas, whose FOXP2 proteins are identical. In fact, these amino acids are identical in *all* mammals tested except for humans. Moreover, statistical analysis of the pattern of DNA variation in and around the gene suggests that natural selection may have had a role in shaping the protein during human evolution. It is therefore tempting (but premature) to suggest that FOXP2 is the evolutionary equivalent of a smoking gun – a glimpse of a critical step in the origin of language.

Pääbo's lab has also pioneered a promising and original approach to identifying other genes that may encode the critical difference(s). Using DNA microarrays, which determine what genes are switched on in a particular tissue (see chapter 8), Pääbo has compared patterns of gene expression – which genes are switched on – in humans, chimpanzees, and macaque monkeys for three different tissues: white blood cells, liver, and brain. As would be expected on the basis of their close relationship, humans and chimpanzees fall out close to each other for both blood cells and liver. However, the pattern of gene expression in the brain tells a totally different story: the human brain is very different from those of the chimpanzee and macaque. Perhaps this is not entirely surprising: most of us would not need a laboratory full of equipment to figure out that human brains are distinct from chimpanzee brains. The research's significance lies instead in its ability to provide us with an inventory of the genes whose expression differs between human and chimpanzee brains. Even that will be only a start at best. It is unlikely that, even once we have a full catalog of the underlying mechanisms, we shall understand precisely *how* they set us apart. Our humanness is likely much more difficult to describe than even a precisely detailed list of controlled molecular events. But in our search for its genetic underpinning we are now at least beginning to assemble a list of suspects.

As I write this, the chimpanzee genome project is nearing completion. When it is done, the DNA making up the 1 percent difference that King and Wilson identified will be revealed. My guess is that they will be proved right: the critical differences will lie not in the genes themselves but in their regulation. Humans, I suspect, are simply great apes with a few unique – and special – genetic switches.

Molecular biology's grandest mission is surely to answer questions about ourselves and our origins as a species. But each human soul yearns to know its own story as well as that of its kind. DNA can provide a more individualized account of ancestry as well. In a sense, written in my DNA molecules is the history of my evolutionary lineage, a narrative that can be viewed at different levels. I can situate the sequence of my mtDNA into Cann and Wilson's human family tree, or I can look in greater detail at my known family's past. My Y chromosome and mtDNA will tell different stories – my mother's side, and my father's.

I was never interested in genealogy. But my family – like many, I suspect – had its own in-house archivist in the form of my aunt Betty, who spent a lifetime worrying about who was related to whom and how. It was she who found that the Watsons – of lowland Scots stock – first appeared in the United States in 1795 in Camden, New Jersey.

And it was she who insisted that some paternal ancestor of mine designed Abe Lincoln's house in Springfield, Illinois. But I've always been more interested in my Irish side, my maternal grandmother's family. My mother's grandparents fled Ireland during the great potato famine of the 1840s, ending up in Indiana, where her grandfather, Michael Gleason, died in 1899, the year my mother was born. On his gravestone it says he had come from a town in Ireland called Glay.

On a visit to Ireland, I tried to find out more about my great-grandfather at the County Tipperary Records Office, whose quarters in Neneagh, twenty miles from Limerick, had formerly been a prison. My sleuthing was singularly unsuccessful. Finding no record at all of "Glay," I could only conclude that name as spelled on the tombstone of my probably illiterate ancestor was fanciful. Thus ended my only brush with genealogical research, until recently. Now that the framework of the human family tree has been laid out by Cann and others, I am keen to see where I fit in. Companies like Bryan Sykes's Oxford Ancestors represent the new face of genealogical research, with high-tech laboratories to replace dusty archives. With a sample of my DNA, Oxford Ancestors has conducted both mtDNA and Y chromosome analysis. Sadly the tests revealed nothing romantic, no exotic ancestry. I really am, as I feared, largely the product of generic Scots-Irish stock. I cannot even blame my more brutish attributes on ancient Viking incursions into my bloodline.

CHAPTER TEN

GENETIC FINGERPRINTING: DNA'S DAY IN COURT

In 1998 Marvin Lamont Anderson, thirty-four years old, was released from the Virginia State Penitentiary. He'd been there for fifteen years, almost all his adult life, convicted of a horrific crime: the brutal rape of a young woman in July 1982. The prosecution had presented an unambiguous case: the victim recognized Anderson from a photograph; she picked him out in a lineup; and she identified him in court. Found guilty on all counts, he was given consecutive sentences totaling over two hundred years.

A clear-cut case. A better defense attorney, however, might have been more effective in countering the prosecution's efforts to stack the deck against the defendant. Anderson was picked up based exclusively on the (white) victim's report to the police that her (black) assailant had boasted of "having a white woman"; so far as the authorities knew, Anderson was the only local black man with a white girlfriend.

Among the mug shots the victim looked at, only Anderson's was a color photograph. And of the men whose pictures she was shown, he alone was placed in the lineup. And although another man, John Otis Lincoln, was shown to have stolen, about thirty minutes before the attack took place, the bicycle used by the assailant, Anderson's attorney failed to call Lincoln as a witness.

Five years after Anderson's trial, Lincoln confessed under oath to the crime, but the trial judge declared him a liar and refused to act. Anderson meanwhile continued to protest his innocence and requested that DNA analysis be done on the physical evidence from the crime scene. But he was told that it had all been destroyed in accordance with standard procedure. It was then that Anderson contacted the lawyers of the Innocence Project, a group that had gained national attention using DNA analysis to establish definitive evidence of guilt or innocence in criminal proceedings. While the Innocence Project worked on Anderson's request, he was released on parole; assuming no violations, he would remain a parolee until 2088, easily the rest of his life.

In the end, Anderson's salvation was the sloppiness of the police technician who had performed the inconclusive blood group analysis on the crime scene material in 1982. She had failed to return the samples to the proper authorities for routine destruction, and so they still existed when Anderson asked for a reexamination. The director of the Virginia Department of Criminal Justice, however, refused the request, arguing it might establish an "unwelcome precedent." But under a new statute, the Innocence Project attorneys won a court order calling for the tests to be performed, and, in December 2001, the results proved categorically that Anderson could not have been the assailant. The DNA "fingerprint" matched Lincoln's. Lincoln has since been indicted and Anderson pardoned by Governor Mark Warner of Virginia.

DNA fingerprinting – the technique that rescued Marvin Anderson from an undeserved life sentence – was discovered by accident by a British geneticist, Alec Jeffreys (see Plate 52). From the earliest days of the recombinant DNA revolution, Jeffreys had been interested in genetic differences among species. His research at Leicester University focused on the myoglobin gene, which produces a protein similar to hemoglobin, found mainly in muscle. It was in the course of this "molecular dissection" that Jeffreys found something very strange: a short piece of DNA that repeated over and over again. A similar phenomenon had been observed in 1980 by Ray White and Arlene Wyman, who, looking at a different gene, had shown that such repeats varied in number from individual to individual. Jeffreys determined that his repeats were junk DNA, not involved in coding for protein, but he was soon to discover that this particular junk could be put to good use.

Jeffreys found that this short stretch of repeating DNA existed not only in the myoglobin gene but was scattered throughout the genome. And although the stretches varied somewhat from one repetition to the next, all of them shared one short, virtually identical sequence of some fifteen nucleotides. Jeffreys decided to apply this sequence as a "probe": using a purified sample of the sequence tagged with a radioactive molecule, he could hunt for the sequence genomewide. With DNA from the genome laid out on a special nylon sheet, the probe would stick down, by base-pairing, wherever it encountered its complementary sequence. By placing the nylon on a piece of X-ray film, Jeffreys could then record the pattern of radioactive spots. When he developed the film from the experiment, he was astonished by what he saw. The probe had detected many similar sequences across a range of DNA samples. But there was still so much variability from one sample to the next that even among ones taken from members of the same family you could tell the individuals apart. As he wrote in the resulting paper in *Nature* in 1985, the "profile provides an individual-specific DNA 'fingerprint.' "

Jeffreys's choice of the term "DNA fingerprint" was quite deliberate. This technology clearly had the power to identify an individual, just like traditional fingerprinting. Jeffreys and his staff obtained DNA samples from their own blood and subjected them to the same procedure. The images on X-ray film, as expected, made it possible to distinguish unambiguously between people. He realized the range of potential uses was extensive:

> In theory, we knew it could be used for forensic identification and for paternity testing. It could also be used to establish whether twins were identical – important information in transplantation operations. It could be applied to bone marrow grafts to see if they'd taken or not. We could also see that the technique [would work] on animals and birds. We could figure out how creatures are related to one another – if you want to understand the natural history of a species, this is basic information. We could also see it being applied to conservation biology. The list of applications seemed endless.

But the procedure's first practical application was stranger than any Jeffreys had anticipated.

In the summer of 1985, Christiana Sarbah was at her wits' end. Two years before, her son, Andrew, had returned to England after visiting his father in Ghana. But at Heathrow, British immigration authorities had refused to admit the boy, though he had been born in Britain and was a British subject. Denying that Sarbah was his mother, they alleged that Andrew was, in fact, the son of one of Sarbah's sisters and was trying to enter the country illegally on a forged passport. After reading a newspaper report about Jeffreys's work, a lawyer familiar with the case asked the geneticist for help. Could this new DNA test prove that Andrew was Mrs. Sarbah's son and not her nephew?

The analysis was complicated by the fact that neither the father nor Sarbah's sisters were available to give samples. Jeffreys prepared DNA from samples taken from the mother and three of her undisputed children. The analysis showed that Andrew had the same father as the other children, and that Sarbah was his mother. Or more specifically, that chances were less than 1 in 6 million that one of her sisters was his mother. The immigration authorities did not challenge Jeffreys's results but avoided formally admitting the error by simply dropping the case. Andrew was reunited with his mother. Jeffreys saw them afterward: "The look of relief on her face was pure magic!"

But would the technique work with blood, semen, and hair, the body tissues typically found at crime scenes? Jeffreys was quick to prove that it could indeed, and soon his DNA fingerprints would gain worldwide attention, revolutionizing forensic science.

On a Tuesday morning in November 1983, the body of a fifteen-year-old schoolgirl named Lynda Mann was found on the Black Pad, a footpath outside the village of Narborough, near Leicester in England. She had been sexually assaulted. Three years passed with no arrest in the case. Then, it happened again: on a Saturday in August 1986 the body of Dawn Ashworth, another fifteen-year-old, was found on Ten Pond Lane, another footpath in Narborough. The police were convinced that the same man had committed both murders and soon accused a seventeen-year-old kitchen assistant. But, while confessing to the Ashworth murder, the suspect denied involvement in the earlier case. So it was that the police consulted Alec Jeffreys to confirm that their suspect had killed both girls.

Jeffreys's fingerprint analysis contained both good and bad news for the authorities: Comparison of samples from the two victims showed that the same man had indeed carried out both murders, as the police believed. Unfortunately (for the police) the same test also proved that the kitchen worker in custody had not murdered either girl, a result confirmed by other experts the police called in. The suspect was released.

With their only lead now blown, and worries rising in the local community, the police took an extraordinary step. Confident that DNA fingerprinting would yet prove the key to success, they decided to request DNA samples from all adult males in and around Narborough. They set up stations to collect blood samples and were able to eliminate a great many candidates by the traditional, and cheaper, test for blood type. The remaining samples were sent for DNA fingerprinting. A good Hollywood version of the story would, of course, have Jeffreys identifying the true killer. And it did happen that way, but not without a further plot twist, also worthy of Tinseltown. The culprit initially eluded the genetic dragnet. When faced with providing the mandatory sample, Colin Pitchfork, pleading a terror of needles, persuaded a friend to furnish a sample in his stead. It was only later, when the friend was overheard telling of what he had done, that Pitchfork was picked up and thus gained the dubious honor of being the first criminal ever apprehended on the basis of DNA fingerprints.

The Narborough case showed law enforcement agencies worldwide that DNA fingerprinting was indeed the future of criminal prosecution. And it would not be long before such evidence was first adduced in an American legal proceeding.

Perhaps the British are, culturally, more accepting of authority, or perhaps recondite molecular mumbo jumbo was just more likely to rub Americans the wrong way, but in any case the introduction of DNA fingerprinting into the United States was highly controversial.

The law has always had difficulty assimilating the implications, if not the very idea, of scientific evidence. Even the most intelligent lawyers, judges, and juries have customarily found it difficult to understand at first. In one famous early instance of forensic courtroom drama, blood-typing had unequivocally ruled out Charlie Chaplin as the father of a child whose mother had slapped a paternity suit on the silent-screen legend. The jury nevertheless ruled in the mother's favor.

American courts had long applied the Frye test as their standard for admissibility of scientific evidence. Based on one of the first trials to introduce forensic proof, it tries to keep out unreliable evidence by requiring that the science on which it is based "must be sufficiently established to have gained general acceptance in the particular field in which it belongs." But being based on a poor understanding of what constitutes well-established science, the test proved an ineffective way of determining the credibility of "expert" testimony. It was not until 1993, in *Daubert vs. Merrell Dow Pharmaceuticals* that the Supreme Court ruled the Federal Rules of Evidence should be used: the judge in a trial should determine whether the proffered evidence is reliable (i.e., whether it can be trusted as scientifically valid).

Nowadays, with Court TV an established part of the television landscape, and with prime-time series focusing on forensic investigations a staple of the networks, it may be hard to appreciate how difficult it was for the American legal system to swallow DNA. Though everyone had been hearing about it since our landmark discovery in 1953, it still had about it an impenetrable scientific aura. Indeed, the field of genetics seemed only more arcane every time the popular media hailed a new advance. Perhaps worst of all was the fact that DNA-supported charges were presented not as dead certainties but as probabilities. And what probabilities they were! With figures like "1 in 50 billion" bandied about to establish the guilt or innocence of the accused, little wonder some questioned the value of lawyers, judges, juries, and expensive trials when a geneticist, wrapped in the authority of science, could settle a case.

But at all events, most trials depend on more than the comparison of two DNA samples. Meanwhile, the acceptance of the new methods progressed slowly but ineluctably. In some sense the cause of broader understanding and acceptance was aided by lawyers who made their name challenging the very cases that depended on DNA evidence. Skilled attorneys like Barry Scheck and Peter Neufeld became as knowledgeable as the experts they were cross-examining. Scheck – short, messy, and pugnacious – and Neufeld – tall, tidy, and pugnacious – gained attention searching for technical flaws in cases presented during the early days of genetic fingerprinting. The two first met in 1977 as colleagues at the office of the Bronx Legal Aid Society, a local center of legal advocacy for the indigent. After growing up in New York City, the son of a successful impresario who managed stars like Connie Francis, Scheck found his political calling when he went to college at Yale, taking part in the national student strike that followed the Kent State shootings in 1970. Ever suspicious of entrenched authority and the abuse of power, he volunteered to assist Bobby Seale's defense team during the Black Panther's trial in New Haven. Peter Neufeld grew up in suburban Long Island, where his mother still lives, not far from Cold Spring Harbor Laboratory. He was no less precocious in his leftward leanings, having been reprimanded in the eleventh grade for organizing antiwar protests.

It was little surprise when the two young bred-in-the-bone social progressives became crusading lawyers manning the barricades of legal aid in New York City – at a tumultuous moment in the life of the city, when rising crime rates made "justice for all" seem to some an ideal endangered in the pursuit of public safety. A decade later, Scheck would be professor at Cardozo School of Law, and Neufeld would be in private practice.

I first met Scheck and Neufeld at an historic conference on DNA fingerprinting held at Cold Spring Harbor Laboratory. The controversy was at its height in part because the forensic technology was being applied more and more broadly despite still being done with Jeffreys's as-yet-unrefined original technique, the arcane-sounding analysis of restriction fragment length polymorphisms, or RFLPs. Inevitably some results were difficult to interpret, and so DNA fingerprinting was being challenged on technical and legal grounds. The Cold Spring Harbor gathering was actually the first occasion on which molecular geneticists – including Alec Jeffreys – would confront the forensic specialists and lawyers now using DNA in the courtroom. The discussions were heated. The molecular geneticists accused the forensic scientists of sloppy laboratory techniques, of simply not doing the testing carefully enough. Indeed, in those days DNA fingerprinting in forensic laboratories was subject to little, if any, regulation or oversight. There were also challenges to the statistical assumptions, likewise unstandardized, used to calculate those imposing numbers suggesting virtual certainty. The geneticist Eric Lander spoke for more than a few concerned participants when he proclaimed bluntly: "The implementation [of DNA finger-printing] has been far too hasty."

These practical problems were typified in a case Scheck and Neufeld were working on in New York. Joseph Castro was accused of murdering a pregnant woman and her two-year-old daughter. RFLP analysis, performed by a company called Lifecodes, had established that a bloodstain on his wristwatch was from the murdered mother. After a sustained examination of the DNA data, however, the expert witnesses of both the prosecution and defense jointly informed the judge in a pretrial hearing that, in their view, the DNA tests had not been done competently. The judge excluded the DNA evidence as inadmissible. The case never came to trial because Castro pleaded guilty to the murders in late 1989.

Despite the exclusion of the DNA evidence, the Castro case helped establish the legal standards for genetic forensics. These were the standards that would be applied in a much more prominent case Scheck and Neufeld were to take on, one that would make DNA fingerprinting a household term in America and indeed everywhere one could find a television: the trial of O. J. Simpson in 1994. The former sports icon was facing a possible death penalty if convicted of the heinous crimes he was charged with by the Los Angeles district attorney: the gory murder of Simpson's ex-wife, Nicole Brown Simpson, and her friend, Ronald Goldman. As part of the legal "dream team" assembled by the accused, Scheck and Neufeld would make critical contributions to Simpson's defense and acquittal. Forensic detectives had collected bloodstains from the crime scene at Nicole Brown Simpson's house, from O. J. Simpson's house, from an infamous glove and sock, and from Simpson's equally infamous white Bronco. The DNA evidence – forty-five blood specimens in all – contributed, according to the prosecution's case, a "mountain of evidence" pointing to Simpson's guilt. But Simpson had in his corner the most skillful mountaineers money could buy. The challenges from the defense came thick and fast, and as the whole world watched on TV, these counterclaims would bring some of the central controversies that had been simmering for years in forensic science up to a full-blown boil.

A decade before the Simpson trial, back in the days when prosecutors first began presenting DNA evidence, and only prosecutors commissioned the application of genetic technology, defense attorneys were quick to raise an obvious question: By what standard could one define a match between a DNA sample found at a crime scene and one derived from blood taken from the suspect? It was a particularly contentious issue when the technology still depended on RFLPs. In this method, the DNA fingerprint appears as a series of bands on an X-ray film. If bands produced by the crime scene DNA were not identical to those produced by the suspect's, just how much difference could be legitimately tolerated before one had to exclude the possibility of a match? Or how same does "the same" have to be? Technical competence came into question as well. Initially, when DNA fingerprinting was done in forensic laboratories without special expertise in handling and analyzing DNA, critical mistakes were not uncommon. Law enforcement agencies understood that if their powerful new weapon were to remain in commission, these questions would have to be answered. A new form of genetic marker – short tandem repeats (STRs) – replaced the RFLP method. The size of these STR genetic markers can be measured very accurately, doing away with the subjective assessment of RFLP bands on an X-ray film. The forensic science community itself dealt with the problem of variable technical competence by establishing a uniform code of procedures for doing DNA fingerprinting, as well as a system of accreditation (see Plate 51).

Perhaps the toughest attacks, however, were launched against the numbers. While prosecutors were given to presenting DNA evidence in terms of dispassionate, seemingly incontrovertible statistics, sometimes, as defense lawyers began to argue, tendentious assumptions had been made in calculating the state's one-in-a-billion margins of certitude. If you have a DNA fingerprint from the crime scene, on what basis do you calculate the likelihood (or, more often, the unlikelihood) that it might belong to someone other than prime suspect A? Should you compare the DNA to that of a random cross-section of individuals? Or, if prime suspect A is, for instance, Caucasian, should your sample be compared only to DNA from other Caucasians (since genetic similarity tends to run higher among members of the same racial group than in a random cross-section of people)? The odds will vary depending on what one deems a reasonable assumption.

And an effort to defend a conclusion founded on the arcane principles of population genetics can backfire, confusing jurors or putting them to sleep. The sight of someone struggling manfully to put on a glove that simply doesn't fit is worth more – much more, experience tells us – than a mountain of statistics.

In fact, DNA fingerprinting evidence presented in the Simpson trial pointed to the accused. A blood drop collected close to Nicole Brown Simpson's body, as well as other drops found on the walkway at the crime scene, were shown with virtual certainty to be his. With an equal lack of doubt, the blood staining the glove retrieved from his home was determined to be a mixture of Simpson's and that of the two victims; the blood found on the socks and in the Bronco proved to match the blood of Simpson and that of his ex-wife.

No, finally, in the eyes of the jury, the undoing of the forensic case against Simpson had less to do with a failure to explain the arcana of population genetics than with the old charge of police incompetence. DNA is such a stable molecule that it can be extracted from semen stains several years old or from bloodstains scraped off sidewalks or from the steering wheel of an SUV. But it is also true that DNA can degrade, especially in moist conditions. Like any type of evidence, however, DNA is only as credible as the procedures for collecting, sorting, and presenting it. Criminal trials always include the formality of establishing the "chain of evidence," verifying that what the police say was found in such-and-such a location did indeed start there before winding up in a Ziploc bag as Exhibit A. Keeping track of molecular evidence, as opposed to knives and guns, can be an especially demanding chore: scrapings from a sidewalk may be visually indistinguishable from scrapings from a gatepost, and the subsequently extracted DNA samples will doubtless look even more alike when placed in small plastic test tubes. Simpson's defense team was able to point to a number of instances when it seemed at least possible, if not probable, that samples had been confused or, even worse, contaminated.

There was, for example, the question of the bloodstain on the back gate of Nicole Brown Simpson's house. This was somehow missed in the early survey of the crime scene and not collected until three weeks after the murders. Forensic scientist Dennis Fung presented a photograph of the stain, but Barry Scheck countered it with another photograph taken the day after the murder, in which no stain appeared. "Where is it, Mr. Fung?" Scheck asked with a rhetorical flourish worthy of Perry Mason. There was no answer. The defense was able raise sufficient doubt in the minds of the jurors about the handling and sources of the DNA samples that the DNA evidence became irrelevant.

As we saw in the last chapter, sample contamination is one of the foremost banes of efforts to establish identity by genetic methods. Because it can yield a DNA fingerprint from even the tiniest sample, the polymerase chain reaction (PCR) is the modern forensic scientist's method of choice for amplifying particular segments of DNA. In the Simpson trial, for instance, crucial evidence included a single blood drop scraped from the sidewalk. But sufficient DNA for PCR can be extracted from cells in the saliva left on a cigarette butt. In fact, PCR can successfully amplify DNA from a single molecule, so if even the slightest trace of DNA from another source – someone handling the samples, for example – contaminates the evidence sample, the results are at best confused and at worst useless.

In the past decade, with the broadening application and acceptance of the DNA fingerprint as proof of identity, the law enforcement community had a flash of inspiration: Doesn't it make sense to DNA fingerprint, well, everyone – at least everyone who might be a criminal? Surely, the argument goes, the FBI should have a central database of DNA records, rather as it does for conventional fingerprints. Indeed, a number of states have passed laws requiring that DNA samples be taken from anyone convicted of a violent felony, like rape or murder. For example, in 1994 North Carolina passed legislation that authorizes taking blood samples from imprisoned felons, by force if necessary. And some of those states have since extended the mandate to cover all individuals who are arrested, whether they are ultimately found guilty of a crime or not.

The outcry from civil libertarians has been intense, and not without reason: DNA fingerprints are not like finger fingerprints. A DNA sample taken for fingerprinting purposes can, in principle, be used for a lot more than merely proving identity: it can tell you a lot about me – whether I carry mutations for disorders like cystic fibrosis, sickle-cell disease, or Tay-Sachs disease. Some time in the not so distant future, it may even tell you whether I carry the genetic variations predisposing me to schizophrenia or alcoholism – or traits even more likely to disturb the peace. Might the authorities, for instance, one day subject me to a more intensive scrutiny than would otherwise be the case simply because I have a mutation in the monoamine oxidase gene that reduces the activity of the enzyme? Some research suggests that this mutation may predispose me to antisocial behavior under certain circumstances. Could genetic profiling indeed become a new tool for preemptive action in law enforcement? Philip K. Dick's 1956 story (which inspired the 2002 movie) "The Minority Report" may not be such far-fetched science fiction as we like to imagine.

Whatever the outcome of the ongoing debate about who should be compelled to provide DNA samples and under what safeguards these ought to be maintained, the fact is that as I write there is a huge amount of DNA fingerprinting going on. In 1990, the FBI established its DNA database, CODIS (Combined DNA Index System), and by June 2002 it contained 1,013,746 DNA fingerprints. Of these, 977,895 are from convicted offenders and 35,851 are forensic crime scene samples for unsolved cases. Since its inception, CODIS has been used to make some 4,500 identifications that would not otherwise have been made.

One major justification for a national database is the potential for making "cold hits." Suppose investigators find some DNA – blood on a broken window, semen on underwear – at the crime scene and a fingerprint is made. Now suppose they have no leads by conventional investigative means, but when the fingerprint is entered into CODIS a match is found. That is what happened in St. Louis in 1996. The police were investigating the rapes of two young girls at opposite sides of the city, and although the two samples of semen revealed under RFLP fingerprint analysis that the same man had committed both crimes, a suspect could not be identified. Three years later, the samples were reanalyzed using STRs and the data compared with the entries in CODIS. In 2001, they found the rapist, Dominic Moore, whose DNA fingerprint was in CODIS because he had confessed to committing three other rapes in 1999.

The interval between a crime and a cold hit can be even more dramatic, and some malefactors have been shocked to face the molecular "j'accuse" of victims long buried. In Britain, fourteen-year-old Marion Crofts was raped and murdered in 1981, long before DNA fingerprinting was in use. Fortunately, some physical evidence was preserved, so it was possible to make a DNA fingerprint in 1999. The authorities and Crofts's bereaved family were disappointed again, this time in learning there was no match in the United Kingdom National DNA Database. In April 2001, however, when Tony Jasinskyj was arrested for assaulting his wife, a DNA sample was taken from him as a matter of routine procedure. When it was entered into the database, a match came up: Jasinskyj was found to be the unknown rapist of twenty years before.

In the United States, crimes like rape have customarily been subject to statutes of limitations in many states. In Wisconsin, for example, a warrant for the arrest of an alleged rapist cannot be issued more than six years after the crime has taken place. Although such statutes may seem devastatingly unfair to victims – after all, does the horror of a crime simply disappear after six years? – they have by tradition served the interests of due process. Eyewitness accounts in particular are notoriously unreliable, and all memories grow hazier over time; statutes of limitations are intended to prevent miscarriages of justice. But DNA is a witness of quite a different order. Samples stored properly remain stable for many years, and the DNA fingerprints themselves lose none of their authority to incriminate.

In 1997, Wisconsin's State Crime Laboratory established a DNA fingerprint registry and that same year the Milwaukee Police Department began reviewing all unsolved rape cases with physical evidence available for possible matching. They found fifty-three, and in six months they had scored eight cold hits against DNA fingerprints from felons already serving time. In one case, the identification was made so late the arrest warrant was issued only eight hours before the statute of limitations kicked in.

Among the cold cases, the State Police Department was also to establish evidence of a serial rapist – three separate assaults, three separate semen samples, the DNA fingerprints of all of them pointing to the same man. With the statute of limitations soon to take effect, Norm Gahn, an assistant district attorney, faced a dilemma. There was not enough time to identify the assailant in the database, but he could not draft a warrant without the suspect's name. Gahn hit on a clever strategy. The Wisconsin criminal code held that in the event a suspect's name was unknown, a valid warrant could be issued on the basis of "any description by which the person to be arrested can be identified with reasonable certainty." Surely, Gahn reasoned, any court would accept a DNA fingerprint as identifying someone by that standard. He made out the warrant: "State of Wisconsin vs. John Doe, unknown male with matching deoxyribonucleic acid (DNA) profile at genetic locations D1S7, D2S44, D5S110, D10S28, and D17S79." Despite Gahn's ingenuity, though, this John Doe still has not been caught.

Meanwhile the first challenge in court of a John Doe DNA warrant came in Sacramento, where one man, called the "Second Story Rapist," was believed to have committed three rapes over several years. Anne Marie Schubert, a local prosecutor, followed Gahn's lead in filing a John Doe DNA warrant just three days before the statute of limitations was to take effect. But she had to satisfy the requirements of her own jurisdiction, in particular the California law requiring that a warrant identify the suspect with "reasonable particularity"; toward this end she specified: "unknown male . . . with said genetic profile being unique, occurring in approximately 1 in 21 sextillion of the Caucasian population, 1 in 650 quadrillion of the African American population, 1 in 420 sextillion of the Hispanic population." Shortly after the warrant was issued, when John Doe's DNA fingerprint was entered into the state database, it turned out to match that of one Paul Eugene Robinson, who had been arrested in 1998 for violating parole. The warrant was amended with "Paul Eugene Robinson" in the place of John Doe and his STR markers, and Robinson was duly arrested. His attorney argued that the first warrant was invalid as it did not name Robinson. Fortunately, the judge upheld the validity of the warrant, remarking that "DNA appears to be the best identifier of a person that we have."

In the wake of the publicity stirred by these successful "John Doe DNA" warrants, many states have amended their rape statutes to permit an exception when DNA evidence is available.

The reach of DNA fingerprinting now even extends beyond the grave. In 1973, Sandra Newton, Pauline Floyd, and Geraldine Hughes, all teenagers, were raped and murdered in South Wales. Twenty-six years later, DNA fingerprints were prepared from samples saved from the crime scenes, but unfortunately the National DNA Database yielded no matches. So, rather than looking for an exact match, the forensic scientists looked for individuals whose DNA fingerprints indicated that they might be related to the murderer. They thus identified a hundred men, furnishing the police with a wealth of leads in light of which to reassess the masses of information that they had collected during the original investigation. Through a combination of state-of-the-art DNA forensics and good old-fashioned detective work they found a trail leading to one suspect, Joe Kappen. The only trouble was that Mr. Kappen had died of cancer in 1991 – what was to be done?

In 1999 Kappen was exhumed and fingerprinted. And the finger-prints indeed matched those from DNA recovered from the three victims. Cancer may have exacted the ultimate price before the law could find him, but at least the girls' families had the long-postponed satisfaction of knowing his name.

DNA fingerprinting has solved mysteries involving bodies much more illustrious than Joe Kappen's. Take the extraordinary story of the Russian royal family, the Romanovs.

In July 1991, a small group of detectives, forensic experts, and police assembled in a muddy, rain-soaked clearing in the forest at Koptyaki, Siberia. Here, in July 1918, eleven bodies had been hurriedly buried. They were the remains of Tsar Nicholas II and Tsarina Alexandra; their son, Alexis, heir to the throne; their four daughters, Olga, Tatiana, Marie, and Anastasia; and four companions – all of whom had been brutally murdered a few days before, Anastasia still holding Jemmy, her pet King Charles spaniel, as she met her end in a hail of bullets. The killers initially tossed the bodies down a mine but, fearing discovery, recovered them the next day before finally burying them in that pit in the forest.

The grave had first been discovered in 1979 thanks to the detective work of Alexander Avdonin, a geologist obsessed with learning the fate of the tsar's family, and the filmmaker Geli Ryabov, who, having earned the privilege of making an official documentary of the Revolution, had gained access to relevant secret archives. In fact, it was a report written by the chief murderer for his bosses in Moscow that led Avdonin and Ryabov to the gravesite. They found three skulls and other bones. But as the chokehold of the Communist Party was then as tight as ever, they rightly realized they would do themselves no favors by drawing attention to the Bolsheviks' butchery of the royal family. They reburied the remains.

With the thawing of the political climate that culminated in the demise of the Soviet Union came the opportunity Avdonin and Ryabov had been waiting for. So it was that picks and shovels were again wielded in the forest clearing.

The exhumed remains – a total of more than one thousand pieces of skull and bone – were taken to a Moscow morgue, where the painstaking process of reassembling and identifying the skeletons began. There was an immediate surprise. The murdered were known to have numbered eleven, six females and five males, but the grave contained the bones of only nine bodies – five female and four male. It was clear from the skeletal remains that the missing bodies were those of Alexis (fourteen at his death) and Anastasia (who had been seventeen).

The claims of identification were viewed with some skepticism, especially as there had been disagreement between the Russian scientists and an American team that had come to assist. And so in September 1992, Dr. Pavel Ivanov brought nine bone samples to Peter Gill's laboratory at the British Forensic Science Service. Gill and his colleague David Werrett had been coauthors of the first paper Alec Jeffreys published in this field and had since established the Forensic Science Service as the UK's premier laboratory for DNA fingerprinting.

Gill had developed a DNA fingerprinting method using mitochondrial DNA (mtDNA), which, as we saw in the analysis of Neanderthal mtDNA, has a special advantage in cases when DNA is old or difficult to obtain: it is far more abundant than the chromosomal DNA from the nucleus.

Gill and Ivanov's first task was the delicate job of extracting both nuclear and mtDNA from the bone samples. The analysis showed that five of the bodies were related and that three were female siblings. But were these the bones of the Romanovs? In the case of the Empress Alexandra at least, an answer could be found by comparing the mtDNA fingerprint from the bones thought to be hers with an mtDNA fingerprint from her grandnephew, Prince Phillip, the Duke of Edinburgh. The fingerprints matched.

It was rather more difficult to find a relative for the tsar. The body of the Grand Duke Georgij Romanov, his younger brother, dwelt in an exquisite marble sarcophagus deemed too precious to open. The tsar's nephew refused to help, still bitter over the British government's refusal to grant his family refuge at the onset of the Revolution. A bloodstained handkerchief was known to exist in Japan, one the tsar had used when he was attacked by a sword-wielding assassin in 1892. Gill and Ivanov secured a narrow strip of it but found that over the years the relic had been contaminated beyond usefulness with the DNA of others. It wasn't until two distant relatives were finally found that the mtDNA fingerprint was confirmed as the tsar's.

But the analysis had yet one more surprise in store: the mtDNA sequences from the presumed tsar and his modern relatives were similar but not identical. Specifically, at position 16,169, where the tsar's mtDNA had a C, that of the two relatives showed a T. And further testing revealed only further complications. The tsar's mitochondrial DNA was actually a mix of two types, both C and T. This unusual condition is called "heteroplasmy" – the coexistence within a single individual of more than one mtDNA type.

A few years later the worries of all but the most committed conspiracy theorists were finally put to rest. The Russian government finally agreed to crack the sarcophagus and provide Ivanov with a tissue sample from Georgij Romanov, the tsar's brother. The grand duke's mitochondria showed the very same heteroplasmy as those found in the bones from the pit. Those bones were without question the tsar's.

But what of the legendary Anastasia, whose skeleton was never recovered from the grave in the forest? There has been no lack of pretenders to the Romanov line, and among these none was more persistent than one Anna Anderson, who asserted for a lifetime that she was the lost grand duchess. She'd first made the claim as early as 1920 and went on to become the subject of many books as well as the film *Anastasia,* in which, played by Ingrid Bergman, she was indeed found to be the grand duchess. When Anderson died in 1984, her identity was still in dispute, but as the claims and counterclaims of her supporters and critics continued, the means for a resolution were at hand.

Anna Manahan (Anna Anderson's married name) had been cremated, making tissue retrieval from her remains impossible. But an alternative source of her DNA was discovered: in August 1970, she had undergone emergency abdominal surgery at the Martha Jefferson Hospital in Charlottesville. Tissue removed during the operation had been sent to a pathology laboratory where it was prepared for microscopy, and where, twenty-four years later, it was still filed away. After an appropriately Byzantine series of court cases over access to the specimen, Peter Gill traveled to Charlottesville in June 1994 and departed with a little preserved slice of Anna Manahan.

The results were crystal clear. Anna Anderson was related neither to Tsar Nicholas II nor to the Empress Alexandra. But in the wake of such a long odyssey, it is perhaps not surprising that some chose to ignore the DNA and believe what they would: the myth that Anna was Anastasia still lives on.

The fate of the Romanovs and Anna Anderson may be the stuff of fairy tales, remote from most of our lives, but DNA fingerprinting is ordinarily applied to grim realities painfully all too close. One of the most awful tasks facing investigators after a violent catastrophe like a plane crash is the identification of bodies. For various reasons – to permit the issuing of a death certificate, for instance – the law requires that it be done. And no one should underestimate the desperate emotional need of families to bury their loved ones with proper ceremony; for most of us, respect for the dead requires the recovery of their remains, however fragmented, and this task depends on positive identification.

In 1972, an American warplane believed to have been piloted by Michael Blassie was shot down during the Battle of An Loc in Vietnam. Remains were recovered from the crash site, but an inadequate forensic examination in 1978 based on blood type and analysis of the bones indicated that they were not Blassie's. The anonymous bones were labeled "X-26, Case 1853," and in a solemn ceremony attended by President Reagan, they were laid to rest in the Tomb of the Unknowns at Arlington National Cemetery. In 1994, CBS News picked up a story by Ted Sampley in the *U.S. Veteran Dispatch,* claiming that X-26 was Blassie. When the subsequent investigation by CBS uncovered evidence corroborating Sampley's claim, Blassie's family petitioned the Department of Defense to examine it. This time mtDNA fingerprints from the unknown's bones were found to match those of Blassie's mother and sister. Twenty years after his death, Blassie came back to St. Louis. Standing beside the gravestone, his mother was able to say, "My son is home. My son is finally home."

The Department of Defense has since established the Armed Forces Repository of Specimen Samples for the Identification of Remains. Blood samples are taken and DNA isolated from all new members of the military, both those on active duty and reservists. By March 2001, the repository contained more than 3 million samples.

I was on my way to my office when I heard that a plane had crashed into one of the World Trade Center towers. Like many others, I assumed initially it was an accident – anything else was unimaginable. But all too soon, when the second plane hit the other tower, it was apparent that a criminal act of the most ghastly kind had been perpetrated against thousands of innocent people. No one who watched that day is likely ever to forget the images of people leaning out of windows high on the towers, or falling to their deaths. And we were not shielded from the tragedy's immediate toll even on the tranquil campus of Cold Spring Harbor Laboratory, thirty miles from Manhattan: two of our staff lost sons that day.

The final loss of life has been reckoned at 2,792 – an extraordinarily low number considering that as many as 50,000 may have been in the towers at the time of the attack. Nevertheless, given an event of such cataclysmic force, one can expect to find few bodies intact, much less alive. And so the search for survivors was transformed with a tragic inevitability into the hunt for remains; a million tons of mangled steel, pulverized concrete, and crushed glass were sifted for any human part they might yield. Some 20,000 were found and taken to twenty refrigerated semi trucks arrayed near the medical examiner's office. Since the beginning of this herculean forensic effort, many identifications have been made using dental records and conventional fingerprints, but as the easy cases are closed, increasingly the load shifts to DNA analysis. For comparison with all genetic traces from the site, relatives have supplied either samples of their own blood or items like the toothbrushes and hairbrushes of the dead, any possession that may have picked up even a few of its owner's cells from which DNA could be extracted. The task of carrying out the DNA fingerprinting has fallen to Myriad Genetics in Salt Lake City and Celera Genomics, both of which are accustomed to analyzing DNA on an enormous scale. But even with the very latest technology, this is a slow and painstaking process.

It is a common human desire to know one's forebears: who they were and where they came from. In the United States, a nation built by generation after generation of immigrants, the longing is especially intense. In recent years, the genealogical craze has been aided by the World Wide Web, which also supplies us with an informal measure of the phenomenon's dimensions: a Google search for "genealogy" yields over 10 million hits (a search for "DNA" gets you only 5 million). By comparing the fingerprints of individuals, DNA makes possible the highly specific sort of genealogical inquiry that Gill and Ivanov carried out to uncover, for instance, Anna Anderson's relationship to the Romanovs (none). But genealogies can also be constructed at a broader level, finding connections by comparing the DNA fingerprint of an individual with those of whole populations.

At Oxford, Brian Sykes used DNA analysis to delve into his own genetic history. Knowing that both surnames and Y chromosomes are transmitted down the male line, he surmised that all men born with the same surname should also have the same Y chromosome – the one belonging to the very first man to take that name. Of course, this linkage of Y chromosome and surname breaks down if a name should arise independently more than once, if men change the family name for one reason or another, or if many boys take the name of a man other than their biological father (a lad secretly sired by the milkman, for instance, would likely wind up with the surname of his mother's husband).

After contacting 269 men called Sykes, Professor Sykes managed to collect 48 samples for analysis. He found that about 50 percent of the Y chromosomes were indeed identical to his own "Sykes" chromosome; the rest bore evidence of conjugal lapses on the part of more than one Mrs. Sykes of generations gone by. Because the origin of the name is documented and can be dated to around seven hundred years ago, it is possible to work out the per-generation rate of infidelity. It averages out to a perfectly respectable 1 percent, suggesting that 99 percent of Sykes wives in every generation managed to resist extramarital temptation.

When Sykes set up a company to market genealogical DNA finger-printing services, one of his first clients was the John Clough Society, whose members trace their ancestry back to a Briton of the same name who emigrated to Massachusetts in 1635. The society even knew that an ancestor of his, Richard, from the Welsh line of the family, had been knighted for his deeds on a crusade to the Holy Land. What they lacked, however, was any historical proof to link their families to those on the other side of the Atlantic. Sykes's company analyzed Y chromosome DNA from the Massachusetts Cloughs and from a direct male descendant of Sir Richard; the two were indeed identical – vindication for the Massachusetts branch. But not all the American Cloughs were as lucky; society members from Alabama and North Carolina were found to be unrelated not only to Sir Richard but to the Massachusetts Cloughs as well.

On *The Montel Williams Show,* or *Ricki Lake,* or *Jenny Jones,* you can see the young women and men looking nervous. The host opens an envelope, gives the couple a meaningful look, and then reads the card. The woman covers her face with her hands and bursts into tears, while the man leaps into the air, pumping his fist. Alternatively, the woman leaps to her feet, pointing triumphantly at the man who remains slumped, shoulders bowed, in his seat. In either case, we have just seen one of the more outlandish applications of DNA fingerprinting – the ultimate in infotainment.

Daytime television may make theater of the subject, but paternity testing is a serious business with a long tradition. Since the beginning of human history, much of one's life – its psychological, social, and legal realities – has depended on the identity of one's father. So, quite naturally, science has been drafted into the service of paternity testing ever since genetic techniques for distinguishing individuals were first developed. Until the advent of molecular genetics, blood itself was the most scientific clue to paternity. The patterns of inheritance were reliable and well understood, but with only a handful of blood groups to test for, the trait's power to discriminate was limited. Practically speaking, a test for blood type has limited power to exclude wrongly accused fathers, and it can never provide definitive affirmation of the right one. If our blood types are not compatible, I am assuredly not your father; but if they are, it's no certain proof that I am – the same will be true of any number of men who have the same blood type I have. Using other markers in addition to the familiar ABO blood group markers improves the resolving power of this kind of test but it still cannot match the statistical muscle of STR-typing: an STR-based genetic fingerprint can establish proof positive of paternity. And in the era of PCR, it is convenient enough to use.

So convenient, in fact, that mail-order paternity testing companies do a thriving business. In some cities huge roadside billboards advertise a local paternity testing service with the none-too-subtle pitch line: "Who's the Daddy?" For a fee, these companies will mail you a DNA sampling kit that includes a swab to scrape some cells from the interior of the mouth. (Samples collected this way would not stand up in court. To be admissible, a DNA fingerprint must be based on a sample collected by a certified lab, which must verify the chain of evidence so as to prevent the sort of genetic switcheroo we saw in the Pitchfork case.) The tissue samples are sent by overnight courier to the testing laboratory, where the DNA is extracted.

The child's DNA fingerprint is compared with that of the mother; any STR repeats present in the child but not in the mother are presumed to have come from the father, whoever he may be. If the fingerprint of a supposed father lacks any of these repeats, he must be excluded. If none are missing, the number of repeats allows us to quantify the likelihood that a match is definitive by the so-called Paternity Index (PI). This measures the chances that some man other than the actual father could have contributed a particular STR, and it varies in relation to how common a given STR is in the population. The PIs for all STRs are multiplied together to give a Combined Paternity Index.

Most paternity tests are, of course, handled with the utmost discretion (unless you happen to be on a talk show), but one recent analysis drew many headlines owing to the great historical interest in the alleged father. It had long been suspected that Thomas Jefferson, third president of the United States and the principal author of the Declaration of Independence, was more than a founding father: he was thought to have had one or more children by his slave Sally Hemings. The first accusation was made in 1802, just twelve years after the birth of a boy, Tom, who later took the last name of one of his subsequent masters, Woodson. In addition a strong resemblance to Jefferson had been widely remarked in Hemings's last son, Eston. DNA was destined to set the record straight.

Jefferson had no legitimate male descendants so it is impossible to determine the markers on his Y chromosome. Instead, researchers took DNA samples from male descendants of Jefferson's paternal uncle, Field Jefferson (whose Y chromosome would have been identical to the president's), and compared them with samples from the male descendants of Tom and Eston. The results showed a distinct Jefferson fingerprint for the Y chromosome, but this DNA fingerprint was not present in the descendants of Tom Woodson. Jefferson's reputation had dodged that bullet. In Eston Hemings's descendants, however, the Jeffersonian Y chromosome signature came through loud and clear. But what the DNA cannot confirm beyond reasonable doubt is the source of that chromosome. We cannot say with certainty whether Eston's father was in fact Thomas Jefferson or some other male in the Jefferson lineage who might also have had access to Sally Hemings. Indeed, some suspicions have been cast on Isham Jefferson, the president's nephew.

Centuries of national reverence, then, are no protection against the harsh revealing light of DNA evidence. Nor, it seems, is any amount of celebrity or money. When the Brazilian model Luciana Morad claimed that Mick Jagger was the father of her son (whom she named Lucas Morad Jagger), the Rolling Stone denied it and demanded DNA testing. Perhaps Jagger was bluffing, hoping that the threat of a forensic denouement would weaken Ms. Morad's resolve and induce her to drop the case. But she did not. The tests were positive, and Jagger found himself legally obliged to contribute to the upbringing of his son. Boris Becker, too, submitted to a paternity test over a girl born to Russian model Angela Ermakova. The tabloids had a field day with stories that the tennis star believed himself the victim of a blackmail scheme contrived by the Russian mafia – the lurid details of how this plot was supposedly perpetrated are best left in the pages of the tabloids. Suffice it to say that when the DNA results were in, the swaggering Becker acknowledged his deed and pledged to support his daughter.

DNA fingerprinting to identify a child's biological relatives has been applied to causes rather more uplifting than those of Messrs. Jagger and Becker. In Argentina, between 1975 and 1983, 15,000 people were quietly eliminated for holding opinions unpopular with the ruling military junta. Many of the children of the "disappeared" were subsequently placed in orphanages or adopted illegally by military officers. Having lost their own children to the regime, the mothers of the disappeared then set about finding their children's children – to reclaim their grandchildren. Las Abuelas (grandmothers) drew attention to their nationwide quest by marching every Thursday in the central square in Buenos Aires. They continue their search to this day. Once a child has been located, genetic fingerprinting methods can be used to determine who is related. Since 1984, Mary-Claire King – whom we encountered earlier grappling with another set of relationships, that between humans and chimpanzees – has provided Las Abuelas with the genetic analysis needed to reunite families torn apart by eight nightmarish years of misrule.

DNA fingerprinting has come a long way since its first forensic applications. It is now a staple of our popular culture, a consumer good for the genealogically curious; a mousetrap in the ongoing spectacle of "gotcha" we play with celebrities and with those ordinary folk who wish only to be on television. But its most serious application remains in the resolution of legal questions involving life and death. The United States is the only nation in the Western world that still imposes the death penalty. Between 1976, when the Supreme Court reinstated capital punishment after a ten-year hiatus, and 2001, 749 convicts were put to death, and by the end of that period there were 3,593 prisoners on death row. It is against this background that we need to examine the work of the Innocence Project, and its founders, Barry Scheck and Peter Neufeld, some of the earliest and staunchest critics of DNA fingerprinting, at least as it was first practiced. Since the early days, Scheck, Neufeld, and other defense attorneys have come to realize that the forensic technology they opposed is actually a powerful tool for justice – more capable, in fact, of exculpating the innocent than of convicting the guilty. Proving innocence merely requires finding a single mismatch between a defendant's DNA fingerprint and that taken from the crime scene; proving guilt, on the other hand, requires demonstrating statistically that the chances of someone other than the accused having the specified fingerprint are negligible.

As of November 2002, the work of lawyers and students in Innocence Projects (there is now a whole network of them, based at law schools throughout the country) has led to the exoneration of 117 wrongfully convicted individuals. In Illinois, six of these mistaken convictions had resulted in death sentences, leading Governor George Ryan to take a remarkable and – given popular support for law-and-order palliatives like capital punishment – politically dangerous step of imposing an indefinite moratorium on executions in the state. In addition, Ryan appointed a special commission to review the handling of capital cases; published in April 2002, this commission's report listed among its strongest recommendations that provision be made to facilitate DNA testing of all defendants and convicts in the state's criminal justice system.

By no means has all DNA testing of those who insist on their innocence led to the overturning of convictions. James Hanratty was convicted of one of the most notorious murders in twentieth-century Britain. He accosted a young couple, shot the man fatally, and raped the woman before shooting her five times and leaving her for dead. Despite his insistence that he'd been miles away when the crime occurred, Hanratty was found guilty and sentenced to hang. In 1962, he became one of the last criminals to be executed in Britain.

Hanratty died proclaiming his innocence, and his family began a posthumous campaign to clear his name. Their efforts became a cause célèbre: they succeeded in compelling the authorities to have DNA extracted from the female victim's semen-stained underwear and from the handkerchief that had masked the assailant's face; both samples were then compared with DNA fingerprints from Hanratty's brother and mother. To their chagrin, it was determined that the crime scene DNA had indeed come from a member of the Hanratty family. Still unsatisfied, the Hanrattys had their black sheep's body exhumed in 2000 in order to retrieve tissue samples for DNA extraction. That more direct analysis showed it was unequivocally Hanratty's DNA on the underwear and the handkerchief. Finally, grasping for straws, the family argued, following the recently successful Simpson defense, that the sample sources had been handled improperly and become contaminated. But the Lord Chief Justice proved less distractable than the Simpson jury. He rejected this claim out of hand: "The DNA evidence establishes beyond doubt that James Hanratty was the murderer."

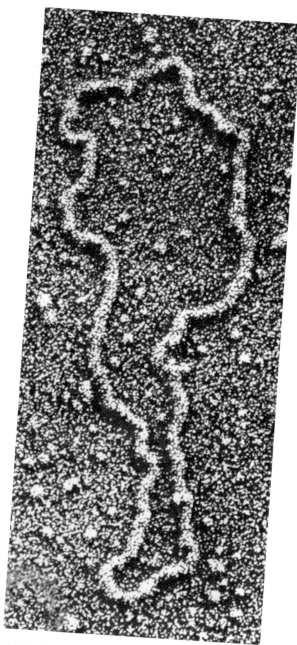

30. A P4 lab
   lethal bugs
   During the
   research

26. *A plasmid as viewed by the electron microscope*

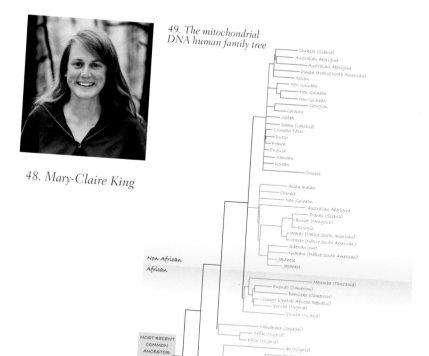

48. Mary-Claire King

49. The mitochondrial
DNA human family tree

Chukchi (Siberia)
Australian Aborigine
Australian Aborigine
Piman (Native North American)
Italian
New Guinean
New Guinean
New Guinean
Georgian
German
Uzbek
Saami (Lapland)
Crimean Tatar
Dutch
French
English
Samoan
Korean
Chinese

Asian Indian
Chinese
New Guinean
Australian Aborigine
Evenki (Siberia)
Buriat (Mongolia)
Khirgiz
Warao (Native South American)
Warao (Native South American)
Siberian Inuit
Guarani (Native South American)
Japanese
Japanese

Non-African
African

Mkamba (Tanzania)
Ewondo (Cameroon)
Bamileke (Cameroon)
Lisongo (Central African Republic)
Yoruba (Nigeria)
Yoruba (Nigeria)

MOST RECENT
COMMON
ANCESTOR
OF ALL
LIVING
HUMANS

Mandenka (Senegal)
Effik (Nigeria)
Effik (Nigeria)
Ibo (Nigeria)
Ibo (Nigeria)
Mbenzele (Pygmy, Central African Republic)
Biaka (Pygmy, Central African Republic)
Biaka (Pygmy, Central African Republic)
Mbenzele (Pygmy, Central African Republic)
Kikuyu (Kenya)
Hausa (Nigeria)
Mbuti (Pygmy, Democratic Republic of Congo)
Mbuti (Pygmy, Democratic Republic of Congo)
San (Bushman, Botswana)
San (Bushman, Botswana)

bacterial cell

recombinant
introduced

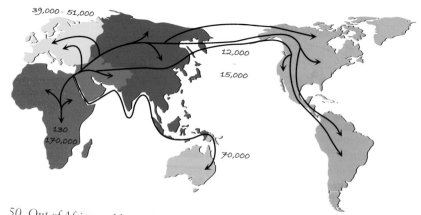

39,000 - 51,000
12,000
15,000
130
170,000
70,000

50. Out of Africa and beyond: Our species originated in Africa and spread out
from there. Estimated colonization dates are based on mtDNA data.

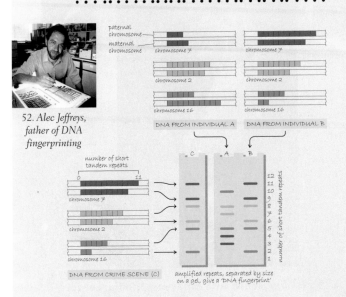

52. Alec Jeffreys, father of DNA fingerprinting

paternal chromosome
maternal chromosome
chromosome 7

chromosome 7

chromosome 2

chromosome 2

chromosome 16

chromosome 16

DNA FROM INDIVIDUAL A

DNA FROM INDIVIDUAL B

number of short tandem repeats

0   11

chromosome 7

chromosome 2

chromosome 16

DNA FROM CRIME SCENE (C)

amplified repeats, separated by size on a gel, give a 'DNA fingerprint'

number of short tandem repeats

51. DNA fingerprinting using STRs. The DNA of two suspects is compared to DNA recovered from the crime scene. The fingerprint of B matches that of the crime scene DNA.

Today short tandem repeats (STRs) have replaced RFLPs as the keys to genetic identification. STRs, in which sequences of two to four bases recur as many as seventeen times, are the segments routinely amplified by PCR. For example, D7S820 is a region on chromosome 7 where the sequence AGAT can occur between 7 and 14 times. It happens that DNA polymerase, the enzyme that copies DNA, does a bad job of copying these repeating chunks of DNA—it tends to get the count wrong—so there is a high mutation rate in copy number of the AGAT sequence at D7S820. To put it another way, there is a great deal of variation in the number of AGAT copies among individual humans. With two copies of chromosome 7 (one from our father, the other from our mother), we typically have a different AGAT repeat count on each—say, 8 on one and 11 on the other. This is not to say, however, that an individual cannot be homozygous for a particular repeat count (e.g., 11 and 11). If we carry out DNA fingerprint analysis on a crime-scene blood sample and find it matches a suspect's fingerprint for D7S820 (say, 8 and 11 repeats), we have one indication of a match but not conclusive proof. After all, many others also have an 8/11 genotype for D7S820. It's therefore necessary to look at multiple regions; the more regions in which the crime scene DNA matches a suspect's, the greater the probability of a match, and the more remote the chances that the crime scene DNA could have come from anyone else. Under the FBI's system, a DNA fingerprint is produced from the analysis of twelve such regions, plus a marker that determines the sex of the individual from whom the DNA sample is derived.

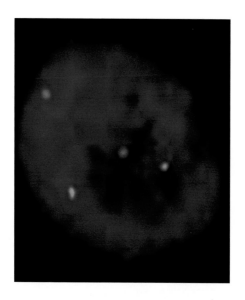

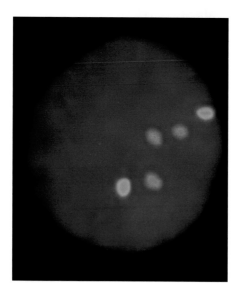

56 and 57. *Fluorescent staining for chromosome number. A cell nucleus (dark blue) is probed for chromosome 10 (light blue) and chromosome 21 (pink). The image on the far left shows a normal karyotype with two copies of each chromosome; in the other, we see a Down karyotype, which has an extra copy of chromosome 21.*

58. *Trofim Lysenko measuring wheat plants in a rare burst of empiricism on a collective farm near Odessa, Ukraine*

59 and 60. The impact of just one gene. At top, a normal mouse mother is highly attentive to her offspring. The mother below, lacking a functional fos-B gene, ignores her newborns.

61 and 62. The cultural wonder that is Homo sapiens. *Two contrasting notions of chic: Paris 1950s and the highlands of Papua New Guinea. Evolutionary psychology seeks the common denominators underlying all our widely divergent behavior.*

Usually the strongest objection to reopening a case comes from the district attorney, who is understandably reluctant to see a hard-won conviction subject to post-trial scrutiny. But sometimes such rigidity can be self-defeating, and if prosecutors have now learned that genetic evidence can nail a case, they should also recognize that DNA may also be the surest way to keep one shut. The example of Benjamin LaGuer illustrates the point. Sentenced in 1984 to forty years in prison for a rape in Worcester, Massachusetts, he never stopped protesting his innocence. Like Hanratty, he attracted a retinue of rich and famous sympathizers, who in 2001 arranged and paid for samples of DNA to be analyzed. The results must have surprised them all: LaGuer was the rapist. One can imagine that a man facing forty years behind bars rightly imagined he had nothing to lose in making such a demand. But ironically, it had taken two years to get the district attorney's office to agree to the DNA fingerprinting. As an editorial in the *St. Petersburg Times* sensibly remarked, "In hindsight, the prosecutor could have wasted less time arguing and gotten the pleasure of saying 'I told you so' much sooner had he consented early on to the DNA test."

Civil libertarians will always object to the broad application of DNA fingerprinting in society as a whole. But it is hard to argue with the social utility of applying the technology to those who, for whatever reason, pass through the criminal justice system; for the chances are, sadly, that those who pass through once will pass through again. Criminological data indicate that those convicted of minor crimes are likely to commit more serious offenses; 28 percent of homicides and 12 percent of sexual assaults in Florida have been linked to individuals previously convicted of burglary. And such patterns of recidivism can be detected among white-collar criminals as well: of twenty-two who had been convicted for forgery in Virginia, ten were linked through DNA fingerprinting to murders or sexual assaults. It would seem prudent to make the corporate bosses of Enron, ImClone, and Adelphia Communications provide DNA samples.

Efforts are under way to broaden DNA fingerprint databases. Recently, the British government has proposed allowing the police to keep DNA samples taken both from acquitted defendants and from those arrested but never charged. The same rule would permit the authorities to keep samples given voluntarily (when, for example, the police test everyone in a location, as they did in Narborough). These changes in collection rules will triple the number of entries in the police database within three years. In the United States, nineteen states now collect DNA samples from all felons, not just those involved in violent crime.

I think everyone should give a DNA sample. It is not that I am insensitive to the concerns about individual privacy or to the potential for inappropriate use of genetic information; as I have said earlier, in my role as the first director of the Human Genome Project, I set aside a substantial chunk of our funding to examine such questions in relation to clinically applied genetic information. But criminal justice is a different matter. Here by my calculation the potential for the greater social good far outweighs the risks of abuse. And since we must all surrender something for the benefit of living in a free society, the sacrifice of this particular form of anonymity does not seem an unreasonable price to pay, provided our laws see to a strict and judicious control over access to databases. Frankly, the remote possibility that Big Brother will one day be perusing my genetic fingerprint for some nefarious end worries me less than the thought that tomorrow a dangerous criminal may go free – perhaps only to do further evil – or an innocent individual may languish in prison for want of a simple DNA test.

But objections to DNA collection in general continue to be heard, and often from the most surprising and far-flung quarters. In both New York City and the Australian state of Tasmania, lawmakers have proposed that the entire police force be fingerprinted. The logic is simple: keep the police on file so their DNA can readily be excluded from any crime they might investigate. Remarkably, the measures were denounced by law enforcement bodies in both jurisdictions: those presumed to be the most law-upholding of citizens, those whose work only promises to be facilitated by the widespread availability of DNA fingerprinting, want no part of it where their own DNA is concerned. My suspicion is that there is something of the irrational at play here. As in the case of genetically modified foods, DNA has in the popular imagination a voodoo quality: there's something scary, mysterious about it. And a lack of understanding of genetic complexities leaves one susceptible to the worst anxieties and conspiracy theories. Once people understand the issues, I hope this hesitation in making the most of a new and powerful beneficial technology will vanish.

Barry Scheck and Peter Neufeld put it well in the preface to their book *Actual Innocence:* "DNA testing is to justice what the telescope is for the stars; not a lesson in biochemistry, not a display of the wonders of magnifying glass, but a way to see things as they really are." What could be wrong with that?

CHAPTER ELEVEN

GENE HUNTING:
THE GENETICS OF HUMAN DISEASE

It was too early in the day for anyone, let alone an impeccably dressed middle-aged woman, to be drunk. But as she swayed unsteadily across the street, drunk is what she seemed, even to the cop on duty near the courthouse, who reprimanded her for creating a public spectacle. In fact, Leonore Wexler wasn't drunk at all. She was beginning to succumb to a ghastly fate that had already destroyed several close relatives before her eyes, a fate she had hoped would pass her by.

Not long thereafter, in 1968, Wexler's ex-husband, Milton, was to celebrate his sixtieth birthday in Los Angeles with their two daughters, Alice, 26, and Nancy, 23. But celebration, as it turned out, was not the order of the day. Milton told his daughters that their mother, 53, was suffering from Huntington disease (HD), a devastating neurological disorder that causes a progressive deterioration in brain function such that those afflicted gradually lose all knowledge of themselves and their loved ones. They also lose control of their arms and legs; at first walking is affected, as in Leonore's case, but as the decline continues patients also experience involuntary, jerky movements. There was no cure and no treatment to delay the relentless slide toward death.

Now Alice and Nancy could make sense of some disquieting facts about their mother's relatives as well as hints she herself had dropped that all was not right in the family. They knew that their uncles, Leonore's three brothers, had all died young; before his end, each had developed the same strange grimace, unsteady walk, and slurred speech. They knew that Leonore's father, their grandfather, Abraham Sabin, had also died young, though Leonore had carefully never mentioned he too had shown those symptoms. Huntington disease, it was becoming clear to them, ran in the family. It was Milton's grim task to answer their immediate question: What was the risk that Alice or Nancy might succumb? "Fifty-fifty," their father told them.

The disease that would afflict Abraham Sabin and his descendants was first identified by George Huntington. Born into a medical family, Huntington grew up in East Hampton, Long Island, where as a young boy he accompanied his father on his rounds. After qualifying as a physician at Columbia University, Huntington returned to the family practice on Long Island for a few years before moving to Pomeroy, Ohio. In 1872, he presented a paper at the Meigs and Mason Academy of Medicine in nearby Middleport entitled "On Chorea." Derived from the Greek word for dance, "chorea" was the name physicians had since the seventeenth century given to illnesses that produced jerky movements in their victims. Late in life Huntington would recount how he had come to be fascinated by the mysterious malady:

> Over 50 years ago, in riding with my father on his rounds I saw my first case of "that disorder," which was the way the natives always referred to

the dreaded disease. I recall it as vividly as though it had occurred but yesterday. It made a most enduring impression upon my boyish mind, an impression which was the very first impulse to my choosing chorea as my virgin contribution to medical lore. Driving with my father through a wooded road leading from East Hampton to Amagansett, we suddenly came upon two women both bowing, twisting, grimacing. I stared in wonderment, almost in fear. What could it mean? My father paused to speak with them and we passed on. Then my Gamaliel-like* instruction began; my medical instruction had its inception. From this point on my interest in the disease has never wholly ceased.

Drawing on his own observations as well as the clinical notes of both his father and grandfather (the original manuscript has annotations penciled in by his father), the young physician's paper offered a masterful description of what became known as Huntington's chorea and is now called Huntington disease. The "chorea" movements, he explained, "gradually increase when muscles hitherto unaffected take on the spasmodic action, until every muscle in the body becomes affected." He noted the attendant mental deterioration: "As the disease progresses the mind becomes more or less impaired, in many amounting to insanity, while in others mind and body gradually fail until death relieves them of their suffering." And he recognized that the disorder was inherited: "When either or both the parents have shown manifestations of the disease, one or more of the offspring invariably suffer from the condition. It never skips a generation to again manifest itself in another. Once having yielded its claims, it never regains them."

Huntington correctly identified the key features of this kind of genetic disorder. He recognized that it affected both males and females

---

*Gamaliel, a famous rabbi and teacher of St. Paul (Acts 22:3), believed in integrating book learning with everyday experience.

and understood that it passed from generation to generation. Each child of a parent with Huntington disease has a 50-50 chance of inheriting it. By the luck of the draw, in some families everyone is affected; in others, none are. If a person does not inherit the abnormal gene from a parent, he or she cannot pass on the gene to the next generation. Today we know Huntington disease is caused by a mutation and since the gene is not preferentially expressed in one sex over the other (i.e., is not sex-linked), we have inferred that the affected gene is on neither the X nor Y sex chromosome. Let's call the normal version of the gene H and the mutant version h. We have two copies of each non-sex chromosome (called "autosomes") and so two copies of the Huntington gene. Individuals with the two copies of the normal gene (HH) are, predictably, disease free. But individuals with two (hh) or even one copy of the mutated gene (Hh) are bound to develop the disease. We call this pattern "autosomal dominant inheritance." ("Dominant" means that only one copy of a mutated gene is sufficient to cause disease – the abnormal gene dominates its normal partner.)

Since it is far likelier that a person will acquire one rather than two copies of the mutant form, most Huntington sufferers are Hh. Such individuals could pass on H or h to their children, yielding a 50 percent chance that a particular child would be affected, just as Milton Wexler told Alice and Nancy.

Back in 1968, not much was known about Huntington disease beyond these facts: it is heritable, and it makes its irreversible progress by killing nerve cells in specific areas of the brain. Milton Wexler resolved that he would take on the terror striking his family: he established the Hereditary Disease Foundation (HDF) to raise money and press for more government funding for Huntington disease research. His daughter Nancy was drawn in as well. While completing a doctorate in psychology at the University of Michigan – her thesis fittingly concerned the psychology of being at risk – she found herself increasingly involved in the affairs of the foundation. In the 1970s, when it became apparent that real progress would depend upon a better understanding of the genetics of the disease, Nancy Wexler began to reinvent herself as a geneticist.

On the shores of Lake Maracaibo, Venezuela, the burden of grinding poverty is compounded by a remarkably high incidence of Huntington disease. If Huntington were to divulge its genetic secrets anywhere, Lake Maracaibo seemed a likely place. In 1979, Wexler began to collect DNA samples and to record family histories with the goal of preparing a genealogy of all affected people. For the geneticist it was a great labor, but for Wexler, the daughter of a Huntington victim with the possibility of the disease in her own future, it was more than that. It involved seeing the familiar in such unfamiliar surroundings: people who lived in tin-roofed wooden huts on poles above the waters of the lake yet walked with that same drunken stagger that had overtaken her mother. Since her first trip to Lake Maracaibo in 1979, Wexler has returned annually to continue the work there. The people she works with have come to call her La Catira for her long blond hair. As Americo Negrette, her Venezuelan colleague and the scientist who first reported the occurrence of Huntington at Lake Maracaibo, describes it, she has made of them an extended family, greeting them each time "without theater, without simulation, without pose. With a tenderness that jumps from her eyes." (see Plate 53).

But tenderness could only mitigate the devastation Huntington disease had visited upon so many. The goal of Wexler's expeditions was ultimately to find the gene responsible for the disease. But how could her Maracaibo genealogies help to identify the culprit? The key lay in advances in human genetics.

If they were to home in on the Huntington gene, Wexler and others interested in genetic disease knew they would have to do for humans what Morgan and his students had started doing for fruit flies more than half a century earlier. As we have seen (in chapter 1), Morgan compared rates at which particular genetic markers – white (as opposed to red) eye color, say, and curly (as opposed to straight) wings – coincided in the offspring of crosses between parents showing various combinations of these traits; from these data he was able to determine how near each other on a chromosome were the genes governing those traits. But human genetics had lagged behind the fruit fly's for two major reasons. First was the impossibility – on moral and practical grounds – of doing the kind of experiments that were still the mainstay of genetic analysis: you can't simply breed two human beings you're interested in and then analyze the progeny two weeks later. Second, even if humans could be crossed at will, they were still lacking in genetic markers. Morgan was able to track a number of simple and obvious differences in appearance caused by specific mutations in individual genes. Humans unfortunately don't possess many easily analyzed traits that are inherited in this simple way; even the canonical example, eye color, turns out to be governed by several genes, not just one. Furthermore, with fruit flies, you can increase levels of genetic variation by subjecting individuals to X rays, or to other mutagenic agents: such options, happily, are not available in dealing with humans. Only with the advent of recombinant DNA did solutions to the two major obstacles present themselves.

In the age of DNA sequencing, genetic markers need no longer be visible, like white eyes in a fruit fly; a variation in the sequence itself will suffice, and you can track such a DNA marker through a family tree – that is, through a number of genetic crosses – simply by analyzing DNA from several generations. The revolution had begun the year before Wexler started her genealogical research. And, as with so many advances in science, a measure of serendipity was involved.

It had become an annual ritual: a small group of graduate students from the University of Utah would accompany their advisers to the Wasatch Mountain ski resort of Alta for an intensive workshop on their research (and, well, a little skiing on the side). Typically, a couple of big-shot scientists from other institutions would be invited, to cast a critical eye over the data presented by each nervous student. In 1978, the big shots included David Botstein from MIT and Ron Davis from Stanford.

David Botstein, it's been noted, "tends to think and talk excessively fast, and often at the same time." Ron Davis is quiet and retiring. That April in Utah, despite their contrasting styles, Botstein and Davis shared an epiphany. As they listened to Mark Skolnick's graduate students discuss genetic disorders traced in the very large pedigrees of Mormon families, Botstein's and Davis's eyes suddenly met as both registered simultaneously the same insight. Though both were experts on yeast, they saw a way to locate *human* genes! What they saw was that cutting-edge recombinant DNA techniques would allow them to apply to humans the very sort of genetic analysis first used by Morgan to study the fruit fly. In fact, DNA markers had already been used to map genes in a number of other species, but Botstein and Davis would be the first to develop the technique's potential in humans.

The technique, called "linkage analysis," determines the position of a gene in relation to the known positions of particular genetic landmarks. The principle is simple: it would be difficult for you, given no other information, to find Springfield on a map of the United States, but if I tell you that Springfield lies about halfway between New York and Boston – two landmarks labeled on the map – then your task is made very much easier. Linkage analysis aims to do this with genes: it establishes links between known genetic markers and unknown genes. It was a very successful method with the fruit fly, but, as we have seen, the dearth of known genetic markers in human beings prevented its application to human diseases – until Botstein and Davis recognized that advances in molecular biology had solved the problem.

The DNA markers that caught their eye were restriction fragment length polymorphisms (RFLPs). They occur when a DNA sequence cut by a particular restriction enzyme in one individual has changed in another so that it can no longer be cut by that enzyme. (Remember that restriction enzymes are sequence-specific: enzyme *Eco*R1 cuts only when it encounters GAATTC. That sequence occurs at a given location in the genome, but through mutation some individuals may have a variant form of that segment – say, GAAGTC. The enzyme will be able to cut only unchanged sequences, not the altered version.) These are naturally occurring differences in DNA sequence; they occur most often in junk DNA, and so there is no functional effect. Still, literally millions of them are scattered through our genome.

In the months following the Alta meeting, Botstein, Davis, and Skolnick, together with Ray White, then at the University of Massachusetts, pursued the RFLP concept. In 1980, a landmark paper that grew out of this collaboration heralded the new age of molecular human genetics. They laid out a clear plan showing how RFLPs could be used, and they worked out the math concerning how many would be needed to ensure that every point in the human genome was within reasonable proximity of at least one RFLP marker – conditions that would in principle permit the mapping of the entire genome. It would be like having enough U.S. cities fixed on a map of North America to allow any unmarked place to be located with respectable accuracy using only information about how close it was to the labeled cities. But, for the genetic map, what was "reasonable" proximity? Botstein and his colleagues calculated that 150 RFLPs spread uniformly across the entire human genome would be enough. The most immediate benefit of the system was a new strategy for identifying genes that cause disease. Using families in which a disorder spanned several generations, they would take DNA samples from both affected and unaffected individuals. Then they would use recombinant methods to test RFLPs one after another, looking for ones that tracked the disease through the families (see Plate 54).

In 1979, before the publication of the paper, White presented these ideas at a Cold Spring Harbor Laboratory conference. He noted that "among the more kosher molecular biologists, there was a lot of bitching and grumbling." What he was hearing was great skepticism as to whether the method would work at all; even those who thought it would couldn't agree on the best way to go about using it. These disagreements came into the open during a later meeting to discuss how RFLP linkage analysis could be used to find the gene involved in Huntington disease.

Nancy Wexler wanted her Lake Maracaibo genealogy to be considered immediately for linkage studies, but Botstein and White thought it was far too early to use RFLP linkage analysis to look for the Huntington gene or any other. They argued that much groundwork needed to be done first – the markers themselves had to be found and mapped – before the technique could be applied for such a specific purpose. In the end, Wexler's determination resulted in a parting of ways: while the Hereditary Disease Foundation pressed on with the hunt for the Huntington gene, Botstein and White pushed for a complete map of the human genome.

The latter goal required finding RFLP markers on every chromosome, and finding enough of them to ensure that at least one was close to every point in the genome. It was soon necessary to make an upward revision of the initial estimate of 150. But, undeterred, academic laboratories like White's began to isolate RFLPs, and soon commercial biotechnology was getting in on the action as well.

In 1983 Helen Donis-Keller, an experienced molecular biologist and in those days David Botstein's wife, established the Human Genetics Department at Collaborative Research, Inc., a Boston-area company. She aimed to produce an RFLP linkage map of the whole human genome, with sufficient markers to locate disease genes on any chromosome. The fruits of the effort were published four years later, in a paper aptly entitled "A Genetic Linkage Map of the Human Genome." The map included 403 loci – many more than Botstein's original estimate – and calculations showed that 95 percent of the genome was within reasonable proximity (or "linked") to a marker. It was a great day for genome mapping, but by 1987 rifts and rivalries were appearing once more among the researchers.

For one thing, there was resentment in academic quarters that Collaborative had incorporated freely available data from university labs while disclosing none of its own. (In this respect, Collaborative was pioneering the best-of-both-worlds strategy that Craig Venter and other would-be genome profiteers were soon to follow in the sequencing sweepstakes.) The French immunologist Jean Dausset, for instance, had been following a somewhat different course. His 1980 Nobel Prize in Physiology or Medicine attracted a generous benefactor, whose substantial gift allowed Dausset to pursue his own strategy for preparing a human linkage map. He realized the task would be much easier if all researchers worldwide were working with a standard set of pedigrees – DNA samples from the same families. So, Dausset created the Centre d'Etude Polymorphisme Humain (CEPH) in Paris to collect pedigrees optimal for genetic analysis: large families with three living generations from which to draw samples. The CEPH collection eventually contained DNA from sixty-one families, including many of the Mormons studied by Ray White, Nancy Wexler's Lake Maracaibo families, and Amish families catalogued by Victor McKusick of the Johns Hopkins Medical School. CEPH made DNA samples from all these families freely available to researchers, with the sole proviso that recipients give their analyses to CEPH for integration into the worldwide database. Collaborative Research took full and fair advantage of this resource.

By far the most serious criticism of the Collaborative map, however, was the patchiness of the distribution of its markers. Chromosome 7 – linked to cystic fibrosis, one of Collaborative's targets – had 63 markers, but on chromosome 14 only 6 were identified. The distance between markers on the marker-poor chromosomes was very much greater than the average for the genome as a whole. Ray White was particularly upset by Collaborative's claims. He himself had found over 470 markers but had been publishing his data chromosome by chromosome as each was filled in with the required density of RFLPs. "We would never have dreamed of making such a publication with our data set, which is substantially larger than theirs, because we still have significant gaps," he remarked, rejecting Collaborative's grandiose claim. Whether the claims were grandiose or not, though, Collaborative's map had proved the feasibility of genome-wide mapping and was a significant advance.

But as we have noted, some, like Nancy Wexler, had seen another path opening up in the wake of the breakthrough 1980 RFLP paper. As efforts to produce a comprehensive map gathered steam, David Housman at MIT was gearing up for what David Botstein had declared to be mission impossible at this stage of the game: to discover the location of the Huntington disease gene. He placed this tall order in the hands of Jim Gusella, who had just completed his Ph.D. in Housman's lab. Now the mapping work would surge ahead on another front.

Botstein's initial pessimism stemmed from the lack of markers: RFLPs looked good on paper, but the work of actually collecting them had only just begun. Indeed, it would take years of effort on the part of White, Donis-Keller, and others for the number of known markers to creep up into the hundreds. Starting out in the dawn of the RFLP era, Gusella had his work cut out for him. By 1982, he had a total of only twelve markers, five he had found himself and seven supplied by others. Wexler meanwhile was back at Lake Maracaibo, trying to fine-tune her genealogy: working out who was married to whom, what children they had, who was whose cousin. Local custom was sometimes a hindrance: some names were quite common, and many individuals were known by more than one. The tree Wexler managed to construct for one family nevertheless wound up with seventeen thousand names on it! Periodically she and her colleagues would set aside a whole day just to collect blood; samples had to be dispatched to Boston all together lest the tropical heat of Lake Maracaibo accelerate the degradation of the DNA.

As for Gusella, he wasn't waiting for the Lake Maracaibo samples. I remember a meeting at Cold Spring Harbor in October 1982 at which he presented his earliest data. With a small Huntington-afflicted family from Iowa as his sample, he had tested just five of his twelve RFLPs, checking each to see whether it correlated with the disease. None did, and I couldn't help thinking that having set out to find a needle in a haystack, he was making rather much of having lifted out a few straws. Only with careful analysis of the whole haystack – the vast genome in its entirety – or, alternatively, with a lot of luck could anyone hope to find what Gusella was looking for. And so when he closed his talk by saying that the "localization of the HD gene is now just a matter of time," I said to myself, "Yes, a *very long* time."

But fortune favors the brave. Gusella returned to his laboratory and tried more RFLP markers. To his astonishment, the twelfth, called G8, seemed to show linkage with Huntington disease in the Iowa family. But the statistical correlation wasn't very strong. And so he eagerly awaited samples from Lake Maracaibo, testing them for G8 as soon as he received them. Now excitement was irrepressible: G8 indeed tracked with Huntington disease. By the summer of 1983, against all the odds, Gusella had discovered a linkage after trying only twelve RFLPs. But this was no ordinary stroke of luck: for the first time, the gene for a human disorder had been located on a chromosome without the helping hand of sex linkage and without any prior knowledge of the illness's biochemical basis. Suddenly a new scientific vista was opening up: it seemed we would finally be able to analyze rigorously all those genetic defects that have plagued our species for as long as it has existed. RFLPs had proved they were indeed an effective tool. And having traced the Huntington disease gene to a manageable portion of the human genome, it was surely just a matter of time before our powerful gene cloning techniques would lead to the isolation of the gene itself.

Huntington disease strikes its terrible blow in adulthood. But genetic disorders that strike in childhood have an added awfulness, afflicting those who have hardly had a chance to live. Following a diagnosis, it is often possible to predict with grim certainty the course of the child's life. Such is the case with Duchenne muscular dystrophy (DMD), a progressive muscle-wasting disease. DMD is a sex-linked disorder: the mutation responsible occurs in a gene carried on the X chromosome. Women may carry the mutation on one of their two X chromosomes, but they are usually protected by the presence of a normal version of the gene on their other X chromosome. It's highly unlikely a female will receive two defective copies since males carrying the mutation almost never survive to have children. If, however, the chromosome with the mutated gene is passed to a son, the boy will develop DMD because he has no other X chromosome to supply a normal copy of the gene. When he is about five years old, his parents will notice he has difficulty getting up from the floor or climbing stairs. By about ten he will need a wheelchair. He will probably die in his late teens or early twenties. DMD is not rare: it affects 1 in every 5,000 male children.

The hunt for genes involved in human disorders is a story dominated less by great research institutions and plucky entrepreneurs than by groups like the Hereditary Disease Foundation, organizations founded by those with firsthand experience of the devastation a particular genetic illness can bring. Led by people with something very precious at stake, these groups are by nature more willing to back risky or novel research, going where universities or biotech companies may fear to tread.

The Muscular Dystrophy Association of America and its counterparts in Europe had long supported laboratory research directed at understanding the basic biology of Duchenne muscular dystrophy. In the late seventies cytogeneticists (who study chromosomes microscopically) provided the first genetic clue. Among the very small number of girls who do develop DMD an abnormality was found on the short arm of one of their X chromosomes, at a location called Xp21. Could this be the location of the DMD gene?

Not long thereafter, Bob Williamson at St. Mary's Hospital Medical School in London initiated RFLP-based searches for both the gene causing cystic fibrosis and the one involved in DMD. His colleague Kay Davies hunted up RFLPs on the X chromosome and tested them for linkage to Duchenne. She was successful, and the clincher was their location: they were in the Xp21 region, just as would have been expected given those strange X chromosomes in the women with DMD.

While the gene hunters pushed ahead trying to isolate the genes involved in Huntington disease and DMD, a revolution of a quieter kind was taking place in the offices of clinical geneticists. From the first, Nancy Wexler and David Housman realized that RFLPs linked to a disease gene could be used not only to localize the gene itself but also as a diagnostic test to determine which members of a particular family were carrying the mutation. They could be used to test even an unborn child. Consider the case of a hypothetical family with DMD. At least one boy will be diagnosed – the "index case" that first reveals the presence of a DMD mutation in the family. His mother, the carrier of a mutated gene, also has one normal copy. Her sisters may also be carriers, so any sons they may have are at risk. Now suppose the mother becomes pregnant once again with a male fetus; the chances are 50-50 that the second son will be affected. But with RFLPs her physician can tell her what fate awaits that fetus if carried to term.

First, the affected son's X chromosome is analyzed to identify the particular RFLPs linked to the DMD gene in this family. Next, DNA is taken from the fetus, either a sample of the placenta or of the amniotic fluid, which contains fetal cells. If the fetus's RFLPs match those of the affected boy, then we can be pretty certain that the unborn fetus will also be affected. Why only pretty certain? As we saw in chapter 1, when egg cells are produced, the chromosome pairs undergo recombination, exchanging DNA: the two copies of chromosome 1 trade with each other, as do the two copies of chromosome 2, the two copies of the X chromosome, and so on. If this swap should occur at a point on the X chromosome between the RFLP markers and the DMD gene, the RFLPs we have found to be associated with the normal version could possibly wind up associated with the mutated (DMD) copy. Experience taught us that with the first RFLPs for DMD this happens about 5 percent of the time, and so RFLP-based diagnosis has only a 95 percent chance of being accurate. This degree of imprecision is an unavoidable consequence of recombination. So while such diagnosis represented a tremendous advance, absolute certainty depended on identifying the gene itself, not merely the markers associated with it.

The key to isolating the DMD gene was a young boy named Bruce Bryer, whose X chromosome was missing a very large piece from the short arm. The piece was so large that Bruce suffered from three other genetic disorders in addition to DMD. In 1985 Lou Kunkel at Harvard Medical School reasoned he could use Bruce's DNA to "fish out" a normal gene from the DNA of an unaffected boy. Bruce's case was special because the disease was caused not by a defective copy of the gene but by its complete absence. Kunkel realized that all of Bruce's DNA should be present in a normal boy's, but whatever sequences the latter had and Bruce lacked would hold the key. Using recombinant methods, Kunkel subtracted Bryer's DNA from the normal DNA and kept the difference – the DNA that should contain the DMD gene. The subtraction didn't work perfectly, but it did work well enough that he could find the DNA pieces he wanted by using genetic markers associated with the Xp21 region.

Tony Monaco, a graduate student with Kunkel, took on the job of determining which, if any, of these Xp21 pieces of DNA might constitute part of the DMD gene itself. The only way to do this was to test each piece against DNA from several unrelated patients with DMD. Monaco hit the jackpot with the eighth try: a sequence called pERT87 was found to be absent in five of his DMD boys. This meant almost certainly that pERT87 was very close to the gene and perhaps even a part of it. Monaco began to isolate other sequences close to pERT87, and these too proved to be missing in the DNA of DMD patients. By 1987, Kunkel's group had isolated the complete gene. Now it could be given a proper name: dystrophin. Even with the genome sequence completed, it still holds the record for largest gene in the human genome, owing mainly to its many large introns.

Immediately the new knowledge was applied to produce foolproof prenatal diagnosis for DMD. And soon scientists discovered that a range of different mutations could impair dystrophin and cause the disease. But it remained unclear what the gene actually did. Would its function give us clues to developing effective therapies for DMD?

The first step was to locate the protein produced by the gene in muscle cells. Eric Hoffman in Lou Kunkel's laboratory found that the dystrophin protein was typically located in muscle cells just below the membrane that encloses the muscle fiber. Further studies have revealed dystrophin's critical role in connecting proteins that make up the muscle cell's interior architecture to a set of molecules that span the cell membrane and interact with other proteins outside the cell. The linking of the interior molecules to those in the membrane somehow secures the cell membrane when muscles contract and relax. Without dystrophin, the membrane suffers damage and the muscle dies cell by cell. Given our new and detailed knowledge of dystrophin and its function, it may seem remarkable that there is still no cure for DMD. This is the central frustration inherent in the current state of the art: genetics has made it possible to identify and understand disease, without yet permitting us in most cases to right the genetic wrong.

Kunkel's approach typifies the modern mapping-based approach to dissecting a disorder. Though now common practice, when Kunkel applied it the method was far enough beyond the bounds of research orthodoxy that the Muscular Dystrophy Association was taking something of a gamble in supporting four years of his efforts – a gamble that paid off handsomely. In the old days you tried to use biochemical analyses of a disease's symptoms to identify the disease gene; these days, following Kunkel, you map the gene, and then interpret the symptoms in the light of the gene's function.

One of the strongest arguments in favor of the mapping approach was that the work produced was useful even before the gene had finally been identified. The hunt for the Huntington disease and Duchenne muscular dystrophy genes yielded genetic markers that could be applied in diagnosis before the genes themselves were found. So it was too with one of the most prevalent genetic disorders, cystic fibrosis (CF). But the hunt for the CF gene would prove particularly notable for two reasons: it marked the first time that a company became involved in mapping a human disease gene, and the first instance of brutal competition among the scientists involved in such an endeavor.

In cystic fibrosis patients, a thick mucus accumulates in the lungs, making it difficult to breathe. The cells lining the tubes of the lungs can't clear out the mucus in which bacteria thrive, producing pulmonary infections. Before antibiotics those afflicted had a life expectancy of just ten years; today survival rates are substantially better. CF is also one of the most common genetic disorders, with about 1 in 2,500 individuals of northern European descent affected. It follows a recessive pattern of inheritance: you need two mutant versions of the gene to be affected. But since as many as 1 in 25 people of northern European descent carry a single mutant version (though are themselves protected because they have one normal copy), there is a relatively high risk that two carriers will get together and both pass it on to their children. It therefore became a medical priority to devise a diagnostic test as soon as this became a realistic goal.

Born in Shanghai, and raised and educated in Hong Kong, Lap-Chee Tsui came to the United States as a graduate student in 1974. Tsui learned his molecular genetics doing research on viruses before moving in 1981 to Manuel Buchwald's laboratory in Toronto to work on cystic fibrosis. Tsui is a quiet, pleasant man who is nevertheless intense and passionate about his goals. Planning to track down the gene via RFLP linkage analysis, he spent the first couple of years finding CF families before starting the painstaking process of testing their DNA with every RFLP he could lay his hands on. But the luck that had smiled upon Jim Gusella in his pursuit of the Huntington disease gene did not favor Tsui: after about a year all he had managed to do was eliminate a lot of RFLPs. He needed more, and was thrilled when Collaborative Research offered to share its RFLP markers with him.

Tsui's Toronto group was not alone in pursuing the CF gene: Bob Williamson in London, who had worked on DMD, also took up the hunt, as did Ray White, now in Utah, attracted by access to the very extensive pedigrees assembled by the Mormon church. These records, the Ancestral File, permit present-day members of the church to make provision for deceased forebears, who lived their lives outside the fold or died before the church was founded in 1830. The aim is to unite families for eternity. Seldom have the needs of religion and genetics been so happily aligned.

But it was the Toronto group that would notch the first success, when it found in 1985 a linkage between one of Collaborative Research's RFLPs and the CF gene. At the time, the location of that RFLP was unknown, but, seeing its potential as a golden egg, Collaborative Research quickly set about locating it. They soon determined that it was on chromosome 7 but did not immediately inform Tsui, their collaborator. Nor did they mention the chromosomal location when they announced the discovery in the November 22 issue of the prestigious journal *Science*. Clearly they were trying to preserve their monopoly on the new information, but secrecy and science often don't mix well: word soon spread on the grapevine that 7 was the place to be.

As Collaborative kept quiet, Williamson and White were just days away from the same discovery. Their own two papers, published in

Nature, Science's British rival, both mentioned that the key RFLPs were on 7. Tsui was incensed: he was about to lose his claim to the linkage discovery thanks to his partners' shenanigans – in science there are no prizes for coming in second – but Helen Donis-Keller persuaded *Nature* to accept a paper from the Toronto-Collaborative team announcing the location. So it was that three papers appeared in the November 28 issue of *Nature,* along with an editorial explaining how it had all come about.

The Toronto-Collaborative partnership did not survive the clash of academic and commercial cultures. Collaborative Research would find that the academic world had become wary of collaborating with them, a situation hardly helped by the crass and not very sturdy claim made by Orrie Friedman, Collaborative's CEO, that "we own chromosome 7." Fortunately, this soap opera saw its final episode in December 1985, when all the research groups agreed to pool their resources in order to test 211 families for linkage to chromosome 7 RFLPs. The results were spectacular. The RFLPs were very close to the gene, within 1 million base pairs – which made them useful in diagnosis, one of the major goals of the CF research.

The next step promised to be even more difficult. Learning that New York is halfway between Washington, D.C., and Boston is better than merely knowing it is somewhere in the United States. But when one must set out on foot from Washington to Boston, looking yard by yard for a sign that reads "Welcome to New York," the clue may seem not so helpful after all. One million base pairs might be close by the standards of linkage analysis, but it is a very long way by the standards of gene cloners who analyze regions one base pair at a time. To go the distance from the two RFLPs nearest the CF gene, Tsui teamed up with Francis Collins, who was then at the University of Michigan and would later succeed me as director of the Human Genome Project.

Collins had developed "jumping" techniques to facilitate the cloning of a gene between a pair of known RFLPs, but he was under no more illusion than Tsui about the magnitude of the problems facing them. After two years of work, they managed to localize the CF gene to a 280,000-base-pair segment of DNA, within which they found the sequence of a gene known to play an important role in human sweat glands, which are dysfunctional in cystic fibrosis patients. It seemed the complete CF gene might finally have been corralled.

The only way to be sure they had got it right was to sequence the cDNA and search for the disease-causing mutations. Given a region 6,500 base pairs long, this was quite a challenge in 1989, and it had to be done twice: once using DNA from a CF patient and once with DNA from a healthy individual. The result, however, was clear-cut: the patient's DNA was missing a stretch of three base pairs, resulting in the absence of just one amino acid in the protein. This one mutation accounts for about 70 percent of CF cases, but over a thousand others found in the CF gene also cause the disease. This multiplicity of harmful variants has greatly complicated the task of DNA-based diagnosis.

Let us now return to Nancy Wexler, David Housman, Jim Gusella, and their colleagues, whom we left back in 1983 at the triumphant moment when a particular RFLP, G8, had been linked to the gene for Huntington disease. If it seemed up until then that they had enjoyed more than their collective share of good luck in locating the HD gene with astonishing speed, the gods were soon to redress the imbalance. Finding the gene had taken a mere three years; isolating it for detailed analysis would take ten years and an international team of 150 scientists. In this case, the region where the gene had been localized was 4 million base pairs long. The Huntington disease geneticists worked hard to narrow that window, but genetic mapping gets more difficult as the genetic distance gets smaller, and finally these efforts were rewarded only with ambiguous data. Imagine the foot journey from Washington to Boston, in search of New York. Now imagine arriving at an intersection in Philadelphia to find a signpost indicating New York in both directions.

Giving up on the contradictory linkage analysis, the Huntington gene hunters devised an alternative strategy, focusing on the region that was most similar among Huntington disease patients. This approach eventually reduced the region to only 500,000 base pairs, and the time had come to turn to gene-cloning techniques. The first results were disappointing: they found three genes in the right-hand half of the region, but none showed any abnormalities in patients with Huntington. Undaunted, they explored the left-hand side and found a single gene, with the prosaic name IT15. Finally, after ten years and many losing lottery tickets, luck had begun to smile on them once more. The gene contained a short sequence, CAG, that repeated over and over again, like the short tandem repeats (STRs) used in DNA fingerprinting. It turned out that unaffected people have fewer than thirty-five CAGs, while people with more than forty will develop Huntington as adults; in the rare instance of more than sixty, a severe form of Huntington develops before the age of twenty. CAG is the genetic code for the amino acid glutamine, so each of the CAG repeats adds an extra glutamine to the protein. In the case of Huntington sufferers, the protein coded by the HD gene – the rather difficult-to-say huntingtin – contains extra glutamines. This difference likely affects the behavior of the protein in brain cells, probably by causing molecules to stick together in gluey lumps within the cell, somehow causing its death.

It had been a tremendous effort by all the laboratories in the Hereditary Disease Foundation's team, and in recognition that it was truly a collaboration, the only name appearing as the author of the article was that of the Huntington Disease Collaborative Research Group. The same strange type of mutation – repeats of the three-base-pair sequence – had already been implicated in three other disorders, remarkably all of them also neurological diseases. We now know of fourteen of these "trinucleotide repeat disorders," but we still are no closer to understanding why brain cells are so susceptible to this kind of mutation.

It may be depressing to know that despite the substantial time it has taken to hunt down their respective genes, these disorders – Huntington, Duchenne, and cystic fibrosis – are, by the standards of geneticists, "simple." They are caused by mutations in a single gene and not much affected by environment. If you have the three-base-pair deletion in both your cystic fibrosis genes or more than forty CAG repeats in one of your Huntington disease genes, you will develop those disorders no matter where you live or what you eat or drink. There is a large number of single-gene disorders – the current genetic disease database lists several thousand – but the majority are extremely rare, each occurring in just a few families.

Much more common are "complex" or "polygenic" disorders, which include many of our most common ills: asthma, schizophrenia, depression, congenital heart disease, hypertension, diabetes, and cancer. These are caused by the interaction of several – perhaps many – genes, each of which alone has only a small effect and perhaps no detectable effect at all. And typically in polygenic disorders, there is a further complication: these sets of interacting genes may create a predisposition to a particular disease, but whether you actually develop a case of it depends on environmental factors. Suppose that you have a set of gene variants that predisposes you to alcoholism. Whether or not you actually become an alcoholic depends on your exposure to the environmental trigger, alcohol. Your fate may be quite different growing up in a dry county in Texas as compared with

Manhattan. The same principle holds for asthma; in a "good" summer, when the pollen and spore counts are low, you may develop no symptoms despite being genetically disposed to the disease.

The complex interplay of genes and environment is nowhere more evident than in cancer. Cancer is fundamentally a genetic disorder caused by mutations in several genes. Each mutation alters one more element in the cell's behavior until it acquires all the characteristics of a fully malignant cell. Cancer mutations arise in two ways. Some are inherited. We have all heard the phrase "it runs in the family," and while some traits described this way – Catholicism for one – are not necessarily heritable, some kinds of cancer are. Still, the disease is so lamentably common that it is not so unusual to have two or even three cases in one family even without a hereditary component. (Geneticists studying "cancer families" therefore apply very strict criteria in deciding whether a cancer is inherited.) Plenty of cancer mutations also arise in the normal course of living. DNA can become damaged owing to errors the enzymes make in the course of duplicating or repairing the genetic molecule, or as a consequence of the side effects of the normal chemical reactions within the cell. And many cancers arise thanks to our own foolishness. Ultraviolet rays in sunlight are potent mutagenic agents to which sun worshippers willingly expose themselves, and cigarettes are a very efficient way to deliver carcinogens straight into your lungs, where they cause lung cancer. Other environmental factors, for instance asbestos in the workplace, have also been shown to promote cancer. The point is that DNA can get damaged quite naturally, but it is up to us to minimize the damage through informed social and personal choices.

In 1974, Mary-Claire King (of human/chimpanzee and Las Abuelas fame) moved to UC San Francisco to work in a laboratory studying breast cancer, where she decided to commit herself to the hunt for a breast cancer gene. At the time, the RFLP linkage approach was still six years away, but King knew that there would be clues in pedigrees, so she set about collecting families. She looked for families in which members had developed breast cancer at an early age, and in which there was also ovarian cancer, reasoning the odds favored a hereditary culprit in such cases. The only genetic markers available to her were protein markers, and after a few years she published her first breast cancer paper, describing unsuccessful tests for linkage with cell surface proteins. This was followed by other papers showing similarly negative results. Naysayers were equally negative: breast cancer is too heavily affected by the environment to permit genetic analysis, they said, referring predictably to needles and haystacks. Undeterred, King continued to refine her data-set, and by 1988, with an analysis of 1,579 families, she thought she had good evidence for a breast cancer gene in these high-risk families.

The medical world was astonished when in 1990 she reported that she had found an RFLP on chromosome 17 linked to breast cancer in a subset of 23 of her families, involving a total of 146 cases of breast cancer over three generations. She checked factors that might have confounded the analysis – perhaps these women had been exposed to more X rays, or they differed from others in their pregnancy histories – but her data held up. There was a gene at chromosome location 17q21 that when mutated greatly increased a woman's risk. King's paper set off a race to isolate the gene itself, called BRCA1 (for Breast Cancer 1), and an ongoing controversy about the commercial exploitation of genes.

Isolating the BRCA1 gene would inevitably be a big event. Even if it was important only in a small subset of high-risk families (i.e., it would only be responsible for a small proportion of all breast cancers), the insights that might come from knowing what the gene did would be cause enough for excitement. King teamed up with Francis Collins, whose gene-hunting credentials were impeccable, but the pair had tough competition. Mark Skolnick, the Utah population geneticist involved in the RFLP linkage breakthrough, formed a company, Myriad Genetics, with Wally Gilbert, whose entrepreneurial spirit had survived his uneasy tenure at the helm of Biogen. Myriad's business plan was to use the power of the Mormon family pedigrees to map and clone genes, and BRCA1 came within their crosshairs very soon. In 1994, a consortium of geneticists from Myriad, the University of Utah, NIH, McGill University, and Eli Lilly beat the rest of the world, announcing what they rather coyly called a "strong candidate" for the BRCA1 gene. They had found it. Everyone involved filed for a patent (although Myriad initially saw to the exclusion of the NIH scientists). In 1997, Myriad's application was approved.

At the moment BRCA1 was being cloned, a different consortium of geneticists, including scientists from Myriad and the Institute for Cancer Research in England, reported they had located a second breast cancer gene, BRCA2, on human chromosome 13. Once again a race began, and within a year the English group claimed success in isolating BRCA2. They knew they had bagged their quarry once they had determined about two-thirds of the gene's DNA sequence and shown it to be defective in six different families. Not to be outdone, Myriad formed yet another consortium, this one comprising institutes in Canada and France; soon they would publish the complete sequence of BRCA2, a very large gene. Of course both Myriad and the Institute for Cancer Research filed patent claims.

It was clear that these were going to be commercially important genes. Mutations in them have very serious consequences for women. The risk of a woman developing breast cancer by age seventy because of a mutated copy of either BRCA1 or BRCA2 can be as high as 80 percent. And it has been established that the same mutations also raise the risk of ovarian cancer to as high as 45 percent. Women in whose families these mutations run need to be informed as early as possible whether they are carrying a defective variant of either gene. There are difficult but potentially life-saving choices to be made: an elective bilateral mastectomy in high-risk women reduces cancer incidence by 90 percent. At the same time, genetic screening can identify individuals in these families who have normal genes; this affords them the comfort of knowing they are not at increased risk.

It sounds like a worthy thing to have brought to market: a genetic test for a very serious disease, a means to help women make informed decisions about their health. Why, then, is Myriad frequently portrayed as exemplifying all that is wrong when commerce is married to science? Myriad now has nine U.S. patents covering BRCA1 and BRCA2, and in 2001 it was granted one in the European Union, one in New Zealand, four in Canada, and two in Australia. In effect, the company now enjoys a global monopoly on these genes and worldwide control over how they are used. It is entirely reasonable that Myriad should make money from testing for BRCA1 and BRCA2 mutations – the company provides a valuable service and has invested a great deal of money to develop the test. But how much money should the company reasonably be making? Today each test costs more than $2,700. At the same time, Myriad restricts academic researchers from using the BRCA gene sequences to develop alternative tests. And information about BRCA mutations gleaned from DNA sequencing among patients enrolled in academic research projects is withheld even from the patients themselves; to do otherwise would be a diagnostic clinical use, infringing the Myriad patents.

Myriad has lately made some concessions. A government deal now permits scientists doing NIH-funded research to use the test at a cut rate of $1,400. But critics have viewed this as a token gesture and remain particularly vocal in Canada and Europe. The European Parliament has passed a resolution expressing "dismay" at the actions of the European Patent Office and instructing the Parliament's staff to prepare challenges to Myriad's patents on both genes. Myriad's French partners in the BRCA2 sequencing – the Institut Curie and the Institut Gustave-Roussy – were particularly incensed about Myriad's BRCA2 patent and have filed a joint complaint to the European Patent Office. In any case, Myriad's monopoly may not be good for patients. The company's test fails to detect all possible cancer-causing changes affecting the gene, so people who test negative for the screened mutations may nevertheless still be at risk. Now when you're tested Myriad has you sign a waiver to the effect that a negative result does not necessarily indicate a clean bill of genetic health. Developing a more comprehensive test is difficult for technical reasons, but more breast cancer labs around the world would likely be trying it right now if not for Myriad's research-stifling patent.

Over the past dozen or so years, linkage analysis has also zeroed in on several other important cancer genes, including ones involved in neurofibromatosis ("Elephant Man disease," not commonly understood to be a form of cancer), colorectal cancer, and prostate cancer. But while effective, the gene-by-gene approach is slow and painstaking, with each study being dependent upon finding appropriate families for analysis. It is here that the Human Genome Project will prove its tremendous value. The DNA and protein micro-arrays we looked at in chapter 8 will furnish cancer-gene hunters with a powerful high-caliber weapon. When I first became interested in cancer research in the 1960s, we knew so little about the underlying genetics and relied on such primitive tools that I turned to viruses that cause cancer in animals. It was my hope that by studying such viruses – which, having very few genes, were manageable even then – we might glean some insight into human cancer. Nowadays, cancer research is no longer confined to viruses; the tens of thousands of genes in actual human tumors are within our powers to map and clone. An enormous wealth of knowledge awaits us as we discover, in greater and greater detail, all the minute biochemical deviations that contribute to turning a normal cell into a cancerous one.

Most linkage analysis studies depend on tracking one's genetic quarry through as large a pedigree as possible. But there is another strategy that looks at small populations with a high incidence of a disorder. And you can't get much smaller than the population of Tristan da Cunha.

A volcanic island rising steeply and inhospitably out of the sea, Tristan da Cunha is a speck of land – just forty square miles – in the middle of the South Atlantic, among the most remote places on the planet. The first permanent settlement was a British garrison established there in 1816 to prevent the French from using the island as a base from which to spring Napoleon from his exile on St. Helena, an island 1,200 miles to the north. Subsequent population growth was sporadic – a few settlers here, a few survivors of shipwrecks there – and as of the unofficial census of 1993, the total was only 301. That year a team from the University of Toronto went to the island to follow up on medical studies done on the islanders in 1961, when the entire population was evacuated to England because the island's dormant volcano had become temporarily active. The most surprising finding had been that about half of the evacuees had a history of asthma.

When the Toronto Genetics of Asthma Program examined 282 of the inhabitants in 1993, they found that 161 (57 percent) showed some symptoms of asthma. The Canadians prepared a genealogy of all the local families; it wasn't overly difficult since all the islanders are descendants of fifteen early settlers and so their lineages are closely interrelated. Asthma was apparently introduced into the island by two women who settled there in 1827. A population like this is a boon for gene hunters: the island is essentially an extended family, and so the genes causing any observable disorder are likely to be the same throughout the population – a best-case scenario for linkage analysis. In a larger, more mixed population, some people's asthma may be caused by one set of genes while that of others is caused by another set. Such heterogeneity is what makes nailing the genetic determinants of complex diseases so difficult.

The Toronto team collected blood samples and prepared DNA, but needed major funding to complete the study. It was then that they heard from Sequana, a company founded to hunt down disease genes. Sequana financed the study, immediately provoking charges that the company was exploiting the islanders, who did not perhaps fully appreciate their role in Sequana's business strategy. Canadian activists calling themselves the Rural Advancement Foundation International claimed that Sequana was "committing an act of biopiracy ... violating the fundamental human rights of the people from whom the DNA samples are taken." Sequana claimed – in a move that was guaranteed to provoke more accusations of "biopiracy" – to have found two genes conferring susceptibility to asthma but refused to reveal them publicly until the European patent application was filed. The genes are located on chromosome 11, and subsequent studies of heterogeneous mainland populations have confirmed a role for chromosome 11 in asthma. It would thus seem that the genetic factors underlying the high incidence of asthma on Tristan da Cunha are not relevant solely to the isolated inhabitants of South Atlantic islands.

The storm over Sequana's "biopiracy" was nothing compared to the hurricane that was to envelop Kari Stefansson and his company, deCODE Genetics, a few years later. Recognizing that it was tedious and inefficient to look for a different Tristan da Cunha–like micropopulation for every disorder, Stefansson reasoned that what he needed was an isolated island but one with a much larger population, among whose members one could look for a number of disease genes at once. It just so happened that Kari Stefansson was born on such an island.

Iceland is about the size of Kentucky but its population of only 272,512 is one-fifteenth that of the Bluegrass State. The island was settled in the ninth and tenth centuries by the Vikings, who brought with them women kidnapped from Ireland during the voyage. Iceland offers several advantages for the enterprising gene hunter. First, the population is very homogeneous, derived almost entirely from the original settlers; there has been very little immigration since Viking days. Second, there are detailed genealogical records going back many generations; many Icelanders can trace their ancestry back five hundred years. This handy resource is supplemented by a detailed register of births begun in 1840 at the University of Iceland. Third, Iceland has had a nationalized health care service since 1914, so the entire nation's medical records are uniformly ordered and readily accessible – at least in principle.

A Harvard neurologist, Stefansson was interested in genetically complex disorders such as multiple sclerosis and Alzheimer disease. Recognizing his own people as a nearly perfect population for genetic research, he devised a project to link the genealogical and medical records to create a database for gene-hunting. Despite the project's worthy purpose, local privacy statutes stood in the way until the Althingi (the Icelandic Parliament, founded in A.D. 930) passed the Law on a Health Sector Database in 1998. The legislation authorized "the creation and operation of a centralized database of non-personally identifiable health data with the aim of increasing knowledge in order to improve health and health services."

In 2000, deCODE was awarded a twelve-year license to build and run the Icelandic Healthcare Database at its own expense in exchange for an annual fee payable to the national government. The genealogical component of the database contains information in the public domain, but access to the medical records database is more restrictive, operating on a basis of "presumed consent" – information about people's health is entered into the database unless they take the initiative to opt out. The genotype component is most restrictive, relying on informed consent – individuals must actively agree to give tissue samples for DNA extraction. Here's the hot spot: although deCODE has a system in place to protect donor privacy, critics argue it is inadequate. Since a person's DNA must be correlated with genealogical and medical records, samples are not taken anonymously; instead the source's identity is encrypted. In theory such encryption could be broken. Word might get around, especially in such a small population, that one family or another is carrying "bad" genes, opening up the possibility of gene-based discrimination. The deCODE project crystallized in microcosm many of the issues concerning genetic privacy that had been discussed rather more hypothetically elsewhere. Nevertheless, despite the controversy, most Icelanders were in favor of the company, viewing it as a means of combining a noble mission – fighting genetic disease – with the happy prospect of serious money swelling the country's small economy. Caught up in the excitement, many Icelanders invested heavily in the company, snapping up shares for as much as sixty-five dollars long before deCODE was officially floated on NASDAQ. But the economic downturn has not been kind to biotech in general, or to deCODE in particular. At the time of writing, shares are worth around two dollars. Many Icelanders have been left to rue those heady *buy!-buy!-buy!* days. Still, deCODE has undeniably brought money into Iceland through its lucrative collaborations with both Hoffmann–La Roche and with Merck. But with the government of Iceland willing to provide a $200 million loan guarantee and the company forced to lay off a substantial proportion of its workforce, the financial reality is that deCODE may yet prove to be less of an economic boon to the country than had been hoped.

The real test of deCODE, however, lies not in the vagaries of the stock market, but in the science it produces. Here unfortunately the commercial imperative of non- or delayed disclosure makes evaluation difficult. The company, we learn via press releases, is now carrying out linkage analysis on forty-six disorders, including asthma, depression, cancer, osteoporosis, and hypertension. It has found linkage markers for twenty-three of these disorders, and it has isolated genes contributing to peripheral blood disease, stroke, and schizophrenia. But with few of the details published in scientific journals, it is too often difficult to distinguish science from hype. Still, deCODE has certainly shown itself capable of making useful contributions: in June 2002, deCODE researchers published a new map of the human genome with significantly higher resolution than that of the old CEPH map. In addition, the long-anticipated publication of the company's research on schizophrenia hints at a scientifically – and commercially – productive future for the company.

Whatever the years to come hold for deCODE, it's clear that its overall approach – a three-way marriage of medical, genealogical, and genetic records for a well-defined population – has great potential. The company is, therefore, not alone in subjecting the population of a whole country to genetic scrutiny. Finland, for example, has a population of 6 million and notable incidence of some thirty-five genetic disorders, some unique, others simply more common there than in other European countries. The Finns have duly attracted considerable interest among human geneticists. Other countries, too, are leaping on the genetic database bandwagon. In April 2002, Britain launched its "Biobank" program, and the government of Estonia is promoting a comparable nationwide effort as well. Large-scale population studies like these will ultimately help us track down even the most elusive of genes.

Human genetics has a long history, starting with our early ancestors' curiosity about how certain characteristics were passed down through generations. But for virtually all of that history, the scientific foundation of the inquiry was weak at best. Charles Davenport's attempts to bolster his eugenics program by searching for the genetic basis of what he called "feeblemindedness" scarcely rate as science. As a measure of just how slow the field was to develop, it's worth noting that for a long time the accepted figure for a fundamental genetic parameter defining our species was wrong. It was not until 1956, three years after the discovery of the double helix, that the number of human chromosomes was determined correctly to be forty-six, and not forty-eight, as had been supposed without question since 1935. But the flood of knowledge let loose in the twenty years since RFLP linkage studies began has, with fantastic speed, created a fertile field from formerly barren ground. Having completed the sequencing of the human genome, we will likely soon find the genes underlying virtually all important genetic diseases. The question then becomes: What do we do with them?

CHAPTER TWELVE

DEFYING DISEASE:
TREATING AND PREVENTING
GENETIC DISORDERS

From the moment he was born, David Vetter never felt the direct touch of another human being. David suffered from an inherited condition called severe combined immunodeficiency disorder (SCID). The failure of his body to develop B and T cells, both crucial elements in the immune response to disease, left him susceptible to the slightest infection.

David's parents knew before he was born that he might have SCID: their firstborn son had died of it. The Vetters and the doctors were ready. They decided early on that if the infant should prove to have SCID, he would be isolated in a germ-free environment until a treatment could be developed – surely it would not be long given the rate of progress in medicine. David was delivered by cesarean section in September 1971 and immediately placed in a sterile incubator. All contact with him was made using latex gloves built into the little chamber. As he grew older he was transferred into larger and larger sterile environments, plastic "bubbles," but one constant remained: the gloves. They would continue to be his only way of feeling anyone or anything in the outside world (see Plate 55).

The hoped-for cure proved elusive. David remained in his bubble, where he drew national attention. NASA tried to help him with a Mobile Biologistical Isolation System, essentially a space suit permitting the boy the freedom to venture beyond the bubble. But a space suit is really only a bubble of a different kind.

Advances in transplantation methods looked promising, and in October 1983, a month after his twelfth birthday, David received a bone marrow transplant from his older sister. Unfortunately, her marrow proved to contain a virus that caused a pernicious lymphoma to develop in David's defenseless system. By February 1984, he had to forsake his bubble and be placed in intensive care. He died soon after, but at least in those final days he was able to experience at last the warmth of human touch.

We can be thankful that SCID is rare, but genetic disorders are surprisingly common among children. In fact, about 2 percent of all babies are born with some kind of serious genetic abnormality. It's estimated that genes are directly responsible for one-tenth of admissions to children's hospitals, and indirectly implicated in about half. David Vetter's case is sadly representative of where our knowledge stands regarding most genetic diseases: we can understand what's wrong and we can diagnose them, but there is relatively little we can do to treat, much less cure, them.

It is interesting to chart the image of SCID in popular culture. In the seventies the condition inspired a made-for-television tearjerker called *The Boy in the Plastic Bubble.* By the nineties, the Bubble Boy had become a figure of fun in the sitcom *Seinfeld.* And in 2001, Disney released a tasteless film reimagining as a series of goofy adventures the life of a boy confined in a bubble on account of an unnamed but unmistakable condition.* This enduring powerlessness of science in the face of such a horrific illness must to some degree account for this trajectory from sentimentalism to farce. But the same powerlessness makes the diseases only harder to bear for the afflicted and their families. Especially with diseases that cause a progressive and inexorable decline, a diagnosis is virtually a death sentence. In the absence of treatment, some would prefer not to know their awful fate, particularly if they have witnessed its ravages in loved ones. In the previous chapter we met Nancy Wexler; with a 50 percent chance of developing Huntington disease, the scourge that had claimed her mother and uncles, Wexler worked so long and so hard at Lake Maracaibo and in genetics laboratories in the United States to track down the genetic culprit. But even though her extraordinary crusade resulted in isolating the gene and identifying the lethal mutations, a cure is still nowhere in sight. And although she has done so much to make a diagnostic genetic test available, Wexler herself has said that she will not be tested – at least, not until a viable treatment seems near. She would prefer to live with a huge uncertainty than discover the truth in a 50-50 gamble: the odds are even that she will face a mental and physical decline leaving her a shell of the dynamic woman she is now.

Sometimes it is almost more unbearable to care for a sufferer than to become one. Carol Carr of Hampton, Georgia, watched her husband, Hoyt, develop Huntington disease in his thirties. His sister Roslyn died of it, and his brother George committed suicide soon after being diagnosed. Carol quit her job and became Hoyt's full-time nurse for the next twenty years, as he fell apart. They already had three

*The Disney ending: The movie's bubble inhabitant turns out to be healthy after all.

sons by the time Hoyt was diagnosed, and when he died in 1995, Carol was already nursing her two eldest, Randy and Andy, as she had her husband – feeding and bathing them, giving them their medicine, helping them to the bathroom. Soon James, her youngest, was also developing symptoms. In despair, Carol reluctantly placed Randy and Andy in a nursing home, where, on June 8, 2002, she shot them both dead. The *New York Times* reported James's opinion that Huntington had killed his brothers long before his heartbroken mother ever pulled the trigger.

Not all genetic diseases are tragedies of medical helplessness. Perhaps the best example to the contrary is the disorder responsible for that strange fine-print warning appearing on some food products, especially soft drinks: "Contains Phenylalanine." Phenylalanine is an amino acid – a common component of proteins – that cannot be processed by people with a genetic disorder called phenylketonuria (PKU).

The story starts in Norway in 1934. A young mother was determined to find out what was wrong with her two children, ages four and seven, both of whom had seemed perfectly normal at birth. The elder was not fully toilet trained and was barely capable of speaking a few words, let alone forming a complete sentence. The case came to the attention of Asbjørn Følling, a biochemist and physician. After conducting a battery of tests, Følling found a biochemical abnormality he linked to their condition: they had too much phenylalanine in their urine. But he also learned that theirs was no isolated case: he discovered thirty-four others in twenty-two families across Norway, and realized that he had stumbled across a genetic disease.

We know now that PKU is caused by a mutation in the gene for phenylalanine hydroxylase, an enzyme that converts phenylalanine to another amino acid, tyrosine. It is a rare disorder, affecting about 1 in 10,000 people in North America, and shows a recessive pattern of inheritance: you must have two mutated copies of the gene, one from each parent, to develop PKU. In the children affected, who lack a functioning enzyme, phenylalanine accumulates in the blood, impairing brain development and leading to severe mental handicap. Prevention is simple: PKU children raised from birth on a diet low in phenylalanine – with minimal protein and no artificially sweetened soft drinks, the two principal sources – grow up normal. Nutrition alone can make the difference between normal brain development and profound disability. Clearly it is important to know a child's PKU status as soon after birth as possible. Robert Guthrie devised a simple diagnostic test for blood levels of phenylalanine and tirelessly promoted its use until it became standard neonatal practice. Since 1966, a heel-prick blood sample has been taken from every newborn and analyzed for phenylalanine levels. Thus without ever examining a single base pair of DNA, the Guthrie test screens for a genetic disease in millions of babies every year. Prior to this testing program, as much as 1 percent of mental retardation in the United States was attributable to PKU; now there are only a handful of cases a year.

The 1950s saw the development of cytogenetics, the study of chromosomes through the microscope. Employed diagnostically, this approach soon revealed that abnormalities in chromosome number – usually one too many or one too few – invariably cause profound dysfunction. These problems stem from an imbalance in the number of genes, a departure from the norm of two of each. Such conditions do not run in families like Duchenne muscular dystrophy (DMD) or cystic fibrosis (CF), but they are still very much *genetic;* they arise spontaneously through accidents in the cell divisions leading to the generation of sperm and egg cells.

The best known is Down syndrome, named for John Langdon Down, who in 1866, as the medical superintendent of a home for the retarded, was first to describe its characteristic clinical features. He noted that 10 percent of the residents of his institution resembled one another: "So marked is this, that when placed side to side, it is difficult to believe that the specimens compared are not children of the same parents." But the first insight into the condition's biological basis did not come until ninety years later, when the French physician Jérôme Lejeune found that children with Down syndrome have three copies of one chromosome, later shown to be chromosome 21. The normal condition, two copies of a chromosome, is called "disomy"; so Down syndrome is known in genetic parlance as "trisomy 21."

The incidence of Down syndrome increases with the age of the mother. At age 20, a woman's chance of producing a Down baby is about 1 in 1,700; but at 35, it jumps to 1 in 400; and at 45 shoots to 1 in just 30. For this reason, many older pregnant women choose to have prenatal diagnosis performed on their fetus to determine whether it might possess 21 in triplicate. The test was first done in 1968, and today it is routinely offered to all pregnant women over 35.

Because the developing fetus must be big enough to withstand safely the extraction of a tissue sample, such diagnosis cannot be performed in the very early stages of pregnancy. Typically it is done in the fifteenth to eighteenth week using amniocentesis, a procedure that entails drawing some amniotic fluid (which naturally contains cells from the fetus). An alternative test, which may be done as early as the tenth week, gathers cells from the chorionic villus, the part of the placenta that attaches to the uterine wall, but this method is less reliable. Because both procedures are mildly risky – amniocentesis results in a 1 percent rate of miscarriage, and chorionic villus sampling in a 2 percent rate – younger women are usually advised to avoid them: the probability that their fetus has a genetic defect is actually lower than the probability that it will be damaged by the procedure. At one time, extracted fetal cells had to be grown in petri dishes before being processed for chromosome analysis. Nowadays, a more rapid diagnosis can be done using fluorescence in situ hybridization (FISH); in this method, a small fluorescent molecule is attached to a stretch of DNA sequence specific to chromosome 21 and introduced to the sample, where it binds to the fetal chromosome 21 DNA. If two fluorescent patches appear in the nucleus of a cell, the fetus is normal; if three, the fetus has Down syndrome (see Plates 56 & 57).

In Britain, 30 percent of Down pregnancies are detected by routinely testing the oldest 5 percent of women bearing children. This method boasts a clear efficiency in simple terms of detections per pound spent (Britain's National Health Service has been subject to such a calculus ever since Mrs. Thatcher's assault on health spending), but what of the remaining 70 percent of Down cases? Down is rarer in the babies of younger mothers, but these women account for the vast majority of all pregnancies. Since the standard tests are statistically not worth their attendant risks, there have been attempts to find alternative, noninvasive, indicators. It turns out that substances detectable in the mother's blood yield useful information. Low levels of alpha-fetoprotein and high levels of chorionic gonadotropin correlate to a significant degree with Down (though they are by no means ironclad indicators of trisomy). Modern practice, then, is to offer younger women the blood test, and, if it suggests the possibility of Down, they are then counseled to undergo amniocentesis or chorionic villus sampling for a definitive diagnosis.

Sadly, today a woman who learns that her fetus has Down syndrome has only two choices: to become the mother of a Down baby or to abort the fetus. It is a painful decision that is made no easier by the variable severity of the affliction. People with Down all share the characteristic facial features identified by Dr. Down – a broad flat face, a small nose, and narrow slanting eyelids* – but they range considerably in IQ, scoring between 20 and 85 (i.e., from severely handicapped to low normal). They are especially prone to a range of ailments, including heart disease (which claims about 15 percent in the first year of life), gastrointestinal anomalies, leukemia, and, with increasing age, cataracts and Alzheimer; but it's also perfectly possible that an individual will have relatively few health problems. With improved care, and better knowledge of medical hazards posed by possession of that extra chromosome, life expectancy has increased substantially: 50 percent of affected individuals today survive into their fifties. Despite typically acquiring over time what most would consider a depressing familiarity with the insides of hospitals, people with Down generally enjoy life, and have brightened many a family. The condition is perhaps tougher on their parents, who must adjust to caring for someone with special medical needs, as well as to the knowledge that their child will, in many ways, never really grow up.

In general, women who learn that they are carrying a Down fetus choose to terminate the pregnancy.† As a result, in countries with routine prenatal screening, the number of Down babies born is

*Dr. Down originally called the disorder "mongolism" on the basis of these characteristics, entitling his 1866 paper "Observations of an Ethnic Classification of Idiots." Subscribing to the racist evolutionary views of his day, he believed that Down represented an evolutionary step backward from the exalted Caucasian state to the "inferior" Mongoloid one. To give him his due, though, he concluded that what he called "retrogression" undermined the claims of those who refused to accept that Caucasians and non-Caucasians were members of the same species.

†In the United Kingdom, 92 percent of fetuses diagnosed with Down are currently aborted. Typically only women who are willing to consider an abortion undergo prenatal testing (there's no point subjecting a fetus to the risks associated with testing if the mother intends to carry the pregnancy to term whatever the result), so we would expect this figure to be high.

declining. Statistically, however, this claim is more complicated than it sounds: the trend toward deferring motherhood – often for professional reasons – has actually increased the ranks of women at risk for a Down pregnancy. In Britain, therefore, the efficacy of screening programs is measured relative to the *expected* number of Down babies given the ages of the women having children that year. We are seeing an ever-decreasing proportion of Down babies; in 1994, for instance, screening programs reduced the incidence of Down by about 40 percent.

Trisomies can also occur for other chromosomes, but these result in abnormalities so severe that the pregnancies abort spontaneously in all cases except trisomies of chromosomes 13 and 18. But children with trisomy 13 seldom live more than a few weeks, and those with trisomy 18 usually die before their first birthday. Chromosomal abnormalities, trisomies included, are probably very common. While many are lethal – a current estimate is that as many as 30 percent of conceptions end in spontaneous abortion, and in about half of these there is some form of chromosomal aberration – some have little or no effect. Alterations may be far less drastic than the loss or gain of an entire chromosome, involving the rearrangement of segments within a chromosome or the transfer of part of one chromosome to another. If there has been a net loss or gain of genetic material, then, as in the case of a whole extra chromosome, the resulting imbalance will usually prove deleterious. Unfortunately, standard cytological analysis of fetal chromosomes can detect only gross imbalances, and yet even minor ones can have disastrous effects.

After struggling to become pregnant for the first time at thirty-seven, Kathleen McAuliffe was relieved to learn that only two chromosome 21s had shown up in her amniocentesis. But what she had not realized was that the test could reveal other chromosomal abnormalities as well. The cytogeneticist had spotted an inversion in the fetus's chromosome 2: it was as though a segment had been popped out of the chromosome, flipped, and reinserted the other way around. The information was not accompanied by any useful advice: there was a chance that the inversion might create a problem – it might, for example, have resulted in a genetic imbalance – but then again it might have no effect. One way to find out more was to look at McAuliffe's own second chromosome, and that of her husband. If either parent had the inversion (i.e., if it was not a spontaneous alteration in their child), one could infer it would have little or no impact since both parents were normal. But neither McAuliffe nor her husband had an inverted chromosome 2, implying that it had arisen de novo in the sperm or egg. What would the inversion do to the baby? McAuliffe suddenly found herself confronting a life or death decision. After agonizing at length, she decided that the uncertainty was too great and she terminated the pregnancy. Despite a specific request that she not be informed of the autopsy results – she was sad and guilt-ridden over the loss of her fetus – by some administrative gaffe the report was sent to her home and she discovered that the fetus had indeed been profoundly abnormal. But this was cold comfort, and McAuliffe still keeps the ultrasound image tucked away in a drawer. Happily, subsequent pregnancies have met with no such complications, and McAuliffe is now blessed with two young children in, as she puts it, "ear-piercing good health."

Genetic knowledge creates ethical dilemmas. McAuliffe had never been warned that her amniocentesis might detect problems other than trisomy 21; perhaps the cytogeneticist overstepped the bounds of duty and should have reported only the results for the test that had been ordered. Certainly there would have been no choice had the clinician used the FISH method, which reveals *only* the number of 21s present. As it grows more sophisticated, genetic testing becomes a Pandora's box, its consequences going far beyond the original issues motivating the test, sometimes spreading to lives beyond those of the tested individuals. Nowhere is this more evident than in genetic testing conducted among families with histories of an inherited condition like DMD, Huntington disease, or cystic fibrosis. In these cases, diagnosis is carried out not by a cytogeneticist but by a molecular biologist, who analyzes not chunks of chromosomes but specified stretches of DNA. DNA is extracted from a sample of tissue, which is obtained from a fetus by amniocentesis, or from a child or adult by taking blood or harvesting cheek cells from inside the mouth with the scrape of a spatula. These days tests usually involve PCR amplification of the critical region – the suspect gene – from the DNA sample, followed by sequence analysis to determine whether or not it carries the mutation. And the test results for any individual may tell us something about the genetic status of his or her relatives.

Let's take, for example, a test for Huntington disease. In a recent case, a man in his twenties came into a genetics clinic to request that he be tested for Huntington. His paternal grandfather had died of the disease, and his father, in his forties, had decided not to be tested, preferring, like Nancy Wexler, to live with 50-50 uncertainty over knowing for sure. Because Huntington strikes relatively late in life, it was possible that the father was in fact carrying the mutation even though symptoms had not yet appeared. The young man knew that the probability of his having the mutation – and therefore the disease in his future – was 1 in 4.* But he wanted to know for sure. The problem is this: if he found out that he did indeed have the mutation, then he must have received it from his father, meaning that his father, too, would definitely develop the disease. The son's quest for genetic

*There is a 1 in 2 chance that the father had received the mutation from the grandfather, and then, if the father has it, another 1 in 2 chance of his having passed it on to the son. The probability for the son is the product of these independent events, a 1 in 4 or 25 percent chance.

knowledge would directly contravene the father's desire to avoid it. A family feud developed, and in the end, only intervention by the young man's mother prevented him from proceeding with the test. His desire to know, she argued, surely paled beside her husband's right to be shielded from what may be a devastating death sentence. This dramatic example illustrates the difference between genetic diagnosis and any other kind. What I might learn about my genes has implications for my biological relatives, whether they care to know or not.

Sometimes the implications may not bear on the present generation but rather on generations to come. Fragile X is the commonest form of inherited mental retardation. (Down syndrome is more frequent, but, occurring spontaneously, it is not usually inherited.) In addition to a low IQ, symptoms typically include a notably long face with an outsize jaw and ears, and a hyperactive, occasionally irritable temperament. Like DMD, it is a sex-linked disorder (the gene responsible is on the X chromosome), but, unlike DMD, it affects females as well as males. One normal copy of the gene is evidently not enough to render the effect of a mutated one negligible; still, women tend to suffer less severe symptoms, and their incidence is 1 in 8,300, compared with 1 in 5,000 for males. Fragile X is caused by a mutation similar to the one responsible for Huntington disease: a DNA triplet, CGG, is repeated over and over again. Normal individuals have about 30 while carriers of fragile X have at least 50 and sometimes as many as 90. For reasons we do not fully understand, the number of repeats tends to increase with each generation; and once there are about 230 CGG triplets, the gene can no longer make mRNA and therefore ceases to function. The condition gets its name from a discernible structural weakness in the X chromosome caused by all these repeats.

As the number of repeats increases from one generation to the next, so the severity of the condition increases, and the age of onset decreases in each family line. The latest descendants in a fragile X pedigree have the largest numbers of repeats and are typically affected earlier in life and more severely than those from whom they inherited the mutation. Geneticists may therefore identify individuals carrying a "premutation" – too few repeats to cause problems at present, but sufficient to result in fragile X in subsequent generations, given the likely expansion next time around. We do not yet know exactly what the protein produced by the affected gene does, but it seems to bind to messenger RNA molecules in the connections – synapses – between nerve cells.

Like ongoing research into Huntington, DMD, and many other genetic afflictions, studies of fragile X have been galvanized by those most directly affected: the families and loved ones of sufferers. FRAXA, the Fragile X Association, has been hugely effective in raising money and in inducing Congress to support fragile X research. Though some scientists may cynically view such groups merely as agencies that offer individuals in dire straits the comforting illusion that they are not entirely powerless, experience shows that dedicated, resourceful, and, above all, motivated organizations like FRAXA sometimes do hold the key to cracking these diseases against the long odds. To those who take the biggest gambles – financial and scientific – sometimes, with luck, go the biggest rewards.

Many women reading this may be asking themselves the question: Why wasn't I tested during my pregnancy for cystic fibrosis, or fragile X, or DMD? Sadly, some of them may even have children with one of these afflictions. In the wake of the genetic revolution that has transformed medical technology, one notes a singularly depressing and senseless fact: the uncoupling of scientific progress and patient care. Actually, it might be more accurate to say that due attention never was paid to coupling them properly in the first instance. In any event, many women are simply not informed of their options, and tests now readily available are hugely underused.

As a head of the Human Genome Project, I made sure to fund efforts to promote understanding of how the knowledge that would soon be pouring out of the sequencing machines would affect, for good or ill, the lives of countless people. Having set aside initially 3 percent of our total budget (and later 5 percent) for this purpose, I appointed Nancy Wexler, the Huntington expert, to run a panel called ELSI charged with exploring the ethical, legal, and social implications of our research. One of ELSI's major initiatives was a series of pilot studies of genetic screening. At a time when every newborn was screened for PKU, it was necessary to ask whether medicine could responsibly fail to offer at least the option of screening for cystic fibrosis, DMD, fragile X, and every other grave human ill that was within the power of science to predict. That was in the early nineties. Today, things have scarcely advanced beyond the pilot stage: small-scale studies are still carried out here or there. The reasons for this paralysis are varied, ranging from dollars-and-cents practicality to profound philosophical disagreements about the essence of human life and dignity. In short, they encompass the gamut of social phenomena, from jockeying for funds to collective soul-searching, that have attended the genetic revolution.

Testing for DMD and Huntington is ordinarily done only in families that already have an affected member. The rationale for the limitation is that these disorders are rare and the tests are costly. This social calculus is debatable, but the same reasoning does not hold in the case of cystic fibrosis, for which testing is nevertheless also limited. Cystic fibrosis, remember, affects about 1 in 2,500 people, making it one of the most prevalent genetic disorders. It is especially common among people of northern European descent. The high incidence seems all the more remarkable when we consider that the underlying defect, which occurs in a gene on chromosome 7, follows a recessive pattern of inheritance, meaning that in order to develop cystic fibrosis one must receive two mutated copies. People with just one are unaffected but being carriers they can pass the mutation on to their children. Epidemiological surveys and calculations tell us that 1 in 25 Americans of European ancestry are carriers.

One difficulty with cystic fibrosis testing is technical, having to do with variability in the underlying defect. One specific form of mutation accounts for about 70 percent of cases: a deletion called ΔF508, which eliminates three bases, CTT.* If just a few other mutations accounted for the remaining 30 percent, then general population screening for CF carriers would not be impractical. But most of the other causative mutations occur in just a single family line and more than a thousand different CF-causing mutations have been discovered to date. What does this mean for population screening? In practice any test could screen for at most twenty-five different mutations, but those twenty-five most common forms would still account for only about 85 percent of all cases. As a result, we'd be missing about one in six mutations – not a very good batting average for a diagnostic. Now, say we have a couple, both of whom have tested negative for CF mutations according to our highly imperfect screen. We could hardly tell them with any confidence that there was no danger of their producing a child with cystic fibrosis. Why bother, the argument goes, with an inconclusive test that typically costs $300?

But, despite the technical difficulties, prenatal screening for cystic

---

*The 70 percent figure applies to people of northern European ancestry, the population in which cystic fibrosis is most common. However, ΔF508 only accounts for around 35 percent of CF mutations in both the African American and Ashkenazi Jewish populations. Ancestral differences like these complicate the design of screening programs.

fibrosis can still identify a large proportion of all affected fetuses. Why isn't it more widely adopted? Paradoxically, cystic fibrosis advocacy groups have played a major part in limiting CF testing to families already affected. Broader testing, it is feared, would divert limited resources away from the ultimate goal of finding a cure. That concern is understandable, particularly at this moment. An estimated thirty thousand Americans have cystic fibrosis. Treatment advances have already extended life expectancies considerably, and it is conceivable that a cure will be in hand before long. Having said that, it would be irresponsible to suggest that a cure is around the corner: babies born today with cystic fibrosis still face the prospect of a lifelong struggle against a debilitating disease. Though curing cystic fibrosis is definitely a top priority, there should still be room to permit an expectant mother to have access, if she wants, to testing. Then, fully informed about the status of her fetus, she has the freedom to make whatever choices she sees fit.

Broader testing is also resisted for less material reasons. There are those who view screening as an admission of defeat, the wrong manner of solution. Advocacy groups are in the business of ensuring that people with the disease feel that they belong to a community and are valued by society – how does one reconcile that mission with testing, which, to put it in bluntest possible terms, is promoting the abortion of affected fetuses?

Cystic fibrosis advocates are anxious that people with CF do not become stigmatized, and they worry that indirectly testing does just that. In fact, there is an unfortunate precedent in the history of genetic testing that haunts all patient advocacy groups. Long before the advent of DNA screening, one of the earliest diagnostics for a genetic disease was developed to detect sickle-cell anemia, which in the United States affects primarily African Americans. As we saw in chapter 3, those with two copies of the mutated "sickle" hemoglobin gene will suffer painful, debilitating symptoms, while those with only a single copy – carriers – will notice little effect.

Following the development of simple blood tests in the 1960s, screening programs were hastily established across the country. Despite the best of intentions they did more harm than good. Screeners generally failed to counsel test subjects properly as to the significance of the test or its results. Many diagnosed as carriers mistakenly assumed that they themselves had the disease; some were even denied jobs or health insurance on the basis of the test; and couples who were at risk of producing sickle-cell children were advised rather heavy-handedly to think twice. The tests – some programs were mandatory – were coercive in effect, suggesting to some that the United States was entering a renaissance of racist eugenics, stigmatizing all who tested positive. The sad irony is that, from a purely medical point of view, the campaign was in fact sensibly conceived: despite advances in treatment, sickle-cell anemia remains a chronic, painful condition. Screening is the best remedy for a disease that is far more easily avoided than confronted, but the first mechanisms designed to eradicate it were so badly managed as to have rightly angered many intended beneficiaries.

Fortunately in 1972 new federal guidelines redesigned the sickle-cell screening program, allowing it to be effective without the widespread concern raised by the initial effort. Harder to repair has been the trust of advocacy groups for genetic diseases in general; the experience of the community of those affected by sickle-cell disease has left them ever leery of screening programs, and the fear of stigma persists, sometimes, alas, at the expense of better public health.

In so many ways, genetic testing, despite its incontrovertible usefulness, proves to be flypaper for controversy. Randi Hagerman, then at Denver Children's Hospital, decided to apply a DNA test for fragile X to children in special education classes in Denver. The reasoning was simple: children whose learning was impaired by that disorder would be better served if they were identified, whereupon their schooling could be tailored to their particular needs. Of 439 students tested, 5 with fragile X mutations were discovered. (A more extensive survey of schools in Holland had found 11 previously undiagnosed cases of fragile X in a group of 1,531 students.)

Perhaps the most interesting part of the Denver study was the response of the parents and guardians to Hagerman's offer. Most recognized the benefit of a diagnosis, both for the potential to improve their child's education and for the identification of the presence of the disorder in the family line. But fully a third refused the test, citing either a certainty that *their* children did not have fragile X or a concern that their children might find the test too stressful. Hagerman has been criticized for her efforts; it was a field day for those who insist upon seeing the menace of a totalitarian genetic future in every attempt at harnessing DNA to address a social problem.

The issue is indeed social as well as personal. The high incidence of the fragile X premutation – it is on perhaps as many as 1 in 200 X chromosomes – may warrant population screening. In the United States, it is estimated that just one reasonably severe case will cost, over a lifetime of nonwork and institutionalization, some $2 million in current dollar value. The ever-increasing challenge of providing affordable health care should itself suggest a potent argument for giving every mother the opportunity to be tested. The logic of hard realities is not lost on smaller countries, where the margins for policy error are not as great. A pilot study in Israel screened 14,334 women; 207 were found to have a premutation. Prenatal diagnosis was made available upon request, identifying five fetuses with extended CGG repeats. The fate of those pregnancies was rightly the choice of the expectant mother: a free society should no more require a woman to abort a fetus with a genetic disorder than it should require her to carry it to term. But not every woman is prepared to raise a disabled child, nor is every woman prepared to terminate a pregnancy on account of the child's foreseeable quality of life. Whatever the individual choice, however, the fact remains that screening can only reduce the incidence of affliction, and that is an unambiguous social good.

Despite the frustrating reluctance to take advantage of genetic screening on a broad scale, the short history of the practice hasn't been entirely one of small-scale pilot studies and damning controversy. There are some happy and illuminating stories to tell about the triumphs of screening programs for genetic disorders in high-risk populations.

Hemoglobinopathies are diseases caused by some malfunction in the hemoglobin molecule. Including the various thalassemias and sickle-cell disease, they are thought to make up the most common class of genetic disorders, with about 4.5 percent of the world's population carrying a mutation for one of them. As we have seen, the sickle-cell gene carried with it antimalarial properties, and so was promoted by natural selection in areas where malaria was prevalent. As a result, the mutation was originally at high frequency only in such parts of the world. The same adaptive advantage accounts for the similar distribution pattern of other hemoglobinopathies as well. Medicine has for some time understood that certain mutations therefore tend to be much more common in some ethnic groups than others, wherever the individuals may now find themselves.

Among the population of Greek Cypriot immigrants in London, thalassemia carriers represent a remarkable 17 percent. In its severe form the condition is the most pernicious of the hemoglobinopathies, resulting in misshapen and sometimes nucleated red blood cells that cause enlargement of the liver and spleen, often leading to death before adulthood. A systematic screening program begun in 1974 by Bernadette Modell of the Royal Free Medical School was welcomed enthusiastically by London's Cypriots, who were only too well aware of the seriousness of the disorder that had long blighted their community. A similar program in Sardinia, also begun in 1974, has dramatically reduced the incidence of thalassemia from 1 in 250 to 1 in 4,000.

Ashkenazi Jews are another group with a bitter awareness of what a deadly mutation can do to a genetically isolated population. Tay-Sachs (TS) is a ghastly disease 100 times more common in this group than in most non-Jewish ones. TS babies are born apparently healthy, but gradually their development slows and they begin to go blind. By about two, they are stricken with seizures. Deterioration continues until they die usually by the age of four, blind and paralyzed. Unlike hemo-globinopathies, whose relative commonness in certain populations can usually be explained by the concomitant adaptive protection against malaria, the high frequency of TS among the Ashkenazim remains a mystery. Perhaps a genetic bottleneck is to blame: the mutation may have been present among the relatively small segment who branched off to become the Ashkenazim during the second Diaspora. A similar phenomenon might also account for why the mutation is also anomalously common among the French Canadians of southwest Quebec as well as the Cajuns in Louisiana: the chance presence of an unfortunate mutation in the small founding populations. An alternative explanation holds that being a carrier of this recessive gene (having one copy of the TS mutation) may confer some resistance to tuberculosis, an advantage perhaps for European Jews who historically tended to live in densely populated urban centers.

The cause of Tay-Sachs was discovered in 1968 when it was recognized that the red blood cells of patients were overloaded with ganglioside GM2. This chemical is an essential component of the cell membrane, and in normal individuals any excess is broken down into related compounds by a key enzyme, which is lacking in TS sufferers. In 1985, Rachel Myerowitz and her colleagues at NIH isolated the gene coding for that enzyme and showed that it was indeed mutated in Tay-Sachs patients.

Thereafter we had the basis for a foolproof prenatal test and a well-defined target population – conditions tailor-made for the implementation of a successful screening program. But prenatal screening effectively offers only one remedy in the event of a positive diagnosis: abortion, which, at least among the observant Orthodox segment of the Ashkenazim, is forbidden. Fortunately, it is also possible to screen prospective parents and so the solution morally acceptable to the devout was a program aimed at couples. Rabbi Yosef Eckstein of New York saw four of his ten children die from Tay-Sachs. In 1985, he established Dor Yeshorim, the "generation of the righteous," a program to carry out TS testing in the local Orthodox Jewish community. Young people are encouraged to take advantage of free-testing days at high schools and colleges. An unusual aspect of this program is its extreme confidentiality: not even those tested are informed whether they are carriers; instead, each is given a code number. Later, when two people are contemplating marriage, they each phone Dor Yeshorim and give their numbers. Only in the event of both being carriers is the status of either partner revealed, together with an offer of counseling. This disclosure on a need-to-know basis is intended to avoid stigmatization of carriers, while still countering the threat of Tay-Sachs.

To date, the Dor Yeshorim program has tested more than seventy thousand individuals and detected more than a hundred couples at risk. Steadily reducing the incidence of Tay-Sachs, it would appear an unqualified success, yet there are those within the Jewish community who find fault with it. Some see coercion in the program's call for all young people to be tested, and intimidation in its strong recommendation that some individuals reconsider their decision to marry. Opponents have labeled Rabbi Eckstein's crusade "eugenics" (a word whose resonance is nowhere more painful than in the Jewish community), but such demagoguery hardly alters the central fact of the matter: the program clearly enjoys strong support within the community it serves, a community that understands the horrors of Tay-Sachs. Indeed, Dor Yeshorim has demonstrated that a screening program can be both effective and culturally responsive, working even in a situation where social mores and religious precepts seem to be at odds with genetic testing in principle.

Prenatal screening offers a stark choice for any woman carrying a fetus that has tested positive for a genetic disorder: to terminate or not to terminate the pregnancy. The fact that amniocentesis cannot be performed until a fetus is at least fifteen weeks old makes the option of termination only more traumatic. At this stage an abortion does not eliminate a featureless ball of cells, but a tiny being – real enough that a parental bond may have already formed with the developing fetus, thanks to the power of ultrasound imaging. Most parents – at least those who do not oppose all abortion on principle – would infinitely prefer to make hard choices presented by genetic testing at an earlier stage of development. Such was the inspiration for the invention of preimplantation diagnosis.

Robert Winston at the Hammersmith Hospital in London is a leading gynecological microsurgeon, an expert in such procedures as the correction of Fallopian tube defects that prevent a woman from conceiving. He has also become one of British television's leading popularizers of science and biomedical research and even finds time, as Lord Winston of Hammersmith, to sit in Parliament advising the government on such matters. By combining two state-of-the-art technologies – in vitro fertilization (IVF)* and PCR-based DNA diagnosis – Winston pioneered a method for checking the genetic status of an embryo before it is implanted in a woman's uterus and begins to develop. After in vitro fertilization, the several conceptuses are grown in the laboratory until each fertilized egg has divided three or four times to produce a ball of eight to sixteen cells. One or two cells are carefully removed from each for DNA extraction and then PCR is used to amplify the relevant sequences to determine in each case whether or not a mutation is present. It is the astonishing capacity of PCR to amplify even the tiniest quantities of target DNA that makes possible this method of ultra-early diagnosis. Parents are then free to implant only those embryos that test negative for genetic disease.

The first preimplantation tests performed in 1989 screened for the sex of the fetus – important information when the risk is for a sex-linked disorder such as DMD. A mother who is a carrier may select only female embryos on the premise that they will not be affected by the disorder, though they may be carriers. It was Winston's colleague, Alan Handyside, and others who subsequently extended pre-implantation diagnosis beyond simple sex determination to the detection of specific mutations: in 1992, they first applied the technique to screen for cystic fibrosis, which is not sex-linked.

As we have seen, despite being sex-linked, fragile X can affect both males and females. The disorder is therefore a natural target for gene-

*IVF is a method of assisted reproduction in which sperm and egg are fused in a laboratory dish. The resulting embryo—rather ominously called a "test-tube baby" in the early days of the technology—is then transferred to the uterus to develop naturally.

specific preimplantation diagnosis, but it still took impassioned parents familiar with the difficulties of raising a fragile X child to mobilize doctors to do it. Debbie Stevenson, a CNBC television news reporter, has a son, Taylor, whose fragile X was only diagnosed after the birth of a second son, James. Though James fortunately beat the 50-50 odds of being affected, the Stevensons were unwilling to trust their third child to fate. They decided to seek preimplantation diagnosis: "Some people think it's unethical to select for healthy embryos," says Debbie Stevenson, "but I think it's better than having to make the wrenching decision about whether to terminate or continue a pregnancy after learning that your baby has a significant disorder." In 2000, at the end of the family's frustrating yearlong search for a lab willing to carry out the procedure, the newest member of the Stevenson family was conceived and, just days later, tested for fragile X. Like James, Samantha is free of Taylor's debilitating disorder.

In our culture human reproductive biology seems an inexhaustible source of controversy, and a procedure involving the manipulation of human embryos for any purpose is sure to become a lightning rod. Preimplantation diagnosis has been no exception. Ethical considerations aside, however, the procedure still has two major drawbacks: it requires a huge commitment on the part of the couple undertaking it, and, like all forms of IVF, it is very expensive. But the method is so powerful in principle, and so much less traumatic than abortion, one can only hope that with time we shall see improved techniques, and with them diminished cost – the usual pattern with developing technology. Preimplantation diagnosis has the potential to become an extremely important weapon in our war on genetic disease.

The disorders we have discussed so far are all "simple" in the genetic sense: they are caused by a mutation in a single gene, and environment has no bearing on whether or not you will get one of these diseases. The situation is rather more complicated in the case of illnesses like cancer, which, as we've seen, may be triggered by a combination of hereditary and environmental influences. But even with cancer, there are some individual genes that have a major effect, regardless of the environment. Though BRCA1, one of the genes implicated in breast cancer, only accounts for about 5 percent of all cases, women with mutations in the gene have an estimated 90 percent chance of developing the disease by the age of sixty.

In the early 1990s, Francis Collins, then at the University of Michigan, joined forces with Mary-Claire King at UC Berkeley in the hunt for BRCA1. They took the standard approach: collecting families, preparing DNA samples, and testing markers, all with a view to homing in on the gene. One family of more than fifty members included multiple cases of breast cancer – a clear instance of an inherited predisposition to the disease. In September 1992, one member of that family – I'll call her Anne – revealed to Barbara Weber, an associate of Collins's, that she had scheduled a bilateral mastectomy the following week, even though there was no sign that she had cancer. Anne had decided that she could no longer tolerate the uncertainty, the question mark hanging over her future, and preferred to take this drastic preemptive step. Weber, however, had concluded from the DNA analysis that Anne was actually not in particular danger: her risk of breast cancer was no greater than that of a woman without a family history of the disease. But this inference was made in the context of a research project and it had been agreed long in advance that as a rule such preliminary data should not be used for clinical diagnosis.

Weber and Collins, however, decided that Anne's plight

outweighed the rule book: they informed Anne that her risk was low, and with great relief she canceled the surgery. But having disclosed their findings to one member of the family, the researchers felt obliged to offer the same benefit to others who asked for it; and so Weber and Collins set up an ad hoc breast cancer genetic counseling program. One family member who also proved to be at no special risk had already undergone a prophylactic bilateral mastectomy five years before. She received the belated diagnosis philosophically: the surgery, she figured, had bought her five years' peace of mind. But had she tested positive for the mutation, the radical course might in fact have bought her more than peace of mind. For years, prophylactic mastectomy had been recommended by clinicians, even though no surgery can feasibly remove all the breast tissue and there were no solid data showing that the measure was saving lives. Today, however, there is proof that the extreme approach does indeed reduce mortality rates among women at high risk; in one group of 639 who had the surgery, only 2, rather than the 20 to 40 statistically expected, actually died of breast cancer. Similarly, removal of the ovaries before age forty (but after a woman has finished having children) reduces the risk of both ovarian and breast cancers. Genetic analysis can give women the power to make decisions that can literally make the difference between life and death.

But the keyhole view into the future that DNA analysis permits can also create opportunities to defeat breast cancer by less extreme means, as another story from the Michigan study reveals. A cousin of Anne's was told that she was in all likelihood carrying the BRCA1 mutation that was devastating her family. Since she had not had a mammogram in years – a fear-based negligence ironically not uncommon in high-risk families – she panicked. Weber scheduled one for later that day and a tiny incipient tumor was found; it was easily removed but would almost certainly have been missed in a routine examination. Self-examination and regular mammography have doubtless saved many lives, but the campaign to universalize these procedures may have had the unintended consequence of creating a false sense of security in some cases. Screening for genetic risk allows us to find those individuals whose imaging examinations merit extra-high scrutiny. Greater risk demands greater surveillance. And in the long run more needles will be found the smaller we make the haystack.

Nancy Wexler, as a member of a Huntington family, and Anne, as a member of a breast cancer family, are both part of a new generation for whom newly available screening tests can provide glimpses of genetic destiny. And as we learn more about the genetic basis of relatively common adult afflictions, from diabetes to heart disease, the biological crystal ball will become ever more powerful, telling the genetic fortunes relevant to us all.

In the past decade, few diseases have struck terror in as many hearts as Alzheimer, which each year draws ever greater numbers into its grip of awful mental and physical debilitation – the disease affects more than 4 million Americans. Family and friends of sufferers may notice first some minor lapses of memory – trouble recalling recent events or finding the right word – which they might hopefully attribute to the ordinary effects of getting on in years. The afflicted may then begin to show evidence of mood swings, also not altogether unnatural among the elderly. But as the disease progresses, the symptoms become more pronounced and unmistakable; the memory loss soon grows unnaturally severe, making the familiar challenges at work and even simple household tasks unmanageable. Speech becomes more labored; sentences go unfinished as the victim loses the train of thought. And the person's awareness of these changes may lead to depression, which in turn intensifies the effect of other increasingly distressing changes in personality. Advanced Alzheimer patients do not know who or where they are; they cannot recognize even their closest kin. With the inexorable erosion of memory and personality, their very essence as individuals is gradually destroyed.

Alzheimer typically first appears at around age sixty, but a rarer form, accounting for about 5 percent of all cases, strikes individuals in their forties. This early-onset form of the disease puts families through the same kind of hell as Huntington disease does, seizing its victims in the prime of life and gradually, relentlessly destroying them. One family with multiple affected members over several generations was described as having been struck by its own "biological Holocaust." Following the argument first advanced by Mary-Claire King in her breakthrough study of breast cancer, that any early-onset version of a disease is likelier than the ordinary form to have a clear genetic basis, most initial Alzheimer research focused on the early-onset form. By 1995 three genes had been found, all of them involved in some way with the processing of amyloid protein deposits, whose accumulation in the brains of patients was noted as early as 1906, in Dr. Alois Alzheimer's original description of the disease. Early-onset Alzheimer is, then, clearly inherited. But what of the more common variety?

Allen Roses at Duke University preferred to ignore the wisdom of the majority, and set out straightaway to tackle the much more familiar late-onset form, which only occasionally runs in families. Ronald Reagan, for instance, who announced his affliction in 1994, lost his brother Neil to late-onset Alzheimer two years later. Their mother had died of it as well.

With training as a neurologist and a background in muscle disorders like DMD, Roses began his search in 1984. His claim in 1990 that a gene on chromosome 19 appeared to correlate with the disease was met by skepticism. Nothing, however, gives Roses more pleasure than an opportunity to prove everyone else wrong. Two years later, he had actually identified the critical gene. It turned out to code for apolipoprotein E (APOE), a protein involved in processing cholesterol. The gene comes in three forms (alleles), APOEε2, APOEε3, and APOEε4, but it was APOEε4 that proved the crucial one: a single copy of that variant increased fourfold one's risk of developing Alzheimer. And individuals with two copies were at a risk ten times greater than that of persons with no APOEε4 allele. Roses found that 55 percent of those with two copies of APOEε4 will have developed Alzheimer by age eighty. Could this correlation be the basis for a genetic test? Probably not. Despite being correlated with the disease, the APOEε4 allele is common and is not a good enough predictor of Alzheimer for testing purposes: though their risk is higher, plenty of people with two APOEε4 alleles never develop Alzheimer. But the use of APOEε4 screens in conjunction with clinical evaluations does improve the accuracy of Alzheimer diagnosis. And perhaps once we understand the correlation in causal terms, the genetic analysis can be refined. Recent work that induced Alzheimer-like symptoms in mice has suggested APOE is involved in the metabolism of the protein that causes nerve cell death in human Alzheimer sufferers.

What of treatment? Most genetic diseases present us with much the same heartrending frustration that comes with Huntington: we know enough to diagnose them, perhaps to evade them, but not to treat them. Happily, there are a few cases in which our genetic understanding has taken us the rest of the way, providing therapies that work. Unfortunately, few of these remedies are as simple and effective as that for PKU, from which a normal life can be retrieved through a few dietary restrictions.

Too often genetic disorders result in the cell-by-cell decimation of particular tissues: muscles in DMD, nerve cells in Huntington and Alzheimer. There is no quick fix to this kind of insidious decay. But though these are early days yet, I think there is a realistic chance that we will eventually be able to treat diseases like these using stem cells. Most cells in the body are capable only of reproducing themselves – a liver cell, for instance, produces only liver cells – but stem cells can generate a variety of specialized cell types. In the simplest case, a newly fertilized egg – the stem cell with maximum potential – will ultimately give rise to every one of the 216 recognized human cell types. Stem cells are accordingly most readily derived from embryos; they can also be found in adults, but such cells tend to lack that embryonic ability to differentiate into *any* cell type. We are beginning to learn how to induce stem cells to produce particular cell types, and someday, I hope, we will be able to replace the lost brain cells in people with Huntington and Alzheimer with new healthy cells. But I caution that we have a long way to go before we have a thorough understanding of the molecular triggers that cause a cell to develop in one direction rather than another. It will take ten years or so of grappling with this fundamental problem in developmental biology before we are in a position to explore properly the therapeutic value of stem cells. I think it would be a tragedy for science, and for all people who may eventually benefit from stem-cell therapy, if research is hindered by religious considerations. Polls consistently show that the majority of Americans favor research using embryonic stem cells, and yet politicians continue to pander to the outspoken religious minority that opposes it. The result is restrictive legislation in the United States that is hampering efforts to develop this potentially valuable technology.

For now, treating genetic disorders does not extend to the whole-sale replacement of cells à la stem-cell therapy, but it may involve replacement of a missing protein. Striking 1 in 40,000 individuals, Gaucher disease is a rare condition resulting from a mutation in the gene for glucocerebrosidase, an enzyme that helps break down a particular kind of fat molecule, which otherwise accumulates harmfully in the body's cells. The disorder can be devastating, with a suite of symptoms including bone pain and anemia. Initial attempts to supply the missing enzyme directly were made as early as 1974. The results were promising but the logistics were nightmarish: the replacement enzyme had to be extracted from human placentas, and twenty thousand placentas were needed to furnish a year's supply for a single patient. A big breakthrough came in the early 1990s, when researchers synthesized a modified form of the enzyme, one that was taken up more efficiently by the cells that most needed it. In 1994, Genzyme, a biotech company, started to produce the modified form using recombinant methods. Treatment of Gaucher does not combat the genetic root of the disorder but rather the effect of the mutation: it provides the patient with the vital protein that the faulty gene cannot.

Righting the genetic abnormality via this biochemical route is evidently feasible and effective. But even given the remarkable efficiency of recombinant methods, the treatment is expensive – $175,000 per year – and the need for continual infusions imposes a burden on patients. Naturally, therefore, geneticists have long dreamed of a practical way to fix the cause of the problem rather than compensate for its effects. The ideal treatment for genetic disorders would be a form of genetic alteration, a correction of the genes that cause the problem. And the benefit of such gene therapy would last the patient's whole life; once fixed, it's fixed for good. There are, at least in principle, two approaches: somatic gene therapy, by which we change the genes within a patient's body cells; or germ-line therapy, by which we alter the genes in a patient's sperm or egg cells, preventing the transmission of the harmful mutation to the next generation.

Such solutions to the ravages of genetic defects might be obvious, but the idea of gene therapy has not met with the warmest of professional or public receptions. Such reactions are not altogether surprising: a culture wary of genetic modification in a corn plant might be expected to be averse to transgenic people – GM humans, if you prefer – however great the potential benefit. And more vociferous objections are made, also not unexpectedly, to the germ-line approach, because of the risk of causing genetic damage when manipulating the DNA. In somatic gene therapy, such damage may be limited in its effect; in germ-line therapy the possibility exists of accidentally producing impaired people. Even its proponents – of whom I count myself one – would never suggest that such a procedure should be carried out until our techniques are good enough for us to be confident that we will not inadvertently cause damage. Many scientists are convinced, though, that we should never attempt germ-line gene therapy. Whether based in ethics or unfounded fears of the unknown, such arguments are ultimately not compelling in my judgment. Germ-line therapy is in principle simply putting right what chance has put horribly wrong. But for now the controversy is academic: germ-line therapy is still way beyond our technical powers. Until it is in reach, we should concentrate our efforts on making somatic gene therapy a powerful tool in its own right.

The first apparently successful gene therapy was carried out by French Anderson, Michael Blaese, and Ken Culver at the National Institutes of Health in 1990. They chose a very rare disorder called adenosine deaminase deficiency (ADA), in which the lack of an enzyme disables the immune system, leaving one as defenseless as

David Vetter, the boy in the plastic bubble. The experimental subjects were two young girls, four-year-old Ashanti DeSilva and nine-year-old Cindy Cutshall.

How do you shoot a new gene into a patient? At that time, retroviruses suggested themselves as the logical weapon of choice. In general, viruses are efficient genetic vectors; they make their living by injecting DNA into other cells. Retroviruses, a special group, have RNA, rather than DNA, as their genetic material. But while most viruses infect a cell, reproduce, and then kill the host cell as the "daughter" viruses escape to infect others, retroviruses are typically kinder and gentler, at least to the host cell: new viral copies are dispatched without destroying it. This does not necessarily mean that a retrovirus is any easier on the host organism; sometimes it is quite the contrary, as demonstrated by the effects of HIV, perhaps the best-known retrovirus. But it does mean that the viral genes – and any extra gene the virus may be induced to ferry – will become a permanent part of the undestroyed cell's genome. Genetic engineering has produced retroviruses that are as safe as possible for gene therapy; stripped of all the viral genes that aren't essential for invading the host cell's genome – and their means for accomplishing this purpose are formidable – retroviruses become the ideal gene vector.

But we are still left with the problem of how to target only the cells affected by the mutation, the ones that need the replacement gene. Today this remains the greatest challenge facing gene therapy: how do you get the good gene into muscle cells to treat DMD, lung cells to treat cystic fibrosis, or brain cells to treat Huntington disease? The choice of the obscure disease ADA was therefore a very sensible one for the first gene therapy trial: the target cells for ADA are readily available – immune-system cells circulating in the blood. Anderson's team was able to extract millions and millions of immune cells from the girls' blood and grow them in petri dishes, where they could be infected with a retrovirus carrying a functional copy of the gene. Once the cell's natural DNA had incorporated the viral genome carrying the replacement gene, the cells were ready to be fed back into the patients' blood.

In September 1990, Ashanti DeSilva was the first to undergo the procedure; Cindy Cutshall's therapy followed four months later. Each received infusions of genetically doctored immune cells every few months. At the same time, each was continued on the non-gene therapy of enzyme replacement, the same manner in which Gaucher disease patients are treated, but in lower doses. This precaution was required by the NIH's Human Gene Therapy Subcommittee, which argued reasonably that it was too dangerous to expose the girls to a new therapy without a safety net. The experiment, though not a perfectly controlled one, did seem to work: the immune systems of both girls improved, and they were better able to fight off minor infections. I personally can attest that Cutshall seemed a very healthy eleven-year-old when she and her family visited Cold Spring Harbor in 1992. Eleven years later, however, the results are not entirely conclusive. DeSilva's immune system is now approaching normal function, but only about one-quarter of her T cells are derived from the gene therapy. Cutshall's blood contains an even smaller proportion of gene-therapy T cells, though her immune system is working well. It is, however, difficult to say exactly how much of the girls' improvement is due to the gene therapy and how much to the continuing enzyme treatment. The result is therefore too ambiguous to be understood without reservation as a clear gene-therapy success.

The Cutshall/DeSilva trials were not the first time the NIH had thrown its weight around in the world of gene therapy. In fact, the Human Gene Therapy Subcommittee was formed at the NIH in 1980 in response to the first gene-therapy experiment ever performed. The trial was a failure and stirred up such a controversy that the government almost moved to strangle the newborn enterprise in its cradle. By all accounts, the man at the center of the storm, Martin Cline, was a clever, ambitious clinician, devoted to the relief of his patients' woes. His special interest was beta-thalassemia, the hemoglobinopathy Bernadette Modell had screened for among London's Cypriot community. After successful animal experiments, Cline applied to the review board of the University of California Los Angeles, where he worked, for permission to try the gene therapy on humans using nonrecombinant DNA. While the application was still being reviewed, an overzealous Cline had arranged to treat two women outside the United States, one in Israel and one in Italy, but he used recombinant genes, whose use was still prohibited under NIH guidelines. On returning to Los Angeles, Cline found that his application had been rejected; the review board ruled that more animal data would have to be presented before an attempt with humans could be sanctioned. Cline had broken just about every rule in the book: not only had he proceeded to treat human subjects without authorization, but he had also used an unequivocally prohibited method. Cline suffered the consequences: he lost his federal funding and was forced to resign as chairman of his department. Gene therapy had lost its first practitioner.

The Cline episode was by no means the last time that scientists attempting gene therapy found themselves in hot water with regulators. Tragically, it took the death of a patient in a gene-therapy trial to bring home the sobering message: gene therapy – that complicated cocktail of viruses, growth factors, and patients – is dangerous. But the message was more than that: because there are so many unknowns in the gene-therapy equation, strict oversight of all procedures involving humans is absolutely necessary. Jesse Gelsinger died both because we do not know enough to predict with complete confidence an individual's response to gene therapy and because scientists took inexcusable shortcuts.

In 1999, Gelsinger, an Arizona teenager, heard about an experiment being conducted by James Wilson, director of the Institute for Human Gene Therapy at the University of Pennsylvania. Gelsinger was suffering from ornithine trans-carbamylase deficiency (OTC), a hereditary impairment of the liver's ability to process urea, a natural product of protein metabolism. Untreated, the disease can be lethal, and though, like PKU, it can be managed with simple medication and an appropriate diet, OTC does leave its victims particularly vulnerable to other ailments. The eighteen-year-old Gelsinger had only a mild case, but a childhood brush with death precipitated by his condition emboldened him to volunteer in the hope of helping to find a cure for himself and others like him. The Pennsylvania therapy aimed to use an adenovirus (a member of the group that causes the common cold) as the vector of the corrected gene. But a few hours after the viruses carrying a normal version of the OTC gene had been injected into his liver, Gelsinger developed a fever. A rampant infection followed, accompanied by blood clots and liver hemorrhaging. Three days after the injection, Jesse Gelsinger was dead.

The teenager's death was a shock not only to his family but to the research community as well. A detailed investigation revealed serious procedural lapses. Most glaring perhaps was this: although two patients had shown signs of liver toxicity earlier in the same study, the cases had gone unreported to any regulatory authority and were never disclosed to the volunteers in the study. Had the Gelsingers been informed, Jesse would likely not have been so quick to volunteer, and he might well be alive today. The tragedy dealt a serious blow to the progress of gene therapy. For a time, the FDA halted all such experiments at the university and at several other programs across the country. Bill Frist of Tennessee, the Senate's sole physician, conducted an investigation into reporting procedures in human trials; President Clinton called for improvement in standards of "informed consent," championing the right of experimental subjects to be apprised of all potential risks. If any good has come from Jesse Gelsinger's death, it is that federal oversight of human trials has been tightened.

The gene-therapy community was still reeling from the shock wave caused by Gelsinger's death when heartening news of a success story came from France. The disorder targeted was SCID, the immune deficiency that condemned David Vetter to life in a bubble. Though bone marrow transplants can effect a cure – the recipient of the first transplant, done in 1968, is still healthy today – the success rate is only about 40 percent and even successful transplants frequently lead to grim complications, as in Vetter's sad case. In 2000 a team under Alain Fischer at the Necker Hospital in Paris carried out gene therapy on two infants who, like David, had been kept in sterile isolation since birth. As with the treatment for ADA, a retrovirus was used to ferry the needed gene into cells extracted from the babies; the cells were then reintroduced. But, in a notable innovation, the French group harvested the cells to be modified from the infants' bone marrow. By using the marrow's immune stem cells rather than ordinary T cells found in the blood, the method, if successful, promised to furnish a self-perpetuating genetic fix. When stem cells reproduce, they increase not only their own numbers but also the numbers of the specialized somatic cells into which they naturally differentiate. Therefore any T cells produced from altered stem cells would also carry the inserted gene, making the repeated infusions of modified cells unnecessary.

And that is exactly what happened: ten months later, T cells containing a working copy of the missing gene were found in both patients, and their immune systems were performing as well as those of any normal children. Fischer's method has since been applied to other SCID children. After a long and not altogether auspicious start, gene therapy had finally notched an unequivocal success. But the champagne celebration did not last long. In October 2002, doctors found that one of the two original patients was suffering from leukemia, a cancer of the bone marrow in which certain types of cell are overproduced. Though it has not been established for sure that the genetic procedure was responsible, the circumstantial evidence is mighty strong. Gene therapy seems to have cured the baby's SCID but caused leukemia as a side effect.

Side effects have always dogged medicine. Drugs may affect more than just their intended target, or surgical procedures may end up causing complications. Though in many ways a departure from conventional medicine, genetic medicine, we now know, is also subject to the same law of unintended consequences. Fischer's SCID treatment probably inadvertently created new problems in the process of fixing the original one. After all, any treatment requiring the insertion of viral DNA into the DNA of patients' cells is inherently risky because the foreign DNA may by chance disrupt the functioning of a critical gene. Because typically the cell in which this has occurred will die, such events usually have no impact. But it is possible that the disrupted gene is one whose elimination does not kill the cell, but rather unlocks its capacity to multiply unchecked: the viral insertion can cause cancer. This seems to be what happened to the SCID baby.

Gene therapy may yet be a long way from delivering the miracles foreseen at the dawn of the genetic revolution. Jesse Gelsinger's death was a severe setback. The leukemia side effect of the SCID treatment, however, is even more damaging. In Gelsinger's case, it seems that unpardonable mismanagement was largely responsible – a problem that has hopefully been fixed by tighter regulation. But there is no ready solution to the side-effect problem. Probably we will have to rely on the depressing calculus that applies in this case: that gene therapy has at least cured a condition, SCID, worse than the one it has caused, leukemia. The good news is that the baby boy in question is apparently responding well to the chemotherapy used to treat the leukemia. However, between them the Gelsinger and leukemia incidents have crystallized many of the difficult issues that need yet to be resolved if somatic gene therapy is to enter the medical mainstream. And I am not so naive as to deny that future trials will likely uncover yet more difficulties. It may be some time before we can claim beyond doubt to have neutralized every conceivable danger, but I nevertheless believe that the potential of this technology to lift the curse of genetic disease is simply too great for medicine to turn away from it.

Your DNA can tell someone a lot about you. As we've seen, if Huntington runs in your family, your DNA can literally reveal your future; soon, depending on whether you possess a particular variant of a gene (or combination of genes), DNA may also speak to your relative risk of succumbing to common killers like heart disease. What version you have of the APOE gene can already serve as a predictor of Alzheimer. But should you worry that this profoundly personal information might be used against you? Not surprisingly, for many Americans the greatest concern is that genetic profiling may one day lead to their being denied health-care insurance.

In 2000 the *American Journal of Human Genetics* published the results of a survey that had asked health insurers whether they would adjust rates to take account of genetic information if it were available to them. Would they, in principle, be prepared to charge more for a customer in perfect health who carried a mutation predisposing him or her to a disorder? About two-thirds admitted they would. The other third were in all likelihood lying. Insurance companies are not philanthropies, but businesses, with shareholders to please. There is no reason to suppose that, left to their own devices, they wouldn't do what they have always done: maximize the premiums of those at risk, and where possible avoid altogether those customers most likely to collect. The same report described a case where an insurance company raised an individual's rates based on a suspicion of a genetic disorder, merely because this individual had requested a diagnostic test for Huntington disease.

As we come to know ourselves at the molecular level, is it inevitable that those of us who have drawn the short straws in the genetic lottery will be made to pay a price, in this way and others? And why presume such abuse would begin and end with insurers? My DNA profile might show that I am likely to have a heart attack or stroke, become an alcoholic, or suffer clinical depression. Might such information cause a prospective employer to think twice about hiring me?

Such questions suggest that *Brave New World* may be upon us well before the twenty-fifth century of Huxley's imagination. DNA is a potent fact of twenty-first-century life – a genie that will never be put back in the bottle. What we allow to be done with it, however, is something we must decide as a democratic society. Unfortunately, in such societies, laws tend to lag behind the need for them: a traffic light isn't typically installed at a dangerous intersection until after a few accidents have occurred. It may require a few horror stories of gross injustice, of individuals made the victims of their own genomes, to motivate the passage of appropriate legislation. What should it look like? Genetic privacy should be a touchstone, but not necessarily the ultimate objective. Balances will need to be struck with society's other priorities, not least the fight against disease, an effort whose progress will more and more depend upon giving medical researchers access to as much genetic data as can be collected from the general population.

While legislation ought not defeat our ambition to exploit the full potential of DNA to alleviate human suffering, or to tell us about ourselves and our origins, or to identify those among us guilty of crimes, it must minimally ensure that no citizen be deprived of civil or human rights on the basis of what might be inscribed in his or her genes.

Meanwhile, it may be a comfort to know that despite a wealth of genetic information already at the industry's disposal and regardless of what they tell pollsters, insurance companies have, on the whole, shown little impulse toward factoring genetic considerations into their calculations for setting rates. The wretched pale skin I inherited has already proved its susceptibility to cancer, but the last time I looked, I still wasn't being charged a higher premium for it. Again, the rationale is business, not charity. Insurers have traditionally set rates using actuarial tables that estimate overall health and longevity mainly on the basis of how we live. I suspect that even if genetic data were universally available, insurers would still find such lifestyle factors – whether one smokes or not, whether one works in a coal mine as opposed to a flower shop – vastly more predictive of one's health risk than the overwhelming majority of relatively subtle differences determined by genetic variation from one person to another. It's indisputable that those whose DNA reveals an unavoidable destiny of debilitation need special protection under the law, but the propensity toward ailments like heart disease and cancer are certain to prove so widespread and complicated as to make them an impractical basis for cost-cutting discriminations. The essential premise of insurance, that payments of the happy many who never have cause to collect will underwrite the relief of the unfortunate few, is not likely to be abolished owing to the accumulation of any amount of genetic information.

But even if our individual rights can be secured against what our DNA may disclose, our peace of mind may not be so easily restored; as Nancy Wexler well knows, genetic knowledge can be a scary prospect. And I agree with her: there is no point in knowing about something that we are powerless to remedy or ameliorate. Alzheimer is a major concern of people my age, but in the absence of proven treatment possibilities, I have no desire to be tested for the presence of the APOEε4 allele. Craig Venter, incidentally, does indubitably have one copy of it. We know this because he insisted on disclosing that the genome sequenced by Celera was his own. And this allele, through its role in cholesterol processing, is associated not only with a higher risk of Alzheimer but of heart disease as well. (APOEε4 is not an asset in any respect.) Stuck with this genetic self-awareness, Venter is wisely trying to respond prophylactically as best he can: he is taking drugs called statins, which lower cholesterol levels and may retard or prevent the onset of Alzheimer. Even without knowledge of my APOE alleles, I'm also taking statins, figuring that a little preemptive medication can't do any harm. If statins are as effective as some claim them to be, we can look forward to many more years of Venter-generated (and, hopefully, Watson-generated) controversy.

Genetic knowledge will remain frightening so long as we remain in the present intermediate stage, possessing in general the power to diagnose but not to cure. But ours is not an unprecedented medical predicament. Think back to the early years of the twentieth century: a diagnosis of infant diabetes was a death sentence. Today, with insulin therapy, such a child can expect to live to a ripe old age. The hope of our research efforts is that one day soon a diagnosis of diseases like Huntington will be transformed in the same way – from death sentence to prescription.

Already, we are in a far stronger position to deal with bad rolls of the genetic dice than we were even twenty years ago; the ever-increasing life expectancies of people with, for example, Down syndrome and cystic fibrosis attest to this progress. But for now our most powerful weapons are diagnostics. The choice of whether to be tested is one best left to each individual or parent, those who will most directly bear the burden of genetic knowledge. In the case of prenatal diagnosis, it is the prospective mother who should make the decisions. That is not to say that others shouldn't participate, but ultimately the choices should lie with the woman: not only is she the one having the baby, but, like it or not, our world is still one in which women are expected to bear the brunt of the day-to-day care of children. Regardless of the specifics of decision making, however, one thing is completely clear to me: over the ages genetic disorders have rained unthinkable misery upon countless families such as Carol Carr's, which has been ravaged by Huntington disease. Testing holds the power to reduce misery by preventing it. Having developed the tests, it would be unconscionable not to make their existence known to those who might want to use them, inexcusable not to make them universally available.

CHAPTER THIRTEEN

WHO WE ARE:
NATURE VS. NURTURE

Growing up, I worried quietly about my Irish heritage, my mother's side of the family. My ambition was to be the smartest kid in the class, and yet the Irish were the butt of all those jokes. Moreover I was told that in the old days signs announcing the availability of jobs often ended with "No Irish Need Apply." I wasn't yet equipped to understand that such discrimination might have to do with more than an honest assessment of Irish aptitudes. I knew only that though I myself possessed lots of Irish genes there was no evidence that I was slow-witted. So I figured that the Irish intellect, and the shortcomings for which it was known, must have been shaped by the Irish environment, not by those genes: nurture, not nature, was to blame. Now, knowing some Irish history, I can see that my juvenile conclusion was not far from the truth. The Irish aren't in the least stupid, but the British tried mightily to make them so.

Oliver Cromwell's conquest of Ireland surely ranks high among history's most brutal episodes. It culminated in banishing the native Irish population to the country's undeveloped and inhospitable western regions like Connaught while the spoils of the more salubrious east were divided up among the lord protector's supporters, who would start to Anglicize the vanquished province. With the incoming Protestants believing the heresies of Catholicism to be a one-way ticket to perdition, Cromwell duly proclaimed in 1654 that the Irish had a choice: they could "Go to Hell or Connaught." At the time, it probably wasn't clear which was worse. Seeing Catholicism as the root of the "Irish Problem," the British took draconian measures to suppress the religion, and with it, they hoped, Irish culture and Irish national identity. The ensuing period of Irish history was thus characterized by a form of apartheid every bit as severe as that so infamously practiced in South Africa, with the principal difference being the basis of discrimination: religion rather than skin color.

Among the "Penal Laws" passed to "Prevent the Further Growth of Popery," education was a particular target. One statute of 1709 included the following provisions:

> Whatever person of the popish religion who shall publickly teach school or instruct youth in learning in any private house, or as an usher or assistant to any protestant schoolmaster, shall be prosecuted.
>
> For discovering, so to lead to the apprehension and conviction of any popish archbishop, bishop, vicar general, Jesuit, monk, or other person exercising foreign ecclesiastical jurisdiction, a reward of 50 pounds, and 20 pounds for each regular clergyman or non-registered secular clergyman so discovered, and 10 pounds for each popish schoolmaster, usher or assistant; said reward to be levied on the popish inhabitants of the county where found.

The British hoped that the Irish young attending British-sponsored Protestant schools would wean themselves off Catholicism. But they hoped in vain: it would take more than oppression or even bounty to prise apart the Irish and their religion. The result was a spontaneous underground educational movement, the "hedge schools," with itinerant Catholic teachers leading secret classes in ever-changing outdoor locations. Often conditions were appalling, as a visitor noticed in 1776: "They might as well be termed ditch schools, for I have seen many a ditch full of scholars." But by 1826, of the entire student body of 550,000 an estimated 403,000 were enrolled in hedge schools. Increasingly a romantic symbol of Irish resistance, the schools inspired the poet John O'Hagan to write:

> Still crouching 'neath the sheltering hedge,
> Or stretched on mountain fern,
> The teacher and his pupils met feloniously to learn.

But if the British had failed in their goal to enforce religious conversion, they had, despite the heroic efforts of the hedge schoolteachers, successfully impaired the quality of education for generations of Irish. The resulting archetype of the "stupid" Irishman would have been more aptly identified as the "ignorant" Irishman, a direct legacy of the anti-Catholic policies of Cromwell and his successors.

And in this way my boyish conclusion was not far off the mark: the so-called curse of the Irish was indeed the result of nurture – development in an environment of substandard educational opportunities – rather than nature – Irish genes. Today, of course, nobody, not even the most bigoted Englishman, can legitimately claim that the Irish aren't as smart as other people. Ireland's modern education system has more than undone the damage of the hedge school era: the Irish population is today one of the best educated on the planet. My youthful reasoning on the subject, however absurdly misinformed, nevertheless taught me a very valuable lesson: the danger of assuming that genes are responsible for differences we see among individuals or groups. We can err mightily unless we can be confident that environmental factors have not played the more decisive role.

This tendency to prefer explanations grounded in "nurture" over ones rooted in "nature" has served a useful social purpose in redressing generations of bigotry. Unfortunately, we have now cultivated too much of a good thing. The current epidemic of political correctness has delivered us to a moment when even the possibility of a genetic basis for difference is a hot potato: there is a fundamentally dishonest resistance to admitting the role our genes almost surely play in setting one individual apart from another.

Science and politics are to a degree inseparable. The connection is obvious in countries like the United States, where a considerable proportion of the scientific research budget depends on the appropriations of the democratically elected government. But politics intrudes upon the pursuit of knowledge in more subtle ways as well. The scientific agenda reflects society's preoccupations, and all too often social and political considerations end up outweighing purely scientific ones. The rise of eugenics, a response on the part of some geneticists to prevailing social concerns of the era, is a case in point. With a scientific basis weak to the point of vanishing, the movement progressed mainly as a pseudoscientific vehicle for the notably unscientific prejudices of men like Madison Grant and Harry Laughlin.

Modern genetics has taken to heart the lessons of the eugenics experience. Scientists are typically careful to avoid questions with overtly political implications and even those whose potential as political fodder is less clear. We have seen, for instance, how such an obvious human trait as skin color has been neglected by geneticists. It's hard to blame them: after all, with any number of interesting questions available for investigation, why choose one that might land you in hot water with the popular press or, worse, earn you an honorable mention in white supremacist propaganda? But the aversion to controversy has an even more practical – and more insidious – political dimension. It happens that scientists, like most academics, tend to be liberal and vote Democratic. While no one can tell how much of this affiliation is principled and how much is pragmatic, it's certainly the case that Democratic administrations are assumed to be invariably more generous toward research than Republican ones.* And so having signed on to the liberal end of the political spectrum, and finding themselves in a climate intolerant of truths that don't conform to ideology, most scientists carefully steer clear of research that might uncover such truths. The fact that they duly hew to the prevailing line of liberal orthodoxy – which seeks to honor and entitle difference while shunning any consideration of its biochemical basis – is, I think, bad for science, for a democratic society, and ultimately for human welfare.

Knowledge, even that which may unsettle us, is surely to be preferred to ignorance, however blissful in the short term the latter may be. All too often, however, political anxiousness favors ignorance and its apparent safety: we had better not learn about the genetics of skin color, goes the unspoken fear, lest such information be marshaled somehow by hatemongers opposed to mixing among the races. But that same genetic knowledge may actually be vitally useful

*Wrongly, it turns out. The stingiest science budget in recent history was Jimmy Carter's.

to people like me, whose Irish-Scots complexion is vulnerable to skin cancer in climes sunnier than Tipperary and the Isle of Skye, where my mother's ancestors hailed from. Similarly, research into the genetics of difference in mental ability among people may raise awkward questions, but that knowledge would also be a boon to educators, allowing them to develop an individual's educational experience with his or her strengths in mind. The tendency is to focus on the worst-case scenario and to shy away from potentially controversial science; it is time, I think, we looked instead at the benefits.

There is no legitimate rationale for modern genetics to avoid certain questions simply because they were of interest to the discredited eugenics movement. The critical difference is this: Davenport and his like simply had no scientific tools with which to uncover a genetic basis for any of the behavioral traits they studied. Their science was not equipped to reveal any material realities that would have confirmed or refuted their speculations. As a consequence, all they "saw" was what they wished to see – a practice that really doesn't merit the name science – and often they came to conclusions manifestly at odds with the truth: for instance that "feeblemindedness" is transmitted as an autosomal recessive. Whatever the implications of modern genetics may be, they simply bear no relation to this manner of reasoning. Now, if we find a certain mutation in the gene associated with Huntington disease, we can be sure that its possessor will develop the disease. Human genetics has moved from speculation to fact. Differences in DNA sequence are unambiguous; they're not open to interpretation.

It is ironic that those who worry most about what unchecked genetics might reveal should lead the way in politicizing the field's most basic insights. Take, for example, the discovery that the history of our species implies that there are no major genetic differences among the groups traditionally distinguished as "races": it has been suggested that as a matter of general practice our society should accordingly cease to recognize the category "race" in any context, eliminating it, for instance, from medical records. The theory here goes that the quality of treatment you receive in a hospital may vary depending on how you identify your ethnicity on the admission form. Racism can surely be found in the ranks of any profession, medicine included. But it's not altogether apparent how much protection your ethnic anonymity on the form will confer once you are face-to-face with a doctor who is a bigot. What is more apparent, however, is the danger of withholding information that might be diagnostically important. It is a fact that some diseases have higher rates of incidence within certain ethnic groups as compared with the human population as a whole: Native Americans of the Pima tribe have a particular propensity to Type II diabetes; African Americans are much more likely to suffer from sickle-cell anemia than Irish Americans; cystic fibrosis affects mainly people of northern European origin; Tay-Sachs is much more common among Ashkenazi Jews than others. This is not fascism, racism, or the unwelcome intrusion of Big Brother. This is simply a matter of making the best possible use of whatever information is available.

For such a young science, genetics has played a central role in a remarkable number of notably ugly political episodes. Eugenics, as we have seen, was partly of the geneticists' own making. The pseudoscience known as Lysenkoism, which flourished in the Soviet Union in the middle of the twentieth century, however, was visited upon genetics from on high – literally: Stalin had plenty to say on the matter. Lysenkoism represents the most egregious incursion of politics into science since the Inquisition.

In the late 1920s, the Soviet Union was still finding its feet. Stalin had won the battle of succession after Lenin's death and was consolidating power. The collectivization of agriculture was under way. And in an obscure agricultural research station in distant Azerbaijan, an uneducated but ambitious peasant was making a name for himself. Trofim Lysenko, born in the Ukraine in 1898, appeared an unlikely choice to oversee Stalin's agricultural revolution. Barely literate, he was working as a minor technician at Gandzha at the Ordzhonikidze Central Plant-Breeding Experiment Station when, in 1927, he was catapulted from obscurity by a visiting *Pravda* correspondent who, perhaps at a loss for good copy, was inspired by the sight of Lysenko: here was the "barefoot professor" solving agricultural problems so that the local "Turkic peasant can live through the winter without trembling at the thought of the morrow." Critically, the article painted Lysenko as a problem solver, not a highfalutin academic: "He didn't study the hairy legs of [fruit] flies, but went to the root of things."

The image of the barefoot professor was irresistible to Soviet apparatchiks. Here was a son of the soil, the true flowering of the Soviet man, of the rural peasant class; his agricultural intuition was surely worth more than all the book learning of the shiftless intellectuals. Not to disappoint, Lysenko was quick to capitalize on his newfound prominence by proposing that winter wheat be "vernalized." Winter wheat is normally planted in the fall; it overwinters as a shoot, with some of the crop perishing, the rest maturing during the spring. Through "vernalization," Lysenko suggested, the losses of winter could be avoided. He claimed that you could fool the wheat seeds into germinating in the spring simply by chilling and wetting them, and that increased yields would be achieved in the bargain (see Plate 58). The definitive experimental demonstration of the method was carried out by none other than Lysenko's father in his own fields. Indeed, the yield was some three times greater than that of conventional unvernalized wheat planted in the same district.

Vernalization did not in fact originate with Lysenko; wherever he may have picked it up, the procedure dates back to the preceding century at least, appearing, for example, in the Ohio agricultural literature of the 1850s. But here Lysenko's lack of education (and therefore ignorance of what had been accomplished elsewhere) stood him in good stead when it came to claiming originality. The same, however, could not be said for every further attempt to apply the method, whose results can vary a good deal depending on local conditions – something the Ohio farmers knew but the barefoot professor apparently did not.

Within a couple of years, beset by failures, Lysenko stopped advocating the vernalization of winter wheat and was pushing instead the vernalization of spring wheat – a ploy worthy of the sharpest Soviet satire, considering that the crop is indeed named after the season in which it is normally planted. Later, his wheat yield policy did another U-turn when Lysenko called for *warming* (instead of cooling) the seed prior to planting. Wheat vernalization was but one of many agricultural nostrums that Lysenko peddled, but it illustrates well his overall strategy. A complete disregard for expert knowledge was de rigueur, as was a refusal to conduct consistent and rigorous tests. Essentially, any idea intuitively appealing to Lysenko was good enough to be implemented. What scientific method he did espouse almost seems inspired by theological reasoning, odd coming from the tool of a godless Communist state: "In order to obtain a certain result, you must want to obtain precisely that result; if you want to obtain a certain result, you will obtain it."

Much more astutely reasoned was Lysenko's careful manipulation of the media. His original brush with fame in *Pravda* had taught him that the state-controlled press was a better venue for scientific self-promotion than the dusty pages of academic or trade journals. In 1929 *Pravda* twice featured the barefoot professor's success with vernalization, each time reporting in loving detail the down-home contribution of Lysenko Senior.

At that point the Soviet Union needed a Lysenko. The "agricultural reorganization," as Stalin preferred to call the collectivization of farms, was proving a catastrophe. Even the official estimates, notorious for their rosy overstatements, painted a grim picture of rural productivity during this period. Lysenko's intuitive quick fixes made him the man of the hour, even if they wound up doing more harm than good the morning after. He embodied an important Bolshevist ideal, *deistvennost* – "action quality." No messing around with grand theories or arcane academic concerns, Lysenko, the can-do barefoot professor, was all about action and solving practical problems.

Lysenko learned quickly how to play the Soviet system. His lectures made no pretense to being scientific in any sense that we would recognize: they consisted instead of ideological rants peppered with all the Marxist-Leninist jargon du jour. Small wonder Stalin was a fan, leading the standing ovation (at a Congress of Collective Farm Shock-Brigade Workers) with cries of "Bravo, Comrade Lysenko!" In return, Lysenko astutely named his latest big idea, a variety of branched wheat, for Stalin. Happily accepting the honor, the generalissimo fortunately never found out that the branched wheat was another bust: though inherently higher in yield, it requires such low-density planting as to more than offset the advantage of its multiple seed heads.

Sucking all of Soviet agriculture into a vast experiment each time he introduced another hopelessly impractical new scheme, Lysenko was ultimately responsible for the starvation of millions. But since Soviet records of the era – especially those kept by Lysenko himself – are woefully self-serving, we will probably never know the actual number of lives sacrificed on the altar of Lysenko's career. Suffice it to say that at the time of Stalin's death in 1953, the availability of meat and vegetables has been estimated by more objective analysts to have been no greater than in the darkest feudal days of Tsar Nicholas II. Lysenko's pernicious influence, however, was not limited to agriculture.

If Soviet farming had its homegrown principles, Soviet science too needed a scientific credo of its own; it pained Lysenko and his followers to imagine the new Soviet man following so meekly in the footsteps of "bourgeois" Western scientists. And Lysenko's wild theories of agricultural development had boiled down to the idea that you could transform any crop so long as you subjected it to the right environment: winter wheat could *become* spring wheat through a simple environmental manipulation. And it was no one-season fix, for, according to Lysenko, such changes would then breed true – acquired traits would be passed on to the next generation. Eventually Lysenko became a full-fledged Lamarckist.* In an unusual fit of experimental enthusiasm, he even commissioned experiments to "disprove" Mendelism – the basis of genetics in the decadent Western tradition. In his mathematical incompetence, Lysenko actually became convinced that the results refuted Mendel's ratios, even when a reanalysis of the data by a distinguished Soviet mathematician showed that in fact the ratios fitted Mendel's predictions exactly. Lysenko, then, was not above doing an occasional experiment, but was never one to countenance a result contradicting even the most outlandish hypothesis.

---

*In 1801 Jean-Baptiste Lamarck first published his theory of inheritance of acquired characteristics, erroneously suggesting that traits acquired during an individual's life could be passed on to its offspring. Flawed though his idea was, Lamarck, unlike Lysenko, was at least trying to base his inferences on observation.

The late 1930s saw a series of debates between Lysenko, backed by what have been described as his "hard core of militant ignoramuses," and the Soviet genetics community – a distinguished group by the standards of the international science of the day. H. J. Muller, one of T. H. Morgan's students (and my professor in graduate school at Indiana University), went to Russia to participate in the great social experiment of Communism and found himself instead embroiled in bizarre, largely stage-managed public discussions about Larmarckian inheritance. In this era of Stalin's purges, political truths carried much more weight than mere scientific ones. To what extent Lysenko directly contributed to the "repression" – to use the preferred Soviet euphemism for Stalin's purges – of the geneticists who spoke against him will probably never be known but, no matter who gave the orders, the fact remains that much of the opposition to Lysenko's Lamarckian ideas simply disappeared as the 1930s drew to a close. Some geneticists heroically stood their ground as outspoken critics. Muller was forced to flee for his life. The doyen of Soviet genetics (and an ardent Soviet patriot), Nikolai Vavilov, was arrested in 1940. He died in prison of malnutrition.

In 1948, it was officially decreed that the debate was over: Mendelism was out, Lysenkoism was in – an absurd and tragic outcome particularly when one considers that it came four years *after* Avery's landmark experiment showing DNA to be the transforming (Mendelian) factor. The Lysenkoite response to the discovery of the double helix, incidentally, was characteristically obscurantist: "It deals with the doubling, but not the division of a single thing into opposites, that is, with repetition, with increase, but not with development." I've no idea what this means, but it seems to be consistent (in its meaninglessness) with Lysenko's other writings on heredity:

In our conception, the entire organism consists only of the ordinary body that everyone knows. There is in an organism no special substance apart from the ordinary body. But any little particle, figuratively speaking, any granule, any droplet of a living body, once it is alive, necessarily possesses the property of heredity, that is, the requirement of appropriate conditions for its life, growth and development.

Darwin was next in line for the Lysenko treatment. The out-of-control peasant-made-good denied the cardinal precept of Darwinism – competition among individuals within a species for access to limited resources – and postulated, as perhaps a good Communist should, that individuals do not compete, but cooperate. He went further, combining his anti-Mendelian and anti-Darwinian views in a bizarre unified theory of the origin of species: given that organisms are molded by their environment, it should be possible, with the right environmental conditions, to transform any one species into any other. Change a warbler's diet to caterpillars, to take his favorite example, and you can produce a cuckoo. Ardent Lysenkoists from around the country were soon writing in with reports of their own transformational successes: viruses turned into bacteria, a rabbit into a chicken. Soviet biology was itself undergoing a transformation of sorts: from science to joke.

Lysenko's rejection of Darwin eventually put him in a position so awkward as to tax even his own formidable skills of political survival. Stalin's final years witnessed the "Great Stalin Plan for the Transformation of Nature." In part, this involved planting a lot of trees to protect the steppes from the vicious east winds, thus moderating the climate in general. It was not a bad idea in principle, but as one might expect, Lysenko had his notions about the best way to grow trees: plant them in a cluster, he argued, and the individual seedlings will *not* compete with each other for sunlight and nutrients but will rather cooperate for the good of the community. In the late forties, armies of peasants fanned out across the steppe planting oak trees in clusters in accordance with the Lysenko method. The result? Intense competition among individual trees, which enfeebled all members of each cluster. By 1956 only 4 percent of the oaks planted were thriving; only 15 percent had even survived. The Ministry of Agriculture withdrew its endorsement of the Lysenko planting protocol, but only after a sum estimated at over 1 billion rubles had been squandered.*

It was a stunning setback, but so entrenched was Lysenko's authority, and so crowded with his protégés were the ranks of Soviet biology, that it wasn't until 1964 that the Kremlin turned its back on him for good. The barefoot professor had managed to persuade Stalin's successor that he was still the man to create a Soviet agricultural miracle. Indeed, when Khrushchev was bundled ignominiously out of office by the Soviet Central Committee (and replaced with Brezhnev), it was rumored that one important reason for the intervention was a general frustration with Khrushchev's continued reliance on Comrade Lysenko. Lysenko himself died in 1976. His family requested that he be buried in the most prestigious Russian national cemetery at the Novo-Devichi convent. The request was denied.

I would not for a moment wish to suggest, by telling the parable of Lysenko, that the fate of Soviet science under that fool's sway is remotely comparable to the state of contemporary Western research in even the most politically overpowering university setting. But the

*The modern dollar equivalent is difficult to compute because the official exchange rates of that era generally reflected Communist wishful thinking rather than financial reality. To put the sum in context, however, it can be noted that in 1956 one-sixth of the Soviet workforce earned an annual wage of around three thousand rubles.

extreme instance should suffice to demonstrate that ideology – of any kind – and science are at best inappropriate bedfellows. Science may indeed uncover unpleasant truths, but the critical thing is that they are *truths*. Any effort, whether wicked or well-meaning, to conceal truth or impede its disclosure is destructive. Too often in our free society, scientists willing to take on questions with political ramifications have been made to pay an unjust price. When in 1975 E. O. Wilson of Harvard published *Sociobiology*, a monumental analysis of the evolutionary factors underlying the behavior of animals ranging from ants – his own particular subject of expertise – to humans, he faced a firestorm of rebuke in the professional literature as well as the popular media. An anti-Wilson book published in 1984 bore a title that said it all: *Not in Our Genes*. Wilson was even attacked physically, when protesters objecting to the genetic determinism they perceived in his work dumped a jug of water on him during a public meeting. Similarly, Robert Plomin, whose work on the genetics of human intelligence we shall presently address, found the American academy so hostile that he decamped from Penn State to England.

Passions inevitably run high when science threatens to unsettle or redefine our assumptions about human society and our sense of ourselves – our identity as a species, and our identities as individuals. What could be a more radical question than this: Does the way I am owe more to a sequence of As, Ts, Gs, and Cs inherited from my parents, or to the experiences I've had ever since my father's sperm and mother's egg fused together many long years ago? It was Francis Galton, the father of eugenics, who was the first to frame the question as one of nature versus nurture. And the implications spill over into less philosophical, more practical areas. Are all math students born equal, for instance? If the answer is no, it may well be a waste of time and money trying to force differential equations down the throats of people like me who are simply not wired to take the stuff on board. In a society built on an egalitarian ideal, the notion that all men are *not* born equal is an anathema to many people. And not only is there a lot at stake, but the issues are very difficult to resolve. An individual is a product of both genes and environment: how can we disentangle the two factors to determine the extent of each one's contribution? If we were dealing with laboratory rats, we could conduct a set of simple experiments, involving breeding and rearing under specified uniform conditions. But, happily, humans are not rats, so illuminating data are hard to come by. This combination of the debate's importance and the near impossibility of satisfactorily resolving it makes for perennially lively argument. But a free society should not shrink from honest questions honestly asked. And what is critical is that the truths we discover are then applied only in ethical ways.

With a lack of reliable data, the nature/nurture debate was entirely subject to the shifting winds of social change. Early in the twentieth century, during the heyday of the eugenics movement, nature was king. But when the fallacies of eugenics became apparent, culminating in its horrific applications by the Nazis and others, nurture began to gain the upper hand. In 1924 John Watson (no relation), the American father of an influential school of psychology called "behaviorism," summarized his nurture-ist perspective as follows:

> Give me a dozen healthy infants, well-formed, and my own specified world to bring them up in and I'll guarantee to take any one at random and train him to become any type of specialist I might select: doctor, lawyer, artist, merchant-chief, and yes, even beggar-man and thief, regardless of his talents, penchants, tendencies, abilities, vocations, and race of his ancestors.

The notion of the child as tabula rasa – a blank slate upon which experience and education can write any future – dovetailed nicely with the liberal agenda that grew out of the sixties. Genes (and the determinism they stood for) were out. Discounting inheritance, psychiatrists preached that mental illness was caused by varieties of environmental stress, an assertion that inspired endless guilt and paranoia among the parents of the afflicted: where did we go wrong? they asked. The tabula rasa remains the paradigm of choice among the politically entrenched defenders of some increasingly untenable views of human development. Among some unregenerate hardliners of the women's movement, for example, the notion of biological – genetic – differences in cognitive aptitudes between the sexes is simply unspeakable: men and women are equally capable of learning any task, period. The fact that men are more common in some fields and women in others is, these theorists would have it, purely a result of divergent social pressures: the male slate is inscribed with one destiny, the female with another, and it begins with our laying that pink blanket on the baby girl and the blue one on the baby boy.

Today we are seeing a swing away from the extreme nurture-ist position embodied by the other Watson. And it is no coincidence that this drift away from behaviorism is coinciding with our first glimpses of the genetics underpinning behavior. As we saw in chapter 11, for years the genetics of humans has lagged behind that of fruit flies and other creatures owing to a lack of genetic markers and the impossibility of doing breeding experiments on people. But since the introduction in 1980 of DNA-based genetic markers the analysis of human traits through the mapping of related genes has advanced by leaps and bounds. Most of the effort has understandably been expended on meeting the most urgent human need genetics can address: diagnosis and treatment of inherited disease. Nevertheless, some efforts have been directed toward nonmedical questions. Robert

Plomin, for example, has used this approach to hunt up genes influencing IQ, taking advantage of an annual gathering in Iowa of superbright schoolchildren from across the nation. With an average IQ of 160, these slightly scary kids were an obvious place to start to look for genes that might affect IQ. Plomin compared their DNA to that of a sample of "normal" kids with average range IQs like yours and mine* – and indeed he found a weak association between a genetic marker on chromosome 6 and stratospheric IQ. Here was reason to suppose that a gene or several genes in that region might in some way contribute to IQ. Of course, any mechanism governing such a complex trait is apt to involve many genes.

In chapter 11, we discussed the difficulty of mapping polygenic traits, like heart disease, that are produced by multiple genes, each with a small individual effect, each mediated by the environment. Behavioral traits generally fall in this category. So far as we know, having the appropriate variant on chromosome 6 would not by itself a genius make: there are doubtless necessary variants in genes as yet undiscovered. And even a solid genetic basis might not get you there unless you were also reared in an environment in which learning and thinking were honored over Nickleodeon. But the discovery, and the acknowledgment, of any molecular basis for intelligence is a breakthrough such as only the genetic revolution could foster.

Before DNA markers were available, the meat and potatoes of behavioral genetics was twin studies. Twins come in two varieties: dizygotic (DZ), meaning two individuals develop from two separate eggs, each fertilized by a different sperm; and monozygotic (MZ), meaning that both come from a single fertilized egg, which in early

*Mine is a respectable, but definitely not stellar, 122. I discovered it by sneaking a quick look at a list on a teacher's desk when I was eleven.

development – usually the 8- or 16-cell stage – splits into two balls of cells. DZ twins are no more genetically similar than any two siblings, but MZ twins are genetically identical. MZ twins are therefore always the same sex, while DZ twins may or may not be. Surprisingly, it hasn't been very long since we first understood this fundamental difference between twin types. In 1876, when Francis Galton first suggested that twins might be useful in determining the relative contributions of heredity and upbringing, he was unaware of the difference (whose basis had been worked out just two years earlier), and assumed wrongly that it was possible for different-sex twins to be derived from a single fertilized egg. From his later publications, however, it's clear that the message eventually got through.

MZ twinning occurs globally in about four of every thousand pregnancies and seems to be nothing more than a random accident. DZ twinning, on the other hand, may run in families, and varies from population to population: a group in Nigeria tops the list with DZ twins accounting for forty pregnancies per thousand, while there are only three per thousand in Japan.

The basic premise of the standard form of twin study is that both members of a same-sex pair of twins, whether DZ or MZ, are raised the same way (i.e., receive similar "nurtures"). Suppose we are interested in a simply measurable characteristic like height. If DZ1 and DZ2 were both raised on the same diet of food, love, and so on, any difference in height between them would be attributable to some combined effect of genetic differences and whatever subtle differences of nurture may have crept in (for instance, DZ1 always finishes her milk; DZ2 never does). But if we follow the same program with MZ1 and MZ2, the fact that these twins are genetically identical eliminates genetic variation as a factor; any differences in height must be a function of *only* those subtle environmental differences. All things being equal, MZ twins will then tend to be more similar in height than

DZ twins, and the extent to which this is true gives us a measure of how much genetic factors influence height. Similarly, the extent to which MZ twins have more similar IQs than DZ twins reflects the effect of genetic variation on IQ.

This kind of analysis is also applicable to the inheritance of genetic diseases. We say that twins are "concordant" if they both have the disease. An increase in concordance when we look from DZ twins to MZ twins would support the claim of a strong genetic basis to the disease: for example, DZ twins are 25 percent concordant for late-onset diabetes (if one twin has it, then there is a one in four chance that the other does too), whereas MZ twins are 95 percent concordant for the disease (if one twin has it, then nineteen out of twenty times the other does too). The conclusion: late-onset diabetes has a strong genetic component. Even here, though, the environment plays an obvious role: if it were not so, we would see 100 percent concordance in MZ twins.

A long-standing criticism of this kind of twin study addresses methodology: MZ twins tend to be treated more alike by their parents than DZ twins. Parents sometimes make a virtual fetish of identicalness: often, for instance, MZ twins are even dressed exactly alike, a habit some weirdly carry into adulthood. This is a legitimate criticism insofar as the more pronounced similarity of MZ twins (as compared with DZs) is interpreted as evidence of genetic influence when in fact it could simply be a reflection of the more precisely similar nurtures shared by MZs. And here is a further wrinkle in the same problem: how do we tell whether a pair of same-sex twins is DZ or MZ? "It's easy," you say. "Just look at them." Wrong. In a small but significant proportion of cases, parents mistake their same-sex DZ twins for MZ twins (and thus tend to subject them to the supersimilar nurture routine – the same frilly pink frock for each); and conversely a small proportion of parents with MZ twins wrongly take them for

DZs (dressing one in frilly pink and the other in bright green). Fortunately, DNA fingerprinting techniques have rescued twin studies from this comedy of errors. The test can determine for sure whether the pair are indeed as they were supposed to be, whether DZ or MZ. The mistaken-identity groups then serve as the perfect experimental control in the analysis: for example, height difference in DZ twins cannot be put down to differences in nurture if the parents were raising them as MZs.

Perhaps no form of twin study holds more popular appeal than analysis of MZs separated at birth. In such cases, the rearing environments are often very different, and so marked similarities are attributable to what the twins have in common: genes. It makes good copy: you see reports of MZs separated at birth who, it turns out, both have red velvet sofas and dogs called Ernest. Striking though these similarities may be, however, chances are they are mere coincidences. There is almost assuredly no gene coding for the red velvet upholstery preference, or for impulses in dog naming. Statistically, if you list a thousand attributes – make and model of car, favorite TV show, etc. – of any two people, you will inevitably find ones that overlap, but in the press these are the ones that get reported, usually in the Believe It or Not column. My coauthor and I both drive Volvo station wagons and appreciate a cocktail or two, but we are most definitely not related.

Popular or no, twin studies have had a checkered history. Part of the ill repute stems from the controversy surrounding Sir Cyril Burt, the distinguished British psychologist who did much to establish the use of twins in studies of the genetics of IQ. After his death in 1971, a detailed examination of his work suggested that some of it was fraudulent – Sir Cyril, some alleged, was not above inventing a few twins from time to time if he needed to bolster sample sizes. The truth of these charges is still debated, but one thing is undeniable: the episode cast a shadow of suspicion over not only twin studies but all attempts to understand the genetic basis of intelligence. In fact, the combination of the Burt affair and hair-trigger political sensitivity to the topic have in effect stifled research by cutting the flow of grant money. No money, no research. Tom Bouchard at the University of Minnesota, a distinguished scientist whose massive 1990 survey of twins reared apart redefined twin studies, had such difficulty raising funds that he was forced to go cap in hand to a right-wing organization that supports behavioral genetics to further its own dubious political agenda. Founded in 1937, the Pioneer Fund counts among its early luminaries Harry Laughlin, the chicken geneticist we encountered in chapter 1 who turned his attentions to humans and entered the vanguard of American scientific racism. The fund's charter was "race betterment with special reference to the people of the United States." That legitimate researchers like Bouchard should be faced with a choice between seeking such a tainted sponsor or seeing their work perish represents a staggering indictment of the federal agencies that fund scientific research. Tax dollars are being allocated according to political rather than scientific merit.

Bouchard's Minnesota twins study revealed that a host battery of personality traits – as measured using standardized psychological tests – were substantially affected by genes. In fact, more than 50 percent of the variability observed in a range of characteristics – the tendency to be religious, to name one – was typically caused by underlying variation in genes. Bouchard concluded that one's upbringing has surprisingly little effect upon personality: "On multiple measures of personality and temperament, occupational and leisure-time interests, and social attitudes, MZ twins reared apart are about as similar as MZ twins reared together." In other words, when it comes to measurable components of personality, nature seems to trump nurture. This lack of impact of upbringing on personality development has even Bouchard scratching his head. Upbringing has little effect, and yet the data still show the environment's considerable effect: MZ twins raised apart are as similar to each other as those raised together, but there are nevertheless differences in both cases between members of a pair. Could there be an aspect of environment distinguishable from upbringing? One suggestion is that variation in prenatal experience, the life of a fetus in utero, may be important; even small differences at this early developmental stage – when, after all, the brain is being assembled – may have a significant impact on who we become. Even MZ twins may find themselves in very different uterine settings courtesy of the natural whims of implantation – the lodging of the embryo in the wall of the uterus – and the development of the placenta. The popular belief that all MZ twins share a single placenta (and therefore have similar uterine environments) is wrong: 25 percent of MZ pairs have separate placentas. Studies have shown that such twins differ more from each other than do pairs who have shared a placenta.

The elephant in the living room of all twin studies is the genetics of intelligence. How much of our smarts is determined by our genes? Everyday experience suffices to prove there is a lot of variation out there. While teaching at Harvard, I became intimately acquainted with the familiar pattern: in any population, there are a few who really aren't too bright, a few who are alarmingly smart, and a vast majority who are middling. The fact that the setting was Harvard, where the population had been preselected in favor of intelligence, makes no difference: the same proportions hold whatever the group. This "bell curve" distribution of course can describe just about any trait that varies in humans: most of us are medium tall, but there are a few super-tall and a few super-short among us. But when used to describe variations in human intelligence the bell curve has demonstrated powers to raise a dust storm of objection. The reason is that in a land of equal opportunity, where we are each free to advance as far as our wits will carry us, intelligence is a trait with profound socioeconomic implications: the measure of it is predictive of how one will fare in life. And so in this matter the nature/nurture debate becomes entangled with the noble aspirations of our meritocratic society. But given the complex interplay of the two factors, how can we reliably judge their respective weights? Smart parents not only pass on smart genes; they also tend to rear their children in ways that foster intellectual growth, thus confounding the effects of genes and environment. This is the reason careful twin studies are so valuable in permitting us to analyze the constituents of intelligence.

Bouchard's study and earlier ones as well have found that as much as 70 percent of the variation in IQ is attributable to corresponding genetic variation: a strong argument for the primacy of nature over nurture. But does this really mean that our intellectual fate is largely sealed by our genes – that education (even our own free will) has little to do with who we are? Not at all. As with all traits, it is nice to be blessed with favorable genes, but there is much that nurture can do to influence the standing of any individual, at least in the bell curve's vast midland, where variations in social circumstance are mainly determined.

Take the case of one group within Japan, the Buraku. They are the descendants of Japanese who by feudal custom had once been condemned to perform society's "unclean" tasks, like slaughtering animals. Despite the modernization of Japanese society, the Buraku remain impoverished and marginalized outsiders, scoring on average ten to fifteen IQ points lower than the national Japanese mean. Are they genetically inferior, or is their IQ simply a reflection of their lowly status in Japan? It would seem to be the latter: Buraku who have immigrated to the United States, where they are indistinguishable from other Japanese Americans, have shown an increase in IQ and over time the fifteen-point gap with their fellows in the homeland disappears. Education matters.

In 1994, Charles Murray and Richard Herrnstein published *The Bell Curve,* arguing that, despite the well-established effect of education, the discrepancies in the average IQ scores of different races may themselves be attributable to genes. It was a profoundly controversial claim, but not as simpleminded as many have supposed. Murray and Herrnstein understood that the combined observations of a genetic basis to IQ and of differences in average IQ among groups do *not* lead directly to the conclusion that genes are responsible for the intergroup differences. Imagine sowing the seeds of a particular plant species in which height varies genetically. Put one set of seeds in a tray with high-grade soil, and another in a tray of poor soil: in both trays we see variation in height; some individuals are taller than others – as expected, given genetic variation. But we also observe that the average height for plants in the tray of poor soil is less than the average for those in the tray of rich earth. The environment, in the form of soil quality, has affected the plants. While genetics is the dominant factor in determining height differences among plants within a tray – all other factors being equal – genetics has nothing to do with the differences seen between the two trays.

Does this same argument apply to African Americans, who lag behind other Americans in measures of IQ? Since poverty rates among African Americans are relatively high, with a large proportion of individuals finding themselves rooted in the relatively poor educational soil of the inner city, environment surely does contribute to their underperformance on IQ tests. Murray and Herrnstein's point, however, was that the discrepancy was so great that environment likely couldn't explain it all. Similarly, environmental factors alone may not account for why, globally, Asians, have on average higher IQs than other racial groups. The idea of measurable variations in average intelligence among ethnic groups is not one, I admit, I want to live with. But though *The Bell Curve*'s claims remain questionable, we should not allow political anxieties to keep us from looking into them further.

There is perhaps no more heartening proof of the role of environment in human intelligence than the Flynn effect, the worldwide phenomenon of upwardly trending IQ, named for the New Zealand psychologist who first described it. Since the early years of the twentieth century, gains have ranged between nine and twenty points per generation in the United States, Britain, and other industrialized nations for which reliable data-sets are available. With our knowledge of evolutionary processes, we can be sure of one thing: we are *not* seeing wholesale genetic change in the global population. No, these changes must be recognized as largely the fruits of improvement in overall standards both of education and of health and nutrition. Other factors as yet not understood doubtless play a role, but the Flynn effect serves nicely to make the point that even a trait whose variation is largely determined by genetic differences is in the end significantly malleable. We are not mere puppets upon whose strings our genes alone tug.

The finding that there is a substantial genetic component to our behavior should not surprise us; indeed, it would be far more surprising if this were not the case. We are products of evolution: among our ancestors, natural selection indubitably exerted a strong influence over all traits that have figured in our survival. The human hand, with its marvelous opposable thumb, is the product of natural selection. In the past, therefore, there must have been varying forms of the hand, with natural selection favoring the version we have today by promoting the spread of the genetic variants underlying it; in this way, evolution ensured that every member of our species would be endowed with this supremely valuable asset.

Behavior, too, has been critical to human survival, and therefore sternly governed by natural selection. Presumably our enthusiasm for fatty and sweet foods evolved this way. Our ancestors were ever pressed to meet their nutritional requirements; therefore the propensity to take full advantage of all energy-rich foods whenever any became available was of huge benefit. Natural selection would have favored any genetic variations that ensured a sweet tooth since those with it survived better. Today those same genes are the scourge of everyone who struggles to keep off the weight in parts of the world with abundant food sources: what was adaptive in our ancestors is now maladaptive.

Ours is a strikingly social species; it is logical, therefore, to infer that natural selection once favored genetic adaptations facilitating social interaction. Not only would gestures, like smiling, have evolved as a means of signaling one's state of mind to other members of the group, but presumably there would also have been strong selective pressures in favor of psychological adaptations permitting one to judge the intentions of others. Social groups are prey to parasitism; there are always individuals who seek to benefit from membership without contributing to the general good. The capacity to detect such freeloaders is vital to the success of a cooperative social dynamic. And though we are no longer hovering in small groups around one fire roasting the communal supper, our gifts for sensing one another's moods and motivations may nevertheless come from those early phases of our development as a social species.

Since the publication of E. O. Wilson's *Sociobiology* in 1975, evolutionary approaches to understanding human behavior have themselves evolved, giving rise to the modern discipline of evolutionary psychology. In this field the search is for the common denominators of our behavior – human nature, the characteristics shared by all of us, whether New Guinea Highlander or Parisienne –

which we seek to understand, trait by trait in relation to some past adaptive advantage conferred by each (see Plate 61 & 62). Some such correlations are simple and relatively uncontroversial: the grasping reflex of a newborn, for instance, strong enough that a baby can use its hands and feet to suspend its full body weight, is presumably a legacy from the time when the ability to cling to a hirsute mother was important for infant survival.

Evolutionary psychology does not, however, limit its scope to such mundane faculties. Is the relatively low representation of women in the mathematical sciences worldwide a universal fact of culture, or might eons of evolution have selected male and female brains for different purposes? Can we understand in strictly Darwinian terms the tendency of older men to marry younger women? With a teenager likely to produce more children than a thirty-five-year-old, might such men be seen as succumbing to the power of evolutionary hardwiring that urges each of them to maximize the number of his offspring? Similarly, do younger women go for wealthy older men because natural selection has operated in the past to favor such a preference: a powerful male with plenty of resources? For now any answers to these questions are mainly conjectural. As we discover more of the genes underpinning behavior, however, I am confident that evolutionary psychology will migrate from its current position on the fringes of anthropology to the very heart of the discipline.

For now the power of genes to affect behavior is more evident in other species, whose nature we can actually manipulate using genetic tricks. One of the oldest, and most effective, of those tricks is artificial selection, which farmers have long used to increase milk yield in cows or wool quality in sheep. But its applications have not been limited to agriculturally valuable traits like these. Dogs are derived from wolves – possibly from wolf individuals that tended to hang around human settlements looking for a scrap and thereby conveniently assisting in garbage disposal. It is thought that they first laid claim to the title of "man's best friend" 10,000 years ago, roughly coinciding with the origin of agriculture. In the brief time since, the anatomical and behavioral diversity engendered by dog breeders has become literally a thing to behold. Dog shows are celebrations of the power of genes, with each breed effectively a genetic isolate – a freeze frame in the spectacular feature film of canine genetic diversity. Of course, the morphological differences are the most striking and fun to consider: the fluff ball called Pekingese; the enormous shaggy English mastiff, which sometimes weighs over 300 pounds; the stretched-out dachshund; the face-flattened bulldog. But it is the behavioral differences that I find most impressive.

Of course, not all dogs within a breed behave alike (or look alike), but typically individuals in the same breed have much more in common with one another than with specimens from other breeds. The Labrador retriever is affectionate and pliant; the greyhound is twitchy; the border collie will round up anything available if a flock of sheep is not to be had; the pit bull, as news reports occasionally remind us, is the canine embodiment of aggression. Some dog behaviors are so engrained as to have become stereotypes. Think of a pointer's elaborate "pointing" stance: that is no "stupid pet trick" taught each individual dog, but a hardwired part of the breed's genetic makeup. Despite their diversity, all modern dogs remain members of one species – meaning that in principle even the most apparently dissimilar pair can produce offspring, as one doughty male dachshund demonstrated in 1972 when he managed to inseminate a sleeping Great Dane bitch. Thirteen "Great Dachshunds" resulted.

While the basis of most behavior is surely polygenic – affected by many genes – a number of simple genetic manipulations in mice reveal that changing even a single gene can have a major behavioral effect. In 1999, Princeton neurologist Joe Tsien used sophisticated recombinant DNA techniques to create a "smart mouse" with extra copies of a gene producing a protein that acts as a receptor for chemical signals in the nervous system. Tsien's transgenic rodent performed better than normal mice in a battery of tests of learning and memory; for example, it was better at figuring out mazes and at retaining that knowledge. Tsien named the mouse strain "Doogie" after the precocious medic in the television show *Doogie Howser, M.D.* In 2002 Catherine Dulac at Harvard discovered that by deleting a gene from a mouse of her own she could affect the processing of chemical information contained in pheromones, the odors mice use to communicate. Whereas male mice will typically attack other males and attempt to mate with females, Dulac's doctored males failed to distinguish between males and females and attempted to mate with any mouse they encountered. Nurturing behavior in mice is also subject to a sex-specific manipulation of a single gene. Females instinctively look after their newborns, but Jennifer Brown and Mike Greenberg at Harvard Medical School found a way to short-circuit this native sense by knocking out the function of a gene called *fos*-B. Otherwise entirely normal, a mouse so altered simply ignores her offspring (see Plates 59 & 60).

Rodents have even given us a glimpse of the mechanistic basis of what in humans we call "love" (in rodents, it's less romantically termed "pair-bonding"). Mouselike voles are common throughout North America. Although they all look pretty much alike, different species have wildly different approaches to life. The prairie vole is monogamous, meaning that couples bond for life, but its close relative, the montane vole, is promiscuous: a male mates and moves on, and over the course of her life the female typically produces litters by several different males. What differences could underlie such widely divergent sexual strategies? Hormones provide the first half of the answer. In all mammals, oxytocin figures in many aspects of mothering: it stimulates both the contractions in labor and the production of breast milk for the newborn; thus it also plays a part in creating the nurturing bond between a mother and her young. Could the same hormone also engender another kind of bond, one between members of a prairie vole couple? In fact it does, together with another garden-variety mammalian hormone, vasopressin, which is primarily known for controlling urine production. But why is the montane vole, which also produces both hormones, such a randy little critter compared with its prairie cousin? The key, it turns out, is in the hormone receptors – the molecules that bind to circulating hormones, initiating a cell's response to the hormone signal.

Focusing on the vasopressin receptor, Tom Insel, a psychiatrist at Emory University, found a major difference between the vole species, not in the receptor gene itself but in an adjacent DNA region that determines when and where the gene is turned on. As a result the distribution of vasopressin receptors in the brain is very different in the prairie vole and the montane vole. But does this difference in gene regulation alone explain why one species is, in human terms, loving and the other cavalier? Apparently it does. Insel and his colleague Larry Young inserted the prairie vole vasopressin gene, complete with its next-door regulatory region, into a regular lab mouse (a species that is typically promiscuous, like the montane vole). Though the transgenic mice did not instantly become romantic pair-bonders, Insel and Young observed a marked change in their behavior. Rather than doing the typical mouse thing of mating with a female and then hightailing it unceremoniously, the transgenic male appeared in contrast tenderly solicitous of the female – in short, the addition of the gene, while not a guarantee for everlasting love, did seem to make the affected mouse less of a rat.

We must not forget that human brain function remains a million miles removed from that of a mouse. No rodent, whether from the mountains or the prairies, has yet produced a major work of art. It nevertheless remains worthwhile to bear in mind that most sobering lesson of the Human Genome Project: our genome and the mouse's are strikingly similar. The basic genetic software governing both mice and men has not changed much over the 75 million years of evolution since our lineages separated.

Unable to target specific genes for inactivation or enhancement as mouse geneticists can, human geneticists must rely on what might be called "natural experiments" – spontaneous genetic changes that affect brain function. Many of the best characterized genetic disorders affect mental performance. Down syndrome, caused by having an extra copy of chromosome 21, results in lowered IQ and, in many cases, a disarmingly sunny disposition. People with Williams syndrome, caused by the loss of a small portion of chromosome 7, also have low IQs but are often preternaturally talented as musicians.

But these are instances in which the mental aspects of a given disorder are effectively by-products of systemwide dysfunction. Thus they tell us relatively little about the particular genetics of behavior. It's a bit like discovering that your computer doesn't work during a power outage. OK, so you know now that it requires electricity, but you've learned very little about the specifics of computer function. To understand the genetics of behavior we need to look at disorders that affect the mind directly.

Among the mental scourges that have attracted close attention from gene mappers, two of the most formidable are bipolar (or manic-depressive) disease (BPD) and schizophrenia. Both diseases have strong genetic components (identical-twin concordance for BPD is as high as

80 percent; for schizophrenia it is close to 50 percent), and both take a devastating toll on mental health worldwide. One in every hundred people is schizophrenic, and the figure is about the same for BPD.

As we have seen, mapping polygenic traits is difficult because each individual gene has only a small incremental effect, and the trait as a whole is often mediated strongly by the environment, as with both these afflictions. But the overall difficulty has also bred a bad habit among researchers: they tend to publish only positive results, failing to notify the field of eliminated possibilities. Compounding the problem is the converse: the understandable but ultimately counterproductive impulse to publish any correlation that appears, after countless other genetic markers have proved a dead end. Ideally, spotting a correlation should only be the beginning of more in-depth analysis to sort out meaningful results from statistical coincidence – after all, if we try enough markers, we should expect occasionally to see, even in the absence of genetic linkage, a correlation produced by chance alone. Too often the pressure to get results leads to premature pronouncements, which must later be withdrawn sheepishly after other groups fail to replicate the finding.

There are additional impediments to the gene hunt when mental illness is the target. Diagnosis, however psychiatric manuals may try to standardize it, is often more an art than a science. Cases may be identified on the basis of ambiguous symptoms, and so a portion of the individuals in a pedigree may be misdiagnosed; these false positives wreak havoc in the mapping analysis. Another complicating factor is that the disorders are defined and diagnosed according to their symptoms, and yet it is likely that a number of genetic causes result in a similar set of symptoms. Thus the genes underlying schizophrenia may differ from one case to the next. Even apparently clear differences between syndromes can prove messy when viewed through the microscope of genetics. Since 1957 we have known that BPD and unipolar disease (the condition characterized by depression alone) are genetically distinct syndromes, but, confusingly, there is some genetic overlap between the two: unipolar depression is much more common among relatives of BPD patients than in the overall population.

Partly for such reasons, the genetic culprits of mental illness have so far proved particularly elusive. A recent study reveals that as many as twelve chromosomes – half the total – have been shown through mapping analysis to contain genes contributing to schizophrenia. It's the same story for BPD, in which genes on ten chromosomes have been implicated. One interesting finding is that there does seem to be some overlap between the gene regions identified in the separate mapping studies of the two maladies. Perhaps, then, there are some genes responsible for the overall organization and structure of our brains. Malfunctions in these genes may be the cause of the delusional or hallucinatory episodes common to both BPD and schizophrenia. The history of this research is full of high hopes brought low. A study will identify a strong correlation in one pedigree, and then subsequent research will fail to generalize the result in other populations. Such was the case in 1987, with a much-publicized study of BPD in the Amish: a promising connection with chromosome 11 did not fare well in follow-up studies. Stanford geneticists Neil Risch and David Botstein have aptly articulated these disappointments:

In no field has the difficulty [of mapping disease genes] been more frustrating than in the field of psychiatric genetics. Manic depression (bipolar illness) provides a typical case in point. Indeed, one might argue that the recent history of genetic linkage studies for this disease is rivaled only by the course of the illness itself. The euphoria of linkage findings being replaced by the dysphoria of non-replication [in other populations] has become a regular pattern, creating a roller coaster-type existence for many psychiatric genetics practitioners as well as their interested observers.

Without denying these difficulties I am extremely hopeful that we are right now entering an era of genetic analysis that will soon take us beyond this irritating game of "now we have it, now we don't." Two innovations hold the key. First, the "candidate gene" approach to finding the genes. With both the complete human genome sequence and a rudimentary functional understanding of many genes finally in hand, we can narrow our search as never before, homing in on genes with functions related to a given disorder. In the case of BPD, for example, a condition apparently connected with a fault in the mechanism by which the brain regulates its concentration of certain chemical neurotransmitters like serotonin and dopamine, we might choose to concentrate on genes that produce neurotransmitters or their receptors. Having chosen our candidate gene, we simply compare its sequence in affected and unaffected individuals to determine whether or not a particular variant might correlate with the disorder. In 2002, Eric Lander's team at MIT's Whitehead Institute surveyed seventy-six BPD candidate genes. Only one – a gene encoding the brain-specific nerve growth factor, the neurochemical tested as a possible treatment for Lou Gehrig's disease (see chapter 5) – proved to correlate with the disorder. But one truly relevant gene can be extremely valuable. The one found resides on chromosome 11, apparently vindicating the original Amish study, which long ago implicated the same region of the chromosome in BPD.

Technological improvements underlie the other reason for my optimism about the hunt for these elusive genes. To detect the subtle effect of a particular gene, we need extra-sensitive statistical analyses, which themselves require very large data-sets. Only with the advent of high-throughput sequencing and genetic-typing technologies have we had the capacity to collect appropriate data for huge numbers of markers from huge numbers of people. Not surprisingly, such industrial-scale genetic analysis is beyond the reach of most academic labs, so we will see biotech companies bankrolled by the pharmaceutical industry come to play an increasingly prominent role in this area. In 2002, two such companies, Genset in France and deCODE in Iceland, identified separate genes implicated in schizophrenia. These discoveries are a major step forward: because we have now fingered actual genes – as opposed to merely mapping an effect to a region of a chromosome – we can study gene function to learn about the biochemical basis of the disorder. Strikingly, both genes are involved in regulating the function of a particular neurotransmitter, glutamate.

With these new approaches – candidate genes and super-powerful genetic mapping – I am confident that we will soon uncover the major genes contributing to BPD and schizophrenia. Hopefully that will lead to improved treatments, as well as to a better understanding of how genes govern the workings of our brain.

For traits about whose neurochemical basis we have no clue, however, the roller-coaster ride of euphoric expectations dysphorically dashed is likely to continue. This has often been the case in studies of nonpathological behavior. Dean Hamer's 1993 analysis of the genetics of male homosexuality provides a case in point. It caused quite a stir when he found a particular region on the X chromosome that seemed to correlate with being gay. If being gay were proven to be as much a function of genes as is, say, skin color, then perhaps antidiscrimination legislation applicable to skin color was equally applicable to gays. Hamer's finding, however, has not withstood the test of time. Nevertheless, I suspect that as we develop more statistically powerful means of analysis (and learn to recognize and discount weaker correlations), we will indeed eventually identify some genetic factors that predispose us to our respective sexual orientations. But this should not be taken as purely determinist conjecture; environment is never to be discounted and a predisposition does not a predetermination make.

My pasty complexion may predispose me to skin cancer but, absent the effects of ultraviolet input from the environment, my genes are merely a matter of potential.

Hamer's other high-profile discovery looks more robust. He looked into the genetics underlying the urge for novelty, one of five key "personality dimensions" identified by psychologists. Do you cower in a corner when your routine gets disrupted? Or do you go out of your way to avoid a rut, subjecting yourself to an ever-changing kaleidoscope of new adventures? These, of course, are the extremes. Hamer's evidence pointed to a slight but significant effect of variation in a gene underlying a receptor for the brain signal molecule dopamine. Some attempts to replicate this result have failed, but others have extended it, finding the same gene implicated in particular types of novelty seeking, including drug abuse.

Violence, too, can be viewed through the lens of genetics. Some people are more violent than others. That's a fact. And violent behavior may be governed by a single gene interacting with environmental factors. This does not, of course, mean that we all carry a "violence gene" (though it's likely that most violent individuals do possess a Y chromosome), but we have identified at least one simple genetic change that can lead to violent outbursts. In 1978, Dr. Hans Brunner, a clinical geneticist at University Hospital in Nijmegen, Holland, learned of a family whose men tended to border on mental retardation and were prone to aggressive episodes. Thirty years earlier, in an attempt to document this "curse," a relative had compiled an extensive dossier on the family's woes. Brunner brought the survey up-to-date. He found eight men in the clan who, despite coming from different nuclear families, evinced similar patterns of violence. One had raped his sister and subsequently stabbed a prison guard; another used his car to hit his boss after being mildly reprimanded for laziness; two others were arsonists.

That only men were affected suggested sex linkage. The inheritance pattern was consistent with a gene likely on the X chromosome, and recessive, meaning that it was typically unexpressed in women, in whom the other (normal) copy on their second X would mask the faulty one's effect. In men, with their single X, the recessive variant was automatically expressed. By comparing the DNA of affected and unaffected members of the family, Brunner and his team duly mapped the gene to the long arm of the X chromosome. In collaboration with Xandra Breakfield at Massachusetts General Hospital, he found that the eight violent men all had a mutated – and nonfunctional – copy of a gene coding for monoamine oxidase. This protein, found in the brain, regulates levels of a class of neurotransmitters called "monoamines," which include adrenaline and serotonin.

The monoamine oxidase story does not end with the eight violent Dutchmen. It turns out to provide an illuminating glimpse of the interaction between genes and the environment, the complex duet of nature and nurture that informs all our behavior. In 2002, Avshalom Caspi and others at London's Institute of Psychiatry examined why some boys from abusive homes grow up normal while others end up antisocial (in the technical sense of having a history of behavioral problems – not in the sense of preferring to keep company with Web pages over people or of tending to be found picking at the canapés in a lonely corner at parties). The survey revealed a genetic predictor of development: the presence or absence of a mutation in the region adjacent to the monoamine oxidase gene, the switch regulating the amount of the enzyme produced. Maltreated boys with high levels of the enzyme were less likely to become antisocial than those with low levels. In the latter case, genes and the environment conspired to predispose the boys to lives punctuated by brushes with the law. Girls are less likely to be affected because, with the gene located on the X chromosome, they must inherit *two* copies of the low-level version rather than one. Girls who do have two copies, however, are likely to have antisocial tendencies similar to those of affected boys. But again the causal relation is nowhere near 100 percent in either boys or girls: growing up abused and having low monoamine oxidase levels in no way guarantees a career in crime.

Among the most surprising discoveries of a monogenic (single-gene) impact on a complex form of human behavior is what the press have dubbed the "grammar gene." As we discussed in the context of human evolution (chapter 9), in 2001 mutations detected by Tony Monaco at Oxford in the FOXP2 gene were found to impair the ability to use and process language. Not only do those so affected have difficulty articulating, but they are stymied by simple grammatical reasoning that poses no trouble for the typical four-year-old: "Every day I wug; yesterday I ____." FOXP2, remember, encodes a transcription factor – a genetic switch – that apparently plays a crucial role in brain development. Rather than exerting a simple direct behavioral impact (like that of monoamine oxidase), FOXP2 affects behavior by shaping the very organ at the center of it all. FOXP2 will prove, I believe, a model for momentous discoveries yet to be made; if I am right, many of the most important genes governing behavior will indeed turn out to be those involved in constructing that most extraordinary of organs, that still supremely inscrutable mass of matter, the human brain. These genes influence us by how they build the exquisite piece of hardware that mediates all we do.

We are as yet in the early days of our attempts to understand the genetic underpinnings of our behavior, both that which we all have in common – human nature – and that which sets us apart, one person from another. But this is a fast-moving area of research; I'm sure that what I've written will be out-of-date by the time this book is published. The future promises a detailed genetic dissection of personality, and it is hard to imagine that what we discover will not tip the scales of the nature/nurture debate more and more in the direction of nature – a frightening thought for some, but only if we persist in being held hostage to a static, ultimately meaningless dichotomy. To find that any trait, even one with formidable political implications, has a mainly genetic basis is not to find something set immutably in stone. It is merely to understand the nature upon which nurture is ever acting, and those things we, as a society and as individuals, need to do if we are better to assist the process. Let us not allow transient political considerations to set the scientific agenda. Yes, we may uncover truths that make us uneasy in the light of our present circumstances, but it is those circumstances, not nature's truth, to which policy makers ought to address themselves. As those Irish children who packed the hedge schools understood very well, knowledge, however awkwardly acquired, is still preferable to ignorance.

OUR GENES AND OUR FUTURE

"The event on which this fiction is founded has been supposed, by Dr. Darwin, and some of the physiological writers of Germany, as not of impossible occurrence."

So begins Percy Bysshe Shelley's anonymous preface to his wife Mary Shelley's novel *Frankenstein*, a story whose grip on the modern imagination has exceeded by far that of anything the poet himself ever wrote. Perhaps no work since *Frankenstein* has so hauntingly captured the terrifying thrill of science at the point of discovering the secret of life. And probably none has dealt so profoundly with the social consequences of having appropriated such godlike power.

The idea of animating the inanimate, and improving upon life as it occurs naturally on earth, had captured the human imagination long before the publication of Mary Shelley's work in 1818. Greek mythology tells of the sculptor Pygmalion, who successfully petitioned Aphrodite, goddess of love, to breathe life into the statue of the beautiful woman he had carved from ivory. But it was during the feverish burst of scientific progress following the Enlightenment that it first dawned upon scientists that the secret of life might be within human reach. Indeed, the Dr. Darwin to whom the preface refers is not the familiar Charles but rather his grandfather Erasmus, whose experimental use of electricity to spark life back into dead body parts fascinated his acquaintance Shelley. In retrospect we know that Dr. Darwin's exploration of what was called "galvanism" was a red herring; the secret of life remained a secret until 1953. Only with the discovery of the double helix and the ensuing genetic revolution have we had grounds for thinking that the powers held traditionally to be the exclusive property of the gods might one day be ours. Life, we now know, is nothing but a vast array of coordinated chemical reactions. The "secret" to that coordination is the breathtakingly complex set of instructions inscribed, again chemically, in our DNA.

But we still have a long way to go on our journey toward a full understanding of how DNA does its work. In the study of human consciousness, for example, our knowledge is so rudimentary that arguments incorporating some element of vitalism persist, even as these notions have been debunked elsewhere. Nevertheless, both our understanding of life and our demonstrated ability to manipulate it are facts of our culture. Not surprisingly, then, Mary Shelley has many would-be successors: artists and scientists alike have been keen to explore the ramifications of our newfound genetic knowledge.

Many of these efforts are shallow and betray their creators' ignorance of what is and is not biologically feasible. But one in particular stands out in my mind as raising important questions, and doing so in a stylish and compelling way. Andrew Niccols's 1997 film *Gattaca* carries to the present limits of our imagination the implications of a society obsessed with genetic perfection. In a future world two types of humans exist – a genetically enhanced ruling class and an underclass that lives with the imperfect genetic endowments of today's humans. Supersensitive DNA analyses ensure that the plum jobs go to the genetic elite while "in-valids" are discriminated against at every turn. Gattaca's hero is the "in-valid" Vincent (Ethan Hawke), conceived in the heat of reckless passion by a couple in the back of a car. Vincent's younger brother, Anton, is later properly engineered in the laboratory and so endowed with all the finest genetic attributes. As the two grow up, Vincent is reminded of his own inferiority every time he tries, fruitlessly, to best his little brother in swim races. Genetic discrimination eventually forces Vincent to accept a menial job as a porter with the Gattaca Corporation.

At Gattaca, Vincent nurtures an impossible dream: to travel into space. But to qualify for the manned mission to Titan he must conceal his "in-valid" status. He therefore assumes the identity of the genetically elite Jerome (Jude Law), a one-time athlete, who, crippled in an accident, needs Vincent's help. Vincent buys samples of Jerome's hair and urine and uses them to secure illicit admission into the flight-training program. All seems to be going well when he encounters the statuesque Irene (Uma Thurman) and falls in love. But a week before he is to fly off into space, disaster strikes: the mission director is murdered and in the ensuing police investigation the hair of an "in-valid" is discovered at the crime scene. An eyelash Vincent has lost threatens not only to dash his desperate dream but to unjustly implicate him by DNA evidence as the director's murderer. Vincent's unmaking seems foreordained, but he evades a nightmarish genetic dragnet until another of Gattaca's directors is found to be the actual murderer. The film's ending is only semi-happy: Vincent will fly off into space but without Irene, who is found to carry certain genetic imperfections incompatible with long space missions. In real life, the two actors who play Vincent and Irene have their futures more under their personal control. Ethan Hawke and Uma Thurman later married and now live in New York City.

Few, if any, of us would wish to imagine our descendants living under the sort of genetic tyranny suggested by *Gattaca*. Setting aside the question of whether the scenario foreseen is technologically feasible, we must address the central issue raised by the film: Does DNA knowledge make a genetic caste system inevitable? A world of congenital haves and have-nots? The most pessimistic commentators foresee an even worse scenario: Might we one day go so far as to breed a race of clones, condemned to servile lives mandated by their DNA? Rather than strive to fortify the weak, would we aim to make the descendants of the strong ever stronger? Most fundamentally, should we manipulate human genes at all? The answers to these questions depend very much on our views of human nature.

Today much of the public paranoia surrounding the dangers of human genetic manipulation is inspired by a legitimate recognition of our selfish side – that aspect of our nature that evolution has hardwired to promote our own survival, if necessary at the expense of others. Critics envision a world in which genetic knowledge would be used solely to widen the gap between the privileged (those best positioned to press genetics into their own service) and the down-trodden (those whom genetics can only put at greater disadvantage). But such a view recognizes only one side of our humanity.

If I see the consequences of our increasing genetic understanding and know-how rather differently, it is because I acknowledge the other side as well. Disposed though we might be to competition, humans are also profoundly social. Compassion for others in need or distress is as much a genetic element of our nature as the tendency to smile when we're happy. Even if some contemporary moral theorists are content to ascribe our unselfish impulses to ultimately selfish considerations – kindness to others seen as simply a conditioned way of promoting the same benefit in return – the fact remains: ours is a uniquely social species. Ever since our ancestors first teamed up to hunt a mammoth for dinner, cooperation among individuals has been at the heart of the human success story. Given the powerful evolutionary advantage of acting collectively in this way, natural selection itself has likely endowed each of us with a desire to see others (and therefore our society) do well rather than fail.

Even those who accept that the urge to improve the lot of others is part of human nature disagree on the best way to go about it. It is a perennial subject of social and political debate. The prevailing orthodoxy holds that the best way we can help our fellow citizens is by addressing problems with their nurture. Underfed, unloved, and uneducated human beings have diminished potential to lead productive lives. But as we have seen, nurture, while greatly influential, has its limits, which reveal themselves most dramatically in cases of profound genetic disadvantage. Even with the most perfectly devised nutrition and schooling, boys with severe fragile X disease will still never be able to take care of themselves. Nor will all the extra tutoring in the world ever grant naturally slow learners a chance to get to the head of the class. If, therefore, we are serious about improving education, we cannot in good conscience ultimately limit ourselves to seeking remedies in nurture. My suspicion, however, is that education policies are too often set by politicians to whom the glib slogan "leave no child behind" appeals precisely because it is so completely unobjectionable. But children *will* get left behind if we continue to insist that each one has the same potential for learning.

We do not as yet understand why some children learn faster than others, and I don't know when we will. But if we consider how many commonplace biological insights, unimaginable fifty years ago, have been made possible through the genetic revolution, the question becomes pointless. The issue rather is this: Are we prepared to embrace the undeniably vast potential of genetics to improve the human condition, individually and collectively? Most immediate, would we want the guidance of genetic information to design learning best suited to our children's individual needs? Would we in time want a pill that would allow fragile X boys to go to school with other children, or one that would allow naturally slow learners to keep pace in class with naturally fast ones? And what about the even more distant prospect of viable germ-line gene therapy? Having identified the relevant genes, would we want to exercise a future power to transform slow learners into fast ones before they are even born? We are not dealing in science fiction here: we can already give mice better memories. Is there a reason why our goal shouldn't be to do the same for humans?

One wonders what our visceral response to such possibilities might be had human history never known the dark passage of the eugenics movement. Would we still shudder at the term "genetic enhancement"? The reality is that the idea of improving on the genes that nature has given us alarms people. When discussing our genes, we seem ready to commit what philosophers call the "naturalistic fallacy," assuming that the way nature intended it is best. By centrally heating our homes and taking antibiotics when we have an infection, we carefully steer clear of the fallacy in our daily lives, but mentions of genetic improvement have us rushing to run the "nature knows best" flag up the mast. For this reason, I think that the acceptance of genetic enhancement will most likely come about through efforts to prevent disease.

Germ-line gene therapy has the potential for making humans resistant to the ravages of HIV. The recombinant DNA procedures that have let plant molecular geneticists breed potatoes resistant to potato viruses could equally well make humans resistant to AIDS. But should this be pursued? There are those who would argue that rather than altering people's genes, we should concentrate our efforts on treating those we can and impressing upon everyone else the dangers of promiscuous sex. But I find such a moralistic response to be profoundly *immoral*. Education has proven a powerful but hopelessly insufficient weapon in our war. As I write, we are entering the third decade of the worldwide AIDS crisis; our best scientific minds have been bamboozled by the virus's remarkable capacity for eluding attempts to control it. And while the spread of the disease has been slowed for the moment in the developed world, huge swaths of the planet tick away as demographic time bombs. I am filled with dread for the future of those regions, populated largely by people who are neither wealthy nor educated enough to mount an effective response. We may wishfully expect that powerful antiviral drugs or effective HIV vaccines will be produced economically enough for them to be available to everyone everywhere. But given our record in developing therapies to date, the odds against such dramatic progress occurring are high. And yet those who propose to use germ-line gene modifications to fight AIDS may, sadly, need to wait until such conventional hopes turn to despair – and global catastrophe – before being given clearance to proceed.

All over the world government regulations now forbid scientists from adding DNA to human germ cells. Support for these prohibitions comes from a variety of constituencies. Religious groups – who believe that to tamper with the human germ line is in effect to play God – account for much of the strong knee-jerk opposition among the general public. For their part, secular critics, as we have seen, fear a nightmarish social transformation such as that suggested in *Gattaca* – with natural human inequalities grotesquely amplified and any vestige of an egalitarian society erased. But though this premise makes for a good script, to me it seems no less fanciful than the notion that genetics will pave the way to utopia.

But even if we allow hypothetically that gene enhancement *could* – like any powerful technology – be applied to nefarious social ends, that only strengthens the case for our developing it. Considering the near impossibility of repressing technological progress, and the fact that much of what is now prohibited is well on its way to becoming practicable, do we dare restrain our own research community and risk allowing some culture that does not share our values to gain the upper hand? From the time the first of our ancestors fashioned a stick into a spear, the outcomes of conflicts throughout history have been dictated by technology. Hitler, we mustn't forget, was desperately pressing the physicists of the Third Reich to develop nuclear weapons. Perhaps one day, the struggle against a latter-day Hitler will hinge on our mastery of genetic technologies.

I see only one truly rational argument for delay in the advance of human genetic enhancement. Most scientists share this uncertainty: can germ-line gene therapy ever be carried out safely? The case of Jesse Gelsinger has cast a long shadow on gene therapy in general. It's worth pointing out, though, that contrary to appearances, germ-line gene therapy should in principle be easier to accomplish safely than somatic cell therapy. In the latter case, we are introducing genes into billions of cells, and there is always a chance, as in the recent SCID case in France, that a crucial gene or genes will be damaged in one of those cells, resulting in the nightmarish side effect of cancer. With germ-line gene therapy, in contrast, we are inserting DNA into a single cell, and the whole process can accordingly be much more tightly monitored. But the stakes are even higher in germ-line therapy: a failed germ-line experiment would be an unthinkable catastrophe – a human being born flawed, perhaps unimaginably so, owing to our manipulation of his or her genes. The consequences would be tragic. Not only would the affected family suffer, but all of humankind would lose because science would be set back.

When gene therapy experiments in mice run aground, no career is aborted, no funding withdrawn. But should gene improvement protocols ever lead to children with diminished rather than improved potential for life, the quest to harness the power of DNA would surely be delayed for years. We should attempt human experimentation only after we have perfected methods to introduce functional genes into our close primate relatives. But even when monkeys and chimpanzees (an even closer match) can be safely gene enhanced, the start of human experimentation will require resolute courage; the promise of enormous benefit won't be fulfilled except through experiments that will ultimately put lives at some risk. As it is, conventional medical procedures, especially new ones, require similar courage: brain surgery too may go awry, and yet patients will undergo it if its potential positives outweigh the dangers.

My view is that, despite the risks, we should give serious consideration to germ-line gene therapy. I only hope that the many biologists who share my opinion will stand tall in the debates to come and not be intimidated by the inevitable criticism. Some of us already know the pain of being tarred with the brush once reserved for eugenicists. But that is ultimately a small price to pay to redress genetic injustice. If such work be called eugenics, then I am a eugenicist.

Over my career since the discovery of the double helix, my awe at the majesty of what evolution has installed in our every cell has been rivaled only by anguish at the cruel arbitrariness of genetic disadvantage and defect, particularly as it blights the lives of children. In the past it was the remit of natural selection – a process that is at once marvelously efficient and woefully brutal – to eliminate those deleterious genetic mutations. Today, natural selection still often holds sway: a child born with Tay-Sachs who dies within a few years is – from a dispassionate biological perspective – a victim of selection against the Tay-Sachs mutation. But now, having identified many of those mutations that have caused so much misery over the years, it is in our power to sidestep natural selection. Surely, given some form of preemptive diagnosis, anyone would think twice before choosing to bring a child with Tay-Sachs into the world. The baby faces the prospect of three or four long years of suffering before death comes as a merciful release. And so if there is a paramount ethical issue attending the vast new genetic knowledge created by the Human Genome Project, in my view it is the slow pace at which what we now know is being deployed to diminish human suffering. Leaving aside the uncertainties of gene therapy, I find the lag in embracing even the most unambiguous benefits to be utterly unconscionable. That in our medically advanced society almost no women are screened for the fragile X mutation a full decade after its discovery can attest only to ignorance or intransigence. Any woman reading these words should realize that one of the important things she can do as a potential or actual parent is to gather information on the genetic dangers facing her unborn children – by looking for deleterious genes in her family line and her partner's, or, directly, in the embryo of a child she has conceived. And let no one suggest that a woman is not entitled to this knowledge. Access to it is her right, as it is her right to act upon it. She is the one who will bear the immediate consequences.

Four years ago, my views on this subject received a very cold reception in Germany. The publication of my essay, "Ethical Implications of the Human Genome Project," in the highly respected newspaper *Frankfurter Allgemeine Zeitung* (*FAZ*), provoked a storm of criticism. Perhaps this was the editors' intent: Without my knowledge, let alone consent, the paper had given my essay a new title devised by the translator as "The Ethic of the Genome – Why We Should Not Leave the Future of the Human Race to God." While I subscribe to no religion and make no secret of my secular views, I would never have framed my position as a provocation to those who do. A surprisingly hostile response came from a man of science, the president of the German Federal Chamber of Medical Doctors, who accused me of "following the logic of the Nazis who differentiate between a life worth living and a life not worth living." A day later, an editorial entitled "Unethical Offer" appeared in the same paper that had published mine. The writer, Henning Ritter, argued with self-righteous conviction that in Germany the decision to end the lives of genetically damaged fetuses would never become a private matter. In fact, his grandstanding displayed a simple ignorance of the nation's law; in Germany today, it is solely the right of a pregnant woman, upon receipt of medical advice, to decide whether to carry her fetus to term.

The more honorable critics were those who argued openly from personal beliefs, rather than exploiting the terrifying specter of the German past. The respected German president, Johannes Rau, countered my views with an assertion that "value and sense are not solely based on knowledge." As a practicing Protestant, he finds truths in religious revelation while I, a scientist, depend only on observation and experimentation. I therefore must evaluate actions on the basis of my moral intuition. And I see only needless harm in denying women access to prenatal diagnosis until, as some would have it, cures exist for the defects in question. In a less measured comment, the Protestant theologian Dietmar Mieth called my essay the "Ethics of Horror," taking issue with my assertion that greater knowledge will furnish humans better answers to ethical dilemmas. But the existence of a dilemma implies a choice to be made, and choice to my mind is better than no choice. A woman who learns that her fetus has Tay-Sachs now faces a dilemma about what to do, but at least she has a choice, where before she had none. Though I am sure that many German scientists agree with me, too many seem to be cowed by the political past and the religious present: except for my longtime valued friend Benno Müller-Hill, whose brave book on Nazi eugenics, *Murderous Science* (*Tödliche Wissenschaft*), still rankles the German academic establishment, no German scientist saw reason to rise to my defense.

I do not dispute the right of individuals to look to religion for a private moral compass, but I do object to the assumption of too many religious people that atheists live in a moral vacuum. Those of us who feel no need for a moral code written down in an ancient tome have, in my opinion, recourse to an innate moral intuition long ago shaped by natural selection promoting social cohesion in groups of our ancestors.

The rift between tradition and secularism first opened by the Enlightenment has, in more or less its present form, dictated biology's place in society since the Victorian period. There are those who will continue to believe humans are creations of God, whose will we must serve, while others will continue to embrace the empirical evidence indicating that humans are the product of many millions of generations of evolutionary change. John Scopes, the Tennessee high school teacher famously convicted in 1925 of teaching evolution, continues to be symbolically retried in the twenty-first century; religious fundamentalists, having their say in designing public school curricula, continue to demand that a religious story be taught as a serious alternative to Darwinism. With its direct contradiction of religious accounts of creation, evolution represents science's most direct incursion into the religious domain and accordingly provokes the acute defensiveness that characterizes creationism. It could be that as genetic knowledge grows in centuries to come, with ever more individuals coming to understand themselves as products of random throws of the genetic dice – chance mixtures of their parents' genes and a few equally accidental mutations – a new gnosis in fact much more ancient than today's religions will come to be sanctified. Our DNA, the instruction book of human creation, may well come to rival religious scripture as the keeper of the truth.

I may not be religious, but I still see much in scripture that is profoundly true. In the first letter to the Corinthians, for example, Paul writes:

> Though I speak with the tongues of men and of angels, but have not love, I have become sounding brass or a clanging cymbal.
>
> And though I have the gift of prophecy, and understand all mysteries and all knowledge, and though I have all faith, so that I could remove mountains, but have not love, I am nothing.

Paul has in my judgment proclaimed rightly the essence of our humanity. Love, that impulse which promotes our caring for one another, is what has permitted our survival and success on the planet. It is this impulse that I believe will safeguard our future as we venture into uncharted genetic territory. So fundamental is it to human nature that I am sure that the capacity to love is inscribed in our DNA – a secular Paul would say that love is the greatest gift of our genes to humanity. And if someday those particular genes too could be enhanced by our science, to defeat petty hatreds and violence, in what sense would our humanity be diminished?

In addition to laying out a misleadingly dismal vision of our future within the film itself, the creators of *Gattaca* concocted a promotional tag line aimed at the deepest prejudices against genetic knowledge: "There is no gene for the human spirit." It remains a dangerous blind spot in our society that so many wish this were so. If the truth revealed by DNA could be accepted without fear, we should not despair for those who follow us.

NOTES

(London: MacMillan, 1892), p. 12.

22 "It is easy": ibid., p. 1.

22 "there is now no": George Bernard Shaw as cited in Diane B. Paul, *Controlling Human Heredity* (Atlantic Highlands, N.J.: Humanities Press, 1995), p. 75.

24 "The Wyandotte is": ibid., p. 66.

26 "family with mechanical": C. B. Davenport, *Heredity in Relation to Eugenics* (New York: Henry Holt, 1911), p. 56.

26 "broad-shouldered, dark hair": ibid., p. 245.

27 "More children": Margaret Sanger as quoted in D. M. Kennedy, *Birth Control in America* (New Haven: Yale University Press, 1970), p. 115.

28 "criminals, idiots": Harry Sharp as cited in E. A. Carlson, *The Unfit* (Cold Spring Harbor, N.Y.: Cold Spring Harbor Laboratory Press, 2001), p. 218.

28 "It is better": Oliver Wendell Holmes as cited in ibid., p. 255.

30 "Under existing": Madison Grant, *The Passing of the Great Race* (New York: Scribner, 1916), p. 49.

31 "America must": Calvin Coolidge as cited in D. Kevles, *In the Name of Eugenics* (Cambridge, Mass.: Harvard University Press, 1995), p. 97.

31 "the farseeing": Harry Laughlin as cited in S. Kühl, *The Nazi Connection* (New York: Oxford University Press, 1994), p. 88.

31 "must declare": Adolf Hitler's *Mein Kampf* as cited in Paul, p. 86.

31 "Those who": Adolf Hitler, *Mein Kampf,* trans. Ralph Manheim (Boston: Houghton Mifflin Company, 1971), p. 404.

31 "law for": Benno Müller-Hill, *Murderous Science* (Cold Spring Harbor, N.Y.: Cold Spring Harbor Laboratory Press, 1998), p. 35.

32 "extra-marital": ibid.

33 "simply the meddlesome": Alfred Russel Wallace as cited in A. Berry, *Infinite Tropics* (New York: Verso, 2002), p. 214.

33 "orthodox eugenicists": Raymond Pearl as cited in D. Miklos and E. A. Carlson, "Engineering American Society: The Lesson of Eugenics," *Nature Genetics* 1 (2000): 153–58.

CHAPTER 2: THE DOUBLE HELIX

36 "Inheritance insures": Friedrich Miescher as cited in Franklin Portugal and Jack Cohen, *A Century of DNA* (Cambridge, Mass.: MIT Press, 1977), p. 107.

46 "stupid, bigoted": Rosalind Franklin as cited in Brenda Maddox, *Rosalind Franklin* (New York: HarperCollins, 2002), p. 82.

54 "Nobody told me": Linus Pauling, interview, as cited at http://www.achievement.org/autodoc/page/pau0int-1

56 "The most beautiful experiment": John Cairns as quoted in Horace Judson, *The Eighth Day of Creation* (New York: Simon & Schuster, 1979), p. 188.

CHAPTER 3: READING THE CODE

68 "That's when I saw": Sydney Brenner, *My Life in Science* (London: BioMed Central, 2001), p. 26.

71 "We're the only two": Francis Crick as quoted in Horace Judson, *The Eighth Day of Creation* (New York: Simon & Schuster, 1979), p. 485.

77 "Without giving me": François Jacob as quoted in ibid., p. 385.

CHAPTER 4: PLAYING GOD

84 "rivaled the importance": Jeremy Rifkin as quoted by Randall Rothenberg in "Robert A. Swanson: Chief Genetic Officer," *Esquire*, December 1984.

88 "from corned beef to": Stanley Cohen, http://www. accessexcellence.org/AB/WYW/cohen/

92 "She made": Paul Berg as quoted at http://www. ascb.org/profiles/9610.html

93 "scientists throughout": Paul Berg et al., "Potential Biohazards of Recombinant DNA Molecules," letter to *Science* 185 (1974): 303.

93 "until the potential": ibid.

93 "our concern": ibid.

94 "the molecular biologists had clearly reached": Michael Rogers, "The Pandora's Box Congress," *Rolling Stone* 189 (1975): 36–48.

98 "I felt": Leon Heppel as quoted in James D. Watson and J. Tooze, *The DNA Story* (San Francisco: W. H. Freeman and Co., 1981), p. 204.

98 "In his cranberry": Arthur Lubow as cited in ibid., p. 121.

99 "In today's": Alfred Vellucci as cited in ibid., p. 206.

100 "Compared to": Watson as cited in James D. Watson, *A Passion for DNA* (Cold Spring Harbor, N.Y.: Cold Spring Harbor Laboratory Press, 2001), p. 73.

104 "You get a nice gold medal": Fred Sanger as quoted by Anjana Ahuja, "The Double Nobel Laureate Who Began the Book of Life," *The Times* (London), 12 January 2000.

CHAPTER 5: DNA, DOLLARS, AND DRUGS

109 "to become": Herb Boyer as quoted in Stephen Hall, *Invisible Frontiers* (New York: Oxford University Press, 2002), p. 65.

118 "a live human-made": *Diamond vs. Chakrabarty et al.* as cited in Nicholas Wade, "Court Says Lab-Made Life Can Be Patented," *Science* 208 (1980): 1445.

132 "I'll *make* it an issue": Jeremy Rifkin as cited in Daniel Charles, *Lords of the Harvest* (Cambridge, Mass.: Perseus, 2001), p. 94.

CHAPTER 6: TEMPEST IN A CEREAL BOX

134 "If man": http://www.nrdc.org/health/pesticides/hcarson.asp

137 "operating outside": Mary-Dell Chilton et al. as cited in Daniel Charles, *Lords of the Harvest* (Cambridge, Mass.: Perseus, 2001), p. 16.

137 "loved the smell": Rob Horsch as quoted in ibid., p. 1.

149 "Put a molecular": Bob Meyer as quoted in ibid., p. 132.

154 "I naively": Roger Beachy, Daphne Preuss, and Dean Dellapenna, "The Genomic Revolution: Everything You Wanted to Know About Plant Genetic Engineering but Were Afraid to Ask," *Bulletin of the American Academy of Arts and Sciences,* Spring 2002, p. 31.

156 "After BSE": Friends of the Earth press release as cited in Charles, p. 214.

158 "this kind": Charles, Prince of Wales, "The Seeds of Disaster," *Daily Telegraph* (London), 8 June 1998.

162 "Where genetically": E. O. Wilson, *The Future of Life* (New York: Knopf, 2002), p. 163.

CHAPTER 7: THE HUMAN GENOME

167 "put Santa Cruz": Robert Sinsheimer as quoted in Robert Cook-Deegan, *The Gene Wars* (New York: W. W. Norton & Co., 1994), p. 79.

167 "DOE's program": David Botstein as cited in ibid., p. 98.

167 "the National Bureau": James Wyngaarden as quoted in ibid., p. 139.

168 "It means": David Botstein as quoted in ibid., p. 111.

169 "an incomparable": Walter Gilbert as cited in ibid., p. 88.

171 "from the very start": James Wyngaarden as quoted in ibid., p. 142.

175 "The revelation": Kary B. Mullis, "The Unusual Origin of the

Polymerase Chain Reaction," *Scientific American* 262 (April 1990): 56–65.

175   "Mullis had": Frank McCormick as quoted in Nicholas Wade, "After the Eureka, a Nobelist Drops Out," *New York Times,* 15 September 1998.

184   "If somebody": William Haseltine as quoted in Paul Jacobs and Peter G. Gosselin, "Experts Fret Over Effect of Gene Patents on Research," *Los Angeles Times,* 28 February 2000.

184   "We'd be entitled": William Haseltine as quoted in ibid.

185   "I was": Francis Collins as quoted in interview, *Christianity Today,* 1 October 2001.

186   "little more": John Sulston and Georgina Ferry, *The Common Thread* (London: Bantam Press), p. 123.

186   "After we'd invested": Bridget Ogilvie as quoted in ibid., p. 125.

189   "Fix it": President Clinton as quoted in Kevin Davies, *Cracking the Code* (New York: The Free Press, 2001), p. 238.

190   "I'd love": Rhoda Lander as quoted in Aaron Zitner, "The DNA Detective," *Boston Globe Sunday Magazine,* 10 October 1999.

190   "an isolated": Eric Lander as quoted in ibid.

190   "I pretty much": Eric Lander as quoted in ibid.

193   "Today, we are": White House Press Release, available at: http://www.ornl.gov/hgmis/project/clinton1.html

CHAPTER 8: READING GENOMES

196   "seeing the surface": Mark Patterson as cited in Kevin Davies, *Cracking the Code* (New York: The Free Press, 2001), p. 194.

207   "Really trust": Barbara McClintock as paraphrased by Elizabeth Blackburn at http://www.cshl.edu/cgi-bin/ubb/library/ultimatebb.cgi?ubb=get_topic;f=1;t=000015

213   "had felt like an outcast": Claire Fraser as quoted in Ricki Lewis, "Exploring the Very Depths of Life," *Rennselaer Magazine,*

March 2001.

213   "Well, young lady": Claire Fraser as quoted in ibid.

213   "We went to": Claire Fraser as quoted in ibid.

223   "a new kind": http://cmgm.stanford.edu/biochem/brown.html

223   "We're toddlers": Pat Brown as quoted in Dan Cray, "Gene Detective," *Time* 158 (20 August 2001): 35–36.

224   "It was like thinking": Pat Brown as quoted in ibid.

228   "Because embryos are beautiful": Eric Wieschaus as quoted in Ethan Bier, *The Coiled Spring* (Cold Spring Harbor, N.Y.: Cold Spring Harbor Laboratory Press, 2000), p. 64

CHAPTER 9: OUT OF AFRICA

234   "It was like": Ralf Schmitz as quoted in Steve Olson, *Mapping Human History* (Boston: Houghton Mifflin, 2002), p. 80.

236   "I can't": Matthias Krings as quoted in Patricia Kahn and Ann Gibbons, "DNA from an extinct human," *Science* 277 (1997): 176–78.

236   "That's when": Matthias Krings as quoted in ibid.

240   "If everyone": Allan Wilson as quoted by Mary-Claire King at http://www.chemheritage.org/EducationalServices/pharm/chemo/readings/king.htm

244   "It turned out": Luigi Luca Cavalli-Sforza as quoted in Olson, p. 164.

255   "If we": Charles Darwin, *The Origin of Species* (New York, Penguin, 1985), p. 406.

CHAPTER 10: GENETIC FINGERPRINTING

268   "having a white woman": Brooke A. Masters, "For Trucker, the High Road to DNA Victory," *Washington Post,* Saturday, 8 December 2001, p. B01.

269   "unwelcome precedent": the Director of the Virginia

Department of Criminal Justice as quoted at http://www.innocenceproject.org/case/display_profile.php?id= 99

271 "profile provides": Alec Jeffreys, Victoria Wilson, and Swee Lay Thein, "Hypervariable 'minisatellite' regions in human DNA," *Nature* 314 (1985): 67–73.

271 "In theory": Alec Jeffreys as quoted at http://www.dist. gov.au/events/ausprize/ap98/jeffreys.html

274 "must be sufficiently": *Frye vs. United States*, 293 F.2d 1013, at 104.

276 "The implementation": Eric Lander, "Population genetic considerations in the forensic use of DNA typing," in Jack Ballantyne, et al., *DNA Technology and Forensic Science* (Cold Spring Harbor, N.Y.: Cold Spring Harbor Laboratory Press, 1989), p. 153.

277 "mountain of evidence": Johnnie Cochran as quoted at http://simpson.walraven.org/sep27.html

280 "Where is": Barry Scheck as quoted at http://simpson. walraven.org/apr11.html

284 "State of Wisconsin": Geraldine Sealey, "DNA Profile Charged in Rape," http://abcnews.go.com/sections/us/DailyNews/ dna991007.html

284 "unknown male": Case Number 00F06871, *The People of the State of California vs. John Doe*, Aug. 21, 2000.

285 "DNA appears": Judge Tani G. Cantil-Sakauye, Case Number 00F06871, *The People of the State of California vs. Paul Robinson*, Motion to Dismiss, reporter's transcript, p. 136, Feb. 23, 2001.

290 "My son": Jean Blassie as quoted in Pat McKenna, "Unknown, No More," http://www.af.mil/news/airman/0998/unknown.htm

298 "The DNA evidence": Lord Woolf, Case Number 199902010 S2, "Regina and James Hanratty," Judgment, May 10, 2002,

paragraph 211.

299   "In hindsight": "DNA testing also proves guilt," editorial, *St. Petersburg Times*, 30 May 2002.

301   "DNA testing": Barry Scheck et al., *Actual Innocence* (New York: Doubleday, 2000), p. xv.

CHAPTER 11: GENE HUNTING

303   "Fifty-fifty": Milton Wexler as quoted in Alice Wexler, *Mapping Fate* (New York: Random House, 1995), p. 43.

303   "Over 50 years": George Huntington as cited in Charles Stevenson "A Biography of George Huntington, M.D.," *Bulletin of the Institute of the History of Medicine* 2 (1934).

304   "gradually increase": ibid.

304   "As the disease": ibid.

304   "When either": ibid.

306   "without theater": Americo Negrette as quoted in Robert Cook-Deegan, *The Gene Wars* (New York: W. W. Norton & Co., 1994), p. 235.

308   "tends to think": ibid., p. 37.

312   "We would never": Ray White as quoted in Leslie Roberts, "Flap arises over genetic map," *Science* 239 (1987): 750–52.

321   "we own": Orrie Friedman as quoted in Richard Saltus, "Biotech Firms compete in Genetic Diagnosis," *Science* 234 (1986): 1318–20.

331   "committing": Rural Advancement Foundation International, at http://www.rafi.org/article.asp?newsid=207

332   "the creation": Althing (Icelandic Parliament); "Law on a Health Sector Database," at http://www.mannvernd.is/english/laws/law.HSD.html

CHAPTER 12: DEFYING DISEASE

341 "So marked": John Langdon Down as quoted in Elaine Johansen Mange and Arthur P. Mange, *Basic Human Genetics* (Sunderland, Mass.: Sinauer Associates, 1999), p. 267

345 "ear-piercing": Kathleen McAuliffe, "The Hardest Choice," at http://blueprint.bluecrossmn.com/topic/hardestchoice

359 "Some people": Debbie Stevenson, "The Mystery Disease No One Tests For," *Redbook,* July 2002: 137.

363 "biological Holocaust": Daniel Pollen, *Hannah's Heirs* (New York: Oxford University Press, 1993), p. 14.

372 "informed consent": U.S. Department of Health and Human Services Press Release "New Initiatives to Protect Participants in Gene Therapy Trials," 7 March 2000. Available at http://www.fda.gov/bbs/topics/NEWS/NEW00717.html

CHAPTER 13: WHO WE ARE

380 "Whatever person": Penal Laws as cited at http://www.law.umn.edu/irishlaw/education.html

381 "They might as well": Arthur Young as cited in Julie Henigan, "For Want of Education: The Origins of the Hedge Schoolmaster Songs," *Ulster Folklife* 40 (1994): 27–38.

381 "Still crouching": John O'Hagan as cited at http://www.in2it.co.uk/history/2.html

386 "barefoot professor": Vitaly Fyodorovich as cited in David Joravsky, *The Lysenko Affair* (Cambridge, Mass.: Harvard University Press, 1970), p. 189.

386 "Turkic peasant": Vitaly Fyodorovich as cited in Valery N. Soyfer, *Lysenko and the Tragedy of Soviet Science* (New Brunswick, N.J.: Rutgers University Press, 1994), p. 11.

386 "He didn't": Vitaly Fyodorovich as cited in ibid., p. 11.

387 "In order": Trofim Lysenko as cited in Joravsky, p. 110.

390 "hard core": ibid., p. 226

390 "It deals": K. Iu. Kostriukova as cited in ibid., p. 247.

390 "In our": Trofim Lysenko as cited in ibid., p. 210.

394 "Give me": John B. Watson, *Behaviorism* (New York: W. W. Norton & Co., 1924), p. 104.

400 "On multiple": Thomas J. Bouchard et al., "Sources of Human Psychological Differences: The Minnesota Study of Twins Reared Apart," *Science* 250 (1990): 223–28.

412 "In no field": Neil Risch and David Botstein, "A Manic Depressive History," *Nature Genetics* 12 (1996): 351–53.

CODA

419 "The event": Percy Bysshe Shelley, introduction to Mary Wollstonecraft Shelley, *Frankenstein* (New York: Oxford University Press, 1969), p. 13.

428 "Ethical Implications": James D. Watson in *Frankfurter Allgemeine Zeitung,* September 26, 2000.

429 "following the logic": Jörg Dietrich Hoppe as cited in Benno Müller-Hill, "Speaking Out in Favor of the Right to Choose" *Frankfurter Allgemeine Zeitung,* December 5, 2000.

429 "value and sense": Johannes Rau as cited in ibid.

429 "Ethics of Horror": Dietmar Mieth in *Frankfurter Allgemeine Zeitung* as cited in ibid.

431 "Though I speak": 1 Corinthians 13: 1–2.

FURTHER READING

CHAPTER 1: BEGINNINGS OF GENETICS

Carlson, Elof Axel. *The Unfit: A History of a Bad Idea.* Cold Spring Harbor, N.Y.: Cold Spring Harbor Laboratory Press, 2002. Discussion of eugenics beginning in biblical times and ending with contemporary clinical genetics.

Gillham, Nicholas Wright. *A Life of Sir Francis Galton: From African Exploration to the Birth of Eugenics.* New York: Oxford University Press, 2001. Engaging recent study of an extraordinary but neglected figure.

Jacob, François. *The Logic of Life: A History of Heredity.* Princeton: Princeton University Press, 1993. Reflections by one of the founders of molecular genetics.

Kevles, Daniel J. *In the Name of Eugenics: Genetics and the Uses of Human Heredity.* New York: Alfred A. Knopf, 1985. Scholarly but readable account of eugenics.

Kohler, Robert E. *Lords of the Fly: Drosophila Genetics and the Experimental Life.* Chicago: University of Chicago Press, 1994. Chronicle of the early days of fruit fly genetics.

Kühl, Stefan. *The Nazi Connection: Eugenics, American Racism, and German National Socialism.* New York: Oxford University Press, 1994.

Mayr, Ernst. *This Is Biology: The Science of the Living World.* Cambridge, Mass.: Harvard University Press, 1997. Fine overview from a biologist who has just celebrated the seventy-fifth anniversary of earning his Ph.D.

Müller-Hill, Benno. *Murderous Science: Elimination by Scientific Selection of Jews, Gypsies, and Others in Germany, 1933–1945.* Translated by Todliche Wissenschaft. New York: Oxford University Press, 1988. Reveals how German scientists and physicians were implicated in Nazi policies and how they resumed their academic positions after the war.

Olby, Robert C. *Origins of Mendelism.* Chicago: University of Chicago Press, 1985.

Orel, Vítezslav. *Gregor Mendel: The First Geneticist.* New York: Oxford University Press, 1996. The most complete biography to date.

Paul, Diane B. *Controlling Human Heredity, 1865 to the Present.* Atlantic Highlands, N.J.: Humanities Press, 1995. A succinct history of eugenics.

CHAPTER 2: THE DOUBLE HELIX

Crick, Francis H. C. *What Mad Pursuit: A Personal View of Scientific Discovery.* New York: Basic Books, 1988.

Hager, Thomas. *Force of Nature: The Life of Linus Pauling.* New York: Simon & Schuster, 1995. An excellent biography of a scientific giant.

Holmes, Frederic Lawrence. *Meselson, Stahl, and the Replication of*

DNA: A History of "The Most Beautiful Experiment in Biology." New Haven: Yale University Press, 2001.

McCarty, Maclyn. *The Transforming Principle: Discovering That Genes are Made of DNA.* New York: W. W. Norton & Co., 1995. Account of the experiments that showed DNA to be the hereditary material by one of the three scientists who carried them out.

Maddox, Brenda. *Rosalind Franklin: The Dark Lady of DNA.* New York: HarperCollins, 2002. Thorough biography that casts new light on Franklin.

Olby, Robert. *The Path to the Double Helix: The Discovery of DNA.* Foreword by Francis Crick. Dover Publishers, 1994. Scholarly historical perspective.

Watson, James D. *The Double Helix: A Personal Account of the Discovery of the Structure of DNA.* New York: Atheneum Press, 1968.

CHAPTER 3: READING THE CODE

Brenner, Sydney. *My Life in Science.* London: BioMed Central Limited, 2001. A rare combination: illuminating *and* funny.

Hunt, Tim, Steve Prentis, and John Tooze, ed. *DNA Makes RNA Makes Protein.* New York: Elsevier Biomedical Press, 1983. Collection of essays summarizing the state of molecular genetics in 1980.

Jacob, François. *The Statue Within: An Autobiography.* Translated by Franklin Philip. Cold Spring Harbor, N.Y.: Cold Spring Harbor Laboratory Press, 1995. Lucid and beautifully written.

Judson, Horace Freeland. *The Eighth Day of Creation: Makers of the Revolution in Biology.* Expanded edition. Cold Spring Harbor, N.Y.: Cold Spring Harbor Laboratory Press, 1996. Classic study on the origins of molecular biology.

Monod, Jacques. *Chance and Necessity: An Essay on the Natural*

Philosophy of Modern Biology. Translated by Austryn Wainhouse. New York: Alfred A. Knopf, 1971. Philosophical musings by a key figure in molecular genetics.

Watson, James D. *Genes, Girls, and Gamow.* New York: Alfred A. Knopf, 2001. Sequel to *The Double Helix.*

CHAPTER 4: PLAYING GOD

Fredrickson, Donald S. *The Recombinant DNA Controversy: A Memoir: Science, Politics, and the Public Interest 1974–1981.* Washington, D.C.: American Society for Microbiology Press, 2001. Account of turbulent times in biomedical research by the then-director of the National Institutes of Health.

Krimsky, Sheldon. *Genetic Alchemy: The Social History of the Recombinant DNA Controversy.* Cambridge, Mass.: MIT Press, 1982. A critic's perspective.

Rogers, Michael. *Biohazard.* New York: Alfred A. Knopf, 1977. Expansion of Rogers's insightful account in *Rolling Stone* of the Asilomar meeting.

Watson, James D. *A Passion for DNA: Genes, Genomes, and Society.* Cold Spring Harbor, N.Y.: Cold Spring Harbor Laboratory Press, 2000. Collection of essays drawn from newspapers, magazines, talks, and Cold Spring Harbor Laboratory reports.

Watson, James D., Michael Gilman, Jan Witkowski, and Mark Zoller. *Recombinant DNA.* New York: Scientific American Books, distributed by W. H. Freeman, 1992. Now out of date but still a sound introduction to the basic science underlying genetic engineering.

Watson, James D., and John Tooze. *The DNA Story: A Documentary History of Gene Cloning.* San Francisco: W. H. Freeman and Co., 1981. The recombinant DNA debate recounted through contemporary articles and documents.

CHAPTER 5: DNA, DOLLARS, AND DRUGS

Cooke, Robert. *Dr. Folkman's War: Angiogenesis and the Struggle to Defeat Cancer*. New York: Random House, 2001.

Hall, Stephen S. *Invisible Frontiers: The Race to Synthesize a Human Gene*. New York: Atlantic Monthly Press, 1987. Tells the insulin-cloning story with verve.

Kornberg, Arthur. *The Golden Helix: Inside Biotech Ventures*. Sausalito, Calif.: University Science Books, 1995. The founder of several companies describes the rise of the biotechnology industry.

Werth, Barry. *The Billion-Dollar Molecule: One Company's Quest for the Perfect Drug*. New York: Touchstone Books/Simon & Schuster, 1995. The story of Vertex, a company typifying the biotech approach to the pharmaceutical business.

CHAPTER 6: TEMPEST IN A CEREAL BOX

Charles, Daniel. *Lords of the Harvest: Biotech, Big Money, and the Future of Food*. Cambridge, Mass.: Perseus Publishing, 2001. Fascinating account of the genetically modified food controversy, emphasizing the business side and focusing primarily on Monsanto.

McHughen, Alan. *Pandora's Picnic Basket: The Potential and Hazards of Genetically Modified Foods*. New York: Oxford University Press, 2000. Spotty introduction to some of the issues, including scientific ones, behind the controversy.

CHAPTER 7: THE HUMAN GENOME

Cook-Deegan, Robert M. *The Gene Wars: Science, Politics, and the Human Genome*. New York: W. W. Norton & Co., 1994. Brilliantly comprehensive account of the origins and early days of the Human Genome Project.

Davies, Kevin. *Cracking the Genome: Inside the Race to Unlock Human DNA.* New York: Free Press, 2001. Continuation of Cook-Deegan's story, bringing it up to the completion of the first draft of the human genome.

Sulston, John, and Georgina Ferry. *The Common Thread: A Story of Science, Politics, Ethics, and the Human Genome.* Washington, D.C.: Joseph Henry Press, 2002. Personal account of research on the worm and of the British end of the Human Genome Project. Sulston's disdain for individuals and companies profiting from the human genome sequence drives his story and his science.

CHAPTER 8: READING GENOMES

Bier, Ethan. *The Coiled Spring: How Life Begins.* Cold Spring Harbor, N.Y.: Cold Spring Harbor Laboratory Press, 2000.

Comfort, Nathaniel C. *The Tangled Field: Barbara McClintock's Search for the Patterns of Genetic Control.* Cambridge, Mass.: Harvard University Press, 2001. A scholarly but approachable account of the life and work of Barbara McClintock.

Lawrence, Peter A. *The Making of a Fly: The Genetics of Animal Design.* Boston: Blackwell Scientific Publications, 1992. Now out of date but still an excellent introduction to the excitement generated when genetics meets developmental biology.

Ridley, Matt. *Genome: The Autobiography of a Species in 23 Chapters.* New York: HarperCollins, 1999. Hugely accessible introduction to modern studies of human genetics.

CHAPTER 9: OUT OF AFRICA

Cavalli-Sforza, L. L. (Luigi Luca). *Genes, Peoples, and Languages.* Translated by Mark Seielstad. New York: North Point Press, 2000. Personal account of human-evolution studies by the field's leader.

Olson, Steve. *Mapping Human History: Discovering the Past Through*

Our Genes. Boston: Houghton Mifflin, 2002. Balanced and up-to-date account of human evolution and the impact that our past has on our present.

Sykes, Bryan. *The Seven Daughters of Eve*. New York: W. W. Norton & Co., 2001.

CHAPTER 10: GENETIC FINGERPRINTING

Massie, Robert K. *The Romanovs: The Final Chapter*. New York: Random House, 1995. The story of the Romanovs' murders and of how DNA fingerprinting established the authenticity of the remains and unmasked impostors.

Scheck, Barry, Peter Neufeld, and Jim Dwyer. *Actual Innocence: Five Days to Execution and Other Dispatches from the Wrongly Convicted*. New York: Doubleday, 2000. From the horses' mouths, an examination of the power of DNA fingerprinting to exonerate the wrongfully convicted.

Wambaugh, Joseph. *The Blooding*. New York: Bantam Books, 1989. Exciting account of the first use of DNA fingerprinting to apprehend a criminal.

CHAPTER 11: GENE HUNTING

Bishop, Jerry E., and Michael Waldholz. *Genome: The Story of the Most Astonishing Scientific Adventure of Our Time – The Attempt to Map All the Genes in the Human Body*. New York: Simon & Schuster, 1990. Still one of the best accounts of the early days of hunting human disease genes.

Gelehrter, Thomas D., Francis Collins, and David Ginsburg. *Principles of Medical Genetics*. Baltimore: Williams & Wilkins, 1998. A short and readable textbook on modern human molecular genetics.

Pollen, Daniel A. *Hannah's Heirs: The Quest for the Genetic Origins of Alzheimer's Disease*. New York: Oxford University Press, 1993.

Captures the thrill of the chase and highlights the awfulness of the disease.

Wexler, Alice. *Mapping Fate: A Memoir of Family, Risk, and Genetic Research.* New York: Random House, 1995. Searingly honest testimony from Nancy Wexler's sister.

CHAPTER 12: DEFYING DISEASE

Davies, Kevin, with Michael White. *Breakthrough: The Race to Find the Breast Cancer Gene.* New York: John Wiley & Sons Inc., 1996. Story of immensely hard work, dedication, ambition, and greed.

Kitcher, Philip. *The Lives to Come: The Genetic Revolution and Human Possibilities.* New York: Simon & Schuster, 1997. Philosophical and ethical discussion about how to use what we have learned of human molecular genetics.

Lyon, Jeff, with Peter Gorner. *Altered Fates: Gene Therapy and the Retooling of Human Life.* New York: W. W. Norton & Co., 1995. Includes a good account of the treatment of the two girls with ADA deficiency.

Reilly, Philip R. *Abraham Lincoln's DNA and Other Adventures in Genetics.* Cold Spring Harbor, N.Y.: Cold Spring Harbor Laboratory, 2000. Essays on topical issues written from the unusually informed perspective of a physician-cum-lawyer.

Thompson, Larry. *Correcting the Code: Inventing the Genetic Cure for the Human Body.* New York: Simon & Schuster, 1994. Account of the development of gene therapy, including the Martin Cline episode.

CHAPTER 13: WHO WE ARE

Coppinger, Raymond, and Lorna Coppinger. *Dogs: A Startling New Understanding of Canine Origin, Behavior, and Evolution.* New York: Scribner, 2001. Overview of the enormous differences, in body and mind, among dogs.

Crick, Francis H. C. *The Astonishing Hypothesis: The Scientific Search for the Soul.* New York: Scribner, 1993. A materialist perspective on the problem of consciousness. Crick concludes that we are "no more than the behavior of a vast assembly of nerve cells and their associated molecules."

Herrnstein, Richard J., and Charles Murray. *The Bell Curve: Intelligence and Class Structure in American Life.* New York: Free Press, 1994. More talked about than read.

Jacoby, Russell, and Naomi Glauberman, ed. *The Bell Curve Debate: History, Documents, Opinions.* New York: Times Books, 1995. Collection of eighty essays about and reviews of *The Bell Curve.*

Lewontin, R. C., Steven Rose, and Leon J. Kamin. *Not in Our Genes: Biology, Ideology, and Human Nature.* New York: Pantheon Books, 1984. The academic left's response to Wilson's *Sociobiology.*

Mendvedev, Zhores A. *The Rise and Fall of T. D. Lysenko.* New York: Columbia University Press, 1969. Firsthand account by a scientist who suffered from the Communist Party's control of Soviet science.

Pinker, Steven. *The Blank Slate: The Modern Denial of Human Nature.* New York: Viking Penguin, 2002.

Pinker, Steven. *How the Mind Works.* New York: W. W. Norton & Co., 1997. Evolutionary psychology outlined by one its most eloquent proponents.

Ridley, Matt. *Nature via Nurture: Genes, Experience, and What Makes Us Human.* New York: HarperCollins, 2003.

Soyfer, Valery N. *Lysenko and the Tragedy of Soviet Science.* Translated by Leo Gruliow and Rebecca Gruliow. New Brunswick, N.J.: Rutgers University Press, 1994. An account from someone who knew Lysenko.

Wilson, Edward O. *Sociobiology: The New Synthesis.* Cambridge,

Mass.: Belknap Press of Harvard University Press, 1975. Proposes an evolutionary explanation for much of our behavior.

ACKNOWLEDGMENTS

This book is one of several strands that together comprise a major effort to commemorate the fiftieth anniversary of the discovery of the double helix. All of the projects – this book, a five-part TV series, a multimedia educational product, and a short film for science museum audiences – are interconnected in many ways. We therefore find ourselves indebted to more than the usual slew of readers, editors, and spouses found in the acknowledgments section of a typical nonfiction book. What follows is a reflection of the size and scope of a sprawling collaborative project.

Throughout, the Alfred P. Sloan Foundation, the Howard Hughes Medical Institute, and the University of North Carolina have been phenomenally generous in their support. With wisdom and good sense, John Cleary and John Maroney oversaw the project's alarmingly complex logistics, ensuring that its many strands never became unraveled.

The television series was produced by David Dugan of Windfall Productions in London under the direction of David Glover and Carlo Masarella. To create the educational components, Max Whitby of the Red Green & Blue Company, also in London, collaborated with a team under Dave Micklos at the Cold Spring Harbor Dolan DNA Learning Center and with genius animator Drew Berry (no relation) at the Walter and Eliza Hall Institute in Melbourne, Australia.

The illustrations for the book were prepared by Keith Roberts of the John Innes Centre in Norwich, England. With his customary flair for combining design with scientific clarity, Keith has, with Nigel Orme, produced a series of illustrations that we feel massively enhance the value of the book. Robin Reardon, the assistant editor at Knopf, managed against all odds to coax us into making deadline after deadline (well, more or less) without once having to resort to physical intimidation. Designer Peter Andersen, also at Knopf, effected the miraculous marriage between the text and the images. Keith, Robin, and Peter were indispensable members of the team.

Many people read versions of the book or of chapters addressing their particular areas of expertise. The following graciously supplied detailed and insightful comments on the manuscript: Fred Ausubel, Paul Berg, David Botstein, Stanley Cohen, Francis Collins, Jonathan Eisen, Mike Hammer, Doug Hanahan, Rob Horsch, Sir Alec Jeffreys, Mary-Claire King, Eric Lander, Phil Leder, Victor McElheny, Svante Pääbo, Joe Sambrook, and Nancy Wexler.

Many others also supplied helpful information and/or images: Bruce Ames, Jay Aronson, Antonio Barbadilla, John Barranger, Jacqueline Barataud, Caroline Berry, Sam Berry, Ewan Birney, Richard Bondi, Herb Boyer, Pat Brown, Clare Bunce, Caroline Caskey, Tom Caskey, Luigi Luca Cavalli-Sforza, Shirley Chan, Francis A. Chifari, Kenneth Culver, Charles DeLisi, John Doebley, Helen Donis-Keller, Cat Eberstark, Mike Fletcher, Judah Folkman, Norm Gahn, Wally Gilbert,

Janice Goldblum, Eric Green, Wayne Grody, Mike Hammer, Krista Ingram, Leemor Joshua-Tor, Linda Pauling Kamb, David King, Robert Koenig, Teresa Kruger, Brenda Maddox, Tom Maniatis, Richard McCombie, Benno Müller-Hill, Tim Mulligan, Kary Mullis, Harry Noller, Peter Neufeld, Margaret Nance Pierce, Naomi Pierce, Tomi Pierce, Daniel Pollen, Mila Pollock, Sue Richards, Tim Reynolds, Matt Ridley, Julie Reza, Barry Scheck, Mark Seielstad, Phil Sharp, David Spector, Rick Stafford, Debbie Stevenson, Bronwyn Terrill, William C. Thompson, Lap-Chee Tsui, Peter Underhill, Elizabeth Watson, Diana Wellesley, Rick Wilson, David Witt, Jennifer Whiting, James Wyngaarden, Larry Young, Norton Zinder.

Thank you all.

All the above did their best to ensure that we got things right. Nevertheless we are wholly responsible for the errors that no doubt remain.

INDEX

Entries referring to key terms or concepts in the text are set in **bold**.

PENGUIN CLASSICS

THE ARABIAN NIGHTS
TALES OF 1001 NIGHTS
VOLUME 3

MALCOLM C. LYONS, sometime Sir Thomas Adams Professor of Arabic at Cambridge University and a life Fellow of Pembroke College, Cambridge, is a specialist in the field of classical Arabic Literature. His published works include the biography *Saladin: The Politics of the Holy War*, *The Arabian Epic: Heroic and Oral Story-telling*, *Identification and Identity in Classical Arabic Poetry* and many articles on Arabic literature.

URSULA LYONS, formerly an Affiliated Lecturer at the Faculty of Oriental Studies at Cambridge University and, since 1976, an Emeritus Fellow of Lucy Cavendish College, Cambridge, specializes in modern Arabic literature.

ROBERT IRWIN is the author of *For Lust of Knowing: The Orientalists and Their Enemies*, *The Middle East in the Middle Ages*, *The Arabian Nights: A Companion* and numerous other specialized studies of Middle Eastern politics, art and mysticism. His novels include *The Limits of Vision*, *The Arabian Nightmare*, *The Mysteries of Algiers* and *Satan Wants Me*.

The Arabian Nights

Tales of 1001 Nights

Volume 3
Nights 719 to 1001

Translated by MALCOLM C. LYONS,
with URSULA LYONS
Introduced and Annotated by ROBERT IRWIN

PENGUIN BOOKS

PENGUIN CLASSICS

Published by the Penguin Group
Penguin Books Ltd, 80 Strand, London WC2R ORL, England
Penguin Group (USA) Inc., 375 Hudson Street, New York, New York 10014, USA
Penguin Group (Canada), 90 Eglinton Avenue East, Suite 700, Toronto, Ontario, Canada M4P 2Y3
(a division of Pearson Penguin Canada Inc.)
Penguin Ireland, 25 St Stephen's Green, Dublin 2, Ireland (a division of Penguin Books Ltd)
Penguin Group (Australia), 250 Camberwell Road, Camberwell, Victoria 3124, Australia
(a division of Pearson Australia Group Pty Ltd)
Penguin Books India Pvt Ltd, 11 Community Centre, Panchsheel Park, New Delhi – 110 017, India
Penguin Group (NZ), 67 Apollo Drive, Rosedale, North Shore 0632, New Zealand
(a division of Pearson New Zealand Ltd)
Penguin Books (South Africa) (Pty) Ltd, 24 Sturdee Avenue, Rosebank,
Johannesburg 2196, South Africa

Penguin Books Ltd, Registered Offices: 80 Strand, London WC2R ORL, England

www.penguin.com

This translation first published in Penguin Classics hardback 2008
Published in paperback 2010

018

Translation of Nights 719 to 1001 copyright © Malcolm C. Lyons, 2008
Translation of 'The story of Aladdin, or The Magic Lamp' copyright © Ursula Lyons, 2008
Introduction and Glossary copyright © Robert Irwin, 2008
All rights reserved

The moral right of the translators and editor has been asserted

Text illustrations design by Coralie Bickford-Smith; images: Topkapi Palace Museum,
Istanbul/The Bridgeman Art Library

Printed in England by Clays Ltd, Elcograf S.p.A.

ISBN: 978-0-140-44940-2

www.greenpenguin.co.uk

Penguin Books is committed to a sustainable
future for our business, our readers and our planet.
This book is made from Forest Stewardship
Council™ certified paper.

Contents

Editorial Note

This new English version of *The Arabian Nights* (also known as *The Thousand and One Nights*) is the first complete translation of the Arabic text known as the Macnaghten edition or Calcutta II since Richard Burton's famous translation of it in 1885–8. A great achievement in its time, Burton's translation nonetheless contained many errors, and even in the 1880s his English read strangely.

In this new edition, in addition to Malcolm Lyons's translation of all the stories found in the Arabic text of Calcutta II, Ursula Lyons has translated the tales of Aladdin and Ali Baba, as well as an alternative ending to 'The seventh journey of Sindbad', from Antoine Galland's eighteenth-century French. (For the Aladdin and Ali Baba stories no original Arabic text has survived and consequently these are classed as 'orphan stories'.)

The text appears in three volumes, each with an introduction, which, in Volume 1, discusses the strange nature of the *Nights*; in Volume 2, their history and provenance; and, in Volume 3, the influence the tales have exerted on writers through the centuries. Volume 1 also includes an explanatory note on the translation, a note on the text and an introduction to the 'orphan stories' ('Editing Galland'), in addition to a chronology and suggestions for further reading. Footnotes, a glossary and maps appear in all three volumes.

As often happens in popular narrative, inconsistencies and contradictions abound in the text of the *Nights*. It would be easy to emend these, and where names have been misplaced this has been done to avoid confusion. Elsewhere, however, emendations for which there is no textual authority would run counter to the fluid and uncritical spirit of the Arabic narrative. In such circumstances no changes have been made.

Introduction

The Christians of medieval Europe believed Asia to be a region of fabulous riches, strange marvels and wise sages. Cannibals and dog-headed men dwelt there and lambs grew from the soil as plants. *The Travels of Sir John de Mandeville*, written sometime between 1357 and 1371, gave an account of the marvels of Asia that was supposedly based on the author's journeyings. However, Mandeville's *Travels* was no kind of *Rough Guide* to Asia, providing reliable information for prospective travellers. It was, rather, a work of entertainment in which interesting facts were mixed in with even more interesting fictions. Some of the wonders conjured up by Mandeville are common to *The Arabian Nights* and to *The Seven Voyages of Sindbad*. These include the giant bird known as the *rukh*, the Amazon warrior women, the Magnetic Mountain, the Fountain of Youth and the earthly paradise.

In later centuries, Galland, Lane and Burton were to use their translations of *The Arabian Nights* as vehicles for instructive glosses and foot-notes about Islamic and Arab manners and customs. But medieval Christian storytellers were not so interested in such things, and they had little sense of the otherness of the Arab world. They did not compose or adapt stories featuring veiled women, harems, eunuchs and camels. There seems to have been no attempt to produce a translation of the *Nights* that might have served any educational purpose. Instead, individual storytelling items were absorbed piecemeal by medieval European romancers and added to their fictional repertoire. Detached from Shahrazad's frame, such plot motifs, images or accessories – for example, the flying carpet – even reached as far as Iceland.

The unfinished 'Squire's Tale' in Chaucer's *Canterbury Tales* (1387–1400) provides one example of how all sorts of bits and pieces were taken from the *Nights* and other Oriental sources, yet the story as a whole is unmistakably European and reveals no interest at all in the real

Orient. In Tartarye (the Mongol lands) there was a great king called Cambyuskan. An envoy from Arabia brought him gifts, including a horse of brass, a mirror, a gold ring and a sword. The mirror and the ring were for the king's daughter Canacee. The mirror allows the viewer to see danger and to detect falsehood in a woman. The sword can cut through armour and deliver wounds that cannot be cured save by the application of the flat of the same sword. The ring permits its wearer to understand the language of birds; hence Canacee is able to hear a female falcon lament about how she has been deserted by a tercel (male hawk). Canacee nurses the bird, which has swooned from grief, and, shortly after this episode, 'The Squire's Tale' breaks off.

It is impossible to know how the story as a whole would have developed further and what part the horse, mirror and sword would have played in it. However, the deserted female falcon features in the *Nights* stories of Ardashir and Hayat al-Nufus and of Taj al-Muluk and Princess Dunya. From the *Nights* tales we can deduce that Canacee, having heard the female falcon's story, will come to distrust all men and rebuff their approaches, until some prince completes the story by showing how the male hawk did not deliberately abandon the female, but was seized by a bigger raptor, such as a kite. Once Canacee understands the full story, she will accept the prince's suit.

The mechanical horse and the magic mirror that feature in 'The Squire's Tale' also have their precursors in *Nights* stories. But this is not the place to track down and examine each and every example of Arab stories and images that appeared in the romances of medieval and Renaissance Europe. *Orlando Furioso*, the mock-heroic epic composed by Ludovico Ariosto and published in 1532, provides a striking example of the adaptation of a *Nights* story, almost certainly via an Italian intermediary source. One of the great classics of European literature, it is set in the time of Charlemagne and recounts the struggles of Orlando (Roland) and others of Charlemagne's paladins against the Saracens and pagans. Their destinies cross with those of distressed damsels, sorcerers and monsters. In Canto 28, an innkeeper recounts to Rodomont the story of two kings, Astolfo and Iocondo, who were betrayed by their wives with a knave and a dwarf respectively. Eventually the kings accept the propensity of women to be unfaithful. Rodomont, having listened to the innkeeper, is forced to accept that there is no limit to women's wiles. Evidently the innkeeper's story is an adaptation of the story with which the *Nights* opens, the tale of how Shahriyar and Shah Zaman were betrayed by their adulterous wives and how, after a sexual encounter with a woman, supposedly kept under guard by a *jinni*, they come to recognize that there is no such thing as a faithful woman.

Arab, Persian and Turkish stories percolated into Europe, carried there perhaps by sailors, merchants and prisoners of war. Spain and Sicily were important as channels of transmission for Arab and Islamic culture, while another region where Muslim and Christian alternately fought one another or lived together in uneasy coexistence was the Balkans. The degree to which *Arabian Nights* stories were known by Balkan and Greek Christians and transmitted by them prior to the publication of Galland's French translation has yet to be properly investigated. But a version of 'Ali Baba and the Forty Thieves' circulated in the Balkans (though 'Ali Baba' was one of Galland's 'orphan stories' for which no Arabic original has been found). *Nights* tales also circulated in Romania, and there was a Vlach version of the story of Hasan of Basra. In 1835, Alexander Kinglake, author of the high-spirited travel masterpiece *Eothen*, set out for the Holy Land and Egypt and, at one stage of his journey, took a Greek boat from Smyrna to Cyprus. One of the things that struck him was the Greek crew's fondness for long stories. These were 'mostly founded upon oriental topics, and in one of them I recognized with some alteration, an old friend of *The Arabian Nights*. I inquired as to the source from which the story had been derived, and the crew all agreed that it had been handed down unwritten from Greek to Greek.' (Kinglake went on to speculate, provocatively and foolishly, that the *Nights* as a whole might have a Greek rather than an Oriental origin.) It is clear that some of the *Nights* stories circulated in oral form in Ottoman-occupied Greece and came via Turkish versions.

Romanians, Bulgarians, Albanians, Greeks and others could have become familiar with *Nights* stories via translations made from Arabic or Syriac. But it is perhaps more likely that they came to the stories in Turkish versions produced during the Ottoman period. Although the main corpus of *The Arabian Nights* was not translated into any European language until the eighteenth century, a substantial section had been translated into Turkish at a remarkably early date by Abdi in 1429 under the title *Binbir Gece* ('Thousand and One Nights'). Several other translations were later made into Turkish and these survive in various manuscripts. One in the British Library, apparently dating from the seventeenth century, has the 'night stories' told by Shahrazad, or rather 'Shehzad' as she features in Turkish. These stories include the hunchback cycle, but the 'night stories' are interleaved with 'day stories' related by another narrator about the great Sufi saint Junayd. There is also the ten-volume Beyani manuscript of 1636, a translation of the *Nights* into Turkish made on the orders of Murad IV. This manuscript was purchased by Galland and brought by him to Paris; it is currently in the Bibliothèque Nationale. It is possible that Galland consulted this Turkish manuscript in order to supplement the material in the fifteenth-century Arabic manuscript he was translating from.

Galland's translation was rather free and stories were edited in order to conform to eighteenth-century French standards of decorum and refinement. He also conceived of the publication of these stories as having a twofold purpose: they would not only give readers instruction about the manners and customs of Oriental peoples, but those readers would 'benefit from the examples of virtues and vices' contained in the stories. His French translation appeared in 1704–17 as *Les Mille et une nuits* and it was in turn swiftly translated into English, German, Italian and most of Europe's leading languages. Adaptations, parodies, pastiches and other works inspired in one way or another by the *Nights* followed its publication. These included Jacques Cazotte's *Les Mille et une fadaises* (1742), Crébillon *fils*'s *Le Sopha* (1742), Denis Diderot's *Les Bijoux indiscrets* (1748), Voltaire's *Zadig* (1748), Samuel Johnson's *Rasselas* (1759), John Hawkesworth's *Almoran and Hamet* (1761), James Ridley's *Tales of the Genii* (1764) and Cazotte's *Le Diable amoureux* (1772). The majority of these publications echoed Galland's earnest purpose in that their narratives offered improving examples of the 'virtues and vices'.

France and, more precisely, Paris in the early eighteenth century had a central role as the arbiter of taste and civilization. In his introduction to *Spells of Enchantment: The Wondrous Fairy of Western Culture* (1991), Jack Zipes, a leading authority on the history of the fairy tale, having noted this, goes on to remark of Galland's translation that 'the literary fairy tale became an acceptable social symbolic form through which conventionalized motifs, characters, and plots were selected, composed, arranged, and rearranged to comment on the civilizing process and to keep alive the possibility of miraculous change and a sense of wonderment'. In the centuries that followed the French publication of the *Nights*, the stories were imitated, parodied and emulated. Some writers imitated the manner; others merely borrowed a few Oriental props or phrases. Words such as 'carbuncle', 'talisman' and 'hieroglyphic' and phrases such as 'Barmecide feast' and 'Aladdin's cave' were part of the common stock of literary bric-à-brac from a cultural attic. In more modern times, writers have played intertextual games with the original stories. Often overt or covert reference to the *Nights* has been used as a kind of literary echo chamber in order to give depth to a more modern story.

For some eighteenth-century authors, the stories of the *Nights* were not moralistic enough and they laboriously 'improved' them; for others in the nineteenth and twentieth centuries, the stories were not erotic enough. From the second half of the nineteenth century onwards, the existence of *The Arabian Nights* was also to serve as a kind of licensing authority, permitting literary fantasy, eroticism and violence. Later, from the mid twentieth century onwards, there have been many attempts by women writers to redress the injustice of Shahriyar's treatment of women and his threat to execute Shahrazad as well as to reply to the fairly pervasive misogyny of the medieval Arab stories.

Translated versions of the *Nights* influenced in different ways such well-known writers as Joseph Addison, Samuel Johnson, Voltaire, William Beckford, Samuel Taylor Coleridge, Marcel Proust, Jorge Luis Borges and John Barth (and I have discussed the nature of the various influences in my *Arabian Nights: A Companion*, 2004). The account of the influence of the *Nights* that follows will concentrate on a small handful of selected examples from British and French literature, but, of course, the influence of the *Nights* spread more widely and any truly comprehensive account of its influence would also need to discuss such figures as the Germans, Johann Wolfgang Goethe, Hugo von Hofmannstahl and Ernst Junger, the Danes Adam Oehlenschläger and Hans Christian Andersen, the Italian Italo Calvino and the Japanese Yukio Mishima. It would also cover the impact of the *Nights* on modern Arabic literature. In the Arabic-speaking world, the *Nights*, because of its colloquial style, frequently incorrect Arabic and occasional bawdiness, used not to enjoy a high reputation. However, from the twentieth century onwards and following the acclamation of Western writers and intellectuals, many of the Arab world's most famous writers have championed the *Nights*, praised the liberating qualities of imaginative fiction and pastiched its themes. They include Tawfiq al-Hakim, Taha Hussein, Jabra Ibrahim Jabra, Naguib Mahfouz and Edwar Kharrat. 'We are a doomed people, so regale us with amusing stories' is the bitter reflection of the narrator in the Sudanese writer Tayeb Salih's *Season of Migration to the North* (originally published in Arabic in 1966). This remarkable novel – one of the finest, perhaps *the* finest ever to have been written in Arabic, about traditional values, colonialism, cross-cultural sexual encounters and much else – draws complex parallels with the *Nights*, its protagonist a modern avatar of Shahriyar, driven to kill the women with whom he sleeps.

In eighteenth-century Britain, the influence the *Nights* exercised on young people, some of whom were destined to go on to become novelists, was redoubled by the many imitations and pastiches that were published. These included John Hawkesworth's aforementioned *Almoran and Hamet* (1761), William Beckford's *Vathek* (1786) and Thomas Gueulette's early eighteenth-century, mock-Oriental Tartar, Moghul and Chinese tales (written in French, but translated into English). But one English mock-Oriental story collection was of particular importance. This was *Tales of the Genii: or The Delightful Lessons of Horam, the Son of Asmar* (1764) by the Reverend James Ridley. In this book, Ridley, who had served as an army chaplain in India, sought to promote 'the true doctrines of morality under the delightful allegories of romantic enchantment'. Those 'true doctrines' were of course Christian and Protestant. Ridley's tales contain a lot of sorcery, magical transformations, genii, richly decorated palaces and all the conventional settings and trappings of the Orient. His heroes and heroines pass through many ordeals and unmask all sorts of enchantments in order to discover virtues that are veiled by appearances. The book, with its heavy freight of Christian doctrine and moralizing, does not read well today, but it was enormously popular in the eighteenth and nineteenth centuries and, as we shall see, it helped form the youthful imaginations of better writers.

Responses to the *Nights* over the centuries were shaped by changes in society and taste. The Romantics were less interested than their predecessors in the moral lessons to be drawn from the Arabian stories, but enthusiastic about the wonders of magic, the exotic, and the sublime qualities of the vast and wild. Moreover, the *Nights* came to be associated with childhood reading and the opening up of the imagination that came from it. 'Should children be permitted to read romances, and relations of giants and magicians and genii?' Coleridge asked (in a letter to Thomas Poole in 1797), before answering: 'I know all that has been said against it; but I have formed my faith in the affirmative. I know of no other way of giving the mind a love of the Great and the Whole.' (The first selection of tales from the *Nights* made specifically for children was published by Elizabeth Newberry in 1791.)

William Wordsworth's *The Prelude or, Growth of a Poet's Mind* (1850) is an autobiographical poem in blank verse, in which the poet searches for past sources of joy and recalls his childhood, including his youthful reading. In the course of this, he praises the authors of pulp fiction: 'Ye dreamers, then, / Forgers of lawless tales! we bless you then'. He believed that Arabian and similar tales were 'eminently useful in calling forth intellectual power' (as expressed in a letter of 1845). In the fifth book of *The Prelude*, he writes:

> A precious treasure had I long possessed,
> A little yellow, canvas-covered book,
> A slender abstract of the Arabian tales . . .

The Brontës – Emily, Charlotte, Anne and Bramwell – all read the *Nights* when young. Ridley's *Tales of the Genii* also exercised a powerful influence on their youthful imaginations and the four children called themselves the 'Genii'. Charlotte and Bramwell constructed an imaginary kingdom called Angria and wrote stories about it, while Emily and Anne composed poems about another imaginary kingdom, Gondal. Both the stories and the poems drew upon Oriental and pseudo-Oriental tales.

'I took a book – some Arabian tales.' This description of what Jane did after a conversation with Mr Brocklehurst at Gateshead is one of several explicit references to the *Nights* in Charlotte Brontë's *Jane Eyre* (1847). To take another example, Rochester's horse is called Mesrour – derived from 'Masrur', the name of the eunuch who accompanies Harun al-Rashid on his nocturnal explorations of Baghdad in the *Nights*. More profoundly, such incidental and trivial references are surely intended to suggest that, in its broadest outline, *Jane Eyre* is patterned on the frame story of the *Nights*. Jane is a kind of reincarnation of Shahrazad, talking and teaching for her future. Correspondingly, Rochester is Shahriyar, an embittered despot who distrusts women (though he is, of course, also a kind of Bluebeard, presiding over a great house with a locked chamber). In fact, Jane refers to him as 'sultan'. The novel is, in short, the story of how the autocratic sultan is tamed by a good woman (subsequently the theme of so many women's romances).

Emily Brontë's *Wuthering Heights* (1847) also contains many explicit references to the *Nights*, as Nelly Dean assumes the role of a latter-day Shahrazad, while Lockwood is a kind of avatar of Shahriyar. Early on in the novel, Catherine pretends to herself that she is like a merchant with his caravan – such as the merchant in the *Nights* story of the merchant and the *jinni* – but whereas the merchant's expedition leads him to a dangerous encounter with the *jinni*, who wants to kill him, Catherine's leads her to meet the demonic Heathcliff. Heathcliff is several times compared to a ghoul. On the other hand, Nelly Dean fancies that Heathcliff might be an Oriental prince in disguise. More important, perhaps, than specific references to the *Nights*, was the licence that the Arabic stories conferred for the wildness and passion that characterizes the storytelling of both *Jane Eyre* and *Wuthering Heights*.

In *Charles Dickens: A Critical Study* (1898), the nineteenth-century novelist and essayist George Gissing wrote about Dickens's novels in the following terms:

Oddly enough, Dickens seems to make more allusions to the *Arabian Nights* than to any other book or author ... Where the ordinary man sees nothing but everyday habit, Dickens is filled with the perception of marvellous possibilities. Again and again he has put the spirit of the *Arabian Nights* into his pictures of life by the river Thames ... He sought for wonders amid the dreary life of common streets; and perhaps in this direction was also encouraged when he made acquaintance with the dazzling Eastern fables, and took them alternately with that more solid nutriment of the eighteenth-century novel.

Dickens, who read the *Nights* as a boy, was delighted by the stories, but his pleasure in Oriental storytelling was also fuelled by his passion for Ridley's tales. As a child, Dickens composed a tragic drama, *Misnar, The Sultan of India*, based one of the stories in *Tales of the Genii*, essentially the story of an Indian prince's struggle to retain his throne against challenges presented by his ambitious brother assisted by seven genii and the illusions conjured up by them.

The power of Ridley's story stayed with Dickens throughout his life. Towards the end of *Great Expectations* (1861), Pip reflects on how, without his foreknowledge, everything is slowly but inexorably moving in such a way to bring catastrophe suddenly upon his head: 'In the Eastern story, the heavy slab that was to fall on the bed of state in the flush of conquest was slowly wrought out of the quarry, the tunnel for the rope to hold it in its place was slowly carried through leagues of rock ...' The reference is to the story of Misnar and his wise counsellor who design a pavilion with a great stone slab set above it to trap and kill two evil enchanters. At one point in the *The Old Curiosity Shop* (1841), Dick Swiveller wakes up in a strange bed: 'If this is not a dream, I have woke up, by mistake, in a dream in an Arabian Night, instead of a London one.' Likewise, David Copperfield, who as a schoolboy is compelled by the domineering Steerforth to tell stories late at night, compares his fate to that of Shahrazad. (And, of course, since Dickens both published many of his novels in serial form and gave readings from them, it would be natural for him to think of himself as a latter-day male version of Shahrazad.) It would be very easy to go on listing overt and covert references to the *Nights* and to Ridley's ersatz version elsewhere in Dickens's works. What is more important is the feel of the *Nights* stories and their impalpable but pervasive influence over Dickens's fantastical plots with their moralizing outcomes. Enigmatic philanthropists cloaked in disguise walk the streets at night following in the footsteps of Harun al-Rashid. Baghdad is reconfigured as London, and the Dickensian city of mysteries and marvellous possibilities teems with grotesque characters who may be distant descendants of the hunchback or of the barber's seven disabled brothers.

In George Eliot's *Daniel Deronda* (1876), in large part a novel about the Jewish people and their future prospects, she nevertheless makes frequent use of references to *The Arabian Nights* in order to heighten the Oriental feel of the novel. In particular, the protagonist, Daniel Deronda, is repeatedly compared to Qamar al-Zaman, the prince who, because of his education, is suspicious of women (and in Lane's translation is so handsome that he is 'a temptation unto lovers, a paradise to the desirous'). Correspondingly, the Jewess Mira is compared to Princess Budur, who is fated to marry Qamar al-Zaman. Throughout the novel, allusions to the *Nights* are used not only to suggest the Oriental, but also the sensual passion that is the theme of so many of its stories.

For a long time, European and Japanese knowledge of *The Arabian Nights* was mediated by Galland's courtly French and there is a sense in which the *Nights* in the eighteenth and nineteenth centuries can be regarded as largely a work of French literature. Its influence on French literary culture was, if anything, more overpowering than its legacy in Britain. In his *Souvenirs d'égotisme* (1832), Stendhal wrote of the *Nights*: 'I would wear a mask with pleasure. I would love to change my name. *The Arabian Nights* which I adore occupy more than a quarter of my head.' It would probably be fruitless to search *Le Rouge et le noir* (1830) or *La Chartreuse de Parme* (1839) for plots or borrowed props.

Nevertheless the *Nights*, and in particular its stress on magical powers, did help shape Stendhal's image of himself as a novelist. Late in life he awarded himself magical powers as a writer, including becoming another person (as all good novelists should strive to do). He wanted to live like Harun in disguise. (He also wished for the ring of Angelica which conferred invisibility in Ariosto's *Orlando Furioso*.) The power to become invisible, to assume another's identity, to read another's mind – all these staples of Islamic occultism and storytelling gave Stendhal metaphors for himself as an observer of humanity and a writer.

The *Nights* influenced the storytelling of Alexandre Dumas *père* at a more obvious and superficial level. His *Le Comte de Monte-Cristo* (1845–6) is a wonderful melodrama of imprisonment, escape, enrichment and revenge, which was first published as a magazine serial. Dumas worked with a vocabulary of Oriental fantasy images that was shared with his readers. In the novel, Edmond Dantès, wrongfully imprisoned as a Bonapartist conspirator, escapes and, having discovered treasure on the island of Monte-Cristo, returns to Paris to take revenge on those who put him in prison.

There are many overt references to the *Nights* in the novel, but the most sustained evocation of its Oriental matter comes in chapter thirty-one, in which a Baron Franz d'Epinay lands on Monte-Cristo on a venture to encounter smugglers or bandits. Having landed, he encounters a group of smugglers who are going to roast a goat. They invite Franz to dine with them, but he can only join them if he is first blindfolded. The place he is led to is compared by his guide to the caves of Ali Baba. On being told that it is rumoured that the cave has a door that only opens to a magic password, Franz exclaims that he has 'definitely stepped off into a tale from the *Thousand and One Nights*'. When the blindfold is removed, he finds himself in a cave furnished sumptuously in an Oriental style.

He is greeted by a mysterious and strikingly pale man who introduces himself as Sindbad the Sailor (but he is, of course, the Count of Monte-Cristo, alias Dantès, a Byronic figure who revels in mystery). Whereupon Franz decides to take the name Aladdin for the evening. The splendid dinner is served by Ali, a Nubian mute. In an evening of Eastern opulence, dinner is followed by hashish paste and, as the hashish takes effect, Franz drifts off into erotic dreams. He wakes on a bed of heather in a cave – transported back from Oriental enchantment to mundane reality.

In the course of the nineteenth century, abridged, bowdlerized and illustrated versions of *The Arabian Nights* proliferated and the *Nights* was in danger of being classified as merely children's literature. The fresh translation from the Arabic into French by Joseph Charles Mardrus was to reverse that trend – at least for some readers. His highly literary version had been produced at the urging of the poet Stéphane Mallarmé and it was published from 1898 to 1904 by *La Revue blanche*, a periodical devoted to Symbolism and modernism. Although Mardrus's translation was inaccurate, unscholarly and somewhat fraudulent, it read well and it was a huge hit. André Gide was one of its leading enthusiasts. In Britain, the poets W. B. Yeats and James Elroy Flecker were among those who found fresh inspiration in Mardrus.

Jean Cocteau, dandy, *enfant terrible*, poet, novelist, artist and filmmaker, was, from his youth onwards, obsessed with the Mardrus *Nights*. He founded a literary and artistic magazine, *Shéhérazade*, which ran from 1909 to 1911. He was particularly fascinated by the figure of Shahrazad, the woman who talks for her life, whom he compared to a snake charmer who plays a flute in front of a cobra in the knowledge that, if the flute's melody should cease, the cobra would strike. But, in some strange way, the young Cocteau, who had passionate fixations for a succession of beautiful women, but who was coming to terms with his homosexuality, seems to have regarded Shahrazad as representing the feminization of the world and therefore a figure to be resisted. His first volume of verse was entitled *La Lampe d'Aladdin* (1909). In the preface, Cocteau wrote: 'I have wandered amid the gloom of life, with the marvellous lamp. Young like Aladdin, walking with fearful step, I have seen fruits, jewels, gleams and shadows. And my heart filled with illusions, I have wept at the difficulty of giving them to an unbelieving world.' Cocteau also furnished a preface to an edition of the *Nights*. Sensuality, drugs, mirages, the Alexandrian Cabala, all such things were meat and drink, or rather opium and alcohol, to Cocteau.

But it was not just writers who were inspired by the Mardrus's representation of the *Nights*. When the Ballets Russes presented their version of Rimsky-Korsakov's *Shéhérazade* in Paris in 1909, they discarded the four-part scenario that Rimsky-Korsakov had provided for his opus. Instead, quite a different story was substituted that was loosely modelled on the opening *Nights* story of Shahriyar's discovery of his wife's infidelity and his bloody vengeance. The story of illicit lust and violent requital in an opulent, lushly coloured setting echoed the general tenor of the Mardrus version of the *Nights*.

In the second half of the twentieth century there was a revival of interest in the *Nights*, particularly among writers who were interested in the opportunities offered for modernist and postmodernist literary experiments. Jorge Luis Borges, John Barth, Italo Calvino and Angela Carter are among those who can be mentioned here. Indeed, Angela Carter was one among many women writers who have been preoccupied with the role of the female storyteller. A. S. Byatt shares this preoccupation. In one of her short story collections, *The Djinn in the Nightingale's Eye* (1994), the title story is an account of the encounter of an ageing narratologist (an academic expert on storytelling) with a djinn in an hotel room in modern Ankara. The djinn offers her three wishes. She wishes to be as she was when she last felt good about her body. Secondly she wishes that the djinn should love her. Finally, she wishes that the djinn should have his freedom. 'The Djinn in the Nightingale's Eye' is a story about how stories will outlive us and it plays with conventions of the fairy-tale genre, using the fairy-tale form as a vehicle for the exploration of how fairy tales work.

There is no limit to the messages that can be drawn from various readings and retellings of the stories of *The Arabian Nights*. When one reads these stories, one has the impression that one has entered the engine-room of all stories, where all the possible plots have been stripped down to their essential elements. One senses also that, just as Sindbad or the one-eyed princes are the stories they tell about themselves, so we readers are the sum of our own stories. As the heroine of 'The Djinn in the Nightingale's Eye' reflects: 'These tales are not psychological novels, are not concerned with states of mind or development of character, but bluntly with Fate, with Destiny, with what is prepared for human beings.' Not only literature, *The Arabian Nights* is and continues to be one of the most inspiring sourcebooks of literature ever created.

Robert Irwin
London

The Arabian Nights

Nights 719 to 1001

A story is also told, O fortunate king, that in the city of Shiraz there was a great king whose name was al-Saif al-A'zam Shah. He was a childless old man and so he gathered together wise men and doctors and told them: 'I am old and you know my position, the state of my kingdom and how it is governed. I am afraid of what will happen to my subjects when I am dead, as I have no son yet.' 'We shall prepare some drugs to help you, if Almighty God wills it,' they told him, and when he took what they produced and lay with his wife, she conceived with the permission of Almighty God, Who says to something: 'Be' and it is. When the months of her pregnancy had been completed, she gave birth to a boy as beautiful as the moon who was given the name Ardashir. He grew up studying science and literature until he reached the age of fifteen.

In Iraq there was a king called 'Abd al-Qadir who had a daughter as beautiful as the full moon when it rises, named Hayat al-Nufus. This girl had such a hatred of men that no one could mention them in her presence and, although sovereign kings had asked her father for her hand in marriage, when one of them approached her, she would always say: 'I shall never marry and if you force me to do that, I shall kill myself.' Prince Ardashir heard about her and told his father that he would like to marry her. The king was sympathetic when he saw that his son was in love, and every day he would promise to get Hayat al-Nufus for him as a wife. He sent his vizier off to ask her father for her hand, but the request was refused. When the vizier got back and told his master of his failure, the latter was furiously angry and exclaimed: 'Does someone in my position send a request to a king to have it refused?' He ordered a herald to proclaim that his troops were to bring out their tents and equip themselves as best they could, even if they had to borrow the money for their expenses. 'I shall not draw back,' he said, 'until I have ravaged the lands of King 'Abd al-Qadir, killed his men, removed all traces of him and plundered his wealth.'

When Ardashir came to hear of this, he rose from his bed and went to his father. After kissing the ground before him, he said: 'Great king, do not put yourself to any trouble over this . . .'

Night 720

Morning now dawned and Shahrazad broke off from what she had been allowed to say. Then, when it was the seven hundred and twentieth night, SHE CONTINUED:

I have heard, O fortunate king, that when Ardashir came to hear of this, he went to his father, kissed the ground before him and said: 'Great king, do not put yourself to any trouble over this by sending out these champions and this army and spending your money. You are stronger than 'Abd al-Qadir and, if you send this army of yours against him, you will certainly be able to ravage his lands, kill his fighters and seize his wealth. He himself will be killed and when his daughter hears that she is responsible for his death and the deaths of his men, she will kill herself and because of her I shall die, as I could not live on after her.' 'What do you advise, then, my son?' his father asked, and Ardashir told him: 'I shall set out myself on my own errand. I shall dress as a merchant and find some way of reaching the princess, after which I shall look to see how I can get what I want.' 'Is this what you choose?' asked the king, and when Ardashir said that it was, the king summoned his vizier and said: 'Go with my son, the fruit of my heart; help him in his quest; protect him and use your sound judgement to guide him.' He then gave Ardashir three hundred thousand gold dinars, gems, ring stones and jewellery, as well as other goods, treasures and the like.

Ardashir went to his mother, kissed her hands and asked for her blessing, which she gave him. She then got up quickly and opened her treasure chests, from which she produced for him valuables such as necklaces, jewellery, robes and rarities, together with all kinds of other things, including relics of former kings that had been stored away and were past all price. The prince then took with him such mamluks, servants and beasts as would be needed on the journey and elsewhere, and he and the vizier, together with their companions, dressed as merchants.

When he had taken leave of his parents, his family and his relations, he and his party set out across the desert wastes, travelling night and day. Finding the way long, he recited the following lines:

My passionate longing and lovesickness increase;
There is none to help me against the injuries of Time.
I watch the Pleiades and Arcturus when they rise,
As though in the fervour of my love I worshipped them.
I look for the morning star, and when it comes
I am stirred by passion and my ardour grows.
I swear that I have never changed your love for hate,
And that I am a lover who is left without sleep.
Matching the greatness of my hopes is my increasing weakness;
You have gone, and I have small endurance and few helpers.
But still I shall endure until God reunites us,
To the chagrin of all my envious foes.

When he finished his recitation he swooned away for a time and the vizier sprinkled him with rosewater. 'Be patient, prince,' he told him on his recovery, 'for with patience comes relief, and here you are on your way to what you desire.' The vizier continued to soothe and console him until he regained his composure, and he and his companions pressed on with their journey. Again, however, disheartened by its length, he thought of his beloved and recited:

She has long been absent; my cares and my distress increase;
My heart's blood is aflame with fires of love.
The passion that afflicts me has made my hair turn white,
And tears pour from my eyes.
You are my desire, the goal of all my hopes;
I swear by the Creator, Who fashioned branch and leaf,
I have endured for you, who are my hope,
Such passion as no other lover could endure.
Ask the night to let you know of me,
Whether through all its length my eyelids ever closed.

When he finished his recitation he wept bitterly and complained of the violence of his passion, and again the vizier soothed and consoled him, promising him that he would win through to his goal.

After a few more days of travelling, they came after sunrise within sight of the White City and the vizier told the prince: 'Good news, prince.

Look, for this is the city that you have been making for.' In his delight
the prince recited:

My two companions, I am a lover sick with passion;
The ardour of my love is always with me.
I moan like a parent who has lost a child, sleepless through grief.
In the dark of night there is none who may pity my love.
The winds that blow here from your land
Bring coolness to my heart.
The tears shed by my eyes are like rain clouds,
And, in their sea, my heart is left to float.

The prince and the vizier then entered the city and asked where wealthy
merchants stayed. They were directed to a *khan* where they took lodgings
and in which they hired three storerooms. When they had been given the
keys, they opened these up and stored their goods and possessions in
them, after which they rested. The vizier, who had begun to consider
how the prince might set about his quest . . .

Night 721

Morning now dawned and Shahrazad broke off from what she had been
allowed to say. Then, when it was the seven hundred and twenty-first
night, SHE CONTINUED:

I have heard, O fortunate king, that the prince and the vizier stopped
at the *khan*, put their goods in the storerooms and lodged their servants
there. They stayed until they were rested, after which the vizier, who
had begun to consider how the prince might set about his quest, then
said to him: 'Something has occurred to me which, I think, may serve
your purpose, if Almighty God wills it.' 'Your counsel is always good,'
the prince told him, 'so do whatever comes to your mind, may God
guide you.' The vizier explained: 'I propose to hire a shop for you in the
drapers' market, in which you can sit. Everyone, high and low alike,
finds the need to go to the market and I think that, if people see you
sitting in your shop, they will be drawn towards you and this will help
you to get what you want. For you are a handsome young man and your
good looks will serve as an attraction.' 'Do whatever you want,' the
prince told him.

The vizier got up immediately and put on his most splendid clothes, as did the prince, and he put a purse containing a thousand dinars in his pocket. The two of them went out, and the people who looked at them as they walked around the city were astonished by the good looks of the prince, exclaiming: 'Praise be to God, Who created this young man from "a despicable drop".* Blessed be God, the best of creators'.† There was a great deal of talk about him, some quoting from the Quran: 'This is no mortal man but a noble angel,'‡ while others were asking: 'Has Ridwan, the guardian of Paradise, let him out of its gate by mistake?' People followed the two to the drapers' market, which they entered and then stopped. A venerable and dignified *shaikh* came up and, after they had exchanged greetings, he asked if there was anything they wanted so that he might have the honour of satisfying their needs. The vizier asked him who he was, to which he replied that he was the market superintendent. The vizier then said: '*Shaikh*, you must know that this young man is my son and I want to take a shop for him in this market so that he may sit here and learn the techniques of trading and how to behave as a merchant.' 'To hear is to obey,' the superintendent said, and there and then he produced the key of a shop and ordered the salesmen to sweep it out, which they did.

When it had been cleaned, the vizier sent for a high cushioned seat stuffed with ostrich feathers on which was placed a small prayer mat and whose borders were embroidered with red gold. He also had a cushion fetched, before bringing in as much of the materials that they had taken with them as would fill the shop.

The next day, the young prince came to open the shop. He sat down on the seat with two splendidly dressed mamluks standing before him, while two handsome Abyssinian slaves stood at the lower end. The vizier advised him that, to help him achieve his goal, he should keep his secret from the townspeople, and after telling him to let him know day by day everything that happened to him in the shop, the vizier went off and left him sitting there like the moon when it is full. News of the prince's beauty spread, and people would come to him in the market without wanting to buy anything but merely to admire his grace and his symmetrical form, praising Almighty God for the excellence of His creation. Such were the crowds that it became impossible for anyone to pass through the market, while the prince himself was taken aback by his admirers and kept looking from right to left, hoping to make the acquaintance of

* Quran 32.7, 77.20.   † Quran 23.14.   ‡ Quran 12.31.

someone with a connection to the court who might be able to tell him about the princess.

He became depressed by his failure to do this, although every day the vizier encouraged him to hope for success. When this had been going on for a long time, one day as he was sitting there, a respectable, dignified and sedate old woman arrived, dressed as a person of piety and followed by two slave girls, beautiful as moons. She stopped by the shop and, after having looked at the prince for some time, she exclaimed: 'Praise be to God, Who has perfected the creation of that face!' She then greeted him, and when he had returned her greeting, he gave her a seat beside him. 'From what land do you come, you handsome man?' she asked, and he replied: 'From Indian parts, mother, and I am here to look around the city.' 'You are an honoured visitor,' she told him, before going on to ask what goods and materials he had and asking him to show her something beautiful that would be suitable for royalty. 'If it is something beautiful that you want to be shown,' the prince said, 'I have things to suit all customers.' 'My son,' she told him, 'I want something expensive and elegant, the finest that you have.' He said: 'You will have to tell me for whom you want the goods so that I can show you something to match the client.' 'That is true,' she agreed, and she went on to tell him that she wanted something for her mistress, the princess Hayat al-Nufus, daughter of 'Abd al-Qadir, the ruler of the country. On hearing this, the prince was wild with joy and his heart fluttered. He gave no instructions to his mamluks or his slaves, but reaching behind his back he brought out a purse containing a hundred dinars, which he passed to the old woman, saying: 'This is to pay for your laundry.' Then from a bundle he drew out a dress worth ten thousand dinars or more and said: 'This is one of the things that I have brought to your country.' The old woman looked admiringly at it and asked: 'How much is this, you master of perfection?' 'I make no charge for it,' he said, and she thanked him before repeating the question. He insisted: 'By God, I shall take nothing for it. If the princess will not accept it, you can have it as a gift from me. I thank God, Who has brought us together, and if some day I happen to need something, I may find in you someone to help me get it.'

The old woman admired his eloquence, generosity and good manners. She asked him his name, and when he told her that it was Ardashir, she exclaimed: 'By God, this is a remarkable name, and one that is given to princes, although you yourself are dressed as a merchant's son!' He said: 'It was because of my father's love for me that he gave me this name, and it has no other significance.' The old woman found this remarkable and pressed him to accept payment for the dress, but he swore that he would take nothing. 'My dear,' she told him, 'you must know that truth is the greatest of all things. There must be some reason behind this generosity that you have shown me, so tell me about your hidden purpose, as it may be that I can help you to get what you want.'

The prince now put his hand in hers and, after pledging her to secrecy, he told her the whole story of his love for the princess and what he was suffering because of her. 'You are now telling the truth,' she said, shaking her head, 'but, my son, the wise have a proverb that runs: "If you want to be obeyed, don't ask for the impossible." You are called a merchant, and even if you had the keys to treasure hoards, this is still how you would be described. If you want to climb the next step on the ladder, try to marry the daughter of a *qadi* or of an emir, for why must you look for the hand of the daughter of the leading sovereign of the age? She is a virgin who knows nothing of the world and has never seen anything in her whole life except for the palace where she lives, but for all her youth she is intelligent, sensible and shrewd. She has a sound mind, acts with propriety and shows penetration in her judgement. She is her father's only child and is dearer to him than his own life. He comes to visit her every morning and everyone in the palace is afraid of her. Don't imagine that anyone can talk to her about love, as there is no way in which that could be done. I love you, my son, with my heart and body, and I wish you could be with her, but all I can suggest is something that might cure your heartache and which I would give my life and my wealth to bring about for you.' 'What is that, mother?' he said and, in reply, she told him to ask her to find him the daughter of a vizier or of an emir, promising: 'If you ask me that, I will do it for you, but nobody can cover the distance between earth and heaven in a single bound.' The prince replied with courtesy and good sense: 'You are an intelligent woman, mother, and you know how things fall out. Does a man with a headache tie a bandage round his hand?' When she said no, he went on: 'My heart will accept no one but the princess, and it is only her love that kills me. If I don't find any helper to lead me to her, I am a dead man. I implore you in God's Name, mother, to pity me, stranger as I am, and the tears that I shed.'

Night 722

Morning now dawned and Shahrazad broke off from what she had been allowed to say. Then, when it was the seven hundred and twenty-second night, SHE CONTINUED:

I have heard, O fortunate king, that the prince said: 'I implore you in God's Name, mother, to pity me, stranger as I am, and the tears that I shed.' 'By God, my son,' she told him, 'my heart is cut to pieces by what you say, but I have no way of bringing this about.' 'In your kindness,' he said, 'would you take this note of mine and kiss her hands as you bring it to her?' As she felt sorry for him, she told him to write whatever he wanted for her to take to the princess. When he heard that he was overjoyed, and after calling for an inkwell and paper he wrote the following lines to her:

Hayat al-Nufus, grant the fulfilment of his hope
To a lover who, without you, is doomed to pine away.
The pleasures of life used to be mine to enjoy,
But now I am distracted and confused.
And all night long I stay awake,
With cares my only comrades.
Have pity on the unfortunate, tormented lover,
Whose longing ulcerates his eyelids;
And when the true dawn breaks,
It finds him drunken with the wine of love.

When he had finished writing this, he folded it, kissed it and handed it to the old woman. He then stretched out a hand to his money-box and brought out another purse, containing a hundred dinars, which he gave to her, telling her to distribute it among the slave girls. She refused and told him: 'My son, it is not for anything like this that I am ready to help you.' He thanked her but insisted, and she took the money from him, kissed his hands and went off.

When she came to the princess, she said: 'My lady, I have brought you something from one of the townsfolk, and on the face of the earth there is no one more handsome than the young man who has sent it.' 'Where does he come from, nurse?' the princess asked, and the old woman said: 'From Indian parts, and he has given me this dress embroidered with gold and studded with pearls and gems, which is worth the kingdoms of

Chosroe and Caesar.' When the dress was brought out, it filled the whole palace with radiance because of the beauty of its workmanship and the number of gemstones and jewels it contained, astonishing everyone there. The princess inspected it and found that its price could be no less than a full year's worth of the revenues of her father's kingdom. 'Nurse,' she asked, 'does this dress come from the young man or from someone else?' 'From him,' the nurse replied, and the princess went on to ask: 'Does he come from our city or is he a stranger?' 'He is a stranger, lady, who has only just come here. He is a man with followers and servants. He has a handsome face and a well-built figure, and is also generous and open-hearted. The only person whom I have ever seen who surpasses him in beauty is you yourself.' 'There is something strange here,' the princess said. 'How does this dress, past all price, come to be in the possession of a merchant and how much did he tell you he wanted for it?' 'By God, my lady,' replied the old woman, 'he said nothing about its price. Instead, he said that he would not take any money for it and that he was sending it to you as a gift, as no one else was fit to wear it. He handed back the gold that you had sent with me, swearing that he would not take it, and he added that if you did not accept the dress, I was to keep it myself.' 'By God,' the princess exclaimed, 'this is supreme generosity, but I am afraid that it may lead him into difficulties. Why didn't you ask him whether there was anything he needed that we could do for him?' 'That is exactly what I said to him,' the old woman replied, 'and he agreed that there was something, but instead of saying what it was, he gave me this note and told me to hand it to you.'

The princess took the note from her, opened it and read it to the end. Her mood changed; she lost her self-control and, turning pale, she said to the old woman: 'How is this dog to be answered who speaks like this to the king's daughter, and what connection is there between him and me that allows him to write to me? I swear by the Almighty God, the Lord of Zamzam and the Hatim,* that were it not for my fear of Him, I would send men to pinion this fellow, slit his nostrils, cut off his nose and his ears to make an example of him, and then crucify him on the gate of the market where he has his shop.'

When the old woman heard this, she turned pale and shivered. For a time she was unable to speak, but then she took heart and said: 'Well and good, my lady, but what is in the note to upset you? Is it anything

---

* Zamzam is a sacred well in Mecca; the Hatim is a piece of sacred ground near the Ka'ba.

but a petition that he has sent you complaining of poverty or injustice in the hope that you may be generous to him or right his wrongs?' 'No, by God, nurse,' the princess replied. 'What is in it is a poem objectionably expressed. There are only three possibilities: the dog may be mad and have lost his wits; he may be trying to commit suicide; he may hope to have his way with me through the help of some powerful and mighty sultan or he may be sending me filthy verses to seduce me because he has heard that I am one of the town prostitutes who spends a night or two with anyone who asks her.' 'By God, you are right, my lady,' the old woman told her, 'but there is no need for you to concern yourself about this ignorant dog. Here you are in this lofty and secure palace, so high that not even birds can fly over it and which the wind cannot pass, while he wanders about helplessly. Write him a letter filled with every possible kind of reproach and with the direst of threats, promising to have him put to death. Say: "Where did you find out about me so that you wrote to me, dog of a trader, who spends his life wandering through deserts hunting for money? By God, unless you wake up and recover from your drunkenness, I shall have you crucified on the gate of the market where you have your shop."' The princess told her: 'I'm afraid that if I write to him, it may encourage him,' but the old woman replied: 'How could a man of his status and position be encouraged to think of you? It is to stop him having any hopes of you and to frighten him that you should write to him.'

She continued to use her wiles on the princess, who at last had an inkstand and paper fetched and wrote the following lines to Ardashir:

You claim to be in love and to suffer sleeplessness,
Passing your nights among the cares of passion.
Do you hope for union with the moon, deluded fool,
And can a man get what he wishes from the moon?
I give you advice to which you should attend:
Stop, for you are between death and danger.
If you put this request to me again,
Your punishment will bring you harm on harm.
Behave yourself with sense, wisdom and intelligence.
I give you this advice in poetry and in what I say.
I swear by Him, Who created all things from the void,
And adorned the face of heaven with bright stars,
If you say once more what you have said just now,
I shall nail you to the trunk of a tree.

She folded the note and handed it to the old woman, who took it off to Ardashir's shop and gave it to him . . .

Night 723

Morning now dawned and Shahrazad broke off from what she had been allowed to say. Then, when it was the seven hundred and twenty-third night, SHE CONTINUED:

I have heard, O fortunate king, that the old woman took the note from the princess and brought it to Ardashir in his shop, saying: 'Read her reply. You should know that when she read your note she was furiously angry, but I continued to soothe her until she wrote back to you.' The prince took the letter gladly, but when he had finished reading it and had grasped its contents he shed bitter tears. The old woman felt sorry for him and said: 'My son, may God not bring tears to your eyes or sorrow to your heart. What can be more courteous than for her to reply to you, after you had acted as you did?' 'Mother,' he said, 'what more subtle approach can I try? She writes threatening to have me killed or crucified and tells me never to write to her again. By God, I would prefer to die rather than to go on living, and I ask you in your kindness to take this note and bring it to her.' 'Write,' she told him, 'and I shall see that you get an answer. I shall risk my life to see that you reach your goal, even if I have to die for your sake.'

Ardashir thanked her, kissing her hands, and he then wrote the following lines to the princess:

You threaten to kill me because I love you;
This would bring me relief, and death is our fate.
For the lover, death is easier to bear
Than long life, when he is rejected and rebuffed.
If you visit a lover who has few to help him,
Remember that those who strive for good win thanks.
Do whatever it may be that you intend;
I am your slave, and the slave is a man in bonds.
What can I do, who cannot endure without you,
And how can there be endurance when the heart is constrained to love?
Show mercy to one who is sick with love for you,
For all who love the noble are to be excused.

When he had folded the note he gave it to the old woman, presenting her with two purses, each containing a hundred dinars. She was reluctant to accept, but took them when he conjured her in God's Name, and she swore that in spite of enemies she would get him what he wanted. She then went back to Hayat al-Nufus and gave her the letter. 'What is this, nurse?' the princess said. 'We have begun a correspondence, with you going to and fro, and I am afraid that word will get out and we will be put to shame.' 'How could that be, lady,' the old woman asked, 'and who would dare say such a thing?'

The princess then took the letter and after reading it and grasping its contents, she struck one hand against the other and exclaimed: 'This is disastrous, and we don't even know where this young man has come from!' 'My lady,' the old woman said, 'for God's sake write him a letter in harsh terms threatening to have his head cut off if he writes again.' 'I know that that will not end the matter and it would be better not to write at all,' said the princess, adding: 'And if the dog is not stopped by my earlier threat, then I shall have his head cut off.' 'Write and tell him how things stand,' insisted the old woman, and so the princess called for an inkwell and paper and wrote the following threatening lines:

You, who are ignorant of the blows of fate,
Whose heart longs for my union,
Think, fool: can you reach to the sky,
And join the bright moon at its full?
I shall make you taste fire with inextinguishable flames,
And the destructive swords will leave you dead.
Between you and your goal, my friend, is a vast space,
With hidden perils whitening the parting of the hair.
Take my advice; give up this love of yours,
Abandoning a goal which you will not attain.

She folded the letter and handed it to the old woman, who did not know what to make of it but who took it and went off to give it to the young prince. He took it and read it before bending towards the ground, saying nothing but tracing lines with his finger. 'Why don't you say something or answer, my son?' the old woman asked him. 'What can I say, mother,' he replied, 'when she threatens me and treats me with even greater harshness and aversion?' 'Write to tell her what you want,' she replied, 'and I shall defend you. Be of good heart, for I shall see that the two of you are united.' He thanked her for her kindness, kissed her hands and wrote these lines to the princess:

How strange that there is a heart that does not soften
For a passionate lover, who longs for union with his love,
With eyelids wounded by tears
When covered by the gloom of sombre night.
Give generously; show mercy and charity
To one emaciated by love, who leaves his friends.
All night long he knows no sleep,
Consumed by fire, drowned in a sea of tears.
Do not cut from my heart its hopes,
Wretched and troubled as it is, throbbing with love.

He folded the letter and, after handing it to the old woman, he presented her with three hundred dinars, saying: 'This is to help you wash your hands.' She thanked him, kissed his hands and went off to give his letter to the princess. When the princess had finished reading it, she threw it away, jumped to her feet and, wearing golden slippers studded with pearls and gems, she walked to her father's palace with a vein standing out so angrily between her eyes that no one dared question her. When she reached the palace and asked for the king, her father, his slave girls and concubines told her that he had gone out hunting. She went back to her own quarters like a ravening lioness and it was only after three hours that she addressed a word to anyone, by which time her face had cleared and her mood had softened.

When the old woman saw that she had recovered from her vexation and anger, she came forward and kissed the ground in front of her, before asking where she had gone. 'To my father's palace,' replied the princess. 'But could no one have run the errand for you?' the old woman asked, to which the princess replied: 'I went to tell my father what this dog of a merchant had done to me and to get him to use his authority to arrest him and everyone else in his market and to crucify them over their shops, as well as to expel every single other foreign merchant from our city.' 'Was that the only reason you went to your father?' the old woman asked, and the princess said: 'Yes, but I didn't find him there. I discovered that he had gone out hunting, but I shall wait for him to get back.' 'I take refuge with God, the All-Hearing, the Omniscient,' the old woman said. 'You, my lady – praise be to God – are the most intelligent of people, so how could you tell the king nonsense like this that should never be spread about?' 'Why is that?' the princess asked, and the old woman explained: 'Suppose that you had found the king in his palace and told him this story. He would have sent to have the merchants hung over their shops. People looking at them would have asked the reason for that, and the answer would be that they had tried to debauch the princess.'

Night 724

Morning now dawned and Shahrazad broke off from what she had been allowed to say. Then, when it was the seven hundred and twenty-fourth night, SHE CONTINUED:

I have heard, O fortunate king, that the old woman told the princess: 'Suppose you had told that to the king and he had ordered the merchants to be hung. Would not the people, looking at them, have asked the reason, and the reply would have been that they had tried to debauch the princess. There would be various stories. Some would say that the princess had left her palace and stayed with them for ten days until they had had enough of her, while others would give different versions, but honour, my lady, is like milk which is spoiled by the smallest bit of dust, or like glass which, when it has been cracked, cannot be mended. So be careful not to tell your father or anyone else about this affair lest you be dishonoured, for it will do you no good to tell people anything. Weigh this with your own superior intelligence, and if you find that it is not sound, then do what you want.' When the princess heard this, she thought it over and found the advice to be excellent. 'What you say is right, nurse,' she told the old woman, 'but I was blinded by anger.'

'Almighty God will approve of your intention not to tell anyone,' the old woman said, 'but there is something still to be done, for we must not ignore the shamelessness of this dog, the vilest of merchants. Write to him in these terms and tell him: "Had I not found the king to be absent, I would immediately have ordered you and all your neighbours to be crucified. But you will not escape, and I swear by Almighty God that if you say this kind of thing again, I shall remove all trace of you from the face of the earth." Use harsh language to him to check him and rouse him from his heedlessness.' 'Will this kind of talk make him turn back?' the princess asked, to which the old woman replied: 'How can he fail to do this after I have talked to him and told him what has happened?'

So the princess called for an inkwell and paper and wrote him the following lines:

You have fixed your hopes on union with me,
Aiming to achieve your goal.
It is his own folly that destroys a man,
And what you seek from me will bring disaster.
You are not a man of might with many followers,
Nor are you a sultan or his deputy.
Had an equal of mine acted like this,
The terrors of war would have whitened his hair.
But now I shall forgive your fault,
So that you may repent of what you did.

She gave the letter to the old woman and said: 'Nurse, stop this dog from pursuing me lest I commit a sin by having his head cut off.' 'My lady,' the old woman promised, 'I shall leave him no side on to which to turn.' Then she took the letter and brought it to Ardashir. They exchanged greetings and he took it and read it, shaking his head and reciting the formula: 'To God we belong and to Him do we return.' He went on: 'What am I to do, mother? I am too weak to endure any more.' 'Be patient, my son,' she told him, 'and it may be that God will now bring something about. Write down what is in your heart and I shall fetch you a reply. Take heart and be comforted, for, if God Almighty wills it, I shall very certainly bring the two of you together.'

Ardashir blessed her and wrote a letter in which were the following lines:

There is no one to help me in my love;
The tyranny of passion brings about my death.
Within my entrails I have to endure its fires
By day, and in the night I have no rest.
How can I give up hope of you, the goal of my desires?
I am content with the passion that I feel.
I pray to the Lord of the throne for satisfaction,
As love for the beautiful is destroying me.
I ask Him for the happiness of instant union,
For I am smitten by the terrors of desire.

He folded the letter and handed it to the old woman together with a purse containing four hundred dinars, which she took. She then went back to pass the letter on to the princess, who, for her part, refused to take it and asked what it was. When the old woman told her: 'My lady, this is the reply to the letter that you sent to that dog of a merchant,' the princess said: 'Didn't you forbid him to do this, as I told you?' 'Yes,' replied the old woman, 'and this is his reply.' So the princess took the letter from her and read it through to the end, before turning to the old woman and saying: 'What is the result of the message you gave him?' 'Didn't he say in his reply that he had abandoned his presumption, asking to be excused for his earlier behaviour?' the old woman asked. 'No, by God; he went even further,' the princess told her, at which the old woman said: 'Write him another letter and then see what I shall do to him.' 'I have no need to write or to reply,' said the princess, but the old woman told her: 'You have to write so that I can speak harshly to him and cut off his hopes.' 'You can cut off his hopes without taking him a letter,' the princess pointed out, but the old woman insisted that, in order to do this, she must go with a letter.

The princess then called for an inkwell and paper and wrote these lines to Ardashir:

I have long reproached you but this has not held you back.
How many lines of verse have I written to forbid you?
Hide your love; never let it show.
Disobey me and I shall not protect you.
If you protest your love once more,
The messenger of death will call your name.
Soon you will find the storm winds blowing;
The desert birds will flock around your corpse.
If you act well again, you will succeed,
But look for obscene filth and this will bring your death.

When she had finished writing she threw the paper away angrily, but the old woman picked it up and brought it to Ardashir, who took it from her. When he had read it through, he realized that the princess had not softened and that he would never get to her, as she was even angrier with him. He thought of writing a reply to call down a curse on her, so he composed the following lines:

Lord, by the five planets, I implore you: save me
From one whose love has brought me suffering.
You know the fires of my love

And my lovesickness for one who shows no mercy.
She has no pity for what afflicts me;
She oppresses my weakness and wrongs me.
I am lost, overwhelmed by separation from her,
And there is no one who will help my cause.
How often do I pass the dark of night in tears,
Both secretly and openly lamenting this!
I find no way through which to forget your love.
How could I, when passion has destroyed my patience?
Bird that tells of her parting, bring me news.
Is she safe from the disasters and distress of Time?

When he had folded the letter and passed it to the old woman, he gave
her a purse containing five hundred dinars and she brought the letter to
the princess. The princess read it and, having grasped its contents, she
threw it away and exclaimed: 'You evil old woman, tell me why you
have done all this and why you have schemed to present this man to me
in a good light? You have made me write letter after letter, and while
you have been carrying these to and fro, you have led people to start
talking about the correspondence we have been exchanging. Every time
you say: "I shall see that he does you no more harm and stop him from
talking about you," but you only do this to get me to write him another
letter so you can act as a go-between, coming and going until you have
destroyed my honour.' She cursed the old woman and told the eunuchs
to seize her and beat her, which they did until she fainted, with blood
pouring from all parts of her body. The princess then told the slave girls
to remove her, and they dragged her away by the legs to the furthest part
of the palace. One of them was ordered to stand by her head and, when
she had recovered consciousness, to say: 'The princess has sworn an oath
that you are not to be allowed to come back to the palace or to enter it,
and if you do, you are to be put to death without fail.'

When the old woman came to her senses again, the slave girl passed
on the princess's message, and she replied: 'To hear is to obey.' Then the
girls brought a basket chair for her and told a porter to carry her back
to her own house, which he did. Then they sent a doctor, who on their
instructions treated her gently until she had recovered. She now rode off
to Ardashir, who had been filled with grief at her non-appearance, as he
was longing to hear her news. When he saw her coming, he jumped to
his feet to meet her and, finding that she was showing signs of weakness,

he asked her how she was. She told him everything that had happened
to her with the princess. He was distressed to hear her tale and struck
one hand against the other, exclaiming: 'I find this hard to bear!' He
then went on to ask why it was that the princess had such a hatred for
men, and she told him: 'My son, you must know that the princess has a
garden unsurpassed in beauty throughout the whole world. One night,
while she was happily sleeping there, she had a dream in which she saw
that a hunter had set up his nets and spread grain around them, before
sitting down at a distance to see what game would fall into them. In her
dream the princess had gone there and before long she saw birds gathering to pick up the grain. A cock bird was trapped fluttering in the net,
and the others, including his mate, took flight and left him. Before long,
however, his mate came back to him and went up to the net to test the
part of the mesh in which his foot was trapped. She continued to work
at this with her beak until she managed to cut through it and free him.

'All the while the hunter was sitting snoozing, but when he woke up
he noticed what had happened to his net, mended it, scattered more
grain and sat down again at a distance from it. After a while the birds
gathered round it again, among them being the cock bird and his mate.
When they went up to gather the grain, this time it was the female who
was trapped, and as she fluttered in the net all the others flew off,
including her mate whom she had rescued, and, although the hunter had
fallen deeply asleep, the bird did not return. The hunter slept for a long
time, but when he woke to find the bird in his net, he got up, went to it
and, after freeing its legs from the net, he cut its throat. The princess
awoke in panic and said: "This is how men act with women. The woman
has pity for the man and risks her life for him when he is in difficulties,
but then when in accordance with God's decree she herself is in trouble,
he leaves her, making no attempt to free her, and the good she did him
is left unrequited. May God curse all those who put their trust in men,
for men will not acknowledge the services that women do for them."
She has hated men ever since.'

Ardashir now asked the old woman: 'Mother, does the princess not
go out at all?' 'No, my son,' she told him, 'except that she has one of
the finest pleasure gardens to be found in this age, and every year in the
fruiting season she goes to enjoy herself there for one day, returning at
night to her palace. She gets to the garden by a private door which leads
into it, and I propose to tell you something which, God willing, may be
of service to you. It is now one month before the time of her visit to see

the fruits, and I advise you to go straight away to the gardener who looks after the orchard and make friends with him. He doesn't let any living soul enter it, because it adjoins the princess's palace, but I shall let you know two days before she goes there. You must have made a habit of entering it, and you will then have to go in and contrive to spend the night there, so that when she comes you will be in some hiding place.'

Night 725

Morning now dawned and Shahrazad broke off from what she had been allowed to say. Then, when it was the seven hundred and twenty-fifth night, SHE CONTINUED:

I have heard, O fortunate king, that the old woman told Ardashir that the princess used to visit the garden and that she would let him know two days before the princess's visit. He was to be in a place of concealment. The old woman went on: 'Then, when you catch sight of her, come out, and at the sight of you she will fall in love with you, for love conceals everything. You must know, my son, that your appearance will fascinate her because you are a handsome man. So console yourself and be of good heart, as I shall very certainly bring the two of you together.'

Ardashir kissed her hand and, after thanking her, he gave her three lengths of Alexandrian silk and three of variously coloured satins. With each of these he added extra material for shirts and trousers, as well as a kerchief to wear around the head and Baalbaki cloth to provide linings, so that there were three full sets of clothes for her, each more beautiful than the others. He also gave her a purse of six hundred dinars, saying: 'This is for the sewing.' She took all this and then said: 'My son, would you like to know how to get to my house and to tell me at the same time where you live?' He agreed to this and sent a mamluk with her to find the way to her house and to show her his.

When she had gone he got up, and after telling his servants to close the shop he went back to the vizier and reported everything that had happened in his encounters with the old woman from start to finish. The vizier listened but then said: 'My son, if Hayat al-Nufus comes to the orchard but is not prepared to respond to your advances, what are you going to do then?' 'The only thing that I could do,' Ardashir replied, 'would be to move from words to deeds and risk carrying her off from among her eunuchs. I would have to take her up behind me on my horse and then ride out into the desert. Were I to escape, I should have got what I wanted, and if I perish, then I shall have done with this worthless life.' 'My son,' exclaimed the vizier, 'how are you going to survive if this is your idea of wisdom? How are we going to make the journey when we are so far from home, and how could you do this kind of thing to one of the kings of the age, who has a hundred thousand riders at his command? We could not be sure that he would not send out some of them to cut us off. This is not a good idea, and no sensible man would act like that.' Ardashir said: 'What do you, in your prudence, then advise me to do, vizier? For my death is certain.' 'Wait till tomorrow when we can take a look at this orchard and see what the position is, as well as seeing how we get on with the gardener,' the vizier told him.

The next morning, the two of them got up and the vizier took a thousand dinars with him in his pocket. They walked to the orchard, which they found enclosed with high, strong walls and containing many trees, numerous streams and fine fruits. With the flowers spreading their perfumes and the birds singing, it was like one of the gardens of Paradise. An old man was sitting on a bench inside the gate. When he saw them, noticing their air of dignity, he got to his feet and, after they had exchanged greetings, he asked them politely whether there was anything he could do for them. The vizier told him that they were strangers and added: 'We are finding the heat excessive and our lodgings are far away at the end of the city. Would you be kind enough to take these two dinars and buy us something to eat and then open the gate of this orchard for us so that we can sit in a shady spot by cool water until you bring us the food? Then we can all eat together, and when we are rested, we shall go on our way.' He put his hand in his pocket and brought out the two dinars, which he handed over to the gardener. This man, who had never before in all his seventy years seen anything like that in his hand, was overjoyed. He got up immediately to open the gate and let in his visitors, finding a seat for them under a shady fruit tree. 'Sit here,' he told them, 'but don't go into the garden itself as there is a private door in it that connects with the palace of Princess Hayat al-Nufus.'

Ardashir and the vizier assured him that they would not move from where they were, and he then went off to buy the food they had asked for. When he came back some time later, he was carrying on his head a roasted lamb together with bread. They ate, drank and talked for a while, after which the vizier got to his feet and started looking from right to left through the orchard. In it he noticed a lofty pavilion whose walls were peeling with age and whose pillars were broken. 'Shaikh,' he asked, 'is this garden your property or do you rent it?' 'It is not my property, sir,' the man replied, 'and I am not its tenant but its watchman.' 'How much do you earn?' asked the vizier, and the man told him that he got one dinar a month. 'That is not fair,' said the vizier, 'particularly if you have a family.' 'By God, sir,' he replied, 'there are eight children as well as myself.' 'There is no might and no power except with God, the Exalted, the Omnipotent!' the vizier exclaimed. 'I share your cares, poor man. What would you say of someone who was willing to help you for your family's sake?' The gardener replied: 'Sir, whatever good you do will be stored up as treasure for you with Almighty God.' 'Know then, shaikh,' said the vizier, 'that this orchard is a pleasant place but the pavilion there is old and run-down. I would like to repair it, plaster it and give it a good coat of paint until it becomes the most beautiful thing in the orchard. When the owner comes and discovers that it has been repaired and is now in a fine state, he is bound to ask questions. When he does, tell him: "This was my doing. I saw that it was in such a state that no one could use it or sit there as it was falling into ruins, and so I spent money on having it repaired." When he goes on to ask where you got the money from, tell him that you spent your own savings in order to find favour with him and in the hope of being rewarded. He will then have to repay you for what has been spent on the place. Tomorrow I shall get builders, plasterers and painters to do the job and I shall give you what I have promised.' He took from his pocket a purse containing five hundred dinars and said: 'Take this and spend it on your family, telling them to pray for me and for my son here.' When Ardashir asked him why he was doing this, the vizier said: 'You will see what happens.'

Night 726

Morning now dawned and Shahrazad broke off from what she had been allowed to say. Then, when it was the seven hundred and twenty-sixth night, SHE CONTINUED:

I have heard, O fortunate king, that the vizier gave the five hundred dinars to the gardener and said: 'Take these and spend them on your family, telling them to pray for me and for my son here.' When the gardener saw the gold, he nearly went out of his mind and he threw himself at the vizier's feet to kiss them, calling down blessings on him and on his 'son'. When they were about to leave, he said: 'I shall be waiting for you tomorrow, and may Almighty God never part us night or day.' The vizier went off the next day and summoned the master builder. When he came, the vizier took him to the orchard, where the gardener was delighted to see him. The vizier handed him money for provisions and for all that the workmen needed for the job, after which they set about rebuilding, plastering and painting. He told the painters to listen to his instructions in order to understand what it was he wanted. 'Know,' he said, 'that I once had an orchard like this, and one night when I was sleeping there I had a dream in which I saw a hunter laying nets, around which he scattered grain. When the birds gathered there to pick this up, a cock bird fell into the net, at which all the others, including his mate, were frightened away. Then, after a time, his mate came back by herself and worked at the part of the mesh holding his foot until she had freed him and he flew off. The hunter meanwhile had been asleep, and when he woke up to find a hole in his net, he repaired it and scattered more seed, after which he sat down at a distance, waiting for something to fall into the trap. The birds came back to pick up the grain and this time it was the female who was trapped. The others, including her mate, flew off and left her, and when he did not come back, the hunter got up, took her and cut her throat. As for her mate, however, when he flew off with the others, he fell victim to a bird of prey which killed him, drank his blood and ate his flesh. I want you to paint the whole scene as well as you can, just as I have described it to you, using this orchard with its walls, trees and birds as a model in your painting. Paint a picture of the hunter with his nets and of what happened to the cock bird when it was carried off by the raptor. If you follow my instructions and I am pleased with the result, you will be happy with the bonus that I shall give you on top of your wages.'

When the painters heard what he had to say, they did their very best and produced their finest work. When it was all finished, they showed it to the vizier and he was so delighted to see an exact portrayal of the dream that he had described to them that, after thanking them, he rewarded them with the greatest generosity. Ardashir then came as usual, but he did not know what the vizier had been doing, and when he came and looked at the pavilion, he was astonished to see a picture of the orchard with the hunter and his nets, the other birds, and the cock bird in the talons of the raptor which had killed it and drunk its blood and eaten its flesh. He went back to the vizier and said: 'Prudent man, today

I have seen a marvel which, were it written with needles on men's eyeballs, would serve as a lesson for all who can learn.' When the vizier asked what this might be, Ardashir replied: 'I told you, didn't I, of the princess's dream which led her to hate men?' 'You did,' said the vizier, and Ardashir went on: 'I have just seen it painted there so vividly that I might be looking at the real thing, and I have discovered a detail which the princess didn't know and had not seen, something on which I can rely to get me to my goal.' 'What is that, my son?' asked the vizier, and Ardashir told him: 'I have found that when the cock bird abandoned its mate which was caught in the net, and did not go back, it had itself been seized and killed by a raptor, which drank its blood and ate its flesh. I wish that the princess had seen the whole dream and followed the tale to its end. She would have discovered that the cock bird did not come back to free its mate because it had been carried off by a bird of prey.' 'By God, O fortunate prince,' exclaimed the vizier, 'this is certainly a remarkable thing!'

Ardashir continued to admire the painting and to regret the fact that the princess had not seen the whole dream, saying to himself: 'I wish that she had seen it all or could see the whole of it again, however confused it might be.' The vizier then said: 'You asked me why I had had this work done here, and I told you that you would see the result of it. This is the result, and I am responsible for it as I told the painters to paint the dream and to show the cock bird in the talons of the bird of prey which killed it, drank its blood and ate its flesh. When the princess comes down to look at this, she will see not only what she dreamt, but also how the cock bird was killed. This will lead her to excuse it and to abandon her hatred of men.'

When Ardashir heard this, he kissed the hands of the vizier, thanked him for what he had done and exclaimed: 'A man like you should be the vizier of the greatest of kings! By God, if I reach my goal and return happily to the king, my father, I shall tell him about this so that he may show you even greater favour, ennoble you further and listen to what you say.' The vizier kissed his hand, after which the two of them went to the old gardener and said: 'Look at how beautiful this place is.' 'That is thanks to the good fortune that you have brought,' the man replied, and they then emphasized that, when the owners asked him who had repaired it, he was to say that he had paid for it with his own money in order that he might receive a generous reward. 'To hear is to obey,' he told them, and after that Ardashir remained his constant companion.

So much for the prince and the vizier, but as for Hayat al-Nufus, she was delighted when Ardashir's letters stopped coming and when the old woman stayed away from her, as she thought that her young suitor must have gone back to his own country. Then, one day, a covered dish came to her from her father and when she removed the cover she found ripe fruits. 'Has the season for these come?' she asked, and when her attendants told her that it had, she said that she wanted to make her preparations for a pleasure trip to the orchard.

Night 727

Morning now dawned and Shahrazad broke off from what she had been allowed to say. Then, when it was the seven hundred and twenty-seventh night, SHE CONTINUED:

I have heard, O fortunate king, that when her father sent fruits to the princess, she asked whether they were now in season. On being told that they were, she said that she wanted to make her preparations for a pleasure trip to the orchard.

'What an excellent idea, my lady!' her slave girls exclaimed. 'We have been longing to go there.' 'But what are we going to do?' the princess asked, explaining: 'Every year it has been my nurse who has shown us over it and explained the differences between the various branches, but I have had her beaten and banished from my presence. I am sorry for what I did to her, as she was, after all, my nurse and I owe her a debt for having raised me, but there is no might and no power except with the Exalted and Almighty God.' When the slave girls heard her say this, they all got up, kissed the ground in front of her and implored her in God's Name to pardon the old woman and to command her to come back. 'By God, I have made up my mind to do that,' the princess told them, 'but which of you will go to her and take the splendid robe that I have prepared for her?' Two of the seniors, particular favourites of the princess, came forward, one called Bulbul and the other Sawad al-'Ain, both graceful and pretty girls. They volunteered to go, and the princess told them: 'Do as you like.' So they went to the old woman's house, knocked on the door and went in to see her. When she recognized them she welcomed them with open arms, and when they had sat down they said: 'Nurse, the princess has forgiven you and taken you back into favour.' 'I shall never go back,' the old woman exclaimed, 'even if this brings destruction on me! Have you forgotten how I was punished in front of friends and enemies alike, when my clothes were stained with blood and I was beaten so harshly that I almost died? Then I was dragged off by the legs like a dead dog and thrown out of the door. By God, I shall never go back to her or rest my eyes on her again.' The girls protested: 'Don't let our efforts be wasted. Where is your courtesy to us? Are you thinking over possible visitors and do you want someone to come who is of a higher standing with the princess?' 'God forbid,' the old woman replied. 'I know that I am your inferior, but the princess raised me to a position of such importance among her slave girls and eunuchs that if I were angry with even the greatest of them, such a one would almost die of fear.' 'Things are still the same,' the girls said. 'Nothing has changed and your position is even stronger than before, as the princess has humbled herself before you and asked for a reconciliation without an intermediary.' 'By God,' replied the old woman, 'had you not come to me, I would not have gone back to her even if she had ordered my execution.'

The girls thanked her and she got up immediately, put on her outdoor clothes and set out with them. They brought her to Hayat al-Nufus, who got to her feet at the sight of her nurse, and she, for her part, called on the Name of God and asked: 'Was the fault mine or yours, princess?' 'The fault was mine,' the princess replied, 'and so forgiveness and reconciliation must come from you. By God, nurse, I value you highly and I owe you a debt for having raised me, but you know that the Sublime and Exalted God has allotted four things to his creation: character, life, sustenance and their decreed fate. It is not in human power to ward off destiny. I could not control myself nor regain my balance, but I am sorry for what I did.' At that, the old woman recovered from her anger and came forward and kissed the ground before the princess, who called for a costly robe of honour. This she put over the shoulders of her delighted nurse as the eunuchs and slave girls stood in front of her.

When this had been done, the princess asked about the fruits in the orchard and the old woman replied: 'By God, my lady, I have seen most kinds of fruits in the town, but I shall go and look into the matter today and come back to you with an answer.' She then left, having been treated with the greatest honour, and went to Ardashir. He welcomed her joyfully, embraced her and relaxed, taking her arrival as a good omen, after a long period of expectation. She gave him an account of her encounter with the princess, and told him that she was intending to go to the orchard on such-and-such a day.

Night 728

Morning now dawned and Shahrazad broke off from what she had been allowed to say. Then, when it was the seven hundred and twenty-eighth night, SHE CONTINUED:

I have heard, O fortunate king, that when the old woman went to Ardashir, she told him of her encounter with the princess and that she was intending to go to the orchard on such-and-such a day. 'Have you done what I told you to do with the gardener and have you treated him generously?' she asked. 'Yes,' said Ardashir, 'he has become my friend; his path is mine and he would be delighted if there was anything I needed from him.' He then told her of how the vizier had had a painting made of the princess's dream, with the hunter, the nets and the raptor. She exclaimed in joy at this. 'By God, you should give your vizier a place in the centre of your heart, for what he has done shows the soundness of his judgement and he has helped you to achieve what you want. Now get up immediately and go to the baths, after which put on your most splendid clothes, as nothing that we can do can have more effect than this. Next go to the gardener and find some way of making him allow you to spend the night in the orchard, as even if he were presented with enough gold to fill the whole earth, he would never let anyone go into it. When you do get in, hide away out of sight and stay hidden until you hear me say: "God, Who hides His favours, save us from what we fear." Then come out from your hiding place and show your beauty but remain in the shelter of the trees. This beauty will put the moon to shame, and when Princess Hayat al-Nufus sets eyes on you, her heart and body will be filled with love. You will achieve your goal and your cares will vanish.' 'To hear is to obey,' Ardashir said, and he then brought out a purse containing a thousand dinars which she took before going off.

Ardashir himself got up immediately and, after a luxurious visit to the baths, he put on the finest of his royal robes with a sash studded with precious stones of all kinds and a turban embroidered with threads of red gold and set with pearls and gems. His cheeks were rosy, his lips were red, his eyelids could exchange loving glances with gazelles and he swayed like a drunken man. His whole body was filled with beauty and his pliant form put the branches to shame. He put a purse with a thousand dinars in his pocket and then went to the orchard, where he knocked on the gate. The gardener answered his knock and, on opening the gate, was delighted to see him and greeted him with the greatest courtesy. Seeing that he looked gloomy, the gardener asked him how he was, and Ardashir told him: 'You must know, *shaikh*, that I have always been well treated by my father and before today he never laid a hand on me. Today, however, we had a quarrel and he abused me, struck me in the face and threw me out after beating me with a stick. I can think of no one to befriend me and I am afraid of the treachery of time, for, as you know, the anger of parents is no small thing. I have come to you, uncle, because my father knows about you and I want you to be kind enough to let me stay in the orchard until the end of the day or even to spend the night there until God reconciles the two of us.'

The gardener was distressed to hear about the quarrel and said: 'Sir, would you allow me to go as a peacemaker to intervene with your father?' 'Uncle,' replied Ardashir, 'you should know that my father is a man of difficult temper, and if you approach him about a reconciliation while he is still burning with anger he will pay no attention to you.' 'To hear is to obey,' said the gardener, adding: 'But come to my house with me and if you spend the night with my children and the rest of my family, no one can disapprove.' Ardashir explained that he was so angry that he had to be alone. 'I would find it hard if you slept alone in the orchard when I have a house,' the gardener told him, but Ardashir replied: 'I want to do this so that my mood may pass, and I know that this is the way to get my father's approval and win back his affection.' 'If this is how it has to be,' the gardener said, 'I shall get you a mattress to lie on and something to cover you.' 'There is nothing wrong with that,' agreed Ardashir, and the gardener got up, opened the orchard gate and fetched the mattress and the blanket, not knowing that the princess was about to come there.

So much for Ardashir, but as for the old woman, she went to the princess and told her that the fruits were ripe on the trees, and the princess invited her to come with her the next day to look at them, adding that she was to warn the gardener of their visit. So the old woman sent word to say: 'The princess will be with you in the orchard tomorrow. See that you leave no one there to water the trees and that there are no seasonal workers. No one else at all is to be allowed in.'

When these instructions came to the gardener, he saw to it that the water channels were flowing properly and then went to Ardashir and said: 'The princess is the owner of this orchard. I have to excuse myself to you, as the place is yours and I live only through your kindness, but

I am not my own master and I have to tell you that the princess intends to come here early tomorrow morning. I have been ordered to let no one into the garden who might look at her. Would you be good enough to leave it today, for she will only stay until the afternoon and then you can have it for months and years and ages.' 'It may be that we have done you some harm, *shaikh*,' suggested Ardashir, at which the gardener protested: 'No, by God, master, you have done me nothing but honour.' 'If that is so,' Ardashir replied, 'nothing but good will come to you from me, and I shall hide in the garden without being seen by anyone until the princess goes back to her palace.' The gardener said: 'Master, if the princess catches sight of the shadow of a man, she will have my head cut off.'

Night 729

Morning now dawned and Shahrazad broke off from what she had been allowed to say. Then, when it was the seven hundred and twenty-ninth night, SHE CONTINUED:

I have heard, O fortunate king, that the gardener said that if the princess saw the shadow of a man, she would have the gardener's head cut off. Ardashir reassured him that he would not let anyone at all see him, and added: 'I am sure that you must be short of money to spend on your family.' He then reached for his purse and brought out five hundred dinars, saying: 'Take this and use it for your family expenses, so that you need no longer be worried about them.' The sight of the gold convinced the man that the risk was not too great, and after impressing on Ardashir that he was not to show himself in the garden, he left him sitting there.

So much for the gardener and the prince, but as for the princess, early next morning, when her eunuchs came in, she ordered the private door to be opened which led to the orchard with the painted pavilion. She put on a regal dress studded with all types of pearls and gems and under it she wore a delicate chemise set with rubies, while beneath all this was what cannot be described, leaving the heart bewildered and giving the courage of love to the coward. On her head was a crown of red gold encrusted with pearls, and her golden slippers were adorned with fresh pearls, together with all kinds of precious stones. Resting her hand on the shoulder of her nurse, she gave the order to leave by the private door.

When the old woman looked at the garden she found it full of eunuchs and slave girls, who were eating the fruits, muddying the streams and enjoying themselves happily as they played around. 'You are a wise and intelligent lady,' she told the princess, 'and you know that you don't need all these servants here. If you went outside your father's palace, you would have to have them to preserve your dignity, but you are coming out by a private door into the garden where none of God's creation can lay eyes on you.' The princess agreed and asked what should be done about it. 'Tell the eunuchs to go,' said the old woman, adding, 'and I only tell you this out of respect for the king.' When the princess had given the order, the old woman pointed out: 'There are still some of them left, doing damage here. Send them away and keep no more than two slave girls with you so that we can relax with them.'

When the old woman saw that the princess was happy and at ease, she said: 'Now we can enjoy ourselves, so come with me into the garden.' The princess put her hand on her shoulder and went out through the private door with two slave girls leading the way. She was laughing at them and swaying in her robes, while the old woman went ahead to show her the trees and give her fruits to taste as she moved from place to place, leading her on until she came to the pavilion. When the princess looked at this, she saw that it had been restored and she exclaimed: 'Look, nurse! The pillars have been restored and the walls plastered.' 'By God, lady,' the old woman replied, 'I did hear that the gardener had got some materials from a group of merchants and had sold them before spending the money on bricks, lime, gypsum, stones and so on. I asked him about that and he told me that he had used all this to repair the ruined pavilion. The merchants had asked him to repay what he owed them and he told them to wait until the princess came to the garden and saw and admired the restoration work. He would then take whatever she was pleased to give him and hand them their share. I asked him what had prompted him to do this and he said: "I saw that the pavilion was collapsing, its supporting pillars had fallen, its white surface was flaking and no one had the decency to repair it. I pledged my own credit to borrow money, hoping that the princess would act as befits her." I told him that you were full of generosity and eager to repay debts, for it was in the hope of your favour that he did all this.' 'By God,' the princess said, 'he acted as an honourable and a generous man.'

The princess now told the old woman to call her treasurer, and when he came immediately, she ordered him to give the gardener two thousand dinars. The old woman sent a messenger, who went to fetch him and said: 'Obey the summons of the princess.' When he heard this, the gardener was unnerved and trembled, thinking to himself that the princess must have caught sight of Ardashir and that this was his unluckiest day. He went home and told his wife and children what had happened, giving them his last instructions and saying goodbye as they wept over him. He then walked away, and when he came before the princess, his face was the colour of Indian saffron and he was on the point of total collapse. The old woman realized this and spoke quickly, telling him to kiss the ground in gratitude to Almighty God and to call down blessings on the princess. 'I told her what you did to repair the ruined pavilion,' she said, 'and she has been pleased to present you with two thousand dinars in return for this. Take the money from the treasurer, invoke a blessing on her and kiss the ground in front of her before going on your way.' On hearing this, the gardener did as she told him, before going home with the two thousand dinars to the delight of his family, who called down blessings on the author of this good fortune.

Night 730

Morning now dawned and Shahrazad broke off from what she had been allowed to say. Then, when it was the seven hundred and thirtieth night, SHE CONTINUED:

I have heard, O fortunate king, that the gardener took the two thousand dinars from the princess before going back home. His family were delighted and called down blessings on the author of this good fortune.

So much for them, but as for the old woman, she said to the princess: 'How beautiful this pavilion has become! I have never seen anything whiter than its plaster or finer than its painting, but do you suppose that he tackled both the outside and the inside, or whitened the outside and left the inside black? Come in with me and have a look.' She went in, followed by the princess, and they discovered that the interior was most splendidly painted and embellished. The princess glanced from right to left until she came to the top of the room, where she took a long look. The old woman realized that she had caught sight of the representation of her dream, and so she took away the two slave girls lest they distract her.

When the princess had finished looking at the painting, she turned in wonder to the old woman, striking one hand against the other. 'Nurse,' she said, 'come and see something so remarkable that, were it written with needles on men's eyeballs, it would serve as a lesson for all who can learn.' The old woman asked what it was, and the princess said: 'Come up to the top of the room, look and then tell me what you see.' The old woman went into the room, where she studied the painting of the dream, and when she went out she expressed astonishment and said: 'My lady, this is a picture of the orchard with the hunter and his net, together with everything that you saw in your dream. It turns out that the cock bird had no chance to go back to rescue his mate from the hunter's net, as I saw that it was clutched in the talons of a raptor, which had killed it, drunk its blood and torn and eaten its flesh. This was why it could not get back in time to save her, but the real wonder is how the dream came to be painted, for you yourself could not have done it even if you had wanted to. By God, this is something marvellous which deserves to be recorded in histories. It may be that the angels, whose task it is to look after mankind, realized that we had wrongly blamed the cock bird for not returning, and produced this in order to establish that it had an excuse, as I have just seen its corpse in the raptor's talons.' The princess agreed, saying: 'The bird was a victim of fate, and we wronged it.' 'Adversaries will meet before the presence of Almighty God,' said the old woman, 'but as far as we are concerned, the truth has been made clear and we have been shown that the bird was to be excused. Had it not been seized and killed by the raptor, which drank its blood and ate its flesh, it would not have been slow to return. It would have gone back and saved its mate from the net, but there is no escape from death. Men are special in that they will starve themselves and feed their wives, strip themselves to clothe their wives, anger their own families to please them, and disobediently refuse to their parents what they give to their wives. A wife knows her husband's hidden secrets and cannot bear to be parted from him for a single hour. If he is away for a night she cannot sleep; no one, not even her parents, is dearer to her than him, and as they sleep in one another's arms, his hand is beneath her neck and hers beneath his, as the poet says:

> I pillowed her in my arms and spent the night with her;
> "Be long," I told the night, "for the full moon has risen."
> O for a night whose like God never made!
> It started sweetly only to end in bitterness.

He will then kiss her and she will kiss him. There was a king whose wife sickened and died and who had himself buried alive with her, being content to die himself out of his love for her and because of the strength of their mutual affection. Similarly there was another king who died of an illness, and when her family wanted to bury him, his wife told them to let her bury herself alive with him, saying: "Otherwise I shall kill myself and the responsibility for that will be yours." When they realized that she would not change her mind, they let her be and she threw herself into his grave because of the strength of her love for him.'

The old woman went on telling the princess stories of men and women until her hatred of men left her, and when the old woman recognized that the princess had a newly awakened inclination towards men, she said: 'It is time that we looked around the garden.' So they left the pavilion and walked among the trees. Ardashir turned and caught sight of the princess, and when he saw her figure, her symmetrical form, her rosy cheeks, dark eyes, remarkable gracefulness and the splendour and perfection of her beauty, he was lost in astonishment. As he fixed his eyes on her, passion overcame his reason and his love passed all bounds. He wanted to serve her with every limb of his body, and so fiercely did the fire of love blaze within him that he lost consciousness and fell to the ground in a faint. When he recovered, he found that she had gone out of sight among the trees.

Night 731

Morning now dawned and Shahrazad broke off from what she had been allowed to say. Then, when it was the seven hundred and thirty-first night, SHE CONTINUED:

I have heard, O fortunate king, that while Ardashir was hiding in the garden, the princess and the old woman came down and walked among the trees. The violence of his passion made him faint at the sight of her, and when he recovered he found that she had gone out of sight among the trees. He heaved the deepest of sighs and recited these lines:

When my eyes saw the uniqueness of her beauty,
My heart was rent by passionate love.
I found myself thrown on the ground,
But the princess knew nothing of my suffering.

Swaying, she captivated the heart of her love's slave;
For God's sake, be merciful; have pity on my passion;
Lord, bring close the time of union; let me enjoy
Her, who is my heart's blood, before I travel to the grave.
I shall give her ten kisses, then ten and then another ten,
Gifts to her cheek from a lover emaciated and distressed.

The old woman continued to show the princess around the orchard until she came to Ardashir's hiding place, where she exclaimed: 'God, Who hides His favours, save us from what we fear!' On hearing his cue, Ardashir came out, walking proudly and haughtily among the trees. His figure put the branches to shame, his forehead was pearled with sweat and his cheeks were like the evening sun. Praise be to God, Whose glory shines over His creation! The princess turned and, catching sight of him, she stayed for a long time looking at him and taking note of his beauty and grace, the symmetry of his form, his eyes, which could exchange loving glances with gazelles, and his figure, which would shame the branches of the *ban* tree. His eyes shot arrows at her heart, bemusing her and robbing her of her wits. 'Where did this handsome young man come from, nurse?' she asked. 'Where is he?' the old woman replied, and the princess said: 'Close by, among the trees.' The old woman began to look right and left as though she knew nothing about the matter, and then exclaimed: 'Who told him how to get here?' 'Who can tell us about him?' the princess asked, adding: 'Glory be to God, Who created men.'

She then asked the old woman: 'Do you recognize him?' and the woman replied: 'This is the young man whose letters I carried to you.' The princess, drowning in the sea of love and consumed by the fire of longing, said: 'Nurse, how handsome he is and how attractive. I don't believe that on the face of the earth there can be any better looking man.' The old woman, recognizing that love had mastered her, said: 'Didn't I tell you, my lady, that he was a handsome young man with a lovely face?' 'Nurse,' the princess went on, 'the daughters of kings know nothing about the affairs of the outside world or the qualities of its inhabitants as they have no association with them in either giving or taking. So tell me how I can get to him or arrange to meet him face to face. What should I say to him and what would he reply?' 'What can I possibly do now, when, thanks to you, we are left here in a state of confusion?' the old woman asked. 'I tell you, nurse, that if no one else has ever died of love, I shall die of it, and I am sure that this will happen straight away, thanks to the fire of my passion.'

When the old woman heard this and saw how passionately she loved Ardashir, she said: 'My lady, there is no way in which he can come to you, and because you are so young you cannot be expected to go to him. So come with me and I shall go ahead of you until you get to him. I shall do the talking, so that you need not feel ashamed, and then in the twinkling of an eye the two of you will find yourselves on friendly terms.' 'Go on in front of me,' the princess told her, 'for God's decree cannot be turned back.'

The two of them set off and came to where Ardashir was sitting, looking as beautiful as the moon at its full. When they got to him, the old woman said: 'Young man, look and see who stands before you. This is Hayat al-Nufus, the daughter of the king of the age. In recognition of her status and the honour that she has done you by walking here to you, get up out of respect for her and remain on your feet.' Ardashir rose to his feet immediately, and when the two looked into each other's eyes, they both became intoxicated but not through wine. The longing and love that the princess felt for Ardashir grew stronger; she opened her arms, he opened his, and they embraced in a paroxysm of desire until, overwhelmed by love and passion, they both fainted and collapsed on the ground. There they remained for so long that the old woman, fearing a scandal, brought them into the pavilion and sat there at the door. She told the slave girls to take the chance to enjoy themselves, telling them that the princess was asleep, and this they went back to do.

When the two lovers regained consciousness, they found themselves inside the pavilion. Ardashir said to the princess: 'Tell me in God's Name, queen of beauties, is this a dream or a vision seen in sleep?' The two embraced, drunk with love and complaining of the torment of passion. Ardashir then recited:

The sun rises from her bright face,
With the glow of its setting in her cheeks.
When onlookers see this face,
The evening star hides away in shame.
When lightning flashes from her teeth,
Dawn breaks and clears away dark gloom,
And as her body bends and sways,
The leafy branches of the *ban* tree are jealous.

I need no more than to catch sight of her;
May the Lord of mankind and of the dawn guard her.
She lent the moon part of her beauty;
The sun has tried to rival her but failed.
How can the sun have sides that sway?
Where can the moon find such beauty of shape and nature?
Who can blame me, when my whole being is in love with her,
Whether in its separate parts or in its whole?
She has conquered my heart with a single glance,
And what protection is there for lovers' hearts?

Night 732

Morning now dawned and Shahrazad broke off from what she had been allowed to say. Then, when it was the seven hundred and thirty-second night, SHE CONTINUED:

I have heard, O fortunate king, that when he had completed these lines, the princess clasped him to her breast, kissing him on the mouth and between the eyes. He regained his spirit and started to complain to her of what he had suffered through love and the tyranny of passion, the extent of his longing and his infatuation, and what the hardness of her heart had caused him to endure. When she heard this, she kissed his hands and his feet and uncovered her head, showing the darkness of her hair through which the full moons shone. 'My darling,' she said, 'and the goal of my desires, I wish that there had never been a day on which I rejected you, and may God ensure that it never returns.' They embraced each other tearfully, and the princess then recited:

You, who shame the moon, and the sun by day,
You have caused the beauty of your face to kill me wrongfully,
With the sword of a glance that pierces my entrails.
Where can one flee from such a sword?
The bow of your eyebrows has shot at my heart
With an arrow of burning love.
Your cheeks are for me a garden of fruits,
And how is my heart to endure without them?
Your swaying form is like a flowering branch;
The fruits it bears are there for me to pick.

You captured me by force, and left me without sleep;
Through love for you I have cast off all shame.
May God aid you with radiant light,
Making the far things close and bringing near a meeting.
Pity a heart seared by your love,
Of an emaciated lover whose refuge is in your nobility.

When she had finished these lines, she was swamped by a wave of passionate love and shed floods of tears. Ardashir, himself consumed and distressed by his love for her, went up to her and kissed her hands, shedding tears of his own. The two of them continued to exchange reproaches, intimate talk and poetry, but no more than this, until the call to the afternoon prayer, when they had to think of parting. 'Light of my eyes and breath of my life,' the princess said, 'this is the time we must part, but when shall we meet again?' Ardashir, whose heart had been pierced by this, exclaimed: 'By God, I do not like this talk of parting!' She then left the pavilion, and when he turned to look at her he found her uttering moans that would melt the rocks and shedding tears like rain. Drowning in a sea of the disasters of love, he recited these lines:

Heart's desire, my concern has grown
Through my great love for you. What shall I do?
Wherever your face appears, it is like dawn;
The colour of your hair is that of night.
Your figure is that of a pliant bough,
Swaying in the breath of the north wind.
Your glances are those of a gazelle,
Looked at by noble men.
Your waist is slender and your buttocks heavy,
The weight of one matched by the slimness of the other.
From your mouth comes the sweetest wine,
With pure musk and cool water.
Gazelle girl of the tribe, end my distress
And, of your bounty, send your phantom as a visitor.

When the princess heard how he described her, she came back and embraced him with a heart burning with the fire of separation that could only be extinguished by a kiss and an embrace. She quoted the proverb that says: 'It is patience that lovers must show and not its absence,' adding: 'I must think of some way for us to meet.' Then she left after saying goodbye to him, and, although such was her love that she had no idea where she was walking, she went on until at last she threw herself down in her own room. Ardashir, for his part, felt such an increase of longing and love that he was unable to taste the sweetness of sleep, while the princess could not eat and, unable to endure her plight, she fell sick.

The next morning she sent for her nurse, who arrived to find that the condition of her mistress had changed. 'Don't ask how I am,' the princess told her, 'for you are responsible for all this. Where is my heart's darling?' 'When did he leave you?' the old woman asked. 'Has he been away for more than this one night?' The princess answered: 'How can I bear to be parted from him for a single hour? Go and find some means of bringing us together quickly, for I am almost about to breathe my last.' 'Have patience, lady,' the old woman told her, 'and I shall produce a subtle and undetectable scheme.' The princess replied: 'I swear by the Omnipotent God that if you don't fetch him today, I shall go and tell the king that you have corrupted me and he will cut off your head.' 'For God's sake, give me time, as this is a serious matter,' the old woman implored, and she continued to beg humbly until she got the princess to wait for three days. The princess then said: 'These three days have been like three years for me, nurse, and if a fourth day goes by without your having brought him to me, I shall see to it that you are killed.'

The old woman left her presence and went back to her house. Then, on the morning of the fourth day, she collected the bridal hairdressers of the town and asked them to provide her with the best cosmetics in order to adorn, paint and tattoo a young virgin. When they had brought the finest of these as she had asked, she sent for Ardashir and on his arrival she opened a chest and took out a package containing a woman's dress worth five thousand dinars, together with a headband set with all kinds of jewels. 'My son,' she said, 'do you want to meet Hayat al-Nufus?' When he said yes, she took out tweezers with which she removed the hair from his face before dying it with kohl. Next she stripped him and painted his arms from finger to shoulder and his legs from the insteps to the thighs, after which she tattooed the rest of his body until he had become like a red rose lying on a slab of marble. Then, after a brief pause, she washed and cleaned him, before bringing out a chemise and drawers, over which she made him wear the regal dress she had ready, together with the headband and a veil. She taught him how he should walk, telling him to lead with the left leg and draw back the right, and, following her orders, he walked in front of her like a houri coming from Paradise.

'Steel yourself,' she told him, 'for you are going to the king's palace and there are bound to be guards and eunuchs at the gate. If you seem nervous of them or give them grounds for suspicion, they will investigate you and find you out. That would be disastrous and would lose us our lives, and so, if you cannot do it, tell me.' 'This does not frighten me,' replied Ardashir. 'There is no need for you to worry and you can be at ease.' She then went out and walked on ahead of him until they got to the palace gate, which was crowded with eunuchs. She turned to see whether he was showing signs of fright or not, and discovered that he appeared entirely unmoved.

On her arrival, the chief eunuch looked at her and recognized her, but he could see that she was followed by a 'girl' of indescribable beauty. He said to himself: 'The old woman is the princess's nurse, but no one in our land can match the figure of the girl who is following her, or come near her in point of beauty or grace, except for the princess Hayat al-Nufus and she is kept in seclusion and never goes out. I wish I knew how she comes to be out in the street, and I wonder whether she did this with the king's permission or without it.' He got to his feet to investigate and was followed by some thirty eunuchs. When the old woman saw this, she was taken aback and recited the formula: 'We belong to God and to Him do we return,' adding: 'Now we are dead – there is no doubt about it.'

Night 733

Morning now dawned and Shahrazad broke off from what she had been allowed to say. Then, when it was the seven hundred and thirty-third night, SHE CONTINUED:

I have heard, O fortunate king, that when the old woman saw the chief eunuch coming towards her with his followers, she was terrified and recited the formula: 'We belong to God and to Him do we return,' adding: 'Now we are dead – there is no doubt about it.'

The chief eunuch heard this and became suspicious, knowing that the princess was harsh in her dealings and that her father was under her thumb. So he said to himself: 'It may be that the king ordered the princess's nurse to take her out on some errand and did not want anyone to know about it. If I interfere with her, she will be furious with me and will say: "This eunuch confronted me in order to find out what I was doing," and she will try her best to have me killed. There can be no need to do anything about this.' So he went back to the palace gate, followed by the other thirty, and they drove the bystanders away, letting the nurse enter in safety, as they stood by respectfully and exchanged greetings with her.

Ardashir followed her in and together they went through one gate after another, passing by all the guards under the protection of God, the Veiler, until they came to the seventh gate. This was the gate of the main palace where the king's throne was and which gave access to the apartments of his concubines, the harem and the pavilion of the princess. Here the old woman stopped and said: 'My son, glory be to God, Who has brought us this far, but it is only at night that you can meet the princess, as night is a cover for the fearful.' 'That is true,' he agreed, 'but what are we to do?' 'Hide in the darkness here,' she said, showing him a cistern, in which he sat down. She then went off somewhere else, leaving him there until the end of the day, when she came back and took him out. Together they went through the gate of the pavilion and walked on until they got to the apartments of Hayat al-Nufus. The old woman knocked on the door and a little girl came out to ask who was there. 'It is I,' said the old woman, and the girl went back to ask whether her mistress would admit her nurse. 'Open the door for her and let her come in with whomever she has with her,' the princess told her, and so the two visitors entered.

When they came forward, the old woman turned to Hayat al-Nufus and saw that she had prepared the room, arranging rows of candles and covering the benches and the raised floors with carpets. Cushions had been placed there; candles had been lit in chandeliers of gold and silver; food, fruits and sweetmeats had been set out and the room was perfumed with musk, aloes and ambergris. The princess was sitting surrounded by lamps and candles, but the radiance of her face outshone them all. She looked at the old woman and asked: 'Nurse, where is my heart's darling?' 'My lady,' the old woman answered, 'I have not met him or seen him, but I have brought you his sister.' 'Are you mad?' exclaimed the princess. 'I don't need his sister. When someone has a wounded head, does he bandage his hand?' The old woman replied: 'No, by God, my lady, but nevertheless take a look at her and, if she pleases you, let her stay with you.' At that she unveiled Ardashir's face and, when the princess recognized him, she got to her feet and clasped him to her breast while he clasped her to his before they both fell to the ground unconscious. When they had stayed like this for a long time, the old woman sprinkled them with rosewater and they recovered.

Hayat al-Nufus now kissed Ardashir on the mouth more than a thousand times and recited these lines:

> My heart's darling visited me in the dark;
> I rose to honour him until he took his seat.
> I said: 'You, who are my wish and my desire,
> Did you not fear the guards as you came here by night?'
> 'I did,' he said, 'but it is love that holds my life and soul.'
> We clung together and embraced;
> For here is safety, with no guards to dread.
> We got up with our honour still unsoiled,
> Shaking out robes that had remained unstained.

Night 734

Morning now dawned and Shahrazad broke off from what she had been allowed to say. Then, when it was the seven hundred and thirty-fourth night, SHE CONTINUED:

I have heard, O fortunate king, that when Hayat al-Nufus met her beloved in the palace, they embraced and she recited appropriate lines of poetry. When she had finished her recitation, she said: 'Am I really seeing you here as my companion and friend?' The pangs of love and passion took such a hold on her that she almost went out of her mind with joy and recited:

> I would ransom with my life the one who comes by night,
> For whose return I had been waiting.
> I was startled by the soft sound of his weeping,
> And told him: 'Come, for you are welcome here.'
> I left a thousand kisses on his cheek,
> Embracing him a thousand times, hidden as he was.
> I said: 'I have obtained my hope;
> Praise be to God, Who made this hope come true.'
> We passed the night as we wished, in delight,
> Until the dark was cleared away by dawn.

When morning came, she hid Ardashir away in her apartments and it was not until nightfall that she brought him out, after which they sat and talked. He said: 'I plan to go back to my own country and then tell my father about you so that he may send his vizier to your father to ask him for your hand.' 'My darling,' she objected, 'I am afraid that, if you go home, you may be distracted and forget your love for me or that your father may not agree to what you suggest. Then there would be nothing for me to do but die. The best thing is for me to hold you here, so that we can see each other until I can think of some scheme to allow us to leave the palace together on the same night, after which we can go to your country, as I no longer have any hopes of my own people and despair of them.' 'To hear is to obey,' said Ardashir, and they went on drinking wine.

One night, they enjoyed their wine so much that they did not settle down to sleep until dawn. As it happened, a king had sent gifts to Hayat al-Nufus's father, among them being a necklace comprising twenty-nine unequalled gems that were worth more than the treasuries of kings. 'The only person worthy of this is my daughter, Hayat al-Nufus,' the king said, and he turned to a eunuch whose molar teeth the princess had had pulled out for some reason. Having summoned him, the king told him to take the necklace to his daughter. He was to tell her that it was a present beyond all price, sent to her father from a king, and that she was to put it round her neck.

The eunuch took it, saying to himself: 'May Almighty God ensure that this is the last thing she wears in this world, as she deprived me of the use of my teeth.' When he came to the door of her apartments, he found it locked with the old woman sleeping beside it. He roused her and she woke with a start and asked him what he wanted. When he told her that the king had sent him on an errand to his daughter, she said: 'The key is not here. Go away until it can be fetched.' But when he objected that he could not return to the king, she set out herself to get the key but then became afraid and decided to save her own life. When she was slow to return, the eunuch became afraid that the king would think that he was taking too long and so he rattled the door and shook it. The hasp of the lock broke and the door opened. He went in and kept on until he passed through the seventh door and came to the princess's room, which he found spread with splendid carpets and arranged with candles and wine flasks. This surprised him and he walked up to the dais, which was covered by a silken curtain, surmounted by a net of jewels. He drew the curtain and discovered the princess asleep with a young man in her arms who was even more beautiful than she. He glorified God, Who had created this young man from 'a despicable drop',* before adding: 'These are fine doings for a girl who hates men. How did she manage this? I think it must have been because of him that she had my teeth pulled out.'

He put the curtain back in its place and was making for the door when the princess woke up in a panic and, seeing the eunuch, whose name was Kafur, she called to him but got no reply. She came down from the dais and caught up with him before taking hold of the skirt of his gown, putting it on her head, kissing his feet and exclaiming: 'Conceal what God has concealed!' 'May God not shelter you,' he replied, 'nor anyone who gives you shelter. You had my teeth pulled out, saying that no one was to tell you anything about the qualities of men.' He freed himself from her grip and ran out, bolting the door behind him and posting a eunuch to guard it. He then went to the king, who asked him whether he had handed over the necklace to the princess. 'By God,' he replied, 'you deserve better than this.' 'What has happened? Tell me and be quick,' the king said. 'I can only do this in private,' objected the eunuch, but the king insisted that he tell him there and then. 'Guarantee my safety,' said the eunuch, and the king gave him a kerchief as a token of this. The eunuch then said: 'Your majesty, I went to see Hayat al-Nufus and found her asleep in a luxuriously furnished room with a young man in her arms. I locked the door on them before coming to you.'

When the king heard this, he jumped up, grasped a sword in his hand and shouted to the chief eunuch, telling him to take his followers and go to fetch Hayat al-Nufus and the man with whom she was sleeping on the dais, and to cover them both up.

Night 735

Morning now dawned and Shahrazad broke off from what she had been allowed to say. Then, when it was the seven hundred and thirty-fifth night, SHE CONTINUED:

I have heard, O fortunate king, that the king ordered the eunuch to take his followers, go to Hayat al-Nufus and bring her to him together

* Quran 32.7, 77.20.

with the man who was with her. The eunuchs went off, and on entering the room they discovered Hayat al-Nufus dissolved in tears, as was Ardashir. The chief eunuch told them both to lie down on the couch again, and, fearing for his life, she told Ardashir: 'This is no time to disobey.' The two of them lay down and were carried off to the king. When they were uncovered, Hayat al-Nufus sprang to her feet, and the king, looking at her, was about to strike off her head when Ardashir forestalled him by throwing himself on his chest and exclaiming: 'The fault is not hers but mine, so kill me first!' The king was on the point of striking him when Hayat al-Nufus threw herself at her father, calling out: 'No, kill me and not him, for he is the son of the great king, the ruler of the whole land throughout its length and breadth.' When the king heard this, he turned to his grand vizier, a man who united all the vices, and asked him for his advice. 'What I say,' the vizier replied, 'is that those who are involved in this affair must necessarily lie, and the only thing to do is to put them to all kinds of tortures before cutting off their heads.'

The king summoned his executioner, who came with his assistants. 'Take this scoundrel and cut off his head,' he said, 'and then do the same to this harlot before burning their bodies. Do not ask me about this again.' The executioner put his hand on the princess's back in order to take her, but the king shouted at him and threw something that he was holding in his hand, almost killing him. 'Dog,' he cried, 'when I am angry, should you be gentle? Take her by the hair and pull her so that she falls on her face.' The executioner did as he was told and dragged both her and Ardashir on their faces to the place of execution. He cut a piece from the bottom of his robe and tied it over Ardashir's eyes before drawing his sharp sword. He left Hayat al-Nufus until second, in the hope that someone might intercede for her, and busied himself with Ardashir, swinging his sword three times, while all the soldiers wept and prayed to God for intercession. He was raising his sword in his hand when suddenly a dust cloud could be seen filling the horizon.

The reason for this was that, when news of Ardashir had been slow in reaching his father, he had mustered a huge army and set out in person to look for his son. As for King 'Abd al-Qadir, at the sight of the dust he asked what was happening and what the dust meant. The grand vizier left to investigate and discovered an innumerable and inexhaustible force like a swarm of locusts filling the mountains, valleys and hills. He came back to tell the king of this and the king ordered him to go and learn more about them and to discover why they had come. 'Find out about their leader,' the king said. 'Give him my greetings and ask him why he has come. If he is here on some errand, we shall help him; if he wants to take revenge on some king, we shall ride with him; and if he wants a gift, we shall give him one. For this is an enormous force and there is reason to fear that it may attack our lands.'

The vizier went off and continued to walk between tents, troops and camp servants from first light until nearly sunset, when he came across guards with gilded swords, and star-spangled tents. He then found emirs, viziers, chamberlains and deputies and went on till he came to the commander-in-chief, whom he found to be a mighty king and whose officials, on seeing him, shouted: 'Kiss the ground, kiss the ground!' The vizier did so and then rose, but they shouted a second and then a third time until, when he lifted his head and tried to get up, such was the awe inspired in him by the king that he fell full length on the ground. When at last he stood before the king, he said: 'May God lengthen your days, add glory to your rule and exalt you, O fortunate king. King 'Abd al-Qadir sends you his greetings, kisses the ground before you and asks what important business has brought you here. If you are looking for vengeance on some king, he will ride out under your command, or if there is anything he can do to further any purpose you may have, he is at your service.' The king told him: 'Messenger, go back to your master and tell him that the great king has a son who has long been absent. News of him has been slow to come in and he seems to have vanished without trace. If he is in this city of yours, his father will take him and leave you, but if anything has happened to him or if harm has come to him at your hands, he will ravage your lands, plunder your wealth, kill your men and enslave your women. Go quickly to tell this to your master before harm befalls him.' 'To hear is to obey,' said the vizier, and he was about to turn away when the chamberlains shouted to him: 'Kiss the ground, kiss the ground!' He kissed it twenty times and by the time he rose he was almost at his last breath.

When he left the king's audience chamber, he went off in a state of concern about this foreign king and the size of his army until, by the time he got back to 'Abd al-Qadir, such was his fear that he had lost his colour and was trembling. He gave an account of what had happened to him . . .

Night 736

Morning now dawned and Shahrazad broke off from what she had been allowed to say. Then, when it was the seven hundred and thirty-sixth night, SHE CONTINUED:

I have heard, O fortunate king, that when the vizier returned from the great king to 'Abd al-Qadir, he gave an account of what had happened to him, and 'Abd al-Qadir, who was anxious and fearful both for his own sake and for his people, asked him who the son of this king might be. 'His son,' replied the vizier, 'is this man whose execution you ordered, and we owe praise to God, Who ensured that this was not done in a hurry, as otherwise his father would have ravaged our lands and plundered our wealth.' The king said: 'See what bad advice you gave when you told me to have him killed. So where is he, this son of the great king?' 'It was you who gave the order for his execution,' the vizier said, and the king in a state of confusion gave a cry wrenched from the depths of his heart and his head, calling out: 'Damn you, bring the executioner lest he kill him.' The man was fetched immediately, and when he had come he said: 'King of the age, I cut off his head as you ordered.' 'Dog,' the king told him, 'if this is true, I shall send you to join him.' 'You told me to kill him and not to ask you about it again,' objected the man, but the king said: 'I was angry, but tell me the truth before you get yourself killed.' At that point the man said that the prisoner was still alive, and the king, delighted and relieved, ordered him to be brought before him.

When Ardashir was fetched, the king rose to his feet and kissed him on the mouth. 'My son,' he said, 'I ask pardon from Almighty God for what I did to you and I ask you not to say anything that might lower my standing in the eyes of the great king, your father.' Ardashir asked where his father was, and the king told him: 'He has come because of you.' 'I swear by the respect in which you are held,' Ardashir said, 'that I shall not leave your presence until I have cleared my honour and the honour of your daughter from what you have imputed to us. She is a virgin. Summon the midwives to examine her in front of you. If you find that her maidenhead has been lost, then I give you leave to kill me, but if not, then proclaim my innocence and hers.' The midwives were called and on inspection they confirmed that the princess was a virgin. They told this to the king and asked for a reward, which he gave them, taking off and giving away his own robes, as well as presenting gifts to everyone in the harem. Scent bowls were produced and all the state officials were perfumed to their great delight. The king embraced Ardashir, treating him with the greatest honour and respect, and on his orders his particular favourites among the eunuchs escorted him to the baths. When Ardashir came out, the king presented him with a splendid robe of honour as well as a jewelled crown and a sash of silk ornamented with red gold and studded with pearls and gems. He mounted him on a splendid horse with a golden saddle set with jewels and ordered his state officials and dignitaries to form an escort to take him to his father, instructing him to say to the great king: 'King 'Abd al-Qadir is at your service and will obediently follow all your instructions.' Ardashir promised to do this.

Ardashir then took leave of 'Abd al-Qadir and set off to visit his father, who was overcome with delight when he saw him, jumping to his feet and advancing to embrace him, as joy spread throughout the whole army. He summoned every one of his viziers and chamberlains, as well as all his troops and their officers, who kissed the ground before Ardashir in joy at his arrival. This was a day of great rejoicing. Ardashir allowed his escort and others from 'Abd al-Qadir's city to inspect his father's army without hindrance, so that they might note its size and the greatness of his power. All those who had earlier visited the drapers' market and seen him sitting in his shop were amazed at how someone of such honourable and dignified rank could have been content with such a position, although it was his love for the princess that had forced him to it.

News spread through the city about the huge size of his army, and word of this reached Hayat al-Nufus. She looked from the castle roof towards the mountains and saw them filled with armed men, while she herself was still being held prisoner until it was known whether the king would be content to release her or whether she would be condemned to death and burned. The sight of the army and the knowledge that it belonged to Ardashir's father made her fear that Ardashir might be distracted by this and forget about her, leaving her to be put to death by her father. So she sent off a slave girl who was acting as her attendant in her apartments, telling her: 'Go to Ardashir, the king's son, and don't be afraid. When you reach him, kiss the ground before him and tell him who you are. Then say: "My mistress sends you her greetings. She is being held a prisoner in the palace on her father's orders, either to be released or to be put to death. She asks you not to forget her or to abandon her, for today you are in a position of power and no one can disobey your commands, whatever they may be. If you think of getting her father to free her and taking her to join you, this will be an act of kindness, as it is because of you that she has had to endure these sufferings. But if you do not want to do that, now that you have got what you want from her, speak to your father, the great king, so that he may intercede for her with her father and not leave before she has been freed. He should get her father to pledge that he will not harm her or have her killed. This is the end of her message, may God not grieve her by your absence. Peace be on you."'

Night 737

Morning now dawned and Shahrazad broke off from what she had been allowed to say. Then, when it was the seven hundred and thirty-seventh night, SHE CONTINUED:

I have heard, O fortunate king, that Hayat al-Nufus sent a slave girl to Ardashir. The girl went to Ardashir and gave him the princess's message. When he heard it, he wept bitterly and said: 'Know that Hayat al-Nufus is my mistress and I am her slave, the captive of her love. I have not forgotten what passed between us or the bitterness of parting. Kiss her feet and tell her that I shall talk about her to my father, who will not be able to refuse me, and he will send his vizier, who asked the king for her hand on the first occasion. If your father consults you about this, do not say no, for I shall not return to my own country without you.' The girl went back and passed this message to the princess, after she had kissed her hands, and, when the princess heard it, she wept for joy and gave praise to Almighty God.

So much for her, but as for Ardashir, that night when he was alone with his father and was asked about himself and what had happened to him, he told him the whole story from beginning to end. 'What do you want me to do, my son?' his father asked. 'If you want 'Abd al-Qadir destroyed, I shall ravage his lands, seize his wealth and have his women raped.' 'I do not want that, father,' replied Ardashir, 'for he has not done anything to me to deserve that. What I want is to marry the princess, so would you be good enough to prepare a valuable gift and send it to her father with your vizier, the master of sound counsel?' 'To hear is to obey,' replied his father, and from the treasures that had been stored away from early times he chose the most valuable pieces and showed them to Ardashir, who approved of his choice. He then summoned his vizier, telling him to take these to 'Abd al-Qadir and to request the princess's hand in marriage to his son, asking him to accept the gift and to send back a reply.

The vizier set off to go to 'Abd al-Qadir, who had been filled with gloom since he parted from Ardashir, being preoccupied with fears that his kingdom would be ravaged and his estates seized. He was then suddenly confronted by the vizier, who greeted him and kissed the ground before him. He rose to his feet and greeted his visitor courteously, but the vizier quickly prostrated himself at his feet and kissed them, exclaiming: 'Pardon, king of the age, one in your position does not stand up for the likes of me, the least of the slaves of slaves. Know that the prince has talked with his father and told him of some of the goodness and generosity with which you treated him, which has earned you the king's gratitude. He has sent me, your servant who stands before you, with a gift for you and he greets you and sends you an honourable salutation.'

The king had been in such a state of fear that he did not believe his ears until the gift was presented to him, but he then discovered that it was past all price and that no other ruler among the kings of the earth could have produced its like. This made him feel diminished by comparison, and he rose to his feet, praising and glorifying Almighty God, as well as expressing his gratitude to Prince Ardashir. The vizier then said: 'Noble king, listen to my words and know that the great king has come to you, choosing to establish a relationship with you. My mission is to seek and request the hand of your daughter, the sheltered lady, the hidden jewel, Hayat al-Nufus, in marriage to the king's son, Ardashir, and if you are pleased to accept, then the two of us can come to an agreement about her marriage portion.' When the king heard that, he said: 'To hear is to obey. For my own part I have no objection and would like nothing better. My daughter, however, is of legal age and is in control of her own affairs. I shall refer the matter to her and she can choose for herself.' Turning to the chief eunuch, he said: 'Go to my daughter and tell her about this.' 'To hear is to obey,' the eunuch said and he walked to the harem quarters, where he came to the princess, kissed her hands and, after telling her what the king had said, asked: 'What is your reply to this?' 'To hear is to obey,' she answered.

Night 738

Morning now dawned and Shahrazad broke off from what she had been allowed to say. Then, when it was the seven hundred and thirty-eighth night, SHE CONTINUED:

I have heard, O fortunate king, that when the chief eunuch of the harem told Hayat al-Nufus that her hand had been requested in marriage to the son of the great king, she replied: 'To hear is to obey.' When the chief eunuch heard this, he went back to tell the king of his daughter's answer. The king was delighted and called for a splendid robe of honour, which he presented to the vizier, together with ten thousand dinars, telling him to inform Ardashir's father that Hayat al-Nufus had accepted the proposal and to ask for his permission to visit him. 'To hear is to obey,' said the vizier, who then left 'Abd al-Qadir's court and returned to pass the messages to his master. The great king was pleased, and as for Ardashir he was almost out of his mind with joy, being filled with contentment and happiness. Permission was granted to 'Abd al-Qadir to come to visit the king, and on the following day he rode out and was met on his arrival by the king himself, who greeted him and gave him a place of honour. The two kings sat, while Ardashir remained standing in front of them, and an orator, one of 'Abd al-Qadir's close associates, rose to deliver an eloquent address in which he congratulated Ardashir on having achieved his wish to marry the sovereign of princesses. When he had sat down, the great king ordered a chest filled with pearls and gems to be produced, together with fifty thousand dinars, and he told 'Abd al-Qadir that he was acting as his son's representative in the matter of the marriage settlements. 'Abd al-Qadir acknowledged the receipt of the bride price, which included the fifty thousand dinars for the wedding celebrations. The judges and notaries were then summoned and a marriage contract drawn up between Hayat al-Nufus, daughter of King 'Abd al-Qadir, and Ardashir, son of the great king. This was a momentous day, delighting all lovers and angering those who were filled with jealous hate. There were banquets and receptions, and after that Ardashir lay with his bride and discovered her to be an unpierced pearl and a filly that no one else had ridden, a precious jewel well-guarded and a gem that had been hidden away, a fact that he passed on to her father.

The great king then asked Ardashir whether there was anything else that he wanted before they left. 'Yes, O king,' he replied, 'I want to be revenged on the vizier who maltreated us and on the eunuch who lied about us.' His father instantly sent to 'Abd al-Qadir to demand the two, who were then delivered to him, and when they appeared before him he ordered them to be hung over the city gate. After a further brief delay, 'Abd al-Qadir was asked to allow his daughter to make her preparations for travelling. He provided for her needs and she was mounted on a carriage of red gold studded with pearls and gems and drawn by noble horses. She took with her all her slave girls and eunuchs, and her nurse, who had returned after her flight to take up her usual position again. The great king and Ardashir mounted, and 'Abd al-Qadir and his whole court rode out to say goodbye to his son-in-law and his daughter on what was counted as one of the best of days. When they reached open country, the great king insisted that 'Abd al-Qadir turn back, which he did after first clasping the king to his breast, kissing him between the eyes, thanking him for all his goodness and entrusting his daughter to his care. When he had said goodbye to the king and to Ardashir, he went back and embraced his daughter, who kissed his hands, and the two of them wept as they parted.

While 'Abd al-Qadir went back to his capital, Ardashir, together with his wife and his father, went on to his own country, where they held a second wedding feast. There they remained enjoying the pleasantest of lives in joy and luxury until they were visited by the destroyer of delights, the parter of companions, the ravager of palaces and the filler of graveyards. This is the end of the story.

A story is also told, O fortunate king, that long ago in Persia there was a king named Shahriman who lived in Khurasan. In spite of the fact that he had a hundred concubines, none of them had ever in his life provided him with either a son or a daughter. One day, as he thought this, he grew melancholy, reflecting that the greater part of his life was over and there was no son to inherit the kingdom after his death in the way that he himself had inherited it from his father and forefathers. He was filled with sorrow and deep distress, but then as he was sitting in his palace one of his mamluks came to him and said: 'Master, at the gate is a merchant with as beautiful a slave girl as I have ever seen.' 'Bring them to me,' the king ordered, and when they entered he looked at the girl and discovered that she was like a Rudaini spear and was wrapped in a silken shawl embroidered with gold. When the merchant unveiled her face, the room was illumined by her beauty. She had seven locks of hair hanging down to her anklets like ponies' tails; her eyes were darkened with kohl; she had heavy buttocks and a slender waist and could cure the illnesses of the sick and quench the fires of thirst, as has been described by the poet in these lines:

I fell in love with her in the perfection of her beauty,
Crowned, as it was, by calmness and by dignity.
She was neither too tall nor yet too short;
Her waist-wrapper was too narrow for her buttocks;
Her figure struck a balance between extremes of height,
And could not be criticized for either fault.
Her hair hung down before her anklets,
While her face was always bright as day.

The king, admiring the sight of her graceful beauty and her symmetrical form, asked the merchant her price. 'Master,' he replied, 'I bought her from her previous owner for two thousand dinars; I have spent three years travelling with her and this has cost me another three thousand up to the time that I arrived here, but I give her to you as a present.' The king ordered that the merchant be given a splendid robe of honour together with the sum of ten thousand dinars. He took these and left after kissing the king's hands and thanking him for his favour and generosity.

The king then handed over the girl to the maids, telling them to see to her and to adorn her, as well as to furnish apartments for her and bring her to them, while his chamberlains were instructed to fetch her everything that she might need. The king's capital was known as the White City, and thanks to the fact that his realm lay by the sea, the apartments to which the girl was brought had windows overlooking them.

Night 739

Morning now dawned and Shahrazad broke off from what she had been allowed to say. Then, when it was the seven hundred and thirty-ninth night, SHE CONTINUED:

I have heard, O fortunate king, that when the king took the slave girl, he handed her over to the maids, telling them to see to her. They took her to her apartments, and the king instructed his chamberlains to fetch everything she might need and then lock all the doors. The apartments to which they escorted her had windows overlooking the sea.

When the king went in to visit her, she said nothing and took no notice of him, leaving him to imagine that the people with whom she had been had not taught her manners. Then he looked at her outstanding beauty and grace, her symmetrical figure and her face that was like the full moon or the sun shining in a clear sky, and in his admiration for these qualities he praised God, the Omnipotent Creator. He went up to her, sat beside her and then clasped her to his breast, seating her on his lap and sucking the saliva of her mouth, which he found sweeter than honey. On his orders, all kinds of the most delicious foods were brought in, which he ate himself and morsels of which he gave to the girl until she had had enough. During all this time she uttered no single word and, although he started to talk to her and ask her her name, she remained silent, saying nothing and making no reply, with her head bent towards the ground. It was only her great beauty and attractiveness that saved her from the king's anger, and he said to himself: 'Praise be to God, Who created her – how lovely she is! She does not speak, but perfection belongs to the Almighty alone.' He asked the slave girls whether she had said anything and they told him that, from the time of her arrival until then, they had not heard a word from her. Some of these girls, together with a number of the concubines, were then told to sing to her and relax with her in the hope of breaking her silence. They played different musical instruments in front of her and indulged in various games and so on, until everyone there was moved to delight by their songs, but the girl herself was a silent spectator and neither laughed nor spoke, to the distress of the king. He then dismissed the other girls and remained alone with her. He took off his own clothes and undressed her, discovering that she had a body like a silver ingot. He fell deeply in love with her and deflowered her, being delighted to discover that she was a virgin. 'By God, it is remarkable,' he said to himself, 'that the merchant should have left untouched a beautiful girl with such a lovely figure.'

He then became entirely devoted to her, disregarding everyone else and abandoning all his concubines, while the year that he then spent with her passed like a single day. She had still said nothing, and one day, when his passionate love had grown even stronger, he told her: 'Heart's desire, I love you deeply and for your sake I have forsaken all my slave girls and concubines, as well as every other woman, since you are the only thing that I now want in this world. I have waited patiently for a whole year, and I now ask Almighty God of His grace that He may soften your heart towards me so that you may speak to me or, if you are dumb, then make a sign to tell me, and I will give up the thought that you might ever talk. My hope is that Almighty God will grant me a son by you to inherit the kingdom after my death, for I am entirely alone and in my old age I have no heir. I implore you in God's Name to answer me if you love me.'

The girl looked down thoughtfully towards the ground and then she raised her head and smiled at the king, to whom it seemed as though lightning had filled the room. 'Great king and strong lion,' she said, 'God has answered your prayer, for I am carrying your child and am near the time of delivery, although I do not know whether it will be a boy or a girl. Had I not conceived, I would not have spoken a word to you.' When the king heard this, he beamed with joy and delight and kissed her head and her hands in his gladness, exclaiming: 'Praise be to God, Who has granted me my wishes, as, firstly, you have spoken to me and, secondly, you have told me that you are carrying my child!'

The king then left her and went out to sit on his royal throne in a state of great happiness, and he ordered his vizier to distribute a hundred thousand dinars as alms to the poor, the wretched and the widows, among others, in gratitude to Almighty God, an order which the vizier carried out. The king then returned to the girl and sat with her, holding her to his breast. He said: 'My mistress, whose slave I am, why did you stay silent? You have been with me day and night, waking and sleeping for a whole year now, and during all this time you have not spoken a word until now. What was the reason for this?'

'You must understand, O king of the age,' replied the girl, 'that I was a poor broken-spirited stranger, having been parted from my mother, my brother and my family.' The king listened to this and grasped her point but objected: 'When you say that you are poor, this is not true, as my kingdom, my goods and all that I have are at your service, while I myself am your slave, but when you say that you have been parted from mother, brother and family, tell me where they are and I shall send to fetch them to you.' She replied: 'Know, O fortunate king, that my name is Julnar of the sea. My father was one of the rulers of the sea and when he died he left his kingdom to us, his family. We held it until another king moved against us and took it from us. I have a brother named Salih, while my mother is one of the sea people. I quarrelled with my brother and swore that I would throw myself away on a landsman, so I came out of the sea and sat by moonlight on the shore of an island. A man passed by and took me off to his house, where he tried to seduce me, but I struck him on the head and almost killed him. He then took me out and sold me to the merchant from whom you bought me, a good and pious man, religious, honest and honourable. Had you not truly loved me and preferred me to all your concubines, I would not have stayed with you for a single hour and would have thrown myself into the sea from this window and gone off to my mother and my people. But I was ashamed to go to them carrying your child, for whatever oath I took they would have suspected my virtue if I told them that a king had bought me for money, devoted himself to me and preferred me to his wives and everything else he owned. This is my story.'

Night 740

Morning now dawned and Shahrazad broke off from what she had been allowed to say. Then, when it was the seven hundred and fortieth night, SHE CONTINUED:

I have heard, O fortunate king, that when King Shahriman questioned Julnar of the sea, she told him her story from beginning to end.

When the king heard all this, he thanked her, kissed her between the eyes and said: 'By God, my lady and the light of my eyes, I cannot bear to be parted from you for a single hour, and if you leave me I shall die on the spot, so how are things going to be?' 'Master,' she replied, 'it is nearly time for my child to be born and my family will have to be there to look after me, as women from the land don't know how sea women give birth and vice versa. Then, after they have come, we shall be reconciled with each other again.'

The king asked her: 'How can they walk in the sea and not get wet?' She told him: 'We walk in the sea in the same way you walk on the land, and this is thanks to the blessing conferred by the names written on the seal ring of Solomon, son of David, on both of whom be peace. When my family and my brothers come, I shall tell them that you paid money to buy me, after which you treated me with kindness and courtesy, something that you must confirm for them. They will see your grandeur with their own eyes, and realize that you are a king and the son of a king.' 'Do what you want, as it occurs to you,' said the king, 'for I shall obey you in everything you do.' 'Know, king of the age,' she explained, 'that when we walk in the sea our eyes are open and we can look at what is there, as well as at the sun, the moon, the stars and the sky, which we can see just as though we were on dry land, and that does us no harm. You must know also that there are many different races in the sea, as well as different varieties of all the types of creatures that are to be found on land. Furthermore, what the land contains is tiny in comparison with what is in the sea.' The king was astonished by what she had to say.

She then produced from the palm of her hand two bits of Qumari aloes wood and, taking one of them, she threw it into a brazier that she had lit. She whistled piercingly and started to speak some unintelligible words. As the king watched, a large cloud of smoke rose up and Julnar said: 'Master, hide yourself away in a small room so that I may show you my brother, my mother and the rest of my family without their seeing you. I want to summon them, and here and now you shall see a marvel and wonder at the different shapes and strange forms that God Almighty has created.' The king got up immediately and went into a small room from where he could see what she was doing. She started to burn incense and to recite spells, until the sea began to foam, and from the disturbed water out came a splendidly handsome young man like the full moon, with a radiant forehead, ruddy cheeks and teeth like pearls. He was very like his sister and the following lines would spring automatically to mind:

Once every month the moon becomes full,
But the beauty of your face is perfect every day.
The heart of a single zodiac sign holds the full moon,
But the hearts of all mankind serve as your resting place.

He was followed out of the sea by a grey-haired old lady, who was accompanied by five girls as beautiful as moons, all bearing a resemblance to Julnar. As the king could see, every one of them was walking on the surface of the sea as they approached her, and when she saw them close to her window, she got up and went to meet them with joy and delight. When they saw her and recognized her, they came in and embraced her, raining down tears. 'Julnar,' they said to her, 'how could you have abandoned us for four years? We didn't know where you were and the pain we suffered because of your loss made our lives miserable; there was no single day on which we could enjoy food or drink, and such was our longing for you that we wept night and day.'

Julnar kissed her brother's hand and those of her mother and her cousins. They sat with her for a time, asking her about herself, what had happened to her and how she was. In reply she said: 'Know that when I parted from you and came out of the sea, I sat on the shore of an island and a man took me and sold me to a merchant, who brought me to this city and sold me to its king for ten thousand dinars. The king received me with honour, abandoning all his concubines, his women and his favourites for my sake and occupying himself with me to the exclusion of all his other affairs and the business of his city.'

When Julnar's brother heard her story, he thanked God for having reunited them and added: 'I want you to get up and come with us to our own country and our people.' On hearing that, the king was distraught lest she agree to this suggestion and, passionately in love with her as he was, he would not be able to stop her. He remained dismayed and fearful at the thought of losing her, but, for her part, when she heard what her brother had to say, she told him: 'By God, brother, the man who bought me is the ruler of this city, a great king, an intelligent and noble man of vast liberality. He has shown me honour, and he is a chivalrous and a wealthy man, who has neither sons nor daughters; he has been good to me; he has treated me generously, and from the day that I came here until now I have heard no harsh word from him to cause me distress. On the contrary, he has always been kind to me, consulting me in all that he does, and with him I have enjoyed the pleasantest of lives and the greatest of benefits. Further, if I leave him, he will die as he cannot bear to be parted from me for a single hour, and I too would die if I left him, so deeply do I love him because of the goodness he has shown me during the time I have been with him. Were my father still alive, I would not enjoy the same standing with him as I do with this great and powerful king. You can see that I am carrying his child, and I praise God, Who made me the daughter of the king of the sea and has given me as a husband the greatest of the rulers of the land. He did not abandon me but gave me a good exchange . . .'

Night 741

Morning now dawned and Shahrazad broke off from what she had been allowed to say. Then, when it was the seven hundred and forty-first night, SHE CONTINUED:

I have heard, O fortunate king, that Julnar told her story to her brother and said: 'The Almighty did not abandon me but gave me a good exchange, and as the king has no son and no daughter, I ask Almighty God to let me give birth to a boy, who may inherit from his powerful father all that He gave him in the way of buildings, palaces and possessions.' Julnar's brother and her cousins were pleased by what she had to say, and they said: 'Julnar, you know how deeply we love you, for, as you must already know, you are the dearest of all people to us. You must believe that all we want for you is that you should enjoy a life of ease, without distress or drudgery. If you are not free from worry, then come back with us to your own land and to your family, but if you are and you are both well regarded and happy here, then this is all we hope for, and we can want nothing more for you.' 'By God,' she replied, 'I am enjoying all possible comfort, happiness and dignity, and everything I wish for is mine.' When the king heard what she had to say, he was pleased and relieved; he was grateful to her and loved her even more from the bottom of his heart, as he realized that she returned his love and wanted to stay with him so that she might see his child.

Julnar then gave orders to her slave girl for tables laden with various types of food to be produced, after she herself had supervised the preparations in the kitchen, and then these, together with sweetmeats and fruit, were brought in by the slave girls. When she and her family had eaten, they said to her: 'Julnar, your husband is a stranger to us and we have entered his palace without his leave. You have told us that you are grateful for his kindness, but he does not know about us and although we have eaten his food that you brought us, we have not seen him or met him and he has not seen us. He has not come to join us and has not eaten with us so as to establish the tie of bread and salt.' At that they all stopped eating, and they were so angry with her that what looked like flaming torches came out of their mouths.

When he saw this, the king became distraught with fear, but Julnar, having first got up to mollify them, went to the room where he was and said: 'Master, did you see and hear how I thanked you and praised you to my family, and did you hear them say how they wanted to take me home with them?' 'I both heard and saw,' the king told her, adding: 'May God reward you on my behalf, for I never knew how much you loved me before this blessed hour, a love which I cannot doubt.' 'Master,' she then said, 'one good deed deserves another. You have treated me well and showered great favours on me; you love me deeply, as I can see; you have done everything that you can for me and you have preferred me to all others whom you love and desire. So how could I be content to part from you and leave you? How could I do that after all the kindness that you have shown me? But now I would like you to be good enough to come and greet my family, so that after you have seen each other there may be a bond of sincere affection between you. For you must know, king of the age, that my brother, my mother and my cousins have a great love for you, because I told them how grateful I am to you. They said that they would not leave me and return home until they have met you and greeted you, as they want to see you and make friends with you.' 'To hear is to obey,' said the king, 'for that is what I want myself.'

He then got up and went to Julnar's family, whom he greeted in the friendliest manner, and they, for their part, sprang to their feet and met him with the greatest courtesy. He sat with them in the palace and ate at the same table, after which they stayed with him for thirty days. At the end of this period they wanted to set off for home, and after they had asked for his permission and for that of Queen Julnar, they left, having received the most honourable treatment. It was then that Julnar reached the term of her pregnancy, and when she had gone into labour she gave birth to a boy like a full moon. The king was overjoyed, as he had no other children, male or female, and celebrations were held, the city being adorned with decorations for seven days among general pleasure and rejoicing.

On the seventh day, Julnar's mother came back with her brother and her cousins, having heard of her delivery.

Night 742

Morning now dawned and Shahrazad broke off from what she had been allowed to say. Then, when it was the seven hundred and forty-second night, SHE CONTINUED:

I have heard, O fortunate king, that when Julnar had given birth, her family returned. They were met by the king, who was pleased by their arrival and who told them: 'I said that I would not name my son until you arrived to choose his name, wise as you are.' They all agreed to call him Badr Basim and he was shown to his uncle, Salih, who took him in his arms. Salih then left the others, and after walking to and fro in the palace he left it and went down to the sea. He walked on into it until he was out of the king's sight, and when the king saw that his son had been taken away from him and plunged into the depths of the sea, he began to weep and sob in despair. Seeing this, Julnar said: 'Do not be afraid or grieve for your son, king of the age, for I love him more than you do. He is with my brother, and there is no need to concern yourself about the sea or to fear lest he be drowned, for if my brother knew that any harm would come to the child he would not have acted as he has. God willing, he is just about to bring him back safe and sound.'

Soon there was a disturbance in the sea and from the frothing water emerged Salih with the uninjured child. Carrying him in his arms he flew out to rejoin the others, and the child meanwhile made no sound, while his face was as radiant as the full moon. Looking at the king, Salih said: 'Were you, perhaps, afraid that harm might come to your son when I took him down with me into the sea?' 'I was indeed, sir,' the king replied, 'and I never thought that he would come out alive.' 'King of the land,' said Salih, 'we rubbed his eyes with a type of kohl we know of, and recited over him the names engraved on the signet ring of Solomon, son of David, on both of whom be peace, as this is what we do with our own newborn babies. You need never fear that he will drown or suffocate or that any sea will be dangerous for him, for, as you walk on the land, so we walk in the sea.' From his pocket he then brought out a sealed case with writing on it. He broke the seal and spread out the contents – strings of gems of all types, sapphires and so on, together with three hundred emerald rods* and three hundred pierced jewels, each as large as an ostrich egg, gleaming more brightly than the sun and the moon. 'King of the age,' he said, 'these gems are a gift from me to you. We had not brought you anything because we had had no news of Julnar and did not know where she was, but now that we see your relationship to her and we have become one family, we make you this present, and after every few days, God willing, we shall bring you the same again, because there are more of these jewels with us than there are pebbles on the land. We know which of them are good and which are bad, and we know how to get to them and where they are to be found, and so it is easy for us to obtain them.'

The king looked at the jewels in astonishment and amazement, exclaiming: 'By God, a single one of these is worth as much as my kingdom!' He thanked Salih, and then, looking at Julnar, he said: 'Your

---

* These 'emerald rods' were explained by Reinhart Dozy as a cluster of emeralds set together in the form of a branch.

brother has made me feel ashamed by giving me a more splendid present than anyone on earth could produce.' Julnar, for her part, thanked Salih, who said: 'King of the age, you have a prior claim on us, and it is right that we should thank you because of your kindness to my sister, and because we have entered your palace and eaten your food. The poet has said:

> Had I shed tears of love for Su'da before she wept,
> I would have cured my soul before having to repent.
> But she wept first, prompting my tears,
> And I said: "The merit belongs to the first to weep."'

He continued: 'King of the age, even if we served you as best we could for a thousand years, we still could not repay you, and that would be little enough to do for you.'

The king thanked him profusely, after which Salih and his mother and his cousins stayed there for forty days. At the end of this period, he came and kissed the ground before the king, and when the latter asked what he wanted, he said: 'King of the age, you have shown us kindness, and I want you to be good enough to give us leave to go, as we long for our own folk, our lands, our relatives and our home. I shall never cut myself off from your service or from my sister and my nephew, and I swear by God that I do not want to leave you, but what can we do? We were brought up in the sea and have no liking for the land.' When the king heard what he had to say, he got to his feet to take leave of him, as well as of his mother and his cousins. They all shed tears of parting, but said: 'We shall be with you again soon and will never cut ourselves off from you.' They flew off, making for the sea, into which they plunged and were lost from sight.

Night 743

Morning now dawned and Shahrazad broke off from what she had been allowed to say. Then, when it was the seven hundred and forty-third night, SHE CONTINUED:

I have heard, O fortunate king, that when Julnar's relatives had said goodbye both to the king and to Julnar, they shed tears of sorrow at parting, but then flew off, plunged into the sea and were lost from sight.

The king treated Julnar well, showing her even greater respect; the child was raised successfully and his uncle, his grandmother, his aunt and his mother's cousins would come to the palace every few days and stay with him for a month or two before returning home. As he grew older, the child grew ever more handsome and graceful, until by the time he was fifteen he was unique in his perfection and the symmetry of his figure. He had studied writing, reading, history, grammar and philology, as well as archery; he could manage a lance and a horse, and had learned everything else that princes need to know. There was no one among the children of the citizens there, either male or female, who did not talk about his beauty, for this was so marked that the poet's lines could be applied to it:

> Down has written with ambergris on pearl,
> Inscribing an apple with twin lines of jet.
> When he looks, there is death in his languorous eyes,
> And drunkenness is in his cheeks, not in the wine.

Another wrote:

> Down spread upon the surface of his cheek,
> Like embroidery, and there it thickened.
> It was as though a lamp hung there suspended
> On chains of ambergris beneath the dark.

One day the king, who was exceedingly fond of his son, Badr Basim, summoned the emirs, viziers, state officials and the leading men of his kingdom to make them take a solemn oath that, in succession to him, they would take Badr as their king. They swore to this gladly, as the king was a man who treated his people well, spoke pleasantly and united in himself all virtues, while consulting the general good in what he said. On the following day he rode out, accompanied by the state officials and the other emirs, together with all the troops. They paraded through the city before turning back and then, when they were near the palace, the king dismounted as an act of deference to his son, who went on ahead, with each of the emirs and the officials taking turns to carry the royal saddle cloth before him. When they reached the entrance hall of the palace, Badr dismounted from his horse and was embraced by his father as well as by the emirs. They sat him on the royal throne, with his father and the emirs standing before him.

Badr now gave his decisions to his people, deposing unjust officials and replacing them with honest men. He continued exercising his authority until it was almost noon, when he got up from the throne and went to visit his mother, wearing the crown and looking as radiant as the moon. When his mother saw him standing there with the king in front of him, she got up to kiss him and congratulate him on having taken power, wishing both him and his father a long life and victory over their enemies. He sat and relaxed with his mother, until in the afternoon he rode out to the exercise ground, preceded by the emirs, and there he joined his father and his officials in practising with his weapons until evening. Then everyone there went on ahead as he returned to the palace.

Every day he would follow his visit to the exercise ground by sitting in judgement among the people, giving fair treatment to the powerful and the poor alike. He continued this practice for a whole year, and after that he started to ride out hunting, touring the lands and districts that were under his control, proclaiming security and peace and acting as a king. He was unequalled in his time for his renown and courage, as well as the justice with which he treated his subjects.

It happened that one day the king, Badr's father, fell ill, and the throbbing of his heart alerted him to the fact that he was about to move to his eternal home. When the illness had become so grave that he was on the point of death, he summoned his son and instructed him to look after his subjects, his mother, his officials and all his retainers. These, in turn, he made bind themselves again with an oath to obey Badr, and a few days later he died and was taken to a merciful God. He was mourned by his son and by Julnar, his wife, as well as by the emirs, viziers and officials of state. They built a tomb for him and buried him there, after which they sat in mourning for a whole month. Salih arrived with Julnar's mother and her cousins to share in the mourning, saying: 'Julnar, although the king is dead, he has left an excellent son and no one dies who leaves behind one like this. This young man has no equal; he is a savage lion . . .'

Night 744

Morning now dawned and Shahrazad broke off from what she had been allowed to say. Then, when it was the seven hundred and forty-fourth night, SHE CONTINUED:

I have heard, O fortunate king, that Julnar's brother, Salih, together with her mother and her cousins, told her that, although the king was dead, he had left behind a son without equal, a savage lion and a shining moon. The officials of state and the leading men came to Badr, the new king, and said: 'There is nothing wrong with grieving for your father, but mourning is something that is only suitable for women, so do not let yourself or us become preoccupied with sorrow for your father. He has died, leaving you to succeed him, and no one who has left behind someone like you is really dead.' They spoke gently to him, consoling him, and they then escorted him to the baths. When he came out, he put on splendid clothes embroidered with gold and studded with gems and sapphires, and, after placing the kingly crown on his head, he took his seat on his royal throne and judged the affairs of his people, giving weak justice against the strong and helping the poor to their rights over the emirs. In this way he gained the affection of his people.

This state of affairs continued for a whole year, with his family from the sea visiting him at short intervals. He led a pleasant and a comfortable life and this went on for a long time until it happened that one night his uncle came to visit Julnar. He greeted her and she got up to embrace him, after which she made him sit by her side and asked him how he was and about her mother and her cousins. He assured her that all was very well with them and that the only thing they lacked was the sight of her. She had food brought for him and after he had eaten they began to talk and the conversation turned to Badr, his beauty and grace, his symmetrical figure, his mastery of horsemanship, his intelligence and good manners. Badr himself happened to be resting there, propped on his elbow, and when he heard his mother and his uncle talking about him, he listened to what they were saying while pretending to be asleep. Salih said to Julnar: 'Your son is seventeen years old but, as he has no wife, I am afraid that something may happen to him and he will leave no heir. So I want to marry him to one of the sea princesses, his equal in beauty and grace.' 'Tell me their names,' said Julnar, 'for I know them.' Salih started to list the princesses one by one, but Julnar kept saying: 'She will not do for my son, and I shall only marry him to someone as beautiful and graceful as he is, who is his match in intelligence, piety, good manners, generous qualities, sovereignty and noble birth.' Salih said: 'I don't know of any other princess. I have given you more than a hundred names, but you don't approve of any of them.'

He then told her to check to see whether Badr was really asleep or not. When she had touched him and found that it looked as though he was, she confirmed this to Salih and added: 'So what do you have to say and why do you want to find whether he is asleep or not?' 'I have thought of a princess who would suit him,' he said, 'but I would be afraid of mentioning her if he were awake lest he fall in love with her, and then as he might not be able to reach her, he would suffer as would we and his ministers. This would be a cause of concern for us, as the poet has said:

> Love starts as a drop of water,
> But when it takes hold, it is an ocean.'

'Tell me about this girl,' said Julnar, when she heard this. 'What is her name? For I know the sea princesses and the other girls as well, and if I see that she is a suitable bride for Badr, I shall ask her father for her hand, even if I have to spend all that I have in order to get her. Don't be afraid to tell me, as he is asleep.' Salih objected: 'I'm afraid that he may be awake, and as the poet has said:

> I loved her when she was described to me;
> At times the ear falls in love before the eye.'

'You can speak without fear, but keep it brief, brother,' Julnar told him. He then said: 'Sister, the only suitable girl for your son is Princess Jauhara, the daughter of King Samandal, who is his equal in beauty, grace, splendour and perfection. Nowhere on land or sea is there anyone more charming or sweeter tempered than she; she is a lovely girl with a good figure, red cheeks, a radiant brow, teeth like pearls, dark eyes, heavy buttocks, a slender waist and a beautiful face. If she turns, she puts to shame the wild cows and the gazelles; she sways as she walks and the branch of the *ban* tree is filled with jealousy, while, when she is unveiled, she shames the sun and the moon, captivating all who see her. Her lips are sweet to kiss and she is soft to embrace.'

When Julnar heard what her brother had to say, she exclaimed: 'By God, that is true! I have seen her many times, for when we were small she was my companion, but now, as we live so far apart, our acquaintanceship has lapsed and I have not set eyes on her for eighteen years. She and she alone is worthy of my son.' Badr heard and understood all that they had to say from start to finish about Princess Jauhara. What he heard made him fall in love with her, and although he kept up the pretence of being asleep, the fire of love blazed up in his heart and he drowned in a restless sea without a shore.

Night 745

Morning now dawned and Shahrazad broke off from what she had been allowed to say. Then, when it was the seven hundred and forty-fifth night, SHE CONTINUED:

I have heard, O fortunate king, that when Badr heard what his uncle, Salih, and his mother, Julnar, had to say about the daughter of King Samandal, the fire of love blazed up in his heart and he drowned in a restless sea without a shore.

Salih now looked at his sister and said: 'By God, there is no stupider man than her father among the kings of the sea and no one more violent by nature. So don't tell your son anything about the girl until we can ask her father for her hand. If he agrees, we can praise Almighty God, but if he turns us away and refuses to marry her to your son, then we can let the matter rest and ask for the hand of someone else.' 'That is a good idea,' replied Julnar and the two of them said no more. Salih spent the night with his sister, but as for Badr he was on fire for love of Princess Jauhara, and although he concealed the matter and said nothing about her to his mother or his uncle, he was roasting on the coals of passion.

The next morning he went with his uncle to the baths, and after they had washed and come out, they drank. Food was brought and Badr ate together with his mother and his uncle, until, when they had had enough, they washed their hands. Salih then got to his feet and asked the other two for permission to leave, explaining that he wanted to go back to his mother since he had stayed with them for a number of days and she would be concerned about him and expecting him back. 'Stay with us for today,' Badr told him, and when he had agreed to do this, Badr told Salih to come into the garden with him. The two of them went out and strolled around inspecting the garden, until Badr sat down under a shady tree with the intention of taking a restful nap. He then remembered what Salih had said about the beauty of the princess, and, bursting into a flood of tears, he recited these lines:

The fire is kindled and blazes in my heart
As my entrails burn.
Were I asked which I would prefer, to see them,
Or to enjoy a drink of cold water, I would choose them.

With further complaints, moans and tears he recited:

Who will protect me from a human gazelle,
Whose face is like the sun, or even lovelier?
Love had not troubled my heart before,
Until it blazed with passion for the daughter of Samandal.

When Salih heard this, he struck his hands together and recited the formula: 'There is no god but God; Muhammad is God's Prophet and there is no might and no power except with God, the Exalted, the Omnipotent.' He then asked: 'My son, did you hear what your mother and I said when we were talking about Princess Jauhara and describing her?' 'Yes, uncle,' Badr replied, 'and I fell in love when I heard what you said. So deeply is my heart attached to her that I cannot endure to be without her.' 'Your majesty,' said Salih, 'let us go back to your mother and tell her what has happened, and then I shall ask her permission to take you with me so that I may woo the princess for you. When we have said goodbye to your mother, you and I will come back, for I am afraid that, if I were to take you and go off without her leave, she would quite rightly be angry with me, as I would be responsible for parting the two of you, just as I was responsible for her leaving us. The city would have no king and no one to lead its citizens and look after their affairs. The kingdom would be ruined and you would lose your throne.'

On hearing this, Badr said: 'Uncle, you must know that were I to go back to my mother and consult her on this she would never allow me to go, and so I shall not do this.' He then burst into tears in front of his uncle and said: 'I shall go off with you without telling my mother, and after that I shall come back.' When Salih heard this, he was at a loss to know what to do, and he exclaimed: 'I ask for help from Almighty God in every circumstance!' He saw the state his nephew was in, and, realizing that he did not want to return to his mother but was going to leave with him, he took from his finger a ring inscribed with some of the Names of Almighty God and gave it to him. 'Put this on your finger,' he said, 'and it will save you from drowning and from other disasters, as well as from any harmful sea beasts and fishes.'

Badr took the ring from his uncle and, after he had put it on his finger, the two of them dived into the sea.

Night 746

I have heard, O fortunate king, that after Badr and his uncle, Salih, had dived into the sea, they travelled on to Salih's palace, where Badr's grandmother was seated with her relatives. The two entered and kissed their hands, and when Badr's grandmother saw him, she got up to greet him, embracing him and kissing him between the eyes. 'This is a happy arrival, my son,' she said, before going on to ask: 'How was your mother, Julnar, when you left her?' 'Well and in good health,' he replied, 'and she sent her greetings to you and to her cousins.' Salih then told his mother of what had happened between him and Julnar and that, because of what he had heard, Badr had fallen in love with Princess Jauhara, the daughter of King Samandal.

When he had explained the whole story from beginning to end, he went on to say that the reason why Badr had come was to ask Samandal for Jauhara's hand and then to marry her. On hearing this, his mother became furiously angry with him and, disturbed and distressed, she exclaimed: 'My son, you were wrong to mention Jauhara in front of your nephew, for you know how stupid Samandal is and that he is a savage tyrant with little intelligence, who is reluctant to give his daughter to any of her suitors. The other sea kings have asked him for her hand, but he has refused and has not been willing to give her to any single one of them. In rejecting them, he told them that they were no match for her in beauty, grace or anything else. I am afraid that if we approach him he will turn us away as he turned away the others and, honourable as we are, we will come back disappointed.' On hearing this, Salih asked: 'What can we do then, mother, since it was when I mentioned her to my sister, Julnar, that Badr fell in love with her? We have to ask for her hand, even if it costs me all that I own, for he claims that, unless he marries her, he will die of love for her.'

He went on to tell his mother that Badr was more beautiful and graceful than Jauhara, and that, since Badr had succeeded his father as king of all the Persians, he was the only suitable husband for her. 'I propose to take sapphires and other gems and bring a suitable present to Samandal before asking him for Jauhara's hand. If he objects, telling us that he is a king, then we can say that Badr is a king and the son of a king; if he points to her beauty, Badr outdoes her in this, and as for the size of the kingdom, Badr rules over wider lands than she and her father, and has more troops and guards and a larger state. I must do my best to settle this affair for my nephew, even if it costs me my life, for I was the cause of the whole affair and, since it was I who threw him into the sea of love, it is up to me to try to get him married to the princess, with the help of Almighty God.' 'Do what you want,' said his mother, 'but take care not to speak rudely to Samandal, as you know how stupid and violent he is, and I am afraid he might strike you, as he pays no regard to anyone's rank.'

'To hear is to obey,' said Salih, and he then got up and took two sacks filled with sapphires and other gems, together with emerald rods and precious stones of all kinds, which he gave to his servants to carry. He and Badr then set out with these treasures, making for the palace of King Samandal. He asked leave to approach the king, and when this had been granted, he went in and kissed the ground before him, greeting him with the greatest courtesy. For his part, on seeing him the king rose to greet him and treated him with all possible respect. When he had sat down, as he had been told to do, the king said: 'This is a happy arrival, Salih, for we have been lonely without you. Tell me what has brought you here, so that I may do whatever it is that you want.'

Salih got up and then kissed the ground a second time before saying: 'King of the age, I address my need to God and to the magnanimous king, the great lion, news of whose fame is carried by the caravans and the renown of whose bounty, generous deeds, forgiveness, mercy and benevolence has spread throughout all regions and lands.' He then opened up the two bags, and when he had removed from them the jewels and what else was there, he spread them out in front of Samandal, saying: 'King of the age, perhaps you would be kind enough to gratify me by accepting this gift of mine.'

Night 747

Morning now dawned and Shahrazad broke off from what she had been allowed to say. Then, when it was the seven hundred and forty-seventh night, SHE CONTINUED:

I have heard, O fortunate king, that Salih presented his gift and said: 'O king, perhaps you would be kind enough to gratify me by accepting this gift of mine.' 'Why have you given me this?' asked Samandal. 'Tell me your story and let me know what you need, for if I can, I shall at once satisfy it for you without putting you to any trouble, but if I cannot, then God does not force anyone to do what is beyond his power.'

Salih rose again and, having kissed the ground thrice, he said: 'King of the age, you can indeed grant my request, as this is something that is within your power and under your control. I am not asking you to do anything difficult and I am not mad enough to talk to you about something that you cannot do, for a wise man has said: "If you want to be obeyed, then ask for what is possible." What I have come to seek is something that the king, may God preserve him, is able to grant.' 'Tell me what you want; explain the matter and then make your request,' the king told him, and Salih then said: 'King of the age, know that I have come to ask for the hand of the unique pearl, the hidden jewel, the princess Jauhara, your daughter.'

When the king heard this, he laughed scornfully until he fell over backwards. Then he said: 'Salih, I used to think of you as someone of sense, an excellent young man who acted with understanding and said nothing that was not reasonable. What has happened to your intelligence and induced you to undertake so grave a business and risk such great danger, leading you to ask for the hand of the daughter of a king who rules over lands and climes? Have you reached such a position, coming at last to so high a rank, or has your mind weakened to such an extent that you dare say this to my face?' 'May God bring the king good fortune,' Salih replied. 'I am not asking this for myself, although if I did, I am at least her equal, for, as you know, my father was one of the kings of the sea, even though today it is you who are our ruler. I am asking for the princess's hand on behalf of King Badr Basim, ruler of the regions of Persia, whose father was King Shahriman, whose power you know. You claim to be a great king, but King Badr Basim is greater; you claim that your daughter is beautiful, but King Badr Basim is more beautiful with a more handsome appearance and of a more famous lineage. He is the champion horseman of his time, and if you agree to my request, king of the age, you will be putting something where it belongs, while if you stand on your dignity and refuse, you will not be doing us justice or treating us fairly. As you know, your majesty, Princess Jauhara, your daughter, has to marry, for the sage has said: "The only choice for a girl is between marriage and the tomb." If you intend to find a husband for her, my nephew has a better right to her than anyone else.'

Samandal, on hearing this, fell into such a rage that he almost lost his mind and came close to expiring. 'Dog,' he cried, 'does a man like you dare to speak to me like this and to mention my daughter's name in a public gathering? You say that the son of your sister Julnar is her equal, but who are you and who is your sister? Who is her son and who is his father to lead you to say this and to address me in this way? Compared to her, are you all anything more than dogs?' He then shouted to his servants: 'Cut off the head of this scoundrel!' The servants unsheathed their swords and made for Salih, who ran back to the palace gate. When he got there he found more than a thousand of his cousins, relatives, clansmen and servants, all in full armour, holding spears and gleaming swords, who had been sent by his mother to help him. On seeing the state that he was in they asked him what had happened, and after he had told them his story, realizing how stupid and violent Samandal was, they dismounted and entered his court with drawn swords. They found him, still enraged against Salih, seated on his throne. He had not noticed them and his servants, pages and guards were unprepared, but when he saw them brandishing their naked swords, he called out to his followers: 'Damn you, cut off the heads of these dogs!' Before long, however, his followers were routed and fled, after which Salih and his relatives laid hold of him and tied him up.

Night 748

Morning now dawned and Shahrazad broke off from what she had been allowed to say. Then, when it was the seven hundred and forty-eighth night, SHE CONTINUED:

I have heard, O fortunate king, that Salih and his relatives tied up King Samandal.

When Jauhara woke to find that her father had been captured and his guards killed, she left the palace and fled to an island, where she made for a high tree and hid herself in its topmost branches. During the battle some of Samandal's pages had fled and Badr, seeing them, asked what had happened to them. They told him, and when he heard that Samandal had been captured, he fled in fear for his own life, saying to himself: 'It was I who was the cause of this disturbance, and it is I for whom they will look.' So he turned tail to look for safety, without knowing where he was going. As had been predestined throughout eternity, he reached the island where Jauhara had taken refuge and, going to her tree, he threw himself down like a dead man to get some rest, without realizing that there can be no rest for the pursued and that no one knows what fate has hidden in the future.

As he lay there, he looked up at the tree and his eyes met those of Jauhara. Gazing at her, he saw that she was like a shining moon and he exclaimed: 'Praise be to the Creator of all, the Omnipotent, Who has formed this marvellous shape! Praise be to God, the Glorious, the Creator, the Maker, the Fashioner! By God, if my suspicions are right, this must be Jauhara, the daughter of King Samandal, and I suppose that when she heard about the battle she must have run away to this island and hidden herself at the top of the tree. If it is not her, it is someone even more lovely.'

He then started to think things over and said to himself that he should go and take hold of her, before asking her about herself. Then, if she turned out to be Jauhara, he could ask her to marry him, which was what he wanted. So he got up and said: 'You who are the goal of desire, who are you and what has brought you here?' Jauhara looked at him and saw that he was like a full moon appearing beneath a dark cloud, a slender youth with a lovely smile. She told him: 'Excellent young man, I am Princess Jauhara, the daughter of King Samandal. I have fled here because Salih and his army fought with my father, killed some of his men and captured others, together with my father himself. It was because of this that I ran away in fear for my life.' Having repeated this, she added that she did not know what had happened to her father.

When Badr heard what she had to say, he was astonished at this remarkable coincidence and said to himself: 'There is no doubt that I have got what I want thanks to the capture of her father.' So he looked at her and said: 'Come down from the tree, lady, for love for you is killing me and your eyes have captured me. All this fighting was because of my wish to marry you, and you must know that I am Badr Basim, king of Persia, and that Salih, who is my maternal uncle, approached your father to ask for your hand. It was because of you that I left my kingdom, and it is an extraordinary chance that we should have met here. Come down to me so that you and I may go to your father's palace, where I shall ask my uncle to free him, and we can then be married lawfully.'

Jauhara listened to this but said to herself: 'It is thanks to this miserable wretch that all this happened, that my father was captured and his chamberlains and followers killed, while I myself have been driven from my palace and forced to come here. I shall have to think of some scheme to protect myself from him, for otherwise, if I fall into his power, he will take what he wants, as he is a lover and whatever lovers do is not held to be a fault.' So she tricked him with soft words and he did not realize the deception that she had in mind. 'Master and light of my eyes,' she said, 'are you really King Badr Basim, the son of Queen Julnar?' 'Yes, my lady,' he told her.

Night 749

Morning now dawned and Shahrazad broke off from what she had been allowed to say. Then, when it was the seven hundred and forty-ninth night, SHE CONTINUED:

I have heard, O fortunate king, that Jauhara said to Badr: 'Master, are you really King Badr Basim, the son of Queen Julnar?' 'Yes, my lady,' he told her. She said: 'If my father was looking for someone more handsome than you or with finer qualities, then may God destroy him, take away his kingdom, give him no comfort and leave him in exile. By God, he is a stupid blunderer, but don't blame him for what he did. If you love me an inch, I love you a yard. I have become entangled in your love and am one of your victims, as your own love has moved to me, leaving you with only a tenth of what I have.'

She climbed down from the treetop, went up to him and embraced him, clasping him to her breast and beginning to kiss him. When he saw what she was doing, the passion of his love for her increased because he was sure that she loved him. So he returned her embraces, kissing her and exclaiming: 'Princess, I swear by God that my uncle Salih never told me a quarter of a tenth of your beauty, nor a quarter of one carat from among its twenty-four.' She clasped him again to her breast, but then, after muttering some unintelligible words, she spat in his face and said: 'Leave this mortal shape and take the form of a lovely bird with white feathers and a red beak and feet.' Before she had finished speaking, Badr had turned into a beautiful bird, and after shaking himself, he stood up on his feet and remained staring at her.

Jauhara had a slave girl named Marsina at whom she now looked and said: 'By God, if I were not afraid because his uncle is holding my father prisoner, I would kill him, may God give him no good reward. What ill fortune he has brought here with him, for all this business is thanks to him!' She then told the girl to take him to the Waterless Island and leave him there to die of thirst. The girl took him there and was about to abandon him and come back, when she said to herself: 'By God, no one as beautiful and graceful as this deserves to die of thirst,' and so she took him from there to a wooded island with fruits and streams, where she left him. She then returned to her mistress and said that she had deposited the bird on the Waterless Island.

So much for Badr, but as for his uncle Salih, when he had killed Samandal's guards and servants and had taken the king himself as a prisoner, he looked for Jauhara, the king's daughter, but failed to find her. He then rejoined his mother in his palace and asked her where his nephew Badr was. 'By God, my son,' she said, 'I have no idea, and I don't know where he went, but when he heard that you had been fighting with King Samandal he took fright and fled.' Salih was distressed to hear this and said: 'We neglected him. I'm afraid that he may die or fall in with one of Samandal's men or even with Princess Jauhara herself. This will bring shame on us as far as his mother is concerned, and no good will come to us from her, for I took him without her permission.' He sent out guards and scouts towards the sea and in other directions to try to track Badr, but they could find no news of him and had to come back and report their failure. This added to Salih's distress and anxiety for his nephew.

So much for Badr and his uncle Salih, but as for his mother, Julnar, the sea queen, after her son had gone down with Salih, she waited for him, but he did not come back and news of him was slow in coming. After having sat waiting for some days, she got up, plunged into the sea and went to her mother, who, on seeing her, rose to kiss and embrace her, as did her cousins. She then asked about Badr, and her mother told her: 'He came with his uncle, who took sapphires and other gems and set off with them, accompanied by your son, to King Samandal, whom he asked for the hand of his daughter. Samandal refused and spoke roughly to your brother. I had sent him something like a thousand riders, and there was a fight between them and Samandal in which God gave victory to your brother. Samandal's guards and his troops were killed and he himself was captured. Badr heard of that and he seems to have feared for his own life, as he fled from us through no choice of ours. Since then he has not come back and we have heard no news of him.' Julnar then asked about Salih and was told by her mother that he was seated on the royal throne in Samandal's palace and had sent out scouts in all directions to search both for Badr and for Princess Jauhara.

Julnar was filled with sorrow for her son when she heard her mother's news, and she was furiously angry with her brother Salih for having taken Badr off into the sea without her permission. She told her mother: 'I am afraid for our kingdom because I came to you without telling anyone there, and if I am slow to return I fear that things may go wrong and that we may lose control. I think that the right thing for me to do is to go back and govern the kingdom until God settles the affair of my son, but you are not to forget about him or neglect to search for him. If he comes to harm, I shall very certainly die, as he is my only link with this world and his life is my only pleasure.' 'We shall do this willingly, my daughter,' her mother replied, 'and you need not ask how saddened we are to be separated from him when he is away.' While her mother sent out search parties, Julnar herself returned to her kingdom in sorrowful tears and in a state of great distress.

Night 750

Morning now dawned and Shahrazad broke off from what she had been allowed to say. Then, when it was the seven hundred and fiftieth night, SHE CONTINUED:

I have heard, O fortunate king, that Julnar left her mother and returned to her kingdom in a state of great distress. So much for her.

As for Badr himself, after Jauhara had cast a spell on him she had ordered a slave girl to take him off and abandon him to die of thirst on the Waterless Island. The girl, however, disobeyed her and left him on an island that was green and well wooded, with fruits which he ate and streams from which he drank. He stayed there in his bird shape for a number of days and nights, not knowing where to go or how to fly. Then, one day, a hunter came looking for something to catch and eat. He caught sight of the Badr bird with its white feathers and red beak and legs, which captivated the eye and enchanted the heart. The sight of it delighted the hunter and he said to himself: 'This is a lovely creature and its beauty and form are such as I have never seen in any other bird.' He threw his net and caught it, after which he went to the local town, telling himself that he would sell it for a good price. One of the towns-people met him and asked how much he wanted for it. 'If I sell it to you, what will you do with it?' the hunter asked, and the man replied: 'I'll kill it and eat it.' 'Who could bring himself to treat it like that?' asked the hunter. 'I intend to present it to the king, who will give me more than I would get from you and, far from killing it, will enjoy looking at its beauty, for in all my life as a hunter I have never seen anything like it on land or sea. As for you, even if you wanted it, the most that you would give me for it would be a single dirham and so I swear by Almighty God that I shall not sell it.'

The hunter then took the bird to the palace and the king, struck by its beauty and the redness of its beak and legs, sent a eunuch to buy it from him. The eunuch asked if he was prepared to sell and the man said: 'No, but I shall give it to the king as a present from me.' The eunuch took the bird and went to tell the king what the man had said, and the king accepted it, giving the hunter ten dinars. He took these and, after having kissed the ground, he went off, leaving the eunuch to take the bird to the palace, where he hung it up in a fine cage with food and water. When the king came he asked for it to be brought to him so that he could admire its beauty, but when the eunuch brought it and set it in front of him, he could see that it had eaten none of its food. 'By God,' said the king, 'I don't know how to feed it as I don't know what it eats.' He then ordered food to be brought, and when the tables had been set before him he sampled it. When the bird saw the meat and the other foodstuffs, together with the sweetmeats and the fruit, it started to eat from all the dishes that were in front of the king. When he saw what it ate, the king was astonished and amazed, as was everyone else there. He told his entourage of eunuchs and mamluks: 'Never in my life have I seen a bird eating like this one,' and he ordered that his wife be brought to look at it. The eunuch was sent to fetch her and, when he saw her, he said: 'Mistress, the king wants you to come and look at the bird that he has bought. When we brought in food, it flew out of its cage and settled on the table, after which it ate from all the dishes that were there. Come and look, mistress, as it is a beautiful creature, one of the marvels of the age.'

When the queen heard this, she came quickly, but after she had studied the bird carefully, she covered her face and turned back again. The king followed her and asked her why she had veiled herself when there was no one there apart from the slave girls and eunuchs who were in her service, as well as her own husband. 'Your majesty,' she answered, 'this is no bird but a man like you.' 'That cannot be true!' he exclaimed. 'You must be joking. How can it be anything but a bird?' 'By God, I am not joking with you and I have told you nothing but the truth,' she said, adding: 'This is King Badr Basim, the son of Shahriman, king of Persia, and of Julnar, the sea queen.'

Night 751

Morning now dawned and Shahrazad broke off from what she had been allowed to say. Then, when it was the seven hundred and fifty-first night, SHE CONTINUED:

I have heard, O fortunate king, that the queen told the king: 'This is no bird but a man like you, and he is Badr Basim, the son of King Shahriman and Julnar, the sea queen.' 'How did he come to be in this shape?' her husband asked, and she explained: 'Princess Jauhara, the daughter of King Samandal, put a spell on him.' Then she told him everything that had happened to Badr from beginning to end, explaining that he had asked her father for Jauhara's hand and had been refused and how his uncle Salih had then fought with Samandal and got the better of him before capturing him. The king was astonished to hear all this, but, as she was the leading sorceress of her age, he said: 'I implore you to release him from this spell and free him from his torture. May God cut off Jauhara's hand. How foul and irreligious she must be and how full of deceit and trickery!' The queen told him: 'Say to him: "Badr, go into this closet."' The king did that, and when Badr heard the order, he obeyed. The queen then veiled her face and went into the closet with a cup of water in her hand, over which she recited some unintelligible words. Then she said: 'I conjure you by the virtue of these great names, by the noble verses of the Quran and by the truth of God, the Exalted, the Creator of the heavens and the earth, Who brings the dead to life, Who apportions men their livelihood and allots them their life spans, leave your present shape and return to the form in which God created you.' Before she had finished speaking, Badr shook himself and resumed his human form, appearing before the king as a young man unsurpassed in beauty by any of the inhabitants of the earth.

For his part, when Badr saw what had happened, he exclaimed: 'There is no god but God and Muhammad is the Prophet of God! Praise be to the Creator of all things, Who decrees for men their livelihoods and their allotted spans.' Then he kissed the king's hands, praying that he be granted long life, and in return the king kissed his head and said: 'Badr, tell me your story from beginning to end.' Badr did this, concealing nothing, and the astonished king said: 'Now that God has freed you from this sorcery, what plan do you have and what do you want to do?' 'King of the age,' Badr replied, 'would you be kind enough to prepare a ship for me, manned by a crew of your servants and supplied with the necessary provisions? I have been away for a long time and I am afraid of losing my kingdom. I doubt whether my mother will still be alive, thanks to my absence, as I think it likely that grief for me will have killed her. She does not know what happened to me or whether I am alive or dead, and so I would ask you to complete your kindness to me by granting my request.' The king, taking note of Badr's beauty and eloquence, agreed to this, exclaiming: 'To hear is to obey!' He got a ship ready for him, provisioned it and detailed a number of his servants to go with him.

After Badr had taken his leave and embarked, the ship put to sea and enjoyed favourable winds for ten consecutive days. Then, on the eleventh day, the sea became very rough, the ship rising and falling out of its crew's control, and it continued to be tossed to and fro by the waves until it approached a reef, on which it struck. The ship itself was smashed and all on board were drowned except for Badr, who had come close to death but had managed to climb on to a plank. He had no idea where he was going and he had no means of controlling the plank as it took him wherever wind and wave carried it. This went on for three days and on the fourth it was washed up on a shore where Badr discovered a city gleaming like the whitest of doves. It was built on a peninsula that ran out to sea and was tall and well constructed, with lofty walls against which the sea broke.

Badr was delighted by the sight of this peninsula, as he was half-dead of hunger and thirst. He got off his plank with the intention of going up to the city, but, as he did so, herds of mules, donkeys and horses, in numbers like grains of sand, approached and began to strike at him, preventing him from getting there. So he swam around to the far side of the city, and when he landed he was astonished to find that there was no one there. 'Who do you suppose owns this city,' he asked himself, 'as it has no king and no inhabitants? Where did the mules, donkeys and horses come from that stopped me reaching it?' He was thinking about this as he walked along with no notion where he was going when he caught sight of an old greengrocer, with whom he exchanged greetings. The man looked at him, and, on seeing how handsome he was, he said:

'Where have you come from, young man, and what has brought you here?' Badr astonished him with his story, which he told from beginning to end, and the old man then asked: 'My son, did you see anyone as you came?' 'Father,' said Badr, 'I was surprised to discover that there was no one in the city.' 'Come into my shop,' the man said, 'lest you be killed.'

Badr went in and sat down in the shop while his host got up and fetched him food, telling him to come to the inner room and exclaiming, to his great alarm: 'Glory be to God, Who has saved you from this she-devil!' After having eaten his fill, Badr washed his hands and, looking at the old man, he asked: 'Master, what are you talking about? You have made me frightened of the city and its people.' 'My son,' replied the other, 'you must know that this is a city of magicians and that its queen is a witch, a she-devil, a soothsayer and a mistress of magic, wiles and treachery. The horses, mules and donkeys that you can see were all men like you and me, who came here as strangers. Any handsome young man like you who arrives here is taken by this infidel sorceress, who keeps him with her for forty days and after that she transforms him into a mule or a horse or a donkey, like those that you saw by the shore.'

Night 752

Morning now dawned and Shahrazad broke off from what she had been allowed to say. Then, when it was the seven hundred and fifty-second night, SHE CONTINUED:

I have heard, O fortunate king, that the old greengrocer told Badr about the sorceress queen and explained that she had transformed all the inhabitants of the city, going on: 'When you wanted to land, they were afraid that you might be enchanted as they had been, and it was out of pity for you that they were trying to tell you by their gestures not to land lest the sorceress see you and treat you as she had treated them.' He went on to say that the sorceress had taken the city from its people by magic and that her name was Queen Lab, a word meaning 'calculation of the sun'. Badr was terrified when he heard this and began to tremble like a reed in the wind. 'I have only just managed to escape from a misfortune brought on me by magic, before fate delivers me to an even worse place.'

The old man saw that he was brooding about his circumstances and his experiences and that he was terribly afraid. 'My son,' he said, 'come and sit on the doorstep of the shop and look at these creatures, their different coverings and species, and the enchantment from which they suffer. There is no need to be frightened, because the queen and all the townsfolk are on friendly terms with me. They look after me and cause me no alarm or distress.' When Badr heard that, he came out and sat by the shop door as the people passed by, looking at the sights and the innumerable creatures that were there. When the people saw him, they approached the old man and asked: 'Is this a captive of yours whom you have taken recently?' 'This is my nephew,' he told them. 'I heard that his father had died and so I sent to him and brought him here to quench the fire of my longing.' They said: 'He is a handsome young man and we are afraid that Queen Lab will trick you and take him from you, as she has a liking for pretty boys.' 'The queen will not go against what I tell her,' replied the old man, 'for she looks after my interests and is fond of me. When she finds out that he is my nephew, she will not do anything to harm him or cause me concern about him.'

Badr stayed for some months with the old man, who supplied him with food and drink and had deep affection for him. Then, one day, while he was in his usual place by the shop door, he was suddenly confronted by a thousand eunuchs with drawn swords in their hands, variously clad, but with jewel-studded belts around their waists. They were riding on Arab horses with their Indian swords hanging from baldrics, and after they had come to the old man's shop, they greeted him and then went off. They were followed by a thousand slave girls as beautiful as moons, who wore smooth silks embroidered with gold and set with jewels of all kinds. They were all carrying spears and in the middle of them was one riding on an Arab mare with a saddle of gold, studded with sapphires and other gems. They too came to the shop and greeted the old man before leaving. Next came Queen Lab herself in a great procession, and she rode up to the shop, where she saw Badr sitting like the full moon.

She was astounded and taken aback by his beauty and became infatuated with him. As a result, she came to the shop, dismounted and sat down beside him. 'Where did this handsome youth arrive from?' she asked the old man. 'He is a nephew of mine who has only recently come,' the man told her. 'Let him spend the night with me so that we can talk,' she said. 'Do you agree not to put a spell on him if you take him from me?' he asked, and when she said yes, he asked her to swear an oath for him. She swore not to harm him or enchant him, and then on her orders a fine horse was brought for him, saddled and provided with a golden bridle, all of its trappings being made of gold and set with jewels. She then presented the old man with a thousand dinars for his own use.

She now went off with Badr, who was looking like a full moon on its fourteenth night, and as he rode with her, all those who saw how handsome he was felt sorry for him, saying: 'By God, this young man does not deserve to be enchanted by this damned woman.' Badr heard what they were saying but he kept quiet, having entrusted his affairs to Almighty God. They rode on to the palace . . .

Night 753

Morning now dawned and Shahrazad broke off from what she had been allowed to say. Then, when it was the seven hundred and fifty-third night, SHE CONTINUED:

I have heard, O fortunate king, that Badr, Queen Lab and her followers rode on to the palace door, where the emirs, the great ministers of state and the eunuchs dismounted. The queen instructed her chamberlains to order all the officials to disperse, which they did after having kissed the ground. Accompanied by her eunuchs and slave girls she then entered her palace, a building the like of which Badr had never seen. Its walls were of gold and in the centre of it was a huge lake with a plentiful supply of water set in an extensive garden. In the garden, Badr saw birds whose songs were of all kinds, joyful and sad, while they themselves were of various shapes and colours. At these indications of great power he exclaimed: 'Praise be to God, Who through His generosity and clemency provides for those who worship other gods!'

The queen took her seat on an ivory couch with splendid coverings by a window that overlooked the garden. As Badr sat down beside her, she kissed him and clasped him to her breast before telling her slave girls to fetch a table of food. The table they brought in was made of red gold set with pearls and other jewels and laden with all kinds of food. The two of them ate their fill, and when they had washed their hands, the slave girls fetched vessels of gold, silver and crystal, together with flowers and bowls of dried fruits. On the queen's orders, ten singing girls as beautiful as moons then entered with various musical instruments in their hands. She herself filled a goblet with wine and drank it, after which she filled another and passed it to Badr. He took it and drank and they went on drinking like that until they had had enough. The queen then told the girls to sing, and the melodies they produced made it seem to Badr that the palace itself was dancing with joy. He became light-headed, and in his delight he forgot that he was a stranger. 'This queen is a lovely young woman,' he told himself. 'I shall never leave her, for this kingdom of hers is bigger than mine and she is more beautiful than Jauhara.'

He went on drinking with her until evening, when the lamps and the candles were lit and incense released. They both drank until they were drunk, to the accompaniment of the songs of the singing girls, and when the queen was drunk she got up and lay down on a couch, dismissing her girls, and she then told Badr to lie down beside her. He slept with her, enjoying the greatest pleasure until morning came.

Night 754

Morning now dawned and Shahrazad broke off from what she had been allowed to say. Then, when it was the seven hundred and fifty-fourth night, SHE CONTINUED:

I have heard, O fortunate king, that the queen then woke and took Badr with her to the palace baths, where they bathed. When they left, she produced for him the finest of clothes, and when they had drunk from goblets fetched, on her instructions, by slave girls, she took him by the hand and they both sat down on chairs. She ordered food to be brought and after they had eaten they washed their hands. The slave girls then fetched goblets and fruits, fresh as well as dried, together with flowers, and Badr and the queen went on eating and drinking as the girls sang a variety of melodies until evening came. They continued to enjoy themselves in this way for forty days, after which the queen asked: 'Badr, which is the more pleasant place, this or your uncle's shop?' 'By God, your majesty,' Badr replied, 'it is more pleasant here, as my uncle is a poor man who sells beans,' at which she laughed and the two of them passed the night in enjoyment until morning.

When Badr woke up, he failed to find the queen lying beside him and he wondered where she could have gone. He was disturbed by her absence and he did not know what he should do. A long time passed and when she still had not come back he put on his clothes, saying to himself: 'Where can she have got to?' He searched for her unsuccessfully and then he told himself that she might have gone to the garden. He went there himself, and there beside a flowing stream he caught sight of a white bird with other birds of various colours on top of a tree that was growing by the bank. As he looked, he saw a black bird fly down from the tree and start to feed the white bird with its beak as doves do, before treading it thrice. After a while the white bird changed into human shape, and when Badr looked closely he saw that this was Queen Lab. He realized that the black bird must be a man under a spell and that the queen, being in love with him, had transformed herself into a bird in order to copulate with him. This made him jealous and because of the bird he became angry with the queen.

When he got back to his room he lay down on his bed, and after a time the queen came and started to kiss him and to joke with him, but he was too angry to say a single word. She realized what he was feeling and was sure that he must have seen her when she had turned into a bird, to be trodden by her mate, but she gave nothing away and kept this secret. After Badr had satisfied her lust, he asked her to allow him to go to his uncle's shop, pointing out that he had not seen him for forty days and was longing for a meeting. 'Go to him,' she said, 'but come back to me quickly, for I cannot bear to be parted from you for a single hour.' 'To hear is to obey,' he replied, and he then mounted and rode to the old man's shop. The old man got up to embrace him and then asked: 'How are you getting on with the infidel woman?' 'All was going well,' Badr replied, 'but this last night she went to sleep beside me and when I woke up I couldn't find her. I put on my clothes and went around looking for her until I came to the garden.' He then explained about the stream and the birds that he had seen on the treetop. When the old man heard about this, he said: 'Be on your guard against her. You must know that the birds on the tree were all young strangers who became her lovers and whom she then changed by magic into birds. The black bird was one of her mamluks of whom she was particularly fond, but he turned his attention to one of the slave girls and she transformed him by magic into his present shape.'

Night 755

Morning now dawned and Shahrazad broke off from what she had been allowed to say. Then, when it was the seven hundred and fifty-fifth night,
SHE CONTINUED:

I have heard, O fortunate king, that Badr told the old greengrocer all about Queen Lab and what he had seen her do. The man explained that all the birds on the tree were young strangers, whom she had transformed by magic. The black bird had been one of her mamluks, whom she had turned into that shape by a spell. 'Then, whenever she wants him,' he went on, 'she turns herself into a bird so that he can tread her, as she is still deeply in love with him. Now that she has realized that you know about this, she will harbour a grudge against you and try to harm you. But no harm will come to you as long as I am here to protect you. There is no need to be afraid, for I am a Muslim named 'Abdallah, and there is no greater sorcerer in this age, although I only use magic when I am forced to it. I have often frustrated the spells of this damned witch and saved people from her. She does not worry me as she has no power over me, but, on the contrary, she is terrified of me, as are all the other sorcerers of her kind in the city. They are all co-religionaries of hers, worshipping fire in place of the Omnipotent God. Come back to me tomorrow and tell me what she did to you, for she will try her best to destroy you tonight, and I will tell you what to do to in order to escape from her wiles.'

Badr said goodbye to the old man and went back to the queen, whom he found sitting and waiting for him. When she saw him she got up to welcome him, before making him sit and fetching him food and drink. The two of them ate until they had had enough, and, after they had then washed their hands, she ordered wine to be brought. When it had come, they started to drink and went on until midnight, as she kept leaning over to pass him wine cups until he became drunk and incapable. When she saw this, she said: 'In God's Name, I conjure you by the object of your worship to tell me, if I ask you about something, will you give me a truthful answer?' 'Yes,' replied the drunken Badr, and she went on: 'You did not find me there when you woke up and so you looked for me and came to me in the garden, where you saw me in the form of a white bird and you saw the black bird which flew down on me. I shall tell you the truth about this bird, which was one of my mamluks, whom I loved dearly. One day he eyed one of my slave girls, and in my jealousy I changed him by magic into the shape of a bird and killed the girl. Now, however, I find that I cannot do without him for a single hour and whenever I want him I turn myself into a bird and go to him so that he can tread me and possess me as you saw. Is this not why you are angry with me? Although I swear by fire, light, dark and heat that I love you

even more, and that you are all that I want from the world.' Badr, still
in his drunken state, said: 'You are right in thinking that it was this and
only this that made me angry.'

She then clasped him to her breast and kissed him, pretending to be
in love with him, and she then went to sleep while he slept beside her.
Halfway through the night she got out of bed. Badr was awake but
pretended to be sleeping, and he stole a glance at her to see what she
was doing. From a red bag she took something red which she scattered
in the middle of the palace, where it turned into a flowing stream like a
sea. Next she took a handful of barley, sowed it on the soil and, when
she had watered it from the stream, the grains became stalks with ears,
which she took and ground into flour. She put this away somewhere and
then went back to sleep beside Badr until morning.

In the morning, Badr got up, washed his face and then asked the
queen's permission to visit the old greengrocer. When she had allowed
him to go, he went off and told the man what she had done and what
he had seen. On hearing this, the old man burst out laughing and said:
'By God, this infidel witch has laid a trap for you, but you have no need
to worry about her.' He then brought out a *ratl*'s weight of barley gruel
and told Badr to take it with him. 'When she sees you,' he went on, 'she
will ask what it is and what you are doing with it. Tell her: "Eat it, as
you can never have too much of a good thing." She will then produce
sawiq of her own, which she will tell you to sample. You must pretend
to do that, but instead you must eat mine, taking care not to take a single
grain of hers. Were you to do that, you would be under her spell; she
would bewitch you and when she said: "Leave your human shape," you
would find yourself transformed into whatever shape she might choose,
but if you don't eat any of it, her spell will not work and no harm will
come to you. She will be covered with embarrassment and will tell you:
"I was only joking," while professing love and affection for you, but this
will merely be hypocrisy and trickery on her part. You yourself should
make the same pretence of love for her and say: "My mistress and the
light of my eyes, try this dish of mine and see how tasty it is." If she eats
even one grain, take some water in your hand and dash it in her face,
telling her to quit her mortal shape for any other that you want. Then
leave her and come back to me so that I can arrange something for you.'

Badr said goodbye to him and went up to the palace. When he came
to the queen, she welcomed him, getting up and kissing him. 'You have
been slow in coming back to me,' she complained, and he told her: 'I

was with my uncle, who gave me some of this *sawiq* to eat.' 'I have some that is better,' she said, and she then put his on one plate and her own on another before saying: 'Try some of this, for it is tastier than yours.' He pretended to do this, and when she thought that he had eaten it she took some water in her hand and sprinkled it over him before saying: 'Leave this shape, you miserable wretch, and become an ugly one-eyed mule.' When she saw that, far from being transformed, he was still exactly as he had been, she got up, kissed him between the eyes and said: 'My darling, I was playing a joke on you. Don't be angry with me.' 'By God, lady,' he replied, 'my feelings towards you have not changed at all. I am sure that you love me, so eat some of this food of mine.' She took a mouthful and ate it, but when it settled in her stomach she had convulsions and he took some water in his hand and sprinkled it over her face, saying: 'Quit your human shape and become a dappled mule.' When she looked at herself and found that this was what she was, tears began to roll down her cheeks and she started to rub at them with her legs. Badr fetched a bridle but she refused to accept it, and so he left her and went back to the old man to tell him what had happened.

The old man now went and fetched another bridle for him, telling him to take it and use it on her. He brought it back to her, and when she saw him she went up and he put it in her mouth. After that, he mounted her and left the palace to return to 'Abdallah. He, on seeing the mule, went up to her and said: 'May Almighty God pay you back, you damned woman.' He then told Badr: 'My son, you cannot stay here any longer, so mount her and ride off to wherever you want, but take care not to hand over the bridle to anyone.'

Badr thanked him, took his leave and left. When he had ridden for three days, he came in sight of a city and was met by an old man with a handsome head of white hair who asked him where he had come from. 'From the city of the witch,' Badr told him, at which the man offered him hospitality for the night. He accepted and as he was accompanying his host along the road, he came across an old woman who wept when she saw the mule and exclaimed: 'There is no god but God! This is very like my son's mule which, to my sorrow, has just died. I implore you in God's Name, master, to sell yours to me.' 'By God, mother, I cannot do that,' he told her, but she urged him not to refuse her, saying: 'If I don't buy it for him, my son will very certainly die.' She went on and on pressing him, until at last he said: 'I shall only sell it for a thousand dinars,' thinking to himself: 'Where will an old woman like this get a thousand dinars?' At that, however, she produced the money from her belt and when Badr said: 'I was only joking, mother, and I cannot sell the mule,' the old man looked at him and said: 'My son, in this town no one tells lies, for any liar found here is put to death.'

Badr dismounted . . .

Night 756

Morning now dawned and Shahrazad broke off from what she had been allowed to say. Then, when it was the seven hundred and fifty-sixth night, SHE CONTINUED:

I have heard, O fortunate king, that Badr dismounted and handed the mule over to the old woman. She removed the bit from its mouth and then took some water in her hand, which she sprinkled over it, saying: 'Daughter, leave this shape and resume the one that you had before.' At that, the mule instantly became a woman again and she and her rescuer went up to each other and embraced. Badr realized that this must be Lab's mother and that he had been tricked. He was about to take to his heels when the old woman gave a loud whistle and an *ifrit* like a huge mountain appeared in front of her. Badr stopped still in fear and the old woman mounted on the *ifrit*'s back, taking up her daughter behind her and setting Badr in front of her. The *ifrit* flew off with them and no more than an hour had passed before they had returned to the queen's palace.

When she had taken her seat on her royal throne, she turned to Badr and said: 'Miserable wretch, here I am again with my wishes granted, and I shall now show you what I shall do with you and the old green-grocer. How many favours have I shown him, but still he tries to harm me, and it was only thanks to him that you got what you wanted.' She then took water, sprinkled him with it and said: 'Quit your shape and become the ugliest of all birds.' Instantly Badr was transformed into a repulsive-looking bird and the queen put him in a cage, allowing him neither food nor drink. He was seen, however, by a slave girl who took pity on him and began to feed him and supply him with water without the queen's knowledge.

One day this girl found that her mistress was not paying attention to her, and so she went off to tell the old greengrocer what had happened, saying: 'Queen Lab intends to kill your nephew.' He thanked her and said: 'I am going to have to take the city from her and put you in her place as its queen.' He gave a loud whistle, at which a four-winged '*ifrit* appeared. 'Take this girl,' the old man said, 'and carry her to the city of Julnar, the sea queen, and her mother, Farasha, for they are the finest magicians on the face of the earth.' He then told the girl: 'When you get there, tell the two of them that King Badr Basim is being held prisoner by Queen Lab.' The '*ifrit* took up the girl and no more than an hour later he set her down in Julnar's palace. She came down from the roof and entered Julnar's presence, where she kissed the ground and told her all that had happened to her son from start to finish. Julnar got up, treated her with honour and thanked her, after which she had drums beaten throughout the city to spread the good news, announcing to the citizens and the state officials that King Badr had been found.

Julnar and her mother, Farasha, together with Salih, her brother, then summoned all the tribes of the *jinn* and the armies of the sea, since after the capture of Samandal the *jinn* kings obeyed them. They then flew off through the air and alighted at Lab's city, where they sacked the palace and in the twinking of an eye had killed all the infidels that they found there. Julnar then asked the girl where her son was, and the girl fetched the cage and put it down in front of her. Pointing at the bird inside it, she said: 'Here he is.' Julnar brought him out of the cage and then, taking some water in her hand, she sprinkled him with it and said: 'Leave this shape for the one you had before.' Before she had finished speaking, the bird shivered and became a man again. Julnar, seeing her son in his original shape, embraced him, as did Salih, his uncle, Farasha, his grandmother, and his cousins, who started to kiss his hands and feet, while he, for his part, shed floods of tears. Julnar sent for the old greengrocer and thanked him for his goodness to her son, before marrying him to the girl whom he had sent to bring her the news of Badr. The marriage was consummated, and Julnar then appointed 'Abdallah as king of the city. She summoned the Muslims who remained there and made them take an oath of allegiance to 'Abdallah and vow to obey and serve him, which they agreed to do.

When Julnar and the others had taken their leave of 'Abdallah they returned to their own city, and when they reached the palace they were met by the citizens with gladness and rejoicing. As an expression of their delight at the return of Badr, their king, the city was adorned with decorations for three days, after which Badr told his mother: 'It only remains for me to get married, so that we may all remain together.' 'That is a good idea,' she agreed, 'but wait until we can ask which princess would make you a suitable bride.' Farasha, Badr's grandmother, together with his uncle Salih and his cousins, all promised to help, and each one of them went out on this quest throughout the lands. Julnar herself sent her slave girls, riding on *'ifrits*, with orders to look at all the pretty girls that were to be found in every city and every royal palace. When Badr saw the trouble that they were taking in the search, he told his mother to stop, adding: 'No one will satisfy me except Jauhara, the daughter of King Samandal, for she is a jewel, like her name.'* When she knew what he wanted, Julnar had Samandal brought before her, after which she sent for Badr and told him that Samandal was there. Badr went into the room, and when Samandal saw him coming he got up, greeted him and welcomed him. Badr then asked him for his daughter's hand and Samandal replied: 'She is at your service as your slave to command.'

Samandal now sent a number of his companions back to his own country with orders to fetch his daughter and to tell her that her father was with King Badr. Off they flew and before an hour had passed they had returned, bringing the princess with them. When she saw her father, she went up to him and embraced him. For his part, he looked at her and said: 'Daughter, know that I have married you to this great king, the mighty lion, King Badr Basim, son of Queen Julnar, for he is the handsomest man of his time, the highest in rank and the noblest. He is the only suitable husband for you and you are the only suitable wife for him.' 'I cannot disobey you, father,' she said, 'so do what you want. My worries and distress are over and I am one of his servants.'

At that, the *qadis* and notaries were brought forward and a marriage contract drawn up between King Badr, son of Julnar the sea queen, and the Princess Jauhara. The city was adorned with decorations, drums were beaten to spread the good news, all prisoners were released and the king provided clothes for widows and orphans as well as presenting robes of honour to the officers of state, the emirs and the grandees. There was a great festival with banquets and celebrations that continued night and day for ten days. Jauhara was displayed to Badr in nine different robes, after which he gave a robe of honour to King Samandal and returned him to his own country, his family and his relations. They remained enjoying the pleasantest of lives and the most delightful of times, eating, drinking and living in comfort until they were visited by

* Jauhar is Arabic for 'jewel'.

the destroyer of delights and the parter of companions. This is the end of their story, may God have mercy on them all.

A story is told, O fortunate king, that in the old days there was a Persian king named Muhammad ibn Saba'ik, who ruled over Khurasan. Every year he would launch raids on the lands of the unbelievers in Hind, Sind, China, Transoxania and other foreign parts. He was a just and courageous ruler, characterized by generosity, and he had a fondness for social gatherings, tales, poems, histories and stories, narratives and conversations. Anyone who could remember a remarkable tale and tell it to him would be rewarded, and it is said that when a stranger came with this kind of thing, and the king approved of it, he would be presented with a splendid robe of honour, given a thousand dinars and provided with a horse, saddled and bridled. After having been clothed from top to toe, he would be given great gifts to take away with him.

It happened on one occasion that an elderly man arrived and won the king's favour with a remarkable story. The king ordered him to be given a splendid reward, part of which consisted of a thousand Khurasanian dinars and a horse with all its trappings. News of his generosity now spread throughout all lands and his reputation reached a merchant named Hasan, a learned man, liberal and generous and an excellent poet. The king had as a vizier an envious and malevolent person who had no fondness for anyone, rich or poor. He was jealous of anyone who came to the king and was given a reward, and he used to say that the king's generosity wasted money and was ruining the kingdom. As the king continued to act like this, what the vizier said was prompted solely by envy and hatred.

As it happened, the king heard of Hasan the merchant and sent to have him brought to his court. When he arrived, the king told him: 'My vizier is opposed to the way in which I give money to poets as well as to those who entertain me with their conversation and who tell stories and quote verses. I want you to tell me a story that is both pleasant and remarkable, such as I have never heard before. If I approve of it, I shall present you with many lands, together with their castles, and add to your fiefs, and I shall give you authority over the whole of my kingdom, appointing you as my grand vizier, where you will sit at my right hand, delivering judgements to the people. But if you fail in this, I shall seize all your goods and expel you from my country.' 'To hear is to obey, your majesty,' Hasan answered, 'but your servant asks you for a year's

delay and he shall then produce for you a story the like of which you
have never heard in your life and whose equal or superior no one else
has ever been told.' 'I grant you a full year,' the king said, and he then
called for a splendid robe of honour, which he gave to the man, saying:
'Stay at home; don't ride out or go to and fro for a whole year until you
bring me what I have asked for. If you do this, you will enjoy my special
favour and you will be glad to learn that I shall keep my promise, but if
you don't, we shall have nothing more to do with each other.'

Night 757

Morning now dawned and Shahrazad broke off from what she had been
allowed to say. Then, when it was the seven hundred and fifty-seventh
night, SHE CONTINUED:

I have heard, O fortunate king, that the king, Muhammad ibn Saba'ik,
said: 'If you do this, you will enjoy my special favour and you will be
glad to learn that I shall keep my promise, but if you don't, we shall
have nothing more to do with each other.' After kissing the ground in
front of the king, Hasan went out. He selected five of his best mamluks,
all of whom could read and write and were excellent, intelligent and
cultured men. He gave each of them five thousand dinars and told them:
'It was for a day like this that I brought you up, so come to my rescue
by helping me to do what the king has asked for.' 'What do you want
to have done?' they asked. 'For we would ransom you with our lives.' 'I
want each of you to go to a different region,' he told them, 'where you
are to question men of learning, culture and excellence, together with
the tellers of strange stories and remarkable tales. Track down for me
the story of Saif al-Muluk and then bring it to me, and if you find anyone
who has it, tempt him to sell it, and give him whatever he wants in the
way of gold or silver, even if he asks you for a thousand dinars. Pay
what you can and promise him the rest, but bring it to me. Whichever
of you finds the story and fetches it for me will receive splendid robes
and generous favours and he will be the dearest of men to me.'

Hasan told one of the five to go to Hind and Sind, and the districts
and regions there; another was to visit Persia, China and their regions;
a third was to go to the districts of Khurasan; a fourth was told to visit
all the quarters, regions, territories and borders of the Maghrib, while
the fifth was to travel through all parts of Syria and Egypt. Hasan chose

an auspicious day for their departure and told them to do everything they could to get what he wanted without slacking in their quest, even if it cost them their lives. They took their leave of him and set off, each going where they had been told. After an absence of four months, four of them had found nothing in their searches and these went home, to Hasan's distress, to tell him that they had looked through cities, lands and regions searching for the story he wanted, but had discovered no trace of it.

The fifth mamluk had gone to Syria, where he reached the city of Damascus. He found this to be a pleasant and secure place, with trees, rivers and fruits, where the birds sang the praises of the One Almighty God, the Creator of night and day. For some days he stayed there, making enquiries about what his master wanted but without getting any answers. He was on the point of leaving to go somewhere else when a young man ran past and tripped over the edge of his cloak. 'Why are you running like this? Where are you off to?' the mamluk asked. The young man told him: 'There is an excellent *shaikh* here who takes his seat every day at about this time and who tells splendid stories, tales and legends that have never been heard before. I am running to get a place close to him, but there are so many people that I'm afraid I won't find one.' 'Take me with you,' the mamluk said, to which the man replied: 'Move fast, then,' and the mamluk closed his door and hurried on to the place where the storyteller sat with his audience. He found him to be an old man with a handsome face who was seated on a chair telling his stories to the people.

The mamluk sat down close to him to listen, until at sunset, when the storyteller had finished and the audience had dispersed, he went up and greeted him. The storyteller returned his greeting with the greatest courtesy, and the mamluk then said: 'You are a fine man who deserves respect and your stories are excellent. I want to ask you about something.' 'Ask whatever you want,' the storyteller replied, and the mamluk then asked him if he knew the tale of Saif al-Muluk and Badi' al-Jamal. 'Where did you hear about this? Who told you?' the man asked, and the mamluk assured him that no one had told him, going on to explain: 'I have come from a distant land to look for this story, and if you have it and are kind enough to pass it to me, I shall give you whatever you ask. For this would be an act of kindness and generosity on your part, and were I able to pay you for it with my life, I would be happy to do so.' 'There is no need for you to be so concerned,' the man told him, 'for you can have it. But this is not something which is to be told in public, and I would not hand it over to everyone.' 'By God, master,' the mamluk begged him, 'do not grudge it to me but ask me whatever you like for it.' 'If you want it,' said the storyteller, 'I shall let you have it for a hundred dinars, but on five conditions.'

When the mamluk realized that the *shaikh* really had the story and was prepared to give it to him, he was delighted and said: 'I shall give you the hundred dinars as its price and an extra ten for yourself, and I accept what you say about the conditions.' 'Go and fetch the gold,' the other replied, 'and then you can take what you have come for.' The mamluk got up, kissed the *shaikh*'s hand and went back delightedly to his lodgings, after which he got a hundred and ten dinars and put them in a purse that he had with him. The next morning, after he had risen and dressed, he fetched the purse and brought it to the *shaikh*, who took the money and then went into his house, taking the mamluk with him. He gave him a seat, providing him with an inkstand, pen and paper, and he then produced a book for him and said: 'Copy from this book the story that you have been looking for, the legend of Saif al-Muluk.'

The mamluk sat down and copied the story until he had transcribed it all, and the *shaikh* read it over and confirmed the accuracy of the copy. The *shaikh* then said: 'Know, my son, that the first of my conditions is that you are not to recite this story in public, or to tell it to women or girls, or to slaves or foolish men, or to boys. It must only be told to emirs, kings, viziers, scholarly commentators and so on.' The mamluk agreed to this, kissed the *shaikh*'s hands, took his leave of him and then went off.

Night 758

Morning now dawned and Shahrazad broke off from what she had been allowed to say. Then, when it was the seven hundred and fifty-eighth night, SHE CONTINUED:

I have heard, O fortunate king, that Hasan's mamluk transcribed the story from the book of the Damascene *shaikh* and was told his conditions, after which he took his leave of him and went off.

He set out that same day in a state of great delight and pressed on with his journey, full of joy at having got hold of the story, until he reached his own country and sent on his servant to tell Hasan that he was back safely with what he had been looking for. By the time he had reached Hasan's city and sent him the good news, only ten days were left before the end of the time limit agreed between Hasan and the king. The mamluk went to see Hasan, who was filled with joy when he heard what had happened, and to whom, after having rested in his own room, he handed the book in which he had copied the story of Saif al-Muluk and Badi' al-Jamal. At the sight of it, Hasan presented him with all the robes he was wearing, together with ten fine horses, ten camels, ten mules, three slaves and two mamluks. He then took the story and copied it out clearly in his own hand before taking it to the king. 'Your august majesty,' he said, 'I have brought you the account of a splendid legend, the like of which no one has ever heard.' On hearing this, the king immediately ordered that all the wise emirs, excellent scholars, men of culture, poets and the intelligent should be brought to him. When Hasan the merchant had taken his seat, he read out the story to the king, who was full of admiration and approval, as was the entire audience. They scattered gold, silver and jewels over Hasan, and the king ordered him to be presented with a magnificent robe of honour and gave him a large city, complete with its castles and estates. He appointed him as one of his principal viziers and sat him at his right hand. His scribes were ordered to copy out the story in letters of gold so that it might be stored in his private treasury. Whenever he felt depressed, he would have Hasan fetched to read it out. THE STORY RAN AS FOLLOWS:

In the old days there was a king of Egypt named 'Asim ibn Safwan, a generous and liberal man, characterized by dignity and gravity, who ruled over many lands, castles and fortresses, with armies and troops at his command. He had a vizier named Faris ibn Salih, and he, together with all his people, were fire worshippers who took the sun as god in place of the Almighty, Exalted and Omnipotent God. The king had become a very old man, weakened by age, illness and decrepitude, as he had lived for a hundred and eighty years. He had no children, sons or daughters, and this was something that caused him distress night and day.

One day, he happened to be sitting on his royal throne with the emirs, viziers, officers and state officials present as usual to serve him, each in their proper position according to their rank. Whenever an emir came in accompanied by one son or two, the king would feel envious of him and say to himself: 'Every one of these men is happy and glad of his children, but I have no son and soon I shall die and have to leave behind my kingdom, my throne, my estates, my treasuries and my wealth. Strangers will take them; no one will remember me or mention my name anywhere in the world.' He plunged into the depths of thought, and such were his cares and sorrows that he burst into tears and came down from his throne to sit weeping in humiliation on the ground. When the vizier and the other state leaders present saw what he was doing, they called out to the people, telling them to go back and stay quietly at home until the king had recovered. When they had gone, no one was left there apart from the king and the vizier.

After the king had regained control of himself, the vizier kissed the ground in front of him and asked him why he had wept. 'Who,' he asked, 'of the kings, lords of castles, emirs or ministers of state has acted against you? If you tell me who has thwarted you, we shall join forces against him and take his life.' The king said nothing and did not raise his head, and so the vizier again kissed the ground before him and said: 'King of the age, I am your son and your slave and it is you who nurtured me. If I don't know the cause of your present grief and distress, how can anyone else discover it and take my place in front of you? So tell me why it is that you are weeping.' Still the king said nothing, neither opening his mouth nor raising his head. He continued to shed tears, crying out aloud, lamenting even more bitterly and moaning. The vizier waited patiently for him to stop but then said: 'If you don't tell me the reason for this, then, rather than see you so distressed, I shall immediately kill myself before your eyes.' At that point King 'Asim did lift his head, and after he had wiped away his tears, he said: 'My good counsellor, leave me to my grief, for the sorrows in my heart are enough for me.' The vizier insisted: 'If you tell me the reason for your tears, it may be that God will let me rescue you from your affliction.'

Night 759

Morning now dawned and Shahrazad broke off from what she had been allowed to say. Then, when it was the seven hundred and fifty-ninth night, SHE CONTINUED:

I have heard, O fortunate king, that the vizier asked King 'Asim to tell him why he was weeping as it might be that God would allow him to rescue him. 'Vizier,' the king said, 'it is not because of money or horses or anything else of the kind that I weep, but I am an old man and have lived for almost a hundred and eighty years without having any children, male or female. When I am dead and buried, no trace of me will remain; my name will be lost; strangers will take my throne and my kingdom and no one will remember me any more.' 'King of the age,' replied the vizier, 'I am a hundred years older than you, and I too have never had any children. This causes me pain and distress night and day, but what can we do, you and I? I have, however, heard accounts of Solomon, the son of David, on both of whom be peace, and that he has a mighty Lord Who is able to do all things. I think that I should go to him with a present to see if he would ask his Lord to provide each of us with a son.' He then made his preparations for the journey and set off to bring Solomon a splendid gift.

So much for him, but as for Solomon himself, the Glorious and Almighty God sent him a divine revelation to tell him: 'Solomon, the king of Egypt has sent you his grand vizier with such-and-such precious gifts. Send out your own vizier, Asaf, son of Barkhiya, to give him an honourable reception and to provide provisions for him where he camps. When he appears before you, tell him: "The king has sent you with such-and-such a request and you yourself have such-and-such a need." After that, expound the true faith to him.' So Solomon instructed Asaf to take an escort with him as a sign of respect to the visitors on their arrival and to have splendid provisions ready for them where they camped.

Asaf went to meet the Egyptian vizier, Faris, whom he greeted, receiving both him and his escort with the greatest courtesy as well as supplying them with food and fodder for their camps. 'Welcome to the guests who have arrived,' he said, adding: 'You will be happy to hear that what you need will be granted you, and so you can gladly relax.' 'Who told them about this?' said Faris to himself, and he then put the same question to Asaf. 'It was Solomon, the son of David, on both of whom be peace, who told us,' Asaf replied. 'But who told our master Solomon?' asked Faris, and Asaf said: 'He was told by the Lord of heaven and earth, the God of all creation.' 'This must be a mighty god indeed!' exclaimed Faris. 'Do you not worship him?' Asaf asked, and when Faris told him that they worshipped the sun, he said: 'Vizier Faris, the sun is a star, one of those which the Glorious and Almighty God created. It could not possibly be a god itself, as sometimes it can be seen and at other times it is hidden, whereas our Lord is always present and never absent, and He has power over all things.'

After a short journey the visitors came to the land of Saba and approached Solomon's throne. On Solomon's instructions his armies of men, *jinn* and others were drawn up in ranks along their route, and in these ranks were all the sea beasts together with elephants, leopards and lynxes, with the various classes of each species having their separate places. The same was true of the *jinn*, all of whom appeared without concealment in their various terrifying forms. As the two lines stood there, the birds spread their wings over them to shade them, singing to each other, every one with his own tuneful call. When the Egyptians arrived, they were too awestruck to dare advance, but Asaf said: 'Walk between them without fear, for they are Solomon's subjects and not one of them will harm you.' He himself led the way and everyone else followed, including Faris's escort, frightened as they were.

They went on until they came to the city, where they were lodged in the guest quarters and treated with the greatest honour, with splendid guest provisions being supplied to them over a period of three days. They were then brought before Solomon, the prophet of God, upon whom be peace, and when they had entered and were about to kiss the ground in front of him, he stopped them and said: 'No man should prostrate himself to any but the Great and Glorious God, the Creator of earth, heaven and every other thing. Any of you who wish to stand may do so, but not as a sign that you are at my service.' They obeyed and, while the vizier Faris sat down, as did a number of his servants, some of their juniors remained on their feet deferentially. When the rest were comfortably seated, tables of food were produced and everyone ate their fill, after which Solomon invited Faris to say what it was that he wanted. 'Speak and hide nothing about why you are here,' he said, 'since you have come for a specific purpose,' adding, 'and I will tell you what it is. 'Asim, the king of Egypt, has become old, decrepit and weak, but Almighty God has not provided him with either a son or a daughter. This has caused him grief and distress night and day, and one day when he was seated on his royal throne he noticed that, as the emirs, viziers and the principal officers of state came in to present their services to him, some had with them one son, others two and others three. He thought about this and exclaimed in his grief: "Who will take over my kingdom after my death? Will it be a stranger and will I be forgotten as if I had never existed at all?" This plunged him into so deep a sea of care and he was so distressed that his eyes flooded with tears and, covering his face with his kerchief, he wept bitterly. He got up from his throne and sat on the ground weeping and wailing, while only God Almighty knew what was in his heart.'

Night 760

Morning now dawned and Shahrazad broke off from what she had been allowed to say. Then, when it was the seven hundred and sixtieth night, SHE CONTINUED:

I have heard, O fortunate king, that the prophet of God, Solomon, son of David, on both of whom be peace, told the vizier Faris of how the king had wept in sorrow and of all that had passed between them from beginning to end. 'Is what I have told you true, vizier?' Solomon asked. 'Prophet of God,' replied Faris, 'it is entirely true, but when the king and I were talking about this there was no one else there and no one knew about it, so who told you all this?' Solomon said: 'I was told by my Lord, Who knows what is meant by the surreptitious glance and Who is familiar with the secrets hidden in men's breasts.' Faris said: 'Prophet of God, this is a generous and exalted Lord, with power over all things,' after which he and his escort accepted Islam.

Solomon then told him: 'You have brought such-and-such with you in the way of gifts and presents.' Faris agreed, and Solomon said: 'I accept all of them from you, but I then make them over to you. So you and your companions must rest in your present lodgings until you have recovered from the fatigues of your journey, and then tomorrow you will fulfil and complete your mission through the will of Almighty God, the Lord of earth and heaven, the Creator of all things.' Faris went back to his lodgings, and when he returned to Solomon the next day, he was told: 'When you go back and rejoin King 'Asim, you must go out together to climb a certain tree and sit there quietly. Between the midday and the afternoon prayers, when the noon heat has worn off, go down to the foot of the tree and there you will find two snakes emerging, one with the head of an ape and the other with the head of an 'ifrit. Shoot arrows at them and when you have killed them, take their bodies and cut off one span from the head downwards and another from the tail upwards and throw these bits away. Then take what is left of their flesh, cook it thoroughly and give it to your wives to eat. Sleep with them that same night, and with the permission of Almighty God they will conceive male children.' Solomon next brought out a seal ring, a sword and a package

containing two cloaks set with gems and told Faris that when the two
boys had grown to man's estate they were each to be given one of these
cloaks. Then he said to him: 'In the Name of God: the Almighty has
granted your request and all that remains for you is to set off with His
blessing, for your king is spending day and night waiting for you with
his eyes fixed on the road.'

Faris went up to Solomon, took his leave of him and left, after having
kissed his hands. He travelled throughout the rest of that day, happy at
having fulfilled his mission, and after that he pressed on, night and day,
until he got near Cairo. He sent one of his servants to bring the news to
the king, who, when he heard of his vizier's return and of the success of
his quest, was overwhelmed with joy, a joy shared by his close associates,
his ministers of state and all his troops. They were particularly delighted
by the vizier's safe return. For his part, when he met the king, Faris
dismounted, kissed the ground in front of him, gave him the good news
of his complete success and expounded to him the faith of Islam. The
king accepted this and then said: 'Go home to rest tonight and for
another week. After that, go to the baths and then come back to me, so
that I can tell you what plans we have to make.' The vizier kissed the
ground and left for his house with his retinue, his pages and his servants.

He rested for eight days, after which he set off back to the king and
told him the whole story of his encounter with Solomon. 'You must
come alone with me,' he said, and so the two of them took two bows
and two arrows, climbed the tree and sat there quietly until it was past
noon. They stayed where they were until it was almost time for the
afternoon prayer before climbing down and then, when they looked,
they could see two snakes emerging from the foot of the tree. Staring at
them, the king was struck by the fact that they were wearing collars of
gold, and he felt a liking for them. 'By God, this is a marvel!' he exclaimed. 'Let us catch these collared snakes and put them in a cage so
that we may enjoy looking at them.' But the vizier said: 'These two
creatures were created by God for the benefit that they can bring. So you
must shoot one and I shall shoot the other.' Both of them then shot and
killed the two, after which they cut off one span from their heads
downwards and another from their tails upwards and threw the pieces
away. The rest they took back to the royal palace, where they summoned
the cook and told him to cook the flesh well with onion sauce and spices,
put it in two bowls and then bring these to them without delay at a time
they specified.

Night 761

Morning now dawned and Shahrazad broke off from what she had been allowed to say. Then, when it was the seven hundred and sixty-first night, SHE CONTINUED:

I have heard, O fortunate king, that when the king and the vizier had given the snakes' flesh to the cook, they told him to cook it, put it into two bowls and bring these to them without delay. The cook took the flesh to the kitchen and cooked it expertly, adding large quantities of onion sauce, after which he ladled it out into two bowls which he brought to the king and the vizier. They took one each and gave it to their wives, with whom they then spent the night, and through the will and power of the Glorious and Almighty God both women conceived immediately. The king spent the next three months in a state of confusion, saying to himself: 'I wonder whether this can be true or not.' Then one day, while his wife was sitting, the child moved in her womb and she realized that she must be pregnant. She felt a pain; her colour changed and, sending for her chief eunuch, she told him to go to the king, wherever he might be, and to give him the good news that the signs of her pregnancy were now clear and that the child had moved in her womb. The eunuch hurried off joyfully and found the king alone, with his hand to his cheek, brooding over the matter. He went up to him, kissed the ground in front of him and told him that his wife was pregnant. On hearing this, the king leapt to his feet, and such was his delight that he kissed the hand and head of the eunuch and stripped off his own robes to present them to him. He then told everyone present: 'Let whoever loves me make a present to this man,' and what they then gave the eunuch in the way of money, jewels of all sorts, horses and mules, as well as orchards, was more than could be counted.

At that point the vizier came in and said: 'King of the age, just now I was sitting alone at home, preoccupied and worried about our problem, saying to myself: "I wonder if this is true and if my wife is pregnant or not." Just then, in came a eunuch to tell me that she was and that the child had moved in her womb and her colour had changed. I was so glad that I stripped off all the robes that I was wearing and presented them to him, together with a thousand dinars, and I appointed him chief eunuch.' 'Vizier,' said the king, 'the Blessed and Exalted God has shown us favour through His grace and goodness, treating us generously,

granting our requests and revealing to us the true religion. In His bounty
He has removed us from darkness to light, and as a result I want to bring
relief and joy to my people.' 'Do as you want,' the vizier replied, and the
king then told him to go immediately to free all the criminals and debtors
who were being held in prison, while adding that anyone who later
committed a crime would be punished as he deserved. There was to be
a three-year remission of taxes, and kitchens were to be set up within
the circuit of the city walls where cooks were to hang up various types
of cooking pots and to cook food of all kinds night and day. The citizens
and those from surrounding districts, far and near, were to be allowed
to eat and drink there and to take food back home. A festival was to be
held, and the city adorned with decorations for seven days, and shops
were not to close by night or by day.

The vizier left immediately and carried out the king's orders. The
city, the citadel and the towers were adorned with the most splendid
decorations; everyone put on their finest clothes and they ate, drank,
played and enjoyed themselves until the night came when, after the
completion of the term of her pregnancy, the queen gave birth. The
king gave orders that all scholars, astronomers, men of culture, leading
savants and astrologers who were in the city, together with the men of
learning and experts in divination, were to be assembled. They sat waiting for a glass ball to be thrown at the window, which was to serve as a
sign for the astrologers and the other senior figures gathered there. They
all stayed there expectantly until the queen gave birth to a boy like
the disc of the moon when it becomes full. They instantly began their
calculations, noting the star under which the child was born and the
time of his birth and working out dates. On being summoned, they all
kissed the ground and gave the king the good news that the child would
be blessed and fortunate, but they added: 'At the start of his life he will
encounter a danger which we are afraid to mention to your majesty.' He
assured them that they could speak without fear, and they said: 'Your
majesty, this child will leave this country and travel abroad; he will be
shipwrecked and fall into difficulties; he will be captured and find himself
in trouble, as there are many hardships for him to face. But he will win
free, reach his goal and spend the rest of his life at ease, ruling over
peoples and lands and exercising power in spite of his jealous enemies.'
'There is obscurity here,' said the king, 'but whatever good or evil
Almighty God has allotted to His servants in the book of fate will come
about, and from one day to the next he is certain to experience many

joys.' So he paid no attention to what they said, and after they and everyone else there had been given robes of honour, they all left.

At that point the vizier came in, full of joy, and kissed the ground in front of the king. 'Good news, your majesty,' he said, 'for my wife has just given birth to a boy like a sliver of the moon.' 'Bring him here,' said the king, 'so that the two children can be brought up together in my palace, and your wife can stay with mine so that they may both look after them.' The vizier fetched his wife and his son, and the two children were entrusted to nurses, both dry and wet. Seven days later they were brought before the king, and his attendants asked what names he was going to give them. 'Name them yourselves,' he said, but they objected that this was the father's duty, at which the king said: 'Name my son Saif al-Muluk after my grandfather, and name the son of the vizier Sa'id.' He then presented the nurses with robes of honour and instructed them to treat the children with tenderness and to bring them up as well as possible.

The nurses did their best until, when the children were five years old, they were transferred to the charge of a school instructor who taught them the Quran, as well as the art of writing. When they were ten, they were provided with other teachers to learn how to ride, shoot, use the lance and play polo, as well as to master the arts of horsemanship. By the time they were fifteen, they had become expert in all of these skills and had no match as riders, both of them being accustomed to take the field successfully alone against a thousand.

After they had reached manhood, the king used to be filled with delight whenever he saw them, and then, when they were twenty, he summoned the vizier to a private audience and said: 'Vizier, I have thought of something that I wish to do and I want to consult you about it.' 'Whatever it is, do it,' said the vizier, 'because your ideas are always fortunate.' The king explained: 'I have become a frail old man, stricken in years. I want to retire to a small mosque in order to worship Almighty God, and I want to hand over the governance of my kingdom to my son, Saif al-Muluk. He has become an excellent young man, a fine horseman and a man of intelligence and culture, as well as being modest and having a capacity for leadership. What do you think of this idea?' 'It is an excellent one,' the vizier replied, 'and one which will bring blessings and good fortune. If you carry it through, I for my part will do the same thing, and my son, Sa'id, can be his vizier, for he too is a fine young man with a store of learning and good counsel and the two of them will be together. We shall not neglect them but can help arrange things for them and guide them on the right way.' 'Write letters,' the king instructed him, 'and have the couriers take them to every region, land, fortress and castle in my realm, telling their lords to come to the Elephant Field in such-and-such a month.'

The vizier left immediately and wrote to all the governors, castle lords and the other subjects of the king, ordering them to assemble at the specified time, and he also summoned all the citizens, wherever they lived, to come too. When it was close to the time of the meeting, he ordered servants to erect pavilions in the centre of the field with the most splendid decorations, and to set up the great throne, which was only used by the king on feast days. All these instructions were carried out immediately, and when the throne had been put in place, the officers, chamberlains and emirs filed out, followed by the king himself. He had a proclamation made to the people ordering them in the Name of God to come to the field. Emirs, viziers, lords of provinces and estates all went there, presenting their services to the king in the usual way and taking their places according to their rank, some sitting and some standing, until they had all arrived. Tables were set out on the king's orders and they ate, drank and called down blessings on him. He then told his chamberlains to proclaim that no one was to leave and so they called out: 'No one is to go until he has listened to the king's speech!' The curtains around the throne were raised and the king said: 'Let whoever loves me stay and hear what I have to say.' They had started by being alarmed, but now everyone sat quietly as the king got up and made them swear not to leave their places.

He then began: 'Emirs, viziers and ministers of state, great and small, and all you who are here, do you know that I inherited this kingdom from my fathers and forefathers?' 'Yes, your majesty,' they said, 'we all know that.' He went on: 'You and I used always to worship the sun and the moon until Almighty God supplied us with the true faith, rescuing us from darkness, bringing us into the light and leading us, Glorious and Exalted as He is, to the religion of Islam. As you know, I am now an old man, decrepit and feeble, and I want to sit in a mosque worshipping Almighty God and asking forgiveness for my past sins while my son, Saif al-Muluk, acts as ruler. As you know, he is a fine young man, gifted with eloquence and a knowledge of affairs, intelligent, excellent and just. I want immediately to hand over the kingdom to him and to appoint him as king over you in my place, surrendering my power to him. I shall then retire on my own to worship God, while he takes office and rules

over you. What have you all to say to this?' All present rose, kissed
the ground and expressed their agreement, saying: 'Our king and our
protector, were you to appoint one of your slaves to rule us, we would
follow him obediently, so how much more readily will we obey your
son, Saif al-Muluk. We accept him gladly and willingly.'

At that, the king came down from his place and installed his son on
the royal throne. He took the crown from his own head and put it on
the head of his son, and he fastened the royal belt around his waist.
He then took his seat on the throne alongside him as the emirs, viziers
and principal officers of state rose and kissed the ground before him. As
they stood there, they said to one another: 'Saif al-Muluk is worthy of
the kingdom and has a better right to it than anyone else.' They called
out their prayers for his safety and wished him victory and good fortune,
while for his part Saif al-Muluk scattered gold and silver over the heads
of them all . . .

Night 762

Morning now dawned and Shahrazad broke off from what she had been
allowed to say. Then, when it was the seven hundred and sixty-second
night, SHE CONTINUED:

I have heard, O fortunate king, that King 'Asim installed his son, Saif
al-Muluk, on his throne and all the people prayed that he be given
victory and good fortune. For his part, Saif al-Muluk scattered gold and
silver over the heads of them all, as well as distributing robes of honour
and other gifts.

After a short pause, the vizier Faris got to his feet, kissed the ground
and said: 'Emirs and ministers of state, do you know that I am a vizier
of long standing who held the office before King 'Asim came to the
throne, and that he has now deposed himself and appointed his son in
his place?' 'We know that you inherited the vizierate from your ancestors,' they agreed, and he went on: 'I now depose myself and put my
son, Sa'id, in my place, for he is intelligent, shrewd and well informed.
What do you all say?' They replied: 'Sa'id, your son, is the only suitable
vizier for King Saif al-Muluk, for they are well matched.' At that, Faris
got up, removed his vizier's turban and placed it on his son's head, as
well as putting in front of him the official inkstand. 'He deserves the
vizierate,' agreed the chamberlains and the emirs.

King 'Asim and Faris, his vizier, then threw open the treasuries and distributed splendid robes to kings, emirs, viziers, the great officers of state and to the people as a whole. They provided spending money as well as additional grants, and had new charters and diplomas written and sealed with the seals of Saif al-Muluk and Sa'id, the vizier. Everyone stayed in the city for a week, after which they dispersed to their own parts. King 'Asim now took his son and Sa'id back into the city, and when they reached the palace, the treasurer was ordered to produce the ring, the sword, Solomon's package and the signet. From among these, 'Asim told each of the young men to take whichever he chose. Saif al-Muluk was the first to stretch out his hand and he took the package and the ring, while Sa'id, who followed him, took the sword and the signet. They then kissed 'Asim's hands and returned to their quarters.

When Saif took the package, he did not open it to see what was in it but instead threw it down on the couch which he shared with Sa'id, for they were in the habit of sleeping together. The two of them spread out the bedclothes and lay down together in the candlelit room. When midnight came Saif woke up and, on catching sight of the package beside his head, he said to himself: 'I wonder what is in this gift that my father has given me.' So he took it and picked up a candle, before getting down from the couch, where he left Sa'id asleep. He went to the treasury, where he opened the package, and in it he found a gown of *jinn* workmanship. When he unfolded this and turned it over, he found on its inner surface at the back the picture of an astonishingly beautiful girl worked in gold. The sight of this robbed him of his wits; he fell madly in love with the picture and, after collapsing unconscious on the ground, he began to weep and sob, striking his face and breast and kissing it. Then he recited these lines:

Love at first is a drop of water,
Brought by fate,
But when a man plunges into its depths,
It is too vast to be endured.

He added a few more lines:

Had I known that love was like this,
A robber of lives, I would have been on guard,
But I threw myself into it on purpose,
Not knowing what the end would be.

He kept on sobbing, weeping and striking his face and his breast until Sa'id awoke to find that he was alone on the couch and that there was only a single candle. He wondered where Saif could have gone, and so he took the candle and began to look around the whole palace until he came to the treasury, where he found him. Seeing him weeping so bitterly, he said: 'Brother, what is the reason for this? Tell me what has happened to you.' Saif neither spoke nor raised his head but continued to sob and to strike his chest with his hand. Seeing this, Sa'id said: 'I am your vizier and your brother. We were brought up together, and if you don't tell your secret to me, to whom are you going to tell it?' For some time he continued with his entreaties, kissing the ground, but Saif went on weeping, without turning towards him or saying a single word. Alarmed by this and unable to arouse him, Sa'id left him and fetched a sword, which he brought back to the room, setting its point against his own breast. 'Rouse up, brother,' he said, 'for unless you tell me what has happened to you, I shall kill myself rather than see you in such a state.' At that, Saif looked at him and said: 'Brother, I am ashamed to tell you this.' Sa'id replied: 'I implore you by God, the Lord of lords, the Emancipator, the Cause of causes, the One, the Merciful, the Generous, the Giver, answer me. You need feel no shame before me, for I am your slave, your vizier and your advisor in all matters.' Saif then said: 'Look at this picture.' Sa'id did that, and after he had studied it for a long time, he noticed an inscription at the top of it, spelt out through an arrangement of pearls, which read: 'This is the picture of Badi' al-Jamal, daughter of Shammakh, son of Sharukh, a king of the believing *jinn*, inhabitants of the city of Babel who live in the garden of Iram, the son of 'Ad the Great.'

Night 763

Morning now dawned and Shahrazad broke off from what she had been allowed to say. Then, when it was the seven hundred and sixty-third night, SHE CONTINUED:

I have heard, O fortunate king, that King Saif al-Muluk and Sa'id, son of the vizier Faris, read what was written on the gown and studied the picture of Badi' al-Jamal, daughter of Shammakh, son of Sharukh, a king of the believing *jinn*, inhabitants of the city of Babel who live in the garden of Iram, the son of 'Ad the Great. Sa'id asked Saif: 'Brother, do you know whose portrait this is, so that we can look for her?' and when Saif said that he did not, Sa'id said: 'Come here and read this inscription.' Saif went up, but when he had read and understood what was written on the girl's crown, he called out: 'Ah! ah! ah!' from the bottom of his heart. Sa'id said: 'If the subject of this portrait exists, and if her name is Badi' al-Jamal, and if she is to be found in this world, then I shall begin to search for her quickly and without delay, so that you may get what you want, but for God's sake stop this weeping and allow the officers of state to come in to present their services. Then, later in the morning, summon the merchants, the *faqirs*, the travellers and the poor wanderers and ask them about the city of Babel, for it may be that, through the blessing and aid of the Glorious and Almighty God, one of them may be able to tell us how to get to it and to the garden of Iram.'

In the morning, Saif got up and took his seat on the throne, still clasping the gown, for he could neither sit nor stand nor sleep unless he had it with him. The emirs, viziers, soldiers and ministers of state came in, and when all the court was there, each in his proper place, Saif told Sa'id: 'Go out and tell them that the king is suffering from a disorder and last night he was ill.' Sa'id announced this to the people, and when King 'Asim heard it he was concerned and went to see his son with the doctors and astrologers whom he had summoned. When they had looked at him, they prescribed a draught, but the illness continued for three months. King 'Asim was angry with the doctors and said: 'Damn you, you dogs, can none of you cure my son? If you don't do this immediately, I shall kill the lot of you.' Their leader said: 'King of the age, we know that this is your son and you know that we always do our best when it comes to treating strangers, so how much more are we going to exert ourselves in order to cure your son? In his case, however, the disease is hard to treat, but if you want to know about it, we will tell you what it is.' 'What have you discovered about it?' the king asked, and the leading doctor replied: 'King of the age, your son is in love, and there is no way in which he can reach the object of his love.' The king was furious with them and said: 'How did you come to learn that he is in love, and how could this have happened to him?' 'Ask Sa'id, his brother and vizier,' the doctor answered, 'for he knows about your son's condition.'

At this point the king got up and went alone to the treasury, where he summoned Sa'id and ordered him to tell the truth about Saif's illness. When Sa'id protested that he did not know, the king told the executioner to take him, blindfold him and then cut off his head. In fear for his life

Sa'id asked for pardon, and the king said: 'If you tell me the truth, I shall pardon you.' 'Your son is in love,' Sa'id told him, and when the king asked who it was whom he loved, Sa'id replied: 'The daughter of one of the kings of the *jinn*, whose picture he saw on a gown from the package that you were given by Solomon, the prophet of God.' At that the king got up and went to see his son, Saif. 'My son,' he asked, 'what misfortune has overtaken you? What is this picture with which you have fallen in love and why did you not tell me?' 'I was ashamed to do that, father,' he said, 'and I could not bring myself to tell you or anyone else about it, but now that you know about my love, see what you can do to cure me.' The king replied: 'What is there that can be done? If this were a mortal, we could find some way for you to reach her, but as she is a *jinn* princess, the one person who can do anything about her is Solomon, the son of David, who alone has this power. You, however, my son, should get up immediately, take heart and mount your horse. Then go out to hunt or to exercise yourself; concern yourself with eating and drinking; banish all cares from your heart and I shall bring you a hundred princesses, for you have no need of the daughters of the *jinn*, over whom we have no power and who are not of our race.' 'I am not going to abandon her or to look for anyone else,' said Saif. When his father then asked him what was to be done, he said: 'Summon all the merchants, the travellers and those who have visited foreign lands, and question them, for it may be that God will guide us to the garden of Iram and the city of Babel.'

The king gave orders that every merchant in the city, every foreigner and every sea captain was to be brought to him. When they had come, he asked them about these places, but none of them knew anything about them or could tell him anything; however, as they were dispersing, one of them said: 'King of the age, if you want to know about this, you should go to the lands of China, where there is a huge city, and it may be that someone there will be able to guide you.' At that, Saif asked his father to fit out a ship in which he could sail to China, but his father said: 'My son, won't you take your seat on your royal throne and govern your people, while I go to China and look after the affair myself?' 'This is my responsibility,' Saif replied, 'and no one else can conduct the search like me. Whatever happens, if you give me leave to go, I shall set out and I may be away for some time. If I find any news of the princess, I shall have got what I want, but even if I fail, the journey itself will relieve me, allow me to recover my spirits and stop me from taking the affair too seriously. Then, if I live, I shall come back to you safely.'

Night 764

Morning now dawned and Shahrazad broke off from what she had been allowed to say. Then, when it was the seven hundred and sixty-fourth night, SHE CONTINUED:

I have heard, O fortunate king, that Saif al-Muluk told his father, King 'Asim, to prepare a ship to take him to China so that he might pursue his quest. 'If I live,' he said, 'I shall come back to you safely.' Looking at his son, the king could see that there was nothing for it except to do what he wanted. He gave him his permission to set out and he prepared a fleet of forty ships for him, together with twenty thousand mamluks, not counting servants, providing him with money and treasures as well as all the military equipment that he might need. He prayed for his son's success, health and safety, saying: 'I entrust you to One with Whom everything deposited is safe.' He and Saif's mother then said goodbye to their son, and after the ships had been loaded with water, provisions, arms and troops, they set out and continued on their way until they reached the capital of China.

When the Chinese heard of the arrival of forty ships carrying men, equipment, weapons and treasure, they thought that these must be enemies who had come to fight them and lay siege to their city, and so they shut the gates and prepared their mangonels. On hearing of that, Saif sent out two of his leading mamluks with orders to go to the king of China with the message: 'Saif al-Muluk, the son of King 'Asim, has come to your city as a guest to enjoy the sights of your country for a time, with no intention of fighting against you or of acting in any hostile manner. If you are willing to accept him, he will land, but if not, he will sail back without disturbing you or the people of your city.' When the mamluks arrived at the city and told the people that they were envoys from King Saif al-Muluk, the gates were opened for them and they were escorted to the king.

The king's name was Faghfur Shah and years ago he had been acquainted with King 'Asim. When he heard that his visitor was Saif al-Muluk, 'Asim's son, he presented the messengers with robes of honour and ordered the gates to be opened. Then he had provisions prepared for his guests and went out himself with his leading officials. When he met Saif, the two of them embraced and Faghfur spoke words of welcome, adding: 'I am your servant and the servant of your father; my city is at your disposal and whatever you want will be provided for you.' He sent guest provisions and food to the camps, while Saif himself, together with his vizier Sa'id, his officers of state and all the rest of his troops, moved from the seashore to the city. Drums were beaten to announce the good news, and for forty days they stayed there being generously entertained.

At the end of this time, Faghfur said to Saif: 'How are you, son of my brother, and do you like my city?' 'May God continue to ennoble it by your presence, O king,' replied Saif, after which Faghfur went on: 'You must have come here because of some unexpected need, and whatever you want from my country, I shall provide for you.' 'Mine is a strange story, your majesty,' Saif answered, 'for I have fallen in love with a picture of Badi' al-Jamal.' Faghfur shed tears of pity and asked what it was that he now wanted. Saif told him: 'I would like you to collect all travellers and those who make frequent journeys so that I may ask them about the princess, as one of them may be able to tell me something about her.' So Faghfur sent word to his deputies, chamberlains and assistants, telling them to bring all the travellers who were in their lands. They did what they were told, and a large group of these people was brought together in front of the king. Saif then put his question about the city of Babel and the garden of Iram, but no one there could give him an answer.

He was at a loss to know what to do, but at that point one of the sea captains said: 'Your majesty, if you want to find out about this city and the garden, then you should go to the Indian islands.' So Saif gave orders for his ships to be brought up, and when this had been done, water, food and everything else that was needed was loaded on them. Saif and Sa'id took their leave of Faghfur, and after going on board, they put out to sea. With fair winds they sailed safely and without disturbance for four months, but then one day the wind turned against them, waves came from all directions, rain fell and because of the force of the gale the sea became stormy, causing the ships to dash against one another. They and the smaller boats with them were all wrecked and everyone was drowned with the exception of Saif and a number of his mamluks who were in a little boat. Through the power of Almighty God the wind then died down and the sun came out, but when Saif opened his eyes there was no ship to be seen and nothing else except sky and sea, while he and his companions were alone in the boat. 'Where are the ships and the little boats, and where is Sa'id, my brother?' he asked them. 'King of the age,' they told him, 'there are no ships left, great or small, and none of their crews, for these have all drowned and become food for the fishes.' Saif gave a cry and recited the words that never bring confusion on those who speak them: 'There is no might and no power except with God, the Exalted, the Omnipotent.' He began to slap his face and was on the point of throwing himself into the sea when he was restrained by his mamluks. 'What is the use of that, your majesty?' they asked him. 'You have brought this on yourself, and had you listened to what your father said, it would not have happened. But this has been written in the book of fate from past eternity, according to the will of the Creator of life ...'

Night 765

Morning now dawned and Shahrazad broke off from what she had been allowed to say. Then, when it was the seven hundred and sixty-fifth night, SHE CONTINUED:

I have heard, O fortunate king, that when Saif al-Muluk was about to throw himself into the sea, his mamluks restrained him and said: 'What good will this do? You have done this to yourself, but it was written in the book of fate from past eternity, according to the will of the Creator of life, so that His servants may fulfil their destiny. At your birth the astrologers warned your father that you would have to suffer all these difficulties and there is nothing that we can do except to endure until God releases us from our present distress.' Saif repeated the words: 'There is no might and no power except with God, the Exalted, the Omnipotent' and agreed that there was no way to escape from what God had decreed. He then recited these lines:

By God, I am at a loss, and do not know what to do;
Misgivings assail me, but I do not know their source.
I shall show patience, so that everyone may know
That I endure what is more bitter than aloes,
For in this matter there is nothing I can do.
I entrust my affairs to the Lord God, Who commands.

He was then plunged into a sea of cares and tears showered down his cheeks. He slept for part of the day and then, when he woke up, he asked for something to eat. After he had eaten his fill, the mamluks took away the rest of the food from in front of him, and the boat carried them on, although they did not know in which direction it was heading. For a long time, wind and wave swept them along night and day until their provisions were all exhausted and they had lost all touch with reality and were reduced to extremities because of the pressures of hunger, thirst and distress from which they were suffering. Just then, however, in the distance they caught sight of an island. The wind blew them towards it, and when they had reached the shore and anchored, they disembarked, leaving one of their number as a guard. When they set off to investigate, they discovered quantities of fruits of all kinds from which they ate until they had had enough. Among the trees they then found an odd-looking person seated, with a long face and a white beard and white body, who called to one of the mamluks by name, saying: 'Don't eat those fruits; they are not ripe. Come here to me and I'll give you ripe ones.' The mamluk looked at him and thought that he must be one of his companions who were feared drowned, and that he must have reached the island. He was delighted to see him there and went up to him, without knowing what fate had in store for him and of the destiny written on his forehead. When he came near, the 'man' jumped on him and turned out to be a *marid*, who crooked one of his legs around his neck and let the other hang down over his back. 'Move,' the *marid* told him. 'You can never escape from me, and you will stay here as my donkey.' The mamluk called tearfully to his companions, telling them to save themselves and flee away from the wood. 'One of its people has mounted on my shoulders,' he told them, 'and the others will hunt you down in order to do the same to you.'

When they heard this, they all ran back and embarked on their boat. The *marids* followed them into the sea and called out: 'Where are you going? Come and sit with us so that we can climb up on your shoulders. We will give you food and drink and you will stay as our donkeys.' When they heard this they sailed off as fast as they could until they were some distance away, after which they set a course, entrusting themselves to Almighty God. A month later they came in sight of another island, and when they reached it they found that it too had fruits of various kinds. They busied themselves with eating them and, while they were doing this, they caught sight of something at a distance in the direction that they were going. When they looked from closer at hand, they saw an ugly thing lying there like a column of silver. One of the mamluks kicked it, and it turned out to be a creature with eyes set lengthways and a split head, which had been covered by one of its ears, as, when it wanted to sleep, it would put one ear under its head and shelter itself with the other. The creature snatched up the mamluk who had kicked it and made off with him to the centre of the island, which turned out to be full of cannibal *ghuls*. The mamluk shouted to his companions: 'Save yourselves! This is the island of cannibal *ghuls* and they are going to cut me up and eat me.' On hearing that, the others fled away and embarked on their boat without having collected any of the fruit.

They sailed on for some days before they came in sight of yet another island, and when they reached it they found that it contained a lofty mountain. As they climbed up its slopes, they saw a thick wood and in their hunger they had started to eat its fruits when, before they knew what was happening, a number of terrible-looking creatures came out from among the trees, each of them being fifty cubits high, with projecting tusks like those of elephants. They took Saif and his companions to stand before their king, who was seated on a piece of black felt that had been spread over a rock, and around him were a vast number of blacks, who were standing there in his service. Their captors said: 'We found these birds among the trees,' and, as the king was hungry, he took two of the mamluks, cut their throats and ate them.

Night 766

Morning now dawned and Shahrazad broke off from what she had been allowed to say. Then, when it was the seven hundred and sixty-sixth night, SHE CONTINUED:

I have heard, O fortunate king, that when the blacks captured King Saif and his companions, they brought them to stand before their king, telling him that they had come across these birds among the trees. The king took two of them, cut their throats and ate them. When Saif saw this, he shed tears, fearing for his life, and he then recited these lines:

Disasters have formed a bond of friendship
With my heart's blood, having shunned it earlier,
And the generous man is friendly.
My cares are not all of one kind;
Praise be to God, they come in thousands.

He heaved a sigh and then recited more lines:

Time has afflicted me with disasters;
My heart is covered over by its arrows.
When any one of them strikes me,
Its tip breaks on another that is already there.

When the king heard his tearful lament, he said: 'These birds sing pleasantly and I like their voices, so put them all in a cage.' This was done and the cage was hung above his head so that he could listen to them. Saif and his mamluks stayed there, with food and water being given to them by the blacks, alternately weeping and laughing, talking and staying silent, and the king of the blacks took pleasure in the sounds that they made. Things went on like this for a long time until the king's married daughter, who lived on another island, heard that her father had some sweet-voiced birds. She sent a messenger to ask him for some of them, and he sent the man back with four cages containing Saif himself and three of his mamluks; and when they reached her, she was so pleased with what she saw that she ordered the cages to be hung up above her head.

Saif began to wonder at what had happened to him, contrasting it with his previous grandeur. He wept over his own plight while the three mamluks wept for theirs, and all the time the princess thought that they were singing. It was a habit of hers to show great favour to anyone whom she got hold of from Egypt or some other country, if he pleased her. In accordance with God's decree, when she saw Saif she admired his beauty and grace together with his well-formed figure, and so she ordered that he and his men be treated with respect. One day when she happened to be alone with him, she asked him to lie with her, but he refused and said: 'My lady, I am a stranger here, suffering from the pangs of love, and it is only union with my beloved that can content me.' She tried to seduce him with soft words, but he resisted her and she could find no way at all of getting close to him. Her failure made her angry both with him and with his mamluks, and on her orders they were forced to act as her servants, carrying water and firewood for her. After four years of this, Saif was reduced to despair and he sent a pleading message to the princess, begging her to set him and his companions free so that they might rest from their toil and go on their way. She summoned him and promised that if he agreed to what she wanted, she would free him so that he could return with safety and profit to his own country.

She went on pleading with him and trying to sway him, but when he refused to comply, she turned from him in anger.

As a result, he and his mamluks stayed on the island as her servants, and since the islanders recognized that they were her 'birds', no one there dared do them any harm. She herself had no worries about them, being sure that they could not get away, and they used to leave her for two or three days at a time as they went around collecting firewood from all parts of the island and bringing it to her kitchen. After five years, they were all sitting by the shore one day talking over what had happened to them. Saif, seeing the plight they were in and remembering his mother and father, as well as his brother Sa'id, together with his former grandeur, shed bitter tears and was joined in his lamentations by his mamluks. But they then said: 'King of the age, how long are we going to go on weeping when tears do no good? This is something written by fate on our foreheads in accordance with the decree of the Great and Glorious God. It has been foreordained, and all that we can usefully do is to show patience in the hope that God may free us from this affliction which He has sent us.' Saif asked: 'What can we do to rescue ourselves from this damned woman? I think we will only escape if God saves us by His decree, but it has struck me that we might try to run away so as to free ourselves from this toil.' They said: 'King of the age, where can we go to from this island, for all these parts are full of cannibal *ghuls* and wherever we go they will find us and either eat us or return us here as prisoners to face the anger of the princess?' 'I shall do something for you in the hope that Almighty God may help us escape,' he told them, and when they asked what it was that he proposed, he explained: 'We shall cut some of these tall trees, twist the bark into ropes and use them to bind the trees together so as to form a raft which we can then launch. We can fill it with fruits, make oars for it and then embark in the hope that Almighty God may allow us a release from our misery, for He is Omnipotent. Were He to grant us a fair wind we might reach India and find ourselves free of this damned woman.'

The mamluks delightedly agreed that this was a good plan, and there and then they began to cut logs for the raft and to twist ropes to bind them together. They worked on this for a month, taking back firewood for the princess's kitchen each evening and devoting the rest of the day to the raft.

Night 767

Morning now dawned and Shahrazad broke off from what she had been allowed to say. Then, when it was the seven hundred and sixty-seventh night, SHE CONTINUED:

I have heard, O fortunate king, that Saif and his mamluks cut logs on the island, twisted ropes and lashed together the raft that they had made. When they had finished it, they launched it and, after it had been loaded with fruit from the island's trees, they got ready to sail one evening without having told anyone what they had done. They put out to sea and sailed for four months without knowing where they were going. By this time their provisions had been exhausted and they were tormented by hunger and thirst. The sea then became turbulent with high breakers and they found themselves confronted by a terrible crocodile, which reached out and seized one of the mamluks, whom it then swallowed. When Saif saw this, he burst into tears and he and the surviving mamluk, now alone on the raft, sailed away fearfully from where the crocodile had appeared. They continued on their course until one day, to their delight, they came in sight of an enormous mountain towering into the sky. They were able to make out an island, and they pressed on towards it, happy at the thought of coming ashore. Just then, however, the sea became disturbed and with the disturbance the waves rose and a crocodile raised its head and stretched out to grasp the remaining mamluk, whom it then swallowed.

Saif reached the island alone and after climbing up the mountain he caught sight of a forest, which he entered. He walked among the trees, eating their fruits, but then on the treetops he was terrified to see more than twenty great apes, each bigger than a mule. They came down and surrounded him on all sides, before walking on ahead of him and gesturing to him to follow them. On they went, with Saif behind them, until they arrived at a lofty castle with high buttresses, which they entered. Saif went in after them and was confronted by the sight of rare treasures, jewels and precious stones, such as no tongue could describe. There in the castle was an exceedingly tall young man with no hair on his cheeks. Saif hoped for a friendly reception from this, the only human being in the castle, while the young man, for his part, was astonished to see Saif and asked: 'What is your name? Where have you come from and how did you get here? Tell me your story and don't keep anything back from me.' 'By God,' Saif answered, 'I didn't intend to come here, as this was not where I wanted to go, but all I can do is to move from one place to the next until I reach my goal.' 'And what is your goal?' the young man asked. 'I come from Egypt,' Saif told him. 'My name is Saif al-Muluk and my father is King 'Asim ibn Safwan.' He then gave an account of all that had happened to him from beginning to end, at which the young man rose respectfully and said: 'King of the age, I was in Egypt and I heard that you had set out for China, but this is nowhere near China. This is a strange and remarkable matter.' 'True,' agreed Saif, 'but I then left China for India. We met a gale and in the storm my ships were wrecked.' After telling his story, he ended by saying: 'And now I have reached you here.'

The young man said: 'Prince, you have suffered enough hardships on your journey, and God be thanked Who brought you here. Stay with me so that I may enjoy your company until I die, and you will then be king over the island which extends no one knows how far. The apes are master craftsmen and you will find everything that you want here.' 'Brother,' replied Saif, 'I cannot stay anywhere until I have done what I have to do, even if I have to go round the whole world asking about my goal, in the hope that either God may lead me to it or that I may reach the place where I am fated to die.' The young man then turned and made a sign to an ape, which left for a time before returning with others who had silk aprons tied round their waists. They brought out a table on which they set a hundred dishes of gold and silver filled with foods of all kinds. The apes then stood like servants in front of kings, after which the young man gestured to their chamberlains to sit, which they did, while the ape whose usual duty it was to serve remained standing. They all ate their fill and the cloth was then removed and replaced with bowls and jugs of gold, with which they washed their hands. Something like forty drinking glasses were then brought out, each filled with a different type of drink, and they drank with pleasure, enjoying a pleasant time, while the apes were all dancing and playing as long as the meal continued. Saif was so astonished by this that he forgot about the sufferings he had endured.

Night 768

Morning now dawned and Shahrazad broke off from what she had been allowed to say. Then, when it was the seven hundred and sixty-eighth night, SHE CONTINUED:

I have heard, O fortunate king, that when Saif saw what they were doing and how they were dancing, he was so astonished that he forgot what he had suffered in his exile.

When night fell, candles were lit and placed in candlesticks of gold and silver, after which dishes of fruits, dried and fresh, were brought in, from which Saif and the young man ate what they wanted. Then, when it was time to sleep, bedding was laid down and they slept.

The next morning the young man got up as usual, and, having roused Saif, he told him: 'Put your head out of the window and see what is standing underneath.' When Saif looked, he saw apes filling the wide plain and the whole countryside in numbers known only to Almighty God. Saif asked: 'Why has this enormous crowd collected now?' and the young man told him: 'This is a custom of theirs. All the apes in the island gather together, some after a journey of two or three days, and they come to stand here every Saturday until I wake and put my head out of the window. When they see me, they kiss the ground in front of me and then go off about their business.' He then leaned out of the window and at the sight of him the apes kissed the ground and left.

Saif stayed with the young man for a full month, after which he said goodbye to him and set off. The young man ordered some hundred apes to go with him and they escorted him for seven days until they had reached the furthest point of their islands, where they took their leave of him and went back home. Saif continued alone for four months, traversing mountains, hills, open plains and deserts. He was sometimes hungry and sometimes well fed, as at times he had to live on grasses while at others he could pick fruits from trees. He was beginning to regret what he had done to himself by leaving the young man, and he was thinking of retracing his steps and going back when he caught sight of a dark shape in the distance. 'Is that a town of black stone or something else?' he said to himself, and he made up his mind not to go back until he had found the answer.

When he got near it, he discovered that it was a lofty palace – and it was, in fact, the palace built by Japhet, the son of Noah, which is mentioned by Almighty God in the glorious Quran where He says: 'an abandoned well and a high-built palace'.* Saif sat down by the castle gate, wondering to himself what was inside it and what kings might be living there. 'Who will be able to tell me the truth about this and let me know whether its people are human or *jinn*?' he asked himself. He sat thinking this over for some time without finding anyone who was either going into the palace or coming out of it. So he got up and began to walk on, relying on Almighty God, until he had entered the palace. On his way in he counted seven hallways but still saw nobody, and then on his right he found three doors, while in front of him was another door over which hung a curtain. He came up to this door, raised the curtain with his hand and went in, to find himself in a great room spread with silk carpets at whose upper end was a golden throne on which was seated a girl with a face as radiant as the moon. She was royally dressed like a bride on her wedding night, and beneath the throne were forty spread tables with plates of gold and silver, all filled with costly foods.

When Saif saw the girl he went up to her and, after they had exchanged greetings, she asked him: 'Are you man or *jinn*?' 'I am one of the best of men,' he told her, 'for I am a king and the son of a king.' 'What do you want?' she said, adding: 'Help yourself to this food, and after that you can tell me your story from beginning to end and let me know how you got here.' Saif, who was hungry, sat down and removed the cover from the food, after which he ate his fill from the dishes and then washed his hands. When he had taken his seat beside the girl, she asked him who he was, his name and who had brought him there. 'Mine is a long story,' he said, but she repeated: 'Tell me where you have come from and why, as well as what you want.' He replied: 'It is for you to tell me what you are, what your name is, who brought you here and why you are sitting here alone.' She answered: 'My name is Daulat Khatun; I am the daughter of the king of India and my father lives in the city of Serendib. My father has a large garden whose beauty is unsurpassed anywhere in India or its regions and in it there is a large pool. I went there one day with my slave girls and they and I stripped off our clothes and went into the water, where we played and enjoyed ourselves. Then, before I knew what was happening, something like a cloud swooped down on me and snatched me up from among the others, before flying off with me between heaven and earth. "Do not be afraid or worried, Daulat Khatun," my captor

* Quran 22.44.

kept saying to me, and after a short flight he set me down in this palace. Immediately he changed his shape and turned into a most handsome young man, neatly dressed. "Do you know who I am?" he asked me, and when I said no, he told me: "I am the son of al-Malik al-Azraq, king of the *jinn*. My father lives in the castle of Qalzam and under his command are six hundred thousand *jinn* who can fly through the air and dive down into the sea. I happened to see you as I was passing on my way, and it was because I had fallen in love with you that I came down and snatched you away from your slave girls to bring you to this lofty palace, which is where I live. No one, either human or *jinn*, can come here and it is a hundred-and-twenty-year journey from India, so you can be sure that you will never see the lands of your father and mother again. Stay quietly and contentedly here with me and I shall fetch you whatever you want." Then he embraced me and kissed me.'

Night 769

Morning now dawned and Shahrazad broke off from what she had been allowed to say. Then, when it was the seven hundred and sixty-ninth night, SHE CONTINUED:

I have heard, O fortunate king, that the girl told Saif: 'After the king of the *jinn* had told me this, he embraced me and kissed me, saying: "Sit here and have no fear." He left me for a while and then came back with this table, together with the furnishings and the carpets. Every Tuesday he visits me and stays with me for three days, and then on the fourth afternoon he leaves and does not return until the following Tuesday, when the pattern is repeated. When he comes, he eats and drinks with me, but although he embraces and kisses me, he does nothing more, and I am still a virgin as Almighty God created me. My father's name is Taj al-Muluk, but he can have heard nothing about me or have found any trace of me. This is my story, so now tell me yours.' 'My tale is a long one,' said Saif, 'and I am afraid that if I start to tell it to you, it may go on so long that the *'ifrit* will come back.' 'He left only an hour before you came,' the girl said, 'and he will not return until Tuesday, so relax and tell me all that happened to you from beginning to end.'

'To hear is to obey,' Saif replied, and he began to tell the whole of his story from start to finish. When he came to the mention of Badi' al-Jamal, the girl's eyes brimmed over with tears and she exclaimed: 'I did not think that you would do this, Badi' al-Jamal! Alas for the vicissitudes of time! Don't you remember me and say: "Where has my sister, Daulat Khatun, gone?"' Then she wept even more bitterly, lamenting the fact that Badi' al-Jamal had forgotten her. Saif asked her how the two of them could be sisters since she was human and Badi' al-Jamal was a *jinniya*, and she explained: 'We are foster sisters. The reason for this is that my mother was looking around the garden when she went into labour and it was there that she gave birth to me. Badi' al-Jamal's mother was also in the garden together with her *jinn* servants, and she too went into labour and gave birth to her daughter by its outer edge, where she had alighted. She sent one of her maids to my mother to ask for food and what was needed for the baby's delivery. My mother sent her what she wanted and invited her to come to her. She got up and went to my mother, bringing the baby, whom my mother suckled. The two of them stayed with us in the garden for two months, after which the *jinn* queen set off back to her own country, having given something to my mother and telling her: "Whenever you need me, I will come to you in the middle of this garden."

'Badi' al-Jamal used to come every year with her mother to pass some time with us before going home. If we were all there together as we used to be, I with my mother and you there as well, I would find some way of seeing to it that you got what you wanted, but I am here and they know nothing about me. If they did and if they knew where I was, they would be able to rescue me, but the affair is in the hands of the Glorious and Almighty God, and what can I do?' 'Come away with me,' Saif told her, 'and we can make our escape to wherever He wishes.' 'No, we cannot,' she objected, 'for even if we fled away for the distance of a year's journey, this damned '*ifrit* would reach us within an hour and kill us.' Saif said: 'Then I shall hide somewhere and when he passes by I shall strike him dead with my sword.' 'The only way that you can kill him is by killing his soul,' she told him, and when he asked where his soul was, she replied: 'I have put this question to him many times, but he has never told me. Once he grew angry with me when I pressed him and he said: "How often are you going to go on asking and why do you want me to tell you about my soul?" "Hatim," I said, addressing him by name, "apart from God, I have no one left but you, and as long as I live I shall hold your soul in my embrace. For if I fail to guard it and keep it as my dearest possession, how could I live on after you? But if I know where it is, it shall be mine to protect like my right eye." At that he told me:

"At my birth the astrologers said that I would meet my death at the hands of a mortal prince, and so I took my soul and put it in a sparrow's crop. I shut the sparrow in a casket and put the casket in a box, and the box inside seven other boxes, and the boxes inside seven chests, which I placed in a marble coffer by the shore of this ocean. It is far removed from the lands of men, and no mortal can ever reach it. Now I have told you the secret, but you are not to tell it to anyone else, as it is between you and me."'

Night 770

Morning now dawned and Shahrazad broke off from what she had been allowed to say. Then, when it was the seven hundred and seventieth night, SHE CONTINUED:

I have heard, O fortunate king, that Daulat Khatun told Saif about the soul of the *jinni* who had kidnapped her. She repeated what he had said and that he had told her: 'This is a secret between the two of us.' She continued: 'I said to him: "Whom am I likely to tell? Apart from you I have no visitor to talk to," and then I added: "You have left your soul in a place of the very greatest safety. How could any mortal get there, even if we were to suppose the impossible, and God were to decree something like the astrologers predicted? How could this ever happen?" He replied: "One of them might have on his finger the ring of Solomon, son of David, on both of whom be peace. If he came here and put his hand with the ring on the surface of the water and then said: 'I conjure so-and-so's soul to come out,' the coffer would emerge, and if he broke it, together with the chests and the boxes, the sparrow would come out of the casket. If he then throttled it, I would die."'

At this, Saif said: 'I am a prince and here on my finger is the ring of Solomon, so come to the shore with me to see whether he was telling the truth or not.' So the two of them got up and walked to the sea, where Daulat Khatun stayed on the shore while Saif waded waist-deep into the water. Then he said: 'By the power of the names and the talismans in this ring and by the power of Solomon, on whom be peace, I conjure the soul of the son of al-Malik al-Azraq of the *jinn* to emerge.' At that, there was a disturbance in the sea and out came the coffer. Saif took it and struck it against the water, breaking it and breaking the chests and the boxes.

He took the sparrow from the casket and he and the princess went back to the palace. As they sat down on the throne, a fearsome cloud of dust appeared and something huge flew up, calling out: 'Spare me, prince! Don't kill me but take me as your freed slave and I will bring you to your heart's desire.' 'The *jinni* is here!' cried Daulat Khatun. 'Kill the sparrow lest the damned creature come in and take it from you, after which he will kill first you and then me.' At that, Saif throttled the sparrow, and when it died the *jinni* collapsed by the palace gate and became a heap of black ash. 'We are free from his power,' said Daulat Khatun, 'but what are we going to do now?' 'We must ask help from Almighty God,' replied Saif. 'It was He Who sent us this affliction and He will arrange things for us and help us to escape from our present plight.'

Saif then lifted out some ten of the palace doors, which were made of sandalwood and aloes wood, with nails of gold and silver. He and the princess took ropes of silk that were in the palace and used them to tie the doors together, after which they helped each other to take them down to the sea, where they launched them as a raft. They then moored this raft to the shore and went back to the palace, from which they removed the gold and silver plates as well as sapphires and other jewels and precious stones, together with everything else that was both valuable and light to carry. When they had loaded all this on their raft, they went on board with two pieces of wood that they had fashioned as oars, entrusting themselves to Almighty God, Who does not disappoint the hopes of those who trust in Him and Who suffices them as a helper. They then cast off and let the raft drift out to sea.

For four months they sailed on like this until their provisions were exhausted, and in their distress and despondency they prayed to God for deliverance. During their voyage, when Saif slept he would place Daulat Khatun behind his back, and when he turned over his sword was between them. One night while they were drifting with Saif asleep and Daulat Khatun awake, the raft veered towards the land and came to a harbour where ships were anchored. As Daulat Khatun looked at them, she heard someone talking with the sailors and this turned out to be their leader and captain. When she heard his voice she realized that this must be the harbour of a city and that they had reached civilized parts. Full of joy, she woke Saif and told him to get up and ask the captain the name of the city and of the harbour. He did this happily, addressing the captain as 'brother' and asking the names not only of the city and the harbour but of the local king. 'Fool and idiot,' said the captain, 'if you don't know the harbour and the city, how did you get here?' Saif replied: 'I am a stranger. I was in a merchant ship which was wrecked and lost with all hands, but I managed to climb on a plank and then made my way here. I did not intend any disrespect by my question.' At this, the captain told him that the city was 'Amariya and that the harbour was called Kamin al-Bahrain. When Daulat Khatun heard this, she exclaimed in joy: 'Praise be to God!' and when Saif asked her why, she said: 'Good news! The end of our troubles is close at hand, for the king of this city is my paternal uncle . . .'

Night 771

Morning now dawned and Shahrazad broke off from what she had been allowed to say. Then, when it was the seven hundred and seventy-first night, SHE CONTINUED:

I have heard, O fortunate king, that Daulat Khatun said to Saif: 'Good news! The end of our troubles is close at hand, for the king of this city is my paternal uncle, 'Ali al-Muluk.'

She then told him to check with the captain that this was true, but when he did the captain grew angry again and said: 'You claim to be a stranger who has never been here before in his life, so who told you the king's name?' Daulat Khatun was delighted as she recognized the man as one of her father's sea captains, named Mu'in al-Din. He had come out to search for her when she went missing and, having failed to find her, he had continued to sail around until he came to her uncle's city. She told Saif: 'Say to him: "Captain Mu'in al-Din, come and answer your mistress's summons."' Saif did so and the furious captain, on hearing this, said: 'Dog, who are you and how did you recognize me?' He told one of the sailors: 'Fetch me an ash cudgel so that I can go to this wretch and break his head.' Cudgel in hand, he made towards Saif, but when he looked at the raft he caught sight of something so amazing and delightful that he was completely taken aback. He looked more closely to check what he had seen, and there before his eyes was Daulat Khatun, seated and looking like a sliver of the disc of the moon. 'What have you got there?' he asked Saif, and when Saif said: 'A girl named Daulat Khatun,' the name caused him to fall down in a faint, as he realized that this was his mistress, the king's daughter.

When he recovered consciousness, he left the raft, together with its contents, and set off for the city, where he made for the royal palace and asked for an interview with the king. The chamberlain told the king that Captain Mu'in al-Din had arrived to bring him good news, and the king gave him his permission to enter. When he came in, he kissed the ground and said: 'Good news, your majesty. Your niece, Daulat Khatun, has arrived at the city safe and sound and she is on a raft with a young man as radiant as the moon on the night it becomes full.' The news of his niece delighted the king, who presented the captain with a splendid robe of honour and immediately ordered the city to be adorned with decorations to celebrate her safe return. He sent to have her brought to him, together with Saif, and after greeting them both and congratulating them on their safety, he sent word to his brother to let him know that Daulat Khatun had been found and was with him.

When the messenger reached him, Taj al-Muluk, the princess's father, made his preparations, collected his troops and set out to go to his brother, and when he got there he was reunited with his daughter to everyone's delight. He stayed with his brother for a week, after which he took his daughter and Saif and set off for his own country, Serendib, where Daulat Khatun was reunited with her mother, and feasts were held amid general rejoicing at her safe return. It was a great day, the like of which had never been seen.

The king treated Saif with honour and said: 'Saif al-Muluk, you have done me and my daughter this great service which I am not able to repay, as this can only be done by the Lord of creation. But I want you to take my place on the throne and to rule over the lands of India, for I give you my kingdom, my throne, my treasuries and my servants. All this is a gift from me to you.' Saif got up, kissed the ground before the king, thanked him and said: 'I accept all that you have given me, but I then return it as a gift to you from me. I do not want a kingdom or power, king of the age; all that I want is for Almighty God to bring me to my goal.' 'My treasuries are at your disposal,' the king told him. 'Take whatever you want from them without consulting me, and may God reward you on my behalf with every favour.' Saif repeated that he wanted no share in the kingdom or in its wealth during the course of his quest, but he added that his immediate intention was to see the sights of the city and look at its streets and markets. At that, a fine horse, saddled and bridled, was fetched for him on the king's orders. He mounted and set off for the market, threading his way through the city streets. He was looking from right to left as he rode when he caught sight of a young man who was calling out the price of a cloak that he was selling for fifteen dinars. Saif looked at him closely and saw that he was very like Sa'id, his brother.

The young man was, in fact, Sa'id, but the hardships he had experienced as he wandered through foreign parts had altered his complexion and his appearance. Saif did not recognize him and told his escort to bring him for questioning. Then, when they fetched him, he said: 'Take him to the palace where I am staying and leave him there with you until I come back from my excursion.' The men thought that he had said: 'Take him to prison' and, supposing that the man must be a runaway mamluk belonging to Saif, they took him to prison, fettered him and left him sitting there. When Saif returned from his trip, he went up to the palace, forgetting about Sa'id. No one mentioned him and so he was left in prison, and when the prisoners were taken out to act as building labourers, he was taken with them. He worked with the others for a whole month and became covered in grime, and as he thought things over, he was left to wonder why it was that he had been imprisoned.

For his part, Saif was busy with his pleasures and entertainments until one day as he was sitting he happened to remember Sa'id. He asked his mamluks: 'Where is the mamluk whom you had with you on such-and-such a day?' 'Didn't you tell us to take him to prison?' they answered. 'That's not what I said,' Saif corrected them. 'I told you to take him to my palace.' He then sent his chamberlains to Sa'id, and after they had released him from the fetters that he was still wearing they brought him before Saif. 'Young man,' said Saif, 'from which country do you come?' 'I come from Egypt,' Sa'id replied, 'and my name is Sa'id, son of the vizier Faris.' When Saif heard this, he jumped up from his throne, hurled himself at Sa'id and threw his arms around his neck, shedding floods of tears in his delight. 'Sa'id, my brother,' he cried, 'thank God that you are still alive and that I have seen you. I am your brother Saif.' When Sa'id heard this, he recognized Saif and the two of them embraced tearfully, to the astonishment of all who were present. On Saif's instructions, Sa'id was taken to the baths, and when he emerged he was dressed in splendid clothes, taken to Saif's audience room and seated with him on the throne. When Taj al-Muluk learned what had happened, he was delighted that the two had been reunited; he came himself and all three sat talking over their experiences from start to finish. IT WAS AT THIS POINT THAT SA'ID SAID:

Saif, my brother, when the ship sank and most of the mamluks were drowned, I joined a number of them who were floating on a plank. After a month at sea, in accordance with God's decree, the wind drove us on to an island, where we landed. We were suffering from hunger and so we went in among the trees there and ate their fruit. While we were busy doing this, we were taken by surprise by people like *'ifrits*, who leapt out at us and jumped on our shoulders, telling us to carry them as we had become their donkeys. I asked my rider why he was doing that, but when he heard me speak, he wrapped one of his legs so tightly around my neck that I almost died, while with his other leg he kicked me in the back. I thought that my spine was broken and I collapsed face downwards on the ground as, thanks to hunger and thirst, I had no strength left. When I fell, my rider realized that I must be hungry and so he took me by the hand and led me to a tree which had many fruits resembling pears. He told me to eat my fill, which I did, and then, having no choice in the matter, I began to walk, but I had not gone far before my rider turned around and jumped up on my shoulders. He rode me sometimes at a walk and at others either at a run or a trot, and he kept laughing and saying: 'Never in my life have I seen a donkey like you.'

One day we happened to have collected a quantity of grapes, which we trampled in a pit until it became a large pool. When we went back to it after a while, we found that, thanks to its exposure to the sun, the juice had turned into wine. We stayed there drinking until we got drunk; our faces turned red and we were singing and dancing under the influence of the wine. Our captors asked us why this was, and we said: 'Don't ask. What do you mean by asking us?' But as they insisted that we tell them what we were drinking, we told them that it was pressed grapes. They took us to a valley which we did not know at all, and in it was a seemingly endless quantity of vines, each of whose grape bunches weighed twenty *ratls* and was ready for picking. They told us to pick some of these, and when we had collected a great deal of them, I found a large pit, bigger than a sizeable water tank. We did what we had done earlier, filling it with grapes and then trampling them. When the juice had turned into wine, we told our captors: 'It's ready now, so what do you want to drink it from?' They replied: 'We once had donkeys like you, and when we ate them, their heads were left over, so pour us the drink in their skulls.'

There were about two hundred of the creatures, and when we had poured out wine for them, they fell into a drunken sleep. We said to each other: 'Isn't it enough for them just to ride us? They want to eat us as well, and there is no might and no power except with God, the Exalted,

the Omnipotent! Let us get them thoroughly drunk and then save ourselves by killing them and then escaping.' So we roused them and started
refilling the skulls and passing them the wine. They were complaining
that it was bitter, but we asked them why they said that, explaining that
anyone who called this wine bitter would die the same day unless he
drank ten times. This alarmed them and they told us to give them the
full number of drinks, and after they had taken it all they became
helplessly drunk, and when they were powerless we dragged them off by
their hands. We collected a large number of vine stalks, which we piled
around and on top of the creatures, and then we set fire to them and
stood at a distance to see what would happen to them.

Night 772

Morning now dawned and Shahrazad broke off from what she had been
allowed to say. Then, when it was the seven hundred and seventy-second
night, SHE CONTINUED:

I have heard, O fortunate king, that SA'ID SAID:

When I and the mamluks who were with me set the firewood alight
with the *ghuls* in the middle of it, we stood at a distance to see what
would happen to them. After the fire had died down, we went up and
discovered that they had been reduced to a pile of ashes and we gave
thanks to God for having saved us from them, so giving us the chance
to leave the island. We then went to the coast, where we split up. I
walked on with two of the mamluks until we reached a large and dense
forest. As we were busy eating fruit there, we were confronted by a tall
man with a long beard, drooping ears and eyes like torches, who was
driving a large flock of sheep in front of him. With him were a number
of other people of the same kind, and when he saw us he greeted us
cheerfully. After having spoken words of welcome, he said: 'Come home
with me, and I shall slaughter one of these sheep and roast it for you to
eat.' 'Where do you live?' we asked, and he told us: 'Near this mountain,
so carry on in this direction until you see a cave and then go in. You will
find many other guests like yourselves there, so go and sit with them
until we can prepare a guest meal for you.' We thought that he was
telling the truth, but when we followed his directions and went into the
cave, we discovered that the other 'guests' there were all blind.

When we entered, one of these said: 'I am ill,' while another complained:

'I am weak.' 'What are you saying,' we asked, 'and why is it that you are weak and ill?' At that point they asked us who we were, and when we told them that we were guests, they said: 'How did you fall into the hands of this damned creature? There is no might and no power except with God, the Exalted, the Omnipotent. This is a man-eating *ghul* who has blinded us and intends to eat us.' 'How did he blind you?' we asked, and they said: 'He will blind you now just as he blinded us.' 'How will he do that?' we asked, and they told us: 'He will bring you bowls of sour milk and say: "You must be tired from your journey, so take this milk and drink." Then, when you do, you will become blind just like us.' I said to myself: 'The only way out of this is by a trick,' and so I scooped out a hole in the earth and sat over it. After a time the damned *ghul* came in with bowls of sour milk, one of which he gave to me and the others to my companions. 'You must be thirsty after having come from desert country,' he said, 'so drink this milk while I roast meat for you.' I took the bowl and put it to my lips, but then poured the contents into the hole. I then called out: 'Oh! My sight has gone! I am blind!' and I covered my eyes with my hand and started to weep and cry out, while the *ghul* laughed and told me not to be afraid. In the meantime both my companions had drunk the milk and had gone blind.

The *ghul* immediately went to close the cave's entrance, after which he came up to me and felt my ribs. When he discovered that I was thin with no flesh on my bones, he felt someone else and was glad to find him fat. He then slaughtered three sheep, skinned them, and after skewering their flesh on iron spits that he had brought out, he put these over the fire, and when he had roasted the meat he gave it to my companions and joined them in their meal. Next he brought out a full wine skin, and after drinking he lay face downwards and began to snore. I saw that he was sound asleep and was wondering how to kill him when I remembered the spits. I fetched two of them and after putting them in the fire I waited until they were red hot. I then tucked up my clothes, got to my feet and came up to the damned *ghul* with the spits in my hand and thrust them into his eyes, pressing down on them as hard as I could. He was startled from the rest he had been enjoying and, leaping to his feet, he tried to seize me, blind as he was. He followed me as I ran into the cave, and I called to his blind victims: 'How can I deal with this damned creature?' At that, one of them told me to climb up to a hollow in the wall where I would find a polished sword. 'Take it,' he said, 'and then come back to me and I'll tell you what to do.' I did as he said, took the sword and went back to the man, who told me to aim a blow at the *ghul*'s waist as this would kill him at once. The *ghul*, tired out by running, had gone to kill his blind victims when I ran up behind him and struck at his waist with a blow that cut him in two, but he called out to me: 'If you want to kill me, you will have to strike again.' I was just going to do that when my mentor said: 'Don't strike, for if you do, far from dying, he will come to life and kill us.'

Night 773

Morning now dawned and Shahrazad broke off from what she had been allowed to say. Then, when it was the seven hundred and seventy-third night, SHE CONTINUED:

I have heard, O fortunate king, THAT SA'ID WENT ON:

When I struck the *ghul* with the sword, he called out: 'You have struck me, but if you want to kill me, you will have to strike again.' I was about to do so when the man who had told me where the sword was said: 'Don't do that, for if you do, he will come to life and kill us.' I did what I was told and, as I did not strike him again, the *ghul* died.

At that point, my mentor said: 'Go and let us out by opening up the cave, as it may be that Almighty God will help us to escape from this place.' 'Nothing can go wrong for us now,' I told him, 'so let us rest, slaughter some of these sheep and drink some wine, for it is a long way across the open country.' So we stayed there for two months, eating the sheep and the fruit until one day, as we were sitting by the shore, we happened to catch sight of a large ship out at sea. We gesticulated and shouted to the crew, but they knew that the island was home to a cannibal *ghul* and they were about to sail off in fear. We waved the ends of our turbans at them, and when they sailed closer we started to call out to them. Then one sharp-eyed man on board told the others that we looked like humans and not *ghuls*. They gradually came closer and when they were sure that we were men, they greeted us and, after having returned their greeting, we told them that the damned *ghul* was dead. They thanked us for this, and after we had provided ourselves with fruit from the island we went on board.

For three days we enjoyed a fair wind, but on the fourth a storm blew up, the sky darkened and within an hour the wind had driven the ship against a mountainside, where it broke up, its planks being torn apart.

By the decree of the Great God, I managed to get hold of one of them and I drifted on this for two days. I had a favourable wind and for a time I sat on top of my plank, paddling with my feet, until God brought me safely to shore. Then I got to this city, but as I was entirely on my own as a stranger I did not know what to do. I was suffering the pangs of hunger after my exertions, and so I came to the market and hid myself in order to remove my cloak, telling myself that I would sell it and use the money to buy food, until what God decreed should come about. I had the cloak in my hand and the people were looking at it and bidding for it when you arrived and saw me. You gave orders that I was to be brought to the palace, but your servants took me and put me in prison. Later you remembered me and had me brought here to you. I have now told you what happened to me, and I thank God that we have met again.

Saif and Daulat Khatun's father, Taj al-Muluk, were filled with astonishment when they heard Sa'id's story. Taj al-Muluk assigned pleasant lodgings to his two visitors, and Daulat Khatun herself used to visit Saif and gratefully discuss with him the service that he had done her. Sa'id then told her: 'Princess, what he wants is for you to help him to reach his goal.' 'Yes,' she said, 'I shall do all that I can to see that he does, if this is what Almighty God wills.' She then turned to Saif and told him to rest easy and be comforted.

So much for Saif and his vizier Sa'id, but as for the princess Badi' al-Jamal, news reached her that her sister, Daulat Khatun, had returned to her father and her kingdom, and she said: 'I must go to visit her and greet her, splendidly dressed with ornaments and robes.' So she set off, and when she was near the palace, Daulat Khatun herself met her and greeted her, before embracing her and kissing her between the eyes, while for her part she congratulated the princess on her safe return. The two of them sat talking, and Badi' al-Jamal asked her friend what had happened to her in her enforced exile. 'Don't ask me about that, sister,' said Daulat Khatun. 'What sufferings God's creatures have to endure!' When Badi' al-Jamal pressed her, she went on to explain how she had been held in the Lofty Castle in the power of the son of al-Malik al-Azraq, and she explained the rest of the story from start to finish, telling her about Saif, what had happened to him in the castle, as well as the sufferings and perils that he had endured before getting there, and how he had killed the *jinn* prince. She went on to explain that he had constructed a raft by taking the doors off their hinges, how he had made oars and how he had then got back.

Badi' al-Jamal was amazed and exclaimed: 'By God, sister, this is very remarkable indeed!' Her friend then said: 'I would like to tell you how his story started, but a sense of shame prevents me.' 'Why should you feel shame?' asked Badi' al-Jamal. 'You are my sister and my companion; there are close ties between us. I know that you wish me nothing but good, so why should you be ashamed? Tell me what it is, and don't be bashful or hide anything from me.' So Daulat Khatun said: 'He saw your picture on the robe that your father had sent to Solomon, son of David. Without opening it or inspecting it, Solomon sent it on, among other gifts and treasures, to King 'Asim ibn Safwan, the ruler of Egypt, and he in his turn sent it, still unopened, to his son, Saif al-Muluk. Saif did open it and he was about to put on the robe when he saw your picture and fell in love with it. He then went off to search for you, and it is for your sake that he has endured all these hardships.'

Night 774

Morning now dawned and Shahrazad broke off from what she had been allowed to say. Then, when it was the seven hundred and seventy-fourth night, SHE CONTINUED:

I have heard, O fortunate king, that Daulat Khatun told Badi' al-Jamal that Saif had fallen in love with her thanks to the robe with her picture on it and how, on seeing this, he had left his kingdom and his family for her sake, distracted by love. She explained to her the terrors that he had then endured for her. Badi' al-Jamal blushed modestly at these words but said: 'This can never be, as there can be no union between men and *jinn*.' Daulat Khatun, however, went on describing the prowess of the handsome Saif and what he had achieved, and she continued to praise him and to talk of his qualities, until she ended by saying: 'Sister, for the sake of Almighty God and for my sake, come and talk to him, even if you only say one word.' Badi' al-Jamal told her: 'I am not going to listen to what you say or do what you suggest, and I shall pretend that I did not hear anything.' She felt no love for Saif, however handsome, distinguished or chivalrous he might be, but Daulat Khatun started to implore her, kissing her feet and saying: 'For the sake of the milk on which we were suckled, you and I, and for the sake of what is engraved on the ring of Solomon, on whom be peace, please listen to me, for when I was in the Lofty Castle I undertook to let him see your face. So I implore you in God's Name to show yourself to him once for my sake, and you, for your part, will be able to see him.' She mixed tears with her pleadings and continued to kiss Badi' al-Jamal's hands and feet until she agreed and said: 'For you I shall let him have a single glimpse of my face.'

This made Daulat Khatun happy and, after kissing Badi' al-Jamal's hands and feet again, she went to the main pavilion in the garden and told her slave girls to spread out furnishings, set up a golden couch and provide rows of wine cups. She then went to Saif and Sa'id, who were sitting in their palace, and she gave Saif the good news that he had reached his heart's desire. 'You and your brother must come to the garden,' she told him, 'enter the pavilion and then hide away so that no one there can see you until I bring Badi' al-Jamal.' Saif and Sa'id got up and, following her directions, they went into the pavilion, where they saw food and drink set out, together with a golden couch covered with cushions. When they had been sitting there for some time, Saif found himself overcome by distress and passionate longing at the thought of his beloved. He got up and walked out of the entrance hall of the pavilion, followed by Sa'id, to whom he said: 'Brother, sit down where you were; don't follow me but wait until I come back to you.'

Sa'id did as he was told, but Saif went into the garden, drunk with the wine of passion and bewildered by excess of love. Wracked with longing and overcome by emotion, he recited these lines:

> Badi' al-Jamal, you are the only one for me;
> Have mercy on me, the captive of your love.
> It is for you that I ask; you are my wish and my delight;
> My heart refuses the love of any other.
> I wish I knew whether you know of my tears,
> Which I shed all through my sleepless nights.
> Order sleep to visit my eyelids,
> And it may be I shall see you in my dreams.
> Show sympathy to a passionate lover;
> Save him from being destroyed by your harshness.
> May God add to your joy and your delight;
> All mankind are your ransom.
> At the Resurrection all lovers will collect beneath my banner,
> While all the lovely ladies will be under yours.

He shed tears and then added these lines:

> This marvel of beauty is my only wish;
> She is the secret hidden in my heart.
> If I speak, it is of her loveliness,
> And if I am silent, she is in my inmost thoughts.

After shedding more bitter tears, he went on to recite:

> In my heart a fire burns ever more fiercely;
> You are my desire; my passion is of long standing.
> It is to you and none but you I turn.
> I hope to please you, and the lover is long-suffering.
> Have mercy on one whom love emaciates;
> He is weakened and sick at heart.
> Show pity; be generous and gracious;
> I shall not leave you or ever turn away.

After more tears, he recited:

> Cares visited me together with your love;
> Sleep treated me with harshness similar to yours.
> The messenger told me that you were angry;
> May God ward off the evil of what he said.

Sa'id, finding Saif slow to return, left the pavilion to look for him in the garden. He saw him walking there distractedly, reciting:

> By God, by the Almighty and by the reciter
> Of the *Sura* Fatir from the Quran,
> Whatever beauties I may look at,
> It is your image that stays with me through the night.

After the two of them had met, they started to look around the garden and to eat its fruits.

So much for Sa'id and Saif, but as for Daulat Khatun, she entered the pavilion with Badi' al-Jamal after the eunuchs had carried out her orders, adorning it with splendid decorations of all kinds and preparing a golden couch as a seat for Badi' al-Jamal beside a window that overlooked the garden. When she saw this she sat down there, while the eunuchs produced all kinds of splendid foods. The two princesses ate, with Daulat Khatun feeding her companion on titbits until she had had enough. Then Daulat Khatun called for sweetmeats, and when these had been fetched the two ate their fill and washed their hands. All that was needed for wine to be served was made ready, with jugs and glasses being set out in rows, and Daulat Khatun started to fill the glasses and pass them to Badi' al-Jamal before drinking herself.

It was then that Badi' al-Jamal looked out of the window beside her towards the garden. She saw the fruits and the branches, but then she looked around and noticed Saif as he wandered there, followed by Sa'id, his vizier, and she heard him reciting poetry in floods of tears. This glance of hers was followed by a thousand regrets . . .

Night 775

Morning now dawned and Shahrazad broke off from what she had been allowed to say. Then, when it was the seven hundred and seventy-fifth night, SHE CONTINUED:

I have heard, O fortunate king, that when Badi' al-Jamal noticed Saif wandering in the garden, her glance was followed by a thousand regrets, and she turned to Daulat Khatun, her emotions stirred by wine, and said: 'Sister, who is that young man I see wandering distractedly in the garden, in a state of gloom and sorrow?' Daulat Khatun replied: 'Do I have your permission to bring him here so that we may have a look at him?' 'Yes, if you can,' Badi' al-Jamal told her, and at that Daulat Khatun called out: 'Prince, come here to us, in all your beauty and grace.' Saif, recognizing Daulat Khatun's voice, went up to the pavilion, but when he set eyes on Badi' al-Jamal he fell down in a faint, only recovering after Daulat Khatun had sprinkled a little rosewater over him. He then got up and kissed the ground before Badi' al-Jamal, who was astounded by his beauty. Daulat Khatun introduced him to her, explaining: 'This is the man who rescued me, in accordance with God's decree, and it is he who endured all those hardships for your sake. I want you to look on him with favour.' Badi' al-Jamal laughed and said: 'Why should you suppose that this young man would keep a promise? There is no such thing as love among humans.' Saif exclaimed: 'Princess, I would never break my word and not everyone is the same!' He then shed tears before her and recited:

Badi' al-Jamal, show pity to a wretch
Worn out and afflicted, hurt by a bewitching glance.

By the white beauties of your cheeks,
Combined with the deep red of anemones,
I ask you not to punish with separation one near death;
When you are far away, my body fades.
For this is my desire; this is the goal of hope;
I look for union, if it can ever be.

With still more tears, and overcome by the passion of his love, he greeted
her with these lines:

I bring you greetings from a slave of love;
The noble are always favoured by their like.
Greetings to you and may your image never leave;
May no place where I sit or rest be without you.
I am jealous for your honour; I never speak your name;
All lovers must incline to the beloved.
Do not cut off your favours from your lover;
Grief will destroy him, and he is already sick.
I am frightened as I watch over the bright stars;
Thanks to my passion the night is so long.
I have no patience; there is nothing I can do;
And what words can I use with which to beg?
The peace of God be on you, cruel girl,
A greeting from a distracted lover who endures.

Then, overwhelmed by passion, he added:

If ever I seek out another love,
May I not reach my goal in such a quest.
Who else but you is mistress of such loveliness,
That in it I should find my resurrection?
Heaven forbid that I forget my love,
When I have worn away on you my inmost heart.

When he had finished, he broke down in tears and Badi' al-Jamal said:
'Prince, I am afraid that were I to give myself wholly to you, I would not
find in you any love or affection, for there is little good in humankind
and a great deal of treachery. As you know, Solomon captivated the
queen of Sheba by pretending to love her, but when he saw someone
more beautiful, he left her for his new love.' 'My eye and my soul,' said
Saif, 'God has not created all men the same, and, if He so wills, I shall

either keep faith or die beneath your feet. You will see that my deeds match my words, and it is God who acts as guarantor for what I say.' 'Sit down quietly,' she told him, 'and swear to me by your religion, so that we may make a pact not to betray each other, and if either of us does, then Almighty God will punish the guilty.'

On hearing this, Saif sat down and he and Badi' al-Jamal held hands and swore that neither of them would choose any other lover, human or *jinn*. After a long embrace they shed tears of joy and, overcome by emotion, Saif recited these lines:

> I wept in passionate longing,
> For the love of my inmost heart.
> Her long absence had piled sufferings on me
> And I had not strength enough to reach my goal.
> Grief sapped my firmness
> And showed the censurers some of my distress.
> The width of my patience narrowed,
> Until I had no more strength to endure.
> Do you suppose God will unite us,
> Curing me of my pains and suffering?

After Badi' al-Jamal and Saif had plighted their troth, the two of them got up and walked away. With Badi' al-Jamal was a slave girl who was carrying some food, together with a bottle of wine, and when her mistress sat down, this girl put the food and the wine down in front of her. They had not been waiting long before Saif arrived and was greeted by Badi' al-Jamal. The two embraced . . .

Night 776

Morning now dawned and Shahrazad broke off from what she had been allowed to say. Then, when it was the seven hundred and seventy-sixth night, SHE CONTINUED:

I have heard, O fortunate king, that when Badi' al-Jamal had food and drink fetched, Saif came and she greeted him, and after they had sat there eating and drinking for some time, Badi' al-Jamal told Saif: 'When you come to the garden of Iram, you will see there a large tent of red satin with a lining of green silk. Pluck up your courage and go into it, and there you will see an old lady sitting on a throne of red gold studded with pearls and other jewels. Greet her courteously and modestly and then, if you look at the throne, you will see underneath it a pair of sandals worked from beaten gold and adorned with precious stones. Take these and kiss them before placing them first on your head and after that under your right arm. Then stand silently in front of the old lady with your head bowed. She will ask you where you have come from, how you got there, who told you about the place and why it was that you took the sandals. Say nothing until this maid of mine comes to talk to her and tries to win her sympathy for you by conciliating her, as it may be then that Almighty God will soften her heart towards you and she will grant your request.'

Badi' al-Jamal then called to the slave girl, whose name was Marjana, and said: 'By the love that I have for you, I ask you to do your best to carry out my errand today, and if you do I shall set you free for the sake of Almighty God. I shall honour you as the dearest of my friends and I shall tell my secrets to no one but you.' 'My lady and the light of my eyes,' said the girl, 'tell me what you want done and I shall do it willingly.' Badi' al-Jamal replied: 'Carry this human on your shoulders and bring him to my maternal grandmother in the garden of Iram. Take him to her tent and look after him. When the two of you have entered it and you see him take the sandals and present his services to her, she will ask him where he has come from, how he came, who brought him, why he has taken the sandals and what he wants her to do for him. When this happens, go in quickly and say: "My lady, I brought him. He is the son of the king of Egypt, and it was he who went to the Lofty Castle, where he killed the son of al-Malik al-Azraq, rescued Princess Daulat Khatun and brought her back safely to her father. He was sent off with me and I have brought him here so that you may reward him for bringing you the good news of her safety." Then go on: "I conjure you by God, lady, to tell me whether this young man is handsome." She will say yes, and you are then to say to her: "My lady, he is a man of honour, chivalry and courage; he is king of the realm of Egypt and possesses the full range of praiseworthy qualities." She will ask you what he wants, and you are to say: "My mistress sends you her greetings and asks how long she is to sit at home single and without a husband. A long time has passed, and why is it that you do not find her a husband and marry her off like other girls while both you and her mother are still alive?" My grandmother will say: "If she knows someone who has taken her fancy, tell me about him and I shall fall in with her wishes as far as is possible."

You must then tell her: "My lady, my mistress says: 'You wanted to marry me to Solomon, on whom be peace, and you had my portrait placed on the gown, but he would have nothing to do with me, and he sent the gown to the king of Egypt. He, in turn, gave it to his son, who saw my portrait on it and fell in love with me. He then abandoned the realm of his father and mother, and turning away from worldly delights he wandered around sick with love, enduring the greatest of hardships and perils for my sake.'"'"

Marjana now picked up Saif and, when he had closed his eyes on her instructions, she flew off with him into the air. Some time later she said: 'Open your eyes,' and when he did he saw a garden, this being the garden of Iram. 'Go into this tent,' Marjana told Saif, and when he entered, calling on the Name of God and looking at the garden, he saw an old lady seated on a throne attended by slave girls. He advanced with courtesy and modesty, picked up the sandals, kissed them and then did what Badi' al-Jamal had told him. 'Who are you and where have you come from?' the lady asked, adding: 'Tell me your country and who brought you here. Why did you take these sandals and kiss them? Have you ever told me of a need which I did not fulfil for you?' At that point, Marjana came in and greeted the old lady politely and modestly before repeating what Badi' al-Jamal had told her to say. The lady shouted at her angrily and exclaimed: 'How can there be any agreement between men and *jinn*!'

Night 777

Morning now dawned and Shahrazad broke off from what she had been allowed to say. Then, when it was the seven hundred and seventy-seventh night, SHE CONTINUED:

I have heard, O fortunate king, that when the old lady heard what Marjana had to say, she was furiously angry and exclaimed: 'How can there be any agreement between men and *jinn*!' Saif said: 'I shall make an agreement with you to be her servant, to love her as long as I live, to keep my word and never to look at anyone else. You will see that what I say is no lie but the truth, and that I shall act with chivalry, if Almighty God wills it.' The old lady bowed her head and sat in thought for a time, after which she looked up and said: 'Will you keep to this agreement that you have made, handsome young man?' 'I will,' he replied, 'and I swear this by Him Who raised the heavens and spread out the earth over the waters.' 'If He wills it, then I shall settle the matter for you,' the lady said, 'but for the moment go and look around the garden. You can eat fruits the like of which cannot be found anywhere else in the world, while I send word to my son, Shahyal. When he comes I shall talk to him about this and, God willing, all will turn out for the best, as he will not oppose me or disobey me. You may rest assured that I shall marry you to his daughter, Badi' al-Jamal, and she will be your wife.' When Saif heard this, he thanked her and, after kissing her hands and feet, he left her and went towards the garden. For her part, she turned to Marjana and said: 'Go to look for my son, Shahyal, and when you find out what part of the world he is in, fetch him to me.' Marjana duly went off to look, and when she met Shahyal she brought him back to his mother.

So much for her, but as for Saif, while he was looking around the garden he was seen by five *jinn*, followers of al-Malik al-Azraq. 'Where did this man come from,' they asked one another, 'and who brought him here? It may be that it was he who killed the king's son. Let us see if we can trick him with some questions.' So they started to stroll a little way until they came up with Saif at the edge of the garden and then sat down beside him. 'How well you managed to kill the son of al-Malik al-Azraq and to rescue Daulat Khatun, you handsome young man,' they said, adding, 'and he was a treacherous dog who had duped her. Had God not sent you to her, she would never have escaped. But how did you kill him?' Saif looked at them and said: 'I used this ring on my finger,' so confirming for them that he was indeed the killer. At that point, two of them seized his hands and another two his feet while the fifth put his hand over his mouth lest Shahyal's followers hear his cries and rescue him. They carried him off and flew away with him until they set him down in front of their king, telling him: 'King of the age, we have brought you the killer of your son.' 'Where is he?' the king asked, and the *jinn* replied: 'This is the man.' 'Did you kill my beloved son, the light of my eyes, without due cause and without his having wronged you?' the king asked. 'Yes,' Saif replied, 'I did kill him, but this was because of his unjust and hostile behaviour: he was in the habit of seizing the children of kings and taking them to the Abandoned Well and the Lofty Castle in order to debauch them when he had separated them from their families. I killed him with this ring on my finger, and God hastened his soul to hellfire, an evil resting place.'

When the king knew for sure that Saif was responsible for his son's death, he sent for his vizier and, after having told him that there was no doubt about Saif's guilt, he asked him for his advice and said: 'Shall I have him put to the worst of deaths, or subject him to the most excruciating tortures, or what should I do?' 'Cut off one of his limbs,' advised the grand vizier. 'Have him flogged painfully every day,' said another. A third suggested that he be cut in half. 'Cut off all his fingers and then burn him alive,' said another, while someone else said: 'Crucify him.' Each of them expressed his opinion, but the king had with him an elderly emir who had experience in affairs and a knowledge of past events. This emir now said: 'King of the age, I have something to say to you and you would be well advised to pay attention to what I recommend.' As he was the counsellor of state and its administrative head, the king had been in the habit of listening to him and acting on his advice in all matters. The emir now got to his feet, kissed the ground and continued: 'King of the age, if I give you advice on this affair, will you follow it and give me leave to speak freely?' 'Tell me what you think,' said the king, 'and I guarantee your safety.' So the emir told him: 'If you kill this man now, your majesty, and refuse my advice or fail to grasp what I am saying, this will be a mistake. He is in your power and you are responsible for him as he is your prisoner. You can lay your hands on him whenever you want and do what you want with him. But you should be patient, for he entered the garden of Iram and, as the future husband of Badi' al-Jamal, the daughter of King Shahyal, he has become one of them. A group of your followers seized him and brought him to you, and he did not try to conceal what he had done from them or from you. If you kill him, King Shahyal will look to avenge him on you, and for his daughter's sake he will bring up an army to attack you, which you will not be able to resist.' The king listened to this advice and had Saif put in prison. So much for him.

As for Badi' al-Jamal's grandmother, when she met her son, Shahyal, she sent Marjana to look for Saif but she came back saying that she had not been able to find him in the garden. Marjana then sent for the gardeners and, when she questioned them, they said: 'We saw him sitting under a tree and five of al-Malik al-Azraq's followers sat down and talked with him. Then they picked him up, gagged him and flew away with him.' When the old lady heard this from Marjana, she was not disposed to take it lightly but rose to her feet in a furious rage and said to Shahyal, her son: 'How can you be a king and allow the followers of al-Malik al-Azraq to come to our garden, seize a guest of ours and go off unharmed while you are still alive?' She continued to urge him on, saying that as long as he lived he should never allow such hostile acts. He objected, saying: 'Mother, this human killed a *jinni*, the son of al-Malik al-Azraq, to whom God delivered him. How can I go and attack the king for the sake of the human?' His mother insisted: 'Go and ask him for our guest. If he is still alive and is handed over to you, take him and come back, but if he has been killed, take the king, together with his children, his women and all his followers who are under his protection, and bring them to me alive so that I may cut their throats with my own hand, before ravaging their lands. If you disobey me, I shall hold that you have not repaid me for the milk with which I suckled you, and the nurture that I gave you will prove to have been unlawful.'

Night 778

Morning now dawned and Shahrazad broke off from what she had been allowed to say. Then, when it was the seven hundred and seventy-eighth night, SHE CONTINUED:

I have heard, O fortunate king, that Badi' al-Jamal's grandmother told her son Shahyal to go to al-Malik al-Azraq in order to look for Saif. 'If he is still alive,' she said, 'bring him here, but if the king has killed him, seize him, together with his women and children, as well as everyone who is under his protection, and bring them to me alive so that I may cut their throats with my own hand, before ravaging their lands. If you disobey me, I shall hold that you have not repaid me for the milk with which I suckled you, and the nurture that I gave you will prove to have been unlawful.' At this point the king got up, ordered his troops out and set off for al-Malik al-Azraq, out of respect for his mother and to satisfy her and her loved ones, as well as to bring about something that had been destined from past eternity. He continued to advance with his army until they and the forces of al-Malik al-Azraq met and fought. The latter were defeated and the king's children, old and young, were seized, together with his ministers of state and its principal officers. All these were bound and taken before Shahyal, who asked al-Malik al-Azraq: 'Where is Saif al-Muluk, the human who was my guest?' Al-Malik al-Azraq said: 'Shahyal, you are a *jinni* and so am I. Are you doing this because of a human, the slayer of my son? It was he who killed my dear one, the delight of my soul, so why, then, have you done all this and shed the blood of so many thousand *jinn*?' 'Stop this talk,' replied

Shahyal. 'If he is still alive, produce him and I will free you and all your children whom I have taken, but if you have killed him, I shall cut all your throats.' 'Is he dearer to you than my son?' asked the other, and Shahyal said: 'Your son was an evil-doer who used to kidnap people's children and the daughters of kings, place them in the Abandoned Well and the Lofty Castle and debauch them.' Al-Malik al-Azraq then said: 'Saif is with me, but you must reconcile us with him.' Shahyal did this and presented robes of honour, before drawing up a document of indemnity covering the killing of the prince. After that, Saif was handed over to Shahyal, who entertained them all generously.

After al-Malik al-Azraq and his followers had stayed with him for three days, Shahyal brought Saif back to his mother, who was overjoyed at his recovery, while he himself admired Saif's handsomeness and the perfection of his good looks. Saif told him the story of his adventures from beginning to end, including his encounter with Badiʿ al-Jamal, after which Shahyal said to his mother: 'Mother, as you are content with this business, I shall agree to everything of which you approve. So take Saif back to Serendib and hold a great wedding celebration there, for he is a handsome young man and he has endured many perils for my daughter's sake.'

Badiʿ al-Jamal's grandmother set off with her attendants and brought Saif to Serendib, where they entered the garden belonging to Daulat Khatun's mother, and here Badiʿ al-Jamal caught sight of Saif. They went to meet in the tent, where the old lady told of Saif's encounter with al-Malik al-Azraq and of how he had been imprisoned and had faced imminent death – but there is nothing to be gained by retelling this. Daulat Khatun's father, King Taj al-Muluk, gathered together the grandees of his state and drew up the marriage contract between Badiʿ al-Jamal and Saif, distributing splendid robes of honour and providing banquets for the people. At this point Saif got up and, after having kissed the ground in front of Taj al-Muluk, he said: 'If you will pardon me, your majesty, I have something to ask of you, but I am afraid that you may turn me away disappointed.' 'By God,' exclaimed the king, 'you have done me so great a service that if you asked me for my life I would not refuse you.' Saif then said: 'I would like you to marry Princess Daulat Khatun to my brother Saʿid, so that we may both be your servants.' 'To hear is to obey,' the king said, and he then collected his grandees for a second time and drafted a marriage contract between Daulat Khatun and Saʿid, having it drawn up by the legal scribes. When this had been done, gold and silver were scattered, an order was given for the city to be adorned with decorations, and wedding celebrations were held, after which the marriages of Saif and Badi' al-Jamal and of Sa'id and Daulat Khatun were consummated on the same night.

Saif spent the next forty days alone with Badi' al-Jamal, and then one day she asked him if there was anything he regretted. 'God forbid,' he said. 'I have achieved my goal and there are no regrets in my heart, although I should like to meet my mother and father again in Egypt to see whether they are still well or not.' She then instructed a number of her servants to take him and Sa'id to their families in Egypt. After they had done this and the two had met their parents, they stayed for a week and then, having said their farewells, they returned to Serendib. They continued to come and go whenever they wanted to repeat their visits.

Saif lived with Badi' al-Jamal in the greatest happiness, as did Sa'id with Daulat Khatun, until they were visited by the destroyer of delights and the parter of companions. Praise be to the Living God, Who does not die, the Creator of all things, Who has decreed that they should meet death. He is the First without a beginning and the Last without an end.

This is the end of the story of Saif al-Muluk and Badi' al-Jamal as it has come down to us, but God knows better about the truth of that.

A story is also told that in the old days and former ages there was a rich merchant in Basra who had two sons. God, the All-Hearing and Omniscient, decreed that he should die, leaving behind his wealth, and after his sons had made the necessary preparations and organized his funeral, they divided this inheritance equally between themselves. Each took his share and opened a shop, one being a coppersmith and the other, whose name was Hasan, a goldsmith. While the latter was sitting in his shop one day, a Persian who was walking through the market went past it. He looked at Hasan's work with a knowledgeable eye and admired it, nodding his head, and exclaiming: 'By God, you are an excellent craftsman!' Hasan meanwhile was studying an old book that he was holding, leaving the passers-by preoccupied with his beauty and his well-shaped figure.

At the time of the afternoon prayer, when everyone else had left the shop, the Persian came back and said: 'My son, you are a handsome young man. What is your book? You have no father and I have no son, but I know the best craft in all the world.'

Night 779

Morning now dawned and Shahrazad broke off from what she had been allowed to say. Then, when it was the seven hundred and seventy-ninth night, SHE CONTINUED:

I have heard, O fortunate king, that the Persian came to Hasan, the goldsmith, and said: 'My son, you are a handsome young man. You have no father and I have no son, but I know the best craft in all the world. Many people have asked me to teach it to them and I have never agreed, but I can bring myself to teach it to you and take you as a son. This will serve as protection for you, helping you to ward off poverty, and you will be able to take a rest from what you have to do at present, toiling away with your hammer, charcoal and fire.' 'When will you teach me this, master?' asked Hasan, and the Persian promised to come back the next day in order to turn copper into pure gold before his eyes.

Hasan was delighted and, after saying goodbye to the Persian, he went to his mother and when he had greeted her and shared a meal with her, he told her about him. Hasan was so overwhelmed that he had completely lost his wits, but his mother said: 'What is the matter with you, my son? You should take care not to listen to what people say, particularly when it comes to Persians. Don't do anything that they tell you, for these alchemy teachers are frauds who try to trick people and swindle them out of their money on false pretences.' 'We are poor, mother,' Hasan said, 'and we have nothing that would tempt anyone to try to swindle us. This Persian is a virtuous and quite clearly pious old man, whom God has filled with sympathy for me.' His mother bit back an angry reply, and, for his part, Hasan was preoccupied and could not sleep that night because of his delight at what the Persian had said to him.

The next morning he got up, took the keys and opened up his shop. At that moment he saw the Persian coming, and tried to kiss his feet, but the man would not let him. Instead, he told him to get the crucible ready and to apply the bellows. When he had done this and had lit the charcoal, the Persian asked if he had any copper. 'I have a broken tray,' said Hasan, and the Persian told him to use his shears to cut it into little bits. He did as he was told, and then put the pieces into the crucible, pumping the bellows until the contents became liquid. At that point the Persian put his hand into his turban and removed a twist of paper. When he had unfolded it, he poured half a dirham's weight of what looked like yellow kohl from it into the crucible, telling Hasan to keep on pumping. Hasan did this until the liquid had turned into an ingot of gold.

Hasan was lost in astonishment and joy when he saw what had happened. He took the ingot and turned it over and over before taking a rasp and filing away at it, only to discover that it was indeed pure gold of the very highest quality. Overcome with delight, he bent over the Persian's hand to kiss it, but again the man stopped him and said: 'Take the ingot and sell it in the market. Be quick when you take the money that you get for it, and say nothing.' So Hasan went off to the market and gave his ingot to the auctioneer, who took it from him and, on testing it, confirmed that it was pure gold. He opened the bidding at ten thousand dirhams, but the merchants went on raising their bids until it was sold for fifteen thousand. Hasan took the money and went back home, where he told his mother everything that he had done. When he said that he had learned the art of alchemy, she laughed at him and said: 'There is no might and no power except with God, the Exalted, the Almighty.'

Night 780

Morning now dawned and Shahrazad broke off from what she had been allowed to say. Then, when it was the seven hundred and eightieth night, SHE CONTINUED:

I have heard, O fortunate king, that Hasan, the goldsmith, told his mother what the Persian had done, saying: 'I have learned the art of alchemy.' She exclaimed: 'There is no might and no power except with God, the Exalted, the Almighty!' After that, reluctantly enough, she said no more.

Hasan, in his ignorance, then took a mortar and brought it in to set before the Persian, who was sitting in his shop. 'What do you intend to do with this, my son?' the man asked, and Hasan said: 'We can put it on the fire and make it into gold ingots.' 'Are you mad?' said the Persian, laughing. 'Are you going go to the market with two ingots on one and the same day? Don't you realize that people would become suspicious of us and we would lose our lives? When I teach you this craft, you are only to use it once a year, so that it can keep you going from one year to the next.' Hasan agreed that this was sensible, and he then took his seat in the shop, put charcoal on the fire and set the crucible over it.

'What do you want?' asked the Persian, and Hasan said: 'Teach me this craft.' The Persian laughed again and repeated the formula: 'There is no might and no power except with God, the Exalted, the Almighty,' adding: 'You have not much sense, my son, and you are not fit to study alchemy. Does anyone ever learn it on the open road or in the market? If we start to busy ourselves with it here, people will say: "Those people are practising alchemy." Then the authorities will hear of it, and that will cost us our lives. If you really want to study it, come back to my house with me.'

So Hasan got up, locked his shop and set off with the Persian. On his way, however, he remembered what his mother had said and, thinking it over carefully, he stopped and spent some time looking down at the ground. The Persian turned and laughed to see him standing there. 'Are you mad?' he asked. 'How is it that, when all I am thinking of is to do you some good, you get it into your head that I am going to harm you?' He added: 'If you are afraid of coming to my house with me, I shall go home with you and teach you there.' Hasan agreed to this and when the Persian said: 'Lead the way,' he went on ahead to his own house, with the other following behind him. When Hasan entered, he told his mother that the Persian standing at the door had come with him, and after she had arranged things for them in the house, she went away. Hasan then allowed his guest to enter, and afterwards he himself went to the market with a bowl in his hand to fetch something to eat. When he had got food, he put it in front of the Persian and said: 'Eat, master, so that we may be linked by bread and salt, for Almighty God takes revenge on anyone who breaks this tie.' 'You are right, my son,' said the Persian, smiling, 'and who can know the value of this bond?'

He then came forward and he and Hasan ate until they had had enough, after which Hasan was told to fetch some sweetmeats. He was happy to go back to the market, and he brought the sweetmeats in ten covered dishes. They both ate from these, and the Persian said: 'May God reward you, my son. You are the kind of person whom people take as a friend and to whom they tell their secrets, teaching them what will be useful to them.' He told Hasan to fetch his apparatus, and, scarcely able to trust his ears, Hasan rushed off like a colt let out in the spring, and when he had brought back what was needed from his shop, he set it down in front of the Persian. For his part, the Persian took out a twist of paper and said: 'Hasan, I swear by the tie of bread and salt that if you were not dearer to me than my own son, I would not show you this craft, for the only bit of this elixir I have left is what is in this paper. But look carefully as I mix the drugs and put them in front of you, and bear in mind that you have to add half a dirham's weight of the elixir for every ten *ratls* of copper, and these ten *ratls* will then turn into pure gold.' He added: 'The paper contains three Egyptian ounces of elixir, and when it has all been used up, I shall make you some more.'

Hasan picked up the twist of paper and found that it held a yellow substance finer than the first lot that he had seen. 'Master,' he asked, 'what is this called? Where is it to be found and what is it made of?' The Persian laughed and, hoping to get Hasan into his power, he said: 'Why are you asking questions? Do your work and keep quiet.' He then brought out a bowl from the house, and having cut it into bits, he put these in the crucible and added a small amount from the contents of the paper. The result was an ingot of pure gold, and when Hasan saw this he was so delighted and so preoccupied with the ingot that he could no longer think straight. Quickly the Persian took a packet of *banj* from his turban, one sniff of which would be enough to put an elephant to sleep for twenty-four hours. He cut this up and put it in one of the sweetmeats, after which he said: 'Hasan, you are now a son to me and you have become dearer to me than my money and my life. I have a daughter, and I propose to give her to you as a wife.' 'I am your servant,' replied Hasan, 'and all that you do for me will be recorded by Almighty God.' 'If you wait patiently,' the other told him, 'good will come to you.'

It was at this point that the Persian gave him the drugged sweetmeat. He took it, kissed the Persian's hand and put it in his mouth, with no knowledge of what was in store for him. When he swallowed it, he fell down head over heels and collapsed unconscious. The Persian was delighted at his misfortune and exclaimed: 'You have fallen into my trap, you miserable Arab dog, and I have finally got hold of you after long years of searching!'

Night 781

Morning now dawned and Shahrazad broke off from what she had been allowed to say. Then, when it was the seven hundred and eighty-first night, SHE CONTINUED:

I have heard, O fortunate king, that when Hasan, the goldsmith, ate the sweetmeat which the Persian had given him, he fell unconscious on the ground. The Persian was delighted and exclaimed: 'I have spent many years searching for you, but now I have got you!' He tucked up his clothes, then put Hasan's hands behind his back and tied them to his feet, before fetching a chest whose contents he removed and into which he then locked him. He emptied another chest and in this one he put all Hasan's money, together with both the ingots that he had made. He locked it and then ran out to the market to find a porter, whom he hired to carry both chests out of the city and to put them down on the seashore, after which he went to a ship that was lying there at anchor, ready and prepared for him. The captain was expecting him, and when the sailors caught sight of him, they went over, picked up the chests and loaded them on board. The Persian called out to the captain and the whole crew that he had done what he wanted, and the captain shouted to the crew to raise anchor and unfurl the sails. The ship then got under way, helped by a favourable wind.

So much for Hasan and the Persian, but as for Hasan's mother, she waited until evening without hearing anything or getting any word of him at all. She went home to find the house open and nobody there. The chests and the money had gone, and she realized that, thanks to the decree of fate, she had lost her son. She struck her face, tore her clothes, uttering shrill screams, and began to call out: 'Alas for my son! Alas for the darling of my heart!' Then she recited these lines:

> I can bear it no longer; my restlessness grows worse;
> You have gone; my grief and my distraction multiply.
> By God, I cannot bear your loss;
> How can I resign myself to the ruin of my hopes?
> Where is sleep's pleasure, now that my darling has gone?
> Who can enjoy a life of such disgrace?
> You went, leaving the dwellings and their people desolate,
> Muddying the clear water from which I drank.
> In every hardship I found you my aid;
> You were my glory, my dignity and my helper among men;
> May never a day come when you are out of sight,
> On which I do not see you coming back to me.

She kept on weeping and wailing until dawn, when the neighbours came to ask her about her son. She told them about his meeting with the Persian and, sure that she would never see him again, she wandered around the house shedding tears. As she did so, she caught sight of some lines written on the wall and so she fetched a scholar who read them for her. They ran as follows:

> When I am overcome by drowsiness, a phantom comes at dawn
> While my companions lie in the desert, still asleep.
> But when I wake to see the one who came,
> The camp is empty and the dear one far away.

When she heard this, she exclaimed: 'Yes, my son, the camp is empty and you are too far off to visit!' after which her neighbours took their leave and went away, praying that she might be granted patience and a quick reunion with her son. She continued to weep all night and all day, and in the middle of the house she built a tomb, inscribed with Hasan's name and the date of his loss. From then on she would never leave it.

So much for her, but as for Hasan and the Persian, it turned out that the latter was a Magian who had a great hatred of Muslims, killing any of them who fell into his power. He was a vile and ignoble man, a treasure hunter, an alchemist and an evil-doer, who fitted the poet's description:

> A dog, the son and grandson of a line of dogs;
> What good is in a dog, whose forefathers were dogs?

Other lines also apply:

> A son of ignoble parents, dog, son of a devil,
> A bastard, child of sin, an infidel.

The name of this damned man was Bahram the Magian, and every year he used to kidnap a Muslim and kill him in order to open the way to a treasure.

After he had successfully seized Hasan, he sailed off with him from dawn till dusk, when the ship anchored off shore for the night. At sunrise it put to sea again, and he told his servants to bring him the chest in which Hasan was shut. They did this, and when he opened it, he brought out Hasan and made him inhale some vinegar, while blowing powder into his nostrils. Hasan sneezed, vomited up the drug and opened his eyes. He looked from right to left and found himself on the open sea with the ship under way and the Persian sitting beside him. He realized that the damned man had tricked him and that he had fallen into the trap his mother had warned him about. He recited the words that never bring shame on those who use them: 'There is no might and no power except with God, the Exalted, the Omnipotent; we belong to Him and to Him do we return,' adding: 'My God, be merciful in what You decree for me, and grant me patience to endure the affliction You have brought on me, O Lord of creation.' Then he turned to Bahram and said politely: 'What have you done, my father, and where is the tie of bread and salt and the oath that you swore to me?' 'Dog,' answered Bahram, 'does a man like me recognize such a tie? I have killed nine hundred and ninety-nine young men like you, and you will make up the thousand.' He then shouted at him, and Hasan kept quiet, realizing that he had been struck by the arrow of fate.

Night 782

Morning now dawned and Shahrazad broke off from what she had been allowed to say. Then, when it was the seven hundred and eighty-second night, she CONTINUED:

I have heard, O fortunate king, that when Hasan saw that he had fallen into the hands of the damned Persian, he spoke to him politely, but to no avail, for the man shouted at him. He kept quiet, realizing that he had been struck by the arrow of fate.

At this point Bahram ordered that he should be untied, and he was given a little water. Bahram said with a laugh: 'I swear by fire, light, shade and heat that I never thought that you would fall into my net, but the fire gave me power over you and helped me trap you, so that I might achieve my purpose and take you back as a sacrifice to win its favour.' 'You have betrayed the tie of bread and salt,' said Hasan, at which Bahram raised his hand and struck him so that he fell down, biting the deck with his teeth, with tears running down his cheeks. He fainted and Bahram then told his servants to light a fire. When Hasan asked what he was going to do with it, he said: 'Fire is the mistress of light and of sparks, and it is the fire that I worship. If you join me in this, I will give you half my wealth and marry you to my daughter.' At this, Hasan shouted at him: 'Damn you, you are nothing but an infidel Magian who worships fire in place of the Omnipotent God, the Creator of night and day. This is a disastrous religion.' 'Arab dog,' replied Bahram angrily, 'will you agree to adopt my religion?' When Hasan refused, Bahram got up and having prostrated himself to the fire, he ordered his servants to stretch him out, face downwards. He then set about beating him with a whip of plaited leather until he had cut open the sides of his body. All the while his victim was calling for help, but no one would come to his aid. He raised his eyes towards the Omnipotent God, asking the chosen Prophet to intercede for him and, when he could bear it no longer, with tears running down his cheeks like rain, he recited these lines:

> My God, I must endure the things You have decreed;
> I shall do this, if this is what You want.
> My enemies have power to injure me,
> But through Your grace, You may wipe out the past.

Bahram now told his slaves to allow Hasan to sit up, and to fetch him some food and drink, but when they did, Hasan refused what they offered. Throughout the voyage, Bahram kept torturing him night and day, his heart being hardened by the fact that Hasan put up with the pain and addressed his entreaties to the Great and Glorious God. This went on for three months, but at the end of this time God sent a wind against the ship; the sea darkened and, as the ship was tossed by the storm, the captain and the crew exclaimed: 'By God, all this is thanks to the young man whom the Magian has been torturing for three months, something that He does not allow!' They attacked Bahram and killed his servants and companions, at which, in fear of his life and being certain that he was about to be killed himself, he freed Hasan from his bonds and replaced his rags with other clothes. In order to make his peace with him, he promised to teach him alchemy and return him to his own country, pleading: 'Do not hold against me what I have done to you, my son.' 'How can I ever trust you again?' asked Hasan, to which Bahram replied: 'If there was no sin, there would be no pardon. The only reason that I acted as I did towards you was to test your powers of endurance and, as you know, all things are in the hands of God.'

The crew and the captain were glad that Hasan had been released, and he, for his part, called down blessings on them and gave praise and thanks to Almighty God. At that, the gale died down, the darkness cleared away and they sailed on with a fair wind. Hasan asked Bahram where he was heading, and Bahram told him: 'My son, I am making for the Cloud Mountain, on which is found the elixir used in alchemy,' and he went on to swear by fire and light that he would hide nothing from him. Hasan was happy to hear this, and now that his mind had been set at rest, he began to eat, drink and sleep with Bahram, and to wear his clothes.

The voyage continued for another three months, and after that the ship anchored off a long strip of shore composed of variously coloured pebbles, white, yellow, blue, black and so on. At this point, Bahram got up and told Hasan: 'Come on, disembark. We have reached our goal.' Hasan did as he was told and, after Bahram had told the captain to look after his belongings, the two of them walked off until they were out of sight of the ship. Bahram then sat down and produced from his pocket a little copper drum with a silk-covered plectrum, inscribed in gold with talismanic characters, which he beat. As soon as he had stopped, a dust cloud rose over the plain and Hasan, astonished by this, turned pale with fear. He began to regret having landed with Bahram, but Bahram looked at him and said: 'What is wrong with you, my son? I swear by the fire and the light that you have nothing to fear from me. Were it not for the fact that I cannot reach my goal without you, I would not have got you to leave the ship. You can be assured that all will go well for you, as the dust that you see is raised by our riding beasts, which will help us cross this plain without difficulty.'

Night 783

Morning now dawned and Shahrazad broke off from what she had been allowed to say. Then, when it was the seven hundred and eighty-third night, SHE CONTINUED:

I have heard, O fortunate king, that the Persian said to Hasan: 'This dust is raised by our riding beasts, which will help us cross this plain without difficulty.' Soon afterwards the dust cleared away to show three dromedaries, on two of which Bahram and Hasan mounted, while the third was loaded with their provisions. After a seven-day ride they reached a wide tract of country, and on entering it they caught sight of a dome supported on four columns of red gold. They dismounted from their dromedaries and ate, drank and rested under its shelter, and it was then that Hasan turned and saw something lofty. 'What is that, uncle?' he asked, and Bahram told him: 'It is a palace.' But when Hasan suggested that they should go there to rest and look around, Bahram went off, saying: 'Don't mention that place to me, for my enemy lives there. I am not going to tell you just now what happened between him and me.' He then beat the drum, the dromedaries came up, and both men mounted and rode on for another seven days.

On the eighth day, Bahram asked Hasan what he could see. 'I see a bank of clouds stretching from east to west,' Hasan replied, but Bahram said: 'Those aren't clouds but a huge, towering mountain on which the clouds split, as it is so immensely high that none of them can overshadow its peak. It is this mountain that I am going to, as what we need is there on its summit. This is why I have brought you here, for you will have to fetch it.' At that, Hasan despaired of life and said to Bahram: 'I conjure you by what you worship and by the religion in which you believe to tell me what is this thing for which you have brought me.' Bahram replied: 'Alchemy will not work without a herb which grows on the summit where the clouds split as they pass the mountain. After you have fetched it, I shall show you what the art entails.' Hasan was too afraid to say more than: 'Yes, master,' but, despairing of his life, he wept at being parted from his mother, his family and his country. He regretted not having followed his mother's advice, and he recited these lines:

Consider how your Lord can bring about
The swift release from suffering you desire.
Do not despair when evil fortune catches you;
How many wonders of God's grace misfortune brings!

They travelled on until, when they got to the mountain, they halted under its slopes. Hasan could see a palace there, but when he asked what this was, Bahram told him that it was the home of *jinn, ghuls* and devils. He dismounted, telling Hasan to do the same. He then went up, kissed Hasan's head and said: 'Don't blame me for what I did to you. I shall protect you as you climb up there, but I want you to swear that you will not trick me, whatever you bring back from it. You and I will share and share alike.' 'To hear is to obey,' said Hasan. Bahram then opened a bag from which he took a hand mill and a quantity of grain, which he ground up and then kneaded to form three flat loaves. He lit a fire, baked the loaves and then took out the copper drum, which he struck with the plectrum. The dromedaries came up and after selecting one of them, he slaughtered and skinned it. Turning to Hasan, he said: 'Listen to my instructions, Hasan, my son.' 'Yes,' replied Hasan, and Bahram went on: 'Get into this skin, and I will then sew you up in it and put you down on the ground. Vultures will arrive and carry you off to the summit of the mountain. Take this knife with you, and when they have brought you there, and you find that they have put you down, cut through the skin and come out. The vultures will be afraid of you and will fly away, after which you can look down at me from the summit and talk to me, and I shall be able to tell you what to do.'

Bahram now provided Hasan with the three loaves and a water skin, which he put in the camel skin with him, before sewing it up. A vulture then swooped down and, after picking up the skin, it flew off and deposited its load on the summit. When he realized this, Hasan cut himself free and went away to call down to Bahram. Bahram, hearing his voice, danced with joy and shouted up: 'Go back in the other direction and tell me what you see.' Hasan did as he was told, and what he saw were heaps of dry bones and a large quantity of wood. He told this to Bahram, who said: 'This is what we are looking for. Take six bundles of the wood and throw them down to me, for it is these that we shall use when we are making gold.' Hasan obeyed, and when Bahram saw that he had got the bundles, he shouted back up to Hasan: 'I have got what I wanted from you, you miserable fellow, and now you can stay on the mountain if you want, or else throw yourself down to your death.' He then went away.

Hasan exclaimed: 'There is no might and no power except with God, the Exalted, the Omnipotent! This dog has tricked me.' He sat there bemoaning his fate and he recited these lines:

> When God apportions man his fate,
> Although he may have wit, hearing and sight,
> Fate deafens him and blinds his heart,
> Drawing his wits out, as one pulls out a hair.
> Then, after His decree has come to pass,
> God sends him back his wits, so he may learn his lesson.
> You need not ask: 'How did this happen?'
> All things are bound by the decrees of fate.

Night 784

Morning now dawned and Shahrazad broke off from what she had been allowed to say. Then, when it was the seven hundred and eighty-fourth night, SHE CONTINUED:

I have heard, O fortunate king, that when the Magian had succeeded in getting Hasan to the summit of the mountain and Hasan had thrown down to the Magian what he needed, he cursed Hasan and went off, leaving him behind. Hasan exclaimed: 'There is no might and no power except with God, the Exalted, the Omnipotent! This damned dog has tricked me.' He stood up and, after turning right and left, he went along the summit ridge, convinced that his last hour had come. When he got to the far end, he saw that the mountain was flanked by a blue sea with foaming waves that were clashing together, each the size of an enormous hill. He sat down and, after having quoted appropriate passages from the Quran, he asked God Almighty to grant him either an easy death or else a release from his difficulties. He then recited for himself the Prayer of the Dead and threw himself down into the sea, where, thanks to the power of the Almighty, the waves carried him safely to shore. In his gladness, he praised and thanked God, before starting to walk in search of something to eat. He got as far as the place where he had halted with Bahram, and then, after walking on for a time, he found himself confronted by a huge palace, towering into the sky. This was where, in answer to Hasan's question, Bahram had told him that his enemy lived, and Hasan told himself: 'By God, I must go in, as it may be that here I will find relief from my sufferings.'

When he came to the palace he found the door open, and as he entered he saw a bench in the entrance hall on which were seated two girls as radiant as moons, playing a game of chess on a board that was set out in front of them. One of them looked up at him and called out joyfully: 'By God, this is a human, and I think it must be the one Bahram the Magian brought here this year.' When Hasan heard this, he threw himself down in front of them, burst into floods of tears and said: 'By God, ladies, I am indeed that poor wretch.' The younger of the two said to her elder sister: 'Sister, I call you to witness that I take this man as my brother in accordance with the bond of God's covenant. I shall die with his death, live with his life, rejoice with his joy and grieve at his sorrow.' She went up, embraced and kissed him and then, taking him by the hand, she led him into the palace, accompanied by her sister. She removed the rags that he was wearing and replaced them with royal robes, after which she brought him dishes laden with all kinds of foods.

As the two sisters sat and ate with him, they said: 'Tell us what happened to you with that evil dog of a magician, from the time that you fell into his hands until your escape, and then we shall tell you of our own experience with him from start to finish, so that you may be on your guard against him when you see him.' When Hasan heard this and saw how warmly they welcomed him, he relaxed and recovered his composure. He then started to tell them his story from beginning to end, and when they said: 'Did you ask him about this palace?' he told them that he had, but that Bahram had said: 'I don't like to talk about it, because it belongs to devils and demons.' The girls were furious and asked: 'Did that infidel really say that about us?' Hasan confirmed that he had, and the younger girl, his 'sister', exclaimed: 'By God, I shall see to it that he dies the worst of deaths and rob him of the breath of life!' 'How will you get to him in order to kill him, for he is a treacherous magician?' Hasan asked, and she told him: 'He is in a garden called al-Mushayyad, and I must certainly kill him soon.' Her sister said: 'Hasan has given us a true and accurate account of this dog, and so it is now time for you to tell the whole of our own story to him, so as to fix it in his mind.'

The younger girl began: 'You must know, my brother, that we are princesses and that our father is one of the great kings of the *jinn*, with *marids* as soldiers, guards and servants. The Almighty provided him with seven daughters, all born of the same mother, but such is his boundless stupidity, jealousy and pride that he refused to marry us to anyone. Instead, he summoned his viziers and his companions, and asked them whether they knew of somewhere inaccessible both to men and to *jinn*, with many trees, fruits and streams. When they asked what he wanted with such a place, he told them it was to install his seven daughters in it. "Your majesty," they said, "the palace of the Cloud Mountain would suit them. It was built by one of the rebellious *marids* in the time of our lord Solomon, on whom be peace, and since the death of its builder no one, human or *jinn*, has lived there, as it is cut off and inaccessible. It is surrounded by trees, fruits and streams, and the water that flows around it is sweeter than honey and colder than snow. Whoever suffers from any type of leprosy or any other disease will instantly be cured if he drinks it." When our father heard this, he sent us here with an escort of his troops and he supplied us with everything that we might need. Whenever he wants to ride out, he beats a drum to assemble all his troops, from whom he takes his pick, and the rest disperse. When he wants us to come to him, he orders the magicians among his servants to fetch us, and they come and take us to him so that he can enjoy our company, and we can see to whatever we want done before being returned here. Our five other sisters have gone out to hunt in the countryside here, which is filled with vast quantities of wild game, and two of us take it in turns to stay behind and cook. As it is our turn, this is what my sister and I are doing here now. We used to pray to the Almighty to send us a mortal to keep us company, and praise be to Him for having sent us you. You can relax happily, for no harm will come to you.'

Hasan was delighted and exclaimed: 'Praise be to God, Who led me on the way to safety and touched the hearts of these girls with pity for me.' His 'sister' then took him by the hand and led him to a room where she produced furniture and furnishings such as no one else among God's creation could provide. Some time later, her sisters returned from their hunt; they were delighted to hear about Hasan and went to his room, where they greeted him and congratulated him on his escape. He stayed with them, leading the most pleasant and delightful of lives, and he started to go out hunting with them, slaughtering beasts and enjoying their company. Things went on like this until his health was restored and he recovered from the effects of his sufferings. His body grew strong and he put on bulk and weight because of the good treatment that he received during his stay with the princesses. He would join them as they walked through the sumptuously decorated palace and among the gardens and their flowers. They would do their best to please him and engage him in friendly conversation until all feelings of loneliness left him. They themselves became more and more pleased with him, but the pleasure that he took in them was even greater.

His young 'sister' told the others how Bahram the Magian had called them devils, demons and *ghuls*, and they swore to kill him. In the following year the damned man came back, bringing with him a handsome young Muslim as radiant as the moon, whom he had fettered and tortured viciously. He camped beneath the castle where Hasan had met the princesses, and Hasan himself happened to be sitting under the trees by a stream. At the sight of Bahram, his heart beat fast, he changed colour and he struck his hands together . . .

Night 785

Morning now dawned and Shahrazad broke off from what she had been allowed to say. Then, when it was the seven hundred and eighty-fifth night, SHE CONTINUED:

I have heard, O fortunate king, that at the sight of Bahram, his heart beat fast, he changed colour and he struck his hands together, calling out to the girls: 'For God's sake, sisters, help me kill this damned fellow, for here he is, and he is within your grasp. He has a prisoner with him, a young Muslim from the upper classes, on whom he is inflicting all kinds of painful tortures. I want to give myself the satisfaction of killing him and to win a reward from God by freeing the young man from his tortures. If he is returned to his own land and reunited with his brothers, his family and his friends, that will count as an act of charity on your part and the Almighty will reward you.' 'We shall obey both God and you, Hasan,' they told him, after which they put on mouth-veils, armour and sword belts. They provided Hasan with an excellent horse, armed him fully and gave him a fine sword, and after that they all rode out together. They found that Bahram had slaughtered a camel and skinned it. He was busy punishing the young Muslim and telling him to get into its skin when Hasan came up behind him without attracting his attention. Bahram was dismayed and confounded when Hasan shouted at him and came up, calling out: 'Take your hands off him, damn you, you enemy of God and of the Muslims, you treacherous dog, fire worshipper and follower of the path of evil. Do you worship fire and light and swear by darkness and heat?'

Bahram turned and, on seeing Hasan, he exclaimed: 'My son, how did you escape and who brought you down from the mountain?' 'It was Almighty God Who rescued me,' said Hasan, 'and it is He Who has put your life into the hands of your enemies. You tortured me throughout our journey, you unbelieving infidel, but now you have taken a wrong turning and fallen into misfortune. You have no mother to help you, nor any brother or friend, and there is no firm covenant to protect you. You said: "God takes revenge on whoever betrays the tie of bread and salt," and this was the tie that you yourself betrayed. God has now caused you to fall into my power, and there is no rescue for you near at hand.' 'My son,' said Bahram, 'I swear by God that you are dearer to me than life and than the light of my eyes,' but Hasan advanced on him and quickly struck him such a blow on his shoulder that his sword emerged gleaming from the other side of his body, and God hurried his soul to hell, an evil resting place.

Hasan now took Bahram's bag and opened it, taking out the drum as well as the plectrum, with which he struck it. Quick as lightning, up came the dromedaries, and when Hasan had freed the young man from his fetters, he mounted him on one of them and loaded the other with stores of food and water. 'Go off wherever you want,' he said, and the young man left, having been freed from his predicament by Almighty God through the agency of Hasan. When the princesses saw what Hasan had done to Bahram, they were overjoyed and clustered around him, admiring his courage and strength, thanking him for his action and congratulating him on his safety. 'What you have done is enough to cure the sick and to win the favour of the Great God,' they told him, and he and they went back to the palace.

Hasan continued to live there with the princesses, eating and drinking, playing and laughing, and so much did he enjoy his stay that he forgot about his mother. Then, while he was enjoying the pleasantest of lives with them, suddenly a huge dust cloud rose up from the far side of the plain, darkening the sky. 'Hasan,' the princesses told him, 'go and hide away in your room or, if you prefer, go to the garden and conceal yourself among the trees and the vines. No harm will come to you.' So he went into the palace and locked himself in his room.

When the dust cleared away, a huge army could be seen beneath it, advancing like a raging sea. This had been sent by the girls' father, and when it arrived they provided the newcomers with the best of lodgings and supplied them with guest provisions for three days. After that, they asked their visitors about their mission and were told: 'We have come for you from the king.' 'What does he want of us?' they asked, and the envoys said: 'One of the kings is giving a wedding feast and your father wants you to be there for the celebrations.' When they said: 'How long will we be away from here?' they were told that the journey there and back, together with the stay, would take two months. So they went back into the palace and explained to Hasan what had happened. 'Treat this place as your own,' they said. 'Take your pleasure and enjoy yourself without fear or sorrow. No one can come and so there is no need for you to be disturbed or upset while we are away. Here are the keys of our rooms, which we shall leave with you, but we ask you, brother, for the sake of our fraternal bond, not to open this one door. That is something there is no need for you to do.'

They then said goodbye to Hasan and went off with the escort, leaving him alone in the palace. He soon became depressed and impatient, and as his feelings of gloom and loneliness increased, he bitterly regretted their absence, and spacious as the palace was, he found it too small for him. In his isolation and loneliness he recited these lines, with the princesses in mind:

All this wide space I find too small;
It fills my every thought with gloom.
The dear ones have departed, and my happy days
Are darkened, while my tears flow for their loss.
When they abandoned me, sleep left my eyes,
And all my inmost thoughts were turned to grief.
Will time allow us once again to meet?
Will our delightful evenings come once more?

Night 786

Morning now dawned and Shahrazad broke off from what she had been allowed to say. Then, when it was the seven hundred and eighty-sixth night, SHE CONTINUED:

I have heard, O fortunate king, that when the princesses left him he sat by himself in the palace, depressed by their departure. He took to going out on solitary hunting trips, bringing back slaughtered game and eating on his own. Becoming more and more depressed and disturbed by his loneliness, he started to wander round the palace, investigating all its parts. He opened the rooms of the princesses and, although he found there treasures enough to astound all who looked at them, he could still find no pleasure in any of this because of their absence. A fire of curiosity then began to consume him because of the door which his 'sister' had told him not to approach and never to open. 'She must have told me this because there is something there that she doesn't want anyone to see,' he said to himself, and he swore that he would go and open it to find what was in it, even if what was there was death.

So he took the key and opened the door. Inside there was nothing to be seen in the way of treasure, but at the upper end of the room was a flight of stairs, vaulted over with Yemeni onyx. He climbed the stairs, wondering why he had been told not to do this, and they took him to the palace roof. From there he looked down on fields and orchards, filled with trees and flowers and swarming with wild beasts and birds, which were singing to the glory of Almighty God, the One, the Omnipotent. As he looked at the pleasure gardens, he saw a surging sea with clashing waves, and when he continued to look around, he caught sight of a pavilion raised on four columns. Within it he saw a room set with precious stones of all types – sapphires, emeralds, hyacinth gems and other jewels – and in its walls gold bricks alternated with bricks of silver, together with others comprising sapphires and green emeralds. In the middle there was a pool filled with water and covered with a trellis of sandalwood and aloes wood, with a lattice whose bars were of red gold and green emerald, set with jewels of all kinds, including pearls the size of pigeon's eggs. Beside this pool was a bench of aloes wood studded with pearls and other gems and fronted with tracery of red gold. In it were gemstones and precious metals of various types set opposite one another, and around it birds were hymning the glory of the Almighty in their varied and beautiful songs. Neither Chosroe nor Caesar had ever possessed a place like this, and Hasan was amazed by the sight of it. He sat down to look around, marvelling at the beauty of its construction, the splendour of its various jewels and the excellence of its craftsmanship. He also admired the fields and the birds, as they praised the One Omnipotent God.

As he was studying the monument left by the builder whom He had enabled in His majesty to construct this pavilion, he caught sight of ten birds flying from the landward side towards the palace and the pool. Realizing that they were coming to drink, he hid himself for fear that, if they saw him, they might be frightened off. They first settled on a large and lovely tree and, as they then circled it, Hasan's attention was drawn to the most beautiful of them, a fine large bird around which the others fluttered as though they were its servants. It dominated them, pecking at them with its beak as they fled away from it. Then, as Hasan watched from a distance, they alighted on the bench and each of them used its claws to tear at its skin, before emerging from what turned out to be dresses of feathers. Out of these stepped ten virgins, whose beauty put to shame the splendour of the moon and who, when they were naked, went down into the water, where they bathed, played and enjoyed themselves. Their leader threw them down and ducked them as they tried to escape, but they could never get the better of her.

At the sight of this girl, Hasan lost his wits; he realized that this was why the princesses had told him not to open the door, for such was the leader's beauty and grace, together with the elegance of her figure, that, as she played and sported, splashing water, he fell desperately in love with her. He stood there, regretting that he could not join the bathers, and bemused by the loveliness of their leader. Love for her had captured him in its toils, and while he followed her with his eyes, its fire blazed up in his heart, for 'the soul prompts man to evil'.* He wept tears of longing for her beauty, and his heart flamed with passion, the sparks of whose flame could not be quenched and the effects of whose ardour were not to be concealed.

At this point the girls came out of the pool, while, unseen by them, Hasan was watching, lost in admiration of their beauty, elegance and gracefulness. As he turned, he caught a glimpse of their leader in her nakedness, and between her thighs he could see a large, rounded dome with four pillars, like a bowl of silver or crystal. He remembered the poet's lines:

> I lifted her dress to show her cleft,
> Narrow as my nature and my livelihood.
> I plunged halfway in until she sighed.
> 'Why do you sigh?' I asked, and she said: 'For the rest.'†

On leaving the pool, the girls put on their dresses and their ornaments. The leader's dress was green, and in her loveliness she outshone all the beauties of the world, while the splendour of her face eclipsed the shining moon at its full. She was more supple than a swaying branch and she distracted the mind with sinful thoughts. She was as the poet has described:

> A girl appeared in all her liveliness;
> The sun had borrowed brightness from her cheek.
> She wore a shift of green,
> Green as the branch that hides the pomegranate.
> I asked her what she called her dress,
> And she replied with subtlety:
> 'We pierce our lovers' inmost hearts,
> As a soft breeze blows, which penetrates the seat of love.'

Night 787

Morning now dawned and Shahrazad broke off from what she had been allowed to say. Then, when it was the seven hundred and eighty-seventh night, SHE CONTINUED:

I have heard, O fortunate king, that when Hasan saw the girls coming

---

* Quran 12.53.   † cf. Night 124, which also contains these lines.

out of the pool, the beauty and grace of their leader stole away his wits and he recited those lines. After they had dressed, they sat talking and laughing, still watched by Hasan, who was drowning in the sea of love and astray in the valley of his thoughts. He was telling himself: 'It must have been because she was afraid that I might fall in love with one of these girls that my "sister" told me not to open the door.' He sat there studying the beauty of their leader, who was the loveliest thing God had created in her age, and who surpassed all mankind in her grace. Her mouth was like the ring of Solomon, her hair was blacker than the night of rejection for a wretched lover, her forehead was like the new moon on the festival of Ramadan, her eyes were like those of a gazelle and her shining nose was curved. Her cheeks were like red anemones, her lips were like coral and her teeth were pearls set in a necklace of gold, while her neck was a silver ingot placed above what was like the branch of a *ban* tree. In her belly were folds and nooks, causing the passionate lover to cry out to God, and her navel could contain an ounce of the most fragrant musk. She had plump thighs like marble columns or cushions filled with ostrich feathers, and between them was something roofed and pillared that looked like a large hill or a hare with flattened ears. In the beauty of her form she surpassed the branches of the *ban* tree or the shoots of the bamboo, and she was as the poet described:

She is a girl the moisture of whose mouth is honey,
With a glance more piercing than a sharp Indian sword;
In her movements she shames the *ban* tree's branch,
And when she smiles, the lightning flashes from her teeth.
I compared her cheeks to roses that bloom in line,
But she turned away and said: 'He has no shame
Who likens me to a rose, my breasts to pomegranates.
Have pomegranates a bough that can bear my breasts?
I swear by my beauty, my eyes and my heart's blood,
By the paradise of my union and the hell of my aversion,
If he says this again, I shall forbid to him
The sweetness of my union and burn him with the fire of my rejection.
They say that gardens are adorned with lines of roses,
But these are not the roses of my cheeks, nor is their branch my figure.
If he can find my equal in a garden,
What does he look for when he comes to me?'

The girls continued to laugh and play as Hasan stood watching them, forgetting about food and drink, until it was almost time for the afternoon prayer. Then their leader said to the others: 'Princesses, time is getting on and our lands are far away. We have stayed here long enough, so let us get up and leave for home.' At this they all rose, and when they had put on their feathered dresses, they changed back to birds, as they had been to start with, and then they all flew off together, with their leader in the middle of the flock.

Hasan was left in despair, and when he wanted to go down from his vantage point he was too feeble to get up. Tears poured down his cheeks, and in an excess of passion he recited these lines:

> May I be denied the fulfilment of your pledge
> If, after you have gone, I have enjoyed my sleep.
> Since then I have not closed my eyes,
> Nor taken any pleasure in my rest.
> I seem to see you all the while I dream;
> Would that my dreams were true.
> I need no sleep, but it is dear to me,
> As it may bring you to me in my dreams.

He walked for a little without finding his way, but when he got down from the roof to the bottom of the palace, he crawled to the door of his room, went in and locked it. He then took to his sickbed, neither eating nor drinking but drowning in a sea of cares. He shed tears, lamenting his fate, until morning came, when he recited these lines:

> At evening the birds took flight –
> And one who dies of love has done no wrong.
> While I am able, I shall keep my love concealed,
> But if longing conquers me, it will be shown.
> The phantom of my love, who shines like dawn, came here by night,
> But this long night of passion has no dawn.
> I weep for the absent, while the carefree sleep;
> I am a plaything for the winds of love.
> I am generous with my tears, my wealth and my heart's blood,
> My mind, my soul, and this should profit me.
> The worst of all disasters and distress
> Comes through contention with the lovely girls.
> Union with them is said to be forbidden,

Although to shed the blood of lovers is allowed.
What can the lovesick do but sacrifice his life,
Giving it freely in the game of love?
Longing and passion make me cry to her;
All the distracted lover can do is to sob.

When the sun rose, Hasan reopened the door of the forbidden room and went back to sit opposite the pavilion in the place where he had been on the previous day, but although he stayed until evening waiting for the birds, not one of them came. He wept so bitterly that he collapsed fainting on the ground. On recovering he crawled down to the bottom of the palace, as night fell and his world closed in on him. All night long he wept and lamented his fate, until morning came and the sun rose over the hills and plains. He could neither eat, drink, nor sleep; he found no rest and spent his days in confusion and his nights awake, dazed and drunk with care, thanks to the violence of his passion. He recited the lines of a lovelorn poet:

You shame the gleaming morning sun,
And, without knowing it, humiliate the branch.
Will Time allow you ever to return,
To quench the fires that burn within my heart,
United, as we meet with an embrace,
Your cheek on mine and your neck next to mine?
Who talks of sweetness to be found in love?
Its days are sourer than the bitter aloes.

Night 788

Morning now dawned and Shahrazad broke off from what she had been allowed to say. Then, when it was the seven hundred and eighty-eighth night, SHE CONTINUED:

I have heard, O fortunate king, that when the pangs of Hasan's love increased, he recited poetry to himself while he was alone in the palace, finding no one to console him. While he was suffering from these pangs, he saw a dust cloud rising over the plain and hurried down below to hide himself, as he realized that the palace ladies had come back. Soon afterwards their escort dismounted and spread out around the palace, while the seven princesses came in to take off their weapons and their armour, all except the youngest, Hasan's 'sister'. Instead of disarming, she went straight to Hasan's apartments, and when she failed to find him there, she searched for him until she came across him in a small room. He was weak, emaciated and exhausted; his bones had become brittle; he was pallid, with sunken eyes. All this was because he had taken little food or drink and because of the many tears he had shed in his love for the bird girl.

When his 'sister' saw the state he was in, she was disconcerted and distraught. She asked him what was wrong and what had happened to him, saying: 'Tell me, my brother, so that I may find some way of removing whatever it is that is harming you, and act as your ransom.' Hasan wept bitterly and recited these lines:

When the beloved is not here,
The lover only meets distress and harm.
Passion shows without, while sickness lies within;
Love starts with memory and ends with care.

On hearing this, his 'sister' was filled with admiration for his eloquence, command of language, choice of words and the fact that he had chosen to reply to her in verse. 'My brother,' she asked, 'when did you fall into this state and when did this happen to you, leading you, as I can see, to express yourself in poetry, while you rain down tears? In God's Name, I call on you, by the sanctity of the love there is between us, to tell me what is wrong. Let me know your secret and don't hide from me anything that happened while we were away, as I am distressed and unhappy for your sake.'

Hasan sighed, shed tears like rain and said: 'I'm afraid that if I tell you, you will not help me to get what I want, but will abandon me to die choked with grief.' 'No, by God, brother,' she replied, 'I shall never abandon you, even if it costs me my life.' So he told her what had happened and what he had seen when he opened the door, explaining that the reason for his present distress was that he had fallen in love with the girl whom he had seen, as a result of which for ten days he had tasted neither food nor drink. He then wept and recited the following lines:

Put back my heart where it was in my breast;
Allow my eyes to close in sleep and then leave me.
Do you imagine that Time's passage changed
The covenant of love? May no one yield to change.

The princess sympathized and joined in his tears, in pity for his life of exile. Then she said: 'Take comfort, my brother, for I shall risk my life in your company and sacrifice it in order to bring you pleasure. Even if it ruins me and brings about my death, I shall contrive some scheme to help you achieve your purpose, if God Almighty wills it. But take my advice and keep your secret from my sisters; don't let any one of them know of your condition, lest this lead to my death and to yours, and if they ask you about the door, say that you never opened it. Tell them that you were preoccupied by our absence and in your loneliness you sat by yourself in the palace.' Hasan approved of her advice and kissed her head. He had been afraid of what she might do when she learned that he had opened the door, but at this point he relaxed, his mind at rest. His fear had brought him to the point of death, but now his spirits were restored and he asked for something to eat.

The princess went off tearfully and sadly to her sisters, and when they asked her what was wrong, she told them that she was concerned about Hasan, who was sick and had not eaten for ten days. When they asked why, she told them that it was their absence that had made him lonely, explaining: 'The days that we were away were longer than a thousand years for him, and he is to be excused, for he is a solitary stranger and we left him alone with nobody to keep him company or to cheer him. After all, he is a very young man and it may be that he was remembering his family and his old mother, thinking of her weeping for him night and day in constant grief, whereas while we were with him, we used to divert him.' When the others heard this, they shed tears of sympathy and said: 'By God, he is to be excused.' Then, after they had gone out to dismiss the escort, they went on to greet Hasan. They discovered that he had lost his good looks, that his complexion was pallid and his body emaciated, and they wept in pity for him. They sat with him to keep him company and to console him by talking to him and telling of all the wonders and marvels that they had seen, as well as of the bridegroom's encounter with his bride.

For a whole month they sat with Hasan, treating him with friendship and kindness, but every day his illness grew worse, and they shed bitter tears to see him in such a state, although none of them wept as much as their youngest sister. Then, after a month, the others wanted to go out hunting and, when they had decided on this, they asked her to come with them. She said: 'By God, my sisters, I cannot go with you while my brother is in such a state. Until he gets better and this illness of his leaves him, I shall stay to distract him.' On hearing this, they thanked her for her good nature and told her: 'God will reward you for all you do for this stranger.' Then they left her in the palace with Hasan and rode off, taking provisions for twenty days.

Night 789

Morning now dawned and Shahrazad broke off from what she had been allowed to say. Then, when it was the seven hundred and eighty-ninth night, SHE CONTINUED:

I have heard, O fortunate king, that when the princesses mounted and rode off to hunt, they left their youngest sister sitting with Hasan in the palace. When she was sure that they must be a long way off, the princess went to Hasan and told him to get up and show her the place where he had seen the bird girls. In his delight at this, he willingly agreed, convinced that he would now reach his goal, but when he tried to rise to do this, he found himself unable to walk, and so she had to carry him in her arms. She brought him to the door and opened it for him, after which she carried him up the stairs to the roof of the palace. When they got there, he showed her the place where he saw the bird girls, together with the bench and the pool. She asked him to describe the girls for her and how they had come, after which he told her what he had seen, concentrating in particular on their leader, with whom he had fallen in love. When she heard what he had to say, she recognized the description and became pale and shaken. Hasan asked why, and she told him: 'You must know, brother, that this girl is the daughter of one of the great kings of the *jinn*; her father rules over men and *jinn*, magicians and sorcerers, tribes of *'ifrits*, regions and lands, as well as many islands. He has vast riches and our own father is one of his lieutenants. No one can stand against him because of the size of his armies and the extent of his kingdom and his wealth. He has given his daughters, whom you saw, a tract of land a whole year's journey in length and breadth, encircled by a huge river and inaccessible to both men and *jinn*. He has an army of twenty-five thousand girls, trained to use swords and spears. Any one of them, when mounted and armed, can hold her own against a thousand brave riders, while his seven daughters are a match, and more than a match, for them in courage and horsemanship. This region which I have told you about has been given by the king into the charge of the eldest, the senior of his daughters, who combines bravery and skill as a rider with guile, deceit and magical skill, powers through which she can overcome everyone in her realm, while the others who were with her were her state officials, guards and intimates. Their feather dresses, which they used in order to fly, are the work of *jinn* magicians.

'If you want to lay hold of their leader and marry her, sit here and wait, for they come at the start of each month. When they arrive, hide yourself and make sure that they don't see you, for that would cost us our lives. Pay attention to what I tell you and remember it. Sit near enough to see them without being seen, and when they take off their feather dresses, keep your eyes on the one that belongs to their leader, the one whom you want. Take it, but don't touch anything else, for it is this that allows her to get home, and if you have it, then you will have her as well. But take care not to let her trick you, for she will say: "Whoever you are who stole my dress, give it back, for here am I before you and in your power." If you give it to her, she will kill you, destroy our palaces and kill our father, so be sure you know what you are doing. When the others see that her dress has been taken, they will fly away, leaving her sitting there alone, and then you can go and drag her towards you by her hair. You will have mastered her and she will be yours, but you must keep the feather dress, since while you have it she will be your captive – without it she cannot get back to her own land. When you have got her, carry her to your room, but don't tell her that it was you who took the dress.'

Hasan's mind was set at rest by what he had heard and his dismay and pain left him. He got to his feet and kissed his 'sister's' head, after which the two of them went down from the roof to sleep. Hasan attended to himself until, when morning came and the sun rose, he got up, and climbed to the roof, where he sat until evening. His 'sister' brought him food and drink, as well as a change of clothes, after which he fell asleep, and she continued to help him like this until the first day of the next month. The sight of the new moon raised his expectations and it was at that point that he suddenly caught sight of the birds swooping down like bolts of lightning. He hid himself where he could see without being seen.

The birds settled each in her own place, and then each of them, including their leader, removed their dresses. Hasan was close at hand, and when the leader and her companions went into the pool, he gradually moved forward, keeping under cover and sheltered by God. Not one of the girls saw him as he took the dress, since they were busy laughing and playing with each other. When they had finished, they got out of the pool and put on their feather dresses, but their leader could not find hers. She shrieked, struck her face and tore what she was wearing, at which her companions came up and asked what was wrong. When she told them that she had lost her feathered dress, they too wept, shrieked and slapped their faces, but with the approach of night they could not stay with her any longer and had to abandon her where she was.

Night 790

Morning now dawned and Shahrazad broke off from what she had been allowed to say. Then, when it was the seven hundred and ninetieth night, SHE CONTINUED:

I have heard, O fortunate king, that when Hasan stole the girl's dress, she looked for it but failed to find it and her sisters flew off, leaving her alone. When Hasan saw that they had flown off out of sight, he listened and heard their leader say: 'Whoever you are who has stolen my dress and has left me naked, I ask you to give it back to me and shelter my nudity, may God never cause you to taste distress like mine.' When he heard this, love robbed him of his wits, and so strong was his emotion that he was unable to endure it any longer. He started up and ran at her, seizing her and pulling her towards him. He took her down to the bottom of the palace and put her in his room, throwing her his cloak, as she wept and gnawed at her fingers. He then locked the door and went to tell his 'sister' that he had captured her and that she was sitting crying and biting her fingers. When his 'sister' heard this, she went with him to the room, and, on going in, she found the girl weeping in grief. She kissed the ground in front of her and greeted her, but the girl said: 'Princess, is this the kind of evil treatment that people like you inflict on the daughters of kings? You know well enough that my father is a great king and that all the other kings of the *jinn* are afraid of him and fear his might. He has an invincible array of sorcerers, magicians and diviners, together with devils and *marids*, and his subjects are so numerous that none but God can count them. How is it right for princesses like you to shelter mortals among you and to allow them to look at our affairs and yours, for how could this man have got here otherwise?' Hasan's 'sister' replied: 'Princess, this is a chivalrous man, who has no dishonourable intentions. He loves you, and it is only for men that women were created.

If he were not in love with you, he would not have fallen so ill that he was on the point of death.' She repeated everything that Hasan had said about his love for her and how he had seen the girls fly in and then bathe; how she was the only one of them whom he admired, the others being her servants; and how he had seen her ducking them in the pool without their being able to lay a hand on her.

When she heard this, she gave up hope of escape and Hasan's 'sister' got up and left her, returning later with a splendid robe which she gave her to wear, as well as food and drink, which she shared with her until the girl had taken heart and recovered from her alarm. Hasan's 'sister' kept on talking to her soothingly and gently, saying: 'Have mercy on him who only saw you once but fell victim to your love,' and she went on like this, trying to win her favour with fine words and expressions, as her captive wept. By the time dawn broke, however, the girl had become more cheerful and had stopped weeping, realizing that there was no way out of the trap into which she had fallen. 'Princess,' she said, 'this is what God has decreed for me – that I should live as a stranger, cut off from my country, my family and my sisters – and there is nothing for it but to endure with patience what God has decreed.'

She was then given the finest room in the castle for herself, and the princess stayed with her, consoling and comforting her until she became reconciled to her lot and laughed cheerfully, as the sorrow and distress that she felt at being separated from her family, her homeland, sisters, parents and kingdom left her. At that point, the princess went to Hasan and told him to go to the girl's room and to kiss her hands and feet. He did this and then, after kissing her between the eyes, he said: 'Queen of beauties, life of the soul, delight of the watchers, be at ease, for I have only taken you in order to serve you as a slave until the Last Day, while my "sister" here is your servant. What I want is to marry you in accordance with the ordinance of God and of His Prophet. Then I shall go home, and you and I will live in Baghdad, where I shall buy you slaves, male and female. I have a mother, one of the best of women, who will be at your service; there is no other country better than ours; everything there is better than anything that can be found anywhere else or among any other folk, and its inhabitants are good people with bright faces.'

While he was talking in an attempt to entertain her, and she, for her part, was saying nothing at all, a knock came at the palace gate. When Hasan went out to see who was there, he found that the other princesses had come back from their hunt. He was delighted to see them, and, after his first greeting, they wished him health and safety and he did the same for them. Then they dismounted and, after having entered the palace, each of them went to her own room to take off her worn clothes and put on something better. During their hunt they had caught a large number of gazelles, wild cows, hares, lions, hyenas and other beasts, some of which they slaughtered, while others they kept with them in the palace. Hasan stood among them, with his clothes belted around his waist, doing the butchering for them, as they played around happily, enjoying themselves to the full. When the butchering was finished, they sat down to prepare a meal and Hasan came up to the eldest and kissed her head, after which he went on to kiss the heads of the others, one after the other. 'You are being very respectful to us, brother,' they told him, 'and it is surprising how extremely affectionate you are towards us. But it is we and not you who should be doing this, as you are our superior, you being human while we are *jinn*.'

At this, Hasan began to weep bitterly and they asked: 'What is wrong? What has made you cry? You are filling us with sadness by weeping now, and you seem to be longing for your mother and your own land. If that is so, we shall fit you out for the journey and send you back there to your dear ones.' 'By God,' he told them, 'I do not want to leave you,' and so they asked: 'If that is so, then which of us has caused you such distress?' Hasan was ashamed to say that the reason for this was love, lest they disapprove, so he stayed silent and told them nothing about his condition. It was then that his 'sister' got up and said: 'He has caught a bird from out of the air and he wants you to help him tame it.' The others turned to him and said: 'We are all at your service and we will do whatever you want, but tell us the story and don't keep anything back.' Hasan then said to his 'sister': 'You tell them, for I am too ashamed to talk to them about this.'

Night 791

Morning now dawned and Shahrazad broke off from what she had been allowed to say. Then, when it was the seven hundred and ninety-first night, SHE CONTINUED:

I have heard, O fortunate king, that Hasan said to his 'sister': 'Tell them my story, for I am too ashamed to talk to them about this.' She then told her sisters: 'When we went off and left this poor man alone,

he found the castle oppressive and was in fear of intruders, for you know
how light-witted the sons of Adam are. In his depression and loneliness
he opened the door leading to the palace roof, and he went up on to it
and sat looking down into the valley in the direction of the gate, for fear
of an attack. One day, as he was sitting there, ten birds approached,
making for the palace, and they came down to perch by the pool above
the pavilion. He stared at the most beautiful of the ten, which was
pecking at the others, who could not resist. They then used their claws
to rip open their feathered dresses from the neck down, and, on stepping
out of these, they showed themselves to be girls, as radiant as full moons.
While Hasan watched, they undressed and plunged into the pool, where
they could not resist their leader as she ducked them, she having the
most beautiful face of them all, with the best figure and the finest dress.
They went on enjoying themselves, with Hasan watching, until it was
almost time for the afternoon prayer, when they left the pool and dressed
themselves. Then, having wrapped themselves in their feather dresses,
they flew away. Hasan could think of nothing else; his heart was on fire
with love for their leader, and he regretted that he had not stolen her
dress. So he stayed on the roof waiting for her without food, drink or
sleep, and that went on until the next new moon. Then, as he was sitting
there, he saw them coming again as they had done before. This time,
after they had taken off their feathers and gone down into the pool, he
took the leader's dress, realizing that she could not fly without it. He hid
it, fearing that they might come out and kill him, and after waiting for
the others to fly away, he emerged, seized her and brought her down
from the palace roof.'

Her sisters now asked where the girl was, and she said: 'She is with
him in such-and-such a room.' 'Describe her for us,' they told her, and
she told them: 'She is fairer than the full moon, her face is more radiant
than the sun, her mouth tastes sweeter than honey and her figure is more
graceful than a bough. She has black eyes, a bright face and a luminous
forehead. Her bosom is like a jewel, with breasts like pomegranates, and
her cheeks are like apples. She has a wrinkled belly and a navel like an
ivory casket filled with musk. Her legs are like pillars of marble and she
captures hearts with her dark eyes, her slender waist and her heavy
buttocks, while her speech could cure the sick. She has a lovely figure
and a beautiful smile that shines out like the full moon.' When her sisters
heard this description they turned to Hasan and said: 'Show her to us.'
So, lovesick as he was, he took them to the room where she was, opened

the door and went in ahead of them. At the sight of the captive's beauty they were struck with admiration for her graceful form and her elegance, and, after kissing the ground in front of her, they greeted her and said: 'By God, daughter of the great king, this is a serious business. Had you heard how this mortal has described you as a woman, you would have become his lifelong admirer. He is deeply in love with you, and he is not looking for a dishonourable liaison but for legal marriage. If we knew that girls could do without men, we would stop him, although he came to you himself rather than sending you a messenger. He told us that he burned your dress of feathers, or else we would have taken it from him.'

One of the sisters then, with the girl's consent, acted as her deputy, drawing up a marriage contract between her and Hasan; he then took her hand in his and she was married to him with her consent. The sisters produced a wedding feast suitable for a princess, before bringing the bridegroom to the bride. Hasan then opened the gate, removed the barrier and broke the seal. The passion of his love for his bride increased and multiplied, and when he had reached his goal he congratulated himself and recited the following lines:

> Girl with dark eyes, your figure fascinates;
> Your face sheds drops of beauty;
> The finest picture stands before my eyes,
> Half sapphire, one-third gemstone,
> One-fifth of musk and one-sixth ambergris.
> You are a pearl, but yet more dazzling.
> None of Eve's daughters can compare with you,
> Nor is your equal in the *jinn* of Paradise.
> Torment me if you wish, for this is how love acts,
> But were you to forgive me, this must be your choice.
> You are the world's adornment, and the end of all desires;
> And who can bear to lose the beauty of your face?

Night 792

Morning now dawned and Shahrazad broke off from what she had been allowed to say. Then, when it was the seven hundred and ninety-second night, SHE CONTINUED:

I have heard, O fortunate king, that when Hasan lay with the princess and deflowered her, he experienced the greatest pleasure. His love for her and his passion increased, at which he recited those lines about her.

When the sisters, who were standing at the door, heard these lines, they exclaimed: 'Princess, how can you blame us, now that you have heard the poetry that this mortal recites for love of you?' and the bride herself was filled with happiness and delight. Hasan stayed with her for forty days in a state of joy, pleasure and gladness, on every day of which the sisters provided him with another feast, as well as showering him with favours and rare gifts. He was happy and content, and his bride enjoyed her stay there so much that she forgot her family. At the end of this time, however, Hasan saw in his sleep a vision of his mother, whose bones had become brittle and her body emaciated in her grief for him; she was sallow and her condition had worsened, while he, for his part, was flourishing. When she saw this, she exclaimed: 'Hasan, my son, how can you live at ease and forget me? See what has happened to me after you left! I do not forget you; I shall never stop repeating your name until I die and, so that I may always remember you, I have built a tomb for you in the house. Shall I live to see you back with me, so that we may be together again as we were before?'

Hasan woke up weeping and wailing, with tears running down his cheeks like rain. He became sad and gloomy; he wept constantly and, as he could neither sleep nor rest, his powers of endurance left him. When the princesses came to enjoy themselves with him in their usual morning visit, he paid no attention to them, and when they asked his bride what was wrong, she said that she did not know. They pressed her to ask him, and so she went up to him, and when she had put the question to him, he sighed unhappily and told her what he had seen in his dream. Then he recited these lines:

We are uneasy and are at a loss;
We seek a dear one, but can find no way.
Love's miseries increase for us;
Even its lightness has proved burdensome.

His wife told the princesses what he had said, and when they heard the lines he had recited, they pitied his suffering and told him: 'Please, do what you want. By God, we cannot stop you from visiting your mother, and we shall do everything that we can to help you. On the other hand, you must not cut yourself off from us altogether, and you will have to visit us at least once a year.' 'To hear is to obey,' he replied, and they immediately prepared provisions for him and equipped his bride with ornaments, robes and expensive trappings of all kinds, such as defied description, together with treasures past numbering. Then they beat a drum, at which dromedaries arrived from all directions, and from these they chose enough to carry all the baggage that they had prepared. They provided mounts for Hasan and his bride, and loaded twenty-five chests with gold and fifty with silver. For three days, during which they covered a three-month journey, they accompanied the pair, before saying goodbye to them. When they were about to go back, Hasan's 'sister' embraced him and burst into tears until she fainted. Then, when she recovered, she recited these lines:

> May there never be a day of parting,
> Robbing the eyes of sleep.
> We are separated, you and I,
> And our bodies have lost their strength.

When she had finished, she took her leave of him, impressing on him that, after he had gone home and set his heart at ease by meeting his mother, he must remember to visit her for six months every year. She told him: 'If something worries you, or if you fear some disaster, strike the Magian's drum, and when the dromedaries arrive, be sure to mount and come back to us.' Hasan swore to do that, and he now pressed the princesses to go back. They took their leave and parted sadly, the saddest of all being his young 'sister', who was unable to rest or to endure her loss, and who spent her nights and days in tears.

So much for them, but as for Hasan, he travelled night and day with his bride, crossing plains, deserts, valleys and rough ground in the noonday heat and at early dawn. As God had decreed, they arrived safely at Basra, and they rode through the city before halting their dromedaries at the door of Hasan's house. He sent them off and, as he went up to open the door, he heard the low sound of his mother's broken-hearted weeping. Tortured by the fire of grief, she was reciting these lines:

> How can the sleepless sleep,
> Who lies awake at night while others rest?
> Here was a man with wealth, family and power,
> But he became a solitary stranger.
> Between his ribs there is a burning coal,
> A longing with a violence unsurpassed.

The power of passion has him in its grip;
Strong though he is, his sufferings make him weep.
The love he feels has caused him to appear
Sad and distressed, as witnessed by his tears.

When Hasan listened to his mother's tearful lament, he shed tears of
his own before knocking loudly on the door. 'Who is there?' his mother
asked, and he said: 'Open up.' She did this, and then, when she looked
at him and recognized him, she collapsed in a faint. He looked after her
gently until, when she had recovered, he embraced her and she returned
his embrace and kissed her. Then, as his bride watched them, he brought
all his baggage inside the house, and when his mother had recovered her
composure after the reunion with her son, she recited:

Time has pitied me,
Regretting my long distress.
It has given me my desire,
Removing all my fears.
Its past sins are forgiven,
Including the white parting of my hair.

Night 793

Morning now dawned and Shahrazad broke off from what she had been
allowed to say. Then, when it was the seven hundred and ninety-third
night, SHE CONTINUED:
I have heard, O fortunate king, that as Hasan and his mother sat
talking, she asked him how he had got on with the Persian, and he told
her: 'That was no Persian but a Magian who worshipped fire rather than
the Omnipotent God.' He went on to tell her how the man had taken
him away on a journey and then sewed him into a camel's skin, to be
carried off by birds and put down on the summit of a mountain. He
explained how he had seen a large number of corpses, these being the
victims of the Magian's trickery whom he had abandoned there after
they had served his purpose. He then told her how he had thrown himself
off the mountain into the sea, and how Almighty God had saved him
and brought him to the palace of the princesses. One of them had
adopted him as a brother, and he had stayed there until God had brought
him the Magian, whom he had killed. He then explained how he had

fallen in love with the bird girl and how he had caught her, telling his mother everything that had happened before his marriage.

She was astonished by this account, and she praised God for his health and safe return. She looked at what he had brought in the way of baggage and was delighted when, in answer to her question, he told her what was in it. She then went up to talk to his bride and to entertain her. The sight of the girl's loveliness filled her with astonishment and admiration for her beauty and grace and her well-shaped figure. 'Praise be to God, my son,' she exclaimed again, 'that you have come back safe and sound!' and she sat beside the bride, making her welcome and setting her mind at ease. Early next morning she went to the market, where she bought ten of the finest sets of clothes to be found in the city, together with splendid furnishings, and she gave the robes to her daughter-in-law, decking her out with ornaments of all kinds.

'With all this money we will not be able to go on living in Basra,' she told Hasan, 'for, as you know, we have been poor, and people will suspect that we have been practising alchemy. We had better go to Baghdad, the City of Peace, where we shall be under the protection of the caliph and where you will be able to sit in a shop, buying and selling. If you fear the Great and Glorious God, He will see to it that this wealth serves you well.' Hasan listened to this with approval. He wasted no time in getting up and going out; he sold the house, summoned the dromedaries and loaded them with all his wealth and possessions, after which he mounted his mother and his wife on them and set off. When he got to the Tigris, he hired a boat to take him to Baghdad, to which he transferred the ladies, together with all the goods that he had brought with him. He then went on board himself, and the boat sailed with a favourable wind for ten days until, to the delight of the passengers, they came within sight of Baghdad. When they got to the city, Hasan immediately disembarked and went off to hire a storeroom in one of the *khans*, to which he moved his belongings.

He spent the night in the *khan* and the next morning, having changed his clothes, he approached the market agent, who asked him what he wanted. 'I want a nice, spacious house,' Hasan told him, and when the man showed him what he had in this line, Hasan was attracted by one that belonged to a vizier, which he now bought for a hundred thousand gold dinars, paying over the sum in cash. He then went back to the *khan* where he was staying, and transferred everything to his house, after which he went to the market and bought all the household utensils, furnishings and so on that he needed, together with servants, among them a young black house boy.

For three years he stayed there contentedly, enjoying the most pleasurable and happiest of lives with his wife, who gave birth to two sons, one of whom was named Nasir and the other Mansur. After a while, however, he remembered with longing his *jinn* sisters, recalling how good they had been to him and how they had helped him in his quest. So he went out to the markets, where he bought ornaments, precious materials and dried fruits such as they would never have known or seen before. His mother asked the reason for this, and he told her: 'I have decided to go to see my sisters, who were so good to me and to whose kindness and generosity I owe my good fortune. I want to set off to visit them, but I shall soon be back, if God Almighty wills it.' 'Don't leave me for long, my son,' she said, and he then told her what she had to do with his wife. 'Her feather dress is in a chest buried in the ground,' he explained, 'and you must make sure that she does not take it, for otherwise she will fly off with the children and I shall never hear of them again and so die of sorrow. Take care not to mention it to her, for you must know that she is the daughter of the most powerful of the *jinn* kings, with the largest army and the greatest wealth. She is the queen of her people and her father's favourite. She is also very proud, so wait on her yourself, but don't let her go out of the door or look out of the window or over the wall, for I am afraid that even a breath of wind might harm her. If anything happens to her, I shall kill myself.' 'God forbid that I should not do what you tell me, my son,' his mother replied. 'I should be mad to disobey you, now that you have given me these instructions. You can go off with an easy mind, and, God willing, when you come back safely, she will tell you, when you see her, how I have looked after her. But don't stay away any longer than it takes you to make the journey.'

Night 794

Morning now dawned and Shahrazad broke off from what she had been allowed to say. Then, when it was the seven hundred and ninety-fourth night, SHE CONTINUED:

I have heard, O fortunate king, that when Hasan was intent on going to visit the *jinn* princesses, he instructed his mother, as I mentioned, to look after his wife. As fate had decreed, his wife had been listening to what he said to his mother, although neither of them knew it.

Hasan then went outside the city to beat the drum, and when the dromedaries arrived, he loaded them with twenty bales of Iraqi treasures. He said goodbye to his mother, his wife and his sons, one of whom was a year old and the other two, and after he had repeated his instructions to his mother, he mounted and set off on his journey to the *jinn* princesses. For ten nights and days he rode by valleys, mountains, plains and rough ground before arriving at their palace on the eleventh. He went in carrying the gifts that he had brought for them and, when they saw him, they joyfully congratulated him on his safe arrival. His 'sister' adorned the palace with decorations both inside and out, and when they had accepted his presents, they gave him a room to himself as usual, before asking him about his mother and his wife. He told them that his wife had given birth to two sons, and his young 'sister', delighted to find him well, recited the lines:

I ask the passing wind about you,
For no one else but you comes to my mind.

For three months he stayed with them as an honoured guest, hunting and enjoying a happy and pleasant life. So much for him, but as for his mother and his wife, on the third day after he had left, his wife exclaimed tearfully to his mother that, although she had been there for three years, she had never been to the baths. His mother felt sorry for her and said: 'My daughter, we are strangers in Baghdad and your husband is not in town. If he were, he would look after you, but I myself know nobody here. What I shall do, however, is to heat water for you and to wash your head in the bathroom at home.' 'Lady,' said Hasan's wife, 'had you said this to one of your slave girls, she would have refused to stay with you and demanded to be sold in the market. Men are to be excused, as they are jealous creatures and they imagine that if a woman leaves her house she is going to commit adultery, but not all women are alike. You know that when one of us has set her heart on something, no one can get the better of her, or guard and protect her, or keep her away from the baths or from anywhere else, for she will do exactly what she wants.' She then broke into tears, cursing her fate and lamenting her misfortunes and her exile. Hasan's mother felt sorry for her and realized that what the girl said would have to be done. So she promptly prepared everything that would be needed in the baths and then took her daughter-in-law there. When they had entered and undressed, all the women there began to stare at the younger of the pair, praising God, the Great and Glorious, as they noted the beautiful form that He had created. Others, who were passing the baths, began to come in to enjoy the spectacle, and, as word spread throughout the city, more and more women crowded around, so that, because of the numbers, no one could make their way through the baths. This was something quite out of the ordinary, and, as it happened, one of those who went there was one of the caliph Harun al-Rashid's slave girls, a lute player named Tuhfa. When she saw the crowds of women and girls flocking around the baths and blocking the way through, she asked what was happening and was told about Hasan's wife. She went in to look for herself, and the sight of such beauty and grace left her bemused. She praised God, the Glorious, the Creator of this loveliness, and went no further in but sat there without washing, lost in admiration until the girl had finished her own ablutions and came out wearing a dress that added beauty to beauty. From the steam room she went to sit down among cushions on a rug, but when she turned and saw the women staring at her, she left.

Tuhfa got up and went out with her until she discovered where she lived, after which she left and returned to the caliph's palace. She went to the Lady Zubaida and kissed the ground in front of her. Zubaida then asked why she had been so long at the baths, and Tuhfa told her: 'My lady, I have come across a marvel, the like of which I have never seen before either among men or women. It was this that preoccupied me, and it so bemused and bewildered me that I never even washed my head.' Zubaida asked what the marvel was, and Tuhfa said: 'In the baths I saw a girl with two little boys like two radiant moons. She herself was more beautiful than anyone who has ever been or will be, and there can be no one in the whole world to match her. I swear by your grace, my lady, that if you tell the Commander of the Faithful about her, he will kill her husband and take her from him, as she has no equal among women. I asked about her husband and was told that he was a merchant called Hasan of Basra. When she left the baths, I followed her home and discovered that she is living in the vizier's house with two doors, one facing the river and the other the land. My fear is that if the Commander of the Faithful hears about her, he will break the law by killing her husband and taking her as a wife.'

Night 795

Morning now dawned and Shahrazad broke off from what she had been allowed to say. Then, when it was the seven hundred and ninety-fifth night, SHE CONTINUED:

I have heard, O fortunate king, that when the slave girl of the Commander of the Faithful saw the wife of Hasan of Basra and described her beauty to the Lady Zubaida, she said: 'My lady, I am afraid that the Commander of the Faithful will hear of her. This might lead him to break the law, kill her husband and marry her.' Zubaida exclaimed at that and said: 'Is this girl really so beautiful that the caliph would be prepared to sell his religion for worldly pleasure and break God's law for her sake? By God, I must have a look at her, and if she is not as you have described her, I shall have your head cut off. You whore, in the caliph's harem there are three hundred and sixty girls, one for each day of the year, and not one of them has the qualities that you mentioned.' 'By God, lady,' Tuhfa said, 'this girl has no equal in all Baghdad or among the Persians or the Arabs, for the Great and Glorious God has created no one else like her.'

At this point Zubaida sent for Masrur, who kissed the ground in front of her when he arrived. She told him: 'Go to the house of the vizier with two doors, one facing the river and the other the land, and fetch me quickly and without delay the girl whom you will find there, together with her children and the old woman who is there with her.' 'To hear is to obey,' replied Masrur, and he then left and went to the vizier's house, where he knocked on the door. Out came Hasan's old mother to ask who was there, and when he replied: 'Masrur, the eunuch of the Commander of the Faithful,' she opened the door for him. He came in and, when they had exchanged greetings, she asked what he wanted. He said: 'The Lady Zubaida, the daughter of al-Qasim, the wife of the Commander of the Faithful, Harun al-Rashid, the sixth of the line of al-'Abbas, the paternal uncle of the Prophet, may God bless him and give him peace, summons you and the wife of your son, of whose beauty her women have told her, together with her children.' 'Masrur,' said the old woman, 'we are strangers in Baghdad and my son, the girl's husband, is not here at the moment, but he told me not to go out with her to meet anyone at all. I am afraid that if he comes back and finds that anything has happened, he will kill himself. Please, then, don't try to make me do something that I cannot.' Masrur replied: 'Lady, if I knew that there was anything for you to be afraid of here, I would not make you go, but all that the Lady Zubaida wants is to see your daughter-in-law, before letting her come back. Don't disobey me, for you would regret it, and I shall take you and bring you back here safely, if God Almighty wills it.'

Hasan's mother could not refuse and so she went into the house, prepared her daughter-in-law and brought her out, together with her children. Masrur led the way to the caliph's palace, where he took them to Zubaida. They kissed the ground before her and called down blessings on her, the girl remaining veiled all the while. Zubaida asked her to unveil so that she could see her face, and after kissing the ground again, she did this, revealing a face that would put to shame the moon in the vault of heaven. Zubaida stared, feasting her eyes on her, astonished by her beauty, as the radiance of her face illumined the palace, and all who saw her were so enchanted that no one could say a word. Zubaida got up and, raising the girl to her feet, she clasped her to her breast, before seating her on the couch beside her. On her orders the palace was adorned with decorations, and a most splendid dress, together with a necklace of magnificent gems, were brought and given to the girl to wear.

Zubaida then said: 'Queen of beauties, who has filled my eyes with admiration, what are your skills?' 'My lady,' replied the girl, 'I have a dress of feathers and, were I to wear it in your presence, you would be struck with wonder at the supreme skill that I would display to you, about whose excellence people would talk for generation after generation.' 'Where is this dress?' Zubaida asked her, and she said: 'My mother-in-law has it, so ask her to give it to me.' Zubaida told the old woman to fetch it so that they might all watch the girl's display, after which she would hand it back again, but the old woman said: 'She is a liar. Have you ever seen any woman with a dress of feathers? Only birds have these.' The girl insisted that she did have such a dress and that it was in a chest buried under the storeroom of her house, at which Zubaida removed from her throat a jewelled necklace worth all the treasuries of Chosroe and Caesar. 'Take this, mother,' she said, handing it over to her, 'and then please go and fetch the dress for us to look at, and then you can have it again.' The old woman swore that she had never seen it and did not know where it was, but Zubaida shouted at her and took her key. She summoned Masrur, and when he came, she told him: 'Take this key and go back to the house. Open it and then go to the storeroom' – whose door she described for him – 'in the middle of which is a chest.

Bring it out, break it open and then bring me the dress of feathers that
you will find inside it.'

Night 796

Morning now dawned and Shahrazad broke off from what she had been
allowed to say. Then, when it was the seven hundred and ninety-sixth
night, SHE CONTINUED:

I have heard, O fortunate king, that when the Lady Zubaida took the
key from Hasan's mother and gave it to Masrur, she told him: 'Take this
key and open the storeroom' – which she described for him – 'then take
out the chest, break it open and remove the dress of feathers that is in
it. Then bring this to me.' 'To hear is to obey,' replied Masrur.

He took the key from Zubaida and went off with Hasan's mother,
who was in tears and full of regret that she had let the girl go to the
baths with her, as this had been a trick on the girl's part. She went into
the house with Masrur and opened the storeroom door, after which he
entered and brought out the chest, from which he removed the dress of
feathers. He wrapped it in a towel and brought it to the Lady Zubaida,
who took it and turned it over, admiring how beautifully it had been
made. She handed it to the girl and asked: 'Is this your feather dress?'
'Yes, my lady,' she replied, and she gladly reached out, took it and
inspected it, finding, to her joy, that no harm had come to it and that it
was just as it had been, with no single feather missing. She got up from
beside Zubaida, opened up the dress and slipped into it, holding her
children in her arms. Then, through the power of the Great and Glorious
God and to the amazement of Zubaida and all those who were present,
she turned into a bird. She swayed, walked, danced and played under
the astonished gaze of the spectators. 'Is this good, ladies?' she asked,
speaking clearly, and they replied: 'Yes, indeed, queen of the beauties.
Everything you do is good.' 'But what I am going to do is better,' she
told them, after which she spread her wings and flew up with her children
to perch on top of the domed roof of the palace overlooking Zubaida's
hall. The ladies stared at her and said: 'By God, this is amazingly good
and something that we have never seen before.'

The bird girl was about to fly off back to her own land when she
thought of Hasan and, after telling the ladies to listen, she recited these
lines:

You who have gone from here, hurrying
In your flight to those you love,
Do you think I am at ease with you,
And that you did not darken my life with distress?
When I was captured and fell in the toils of love,
He made his love my prison, though my home was far away.
My dress was hidden, and he felt assured
That I would not ask for it from the One Great God.
He told his mother she must keep it safe,
Hidden in a room, wronging and injuring me.
I heard their conversation and remembered it,
Hoping that benefits would now shower down on me.
I went off to the baths in order that I might
Dazzle the minds of those who looked at me.
The wife of al-Rashid admired my loveliness,
Looking at me from the right and from the left.
I said: 'Wife of the caliph, I possess
A dress of splendid feathers, a source of pride.
If I wore it, the wonders you would see
Would wipe out all distress and disperse cares.'
She asked me where it was, and I replied:
'It is in the house of the one who hid it.'
Masrur hurried to fetch it for her,
And brought it back, gleaming and resplendent.
I took it from his hand and spread it out,
Looking inside it, checking all its buttons.
I put it on and, taking up my children,
I spread my wings and flew away.
Mother of Hasan, tell him when he comes,
If he wants union with me, let him leave his home.

When she had had finished these lines, Zubaida said to her: 'Queen of beauties, will you not come down to us, so that we may enjoy your loveliness? Praise be to God, Who gave you such eloquence and grace.' 'What has gone will not return,' she replied, and then, addressing the mother of poor, miserable Hasan, she said: 'I shall miss you, but if, when your son comes back and finds the period of our separation long, and if he looks for a reunion, as the winds of love and longing blow over him, let him come to the islands of Waq.'

She then flew off with her children, making for her own country, and when her mother-in-law saw this, she burst into tears, slapping her face and sobbing until she fainted. When she recovered, Zubaida addressed her respectfully and said: 'I had no idea that this would happen, and had you told me about her, I would not have interfered with you, but I have only just realized that she is one of the flying *jinn*. If I had known this before, I would never have let her put on the dress or take her children, so do not hold this against me.' Hasan's mother, finding that there was nothing she could do, said: 'I do not hold you responsible.' She left the palace and went back home, where she started to slap her face again until she fainted. Then, after recovering her senses, she expressed her sorrow for the loss of her daughter-in-law and her grandchildren, as well as for the absence of her son, in the following lines:

Your departure on the day of separation made me weep,
Grieving that you had left these lands.
I cried out in distress at the pain of parting,
My eyelids ulcerated by my tears.
After this parting, can there be return?
I can no longer hide my feelings, now that you are gone.
Would that they might keep faith by coming back,
And if they do, my happiness may return.

She got up and dug three graves in the house, over which she wept night and day. When, after Hasan's long absence, her distress and sorrowful longing increased, she recited:

Your image is between my eyelids;
Disturbed or resting, I remember you.
Your love flows through my bones,
As juices penetrate fruit on a tree.
I am sad when I cannot look at you,
And even censurers excuse my sorrow.
Love has reached out to lay its hands on me,
But it is madness that commands this love.
Fear God in what you do to me; have pity;
Your love has made me taste the bitterness of death.

Night 797

Morning now dawned and Shahrazad broke off from what she had been allowed to say. Then, when it was the seven hundred and ninety-seventh night, SHE CONTINUED:

I have heard, O fortunate king, that Hasan's mother began to weep day and night for the loss of her son, his wife and his children. So much for her, but as for Hasan, when he reached the *jinn* princesses, they insisted that he must stay with them for three months. After that they provided him with money, together with five loads of gold, another five of silver and one of provisions. They accompanied him on the start of his journey, until he insisted that they turn back, at which they came up to take their leave of him by embracing him. When the youngest of them did this, she wept so bitterly that she fainted, and when she recovered she recited:

> When will the fire of parting be quenched by your return?
> When will my wish be granted and our old life restored?
> The day of parting brings distress and fear,
> And this leave-taking weakens me more and more.

The next sister then came up, embraced Hasan and recited:

> To leave you is like taking leave of life,
> To lose you is to lose the dearest of my friends.
> When you went off, fire burned my heart,
> While to be near you is like Paradise.

Then the third came up and, after embracing him, recited:

> When we parted, we did not end our leave-taking
> Because we tired of it or for an evil cause.
> For without question, you are my true soul.
> How can one choose to take leave of one's soul?

When the fourth had embraced him, she recited:

> It was the talk of parting made me weep,
> When he told me of this in secret, as he took his leave.
> The pearl that once hung on my ear
> Has now been left to fall down as a tear.

The fifth embraced him and recited:

Do not go, for I cannot bear your loss.
And as you leave, I cannot say goodbye.
I have no power to bear this separation;
There are no tears left for the empty camp.

The sixth recited after embracing him:

The caravan left that took them off from me,
And longing plundered my heart's blood.
I said: 'Had I the power of a king,
I would seize every ship by force.'*

The seventh embraced him and recited:

When you see that it is time to part, endure
And do not be alarmed by this parting.
Look for a quick return,
For with a change of letters, 'farewell' spells 'return'.†

She added two more lines:

I am distressed that we must part,
And I have not the heart to say goodbye.
God knows the reason I do not do this
Is fear that it might melt your heart.

It was then Hasan's turn to say goodbye, and the prospect of parting
made him weep until he fainted. Then, on recovering, he recited:

On the day of parting my tears flowed
Like pearls on threaded necklaces.
As the leader sang to urge his camels on,
Heart, patience and endurance all failed me.
I took my leave and sadly turned away,
Leaving the friendship of familiar haunts.
I went back, though I did not know the way,
Consoled by hopes to see you on return.
Listen, my friend, to what I say of love,
And may your heart not be deaf to my words.
Soul, when you part from them, you part
From all life's pleasure; do not hope for length of life.

* Quran 18.78.   † Arabic: *wada'a* / *'ada.*

He then pressed on, travelling night and day until he reached Baghdad, the City of Peace, the sanctuary of the 'Abbasid caliphate. He went home to greet his mother, knowing nothing of what had happened after he had left on his travels. He found that she had become emaciated, with brittle bones. She had been weeping and wailing sleeplessly for so long that she had become as thin as a tooth-pick and she was not able to reply to him. He dismissed his dromedaries and then, going up to her, he asked her about his wife and children, but she burst into tears and fainted. When he saw the state she was in, he searched the house for his family, but could find no trace of them. Then he looked at the storeroom and the chest and discovered them both open, with the feather dress gone. At this point he realized that his wife must have got hold of it and flown off, taking her children with her. He returned to his mother and, finding that she had recovered consciousness, he asked her about them. She told him tearfully: 'My son, may God reward you well for them. Here are their three graves.' When he heard his mother saying this, he gave a great cry and collapsed unconscious, remaining in that state from the beginning of the day until noon. This piled sorrow on sorrow for his mother, who despaired of his life.

When he eventually recovered, he wept, slapped his face, tore his clothes and wandered aimlessly around the house, before reciting these lines:

As far as I can, I conceal my love,
But fires of passion are not quenched.
It may be that for some this fire is mixed,
But as for me, I drink love neat.

He went on:

Others before me have complained of parting;
Fear of one's love's remoteness haunts the living and the dead,
But never have I seen or heard of passion
To match the love I nurture in my heart.

When he had finished, he drew his sword, went up to his mother and said: 'Unless you tell me the truth, I shall cut off your head and then kill myself.' 'Don't do that, my son,' she said. 'I shall tell you, but sheathe your sword and sit down, so I can let you know what happened.' He did that and, as he sat beside her, she told him the story from beginning to end. 'My son,' she explained, 'I found your wife in tears because she wanted to go to the baths, and I was afraid that when you came back she would complain to you, and you would be angry with me. Had it not been for that, I would not have taken her there, and had the Lady Zubaida not been angry with me, taking the key from me by force, I would not have brought out the dress, even if it cost me my life, but, as you know, my son, no force can match that of the caliphate. When they brought her the dress, she took it and turned it over and over, thinking that there might be something missing. When she discovered that it was undamaged, she was delighted. Then she took the children and tied them to her waist before putting on the dress, and this was after the Lady Zubaida had removed all the ornaments that she herself was wearing and handed them to her as an act of favour and an acknowledgement of her beauty. When she had put on the dress, she shook herself and became a bird, moving through the palace, where everyone watched her, admiring her beauty and grace. She then flew up to the palace roof and looked at me. "If, when your son comes back, he finds the nights of separation long, and if he looks for a reunion, as the winds of love and longing blow over him, let him leave his country and come to the islands of Waq." This is what happened to her while you were away.'

Night 798

Morning now dawned and Shahrazad broke off from what she had been allowed to say. Then, when it was the seven hundred and ninety-eighth night, SHE CONTINUED:

I have heard, O fortunate king, that when Hasan had listened to all that his mother had to tell him about what his wife had done when she flew away, he gave another great cry and fell down in a faint, which lasted until the end of the day. When he recovered, he struck his face and began to writhe on the ground like a snake, as his mother sat weeping at his head until midnight. Then he recovered and, shedding bitter tears, he recited:

Stop to look at the state of the one whom you are leaving;
It may be that after harshness you will turn to pity.
Because of his sickness, you will not recognize the man you see,
As though this was a stranger, quite unknown.
Here is a man slain by his love for you,

Who would be counted dead but for his groans.
Do not think parting to be some light thing;
To the longing lover it is harder to bear than death.

When he finished these lines, he got up and started to wander around the house, sobbing, weeping and lamenting. This went on for five days, during which he tasted neither food nor drink, and although his mother implored him to stop weeping, he would not listen to her but went on with his expressions of grief, paying no attention to her as she tried to console him. Then he recited these lines:

Is this the reward of a consort's love?
Is this the nature of black-eyed gazelles?
Does a honey hive drip from the beloved's lips,
Or do they offer wine for sale?
Tell me the story of those killed by love,
For consolation breathes life for the sad.
You need not come by night for fear of blame,
Nor are you the first whose resolution was ensnared by love.

He continued in this tearful state until dawn, when he fell asleep, and then in a dream he saw his wife weeping sorrowfully. With a cry, he woke with a start and recited:

Your phantom never leaves me for a minute;
It holds the place of honour in my heart.
Only the hope of union lets me live.
I only sleep to see you in my dreams.

In the morning he wept and wailed even more bitterly, and for a whole month he remained tearful and sad at heart, shunning both sleep and food. At the end of this time, it occurred to him that he should visit his *jinn* sisters in the hope that they might help him in his quest to recover his wife. He called up the dromedaries, loaded fifty of them with Iraqi treasures and mounted another. He put his mother in charge of the house, where he had left only a few of his belongings, and left the rest in store. Then he set out for his sisters, hoping for help, and he travelled on until he arrived at their palace on the Cloud Mountain. When he went in to greet them, they were delighted with the gifts that he had brought for them, and congratulated him on his safe arrival. Then they asked: 'But why have you come back so soon, brother? You have not been away for more than two months.' At that, Hasan burst into tears and recited:

> I see my soul troubled by the loss of my beloved,
> Taking no pleasure in the sweetness of life.
> My illness is one with no known cure,
> For none can cure it who was not its cause.
> You robbed me of sweet sleep, and then left me
> To ask the wind about you as it blows,
> Whether it saw you lately, you, my love,
> Whose beauties call my eyes to shed their tears.
> Were you to come close to her land,
> A sweet breath might perhaps revive the heart.

When he had finished these lines, he gave a great cry and fell down unconscious. The princesses sat around him weeping until he had recovered, and he then recited:

> It may be that perhaps Time will turn rein,
> To fetch my love, but it is envious.
> It may fetch happiness and bring me what I need;
> And after what has happened, something else may come.

Having finished, he wept until he fainted, and when he recovered he recited:

> By God, are you content, you, who are now
> My final sickness, as I am content to love?
> Do you abandon me for no good cause?
> Join me and pity your abandoned lover.

On finishing, he wept until he fainted and then, on recovery, he recited:

> Sleep left me to the embrace of sleeplessness;
> My eye is prodigal with its hoarded tears.
> It sheds carnelian drops because of love,
> Whose flow increases more and more with time.
> Lovers, this longing has produced for me
> A fire of love that burns between my ribs.
> When I remember you, each tear that flows
> Has in it lightning mixed with thunderclaps.

When he had finished he wept until he fainted, and when he recovered
he recited:

> Does your approach to love and suffering match my own,
> And is my love for you the same as yours for me?
> May God curse love for this, its bitterness;
> I wish I knew what it can want from me.
> Distant we may be, but your lovely face
> Is there before me everywhere I am.
> My heart is full of memories of your clan,
> And cooing doves stir up my grief.
> Dove, you who call all night long to your mate,
> Increasing longing and so grieving me,
> You left eyelids unwearied by their tears
> For those who have now vanished from my sight.
> I yearn for them in every passing hour,
> While in the dark of night my longing stirs.

His 'sister' heard this and went out to find him lying unconscious. She
shrieked and slapped her face and, on hearing her, her sisters came and
surrounded the prostrate Hasan, weeping over him. It was clear to them
when they saw him that he was afflicted by passionate love and longing,
and when he recovered and they asked him what was wrong, he told
them tearfully what had happened in his absence, when his wife had
taken her children and flown away. They grieved for him and asked him
what she had said when she left. 'Sisters,' he told them, 'she told my
mother: "When your son comes, tell him that if he finds the nights of
separation long and is shaken by the winds of love and longing, he
should come to the islands of Waq if he wants to be reunited with me."'
On hearing this, the princesses thought the matter over, gesturing and
looking at one another, as Hasan watched. For a time they stared down
at the ground, but then they raised their heads and recited the formula:
'There is no might and no power except with God, the Exalted, the
Omnipotent,' adding: 'Stretch out your hand to heaven and, if you can
touch it, then you will be able to reach your wife . . .'

Night 799

Morning now dawned and Shahrazad broke off from what she had been allowed to say. Then, when it was the seven hundred and ninety-ninth night, SHE CONTINUED:

I have heard, O fortunate king, that the princesses said to Hasan: 'Stretch out your hand to heaven and, if you can touch it, then you will be able to reach your wife and your children.'

Tears rained down Hasan's cheeks, soaking his clothes, and he recited:

Her red cheeks and her eyes have stirred my love;
Patience has left me and I cannot sleep.
The cruelty of the soft, fair girls has left me gaunt;
There is no breath of life left here for men to note –
Black-eyed girls moving like the sand gazelles,
Showing such beauty as would ensnare saints.
They walk like the dawn breeze among the fields,
But what their love brings me is sorrow and distress.
One lovely girl among them holds my hopes;
Because of her, my heart is burned by blazing fire,
A soft girl who sways as she walks,
Dawn in her face and twilight in her hair.
It is she who roused my longing, and how many men
Are moved to love by fair girls' cheeks and eyes!

When he had finished, the princesses joined in his tears out of sympathy and concern for him. They gently urged him to show patience, and prayed that he would be reunited with his wife. His 'sister' then went up to him and said: 'Console yourself and take comfort, brother. If you are patient, you will reach your goal, for whoever shows patience and perseverance will get what he wants, since patience is the key to a happy outcome, as the poet has said:

Allow free rein to destiny,
And pass your nights absolved from care.
Within the time your eye can blink,
God changes one condition to another.

Take heart and be resolute, for a child destined to live for ten years will not die when he is nine. The only thing that tears, distress and sorrow produce is sickness, so stay with us until you are rested, and then, if God Almighty wills it, I shall find a way for you to reach your wife and children.'

Hasan, however, shed more tears and recited these lines:

The sickness of my body may be cured,
But my heart's sickness is incurable.
The only remedy for the disease of love
Is for the lover to join the one he loves.

He then sat down beside his 'sister' and she began to talk to him consolingly and to ask him how it was that his wife had come to leave. When he told her, she said: 'By God, brother, I had meant to tell you to burn the feather dress, but Satan made me forget.'

She went on talking gently to him, but after a time he grew even more distressed and recited:

My heart is held by a beloved friend,
And God's decrees are not to be turned back.
In her is all the beauty of the Arabs,
A gazelle, but one that pastures on my heart.
I have few means or patience to confront this love;
I weep, although my tears can do no good.
She is a lovely girl of seven and seven years,
Resembling a moon of fourteen nights.

When the princess saw the state of his lovesickness and passion, she went tearfully and sorrowfully to her sisters and threw herself on them, weeping, kissing their feet and begging them to help Hasan to win back his wife and children, and asking them to promise to find a way of getting him to the islands of Waq. She went on crying until they joined in her tears and comforted her, saying that they would do their best to see that he rejoined his family, if this was the will of Almighty God.

Hasan stayed with them for a whole year, during which he never ceased to shed tears. They had an uncle, a full brother of their father, whose name was 'Abd al-Quddus and who was very fond of the eldest of them, visiting her once a year to see to her affairs. He had been delighted when the princesses had told him the story of Hasan's encounter with the Magian and of how he had managed to kill him. He had given the eldest a bag containing incense, telling her: 'If anything happens to worry or to annoy you, or if there is anything that you need, throw this incense on the fire and call on my name. I will come quickly and settle things for you.' He had said that on the first day of the year, and the eldest princess now told one of her sisters: 'The year is almost over and my uncle has not come. Go and fetch flint and steel and bring me the incense container.' The girl brought this gladly and, after opening it, she got out a pinch of incense and handed it to her sister, who took it and threw it on the fire, calling on her uncle's name. Before it had burned away, a dust cloud appeared at the upper end of the valley, clearing to reveal a *shaikh* riding on an elephant, which was trumpeting beneath him. When he saw the princesses, he started to gesticulate to them with both hands and feet, and when he came up to them he dismounted and embraced them, as they kissed his hands and greeted him.

When he had taken his seat, they started to talk with him and to ask him about his absence. 'I was sitting just now with your aunt,' he told them, 'when I smelt the incense and so I came to you on this elephant.' He then asked his niece what she wanted, and she said: 'We have been longing to see you, uncle. It has been a year since you came and you have not been in the habit of staying away from us for longer than that.' He told them: 'I have been busy, but I was planning to visit you tomorrow.' They thanked him, called down blessings on him and sat talking with him.

Night 800

Morning now dawned and Shahrazad broke off from what she had been allowed to say. Then, when it was the eight hundredth night, SHE CONTINUED:

I have heard, O fortunate king, that the princesses sat talking with their uncle. The eldest then said: 'Uncle, we spoke to you about Hasan of Basra, who was brought here by Bahram the Magian; we told you how he killed Bahram and how, after having endured many difficulties and dangers, he captured and married the daughter of the great king and took her back to his own country.' 'Yes, you did,' agreed her uncle, 'but what happened to him after that?' She replied: 'His wife gave birth to two sons but then betrayed him while he was away and took them back to her own country. She told his mother: "When your son comes back, if he finds the nights of separation long and is shaken by the winds of longing so that he wants a reunion, let him come to the islands of Waq."'

The *shaikh* shook his head and bit his finger, before staring down at the ground and scratching on it. Then he looked from right to left and shook his head again, watched all the time by Hasan, who was there in hiding. 'Answer us, uncle,' said the princesses, 'for you have broken our hearts.' The *shaikh* shook his head at them and said: 'My daughters, this man is wearing himself out and exposing himself to enormous perils and huge dangers, for he cannot possibly reach the islands of Waq.' At that, the princesses called to Hasan, who came up to the *shaikh*, kissed his hand and greeted him. The *shaikh* was glad to meet him and made him sit down by his side, after which the princesses said: 'Uncle, explain to our brother what it is that you meant.' The *shaikh* said: 'My son, give up this painful torment. You could not get to the islands of Waq even if you had with you the flying *jinn* and the planets, since between you and them are seven valleys, seven seas and seven gigantic mountains. How could you get there and who could take you? I urge you in God's Name to go home as soon as possible and not to put yourself to such great trouble.'

When he heard what the *shaikh* had to say, Hasan wept until he fainted, and the princesses sat around him weeping, while the youngest tore her clothes and slapped her face until she too fainted. The *shaikh*, seeing the extent of their passionate concern and grief, was touched by pity and sympathy. He told them to keep quiet and then said to Hasan: 'Take heart. I have news for you that, God willing, you may succeed in your quest. So get up, brace yourself and follow me.' This Hasan did, being filled with joy at the thought of achieving his purpose. He took leave of the princesses and the *shaikh* called to the elephant. When it came he mounted, taking Hasan up behind him, and the two travelled for three days and nights like a lightning flash. They then reached a huge blue-coloured mountain, all of whose rocks were blue, in the middle of which was a cave with a gate of Chinese iron. The *shaikh* took Hasan's hand and helped him to dismount before dismounting himself, and then, after dismissing the elephant, he went up and knocked on the gate. When it opened, out came a hairless black slave, like an *'ifrit*, with a sword in his right hand and a steel shield in his left. At the sight of the *shaikh*, the slave threw these away and went up to kiss his hand. The *shaikh* took Hasan by the hand and when the two of them had gone in, the slave locked the gate behind them.

What Hasan saw was a huge and spacious cave with a vaulted entrance way, and they walked on for a mile before coming to a vast open space in one corner of which were two gates cast in brass. The *shaikh* opened one of them, drawing it shut behind him after telling Hasan: 'Sit here by the gate but take care not to open it and go in until I have entered and come back, which will not take me long.' He then went in and was away for an hour as measured by the sun, after which he emerged with a horse saddled and bridled, whose normal pace was as swift as flight and whose flight no dust could overtake. The *shaikh* brought the horse up to Hasan, telling him to mount it, and he then opened the second gate through which could be seen a wide desert. When Hasan had mounted, the two of them went out through the gate and into the desert. Then the *shaikh* told Hasan: 'Take this letter and ride to where the horse will bring you. When you see him stop at the gate that leads to a cave like this, dismount, put his reins over the saddlebow and then let him go. He will enter the cave, but instead of going in with him you must wait patiently by the entrance for five days. On the sixth, out will come a black *shaikh*, dressed in black, with a long white beard hanging down to his navel. As soon as you see him, kiss his hands and clutch at the edge of his robe, putting it on top of your head, and weeping as you stand in front of him, to make him pity you and ask you what you want. When he does this, hand him this letter. He will take it from you without speaking, and he will then go into the cave, leaving you alone. Stay where you are for five more days without becoming upset, and then on the sixth you can expect someone to come out. If this is the *shaikh* himself, you can be sure that you will get what you want, but if it is one of his servants, you must know that all he wants to do is to kill you, and you have to realize, my son, that whoever is prepared to risk his life is responsible for his own destruction.'

Night 801

Morning now dawned and Shahrazad broke off from what she had been allowed to say. Then, when it was the eight hundred and first night, SHE CONTINUED:

I have heard, O fortunate king, that when Shaikh 'Abd al-Quddus gave Hasan the letter, he told him what would happen to him and added: 'Everyone who risks his life is responsible for his own destruction. If you are afraid, don't expose yourself to this danger, but if not, then do whatever you want, for I have explained the position to you. If you would prefer to go back to your friends, then here is this elephant, which will take you back to my nieces, and they will send you on to your own country. May God see to it that your love for this girl turns out well.' Hasan replied: 'What pleasure could I find in life if I do not reach my goal? By God, I shall not return until I reach my beloved or death overtakes me.'

He then recited these lines:

With the loss of my beloved, my passion grows;
I stand calling out, broken-hearted and abased.
In my longing I kiss the camp's earth,
Which serves only to add to my distress.
God guard the absent ones, whose memory is in my heart;
My pleasure vanished, to be replaced by pain.
They told me to be patient when she went off with her clan,
But as they left, love's fire prompted my sighs,
And her last words filled me with fear:
'Remember me when I am gone; do not forget our days together.'
Where can I find refuge or hope now they are gone?
In them I placed my hope in good times and in bad.
How sadly I returned when they had left,
To the delight of enemies filled with hate!
In my sorrow, this was what I feared;
So, passion, burn more fiercely in my heart.
If dear ones go, no life is left for me;
What joy and what delight would come with their return!
By God, my tears will never cease after their loss,
Following each other in an endless flow.

When Shaikh 'Abd al-Quddus heard this, he realized that words would have no effect on Hasan, who was not going to turn back from his purpose, and the *shaikh* was sure that Hasan would shrink from no risk, even at the cost of his life. So he said: 'You must know, my son, that there are seven islands of Waq, and that they have a vast army, all composed of virgin girls. The inner islands are inhabited by devils, *marids*, sorcerers and various other *jinn* tribes; no one who has gone there to visit them has ever been known to come back. The girl for whom you are looking is the daughter of the supreme king of all the islands, so how will you ever be able to reach her? Listen to my advice, my son, and it may be that God will replace her for you with a better wife.' 'Sir,'

Hasan replied, 'even if I were hacked limb from limb because of my love for her, this love would only increase. I have no choice but to go to these islands in order to see her and my children and, if Almighty God wills it, I shall not come back without them.' 'Go, then, as you must,' said the *shaikh*, and Hasan replied: 'I shall, and I want you to pray to God to help me so that, thanks to Him, I may soon be reunited with my wife and children.'

He then shed tears of longing and recited:

You are my desire; the best of all mankind;
To me you are as dear as ears and eyes.
You have possessed my heart, in which you dwell.
Now you have gone, I stay sunken in gloom.
Do not think I shall ever leave your love,
Although its victim is plunged in distress.
You went, and carried all my joy with you,
As darkness then replaced my happy days.
You left me here in pain, watching the stars,
Pouring down tear drops like the flooding rain.
How long the night is for the restless lover,
Who waits in his ardour for the moon to rise!
Wind, if you pass a camp where they have stopped,
Take them my greetings, for our lives are short.
Tell them what I am left to suffer here,
For my dear ones have not had news of me.

On finishing these lines, he wept bitterly until he fainted. Then, when he had recovered, the *shaikh* reminded him of his mother, telling him not to distress her by his loss, but Hasan replied: 'By God, I shall either come back with my wife or else die.' He shed more tears and recited:

I swear by love that distance has not changed my covenant;
I am not one of those who break their word.
Were I to speak to others of my longing,
All they would say is that this man is mad.
Passion, grief, lamentation and the pain of love –
What is the fate of one who suffers from all these?

When he had finished, the *shaikh* realized that Hasan was never going to change his mind, even at the cost of his life, and so he handed over the letter and told him what to do, explaining: 'In this letter I have recommended you to Abu'l-Ruwaish, the son of Bilqis, daughter of Mu'in, my *shaikh* and teacher, whom both men and *jinn* hold in reverence and fear. So go off with God's blessing.' Hasan rode away with a loose rein and his horse flew faster than lightning. He pressed on for ten days, at the end of which he saw a huge shape, blacker than night, that blocked the horizon from east to west. When he got near it, his horse neighed, and at that other horses, in vast numbers, innumerable as raindrops, collected and started rubbing against it, filling Hasan with alarm. He rode on in the middle of them until he came to the cave that Shaikh 'Abd al-Quddus had described to him. His horse halted by its entrance and Hasan dismounted, looping his bridle over the saddlebow. The horse then entered the cave, while Hasan stayed by the entrance as he had been told by the *shaikh*, wondering how things would turn out, perplexed, confused and not knowing what was going to happen to him.

Night 802

Morning now dawned and Shahrazad broke off from what she had been allowed to say. Then, when it was the eight hundred and second night, SHE CONTINUED:

I have heard, O fortunate king, that when Hasan dismounted, he stood at the entrance to the cave thinking about how the affair would turn out, as he did not know what would happen to him.

For five days and nights he stood there, sleepless and sorrowful, distracted and full of care at having left his family, his own country, his friends and his companions, tearful and sad at heart. He remembered his mother, and thought about what was happening to him, how he had been parted from his wife and children, and of the sufferings that he had endured. He recited these lines:

My heart perishes, but with you is its cure;
Tears from my eyes flood down like mountain streams.
Parting, sorrow, longing and loneliness are here.
I am far from my land and yearning overwhelms me.
I am a passionate lover, overcome
By misery at losing the beloved;
But even if my love brings me disaster,
What noble man has not met such a fate?

Before he had finished these lines, Abu'l-Ruwaish came out to him. He was a black man wearing black clothes, and Hasan recognized him from the description given him by 'Abd al-Quddus. Throwing himself on him, he rubbed his cheeks against his feet, and taking hold of one of these, he set it on his head while bursting into tears. 'What do you want, my son?' asked Abu'l-Ruwaish, at which Hasan held out the letter to him. He took this, and then, without saying anything, he went back into the cave. As he had been told, Hasan sat down by the entrance, in tears. He did not move from there for five days, becoming more disturbed and more frightened all the while. He was unable to sleep, and the pain of separation, allied to sleeplessness, made him tearful and uneasy. He recited these lines:

Praise to the Omnipotent Lord of heaven;
The lover is in distress.
Those who have never tasted love
Do not know the strength of misery.
If I were to check my tears,
I would pour streams of blood.
How many a friend is hard of heart,
In love with misery!
In gentle moods, he blames me for what I do;
I say: 'I am not shedding tears;
I went to put on my cloak,
And then disaster overtook me.'
Even wild beasts lament my loneliness
And the birds of the air weep over me.

He went on weeping until dawn broke, and it was at that moment that Shaikh Abu'l-Ruwaish came out, dressed in white, beckoning him to enter. He went in, and the *shaikh*, taking him by the hand, came with him into the cave. Hasan was filled with joy, being certain that he would get what he wanted, and for half a day he walked on with the *shaikh* until they came to an arched gateway with a steel door. The *shaikh* opened the door and the two of them entered a hallway vaulted with onyx and adorned with gold, through which they proceeded to a large and spacious marble courtyard. In the centre of this was a garden with trees of all kinds, with flowers, fruits and birds on the branches singing the praises of the Omnipotent King. There were four alcoves facing one another and opening off the courtyard, each of which had a sitting place and a fountain, while at the corner of each fountain stood the statue of a golden lion. In every one of these alcoves a *shaikh* was seated on a chair with a great quantity of books in front of him, together with golden braziers in which incense was burning. In front of each *shaikh* were students reading to him from the books, and when Abu'l-Ruwaish and Hasan entered they all rose up respectfully. Abu'l-Ruwaish went up and told the four to dismiss the students, and when they had done this, they came and sat in front of him. They asked him about Hasan, and Abu'l-Ruwaish gestured to him and told him to tell them his story with all the details of what had happened to him from beginning to end. At that, Hasan shed bitter tears and recounted the whole tale.

When he had finished, all the *shaikhs* exclaimed: 'This must be the man whom the Magian got the vultures to carry to the top of the Cloud Mountain in a camel skin.' When Hasan confirmed this, they went up to Abu'l-Ruwaish and said: 'Bahram contrived to get him up to the top of the mountain, but how did he get down and what marvels did he see on the summit?' Abu'l-Ruwaish asked Hasan about this, and he repeated everything that had happened from start to finish, explaining how he had got the better of Bahram and killed him, freeing his victim, and how he had trapped the bird girl, how she had then betrayed him and flown off with his children and what terrors and difficulties he had then experienced. His account astonished everyone there, and they went up to Abu'l-Ruwaish and said: 'Shaikh al-Shuyukh, this youth is the victim of misfortune, but perhaps you may be able to help him to recover his wife and children.'

Night 803

Morning now dawned and Shahrazad broke off from what she had been allowed to say. Then, when it was the eight hundred and third night, SHE CONTINUED:

I have heard, O fortunate king, that when Hasan told his story to the *shaikhs*, they said to Abu'l-Ruwaish: 'This is an unfortunate young man, but perhaps you may be able to help him to recover his wife and children.' Abu'l-Ruwaish replied: 'My brothers, this is a hugely dangerous matter and I have never seen anyone show such contempt for his life as this young man. You know how difficult it is to get to the islands of Waq and that deadly peril has faced all those who make their way there; you know how strong the people there are and what guards they have, while for my part I have taken an oath not to set foot on their soil or to do anything to oppose them. How then can this man reach the daughter of the great king? Who could get him to her or help him in such a business?' 'Shaikh al-Shuyukh,' they replied, 'this is a victim of love who has risked his life and come to you with a letter from your brother, Shaikh 'Abd al-Quddus. It is your duty to help him.' At that, Hasan got up, kissed Abu'l-Ruwaish's foot and, putting the skirt of the *shaikh*'s robe on his head, burst into tears and said: 'I ask you in God's Name to reunite me with my family, even if that costs me my life.' Those who were present joined in his tears and told the *shaikh*: 'Win a reward from God by helping this poor man, and help him for your brother's sake.' 'He is certainly a poor fellow,' the *shaikh* replied, 'and he does not know what awaits him, but I shall help him as far as I can.'

In his joy at hearing that, Hasan kissed his hands and the hands of the other *shaikhs* there, one after the other, asking for their help. Abu'l-Ruwaish took paper and ink and wrote a letter, which he sealed and gave to Hasan. He then handed him a leather bag containing incense and fire-making equipment, such as fire steel and so on. 'Keep this safely,' he told him, 'and when you are in difficulties, use a little of the incense while calling my name, and I shall come and rescue you.' He then ordered one of the others to summon there and then an *'ifrit* from among the flying *jinn*. The *'ifrit* came and Abu'l-Ruwaish asked him his name. 'I am your servant, Dahnash, son of Faqtash,' the *'ifrit* replied. 'Come close to me,' the *shaikh* told him, and when he did, the *shaikh* put his mouth to his ear and said something which made him shake his head. The *shaikh* then told Hasan: 'Get up on his shoulder, my son. When he carries you up into the heavens and you hear the angels in the upper air praising God, do not join in their praise or else both you and Dahnash will be destroyed.' 'I shall not say a word,' Hasan promised, and the *shaikh* went on: 'At dawn on the second day, Dahnash will set you down in a land whose soil is pure white like camphor. You must then walk on your own for ten days until you reach the gate of a city. Go in and ask for the king, and when you meet him, greet him, kiss his hand and give him this letter. Then pay attention to whatever he tells you.' 'To hear is to obey,' Hasan replied.

He and the *'ifrit* then got up, as did the *shaikhs*, who called down blessings on him and instructed the *'ifrit* to look after him. The *'ifrit* carried Hasan on his shoulder and flew up with him into the clouds,

travelling for a day and a night until he could hear the angels praising
God in heaven. On the following morning, he set him down in the land
with camphor-white soil, where he left him and went off on his way.
When Hasan found himself back on the ground alone, he started out,
walking night and day for ten days until he came to the gate of the city,
which he entered. He asked his way and was shown to the king, whose
name, he was told, was Hassun, lord of the Land of Camphor. With
Hassun were enough troops and soldiers to fill the length and breadth
of the land, and when Hasan asked leave to enter, this was granted. He
went in and kissed the ground in front of what he could see was a
majestic king. For his part, the king asked what he wanted and by way
of reply Hasan kissed the letter and handed it to him. He took it and
read it, after which he spent some time shaking his head before telling
one of his courtiers to take Hasan and lodge him in his guest house. The
man did this and for three days Hasan stayed there, eating and drinking
alone apart from an attendant. This man talked to him in a friendly way,
asking him about himself and how he had come to those parts, at
which Hasan told him of his situation and explained everything that
had happened to him.

On the fourth day, the attendant brought him before the king, who
said: 'Hasan, Shaikh al-Shuyukh has explained to me that you are here
because you want to reach the islands of Waq. My son, I am ready to
send you off soon, but in your path are many dangers and waterless
deserts full of perils. If you are patient, however, things may turn out
well, and I shall think of some way of getting you to your goal, if this is
the will of Almighty God. I should tell you, however, that there is a large
force of well-armed Dailami riders here who have tried but failed to get
to the islands of Waq, but for the sake of Shaikh al-Shuyukh, Abu'l-
Ruwaish, I cannot send you back with your purpose unfulfilled. Soon –
in fact very soon – we shall be visited by ships from those islands, and
when this happens, I shall put you on board one of them and tell the
sailors to look after you and see that you get to Waq. You must tell
anyone who asks about you that you are a relation of King Hassun, lord
of the Land of Camphor. Then, when the ship anchors off the islands,
disembark when the captain tells you, and there you will see many
benches set out all along the shore. Choose one for yourself, shelter
underneath it and stay there without moving. When it gets dark, you
will find a band of women crowding around the merchants' goods.
Reach out your hand to the woman under whose bench you are sitting

and ask her to take you under her protection. If she does, you can be sure that you have got what you want and will be able to reach your wife and children, but if she refuses, then mourn your fate, despair of life and be sure that death awaits you. For you must know, my son, that you are risking your life. This is the best that I can do for you . . .'

Night 804

Morning now dawned and Shahrazad broke off from what she had been allowed to say. Then when it was the eight hundred and fourth night, SHE CONTINUED:

I have heard, O fortunate king, that when King Hassun said this to Hasan, and gave him the advice that I have mentioned, he added: 'This is the best that I can do for you, and had the Lord of heaven not watched over you, you would not have got as far as this.'

When Hasan heard this, he wept until he fainted and then, on recovering, he recited these lines:

I have an allotted term of days;
When these are finished, I shall die.
I could defeat a lion in its den,
If my life span were not yet due to end.

When he had finished, he kissed the ground in front of the king and said: 'Great king, how long will it be before the ships come?' The king told him that they would arrive in a month's time and would stay for another two in order to sell their cargoes before setting off back home, as a result of which he could not hope to set off for three whole months. So he told Hasan to go back to his guest house and ordered that he be provided with all that he needed in the way of food, drink and clothes suitable for a king.

Hasan stayed there for a month, after which the ships came and the king and the merchants went out to them, taking Hasan with them. He saw one lying out to sea, on board which were men as numerous as pebbles, their numbers known only to their Creator, with small boats ferrying its goods to the shore.

Hasan stayed there with the others until the goods had been unloaded and its merchants had done their trading. When there were only three days left before it was due to sail, the king had Hasan brought before him, provided him with all that he might need and gave him generous gifts. He then summoned the ship's captain and said: 'Take this young man on board with you without telling anyone, and bring him to the islands of Waq. Leave him there and don't bring him back.' 'To hear is to obey,' the captain said. The king then advised Hasan not to tell anyone on board about himself or to let them know anything of his story. Hasan agreed to this and then took his leave, expressing the hope that the king might enjoy a long life and be victorious over all his envious foes. In return, the king thanked him and wished him a safe and successful outcome to his quest. He then handed him over to the captain, who hid him in a chest which he loaded on a skiff, only putting him on board when everyone else was busy moving their goods.

The ships then set off, and after sailing for ten days, on the eleventh they reached land. The captain arranged for Hasan to disembark, and when he landed he saw benches in such numbers as only God could count. He walked on until he came to one that stood out from all the others, and it was beneath this that he hid himself. Night fell and then a huge number of women arrived like a locust swarm, walking with naked swords in their hands, enveloped in coats of mail. They first busied themselves with an inspection of the trade goods and then sat down to rest. When one of them took her seat on the bench under which Hasan was concealed, he took hold of the skirt of her robe, put it on his head and then threw himself down in front of her, kissing her hands and her feet and shedding tears as he did so. 'Get up, man,' she told him, 'before someone sees you and kills you.' At that, he came out from under the bench, stood up and kissed her hands. 'My lady,' he said, 'I ask for your protection,' adding, with tears: 'Have pity on one who has been parted from his family, his wife and his children, and who has come to meet them at the risk of his life. If you show me pity, be sure that you will win a reward for this in Paradise, but even if you refuse to receive me, I ask you in the Name of the Omnipotent God, the Shelterer, not to give me away.' As he spoke, the merchants started to stare at him, and the woman, who had listened to him and noted his pleadings, felt sympathy for him, realizing that he would not have run the risk of coming there unless it was for something of great importance. So she said to him: 'Take comfort, my son, and be reassured. Go back to where you were hiding beneath the bench and wait till tomorrow night. God's purpose will be fulfilled.'

She took her leave of him and he went back beneath the bench, while the army of women passed the night there until morning, in the light of candles made with a mixture of aloes wood and crude ambergris. When day broke, the ships put in to shore again and the merchants busied themselves with unloading their goods until night fell. Hasan, meanwhile, was still in hiding under his bench, shedding sorrowful tears and not knowing what fate held in store for him. While he was in this state, the woman who had come to trade and from whom he had asked protection arrived and handed him a coat of mail, a sword, a gilded belt and a spear. She then turned away in fear of the army of women, and, on seeing that, Hasan realized that she meant him to wear what she had brought him. He got up, put on the mail coat and fastened the belt around his waist, suspending the sword beneath his armpit and grasping the spear in his hand. He then took his seat on the bench, not forgetting to invoke the Name of Almighty God, to Whom he called for shelter.

Night 805

Morning now dawned and Shahrazad broke off from what she had been allowed to say. Then, when it was the eight hundred and fifth night, SHE CONTINUED:

I have heard, O fortunate king, that when Hasan had taken the arms that he had been given by the young trader whose protection he had asked, she said: 'Sit underneath the bench and don't let anyone find out what you are.' He sat there, continually invoking the Name of God, and asking Him for shelter. While he was there, torches, lanterns and candles marked the arrival of the army of women, and, getting up, he joined in with them, pretending to be one of their number. As dawn approached, they set off back to their camp, accompanied by Hasan, and each entered her own tent. Hasan went into one of these, which turned out to be that of his protectress, who, when she was inside, threw down her weapons and took off her mail coat and her veil. Hasan, who had put down his own weapons, looked at her and discovered her to be a grey-haired old lady with blue eyes and a large nose. She was a calamity, the ugliest of all creatures, with a pockmarked face, eyebrows that had lost their hair, broken teeth, wrinkled cheeks, grey hair, mucus running from her nose and a slobbering mouth. She fitted the poet's description:

In the corners of her face are nine calamities,
Any one of which gives its own view of hell.
In its ugliness this is a hideous face
With the appearance of a snuffling pig.

She was, in fact, unseemly, with scanty hair, like a spotted snake. When she looked at Hasan, she was filled with amazement and said to herself: 'How did this man get here? In what ship did he sail and how is it that he was not killed?' In her astonishment at his arrival she started to ask him about himself, and at that he fell at her feet, rubbing his face against them and weeping until he fainted. Then, when he recovered, he recited:

When will Time allow us to meet,
And reunite us after our parting?
When will I gain from the beloved what I want,
As reproach ends and love remains?
If the Nile flowed like my tears,
It would leave no land unwatered,
Flooding all Egypt, the Hijaz,
With Syria as well, and all Iraq.
You have rejected me, my love;
Be gentle now and promise we shall meet.

When he had finished, he took the skirt of the old woman's dress and put it on his head, weeping and asking her for protection. When she saw his burning passion, together with his grief and distress, she felt pity for him and agreed to shield him, telling him to have no fear. She asked him about himself and he told her everything that had happened to him from beginning to end. His account astonished her and she consoled and heartened him, saying, to his delight: 'Don't be afraid, for you have reached your goal and, God willing, you will achieve what you want.' She then sent to summon the leaders of the army, this being the last day of the month, and when they had come, she said: 'Go and tell all the troops that tomorrow morning everyone must come out. No one is to stay behind, and anyone who does will be put to death.' 'To hear is to obey,' they said, and so they left and gave orders to the entire army to march out the next morning. They then returned to tell her what they had done, and Hasan realized that she must be the army commander, with supreme authority over it. For his own part, he kept his armour on all that day.

The name of his protectress was Shawahi, known as Mother of Disasters. By the time she had finished issuing her orders, dawn had broken and the whole of her army marched out from camp, but she did not go with them. When they had left and the place was deserted, she called for Hasan, and when he came and stood in front of her, she said: 'Why did you risk your life to come here, and what made you willing to face such danger? Tell me the whole truth. Don't be afraid of me and don't hide anything, for I have promised to protect you and have taken pity on you, as I sympathize with your plight. If you tell me the truth, I shall help you get what you want, whatever the cost may be in lives. As you have come to me I shall do you no harm and won't let anyone else in the islands of Waq injure you.'

Hasan repeated his story from start to finish, telling her about his wife and the bird girls, how he had picked her from the ten of them and caught her, how he had then married her and stayed with her until she had given birth to two sons, and how she had then found how to recover her feather dress and had flown off, taking the children with her. He kept nothing back, starting at the beginning and going up to the immediate present, and when the old woman heard the story she shook her head and exclaimed: 'Glory be to God, Who preserved you, brought you here and arranged for you to fall in with me! For had you met anyone else, you would have lost your life and your quest would have failed. It was your sincerity, your love and the depth of your longing for your wife and children that have led you to succeed. If you had not been passionately in love with your wife, you would not have run such risks. I praise God for your safety, and now I must help you to reach your goal and assist you, God willing, to finish your search quickly. Know then, my son, that your wife is in the seventh of the islands of Waq, and from here to there takes seven months of continuous travel by day and night. The route goes from here to a country known as the Land of Birds, where, because of the noise of their cries and the sound of their flapping wings, no one can hear anyone else speak.'

Night 806

Morning now dawned and Shahrazad broke off from what she had been allowed to say. Then, when it was the eight hundred and sixth night,
SHE CONTINUED:

I have heard, O fortunate king, that the old woman told Hasan: 'Your wife is in the seventh island, which is the principal island of Waq, and which lies at a seven-month journey from us. From here we have to travel to the Land of Birds, where, because of the flapping of their wings as they fly, nobody can hear what anyone else says. We have to go through this land for eleven days, travelling day and night, before reaching another that is known as the Land of Beasts. There it is impossible to hear anything because of the noise of the beasts of prey, hyenas and others, the howling of wolves and the roars of lions. There is a twenty-day journey through this land before we get to what is known as the Land of the *Jinn*, where our ears will be deafened and our eyes blinded by their bellowing as they spew flames, with sparks and smoke coming from their mouths, while they exhale noisily and block our path in their arrogance. We shall be deafened, blinded and unable to see or to hear anything, and whoever tries to turn back dies. In that land, a rider must bend his head over his saddlebow and not raise it again for three days. After that we shall be confronted by a huge mountain and a river, both of which are connected to the islands of Waq.

'You must know, my son, that the whole of this army consists of virgin girls and that we are ruled by a woman from the seven islands of Waq. These seven islands extend for a full year's journey for a fast rider. By the bank of the river that I mentioned is another mountain, named Mount Waq, and the name "Waq" is also given to a tree whose branches look like human heads. When the sun rises over them, they all cry out: "Waq, waq! Glory be to God, the Creator!" and when we hear this cry, we know it is sunrise. At sunset they do exactly the same thing, and when they repeat their cry, we realize that the sun has gone. No man can stay with us, reach our country or set foot on it. The queen who rules over it lives a month's journey from this land; everyone there is under her authority; she controls tribes of *jinn*, *marids* and devils, and at her command are so many sorcerers that only their Creator knows their numbers. If you are afraid, I shall send someone to take you to the shore and then fetch someone else to take you on board a ship and bring you back to your own country, but should you prefer to stay with us, I shall not stop you but will look after you until you get what you want, if the Almighty wills it.'

'My lady,' Hasan said, 'I shall never leave you until I am reunited with my wife or I die.' 'There is no difficulty here,' she said, 'so take heart and, God willing, you will achieve your purpose, but I shall have to tell the queen about you so that she may help you to do that.' Hasan called down blessings on her and, kissing her hands and her head, he thanked her for what she had done and for her great generosity. As he went off with her, he thought about the outcome of the affair and the perils that had faced him since he left home. He began to sob and weep, and he recited these lines:

A breeze has blown from where my love has gone
And, as you see, my passion has distracted me.
The night of union is shining dawn;
The day of separation is black night.
How hard it is to take leave of my love!
To leave a dear one is no light affair.
It is only to her that I complain that she is harsh;
I have no true friend among all mankind.
I find no consolation for your loss;
No censurer, himself to blame, helps me forget.
Your beauty and my love are both unique;
You have no match and my heart is deprived.
If someone claims to love you, but is found
To fear reproach, then he is to be blamed.

At that, the old woman ordered the drums to be beaten as a signal for departure and the army moved off. Hasan went with her, plunged in a sea of cares, reciting lines of poetry, and while she urged him to be patient and tried to console him, he would not rouse himself or pay any attention to what she was saying.

They continued on their way until they reached the first of the seven islands, this being the Land of Birds, and when they entered it, such was the shrieking that Hasan thought that the world had turned upside down. His head ached; he could not think; he was both blinded and deafened, and such was his fear that he was certain of death. 'If this is the Land of Birds,' he said to himself, 'what will the Land of Beasts be like?' The old woman laughed when she saw the state he was in, and said: 'My son, if this is what you are like on the first island, what will happen when you get to the others?' Hasan prayed humbly to God, asking Him for help in his affliction and to bring him to his heart's desire. So they went on through the Land of Birds until they emerged from it into the Land of Beasts and, after having passed through this, they got to the Land of the *Jinn*. Hasan was so alarmed by the sight of this that he regretted having come there with his escort, but he prayed to the Almighty for help and went on with them until they had passed through it safely. They then came to the river, on whose bank they halted, pitching their camp under an enormous, towering mountain. The old woman produced for Hasan a couch of marble set with pearls and other gems as well as with ingots of red gold, and this she set down beside the river. Hasan took his seat there and the troops were paraded in front of him for his inspection, after which they pitched their own tents around him and rested for a while. Then, after a meal and a drink, they enjoyed an untroubled sleep as they had come to their own country.

Hasan, wearing a face veil which showed nothing but his eyes, noticed a group of girls walking close to his tent, who then stripped off their clothes and went down into the river. He watched them as they washed and went on to enjoy themselves by playing in the water, unconscious of the fact that he was looking at them. They thought that he must be a princess, but while he looked at them, naked as they were, he had an erection, as he saw between their thighs a variety of different shapes, some smooth and protruding, others fleshy and plump, some with thick lips and others that differed like the various metres of poetry. Their faces were like moons and their hair like night superimposed on day, for these were daughters of kings. As Hasan sat on the couch that the old woman had placed for him, they finished and came out of the river naked, each looking like the moon on the night it becomes full. The old woman then gave orders for the whole escort to gather in front of his tent before taking off their clothes and going into the river to wash, in the hope that he might recognize his wife among them. She kept asking him about them as one group of them followed another, but he always said: 'She is not there.'

Night 807

Morning now dawned and Shahrazad broke off from what she had been allowed to say. Then, when it was the eight hundred and seventh night, SHE CONTINUED:

I have heard, O fortunate king, that as one group of girls after the other came forward, the old woman kept asking Hasan whether his wife was among them, but he kept on telling her that she was not. Among the last to come up was one attended by ten slave girls and thirty servants, all swelling-breasted virgins. They stripped and entered the water with their mistress, who enjoyed herself for some time by throwing them down and ducking them, until they all came out and sat down. They brought their mistress silk towels embroidered with gold and, when she had dried herself, they fetched clothes, outer robes and ornaments of *jinn* workmanship, which she put on before returning to the others with her attendants, swaying as she moved. When Hasan saw her, his heart leapt and he said: 'She is very like the bird girl whom I saw in the pool of my *jinn* sisters' palace, for she too behaved just like that with her attendants.' However, when the old woman asked whether she was his wife, he told her: 'No, by your life, lady, she is not my wife and I have never seen her before in all my days. In fact, among all the girls whom I have seen in this island, there is no one like my wife and no one to match the symmetry of her figure or her beauty and grace.' 'Give me a full description of her,' the old woman said, 'so as to imprint it in my mind, for, as commander-in-chief, I know every girl in the islands of Waq, and if you describe her to me I shall recognize her and work out a way for you to take her.' Hasan replied: 'My wife has a lovely face, an elegant figure and smooth cheeks. She is high-bosomed and dark-eyed, with plump legs, white teeth, a sweet tongue and a witty nature. She is like a pliant branch; her qualities are unequalled; her eyes are dark, while her lips are tender; there is a mole on her right cheek, and on her belly beneath her navel there is a birthmark. Her radiant face is like a rounded moon, her waist is slim while her buttocks are heavy, and her saliva can cure the sick like Kauthar or Salsabil.'*

'Tell me more about her, may God increase your infatuation for her,' said the old woman, and Hasan replied: 'My wife has a beautiful face with smooth cheeks the colour of anemones, a long neck and dark eyes. Her mouth is like a ring of carnelian, with teeth that gleam like lightning and serve in place of wine glass and jug. She has been moulded in beauty, and between her thighs is the caliphal throne, a sanctuary unmatched among the shrines, as the poet said:

> The letters of what has bewildered me are widely known:
> They are four into five, and six into ten.'†

Then he wept and chanted this *mawwal*:‡

---

* Kauthar and Salsabil are waters of Paradise.   † A sexual reference.
‡ A form of popular lyric.

Heart, if the beloved betrays you,
Do not plan to leave, talking of forgetfulness.
If you are patient, you will bury your foes;
God does not disappoint lovers who endure.

He added these lines:

If you want to pass your whole life in safety,
Neither despair and lose heart, nor exult.
Endure without despair or joy,
And in your discontent, recite: 'Have we not opened?'*

The old woman looked down at the ground for a time before raising her head towards Hasan and saying: 'Praise be to the Great God. You have been sent to me as an affliction, and I wish that I had never come across you because, from the description that you have given me, I recognize your wife as the eldest daughter of the great king. She rules over all the islands of Waq, so open your eyes and plan what to do; if you are sleeping, wake up, for you will never be able to reach her, and if you do you will not be able to take her for yourself, as the distance between the two of you is like that between the earth and the sky. Go back quickly, my son, and don't expose yourself to destruction, bringing me down with you, for I don't think that you can have any share in her at all. Return where you came from, lest you cost us our lives.' She said this both for herself and for Hasan, and when he heard it he wept bitterly until he fainted. She continued to sprinkle water over his face until he recovered, but he then started to weep until his clothes were wet with tears, thanks to the great sorrow and distress that her words had brought him. He despaired of life and exclaimed: 'How can I go back, now that I have got here? I never thought that you would not be able to get me what I want, especially since it is you who command the army of girls.' 'For God's sake,' she told him, 'please choose one of these for yourself and I will give her to you in place of your wife, lest you fall into the hands of the rulers, and I shall have no means of saving you. I implore you to listen to me and to pick one of these instead of the princess. You will then get back soon and safely to your own land without involving me in your distress. For, by God, you have brought great misfortune and deadly danger on yourself, from which no one can rescue you.'

Hasan bent his head, shed bitter tears and recited:

* Quran 94.

I told my critics not to blame me;
My eyelids were made only for tears.
They have poured down in torrents on my cheeks
Because of my harsh treatment by my love.
Leave me, emaciated by my passion,
For in this love I love my madness.
Beloved, my longing has increased
For you. Why do you show no mercy?
I pledged a binding pact, but you were harsh;
I was the friend whom you betrayed and left.
The day we parted, when you left
You made me drink a cup of shame.
Heart, melt in passion for my love,
And eyes be prodigal with tears.

Night 808

Morning now dawned and Shahrazad broke off from what she had been allowed to say. Then, when it was the eight hundred and eighth night, SHE CONTINUED:

I have heard, O fortunate king, that the old woman told Hasan: 'For God's sake, my son, listen to me and choose someone other than your wife from these girls, and you can then go back home soon and safely.' Hasan looked down, wept bitterly and recited the lines that have been quoted. When he had finished these lines, he wept until he fainted and the old woman continued to sprinkle his face with water until he recovered. She then went up to him and said: 'Sir, go back home, for if I take you to the city we shall both die. If the queen finds out about this she will blame me for having brought you to her islands, which no mortal man has ever reached. She will kill me for having taken you with me and for having let you see those virgins whom you watched in the water, girls whom no male has touched and with whom no husband has lain.' When Hasan swore that he had not looked at them lustfully, she repeated: 'My son, go home and I shall give you enough wealth, stores and treasures to serve you in place of every woman there is. Listen to me: go back as soon as you can and don't risk your life. This is my advice.'

When Hasan heard this, he burst into tears and rubbed his cheeks against her feet, saying: 'My lady, my mistress, delight of my eyes, how can I go back, now that I have got here, if I have not seen the lady of my quest? I am near her home, and I hope to meet her soon, as it may be that we are destined to be reunited.' Then he recited these lines:

Sovereigns of beauty, pity a prisoner
Captured by eyelids that hold the empire of Chosroe.
Your perfume overwhelms the scent of musk,
The bloom of your beauty has outshone the rose.
The breezes of delight blow where you halt,
And from your camp the east wind spreads its scent.
Censurer, stop bringing me blame and advice;
The advice you bring is nothing but a ruse.
You cannot criticize and blame my love,
When you yourself know nothing of its course.
It was her languid eyes that captured me
With force and violence, driving me to love.
I scatter tears as I compose my verse;
The story of my love is in both verse and prose;
Her rosy cheeks have caused my heart to melt
And all parts of my body are ablaze.
If I am not to talk of this, tell me,
What can I say to ease my burden here?
Mine is a lifelong love for lovely girls,
And this is followed by what God decrees.

When Hasan had finished these lines, the old woman felt pity for him and went up to him and consoled him, telling him to take heart and be comforted. 'You can cease to worry,' she said, 'for I shall join you in risking my own life until you reach your goal or I die.' Hasan took heart, relaxed and sat talking with her until evening, when, as night fell, all the girls dispersed, some going to their homes in the town while others stayed in camp. The old woman took Hasan into the town and provided him with a place where he could lodge on his own, lest anyone find out his secret and tell the queen, who would then kill both him and whoever had brought him. She waited on him herself and told him such alarming stories of the power of the great king, his wife's father, that he burst into tears and said: 'Lady, if I cannot be united with my wife and my children, I would prefer to die, as there is nothing else I want in the world. I am risking my life and shall either reach my goal or die.'

The old woman began to think how to arrange for him to meet his wife and what she could do to help him in his wretchedness, now that he had put his life in danger, heedless of himself and refusing to be deterred from his quest either by fear or by anything else. For, as the proverb says: 'No lover listens to anyone who is not in love.'

The queen of the island on which they were staying was called Nur al-Huda and she had six virgin sisters who lived with their father, the great king, the ruler of the seven islands and the districts of Waq, whose royal seat was in the largest of its cities. Nur al-Huda, his eldest daughter, was the governor of that city which Hasan had reached, and of the other parts of its island. On seeing that Hasan was consumed with longing to be reunited with his wife and children, the old woman went to the palace and, after entering, she kissed the ground before her. Nur al-Huda was in her debt as it was she who had brought up all the princesses, exercising authority over them and being respected by them and highly regarded by the king.

On her entrance, Nur al-Huda got up and embraced her, before making her sit beside her and asking about her journey. 'By God, my lady,' the old woman replied, 'the journey was blessed with good fortune and I have brought back with me a gift which I shall present to you.' She then continued: 'My daughter, queen of the age, it is something wonderful that I have brought and I want to tell you about it so that you can help me with what is needed here.' 'What is it?' asked the queen, at which the old woman told her Hasan's story from beginning to end, trembling all the while like a reed in a storm wind, until eventually she fell down in front of her, saying: 'My lady, someone who was hiding under a bench on the coast asked for my protection and I gave it to him and brought him with me among the army of girls. He was carrying arms so as not to be recognized, and I have brought him into the city.' Then she added: 'I tried to terrify him by telling him of your power and might, but whenever I did this he would weep and recite poetry, saying: "I must find my wife and children or die, and I shall not go back home without them." He has come to the islands of Waq at the risk of his life, and never in all my days have I seen a more resolute or stronger man. He is totally dominated by love.'

Night 809

Morning now dawned and Shahrazad broke off from what she had been
allowed to say. Then, when it was the eight hundred and ninth night,
SHE CONTINUED:

I have heard, O fortunate king, that the old woman told Hasan's story
to Queen Nur al-Huda and said: 'I have never seen a stronger-hearted
man than this, and he is completely dominated by love.' When the queen
heard this and understood what the old woman was saying about Hasan,
she was furiously angry. For a time she looked down at the ground, but
when she raised her head and looked at the old woman, she said: 'You
ill-omened creature, are you so sunk in wickedness that you carry men
with you to the islands of Waq and introduce them to me with no fear
of my power? I swear by the king's head that, were it not for the deference
I owe you for having reared me, I would put the two of you to the vilest
of deaths this very moment, you damned woman, as a lesson to other
travellers, so that no one else might commit a crime of such unmatched
enormity. Go out and bring the man here immediately so that I may
look at him.' The old woman left in a dazed state, not knowing where
she was going. 'It was through Hasan that God has brought this misfortune upon me at the hands of the queen,' she kept saying, and when she
came to him she told him: 'Get up and answer the queen's summons; the
end of your life is near at hand.' So he went off with her, constantly
calling on the Name of the Almighty and saying: 'God, be merciful to
me in Your judgements and save me from this distress.'

The old woman took him before the queen, having coached him on
the way about what to say while talking with her. When he appeared in
front of her he saw that she was wearing a mouth-veil, and after having
kissed the ground before her he recited these lines:

God grant you lasting glory and happiness;
May He confer His boons upon you,
Adding to your magnificence and your splendour,
Aiding you with His power against your foes.

When he had finished, the queen indicated that the old woman was to
act as her spokeswoman so that she might hear his replies. So the old
woman said: 'The queen returns your greeting and asks: "What is your
name? Where have you come from? What are the names of your wife

and children for whom you have come, and what is the name of your country?"' Hasan plucked up his courage and, aided by destiny, he replied: 'Queen of the age, unique among Time's children, my name is Hasan, the mournful, and I come from Basra. My wife's name I do not know, but as for my children, one is called Nasir and the other Mansur.' When the queen heard this, she asked: 'From where did your wife take away her children?' 'From Baghdad,' he said, 'from the caliph's palace.' 'Did she say anything to you as she flew off?' Hasan replied: 'She told my mother: "When your son comes back and is shaken by the winds of longing in the long days of separation, if he wants to be reunited with me, let him come to the islands of Waq."' The queen shook her head and said: 'Had she not wanted you, she would not have told your mother this, for unless she wished for a reunion she would not have said where she was going or invited you to her country.' Hasan went on: 'Empress and ruler over kings and beggars, I have now told you what happened and concealed nothing. I take refuge with God and with you, imploring you not to wrong me but to have pity on me, as God will reward you on my behalf. Help me to be reunited with my wife and children; dispel my sorrow, and comfort and aid me by letting me see my children again.'

With tears of longing and complaint, he then recited these lines:

> I shall do my best to thank you as long as ringdoves coo,
> Even if my task must be left unfulfilled.
> In whatever delights I was involved before,
> I found none where you yourself were not the cause.

For a considerable time the queen looked down at the ground, shaking her head. Then she looked up and said: 'Out of pity for you I have decided to show you every girl in my city and the towns of my island. If you recognize your wife, I shall hand her over to you, but if not, I shall have you executed and crucify you on the door of the old woman's house.' 'I agree to this, queen of the age,' Hasan said, and he then recited these lines:

> You have kindled the passion of my love and then sat back;
> You have left my wounded eyes sleepless and then slept.
> You swore that you would not temporize with me,
> But when you had me in your chains, you played me false.
> I loved you like a child, not knowing what love was;
> If you kill me, I am unjustly slain.

Do you not fear God's anger for a lover's death,
Who spends the night star-watching while the others sleep?
By God, my people, when I die, inscribe
My tombstone with the words: 'Here lies a slave of love.'
It may be that one like me, injured by love,
Will greet me as he passes by my tomb.

When he had finished, he said: 'I accept your condition, and there is no might and no power except with God, the Exalted, the Omnipotent.' At that, the queen ordered that all the girls in the city should come to the castle and parade before Hasan, instructing the old woman to go down there herself to make sure that this was done. The queen then introduced the girls to Hasan in groups of a hundred at a time, until every single girl in the city had been shown to him. He did not see his wife among them, and when the queen asked him, he swore to her that she was not there. The queen became very angry with him and told the old woman to go into the palace and bring out every girl there to show to him. He still could not find his wife, and when he told this to the queen, she called out angrily to her attendants: 'Seize him! Drag him along on his face and then cut off his head, so that no one else may dare to follow what he has done by visiting our country to spy on us and setting foot on our shores.' The attendants did as they were told, throwing the skirt of his own robe over him and blindfolding him. They then stood with drawn swords by his head, waiting for the queen's command, but at that the old woman came up to her, kissed the ground before her and, taking the hem of her robe, she placed it on her own head and said: 'Queen, I implore you by the debt of nurture that you owe me, do not act hastily against him, particularly as you know that the poor man is a stranger who has risked his life to come here. He has endured hardships such as no one else has ever faced before, and the Great and Glorious God has preserved him since he is destined for a long life. He came to the lands that are under your protection because he had heard of your justice, and if you have him killed, reports will spread that you hate strangers and put them to death. At all events, he is in your power, and if his wife is not to be found in your land, he is yours to kill, since whenever you want him, I shall be able to bring him back to you. I only offered him protection because I hoped for generosity on your part in repayment of the debt you owe me for having reared you, and I guaranteed him that you would bring him to his goal because of your justice and sympathy.

Had I not known this about you, I would not have brought him to your city. I said to myself: "The queen will take pleasure in looking at him and listening to his poetry and his agreeable and eloquent words, which are like threaded pearls." He has come to our land and eaten our food, as a result of which we are under an obligation to him . . .'

Night 810

Morning now dawned and Shahrazad broke off from what she had been allowed to say. Then, when it was the eight hundred and tenth night, SHE CONTINUED:

I have heard, O fortunate king, that when Queen Nur al-Huda ordered her servants to seize Hasan and cut off his head, the old woman tried to appease her, saying: 'He has come to our land and eaten our food, as a result of which we are under an obligation to him, especially since I promised him a meeting with you. You know how hard parting is and that it can be fatal, particularly when it involves leaving one's children. Now, since the only woman left here is you, show him your own face.' The queen smiled and said: 'How could this be my husband and have had children by me, so that I should show him my face?' Then, however, she ordered Hasan to be brought before her, and when this was done she unveiled.

Hasan uttered a great cry and fell down unconscious. The old woman did her best to treat him until he recovered, and he then recited the lines:

Breeze coming from Iraq that blows
Through this land where men say: 'Waq,'
Take this message to my dear ones,
That I have tasted the bitterness of love.
Lovers, show me your compassion;
My heart has melted through the pain of parting.

When he had finished these lines, he stood up and, after looking at the queen, he gave so great a cry that the palace almost fell on the heads of everyone in it, before he again collapsed unconscious. The old woman looked after him once more, and when he had recovered she asked him how he was. He said: 'This queen is either my wife or someone exactly like her.'

Night 811

Morning now dawned and Shahrazad broke off from what she had been allowed to say. Then, when it was the eight hundred and eleventh night,
SHE CONTINUED:

I have heard, O fortunate king, that when the old woman questioned him, he said: 'This queen is either my wife or someone exactly like her.' 'Damn you, nurse!' exclaimed the queen. 'This stranger is mad or deranged; he is looking at me with staring eyes.' 'Your majesty,' the old woman replied, 'do not blame him; he is to be excused, for as the proverb has it, there is no cure for the lovesick and he is as insane as a madman.' Hasan then shed bitter tears and recited:

I see the traces of the dear ones, and melt with my longing,
Pouring out my tears in the places where they lived.
I ask God, Who afflicted me by their departure,
To grant me, of His grace, that they return.

He then told the queen: 'By God, you are not my wife, but you have the closest of resemblances to her.' At this, the queen laughed so heartily that she fell over backwards, and then, turning on to her side, she said: 'My dear, go easy on yourself; study me closely; answer my questions and leave aside this madness and confusion, for you will soon find relief.' Hasan replied: 'Ruler of kings and refuge of both rich and poor, the sight of you has robbed me of my wits, for you are either my wife or someone who very closely resembles her. Now ask me whatever questions you want.' She said: 'In what way am I like your wife?' 'My lady,' he answered, 'you are like her in all your qualities of loveliness, grace, charm and elegance, with your shapely figure, the sweetness of your words, your red cheeks, your swelling breasts and your other beauties.'

The queen now turned to the old woman and said: 'Take him back to your house, where he was before, and look after him yourself until I investigate his affair. If he turns out to be a chivalrous man who remains true to the ties of affection, companionship and love, then we must help him reach his goal, especially since he has come to our land and eaten our food, suffering hardships on his journey and facing fearful perils. When you have taken him home, leave him in the care of your servants and come back here quickly. If the Almighty wills it, things will turn out well.'

At that, the old woman took Hasan and brought him back to her house, where she told her slave girls, eunuchs and retainers to look after him, fetching him whatever he might need and leaving nothing undone. She then hurried back to the queen, who ordered her to arm herself and collect a force of a thousand daring riders. When that had been done, she presented herself to the queen and told her that the riders were ready. The queen then ordered her to go to the city of the king, her father, where she was to approach her youngest sister, Manar al-Sana, and tell her: 'Clothe your two sons in the mail coats that their aunt had made for them and send them to her, for she is longing to see them.' The queen went on: 'You are not to tell her anything about Hasan, but when you have got the children, simply give her my invitation to pay me a visit. If she hands over the children and wants to come here with them, you are to bring them to me quickly and leave her to follow at her leisure. You are to take a different route from the one that she will use, travelling night and day and being careful to see that no one finds out what has happened. I swear by everything I hold sacred that if my sister turns out to be Hasan's wife and it becomes clear that the children are his, I shall not prevent him from taking her or stop her going back with him, together with her boys.'

Night 812

Morning now dawned and Shahrazad broke off from what she had been allowed to say. Then, when it was the eight hundred and twelfth night, SHE CONTINUED:

I have heard, O fortunate king, that the queen said: 'I swear to God, by every possible oath, that if he finds his wife, I shall not stop him from taking her. Rather, I shall help him and then aid him to take her back to his own country.'

The old woman believed the queen and did not realize that, in the queen's wickedness, she secretly intended to put Hasan to death if Manar al-Sana turned out not to be his wife and if the children did not resemble him. What the queen said was: 'If I am right in my guess, mother, his wife will turn out to be my sister Manar al-Sana, but God knows better. She fits the description, and all the qualities of outstanding loveliness and grace that he listed belong to no one but my sisters, and in particular, to the youngest.' The old woman then kissed the queen's hand and went back to tell Hasan what she had said. He was deliriously happy, got up and kissed her head, but she said: 'Kiss me on the lips rather than on the head, my son, and this kiss will serve as a titbit to celebrate your safety. Don't shy away, for you can take heart and be comforted and easy in your mind. It was I who was the cause of this meeting, and you can be happy and joyful, with all your fears allayed.'

When she had taken her leave of him and left, he recited these lines:

> There are four witnesses to my love for her –
> Though every case in law needs only two –
> My beating heart, the trembling of my limbs,
> My wasted body and my speechless tongue.

He also recited:

> Were tears of blood shed from my eyes
> Until my end was heralded, two things
> Would not have had a tenth of what was due:
> The prime of youth and parting from my friends.

The old woman took up her weapons and with her thousand armed riders she set out for the island of the queen's sister, travelling for three days until she had got there. On her arrival she went to the princess and, having greeted her, she passed on the greetings of Queen Nur al-Huda, who, she said, was longing to see her and her children, adding that the queen was reproachful because she had not been to visit her. Manar al-Sana said: 'My sister is right. I have been slack about going to see her, but I shall go now.' She gave orders for her tents to be pitched outside the city, and she took with her suitable gifts to present to her sister. Her father, the king, saw the tents from a window in his palace, and when he asked about them, he was told that they had been pitched there on the instructions of Manar al-Sana, who was intending to visit her sister, Nur al-Huda. When he heard that, he provided her with an escort and brought out money, food, drink, gifts and jewels, such as would beggar all description, from his stores. He had seven daughters, all of whom, bar the youngest, were full sisters. Their names in order of age were: Nur al-Huda, the eldest; Najm al-Sabah; Shams al-Duha; Shajarat al-Durr; Qut al-Qulub; Sharaf al-Banat; and Manar al-Sana, the youngest, Hasan's wife, who had a different mother.

The old woman now came forward and kissed the ground in front of Manar al-Sana, who asked: 'Is there anything you want, mother?' She replied: 'Your sister, Nur al-Huda, tells you that, by way of precaution,

you should dress your children in the coats of mail that she had made to
fit them and to send them with me. I shall take them on ahead and bring
her the good news that you are coming.' When Manar al-Sana heard
this, she changed colour and looked down at the ground, remaining like
that for a long time. Then she shook her head and, looking up at the old
woman, she said: 'When you mentioned my children, my heart trembled
and fluttered, for since the day of their birth no one, *jinn* or mortal, man
or woman, has ever seen their faces, and I guard them jealously even
from the breeze when it blows.' 'What are saying, my lady?' asked the
old woman. 'Are you afraid that your sister may harm them?'

Night 813

Morning now dawned and Shahrazad broke off from what she had been
allowed to say. Then, when it was the eight hundred and thirteenth
night, SHE CONTINUED:

I have heard, O fortunate king, that the old woman said to Manar
al-Sana: 'What are you saying, my lady? Are you afraid that your sister
may harm them? Don't be absurd, for even if you wanted to disobey the
queen, you could not, because she would blame you. But your children
are young and you can be excused for being afraid for them, since love
makes one fear the worst. For all that, my daughter, you know my
fondness and affection for you and for your children. I reared you before
they were born, and I shall take charge of them, pillowing them on my
cheek and opening my heart to store them in it. In a case like this I need
no one to tell me what to do. You can be happy and content to send
them off to your sister, and at the most I shall only be one or two days
ahead of you.'

The old woman continued to press Manar al-Sana until she weakened,
being afraid to anger her sister and not knowing what lay hidden for her
in the future. When she had agreed to send off the children, she called
for them, bathed them, made them ready and dressed them in their coats
of mail. When they had been handed over to the old woman, she followed
Nur al-Huda's instructions and took them as swiftly as a bird, by a
different route from that which their mother was about to take. In her
concern for them, she continued to press on with her journey until she
reached Nur al-Huda's city. When she had brought them across the river
and into the city, she took them to their aunt, the queen, who was

delighted to see them, embracing them and clasping them to her breast, seating one of them on her right thigh and the other on her left. Then she turned to the old woman and said: 'Now fetch me Hasan, for I have guaranteed him immunity. He has found shelter here under my protection after having faced perils, hardships and deadly dangers involving ever-increasing distress – but he is still not safe from having to drink the cup of death.'

Night 814

Morning now dawned and Shahrazad broke off from what she had been allowed to say. Then, when it was the eight hundred and fourteenth night, SHE CONTINUED:

I have heard, O fortunate king, that when Queen Nur al-Huda ordered the old woman to produce Hasan, she said: 'He has faced perils, hardships and deadly dangers involving ever-increasing distress – but he is still not safe from having to drink the cup of death.' The old woman replied: 'If I bring him before you, will you reunite him with his children, and if they turn out not to be his, will you pardon him and send him back to his own land?' On hearing this, the queen fell into a rage and said: 'You ill-omened old woman, how long are you going to continue trying to trick me over this stranger, who has had the audacity to come here in order to uncover our secrets and spy on us? Do you suppose that he can come here, look at our faces, stain our honour and then go safely home to bring shame on us in his country and among his people, spreading news of us among kings throughout the world? Traders would then talk about us everywhere, saying: "A human entered the islands of Waq, passing through the country of the sorcerers and magicians, crossing the Land of the *Jinn* and the Lands of Beasts and Birds and then coming home safely." This must never be. I swear by the Creator of the heavens, Who built them, Who laid out the earth, Who created all beings and numbered them, that if these children turn out not to be his, I shall put him to death and cut off his head with my own hands.' She then shouted at the old woman, who collapsed in fear and was seized by the chamberlain and twenty mamluks. 'Go off with this old woman,' the queen ordered them. 'Fetch me the young man whom she has in her house and be quick about it.' They dragged the woman off, pale and trembling, and when she got home and came to Hasan, he kissed her hands and greeted her. For her part, she did not return his greeting but said: 'Get up and answer the summons of the queen. Did I not tell you to go home? I tried to stop you from doing this, but you wouldn't listen. I promised you more wealth than anyone could carry if you would waste no time in leaving, but you refused to obey me or to listen, going against my advice and preferring to bring about your own death as well as mine. You have got what you chose, as your death is near at hand. So get up and answer the summons of this vicious, unjust and tyrannical whore.'

Hasan, sad and fearful, got up despondently, exclaiming: 'Saviour, rescue me! O my God, be merciful to me in the trials that You have decreed for me, and preserve me, You Who are the most merciful of the merciful.' As he set off with the twenty mamluks and the chamberlain, together with the old woman, he despaired of his life, but when he was brought into the presence of the queen, he found his two sons, Nasir and Mansur, sitting on her lap while she kept them amused by playing with them. As soon as he saw them he recognized them, and with a great cry of joy he fell down unconscious . . .

Night 815

Morning now dawned and Shahrazad broke off from what she had been allowed to say. Then, when it was the eight hundred and fifteenth night, SHE CONTINUED:

I have heard, O fortunate king, that when Hasan's eyes fell on his children, he recognized them and with a great cry he fell unconscious on the ground. When he recovered, the children recognized him as he had recognized them, and, moved by natural affection, they got down from the queen's lap and, standing beside him, were prompted by the Almighty to exclaim: 'Father!' The old woman and everyone else there wept out of pity for them, exclaiming: 'Praise be to God, Who has reunited them!' Hasan recovered and embraced them, but then wept again until he fainted a second time, and then, on his recovery, he recited these lines:

I swear that my heart cannot endure this parting,
Even if to stay united means my death.
Your phantom says: 'We meet tomorrow,'
But will my enemies let me live till then?
I swear that since the day I parted from you,

Life, for all its sweetness, has held no joys for me.
Should God decree my death because of love,
It is as the noblest of martyrs I shall die.
A gazelle seeks pasture in the corners of my heart;
But she herself, like sleep, is absent from my eyes.
In a court of law she may deny my murder,
But above her cheeks there are two witnesses.

It was now clear to Queen Nur al-Huda that the boys were, in fact,
Hasan's children and that her sister, Manar al-Sana, must be the wife in
search of whom he had come. She was furious with her sister . . .

Night 816

Morning now dawned and Shahrazad broke off from what she had been
allowed to say. Then, when it was the eight hundred and sixteenth night,
SHE CONTINUED:
I have heard, O fortunate king, that when it was clear to Queen Nur
al-Huda that the boys were, in fact, Hasan's children and that her sister,
Manar al-Sana, must be the wife in search of whom he had come, she
was furious with her and screamed in Hasan's face so that he fainted.
When he recovered, he recited these lines:

You are the nearest to my inmost heart, but even so,
You are removed and stay remote from me.
I swear by God, I have not turned elsewhere,
Enduring patiently the wrongs of Time.
The nights pass by and end, but for my heart
Your love has left sighs, as my passion burns.
I could not bear to leave you for an hour;
How is it, then, that months have now gone by?
I am jealous of the breeze that blows on you –
As I am jealous for the honour of young girls.

Having finished these lines, Hasan again collapsed and when he recovered, he discovered that he had been dragged out of the palace on his
face. He got up, stumbling over the skirts of his robe as he walked and
scarcely able to believe that he had escaped alive from the difficulties he
had faced. This was hard for the old woman, Shawahi, to bear, but she

could say nothing to the queen because of her fury. As for Hasan, he walked away from the palace in a daze, not knowing whether he was coming or going or where he was heading. Wide as it is, the world was too narrow for him; he could find no one to talk to him, befriend him or console him; there was no one to give him advice or to whom he could look for shelter. He was certain that he was going to die; he could not go home as he knew of no guide to go with him. He himself did not know the way and he could never pass by the Land of the *Jinn*, the Land of the Beasts and the Land of the Birds. Despairing of life, he wept over his fate until he fainted and then, on recovering, he thought about his children and his wife, her coming visit to her sister and what would result from their meeting. He regretted having travelled to those lands and the fact that he had refused to listen to any advice, and he recited these lines:

> Let my eyes shed tears for my lost love;
> Consolation is hard to find and miseries have increased.
> I drank unmixed the cup of separation;
> Who has the strength to endure such a loss?
> You unrolled the carpet of reproach between us;
> When will you fold this carpet up again?
> You slept while I was sleepless. When you thought
> That I forgot your love, it was forgetfulness that I forgot.
> My heart longs for your union ardently;
> You are my doctor and you hold my cure.
> Surely you see what shunning me has done?
> I am abased before even the basest men.
> I have concealed your love, but passion spreads the news;
> The fires of love for ever burn my heart.
> So pity me and please be merciful;
> Even in private I have kept my word.
> Will Time ever unite us once again?
> You are my heart's desire and my soul's love.
> My heart is wounded by this parting;
> Send me some news, I beg you, of your clan.

When he had finished reciting these lines, he walked on out of the city and, finding himself by the river, he walked along the bank, not knowing where he was going.

So much for Hasan, but as for his wife, Manar al-Sana, she had been intending to set off on the day after Shawahi had left, but, when she was just about to leave, a chamberlain came from her father and kissed the ground before her.

Night 817

Morning now dawned and Shahrazad broke off from what she had been allowed to say. Then, when it was the eight hundred and seventeenth night, SHE CONTINUED:

I have heard, O fortunate king, that when Manar al-Sana had made up her mind to leave, her father's chamberlain arrived and kissed the ground in front of her. 'Queen,' he said, 'your father, the great king, sends you his greetings and summons you to come to him.' So she got up and went with the chamberlain to find what her father wanted. When he saw her, he made her sit beside him on his couch, before telling her: 'Daughter, last night I had a dream which made me afraid for you, as it seemed to me that this journey of yours would bring you prolonged distress.' She asked him what he had seen, and he said: 'I saw myself going into a treasure chamber filled with vast heaps of money and large quantities of sapphires and other jewels, but among all that treasure and all those jewels the only things that took my fancy were seven gemstones, the finest and most radiant of all that were there. From these I picked the smallest, which was the most beautiful and had the loveliest sheen. I held it in my hand, admiring its beauty, and took it out of the treasure chamber, but outside the door, when I opened my hand and happily turned the gem over, a strange bird, not one of ours but from a distant country, swooped down and snatched it from me, before carrying it back to where it had come from. I was filled with sorrow and distress, and I woke up in a state of panic, still unhappy and saddened by the loss of the gem. When I was awake, I summoned the interpreters of dreams and told them about it. They said: "You have seven daughters and you will lose the youngest of them, who will be taken from you by force against your will." You are the youngest of my daughters, as well as the dearest and most precious of them to me. Here you are, about to go off to your sister, and I do not know what she will do with you. Rather than leaving, you should return to your palace.'

When Manar al-Sana heard what her father had to say, her heart throbbed violently, as she feared for her children. For a time she looked down at the ground, but then, lifting her head and looking at her father, she said: 'Your majesty, Queen Nur al-Huda has made preparations to entertain me as a guest and she is expecting me to come at any minute. It is four years since she saw me, and if I don't go now she will be angry with me. I shall stay with her for a month at most, after which I shall come back to you. What stranger can reach our country and come to the islands of Waq? Who can get as far as the White Land and the Black Mountain and then reach the Land of Camphor and the Castle of the Birds? How could anyone pass the Valley of the Birds, and then the Valleys of the Beasts and the *Jinn* in order to reach our islands? Even if any stranger got here, he would find himself drowning in a sea of mortal perils. There is no need for you to concern yourself or to be worried about my journey, for ours is an inaccessible land.' She went on trying to win him over until at last he gave her his permission to set off.

Night 818

Morning now dawned and Shahrazad broke off from what she had been allowed to say. Then, when it was the eight hundred and eighteenth night, SHE CONTINUED:

I have heard, O fortunate king, that she went on trying to win him over until he gave her his permission to set off. He ordered a thousand riders to escort her as far as the river, where they were to stay until she had reached the city and entered her sister's palace. They were then to wait for her and fetch her back to him. As for Manar al-Sana herself, he told her to stay for two days with her sister and then to hurry home. 'To hear is to obey,' she said, before going out with her father, who then said goodbye to her. His words had made a deep impression on her and she feared for her children's safety, but no defence or precaution can help ward off the blows of fate.

She pressed on with her journey, night and day, until she came to the river and had her camp pitched on its bank. She then crossed with a number of her servants, retainers and viziers and when she got to Nur al-Huda's city, she went to the palace and came into her presence. There she found her children weeping and crying out: 'Father, father!' She joined in their tears, clasped them to her breast and said: 'Have you seen your father? I wish that I had never left him, and had I known that he was still alive I would have taken you to him.' She then recited the

following lines, lamenting her own fate and that of her husband, as well
as mourning her children's tears:

> Dear ones, I am far away and harshly treated,
> But I long for you, wherever you may be.
> My eyes are turned towards your land,
> And my heart yearns for the days I shared with you.
> How many blameless nights we spent together
> As lovers blessed by faithful tenderness!

Then, when her sister saw her clasping her children, Manar al-Sana
said: 'It was I who did this to myself and my children, bringing down
my own house in ruins.' For her part, Nur al-Huda gave her no greeting,
but said: 'Where did you get these children from, you whore? Did you
marry without your father's knowledge or commit fornication? If you
are a fornicator, an example must be made of you, and if you married
without any of us knowing it, why did you abandon your husband and
take your children away, removing them from their father and coming
back here?'

Night 819

Morning now dawned and Shahrazad broke off from what she had been
allowed to say. Then, when it was the eight hundred and nineteenth
night, SHE CONTINUED:

I have heard, O fortunate king, that Queen Nur al-Huda said to her
sister, Manar al-Sana: 'You may have married without our knowledge,
but why did you leave your husband and take your children away from
him, coming back to our country? You tried to hide them away from us,
but do you think that we don't know? The Almighty God, Who knows
all secrets, has revealed it to us, showing us where you stand and
uncovering your shame.' She then ordered her guards to lay hold of
Manar al-Sana, which they did, tying her hands behind her back and
loading her with iron fetters. They then hung her up by her hair and
gave her a painful beating, cutting open her skin, before throwing her
into prison.

Nur al-Huda now wrote to her father, the great king, telling him about
Manar al-Sana and saying: 'A human called Hasan has appeared in our
land, and my sister claims that he is her lawful husband, by whom she

has had two children. She hid them away from us both and told no one anything about what she had done, until this man, Hasan, arrived here. Then she said that she married him and stayed with him for a long time, before taking her children and leaving without his knowledge. She told his mother to tell him that, if he felt a longing for her, he was to come to the islands of Waq. I detained Hasan here and sent the old woman, Shawahi, to tell my sister to come to me with the children, and she made her preparations and came. I had told the old woman to see that the children got to me before their mother, which she did, and then I sent for the man who claimed to be her husband. When he came and saw the children, he recognized them and they recognized him, as a result of which I was sure that they were his, that she must be his wife and that, as what he said was true, he could not be blamed. I realized that the shame and disgrace belongs to Manar al-Sana and I was afraid that our islanders would think of us as being dishonoured. So when the treacherous harlot came to me, I vented my anger on her and gave her a painful beating, hanging her up by her hair. I have now told you about her; it is for you to decide, and I shall do what you tell me, as you will see that this is a matter which brings disgrace both on us and on you. If the islanders hear of it, they will take it as an example of our degeneracy, and so you should send me a reply quickly.'

Nur al-Huda gave this letter to a courier, who took it to the king. When he read it, his anger with Manar al-Sana boiled over and he sent a reply to Nur al-Huda telling her that he entrusted the matter to her, giving her power of life or death over her sister. 'If things are as you say,' he wrote, 'then put her to death and do not consult me again about her.' When the letter reached Nur al-Huda and she had read it, she sent for Manar al-Sana, who was brought before her, bloodstained, tied with her own hair, loaded with heavy shackles of iron and clothed in a hair shirt. She stood there in this ignominious plight and when she saw herself disgraced and humiliated, she thought of her past glory and recited these lines with bitter tears:

Lord, my enemies are trying to destroy me,
Thinking that I shall not escape them.
My hope is in You, that You may confound their plots;
Lord, You are the refuge of all those in fear.

She then indulged in a storm of weeping until she collapsed unconscious. When she recovered, she recited:

Disasters are familiar with my heart and I with them,
Though first I shunned them, for the generous are friendly.
My cares do not come singly;
Praise be to God, they come in thousands.

She then recited more lines:

There is many a misfortune that oppresses us,
But from which God will rescue us.
When the noose was drawn to its tightest,
Contrary to my fears, it was then relaxed.

Night 820

Morning now dawned and Shahrazad broke off from what she had been
allowed to say. Then, when it was the eight hundred and twentieth night,
SHE CONTINUED:

I have heard, O fortunate king, that Queen Nur al-Huda ordered her
sister, Manar al-Sana, to be fetched. She was brought before her with
her hands tied behind her back and then recited the lines that have been
quoted. At this point Nur al-Huda had a wooden ladder brought in,
along which she had her sister laid out on her back, ordering her eunuchs
to tie her to it. When her arms had been stretched out and tied with
ropes, Nur al-Huda had her head uncovered and wound her hair around
the wooden rungs, as all pity had been removed from her heart. When
Manar al-Sana saw herself in such a humiliating and degraded state, she
cried out tearfully, but no one helped her. 'Sister,' she called, 'why have
you hardened your heart against me and why have you no pity for me
or for these little children?' When Nur al-Huda heard this, she became
even harsher and abused her sister, saying: 'You infatuated harlot, may
God show no mercy to anyone who pities you. How can I feel sorry for
you, you traitress?' To this, Manar al-Sana replied as she lay stretched
out: 'I am content to call on the Lord of heaven to refute the slander that
you bring against me. I am innocent of this; I swear by God that I never
committed fornication; my marriage was legal and it is God Who knows
whether I am telling the truth or not. I am full of anger at your harshness
towards me, for how can you accuse me of fornication without knowing
what happened? My Lord will rescue me from your clutches, but if the
charge you have brought against me is true, He will punish me.'

Nur al-Huda paused to think when she heard this, but she then said: 'How dare you address me like this?' and, going up to her, she beat her until she fainted. Water was sprinkled on her face until she revived, but the beating, the bonds and the excessive humiliation she had suffered had impaired her beauty. She recited:

If I am guilty of a crime and have done wrong,
I am sorry for what is past and come to ask your pardon.

This enraged Nur al-Huda even further and she exclaimed: 'Do you dare to recite poetry to me, you whore, trying to excuse the mortal sin that you have committed? I had intended to return you to your husband so that I might see for myself your lustfulness and your lewd looks, as you seem to be proud of your obscene and sinful behaviour.' She told her servants to bring her a switch cut from a palm branch, and when they had done that she tucked up her sleeves and began to beat Manar al-Sana from her head down to her feet. Then she called for a plaited whip, one blow from which would have made an elephant start away at full speed, and she used this to strike her victim's back, belly and all her limbs until she fainted. When the old woman, Shawahi, saw all this, she fled in tears, calling down curses on the queen, who shouted to her eunuchs to fetch her. They ran after her, seized her and brought her back, after which the queen ordered them to throw her down on the ground, instructing her slave girls to drag her out again on her face, which they did.

So much for them, but as for Hasan, he plucked up his courage, got to his feet and walked along the river bank in the direction of the desert. He was perplexed and sorrowful, despairing of life, and in his bemused state he could not distinguish between night and day because of his sufferings. He walked on until he came up to a tree from which he found a piece of paper hanging. He took this in his hand and discovered the following lines written on it:

I arranged your destiny,
While you were still a foetus in your mother's womb.
I filled her with tenderness towards you,
And she held you to her breast.
I shall solve for you
The problem that vexes and distresses you.
Come to Me in all humility;
And I shall help you in this grave affair.

When Hasan finished reading this, he felt certain that he was going to escape from his difficulties and be reunited with his family. He took two steps forward and then found himself in a dangerous desert with no one there to befriend him. Loneliness and fear caused his heart to flutter, and he shivered with the terror which the place inspired in him. He recited:

Breath of the east wind, when you pass by my dear ones' land,
Bring them my greeting in full measure.
Tell them that I am pledged to love,
And that the passion of all lovers is surpassed by mine.
It may be that a breath of pity will blow from them,
And so revive, on the instant, my dried bones.

Night 821

Morning now dawned and Shahrazad broke off from what she had been allowed to say. Then, when it was the eight hundred and twenty-first night, SHE CONTINUED:

I have heard, O fortunate king, that when Hasan read the note, he was certain that he would escape from his difficulties and be reunited with his family. He took two steps forward and then found himself in a dangerous desert with no one there to befriend him. He shed bitter tears and recited the lines that have been quoted. He took another two steps along the river bank and found two small boys, who were descended from a line of sorcerers and diviners. In front of them was a brass rod engraved with talismans, and beside it was a leather cap with three segments, on each of which were worked names and seal inscriptions written in letters of steel. Both the rod and the cap had been thrown down on the ground, as the two boys were quarrelling and fighting. Blood was flowing. 'No one is going to take this rod except me,' one was saying, while the other was insisting: 'I'm going to have it.' Hasan intervened and parted them before asking them the reason for their quarrel. 'Uncle,' they said, 'decide this for us, since Almighty God must have sent you here to give us a fair ruling.' 'Tell me the story,' said Hasan, 'and I shall act as judge.'

They replied: 'We are full brothers. Our father was one of the great sorcerers and he lived in a cave on this mountain. When he died, he left us this cap and rod. Each one of us insists that it is he who must have the rod, so you must decide and settle our quarrel for us.' Hasan listened to this and then said: 'What is the difference between the rod and the cap, and what is their value? As far as I can see, the rod is not worth much and the cap is only worth half that.' 'You don't know their value,' the boys told him, and when he asked them to explain, they said: 'Each of them has a wonderful secret power. The rod is equal to the revenues of all the islands of Waq, together with their districts, and so is the cap.'

Hasan asked them to explain their powers to him and they said: 'These are enormous. Our father spent a hundred and thirty-five years working on them until he had perfected them, incorporating a secret in each of them, so that they could be put to extraordinary uses. He engraved them with a replica of the revolving heavens and used them to cancel the power of every talisman. It was only after he had finished working on them that he fell victim to death, the inevitable fate of all mankind. The secret of the cap is that whoever puts it on his head becomes invisible, and as long as he is wearing it no one can see him. The secret of the rod is that its possessor has authority over seven *jinn* tribes, who are its servants and under its command. When he, whoever it is who holds it, strikes it on the ground, the kings of these tribes abase themselves before him and all their subjects are ready to serve him.'

When Hasan heard this, he said to himself: 'If God Almighty wills it, this rod and this cap will give me the upper hand, and I have a better right to them than these two boys. I shall now try to get them by a trick, so that I can use them to save myself together with my wife and children from this tyrannical queen, and so that we may then quit this gloomy place from which no man can escape in any other way. It may be that this was the reason that God brought me to these boys.' He then raised his head, looked at them and said: 'If you want me to settle the matter, I shall put you to a test, with the winner taking the rod and the loser the cap, for if I distinguish between you in this way I shall know which of you deserves what.' 'Uncle,' they told him, 'we are both ready for your test, so do what you want in order to judge between us.' 'Will you listen to me and obey me?' he asked them, and when they said yes, he explained: 'I shall throw a stone and whoever gets to it first and takes it before the other can have the rod, and the one who comes in second, and fails to take it, will get the cap.' The boys agreed to accept this, and Hasan picked up a stone and threw it as hard as he could so that it went out of sight, with the boys racing each other to get to it as it fell. When they were at a distance, he took the cap himself and put it on, after which he

picked up the rod and went off to investigate whether they had been
telling the truth about the secret powers put in it by their father.

The younger boy got to the stone first, but when he had picked it up
and gone back to where he had left Hasan, he could find no trace of
him. He called out to his brother: 'Where is the man who was acting as
our umpire?' 'I can't see him,' his brother answered, 'and I don't know
whether he has gone up to the highest heaven or down to the bottom of
the earth.' They both looked but could not see him, while Hasan, for his
part, was still standing where they had left him. They abused each other
and then said: 'The rod and the cap are gone and neither of us has them.
This is exactly what our father warned us about, but we forgot what he
said.' They turned back, and no one saw Hasan as he went into the city
wearing the cap and with the rod in his hand. He entered the palace and
went to the room of the old woman, Shawahi, who could not see him
because of the cap. He came up to a shelf that was above her head, on
which were various pieces of glass and china. With a movement of his
hand he swept them on to the ground, at which Shawahi screamed and
struck her face. She put back what had fallen, saying to herself: 'By God,
I think that Queen Nur al-Huda must have sent a devil to do this to me.
I pray to the Almighty to rescue me from her and to save me from her
anger, for if this is how she maltreats her own sister, dear as she is to
her father, beating her and hanging her up by the hair, what will she do
to someone who is not related to her, like me, when she is angry?'

Night 822

Morning now dawned and Shahrazad broke off from what she had been
allowed to say. Then, when it was the eight hundred and twenty-second
night, SHE CONTINUED:

I have heard, O fortunate king, that the old woman, Shawahi Dhat
al-Dawahi, said: 'If Queen Nur al-Huda does this to her sister, how will
she treat a stranger with whom she is angry?' She then said: 'Devil, I
conjure you by God, the Compassionate, the Beneficent, the Mighty
Ruler, Who created both men and *jinn*, and by what is inscribed on the
seal ring of Solomon, son of David, on both of whom be peace, that you
speak to me and answer me.' Hasan replied: 'I am no devil. I am Hasan,
the lovesick, the perplexed,' at which he removed the cap from his head
and became visible to her. She recognized him and, taking him off by

himself, she said: 'What has happened to your wits to make you come here? Go away and hide, for if that whore has tortured your wife, her own sister, what will she do if she comes across you?' She then told him all that had happened to his wife, explaining the distress, punishment and torture that she was suffering, and adding an account of how she herself had been tortured. She went on: 'The queen was sorry that she had let you go and she has sent someone to fetch you back to her, promising him a *qintar*'s weight of gold and offering him the rank that I held at her court. She swore that if she gets hold of you again, she will kill you, together with your wife and your children.' She then burst into tears and showed Hasan the scars left by her treatment at the hands of the queen. Hasan joined her in her tears and said: 'My lady, how am I to escape from this land and from this tyrannical queen, and by what ruse can I rescue my wife and my children and bring them home safe?' The old woman replied: 'Save yourself, poor man,' but he insisted that he had to rescue his wife and his children in spite of the queen. 'How can you do that?' she asked. 'Go off and hide yourself, my son, until Almighty God answers your prayers.'

It was after this that Hasan showed her the brass rod and the cap, and when she saw them she was overcome with joy and exclaimed: 'Glory be to God, Who breathes life into dry bones! By God, my son, you and your wife were as good as dead, but now the two of you, as well as your children, are safe. I know this rod and I knew the man who made it, for it was he who taught me sorcery. He was a magician of power, and he spent a hundred and thirty-five years perfecting both the rod and the cap, but when he had finished, death, the inevitable, overtook him. I heard him telling his two sons: "My boys, these two things will not remain with you. A stranger from foreign parts will come and take them from you by force, but you will not know how he has done it." "Father," they asked him, "tell us how he will manage that," but he said that he did not know. How did you manage to do it?' When Hasan had explained how he had taken the rod and the cap from the boys, she was delighted and said: 'You have the means to rescue your wife and children, but listen to my advice. I cannot stay with this whore after the shameless way in which she mistreated me, and so I shall make my way to the sorcerers' cave in order to stay there and live with them for the rest of my life. So you, my son, must put on the cap and, with the rod in your hand, go to where your wife and children are being held. Then, strike the rod on the ground and call on the servants of the names, and if one

of the chiefs of the *jinn* tribes comes with them, tell him what you want
him to do.'

Hasan took his leave of her and went off to put on the cap. Taking
the rod with him he went to his wife's prison, where he saw her close to
death, stretched out on the ladder and tied to it by her hair. In this
miserable state she was weeping sorrowfully as she could see no way of
escape. She was looking at her children, who were playing under the
ladder, and shedding tears both for them and for her own sufferings,
while enduring the most painful of tortures and beatings. Hasan saw her
in this evil plight and heard her reciting the following lines:

> Nothing remains but a spent breath
> And the faded pupil of an eye,
> A lover whose entrails are consumed
> With fire, but who says nothing.
> Even the malicious pity what they see;
> Alas for one who gets such pity!

At the sight of her torment and degradation, Hasan wept until he fell
unconscious. When he recovered, he saw his children playing while their
mother had fainted through the pain of her sufferings. He took off the
cap and the children called out: 'Father!' at which he put it on again.
When Manar al-Sana recovered her senses, thanks to their shout, all
she could see was the children crying and calling: 'Father!' When she
heard this and saw their tears, she shed tears of her own, distraught
and heartbroken, and in her sorrow and distress she called to them:
'Where are you and where is your father?' Then, remembering the
days when they were together and thinking of what had happened to
her since she left Hasan, she wept so bitterly that her tears furrowed her
cheeks, drowning them in their flood and watering the ground. She had
no hand free with which to wipe them away, and flies feasted on her
skin. Her only relief was in tears and the recitation of poetry, and so she
recited:

> I remembered our leave-taking on the day of parting,
> When, as I returned, my tears flowed down in streams.
> When the camel driver called out to his beasts,
> I found no strength or patience, nor was my heart still mine.
> I went back, though I did not know the way, still sunk
> In the pain and torment of my passionate love,

And as I went, what hurt me most was the malicious glee
Of one who came, feigning humility.
Now that my love has gone, my soul must leave aside
All pleasure in this life, and not hope to survive.
My friend, listen to my tales of love,
And see you pay attention when I speak.
My story is of passion, linked with wonders,
Which I tell like the prince of storytellers.*

Night 823

Morning now dawned and Shahrazad broke off from what she had been
allowed to say. Then, when it was the eight hundred and twenty-third
night, SHE CONTINUED:

I have heard, O fortunate king, that when Hasan came to where his
wife was, he saw his children and heard her reciting the lines that have
been quoted. She then looked right and left to see why the children were
calling to their father, but she could not see anyone and this made her
wonder what had made them talk of him just at that moment.

So much for her, but as for Hasan, when he heard the lines she recited
he wept, tears running down his cheeks like rain, until he fainted. He
then went up to his children and removed the cap, and when they saw
him, they recognized him and again called: 'Father!' Manar al-Sana wept
when she heard this, and exclaimed: 'Nothing can counter the decrees
of God,' while wondering why it was that the children had called out to
their father. She tearfully recited these lines:

The rising moon sheds no light on the camping grounds;
Eyes, do not hold back flooding tears.
They have gone; how can I now endure?
I swear I have no heart and no endurance left.
They have gone, but their place is still within my heart;
After this parting, will there be a return?
What harm would it do if I again enjoyed their company,
And if they pitied me my floods of tears and pain?
The day they left, they made the clouds of my eyes shed moisture –
I wonder that my heart's fire was not quenched.

* The text gives his name, al-Asma'i.

I wanted them to stay, but they would not obey me,
And so they dashed my hopes by leaving me.
I ask my dear ones to come back to me,
For tears enough have now been shed.

Hasan could bear this no longer and took off the cap. Manar al-Sana saw him, and when she recognized him she gave so loud a cry that everyone in the palace was alarmed. 'How did you get here?' she asked. 'Did you come down from the sky or rise up from the earth?' Her eyes brimmed over with tears and Hasan wept with her, but she said: 'This is no time for tears or reproaches. What was fated came to pass; our eyes were blinded, and what had been decreed by God in past eternity has been fulfilled. But wherever you have come from, for the sake of God go and hide yourself, lest someone see you and tell my sister, in which case she will cut my throat and yours as well.' 'My lady, queen of queens,' replied Hasan, 'I have risked my life to get here and I shall either rescue you from your distress and take you and the children back home, in spite of this vicious sister of yours, or else I shall die.'

When Manar al-Sana heard this, she smiled and laughed and then, after shaking her head for a long time, she said: 'My darling, no one is going to be able to rescue me from this plight except for Almighty God. Save yourself; go and don't risk your life. My sister has a huge army which no one can face. Even if you were to take me away from here, how could you escape these islands and the difficulties of these perilous places in order to get back home? On your way here you must have seen marvels, wonders, dangers and difficulties, from which not even the rebel *jinn* could escape. Hurry away; don't burden me with yet another care and another grief, and don't claim that you can rescue me. Who would take me to your country through these valleys, the waterless lands and the perilous places?' Hasan replied: 'Light of my eyes, I swear by your life that I shall not leave here and go off without you.' 'How can you do this?' she asked. 'What kind of a creature are you? You don't know what you are saying, for even if you could command all the various sorts of *jinn* and the sorcerers, not one of them would be able to find a way out of here. Look to your own safety and leave me, as it may be that God will bring about some change.' 'Queen of beauties,' Hasan answered, 'it is to set you free that I have come with this rod and this cap,' and he went on to tell her of his encounter with the two boys.

While he was doing this, Queen Nur al-Huda came in and overheard the conversation. As soon as he saw her, Hasan put on the cap, and she asked her sister: 'Whore, who were you speaking to?' Manar al-Sana replied: 'Who is there here to talk to me except for these children?' At that, Nur al-Huda took the whip and started to beat her, while Hasan stood watching, and she went on until her victim had fainted. Nur al-Huda then ordered her to be taken somewhere else and the slave girls untied her and removed her. Hasan went with them and saw them throw her down unconscious and then stay watching her. When she recovered, she recited these lines:

Our parting filled me with regret,
Regret that brought tears to my eyes.
I swore that if Time reunited us,
I would not talk of 'parting' ever again.
To the envious I say: 'Die of grief,
For, by the Lord, I have achieved my wish.'
My joy has overflowed so far
That its excess has prompted me to weep.
Eye, why are you now so used to shedding tears?
You shed them both in joy and in distress.

When she had finished, the slave girls left her and Hasan removed the cap. She told him: 'All this has happened to me because I disobeyed you and did what you told me not to do, going out without your permission. Please don't blame me for what I did, as you must know that a woman does not know a man's true value until she is parted from him. I have done wrong, but I ask pardon from Almighty God for what I did, and if He reunites us, I shall never disobey you again.'

Night 824

Morning now dawned and Shahrazad broke off from what she had been allowed to say. Then, when it was the eight hundred and twenty-fourth night, SHE CONTINUED:

I have heard, O fortunate king, that Hasan's wife excused herself to him and said: 'Do not reproach me for my disobedience; I ask for pardon from the Omnipotent God.' Hasan, whose heart ached for her, replied: 'You were not at fault; the fault was mine because I went off and left you with someone who didn't know your rank or value. Know, then, heart's darling and the light of my eyes, that the Glorious God has given me the power to free you. So would you like me to take you to the lands of your father, to let you fulfil your destiny there, or will you come to my country as soon as you have been set free?' 'No one except the Lord of heaven can free me,' she said, 'so go back home and give up hoping for this. You don't know how dangerous these parts are, but unless you do what I tell you, you will soon see.' Then she recited these lines:

> I consent to everything you want;
> Why do you turn away so angrily?
> Whatever happened, may our former love
> Never come to an end and be forgot.
> The slanderers stayed with us at our side
> Until we were estranged, and then they went.
> I still think well of you and keep my trust,
> However they provoke me in their folly.
> Our secret will be hidden faithfully,
> Even if the sword of censure is unsheathed.
> I spend my days longing that there may come
> A messenger to say that you are reconciled.

Both she and her children then wept, and, on hearing this, the slave girls came in to find the three of them in tears, but they could not see Hasan there. They themselves shed tears of pity and cursed Queen Nur al-Huda. Hasan himself waited until nightfall, when the guards whose duty it was to watch over Manar al-Sana had gone to bed. He then got up, tightened his belt and, after releasing his wife from her bonds, he clasped her to his breast and kissed her between the eyes. 'How long have we yearned to be reunited in our own land!' he exclaimed, and asked: 'Are we here together in a dream or are we really awake?' He picked up his eldest son while Manar al-Sana carried the younger one, and they left, sheltered by God, but when they had got outside the main part of the palace and had reached the door used to separate it from the queen's quarters, they found it locked. Hasan recited the formula: 'There is no might and no power except with God, the Exalted, the Omnipotent. We belong to Him and to Him do we return.' He and his wife both despaired of life and, striking one hand against another, Hasan called on God, the Dispeller of grief, and exclaimed: 'I took everything into account and worked out its consequences except for this! When day breaks, they will seize us and there is nothing that we can do about it.' He then recited these lines:

When times were good, you thought all would go well,
And did not fear that fate would bring you harm.
Nights were at peace with you, and you were duped;
It is when all is clear that dark clouds rise.

He wept and his wife joined in his tears, lamenting the degradation and suffering that Time had brought on her. Turning to her, Hasan recited:

Time fights against me like an enemy,
Bringing a new misfortune every day.
I look for good; Time brings its opposite,
And if one day is clear, the next is dark.

Then he added:

Time treats me badly, and it does not know
That its misfortunes pale before my glory.
It shows me its hostility, while I
Continue teaching it how to endure.

Manar al-Sana said: 'Our only relief will be to kill ourselves so as to escape from this great misfortune, for otherwise in the morning we shall have to endure painful torture.' While they were talking, a voice spoke from the other side of the door, saying: 'By God, I shall not open this for you, my lady Manar al-Sana, and Hasan, your husband, unless you are ready to do whatever I tell you.' When they heard this they kept quiet and were about to go back to where they had come from, but the voice spoke again and said: 'Why do you stay silent and not answer?' They now recognized that the speaker was the old woman, Shawahi, and so they agreed to do what she told them, adding: 'First open the door, as this is no time to be talking.' 'By God,' the old woman answered, 'I am not going to do that for you unless you swear to take me with you and not to leave me here with this vicious woman. I shall share your fate: if you escape, so shall I, and if not, we shall perish together, for this wicked lesbian despises me and keeps on tormenting me because of the two of you, whereas you, my daughter, know how to value me.'

The fears of Hasan and his wife were calmed when they realized who was there. They reassured Shawahi with the oath that they took, and she opened the door for them. When they went out, they discovered her riding on a Greek amphora of red earthenware, round whose neck was a rope made of palm fibres. The jar was turning round and round beneath her and moving faster than a Nejd colt could gallop. She went on ahead and told them to follow and to have no fear, adding: 'I am mistress of forty types of sorcery, the least of which would allow me to turn this city into a raging sea with clashing waves, and to transform every girl in it into a fish. But although that is something I could accomplish before morning, fear of the high king, together with the duty I owe to the queen's sisters, has stopped me from doing any such harm, as they are supported by *jinn* of various types together with other servants. Soon, however, I shall show you the wonders of my sorcery, so come on, under the blessing and protection of Almighty God.' Hasan and Manar al-Sana were filled with joy, sure now that they were going to escape.

Night 825

Morning now dawned and Shahrazad broke off from what she had been allowed to say. Then, when it was the eight hundred and twenty-fifth night, SHE CONTINUED:

I have heard, O fortunate king, that when Hasan, his wife and the old woman, Shawahi, left the palace, they were sure that they were now safe. When they were outside the city, Hasan took the rod in his hand and, screwing up his courage, he struck it on the ground and summoned those who served the names inscribed on it to come and show themselves to him, in whatever state they were. At that, the earth split open and seven 'ifrits emerged, each with a head that touched the clouds while their feet were still planted on earth. They kissed the ground three times in front of Hasan and, speaking with one voice, they said: 'Here we are, master. What are your instructions for us, for we hear and obey your orders? If you want, we shall dry up the seas and move the mountains.' Hasan was pleased both by what they said and by the speed with which they replied and so, with a brave heart and a firm resolve, he asked them their names, their tribes, their clans and their septs. Having kissed the ground a second time, they replied, speaking again with one voice: 'We are seven kings and, as each of us rules over seven tribes of *jinn*, devils and *marids*, we control forty-nine tribes of these *jinn*, including those who fly, who dive into the sea, the mountain dwellers, those who live in the deserts and the inhabitants of the oceans. Give us what orders you like; we are your slaves and in your service, for whoever has this rod is the master of us all and we must obey him.'

Hasan, his wife and the old woman were overjoyed when they heard this, and Hasan said to the kings: 'I want you to show me the various classes of your followers.' 'Master,' they replied, 'were we to do this, we would be afraid for you and your companions, for there are huge hosts of them and they have a variety of shapes, forms, colours, faces and bodies. Some of us have heads but no bodies, others bodies without heads, while yet others are like wild beasts or lions. If you wish it, however, then we will start by showing you those shaped like wild beasts, but what is it that you want from us now?' At that, Hasan said: 'I want you immediately to carry me, my wife and this virtuous woman to the city of Baghdad.' When they heard this, they hung down their heads, and when Hasan asked them why they did not reply, they all said: 'Lord and master, our master, Solomon, son of David, on both of whom be peace, made us swear never to carry any of the children of men on our backs. From that time on we have never done that or taken them on our shoulders, but what we shall do now is to harness some of the horses of the *jinn* to take you and your companions back to your country.' 'How far is it from here to Baghdad?' asked Hasan, and they said: 'A seven-year journey for a rider who presses his horses.' This astonished Hasan, who asked: 'How did I manage to get here in under a year?' They told him: 'God moved the hearts of His pious servants to pity you; otherwise you would not have got here at all or ever set eyes on these lands. Shaikh 'Abd al-Quddus, who mounted you on the elephant and then on the lucky horse, compressed a three-year journey for a fast rider into three days for you. Then Shaikh Abu'l-Ruwaish handed you on to Dahnash, who in a day and a night carried you over another three-year journey, and this was thanks to the blessing of the Omnipotent God. For Abu'l-Ruwaish is a descendant of Asaf, son of Barkhiya, and he knows the Greatest Name of God. From Baghdad to the palace of the *jinn* princesses is a year's journey, and this makes up the seven years.'

Hasan was amazed by this and exclaimed: 'Praise be to God, Who makes easy what is difficult, Who restores what is broken, Who brings close what is distant and Who abases every stubborn tyrant! It is He Who has helped us out of all our difficulties and has brought me here, causing these *jinn* kings to serve me and reuniting me with my wife and children, so that I don't know whether I'm sleeping or waking, sober or drunk.' He then turned to the *jinn* and asked them: 'If you mount me on your horses, how many days will it take them to get us to Baghdad?'

'It will take less than a year,' they told him, 'but you will first have to face many hardships, difficulties and dangers. You will have to cross waterless valleys, lonely deserts and many other perilous places, and we cannot be sure that you will be safe from the inhabitants of these islands . . .'

Night 826

Morning now dawned and Shahrazad broke off from what she had been allowed to say. Then, when it was the eight hundred and twenty-sixth night, SHE CONTINUED:

I have heard, O fortunate king, that the *jinn* said to Hasan: 'We cannot be sure that you will be safe from the inhabitants of these islands or from the malice of the great king and these sorcerers. It may be that they will get the better of us and take you from us. That would lead us into difficulties, as everyone who heard about it would tell us: "You did wrong. How dare you oppose the great king, removing a human from his land and carrying off his daughter as well?" Had you been on your own, it would have been easy, but God, Who brought you here, is able to take you back to your own land and reunite you soon with your mother. Be resolute; rely on God and have no fear, for we shall be at your service until we bring you home.' Hasan thanked them and said: 'May God reward you well,' after which he told them to fetch the horses quickly. 'To hear is to obey,' they said.

They now stamped on the ground, which split open, and after they had been away for some time, they re-emerged, this time with three horses, saddled and bridled. On each saddlebow was a pair of saddlebags, in one of which was a water bottle, while the other was full of provisions. When the horses had been brought up, Hasan mounted one of them, taking up one of the children in front of him, while Manar al-Sana, with the other, mounted the second horse and the old woman dismounted from her jar and took the third. They rode all night long and then, in the morning, they turned off the road and headed towards the mountain, calling constantly on the Name of God. For the whole day their way took them under the mountain, and while they were travelling Hasan saw what looked like another mountain, a lofty pillar rising into the sky like smoke. He quoted some verses from the Quran and recited the formula: 'I take refuge in God from the accursed Satan.' As he and his party came closer, the black shape became clearer, and when they came up to it, they discovered it to be an *'ifrit*, with a head like a huge dome, hook-like tusks, a throat like a lane, nostrils like jugs and ears as big as leather shields. He had a cavernous mouth, teeth like stone pillars, hands like winnowing forks and feet like ships' masts. While his head was in the clouds, his feet were buried under the surface of the earth.

As Hasan looked at him, the *'ifrit* bent down and kissed the ground before him, saying: 'Don't be afraid of me, Hasan. I am the chief of the inhabitants of this land, which is the first of the islands of Waq. I am a Muslim who believes in the unity of God, and I heard about you and knew that you were on your way here. When I found out about your plight, I wanted to leave the land of the sorcerers and go to some deserted spot, far removed from men and *jinn*, where I could live on my own, worshipping God in solitude for the rest of my life. I would like to go with you, acting as your guide until you leave these islands. I shall only appear at night and you can be easy in your minds about me, for I am a Muslim like you.' Hasan was delighted to hear this, and was sure that he would now be safe. He turned to the *'ifrit* and said: 'May God reward you well. Come with us under the blessing of God.'

With the *'ifrit* leading the way, they went on, talking playfully and happily, and as they relaxed Hasan started to tell his wife everything that had happened to him and the sufferings that he had endured. They rode all night long . . .

Night 827

Morning now dawned and Shahrazad broke off from what she had been allowed to say. Then, when it was the eight hundred and twenty-seventh night, SHE CONTINUED:

I have heard, O fortunate king, that they rode all night long until morning, their horses moving like lightning flashes, and when day broke each of them drank water and reached into the saddlebags to take out something to eat. They then pressed on, led by the *'ifrit*, who took them on an unfrequented track by the seashore. For a whole month they kept on crossing valleys and deserts, and then, on the thirty-first day, they caught sight of a dust cloud filling the horizons and darkening the day. Disquieting sounds were heard and the old woman turned to Hasan, who was pale and confused. 'My son,' she said, 'these are the armies of the islands of Waq; they have caught up with us, and will capture us this very hour.' 'What shall I do, mother?' Hasan asked, and she told him to strike the ground with his rod. He did this and out came the seven kings, who greeted him and kissed the ground before him, saying: 'Do not be afraid or saddened.' Hasan was glad to hear this and exclaimed: 'Well done, lords of the *jinn*! This is your moment.' They told him: 'You, your wife, your children and your companion must go up to the top of the mountain and leave us to deal with these armies. For, as we know that you are in the right and they are in the wrong, God will grant us victory over them.'

Hasan and Manar al-Sana dismounted, together with their children and the old woman, and letting the horses go off, they climbed up the mountain.

Night 828

Morning now dawned and Shahrazad broke off from what she had been allowed to say. Then, when it was the eight hundred and twenty-eighth night, SHE CONTINUED:

I have heard, O fortunate king, that Hasan climbed up the mountain with his wife and children, together with the old woman, after they had let their horses go. Meanwhile, Queen Nur al-Huda advanced, her army formed into two wings while her commanders rode around, drawing up the various squadrons in their ranks. The two hosts then clashed in the heat of battle, with the brave advancing and the cowards turning in flight, while the *jinn* breathed out showers of fiery sparks. When night fell, they parted, dismounted and sat down to rest, lighting their camp-fires. The seven kings went to Hasan and kissed the ground before him, at which he went to meet them, thanked them and prayed to God to help them. When he asked them how they had found Nur al-Huda's army, they told him: 'They will not be able to stand for more than three days against us, for each day we shall get the better of them. We have captured some two thousand of them and killed more than can be counted. So you can take heart and relax.' Then they took their leave of him and went down to see to the safety of their own force. The campfires went on burning until daybreak, when both sides remounted and met with a clash of swords and exchanges of spear thrusts. They spent the next night on horseback, dashing against each other like two seas, with the fire of war blazing between them, and the battle continued to rage until the forces of Waq were beaten, broken, demoralized and routed. Wherever they made for, they were defeated, and so they turned their backs and took refuge in flight. Most of them were killed, while Nur al-Huda herself, together with her state officials and her intimates, were taken prisoner.

In the morning, the seven kings came to Hasan and set up for him a marble throne studded with pearls and other gems. He took his seat on this and the kings set up another for his wife, Manar al-Sana, made of ivory with panels of bright gold. She sat down on this, while the old woman, Shawahi, was seated on a third throne. The prisoners were then brought before Hasan, among them being Nur al-Huda, her hands tied behind her back and fetters on her feet. When the old woman saw her, she said: 'You wicked whore, the only fitting punishment for you will be if we take two starving bitches and two thirsty mares, tie you to the mares' tails and then drive them into the river with the bitches in pursuit so that they may tear your flesh. Some of this flesh will then be cut off and you will be forced to eat it. How could you have done what you did to your sister, you debauched woman, when she was legally married in accordance with the ordinance of God and His Prophet? There is no monkery in Islam; marriage is one of the practices sanctified by God's apostles, on whom be peace, and women were only created for men.' At that, Hasan ordered all the prisoners to be put to death, and the old woman cried out: 'Kill them and spare no one!'

When Manar al-Sana saw the state to which her sister had been reduced, as she stood fettered and bound, she shed tears and said: 'Sister, who has got the better of us and captured us in our own country?'* 'This is a terrible business,' the other replied, 'for this man, whose name is Hasan, has us in his power; God has given him authority over us and over all our dominions, and he has subdued both us and the kings of the *jinn*.' Manar al-Sana said: 'It was by the use of this cap and this rod that God aided him to defeat and capture you.' Nur al-Huda saw that this was true and realized that this was how Hasan had managed to rescue her sister. She continued to ask for mercy until Manar al-Sana felt pity for her and said to her husband: 'What do you intend to do with my sister? She is in your power, but she did not maltreat you yourself in any way that you can hold against her.' 'The fact that she tortured you is

---

* This sentence appears to be derived from a different version of the story.

mistreatment enough,' he answered, but his wife insisted: 'Whatever she did to me was excusable, and if you are going to distress my father by taking me from him, how much worse will it be if he loses her as well?' Hasan said: 'The decision is yours; do what you want.' At that point, Manar al-Sana ordered that all the prisoners be released for the sake of her sister, and when this had been done, her sister too was set free. Manar al-Sana went up and embraced her, and the two of them burst into tears, continuing to weep for some time. 'Do not blame me for what I did to you,' said Nur al-Huda, and her sister replied: 'This was something that had been decreed for me by fate.'

Manar al-Sana then conciliated and reconciled Nur al-Huda with the old woman and with Hasan, and she and her sister then sat on her throne talking to one another. Hasan dismissed the army which had come in answer to the summons of the rod, thanking them for the service they had done him in defeating his enemies. Manar al-Sana gave her sister an account of everything that had happened to her with Hasan, together with all his adventures and the sufferings he had endured for her sake, saying: 'No man who has done all this and shown such strength can be treated with too much respect. Almighty God has added to his might and so allowed him to enter our lands, taking you prisoner and defeating your army, as well as getting the better of our father, the great king, who has dominion over all the kings of the *jinn*.' Nur al-Huda replied: 'What you have said about the marvellous adventures that your husband has endured is true, but was it all for your sake, sister?'

Night 829

Morning now dawned and Shahrazad broke off from what she had been allowed to say. Then, when it was the eight hundred and twenty-ninth night, SHE CONTINUED:

I have heard, O fortunate king, that Manar al-Sana told her sister all about Hasan, and her sister exclaimed: 'By God, too much cannot be said about this man, especially because of his chivalry. Did he do all this for you?' 'Yes,' said Manar al-Sana, and they spent the rest of the night talking until morning came. When the sun rose, Hasan wanted to be off, and so they took leave of one another, and Manar al-Sana said goodbye to Shawahi, after having made up the quarrel between her and Nur al-Huda. Hasan struck his rod on the ground, at which his servants emerged and greeted him, saying: 'Praise be to God that you are now at ease. Tell us what you want done and we shall do it for you in the twinkling of an eye.' He thanked them, asked God to reward them, and told them to saddle him two of the best horses. They did this immediately, and when the horses had been brought up, Hasan mounted one of them, taking his elder son up in front of him, while Manar al-Sana, with the younger child, mounted the other. Both Nur al-Huda and Shawahi also mounted and they all set off on their own ways, with Hasan and his wife turning right and Nur al-Huda and Shawahi turning left.

Hasan, together with his wife and children, travelled on for a whole month before coming in sight of a city surrounded, as they discovered, by trees and streams. They headed for the trees and dismounted in order to rest, but as they were sitting talking they caught sight of a large band of horsemen coming towards them. At this, Hasan got to his feet and went to meet them. They turned out to be led by King Hassun, the lord of the Land of Camphor and the Castle of the Birds. Hasan went up to him, kissed his hands and greeted him, while, for his part, Hassun, seeing who his visitor was, dismounted. He greeted Hasan, and congratulated him delightedly on his safe return, after which the two of them sat down on a carpet spread under the trees. At Hassun's request, Hasan told him the whole story of his adventures from start to finish. Hassun was astonished and said: 'My son, no one who went to the islands of Waq has ever come back except for you. Yours has been an amazing adventure and God be praised that you are safe.'

After this the king remounted and told Hasan to ride with him, which he did, and they went on into the city, where the king returned to his palace and lodged Hasan, together with his wife and children, in his guest house. There they stayed for three days, eating, drinking, making merry and enjoying themselves, after which Hasan asked permission from Hassun to set off on his journey home. This was granted, and he started out with his family, accompanied by the king, who took his leave of him after ten days and set off homewards. Hasan and his family rode on for another whole month, at the end of which they came up to a huge cave whose floor was made of brass. 'Look at this,' Hasan said to Manar al-Sana. 'Do you know what it is?' 'No,' she said, and he then told her: 'It is the home of a *shaikh* named Abu'l-Ruwaish, who did me a very great favour, as it was thanks to him that I made the acquaintance of King Hassun.' He started to tell her about Abu'l-Ruwaish when the *shaikh* himself came out of the cave entrance. At the sight of him, Hasan dismounted and kissed his hands, and the *shaikh* greeted him joyfully, congratulating him on his safe return. He then took him into the cave and the two of them sat down as Hasan began to tell him of his adventures in the islands of Waq. The *shaikh* was filled with amazement and asked him how he had managed to rescue his wife and children, at which Hasan told him the story of the rod and the cap. This astonished the *shaikh*, who said: 'Had it not been for them, my son, you would not have succeeded.' Hasan agreed and as they were talking a knock came at the cave door. Abu'l-Ruwaish went to open it, and there he found Shaikh 'Abd al-Quddus, who had come there on his elephant. Abu'l-Ruwaish went up to greet him, embracing him joyfully and congratulating him on his safety. He then told Hasan to repeat his story to the visitor, and Hasan started to tell him everything that had happened to him from beginning to end until he got to the episode of the rod and the cap.

Night 830

Morning now dawned and Shahrazad broke off from what she had been allowed to say. Then, when it was the eight hundred and thirtieth night, SHE CONTINUED:

I have heard, O fortunate king, that, as they were talking together in the cave, Hasan started to tell his story to Shaikh 'Abd al-Quddus and Shaikh Abu'l-Ruwaish, explaining what had happened to him from beginning to end until he got to the episode of the rod and the cap. It was then that 'Abd al-Quddus said to him: 'My son, as far as you are concerned, you have rescued your wife and your children and so you don't need these any more. It was thanks to us that you reached the islands of Waq, and I did you a favour because of my nieces. So I now ask if you would be good enough to give me the rod and let Shaikh Abu'l-Ruwaish have the cap.' When Hasan heard this, he looked down at the ground, but he was ashamed to refuse, saying to himself: 'These two *shaikhs* did me a very great service. It was thanks to them that I got to the islands of Waq, which I could not have reached without them. I would not have saved my wife and my children, nor would I have got the rod and the cap.' So, when he raised his head, he said: 'Yes, I shall give them to you,' adding, 'but I am afraid of my wife's father, the great king, lest he bring his armies to my country in order to fight me, and without the rod and the cap I should not be able to resist him.' 'Abd al-Quddus said: 'Have no fear, my son. We shall watch over you and protect you there, and we shall drive away anyone who is sent to you by the king. You need not be afraid of anything at all, and you can take heart and rest at ease, for no harm will come to you.'

When Hasan heard that, he felt ashamed of his hesitation and he handed over the cap to Abu'l-Ruwaish, promising to give the rod to 'Abd al-Quddus after he had accompanied Hasan back home. Both *shaikhs* were delighted and they provided indescribable quantities of wealth and treasure for Hasan to take with him. He stayed there for three days and then, when he asked to leave, 'Abd al-Quddus got ready to go with him. When Hasan and Manar al-Sana had both mounted, he whistled, and at the sound a huge elephant came up from the desert, stretching out its legs in a gallop. The *shaikh* caught it, and when he had mounted it, he set off with Hasan and his family while Abu'l-Ruwaish went back into his cave. The travellers went on their way, traversing vast stretches of country, with 'Abd al-Quddus guiding them by easy tracks and accessible passes until they came close to Hasan's own parts. Hasan was delighted to be nearing his mother's home and to have recovered his wife and his children after the perils and difficulties that he had faced. He gave praise and thanks to Almighty God for His favours, reciting these lines:

> It may be that God will soon reunite us,
> And we shall join together in a close embrace.
> I shall tell you of my strange adventures,
> And the pain I felt upon our parting.
> My eyes will be cured by the sight of you,
> For my heart has been stirred by longing.
> I have hidden in my heart what I must say,
> So as to tell it to you when we meet.
> I shall reproach you for what you did,
> But reproach comes to an end and love remains.

When he had finished this poem, he caught sight of the green dome, the fountain and the Green Palace and in the distance he could see the Cloud Mountain. 'I have good news for you, Hasan,' said Shaikh 'Abd al-Quddus, 'for you will spend this night as a guest of my nieces.' Both Hasan and his wife were delighted, and they dismounted by the dome to rest, as well as to eat and drink, after which they remounted and rode on until they were close to the palace. As they approached, the *jinn* princesses came out to meet them. When they had exchanged greetings with their uncle, he told them: 'I have done what your brother Hasan wanted and I have helped him to rescue his wife and children.' The girls went up to Hasan and embraced him joyfully, congratulating him on having come back safe and sound and on having been reunited with his family. This was a feast day for them, but when his 'sister', the youngest of them, embraced him, she shed bitter tears in which he joined because of the length of their loneliness. She complained of the pain of parting and of the sorrow and distress that she had experienced when he left. She recited these lines:

Whoever I looked at, after you had left,
I saw you there before my eyes.
When I shut them, you were there in my sleep,
As though you were between my eyelid and my eye.

When she finished, her tears turned to joy and Hasan said to her: 'It is you, rather than any of your sisters, whom I have to thank in this affair, and may Almighty God aid and assist you.' He followed this with an account of everything that had happened to him on his journey from start to finish, including what he had had to endure and how he had managed to rescue his wife and children. He told her of the wonders he had seen and of his perils and difficulties, explaining how Nur al-Huda had wanted to kill him as well as his wife and children, and how they had only been saved by Almighty God. He followed this with an account of the rod and the cap, of how the *shaikhs* Abu'l-Ruwaish and 'Abd al-Quddus had asked him for them and of how it was only for her sake that he had handed them over. She thanked him and prayed to God to grant him a long life. 'By God,' he said, 'I shall never forget what you did for me from start to finish in this affair.'

Night 831

Morning now dawned and Shahrazad broke off from what she had been allowed to say. Then, when it was the eight hundred and thirty-first night, SHE CONTINUED:

I have heard, O fortunate king, that when Hasan met the princesses he told his 'sister' everything that he had had to endure, and added that he would never forget what she had done for him from the beginning to the end of the affair. The princess then turned to his wife, Manar al-Sana, embraced her and clasped her children to her breast. 'Daughter of the great king,' she said, 'was there no pity in your heart when you took these children from their father, leaving him to suffer pain because of their loss? Did you do this because you wanted him to be killed?' Manar al-Sana laughed and said: 'This was decreed by God, the Glorious, the Exalted, for whoever tricks people will himself be tricked by God.'

The princesses now provided food and drink, and everyone ate, drank and relaxed. Hasan stayed for ten days enjoying himself like this, and then got ready to leave while his 'sister' prepared for him such quantities of money and presents that they cannot be described. She then hugged him to her breast and embraced him by way of farewell, while he, for his part, gestured towards her and recited:

> Consolation is far removed from lovers,
> And it is hard to part from the beloved.
> Harshness and distance bring distress,
> But all love's victims die a martyr's death.
> How long, then, is the lover's night
> Who left his beloved and now stays alone?
> Tears pour down his cheek
> Until he asks whether there are more to flow.

Hasan now delighted Shaikh 'Abd al-Quddus by giving him the rod, and after he had thanked Hasan for it, he took it and went back to his own place. Then Hasan mounted, together with his wife and children, and when they left the palace the princesses went with them in order to say goodbye before returning home. Hasan set off for his own country and for two months and ten days he passed through deserts before reaching Baghdad, the House of Peace. He approached his own house by the postern door which opened on to the desert side, and then knocked on the main door. Ever since he had left, his mother had been unable to sleep and had spent so long lamenting and shedding tears of sorrow that she had fallen ill. She could neither eat nor enjoy sleep, and she wept night and day, constantly calling on Hasan's name. She had despaired of his return, and as he stood at the door he heard her tearfully reciting these lines:

> I call on you, master, to cure your patient,
> Whose body is wasted and whose heart is broken.

If, in your bounty, you grant her union,
The lover will be overwhelmed by his love.
I do not despair of your approach, for God is All-Powerful;
At times what is hard changes and becomes easy.

When she finished, she heard Hasan at the door calling out: 'Mother, time has allowed us to meet again.' She recognized his voice and went to the door, half believing and half in disbelief, but when she opened it and saw her son standing there with his wife and children, she cried out in joy and fell to the ground in a faint. Hasan revived her tenderly and she embraced him and wept, before calling to the servants and slaves and telling them to take all his baggage into the house, which they did. Manar al-Sana and the children entered, and Hasan's mother went and embraced her, kissing her head and her feet and saying: 'Daughter of the great king, if I wronged you I ask pardon from Almighty God.' Then she turned to Hasan and asked him why he had been away for so long, at which he told her the full story of his adventures from start to finish. When she heard all that had happened to him, she gave a great cry and collapsed unconscious on the ground. After he had once again revived her with tender care, she said: 'By God, my son, you lost an opportunity when you gave away the rod and the cap, for had you kept them for yourself, you could have become ruler of the whole world. But, praise be to God, you are safe and so are your wife and children.'

They passed the most pleasant of nights and then in the morning Hasan changed into clothes made of the most splendid material and went off to the market. There he began to buy slaves, male and female, materials, precious ornaments, robes and furnishings, together with utensils of such value that no kings had anything to match them. He also bought houses, orchards and properties together with other things, after which he stayed with his wife, his children and his mother, eating, drinking and savouring pleasures. They continued to enjoy the pleasantest and most luxurious of lives until they were visited by the destroyer of delights and the parter of companions. Praise be to the Sovereign Lord, the Living and Eternal God, Who does not die.

A story is also told that once upon a time in the old days there was in Baghdad a poor wretch of a fisherman named Khalifa, who had never married. One day, he took his net and went out as usual to the river to try to catch some fish before the other fishermen got there. He tucked up his clothes into his waistband, waded into the stream and spread the net out. His first cast brought nothing, and so did the second, but he kept on trying until, when he still had had no success after ten casts, he became despondent. Not knowing what to do, he said: 'I ask pardon from the Omnipotent God, apart from Whom there is no other god, the Living, the Everlasting Lord. I turn to Him in repentance, for there is no might and no power except with Him, the Exalted, the Omnipotent. What He wills comes to pass and without Him nothing takes place. Our daily bread comes from God, the Great and Glorious. What He gives to one of His servants cannot be withheld by anyone else, and what He withholds no one else can grant.' Then, from the depths of his distress, he recited:

When Time afflicts you with misfortune,
Prepare to suffer it and show endurance.
After your hardships, the Lord of all creation,
Thanks to His bounty, will grant you ease of life.

He sat for a time looking down at the ground and thinking over his position, and afterwards he recited:

Be patient, whether Time is sweet or bitter,
And know that God fulfils His purposes.
There is many a night of cares, painful as a boil,
I have endured till a successful dawn.
Misfortunes pass us by and disappear,
And then we never think of them again.

After that, he said to himself: 'I shall make one more cast, relying on God, in the hope that He will not disappoint me.' So he moved forward and hurled the net as far as he could, and then, having twisted the rope to secure it, he waited for a time before hauling it in. He found that it was heavy . . .

Night 832

Morning now dawned and Shahrazad broke off from what she had been allowed to say. Then, when it was the eight hundred and thirty-second night, SHE CONTINUED:

I have heard, O fortunate king, that after Khalifa, the fisherman, had

made a number of casts with his net but had caught nothing, he thought
things over and recited the verses that have been quoted. Then he said
to himself: 'I shall make one more cast, relying on God, in the hope that
He will not disappoint me.' He got up, cast the net and, after waiting
for a time, he drew it towards him. He found that it was heavy, and so
he pulled it in gently until he got it on shore, only to find that what was
in it was a lame, one-eyed ape. When he saw this, he recited the formula:
'There is no might and no power except with God. We belong to Him
and to Him do we return,' adding: 'What is this unlucky and ill-omened
turn of fate? What has gone wrong for me on this blessed day? But all
this happens through the decree of Almighty God.' He then took the
ape, secured it with a rope and went up to a tree overlooking the river
bank. He tied the ape to the tree and, taking a whip that he had with
him, he raised it in the air and was about to bring it down when God
allowed the ape to speak with a clear voice. 'Khalifa,' he said, 'stay your
hand and don't beat me. Leave me tied to this tree and go down to the
river to cast your net, putting your trust in God, for He will make
provision for you.' When Khalifa heard this, he took the net and returned
to the river. He made his cast, loosening the rope and then drawing it
in. This time it was even heavier than before, and he kept on pulling it
in until it came to shore, but what was in it now was another ape,
gap-toothed, its eyes stained with henna and its hands dyed; it was
wearing a tattered waist-wrapper and was laughing. 'Praise be to God!'
Khalifa exclaimed. 'He has changed the river fish into apes!' He went
back to the ape that was tied to the tree and said: 'Look, you ill-omened
creature, see what bad advice you gave me. For it was you and you only
who brought me to this second ape. Since you came to me this morning,
lame and one-eyed as you are, I have become exhausted and worn out,
and here I am, left with no gold and no silver.'

He took the whip in his hand, cracked it three times in the air and was
about to bring it down when the ape appealed to him and said: 'I implore
you in God's Name to spare me for the sake of this companion of mine.
Ask him for what you need, as he will guide you to whatever you want.'
At this, Khalifa decided to spare the first ape and, throwing aside the
whip, he went up to the second one and stood in front of him. 'Khalifa,'
this ape said, 'what I am going to tell you will do you no good unless
you listen carefully. If you do that and obey me without contradiction,
through me you will become a rich man.' Khalifa replied: 'Tell me what
it is that you have to say, so that I may obey you.' The ape said: 'Leave

me tied up here, go back to the river and cast your net, and after that I shall tell you what to do.' So Khalifa took the net, went to the river and made his cast, after which he waited for a while before starting to draw it in. He found it heavy, and when he had pulled it into the bank he found yet another ape in it. This one was red with a blue waist-wrapper, and it had dyed hands and feet, as well as kohl around its eyes. 'Glory be to God, the Omnipotent, the Sovereign King!' exclaimed Khalifa. 'From first to last this has been a blessed day, and the face of the first ape was a good omen, in the way that one can tell what is on a page from its title. This is the day of the apes. There are no fish left in the river and it must have been apes that I came out to catch today. So praise be to God, Who has turned the fish into apes.'

Turning to this third ape, he asked: 'What kind of a thing are you, you unlucky creature?' 'Don't you recognize me, Khalifa?' the ape said, and when Khalifa said no, it went on: 'I am the ape of Abu'l-Sa'adat, the Jewish money-changer.' 'And what do you do?' Khalifa asked, to which the ape replied: 'I say good morning to him at the beginning of the day, and he gets five dinars, and then at the end of the day I say good evening, and he gets another five.' Khalifa turned to the first ape and said: 'Ill-omened beast, see what fine apes other people have, but as for you, you come to me in the morning, lame and one-eyed with your unlucky face, and I become poor, penniless and hungry.' He took his whip, cracked it three times in the air and was about to bring it down on the ape when the third one said: 'Leave him alone, Khalifa. Stay your hand and come to me so that I can tell you what you should do.' Khalifa threw down the whip, went up to the ape and asked: 'What is it that you have to tell me, lord of all the apes?' The ape answered: 'Take your net and cast it into the river, leaving me and these other two apes to sit here with you. Then bring me whatever comes up in the net and I shall tell you something that will delight you.'

Night 833

Morning now dawned and Shahrazad broke off from what she had been allowed to say. Then, when it was the eight hundred and thirty-third night, SHE CONTINUED:

I have heard, O fortunate king, that Abu'l-Sa'adat's ape told Hasan to take his net and cast it into the river. He went on: 'Bring me whatever you find in it, and I shall give you good news.' 'To hear is to obey,' said Khalifa. He picked up his net, folded it over his shoulder and recited these lines:

In my distress I turn for help to my Creator,
Who has the power to make all hard things easy.
Within the blink of an eye, the favour of our Lord
Sets captives free and cures broken men.
All your affairs you should entrust to God;
And all with eyes to see can see His grace.

Then he added these further lines:

You, Who have brought hardship to men,
You cure cares and the causes of distress.
Let me not covet what I shall not get;
How many a covetous man has failed to reach his goal!

When he had finished reciting, he went to the river and cast his net. After he had waited for some time, he drew it up and found in it a large-headed perch with a tail like a ladle and eyes like dinars. Khalifa was delighted to see this, as he had never caught anything like it in his life. Filled with astonishment, he took it to Abu'l-Sa'adat's ape as proudly as if he were king of the whole world. 'What are you thinking of doing with this, Khalifa?' the ape asked. 'And what are you going to do with your own ape?' 'I shall tell you, lord of the apes,' Khalifa replied. 'First of all I shall make sure to kill this damned ape of mine, and I shall then take you in its place and give you whatever food you want each day.' The ape said: 'Now that you have chosen me, I shall tell you what to do and, God willing, this will lead to your advantage. Take note of what I have to say. Get another rope, tie me to a tree and then go out to the middle of the dyke and make a cast into the Tigris. Wait for a little while before drawing in your net, and in it you will find the finest fish that you have ever seen in your life. Bring it to me and I shall tell you what to do after that.'

Khalifa went off straight away, and when he had cast the net and drawn it in he saw that it contained a sheatfish, such as he had never seen before, as big as a lamb and bigger than the perch he had caught earlier. He took it to the ape, who told him to collect a quantity of green grass and put half of this in a basket with the fish on top, and then to use the other half to cover the fish. 'Leave us tied up,' he went on, 'and then take the basket on your shoulder and go into Baghdad. Don't answer anyone who speaks to you or who asks you questions until you get to the money-changers' market. At the top end of this you will find the booth of Master Abu'l-Sa'adat, the Jew, who is the superintendent of the money-changers. He will be sitting on a bench with a bolster at his back; in front of him you will see two chests, one of gold and the other of silver, and with him will be mamluks, slaves and servants. Go up to him and put the basket down in front of him. Then say: "Abu'l-Sa'adat, I went out to fish this morning and invoked your name as I cast my net, at which the Almighty sent me this fish." He will ask you whether you have shown it to anyone else and when you say no, he will take it from you and will give you a dinar. Give this back to him and he will then give you two. Don't accept them, and refuse whatever he offers, even if he is prepared to give you the fish's weight in gold. He will then ask what it is that you do want, and you must tell him: "I will only sell it if you say two things." When he says: "And what are they?" tell him: "Get to your feet and say: 'I call everyone here in the market to witness that I have exchanged my own ape for that of Khalifa the fisherman, and I have exchanged his allotted fortune with mine.' This is the price of the fish, for I have no need of gold." When he has agreed to this exchange with you, I shall come to you every morning and evening, and you will continue to earn ten dinars each day, whereas this lame, one-eyed ape will visit Abu'l-Sa'adat each morning, and God will see to it that every day he will have to pay a fine. This will go on until he is reduced to poverty and has nothing left at all. If you pay attention to what I have told you, you will be guided to prosperity.'

When Khalifa the fisherman heard what the ape had to say, he said: 'I shall accept your advice, king of the apes, but as for this ill-omened ape, may God give him no blessing. I don't know what to do with him.' 'Let him go back into the river,' replied the other, 'and let me go too.' 'To hear is to obey,' Khalifa said, and he went to the apes and released them, after which they entered the water. He then took and cleaned the fish, setting it on top of a bed of grass in a basket and then covering it with yet more grass. With the basket on his shoulder he set off, singing:

You are safe if you entrust your affairs to the Lord of heaven;
Do good throughout your life, and you will have no regrets.
If you associate with dubious men, men will suspect you;
Guard your tongue; do not abuse others, or they will abuse you.

He walked on until he reached Baghdad.

Night 834

Morning now dawned and Shahrazad broke off from what she had been allowed to say. Then, when it was the eight hundred and thirty-fourth night, SHE CONTINUED:

I have heard, O fortunate king, that when Khalifa the fisherman had finished his song, he put the basket on his shoulder and set off, walking on until he reached Baghdad.

As he went into the city, people recognized him and started calling out to him: 'What have you got, Khalifa?' But he paid no attention to them and carried on until he got to the market of the money-changers. He passed the stalls, following the ape's instructions, until he caught sight of the Jew sitting in his booth, surrounded by his servants as though he were a king of Khurasan. Khalifa recognized him, and when he went up to him, the Jew raised his head and looked at him. Then, as their recognition was mutual, the Jew greeted him and said: 'What do you want, Khalifa? If anyone has slandered you or is involved in a dispute with you, tell me so that I may go to the *wali* and see that he gives you your rights.' 'No, no, chief of the Jews,' replied Khalifa, 'nothing of the kind. I left home this morning relying on your good fortune, and when I got to the Tigris I made a cast with my net and brought out this fish.' At that, he opened the basket and put the fish down in front of the Jew, who admired it and exclaimed: 'By the Torah and the Ten Commandments, while I was asleep last night I saw myself in a dream standing before the Virgin, who was telling me that she had sent me a fine gift, and it seems that there can be no doubt that this fish is what she sent.' He turned to Khalifa and asked whether anyone else had seen it, and when Khalifa swore by God and by Abu Bakr al-Siddiq that no one had, the Jew told one of his servants to take it back home and get his wife, Sa'ada, to prepare it, fry it and grill it, so that it was ready for him when he came back after work. Khalifa joined in and said to the servant: 'Go and tell the master's wife to fry part of the fish and grill another piece.' 'To hear is to obey, sir,' the servant replied, and he went off with the fish to the Jew's house.

As for the Jew himself, he held out a dinar to Khalifa and said: 'Take this and spend it on your family.' When Khalifa saw the coin in the palm of his hand, he exclaimed: 'Glory be to the Sovereign King!' as though he had never seen gold before in his life. He took it, but after he had

walked a little way he remembered the ape's instructions and so came
back and threw the coin to the Jew, saying: 'Give me back the fish; it
doesn't belong to you. Are you trying to make a fool of people?' When
he heard this, the Jew thought that Khalifa was joking, and so he handed
over two more dinars in addition to the first one, but Khalifa said: 'Don't
try to be funny, just give me back the fish. Do you really think that I
would sell it for this price?' The Jew reached out for two more dinars
and said: 'Take these five dinars for the fish and don't be greedy.' Khalifa
took the money in his hand and set off with it joyfully. As he looked
admiringly at the gold he was saying: 'Glory be to God! Not even the
caliph of Baghdad can have had luck like mine today.' He walked on
until he had reached the upper end of the market, but then he
remembered what the ape had said and the instructions that he had given
him. So he went back to the Jew and threw the money down. 'What is
wrong, Khalifa?' the Jew asked. 'What do you want? Would you prefer
to change your dinars into dirhams?' Khalifa replied: 'I want neither
dinars nor dirhams, but only that you should give me back the fish,
which doesn't belong to you.' At this, the Jew lost his temper and shouted
at him: 'Fisherman, you bring me a fish that is not worth a dinar but
when I give you five you are not happy. Are you mad? Tell me how
much you want for it.' Khalifa told him: 'I shall only sell it if you say
two things for me.'

When the Jew heard him say this, his eyes started out of his head; he
could scarcely breathe for anger and he ground his teeth, exclaiming:
'Muslim scum, do you want me to abandon my religion because of your
fish? Are you trying to corrupt my faith and my beliefs, which were
followed by my fathers before me?'* He shouted for his servants, and
when they came he told them: 'Take this wretch, beat him on the nape
of his neck and give him a painful thrashing.' The servants set about
beating Khalifa and went on until he fell down under the booth. 'Leave
him to get up,' the Jew ordered, and at that Khalifa sprang to his feet as
though nothing had happened to him. The Jew then said: 'Tell me what
you want in payment for this fish and I shall give it to you, for you didn't
get much good from me just now.' 'You need have no fears for me
because of this beating, master,' Khalifa replied, 'for I can put up with
as many of these as any ten donkeys.' The Jew laughed at this and said:
'For God's sake, tell me what you want and I swear by my religion to

---

* The Jew assumes that Khalifa wants him to recite the Muslim confession of faith.

give it to you.' 'The only price I am prepared to accept for the fish is for you to say two things,' Khalifa replied. The Jew said: 'It seems to me that you want me to become a Muslim.' 'By God, Jew,' Khalifa answered, 'if you do become a Muslim your conversion will not help the Muslims or harm the Jews, and if you stay as an unbeliever this will neither harm the Muslims or benefit the Jews. What I want you to do is to stand up and say: "I call on everyone in the market to witness that I have exchanged my own ape for that of Khalifa the fisherman, and I have exchanged his portion in this world and his luck with mine."' 'If this is what you want, I can do it easily,' replied the Jew . . .

Night 835

Morning now dawned and Shahrazad broke off from what she had been allowed to say. Then, when it was the eight hundred and thirty-fifth night, SHE CONTINUED:

I have heard, O fortunate king, that the Jew said to Khalifa: 'If this is what you want, I can do it easily,' and he got to his feet immediately and used the form of words that Khalifa had proposed. Then, turning to him, he asked: 'Is there anything else that you need from me?' at which Khalifa said no. 'Then goodbye,' said the Jew, after which Khalifa got up straight away and, taking his basket and his net, he went to the Tigris. He made a cast and when he started to draw the net in he found it so heavy that he only managed to pull it in with difficulty, and when he got it out he discovered that it was filled with fish of all kinds. A woman came up to him with a dish and gave him a dinar, in return for which he gave her some of the fish, and she was followed by a eunuch, who also bought a dinar's worth. The stream of customers continued until he had sold his catch for ten dinars. Each day for the next ten days he sold ten dinars' worth of fish, until he had amassed a hundred gold dinars.

Khalifa had a lodging within the merchants' passage, and one night as he lay there to sleep he said to himself: 'Khalifa, everyone knows that you are a poor fisherman and yet you have a hundred gold dinars. Harun al-Rashid, the Commander of the Faithful, is bound to be told about this by somebody, and if he happens to be in need of money he will send to you and say: "I need some dinars; I have been told that you have a hundred of them, so lend them to me." I shall say: "Commander of the Faithful, I am a poor man and whoever told you that I had a hundred dinars was lying. I have not got anything like that at all." Then the caliph will hand me over to the *wali* and say: "Strip off his clothes and beat a confession out of him to make him hand over the money that he has." The right thing to do in order to escape from this dilemma is to start now and punish myself with this whip, so as to get into training for being beaten.' Under the influence of hashish, he told himself to get up and strip off his clothes. He did this straight away and, taking a whip that he had there, he started to aim alternate blows at his leather bolster and his own skin, calling out: 'Oh, oh, I swear to God, master, that this is false and that what they say of me is a lie, for I am a poor fisherman and I have none of this world's goods.'

As he used his whip to strike himself and then the bolster, people heard the noise of the strokes resounding through the night. Among them were the merchants, who asked each other: 'Why do you suppose this poor fellow is screaming, and we can hear the sound of blows landing on him? Maybe he has been attacked by robbers, who are beating him.' The sound of the blows and the screams stirred them all to action and, leaving their houses, they went to Khalifa's lodging, only to find the door locked. 'The robbers may have dropped down on him from the other side of the courtyard,' they told themselves, 'so we shall have to go up on the roofs.' They did this, and then climbed down through the skylight, only to discover the naked Khalifa flogging himself. They asked him what the matter was, and he told them that he was afraid that someone would tell Harun al-Rashid that he had got some dinars and he would then be summoned and asked for them. He would deny that he had them, but, as he explained: 'If I do that, I'm afraid that he will torture me, and so I am beating myself to prepare for what will happen.' The merchants laughed at him and said: 'Stop this. May God not bless you or your dinars, for you have disturbed us tonight and caused us alarm.'

Khalifa then stopped beating himself and slept until morning, but when he got up and was about to go off to work, he thought again about his hundred dinars and said to himself: 'If I leave them here, they will be stolen by thieves, and if I put them in a belt around my waist, someone may see them and then lie in wait for me until I am alone in a deserted spot, when he will kill me and take them. But I have an excellent scheme that will help me.' He got up straight away and sewed a pocket into the collar of his *jubba*; then he tied up the money in a purse and put the purse in the pocket he had made. After all that, he took his net, his basket and his stick and went off to the Tigris.

Night 836

Morning now dawned and Shahrazad broke off from what she had been allowed to say. Then, when it was the eight hundred and thirty-sixth night, SHE CONTINUED:

I have heard, O fortunate king, that Hasan put the hundred dinars in his pocket and then went off to the Tigris with his net, his basket and his stick. When he drew in his net after his first cast it turned out to be empty and so he went to another spot and tried once more, again without success. He went on from place to place, making fruitless casts, until he was half a day's journey from the city. He then told himself: 'Only one more cast, whether it works or not.' So in his pent-up frustration he hurled the net as hard as he could, and the purse with his hundred dinars flew out of his pocket and landed in the middle of the river, where it was carried away by the force of the stream. He threw away his net, stripped off his clothes and left them on the bank, then he waded into the river before diving in after his purse. He kept on diving and coming out again something like a hundred times, until his strength was gone and he had exhausted himself, still without having recovered the purse. Giving up hope of it, he went up on to the bank, but the only things he found there were his stick and net. He looked around for his clothes but could find no trace of them. He said to himself: 'This dirty thief must be the kind of man who, as the proverb says, completes his pilgrimage by copulating with his camel.' He spread out his net and wrapped himself in it. Then, with his stick in his hand and his basket on his shoulder, he started to rush off like a rutting camel, turning right, left, backwards and forwards, with his hair dishevelled as though he were a rebellious *'ifrit* freed from the prison of Solomon.

So much for him, but as for the caliph Harun al-Rashid, he had a friend called Ibn al-Qirnas, a jeweller, who was known by everybody, including the merchants, auctioneers and brokers, to be his agent. No rarity or other valuable piece was sold anywhere in Baghdad before it had been offered to him, and this included both mamluks and slave girls. One day, while he was sitting in his shop, the chief auctioneer came up to him bringing a girl so lovely and graceful, with so elegant a figure, that no one had ever seen her like before. Among her merits was the fact that she was familiar with every science and every art and, in addition to being able to compose poetry, she could play every single musical instrument. Ibn al-Qirnas bought her for five thousand dinars and, after having dressed her in clothes worth another thousand dinars, he took her to the caliph, with whom she spent the night. He tested her in every branch of science and art and found that her skill in all of them was such that she had no equal in her age. Her name was Qut al-Qulub, and she was as the poet has described:

> When she unveils, I look and then must look again;
> But, turning from me, she repels my glance.
> Her neck, when she looks round, is that of a gazelle,
> And, as is said, gazelles turn many times.

But these cannot match the lines:

> Who will fetch me a brunette whose supple body brings to mind
> Long, brown and graceful Samhari spear shafts?
> Her quiet glances and her cheeks of silk
> Preach sermons to the wasted lover's heart.

The next morning the caliph sent for Ibn al-Qirnas, who, on his arrival, was assigned ten thousand dinars in return for the girl, whose name was Qut al-Qulub. The caliph became so obsessed with her that he neglected his cousin and wife, the Lady Zubaida, daughter of al-Qasim, and he abandoned all his concubines, spending an entire month with her, during which time he only went out for the Friday prayer and then returned to her immediately. His ministers of state became concerned by this behaviour and complained to the vizier, Ja'far the Barmecide. Ja'far waited until Friday and he then met the caliph in the mosque and told him all the remarkable stories of love that occurred to him in order to prompt him to reveal his feelings. 'By God, Ja'far,' the caliph told him, 'this is not something that I am doing out of choice, but my heart is so far ensnared in the nets of love that I don't know how to act.' Ja'far replied: 'This concubine of yours, Qut al-Qulub, now belongs to you as one of your servants, and men's souls do not covet what they already possess. There is another point, which is that the thing in which kings and princes take most pride is hunting and taking opportunities to enjoy themselves. That might perhaps distract you and make you forget her.' 'Well said, Ja'far!' exclaimed the caliph. 'Let us go out to hunt immediately.'

When the Friday prayer was over, the two men left the mosque and rode off straight away to hunt . . .

Night 837

Morning now dawned and Shahrazad broke off from what she had been
allowed to say. Then, when it was the eight hundred and thirty-seventh
night, SHE CONTINUED:

I have heard, O fortunate king, that the caliph Harun al-Rashid,
together with Ja'far, rode off to hunt until they reached open country.
Both the caliph and Ja'far were mounted on mules, and as they were
busy chatting to one another, they were outstripped by their escort. In
the burning heat the caliph complained to Ja'far that he was very thirsty
and, gazing around, he thought that he could just make out the figure of
a man on top of a high mound. 'Do you see what I see?' he asked Ja'far,
and Ja'far said: 'Yes, Commander of the Faithful, I can dimly make out
a man there, who must be looking after an orchard or a cucumber bed.
At all events there must be water there.' He added: 'I shall go to him and
fetch some,' but the caliph said: 'My mule is faster than yours, so you
stay here to wait for the others while I go myself, have a drink with the
man and then come back.'

Al-Rashid rode off on his mule, which went like the wind or like water
pouring into a pool, going on without a check until within the blink of
an eye he had reached his goal and discovered that what he had seen
was Khalifa the fisherman. He found Khalifa naked apart from the net
in which he was wrapped; his eyes were as red as firebrands; he was
frightful to look at, bent, with dishevelled hair, and covered with dust,
like an *'ifrit* or a lion. He exchanged greetings with al-Rashid, but he
was still furious and consumed with blazing anger, so when the caliph
asked if he had any water, he said: 'Whoever you are, are you blind or
mad? Can't you see the Tigris there behind this mound?' The caliph
circled round behind the mound and went down to the river, where he
drank and watered his mule. He then went back straight away to Khalifa
and asked him who he was and his trade. Khalifa said: 'This is an even
stranger and odder question than the one that you asked about water.
Can't you see the tools of my trade on my shoulder?' 'You look like a
fisherman,' the caliph said, and when Khalifa said: 'Yes, I am,' the caliph
went on to ask: 'Where is your *jubba*, your cloak, your belt and the rest
of your clothes?' The items that the caliph mentioned exactly matched
the ones that Khalifa had lost, and when he heard the question he
thought to himself that the questioner must be the man who had taken

his clothes from the river bank. So immediately, quicker than a lightning flash, he bounded down from the mound and, grasping the reins of the caliph's mule, he said: 'Man, give me back my things and stop playing games.' 'By God,' said the caliph, 'I haven't seen your clothes and I don't know anything about them.'

The caliph had big cheeks and a small mouth and so Khalifa told him: 'You may be a singer or a flute player by trade, but either you give me my own clothes, or better ones, or else I shall beat you with this stick until you foul yourself with your own piss.' The caliph looked at his stick and saw that he was in Khalifa's power. He said to himself: 'By God, I couldn't endure half a blow from this lunatic beggar's stick,' and as he happened to be wearing a satin gown, he offered this to Khalifa in exchange for his lost clothes. Khalifa took it and, after having turned it inside out, he said: 'My clothes are worth ten times as much as this fancy gown.' 'Put it on,' said the caliph, 'until I can fetch you your own clothes.' Khalifa took it and put it on, but found it too long for him and, as he had a knife tied to the handle of his basket, he took it and cut a third of its length from the bottom of the gown so that it came down to below his knees. He then turned to the caliph and said: 'Tell me, flute player, how much does your master pay you each month for playing to him?' 'Ten gold dinars,' the caliph told him, and Khalifa exclaimed: 'By God, you poor fellow, you make me sorry for you, as this is what I get every day! If you would like to join me in my work I'll teach you how to fish and share the profit with you. You would be working for five dinars a day as my assistant, and I would use this stick of mine to protect you from your master.' 'I accept,' said the caliph, and Khalifa told him: 'Now get down from your donkey and tether him so that you can use him to carry fish, and then come on so that I can start teaching you here and now.'

The caliph dismounted and, when he had tethered his mule, he tucked up the skirts of his gown into his belt. 'Flute player,' Khalifa told him, 'take hold of the net like this, put it over your forearm like this and then throw it into the Tigris like this.' The caliph took heart and, following Khalifa's instructions, he made his cast into the stream, only to find that he could not draw it out. Khalifa came to help him, but even then the net would not come clear. 'You unlucky flute player,' Khalifa said. 'I took your gown in place of my clothes on the first occasion, but this time, if I find that my net has been torn I shall take your donkey to make up for it and I shall beat you until you foul yourself.' 'Let's pull together, you and I,' said the caliph, but it was still only with a great effort that the two of them managed to haul the net up. When they did, they found it filled with fish of all kinds and descriptions.

Night 838

Morning now dawned and Shahrazad broke off from what she had been allowed to say. Then, when it was the eight hundred and thirty-eighth night, SHE CONTINUED:

I have heard, O fortunate king, that when the net emerged, Khalifa the fisherman and the caliph saw that it was filled with all kinds of fish.

'You're an ugly fellow, flute player,' Khalifa told him, 'but if you apply yourself, you could be a great fisherman. The right thing now is for you to get on your donkey and go off to fetch two panniers from the market, while I stand guard over the fish until you come back. Then you and I will load them on the back of your donkey and take them all with us, for I have a pair of scales, weights and everything we need. You won't have to do anything more than to hold the scales and take the cash, for the fish we have here are worth twenty dinars. So hurry to fetch the panniers and don't dawdle.' 'To hear is to obey,' said the caliph.

He then left Khalifa with the fish and rode off happily on his mule, laughing all the while at his adventure until he reached Ja'far. At the sight of him, Ja'far said: 'I suppose, Commander of the Faithful, that when you went for your drink you must have found a pleasant orchard and gone in to enjoy yourself on your own.' When he heard this, the caliph burst out laughing again, at which all the Barmecides rose and, after having kissed the ground before him, said: 'May God prolong your joy and shelter you from sorrow, Commander of the Faithful. Why did you take so long when you went for your drink and what happened to you?' 'I had a most extraordinary and amusing adventure,' he replied, and he then told them about Khalifa the fisherman, how Khalifa had accused him of stealing his clothes, how he himself had then given him his gown and how, finding it too long, Khalifa had cut it down. Ja'far said: 'Commander of the Faithful, I had thought of asking you for the gown, but I shall go straight away to the fisherman and buy it from him.' 'By God,' replied the caliph, 'he ruined it by cutting off a third from the bottom. But this fishing has made me tired, because I made such a large catch, which is there on the river bank with my teacher. He is waiting for me to come back with two panniers and a chopper, after which the two of us are to go to sell the fish in the market and divide the money.' 'I shall bring you a buyer, Commander of the Faithful,' Ja'far promised, and the caliph said: 'I swear by my pure ancestors that whoever brings me one of the fish that my teacher, Khalifa, has in front of him will get a gold dinar for it from me.' As a result, the mamluks of the escort were told to go and buy fish for the caliph and off they went, making for the river bank.

While Khalifa was waiting for Harun to bring him the panniers, the mamluks swooped down on him like eagles, fighting each other to get to him and seizing the fish, which they wrapped in gold-embroidered kerchiefs. 'Without a doubt, these must be some of the fish of Paradise!' exclaimed Khalifa and, taking two in each hand, he waded neck-deep into the water, saying: 'God, I implore you by these fish to let my partner, Your servant the flute player, turn up at this moment.' Just then the chief of all the caliph's black slaves arrived, having been delayed by his horse, which had stopped to urinate on the way, and when he got there, he found that not one fish was left. He looked from right to left and then caught sight of Khalifa standing in the water with his fish. 'Fisherman,' he called, 'come here!' And Khalifa called back: 'Go away and don't make a nuisance of yourself.' But the eunuch went up to him and said: 'Hand over those fish and I'll give you a fair price for them.' 'Are you weak in the head?' Khalifa asked. 'I'm not going to sell them.' The eunuch drew his mace to attack him, but Khalifa said: 'Don't strike, you wretched fellow. Generosity is better than a blow with a mace,' and he threw him the fish. The eunuch took them and wrapped them in his kerchief, but when he put his hand in his pocket he couldn't find even a single dirham. 'Your luck is out, fisherman,' he said, 'for I have no money with me, but come to the caliph's palace tomorrow and ask to be directed to Sandal, the eunuch. You will be shown the way, and when you get to me I'll settle up with you, and you can take the money and go off.' 'This is a lucky day, as was clear from the start,' said Khalifa and, with his net over his shoulder, he walked back to Baghdad. As he went through the market, people could see that he was wearing one of the caliph's robes and they started to stare at him until he turned into his own quarter. It was by the gate of this quarter that the caliph's tailor had his shop, and when he saw Khalifa the fisherman wearing a robe of the caliph's that was worth a thousand dinars, he asked him where he had got it. 'Why are you so inquisitive?' asked Khalifa. 'I got it from a man whom I taught to fish. He became my apprentice and, since I had spared him from having his hand cut off as the penalty for stealing my clothes, he gave me this to make up for them.' The tailor then realized that the caliph must have passed Khalifa as he was fishing and played a joke on him, giving him the robe.

Night 839

Morning now dawned and Shahrazad broke off from what she had been allowed to say. Then, when it was the eight hundred and thirty-ninth night, SHE CONTINUED:

I have heard, O fortunate king, that the tailor realized that the caliph must have passed Khalifa the fisherman as he was fishing and played a joke on him, giving him the robe. Khalifa then went back home.

So much for him, but as for the caliph, he had gone out hunting specifically to distract himself from the influence of his slave girl, Qut al-Qulub. When Zubaida had learned about his obsession, with typical feminine jealousy she took neither food nor drink and could not sleep. She waited for the caliph to go out or leave on a journey so that she might lay a trap for her rival, and when she heard that the caliph had gone hunting she ordered her slave girls to spread furnishings and to adorn the palace with magnificent decorations. She had foodstuffs and sweetmeats set out, and among these was a china dish containing a delicacy of the finest type, which she had drugged with *banj*. On her instructions, a eunuch went to Qut al-Qulub with an invitation for her to eat with the Lady Zubaida. 'The wife of the Commander of the Faithful,' the eunuch told her, 'has taken a purge today, and as she has been told how well you sing, she wants to enjoy some of your artistry.' 'To hear is to obey,' replied the girl, 'for I owe obedience both to God and to the Lady Zubaida.'

She got up immediately, not knowing what the future held in store for her, and, taking the instruments that she needed, she went off with the eunuch and was led to Zubaida. When she entered her presence, she kissed the ground in front of her before standing up and saying: 'Peace be on the unapproachable lady, curtained in majesty, who comes from the stock of the 'Abbasids and the family of the Prophet. May God grant you good fortune and safety throughout the coming days and years.' She then took her place with the slave girls and the eunuchs, as Zubaida looked up. What she saw, when she looked, was a lovely girl with smooth cheeks, breasts like pomegranates, a resplendent forehead, dark eyes with languid lids, and a face of dazzling splendour, radiant as the moon. It was as though the sun rose in the whiteness of her complexion, the darkness of night was in the locks of her hair, the scent of musk was carried on her breath, her beauty caused the flowers to bloom, the moon could be seen shining from her forehead and her figure taught the branches how to bend. She was like the full moon gleaming in the darkness, her eyes spoke of dalliance, her eyebrows were curved like a bow and her lips were fashioned of coral. Her beauty distracted everyone who gazed at her, and all who saw her were bewitched by her glances. Praise be to the One Who created her in so perfect a form. She resembled the subject of the poet's lines:

> When she is angry, you see her victims lying dead,
> But if she is pleased, their spirits soon revive.
> There is sorcery within her glance
> That brings life or death wherever she may please.
> Her eyes take captive all mankind;
> It is as though they all become her slaves.

Zubaida welcomed her and invited her to sit and to entertain her with an exhibition of her artistry, to which she replied: 'To hear is to obey.' She sat down and reached out for her tambourine, an instrument that has been described in the following lines:

> Girl with the tambourine, my heart takes flight in longing,
> Calling out in passion to your beat.
> It is a wounded heart that you have seized,
> And all men long to hear you play.
> Whether your words are grave or light,
> Play what tune you like, for you will please us.
> Cast off restraint in happiness, you lover;
> Get up, dance, bend, give and receive delight.

She struck up a lively air and sang so as to halt the birds in mid-flight and rouse the whole palace. Then, laying aside her tambourine, she took a flute, of which it has been said:

> It has eyes whose pupils point
> With fingers at a harmony without discord.

Another poet has also said:

When the songs reach their desired end,
That is the pleasant time of joyful union.

After having delighted her audience, she put down the flute and took
up a lute, of which the poet says:

Many a moist branch has become a lute, held by a singing girl,
Moving the longing of the noblest and the best.
Such is her skill, she tests it by her touch
With fingers that bring out the perfect chords.

She tightened the strings, adjusted the pegs and then, putting it in her
lap, she bent over it like a mother over her child, as a poet has described:

She brought Arab eloquence out of the Persian string,
Giving intelligence to what has none.
Her lesson was that love can kill,
Destroying the minds of Muslim men –
A lovely girl holding a painted lute
That serves in place of human mouths to speak.
With this she holds in check the course of love,
As a good doctor halts the flow of blood.

She played in fourteen different modes and then sang an entire piece,
distracting the watchers and moving the listeners to delight. She then
recited:

Blessing attended me here with ever-fresh delights;
Good fortune is constant and its happiness never ends.

Night 840

Morning now dawned and Shahrazad broke off from what she had been
allowed to say. Then, when it was the eight hundred and fortieth night,
SHE CONTINUED:

I have heard, O fortunate king, that the slave girl, Qut al-Qulub sang
and played before the Lady Zubaida. After that she gave a display of
juggling tricks and other elegant skills until the Lady Zubaida almost
fell in love with her, and she said to herself: 'My cousin al-Rashid is not

to be blamed for being attracted to her.' After having kissed the ground before the Lady Zubaida, Qut al-Qulub sat down and was brought food and sweetmeats, together with the specially prepared plate. She ate something from this and, when the drugged morsel had settled in her stomach, her head lolled and she collapsed unconscious on the ground. 'Take her up to one of the rooms until I ask for her,' Zubaida told the slave girls, who answered: 'To hear is to obey.' She then ordered a eunuch to make her a chest and bring it to her, and she also gave instructions for a bogus tomb to be constructed and a report to be spread that Qut al-Qulub had choked to death. She warned her close associates that anyone who said that the slave girl was still alive would have their heads cut off.

Just at that moment the caliph returned from hunting, and the first thing he did was to ask about Qut al-Qulub. He was approached by a eunuch who had been coached by Zubaida to tell him, if he asked about her, that she was dead. The eunuch now kissed the ground before him, wished him a long life and told him that Qut al-Qulub had choked on her food and died. 'May God never give you any good news, you evil slave!' exclaimed the caliph, and he went on into the palace, where he heard confirmation of the news from everyone there. 'Where is she buried?' he asked, and they took him to see the fake tomb, telling him that this was her grave. At this sight he uttered a cry, threw his arms over it, burst into tears and recited the following lines:

By God, O grave, has her beauty vanished,
And has that radiant face been changed?
Grave, you are neither garden nor sky,
So how can you hold a bough and a full moon?

He followed this with a fit of weeping, and after a prolonged stay by the tomb he got up and left in the deepest sadness. Zubaida, learning that her scheme had worked, ordered that the chest be brought to her. She had Qut al-Qulub fetched and put inside it, after which she instructed the eunuch to do his best to sell it, laying it down as a condition that the purchaser was to accept it locked. The eunuch was then to give away the purchase price as alms. He took the chest and left to carry out her instructions.

So much for them, but as for Khalifa the fisherman, when day broke he said to himself: 'The best thing I can do today is to go to the eunuch who bought the fish from me and told me to come to see him in the caliph's palace.' So he went out and headed there. When he arrived, he found the mamluks, the black slaves and the eunuchs standing or sitting there, and when he looked more closely, he caught sight of the eunuch who had taken his fish seated with mamluks in attendance on him. One of these shouted to Khalifa and when the eunuch looked to see who it was, he recognized him as the fisherman. Khalifa, seeing this, said: 'I have not failed you, my ruddy friend, and this is how the trustworthy behave.' On hearing this, the eunuch laughed and said: 'By God, fisherman, you are right.' He was about to give Khalifa some money and had put out his hand towards his pocket when there was a great clamour. He looked up to see what was happening, and discovered the vizier, Ja'far the Barmecide, coming from the caliph. He got up and walked over to him, after which the two of them walked to and fro, deep in conversation, for some considerable time.

Khalifa waited for a while without the eunuch turning towards him, and, when this had gone on for too long, he went to stand opposite him at a distance, and waved to him, saying: 'My ruddy friend, let me go.' The eunuch heard him but was ashamed to answer in the presence of Ja'far, as he was engrossed in his conversation and too busy to pay attention to him. So Khalifa called out: 'Hey you, the slow payer, may God disgrace every sluggard and everyone who takes people's property and then makes difficulties for them. I appeal to you, Master Bran-Belly, to give me my due so that I can get off.' The eunuch, on hearing this, was embarrassed because of Ja'far, who, for his part, saw Khalifa waving his hands and talking, although he did not follow what he was saying. He said reproachfully to the eunuch: 'What is this poor beggar asking from you?' 'Don't you recognize him, my lord vizier?' asked the eunuch. 'No, by God,' answered the vizier. 'How could I, when this is the first time that I've seen him?' 'Master,' said the eunuch, 'this is the fisherman whose fish we took from the bank of the Tigris. I hadn't got any of them and I was ashamed to go back empty-handed to the Commander of the Faithful, after all the mamluks had got some. When I got to the bank I found him standing the middle of the stream calling on God and clutching four fish. I told him to give them to me for a fair price, but when he did and I felt in my pocket to give him something, I could find nothing at all. So I told him to call on me at the palace and I would give him something to help relieve his poverty. He came today and I was just reaching out to give him this when you arrived. I got up to present my services to you and this distracted me from him. He has got tired of waiting, and this is why he is standing here.'

Night 841

Morning now dawned and Shahrazad broke off from what she had been allowed to say. Then, when it was the eight hundred and forty-first night,
SHE CONTINUED:

I have heard, O fortunate king, that Sandal the eunuch told Ja'far the Barmecide about Khalifa the fisherman, and concluded by saying: 'This is his story and this is why he is standing here.'

Ja'far, on hearing this, smiled and said: 'Chief of the eunuchs, how is it that this fisherman comes in the hour of his need and you have not settled the matter for him? Don't you know who he is?' 'No,' said the eunuch, and Ja'far told him: 'He is the teacher and partner of the Commander of the Faithful, who is distressed, sorrowful and preoccupied today. It is only this fisherman who will be able to cheer him up, so don't let him go until I consult the caliph. I want to take him to the caliph, whose melancholy God may dispel by his presence, in consolation for the loss of Qut al-Qulub. He might then help the man with a gift, and this would be thanks to you.' 'Do as you wish, master,' said the eunuch, 'and may the Almighty preserve you as a bulwark for the 'Abbasid state and maintain and protect it root and branch.' Ja'far set off back to the caliph, and the eunuch instructed the mamluks to stay with the fisherman. 'How kind of you, my ruddy friend!' Khalifa exclaimed. 'The seeker becomes the sought! I came to look for what was owed me only to be arrested for its arrears.'

When Ja'far entered the caliph's presence, he found him sitting miserably, looking down at the ground and full of care. He was chanting these lines:

> The censurers tell me to forget,
> But what can I do when my heart will not obey?
> How can I endure the loss of a young girl's love,
> When no endurance helps me, now that she has gone?
> I shall not forget how the cup passed between us,
> As the wine of her glance made me sway drunkenly.

On entering, Ja'far said: 'Peace be on you, Commander of the Faithful, defender of the sanctity of the faith, nephew of the master of the apostles, may God bless him and all his family and give them peace.' The caliph looked up and replied: 'Peace be on you, together with God's mercy and blessing.' Ja'far then asked for permission to speak freely, and the caliph said: 'When were you ever not allowed to do this, you who are the master of all viziers? Say whatever you want.' So Ja'far told him: 'When I left your presence on my way home, I saw Khalifa the fisherman, your master, teacher and partner, standing at the gate. He was angry with you and was complaining that he had taught you how to fish and that you went off to fetch him a pair of panniers and never came back. "This is not how partners or teachers should be treated," he was saying. If you want to stay in partnership with him, well and good, but if not, then tell him, so that he can find someone else.'

The caliph smiled at Ja'far's words and his depression left him. 'Is it true that he is standing there?' he asked, and when Ja'far confirmed this, he said: 'By God, I'll do my best to see that he gets what he deserves. If God decrees that he should find misfortune at my hands, so be it, and similarly if He decrees good fortune.' He took a sheet of paper and cut it into pieces before telling Ja'far to write down twenty sums of money, ranging from one dinar to a thousand, together with a number of official posts from the humblest to the caliphate itself, and twenty punishments, from the mildest reproof to death. Ja'far did what he was told, saying: 'To hear is to obey, Commander of the Faithful.' The caliph then said: 'Ja'far, I swear by my pure ancestors and my relationship to Hamza and 'Aqil* that I want you to bring the fisherman and tell him to take one of these bits of paper, on which only you and I will know what is written. Whatever this may be, I shall see that he gets it, even if it is the caliphate, for I would then abdicate and install him in my place without grudging it to him, while if it is a matter of hanging, amputation or death, the same will apply. Now go and fetch him.'

When Ja'far heard this, he said to himself: 'There is no might and no power except with God, the Exalted, the Omnipotent. It may be that this will lead to the poor fellow being killed, and it is I who will be responsible. But as the caliph has sworn an oath, there is nothing for it but to bring the man in, and then it is what God wills that is going to happen.' So he went to Khalifa and took him by the hand in order to bring him to the caliph. For his part, Khalifa went out of his mind with fear, saying to himself: 'What a fool I was to come to this ill-omened slave, the ruddy one, who got me to meet Bran-Belly.' As Ja'far led him on, there were mamluks in front of him and behind him and Khalifa

---

* Hamza was the Prophet's uncle, and 'Aqil was the brother of 'Ali ibn Abi Talib.

said: 'It's not enough for me to have been arrested; now these mamluks are hemming me in to stop me running away.' Ja'far took him through seven halls and then told him that he was standing in the presence of the Commander of the Faithful, the defender of the sanctity of the faith. He then raised the great curtain and Khalifa's eye fell on the caliph seated on his throne, attended by his state officials. On recognizing him, he went up to him and said: 'Greetings, flute player. It was wrong of you to play the fisherman and then go off and not come back, leaving me sitting there to look after the fish. Before I knew what was happening, mamluks rode up on all kinds of beasts and snatched the fish away, leaving me standing there alone, and this was all your fault. If you had fetched the panniers quickly enough, we would have sold the catch for a hundred dinars. I came here to get what was due to me and they arrested me, but who was it who imprisoned you here?' The caliph smiled and then, lifting a corner of the curtain, he leaned out and told him: 'Come here and take one of these bits of paper.' Khalifa said: 'You were a fisherman, and today I see that you have turned into an astrologer, but the jack-of-all-trades is a very poor man.' 'Don't talk so much,' said Ja'far, 'but do what the Commander of the Faithful tells you and take one of these pieces of paper.' So Khalifa advanced and stretched out his hand, saying: 'This flute player will never be my apprentice and come out fishing with me again.' He took a piece of paper and handed it to the caliph, saying: 'Flute player, what does it say my fortune is going to be? Don't keep it secret.'

Night 842

Morning now dawned and Shahrazad broke off from what she had been allowed to say. Then, when it was the eight hundred and forty-second night, SHE CONTINUED:

I have heard, O fortunate king, that Khalifa the fisherman took one of the bits of paper and passed it to the caliph. 'Flute player,' he said, 'what does it say my fortune is going to be? Don't keep it secret.' The caliph took it and passed it to Ja'far, telling him to read out what was on it. When Ja'far looked at it, he exclaimed: 'There is no might and no power except with God, the Exalted, the Omnipotent.' 'What have you seen in it, Ja'far?' the caliph asked. 'Good news, I hope.' But Ja'far told him that, according to the paper, Khalifa was to receive a hundred strokes of the cane. On the caliph's orders this punishment was then carried out, and after the beating Khalifa got up and said: 'Damn this for a game, Bran-Belly! Is imprisonment and beating all part of it?' Ja'far said: 'Commander of the Faithful, this poor fellow has come to the river. How can he be allowed to go back thirsty? I hope that your bounty will allow him to take another slip of paper in the hope that what comes out will allow him to go back with something to relieve his poverty.' 'By God, Ja'far,' the caliph replied, 'if what comes out is death, I shall kill him and it is you who will be responsible.' Ja'far replied: 'If he dies, he will be at rest,' but Khalifa said: 'May God never send you good news. Have I made Baghdad so unpleasant for you that you want to kill me?' 'Take another piece of paper,' Ja'far told him, 'and pray to the Almighty to grant you something good.' So he put out his hand, took one and passed it to Ja'far, who took it, read it and stayed silent. The caliph asked him why this was, and Ja'far told him that, according to the paper, Khalifa was to be given nothing at all. The caliph said: 'He is not going to earn a living here, so tell him to leave my presence.' 'I call on you by your pure ancestors to let him draw a third lot,' Ja'far said, 'and it may be that he will get some money.' 'He can have one more, but nothing else,' replied the caliph, and at that Khalifa reached out for this third piece, on which was written: 'The fisherman is to be given one dinar.' Ja'far said: 'I had hoped to bring you good fortune, but it was God's will that you should get no more than this dinar.' 'One dinar for a hundred strokes – what a good bargain! May God never send you any health,' said Khalifa. The caliph laughed and Ja'far took Khalifa by the hand and led him out.

When he got to the door, Sandal the eunuch caught sight of him and said: 'Come on, fisherman, let me have a share of what the caliph gave you while he was joking with you.' 'Right you are, my ruddy friend,' said Khalifa. 'Do you want to share the hundred strokes of the cane that I got, black skin, together with one single dinar? We can then call it quits.' He threw the dinar at the eunuch and went off with tears running down his cheeks. The eunuch, seeing the state he was in, realized that he had told the truth and called to the servants to bring him back, which they did. He then put his hand into his pocket and brought out a red purse which, when he shook it out, he found contained a hundred gold dinars. 'Fisherman,' he said, 'take this gold in return for your fish and go on your way.' Khalifa was overjoyed and, taking the hundred dinars together with the one the caliph had given him, he went away, forgetting all about his beating.

Almighty God, in order to carry out His decree, caused him to pass by the slave girls' market, where he saw a great circle of people. He wondered what this was and so he went up through the crowd of merchants and others. 'Make way for Captain Trash,' the merchants shouted, and when the crowd allowed him through, he saw an old man standing with a chest in front of him on which a eunuch was sitting. The old man was calling out: 'Wealthy merchants, which of you will take a chance and bid for this chest with its unknown contents, from the palace of the Lady Zubaida, daughter of al-Qasim, the wife of the Commander of the Faithful, Harun al-Rashid? How much are you going to offer, may God bless you?' 'By God,' one of them said, 'this is a risk, but I can't be blamed for saying twenty dinars.' Another offered fifty and they then went on bidding against each other until the price reached a hundred dinars. 'Is there any advance on that?' asked the auctioneer, and at that Khalifa called out: 'A hundred and one.' The merchants laughed when they heard this, thinking that he must be joking, and they told the eunuch: 'Sell the chest to Khalifa for a hundred and one dinars.' 'By God,' the eunuch said, 'I'm not going to sell it to anyone else. Take it, fisherman, God bless you, and hand over the gold.' Khalifa produced it and handed it over to the eunuch, at which the bargain was concluded.

The eunuch gave away the money on the spot as alms and then went back to the palace, where he delighted Lady Zubaida by telling her what he had done. Khalifa tried to carry the chest on his shoulder but it was too heavy for him. So he put it on his head and brought it to the quarter where he lived, before putting it down again. He then sat in exhaustion, thinking of what had happened to him, and he started to say to himself: 'I wish I knew what was in this chest.' Then he opened his house door and managed to bring it in. He tried to open it but failed, which left him wondering why he had been stupid enough to buy it. He told himself that he would have to break it open to see what was in it, but when he found that he couldn't force the lock he decided to leave it until the following day. He wanted to sleep, but as the chest was almost as big as the room, the only place that he could find on which to stretch out was the top of it. An hour later he felt something moving; sleep deserted him and in a blind panic . . .

Night 843

Morning now dawned and Shahrazad broke off from what she had been allowed to say. Then, when it was the eight hundred and forty-third night, SHE CONTINUED:

I have heard, O fortunate king, that after Khalifa the fisherman had been sleeping on top of the chest for an hour, he felt something moving. Sleep deserted him and, in a blind panic, he exclaimed: 'There must be *jinn* in it! Praise be to God, Who stopped me opening it, for otherwise they would have attacked me in the darkness and killed me. That would have done me no good at all.' He tried to go back to sleep, but the chest moved again, this time more violently than before. 'It's happening again,' he said, 'and it is still alarming.' He went to get a lamp but couldn't find one and, as he had no money to buy another, he went outside and called to the people of the quarter. Most of them were asleep, but they woke up when he called, asking him what was wrong. 'Bring me a lamp,' he said, 'because the *jinn* have come to attack me.' They laughed at him, but they did give him a lamp, which he took back to his room. He used a stone to break the lock on the chest, and when he opened it, there was a sleeping girl, looking like a houri of Paradise. She had been drugged, but just at that moment she vomited up the drug, and then recovered and opened her eyes. Finding herself cramped, she moved, and at the sight of her Khalifa jumped forward and said: 'By God, my lady, where have you come from?' Now that her eyes were open, she said: 'Fetch me Narcissus and Jasmine,'* at which Khalifa said: 'I've only got henna plant here.' She regained her senses and looked at Khalifa. 'What are you?' she asked, and then: 'Where am I?' He told her: 'You are in my room.' 'Aren't I in the palace of the caliph, Harun al-Rashid?' she said, at which he exclaimed: 'What is this Rashid, you mad woman? You are my slave girl; I bought you today for a hundred and one dinars and fetched you home while you were asleep in this chest.' When she heard this she asked him his name, to which he answered: 'My name is Khalifa, but how is it that my star is in the ascendant, when this is not what I'm used to?' She laughed and said: 'Enough of that. Have you anything to eat?' 'No, by God,' he told her, 'and there is nothing to drink either. I haven't eaten for two days and I still haven't got a bite.' 'Haven't you

---

* The names of slave girls.

any money?' she asked, and he replied: 'May God preserve this chest. It is this that has bankrupted me, as I paid out all that I had to get it and now I'm penniless.' She laughed at him and said: 'Get up and get me something to eat from your neighbours, for I'm hungry.'

Khalifa got up and went outside, where he shouted to his neighbours. They were roused from sleep and asked him again what was the matter. 'I'm hungry,' he told them, 'and I've nothing to eat.' One of them came down with a loaf, another with a crust, a third with a piece of cheese and a fourth with a cucumber, all of which they put into his lap. He went back in, set them all down in front of her and told her to eat. 'How can I,' she said, laughing, 'when I haven't a jug of water to drink from? I'm afraid that I might choke on a mouthful and die.' 'I'll fill this jar for you,' Khalifa said and, taking it with him, he went to the centre of the quarter and called out to his neighbours. 'What's wrong with you tonight, Khalifa?' they asked, and he said: 'You gave me food, which I ate, but I'm thirsty, so give me something to drink.' One of them fetched a mug, another a jug and a third a bottle, and so, after having filled his jar, he went inside. 'There is nothing else that you can need, my lady,' he said, to which she replied: 'Not for the moment.'

He then asked her to tell him her story, and she said: 'If you don't know who I am, you wretched fellow, I shall introduce myself. I am Qut al-Qulub, the slave girl of the caliph Harun al-Rashid. The Lady Zubaida was jealous of me, drugged me and put me in this chest, but praise be to God that things did not turn out worse. This was thanks to your good fortune, and there is no doubt that you will get enough money from the caliph to make you a rich man.' 'Is that the Rashid in whose palace I was held as a prisoner?' asked Khalifa, and when the girl said yes, he went on: 'By God, I have never seen a meaner man than him, good-for-nothing and stupid flute player that he is. He gave me a hundred strokes of the cane yesterday and presented me with a single dinar, in spite of the fact that I taught him to fish and took him as a partner, only for him to betray me.' 'Don't abuse him like that,' she replied. 'Open your eyes and the next time you see him, remember your manners. Then you will get what you want.' When Khalifa heard this, it was as though he had woken up from sleep, and God removed the veil from his understanding in order to bring him good fortune. He agreed to what she said and then told her to go to sleep, which she did, while he, for his part, slept at a distance from her until morning.

When morning came, she asked him for an inkstand and a sheet of paper. When he had fetched these for her, she wrote to the caliph's friend, Ibn al-Qirnas, the merchant, telling him of the position in which she found herself and how she had been bought by Khalifa the fisherman. She gave the note to Khalifa and told him to go to the jewellers' market, to ask for the booth of Ibn al-Qirnas and then hand him the note without speaking. 'To hear is to obey,' said Khalifa, who then took the note from her hand, went to the market and asked after the booth. When he had been given directions, he went to Ibn al-Qirnas and they exchanged greetings. Ibn al-Qirnas looked at him contemptuously and asked him what he wanted. At that, Khalifa passed him the note, but Ibn al-Qirnas didn't read it, thinking that here was a beggar asking for alms. Instead, he told one of his servants to give Khalifa half a dirham, but Khalifa told him: 'I don't need alms; read the note.' Ibn al-Qirnas now took it, and when he had read it and grasped its contents, he kissed it and put it on his head.

Night 844

Morning now dawned and Shahrazad broke off from what she had been allowed to say. Then, when it was the eight hundred and forty-fourth night, SHE CONTINUED:

I have heard, O fortunate king, that Ibn al-Qirnas read the note, grasped its contents and put it on his head. Then he got to his feet and asked: 'Where is your house, my brother?' 'What do you want with my house?' Khalifa asked. 'Are you thinking of going there to steal my slave girl?' 'No,' Ibn al-Qirnas told him, 'but I shall buy something for the two of you to eat.' At that, Khalifa told him where his house was, and Ibn al-Qirnas said: 'Well done! Good for you, you lucky fellow.' He then told two of his slaves: 'Go with this man to the booth of Muhsin the money-changer and tell Muhsin to give him a thousand dinars in gold. Then hurry back here with him.' Accordingly the slaves took Khalifa off to Muhsin's booth, and at their request he was handed a thousand dinars, which he took back, accompanied by the slaves, to their master's booth. There he found Ibn al-Qirnas mounted on a dappled mule that was worth a thousand dinars, surrounded by mamluks and servants, while standing beside his mule was another just like it, saddled and bridled. Ibn al-Qirnas told Khalifa to mount it, but he said: 'No, by God, I shan't, for I'm afraid that it would throw me.' Ibn al-Qirnas insisted, and when Khalifa came up to mount it, he got on the wrong way round. He clutched at its tail and gave a yell, at which the mule threw him on the ground as the spectators laughed. He got up and said: 'Didn't I tell you that I wasn't prepared to ride this great donkey?'

Ibn al-Qirnas now left him in the market and went off to tell the caliph about Qut al-Qulub, after which he came back and moved her to his own house. It was later that Khalifa went home to see her, only to find all the people of the district crowded together. They were saying: 'Khalifa has got good reason to be terrified today. Where do you suppose that he got the girl from?' One of them said: 'He is a crazy pimp. He may have found her drunk in the street and carried her off to his house, before disappearing when he realized what he had done.' Just at that moment, as they were talking, Khalifa himself came up. 'What kind of trouble are you in, you poor wretch?' they asked. 'Don't you know what has happened to you?' 'No, by God,' he told them, and they said: 'Mamluks came just now and took that slave girl whom you stole. They looked for you, but couldn't find you.' 'How could they take her?' he said, to which one of the crowd replied: 'Had they come across you, they would have killed you.' Khalifa, however, paid no attention to them but ran back to Ibn al-Qirnas's booth. Finding him mounted on his mule, Khalifa said: 'By God, this was not fair. You managed to distract me and then sent your mamluks to take my slave girl.' 'Idiot,' said Ibn al-Qirnas, 'come here and keep quiet.' He then took him to an elegantly built house where, on entering, he found Qut al-Qulub seated on a golden throne, surrounded by ten slave girls as lovely as moons. When Ibn al-Qirnas saw her, he kissed the ground before her and she asked him: 'What have you done with my new master, who bought me with everything he had?' 'My lady,' he told her, 'I gave him a thousand gold dinars,' and he went on to tell her what had happened from start to finish. She laughed and said: 'Don't blame him, for he is only a common man,' and she added, 'and here is another thousand dinars as a gift to him from me, while, God willing, he will get enough from the caliph to make him rich.'

While they were talking, a eunuch, sent by the caliph, came to look for Qut al-Qulub as, when he had learned where she was, he could not bear to be without her and so had given orders that she was to be fetched. As she set off to meet him, she took Khalifa with her. Then, when she reached the palace and came into the caliph's presence, she kissed the ground before him, while he, for his part, got up and, after greeting and welcoming her, he asked her how she had got on with her purchaser. 'He is a man called Khalifa the fisherman,' she told him, 'and he is standing at the door. He told me that he has an account to settle with you, Commander of the Faithful, because of your partnership in the business of fishing.' 'He's there, is he?' the caliph asked, and when Qut al-Qulub confirmed that he was, the caliph ordered him to be brought in. On entering, he kissed the ground before the caliph and prayed to God to prolong his glory and prosperity. This surprised the caliph, who laughed at him and then asked: 'Were you, in fact, my partner yesterday, fisherman?' Khalifa understood what he meant and, plucking up his courage, he replied: 'I swear by God, Who appointed you as the successor to your cousin, the Prophet, that my only acquaintanceship with this girl was one of sight and speech.' Then, as the caliph laughed, he told the full story of his adventures from beginning to end, giving an account of his encounter with the eunuch, and how the man had given him a hundred dinars in addition to the one dinar that he had got from the caliph, how he had then gone to the market and how he had bought the chest for a hundred and one dinars without knowing what was in it. When he had finished, the caliph laughed in relief and said: 'You have returned to me what was mine and I shall give you whatever you want.' Khalifa stayed silent and so the caliph presented him with fifty thousand dinars in gold, a splendid robe of the sort that was worn by great rulers, a mule, and black slaves to wait on him. While Khalifa had become like one of the kings of his age, the caliph for his part was overjoyed at the return of his slave girl.

He knew that Zubaida had been responsible for all this . . .

Night 845

Morning now dawned and Shahrazad broke off from what she had been allowed to say. Then, when it was the eight hundred and forty-fifth night, SHE CONTINUED:

I have heard, O fortunate king, that the caliph was delighted at the return of Qut al-Qulub. He knew that the Lady Zubaida had been responsible for all this and he was so angry with her that for some time he abandoned her, neither visiting her nor showing any fondness towards her. When she realized this, she was so worried that her rosy complexion turned pale, and when she could bear it no longer she sent him a message acknowledging the wrong that she had done and apologizing. She recited the following lines:

I long for the goodwill you used to show
To cure my distress and treat my grief.
Pity me, lord, because of my great love;
You have already punished me enough.
I can bear it no longer, now that you have gone;
You have clouded a life that used to be serene.
When you are faithful to your covenant, I live,
But if you do not grant me this, I die.
Forgive me, wrong though I was in what I did;
How dear the lover is when he forgives.

When her message reached the caliph, he read it and realized that she had confessed and had sent to ask his forgiveness for what she had done. Recalling the words of the Quran: 'God forgives all sins, for he is the Merciful, the Forgiving,'* he delighted her by sending a reply offering reconciliation and forgiveness.

As for Khalifa, the caliph provided him with a monthly grant of fifty dinars, and arranged for him to enjoy high position, rank and dignity at his court, so that, as he left after having kissed the ground, he strutted out proudly. When he got to the door, the eunuch who had given him the hundred dinars recognized him and asked: 'Where did you get all this from, fisherman?' Khalifa told him everything that had happened to him from beginning to end, and the eunuch was delighted to think that it was he who was responsible for Khalifa's new-found wealth. He said: 'Aren't you going to give me something from what you have got?' at which Khalifa reached into his pocket, brought out a purse containing a thousand dinars in gold and handed it to him. But the eunuch said: 'Keep your money, God bless you,' being filled with admiration for his sense of honour and the generosity that he showed, after having been so poor.

Khalifa then left him, mounted on his mule with his slaves holding on to its crupper, and as he rode to the *khan* he was the cynosure of all eyes, while the people wondered at his newly acquired splendour. When he had dismounted, they came up to ask how this had happened and he told them the entire story. Later he bought a fine house on which he lavished money until it reached a pitch of perfection. When he had moved in there, he used to recite the lines:

* Quran 39.54.

Look! This is like a house in Paradise;
It banishes care and can cure the sick.
It has been raised up high
As a lasting home for everything that is good.

When he was settled in it, he wooed the pretty daughter of one of the
leading citizens of the town, and after the marriage had been consummated, he enjoyed a sociable life of ever-increasing good fortune and
contentment, with blessings piled on blessings and in complete happiness.
Seeing himself in this position, he gave thanks to Almighty God for the
favours and benefactions showered on him, praising Him in his gratitude
and chanting the lines:

Praise be to You for the succession of Your favours,
You, Whose endless bounty encompasses all things.
I offer You my praise; accept it, Lord,
For I am mindful of Your liberality.
You have showered on me Your favours and Your gifts
Through Your good grace, and I give thanks to You.
Mankind drinks from the ocean of Your bounty,
And it is You Who aids them in distress.
You have granted me an abundance of good things,
And through You all my sins have been forgiven,
Thanks to the one who came as a mercy to all,
A noble Prophet, truthful and pure of heart,
On whom be blessings and the peace of God.
Blessings be on his helpers and his family, while the pilgrims come,
And on his noble companions, men of power,
For all time, while the birds still sing on trees.

Khalifa used to frequent the society of the caliph as a favoured visitor,
and the caliph treated him with the greatest kindness and generosity. He
continued to enjoy prosperity, happiness and grandeur, with ever-
increasing comforts, rank, pleasure and delight, until he was visited by
the destroyer of delights and the parter of companions – praise be to the
Glorious and Everlasting Lord, the Eternal One Who never dies.

A story is told that in the old days there was a merchant named Masrur
who, in addition to possessing great wealth, was one of the most handsome people of his time. He lived in easy circumstances and enjoyed

taking his pleasure in gardens and orchards, while also delighting in the love of beautiful women. One night it happened that while he was asleep he dreamt that he was in the most beautiful of gardens where there were four birds, among them being a white dove that glistened like polished silver. He was struck with admiration for this dove and conceived a passion for it, but then he saw in his dream that a huge bird swooped down on it and snatched it from his hand, to his great distress. At that point he woke up and, being unable to find the dove, he experienced pangs of longing for it until morning came. 'I must go off today and find someone to interpret this dream for me,' he told himself.

Night 846

Morning now dawned and Shahrazad broke off from what she had been allowed to say. Then, when it was the eight hundred and forty-sixth night, SHE CONTINUED:

I have heard, O fortunate king, that when Masrur, the merchant, woke, he experienced pangs of longing and said to himself: 'I must go off today and find someone to interpret this dream for me.' He got up, but although he walked in various directions until he was a long way from home, he failed to discover anyone who could do this for him. He was on his way back when it occurred to him to turn off to the house of a wealthy merchant. When he got there, he heard the following lines being recited in a plaintive voice from a broken heart:

> The east wind blows from her abandoned camp
> With scent that cures sick-hearted lovers.
> I halted with my question by the worn-out ruins,
> Where nothing but dry bones answered my tears.
> I asked the breeze: 'For God's sake let me know,
> Will its delights ever return again?
> Shall I enjoy the favour of a fawn whose soft form has led me astray,
> Wasting my body, with slumberous, languid eyes?'

When Masrur heard the voice, he looked inside the door and saw the loveliest of gardens, within which there was a curtain of red brocade, studded with pearls and other gems. Behind this were four girls, among whom was another between four and five foot tall like a rounded moon gleaming at its full. Her eyelids were darkened with kohl beneath joining eyebrows; her mouth was like Solomon's ring; her lips and her teeth were pearls and coral; and her beauty and grace, together with the symmetry of her figure, were such as to rob all who saw her of their wits.

At the sight of her, Masrur entered the house and went on as far as the curtain, at which point the girl raised her head and looked at him. He greeted her and, speaking sweetly, she returned his greeting and then, as he looked closely at her, he lost both his wits and his heart. He gazed at the garden, which was filled with jasmine, gillyflowers, violets, roses, oranges and all kinds of scented herbs. All the trees were adorned with fruit, and water was flowing down from four alcoves set facing each other. Masrur looked at the first of these and saw that round it was inscribed in letters of vermilion the following lines:

> House, no sorrow has ever entered you,
> Nor has your owner been betrayed by Time.
> How good a refuge you provide for guests
> Left comfortless elsewhere.

On the second alcove he saw spelt out in letters of red gold:

> House, may you be clothed in fortune
> As long as birds sing on the garden trees.
> May all your air be scented, while in you
> Love finds its consummation.
> May those who live here enjoy fame and happiness,
> While planets circle in the upper heaven.

The inscription around the third alcove was picked out in lapis lazuli and read:

> Remain in glory and good fortune, house,
> As long as nights are dark and the stars shine.
> Happy are those who come within your gate,
> For you shower fortune on your visitors.

Round the fourth alcove, Masrur saw written in yellow ink:

> Here is a garden and a pool –
> A pleasant place to sit, and a forgiving Lord.

There were ringdoves, pigeons, nightingales and turtledoves in the garden, each with its own song, while the lovely girl with her shapely figure swayed in a way that would captivate all who saw her. She now asked Masrur: 'Man, what has brought you to what is not your own house and to girls who are strangers to you, without permission from the owners?' He said: 'My lady, when I saw this garden, I was struck by the beauty of its greenery, the fragrance of its flowers and the songs of its birds, and so I came in to enjoy the sight of it for a time before going on my way.' At that, she spoke words of welcome, and when Masrur heard this and saw the coquetry of her glance and the elegance of her form, he became bewildered both by her beauty and grace and by the charm of the garden and the birds. His wits left him and he recited:

> She appears as a moon in unmatched loveliness
> Among the hills where scented breezes blow,
> With myrtle, eglantine and violets,
> Whose fragrance is diffused among the branches.
> The garden here is perfect in its beauty;
> All flowers are here and every kind of branch.
> The moon unveils itself beneath their shade,
> And here birds sing their sweetest songs.
> The ringdove, nightingale and turtledove
> Join with the bulbul to arouse my grief.
> Passion has halted helpless in my heart,
> Bewildered by her beauty, like a drunken man.

After the girl, whose name was Zain al-Mawasif, had listened to these lines she gave him a glance that robbed him of his wits and was followed by a thousand regrets. She replied to his verses with these lines:

> Do not look for union with the beloved;
> Abandon all the hopes to which you cling.
> Leave this aside, for you can never bear
> To be rejected by the one you love.
> My glances harm the lover, and the words
> That you have spoken are of no avail.

In spite of what he had heard, Masrur showed patience and endurance, concealing his feelings for the girl and telling himself: 'Patience is the only cure for misfortune.' They waited there until nightfall, when Zain ordered food to be brought. A table was set before her and Masrur on which were various dishes, such as quails, young pigeons and mutton. When they had both eaten their fill, she ordered the food to be removed; when this had been done they were brought the wherewithal to wash their hands, and next she had candlesticks fetched, in which were set candles scented with camphor. 'I feel depressed tonight as I'm suffering from fever,' she said. 'May God enliven you and dispel your distress,' said Masrur, and she then told him that she was an experienced chess player and asked whether he knew anything about the game. 'Indeed I do,' he said, and she then produced a chessboard of mixed ebony and ivory, the squares marked with gleaming gold, while the pieces were made of pearls and sapphires.

Night 847

Morning now dawned and Shahrazad broke off from what she had been allowed to say. Then, when it was the eight hundred and forty-seventh night, SHE CONTINUED:

I have heard, O fortunate king, that Zain ordered the chessmen to be brought to her. Zain turned to Masrur, who was bemused by the sight of this, and asked him whether he wanted the red or the white pieces. 'Mistress of beauties and glory of the dawn,' he replied, 'you must take the red, because their loveliness complements yours, and leave me the white.' She agreed to that and, after setting the red pieces opposite the white, she stretched out her hand to one of them in order to make the first move. Masrur looked at her fingers, which were as soft as dough, and was astonished both by their beauty and by her fine qualities. She turned to him and said: 'Don't be so taken aback but show patience and firmness.' 'Lady, whose beauty shames the moon,' he replied, 'how can a lover who looks at you show patience?' While he was still in this state she checkmated him and won the game. Then, realizing that he was maddened by love, she said: 'I shall only play with you for a fixed bet.' 'To hear is to obey,' he answered, and she went on: 'We must swear to each other that neither of us will cheat the other.' When they had done this, she told him: 'If I beat you I shall take ten dinars from you, and if you beat me I shall give you nothing at all.'* Masrur thought that he would win and he said: 'Don't break your word, lady, for I see that you are the stronger player.' 'I agree,' she replied, and they started to play, pushing forward pawns and backing them with their queens, linking them with castles and allowing the knights to move forward.

---

* It appears later that she was staking herself.

Zain was wearing a scarf of blue brocade, which she took from her head, and she rolled back her sleeve to show a wrist like a pillar of light. Passing the palm of her hand over the red pieces she told Masrur to take care, but at the sight of such elegance and grace he lost his wits and, bewildered and dazzled, when he stretched his hand out for the white pieces, it touched the red. 'Where is your intelligence, Masrur?' she asked him, pointing out that the red pieces were hers and that his were the white. 'No one can look at you and keep control of his reason,' he replied, and, seeing the state he was in, she took the white pieces from him and gave him the red. He played with the red and lost, and as they continued to play and as she continued to win, he paid over ten dinars after each game. Seeing how deeply in love with her he was, she said: 'You will never get what you want, Masrur, unless you beat me, for this was the condition. I am not going to go on playing with you unless we put a hundred dinars on the game.' He agreed willingly and so she continued to play and to beat him, and each time he paid over a hundred dinars.

This went on until morning without his winning a single game. He then got to his feet and when she asked him what he was thinking of doing, he said: 'I'm going back home to fetch more money so that perhaps my hopes may come true.' 'Do as you think best,' she told him, and so he went home and fetched all the money he had. Then, when he returned, he recited these lines:

I saw a bird that passed me in a dream
In a pleasant garden where the flowers smiled.
When it appeared, I caught it;
It is for you to show the interpretation of the dream was true.

After he had brought all his money to her, he started to play against her, but she kept on winning and he was unable to take a single game. This lasted for three days, by the end of which she had won all his money, and then she asked: 'What do you want to do now?' 'I'll play you for my perfume shop,' he said, and when she asked what it was worth, he told her: 'Five hundred dinars.' He played five games and lost them all, and then he staked his slave girls, his properties, his orchards and his buildings, all of which she won, leaving him with nothing. She turned to him and asked whether he had anything else to stake, to which he replied: 'By Him Who made me fall into the toils of your love, I have no money left and nothing else, great or small.' 'Masrur,' she told him, 'something that began with contentment should not end with regret, and if it is regret that you feel, then take back your money, leave me and go off. I shall not hold you to our agreement.' 'By God, Who decreed that this should happen to us,' exclaimed Masrur, 'if you wanted to take my life, that would be a small price to pay for contenting you, who are my only love!' So she told him to fetch the *qadi* and the notaries and to write a deed making over to her all his possessions and properties. 'Willingly,' he said and he got up instantly, and when he brought the officials to her, the *qadi* was bemused and bewildered by the beauty of her fingers. 'My lady,' he said, 'I shall only draw up this deed on condition that you are the purchaser of the properties, the slave girls and the other possessions, which are then to be in your hands and at your disposal.' 'We have agreed on that,' she told him, 'so draw me up a deed to say that the property of Masrur, his slave girls and all that he owns are to be transferred to the ownership of Zain al-Mawasif at a total price specified.' The *qadi* drew up the deed and the notaries added their signatures to it.

When Zain got this deed . . .

Night 848

Morning now dawned and Shahrazad broke off from what she had been allowed to say. Then, when it was the eight hundred and forty-eighth night, SHE CONTINUED:

I have heard, O fortunate king, that when Zain got this deed stipulating that everything that Masrur had was to be transferred to her, she said to him: 'Now go away,' but her slave girl Hubub turned to him and told him to recite some poetry. He produced the following lines on the subject of chess:

> I complain of Time and of what has come to me;
> I complain of loss, of chess and of a glance that led to love –
> Love for a delicate and tender girl,
> Who has no match among mankind, female or male.
> She notched an arrow, aiming her eyes at me,
> And ordered out such armies as would conquer every foe,
> Red and white, with knights clashing in combat.
> She challenged me, saying: 'Be on your guard.'
> She left me straying when she reached out her fingers,
> In the gloom of night, dark as her hair.

I could make no move to rescue my white men,
As passion caused my tears to fall in floods.
Pawns, rooks and queens might charge,
But the white army turned back in defeat.
She struck me with an arrow from her eyes,
An arrow that pierced through my heart.
She offered me a choice between the two armies;
I chose the white on which to place my stake.
I said: 'I want the white men; they suit me,
While as for you, you can command the red.'
She played me for a stake I had accepted,
But I could never win my goal of pleasing her.
Alas for my heart, my longing and my sorrow;
I could not reach a girl fair as the moon.
My heart does not burn with regret or sorrow
For my estates while I can fondly look on you.
I am perplexed, bewildered and afraid,
Blaming Time for what it has brought down on me.
She asked: 'Why so distressed?' I said:
'Can a wine drinker sober up when drunk?'
Her figure stole my heart away, and she would be
Human, if her heart were not of stone.
I led myself to hope and said: 'Today
I'll win her through my bet – no need for fear or caution.'
My heart went on and on coveting this,
Until I found myself completely destitute.
Can the lover draw back from his harmful love,
Even if he is drowning in the seas of passion?
The slave of love has no two coins to rub together,
A prisoner to love's longing, who has failed to reach his goal.

When Zain heard this she admired his eloquence, but said: 'Give up this madness, Masrur. Come to your senses and go on your way, for you have wasted all your wealth and your property on playing chess. You have not got what you wanted and there is no way in which you can succeed.' He turned to her and said: 'My lady, ask me for something and I shall fetch you whatever it may be and lay it before you.' 'But you've no money left,' she pointed out, at which he said: 'Goal of my hopes, I may have no money but people will help me.' 'Does the giver ask for gifts?' she queried, to which he answered: 'I have relatives and friends who will give me whatever I ask for.' So she said: 'I want from you four containers of pungent musk, four of combined musk and ambergris perfume, four *ratls* of ambergris, four thousand dinars and four hundred robes of embroidered royal brocade. If you bring me all this, I shall grant you union.' 'This will be easy for me, you who put the moon to shame,' Masrur told her, and then he left in order to fetch what she had asked for.

Zain sent Hubub, her slave girl, after him to see what his standing was with the people whom he had mentioned to her. As he was walking through the city streets he happened to turn and catch sight of her some way away. He waited until she caught up with him, and then asked her where she was going. She explained why her mistress had sent her after him, and reported everything that she had said. 'By God, Hubub,' he said, 'I have nothing at all in the way of money.' 'Why did you make her a promise, then?' she asked, and he replied: 'How many promises remain unfulfilled? There must always be delays in love.' When Hubub heard that, she told him to take heart, promising that she herself would help him to his goal.

She then walked back to her mistress and said: 'By God, my lady, Masrur is valued and respected in the community.' 'No one can resist what the Almighty has decreed!' exclaimed Zain, adding: 'This man did not find me merciful, in that I took his money and gave him no affection in return or sympathized with him when he wanted me, but if I do agree to what he wants, I'm afraid that word may get out.' Hubub replied: 'It's not easy for us to ignore his present plight and the way you took his money. You have no one here with you apart from me and Sukub and, as we are your slave girls, which of us could say anything about you?' Zain stared down at the ground and her girls said: 'Lady, we think that you should send him a gracious message and not leave him to ask favours from some ignoble man, for it is bitter to have to beg.' Their mistress accepted their advice and, after calling for an inkwell and paper, she wrote these lines:

Be glad, Masrur; union is near at hand;
Do not delay but come when the night is dark.
Do not go begging to base men for money.
I was drunk but now I have regained my wits.
I shall return you all your wealth,

And add to that the gift of union,
Because you have been patient and with grace
Accepted the beloved's unjust tyranny.
Come quickly to enjoy the happiness of love.
Do not be careless lest my family learn.
Hurry to me; do not delay, and taste
The fruits of union while my husband is away.

She folded the letter and gave it to Hubub, who took it off to Masrur. She found him tearfully reciting these lines:

A lovesick breeze has blown over my heart,
Captivating it through too much passion.
With the beloved's absence my love has grown
And my eyes have shed an increased flow of tears.
Were I to show the doubts that have beset me
To solid rock, how quickly it would soften.
Shall I ever find something for my delight,
And enjoy reaching the goal for which I hope?
Will the nights of rejection, which followed parting, end,
Curing the wound she left within my heart?

Night 849

Morning now dawned and Shahrazad broke off from what she had been allowed to say. Then, when it was the eight hundred and forty-ninth night, SHE CONTINUED:

I have heard, O fortunate king, that Masrur, overcome by passion and the intensity of his longing, started to recite poetry. When Hubub heard him repetitively chanting these lines, she knocked on his door and he got up and opened it for her. She went in and gave him the letter, which he took and read. 'What news do you have of your mistress?' he asked, but she said: 'There is no need to answer that, thanks to what is in the letter, and you are a man of sense.' Masrur was overjoyed and recited:

A letter has come whose contents bring me joy;
Would that I could preserve it in my heart.
I kissed it and my longing was increased;
It is as if within it was love's pearl.

He then wrote a reply and gave it to Hubub, who took it back to her mistress, and when she got there she started to enlarge on his attractions and to talk of his various qualities and his noble nature, for she had decided to help bring the two of them together. 'He is slow in coming,' said Zain, but Hubub assured her that he would soon be there, and before she had finished speaking he had arrived and knocked on the door. She opened it for him and took him in to her mistress, who welcomed him and made him sit down beside her.

She now told Hubub to bring Masrur a splendid robe, at which she fetched one that was adorned with gold. Zain took this and put it on Masrur, while she, for her part, chose another magnificent one, which she put on herself. On her head she placed a chaplet of fresh pearls fastened with a band of brocade, itself studded with pearls, sapphires and other gems. Beneath it she allowed two plaits to hang down, each adorned with a ruby and picked out with gleaming gold; the hair that she now let down was as dark as night, and she was scented and perfumed with aloes, musk and ambergris. 'May God guard you from the evil eye,' said Hubub, and then, as her mistress walked proudly, swaying from side to side, she recited these remarkable lines:

She shames the *ban* tree's branches as she walks,
And lovers are left powerless by her glance.
A moon is framed in the darkness of her hair,
And a sun is shining among her black locks.
Blessed is the man who spends the night beside such beauty,
And dies swearing an oath 'by your life'.

Zain thanked her and then went up to Masrur like a glorious full moon. He sprang to his feet and said: 'Unless I am deceived, this is no mortal woman but one of the houris of Paradise.' She called for a table to be brought, round the edges of which the following lines were inscribed:

Turn aside your spoons to the spring camp of the bowls;*
Enjoy roasts of all kinds and young partridges.
There are quails I never cease to love,
And other costly young fowls, as well as chickens.
How splendid are the kebabs that bloom so red,
With vegetables dipped in bowls of vinegar.

* A parody on an introductory line of the conventional ode of classical Arabic.

How tasty is rice cooked in milk, in which
Women plunge their hands up to their bracelets.
How I sigh for two kinds of fish,
With two loaves of bread that have been well baked.

They ate, drank and enjoyed themselves, after which the food was
cleared away and replaced with wine. As the wine circulated between
them, they grew happy, and when Masrur filled the glass he addressed
Zain as 'my mistress whose slave I am', and recited:

I wonder that my eye can look its fill
On the beauty of a girl of shining loveliness.
She has no equal in her age,
In elegance and graceful qualities.
The *ban* tree's branch envies her suppleness
When she advances, well poised in her robe;
Her radiant face puts the full moon to shame;
The parting of her hair is like its gleaming crescent.
Wherever she moves, her perfume scents the breeze
That moves across the plains and hills alike.

When Masrur had finished these lines, Zain said to him: 'Masrur,
whoever holds to his religion and has eaten my bread and my salt must
acknowledge what he owes me. So forget what has happened and I shall
give back your property and everything that I took from you.' 'My lady,'
he replied, 'I would not hold you to this, even if you had not kept to the
terms of the oath that we swore to each other. I shall go off and become
a Muslim.' At that point, Hubub said: 'My mistress, learned as you are,
you are still young, and I call on the Almighty to intercede for me with
you. Unless you oblige me by doing what I say, I shall not sleep in your
house tonight.' Zain promised to do what she wanted and then told her
to go and prepare another room. This she did, providing decorations
and supplying the finest of her mistress's preferred perfumes. Food was
prepared and wine brought in, which then was passed between Masrur
and Zain. As they were enjoying themselves . . .

Night 850

Morning now dawned and Shahrazad broke off from what she had been
allowed to say. Then, when it was the eight hundred and fiftieth night,
SHE CONTINUED:

I have heard, O fortunate king, that Zain told Hubub, her slave girl,
to prepare another room. Hubub got up and provided fresh supplies of
food and wine. The wine cup passed around between them and, as they
were enjoying themselves, Zain said: 'Masrur, the time has come for
close embraces and, if you really love me, recite me some original verses.'
So Masrur recited:

> I am a captive in whose heart rages a fire,
> As parting cuts the tie uniting us.
> I love a girl whose form lies in my heart
> And whose soft cheek has robbed me of my reason.
> Her eyebrows join over dark eyes,
> And when she smiles her teeth gleam like a lightning flash.
> She is fourteen, and now the tears
> I shed for love of her are all dyed red.
> I saw her between a garden and a stream,
> With a face outshining the moon in heaven's vault.
> I stood in awe of her, a prisoner,
> And said: 'Peace be on you, lady of the sanctuary.'
> She willingly returned the greeting that I gave,
> With courteous words like pearls upon a string.
> But when she came to know from what I said
> The goal I aimed at, then her heart was deaf.
> She told me this was folly, but I said:
> 'Do not blame a man who is in love.
> If you accept me now, this is a simple matter;
> You are the loved and I the slave of love.'
> When she saw what I wanted, she said smilingly:
> 'I swear by God, Who made both earth and heaven,
> I am a Jewess, following the strictest code
> Of my religion, while you are a Christian.
> How can you look for union, when you do not share my faith?
> If you seek this, regret is bound to come.

Is it allowed by love to juggle with two faiths,
So that someone like me would be wounded by blame,
As a religious outcast, while, as for you,
You would have sinned against my faith and yours?
If you love me, become a Jew for love,
Keeping yourself from all unions but mine.
Now, by the Gospel, take a solemn oath
To hide away the secret of our love,
While I swear by the Torah faithfully
That I shall keep the covenant we have made.'
I swore by my religion and its law,
And made her take the same most solemn oath.
I said: 'Goal of desire, what is your name?'
'Zain, lady of the sanctuary,' she said.
'Zain,' I called out aloud to her,
'Your love has occupied my heart, enslaving me.'
I saw the beauty underneath her veil
To my distress, as I fell deep in love.
She sat behind a curtain, and my humble plea
Came from an ardent heart controlled by love.
Then, when she saw how deep my passion was,
She showed me there unveiled her laughing face.
The winds of union blew for us, as musk
Diffused its scent over her neck and wrist,
And the whole place was perfumed, as I kissed
The sweet wine of her smiling mouth.
Like a *ban*-tree branch, she swayed beneath her gown,
And what had been refused me was allowed.
We passed the night joined in close embrace,
Clasping and kissing, while I sucked red lips.
The splendour of the world is that the one you love
Should be beside you, yielding to your will.
Then, when dawn broke, she rose and took her leave,
With a fair face that puts the moon to shame.
As she recited her farewells, her tears
Were scattered, then collected on her cheeks.
Never in all my life shall I forget our pact,
The beauty of the night and our most solemn oath.

This delighted Zain, who complimented Masrur on his skill, adding: 'May your enemies perish.' She went to her room and called Masrur, who went in and embraced her, hugging and kissing her until she had granted him what he had thought he would never get. He was delighted by the sweetness of union and Zain told him: 'Your property is now lawfully yours again and not mine, as we have become lovers.' So she returned everything that she had taken from him, after which she asked: 'Have you got a garden where we can go to enjoy ourselves?' He told her that he had one of unparalleled beauty, and then he went home to tell his slave girls to get ready a splendid meal and to prepare a handsome room, setting out a magnificent candelabrum. After this he invited Zain to come, and, when she had arrived with her slave girls, they ate, drank and enjoyed themselves cheerfully as the wine circulated. Then, when the lovers were alone, Zain told Masrur: 'An elegant poem has come to my mind which I would like to sing to you, accompanying myself on the lute.' She took a lute, tuned it, passed her hands over the strings and, pitching her voice at the right note, she sang these lines:

> The strings have filled me with delight;
> Sweet-tasting is our early morning wine.
> Love uncovers the lovesick heart
> And shows itself, tearing aside the veils,
> Accompanied by finely flavoured wine
> Like the sun uncovered in the hands of moons.*
> On a night that brought us such a joy
> As served to wipe away all sombre cares.

When she had finished, she said: 'Now recite some of your own lines, Masrur, and let me enjoy the fruits of your eloquence.' So he recited:

> We rejoice as a full moon passed around the wine
> To the lute music in our garden here,
> As the doves sing upon the bending boughs
> At early dawn. Here is the goal of all desire.

When he had finished, Zain said: 'If you are so deeply in love with me, recite some verses about what has happened to us.'

---

* The moons are the beautiful wine-pourers.

Night 851

Morning now dawned and Shahrazad broke off from what she had been allowed to say. Then, when it was the eight hundred and fifty-first night, SHE CONTINUED:

I have heard, O fortunate king, that Zain said to Masrur: 'If you are so deeply in love with me, recite some verses about what has happened to us.' Masrur willingly agreed and recited:

> Stop and hear what it was that came to me
> Thanks to the love of this gazelle.
> She shot me with an arrow,
> Attacking me with her glance.
> I was seduced by passion;
> Love left me powerless.
> I fell in love with a coquette,
> Veiled from me by arrow heads.
> I saw her in a garden,
> A girl with a graceful form.
> I greeted her and when she heard
> My words, she greeted me.
> I asked her name, and she told me:
> 'As befits my beauty,
> I am called Zain.'*
> 'Pity my state,' I said.
> 'I am consumed by passion;
> There was never a lover to match me.'
> She said: 'If you do love me
> And hope for union with me,
> I want huge sums of money,
> More than all reckoning.
> I want you to give me garments
> All made of costly silk,
> And four *qintars* of musk
> In return for a night of love,
> Together with pearls and carnelian,
> All of the costliest,

* A name meaning 'beauty'.

As well as silver and gold
By way of finery.'
I showed the virtue of patience
In spite of my distress,
And so she granted me union
On a night of the crescent moon.
If any wish to blame me,
I say: 'Listen to me:
She has long locks of hair, night-coloured;
There are roses in her cheeks
That bloom like kindled fire.
Beneath her eyelids is a sword,
And arrows are shot by her glance.
Within her mouth is wine,
While her saliva is cool water.
Her teeth are rows of pearls
Arranged in order in her mouth.
She has the neck of a gazelle,
Beautiful in its perfection.
Her bosom is marble white,
With breasts like hills.
The creases of her belly
Are perfumed with precious scent,
While below is something
That is the goal of my desire,
Well fleshed and plump, my masters,
Like a king's throne, to which I bring my case.
Between two pillars you will find raised benches,
While its description astounds the minds of men.
There are two large lips that part as in a mule,
Like a red eye, and a bulge like a camel's lip.
When you approach it, intending to do the deed,
You find a warm encounter, full of lusty strength.
It leaves all brave opponents with no more urge to fight,
And at times you have to meet it with a beard to play for time.
The one who tells you this is a handsome and splendid man,
Like Zain in the perfection of her beauty.
I came to her by night and what I won was sweet,
For that one night I spent with her surpassed all other nights.

When morning came, she rose with a face like a crescent moon,
Her body swaying like a long spear shaft.
She took her leave of me and said: 'When will such nights return?'
'Light of my eyes,' I said to her, 'whenever you wish, come.'

Zain was filled with delight and the greatest of pleasure by this poem, but she then said: 'Dawn is near, Masrur, and I shall have to go for fear of being disgraced.' He agreed and got to his feet to escort her to her house, after which he went back home to spend the rest of the night thinking about her beauty. In the morning he prepared a splendid present, which he brought, and then sat there with her. Things went on like this for a number of days, during which the two of them enjoyed the pleasantest and most delightful of lives, but at that point Zain got a letter from her husband in which he said that he would soon be back with her. 'May God grant him neither safety nor life,' she said to herself, 'for if he comes he will spoil everything. How I wish I were sure that he would never return.'

When Masrur arrived, he sat talking with her as usual until she told him that her husband had sent word that he would soon be back from his travels. 'What are we going to do,' she asked, 'as neither of us can do without the other?' He said: 'I don't know what will happen, but you know more about your own husband's character, and, in particular, you are a very intelligent woman who can produce tricks that no man could think up.' She replied: 'He is a difficult man who guards his household jealously. When you hear that he has got back, go and greet him and sit down beside him, after which you should tell him that you are a perfume seller. Then buy some perfumes from him and go back to him again and again, holding long conversations with him and doing whatever he says. It may be then that I shall be able to contrive something that will look like a chance encounter.' 'To hear is to obey,' Masrur replied, and he left her with his heart ablaze with love.

When Zain's husband did come home, she showed pleasure at his arrival and greeted him warmly. When he looked at her, he saw that she was looking pale. This was because she had used the feminine trick of washing her face with saffron. He asked her how she was and she told him that, after he had gone off on his travels, both she and her slave girls had been sick; they had all been concerned because he had been away for so long, and she began to complain to him of the miseries of separation, shedding floods of tears and saying: 'If you had someone with you I would not be so concerned, and so I implore you in God's Name not to go off again unaccompanied. Don't leave me without news of what you are doing, and then I may be easy in my mind.'

Night 852

Morning now dawned and Shahrazad broke off from what she had been allowed to say. Then, when it was the eight hundred and fifty-second night, SHE CONTINUED:

I have heard, O fortunate king, that Zain said to her husband: 'Don't go off again unaccompanied. Don't leave me without news of what you are doing, and then I may be easy in my mind.' He agreed to this willingly, saying that it was a good and sensible idea and promising to do what she wanted.

He then went off to his shop with some of his goods and opened it up for business. As he was sitting there, along came Masrur, who greeted him, sat down beside him and started talking to him. After a long conversation, Masrur produced a purse, opened it and took out some gold, which he passed to the man, saying: 'Give me in exchange some perfumes that I can sell in my shop.' 'To hear is to obey,' the man replied, and he supplied Masrur with what he wanted. After that, Masrur paid him a number of visits over a period of days until the man turned to him and said: 'I am looking for an associate in my business.' 'That is what I am wanting myself,' replied Masrur, adding: 'My father was a Yemeni merchant who left me a large sum of money, which I'm afraid I may lose.' Zain's husband turned to him and said: 'If you would like to join me, I would be your companion and your friend, whether we go on our travels or stay here, and I would teach you how to buy and sell, make money and spend.' Masrur said that he would welcome this and the man took him home, where he left him sitting in the hall while he went to Zain, his wife, and told her that he had found a partner and had invited him as a guest. He asked her to provide him with a lavish meal, and she, for her part, realizing that this must be Masrur, gladly prepared a splendid feast with excellent food, delighted that her scheme had worked.

When Masrur came to their house, her husband told her to come out with him to greet the visitor, but she made a show of anger and said: 'Do you want to produce me before a stranger and a foreigner? God forbid! Even if you were to cut me in pieces, I would not show myself in front of him.' 'Why should you be shy of him,' he asked, 'when he is a Christian and we are Jews? He and I are going to be companions.' She said: 'I don't want to appear before a stranger whom I have never seen before and whom I don't know.' Her husband, who thought that she was being sincere, continued to press her until she rose, wrapped herself up and brought the food out to Masrur, whom she welcomed. He stared at the ground, pretending to be embarrassed, and, on seeing this, his host was sure that he must be an ascetic.

When they had eaten their fill, the food was removed and the wine produced. Zain sat in front of Masrur and they exchanged glances until evening, when Masrur went back home with a fire burning in his heart, leaving his host to reflect on his courtesy and handsomeness. When night fell, his wife brought him his supper as usual. In his house he had a nightingale which had been in the habit of coming to flutter over his head when he sat down to eat, sharing his meal with him. This bird had become friendly with Masrur and had done the same with him, but when Masrur had gone, it no longer recognized his master on his return and would not come near him, leaving him to wonder what the reason for this might be.

As for Zain, she could not sleep for thinking of Masrur and this went on for a second and a third night. Her husband, realizing that something was wrong, observed her distraction and became suspicious. Halfway through the fourth night, he woke up and found her calling Masrur's name as she lay sleeping in his arms, but he concealed his suspicions. In the morning he went off to his shop, and as he was sitting there Masrur arrived. When they had exchanged greetings, the man said: 'Welcome, brother. I was wanting to see you.' For a time they sat talking and then the man said: 'Come home with me so that we may draw up our pact of brotherhood.' Masrur agreed willingly, and when the two of them arrived at the house, the Jew told Zain that he was there and that they were intending to form a trading partnership based on a bond of brotherhood. He asked her to get a handsome room ready for them and to be there herself as an observer when they swore to the bond. 'For God's sake,' she exclaimed, 'don't produce me in front of this stranger! There is no point in my being there.' He said no more to her but told the slave girls to bring in food and drink and he then called for the nightingale, which perched on Masrur's lap, ignoring its master.

It was now that the Jew asked Masrur his name, and when he told him, he realized that this was the name that his wife had been babbling all night long in her sleep. He looked up and saw her gesticulating and using her eyebrows to make signs to Masrur. Understanding that he had been tricked, he told Masrur to wait while he went to fetch his cousins to attend the ceremony. 'Do as you please,' Masrur told him, and so the Jew got up and left the house, but then went round behind the room where they had been sitting . . .

Night 853

Morning now dawned and Shahrazad broke off from what she had been allowed to say. Then, when it was the eight hundred and fifty-third night, SHE CONTINUED:

I have heard, O fortunate king, that the Jew said to Masrur: 'Wait for me to fetch my cousins to attend this ceremony.' He then went round behind the room where they had been sitting, and stood where he could see the two of them through a convenient window without them seeing him. Zain asked Sukub, her slave girl: 'Where has your master gone?' When Sukub said that he had left the house, Zain told her to lock the door, bar it with an iron bolt and, when he knocked, she was to let her know before opening it for him. 'I shall do that,' replied Sukub, but all the while the Jew was watching what was going on.

Zain took a wine cup and brought it to Masrur, having flavoured it with rosewater and crushed musk. He got up and went to meet her, saying: 'By God, your saliva is sweeter than this.' She started pouring wine for him and he poured it for her, after which she sprinkled him from top to toe with rosewater until the whole room was perfumed. Her husband, watching all this, was astonished by the strength of their mutual affection, and the sight filled him with furious rage and passionate jealousy. He went to the door and, finding it locked, he knocked loudly and angrily. Sukub said: 'Mistress, the master has come.' 'Open the door for him,' said Zain, 'although I wish God had not brought him back safely.' So Sukub went to the door and opened it, and when he asked her why she had locked it, she told him: 'When you were away we always kept it locked; it was never opened night or day.' 'Well done,' he said, 'I approve of that.'

He then went to Masrur, concealing his feelings with a laugh, and said: 'Let's postpone taking our oath of brotherhood until another day.' 'To hear is to obey,' Masrur replied. 'Do as you want.' He then went back home, leaving the Jew to brood about the matter. He was at a loss to know what to do and, being filled with gloom, he said to himself: 'Even the nightingale doesn't know me and the slave girls lock the door in my face and turn to someone else.' In his downcast state he began repeating these lines:

> Masrur has enjoyed a time of pleasure,
> With days of delight, while my own life slips away.
> Time is the enemy of my love,
> And my heart is burned with ever fiercer fires.
> The happy days of my delight in her are gone,
> But I am still held helpless by her beauty,
> And as my eyes have seen her loveliness,
> So is my heart still captured by her love.
> Once she was glad to let me quench my thirst
> With pure wine from the sweetness of her mouth.
> My nightingale has left me. Why?
> What makes it yield to someone else's love?
> My eyes have seen strange sights, of such a kind
> As will arouse me when I try to sleep.
> I saw my darling squandering my love,
> While my pet bird refused to fly to me.
> Now, by the God of all created things,
> Who carries out His wishes among men,
> I shall take vengeance on this evil man
> Who in his folly has approached my wife.

When Zain heard these lines, she shuddered and turned pale. 'Did you hear that?' she asked her slave girl, and the girl replied: 'Never in my life have I heard him recite poetry like that, but let him say what he wants.' For his part, when the Jew was sure that his suspicions were correct, he began to sell all his possessions, telling himself that unless he took his wife away from her own country the two lovers would never recover from their infatuation. Then, when he had sold everything, he read out a letter which he pretended had come from his cousins but which, in fact, he had written himself, purporting to be an invitation to himself and his wife to come and visit them. 'How long are we going to stay with them?' Zain asked, and when he told her twelve days, she agreed and went on to say: 'Shall I take some of the maids with me?' 'Take Hubub and Sukub,' he told her, 'but leave Khatub here.' He prepared a fine howdah for them, having made up his mind to leave.

Zain sent a message to Masrur to say: 'If I don't come at the time we arranged, then you will know that my husband has succeeded in parting us by a trick. Don't forget the oaths by which we are bound, but I am afraid of his cunning and guile.' As her husband continued his preparations for the journey, she began sobbing and weeping and could find no rest by day or night. When he said no word of disapproval at the sight of this, she realized that he was determined to go and so she packed up all her belongings and left them with her sister, telling her what had happened and saying goodbye to her. She left her sister's house in tears, and when she got home she found that her husband had brought up the camels and was starting to load them, having got ready the best of them for her. She realized that parting from Masrur was inevitable, and she was at her wits' end, but when her husband went off on some errand, she went to the first door and wrote the following lines on it . . .

Night 854

Morning now dawned and Shahrazad broke off from what she had been allowed to say. Then, when it was the eight hundred and fifty-fourth night, SHE CONTINUED:

I have heard, O fortunate king, that when Zain saw that her husband had brought up the camels and she realized that they were about to set off, she was at her wits' end. It so happened that her husband went out on some errand, so she went to the first door and wrote the following lines on it:

Dove, nesting on the house, carry a greeting
From a lover to a beloved at their parting.
Tell him of my continued sadness
And sorrow for the happy days gone by.
I shall not cease to be enslaved by love,
And full of grief for all our past delights.
We passed a time of joy and happiness,
Savouring union both by night and day,
Only awakening when we heard the call
With which the crow announced that we must part.
We have moved off; the lands are desolate –
Would it had been that we had never left.

She then went to the second door and wrote on it:

> I ask whoever passes here to look and see
> My darling's beauty shining in the dark.
> Tell him the memory of our union makes me weep,
> And that my flow of tears will never end.
> If you cannot endure what fate has brought,
> Then scatter dust and ashes on your head.
> Go on a journey to the east or west,
> But live in patience, for God's will is done.

When she passed on to the third door, she wept bitterly and wrote:

> Gently, Masrur! If you come to her house,
> Pass by the doors and read the writing there.
> If you stay true, do not forget this bond;
> How many pains and pleasures has she had to taste!
> By God, Masrur, do not forget how near she was;
> Her joys and pleasures have all left with you.
> Weep for the joys of union that has gone,
> When she would lower the curtain as you came.
> For my sake, travel to the furthest parts,
> Plunge in the sea and cross the continents.
> Our nights of union are now past and gone,
> Their light quenched by the gloom of our parting.
> God bless those happy days, when we
> Picked flowers in the gardens of desire.
> Would that they had stayed, as I had hoped,
> But God decreed that we should come to water and then go.
> Will Time ever unite us once again?
> For I would then fulfil my vows to God.
> Be sure that our affairs are in the hands of One
> Who writes our destinies upon our brows.

She wept bitterly and went back into the house sobbing and crying. Then, remembering what had passed, she exclaimed: 'Praise be to God, Who has decreed this for us!' But as the sorrow that she felt for leaving home and parting from her lover increased, she recited:

> God's peace be on you, empty house,
> Where past delights have now come to their end.

Continue your lament, dove of the house,
Which stands deserted by its moon-like girls.
Gently, Masrur! Weep for my loss,
For with your loss my eyes have lost their light.
I wish you could have seen me as I left,
With burning heart adding to fiery tears.
Do not forget the garden where we made our pledge,
Within whose shade we were united and concealed.

She then went to her husband, who placed her in the howdah that he had made for her. When she was mounted on the camel's back, she recited:

God's peace be on you, empty house,
How long, how great was our enjoyment there!
I wish that in your shelter I had spent
My whole life until, still in love, I died.
Distance and longing for my land distresses me;
This is my passion, but I do not know its end.
Shall I ever again see a return to it,
Bringing back the pleasure that was there before?

'Don't be sad at leaving your house, Zain,' her husband said, 'for you will soon return.' He started to soothe and humour her, but when they had left the city with the road stretching in front of them, and she had indeed been parted from Masrur, she found that hard to bear.

While all this was happening, Masrur was sitting at home thinking about himself and his beloved. With a presentiment that they were going to be separated, he jumped quickly to his feet and went to Zain's house, where he found the door shut. He caught sight of the lines that she had written on the outer door, and when he had read them he fell on the ground unconscious. On recovering, he went on to the second door and then to the third, reading the inscriptions. When he had finished, his longing and passionate love increased and he hurried off until he caught up with Zain's party. He saw her at the end of the file, with her husband riding at its head because of the goods he had with him. He caught hold of her howdah, shedding tears of sorrow because of the pain of parting and reciting these lines:

I wish I knew for what fault I am wounded
With arrows of rejection over these long years.

Heart's desire, I came to your house one day,
Suffering from ever-greater pangs of love.
I found the house deserted and forlorn
And I bewailed your absence with sad groans.
I asked the wall about the ones I sought:
Where had they gone, who have my heart in pawn?
It told me they had ridden off,
Leaving my passion hidden in my heart.
But on the wall they had left lines to show
That they were keeping faith with what they pledged.

When Zain heard these lines, she realized that this was Masrur . . .

Night 855

Morning now dawned and Shahrazad broke off from what she had been
allowed to say. Then, when it was the eight hundred and fifty-fifth night,
SHE CONTINUED:

I have heard, O fortunate king, that when she heard these lines, she
realized that this was Masrur, at which both she and her maids burst
into tears. Then she said: 'For God's sake, Masrur, go away and leave
me, lest my husband catch sight of us both.' When Masrur heard this he
fainted, but when he had recovered, the two of them said goodbye to
each other and he recited these lines:

At dawn, the leader called to the caravan
As the morning breeze carried his words.
At his shout, the beasts were harnessed and the travellers
Moved off, pressing ahead along the track,
Filling the land around them with their scent
And moving fast along the valley floor.
Though they have gone, they still possess my love
And they have left me following their trail.
My friends, I never meant to part from them
Until I had bedewed the whole earth there with tears.
Now they are gone, leaving me in distress,
A parting that has left wounds in my heart.

Masrur continued to stay with the riders, sobbing and weeping, while Zain, fearing disgrace, pleaded with him to go back before day broke. He went up to her howdah and took a second farewell of her, before collapsing in a prolonged faint. When he recovered he saw the riders moving ahead of him, and, turning towards them and sniffing the south wind, he chanted these lines:

> When the wind tells him the beloved is near,
> The longing lover complains of pangs of love.
> A breeze blew over him at early dawn,
> But when he woke, the beloved was far away.
> Emaciated, he lies on his sickbed,
> Weeping tears of blood that fall in torrents.
> I mourn for neighbours who rode off, taking my heart with them,
> Carried among the riders who urge on their beasts.
> When any breath of wind says she is near,
> I see her image imprinted on my eye.

Full of longing, Masrur went back to Zain's house and, on discovering it empty and deserted, he wept until his clothes were sodden and he fell in a faint, coming near to the point of death. On his recovery, he recited:

> Spring camp, have pity on my abject state,
> My wasted body and my pouring tears.
> Let the breeze spread its fragrance over me
> To cure my sorrow and my suffering.

When he got home he was bewildered and tearful, and he stayed like that for a period of ten days.

So much for him, but as for Zain, she realized that she had been tricked after her husband had travelled on with her for ten days. When he halted at a city, she wrote Masrur a letter and gave it to her slave girl, Hubub, saying: 'Send this to Masrur so that he may know how I have been tricked and betrayed by the Jew.' Hubub took the letter and sent it on to Masrur, and when he got it, he watered the ground with his tears. He then sent a reply to Zain, which he ended with these lines:

> Which way leads to the doors of forgetfulness,
> And how can one forget, who is burned by fires of love?
> How sweet were the days of love that have now gone –
> I wish that some were ours still to enjoy.

When the letter reached Zain she took it and read it, after which she passed it to Hubub, telling her to keep the news secret. In spite of this, her husband found out about the correspondence and removed her and her maids to another city twenty days' journey away.

Masrur, meanwhile, could not enjoy sleep, stay at rest or show patience. Then, one night, when he had closed his eyes, he saw Zain in a dream visiting him in the garden, but as she started to embrace him, he woke up. Such was his dismay when he could not see her that he became demented and in his passion he recited:

> Greetings to one whose phantom visited me in sleep,
> Stirring up desire and adding to my passion.
> I started up from sleep filled with a longing,
> Roused by the vision shown me in my dream.
> Can dreams of the beloved turn out true,
> And cure the burning sickness of my love?
> At times she passed me wine and then hugged me;
> At other times she soothed me with sweet words.
> Then she rebuked me in the dream, until my eyes
> Were filled with bloody tears, but later, at the end,
> I sucked the nectar from her dark red lips,
> Which was like wine flavoured with finest musk.
> What we did in that dream enraptured me,
> As I received from her my heart's desire,
> But when it ended, all that I could find
> Left by the phantom were the pangs of love.
> I woke up like a madman after it,
> And was still drunk in the evening without wine.
> I ask you, wind, to take for me
> Greetings of longing and my salutation.
> Tell them that Time has poured the cup of death
> To the friend with whom they made a covenant.

He then set off for Zain's house, shedding tears all the way until he reached it. When he looked at it he found it empty, but then he saw Zain's image appearing as though she herself was standing there in front of him. The fires of love flared up; sorrow was piled on sorrow and he fell down unconscious.

Night 856

Morning now dawned and Shahrazad broke off from what she had been allowed to say. Then, when it was the eight hundred and fifty-sixth night,
SHE CONTINUED:

I have heard, O fortunate king, that when Masrur saw Zain embracing him in a dream, he was delighted, but then he woke up and went to her house. He found it empty and this distressed him so much that he fell down unconscious. When he recovered, he recited:

> A perfumed scent wafted to me from them,
> And this increased my passion as I left.
> I try in my distress to cure desire
> In a spring camp deserted by my friends.
> Parting and the pains of love have made me ill,
> Reminding me of what is past and gone.

As he finished he heard a raven croaking at the side of the house, and he exclaimed: 'Glory to God! It is only beside deserted houses that ravens croak,' and then, sighing in regret, he recited:

> Why does the raven cry for the beloved's house,
> While fires rage and consume my inward parts?
> I weep for a time of love that has now gone;
> My heart is lost within a deep abyss.
> I die of passion, burned by the fire of love,
> Writing a message no one can deliver.
> My body wastes away; my love has left;
> I wonder, will she ever come again?
> Breath of the east wind, if you visit her at dawn,
> Halt by her house and carry her my greeting.

Zain had a sister named Nasim, who had been watching Masrur from a high vantage point. When she saw the state he was in, she was distressed and, shedding tears, she recited these lines:

> How often do you come back here to weep,
> While the house itself grieves and laments its builder!
> Here there was joy, before the loved ones left,
> And here the suns shone brightly.

Where are the moons that used to rise?
Fate's changes have erased their splendour.
Forget past friendships with the lovely girls;
Some future time may bring them back again.
It was because of you the people left,
And otherwise no raven would perch here.

On hearing this and understanding the point of the poem, Masrur
wept bitterly. Nasim, who knew how deeply he and her sister were in
love, now addressed him and pleaded with him to stay away from the
house lest people notice him and think that it was because of her that he
had come. 'My sister has gone,' she said. 'Do you want to make me go
as well? You know that had it not been for you the place would not be
deserted, so forget her and let her be, for the past is past.' At that, Masrur
wept again and said: 'Nasim, if I had wings I would fly to her in longing,
so how can I forget her?' 'The only thing you can do is to endure,' she
told him, but he said: 'For God's sake, please forward her a letter,
pretending that it comes from you, and pass on the reply to me to cure
my grief and put out the fire in my heart.' She willingly agreed to this
and fetched an inkstand and paper, after which Masrur began to describe
the intensity of his longing and the pain inflicted on him by separation.
'This letter,' he dictated, 'comes from a grieving lover made wretched by
parting, who can find no rest by day or night. So copious are his tears
that his eyelids are ulcerated and his heart is consumed by sorrow.
Thanks to his endless misery he is as restless as a bird that has lost its
mate and is on the point of death. I grieve for our parting and for the
intimacy we once enjoyed. My body has wasted away, my tears flood
down and the whole world, with its mountains and plains, is too narrow
for me. In the ardour of my love I recite these lines:

My passion for those dwellings still remains,
And my longing for their people grows and grows.
I have sent to you the story of my love;
It was the cup of love fate poured for me.
When you set off, leaving behind your lands,
Tears in their torrents flooded from my eyes.
Caravan leader, turn aside with my well-guarded love;
Ever more fiercely fire burns in my heart.
Carry my greetings to my dear one and say:
It is only her red lips that can enchant her lover.

Time has destroyed him, breaking their fellowship,
As the arrow of separation wounds his heart.
Tell her of the love and passion that I feel,
Now she has gone and we can never meet.
I take an oath by the love that I bear
That I shall always keep the pact I made with you,
Unswerving in my love and unforgetting –
How could the longing lover ever forget?
I send my greetings and my salutations,
Perfumed within these pages with the scent of musk.'

Nasim admired his eloquence, his fine sentiments and the delicacy of
his poetry. She felt sympathy for him and sealed the letter with pungent
musk, perfuming it with a mixture of *nadd* and ambergris. Then she
gave it to a merchant, telling him not to hand it to anyone except her
sister or Hubub, her slave girl, to which he agreed. When it reached
Zain, she realized that it must have been dictated by Masrur, recognizing
that the elegance of its sentiments was an expression of his inner feelings.
She kissed it and placed it over her eyes, continuing to shed tears until
she fainted. Then, when she had recovered, she called for an inkstand
and paper and wrote a reply in which she told of her passionate longing
and the yearning that she felt for her lover, and complained of the
condition to which love had reduced her.

Night 857

Morning now dawned and Shahrazad broke off from what she had been
allowed to say. Then, when it was the eight hundred and fifty-seventh
night, SHE CONTINUED:

I have heard, O fortunate king, that when Zain wrote a reply to
Masrur, she began: 'This letter is to my lord and master, the companion
of my inmost thoughts and my secret words. I cannot sleep and cares
increasingly prey on me. I cannot endure the loss of one whose beauty
outshines both sun and moon. Longing has made me restless, while
passion is killing me. I cannot escape from this as I am doomed. You are
the splendour of the world, life's ornament. How can the cup of fate
taste sweet for one whose breath is stifled, and who is to be numbered
neither among the living nor the dead?' Then she recited:

Your letter, Masrur, has stirred up distress;
By God, I cannot do without you or forget you!
When I read it, my limbs were filled with yearning;
A constant flood of tears flowed from my eyes.
Were I a bird, I would fly off in the dark of night.
You left and there is now no manna of forgetfulness for me.
You left, and I may now no longer live,
For I cannot endure the fire of separation.

She sprinkled the letter with crushed musk and ambergris, sealed it and sent it off with a merchant, telling him to give it only to her sister, Nasim. When it reached Nasim, she passed it on to Masrur, who kissed it, placed it over his eyes and wept until he fell unconscious.

So much for them, but as for Zain's husband, when he found out about this correspondence, he took her away, together with her slave girl, travelling from one place to another until she said to him: 'Where are you going with us, so far from our own land?' 'I am taking you on a year's journey,' he told her, 'so that no more letters will reach you from Masrur. I see how you have taken all my wealth and given it to him, but I shall get back everything that is missing from you, and I wonder what good he will do you or whether he will be able to rescue you from me.' He then went to a blacksmith and had three sets of iron shackles made for Zain and her two slave girls. When he had fetched the shackles, he stripped the women of their silk clothes and gave them hair cloth to wear instead, which he scented with sulphur. Then he brought in the blacksmith and told him to put the shackles on their legs. The first one the man approached was Zain herself, and when he saw her, he lost his head, bit his fingers and fell madly in love. 'What have they done wrong?' he asked the Jew, who told him: 'They are my slave girls and they stole my money and ran away from me.' The blacksmith cursed him and said: 'By God, even if this woman had committed a thousand crimes a day and was brought before the chief *qadi*, he would not blame her. Further, she shows no sign of being a thief, and you cannot put chains on her legs.' He asked the Jew to spare her this, and interceded for her. When Zain saw what he was doing, she said to her husband: 'For God's sake, please don't make me come out in front of this strange man.' 'How was it that you came out in front of Masrur?' he asked, and she made no reply. He did, however, accept the blacksmith's intercession by giving her light shackles, unlike those of the two slave girls, which were heavy, as she had a tender body which could not endure rough treatment. She and her slave girls had to go on wearing hair cloth night and day until they became pale and emaciated.

As for the blacksmith who had fallen so deeply in love with Zain, he went home in great distress and recited these lines:

> Smith, may your right hand wither, for it fixed
> Those fetters round the sinews of her legs,
> And you have chained a tender lady's feet –
> A mortal, but one formed of wonder.
> If there was justice, the anklets round her feet
> Would not be iron, but would be made of gold.
> Her loveliness would move the chief *qadi* himself,
> And he would place her in the highest rank.

As it happened, the chief *qadi* was passing the blacksmith's house as he was reciting this poem. He sent for the man and, when he came, he asked who the subject of his poem might be, with whose love he appeared to be obsessed. The blacksmith stood in front of him, kissed his hand and, after wishing him a long life, described Zain in detail, mentioning her grace, beauty, symmetrical form and the perfection of her elegance, with her lovely face, slender waist and heavy buttocks. He then told the *qadi* how she had been humiliated by being imprisoned in chains with not enough to eat. For his part, the *qadi* told him to tell him where she was and to bring her to him so that he could do her justice. 'She is now your responsibility,' he told the smith, 'and if you don't show her to me, God will punish you on the Day of Judgement.' 'To hear is to obey,' the blacksmith said.

He set off straight away to Zain's house, whose door he found locked, and there he heard a melodious voice that came from a distressed heart. This belonged to Zain, who was just then reciting these lines:

> I was once in my own land, united with my lover,
> While love filled for me the cup of happiness.
> We had our share of pleasure in our love;
> Morning and evening, we faced no distress.
> We passed a time of happiness, with wine,
> The lute, the zither giving us delight.
> But then Time parted us, breaking our comradeship;
> Love left and with it went our times of joy.

I wish the unlucky raven had been forced away,
And that the dawn of love's reunion had appeared.

When the blacksmith heard this, he shed tears like raindrops pouring
from a cloud. He knocked on the door, and when the girls asked: 'Who
is there?' he said that he was the blacksmith. He told them what the *qadi*
had said to him and that this *qadi* wanted them to appear before him
and present their complaint, so that he could see to it that they got
their rights.

Night 858

Morning now dawned and Shahrazad broke off from what she had been
allowed to say. Then, when it was the eight hundred and fifty-eighth
night, SHE CONTINUED:

I have heard, O fortunate king, that the blacksmith told Zain what the
qadi had said to him and that he wanted them to appear before him and
present their complaint, so that he could see to it that they got their rights.

Zain answered: 'How can we go to him when we are behind a locked
door with shackles on our legs, while the keys are with the Jew?' 'I shall
make keys for the locks and open both the door and the shackles,' the
smith said, and when Zain asked who would tell them how to recognize
the *qadi*'s house, he promised to describe it for them. Then Zain said:
'How can we go to him wearing hair cloth smelling of sulphur?' and he
had to assure them that the *qadi* would not hold it against them if they
appeared like that. He immediately made keys for the locks and unlocked
both the door and the shackles, which he removed from their legs, before
ushering them out and pointing the way to the *qadi*'s house. Hubub,
Zain's slave girl, then took off the hair cloth that her mistress was
wearing, escorted her to the baths and, when she had been washed,
clothed her in robes of silk, after which her colour returned.

By great good fortune, her husband happened to be at a banquet given
by one of the merchants and so Zain was able to deck herself in all her
finery, after which she made her way to the *qadi*'s house. When he caught
sight of her, he got to his feet and greeted her in the smoothest and most
agreeable of terms, while, for her part, she pierced him with the arrows
of her glances. 'May God prolong the life of our lord, the *qadi*,' she said,
'and use him to help those who look for justice.' She went on to tell him

of the generous treatment she had received from the blacksmith and also of the astonishing way in which the Jew had tormented her, bringing her and her slave girls ever nearer to the point of death, with no prospect of escape. When the *qadi* asked her name, she told him: 'I am called Zain al-Mawasif, and this slave girl of mine is Hubub.' 'Your name fits you and matches its meaning,' said the *qadi*, at which Zain smiled and covered her face. 'Zain,' he then asked, 'do you have a husband or not?' She told him that she had no husband, after which he said: 'And what is your religion?' 'The religion of Islam, the creed of Muhammad, the best of men,' she replied. He asked her to swear to this by the *shari'a*, with its signs and admonitions, and she took the oath and recited the confession of faith. 'How did you come to spend your youth with this Jew?' he asked. She replied: 'May God graciously prolong your days, fulfil your hopes and set the seal of virtue on your deeds. You must know that on his death my father left me fifteen thousand dinars, which he deposited with the Jew. This was to be used for trading, with the profits to be split between him and me, while the capital was protected by a legal agreement. When my father died, the Jew wanted me and asked my mother for my hand in marriage. "How can I make her abandon her religion and turn her into a Jewess?" my mother said, adding: "By God, I shall tell the authorities about you." The Jew was alarmed by this and, taking the money, he fled to Aden. After we heard where he was, we went in search of him, and when we met him he told us that he was there to trade and that he was buying up large quantities of goods. We believed him, and he continued to pull the wool over our eyes until he managed to imprison us, fetter us and subject us to the worst of tortures. We are strangers here to all except Almighty God and our master, the *qadi*.'

When the *qadi* heard this, he asked Zain's slave girl, Hubub: 'Is this your mistress? Are you strangers and has she no husband?' Hubub said yes to all this, and the *qadi* then went on: 'Marry me to her, and I swear to free all my slaves, to fast, to go on pilgrimage and to give my goods as alms if I don't revenge you on this dog and pay him back for what he has done.' 'To hear is to obey,' Hubub replied, after which the *qadi* said: 'Go off now, you and your mistress, in good heart, and tomorrow, God willing, I shall send for this unbeliever and extort your just dues from him. You will be amazed to see the tortures that I shall inflict on him.' Hubub called down blessings on him and went away, leaving him suffering from the burning pangs of passionate love.

When she and Zain had left his house, they asked the way to the house of the second qadi* and when they had been directed to it, they presented themselves to him and told him the same story, repeating the process with the third and the fourth. All four listened to the case and all four wanted Zain and asked her to marry them. She said yes to each one of them, and none of them knew about the others, while the Jew knew nothing at all, as he had stayed in the house where the banquet had been held. The next morning, Hubub got up and dressed her mistress in her finest robes before taking her before the four qadis in the courtroom. When she saw them there, she lifted her veil to show her face and then greeted them. They returned her greeting, each one recognizing who she was. One of them was writing something and the pen fell out of his hand; another was speaking but began to stutter; while a third made a mistake in the figures he was adding up. 'Lady of grace and beauty,' they said, 'take heart, for we shall see that your rightful dues are restored to you and ensure that you get what you want.' She called down blessings on them, took her leave and went off.

Night 859

Morning now dawned and Shahrazad broke off from what she had been allowed to say. Then, when it was the eight hundred and fifty-ninth night, SHE CONTINUED:

I have heard, O fortunate king, that the qadis said to Zain: 'Lady of grace and beauty, take heart, for we shall see that you get what you want.' She called down blessings on them, took her leave and went off.

While all this was going on, the Jew was with his friends at the banquet and knew nothing about it, while Zain, for her part, was calling on the magistrates and notaries to help her against what she described as a suspected unbeliever and to save her from painful torture. She tearfully recited the following lines:

Eyes, pour a flood of tears;
It may be these will cure my sorrow.
I used to wear embroidered silks,
But now my dress is that of a monk,
Smelling of sulphur – very different, this,
From the nadd perfume and sweet basil that I used.

* There was one for each of the four schools of law.

If you were to know of my plight, Masrur,
You would not tolerate the shame of my disgrace.
Hubub is a chained captive in the power
Of an unbeliever who scorns the One Just God.
I have renounced the creed and customs of the Jews,
For now mine is the highest of all faiths.
I bow before God, the All-Merciful,
And follow what Muhammad has laid down.
Masrur, do not forget the love we shared;
Preserve the covenant to which we swore.
It was because of you I changed my faith,
But I concealed the passion that I felt.
Come quickly if you love me, keeping faith,
Like a true man, and do not hesitate.

She wrote out these lines in a letter that she now sent to Masrur, with an account of everything that the Jew had done to her from beginning to end. She then folded it up and gave it to her slave girl, Hubub, telling her to keep it in her pocket until they could send it to Masrur. Just at that moment the Jew came in and, seeing their joyful expressions, he asked why they were so happy, adding: 'Have you had a letter from your friend Masrur?' Zain said: 'Our only helper against you is God, the Glorious, the Almighty, and it is He Who will rescue us from your injustice. If you don't send us back to our own country, tomorrow we shall take you before the governor of this city and the *qadi*.' 'And who was it who removed the shackles from your legs?' asked the Jew, and he went on: 'I am going to have shackles weighing ten *ratls* made for each of you and then I shall parade you around the city.' 'God willing,' Hubub told him, 'everything that you intend to do to us will happen to you. You have removed us from our own land and tomorrow we shall all stand before the governor.'

They went on in this way until morning, when the Jew got up and went to the blacksmith to have more shackles made for the women. Zain and her slave girls, however, went to the courtroom, and when she went in and saw the *qadis* there, she greeted them and they all returned her greeting. The chief *qadi* said to his entourage: 'This girl is like the Prophet's daughter; everyone who sees her loves her and abases himself before her beauty and grace.' He then gave her an escort of four officials, themselves of the stock of the Prophet, telling them to fetch the accused Jew and to treat him roughly.

So much for Zain, but as for the Jew, when he went home with the shackles he had had made, he was taken aback not to find the women there. It was then that the *qadi*'s men laid hold of him, and after giving him a severe beating, they dragged him face downwards to the *qadi*. When the *qadi* saw him, he shouted into his face: 'Damn you, enemy of God, have you gone so far in wickedness as to do what you did, removing these women from their own land, stealing their money and trying to convert them to Judaism? How dare you try to turn Muslims into unbelievers?' 'Sir,' said the Jew, 'this is my wife.' When the *qadis* heard this, they all cried out: 'Throw this dog to the ground, strike him on the face with your shoes and give him a painful beating, for he has committed an unpardonable offence.' The guards stripped off his silk clothes, dressed him in hair cloth and threw him on to the ground. They then pulled out all the hairs of his beard and struck him painfully on the face with their shoes before mounting him backwards on a donkey with its tail in his hand and parading him round the whole city, ringing bells as they went. He was in a state of abject humiliation when they brought him back to the *qadi*, who, with his three colleagues, condemned him to have his hands and feet cut off and then to be crucified. The damned Jew was out of his mind with fright and he asked: 'What do you want from me, sirs?' They told him: 'Confess that this girl is not your wife, that the money you have is hers and that you unjustly removed her from her own country.' He agreed to this and an official record was made of his confession, after which the money was taken from him and passed to Zain, who was also given the record of what had been said.

Zain then left, and everyone who saw her was bemused at the sight of such loveliness, while each of the *qadis* believed that it was to him that she would entrust herself. When she reached home she got ready everything that she might need and then waited for nightfall before setting out in the dark with her slave girls, taking with her any valuables that were light to carry. She travelled on for three days and nights, while as for the *qadis*, after she had left the courtroom they gave orders for her husband, the Jew, to be put in prison . . .

Night 860

Morning now dawned and Shahrazad broke off from what she had been allowed to say. Then, when it was the eight hundred and sixtieth night, SHE CONTINUED:

I have heard, O fortunate king, that the *qadis* gave orders for her husband, the Jew, to be put in prison, and the next morning they and the notaries waited for Zain to come back. When she failed to visit any of them, the one whom she had first approached said: 'I want to take a trip outside the city today as there is something I need to do there.' He got on his mule and, accompanied by his servant, he began to go to and fro through the city streets, looking for Zain but failing to find any news of her. While he was engaged on this hunt he came across his three colleagues, each of whom believed that Zain was pledged to meet him and no one else. He asked them why they were riding round the streets, and when they told him he realized that they were in the same position as he was himself and were asking the same question. They joined forces in their hunt, and when they still could find no news they each returned home and took to their beds, pining away through love.

Then the chief *qadi* remembered the blacksmith and sent for him. When he arrived, the *qadi* said: 'Smith, do you know anything about that girl whom you directed to my house? By God, if you don't tell me where she is, I shall have you whipped.' On hearing this, the smith recited these lines:

She who holds me subject to her love
Possesses every beauty that there is.
Her eyes are of a fawn; she exhales ambergris;
She is a sun, a rippling pool, a pliant branch.

Then he said: 'By God, master, since she left your gracious presence I have not set eyes on her. She has taken over my heart and my mind and I can talk and think about nothing else, but when I went to her house I couldn't find her and I could discover no one who could tell me anything about her. It is as though she has plunged into the depth of the sea or been swept up into the sky.' When the *qadi* heard this he heaved so deep a sigh that he almost expired, and he exclaimed: 'I wish to God that I had never seen her!'

The smith left and the *qadi* took to his bed, wasting away because of his love for Zain, as did his colleagues, together with the notaries. The doctors visited them time after time, but they were not suffering from a disease that any doctor could cure. The leading citizens then came to visit the chief *qadi*, and after they had greeted him and asked how he was, he sighed and revealed his secret, reciting these lines:

Do not increase my suffering by blaming me;
Excuse a judge whose writ runs among men.
The censurers of my love will soon forgive me;
He who is slain by love should not be blamed.
I was a judge when fortune favoured me,
Rising in rank by means of what I wrote,
Until I suffered from a fatal wound,
Shot from the eye of one who came to shed my blood.
She came complaining that she had been wronged,
A Muslim girl, whose teeth were like a string of pearls.
I saw her face unveiled, and there I found
A full moon shining in the dark of night –
A radiant face, a wonderful smiling mouth,
Covered in loveliness from head to toe.
By God, I never saw a sight like this
In any race from among all mankind.
She made me a fine promise, telling me:
'Judge of the peoples, I keep to my word.'
This is my plight and this is my distress;
High-minded men, do not ask me about my pains.

On finishing these lines, he wept bitter tears and then with a groan he died. When his visitors saw that, they washed his body and covered it with a shroud before praying over it and burying it. The following lines were inscribed over his tomb:

A lover's perfection can be seen in those
Who lie slain by the beloved's cruelty.
Here lies a universal judge,
Feared by the sword imprisoned in its sheath.
But love decreed his fall; never before
Was a master seen abased before his slave.

They then left him to God's mercy and went with the doctor to the second *qadi*, but there was nothing wrong with him that they could discover, nor any pain that needed a doctor's care. They asked him how he was and what was preoccupying him, but when he told them of his love, they criticized him harshly. He replied by reciting these lines:

Her love afflicts me and one like me cannot be blamed;
I have been struck by an arrow from an archer's hand.
There came to me a woman named Hubub,
A mature woman, counting her life in years,
But with her was a girl child, with a face
Outshining the full moon in the dark of night.
She showed her loveliness as she complained
With tears that fell in torrents from her eyes.
I listened to her words and looked at her,
Struck by a wasting sickness as she smiled.
Wherever she has gone, she has my heart,
Leaving me pledged as hostage to her love.
This is my story; now lament my fate
And make my servant here judge in my place.

Then, with a groan, he died, and after they had laid out the corpse, they buried him and left him to God's mercy. The same thing happened when they visited the next two *qadis*, both of whom they found to be sick with love, as was also the case with the notaries, for everyone who had seen Zain died of love for her, or, if they did not die, they lived suffering from the pangs of love . . .

Night 861

Morning now dawned and Shahrazad broke off from what she had been allowed to say. Then, when it was the eight hundred and sixty-first night,
SHE CONTINUED:

I have heard, O fortunate king, that the citizens found all the *qadis* and the notaries sick with love for Zain, for everyone who had seen her died of love for her, or, if they did not die, they lived suffering from the pangs of love, may God have mercy on them all.

So much for them, but as for Zain, she pressed on with her journey for some days until she had covered a considerable distance. Then, while travelling with her slave girls, she happened to pass by a monastery standing near the road, which was the seat of an abbot named Danis, who lived there with forty monks. When this man caught sight of her beauty, he came down to invite her in, saying: 'Rest with us for ten days before going on your way.' So she and her maids halted there, but when he looked at her loveliness, the abbot's faith was corrupted and he became infatuated by her. He started to send monks to her, one after the other, carrying messages to win her favour, but all those he sent fell in love with her themselves and tried to seduce her. She excused herself and refused, but the abbot continued to send his monks until all forty of them had gone to her. They had all fallen in love with her at first sight, and had used all their blandishments in their seduction attempts, without ever mentioning Danis's name. She continued to refuse, answering them in the harshest of terms.

When Danis could not bear the increasing pangs of love any longer, he said to himself: 'The proverb says that nothing can scratch my body better than my own fingernail and that when it comes to getting what I want, it is my own feet that move best.' So he got up and prepared a splendid meal which he brought to Zain on the ninth of the ten days that he had agreed she should stay with him. 'I ask you in God's Name to please accept the best of our food,' he said to her, and in reply, as she reached out her hand, she invoked God, the Compassionate, the Merciful. When she and her companions had eaten he said: 'I would like to produce some lines of verse for you.' With her permission, he then recited:

My heart was conquered by your glances and your lovely cheeks;
Your love is the sole theme of all I write in prose or verse.
Will you reject a passionate lover, sick with love,
Who struggles with his love even in his dreams?
Do not reject me, cast down as I am by love;
Because of love's delight I have neglected all I should do here.
You thought it right to shed your lover's blood;
Hear my complaints; have pity on my state.

When Zain heard this, she produced these lines in reply:

You want me, but do not be led astray by hope,
And give up your pursuit of me, O man.
You should not covet what you will not get;
For what is joined to hopes like these is fear.

On hearing this, the abbot retired to his cell to think things over. He did not know what he should do about Zain, and he passed the night in a wretched state. As for Zain, when it was dark she told her slave girls to get up and go off with her, telling them: 'We cannot cope with these forty monks, each one of whom has tried to seduce me.' They agreed willingly, mounted and rode out of the monastery gate.

Night 862

Morning now dawned and Shahrazad broke off from what she had been allowed to say. Then, when it was the eight hundred and sixty-second night, SHE CONTINUED:

I have heard, O fortunate king, that Zain and her girls left the monastery by night. As they went on their way they came across a caravan, which they joined, and they discovered that this had come from Aden, where Zain herself had lived. She overheard people talking about her and saying that the judges and the notaries had all died of love for her, and that those whom the citizens had appointed to replace them had released her husband from prison. 'Did you hear that?' asked Zain, turning to her slave girls, to which Hubub replied: 'If the monks, for whom it is a religious duty to abstain from women, became infatuated with you, then what about the *qadis*, whose creed dictates that "there is no monkery in Islam"? Take us back home before people find out about us.' They then pressed on with their journey.

So much for them, but as for the monks, they went to greet Zain in the morning, only to find her room empty. This left them sick at heart, and the first of them tore his clothes and recited:

Come here to me, my dear companions,
For I shall leave you soon and go.
Within my breast are burning pains
And pangs of love that are fatal to my heart.
The cause here is a girl, who came to us,
Matched only by the full moon in the sky.
She went, leaving me slain by her loveliness,
Shot through by arrows in my vital parts.

A second monk recited:

You, who have taken off my heart, be kind
To this poor wretch and turn back once again.

They went and with their going went my rest;
They are far off, but I can hear their voices still.
Distant they may be, but I wish that they
Would come back in my dreams to visit me.
They took my heart with them when they rode away,
Leaving my body drowned in floods of tears.

A third then recited:

My heart, my eyes, my ears set you on high;
My heart and all my body is your shelter.
Your name is sweeter than honey in my mouth,
And moves like life itself between my ribs.
I am as thin as a tooth-pick thanks to you,
And you have drowned me in the tears of love.
If I can see you in a dream, it may be then
That my cheeks can recover from this flood.

The fourth recited:

My tongue is dumb; my words are few;
Love is my illness; love is my distress.
You were a full moon, rising in the sky,
And this increased the ardour of my love.

Then the fifth recited:

I love a shapely, graceful moon,
Whose slender waist complains of any hurt.*
Her mouth holds what is like the purest wine;
Her heavy buttocks distract all mankind.
My heart is burning with the fire of love,
And while men talk at night, the lover dies.
Tears are like blood-red jewels on my cheek,
Down which they pour in showering rain.

The sixth recited:

You, whose rejection has destroyed my life,
Branch of the *ban* tree, rising like a star,
You, whom I love, have caused my misery,

---

* The skin is so sensitive that it is easily bruised.

Burning me with the fire of rosy cheeks.
Who can compare with a once pious lover
Who, thanks to you, cannot perform his prayers?

Then the seventh recited:

She jailed my heart, but then released my tears,
Destroying patience but renewing love,
Sweet-natured, bitter when she turns away,
Shooting my heart with arrows when we meet.
You who blame me, stop and turn away;
No one believes you when you talk of love.

All the other monks followed their example, shedding tears and reciting verses, but none of them wept and wailed as bitterly as the abbot who had so fruitlessly pursued Zain. He then started to chant these lines:

Endurance left me when the loved one went;
She parted from me, whom I so desired.
Guide of the caravan, do not press on;
They may show kindness and come back to me.
Sleep shunned my eyes the day they started out;
My sorrows were renewed, my pleasures left.
It is to God that I complain of love
Which has so wasted and so weakened me.

When all their hopes had been shattered, they agreed to keep a picture of Zain in the monastery and they kept to this agreement until they were visited by the destroyer of delights.

So much for them, but as for Zain, she continued to travel in search of her beloved Masrur until she got back home. After having opened the doors and entered the house, she sent word to Nasim, her sister. Nasim was overjoyed and fetched furnishings for the house and rich materials which she gave Zain to wear. Curtains were hung down over the doors, and the whole place was filled to the fullest extent with aloes, *nadd*, ambergris and pungent musk. Zain wore her finest dress with her most splendid ornaments, and all the while, as this was going on, Masrur, who did not know that she had come back, was still careworn and very sorrowful.

Night 863

Morning now dawned and Shahrazad broke off from what she had been allowed to say. Then, when it was the eight hundred and sixty-third night, SHE CONTINUED:

I have heard, O fortunate king, that when Zain got back home, her sister brought her furnishings for the house and rich materials which she gave Zain to wear. While all this was going on, Masrur, who did not know that she had come back, was still careworn and very sorrowful.

Zain sat talking with those of her slave girls who had not gone with her on her journey, to whom she told the story of her adventures from beginning to end. She then turned to Hubub and gave her money with which to buy food for her and the others. It was after both food and drink had been fetched and Zain and the others had eaten and drunk that she told Hubub to go to find Masrur and to see what kind of a state he was in.

Masrur, for his part, could find no rest or endure with patience. When passion and lovesickness got the better of him, he would try to console himself by going to Zain's house, kissing the wall and reciting lines of poetry. As it happened, he had gone to the place where he had said goodbye to Zain, and he was reciting:

I tried to hide my feelings for her, but they still showed,
My eyes exchanging sleep for sleeplessness.
When cares enslaved my heart, I called:
'Time, do not spare me and let me not live.
Flanked as I am by danger and distress.'
If love were ruled by justice,
Sleep would not have been exiled from my eyes.
Have pity, masters, on a lovesick man,
And mourn a chieftain of his clan, abased
By the law of love, a rich man now made poor.
I do not follow those who blame my love for you,
Blocking my ears and telling them they lie.
I have kept faith with those I loved.
They said: 'You love one who has left,' and I said: 'Yes.
Say nothing, for fate blinds men's eyes.'

He went back home and sat in tears until he fell asleep. Then, in his dream, he saw Zain back in her house. He woke up weeping, and set off there, reciting:

Can I forget her, whose love captured me,
While my heart burns like coals upon the fire?
God, I complain to You of my absent love,
And of misfortunes heaped on me by Time.
When shall we meet, O goal of my desire,
And be united, O my rising moon?

As he recited this last line on his way through the street where Zain lived, he detected a whiff of the purest perfume, which so roused his emotions that his heart almost left his breast thanks to the burning passion swelling within him. It was just at that moment that he saw Hubub coming from the head of the street to carry out her mistress's errand. As soon as she caught sight of him, she came up to greet him, and when she had given him the good news of Zain's arrival and that she had sent for him, his delight knew no bounds. Hubub took him back with her, and when Zain saw him, she got down from her couch and they exchanged kisses and embraces, continuing until they fell unconscious and remained so for a long time thanks to the ardour of their reunited love.

When they had recovered, Zain told Hubub to fetch a jug of sugared water and another of lemon juice, and when these had been brought the lovers continued eating and drinking until nightfall while they talked over all that had happened to them from start to finish. Zain pleased Masrur by telling him that she had converted to Islam and he followed her example, as did her slave girls, turning in repentance to Almighty God. In the morning she sent for the *qadi* and the notaries, to whom she said that she was again a single woman and that, as she had waited for the time prescribed by law since leaving her former husband, she now wanted to marry Masrur. The marriage contract was drawn up and the wedded couple enjoyed the most delightful of lives.

So much for the two of them, but as for Zain's ex-husband, the Jew, after he had been freed from prison, he set off back home. When he was within a three-day journey of the city where Zain was living, she heard about this and sent for Hubub. 'Go and dig a grave in the Jewish cemetery,' she told her, 'placing scented herbs on top of it and sprinkling water around it. If the Jew comes and asks about me, tell him that I died twenty days ago because of the way he mistreated me. If he then tells you to show him my grave, take him there and find some way of burying him alive.' 'To hear is to obey,' said Hubub. They then removed all the furnishings of the house and stored them in a small room, while Zain herself moved to Masrur's house, where he and she stayed, eating and drinking, for the next three days.

So much for them, but as for the Jew, when he came to the house, he knocked on the door. 'Who is there?' called Hubub. 'Your master,' he replied, at which she opened the door. He could see tears running down her cheeks and he asked her why she was crying and where her mistress was. 'My mistress is dead thanks to the harshness with which you treated her,' Hubub replied, leaving the Jew taken aback and shedding bitter tears. He asked about her tomb and she took him to the cemetery and showed him the grave that she had dug. In floods of tears he recited:

> If eyes shed tears of blood till they were almost drained,
> Two things would still not get a tenth of what is due:
> Youth's passing and the parting from one's love.

Then, with still more tears, he continued:

> I cannot bear the sorrow that I feel;
> I die of grief now that my love has gone.
> Distress has overcome me since that time,
> And what I did myself has rent my heart.
> I wish my secret had always been kept,
> And I had not revealed the passion in my heart.
> I once enjoyed a pleasant life of ease,
> But since that time, distress has humbled me.
> Hubub, what sorrows you have roused in me,
> Telling me that my chief support in life is dead!
> I wish that I had never parted from my wife,
> And that which robbed me of my life had never been.
> I now regret and blame myself for faithlessness
> Because I wronged the pillar of my life.

When he had finished reciting these lines, he went on weeping and wailing until he fell down unconscious. As soon as that happened, Hubub dragged him off and put him into the grave. He was still alive but in a bemused state, and so she filled up the grave and went back to bring the news to her mistress. Zain was overjoyed and recited:

Time swore that it would always bring me grief;
Do penance, Time, because your oath was false.
The censurer is dead; the lover comes;
Come quickly to the house of all delights!

She and Masrur stayed together, eating, drinking and enjoying pleasures, pastimes and amusements until they were visited by the destroyer of
delights, the parter of companions and the slayer of sons and daughters.

A story is also told that once upon a time, in the old days, among the
leading Cairene merchants was a man named Taj al-Din, a trustworthy
and freeborn citizen. He had a passion for travelling and loved to roam
through deserts and open spaces, traversing plains and rough country,
and visiting islands in various seas to search for profit. He owned black
slaves and mamluks, as well as eunuchs and slave girls, and he was a
man who frequently exposed himself to danger, enduring such perils on
his travels that would turn the hair of small children white. He was not
only the wealthiest merchant of his age but the most articulate; he owned
horses, mules and camels, including Bactrian camels, bales and bags,
trade goods, cash and materials that were nowhere else to be found –
muslins from Homs, robes from Baalbak and Merv, fine silks, cloth from
India, tassels from Baghdad and burnouses from north Africa. He had
Turkish mamluks, Abyssinian eunuchs, slave girls from Rum and Egyptian pages. So great was his wealth that even the sacks in which he kept
his goods were made of silk. He was also an extremely handsome man,
with a swaggering stride, but sympathetic and accommodating. He was
as a poet has described:

I saw the lovers of a merchant quarrelling. He asked: 'Why?'
'Merchant,' I said, 'this is because of your fine eyes.'

Another has produced an excellent description that hits the mark:

A merchant came to visit me and left my heart
Bewildered by his glance. He said:
'Why are you so confused?' and I replied:
'My friend, the reason lies within your eyes.'

This merchant had a son named 'Ali Nur al-Din, who was like a moon
that comes to the full on the fourteenth night, remarkable for his beauty
and grace, as well as for the elegance of his shapely figure. It happened

that one day he was sitting as usual in his father's shop, buying and selling, receiving cash and paying it out. He was surrounded by the sons of other merchants, among whom he stood out like a moon amid the stars, with a radiant forehead, rosy cheeks shadowed by down, and a body like marble. He was as the poet has said:

A handsome boy told me to describe him;
I said: 'You carry off the prize for beauty.'
And then I summed it up and said:
'There is no part of you that is not beautiful.'

Another has said:

A mole is on the surface of his cheek,
Like ambergris upon a marble slab.
His eyes unsheathe their swords and call
To all love's rebels: 'God is great.'

His companions invited him to come with them, saying that they wanted him to accompany them on a pleasure trip to a certain garden. 'Wait till I ask my father,' he replied, 'for I can't go without his permission.' While they were talking, his father arrived and Nur al-Din, on seeing him, told him of the invitation and said: 'Do I have your permission to go?' 'Yes,' his father replied, and he then gave Nur al-Din some money, saying: 'Off you go with them.' The other youths rode donkeys and mules and Nur al-Din, on a mule of his own, went with them to a garden where there was everything to delight the soul and please the eye. Its wall was high and solidly built with a vaulted gateway like an arched hall and a sky-blue door like one of the portals of Paradise. The name of the gatekeeper was Ridwan and over his gate were set a hundred trellises of vines of all colours, coral-red, black like the noses of Negroes and white like pigeons' eggs. In the garden were peaches, pomegranates, pears, plums and apples, all of various colours, growing singly or in pairs . . .

Night 864

Morning now dawned and Shahrazad broke off from what she had been allowed to say. Then, when it was the eight hundred and sixty-fourth night, SHE CONTINUED:

I have heard, O fortunate king, that when the merchants' sons entered the garden they discovered all that lips or tongues could desire, including grapes of various colours, growing singly or in pairs, as the poet has described:

Grapes that taste like wine, black as a raven's wing,
Gleaming between the leaves like dyed fingers of girls.

Another has written:

Grape bunches hanging from the branch,
Slender as my wasted form,
Like honey, or like water in a jug,
Turning from sourness into wine.

When the visitors got as far as the garden arbour they found Ridwan seated, looking like his namesake, the guardian of Paradise, while on the door of the arbour they saw these lines inscribed:

May God send rain to a garden whose grape bunches hang low,
And where the boughs bend, weighted down with sap.
The branches dance when the east wind blows
And the rains speckle them with moist pearls.

Inside, there was another inscription, which ran:

Come with us to a garden, friend,
That cleans the heart of rusty care.
There the breeze trips over its own skirts,
With all its flowers laughing up their sleeves.

In the garden there were not only fruits of all descriptions but birds of every kind, ringdoves, bulbuls, plovers, turtledoves and pigeons, singing on every branch. In the channels, running water rippled with the reflection of flowers and delicious fruits, as the poet has described:

The breeze passed by the branches and they seemed
Like girls who stumble in their lovely robes.
While the streams were like swords,
Unsheathed by horsemen from their scabbards.

Another has written:

Under the branches the stream lies stretched out,
With their reflections mirrored in its heart.

But when the jealous breeze sees what is there,
It blows so as to stop them coming close.

The trees of the garden bore every sort of fruit, growing in pairs, among them pomegranates like silver balls, as the poet has well described them:

Two pomegranates with thin outer skins,
Like virgin's breasts when they stand prominent.
When they are peeled, inside them can be seen
Rubies to dazzle all who look at them.

Another has written:

If you explore inside these rounded fruits
You find red rubies dressed in splendid robes.
The pomegranate that I see is like
A virgin's breast or like a marble dome.
What it contains cures sickness and brings health;
Traditions of the Prophet mention it,
And God the Glorious in His Holy Book
Adorns it with His eloquence.

Also in the garden were sugar-sweet apples, apples with a flavour of musk, as well as *damani* apples, such as to astound those who looked at them, as the poet has described:

An apple of two colours, looking like
The cheeks of lover and beloved both combined –
Its opposite hues were there upon a branch,
One dark, one bright, for me to wonder at,
As though a spy had startled their embrace,
Leaving one red with shame, the other pale with love.

There were almond-apricots, camphor-flavoured apricots, as well apricots from Jillian and 'Antab, as the poet has described:

The almond-apricot is like a lover
Losing his wits when the beloved comes.
It is like him in his lovelorn state,
Pale on the outside, with a broken heart.

There are other excellent lines:

Look at the apricots in bloom –
Gardens whose splendour lights the eyes.
They are like shining stars
On flowering branches, gleaming in the leaves.

Among its other fruits the garden had plums, cherries and grapes, such
as could cure illnesses and dispel giddiness and yellow bile, as well as
figs whose colours varied from red to green, amazing all who saw them.
They were as the poet has written:

Figs showing white and green among the leaves
Were like young Rumis on the castle walls,
Posted as lookouts in the dark of night.

It was well said by another:

Welcome the figs piled up upon the plates
Like bags of food drawn shut without a ring.

Give me a tasty fig looking as lovely as we know it is,
And when you taste it, you will find it gives
A scent of camomile and taste of sugar.
When spread out on a plate figs seem to be
Balls that have been fashioned from green silk.

How well another wrote:

I am accustomed to eat figs
And not the other fruit they so enjoy.
They asked me why this was and I replied:
'Unlike the sycamore, figs are not shared.'*

Even better are the lines:

I prefer figs to every other fruit
Hanging when ripe on a luxuriant branch,
Like an ascetic who, when clouds drop rain,
Sheds his own tears for fear of God on high.

There were pears from al-Tur, Aleppo and Rum, of various colours,
growing singly or in pairs . . .

---

* Sexual double entendre, referring to male and female. See Burton 3.302.

Night 865

Morning now dawned and Shahrazad broke off from what she had been allowed to say. Then, when it was the eight hundred and sixty-fifth night, SHE CONTINUED:

I have heard, O fortunate king, that when the merchants' sons entered the garden, they found those fruits that we have mentioned, as well as pears from al-Tur, Aleppo and Rum, growing singly or in pairs, varying from green to yellow, and filling those who looked at them with wonder, as the poet has described:

> How good for you are pears like pallid lovers,
> Looking like virgins in an inner room
> Before whose faces curtains are let down.

Also to be found there were *sultani* peaches whose colours varied from yellow to red as the poet has described:

> The peach in the garden is coloured darkest red,
> Its kernel yellow gold while its face is dyed with blood.

There were the sweetest of green almonds, like palm cores, with their kernels hidden beneath three coverings, as they were created by God the Giver, and as they were described in the lines:

> Three coverings protect a juicy body,
> Different in shape, but all the work of God.
> They threaten its destruction night and day,
> Imprisoning it, though it has done no wrong.

Other good lines are:

> Do you not see the almond that appears
> When someone picks it from the branch?
> When it is peeled, it shows its core,
> Just like a pearl within an oyster shell.

Even better than these are the lines:

> How fine are these green almonds,
> The smallest of which fill the hand.
> Their nap looks like a young boy's down;

Some have single kernels and some double,
Like pearls stored safely within chrysolite.

Other good lines are:

Nothing is as lovely as almonds at blossom time,
Their white heads matching their young down.

Also in the garden was the lotus fruit, variously coloured and growing from the same root or from several. The poet has described it in these lines:

Look at the lotus fruit arranged upon the branch,
Like splendid apricots gleaming on reeds,
Yellow fruit hanging there like golden bells.

Other good lines are:

The lotus tree has every day a different form of loveliness,
Showing fruits like golden bells fixed to its boughs.

There were oranges coloured like *khalanj* wood, as described by one passionate poet:

Red, filling the hand and splendid in their beauty –
Their outer parts are fire, their inner, snow.
This snow, remarkably, can never melt,
Just as the fire has never any flame.

Another expressed it well in the lines:

If you look carefully, you see the orange fruits
Like women's cheeks when, in their finery,
They deck themselves for feasts in silk brocade.

Another did equally well, writing:

Hills where the orange grows, when breezes blow
And branches sway, are like the cheeks,
Resplendent in their loveliness, that touch,
At times of greeting, the cheeks of those they greet.

Of the same quality are these lines:

We told a fawn: 'Describe for us
Our garden and the oranges it has.'

He said: 'This garden of yours is my face;
Whoever picks its oranges picks fire.'

In the garden were citrons, the colour of gold, growing high up and dangling down among the branches like golden nuggets. The passionate poet has written:

Do you not see the citron trees bearing fruit,
Until you fear that they may bend and break?
When the breeze passes, they are like
A bough weighed down with bars of gold.

There were also *kabbad* citrons hanging on the branches, like the breasts of gazelle-like girls, fulfilling all desires, as the poet has well said:

I saw a *kabbad* citron in the garden
On a tender branch like a young girl's form.
Swaying in the wind, it was like a golden ball,
On a polo stick made out of chrysolite.

There were also sweet-smelling lemons like hen's eggs, that were splendidly yellow when ripe and beautifully scented for those who picked them, as a poet has described:

Do you not see the lemon that hangs there,
Gleaming and catching the eyes of those who look,
Like a hen's egg stained by hand with saffron?

Not only did the garden have fruits of all kinds, but there were also aromatic plants, vegetables and sweet-smelling flowers, such as jasmine, henna blossom, pepper, spikenard, roses of every kind, plantain and myrtle, and all sorts of scented plants.* This was, in fact, a garden so unparalleled in beauty that it looked like a piece of Paradise. Any sick man who entered it would leave like a raging lion, and no tongue can describe its wonders, which exist nowhere else except in Paradise itself. This was not surprising, seeing that the name of its doorkeeper was Ridwan, although how great a difference was there between him and the guardian of Paradise!

After the young merchants had looked around the garden, they took their seats in an alcove, with Nur al-Din in the middle.

---

* See Burton 8.273 for annotation of this passage.

Night 866

Morning now dawned and Shahrazad broke off from what she had been allowed to say. Then, when it was the eight hundred and sixty-sixth night, SHE CONTINUED:

I have heard, O fortunate king, that the merchants' sons took their seats in the alcove, with Nur al-Din in the middle. He sat on an embroidered leather mat, leaning against a rounded pillow made of grey fur and stuffed with ostrich down. His companions handed him a fan of ostrich feathers, inscribed with the lines:

The scented breath of the fan brings back memories of happy times,
Always presenting sweetness to the face of the noble youth.

They all then took off their turbans and their outer robes, and sat talking together and indulging in the give and take of discussion, while all the time every eye was fixed on the handsome Nur al-Din. After they had been sitting like this for some time, a black slave arrived carrying on his head a tray of food together with dishes of china and crystal, following instructions left by one of the young men with his family before he had started out. The food comprised the flesh of all kinds of creatures, whether they walked on land, flew through the air or swam in the sea, including sandgrouse, quail, young pigeons, sheep and the most delicate kinds of fish. When all this had been set before them, the young men came up and ate their fill, and when they had finished eating they washed their hands in pure water with soap scented with musk, and then dried them on towels of brocaded silk. To Nur al-Din they presented a towel embroidered with red gold, on which he wiped his hands. Coffee was then served and each youth drank as much as he wanted.

As they sat and talked, the gardener went off and came back with a basket filled with roses which he offered to them. 'That is good,' they said, 'for roses in particular are not to be turned away.' 'Yes,' agreed the gardener, 'but it is a custom here that we only give away roses in exchange for sociability, and whoever wants some must produce a poem to match the occasion.' There were ten young men there with Nur al-Din, and when one of them had agreed to the condition and promised to produce a suitable poem, the gardener handed him a bunch of roses. He took them and then recited:

I love the rose and never weary of it;
Among the army of sweet-scented flowers
The rose is the commander.
They may boast proudly when it is not there,
But when it comes, it puts them all to shame.

A second was then given a bunch and he recited:

Here, my master, is a rose
With a scent reminding you of musk,
Like a girl seen by her lover,
Who hides her head behind her sleeve.

A third, on taking his bunch, recited:

The sight of a precious rose delights the heart,
While its perfume recalls the scent of *nadd*.
The branch holds it delightedly among its leaves,
Like the kiss of a mouth that never turns away.

The fourth then took a bunch and recited:

Do you see the rose bushes,
With marvellous beauties placed among the branches?
They look like rubies, set about
With chrysolite, to which is added gold.

When the fifth was given his roses, he recited:

Branches of chrysolite loaded, instead of fruit,
With ingots of pure gold.
When drops of dew fall from their leaves
They are like tears that languorous eyes have shed.

The sixth, taking his bunch, recited:

Rose, you combine the finest beauty
And subtle secrets given you by God.
This is like the beloved's cheek
On which the longing lover scatters a dinar.

The seventh took his bunch and recited:

I asked the rose: 'Why are your thorns so quick
To wound all those who touch you?' and it said:

'My armies here are all the scented flowers;
I lead them and my weapon is the thorn.'

The eighth recited, on taking his roses:

God guard the yellow rose,
Splendid and glistening like gold.
Its lovely branches have borne fruit,
Producing shining suns.

With his roses the ninth recited:

Bushes of yellow roses introduce
Joy in the hearts of every slave to love.
How strange it is that what has been
Watered by silver should produce pure gold.

Finally, the tenth took his bunch and recited:

Do you not see the armies of the rose
Rising resplendently in red and yellow?
These roses with their thorns are like
Spearheads of emerald set in a golden shield.

When they all had their roses in their hands, the gardener fetched wine,
setting before them a porcelain tray embellished with red gold and reciting:

The gleam of dawn has spoken, so pour wine –
Old wine that turns a wise man to a fool.
So fine and clear it is, I cannot tell
Whether I see it in the glass or see the glass in it.

He filled a wine cup, drank it down and then passed the wine around
until it reached Nur al-Din, but when the cup was filled and passed to
him, Nur al-Din said: 'I must tell you that this is something about which
I know nothing. I have never drunk wine as this is a great sin, forbidden
in His Book by Almighty God.' 'Sir,' replied the gardener, 'if the only
reason you don't drink is because it is sinful, remember that God, the
Great and Glorious, is generous, forbearing, forgiving and merciful. He
forgives even the greatest of sins and His mercy encompasses everything.
May He forgive the poet who wrote:

Do what you want, for God is generous;
Nothing will hurt you, even if you sin.

But there are two things that you must not do –
Join any other god to God or do the people harm.'

One of his companions then pressed him to drink; another swore to divorce his own wife if he did not; while a third stood there in front of him. Nur al-Din was embarrassed and so he took the cup from the gardener's hands but after taking a mouthful he spat it out, exclaiming: 'This tastes bitter!' The young gardener explained: 'Unless it were bitter, it would not have the advantages that it has. Don't you know that every sweet thing you take as medicine tastes bitter? Wine has many uses. It helps you to digest food, dispels care, cures wind, purifies the blood, clears the complexion and restores the body. It emboldens cowards and encourages copulation, but it would take too long to list every one of its advantages. A poet has written:

The mercy of God encompassed me as I drank;
What the glass brought me cured my sufferings.
I am not deceived and realize the sin,
But as God said, drink has advantages for men.'*

He then got up straight away and opened a closet in the alcove from which he took a loaf of refined sugar. He broke off a large piece and put it in Nur al-Din's cup, saying: 'If you are afraid of drinking the wine because it is too bitter, then take it now that it is sweet.' Nur al-Din took the cup and drank it, after which one of his companions filled another and presented it to him, saying: 'I am your slave, Nur al-Din.' A second said: 'I am one of your servants'; a third: 'Drink this for my sake'; and a fourth: 'Take this to oblige me.' This went on until all ten of them had got him to drink. The wine was unfamiliar to his stomach, as never in his life had he tasted it before that moment. As it rose to his head, he became drunk and, rising to his feet, he said, thickly and stumblingly: 'My friends, you are fine, your words are fine, the place is fine, but we need some pleasant music, for it is better not to drink at all than to drink without music, as the poet has said:

Pass the wine round in large cups or in small,
And take it from the hand of a radiant moon,
But do not drink without music, for I see
That even horses are whistled to when they drink.'

* Quran 2.216.

At that, the young gardener got up and rode off on one of the visitors' mules, coming back later with a Cairene girl like a fresh sheep's tail, a pure silver ingot, a dinar on a china dish or a gazelle in the desert. Her face put the radiant sun to shame; her languid eyes were full of magic; while her eyebrows were like bent bows. Her cheeks were coloured like roses, her teeth were pearls and her kisses tasted of sugar. Her breasts were of ivory and her belly slender with wrinkled folds. She had buttocks like stuffed cushions and thighs like Syrian columns, with between them what looked like a purse tucked away in a folded package. She was as the poet has described:

Had the polytheists seen her face,
They would have worshipped her in place of all their idols,
And if a monk had seen her in the east,
He would have turned and bowed down as a westerner,
While if she spat into the salt sea,
Her spittle would turn all its water fresh.

Another poet has written:

Dark-eyed and lovelier than the moon,
She is a gazelle that hunts down lion cubs.
Her tresses, black as night, hang down,
Making a tent of hair fixed with no pegs.
Fires kindled by her rosy cheeks are fed
By the melted hearts and livers of her lovers.
The beauties of the age would rise and bow to her,
Saying: 'She who is foremost must receive the prize.'

How well another has written:

Three things kept her from coming to see me,
Through fear of spies, the envious and the bitter.
These were her radiant forehead, rustling ornaments,
And the scent of ambergris her limbs diffuse.
Though she may use her sleeve to hide her face,
And take off all her ornaments,
What can she do about the scent she spreads?

The girl was like a moon on its fourteenth night; she was wearing a blue dress with a green veil which hung over a forehead so radiant as to astonish all who saw her, leaving the intelligent bereft of their wits.

Night 867

Morning now dawned and Shahrazad broke off from what she had been allowed to say. Then, when it was the eight hundred and sixty-seventh night, SHE CONTINUED:

I have heard, O fortunate king, that the gardener brought the girl whom I have described. Her beauty, grace and elegant figure fitted the poet's description:

> She came in a dress of azure blue,
> The colour of the sky.
> I saw within her gown
> A summer moon rising in a winter's night.

Other excellent lines are:

> She came veiled, and I said: 'Unveil,
> To show the lustre of the shining moon.'
> She said: 'I fear disgrace.' I said: 'Enough;
> Do not be startled by what Time may bring.'
> She moved the veil of beauty from her cheeks,
> And drops of crystal fell upon a gem.
> I thought of pressing a kiss on her cheek
> So that she might complain of me on the Last Day.
> Ours would be then the first trial of two lovers
> Brought on that day before Almighty God,
> And I would ask the case to be prolonged
> So I might for longer look upon my love.

The young gardener said to the girl: 'Mistress of the beauties and of every star that shines, you must know that we have brought you here to drink with this good-natured young man, Nur al-Din, who is on his first visit to us.' 'I wish you had told me about that before,' she replied, 'as there is something that I could have brought with me.' 'I'll go and fetch it,' he said, and she replied: 'As you like.' At his request, she gave him a kerchief as a token, and he left in a hurry, coming back some time later with a green satin bag with two matching bands of gold. The girl took it from him, untied it and shook out the contents, which turned out to be thirty-two pieces of wood. She fitted these together, male to female and female to male, and when she set up what had been fitted together, revealing her wrists as she did so, it turned out to be a highly polished Indian lute. She bent over it like a mother over her child, and when she ran her fingers over the strings, they produced a plaintive sound as if the lute were longing for its one-time home, remembering the waters that nourished it, the earth from which it had grown, the carpenters that had shaped it, the polishers who had oiled it, the merchants who had imported it and the ships in which it had been carried. The mournful notes that it produced and multiplied made it seem as though the girl was asking it about all that and in its wordless state it was answering with the lines:

I was once a tree on which the bulbuls lived;
I swayed with love for them while my leaves were green;
I learned from them as they perched on me and moaned,
And through that sound my secret was made known.
A woodsman felled me, though I had done no wrong,
Making me into a slender lute, as you see now.
But fingers touching me make me reveal
That, although dead among mankind, I still endure.
Because of this, all those who drink together
Hear my lament and then are drunk with love.
God softens every heart to me, and among guests
I am promoted to the highest place.
The loveliest of girls embrace my form,
Gazelles who gaze with dark and languid eyes.
May God not separate the passionate lovers;
May they not live who turn away their loves.

After an interval of silence, the girl took the lute on her lap and, bending over it like a mother with her child, she tried a number of different modes before returning to the first and reciting:

If she turned back and visited the lover,
The load of longing would lift from his heart.
The nightingale on the bough disputes with him,
Like a lover parted from the one he loves.
Wake, for on nights of union the moon shines,
As though these lovers' meetings were the light of dawn.
Today our envious critics pay no heed,
As the lute strings call us to delights.

Four things are joined here for us to enjoy,
Myrtle and roses, gillyflowers and other blooms,
And here today another four bring luck,
Lover and friend, as well as wine and wealth.
Enjoy your luck, for pleasures in this world
Do not remain and all that does are tales.

When Nur al-Din heard the girl reciting these lines, he fell so deeply in love as he looked at her that he could scarcely control his feelings. She, for her part, returned these feelings, as when she looked at all the young merchants who were there and at Nur al-Din, he seemed to her to be a moon among stars, with his soft words, caressing manners, perfectly formed figure, beauty and grace. He was more delicate than the breeze and finer than the water of Paradise, as described in these lines:

I swear by his cheeks and by his smiling mouth,
By the bewitching arrows that he shoots,
By the soft folds of his body and the shafts of his glance,
The whiteness of his forehead and the blackness of his hair,
Eyebrows that exile all sleep from my eyes,
Lording it over me, commanding and forbidding,
By the scorpion locks let loose over his temples,
That do their best to kill the lover when he leaves,
By the roses of his cheeks, their myrtle down,
The carnelian of his mouth, his pearly teeth,
By his figure like a fruitful bough
With pomegranates blooming on his breast,
Haunches that quiver as he moves or rests,
And by his slender waist,
His silken dress, his ready wit,
By all the beauties that he has combined.
I swear by these, his breath lends scent to all perfume,
And it is thanks to him the breeze is sweet.
The sun has not his radiance, and the new moon
Is nothing but the paring of his fingernail.

Night 868

Morning now dawned and Shahrazad broke off from what she had been allowed to say. Then, when it was the eight hundred and sixty-eighth night, SHE CONTINUED:

I have heard, O fortunate king, that when he heard the girl reciting her lines, Nur al-Din was filled with admiration and, reeling with drunkenness, he started to praise her with the following lines:

> A lute girl led my heart astray,
> Drunk as I was with wine.
> Her lute strings spoke to us and said:
> 'God has endowed us with a voice.'

When Nur al-Din had spoken, the girl looked at him with the eye of love. Her passion for him increased as she admired his graceful beauty and his elegant, well-shaped figure. Unable to restrain herself, she cradled her lute a second time and recited these lines:

> He blames me when I look at him,
> And leaves me, though my life is in his hands.
> He keeps me, knowing what is in my heart
> As though through inspiration sent by God.
> I drew his portrait in my palm,
> Telling my eyes they are to weep for him.
> For him these eyes can find no substitute,
> Nor does my heart allow me to endure.
> Heart, I have torn you from within my breast
> As you are one who envies me my love.
> I tell my heart to turn away from him,
> But all the while it is to him it turns.

Nur al-Din admired the fluency, sweetness of diction and eloquence of expression shown in these fine lines and he was overcome by the passion of his love. Unable to restrain himself, he leaned towards her, clasping her to his breast, as she, for her part, accommodated herself to his embrace. Abandoning herself totally to him, she kissed him between the eyes while he clasped her to him and kissed her mouth, playing at kisses with her like a dove feeding its young, while she turned towards him and did the same. The other young men got up in some confusion, and at that point Nur al-Din shamefacedly let go of the girl. She took the lute and after trying out a number of modes, she reverted to the first and recited these lines:

He is a moon who draws a sword from his eyelids
When he bends, and his eyes mock the gazelles.
He is a king whose armies are rare beauties,
And in the press of war his form is like a spear.
Were his heart as tender as his waist,
He would not bring himself to wrong a lover.
Hard heart and slender waist –
Why have you not exchanged one for the other?
You who blame me for loving him, excuse me now;
Yours is his lasting beauty, mine the transient.

Impressed by the gracefulness of these splendid lines, Nur al-Din turned to her in delight, unable to control his admiration, and recited:

I thought of her in fancy as the sun,
But in my heart her blazing fire still burns.
Would it harm her to greet me with a sign,
Were she to signal with her fingertips?
My critic saw her face and lost his way
Among the beauties that show beauty's self.
He said: 'Is it for her you find yourself lovelorn?
You are to be excused.' I said: 'It is for her.'
She aimed at me on purpose with her glance,
Having no pity on a stranger, humble and distressed.
She stole my heart away, leaving me as a slave,
Weeping and wailing all my days and nights.

When Nur al-Din had finished, the girl, filled with admiration for his eloquence, took the lute, and after playing through all the modes again with a most delicate touch, she recited these lines:

Life of my soul, I swear now by your face
That in despair or hope I'll never turn from you.
You may be cruel, but your phantom visits me;
Though you are out of sight, your memory stays as a friend.
My eyes are lonely thanks to you, and you must know
That only in your love can I find happiness.

Your cheeks are roses, your saliva wine;
Why should you grudge them to me when we meet?

Nur al-Din, moved by delight and admiration, replied to her with lines
of his own:

When she unveils to show the sun's face in the dark,
The full moon in the heaven hides away.
And when the eyes of morning see her hair,
Its parting looks for refuge in the dawn.*
Channel my tears that flow in one long stream
Along the quickest route, to water what I say with love.
To many a girl aiming her arrows, I have said:
'Take care when shooting; you alarm my heart.'
My tears can be compared to the Nile flood,
Whereas your love for me is merely feigned.†
She said: 'Bring me your wealth.' I said: 'Here, take it all.'
She said: 'Give me your sleep.' I said: 'Remove it from my eyes.'

On hearing Nur al-Din's eloquent words, the girl was bemused and
ecstatic; her whole heart was filled with love for him, and clasping him
to her breast she started to kiss him like a dove feeding its young. He
returned her kisses one after the other, but whoever is the first to do
something gets the credit. When she had finished kissing, she picked up
the lute and recited:

The censurer reproaches us. Alas!
Shall I complain of him or to him of my restlessness?
You have abandoned me, although I never thought
That you, who are all mine, would spurn my love.
I used to censure lovers for their love,
But now I show your critics how you humbled me.
While yesterday I would blame all who loved,
Today I find excuses for their pains.
If you part from me, leaving me to grieve,
I call on God to help me, in your name.

When she had finished, she added these lines:

* Quran 113.1.
    † 'Feigned': the Arabic word, *malaq*, refers to land alongside the Nile.

The lovers say: 'Unless he pours for us
The pure wine of his kisses, we shall pray
To God Almighty that He answer us,
And all of us will call our love by name.'

On hearing this poem, Nur al-Din admired the girl's eloquence and
thanked her, infatuated by her grace. When she heard the praise that he
lavished on her, she immediately got to her feet and stripped off all her
clothes and her jewellery, throwing them aside, after which she sat on
his knees and kissed him between the eyes and on the mole on his cheek,
presenting him with everything she had been wearing . . .

Night 869

Morning now dawned and Shahrazad broke off from what she had been
allowed to say. Then, when it was the eight hundred and sixty-ninth
night, SHE CONTINUED:

I have heard, O fortunate king, that the girl presented him with everything she had been wearing, telling him: 'Heart's darling, the value of
the gift depends on that of the giver.' Nur al-Din accepted what she gave
him, but then returned it to her, kissing her on the mouth, the cheeks
and the eyes. When they had finished – and everything comes to an end
except for the Everlasting and Eternal God, Who feeds both the peacock
and the owl – Nur al-Din got up to leave. 'Where are you going, master?'
asked the girl, and he said: 'To my father's house.' The others swore that
he should spend the night with them, but he refused and, mounting on
his mule, he rode back home.

'Why have you been away so long, my son?' his mother asked, adding:
'Both your father and I have been worried and concerned by your
absence.' When she came up to kiss him on the mouth, she detected the
smell of wine and said: 'My son, you have always said your prayers and
worshipped God, so how is it that you have started to drink wine,
disobeying the Omnipotent Lord of creation?'

As she was speaking, Nur al-Din's father, Taj al-Din, came in and
when Nur al-Din threw himself on his bed and fell asleep, his father
asked what was wrong with him. 'The air of the garden has given him a
headache,' his mother explained, but when his father went up to him to
greet him and ask about his headache, he too smelt the wine. Not being

fond of wine drinkers, he exclaimed: 'Damn you, my son, have you
become stupid enough to drink wine?' When Nur al-Din, still in his
drunken state, heard what his father said, he raised his hand and struck
at him, the blow falling, as was fated, on his right eye. The eye was
dislodged on to his father's cheek and he collapsed, remaining unconscious for some time, with rosewater being sprinkled over him. When he
recovered he proposed to beat Nur al-Din, and although his wife stopped
him, he swore to divorce her if he did not have the boy's right hand cut
off next morning.

His wife was distressed to hear this and, fearing for her son, she
continued to coax her husband until he fell asleep, in an attempt to
conciliate him. She then waited until moon-rise, when she went to Nur
al-Din, who had now sobered up and asked him: 'What is this foul thing
you did to your father?' 'What did I do?' he asked her and she told him
that he had knocked his father's eye out on to his cheek with his hand.
'He has sworn to divorce me,' she said, 'if he does not cut off your right
hand in the morning.' Nur al-Din regretted what he had done at a time
when regret was of no use, as his mother went on to tell him. 'You will
have to get up straight away,' she added, 'and try to save yourself by
running away. When you leave, go to one of your friends and stay in
hiding, waiting to see what God will do, for it is He Who is the source
of change.' She opened her money-box and took out a purse containing
a hundred dinars. 'Take these,' she said, 'and use them for your needs.
Then when you have finished them, get word to me and I shall send you
more. At the same time, let me have your news secretly and it may be
that God will decree an end to your misfortunes so that you will be able
to come back home.' She then said goodbye to him, shedding floods
of tears.

Nur al-Din took the purse from his mother and was about to leave
when he caught sight of a larger purse which his mother had overlooked
lying beside the money-box. This one had in it a thousand dinars and
Nur al-Din picked it up, tying both purses round his waist, after which
he left the lane and set off before daybreak in the direction of Bulaq.
With the coming of morning, all His creatures rose to proclaim the unity
of God, the Opener of ways, each going about his business to acquire
what had been allotted to him. Nur al-Din, for his part, reached Bulaq
and began to walk along the river bank, where he saw a ship with its
four anchors fixed to the shore, along whose gangway passengers were
boarding and disembarking. The crew were standing there, and when he

asked them where they were going, they told him that they were bound
for Alexandria. He asked them to take him with them, and they said:
'You will be very welcome, you handsome young man.' He went off
straight away to the market, where he bought what he needed in the way
of food, bedding and coverings. When he got back, he found the ship
ready to sail, and not long after he had boarded, it set off and continued
on its way until it reached Rosetta.

On his arrival at Rosetta, Nur al-Din saw a small boat bound for
Alexandria. He went on board and crossed the water until he came to
what was known as the Jami' Bridge. Here he left the boat and entered
the city by the Lotus Gate, sheltered from sight by God so that he passed
unnoticed by those who were standing there. He walked on into the
city . . .

Night 870

Morning now dawned and Shahrazad broke off from what she had been
allowed to say. Then, when it was the eight hundred and seventieth
night, SHE CONTINUED:

I have heard, O fortunate king, that Nur al-Din entered Alexandria
and found it to be a place with strong walls and lovely parks, enjoyable
for its inhabitants and attractive to incomers. The cold of winter had
retreated from it; spring had brought its roses; flowers were in bloom;
trees were in leaf, with ripening fruits; and water was pouring through
the streams. The city itself had been well planned and laid out; it was
peopled by the best of men and when the gates were closed everyone
there was safe. It has been described in the following lines:

> One day I said to an eloquent friend,
> 'Describe Alexandria.' 'A pleasant port,'
> He said, and when I asked:
> 'Can one make a living in it?' he replied:
> 'If the wind blows.'

Another poet has written:

> Alexandria is a mouth that is sweet to kiss.*
> Its union brings delight, if the crow of parting does not croak.

---

* 'Mouth' and 'port' can be covered by the same word in Arabic – *thaghr.*

Nur al-Din walked on, passing the merchants' market, the market of the money-changers, the markets of the nut sellers and the fruiterers and the market of the apothecaries. He was filled with admiration for the city, whose appearance matched its royal name. While he was in the perfume market an old man came down from his shop to greet him, before taking him by the hand and leading him to his house. Nur al-Din saw an attractive lane – the first part swept out and sprinkled with water, the other end paved with marble – in which a pleasant breeze was blowing and which was shaded by leafy trees. There were three houses there and at the upper end of the lane was a fourth with towering walls, whose foundations were set in water. The square in front of it had been swept and sprinkled, while the visitor was met by the scent of flowers and fanned by a breeze, as though here was the garden of Paradise.

The old man took Nur al-Din to the fourth house and fetched food, which the two of them then ate. When the old man had finished, he asked Nur al-Din when he had arrived from Cairo. 'This last night,' Nur al-Din told him and when the old man went on to ask his name, he replied: ''Ali Nur al-Din.' 'Nur al-Din, my son,' the old man said, 'I swear to divorce my wife three times if you leave me during your stay here, and I will give you a place of your own where you can live.' 'Tell me more about yourself, sir,' said Nur al-Din, and the man replied: 'Some years ago I took merchandise to Cairo and when I had sold it I bought some more, but I needed another thousand dinars to complete the purchase. It was your father, Taj al-Din, who weighed this sum out for me although I was a stranger to him, and he did not even get me to sign a receipt for the money. Instead, he waited until I got back here, when I sent off a servant to take it to him, together with a gift. I saw you when you were a little boy and, God willing, I shall repay you for some of what your father did for me.'

When Nur al-Din heard this, he smiled happily, before producing the purse with the thousand dinars, which he passed over to the old man, asking him to keep it as a deposit until he had bought some merchandise with which to start trading. He then stayed in Alexandria for some days, looking round the streets, eating, drinking and enjoying himself until he had used up the hundred dinars that he had with him as spending money. At this point, he went back to the old apothecary to take some of his thousand dinars to use for expenses, and as the man was not there in his shop Nur al-Din sat down to wait for his return. While he was sitting

there watching the merchants and looking right and left, he caught sight
of a Persian riding into the market on a mule, behind whom sat a girl
like an ingot of pure silver, a fish in a fountain or a gazelle in the desert.
Her face would put the radiant sun to shame; she had bewitching eyes,
breasts of ivory, pearly teeth, a slender belly and curving sides, each leg
like the fat tail of a sheep. She combined beauty, grace, slenderness and
shapeliness, as has been described in these lines:

> It is as though she was created by her own wish,
> With all the glamour of beauty, neither tall nor short.
> Her cheeks would make the roses blush for shame,
> Her figure makes fruit ripen on the branch.
> She appears as the full moon; her breath is scented musk;
> Her figure is a branch; she has no match among the human race.
> She has been fashioned as a pearl and, in her beauty,
> She seems to have a moon in every limb.

The Persian dismounted and after helping the girl down from the
mule, he called for the auctioneer and told him to take her and find a
buyer for her in the market. The man went off with her into the middle
of the market and then, after a brief absence, he came back with a chair
of ebony embellished with ivory. He put this on the ground and got the
girl to sit on it before removing her veil to show a face like a Dailami
shield or a gleaming star. In her surpassing loveliness she was like a full
moon on the fourteenth night, as the poet has said:

> The foolish moon challenged her loveliness
> And was eclipsed and split in anger.
> If the *ban* tree is compared to her figure –
> May her hands perish who has become a wood carrier.*

How excellent are the lines of another poet:

> Say to the pretty girl in the gilded veil:
> 'What have you done to the pious ascetic?
> The radiance of the veil and of your face beneath it
> Between them rout the armies of the dark.'
> If I should steal a glance at her cheek,
> A watching guard will hurl a star at me.

* Quran 111.4.

'How much will you offer me for this pearl brought up by the diver, this gazelle that has escaped the hunter?' the auctioneer called to the merchants. One of them offered a hundred dinars, another two hundred and a third three, after which the bidding went on until it reached nine hundred and fifty dinars. There it was halted to wait for the vendor's acceptance.

Night 871

Morning now dawned and Shahrazad broke off from what she had been allowed to say. Then, when it was the eight hundred and seventy-first night, SHE CONTINUED:

I have heard, O fortunate king, that the merchants went on bidding for the girl until, when the total had reached nine hundred and fifty dinars, the auctioneer went to the Persian, the girl's master, and said: 'There is an offer of nine hundred and fifty dinars for your slave girl. Are you willing to sell and take the cash?' 'Is she happy with that, for I don't want to displease her?' the Persian asked, explaining: 'I fell ill on my journey here and she looked after me so well that I swore only to sell her to someone she wanted, leaving the decision to her. So go and ask her. If she is content, sell her, but if she says no, then don't.' The auctioneer went up to her and said: 'Queen of the beauties, know that your master has left the matter of your sale in your own hands. Nine hundred and fifty dinars have been bid for you, so do I have your consent to sell you?' She replied: 'Show me who it is who wants to buy me before concluding the sale.' At that, the auctioneer took her up to a decrepit old merchant, and having taken a long look at him, she turned to the auctioneer and said: 'Are you mad or weak in the head?' He asked why she had said that to him and she replied: 'Does God allow you to sell someone like me to this decrepit old man? He is like the man who said of his wife:

My pampered wife is angry,
Having called me to do what could not be done.
She says: "If you don't take me properly,
Don't blame me when you are cuckolded.
Your penis is as soft as wax
And when I rub it with my hands it weakens."

Another poet has written:

> My penis falls ignobly asleep
> Whenever I have union with my love,
> But when I am alone at home, it tries
> To go to battle and to thrust alone.

Another has written:

> I have an evil penis, ill behaved,
> Maltreating one who shows it most respect.
> For when I sleep, it stands, and it sleeps when I stand;
> God, do not pity those who pity it.'

The fury of the old merchant when he heard this obscene abuse could not have been surpassed and he said to the auctioneer: 'You unlucky fellow, this is an ill-omened slave girl you have brought to the market in order to make fun of me among my colleagues.' The auctioneer took her away and said: 'Lady, don't forget your manners. This man whom you insulted is the *shaikh*, the market superintendent whom all the other merchants consult.' But the girl only laughed and recited:

> It is the rightful duty of the rulers in our time
> To hang the *wali* over his own door,
> And beat the market foreman with a whip.

She demanded to be sold to another buyer, saying that if she were sold to the old man, he would become ashamed and might sell her to someone else, reducing her to the status of a menial. 'It is not for me to sully myself with menial service,' she explained, 'and, as you know, I am in charge of my own sale.' 'To hear is to obey,' the auctioneer replied, after which he took her off to another of the leading merchants and asked: 'My lady, shall I sell you for nine hundred and fifty dinars to Sharif al-Din?' The girl looked at him and saw that he was an old man, this time with a dyed beard, and so again she asked the auctioneer if he was mad or weak in the head to think of selling her to another old man nearing his end. 'Am I a piece of waste or a rag,' she said, 'that you parade me around from one old man to another? Each one of these is like a crumbling wall or an *'ifrit* struck down by a star. The first appeared to be echoing the words of the poet:

> I asked to kiss her mouth, but she said:

"No, by the Creator of all things from nothing.
There is nothing that I want from white hairs;
Is my mouth to be stuffed with cotton while I am still alive?"

There are other good lines:

They say: "White hair is dazzling radiance,
Adding to faces dignity and light."
Until the parting of my hair shows white
I do not want to lose what should be dark.
Were the beard of the white-haired man a book,
On Judgement Day he would not want it white.*

Even better are the lines:

An immodest guest visited my head;
A sword blow would be kindlier.
Be off, unpleasing whiteness. To my eyes
You are more gloomy than the dark of night.

As for the second man, he is a source of suspicion and disgrace in that he tries to blacken his white hairs by using dye, which is the foulest of lies. It is as though he was reciting:

She said: "You dye your hair." I said to her:
"I hide it from you, who are my ears and eyes."
She laughed and said: "How very strange it is
That your deceit extends into your hair."

How well the poet has written:

You dye your white hairs black,
So that your youth may stay with you.
Try, then, to dye my fortune black.
I guarantee you that it will not change.'

When the man with the dyed beard heard these words, his anger knew no bounds and he said to the auctioneer: 'Unfortunate man, this is an impudent slave girl you have brought to market today. She does nothing but abuse everyone here, one after the other, ridiculing them in boastful poems.' He then came down from his shop and struck the auctioneer in the face, after which the man took her away. 'Never in my life have

---

* Good deeds are coloured white in the Book of Life, which records all lives.

I come across a more shameless girl than you,' he told her angrily. 'You have managed to ruin both of us today, as, thanks to you, all the merchants are angry with me.' While they were going along the road, another merchant, Shihab al-Din by name, happened to catch sight of them and promptly added ten dinars to the price that had been offered. The auctioneer asked her if she would agree to the sale and she said: 'Let me look at him and let me ask him whether he has a certain thing in his house. If he has, I am willing to be sold to him, but if not, no.' The auctioneer left her standing there and went up to Shihab al-Din and told him that the girl was willing to be sold to him if he could answer her question about something that he might or might not have in his house. He went on: 'You heard what she said to your colleagues . . .'

Night 872

Morning now dawned and Shahrazad broke off from what she had been allowed to say. Then, when it was the eight hundred and seventy-second night, SHE CONTINUED:

I have heard, O fortunate king, that the auctioneer said to the merchant: 'You heard what she said to your colleagues and, by God, I'm afraid that if I bring her to you she will treat you as she treated the others and I shall be humiliated in front of you. But if you give me your permission, I shall fetch her to you.' 'Bring her,' said Shihab al-Din, at which the auctioneer went off and fetched her. After having looked at him, she said: 'Sir, do you have at home any cushions stuffed with squirrel fur?' 'Yes, queen of the beauties,' he told her, 'I have ten of them, but what, in God's Name, do you want to do with them?' She said: 'I shall wait until you are asleep and then put them over your mouth and your nose until you are dead.' Then she turned to the auctioneer and said: 'You worthless fellow, you must be mad. You showed me just now to two old men, each of whom had two faults, and then you go on to show me to this Shihab al-Din, who has three. The first is that he is too short; the second is that he has a long nose; while, thirdly, he has a long beard. As the poet has said:

Never among all mankind have I seen or heard of a man like this.
His beard is a cubit long, his nose a span but he stands no more than
a finger high.

Another poet has said:

> He has the minaret of a mosque in his face,
> Thin as a finger in a signet ring.
> If all mankind entered his nose,
> The world would be emptied of its folk.'

When Shihab al-Din heard this, he came down from his shop and seized the auctioneer by the collar. 'You miserable fellow,' he said, 'how dare you bring us a slave girl to insult and abuse us, one after the other, bragging away in poems like these!' At that, the auctioneer removed the girl and told her: 'By God, in all the time that I have been doing this job I have never come across a ruder girl than you or one who has brought me worse luck. You have ruined me today and the only profit that I have got from you has been blows on my neck and people seizing my collar.'

He then took her to a merchant named 'Ala al-Din, who had a number of slaves and servants, and asked whether she was prepared to be sold to him. When she looked at him she saw that he was a hunchback and she pointed this out to the auctioneer, quoting the lines:

> His shoulders are low and his spine is long,
> Like a devil crouching to avoid a star.
> It is as though he tasted the first blow of the whip,
> And is left to wonder when the next will fall.

She quoted another poet:

> The hunchback on his mule
> Is a general figure of fun.
> He ducks as they laugh. Don't be surprised
> When the mule that he is riding bolts away.

And another:

> There are plenty of hunchbacks made even uglier
> By their deformity so that eyes spit them out,
> Shrunk like dried-up branches
> That Time has bent through the fruit that they have borne.

The auctioneer rushed up to her and took her off to yet another merchant, asking if she was prepared to be sold to him. She found, on looking at him, that this one was bleary-eyed and she asked the auctioneer how he could sell her to a man like that, quoting the lines:

The inflammation of his eyes
Destroys his powers.
Come, look and see
The mote that spoils his sight.

The auctioneer then took her to someone else and put the same question to her. This time she saw that the merchant had a long beard and said: 'Damn you, this is a ram with his tail sprouting from his throat. How could you possibly sell me to him, you wretched fellow? Haven't you heard that a long beard is a sign of stupidity, and the longer it is, the stupider is its owner? This is well known to all people of intelligence, as the poet has said:

If a man's long beard adds to his dignity,
It is his lack of wit that gives it extra length.

Another poet has written:

I have a friend with a beard;
God lengthened it, but all to no avail.
It is like a winter's night –
Long, dark and cold.'

It was after this that the auctioneer took her back, and when she asked where he was going with her, he said: 'Back to your master, the Persian, because I have had enough trouble today, thanks to you. You have lost money both for me and for your master by your bad manners.' At that point, as she was looking round the market in all directions, right, left, forwards and backwards, as fate would have it her eye fell on Nur al-Din the Cairene. She could see that he was a handsome fourteen-year-old, with smooth cheeks and a slim figure; he was remarkable for his beauty, grace, elegance and amorous disposition; he looked like a full moon on the fourteenth night, with his radiant forehead, rosy cheeks, neck like marble, pearl-like teeth and saliva sweeter than sugar. He fitted the description given in the lines:

Moons and gazelles came out to match his beauty,
I told them: 'Stop; go slow;
Do not compare yourselves with him, gazelles,
And, moons, do not go to such trouble all in vain.'

There are other excellent lines:

There is a slim youth through whose hair and brow
Mankind is both in darkness and in light.
Do not find fault with the mole upon his cheek;
Anemones all have a speck of black.

When the girl looked at him, she lost her self-control and so deep was the impression that he made on her, she immediately fell in love with him.

Night 873

Morning now dawned and Shahrazad broke off from what she had been allowed to say. Then, when it was the eight hundred and seventy-third night, SHE CONTINUED:

I have heard, O fortunate king, that the girl fell in love with him. She turned to the auctioneer and asked: 'That young merchant who is sitting with the others, wearing a cotton mantle – would he not bid a bit more for me?' 'Queen of the beauties,' the man replied, 'he is a stranger from Cairo, where his father is a leading merchant, the doyen of them all. He has only been here for a short time, staying with one of his father's friends, but he made no bid for you, large or small.' On hearing this, the girl removed a valuable sapphire ring from her finger and told the auctioneer to take her to the handsome young man, saying: 'If he buys me, you can keep this ring for yourself in return for the trouble I have caused you today.' This pleased the auctioneer, who took her off to Nur al-Din. When she looked at him she saw that in his beauty and slender, well-proportioned form he was like the full moon, as a poet has described:

Beauty's pure essence is found in his face,
While arrows are shot by his glance.
The lover is choked for whom he pours
Bitter rejection, while union with him is sweet.
Perfection of perfection of perfection
Are his white complexion, his stature and my love.
The fastenings of his robes
Are buttoned round the collar of a crescent moon.
His eyes, his twin moles and my tears
Are triple nights;
His eyebrows, his appearance and my body

Are triple crescent moons.
His eyes pass lovers wine;
For all its bitterness, I find it sweet.
He gave cool water to me for my thirst
On the day of union, from his smiling mouth.
That he should shed my blood and strike me dead
Is three times lawful for a youth like him.

The girl looked at him and said: 'Sir, I ask you in God's Name to tell
me, am I not beautiful?' 'Queen of the beauties,' he replied, 'is there
anyone lovelier than you in this world?' She then asked: 'When you saw
all the other merchants bidding against each other for me, why did you
stay silent, saying nothing and not adding a single dinar to the bidding
as though you didn't admire me?' He replied: 'Had I been at home I
would have given all the wealth I own in order to buy you.' 'I don't ask
you to buy me against your will,' she said, 'but even if you do not buy
me it would be a comfort to me if you added something to my price so
that the merchants might say: "Had the girl not been beautiful, this
Cairene trader would not have raised her price, as Cairenes know about
slave girls."' Nur al-Din was embarrassed by this and, with a flushed
face, he asked the auctioneer how much had been offered for her. The
man told him that the bidding had reached nine hundred and fifty dinars,
without his own commission, while the state dues were for the vendor
to settle. 'Let me have her for a thousand, including both price and
commission,' said Nur al-Din, at which the girl left the auctioneer and
ran up to him, saying: 'I sell myself to this handsome young man for a
thousand dinars.' For his part, Nur al-Din stayed silent, while someone
standing nearby said: 'The deal is concluded'; another said: 'He is worthy
of her'; a third said: 'Only a damned son of the damned would bid and
not buy'; while a fourth added: 'They are well matched.'

Before Nur al-Din knew what was happening, the auctioneer had
fetched the qadis and the notaries, who drew up a document containing
the bill of sale, which they passed to Nur al-Din saying: 'Take your slave
girl and may God bring you blessings with her, for she would suit no
one but you and you would suit her and her alone.' The auctioneer then
recited these lines:

Good fortune came to him, trailing its skirts;
It fitted only him; he fitted only it.

Nur al-Din, feeling embarrassed in front of the merchants, got up immediately and weighed out the thousand dinars that he had deposited with his father's friend, the apothecary. Then he took the girl and brought her to the house where this man had lodged him. When she entered she saw a worn-out carpet and an old mat, and she said: 'Do you have so low an opinion of me, master, that you don't think me worthy of being taken to your own house where you keep your possessions? Why have you not taken me to your father?' 'By God, queen of the beauties,' he replied, 'this is where I am living. It belongs to an old apothecary, a native of this place, who vacated it for me and let me live here, for, as I told you, I am a stranger and come from Cairo.' She said: 'The meanest of houses will do for me until you get back home, but please, master, go off and fetch me some roast meat and wine, as well as fruits, fresh and dried.' Nur al-Din had to explain that the thousand dinars which he had paid over in order to buy her was all the money that he had and, now that it was gone, he had nothing left at all as he had spent his last few dirhams the day before. She asked: 'Don't you have any friend in this city from whom you could borrow fifty dirhams? If you could bring the money to me, I would tell you how to spend it.' 'The only friend I have here,' he told her, 'is the apothecary.'

He went straight away to this man and exchanged greetings with him. 'What did you buy today with your thousand dinars?' the man asked and when Nur al-Din told him that he had bought a slave girl, he exclaimed: 'Are you mad, my son, that you should spend all this money on a single slave? Do you know of what race she was?' When Nur al-Din said that she was a Frank . . .

Night 874

Morning now dawned and Shahrazad broke off from what she had been allowed to say. Then, when it was the eight hundred and seventy-fourth night, SHE CONTINUED:

I have heard, O fortunate king, that when Nur al-Din told the apothecary that the girl was a Frank, he said: 'Here in this city the best Frankish slaves fetch a hundred dinars. You have been tricked, but if you love her, spend the night with her, enjoy her and then tomorrow morning you can take her back to the market and sell her. Even if you lose two hundred dinars, think of it as a shipwreck or an attack by highwaymen.'

'That is sound advice, uncle,' Nur al-Din replied, 'but, as you know, I only had the thousand dinars which I have spent to buy her, and I have not a single dirham left. Would you be so very kind as to lend me fifty dirhams as spending money to use until tomorrow, when I can repay you from the price that I get for her?' The old man agreed willingly and weighed out the money before saying: 'My son, you are young and this is a pretty girl. You may become attached to her and not find it easy to sell her. As you have no money, when you have spent this fifty dirhams you can come back to me and I shall make you another loan, once, twice, thrice and up to ten times, but if you come back after that, I shall not give you the greeting Islam requires and my friendship with your father will be at an end.'

Nur al-Din took the money that the old man produced and brought it back to the girl. 'Go to the market straight away,' she told him, 'and fetch me twenty dirhams' worth of silk in five different colours and then spend the remaining thirty on meat, bread, fruit, wine and scented flowers.' Nur al-Din did what he was told and when he came back with what she wanted, she rolled up her sleeves and cooked the food expertly. After she had served it up to him they both ate their fill, whereupon she brought out the wine. They drank and she continued to fill his glass and to entertain him until he fell into a drunken sleep. She then immediately got up and removed from the bundle of her belongings a bag of leather from al-Ta'if, which she opened. She then took out two needles and sat working until she had finished making a beautiful sash, which she smoothed out and cleaned before folding it and putting it away under a cushion.

She then got up and took off her clothes before lying down beside Nur al-Din. As she squeezed up against him, he woke up to find by his side a girl like an ingot of pure silver, softer than silk and more succulent than the fat tail of a sheep. She caught the eye more clearly than a banner and was more splendid than a russet camel; she was five feet tall with rounded breasts and eyebrows like bows; her eyes were like those of gazelles and her cheeks like red anemones. She had a slender, dimpled belly and a navel that could accommodate an ounce of frankincense; her thighs were like pillows stuffed with ostrich down. Between them was something that would tire out tongues to describe and at whose mention tears would pour down. It was as though the poet had been describing her when he wrote:

Her hair shows night and her parting the dawn;
Her cheeks produce roses and her saliva wine.
Union with her is paradise while separation is hellfire;
Her teeth are pearls; her face is a full moon.

How well another has written:

She appeared as a moon, bending like a *ban* tree's branch,
Diffusing the scent of ambergris, with the eyes of a gazelle.
It is as though sorrow loves my heart,
Achieving union when my beloved leaves.
Her face is lovelier than the Pleiades
With a brow more radiant than the crescent moon.

Another poet has produced these lines:

They show the crescent, then the full moon, when unveiled,
Bending as branches, with the eyes of wild calves.
The beauty of these eyes, darkened with kohl,
Would make the Pleiades wish to lie beneath her feet.

Nur al-Din wasted no time in turning to her, clasping her to him and sucking first her lower and then her upper lip, before putting his tongue between them. He then mounted her and found her to be an unpierced pearl and a filly whom no one else had ridden. He took her virginity and achieved union, leaving the two of them joined in an unbreakable and inseparable love. He continued to press kisses on her cheeks like pebbles falling into water, thrusting as though with a spear in the heat of battle. He was a man who longed to embrace lovely girls, sucking their lips, loosing their hair, clasping their waists, biting their cheeks and mounting on their breasts. All this went on to the accompaniment of Cairene movements, Yemeni writhings, Abyssinian moans, Indian languor and Nubian passion, the pretended anger of peasant girls, Damiettan groans, the heat of upper Egypt and Alexandrian coolness. All these qualities were combined in this girl, together with her extravagant beauty and coquetry. She was as the poet has described:

Never in time shall I forget this girl,
Nor turn to one who does not bring her near.
In her appearance she is the full moon;
Praise be to Him Who formed and fashioned her.
By loving her I may have greatly sinned,

But how can I repent while I still live in hope?
She left me saddened, sleepless, sick with love;
Her qualities have left my heart perplexed.
She spoke a verse that nobody can grasp
Except a lover trained in poetry:
'No one who has not felt it knows what longing is,
And only lovers know of love.'

Nur al-Din spent a night of joy and pleasure with her until morning . . .

Night 875

Morning now dawned and Shahrazad she broke off from what she had been allowed to say. Then, when it was the eight hundred and seventy-fifth night, SHE CONTINUED:

I have heard, O fortunate king, that Nur al-Din and the girl spent a night of joy and pleasure until morning, wrapped, as the two of them were, in the closest of embraces, secure from the misfortunes of Time. There was nothing to distress them and they had no fear that any scandal would be attached to their union. This was as the admirable poet has written:

Visit her with no thought of what the envious say;
An envious man is no help to a lover.
God has produced no finer sight than that
Which shows two lovers on a single bed,
Embracing one another in content,
Pillowing each other with their wrists and arms.
For when hearts are united in their love,
It is cold iron on which their critics strike.
You who blame the lovers for their love,
Have you the power to mend an ailing heart?
If in your lifetime you find one true friend
How good this is! Live for this friend alone.*

When Nur al-Din woke in the morning, he discovered that the girl had fetched water, and so the two of them performed the ritual ablution. When he had completed the morning prayer, she brought him what food

* cf. Night 22.

and drink there was, and, after he had had his meal, she put her hand under the cushion and brought out the sash that she had made the night before. She passed it to him, telling him to take it, and when he asked where she had got it from, she said: 'This is the silk that you bought yesterday for twenty dirhams. Take it to the Persian market and get the auctioneer to put it up for sale. It must be sold for twenty dinars, cash down.' 'Queen of the beauties,' said Nur al-Din, 'can something that cost twenty dirhams be turned in a single night's work into what is worth twenty dinars?' 'You don't know how valuable this is, master,' said the girl, 'but take it to market and give it to the auctioneer. Then, when he calls for bids, you will see what price it fetches.'

At that, Nur al-Din took the sash from her and gave it to the auctioneer in the Persian market, telling him to put it up for sale. He then sat down on a shop bench and after a while the man came back and told him to come and collect the purchase price of the sash, which had fetched a total of twenty dinars in cash. Nur al-Din was astonished and delighted when he heard this and he got up to take the money, scarcely believing that it could be true. As soon as he had taken it, he went off and spent it on more silks of various colours, out of which the girl could make more sashes, and when he got back home and handed over the silks, he told her to get to work. He also asked her to teach him so that he could work with her, saying: 'Never in all my life have I come across a better craft than this or any that produces a better return. By God, it is a thousand times better than trade.' The girl laughed and told him to go to his friend, the apothecary, to borrow another thirty dirhams from him, adding that on the following day he could use the money from the sash to repay both this new loan and the fifty dirhams that he had borrowed earlier.

Nur al-Din went to his friend and said: 'Uncle, please lend me thirty dirhams and tomorrow, God willing, I shall repay you the eighty that I owe you in one lump sum.' At that, the man weighed out the money, and Nur al-Din took it with him to the market where, as before, he bought meat, bread, fruit both fresh and dried, as well as scented flowers, all of which he took back to the girl, whose name was Miriam, and who was known as the sash-maker. She took the meat and at once produced a splendid meal which she placed before her master. Then she set out the wine and they both began to drink, she filling his glass and he hers. When it had gone to their heads, being struck by his gracefulness and courtesy, she recited these lines:

I greet the slender youth with a glass of wine,
In which there is a lingering scent of musk.
'Is the wine pressed from your cheeks?' I asked, but he said: 'No;
When ever did pressed roses produce wine?'

The two of them continued to entertain each other, with Miriam
giving him more wine and then asking him to fill a cup of cheer for her.
When he laid his hand on her, she would coquettishly stop him, and as
the wine had made her seem even lovelier, he recited:

There is many a slender wine-loving girl who has told her lover
Who was afraid to bore her, as they sat in enjoyment:
'Unless you pass the wine and pour it out for me,
You will have to do without me tonight.' In fear, he filled the glass.

Things went on like this until Nur al-Din fell into a drunken sleep, at
which point Miriam got up at once and set to work as before on the
next sash. When it was finished, she packed it tidily in paper and then
undressed and lay down beside Nur al-Din.

Night 876

Morning now dawned and Shahrazad broke off from what she had been
allowed to say. Then, when it was the eight hundred and seventy-sixth
night, SHE CONTINUED:

I have heard, O fortunate king, that when Miriam had finished working on the sash, she packed it tidily in paper, then undressed and lay
down beside Nur al-Din. They slept together until morning and then,
when Nur al-Din had got up and performed the ritual ablution and
prayer, she gave him the sash and told him to take it to the market and
sell it for twenty dinars, the price fetched by the previous one. He went
off and did this, after which he visited the apothecary and paid back the
eighty dirhams, thanking him for his kindness and calling down blessings
on him. 'Have you sold the girl, my son?' the man asked, and Nur al-Din
exclaimed: 'Are you trying to bring down a curse on me? How can I sell
the soul from out of my body?' He told the whole story of what had
happened to him from beginning to end, to the great delight of the
apothecary, who said: 'By God, my son, you have made me very happy
and, God willing, you will always enjoy good fortune. This is what I

want for you because of my fondness for your father and my continued association with him.' Nur al-Din then left him and went straight back to the market, where, as usual, he bought meat, fruit, wine and everything else that he needed, before returning to Miriam.

For a whole year, the two of them spent their time in loving companionship, eating, drinking and enjoying themselves. Every night, Miriam would make a sash and in the morning Nur al-Din would sell it for twenty dinars. He would use this to buy what they needed and whatever was left over he would give her to keep for when it might be needed. At the end of the year, she told him that when he sold the next day's sash, he was to get her silks of six different colours, explaining: 'I have it in mind to make you a mantle to wear over your shoulders such as no merchant's son or prince has ever been lucky enough to own.' Nur al-Din went off to the market, sold the sash and did what Mirian had told him, after which for a whole week she sat working at the mantle every evening after she had finished the daily sash. This went on, one piece at a time, until she had completed it, and then she gave it to Nur al-Din, who put it over his shoulders and started to walk through the market. Merchants, ordinary folk and the leading citizens all stood in rows to admire not only his beauty but the exquisite workmanship of his mantle.

It happened that one night Nur al-Din woke from sleep to discover Miriam weeping bitterly and reciting these lines:

The time of parting from the beloved is near;
Alas, alas that we should have to part!
My heart is broken. Woe is me
For nights of joy that have now gone!
The envious man will look at us
With eyes of malice and will reach his goal.
Nothing can harm us more than envious eyes,
Eyes of the slanderers and the eyes of spies.

When Nur al-Din asked her why she was crying, she said: 'Because the pain of parting has struck home to my heart.' 'Queen of the beauties,' he replied, 'who is going to part us when I am now the dearest of all men to you and my love for you is unsurpassed?' 'I love you far more than you love me,' she told him, 'but sorrow follows whoever thinks of himself as Time's favourite. The poet has well expressed it in the lines:

You may think well of Time when Time is kind,
With no fear of the evils fate will bring.
Time keeps you safe, but only to deceive,
And then, in a clear sky, the clouds arise.
A myriad stars are found within the heavens,
But only sun and moon suffer eclipse.
How many shoots are there on earth, both green and dry,
But we throw stones only at trees with fruit.
Corpses float on the surface of the sea,
While in the lowest depths are found the pearls.'

Then she said: 'Nur al-Din, my master, if you want to make sure that
we do not part, then be on your guard against a Frank who has lost his
right eye and who limps on his left leg, an old man with a dusty complexion and a thick beard. It is he who will bring about our parting; I
noticed that he is here in Alexandria and I think that he has come to
hunt for me.' 'Queen of the beauties,' Nur al-Din replied, 'if I catch sight
of him, I shall make an example of him and kill him.' 'Don't do that,'
Miriam told him. 'Don't speak to him, trade with him or have any
dealings with him; do not sit with him, walk or talk with him, and never
say: "Peace be upon you." I pray to God to save us from his malice and
his guile.'

The next morning, Nur al-Din took the sash that Miriam had made
and went to the market, where he sat on the bench of a shop chatting
with the other young merchants. Feeling drowsy, he lay down on the
bench to sleep and just at that moment while he was asleep the Frank
whom Miriam had described passed through the market with seven of
his compatriots. He saw Nur al-Din dozing on the bench with his face
covered by his mantle, the fringe of which was in his hand. The Frank
sat down beside him and took this fringe, turning it over and over. He
had been doing this for some time when Nur al-Din, sensing his presence,
woke up to find the very man whom Miriam had described sitting by his
head. He gave a great cry and the Frank, alarmed by this, said: 'Why are
you shouting at me? Have I taken anything of yours?' 'By God, you
damned fellow,' said Nur al-Din, 'had you done that I would have taken
you to the *wali*.' The Frank then said: 'Muslim, I conjure you by the
tenets of your religion to tell me where you got this mantle from.' 'It is
my mother's work,' replied Nur al-Din . . .

Night 877

Morning now dawned and Shahrazad broke off from what she had been allowed to say. Then, when it was the eight hundred and seventy-seventh night, SHE CONTINUED:

I have heard, O fortunate king, that when the Frank asked who had made the mantle, Nur al-Din replied: 'It is my mother's work, she made it for me with her own hands.' 'Will you sell it to me for cash?' asked the Frank. 'God damn you,' said Nur al-Din, 'I shall not sell it to you or to anyone else. It is the only one that my mother made and she made it for me.' 'Sell it to me,' said the Frank, 'and I'll pay you five hundred dinars on the spot. She may have made this one for you, but she can make you another even finer.' 'I shall never sell it,' Nur al-Din insisted, 'as there is nothing in this city to match it.' 'Would you not sell it for six hundred dinars of pure gold?' asked the Frank and he went on offering more and more until he got to nine hundred dinars. 'God will provide for me without my having to sell it,' Nur al-Din told him, adding: 'This is something that I shall never do even for two thousand dinars or more.' The Frank went on trying to tempt him with money until, when he had offered a thousand dinars, a number of other merchants who were there said: 'We will sell it to you for him, so hand over the money.' Nur al-Din repeated that he would not sell, but one of the merchants said: 'My son, you must realize that this mantle is worth a hundred dinars at the most, and that is only if you can find a willing buyer. This Frank is paying a lump sum of a thousand dinars and so you will have made nine hundred more than its true value. What more profit do you want than this? My advice is that you should sell it and take the money. You can tell whoever made it to make you another, and perhaps an even better one, and meanwhile you will have made a thousand dinars from this damned Frank, the enemy of our religion.' Feeling embarrassed by the merchants, Nur al-Din concluded the sale and the Frank paid over the money there and then.

He was about to go off back to Miriam to tell her about the Frank, when the Frank told the others to stop him, explaining: 'You and he are my guests tonight, for I have a cask of old Rumi wine, a fat sheep, fruits, fresh and dried, as well as scented flowers. All of you, without exception, must keep me company this evening.' The others then said to Nur al-Din: 'We would like to have you with us on an evening like this so that we can talk together. It would be very kind of you if you were to go with us as guests of this generous Frank.' Swearing to divorce their wives if he refused, and forcibly keeping him from going home, they got up at once and closed their shops. Then, taking Nur al-Din with them, they went off with the Frank to a pleasant and spacious room with two raised platforms. Here they were given seats by their host, who set before them a tray of curious and remarkable workmanship, on which were portrayed heart-breakers and the broken-hearted, lovers and their beloveds, those who asked favours and those who were asked for them. On this were placed precious bowls of china and crystal, all filled with expensive fruits, fresh and dried, as well as with flowers. He brought out his cask of old Rumi wine, and when the fat sheep had been slaughtered on his orders, he started to roast meat and to feed his guests. He poured them wine and made signs to get the merchants to see that Nur al-Din drank. They, for their part, plied him with drink until he was helplessly drunk.

When the Frank saw the state that he was in, he said: 'You have done me a favour by coming here this evening and you are very welcome indeed.' He then came up and sat down beside him, taking the opportunity to have some private words with him before going on to ask him: 'Would you sell me the slave girl whom you bought a year ago for a thousand dinars in the presence of these merchants and for whom I'm prepared now to pay five thousand dinars, giving you a profit of four thousand?' Nur al-Din refused, but the Frank continued to tempt him, pouring wine for him and offering him more and more money as an inducement until when he had got to ten thousand dinars, Nur al-Din, in his drunkenness, said in front of the others: 'I sell her to you; produce the money.' The delighted Frank called the merchants to witness the deal and they passed the rest of the night eating, drinking and enjoying themselves until morning.

The Frank then called on his servants to fetch the money, and when they had brought it he counted out ten thousand dinars in cash for Nur al-Din, saying: 'Take this as the price of your slave girl whom you sold to me last night in the presence of these Muslim merchants.' 'Damn you,' said Nur al-Din, 'I sold you nothing; you are lying to me and I have no slave girls.' 'You did sell her to me and these merchants will bear witness that the sale was concluded,' the Frank said, and the merchants all agreed with that, confirming that they could testify that he had sold the girl for ten thousand dinars. 'Take the money and hand over the girl,' they told him, 'and may God give you a better one in exchange. Do you deny that you bought her for a thousand dinars and that for a year and a half you have enjoyed her beauty, having the pleasure of her companionship and her favours every day and night? Then, after all that, she has brought you a profit of nine thousand dinars over and above her original price, in addition to the daily sash that she made, which you have been selling for twenty dinars. After all this, are you going to deny having agreed to the sale and turn down the profit as being too little, though what more could you want or hope to gain? You may have been in love with her, but after all this time you must have had enough of her. Take the money and buy a more beautiful slave, or else we can marry you to one of our daughters for a dowry less than half this price. Not only will she be lovelier but you will have the remainder of the money as capital.' The merchants continued to cajole Nur al-Din and trick him into agreeing, until at last he took the money, and immediately the Frank summoned the *qadis* and the notaries, who drew up a document certifying the sale by Nur al-Din of the slave girl Miriam, the sash-maker.

So much for Nur al-Din, but as for Miriam, she had sat waiting for her master all day until evening and then from evening until midnight. When he did not come back, she began to weep bitterly in her distress and, hearing this, the old apothecary sent his wife to her. The woman, finding the girl in tears, asked her what was wrong. 'Mother,' said Miriam, 'I have been sitting waiting for my master, Nur al-Din, but up till now he hasn't come. I'm afraid that someone may have found some way of tricking him into selling me.'

Night 878

Morning now dawned and Shahrazad broke off from what she had been allowed to say. Then, when it was the eight hundred and seventy-eighth night, SHE CONTINUED:

I have heard, O fortunate king, that Miriam said to the apothecary's wife: 'I'm afraid that someone may have found some way of tricking my master into selling me.'

Her visitor reassured her: 'Lady Miriam, even if your master were offered enough gold to fill this room, he would not sell you, as I know how much he loves you. It may be that some people have come from his father in Cairo and he has prepared a banquet for them in the place where they are lodging, being ashamed to bring them here, as the place would be too small for them. They might also not have been of sufficiently high status to be invited home or he might not have wanted to let them know about you, and so he would have stayed with them all night. God willing, he will come back to you safe and sound. Don't trouble yourself with cares and anxieties, for this must be why he has not come to you tonight. I shall stay with you overnight and comfort you until he arrives.' She then tried to distract Miriam and console her by talking all through the night.

When morning came, Miriam saw her master, Nur al-Din, turning into the street, followed by the Frank, who was surrounded by a group of merchants. When she saw him she trembled and turned pale, shaking like a ship caught in a gale at sea. The apothecary's wife asked her why she had changed and become so pale and faded. 'By God, lady,' said Miriam, 'my heart tells me that I am to be parted from my lover and removed far from him.' Then, with deep sighs, she recited these lines:

> Do not let yourself be parted,
> For parting's taste is bitter.
> When the sun sets,
> It changes colour thanks to this same pain,
> While, at its rising,
> Joy of reunion causes it to gleam.

She then sobbed even more bitterly, convinced that parting was at hand. 'Didn't I tell you,' she said to the apothecary's wife, 'that my master, Nur al-Din, must have been tricked into selling me? I am quite sure that he must have sold me last night to this Frank, in spite of the fact that I warned him against the man. But precautions are of no use against fate, and it is clear that what I said is true.' While she and the apothecary's wife were talking, in came Nur al-Din, and, as Miriam could see, he had changed colour; he was trembling and his face showed traces of sorrow and regret. 'Nur al-Din, my master, it looks as though you have sold me,' said Miriam, at which he shed bitter tears, groaned and sighed deeply. Then he recited these lines:

> As this was fated, caution does no good;
> I was at fault, but fate makes no mistakes.
> When God wills it that something should occur,
> A man may be wise and may have ears and eyes,
> But he will find himself both deaf and blind,

While his intelligence is plucked out like a hair.
Then, after God's decree has been fulfilled,
His wits will be restored so that he may take note.
When it has happened, do not ask how it took place;
All happenings are brought about by fate.

Nur al-Din now excused himself to Miriam, saying: 'By God, lady, the pen of fate has written what God decrees. These people tricked me into selling you and I was taken in and agreed to the sale, committing a terrible wrong in respect of you. But it may be that God, Who has decreed our parting, will grant us reunion.' 'I warned you against this,' said Miriam, 'for this was what I suspected would happen.' Then she clasped him to her breast, kissed him between the eyes and recited these lines:

I swear by your love never to forget my love for you,
Even if I lose my life through passionate longing.
Each day and night I weep and wail
As the ringdove laments on a tree in the sandy waste.
Parting from you has ruined my life, beloved;
For, now you have gone from me, we cannot meet.

While the two lovers were talking, the Frank came in and went up to kiss Miriam's hands. She promptly slapped him on the cheek with the palm of her hand, saying: 'Keep away from me, damn you. You went on following me until you managed to trick my master, but, God willing, all will turn out well.' The Frank, who had been taken aback by her reaction, laughed and excused himself to her, saying: 'Lady Miriam, what have I done wrong? It was your master, Nur al-Din, who was content to sell you of his own free will. I swear by the Messiah that, had he loved you, he would not have gone this far, and had he not had his fill of you, he would never have sold you. As a poet has said:

When I have had enough of someone, let him leave,
And if I speak his name again, I'm in the wrong.
The wide world is too small for me
That I should want one who does not want me.

This Miriam was the daughter of the king of Ifranja, a large city – the home of many marvels, crafts and different plants – resembling Constantinople. The story of how she came to leave her father's capital is a strange and remarkable one, which I shall now set out in its proper place in order to delight the audience.

Night 879

Morning now dawned and Shahrazad broke off from what she had been allowed to say. Then, when it was the eight hundred and seventy-ninth night, SHE CONTINUED:

I have heard, O fortunate king, that the reason why Miriam had left her father and mother was a remarkable one. She had been brought up in the height of luxury by her parents, and had studied rhetoric, the art of writing and arithmetic, as well as horsemanship and manly pursuits. She had become expert in all the crafts, including embroidery, sewing, weaving, the making of sashes, trimming, coating silver with gold and gold with silver, and she had mastered the arts of both men and women so as to become unique and unparalleled in her age and time. The Great and Glorious God had granted her such beauty, grace, elegance and perfection that she outshone all her contemporaries. The kings of the islands had asked her father for her hand, but he rejected them all, for such was his love for her that he could not bear to be parted from her for a single hour. He had many sons but, being his only daughter, she was dearer to him than they were.

One year it happened that Miriam fell so gravely ill that she was on the point of death, and she vowed to herself that, if she recovered, she would make a pilgrimage to a certain island monastery, which was held in the greatest respect by the Christians and to which they made votive offerings when in search of a blessing. Later, when she did recover, she wanted to fulfil her vow, and her father, the king, sent her to the monastery in a small ship, with an escort of girls, daughters of the leading citizens, as well as of a number of knights. Near the monastery they were attacked by a Muslim ship, manned by fighters in the holy war, who captured everyone on board, together with its cargo of wealth and treasures. The Muslims sold their spoils in Qairawan and it was there that Miriam fell into the hands of a Persian merchant. This man was impotent and had no dealings with women, never exposing himself to any one of them. He kept Miriam as a servant. He then fell so seriously ill that he almost died. During his long illness, which lasted for months, Miriam did all that she could to look after him until he was cured, in accordance with the will of God. Mindful of the tenderness and compassion with which she had treated him and of the service that she had done him, he wanted to return the favour to her, and as a result he promised to grant her a wish. 'Master,' she told him, 'what I want is that you should only sell me to a man of my own choice whom I could love.' He delighted her by agreeing to this and promising to leave the matter of her sale in her own hands. He had offered her conversion to Islam and, when she accepted it, he taught her the rites of worship. During that period, while he instructed her in her new religion, together with its duties, she learned the Quran by heart as well as acquainting herself with a number of the fields of jurisprudence and the traditions of the Prophet. When he then took her to Alexandria, he allowed her to choose a buyer for herself, putting her in charge of her own sale, as I have said, and I have reported how it was that she came into the hands of Nur al-Din.

This is the story of how Miriam came to leave her own country, but as for her father, the king of Ifranja, he had suffered a great shock when he heard what had happened to her and her companions. Ships and men were sent out to track her down, but although they searched throughout the Muslim islands, they found no trace of her and were forced to return with sad and gloomy tidings. The king in his distress now sent his grand vizier to search for her, this being the man who had lost his right eye and was lame in his left leg, a stubborn and tyrannical master of guile and deception. His orders were to scour all the lands of the Muslims in the hunt for Miriam and to buy her back, even at the cost of a whole shipload of gold. His search led him through islands and cities but it was not until he came to Alexandria that he received any news of her, for there, in answer to his questions, he was told that the girl was with 'Ali Nur al-Din, the Cairene. This was followed by his encounter with Nur al-Din, whom he tricked into selling her, as I have described, the clue coming from the fine workmanship of the mantle, which no one but Miriam could have produced. He had then given instructions to the merchants, making an agreement with them to recover her by guile.

Now that she was with him, she spent her time weeping and wailing. He said to her: 'Lady Miriam, stop weeping so sadly and come back with me to your father's city, the seat of your royal splendour, your own native land, where you will be among your own servants and attendants. You must leave behind the degradation of your exile. As for me, I have had enough of wearisome travelling, not to mention the expense, for I have been journeying and spending money for almost a year and a half now, as your father told me to buy you even if it cost a shipload of gold.' He then began to kiss her feet, prostrating himself before her, but the more he went on kissing her hands and feet as a gesture of courtesy, the angrier she became with him. 'Damn you,' she said. 'God Almighty will not allow you to get what you want.'

The servants now brought her a mule with a decorated saddle, on which they mounted her, raising over her head a silken canopy supported on shafts of gold and silver. Surrounded by an escort of Franks, she was taken to the Sea Gate of Alexandria, where they placed her in a small skiff and rowed her out to a large ship. When she was on board, on the orders of the one-eyed vizier, the crew hauled up the mast, hoisted the sails and raised their colours, spreading awnings of cotton and linen and manning their oars. As the ship sailed off, Miriam stood looking back at the city until it had sunk from sight, at which, standing alone, she shed bitter tears . . .

Night 880

Morning now dawned and Shahrazad broke off from what she had been allowed to say. Then when it was the eight hundred and eightieth night, SHE CONTINUED:

I have heard, O fortunate king, that when the king of Ifranja's vizier sailed off with Miriam, she stood looking towards the city until it was out of sight and then she wept, sobbing and reciting these lines:

Home of the beloved, can there be a return?
I have no knowledge of what God will do.
Swift ships have parted us, and my eyes
Are wounded and now dimmed by tears.
I have lost a dear one, the goal of my desire,
Who cured my sickness and removed my pain.
My God, act as his helper in my place,
For nothing left in trust with You is lost.

Whenever she thought of him, Miriam would weep and wail, and although the Frankish knights tried to soothe her, she refused their comfort, being wholly under the influence of her passionate love. Shedding more tears, she lamented and recited these lines:

The tongue of love speaks to you from my heart,
And tells you that I am in love with you.
My heart melts on the burning coals of love,
And flutters, wounded, now that you have gone.
I cannot hide that which has wasted me,
As tears, wounding my eyelids, flow in floods.

Miriam stayed in this state for the whole of the journey, unable to rest and with no powers of endurance.

So much for her and the lame, one-eyed vizier. As for Nur al-Din, when Miriam had sailed off, he too suffered the same symptoms, as the world closed in on him. He went to the apartment where he had lived with Miriam and to him it seemed dark and gloomy; he looked at the equipment she had used to produce her sashes and at the clothes she had worn. Clasping these to his breast, with tears streaming from his eyes, he recited these lines:

We have parted. Shall we ever be rejoined?
I turn to and fro, overwhelmed by my grief.
What is past and gone will never come any more;
Shall I ever again enjoy union with my love?
Do you suppose that God will reunite us,
And will she bear in mind the covenants of love?
Will she preserve what I have squandered in my foolishness,
Keeping the ties that used to bind us two?
I am a dead man now that she has gone,
But will it be that she approves my death?
What use is it to me to show my sorrow?
For as this grows, I waste away for love.
Heart, feel more passion, and eyes, shed more tears
Until within you nothing more is left.
My love is far away, my patience gone;
I have few helpers and my pains increase.
I pray to the Almighty that He may
Bring back to me that union I once had.

Then, shedding still more tears, he looked at every corner of the room and recited:

I see her traces and melt with desire,
Pouring my tears out where she used to live.

I pray God, Who decreed that we should part,
That one day He may bring her back again.

He then immediately got up, locked the house door and ran down to the shore, where he gazed at the place where Miriam's ship had lain at anchor. Tearfully and with deep sighs, he recited:

I greet you, without whom I cannot live;
Though you are far away, you are still near my heart.
I yearn for you with every passing hour,
As thirsty men long to relieve their thirst.
You hold my sight, my hearing and my heart;
Sweeter than honey is your memory.
Sorrow consumed me when you sailed away
Upon that ship which parted you from me.

He wept, wailed, groaned and lamented, calling out: 'Miriam, Miriam, was it in a dream or a confused fancy that I saw you?' Becoming ever more distressed, he recited:

We parted. Shall I see you once again,
And hear you call from somewhere near at hand?
Once we were happy here. Shall we again
Live here, each having what the heart desires?
Take my bones with you as you go,
And where you halt, give them a burial.
Had I two hearts, I could use one to live
And leave the other filled with love for you.
If I were asked what do I wish from God,
I would reply: 'His favour, and then yours.'

While Nur al-Din was in this state, weeping and calling on Miriam's name, an old man landed from a boat and came towards him. He found him tearfully reciting these lines:

Miriam the beautiful, come back. My eyes
Are rain clouds, shedding floods of tears.
Those who blame me can tell you that my lids
Are drowning in the whiteness of my eyes.

'My son,' said the old man, 'it seems to me that these tears must be for the girl who sailed off yesterday with the Frank.' On hearing this,

Nur al-Din collapsed and remained unconscious for a time. Then, when he had recovered, he wept bitterly and recited:

> After this parting can I hope for union?
> Will pleasure in our perfect love return?
> The fire of love burns fiercely in my heart
> And I am troubled by what slanderers say.
> I spend the day astonished and confused;
> At night I hope her phantom may return.
> By God, I never shall forget my love;
> How could I, though the slanderers weary me?
> She is a delicate and slender girl,
> Whose eyes shoot arrows through my heart.
> Her form is like a branch of the *ban* tree;
> Her beauty puts the shining sun to shame.
> Were it not for my fear of God on high,
> I would exalt her beauty as divine.

The old man looked at Nur al-Din and, noting how handsome he was, together with his well-shaped figure, his eloquence and his attractive manners, he felt sorry for him and was moved by pity for his condition. He himself turned out to be the captain of a ship that was on its way to Miriam's city with a hundred Muslim merchants on board. 'Be patient,' he told Nur al-Din, 'and there will be a happy outcome, as, God willing, I shall see that you get to the girl.'

Night 881

Morning now dawned and Shahrazad broke off from what she had been allowed to say. Then, when it was the eight hundred and eighty-first night, SHE CONTINUED:

I have heard, O fortunate king, that the captain promised Nur al-Din that, God willing, he would bring him to the girl. Nur al-Din asked when he was proposing to sail, and the captain said: 'In three days' time, and all will be well.' Nur al-Din was overjoyed to hear this and thanked the man for his kindness. Then, remembering the days when he and Miriam, unmatched in beauty, were together, he recited these lines:

Will God, the Merciful, unite us once again?
Companions, will I reach my goal or not?
Will changing time bring you back here once more,
So that my eyes may hold you as a miser's hoard?
Were it for sale, I'd buy union with you
With my own life, but it is dearer still.

He went at once to the market where he bought all the provisions and
the equipment that he needed for the voyage, after which he returned to
the captain. 'What have you got with you?' the captain asked, and Nur
al-Din explained that he had brought food and whatever else he might
need. The captain laughed and said: 'My son, do you think you are off
on a sight-seeing trip to Pompey's Pillar? If we have fair winds and calm
conditions, it will take two months to get you to your destination.' He
then took some money from Nur al-Din and went himself to the market
and bought as much as would be needed on the journey, and he also
filled a cask with fresh water.

For three days, Nur al-Din stayed on the ship until the merchants had
completed all their preparations and come on board. The captain then
set sail, but after they had been at sea for fifty-one days, they were
intercepted by pirates, who plundered the ship and took captive all on
board. These, including Nur al-Din, were taken to the city of Ifranja and
shown to the king, who ordered them to be imprisoned. Just as they
were being taken off to prison, the ship carrying Miriam and the one-eyed
vizier arrived at the city and the vizier went up to the king to give him
the good news of his daughter's safe return. Drums were beaten in
celebration and the city was adorned with splendid decorations, while
the king himself rode down to the shore with all his troops and state
officials to meet her.

When her ship had anchored, Miriam landed and was embraced by
her father. They exchanged greetings and a horse was brought forward,
which she mounted. On her arrival at the palace, she was greeted by her
mother, who embraced her and then asked her about herself, whether
she was still a virgin, as she had been when she left, or whether she was
now a woman without a husband. 'Mother,' said Miriam, 'how can
someone who has been sold from one merchant to another in the lands
of the Muslims and who has been in the power of others stay a virgin?
The merchant who bought me threatened to beat me and then raped and
deflowered me, after which he sold me to someone else, who, in turn,

sold me to another.' When her mother heard this, the light turned to darkness in her eyes and she repeated what she had been told to her husband, who found the enormity of it hard to bear. He consulted his officials and officers and they said: 'Your majesty, Muslims have defiled her and she can only be purified if you cut off the heads of a hundred of them.' At that, the king ordered the Muslim prisoners to be fetched from his dungeon and they were all brought to him, including Nur al-Din. He ordered their heads to be cut off, and the first to be executed was the ship's captain, after which the others were killed, one after the other, until the only person left was Nur al-Din. A strip was cut from the end of his robe and used as a blindfold, after which he was led to the execution mat. Then, just as the executioners were about to strike off his head, an old woman came up to the king and said: 'Your majesty, you vowed that if God restored your daughter Miriam to you, you would assign five Muslim prisoners to each church to help with its maintenance. Now that your daughter has come back to you, you must keep the promise that you made.' The king said to her: 'Mother, I swear by the Messiah and the true faith that the only prisoner I have left is this one whom they are about to kill. But you can take him with you to help you in the service of the church until I get more, after which I will send you another four. Had you got here before the others were killed, I would have given you as many as you wanted.'

The old woman thanked the king for his generosity and wished him long life and continuing glory and prosperity, after which she wasted no time in going to Nur al-Din and removing him from the execution mat. When she looked at him she found him to be a graceful and elegant young man with delicate skin and a face like the moon when it comes to the full on the fourteenth night. She took him off to the church and then told him: 'Take off these clothes of yours, my son, as they are only suitable for those who serve the king.' She fetched him a *jubba* and a hood of black wool as well as a broad belt, dressing him in the robe, placing the hood over his head and fastening the belt around his waist. She then told him to start to work in the church. He had been doing this for seven days when the old woman came back to him and said: 'Muslim, get your silk robes and put them on again. Then take these ten dirhams and go off immediately. You are to have a holiday today and you mustn't stay here an hour longer lest this bring about your death.' Nur al-Din asked her what was happening and she told him: 'You must know, my son, that the king's daughter, Miriam the sash-maker, wants to come to the church to acquire a blessing by visiting it and to make an offering to it in thankfulness for her safe return from the lands of the Muslims, fulfilling what she vowed to do if the Messiah rescued her. She is bringing with her four hundred girls, each of unblemished beauty, among them being the vizier's daughter and the daughters of the emirs and the state officials. They will be here within the hour, and were they to catch sight of you here they would cut you to pieces with their swords.'

Nur al-Din put on his own clothes and, taking the ten dirhams, he went off to the market and started to look around the streets in order to familiarize himself with the various districts of the city as well as with its gates.

Night 882

Morning now dawned and Shahrazad broke off from what she had been allowed to say. Then, when it was the eight hundred and eighty-second night, SHE CONTINUED:

I have heard, O fortunate king, that Nur al-Din put on his own clothes, took the ten dirhams and went out into the market. He was away for some time, familiarizing himself with the various districts of the city. Having done this, he went back to the church, where he saw the arrival of Princess Miriam, with her four hundred swelling-breasted virgins, beautiful as moons, among them being the vizier's daughter and the daughters of the emirs and state officials. She was walking between them like a moon among stars, and when Nur al-Din set eyes on her he could not stop himself from uttering a heartfelt cry: 'Miriam, Miriam!' As soon as the girls heard this shout, they rushed at him, drawing swords that gleamed like lightning bolts. They were about to kill him on the spot when Miriam turned to look at him. She had no difficulty in recognizing him, telling the others: 'Leave this young man alone. He must be mad for you can see the signs of madness clearly enough on his face.' Nur al-Din heard what she said and bared his head; with staring eyes he dangled his hands, twisted his feet and frothed at the corner of his mouth. 'Didn't I tell you he was mad?' Miriam said, adding: 'Bring him to me but keep away from him yourselves, so that I can hear what he says, as I know Arabic, and I might then be able to investigate his case and see whether his madness can be cured or not.'

The girls brought Nur al-Din up to Miriam and then stood aside.
Miriam said to him: 'Have you come here for my sake, risking your life
and pretending to be mad?' 'My lady,' Nur al-Din replied, 'haven't you
heard the words of the poet:

> They said: "Your love has made you mad," but I replied:
> "Life's pleasure is restricted to the mad."
> Compare my madness with its source, and if
> You find her worth my madness, I should not be blamed.'

'By God, Nur al-Din,' she told him, 'you brought this on yourself. I
warned you before it happened, but you paid no attention and followed
your own fancy. What I told you was not the result of some inspiration
nor did it come from a study of physiognomy or from a dream, but it
was thanks to what I had seen with my own eyes. For I had caught sight
of the one-eyed vizier and I knew that he would not have come to
Alexandria unless he was looking for me.' Nur al-Din answered: 'God
is the only refuge from mistakes made by the intelligent.' Then, carried
away by passion, he recited these lines:

> Pardon the fault that I made when I slipped;
> The slave is treated generously by his masters.
> It is punishment enough for the wrongdoer
> That he repents when his repentance does no good.
> For me confession is my penalty;
> Where is the generous mercy that I need?

It would take too long to describe how Nur al-Din and Miriam continued to reproach each other, each telling the other what had happened
and reciting poetry, with seas of tears flooding down their cheeks as they
complained of the violence of their love and the pain it had inflicted on
them. This went on until they could say no more; the day was over and
darkness had fallen. All the while Miriam was wearing a green robe
embroidered with red gold and set with pearls and other gems, which
enhanced her beauty and gracefulness. How well the poet has described
this in his lines:

> She appeared like a moon wearing green robes,
> Buttons undone, with loosened locks of hair.
> 'What are you called?' I asked, and she replied:
> 'I am she who burns the hearts of lovers over coals;

I am the gleaming silver and the gold,
Which frees the captive from captivity.'
I said: 'Your harshness makes me pine away.'
She said: 'Do you complain to me, whose heart is stone?'
I said: 'Although it may be stone, yet God
Has caused pure water to flow out of stone.'

As night darkened, Miriam returned to the girls and asked them whether they had shut the church gate and when they said that they had, she went off with them to what was known as the Chapel of the Virgin Mary, the Mother of Light, as the Christians believe that the power of her inner spirit is to be found there. The girls circumambulated the whole church to acquire a blessing, and when they had finished their pious visit, Miriam turned to them and said: 'I want to go into the church alone to seek a blessing, since, thanks to my long absence in the lands of the Muslims, I have felt a longing for it. As for you, when you have finished your visit, sleep where you like.' 'Very well,' they replied, 'and you also do as you like.' So they left her in the church and went off to sleep. When they were out of sight, Miriam went to look for Nur al-Din and found him sitting waiting for her, as though he was on hot coals. At her approach, he got up to kiss her hands, and she then sat down and made him sit beside her. Next she stripped off her ornaments, her robes and the fine linen she was wearing before clasping him to her breast and setting him on her lap. The two of them went on kissing, embracing and dancing to the tunes of love, exclaiming: 'How short is the night of union and how long the day of separation!' They then recited the lines of the poet:

Time's virgin, night of union,
The glory of the splendid nights,
You brought me morning in the afternoon;
Are you the kohl around the eyes of dawn
Or sleep that visits eyes that are inflamed?
But as for you, our parting's endless night,
Your end is always joined to its beginning,
A vicious circle with no end;
Before it goes, the Resurrection comes;
For resurrected lovers are still dead,
Killed through rejection by the ones they loved.

While they were sharing the delights of that splendid night, up above
on the roof of the church one of the servants of the saint tolled the bell
as a signal for the dawn service. This was as the poet has described:

> I saw him as he tolled the bell, and said:
> 'Who taught a fawn how to do such a thing?'
> And then I asked my soul: 'Which pains you more,
> The bell's stroke or the stroke of parting? Judge.'

Night 883

Morning now dawned and Shahrazad broke off from what she had been
allowed to say. Then, when it was the eight hundred and eighty-third
night, SHE CONTINUED:

I have heard, O fortunate king, that Miriam and Nur al-Din continued
in pleasure and delight until the bell-ringer climbed to the roof of the
church and tolled the bell. At that, Miriam got up immediately and put
her clothes and her ornaments back on. Nur al-Din was grieved and
distressed and, shedding tears, he recited these lines:

> I kissed and kissed again the tender, rosy cheek,
> With love-bites that exceeded all the bounds,
> But when we were in bliss, while watchers slept,
> With eyes that fluttered and then closed,
> The bells were struck by those who served
> In place of a muezzin, summoning to prayer.
> She got up quickly to put on her clothes,
> For fear a shooting star might strike her down,*
> Saying: 'You are my wish and my desire,
> But now the pallor of the dawn has come.'
> I swore if ever I was in command
> And was a sultan with the power to rule,
> I would raze all the churches to the ground,
> Killing whatever priests there were on earth.

Miriam held him to her breast and, after kissing his cheek, she asked
him how many days he had been in the city. 'Seven,' he told her, and she
went on to ask: 'Have you gone through it and do you know its streets

---

* A reference to her supernatural beauty.

and passageways, as well as the position of the landward and seaward gates?' He told her that he did and she asked: 'Can you find your way to the chest where the votive offerings are stored in the church?' When he said that he could, she went on: 'As you know all that, at the end of the first third of the night to come go to the chest and, when you have picked out what you want from it, open the church door leading to the alley that goes down to the sea. There you will find a small boat manned by ten sailors. When he sees you, the captain will hold out his hand to you, and when you take it he will help you to board. Stay there with him until I come, but be very, very careful not to fall asleep earlier lest you regret it when regret will do no good.' She then said goodbye to Nur al-Din and, after leaving him, she roused her maids and the other girls from their sleep. They accompanied her as she went to knock on the church door, which was opened by the old woman, and when she went out she found her servants and her escort standing there. They brought her a dappled mule and after she had mounted it, they lowered a silken curtain over her. While the knights took the mule's reins and the girls followed behind, the men-at-arms surrounded her with drawn swords in their hands to escort her back to the palace.

So much for Miriam, but as for Nur-al-Din, he stayed concealed by the curtain behind which he and Miriam had sheltered until day broke. When the door of the church was opened, he mingled with the crowds coming in, and then went up to the old woman. 'Where did you sleep last night?' she asked him and he said: 'Somewhere inside the city, as you told me.' 'That was right, my son,' she said. 'For had you stayed here in the church, the princess would have had you put to a most shameful death.' Nur al-Din replied: 'Praise be to God, Who saved me from the evil of this night,' and he went about his work in the church until the day ended and it became dark. He then went to the offertory chest and removed such precious gems as were easy to carry, after which he waited until the end of the first third of the night before going to the door leading to the sea lane. He prayed to God for shelter as he opened it and he then walked out into the lane and on down to the sea.

Not far from the gate he found a ship anchored by the shore, whose captain, a well-favoured old man with a long beard, was standing amidships with the ten members of his crew there in front of him. Nur al-Din gave him his hand, as Miriam had instructed him, and the man pulled him from the shore into the waist of the ship. He then ordered the crew to raise the anchor and put out to sea before daybreak. 'How can we do that, captain?' one of the crew asked. 'The king told us that he wanted to sail out this coming day to reconnoitre, as he is afraid that Muslim pirates may try to capture Princess Miriam.' 'Damn you all!' the captain shouted at them. 'Do you dare disobey me and bandy words instead?' He drew his sword and thrust the point into the throat of the man who had spoken so that it came out gleaming on the other side of his neck. 'Why did you kill him? What wrong had he done?' asked another, and again the captain grasped his sword and cut off the man's head, after which he struck down the rest, one after the other, until he had killed all ten and thrown their bodies into the sea. Then he turned to Nur al-Din and terrified him by shouting loudly: 'Get down and remove the mooring post.' Nur al-Din, afraid that the captain might strike him with his sword, jumped ashore, removed the post and went back on board quicker than a flash of lightning.

The captain began to give him orders, telling him to follow such-and-such a course and to watch the stars, and Nur al-Din, alarmed and frightened, did all that he was told. When he had raised the sail, the ship headed off with the two of them through the tumultuous waves of the sea . . .

Night 884

Morning now dawned and Shahrazad broke off from what she had been allowed to say. Then, when it was the eight hundred and eighty-fourth night, SHE CONTINUED:

I have heard, O fortunate king, that when the captain had hoisted the sail, the ship, with Nur al-Din on board, headed off through the tumultuous waves with a fair wind. All the while, Nur al-Din kept his hand on the yard, plunged in thought and submerged in care, as he did not know what the future held for him. Every time he looked at the captain he felt a tremor of fear, as he had no notion of where the man was making for, and he remained careworn and anxious until day broke. Then, as he watched, he saw the captain take hold of his long beard and give it a tug. Off it came, and when Nur al-Din looked more closely he could see that it was a false beard that had been stuck on, and on studying the captain's face, he realized that here was his beloved, the darling of his heart, Princess Miriam. She had pulled off this trick by killing the captain and then removing the skin from which his beard grew and attaching it

to her own face. Nur al-Din was astounded by this courageous action
and by her strength of heart.

Unable to contain his joy, he exclaimed: 'Welcome, you who are the
object of my desire, my wish and the goal of my quest.' Longing and
delight stirred in him as he felt certain that his hopes had been fulfilled,
and he chanted tunefully the following lines:

Say to those who know nothing of my love
For a beloved whom they cannot reach,
'Question my clan about my love,
For sweet and tender are my songs of love
For one who lives within my heart.'

When I remember her, all sickness leaves
My heart and all pains cease.
My longing and the passion of my love
Increase when it brings sadness to my heart,
And all men talk of this.

I shall accept no blame on her account,
Nor shall I ever seek forgetfulness.
Love has afflicted me with such distress
That lighted coals are burning in my heart
With fiery heat.

I wonder why my sickness was revealed,
As I remained through the dark night awake.
Why with her harshness did she seek my death,
Thinking it right to kill me for my love,
And, with injustice, being just?

Who was it, do you think, gave you advice
To treat so harshly someone who loved you?
By my own life and by God, Who made you,
I swear that what the critics say of you
Is nothing but a lie.

May God not cure the sickness of my love
Or rid my heart of this, its burning thirst.

I never shall complain I'm tired of love
Nor ever take another in your place.
Torture or join me, as you will.

You may still shun me and cause me distress,
But yet your love is fixed within my heart.
If you are angry or if you are pleased,
Do what you want with me, your slave,
Who offers you his life.

This poem filled Miriam with admiration and, after thanking him, she said: 'Someone who is in such a state must play a man's part and not act like a contemptible coward.' She herself was stout-hearted and a skilled sailor on the open sea, knowing all the wind shifts and what course to steer. 'Had you kept up your deception much longer,' Nur al-Din told her, 'I would have died of fear and terror, especially when added to this was the fire of my love and longing and the painful torment of parting.' Miriam laughed and got up immediately to bring out some food and drink, after which they ate and drank with pleasure and delight. She then produced sapphires and other gems, precious stones, valuable treasures and various gold and silver objects. All these, which she showed to the delighted Nur al-Din, were things which in addition to their value were easy to carry, and which she had removed from her father's palace and his treasuries.

With a moderate wind the ship kept its course and the two lovers sailed on until they came in sight of Alexandria and could make out its distinguishing features, both old and new, including Pompey's Pillar. When they reached harbour, Nur al-Din quickly jumped out of the ship and fastened the mooring rope to one of the fullers' stones. He took some of the treasures that Miriam had brought with her and told her to wait in the ship until he could bring her into the city in suitable style. She agreed, but added: 'Be quick, for delay brings regret.' He assured her: 'There will be no delay,' and so she sat there in the ship while he went off to the house of the apothecary, his father's friend, to borrow from his wife a veil, an outer garment, boots and a shawl of the kind worn by Alexandrian ladies. What he had not calculated on, however, in his ignorance were the shifts of that master of marvels, Time.

So much for Nur al-Din and Miriam, but as for Miriam's father, the king of Ifranja, he had looked for his daughter in the morning and, having failed to find her, he asked her maids and her servants where she was. They told him: 'She went out to the church last night, but we know nothing about what happened to her after that.' While he was talking to them, suddenly the whole place resounded to the sound of two loud cries coming from below the palace. When he asked what was wrong, he was told that ten dead men had been discovered on the seashore; his ship had gone; the door leading from the church to the sea lane had been found open and the prisoner employed in the church had disappeared. 'If my ship has gone from its berth, then my daughter must certainly be on board,' said the king.

Night 885

Morning now dawned and Shahrazad broke off from what she had been allowed to say. Then, when it was the eight hundred and eighty-fifth night, SHE CONTINUED:

I have heard, O fortunate king, that when the king was told that his daughter Miriam had gone and that his ship had been lost, he exclaimed: 'If the ship has gone, then there is no doubt that Miriam must be on board.'

He lost no time in calling for the harbour master, to whom he said: 'I swear by the Messiah and the true religion, if you do not immediately take an armed crew and overtake my ship, bringing it back with whoever is on board, I shall make an example of you by putting you to the foulest of deaths.' He followed this by shouting at the man, who left his presence trembling and went to the old woman who looked after the church. 'Did the prisoner who was here with you ever say anything about where he had come from?' 'He used to say that he was from Alexandria,' she told him, and at that the man went straight to the harbour and shouted to his crew to get ready to make sail.

They followed his orders and sailed night and day until, just as Nur al-Din disembarked, leaving Miriam in the ship, they came in sight of Alexandria. The Franks, among whom was the one-eyed, lame vizier who had bought Miriam from Nur al-Din, saw the ship moored there and recognized it. They tied up their own at a short distance from it and approached it in one of their small boats with a draught of two cubits, carrying a hundred armed men, among whom was the vizier, an obdurate tyrant, a rebellious devil and a wily thief, whom no one could outwit, like Abu Muhammad al-Battal. They rowed up to the ship to launch a concerted attack, but the only person they found on board was Princess Miriam. Having seized both her and her ship, they landed, but after a long wait they went back on board, having got what they wanted without fighting or even having to draw their swords. They then turned back, making for the lands of Rum, and after a trouble-free voyage with a fair wind they reached the city of Ifranja and brought Princess Miriam to her father.

The king was sitting on his throne and when he saw her he exclaimed: 'Traitress, how did you come to abandon the religion of your ancestors and the protection of the Messiah, on whom we rely, to follow Islam, the faith of the wanderers, which has raised the sword against the Cross and the idols?' 'This was no fault of mine,' replied Miriam, 'for I went out at night to look for a blessing by visiting the shrine of the Virgin Mary, and Muslim pirates took me unawares. They gagged and bound me before putting me in a ship and sailing off with me to their own country. I did my best to deceive them and I talked with them about their religion until they released me from my bonds, and I could scarcely believe it when your men found me and set me free. I swear by the Messiah, by the true religion, by the Cross and the Crucified, that I was overjoyed to be saved from them and that my rescue from captivity brought me relief and happiness.' 'You shameless whore,' her father replied, 'by what is established in the Gospel sent by Him Who reveals what is forbidden and what is permitted, I am going to have to make a most terrible example of you and put you to the ugliest of deaths. Wasn't the fact that you managed to deceive me on the first occasion enough for you that you should come and lie to me again?'

He gave orders that she was to be crucified on the palace gate, but at that moment the one-eyed vizier, who had long been in love with her, intervened and said: 'Don't kill her, your majesty, but marry her to me and, eager though I am to have her, I shall not lie with her until I have built her a palace of solid stone too high for any Muslim thief to climb. Then, when it is finished, I shall cut the throats of thirty Muslims at its gate as a sacrifice to the Messiah from me and from her.' The king granted him his wish and the priests, monks and knights were given permission to celebrate the wedding. After this had been done, with the king's consent workmen started to construct a lofty palace suitable for the princess.

So much for Princess Miriam, her father and the one-eyed vizier, but as for Nur al-Din and the old apothecary, after Nur al-Din had gone to his house and borrowed the shawl, veil, boots and outer clothes in the Alexandrian style from his wife, he went back to look for Miriam's ship only to find the mooring empty and the bird flown.

Night 886

Morning now dawned and Shahrazad broke off from what she had been allowed to say. Then, when it was the eight hundred and eighty-sixth night, SHE CONTINUED:

I have heard, O fortunate king, that when Nur al-Din discovered the place empty and realized that Miriam must be far away, with a sorrowful heart and shedding floods of tears, he recited:

Su'da's phantom came by night, alarming me
At early dawn in the desert where my companions slept,
But when its visit roused me as I lay,
There was only emptiness to see, with loved ones far away.

As he walked along the seashore, looking right and left, he saw a crowd of people who were saying: 'O Muslims, is Alexandria no longer inviolable that Franks can come in, snatch up citizens and go back to their own lands without any trouble, and without being pursued by any Muslim or *ghazi*?' 'What has happened?' asked Nur al-Din and they told him how a Frankish ship with an armed crew had just attacked the harbour and seized a ship that was anchored there together with those who were in it, before sailing back safe and sound to their own lands. On hearing this, Nur al-Din collapsed unconscious. When he had recovered they asked him what was wrong and he told them his story from beginning to end. This led them all to abuse him and to say: 'Why did you have to find a shawl and a veil before getting the girl to disembark?' Every one of them said something hurtful, although there were some who objected: 'Let him be; what has happened is punishment enough.' He was attacked on all sides with a stream of reproach and blame until he lost consciousness again.

While this was going on, the old apothecary arrived and, on seeing the crowd, he went up to them to find out what was happening. He discovered Nur al-Din lying there unconscious and sat down by his head to try to revive him. When Nur al-Din had recovered, he asked: 'What is the matter with you, my son?' and Nur al-Din replied: 'Uncle, after suffering many hardships I fetched back my lost slave girl in a ship from her father's city. When I got here, I moored the ship with a line to the shore and left the girl in it, while I myself went to your house to get from your wife the things that she would need before I could bring her into the city. Then the Franks came and seized the ship together with the girl before going back unmolested to their own ships.'

When the old apothecary heard this, the light turned to darkness in his eyes and he was filled with sorrow for Nur al-Din. 'My son,' he said, 'why didn't you take her from the ship to the city without a shawl? But it is no use saying anything now. Come with me to the city, and it may be that God will provide you with a more beautiful slave girl to console you for her absence. Praise be to Him Who inflicts no loss without allowing us to profit from it; union and separation are in His hands, Almighty is He.' 'Uncle,' Nur al-Din replied, 'I can never forget Miriam nor shall I abandon the search for her, even if this brings about my death.' The old man asked him what he intended to do and Nur al-Din told him: 'I shall go back to the lands of Rum and enter the city of Ifranja at the risk of my life, whether for good or for ill.' The old man said: 'It is a current proverb, my son, that "the pitcher can go too often to the well". They may not have harmed you the first time, but now they may kill you, especially as they will be able to recognize you easily enough.' 'Uncle,' Nur al-Din replied, 'let me meet a quick death, thanks to my love, rather than abandon her and be killed by hopeless endurance.'

As chance would have it, a ship was lying in the harbour ready to sail. Its passengers had finished their business and the mooring stakes had been removed just as Nur al-Din came on board. The passengers then enjoyed a pleasant voyage with fair winds for a number of days when they came across a Frankish ship that was cruising at sea and capturing every vessel that came in sight, because the Franks were still afraid lest Muslim pirates seize the princess. Whenever they took a ship they would bring all those who had been on board to the king of Ifranja and he would have them killed because of the vow that he had made in respect of his daughter Miriam. Nur al-Din's ship was sighted by these Franks, who took it, capturing everyone on board. They were brought before the king and when he discovered that there were a hundred of them, all Muslims, he ordered them, Nur al-Din included, to be put to death on the spot. The executioner despatched the others, leaving Nur al-Din to the last out of pity for his youth and his graceful form. When the king saw him he recognized him and said: 'Are you not Nur al-Din who was here with us once before?' 'I have never been here and my name is Ibrahim not Nur al-Din,' the young man said, but the king told him: 'That is a lie. You are the Nur al-Din whom I gave to the old woman in charge of the church to help her look after it.' Nur al-Din continued to insist that he was Ibrahim and the king said: 'When the old woman comes and looks at you, she will be able to tell whether you are Nur al-Din or not.'

At that moment, the one-eyed vizier who had married Miriam came in and kissed the ground before the king. 'I have to tell you, your majesty,' he said, 'that the palace has now been finished. As you know, I made a vow to the Messiah to mark its completion by sacrificing thirty Muslims at its door and I am here to ask you for them so that I may fulfil my vow. I guarantee to treat this as a loan and I shall give you as many again when other Muslims fall into my hands.' 'I swear by the Messiah and the true religion,' the king answered, pointing at Nur al-Din, 'that this is the only one I have left. Take him and cut his throat straight away and then, when I get more, I shall send you the others that you need.' At that, the one-eyed vizier got up and took Nur al-Din off with him to his palace in order to slaughter him on the threshold, but when he got there the painters said to him: 'Master, we still have two days' worth of work to do. Wait for us and put off killing this prisoner until we have finished painting and then maybe you will get the other twenty-nine and will be able to fulfil your vow in a single day by butchering them all.'

The vizier ordered Nur al-Din to be locked up . . .

Night 887

Morning now dawned and Shahrazad broke off from what she had been allowed to say. Then, when it was the eight hundred and eighty-seventh night, SHE CONTINUED:

I have heard, O fortunate king, that the vizier ordered Nur al-Din to be locked up and he was taken in fetters to the stables, hungry, thirsty and feeling sorry for himself, with death staring him in the face.

As had been fated, the king had two stallions, full brothers, one called Sabiq and the other Lahiq, either of which emperors would long to possess. One was pure grey, while the other was dark as night. All the island kings had promised anyone who could steal one of these for them as much red gold, pearls and other gems as he might want, but no one had yet succeeded. One of the horses had been suffering from an eye disease which had turned the whites of his eyes yellow, and although the king had summoned all his experts, none of them could cure it. Miriam's husband, the one-eyed vizier, came in and found the king distressed by this, and, wishing to relieve his mind, he said that if the horse were handed over to him, he would arrange for its cure. The king gave it to him and he transferred it to the stable where Nur al-Din was being held, but on being parted from its brother it neighed and whinnied so loudly that it alarmed everyone who heard it. The vizier, realizing what had caused this, went to tell the king. 'If this animal cannot bear being parted, how can rational creatures endure it?' the king wondered, and he instructed his servants to take the other horse to join its brother in the vizier's stable. He told them to tell the vizier that the horses were a present to him in order to gratify Princess Miriam.

Nur al-Din, fettered and shackled, was lying in the stable when he noticed a film over the eyes of one of the stallions. As he knew something about horses and how to treat their ailments, he said to himself: 'By God, this is my chance. I shall go and tell the vizier that I can cure the stallion. This will be a lie for I shall produce something that will destroy the eye, as a result of which he will kill me and I shall be freed from this degraded life.' He waited for the vizier to come to the stable to inspect the stallions and when he did, Nur al-Din asked him: 'Master, what reward would you give me if I were to cure this stallion and give him something to heal his eye?' 'By my head,' swore the vizier, 'I swear that if you do this, I shall spare your life and grant you any wish.' Nur al-Din asked to be freed from his chains, and when this had been done on the vizier's orders, he took a piece of unshaped glass, ground it up, and then added unslaked lime, which he mixed with onion juice. He applied the mixture to the horse's eyes and covered it with a bandage, telling himself again that the horse's eyes would be destroyed and that he himself would be killed and released from his ignominy.

That night he went to sleep with a carefree heart. He had prayed to Almighty God, saying: 'Lord, You know what I need and so I do not have to ask.' The next morning, when the sun rose over the hills and valleys the vizier arrived at the stables and removed the bandage from the stallion's eyes. After he had inspected them, he discovered them to be in the best possible condition, thanks to the power of God, the Opener of ways. 'Muslim,' he said to Nur al-Din, 'I have never seen anyone in the world as knowledgeable as you and I swear by the Messiah and the true faith that you have astonished me, as all the experts in the country were unable to cure this horse.' He went up to Nur al-Din, released him from his bonds with his own hands and, after giving him a magnificent robe, he appointed him master of the horse, with a salary and allowances, together with lodgings over the stables.

Nur al-Din sat there for some days, eating, drinking and enjoying himself while giving instructions to the grooms as to what to do and what not to do. If any of them went off without having fed the horses that were tethered in the stables and for which he was responsible, he would throw them down and give them a painful beating before having their legs put in iron fetters. The vizier was delighted with him, and, in his delight, he had no inkling of what was going to happen. Every day Nur al-Din would go down to visit the two stallions and stroke them with his hand, as he knew how much the vizier prized them and how fond he was of them.

This vizier had a very beautiful virgin daughter, like a roving gazelle or a swaying branch. In the new palace that he had built for Princess Miriam there was a window that overlooked the vizier's house and Nur al-Din's quarters. It happened that one day this girl was sitting by this window when she heard him singing. He was consoling himself for his misfortunes by reciting these lines:

> You blame me, you who enjoy happiness
> And are proud of its pleasures,
> But if Time brought affliction to you,
> So that you tasted bitterness, then you would say:
> *'Alas for the distress of love*
> *Whose fire consumes my heart.'*

> You may be safe now from love's treachery,
> And from the unjust way in which it ends,
> But do not blame those whom it leads astray,
> Until they too exclaim in passion:
> *'Alas for the distress of love,*
> *Whose fire consumes my heart.'*

> Forgive the lovers in their sorry state
> And do not help whoever censures them.
> Take care lest you be bound by the same cord

And taste the bitter torture that they feel.
Alas for the distress of love
Whose fire consumes my heart.

I was once counted as a pious man,
Passing my nights free from all cares,
In ignorance of love and sleeplessness,
But then love summoned me before his throne.
Alas for the distress of love
Whose fire consumes my heart.

No one can know how love humiliates,
Except the victims of its long disease,
Whom it has robbed of their intelligence,
Making them taste the bitterness of its draught.
Alas for the distress of love
Whose fire consumes my heart.

How many a lover lies awake at night,
With sweet sleep banished from his eyes,
As streams of tears pour down,
Flooding his cheeks because of what he feels.
Alas for the distress of love
Whose fire consumes my heart.

How many desperate lovers are there here
Whose passion robs them of all sleep!
They pine away through the disease of love,
Until they cannot even dream.
Alas for the distress of love
Whose fire consumes my heart.

How slender is my patience; how wasted are my bones;
My tears flow downwards like red dye.
How thin and bitter is the taste
Of what I had been used to find so sweet.
Alas for the distress of love
Whose fire consumes my heart.

Sad is the fate of those who love like me,
Lying awake throughout the dark of night,
Drowning in seas of cruel love,
Complaining of its ardour and its sighs.
Alas for the distress of love
Whose fire consumes my heart.

Who has not felt the sufferings of love,
Or saved himself from even its simplest wiles?
Is there a life that can be free of it,
And where is one whom it has left at rest?
Alas for the distress of love
Whose fire consumes my heart.

Guide, Lord, all those whom love afflicts;
Support them, You who are the best support.
Allow them to display their steadfastness
And grant them grace in all they must endure.
Alas for the distress of love
Whose fire consumes my heart.

When Nur al-Din had come to the end of this poem, the vizier's daughter said to herself: 'By the Messiah and the true faith, this Muslim is a handsome young man and there can be no doubt that he is a lover who has been parted from his love. I wonder if she is as fair as he is and whether she feels the same pangs of love. If she matches him, then it is right for him to shed tears and to complain of his passion, but if not, then he has wasted his life in vain regrets and is denying himself the taste of pleasures.'

Night 888

Morning now dawned and Shahrazad broke off from what she had been allowed to say. Then, when it was the eight hundred and eighty-eighth night, SHE CONTINUED:

I have heard, O fortunate king, that the vizier's daughter said to herself: 'If his beloved is beautiful, then it is right for him to shed tears, but if not, then he has wasted his life in vain regrets.'

Princess Miriam the sash-maker, the wife of the vizier, had been moved to the palace the day before, and the vizier's daughter, knowing that she was melancholy, decided to go to visit her and tell her about the young man and the poem that she had heard him recite. She was still thinking about doing this when Miriam herself sent a message asking her to come and entertain her with conversation. She went and found Miriam in a gloomy state, weeping, with floods of tears flowing down her cheeks. In an attempt to control them, she was reciting these lines:

My life has passed, but passion's life remains;
The violence of my longing grants me no relief.
The pain of parting makes my heart dissolve,
Yet I still hope the old days may return
To reunite us as we were before.
Spare me this blame, for I have lost my heart;
Longing and grief have made me waste away.
Shoot no rebuking arrows at this love;
Nothing exceeds a lover's misery,
But still love's bitterness is sweet to taste.

Her visitor asked why she was so gloomy and distracted and, on hearing this, Miriam recalled the delights that she had once enjoyed, reciting these lines:

I make myself accustomed to endure his loss,
Shedding my tears in endless streams.
Perhaps the Lord will send me some relief,
Who folds up joy within adversity.

'Don't be so unhappy,' the vizier's daughter told her, 'but come with me now to the window, for in our stables there is a well-spoken young man, handsome and elegant, who seems to be a lover parted from his beloved.' 'What makes you think that?' asked Miriam, to which the girl replied: 'Queen, I recognize it because of the poetry that he keeps on reciting at night and throughout the day.' 'If what she says is true,' Miriam said to herself, 'this could be a description of poor, unhappy Nur al-Din. Do you suppose that he can be the man she is talking about?' Overwhelmed by a surge of passion, she got up straight away and went with the girl to look out of the window. There she saw her beloved master, Nur al-Din, although she had to look at him closely to make sure of this, for love, together with the fires of passion, the pain of separation and despairing longing, had made him haggard and emaciated. He then started to recite these lines:

> My heart is enslaved and my eyes shed tears
> In floods that no rain cloud can match.
> I weep; I cannot sleep; passion torments me;
> I lament; I languish for the one I love;
> This love consumes me with regret and anguish.
> These are the eight afflictions I endure,
> And they are followed by another ten –
> So stop and listen as I count them out:
> Memory, cares, sighing, a wasted frame,
> Excess of longing and a mind obsessed,
> Tormented, exiled, sick with love.
> Regret is mixed with joy when I see her.
> I am too weak to bear the force of love,
> And when endurance leaves me, pain arrives.
> The pangs of passion wrack my heart still more,
> If you should ask what is it that I feel.
> My tears spread fires within my inner heart,
> A heart whose fire will never cease to burn.
> I drown within the flood of these same tears,
> While passion's flames consume me in love's hell.

When Miriam saw the speaker and heard these eloquent and remarkable lines, she was sure that this must be Nur al-Din, but she kept it hidden from the vizier's daughter, to whom she said: 'By the Messiah and the true faith, I did not think that you knew about my sadness.' She then got up immediately and left the window, returning to her own quarters, while the vizier's daughter went about her own affairs. Miriam waited for a while and then came back and sat down by the window, from which she set about watching Nur al-Din and admiring his delicate grace. He was like the full moon on the fourteenth night, but he was constantly sighing and shedding tears as he remembered past times, and he recited these lines:

> I hoped for a union with my love that I shall never win;
> In place of this, the bitterness of life is mine.
> My tears are like the flooding sea;
> I try to check them when I meet my censurers.

Woe to the one who prayed that we should part;
If I could get his tongue, I'd cut it out.
The days cannot be blamed for what they did,
Although they mixed a bitter draught for me.
Whom should I go to meet apart from you,
After I left my heart within your courts?
Who takes my part against an unjust judge?
I go to her for judgement and she wrongs me more.
I gave her my soul to keep as her own,
Only to have her squander it and me.
I spent my life on love for her; I wish
I might be granted union in return.
O fawn, who dwells within my inner heart,
The estrangement I have tasted is enough.
Within your face the forms of beauty meet,
Robbing me of all power to endure.
Affliction settled with her in my heart;
I am content with what has settled there.
My tears flow downwards like a flooding sea,
But I can find no other path to tread.
I am afraid that I may die of grief
And never reach the goal for which I strive.

Inspired by these lines spoken by the poor parted lover, and with tears in her eyes, Miriam herself recited the following lines:

I wished for my beloved, but when we met,
I was confused and could not speak or look at him.
I had got ready volumes of reproach,
But at our meeting I could find no words.

On hearing this, Nur al-Din recognized Miriam. He then burst into tears and exclaimed: 'By God, that is Miriam's voice! There is no doubt about it and no need to guess.'

Night 889

Morning now dawned and Shahrazad broke off from what she had been allowed to say. Then, when it was the eight hundred and eighty-ninth night, SHE CONTINUED:

I have heard, O fortunate king, that when Nur al-Din heard her reciting, he said to himself: 'This is Miriam's voice. There is no doubt about it and no need to guess.' But then he added: 'I wonder whether I am right and whether it really is Miriam or someone else.' In an access of sorrow, he moaned and recited these lines:

> He who blames me for my love saw me
> With my beloved in an open place.
> When we met there I spoke no word of blame,
> Though blame can often serve as sorrow's cure.
> He asked me: 'Why is it you do not speak,
> Giving her an answer that might hit the mark?'
> So I replied to him: 'Suspicious man,
> How little do you know of lovers' ways!
> The mark of a true lover is that he
> Stays silent when he meets the one he loves.'

When Nur al-Din had finished, Miriam called for an inkstand and paper and wrote a letter, starting with the Name of God and continuing: 'May God grant you His peace, His mercy and His blessings. This is to let you know that Miriam, your slave girl, gives you her greetings. She is filled with longing for you and is sending you this message. As soon as it reaches you, get up at once and take the greatest care to follow her instructions, making sure that you don't disobey them or fall asleep. Wait for the end of the first third of the night, one of the most auspicious of moments, and the only thing you have to think of then is how to saddle the two stallions and bring them out of the city. If anyone asks you where you are going, say that you are going to exercise them. No one will then stop you, for the citizens rely on the fact that the gates are locked.'

She wrapped the note in a silk kerchief and threw it down to Nur al-Din from the window. He took it, read it and grasped its contents, recognizing Miriam's handwriting, after which he kissed it and placed it between his eyes. Then, remembering the sweetness of their union, he shed tears and recited these lines:

Your letter reached me in the dark of night,
Stirring up longings which have wasted me away,
Bringing back memories of when we were as one.
But God sent parting to afflict me – praised be He!

When it grew dark, he busied himself with getting the stallions ready. He waited until the first third of the night had passed and went immediately to saddle the horses with the finest saddles before leading them out through the stable door, which he locked behind him. He then took them to the city gate, where he sat waiting for Miriam.

So much for him, but as for Miriam herself, she went at once to the chamber that had been prepared for her in the palace, and there she found the one-eyed vizier sitting propped up against a cushion stuffed with ostrich down. He was too ashamed to stretch out his hand towards her or even to address her, and when she saw him she uttered a silent prayer to God that he might not have his way with her and defile her purity. She approached him with pretended affection and, sitting down beside him, spoke gently to him, saying: 'My master, why do you turn from me? Is this pride on your part or coquetry? It is commonly held that, when it comes to greeting, the one who is seated should begin by greeting the one who stands. If you will not come and speak to me, I shall go and speak to you.' 'Yours is the grace and favour, empress of the whole earth,' replied the vizier, 'and I am merely one of the least of your servants. Such is your rank that I would be ashamed to take the liberty of addressing you, you who are the incomparable pearl. I lay my face on the ground before you.' 'No more of such talk,' said Miriam, 'but fetch us food and drink.' At that, the vizier called out his orders to the slave girls and eunuchs, and they produced a table spread with all manner of creatures that walked, flew or swam in the sea, including sandgrouse, quails, young pigeons, lambs and fatted geese, together with roast chickens and many other types of dishes. Miriam reached out and not only took food for herself but used her fingers to feed morsels to the vizier, kissing him on the mouth.

The two of them ate their fill and when they had washed their hands the food was removed and replaced with wine. Miriam filled her cup and, after drinking, she poured wine for the vizier, attending to his needs so well that he became almost ecstatic with joy. As he relaxed and his wits succumbed to the influence of the wine, Miriam drew out from her pocket a pill of pure Maghribi *banj*, so strong that were an elephant to catch a fleeting whiff of it, it would sleep from one year to the next. She had kept it in readiness for this moment and she now crumbled it into the vizier's wine cup without him noticing it, before refilling the cup and handing it to him. He was delighted, scarcely believing that she had given it to him, and took it and drank, but no sooner had it settled in his stomach than he collapsed prostrate on the ground. Miriam got to her feet and fetched two large pairs of saddlebags, which she filled with such valuable items as were easy to carry, including gems, sapphires and precious stones. Dressed and armed as a warrior, she took with her some food and drink as well as splendid clothes, fit for a king, to delight Nur al-Din, and weapons. Hoisting the saddlebags over her shoulders, for she was both strong and brave, she left the palace to look for him. So much for her, but as for Nur al-Din . . .

Night 890

Morning now dawned and Shahrazad broke off from what she had been allowed to say. Then, when it was the eight hundred and ninetieth night, SHE CONTINUED:

I have heard, O fortunate king, that when Miriam, who was both strong and brave, left the palace she set off to meet Nur al-Din. So much for her, but as for Nur al-Din, the poor lover, as he sat by the city gate waiting for her, with the stallions' halters in his hand, God caused him to fall asleep – praise be to Him Who never sleeps. At that time, the island kings had been offering money to induce thieves to steal either both or one of the stallions. It was to a black horse thief who had been brought up in the islands that they had promised a huge sum of money for the theft of one of the stallions, while if he managed to steal them both, they had promised to give him a whole island as well as magnificent robes of honour. For a long time this slave had been scouting through the city in disguise, but while the stallions were in the king's stables he had not been able to take them. Then, to his delight, the king had given them to the one-eyed vizier, to whose stables they had been transferred. This excited the thief's hopes and he swore to himself by the Messiah and the true faith that he would steal them.

On the night in question, he had made for the stables to carry out his plan, but on his way there he happened to notice Nur al-Din lying asleep with the stallions' halters in his hand. He slipped the halters from their heads and was about to mount one and drive the other before him when Princess Miriam came up, carrying the saddlebags over her shoulder. Mistaking him for Nur al-Din, she gave him one of the pairs of saddlebags which he loaded on to one of the stallions, and when she passed him the other, he set it on the back of the second stallion. All the while he kept silent, so that she still thought that he was Nur al-Din. He continued to say nothing until, when they were outside the city, she asked: 'Nur al-Din, my master, why don't you say something?' The slave then turned towards her angrily, saying: 'What are you talking about, girl?' On hearing his barbarous accent, Miriam realized that this was not Nur al-Din's voice and when she looked up at him she saw that he had nostrils as big as jugs. The light turned to darkness in her eyes and she said: 'Who are you, *shaikh* of the children of Ham, and what is your name?' 'Base-born woman,' he replied, 'my name is Mas'ud and I steal horses while people lie asleep.' She said nothing, but drew her sword immediately and struck him a blow on the shoulder, cutting right through the tendons of his neck. He collapsed in a bloody heap on the ground and God hurried his soul to hellfire, an evil resting place.

Miriam now took the two stallions, riding one, with her hand on the other, as she retraced her steps in search of Nur al-Din. She found him lying asleep in the place where she had promised to meet him, with the halters still in his hand. He was snoring as he lay there, unable to distinguish his hands from his feet. Miriam dismounted and struck him with her hand, making him start up in fear. 'My lady,' he said, 'praise be to God that you have got here safely!' 'Get up,' she told him, 'and mount this stallion quietly,' after which they both rode out of the city. When they had been riding for some time, Miriam turned to Nur al-Din and said: 'Didn't I tell you not to fall asleep, for sleepers never prosper?' He replied: 'My lady, I only slept because I was so happy that you had promised to come to me, so tell me what happened.' She told him the story of the slave from start to finish, and Nur al-Din gave thanks to God for her safety.

The two of them pressed on with their journey, entrusting their affair to God the Gracious and Omniscient and talking together until they came to the corpse of the slave whom Miriam had killed and whom Nur al-Din could see lying on the ground like an *'ifrit*. Miriam told him to strip off the man's clothes and to take his weapons, but he objected: 'By God, I cannot dismount and stand over him or go anywhere near him.'

Astonished at the man's physique, he thanked Miriam for what she had done, admiring her courage and her strength of heart. They then went on their way and continued to ride hard for the rest of the night until dawn broke and sunlight spread over the hills and valleys. They had reached a broad green meadow in which gazelles were frisking and every part of it was filled with different kinds of fruit and speckled with flowers until it looked like the belly of a snake. The birds were busy there and streams followed their various courses. It was a scene such as the poet has well and accurately described:

> There is a valley, dark red in the burning heat,
> Where common plants grow up in double quantities.
> We stopped beneath its branching trees, which bent over us
> Like a nursing mother over a young child.
> To quench our thirst it gave cold water,
> Sweeter than wine for those who drink together,
> Warding off the sun wherever it sought us out,
> Refusing it entry, but allowing the soft breeze.
> Its pebbles are like virgins' ornaments,
> Whose touch is that of pearls upon a string.

Similarly, another poet has written:

> With birdsong and the sounding stream
> It stirs the lover's longing in the early dawn.
> Like Paradise, it holds within its flanks
> Shade, fruits and streams of running water.

Miriam and Nur al-Din halted there to rest . . .

Night 891

Morning now dawned and Shahrazad broke off from what she had been allowed to say. Then, when it was the eight hundred and ninety-first night, SHE CONTINUED:

I have heard, O fortunate king, that Miriam and Nur al-Din halted there, eating the fruits and drinking from the streams, while the stallions were turned loose to graze and to drink. The two lovers sat talking, and, as they told of their own experiences, they each complained to the other of the pain and hardship that separation had inflicted on them. But as they were sitting there a dust cloud arose, filling the horizon, and they heard the neighing of horses and the clink of arms.

The reason for this was that the king, having married his daughter to the vizier, had intended to pay them the customary visit on the morning after the consummation of their marriage. He had got up and taken with him silks, as well as gold and silver to scatter for the eunuchs and the bride's attendants to pick up. Accompanied by a number of servants he had walked on until, when he got to the new palace, he discovered the vizier stretched out senseless on the carpet. He then searched throughout the palace but was distressed to find no trace of his daughter. Preoccupied by this and unable to think clearly, he called for hot water, virgin vinegar and frankincense and when they had been fetched, he compounded them and poured the mixture into the nostrils of the vizier. When he then shook him, the drug came out from his stomach, looking like a piece of cheese. After the king had repeated the process, the man came to his senses and the king asked him about himself and about Miriam, his daughter. 'Great king,' said the vizier, 'the only thing that I remember is that she handed me a wine cup. It is only now that I have regained my senses and I know nothing about what has happened to her.'

When the king heard what the vizier had to say, the light turned to darkness in his eyes; he drew his sword and struck the vizier a blow on the head from which the blade emerged gleaming through his teeth. He then sent an immediate summons to the servants and grooms, and when they came he asked them for the two stallions. 'Your majesty,' they told him, 'they both went missing last night, as did our chief, and this morning we found all the doors open.' 'By my religion and my faith,' exclaimed the king, 'it can only have been my daughter who took them, together with the prisoner who was acting as a servant in the church. It was he who went off with her before, and although I recognized him well enough, it was this one-eyed vizier who saved him from me. He has now been repaid for what he did.' The king then immediately called for his three sons, brave heroes, each of whom could take the field alone against a thousand riders. He told them to mount and he himself rode out with them, accompanied by his leading knights, officers and officials. They had followed the tracks of the fugitives and now caught up with them in the valley.

When Miriam saw her pursuers, she got up, mounted her horse, girt on her sword and picked up her weapons. Then she asked Nur al-Din: 'How do you feel and have you the heart for a fight?' He replied: 'I can

no more stand firmly in combat than a peg can stay firmly fixed in bran.'
and he recited these lines:

Miriam, spare me painful rebuke,
And do not seek my death and lengthy suffering.
How can I ever be a warrior,
I whom am frightened by a croaking crow?
The mere sight of a mouse fills me with fear,
And terror makes me soil my clothes.
I only like to thrust in privacy,
When the vagina knows the penis' might.
This is the soundest counsel and all else,
Apart from that, must be considered wrong.

When Miriam heard this, she smiled laughingly and said: 'Stay where
you are, master, and I shall protect you from them, even if there are as
many of them as there are grains of sand.' She got ready at once, mounted
and then dropped her reins, while turning her spearhead against those
of her foes. Beneath her the stallion bounded off like a gust of wind or
like water spurting from a narrow pipe. She was the bravest warrior of
her age and unique in her time, for ever since she was a little girl her
father had taught her how to ride and to plunge into the waves of battle
even in the dark of night. To Nur al-Din she said: 'Mount and ride
behind me, and if we are defeated, take care not to fall, for nothing can
overtake that horse of yours.'

The king had recognized his daughter when he saw her and, turning
to his eldest son, he said: 'Bartaut Ra's al-Qillaut, there can be no doubt
at all that this is your sister, Miriam, who has come to attack us. Ride
out against her, but if you get the better of her, don't kill her until you
have given her the chance to turn back to Christianity. If she returns to
her former faith, then bring her here as a prisoner, but if she refuses, kill
her as brutally as you can. Make a most terrible example of her, and do
the same to that damned fellow who is with her,' 'To hear is to obey,'
said Bartaut, who immediately set out to charge his sister. She, for her
part, rode against him and when she had come close to him, he said:
'Miriam, is it not enough for you to have abandoned the faith of your
fathers and forefathers that you have taken up the religion of the
vagrants, that is, Islam? I swear by the Messiah and the true faith that if
you do not return to the religion of your royal ancestors and follow the
best of paths, I shall put you to the worst of deaths and make the

most terrible example of you.' On hearing her brother say this, Miriam laughed and said: 'It is not likely that what is past will return or that the dead come back to life. I shall give you the bitterest of drinks, for I swear by God that I shall never give up the religion of Muhammad, the son of 'Abd Allah, who has shown the way to all mankind. This is the true faith and I shall never abandon his guidance even if I have to drain the cup of death.'

Night 892

Morning now dawned and Shahrazad broke off from what she had been allowed to say. Then, when it was the eight hundred and ninety-second night, SHE CONTINUED:

I have heard, O fortunate king, that Miriam told her brother: 'Far be it from me to abandon the religion of Muhammad, son of 'Abd Allah, who has brought right guidance to all mankind through the true faith, even if I have to drain the cup of death.'

When the damned Bartaut heard what his sister had to say, the light turned to darkness in his eyes as he thought how monstrous this was. A furious fight broke out between them as, unflinching before its violence, they swept through the length and breadth of the valley, the dazzling focus of all eyes. For a long time they manoeuvred and struggled, but whatever trick of warfare Bartaut tried against his sister she countered, thanks to her skill, proficiency and expert horsemanship. So long did they fight that a dust cloud formed above their heads until they were lost to sight. Miriam exerted herself to block all Bartaut's attacks until he grew tired and demoralized. When he had weakened, she struck him on the shoulder with her sword; the gleaming blade severed the tendons of his neck and God hastened his soul to hellfire, an evil resting place.

Miriam then circled the battlefield, challenging her foes to come and fight and calling out: 'Will anyone meet me in battle? Let no one ride out today who is sluggish or weak; I only want heroes from among the enemies of religion so that I may pour them a drink of ignominious punishment. Idolaters, tyrannical unbelievers, this is a day of glory for the faithful and of ignominy for those who do not believe in the Merciful Lord.' When the king saw that his eldest son had been killed, he struck his face, tore his clothes and called to his middle son: 'Bartus Khara' al-Sus, ride out as fast as you can to meet your sister, Miriam. Avenge

Bartaut, your brother, and bring her to me as a miserable, wretched prisoner.' 'To hear is to obey, father,' Bartus replied, and he rode out to attack her. For her part, she charged against him and the two of them fought a duel that was even more furious than the one before. Bartus then, finding that he was unable to withstand her, tried to ride off, but she was too strong for him, as every time he tried to flee she would close with him and press him until eventually she struck him a blow on the neck, causing the sword to emerge gleaming from the upper part of his chest.

When she had sent him to join his brother, she again circled around the field, calling: 'Where are the bold riders? Where is the one-eyed, lame vizier, the follower of a crooked faith?' At that the king, wounded at heart and bleary-eyed with tears, called out to his youngest son: 'Fasyan Salh al-Sibyan, go out to fight your sister and avenge your brothers. Attack her, whether you win or lose, and if you get the better of her, put her to the vilest of deaths.' At that, Fasyan rode out against Miriam, who met him with her skill and expertise, allied, as it was, to courage, horsemanship and experience of war. 'Damn you,' she called out, 'enemy of God and of the Muslims that you are. I will send you to join your brothers, and miserable is the dwelling place of the unbelievers.' She drew her sword again and struck him a blow that cut through his neck and both arms. He was despatched to join his brothers and God hurried his soul to hellfire, an evil resting place.

When the knights and the riders who were with the king saw that his three sons, the bravest champions of their age, had been killed, they were filled with fear and awe of Miriam. They bent their heads towards the ground, convinced that they were about to face ignominy and total destruction. With their hearts consumed by the fires of rage, they turned tail and took refuge in flight. As for the king, having seen his sons killed and his troops routed, he was perplexed and dismayed. In his anguish he said to himself: 'Miriam has treated us with scorn, but if I risk my own life and go out to meet her in single combat, she may get the better of me and put me to the most disgraceful of deaths, making an evil example of me, in the same way that she killed her brothers. She has no longer anything to hope for among us nor do we want her to return, and so I think it best to preserve my dignity and to go back to my city.' So he dropped his reins and rode off.

When he was settled again in his palace, the loss of his three sons, the rout of his men and the stain on his honour caused anger to blaze up in his heart. Within half an hour, he had summoned his officials and the leaders of his state and complained to them of how Miriam had killed her brothers and of the grief and distress she had inflicted on him. He then asked them for their advice and they were all agreed that he should write to the Muslim caliph, Harun al-Rashid, letting him know what had happened. Accordingly, he sent a letter that, after the preliminary greeting, ran as follows: 'My daughter, Miriam the sash-maker, was seduced by a Muslim captive named 'Ali Nur al-Din, son of Taj al-Din, the Cairene merchant, who turned her against me. This Nur al-Din took her off by night, making for his own country. I ask it as a favour on the part of the Commander of the Faithful that he send orders to all the lands of the Muslims to have her arrested and returned to me with a reliable servant.'

Night 893

Morning now dawned and Shahrazad broke off from what she had been allowed to say. Then, when it was the eight hundred and ninety-third night, SHE CONTINUED:

I have heard, O fortunate king, that the king of Ifranja wrote to the Commander of the Faithful, Harun al-Rashid, the caliph, asking his help in the search for his daughter, Miriam, and asking as a favour that he write to all the lands of the Muslims to have her arrested and returned to him with a reliable servant. In return for his help in this matter, the king promised the caliph in his letter that he would make over to him half the city of Rome, in which he could build mosques for the Muslims and whose tax revenues would be paid over to him.

When this letter had been written on the advice of the state officials, the king folded it and summoned the man whom he had appointed to replace the one-eyed vizier. He gave instructions that it was to be sealed with the royal seal as well with the seals of the state officials, who were to add their signatures to it. He promised the new vizier that if he brought Miriam back, he would be given the estates of two emirs and a robe of honour with a double fringe. He then handed him the letter and told him to take it to Baghdad, the House of Peace, and hand it personally to the caliph.

The vizier set off and, after journeying across valleys and deserts, he reached his destination. Having paused for three days to rest, he asked directions to the caliph's palace and was shown where it was. When he got there he asked permission to enter the caliph's presence and, when this had been granted, he went in and kissed the ground before him. He then handed him the letter from the king of Ifranja, together with gifts and remarkable treasures such as befitted him, and when the caliph had read it and grasped its contents, he immediately ordered his viziers to send messages to every Muslim country. This they did, giving the names and descriptions of Miriam and Nur al-Din, whom they described as fugitives, adding that whoever found them was to arrest them and send them on to the caliph. They were also warned not to show any slackness, carelessness or neglect. The letters were sealed and sent off with couriers to the provincial governors, who quickly followed their instructions and began to organize searches in their lands for a couple who fitted the description they had been given.

So much for the rulers and their followers, but as for Nur al-Din and Miriam, when her father and his men had been routed, they rode off immediately. Protected by the Sheltering God, they reached Syria and arrived at Damascus, but the caliph's instructions had got there ahead of them and the governor knew that, if he found them, he was to arrest them in order to send them to his master. So when they went into the city, they were approached by his agents and asked for their names. They returned a truthful answer and gave a full account of everything that had happened to them, as a result of which the agents, realizing who they were, arrested them and took them to the governor. He in turn sent them on to the caliph in Baghdad, and when they got there their escort asked leave to bring them before the caliph. When this had been granted, they entered, kissed the ground before him and said: 'Commander of the Faithful, this is Miriam, the daughter of the king of Ifranja, while the other is Nur al-Din, the son of Taj al-Din, the Cairene merchant, the prisoner who seduced her and turned her against her father before stealing her away from her own country and escaping with her to Damascus. We discovered the two of them there as soon as they entered the city, and when we asked for their names they told us the truth. After that we brought them here before you.'

The caliph looked at Miriam and discovered her to be slender and eloquent, one of the most beautiful women of her time and unique in her age, combining sweetness of tongue with fortitude and a strong heart. She approached him and having kissed the ground before him, she prayed for the continuance of his glory and fortune and that he might avoid all afflictions and disasters. The caliph was struck not only by the loveliness of her figure but by her honeyed words, together with the quickness of her replies. 'Are you Miriam the sash-maker, daughter of the king of Ifranja?' he asked, and she replied: 'Yes, Commander of the Faithful, imam of the monotheists, protector of the faith and descendant of the Lord of the prophets.' The caliph turned to Nur al-Din and saw that he was a handsome and shapely young man like the moon on the night that it becomes full. 'Are you Nur al-Din, the captive, son of Taj al-Din, the Cairene merchant?' he asked. 'I am, Commander of the Faithful, the support of those who seek you,' Nur al-Din answered. 'How was it that you ran off with this girl from her father's kingdom?' the caliph went on, and Nur al-Din told him the whole story from beginning to end. When he had finished, the caliph was filled with the most pleasurable astonishment and he exclaimed: 'How many sufferings men must endure!'

Night 894

Morning now dawned and Shahrazad broke off from what she had been allowed to say. Then, when it was the eight hundred and ninety-fourth night, SHE CONTINUED:

I have heard, O fortunate king, that when Caliph Harun al-Rashid had asked Nur al-Din for his story and had been told everything that had happened to him from beginning to end, he exclaimed in astonishment: 'How much men have to endure!' Turning to Miriam, he said: 'Know that your father, the king of Ifranja, has written to me about you. What have you to say?' She replied: 'You are God's regent on earth, maintaining the traditions and precepts of His Prophet, may He perpetuate His favour towards you and preserve you from all afflictions and disasters. You are His regent and I have entered into your religion, because it is this that is sound and true, having abandoned the faith of the unbelievers who forge lies against the Messiah. I have come to believe in the Gracious God and in the message brought by his compassionate Prophet. I worship God, the Sublime and Almighty; I acknowledge His unity and I prostrate myself humbly before Him, proclaiming His glory. In your presence, I repeat: "I bear witness that there is no god but God and that Muhammad is the Apostle of God, sent to bring right guidance and the true faith, which is to be given victory over all other faiths, however unwilling the polytheists may be."* Do you find, Commander of the Faithful, that you can accept the message sent you by the king of the heretics and return me to the lands of the unbelievers who associate other gods with the Omniscient Lord, who magnify the Cross, worship idols and believe in the divinity of Jesus, who was a created being? If you do that to me, you who are God's caliph, I shall cling to the skirts of your robe on the Day of Judgement and complain of you to your ancestor, the Apostle of God – may God bless him and give him peace – "on a day when no help is to be found in wealth or children and only those who come to God with pure hearts can be saved".'† The caliph replied: 'God forbid that I should ever do that! How can I send back a Muslim woman who believes in the unity of God and in His Apostle against the commands of God and His Apostle?'

Miriam repeated the formula: 'I bear witness that there is no god but God and that Muhammad is the Apostle of God,' and the caliph said: 'May God bless you, Miriam, and guide you further towards Islam. Since you have become a Muslim, believing in the unity of God, I have an obligation towards you not to wrong you in any way, even if I were to be offered in exchange for you as many jewels and as much gold as would fill up the earth. You can be at ease, happy and content, with your mind at rest. Are you willing to take this young man, 'Ali the Cairene, as a husband, while you become his wife?' 'Certainly I am willing,' Miriam replied, 'for he used his wealth to buy me and he showed me the greatest kindness, crowning this by risking his life for me many times.'

The caliph then arranged for her marriage to Nur al-Din, providing her with a dowry and summoning the *qadi*, the notaries and the principal officers of state, so that her wedding day, when the marriage contract was drawn up, was a memorable one. The caliph then turned at once to the vizier of the king of Ifranja, who was present at the time, and said: 'Did you hear what she said? How can I return her to her father, who is an unbeliever while she is a Muslim, believing in the unity of God? He might mistreat her and be harsh with her, more particularly since she killed his sons, and I would have to bear the burden of this sin on the Day of Resurrection. God Almighty has said: "God will not allow the unbelievers any way to overcome those who believe."‡ Go back to your king and tell him to give up the matter and to abandon his hopes.' The vizier, who was a stupid man, replied: 'Commander of the Faithful, I

---

* Quran 9.33.   † Quran 26.88.   ‡ Quran 4.140.

swear by the Messiah and the true faith that I cannot return without Miriam, even if she is a Muslim, for if I do, he will kill me.' 'Take this damned man and kill him,' the caliph ordered and he recited:

This is the reward of disobedience
For those who disobey their betters.

He had given orders for the man's head to be cut off and his body burned, but Miriam said: 'Commander of the Faithful, do not stain your sword with the blood of this damned man.' She then unsheathed her own sword and struck the vizier's head from his body and his soul went to the fires of hell, an evil resting place, while the caliph was left to admire the strength of her arm and her strong-mindedness.

He presented Nur al-Din with a splendid robe of honour and gave him and Miriam their own apartments in the palace, providing them with salaries, allowances and rations. On his instructions, all that they needed in the way of clothes, furnishings and valuable utensils were brought to them. For a time, they stayed in Baghdad, enjoying the pleasantest and most luxurious of lives, until Nur al-Din felt a longing for his mother and father. He put the matter before the caliph and asked his permission to return to his own country in order to visit his relatives. The caliph summoned Miriam and when she appeared before him, he gave leave to Nur al-Din to set off, presenting him with gifts and objects of value, while enjoining the two of them to look after each other. He had messages sent to the emirs of Cairo, the guarded city, as well as to its men of learning and its dignitaries, instructing them to see to the interests of Nur al-Din, his parents and his wife, and to show them the greatest respect.

When news of this reached Cairo, Taj al-Din and his wife were overjoyed by the prospect of their son's return. The city dignitaries, the emirs and the state officials came out to meet him as the caliph had instructed, and this was a memorable day of great joy, marked by the meeting of lovers and the conclusion of their search. Each day, one of the emirs provided a banquet as they showed their pleasure at the arrival of Nur al-Din and Miriam, showering them with honours upon honours. When Nur al-Din was reunited with his parents, all three were delighted and care and sorrow left them, while Miriam was welcomed with joy and respect. Emirs and leading merchants alike gave them gifts and presents, as each day brought further pleasure and joy surpassing that of the great festival.*

* 'Id al-Adha.

So it was that they continued to live for a time in pleasure and happiness, surrounded by the delights of luxury, eating, drinking and enjoying themselves, until they were visited by the destroyer of delights, the parter of companions, the ravager of houses and palaces, the filler of graves. Death then removed them from this world and numbered them among its own. Praise be to the Living God in Whose hands are the keys to this world and the next.

There is also a story told by the emir Shuja al-Din, governor of Cairo, WHO SAID:

I passed a night in the house of a man in Upper Egypt, who entertained me generously. Although he himself was elderly and very dark in colour, he had three small children who were reddish white. When I asked him about this difference in colour, he told me: 'Their mother was a Frankish woman whom I took as a captive, and I had a strange experience with her.' 'Would you do me the pleasure of telling me about this?' I asked. HE AGREED AND SAID:

You must know that I once sowed a crop of flax here, and by the time I had harvested it and combed it out, I had spent five hundred dinars on it. When I tried to sell it, that was as much as I could get for it and people said: 'Take it to Acre, for there you may get a good profit.' So I went to Acre, which at that time was held by the Franks, and there I sold part of my crop with payment to be deferred for six months. While I was doing my selling, a Frankish lady passed by me, unveiled, as is their custom when they go to market. She came up to me in order to buy some flax and I was astounded by what I saw of her loveliness. I quoted a low price for what I sold her, and she accepted it and went off, only to come back again some days later, when I asked an even lower price. Realizing that I had fallen in love with her, she returned again and again. She was in the habit of walking with an old woman, and it was to this companion of hers that I told my love, asking whether she could think of some way for me to get what I wanted. The old woman promised to help, but added: 'This must be kept as a secret between the three of us, you, me and her, and furthermore it is going to cost money.' I told her: 'If it cost me my life, this would not be too high a price to pay for a meeting.'

Night 895

Morning now dawned and Shahrazad broke off from what she had been allowed to say. Then, when it was the eight hundred and ninety-fifth night, SHE CONTINUED:

I have heard, O fortunate king, that when the old woman agreed to help the man, she said: 'This secret must not go beyond the three of us, me, you and her, and it is going to cost money.' He replied: 'If it cost me my life, this would not be too high a price to pay for a meeting.' HE WENT ON:

It was agreed that the girl should come to me and I should hand over fifty dinars, which I did. When the old woman had taken them, she said: 'Get a place ready for her in your house, and she will come to you tonight,' and so I went out and got ready what I could by way of food, drink, candles and sweetmeats. My house overlooked the sea and, as it was summer, I spread bedding down on the roof. The girl came and after we had eaten and drunk, we lay down at night under the stars with the moon shining on us, as we looked at the reflection of the stars in the sea. It was then that I said to myself: 'Are you not ashamed before the Great and Glorious God that you, a stranger, in the open air beside the sea, should disobey God's commandment with a Christian woman, earning yourself the punishment of hellfire? My God, I call You to witness that I shall keep away from her tonight because of the shame that I fear before You and out of fear of Your punishment.' So I went to sleep until morning, and the girl got up angrily in the early dawn and returned home.

For my part, I went off to sit in my shop and she then passed me, accompanied by the old woman, who was herself furious. The girl was looking lovely as the moon so that I almost died of desire and said to myself: 'Who are you to give up a girl like this? Are you an ascetic like al-Sari al-Sakati, Bishr al-Hafi, Junaid al-Baghdadi or al-Fudail ibn Iyad?' So I caught up with the old woman and told her to bring the girl back to me. 'By the Messiah,' swore the old woman, 'she will only come back to you for a hundred dinars.' I promised her the money and when I had handed it over, the girl visited me again, but when she did, the same thoughts crossed my mind and I kept myself from touching her, abandoning her for the sake of Almighty God. When I walked off again to my shop, the old woman passed me in a state of fury and when I asked her again to fetch me the girl, she exclaimed: 'I swear by the Messiah that it will cost you five hundred dinars if you ever want to have the pleasure of her company in your house again; otherwise you can die of grief.' I shuddered at the thought and made up my mind to save myself from this fate by paying over all the money that I had got from my flax. Before I realized what was happening, however, I heard a herald proclaim: 'Muslims, the truce between us is at an end. Those of you who are here have a week to finish your business, after which you must leave for your own lands.' This was the end of my affair, and I set about collecting the cash for the flax that I had sold on deferred terms, and bartering what remained of it. I then left Acre, taking with me some excellent trade goods, but I was still deeply in love with the Frankish girl, who had taken both my heart and my money.

After leaving, I went to Damascus where I got the highest of prices for the goods that I had brought from Acre, as the flow of trade had been interrupted by the ending of the truce. As a result, through the gracious favour of Almighty God, I made an excellent profit. Then, in order to cure my longing for my Frankish girl, I started to trade in captive slave girls and carried on in this line of business. I was still engaged in it three years later when Saladin fought his well-known battles with the Franks. God gave him the victory over them so that, with His permission, he captured all their kings and conquered the lands of the coast. It then happened that a man came up to ask me to supply a slave girl for Saladin, and I showed him a beautiful girl whom I had with me, whom he bought from me for a hundred dinars. He paid over ninety in cash, but the remaining ten could not be found in the treasury that day, because Saladin had spent all his money on fighting the Franks. When he was told about that, Saladin said: 'Take the man to where the captives are being kept and let him choose one of the Frankish girls, whom he can have in exchange for the ten dinars that are owed him.'

Night 896

Morning now dawned and Shahrazad broke off from what she had been allowed to say. Then, when it was the eight hundred and ninety-sixth night, SHE CONTINUED:

I have heard, O fortunate king, that Saladin said that the man could pick one of the Frankish girls in exchange for the ten dinars that he was owed. HE WENT ON:

They took me to the captives and when I had inspected them all, I saw the girl with whom I had fallen in love. She had been the wife of a Frankish knight and I had no difficulty in recognizing her. 'Give me that one,' I said, and I took her off to my tent, where I asked whether she knew me. When she said no, I told her: 'I was the man who was trading in flax when I came across you. You took my money and told me that it would cost me five hundred dinars to see you again, but now I have got you as my own property for ten dinars.' She replied: 'This is thanks to the secret of your true faith, and I bear witness that there is no god but God and that Muhammad is the Apostle of God.' She became a genuine convert to Islam, and I promised myself that I would not sleep with her until I had freed her and had told the *qadi* about it. So I approached Ibn Shaddad, and when I had let him know what had happened, he drew up a marriage contract for me. It was after that that I slept with the girl and she became pregnant.

The army then moved off and a few days after we had got to Damascus a messenger came from Saladin, asking for all the prisoners and captives, because of an agreement that had been made between the kings. All the others, men and women alike, were handed back and the only captive left was the girl who was with me. When it was pointed out that the wife of a certain knight had not been produced, the authorities made enquiries about her and they carried on searching until they were told that she was with me. They told me to hand her over and I went to her with my colour changed through grief. When she asked me what was wrong and what had happened, I explained to her: 'A messenger has come from Saladin to collect all the captives, and I have been asked to return you.' 'Don't worry,' she said, 'but take me to Saladin, for I know what to say to him.' So I went with her into Saladin's presence and there, seated on his right, was the messenger of the Frankish king. I said: 'Here is the woman who is with me.' Saladin and the Frank asked her: 'Do you want to go to your own country and to your husband, as God has now freed you and the others from captivity?' She said to Saladin: 'I have become a Muslim and, as you can see from my belly, I am pregnant. I can be of no use to the Franks.' 'Which do you prefer,' asked the Frankish envoy, 'this Muslim or your husband, the knight?' She repeated what she had said to Saladin, and he asked the Franks who were there with him whether they had heard her reply. When they said yes, he told me to take my wife and go, but then he quickly sent after me to say: 'Her mother entrusted me with something to take to her, saying: "My daughter is a captive and has no clothes, so I want you to bring her this chest." Take it and hand it over to her.' I took it back home and gave it to her and when she opened it she found in it all her materials together with the two purses containing fifty and a hundred gold dinars respectively, which, I saw, were still tied up exactly as I had left them, and I gave thanks to Almighty God. These are my children by her and she herself is still alive and has prepared this meal for you.

'I was astonished by his story and by the fortune that he had enjoyed,' concluded Shuja al-Din. 'God knows better.'

A story is also told that in the old days there was a well-to-do Baghdadi who had inherited a large sum of money from his father. He was in love with a slave girl whom he had bought, and she, for her part, returned his love. He continued to spend money on her until all his wealth had gone and he was left with nothing. He tried to find ways of making a living, but failed. In the days of his prosperity, he had been in the habit of attending meetings of expert singers and had himself become very proficient in this art. One of his friends, from whom he had asked advice, told him: 'I don't know of anything better for you to do than to sing in partnership with your slave girl, for in this way you would get a great deal of money and be able to eat and drink.' Neither the man himself nor the girl approved of this, and she told him that she had another idea. When he asked what it was, she said: 'If you sell me, then both you and I can escape these hardships. I shall be well off, for only a wealthy man will be able to afford someone like me, and I shall then be able to find a way of coming back to you.'

The Baghdadi took her to the market, where the first to see her was a Hashimite from Basra, an educated, cultured and generous-hearted man, who bought her for fifteen hundred dinars. THE BAGHDADI SAID:

When I had taken the money, I was sorry for what I had done and both I and the girl burst into tears. I wanted to cancel the sale, but the Basran would not agree, and so I put away the money in a purse. My house was desolate without her; I didn't know where to go and I wept, slapped my face and sobbed in a way that I had never done before. Then I entered a mosque and there I sat in tears, so bewildered that I didn't know what I was doing, and eventually I fell asleep with the purse beneath my head like a pillow. While I was sleeping, a man pulled it away from under me and ran off. I woke in alarm and, finding my purse gone, I was going to run after the thief, only to fall flat on my face, as my feet had been tied with a rope. I began to weep and strike myself, exclaiming: 'My soulmate has gone, and I have lost my money!'

Night 897

Morning now dawned and Shahrazad broke off from what she had been allowed to say. Then, when it was the eight hundred and ninety-seventh night, SHE CONTINUED:

I have heard, O fortunate king, that when the young man lost the purse, he said: 'My soulmate has gone and I have lost my money.' HE WENT ON:

In my despair, I went to the Tigris and, after covering my face with my clothes, I threw myself in. Bystanders saw what was happening and said: 'Some great sorrow must have made him do this,' and they jumped in after me and brought me out. When they asked, I told them what had happened to me and they expressed regrets, but then one of them, an old man, came up to me and said: 'You may have lost your money, but why should you take your own life and consign yourself to hellfire? Come with me and show me to your house.' I did so, and when we got there, he sat with me for a while until I grew calmer, and he then left after I had thanked him. When he had gone, I was again on the point of killing myself, but I thought of the next world and of hellfire, and so I fled from my house to visit one of my friends. I told him what had happened to me and he shed tears of pity for me and gave me fifty dinars. 'If you take my advice,' he said, 'you will leave Baghdad straight away and use these dinars as spending money until you stop being obsessed by your love for your slave girl and can console yourself for her loss. Your family were secretaries; you write a good script and you are a very cultured man. So approach whatever governor you please and throw yourself on his compassion, so that God may perhaps reunite you with your love.'

When I heard what he had to say, my resolution was strengthened and some of my cares left me. I made up my mind to go to the territory of Wasit, where I had relatives, and so I went down to the river bank and found a ship moored there which its crew were loading with goods and precious materials. I asked them to allow me to come with them, but they told me that the ship belonged to a Hashimite and that they could not take me, dressed as I was. I tempted them with what I offered as passage money, and they said: 'If you must come, strip off those fine robes of yours, dress as a sailor and sit with us as though you were one of the crew.' So I went back and bought some sailor's clothes, and when I had put them on, I returned to the ship, which was about to leave for Basra. I settled in with the crew, and not long afterwards who was it I saw but my former slave girl, attended by two maids. The vexation that I had been feeling left me, and I told myself that I would be able to look at her and listen to her singing all the way to Basra. Very soon afterwards, the Hashimite came aboard with his party, and the ship set off downriver. Food was produced and he and the girl ate together, while all the others had their meal in the waist of the ship.

The Hashimite now said to the girl: 'How long are you going to go on refusing to sing and staying sad and tearful? You are not the first person to be parted from a lover.' This made me realize that she was suffering because of her love for me. The Hashimite then had a curtain lowered over the part of the ship where she was sitting, after which he called up the people who were near me and sat with them in front of it. I asked about them, and was told that these were members of his family. He then produced wine and dried fruits for them and they all kept on urging the girl to sing, until she called for her lute, tuned it and started to sing these lines:

The company has left at night with my beloved,
Pressing on through the darkness with her for whom I wish.
When the camels have gone, the lover is left
With burning embers lighted in his heart.

She was then overcome by tears and threw away the lute, interrupting her song, to the distress of the company. I myself collapsed in a faint, and as people thought that I must be possessed, one of them started to recite Quranic verses in my ear. The others continued to coax the girl to sing, until she tuned her lute and began:

I stood lamenting those who rode off on their laden camels;
Though they had left on a distant journey, they still stayed in my heart.
By the deserted campsite I stood asking about my friends,
But the place was empty and the dwellings desolate.

She then fainted away and people began to weep, while I myself again fell down unconscious, to the alarm of the sailors, leading one of the

Hashimite's servants to ask: 'How is it that you took this madman on board?' They told each other: 'We can land him at some village or other and get rid of him.' The thought of this greatly distressed and pained me, but I summoned up all my powers of endurance and told myself that the only way in which I could avoid this was by letting the girl know that I was on board, so that she could stop them from throwing me off.

We sailed on until, on coming near a village, the ship's master invited everyone to go ashore. They disembarked, this being in the evening, but I stayed on board, went behind the curtain and took the lute, altering the tuning of its strings until it was adapted to the mode that the girl had learned from me. Then I went back to my place in the ship . . .

Night 898

Morning now dawned and Shahrazad broke off from what she had been allowed to say. Then, when it was the eight hundred and ninety-eighth night, SHE CONTINUED:

I have heard, O fortunate king, that THE YOUNG MAN SAID:

I then went back to my place in the ship before the others returned from the shore. Both land and river were bathed in moonlight, and the Hashimite said to the girl: 'For God's sake, don't spoil our pleasure.' She picked up the lute, but on touching it she gave such a groan that they thought that she was dead. Then she exclaimed: 'By God, my teacher is here with us on this ship!' The Hashimite swore: 'Were he here, I would get him to join our company, as this would perhaps comfort you and allow us to enjoy your singing, but it is not likely that he can be on board.' She told him: 'I cannot play the lute or improvise on the airs when my teacher is here.' He said that he would question the crew. 'Go ahead,' she replied and so he asked them: 'Have you taken anyone on board with you?' They said no, and, as I was afraid that the questioning might end there, I laughed and said: 'Yes, I am her instructor and I taught her when I was her master.' 'By God, this is my master's voice!' she exclaimed, and the Hashimite's servants came and took me to him. He recognized me when he saw me and asked me what was wrong and what had reduced me to this state, so I told him all that had happened.

The tears that I shed were accompanied by loud sobbing from the girl behind the curtain, and the Hashimite, for his part, wept bitterly, together with his relatives, out of sympathy for me. 'By God,' he said,

'I have not approached her or slept with her and not until today have I heard her sing. I am a man to whom God has granted riches and I only came to Baghdad to listen to singing and to get the payments granted me by the Commander of the Faithful. When I had done both these things, I was intending to leave for home, but I told myself that I wanted to hear some more Baghdadi singing and so I bought the girl, not knowing that the two of you were in love. I call God to witness that when I get to Basra I shall free her, marry her to you and give you an allowance that will be more than enough for you. This is on condition that whenever I want to listen to singing, she will sing to me from behind a curtain that will be hung for her, and you will be included among my relations and companions.' I was delighted by this and the Hashimite then put his head through the curtain and asked the girl whether she would agree, at which she started to call down blessings on him and to thank him.

He then summoned a servant and said: 'Take this young man by the hand, remove the clothes that he is wearing and replace them with splendid robes, before perfuming him and bringing him back to me.' The servant carried out his master's orders, and he in his turn poured out wine for me as he had done for the girl. She then started to sing melodiously:

They reproached me for shedding tears
When the beloved came to say farewell.
They had never tasted the pain of parting
Or felt the fire of grief that burned between my ribs.
Only the wretched lover knows what love can be,
Who waits dejectedly among those lands.

They were all delighted and I took the lute from her, struck up the finest of airs and recited:

If you want favours, ask the generous,
The man brought up in wealth and luxury.
By doing so you will inherit splendour,
While from the mean you win a legacy of shame.
If you must be brought low, then face your fate
By bringing your petitions to great men.
To sing their praises does not humble you;
That only happens when you praise the mean.

The audience were delighted with me, and their delight continued and increased as the girl and I took it in turns to sing until we landed again.

The ship anchored and everyone, including me, went ashore. By that time I was drunk and, on sitting down to relieve myself, I fell asleep. All the others went back on board and the ship started off downriver, taking them to Basra, as they were too drunk to realize what had happened to me.

I had given all my money to the girl and had nothing left. It was only the heat of the sun that woke me and when I got up and looked around there was no one in sight. I had forgotten to ask the Hashimite his name, that of his family or the whereabouts of his house in Basra. I didn't know what to do, and the joy that I had felt on meeting the girl vanished like a dream. Things went on like this until a large boat passed by, on which I embarked and which took me down to Basra. I had no acquaintances there; I didn't know where the Hashimite lived and so I went to a greengrocer, from whom I borrowed ink and paper . . .

Night 899

Morning now dawned and Shahrazad broke off from what she had been allowed to say. Then, when it was the eight hundred and ninety-ninth night, SHE CONTINUED:

I have heard, O fortunate king, that the girl's original owner, the Baghdadi, entered Basra but did not know where the Hashimite's house was. HE SAID:

I went to a greengrocer, from whom I borrowed ink and paper before sitting down to write, impressing the greengrocer with the elegance of my script. He saw how dirty my clothes were and asked me about myself. When I told him that I was a poor stranger, he offered to lodge me with him, paying me half a dirham each day and providing me with food and clothing if I would keep the accounts of his shop. I agreed to this and I stayed with him, looking after his business and controlling both his takings and his expenses. After a month, he was grateful to find that the former were increasing and the latter diminishing, and he increased my daily rate of pay to a dirham. When a year had passed, he invited me to marry his daughter, offering to take me as a partner in his shop. I agreed to this, consummated the marriage and stayed in the shop, but for all that I was broken-hearted and could not hide my sorrow. The greengrocer used to invite me to drink with him, but I felt too gloomy to accept.

Things went on like that for two years. Then, while I was in the shop, a group of people passed by, carrying food and drink, and I asked the greengrocer what was happening. He said: 'This is a day on which pleasure-seekers, well-to-do young men, along with musicians and enter-tainers, go out to the river bank to picnic among the trees that line the Ubulla canal.' I felt inclined to go and see this, telling myself that among the people there I might meet my beloved. I told the greengrocer what I wanted, and he agreed to it and prepared food and drink for me, which I took off to the canal. There the people were dispersing and I had decided to go with them when, all of a sudden, I caught sight of the captain of the ship which had carried the Hashimite and the girl. He was coming along the canal, and when I called out to him, he and his companions recognized me and came up to me. 'What, are you still alive?' they exclaimed, embracing me and asking what had happened. When I told them my story, they said: 'We thought that you were so drunk that you must have drowned.' I asked them about the girl and they said: 'When she found out that you had been lost, she tore her clothes, set fire to her lute and started to sob and slap herself. Then, when we got to Basra with the Hashimite, we told her to stop weeping and grieving, but she said: "I shall wear black, build a tomb beside the house and stay there, never singing again." She was allowed to do this, and she is still in this state.'

They took me with them and when we got to the house, there she was by the tomb. She caught sight of me and gave so deep a groan that I thought that she had died. I folded her in a long embrace and the Hashimite told me to take her. I agreed, but said: 'Free her first, as you promised, and then marry her to me.' He did this, and he provided her with valuable possessions, a large quantity of clothes, furnishings and five hundred dinars in cash. 'This,' he told me, 'is what I propose to give the two of you as a monthly allowance, on condition that you become one of my drinking companions and that the girl sings.' He gave us a house of our own, and on his instructions everything that we could need was brought there, so that when I went to it I found it luxuriously furnished. After I had moved the girl into it, I went back to the green-grocer and told him everything that had happened to me. I asked him to allow me to divorce his daughter for no fault of her own, and I handed back her dowry, as well as everything else I owed her. After staying with the Hashimite for two years, I had become a rich man and the lifestyle that I had enjoyed with the girl in Baghdad was restored to me. God in His generosity had rescued us and had showered us with His favours, enabling us, after what we had endured, to reach our goal. Praise be to Him in the beginning and the end, for God it is Who knows better.

A story is also told that in the old days there was a great king in India, a tall, handsome man with a fine figure and a noble character, who was generous to the poor and who treated with affection his subjects and all those who lived in his realm. His name was Jali'ad, and he ruled over seventy-two subject kings; in his dominions there were three hundred and fifty *qadis*; he had seventy viziers; and every ten of his soldiers was under the command of an officer. His chief vizier was a man named Shimas, aged twenty-two, who was well formed and good-natured, pleasantly spoken, an intelligent conversationalist, skilled in affairs, and who in spite of his youth was a wise and a prudent administrator, with a sound knowledge of every branch of learning and culture. The king was very fond of him, favouring him because of his powers of expression, rhetorical skills and grasp of administration, and because God had endowed him with pity and compassion for his subjects. The king himself was a just ruler who protected his people and continued to treat great and small alike with kindness, looking after them as they deserved, aiding them with gifts, providing them with safety and security, and lightening the general burden of taxation. In his fondness for them, whatever their rank, he was generous and sympathetic, following a virtuous course that none of his predecessors had matched. In spite of all that, however, both he and his subjects were distressed by the fact that Almighty God had not provided him with a son.

One night, as he lay in bed worrying about what would happen to his kingdom, he fell asleep and dreamt that he was pouring water over the roots of a tree . . .

Night 900

Morning now dawned and Shahrazad broke off from what she had been allowed to say. Then, when it was the nine hundredth night, SHE CONTINUED:

I have heard, O fortunate king, that the king dreamt that he was pouring water over the roots of a tree around which stood many other trees. Suddenly fire leapt from his tree and burned up all the others, and at that he woke in alarm. He summoned a servant and told him to hurry off to fetch Shimas, the vizier, as quickly as possible. The servant took the message to Shimas, explaining that the king had woken from sleep in a state of fear and adding: 'He sent me to tell you to hurry to him.' When Shimas heard this, he got up straight away and, on coming into the king's presence, he found him sitting on his bed. He prostrated himself before him, praying God to prolong his glory and prosperity, and saying: 'May God bring no sorrow on you, your majesty.' He then asked: 'What was it that alarmed you tonight, making you send for me in such a hurry?' The king told him to sit down, after which he began to tell him what he had seen. 'I had a terrifying dream tonight,' he explained, 'in which I was watering the roots of a tree that stood surrounded by others. As I was doing this, fire leapt from the roots of my tree and burned all the rest. This filled me with a feeling of panic and so, when I woke up, I sent for you because I know that, thanks to the width of your learning and the depth of your understanding, you know how to interpret dreams.'

For a time, Shimas looked down, but then he smiled and the king said: 'Tell me the truth, Shimas, and keep nothing back.' Shimas replied: 'God Almighty has granted you a gift to gladden you, and the outcome of this dream will bring happiness, as He will provide you with a son who, after you have lived a long life, will inherit your kingdom. There is something else in the dream, but I don't want to explain it now, as this is not a suitable time for it.' What he said pleased and delighted the king; his fear left him and he recovered his spirits. 'If your interpretation is correct and things turn out like this,' he said, 'then, when the time is ripe, complete it for me, as what cannot be explained now must be explained at the proper time so that my joy may be complete. All that I look for in this is the approval of God, the Glorious and Exalted.'

Shimas, realizing that the king was determined to know the full interpretation of his dream, had produced a pretext with which to excuse himself. The king then summoned the astrologers and all the interpreters of dreams who were to be found in his kingdom, and when they had come before him, he told them of the dream and asked them for a true interpretation. One of them came forward and asked him for leave to speak. When this had been granted, he said: 'Your majesty, Shimas, your vizier, was not unable to interpret your dream but, as he was reluctant to disturb you, he didn't tell you everything. If you let me speak, I shall do so.' 'Don't hesitate, interpreter,' the king replied, 'but tell me the truth.' At that, the man said: 'You must know, your majesty, that you will have a son who will inherit your kingdom from you when your long life comes to an end, but he will not follow the same course as you in his dealings with his subjects. Rather, he will do the opposite of what you have laid down; he will treat them unjustly, and he will suffer the fate of the mouse when confronted by the cat.' 'What is the story of the cat and the mouse?' the king asked.

After praying that God might prolong the king's life, THE INTER-PRETER BEGAN:

One night, the cat went out to hunt in a garden, but could find nothing and, as the weather was cold and rainy, it became weak and started to try to think of some trick that might help it succeed. As it was patrolling the garden, it noticed a nest at the foot of a tree. It went up and started to sniff and to purr, until it discovered that inside the nest there was a mouse. It did its best to get in so as to catch the mouse, but when the mouse realized what was happening, it turned its back to the cat and started to scrabble with its forepaws and hindpaws in order to block the nest entrance with earth. At that, the cat began to speak in a weak voice, saying: 'Brother, why are you doing this? I have come to you for refuge, hoping that you will have pity on me and allow me to shelter in your nest tonight. I am weak and old; my strength has gone and now that I have got into this garden I can scarcely move. How often have I prayed for death in order to find rest, and here am I prostrate at your door, worn out by the cold and the rain. I appeal in God's Name to your generosity, imploring you to take me by the paw and bring me in, so I may take refuge in the entrance hall of your nest. I am a poor stranger, and it is said that whoever shelters any such person in his house will be lodged in Paradise on the Day of Judgement. It is right that you, my brother, should gain the reward for helping me by allowing me to stay with you all night until morning, when I shall go on my way.'

Night 901

Morning now dawned and Shahrazad broke off from what she had been allowed to say. Then, when it was the nine hundred and first night, SHE CONTINUED:

I have heard, O fortunate king, that the cat asked the mouse to be allowed to spend the night with it before going off on its way. When the mouse heard what the cat had to say, it replied: 'How can you come into my nest, you who are my natural enemy and who live off my flesh? I am afraid of your treachery, as this is a characteristic of yours, and you are bound by no agreements. There is a saying that no fornicator can safely be trusted with a beautiful woman, no destitute man with money and no fire with wood. It wouldn't be right for me to trust you with my life, as it is said that in the case of a natural enemy, the weaker he becomes, the stronger is his enmity.' In the fading voice of one who is in the worst of states, the cat answered: 'I can't deny that what you say is true, but I ask you to overlook the fact that in the past nature did indeed make us enemies, for it is said that whoever forgives another created being like himself will be forgiven by his Creator. I used to be your enemy, but today I look for your friendship, and there is another saying that if you want your enemy to become your friend, do good to him. Brother, I will swear to you by God that I shall never harm you, and, in fact, I haven't the strength to do so. Put your trust in God; do a good deed and accept my pledge and covenant.' 'How can I accept the pledge of one in whom hostility towards me is ingrained, and who is in the habit of treating me with treachery?' asked the mouse. 'If this was not a blood feud, it would be a simple matter, but by nature we are mortal foes. There is a saying that whoever trusts his enemy with his life is like a man who puts his hand into a snake's mouth.' The cat, growing angry, said: 'I am wretched; my spirit is weak and my death throes have come upon me. Soon I shall die on your doorstep and it is you who will be responsible, because you could have saved me. This is my last word.'

The mouse was filled by the fear of Almighty God and its heart was moved by pity. It said to itself: 'Whoever wants the Almighty's help against his enemy, should show pity and do good to him. In this affair I place my trust in God and I shall save the cat's life in order to gain a heavenly reward.' So it went out and pulled the cat into its nest, where it stayed until it had rested and grown stronger, having partially recovered. But it then started to complain of its weakness, the loss of its strength and its lack of friends. The mouse spoke gently in order to comfort it, going close to do what it could for it, while the cat crept up to guard the entrance to the nest lest the mouse use it to escape. For its part, the mouse, wanting to go out, went near the cat, as it had become used to doing, but this time the cat seized it and, taking it in its claws, it started to bite it and toss it to and fro. It would take it in its mouth, lift it off the ground and then throw it down again before running after it, grabbing it with its teeth and then tormenting it.

The mouse cried out for help and prayed to God to rescue it, while reproaching the cat and saying: 'What about the pledge that you gave me? Where are the oaths that you swore? Is this how you repay me for letting you into my nest and trusting you with my life? How right they were who said: "He who accepts a pledge from his enemy is not looking for self-preservation," and "Whoever entrusts himself to his enemy is responsible for his own death." But my reliance is on my Creator, Who will save me from you.' While this was going on and the cat was about to pounce on it and kill it, a hunter arrived, accompanied by fierce hunting dogs, one of which passed by the entrance to the nest. It heard the sounds of a great struggle and, thinking that a fox must be catching something there, it burrowed down to get it until it came across the cat and dragged it up. The cat, finding itself between the dog's paws, was so concerned for itself that it dropped the mouse, alive and uninjured, while the dog tore it to pieces before pulling out the corpse and throwing it down. The two of them thus proved the truth of the saying: 'Whoever shows mercy will, at the last, receive mercy and whoever does wrong will himself quickly be wronged.'

'This, then, your majesty, is what happened to the cat and the mouse,' said the interpreter. 'It shows that no one should break faith with one who trusts him, while whoever acts with treachery will suffer the same fate as the cat; whatever measure you use will be meted out to you and whoever has recourse to what is good will receive his reward. Do not grieve or be distressed, because it may be that, after having acted unjustly and recklessly, your son will turn back to your own virtuous ways. Your learned vizier, Shimas, would have preferred not to keep anything back in the revelation that he made to you. That would have been the right course for him to follow, but it is said that the most learned people, who take the greatest delight in what is good, are the most timorous of all.'

The king accepted this point and ordered a great reward to be given to the sages. He then dismissed them before rising and returning to his own apartments where he sat thinking about the outcome of the affair. When night fell he went to his favourite and most honoured wife and slept with her. Four months later, this woman was delighted to feel the foetus move within her womb, and she told the king, who said: 'My dream has come true and I turn to God for aid.' He then lodged her in the finest apartments and treated her with the greatest distinction, showering her with benefits and loading her with quantities of gifts. He followed this by sending a servant to fetch Shimas, and when Shimas had come, the king told him joyfully that his wife was pregnant, saying: 'My dream has come true and my hopes have been fulfilled. It may be that the child will be a boy and that he will inherit my throne. What have you to say about this, Shimas?' Shimas remained silent and made no reply. 'Why don't you share my joy and why don't you answer me?' asked the king. 'Can it be that the news is not to your liking?' At that, Shimas prostrated himself before the king and said: 'May God prolong your majesty's life. What use is it to seek the shade of a tree if fire is going to break out of it? What pleasure is there in drinking unmixed wine if it is going to choke you, and what is the point of quenching one's thirst with fresh, cool water if you are going to be drowned in it? I serve God and you, your majesty, but there are three things about which an intelligent man should not speak until they have reached their conclusion. One should not talk about a traveller until he has come back from his journey, about a man fighting a war until he has conquered his enemy, and about a pregnant woman until she has given birth . . .'

Night 902

Morning now dawned and Shahrazad broke off from what she had been allowed to say. Then, when it was the nine hundred and second night, she continued:

I have heard, O fortunate king, that the vizier Shimas told the king: 'There are three things about which an intelligent man should not speak until they have reached their conclusion.' He went on: 'You must know that whoever talks about what is unfinished is like the ascetic who had butter poured over his head.' The king asked what this story was, AND SHIMAS TOLD HIM:

An ascetic lived with one of the nobles of a certain city, who gave him as a daily allowance three loaves of bread and a small quantity of butter and honey. Butter was expensive there and the ascetic continued to collect his ration and store it in a jar. When it was full, he hung it over his head, as he was anxious to protect it. Then, one night, while he sat on his bed with his stick in his hand, he happened to think about butter and how dear it was, and he said to himself: 'I shall sell all this butter and use the money to buy a ewe, which I shall share with a farmer. In the first year, it will produce a male and a female, and the following year a female and a male. These, in their turn, will go on producing males and females until, when they have become a large flock, I shall separate off my own holding and sell what I want. Then I shall buy such-and-such a piece of land, set out a garden there and build a large villa. I shall get myself clothes of all kinds and buy slaves of both sexes, before marrying the daughter of So-and-So, the merchant. The wedding feast will be the finest ever seen; I shall slaughter beasts and provide splendid foods, sweetmeats, sugar-coated delicacies, and so on. There will be entertainers, skilled performers and musical instruments, together with flowers, scented herbs and aromatic plants of all kinds. I shall invite rich and poor, men of learning, leaders and officers of state. I shall produce food and drink of all kinds and give everyone what he asks for, getting a crier to proclaim that every wish shall be granted. After that, I shall unveil my bride and lie with her, enjoying her beauty and loveliness, eating, drinking and taking my pleasure. I shall tell myself that I have now got all I wished for and can take a rest from asceticism and the service of God.

'Later, my wife will become pregnant and to my delight she will bear me a son. I shall give banquets in his honour and pamper him as I bring him up, teaching him philosophy, literature and mathematics, until his name becomes well known and I can boast of him in the assemblies. He will not disobey me as I shall instruct him to do good and order him to refrain from fornication and evil actions, recommending him to follow the path of piety and good works. He will be given handsome and splendid presents, and if I see that he remains obedient, I shall give him more. If, on the other hand, I find him inclined to disobey me, I shall bring down this stick on him.' At that, the man raised the stick in order to strike his imaginary son, but it collided with the butter jar that was above him. Its fragments showered down on him and the liquid butter flowed over his head, his clothes and his beard, making a spectacle of him.

'This goes to show, your majesty,' said Shimas, 'that one should not talk about anything until it actually happens.' 'What you say is true,' the king told him, 'and you are an excellent vizier because you speak the truth and give good advice. I value you as you would wish, and you are always acceptable to me.' Shimas prostrated himself in reverence both to God and to the king, and prayed for the king's continued prosperity, saying: 'May God prolong your days and exalt you. Know that I conceal nothing from you, secretly or openly; what satisfies you satisfies me and what angers you angers me; my only joy comes from your gladness and I cannot sleep at night if you are displeased with me. It is through my generous treatment at your hands that Almighty God has provided me with all manner of good things and I pray to Him that He send His angels to guard you and that, when you meet Him, He may reward you well.' He then rose and went off, leaving the king delighted with his speech.

Some time later, the king's wife gave birth to a boy, and when the good news was brought to him, he was overjoyed and poured out his thanks to God, saying: 'Praise be to God who has given me a son after I had despaired, for He is compassionate and merciful to His servants.' He sent messages throughout his realm to tell his subjects what had happened and to summon them to his palace, where the emirs, leaders and officials of state were all gathered. Drums were beaten to spread the joyful news throughout the kingdom and people flocked to the city from all sides, including men of learning such as scholars, philosophers, literary experts and doctors. They all presented themselves to the king and each of them was given his proper place. The king then ordered his seven principal viziers, whose leader was Shimas, to speak one after the other, each dealing with the matter in hand as his wisdom suggested.

Shimas, the chief vizier, was the first to ask leave to speak, and when this had been granted he began: 'Praise be to God, Who has raised us from non-existence into existence and Who has graciously supplied His servants with just and righteous kings, whom He has set as rulers in order that they may perform good deeds, and through whom He has provided sustenance for His people. In particular, we thank Him for our own king, who has restored our dead land to life thanks to the favours that He has granted us, as it is through His preservation of our king that we have enjoyed prosperity, tranquillity and justice. What other king has done for his subjects what ours has done for us, looking after our interests, preserving our rights, establishing justice among us, never neglecting us and protecting us from wrongs? It is a mark of God's grace to a people that their king should concern himself with their affairs and guard them from those foes of theirs, whose chief aim is to defeat and conquer them. Many people present their sons to kings as servants to act for them in place of slaves, so that they may ward off enemies, but in our own case no enemy has set foot in our lands during the reign of our king. This is the ultimate blessing and the height of felicity, so great that it passes the bounds of description. You, your majesty, are truly deserving of this great favour and, as we shelter under the shadow of your wing, may God reward you well and prolong your life. In times past, we have been urgent in our pleas to Almighty God that He should graciously hear our prayers, preserve you for us and grant you a virtuous son to comfort you. These prayers have now been answered, blessed and exalted is He . . .'

Night 903

Morning now dawned and Shahrazad broke off from what she had been allowed to say. Then, when it was the nine hundred and third night, SHE CONTINUED:

I have heard, O fortunate king, that the vizier Shimas told the king: 'Almighty God has heard our prayers and answered them, blessed and exalted is He for giving us prompt relief, as He did to the fish in the pool.' When the king asked him about the story of the fish, SHIMAS SAID:

You must know, your majesty, that in a certain place there was a pool which contained a number of fish. It happened that there was a shortage of water, and bit by bit the pool started to shrink until the fish were on the point of death, as there was not enough water to support them. 'What is going to happen to us?' they wondered. 'How can we manage and whom can we consult about finding a way to save ourselves?' At that, the oldest and most intelligent of them said: 'There is nothing we can do to rescue ourselves except to pray to God. But let us ask advice from our senior, the crab, and go to see what he has to say, as he knows more than we do about the truth of things.' The other fish approved of this and they all went to the crab, whom they found resting in his hole, entirely ignorant of what was happening to them. They greeted him and said: 'Master, are you not concerned about us, you who are our leader and our chief?' The crab returned their greeting and asked them what was wrong and what they wanted, at which they told them their story, explaining how the drought was affecting them and how they would perish were the water to dry up. They added: 'We have come to you expecting that you may be able to advise us how to save ourselves, as you are our senior and you know more than all of us.'

For a time the crab looked down and then it said: 'There is no doubt that you are short of intelligence as you despair of the mercy of Almighty

God and His power to sustain all His creatures. Don't you know that the Blessed and Glorious Creator looks after all His worshippers without counting the cost, and allots them their livelihoods before creating them? To each He has given a fixed term of life and apportioned a livelihood through His divine power, so why should we be anxious about what will happen in the future according to His decree? In my opinion, there is nothing better to do than to pray to Him, while ensuring that we go to Him with a clear conscience, both secretly and in the open, calling on Him to save us and to rescue us from our difficulties. He will never disappoint the hopes of those who put their trust in Him or reject the petition of those who seek His favour. When we have set our own affairs in order, then all will be well and we shall enjoy His good grace. Winter will come and our land will be flooded thanks to the prayers of the virtuous among us, as God will not destroy what He has built up so well. So my advice is that we should wait in expectation of what God will do for us. If we die, as is our common fate, then we shall be at rest, and if we have to flee, then we shall leave our own country and go wherever God wants us to go.' All the fish replied with one voice: 'What you have said is true, master; may God reward you well.' They each returned to their own place and within a few days God sent them a violent rainstorm which filled their pool higher than it had been before.

'In the same way, your majesty, we had despaired of your having a son,' continued Shimas. 'But God has graciously granted this blessed child as a favour to us and to you, and we pray to Him that blessings will continue to attend him, that he may be a source of comfort and prove a virtuous successor to you and that he may provide for us as you have done. For God does not disappoint those who search for Him, and no one should despair of His mercy.'

The second vizier now rose and exchanged greetings with the king. He then said: 'A king cannot be called a king unless he is bountiful and just, a good and generous ruler, who treats his subjects well, maintaining the laws and customs with which they are familiar. He should establish justice among them, avoiding bloodshed and protecting them from harm. He should be marked out by his constant attention to the poor; he should aid both high and low alike, giving them their rightful dues, so that they may all call down blessings upon him and obey his commands. There can be no doubt that a king like this will be beloved by his subjects and that, having risen to the highest rank in this world, in the next he will obtain glory and win the approval of his Creator. Your majesty, we, your servants, acknowledge that you have all these qualities that have been mentioned. As the saying goes, the best thing of all is for a king to be just, wise and experienced, a skilful doctor for his people, acting in accordance with his knowledge. We are now privileged to share in this happiness, whereas earlier we had despaired of your having a son to inherit your throne. But God, glory be to His Name, has not disappointed your hopes and has answered your prayer, because you were content to entrust your affairs to Him. How well have your hopes turned out. Indeed, what has happened to you is like what happened to the crow and the snake.' When the king asked about the story of the crow and the snake, THE SECOND VIZIER SAID:

You must know, your majesty, that a crow and his mate enjoyed the easiest of lives in a tree until the time came for them to hatch out their eggs in the summer. Then a snake came out of its hole, went to the tree and, by clinging to the branches, made its way up until it reached the crow's nest, in which it installed itself and where it stayed all summer long. The exiled crow could find no opportunity to return or any place of shelter until, when the heat had passed, the snake went back to its hole. The crow told its mate: 'We should give thanks to Almighty God, Who has rescued us and saved us from this destructive creature, even though we have been deprived of our brood this year. God has not cut off our hopes and we must be grateful to Him for having granted us both safety and health. We can only rely on Him, and if it is His will that we survive until next year, He may recompense us with another hatching.'

When the hatching season came round, the snake emerged from its hole again and went to the tree, but when it was on one of the branches, making for the nest as it had done before, down swooped a kite, which struck it on the head and then tore at it. It fell to the ground unconscious and ants came and ate it. The crow and its mate enjoyed safety and tranquillity, and they produced many young for which, as well as for their safety, they offered thanks to God.

'Similarly, your majesty,' continued the second vizier, 'it is for us to thank God for the favour that He has shown both to you and to us through the birth of this blessed and fortunate child, coming after a period of despair and hopelessness. May God reward you well and bring this matter to a happy end.'

Night 904

Morning now dawned and Shahrazad broke off from what she had been allowed to say. Then, when it was the nine hundred and fourth night, SHE CONTINUED:

I have heard, O fortunate king, that the second vizier ended his remarks by saying: 'May God reward you well and bring this matter to a happy end.'

The third vizier then rose and said: 'Just king, I bring you happy news of good fortune in this world and reward in the next, as whoever is beloved by the people of the earth is loved by the inhabitants of heaven. Almighty God has given you as your portion love, which He has set in the hearts of your subjects, and to Him be thanks and praise both from us and from you so that He may increase his favours to you and, through you, to us. Know, your majesty, that no man has the power to do anything except by the decree of the Almighty. It is He Who is the giver and the end of every good thing that a man possesses, and He divides his favours among his servants as He wishes. To some He presents numerous gifts, while others He keeps busy by making them earn their daily bread. Some He appoints as leaders, while others, out of their desire for Him, are abstemious in the things of this world. He has said: "I bring both injury and advantage; I cure and make sick; riches and poverty are in My gift; and I bring both life and death. I hold everything in My hands and all things return to Me." We all owe Him our thanks and as for you, your majesty, you are one of those who are marked out by good fortune and piety, as it is said that the most fortunate of the pious are those for whom God has combined the good things of both this world and the next and who are contented with what God has allotted to them, thanking Him for the rank which He has assigned to them. Whoever transgresses and looks for something that God has not decreed, either to his advantage or to his disadvantage, is like the wild ass and the jackal.' When the king asked him about this story, HE SAID:

Know, your majesty, that a jackal used to leave his lair every day to look for food. One evening when he was on a mountain and thinking of going home, he came across another jackal walking there. The two told each other about their hunting, and the second one said: 'Yesterday I came across the body of a wild ass. I was hungry as I hadn't eaten for three days, and in my delight I gave thanks to Almighty God for what

He had given me. I got through to its heart and ate my fill before going home, and although that was three days ago and I've found nothing to eat since then, I'm still full.'

This made the first jackal envious and he told himself that he too must get an ass's heart. He ate nothing for some days until he became emaciated, and he was so weak and feeble that he had reached the point of death as he lay in his lair. Then one day while he was there two hunters walked by in search of game. They had come across a wild ass and had spent the whole day tracking it until one of them managed to shoot it with a forked arrow, which passed through it and struck its heart, killing it just opposite the jackal's lair. When the hunters came up and found it dead, they tried to draw out the arrow, but only succeeded in removing the shaft, while the forked tip remained inside the ass.

In the evening, the jackal left his lair, suffering from weakness and hunger, and he was overjoyed and exultant to find the carcass lying at his door. 'Praise be to God,' he exclaimed, 'Who has brought me what I wanted so easily and without trouble, just when I had no hope of getting a wild ass or anything else! It may be that God brought it for me to my lair and caused it to fall here.' The jackal then leapt on to the body of the ass, tore open the belly and put his head inside, after which he twisted his muzzle around in the entrails until he found the heart, which he took in his teeth and then swallowed. When this was in its throat, the forked head of the arrow stuck in his neckbone and as he could neither dislodge it into his stomach or expel it from his throat, he was certain that he was doomed. 'It is true,' he said to himself, 'that no creature should seek for himself more than God has allotted him. Had I been satisfied with that, I would not have come to this end.'

'So it is, your majesty,' continued the third vizier, 'that every man should be content with what God assigns to him, thanking Him for His favours and not despairing of His goodness. For because of your virtuous intentions and the good deeds that you have performed, God has given you a son, after we had despaired of this. We pray to Him that He will grant this child long life and continuing happiness and bless him as your successor, so that he may keep the covenants that you have made when, at the end of your own long life, he follows you on to the throne.'

The fourth vizier then rose and said: 'A king should be intelligent, a master of all branches of learning . . .'

Night 905

Morning now dawned and Shahrazad broke off from what she had been allowed to say. Then, when it was the nine hundred and fifth night, SHE CONTINUED:

I have heard, O fortunate king, that when the fourth vizier rose, he said: 'A king should be intelligent, a master of all branches of learning, the rules of governance and administration, pure in his intentions and treating his subjects with justice. He must show honour and respect where it is due and, when he is in a position of power, he should display clemency where it is necessary. He must look after the interests of both governors and the governed, lightening their burdens, giving them favours, refraining from shedding their blood, attending to their needs and keeping his promises to them. Such a king rightfully acquires happiness both in this world and the next; he is protected from those who seek to harm him; he is helped to establish his rule; he is given victory over his enemies; his hopes are fulfilled and God overwhelms him with favours, granting him success because, thanks to the gratitude he has shown, he has won the right to divine protection. Any king who follows the opposite path will find himself and his subjects involved in constant disasters and misfortunes, since the injustice that he shows both to strangers and to his own people will involve him in the same situation as that of the wandering prince.' When the king asked about this, THE VIZIER SAID:

Know, your majesty, that there was in the west an unjust, oppressive, tyrannical and despotic king who paid no regard to the interests of his own people or of those who entered his realm. In the case of these latter, his officials would seize four-fifths of their money, leaving them with no more than the remaining fifth. Almighty God had decreed that the king should have a son who was both fortunate and sustained by divine providence. When this prince saw how crooked was the way of this world, he abandoned worldly things and left, while still young, as a wanderer devoted to the service of Almighty God. He rejected the world and all it contained, roaming in obedience to God through both desert wastes and cities.

One day he came to a city where the guards stopped him and searched him. They found that all that he had was a pair of robes, one new and the other old. They took the new one away from him and left him the old, treating him with contempt and derision. He complained of their injustice, pointing out that he was a poor wanderer and asking: 'What good is this robe to you?' He then added: 'Unless you give it back, I shall go to the king and lodge a complaint against you.' 'We have been following the king's orders, so do what you like,' they told him. The prince went to the palace, but when he tried to enter the chamberlains stopped him and so he went back and said to himself: 'I shall watch until the king comes out and then I shall complain to him of my treatment.' At that point he heard a soldier saying that the king was on his way and so he edged forward little by little until he was standing in front of the gate. Then, before he knew it, out came the king. Standing in front of him, the prince first prayed that he might be victorious and then complained to him of how he had been treated by the guards. He explained that he was a man of religion who had abandoned worldly things and had gone out to seek the approval of God as a wanderer through the lands, adding that whatever town or village he entered, those who met him had always shown him generosity, according to their means.

He went on: 'When I came to this city I hoped that the people would treat me as other wanderers are treated, but your men blocked my way, stripped me of one of my two robes and gave me a painful beating. Look into my case; help me and restore my robe to me, and then I shall not stay an hour longer here.' The unjust king replied: 'Who told you to come here, ignorant as you are of the customs of the king?' 'After I have got my robe back, you can do what you want with me,' said the prince. On hearing this, the king lost his temper and said: 'You fool, we took your robe in order to humble you, but now that you have made such a disturbance in my presence, I shall take your life.' He was imprisoned on the king's orders and, on finding himself in custody, he regretted having answered the king back, blaming himself harshly for not having let the matter rest and so escaping with his life.

At midnight, he got to his feet and prayed at length, saying: 'O God, the Just Ruler, You know my plight and what the unjust king has done to me. I, Your wronged servant, pray that, of Your abundant mercy, You may rescue me from his hands and send down Your vengeance on him, since You do not ignore the actions of the evil-doers. If You know that he has wronged me, then punish him this night and visit him with Your chastisement. You are the Righteous Ruler, the Succour of the afflicted and Yours is the power and the might until the end of time.' When the gaoler heard the poor prisoner's prayer, all his limbs trembled and, while he was in this state, fire broke out in the king's palace and burned everything there, including the prison door. The only survivors were the gaoler and the wandering prince, who both left and travelled to another city, while the city of the unjust king was totally consumed because of his wickedness.

'As for us, O fortunate king,' continued the fourth vizier, 'we call down blessings on you morning and evening; we thank the Almighty for having graciously given you to us and we live in tranquillity thanks to your just and excellent conduct. The lack of an heir to inherit your throne had deeply distressed us as we feared that you might be succeeded by some other kind of king, but now God has been gracious to us, removing our anxiety and cheering us by the gift of this blessed child. We pray that God will make him a virtuous successor to you, and grant him glory and lasting fortune, prolonging his benefits.'

The next to rise was the fifth vizier, who said: 'Blessed be God, the Omnipotent . . .'

Night 906

Morning now dawned and Shahrazad broke off from what she had been allowed to say. Then, when it was the nine hundred and sixth night, SHE CONTINUED:

I have heard, O fortunate king, that the fifth vizier said: 'Blessed be God, the Omnipotent, the Giver of all good and splendid gifts, Who we know for certain is gracious to those who are grateful to Him and who abide by His religion. You, O fortunate king, are characterized by splendid virtues, and your just and equitable treatment of your subjects wins the favour of Almighty God. For that reason He has exalted you, filling your days with happiness and giving you the good gift of this fortunate child, after we had been in despair. As a result, ours is perpetual joy and unbroken happiness, since before that your lack of a son had left us in great concern and distress. We thought of your justice and kindness to us and feared that God might decree your death without there being an heir to inherit your throne. In that case, our opinions would have been divided; there would have been discord among us and we would have suffered the fate of the crows.' When the king asked for the story of the crows, THE VIZIER SAID:

Know, O fortunate king, that in a certain wilderness there was a wide valley containing streams, trees and fruits, where birds sang the praises of the One God, the Omnipotent, the Creator of night and day. Among these birds was a flock of crows who enjoyed the most pleasant of lives under the rule of one of their number who treated them with kindliness and sympathy. They lived secure and tranquil lives and they cooperated so well with one another that no other bird was able to get the better of them. It happened, however, that their leader suffered the common fate of all creatures, and died. They were plunged into mourning for him, and what added to their sorrow was the fact that among them there was no other bird of his qualities to take his place. They held a meeting to consult as to who would be a good leader. One faction chose a candidate and claimed that he would make a suitable king, but others disagreed and did not want him. As a result there were splits, wrangling and great dissension, until at last they came to an agreement that, after a night's sleep, none of them should go out early to look for food but they should all wait for morning. Then, when the sun had risen, they should gather in a given place and whichever of them could outdistance the others in flight would be the one ordained by God to be chosen as their king. 'We shall appoint him over us and entrust him with our affairs,' they said, and they all agreed to abide by this arrangement. At that point, however, a hawk flew up and they said: 'Father of good, we have chosen you as our leader to see to our affairs.' The hawk agreed to this and told them: 'If Almighty God wills, you will get great benefits from me.'

After they had made the hawk their king, he and they would fly off each day and he would single out one of them, strike him down and then eat his brains and eyes, leaving the rest of the body. This went on until the crows noticed what he was doing and, seeing that most of them had already been killed, they were sure that they were going to be destroyed. 'What are we going to do,' they asked each other, 'now that most of us are dead? It is only after our leaders have been killed that we have woken up to what is happening. We have to save ourselves.' So the next morning they flew away from the hawk and dispersed.

'We have been faced with the fear that the same thing might happen to us and that we might have a different kind of king,' continued the fifth vizier. 'But God has now granted us this grace, as your face is turned towards us. We are now confident that all will be well and that we shall live in unity, peace, security and safety in our native land. Blessed be the Great God and to Him be our grateful thanks and our finest praise! His blessings have rested on our king and on us, his subjects, and He has provided both us and him with supreme felicity, establishing his good fortune and making him the happiest man of his time.'

The sixth vizier rose next and said: 'Your majesty, God has granted you supreme happiness in this world and the next. The ancients said that whoever prays, fasts, does his duty by his parents and is just in his judgements will find that God is content with him when they meet. You have acted justly as our ruler and have been fortunate in your actions, and so we pray that the Almighty will reward you well, repaying you for the good that you have done. I have heard my wise colleague say that we had feared to be deprived of our good fortune if you died or were succeeded by a different sort of ruler, as there might be dissension among us, leading to disaster. That being so, it was our duty to address our prayers to God in order that He might provide you with a fortunate son to succeed to your throne. However, no man knows how what he wishes and desires in this world is going to turn out, and so we should not ask God for something whose outcome we cannot predict. This may do more harm than good, as it may destroy the man who asked for it and he may suffer the fate of the snake charmer, his wife and children, together with his household.'

Night 907

Morning now dawned and Shahrazad broke off from what she had been allowed to say. Then, when it was the nine hundred and seventh night, SHE CONTINUED:

I have heard, O fortunate king, that the sixth vizier told the king that no one should ask God for something whose outcome he could not predict as it might do him more harm than good and might lead to his destruction, as happened in the case of the snake charmer, his wife, children and whole household. The king asked what this story was and THE VIZIER WENT ON:

You must know, your majesty, that there was a snake charmer who used to train snakes, this being his profession. He had a large basket in which he kept three of them, without the knowledge of his family, and every day he would go out and take them round the city in order to gain a living for himself and his family. Then in the evening he would go home and put them back in the basket secretly, repeating the process the following day. This was his invariable custom, and nobody in the house knew what the basket contained. Then it happened that on one occasion, when he came home as usual, his wife asked him what was in it. 'What do you want with it?' he asked her. 'Don't you have enough and more than enough in the way of provisions? Be content with what God has allotted to you and don't ask for anything else.' His wife stayed silent, but she told herself that she would have to search the basket and find out what was there.

Having made up her mind to do this, she told her children, insisting that they ask their father what was in the basket and press him to tell them. The children got the idea that there was something to eat in there, and every day they would ask their father to show them, but he would put them off, and although he tried to keep them happy, he would tell them not to keep asking. This went on for some time, and all the while their mother was urging them to find out, until eventually they agreed with her that they would neither eat nor drink with their father until he did as they asked and opened the basket for them. After they had made up their minds, the snake charmer came back one night with a large quantity of food and drink. He took his seat and called them to come and eat with him, but they refused, with a show of anger. He tried to cajole them with soft words, saying: 'Look, what is it you want me to bring you, food, drink or clothes?' 'The only thing we want you to do, father,' they replied, 'is to open the basket so that we can look inside, and if you don't we shall kill ourselves.' 'There is nothing in it to your advantage,' he told them. 'Indeed, it will be harmful for you to open it.' This made them even angrier, and when their father saw that, he threatened to beat them if they did not stop. They became still angrier and kept trying to press him. At that, he lost his temper and took up a stick with which to beat them, while, for their part, they ran away from him into the house. He had not hidden the basket, which was sitting there, and his wife, seeing that his attention was taken up with the children, opened it quickly to look inside. Out came the snakes. First they bit the woman and killed her, and then they went around the house killing old and young alike, apart from the snake charmer himself, who left the house and went away.

'If you look into this story, O fortunate king,' continued the sixth vizier, 'you will see that a man should only ask for what the Almighty has not refused him, and he should be content with what has been decreed for him by God's will. As for you, in addition to the depth of your learning and the soundness of your understanding, God has granted you comfort and consolation by providing you with a son after you had despaired, and we ask Him that this son may prove a just successor to you, winning the approval both of God and of his subjects.'

The seventh vizier then rose and said: 'Your majesty, I know for certain the truth of what my colleagues, these wise and learned viziers, have said in your presence as they described your justice, the excellence of your conduct and how you are distinguished from and superior to all other kings. It is our duty to acknowledge this and I myself give thanks to God, Who has entrusted you with His favour and granted you, by His mercy, responsibility for the prosperity of this realm. He has aided both you and us, leading us to add to His praises because of your presence among us. For as long as you are here with us, we fear no injustice, we do not find ourselves wronged and we have no weaknesses that can be used by an aggressor. It is a common saying that the most fortunate of subjects are those who have a just king and the most wretched those whose king is unjust. Another such saying states that it is better to live with ravening lions than with a tyrannical ruler. We give continual praises to Almighty God for having granted us your presence and for having provided you with the blessing of a son, after that had been despaired of and you had grown old. The greatest gift in this world is a virtuous son, and it is said that a childless man leaves nothing behind and will never be remembered. This fortunate child has been granted to you because of your adherence to justice and because of the sincerity of your devotion to God. You have pursued a pious course and shown the virtue of patience, as a result of which the child has come as a divine gift both to us and to you. This is like what happened in the case of the spider and the wind. When the king asked what this story was . . .

Night 908

Morning now dawned and Shahrazad broke off from what she had been allowed to say. Then, when it was the nine hundred and eighth night, SHE CONTINUED:

I have heard, O fortunate king, that when the king asked what this story was, THE VIZIER SAID:

You must know, your majesty, that a spider attached itself to a high gate that stood in isolation, and there it made a home for itself where it lived in safety, thanking Almighty God for having provided it with so pleasant a spot where it would be safe from reptiles. It continued to live like this for some time, giving thanks to God for its quiet life and its constant supply of food. God then decided to put it to the test by expelling it, in order to try out both its gratitude and its patience. He sent out a storm wind from the east which carried it off, web and all, and threw it into the sea, whose waves dragged it back to shore. At that, it gave thanks to God for its safety and started to reproach the wind, saying: 'Wind, why did you do that to me? What good did it do you to move me here from where I was safe and at my ease on top of that gate?' 'Stop blaming me,' said the wind, 'and I shall take you back and bring you to your old place where you were before.' The spider waited patiently and hopefully, but the wind went round to the north and did not keep its promise. Then a south wind came which picked it up and whisked it back in the direction of its old home. When it passed the gate, it recognized it and clung on to it.

'We address our prayers to God,' continued the seventh vizier, 'Who has rewarded the king for his patient endurance of his solitary state, granting him this son after he had despaired and grown old, and not removing him from this world until He had comforted him with the gift of an heir, together with that of kingship and rule, showing mercy to the king's subjects and giving them His grace.'

The king then said: 'To God be thanks surpassing all other thanks, and gratitude surpassing all other gratitude. There is no god but God, the Creator of all things, the radiance of Whose signs shows us the glory of His power. It is He Who grants kingship and rule throughout the lands to those whom He wishes from among His servants, picking those He wants to stand in His place, acting as His deputies among His creation. He orders these rulers to behave with justice and equity, establishing the laws and customs of religion and acting rightly and properly in the affairs of their subjects in accordance with His wishes and theirs. Whichever of them follows God's commands is fortunate in his obedience, and God protects him from the terrors of this world and gives him the best of rewards in the world to come, as He does not neglect the recompense due to those who do good. Those who act against God's orders are gravely mistaken in their disobedience to Him and in their choice of this world over the next. In this world they will leave no imprint and in the world to come they will have no portion, for God does not delay the punishment of the unjust and corrupt, nor does He neglect any of His servants.

'These viziers of mine have said that it is thanks to my justice and good conduct towards them that God has granted good fortune to me and to them, as our gratitude has deserved this increase in His favours. Each one of them has spoken on this subject as God inspired him, and has used the greatest eloquence in thanking and praising Him for His favours and His grace. I too thank Him, as I am a slave under His command. He holds my heart in His hand, my tongue is in His service, and I am content to accept whatever He decrees for me and for His other servants, whatever this may be. Each vizier said what came into his mind on the subject of my son, describing the renewal of God's grace to me at a time when I had reached an age when despair was uppermost and assurance had weakened. Praise be to Him Who has rescued me from deprivation, for a change of rulers is like the change of night and day. This was a great favour to my subjects and to me, and so I praise the Almighty Who has given us this boy as an obedient servant of His and set him on high as the heir to my throne. I pray to Him that, out of His generosity and clemency, He may make the boy fortunate in his actions and successful in the performance of good deeds, as a just ruler of his subjects, preserving them from destructive aberrations by his generous bestowal of favours.'

When the king had finished speaking, the men of learning and wisdom rose and prostrated themselves to God, giving thanks to the king and kissing his hands before leaving, each to his own home. The king then went to his own apartments to look at his son, whom he blessed and named Wird Khan.

When Wird Khan was twelve years old, his father wanted him to be instructed in the various branches of learning. He had a palace built for him in the middle of the city and in which there were three hundred and sixty rooms. The prince was installed there and three learned sages were appointed with orders to apply themselves night and day to teaching him. They were to sit with him for one day in each of the rooms, and they were to take pains to see that no branch of knowledge was omitted from his curriculum, so that he might turn out to be a master of them all. On the door of each of the rooms they were to write down what it was that they were teaching him, and every seven days they were to bring a report of what it was that the boy had learned.

After these teachers had been introduced to the prince, they did not relax their efforts to instruct him, by night or by day, neither did they keep back from him anything that they knew. As for the prince, he displayed intelligence, soundness of understanding and a capacity for absorbing knowledge such as had never been seen before. The weekly report that the king was brought of his son's progress showed him that he had been provided with extensive knowledge and refinement. The teachers told him: 'We have never seen anyone endowed with such intelligence; may God bless you through him and allow you enjoyment of his life.'

When the prince was twelve years old,* such was his mastery of the best part of every branch of learning that he surpassed all the wisest sages of his time. His teachers came to the king and said: 'You can be happy, your majesty, with this fortunate son of yours, whom we have brought to you after he has mastered every branch of knowledge and outstripped all the learned men of this age.' The king was delighted to hear this and in his gratitude he prostrated himself before the Great and Glorious God, exclaiming: 'Praise be to God for His favours that no man can count!' He then summoned Shimas and told him: 'Shimas, the teachers have come to tell me that this son of mine has mastered every branch of learning, all of which they have taught him, until he has come to surpass all previous generations. What have you to say to that?' Shimas prostrated himself before God and the king, whose hand he kissed before saying: 'Even if a ruby is embedded in the solid rock of a mountain, it must still gleam like a lamp. Your son is a ruby and his youth does not prevent him from being wise. Praise be to God for the gifts with which He has entrusted him. Tomorrow, God willing, I shall question him and examine him on what he knows in an assembly at which I shall gather together the leading scholars and emirs . . .'

Night 909

Morning now dawned and Shahrazad broke off from what she had been allowed to say. Then, when it was the nine hundred and ninth night, SHE CONTINUED:

I have heard, O fortunate king, that when King Jali'ad had heard what

* The text shows corruption in this passage.

Shimas had to say, he ordered all the scholars, men of intelligence and excellence and the skilled philosophers to assemble the next day in the royal palace. When those whom he had summoned had all presented themselves at the king's gate, they were given permission to enter. Shimas came up and kissed the hands of the prince, who, in turn, prostrated himself to him. At this, Shimas said: 'The lion cub should not prostrate itself to any of the beasts, nor should light associate itself with darkness.' 'When the cub sees the king's vizier, it prostrates itself before him,' replied the prince.

Shimas then said: 'Tell me, what is the eternal absolute? What are its two forms of being and, of these, which one is permanent?' 'The eternal absolute is the Great and Glorious God,' the prince replied, 'as He is the First without beginning and the Last without end. His two forms of being are this world and the next, and of these what is permanent is the bliss of the world to come.' 'That is true,' said Shimas, 'and I accept your answer, but I would like you to tell me how you know that one of these forms is the present world and the next the world to come.' The prince replied: 'That is because the present world was created, but not from anything that pre-existed, and so it has to be explained as the first form of being. But it is an accident, quick to pass away, and as it involves the need for actions to be rewarded or punished, this calls for the re-creation of the past. The world to come, then, must be the second form of being.' 'True,' said Shimas, 'and I accept your answer, but I would like you to tell me how it is that you know that the delights of the world to come represent the permanent form of being.' The prince said: 'I know that because that is where actions are rewarded, and it has been prepared by the Everlasting God.'

'Tell me then,' said Shimas, 'whose actions in this world are the most praiseworthy?' 'Those who prefer the next world to this,' replied the prince. 'And who is it who does so?' asked Shimas. 'Whoever knows that he is living in a finite world, that he has been created in order to die, that after death he will be judged and that, were immortality possible in this world, he would still not prefer it to the world to come,' replied the prince. 'Can there be a future world without this one?' asked Shimas, to which the prince replied: 'There can be no future world for anyone who has not lived here. I think that this world, its people and the world to which they are going are like an estate and its workers, whom an emir has housed in a narrow building that he has had constructed for them. He has assigned each of them a task, with a fixed time attached to it, and put them in charge of an overseer. Whichever of them completes his task is removed from that building by the overseer, while if anyone fails to do it in time, he is punished. While they are in this position, honey oozes through chinks in the building and when they eat it and taste its sweetness, they slacken off in their work, giving it up and putting up with their narrow quarters and their affliction, together with the prospect of the punishment that they must face, because they are content with that small measure of sweetness. Then, when their time is up, the overseer removes them from the building. We know that this world bewilders our eyes and that fixed terms are appointed for its people. Whoever discovers its brief delights and occupies his time with them will perish because he has preferred the things of this world to those of the next, but whoever prefers the things of the next world to those of this, and pays no attention to its small sweetness, will gain his reward.'

Shimas said: 'I hear and approve of what you have said about this world and the next, but I see that both of them have power over a man and he has to try to please them both, in spite of the fact that they are at variance with one another. If he concerns himself with seeking his livelihood, that will injure his soul in the next world, while by concentrating on the next world he will harm his body, and there is no way in which he can satisfy these two opposites.' THE PRINCE SAID:

To acquire a livelihood in this world is to be strengthened for the next. I see these worlds as two kings, one just and one unjust. The unjust king's land contained trees, fruits and plants, but he allowed no merchants to enter without seizing their money and their goods. They put up with that because of the living that they could make from the fertility of the soil. The just king provided one of his subjects with a large amount of money and sent him out with instructions to go to the territory of the unjust king in order to buy gems. When the man had set off with the money and had reached that land, the unjust king was told that a wealthy merchant had arrived, wanting to buy gems. The king sent for him. 'Who are you?' he asked. 'Where have you come from? Who has brought you here and what do you want?' The man told him the name of his country and explained: 'My king has given me money and told me to buy gems for him from here, and I have come in obedience to his command.' The king replied: 'Damn you, don't you know how I treat people here? Every day I seize what they have, and so how is it that you have been here for some time with this money of yours?' The merchant replied: 'None of it is mine. It has been entrusted to me as a deposit to be returned to its owner.' 'I shall not let you make a living here until you ransom yourself with the entire sum,' the king told him . . .

Night 910

Morning now dawned and Shahrazad broke off from what she had been allowed to say. Then, when it was the nine hundred and tenth night, SHE CONTINUED:

I have heard, O fortunate king, that the unjust king said to the merchant who was wanting to buy jewels from his country: 'I shall not let you make a living here until you ransom yourself with all this money that you have with you; otherwise you will be killed.'

The man said to himself: 'I have fallen between two kings. I know that this one is unjust to all who are in his lands, and if I do not satisfy him I shall certainly be put to death and my money seized without my having performed my errand. On the other hand, if I give him all the money, then the king to whom it belongs will undoubtedly kill me. The only thing that I can do is to win the favour of the unjust king by giving him a little bit of the money and so protect myself and preserve the rest of it. This land is fertile enough to allow me to feed myself until I can buy the gems I want. In this way, I shall have pleased the unjust king with my gift and, then, having taken my fair share from his land, I shall take my master what he wants. Thanks to his justice and forbearance, I can hope that he will not punish me for what this king has taken, especially if it is only a small amount.' Then he called to the king and said: 'Your majesty, I wish to ransom myself and protect this money by giving you a small part of it to cover the time between my entering your country and leaving it.' The king accepted his offer and let him go free, after which the merchant used the rest to buy gems before returning to his master.

'The just king represents the world to come,' continued the prince, 'while the gems found in the land of the unjust king are good deeds and pious actions. The man with the money is like someone who looks for the things of this world, while the money he has with him represents his life. When I recognized this, I realized that whoever seeks a livelihood in this world should not let any day pass without looking towards the next world. He will then have satisfied this world with what he has taken from the fertile land as well as satisfying the next world by devoting part of his life to searching for it.'

Shimas said: 'Now tell me whether the body and the soul share the same rewards and punishments, or is it only the part which lusts and sins that is punished?' The prince replied: 'The inclination to lust and sin can lead to reward in cases where the soul refrains from these and repents of them. The affair is in the hands of the man, who can do what he wishes, and things can be distinguished by their contraries. The body must have something on which to live; there can be no body without a soul and the purification of the soul is accomplished by the sincerity of its intentions in this world and by its turning to what will be of advantage to it in the world to come. These two worlds are like two horses running in a race, two foster brothers or two business partners, and in discriminating between the sum total of men's deeds, one must look at their intentions. Similarly, body and soul share both in actions and in rewards and punishments.' HE WENT ON:

They are like the blind man and the cripple whom the owner allowed into his orchard, telling them not to do any harm or damage to it. When the fruits ripened, the cripple said to the blind man: 'Come! I can see delicious fruits, but although I want them, I can't get up to eat them. There is nothing wrong with your legs, so do you go and bring me something that we can eat.' The blind man cursed him for mentioning the fruits, saying: 'I knew nothing about them and I can't take them as I can't see them. What can we do to get hold of them?' While they were thinking about this, the overseer, a wise man, arrived, and the cripple told him that he wanted fruit, but as it was obvious that he was crippled and his companion blind, he asked what they should do. The overseer said: 'Surely you remember that the owner made an agreement with you not to do anything that would damage his orchard? Give up any thought of this and don't do it.' But the two insisted that they must have a share of the fruit and pressed him to tell them how to get it. As they were not to be dissuaded, the overseer suggested that the blind man should get up and carry the cripple on his back to the tree whose fruits attracted him. When he was up there, he could pick what was within reach.

The blind man followed his advice and the cripple directed him how to get to the tree, from which he, in his turn, picked what he wanted. The two of them went on doing that until they had stripped all the trees in the orchard, but at that point the owner arrived and said: 'Damn you, what have you done? Didn't I make an agreement with you that you were not to do any harm here?' They said: 'You know that there is nothing that we can do to cause harm, as one of us cannot stand, being a cripple, while the other is blind and cannot see what is in front of him, so what could we have done that is wrong?' 'Do you think that I don't know how you managed this and how you have stripped my orchard?' said the owner. 'You, the blind man – I know perfectly well that you carried the cripple on your back and guided him to the tree.' Then he took hold of them and gave them both a painful beating before expelling them from the orchard.

'The blind man is like the body, which can only see through the soul,' continued the prince, 'while the cripple is like the soul which depends on the body in order to move. The orchard is like the works that God's servants perform in order to win their reward, while the overseer is the intellect which orders us to do good and forbids us to do evil. So body and soul share in both rewards and punishments.'

'That is true,' said Shimas, 'and I accept what you say, but tell me which of the scholars do you think deserves most praise.' 'Those who have a knowledge of God,' the prince replied, 'and who are helped by this knowledge.' 'Who are they?' asked Shimas, and the prince told him: 'Those who seek the approval of their Lord and avoid angering Him.' 'Which is the best of them?' Shimas asked. 'He who knows most about God,' came the reply. 'Which learns most from experience?' Shimas asked, and the prince replied: 'He who is patient in his actions, thanks to his knowledge.' 'Tell me,' Shimas then said, 'which are the most tender-hearted?' The prince replied: 'Those who make the most preparations for death and recite God's Name and have the least hope. For whoever opens his soul to the attacks of death is like a man who looks into a clear mirror; he recognizes the truth and the mirror becomes clearer and gleams more brightly.' 'What are the best treasures?' Shimas asked, and the prince replied: 'The treasures of heaven.' 'And of these which is the best?' 'The glorification of God and His praise.' 'Which is the best earthly treasure?' 'Good deeds.'

Night 911

Morning now dawned and Shahrazad broke off from what she had been allowed to say. Then, when it was the nine hundred and eleventh night, SHE CONTINUED:

I have heard, O fortunate king, that when the vizier Shimas asked the prince what was the best earthly treasure, he replied: 'Good deeds.' 'That is true,' said Shimas, 'and I accept what you say, but now tell me about three different things, knowledge, judgement and intellect, and about what combines them.' 'Knowledge,' said the prince, 'comes from study, judgement from experience and intellect from thought, and they are brought together and established in the mind. Whoever combines all three is perfect, and if he unites them with the fear of God he is following the right path.' 'True,' said Shimas, 'and I accept what you say, but tell me about the learned man of sound judgement with lively intelligence and superior intellect – can desire and lust cause a change in these qualities?' The prince replied: 'When desire and lust affect a man, they alter his learning, his understanding, his judgement and his intellect. He is like the rapacious eagle who was clear-sighted enough to avoid snares by staying in the upper air. While he was soaring there, he noticed a fowler setting out his nets and finishing by baiting the trap with meat. The eagle saw the meat and was overcome by so passionate a desire that he forgot the trap that he had seen and the evil fate that attended birds who fell into it. He swooped down on the meat and found himself entangled in the net. When the fowler came, he was amazed to find the eagle trapped and said: "I set out my net in the hope that a pigeon or some other puny bird might fall into it, but how has it caught this eagle?"

'It is said that when an intelligent man is tempted by passionate desire to do something, he thinks about its aftermath and rejects the prompting of desire, subduing it with his strength of mind. When it is urging him to act, he must use his reason as a skilled horseman on a wild horse, who has to tug hard on the bridle to straighten it so that it goes where he wants. The fool has neither knowledge nor judgement; his affairs are in a state of confusion and he is under the control of his passions, acting as they dictate. As a result, he is numbered among those who perish and there is nobody more miserable than he.'

'That is true,' agreed Shimas, 'and I accept what you say, but tell me in what circumstances knowledge is useful and how reason can ward off the evil consequences of passion.' The prince replied: 'That happens when a man uses reason and knowledge in his quest for the next world, a quest on which they can help him. When it comes to a search for worldly things, he should only use them to gain his daily bread and to save himself from harm, while otherwise they should only be kept for actions connected with the world to come.' 'Tell me,' Shimas then asked, 'on what should a man concentrate his attention?' 'Good works,' came the reply, at which Shimas went on: 'If he does that, it will distract him from earning his living, so what is he to do about the daily bread which he cannot do without?' 'There are twenty-four hours in the day,' replied the prince, 'and of these he must set aside one portion for earning his living, another for rest and relaxation, and the rest in a search for knowledge. An intelligent man without knowledge is like barren ground which has not been worked or planted, where plants do not grow. Nothing can be done with it, and it will produce no useful fruits, while if it has been prepared and planted, then it will yield a good crop. Similarly, no good will come from an ignorant man until knowledge has been implanted in him, for then he will bear fruit.'

'Tell me about the position of knowledge where there is no reason,' said Shimas. The prince replied: 'This is like the knowledge of an animal which knows its feeding times and the times for it to drink and to be awake, but which has no powers of reason.' 'That is a brief answer,' Shimas told him, 'but I accept what you say. Now tell me how I must guard myself against the power of the ruler.' The prince replied: 'By allowing him no opportunity to attack you.' 'But how can I do that,' Shimas asked, 'seeing that he has authority over me and controls my affairs?' 'His authority over you is confined to the duty that you owe him,' the prince said, 'and if you give him this, he has no further power over you.' 'What is it right for a king to expect from his vizier?' asked Shimas. 'Advice, diligence both in private and public, sound judgement and the keeping of secrets,' said the prince, adding: 'The vizier must not hide from his king anything that the king should know and he should not neglect to carry out whatever has been entrusted to him, while trying to please his master in whatever way possible without angering him.'

'Then tell me,' Shimas said, 'how should a vizier act with a king?' The prince replied: 'If you want to be safe from him, then both listen to him and tell him more than he expects from you. Only ask him for what your standing with him entitles you to, and take care not to assume for yourself a position of which he may think you unworthy, since he will

take that as an act of insolence on your part. If you do, being misled by
his forbearance, you will be like the hunter who used to kill animals for
their skins, while he would throw away their flesh. A lion would come
up and eat the carcasses, and after repeated visits it came to be on
friendly terms with the hunter, who started to throw it meat and rub its
back, while it wagged its tail. When he saw that the lion appeared quiet,
friendly and humble, the hunter said to himself: "This lion has submitted
itself to me and I am its master. I think that I should get on its back and
then skin it, as I have done with the other beasts." In his greed he dared
to jump on it, and when it saw what he had done, in its fury it raised its
paw and struck at him so that its claws penetrated into his entrails. It
then threw him to the ground and tore him to pieces. From this you can
tell that a vizier should take account of what he sees to be his standing
with the king and not presume on the excellence of his own judgement,
lest the king change his opinion of him.'

Night 912

Morning now dawned and Shahrazad broke off from what she had been
allowed to say. Then, when it was the nine hundred and twelfth night,
SHE CONTINUED:

I have heard, O fortunate king, that the young son of King Jali'ad said
to Shimas the vizier: 'A vizier should take account of what he sees to be
his standing with the king and not presume on the excellence of his own
judgement, lest the king change his opinion of him.' 'Tell me then,' said
Shimas, 'how can a vizier distinguish himself in the sight of his king?'
The prince answered: 'He can do this by carrying out the confidential
tasks with which he is entrusted by means of his good advice, sound
judgement and obedience to orders.' Shimas said: 'You have said that a
king has a right to expect that a vizier should avoid trying to anger him,
acting in a way that will gain his approval and concerning himself with
the tasks entrusted to him. This is certainly right, but tell me how a vizier
should act when what pleases his king is injustice, wrongdoing and
oppression. When he is burdened by having to associate with a king like
that, what is he to do? If he tries to turn him from his passions and his
evil judgements, he will fail, while if he falls in with these passions and
approves of his judgement, his will be the responsibility and he will
become an enemy of the people. What have you to say to that?' The

prince replied: 'What you said about responsibility and guilt only applies if the vizier follows the king in his wrongdoing. When the king asks his advice in a case like this, it is his duty to point out the way of justice and fairness, cautioning against injustice and oppression while instructing his master how to treat his subjects well. He should encourage him to do this by pointing to the rewards that he will receive and warning him of the punishment that will otherwise be his. If the king listens and accepts this, the vizier will have achieved his object, but if he does not, the vizier's only recourse will be amicable separation, as this will bring relief to both of them.'

Shimas then said: 'Tell me what it is that the king owes his subjects and what do they owe him?' The prince said: 'The king's subjects must apply themselves with sincerity to carrying out his orders and obeying him in a way that will satisfy not only him but also God and His Prophet. For his part, the king is obliged to guard their wealth and protect their women, while they must listen to him and obey him, sacrificing their lives for him, paying him his dues and praising him for his justice and good conduct.' 'You have given me a clear answer about the rights of the king and of his subjects, but tell me whether there is anything that you have not yet mentioned that his subjects are owed by their king.' 'Yes,' answered the prince, 'the rights owed by the king to his subjects are more binding than those that they owe to him, in that a failure on his part to respect what they are owed is more destructive than the reverse. It is only because of such a failure on the king's part that he will face destruction, together with the loss of his realm and his prosperity. Three things are necessary to anyone who comes to the throne: he must see to it that religion prospers; that his subjects prosper; and that his administration is properly carried out. If he sticks to these three things, he will remain in power.'

'Tell me, then,' said Shimas, 'how he may successfully see to the prosperity of his subjects?' 'By giving them their due,' the prince replied, 'as well as by maintaining their traditions and by employing learned scholars to teach them. He must also establish fair dealing among them, refrain from shedding their blood or seizing their wealth, and he must lighten their burdens and strengthen their armies.' Shimas asked what it was that the king owed his vizier, to which the prince replied: 'There are three reasons why what the king owes the vizier is more important than what he owes anyone else. The first of these lies in the misfortunes that he will suffer if the vizier gives bad advice, together with the advantages that he will share with his subjects if the advice is sound. The second point is that people must be made to see the high regard in which the vizier is held by the king in order that they may treat him with veneration, respect and deference. The third point is that when the vizier sees that he is appreciated by the king and his subjects, he will protect them from misfortune and see to it that they have what they want.'

Shimas said: 'I have heard and accepted all that you have told me about the qualities of kings, viziers and subjects. Now tell me about the need to keep oneself from telling lies, making foolish remarks, attacking the honour of others and talking exaggeratedly.' The prince replied: 'A man should only say what is good, and he should not talk about what does not concern him. He should never indulge in slander or pass on what he has heard from someone else to the man's enemy. He must not try to injure either friend or foe with the ruler, and when he hopes for good or fears harm, he should look only to the Almighty God, the true source of both. He should not talk of anyone's defects or speak in ignorance, for God will hold him responsible and guilty, and people will hate him. Words are like arrows which, when shot, cannot be recalled. He should be careful not to entrust a secret to anyone who might disclose it, for, although he may have been confident that it would be kept hidden, its disclosure will bring him harm. His secrets should be kept even more carefully from his friends than from his enemies, while for everybody the keeping of secrets is a matter of good faith.'

'Tell me now,' Shimas said, 'about good nature and how it is shown in relation to family members and relatives.' The prince told him: 'Goodness of nature is the only source of comfort to mortal men, and a man must deal fairly with the rest of his family and with his brothers.' 'What is it that he owes his family?' asked Shimas, and the prince replied: 'He must show deference to his parents, speaking to them pleasantly and treating them gently, with honour and respect. To his brothers he owes good advice, gifts of money and help in their affairs. He should share in their gladness and pass over any faults that they may commit, while in return, when they see him acting in this way, they will give him their best advice and offer their lives for him. When you are sure that you can trust your brother, give him your love and help him in all that he does.'

Night 913

Morning now dawned and Shahrazad broke off from what she had been allowed to say. Then, when it was the nine hundred and thirteenth night, SHE CONTINUED:

I have heard, O fortunate king, that when the vizier Shimas put those questions to the young son of King Jali'ad, he answered them. Shimas then said: 'I see that "brothers" are of two kinds, one in whom you can trust and the other social "brothers". What you owe to the former is what you have described, but I now ask you about the latter.' The prince answered: 'From these social "brothers" you get enjoyment, good nature, pleasant conversation and enjoyable society. You should not deprive them of the pleasure that you can give to them. Rather, give them what they give you and treat them as they treat you, with cheerfulness and pleasant words. In this way, your life will be rendered happy and what you say will be acceptable to them.'

'I recognize the truth of all this,' said Shimas, 'but now tell me about the livelihoods allotted by the Creator to His creation. Are these apportioned to each man and each beast until the end of their lives, and in that case what leads us to undergo hardships in the search for what we know we are destined to receive even without any effort on our part, while if it is not destined for us we shall never acquire it, however hard we try? So should we stop trying, rely on God and rest both body and soul?' The prince replied: 'We can see that everyone has allotted to him a fixed livelihood and an appointed life span. But there is a way to achieve this livelihood and there are means to do so. Whoever looks for it may find comfort should he abandon his quest, but nevertheless he has to pursue it. There are two types of seeker, one successful and one who fails. The successful finds comfort twice over, both in acquiring his livelihood and in the fact that the accomplishment of his goal is praiseworthy. The comfort of the unsuccessful is found in three things: firstly the fact that he is prepared to seek his living; second in that he is not a burden to others; and thirdly in that he cannot be blamed.' 'Tell me,' Shimas said, 'how one can look for a livelihood.' The prince replied: 'Any man can consider lawful what the Great and Glorious God has permitted and unlawful what He has prohibited.' The discussion between the two of them ended here.

At this point, Shimas and the other scholars present rose and prostrated themselves before the prince, honouring him and heaping praises on him. His father clasped him to his breast and seated him on the royal throne, exclaiming: 'Praise be to God, Who has given me a son to delight me in my life!' For his part, the prince now said, addressing Shimas and the other scholars: 'Learned master of spiritual questions, even if God had given me only a small quantity of knowledge, I would still have understood what you meant by accepting my answers to your questions, whether I was right or wrong, as it may be that you deliberately overlooked my mistakes. Now I should like to ask you about something which I am unable to understand. It is out of my reach and I can find no words to cover it, since for me it is as hard to grasp as clear water is hard to see in a black bowl. I would like you to explain it to me so that in the future no one else in my position may find it as obscure as I have discovered it to be. For as God has made life to derive from sperm and strength from food, and as the sick are cured by the doctor's treatment, so He has arranged for the ignorant to be cured by the knowledge of the wise. Listen then, to what I have to say.' 'You are a man of luminous intellect and a sound questioner,' replied Shimas. 'All the scholars have testified to your merit, thanks to your powers of discrimination and the excellence with which you have replied to the questions that I put to you. I know that whatever the question may be that you want to put to me, you yourself will be able to explain the matter more accurately and more reliably, as the knowledge that God has given you surpasses what He has given to anyone else. Tell me, then, what is it that you want to ask?'

The prince said: 'Tell me about the Omnipotent Creator. From what did He construct His creation, when nothing had existed before, while in this world we see nothing that has not been created from something else? He is able to create something from nothing, but in spite of His omnipotence, it is His will not to do this.' Shimas replied: 'Potters and other craftsmen cannot create something from nothing, as they are themselves created beings, but as for the Creator, Whose marvellous art has fashioned the world, if you want to reach an understanding of His creative power, Blessed and Exalted as He is, then consider carefully the various things that He has created. Here you will find indications and signs of the perfection of this power that show Him capable of creating things from nothing. He has produced them from absolute non-existence, in that their constituent elements did not exist before. That is made clear and apparent to you beyond any shadow of doubt by the sequence of night and day, which are His signs. When day fades to be replaced by night, it is hidden away and we do not know where it has settled. Similarly, when the forlorn darkness of night has left, day comes and we do not know where night has gone. The sun shines on us, but we do not know where it hides its radiance, and when it sets we do not know where it can be found after its setting. There are many other examples of such actions on the part of the Creator, glory be to His Name, that bewilder intelligent men from among His creatures.' The prince said: 'Man of learning, what you have told me of the Creator's power cannot be denied. But tell me, by what means did He bring His creation into being?' Shimas replied: 'Created things were brought into existence by His word that existed before time, and this was the source of all creation.' The prince said: 'Then God, the Omnipotent, Whose Name is glorious, willed the existence of creation before it had come into being?' Shimas replied: 'It was through His will that He created things by His word, and had that word not been spoken, these things would not have come into being.'

Night 914

Morning now dawned and Shahrazad broke off from what she had been allowed to say. Then, when it was the nine hundred and fourteenth night, SHE CONTINUED:

I have heard, O fortunate king, that when Shimas asked the questions that have been listed, the young man answered them. Shimas then continued: 'My dear son, anyone who gives you a different answer to this question is twisting what has been revealed to us in the doctrines of our religion, and distorting its truths. Among these distortions is the claim that the word itself has power – I take refuge with God from such a belief. When we say that the Great and Glorious God brought creation into being by His word, we mean that both in His essence and His attributes God is one, and not that His word has its own power. Power is one of the attributes of God in the same way that speech and the other qualities of perfection are attributes of His, exalted be He Whose power is glorious. He cannot be described as being separate from His word nor is His word separate from Him. He created all things by His word and without this nothing was created. Thus the word of truth was the agent of creation, and so we are created by truth.'

The prince said: 'I understand what you have said about the Creator and the power of His word, and I can grasp and accept that. But I heard you say that the word of truth was the agent of creation. Truth is the opposite of falsehood, but where does falsehood come from and how can it occur in opposition to truth, such that created beings find the two confused and require some way of distinguishing between them? Does the Great and Glorious Creator love this falsehood or hate it? If you say that He loves truth and has used it to produce His creation, and that He hates falsehood, how did what He hates find a way to oppose the truth that He loves?' Shimas replied: 'When God used truth to create mankind, man had no need of repentance until falsehood was able to attack the truth by which he was created. This was thanks to the capacity with which God had endowed him, this capacity being his will and what is known as the acquisitive instinct. When this happened, falsehood became confused with truth, due to this capacity of man's, which is the part that acts according to his own choice and which is combined with the weakness of his nature. God then created repentance for him, in order to remove falsehood from him and establish him in the way of truth. He also created punishment for him, in case he persisted in attaching himself to falsehood.'

The prince said: 'Tell me why it is that falsehood has managed to insert itself into the path of truth so as to become confused with it, and how was it necessary for there to be punishment so that man requires repentance?' Shimas answered: 'When God created man through truth, he instilled in him a love of truth, and there was no such thing as punishment or repentance. Things remained like this until God added the soul to the human compound and, while this represented man's perfection, the soul was naturally inclined to lust. It was from this that the confusion of falsehood with truth occurred in human nature and it is for this reason that man has a natural fondness for falsehood. When he proceeds along this path, disobedience to God leads him to deviate from truth and, in doing so, he plunges into falsehood.' 'So it is through man's disobedience to God that truth is attacked by falsehood?' the prince asked. 'That is so,' Shimas replied, adding: 'God loves man, and because of the extent of this love He has created man to have a need for Him, this being the essence of truth. It may happen, however, that a man slackens in his quest for God, thanks to his soul's inclination towards lusts, and he goes the opposite way, turning to falsehood through his disobedience to God and so deserving punishment. If he banishes falsehood by means of repentance and reverts to the love of truth, then he merits reward.'

'Tell me,' said the prince, 'about the origin of this opposition to God's wishes, when all mankind is ultimately descended from Adam, whom God created through truth. How was it that he allowed disobedience to enter into the soul which God had added to his composition, disobedience which was later to be linked with repentance, so that as a result there were to be rewards and punishments? We see that some people persist in opposing God, inclining to what runs contrary to His wishes in contradiction to what is required by that love of truth, which is the basis of their creation, and so they earn themselves the wrath of their Lord. Others steadfastly seek the approval of their Creator through obedience, and because of this they merit mercy and reward. But what is it that causes this difference between them?' Shimas replied: 'It was because of Iblis that the creatures God had created were first tempted to disobey Him. Iblis had been the noblest of God's creation from among angels, mortals and *jinn*, and love was innate in his nature, as this was the only thing that he knew. However, when he found himself in this unique position, he became proud, haughty and vainglorious, thinking that faith and obedience to his Creator's commands were beneath his dignity. As a result, God made him inferior to all other created beings, removing him from His love and leaving him to dwell by himself in disobedience. He knew that disobedience was hateful to God, and when he saw that Adam adhered to truth, love and obedience to God, he became envious and used his wiles to turn him away from truth so as to find a partner in falsehood. Adam was punished because he had inclined to disobedience, which his enemy had made to appear attractive, and because he had been led to follow his own lusts, in disobedience to the injunctions of his Lord, thanks to the incursion of falsehood. When God, Whose Names are holy and to Whom all praise is due, saw how weak man was and how quickly he could be tempted to turn to his enemy and abandon truth, in His mercy He made repentance the means whereby he might extricate himself from the abyss of his inclination to disobedience. Armed with repentance he could overcome Iblis, his enemy, together with his armies, and return to the truth which was implicit in his nature.

'When Iblis saw that God had allowed him freedom of action for a prolonged period, he was quick to attack man, scheming against him in order to remove him from God's grace, so that he might be associated with him in the divine wrath that he and his followers had deserved. God, however, endowed man with the capacity to repent, ordering him constantly to follow truth and forbidding him to disobey or oppose Him. He revealed to him that he had on earth an enemy who would never stop attacking him, by night or by day. For this reason, if a man clings to the truth, whose love is inborn in him, he merits a reward, but if his soul, in its inclination towards lusts, gets the better of him, then punishment awaits him.'

Night 915

Morning now dawned and Shahrazad broke off from what she had been allowed to say. Then, when it was the nine hundred and fifteenth night, SHE CONTINUED:

I have heard, O fortunate king, that the young man asked Shimas the questions that have been listed, and he answered them. The prince then said: 'Tell me, how is it that created beings can have the power to oppose their Creator, when He is omnipotent, as you have described, while no one can overcome Him or act against His will? He must surely be able to see to it that they do not disobey Him and ensure that they always love Him.' Shimas replied: 'Almighty God, glory be to His Name, is just and treats those who love Him fairly and with mercy. He has shown them the path to what is good and given them the ability to do what good they want, whereas if what they do is contrary to this, then their disobedience will destroy them.' The prince asked: 'If it is the Creator who has given them the ability to do what they choose, why does He not stop them if what they choose is wrong, and so turn them back to the truth?' 'That is because of His great mercy and the splendour of His wisdom,' replied Shimas. 'As, in the past, He was angry with Iblis and did not show him mercy, so to Adam He showed mercy by allowing him to repent, so that he might escape the divine wrath and regain favour.' 'This is the exact truth,' the prince agreed, 'because it is God the Creator who rewards everyone for his actions and there is no other creator apart from Him, Who has power over all things.' He then continued: 'Did He create what He loves and what He does not love, or did He only create what He loves?' 'He created everything,' Shimas told him, 'but He only favours what He loves.'

'Tell me about two things,' the prince then asked, 'one of which is pleasing to God and brings a reward to whoever performs it, and the other, being hateful to God, brings punishment.' Shimas said: 'Explain to me what these two things are and clarify them for me, so that I may be able to tell you about them.' 'The two are good and evil,' replied the prince, 'which are compounded in our bodies and souls.' Shimas told him: 'Wise prince, I see that you know that good and evil result from actions performed by the body and the soul. Of the two, good is called good because God approves of it, and evil is so named because it earns God's wrath. You must know God's will and please Him by doing what is good, as this is what He has commanded us to do, while He has forbidden us to do evil.' The prince said: 'I see that these two things, good and evil, are performed by what are recognized in the body as the five senses. These comprise the centre of taste, from which comes speech, together with hearing, sight, smell and touch. I would like you to tell me whether these five were all created for good or for evil.' 'The explanation of the point that you have queried,' Shimas replied, 'can be clearly set out and should be stored in your mind and poured into your heart. God, the Blessed and Exalted, created man through truth and installed love for truth in his nature. Nothing was created except through His exalted power, which influences every happening, and what is to be attributed to Him can only be the command to act with justice, fairness and beneficence. He created man to love Him but endowed him with a soul that is naturally inclined towards lust, as well as giving him the ability to follow his inclinations, while his five senses can lead him either to Paradise or to hell.' 'How is that?' the prince asked, and Shimas said: 'The tongue was created for speech, the hands for work, the feet for walking, the eyes for seeing and the ears for hearing. He gave each one of these senses a capacity, stimulating them to act and to move, while ordering them to act only in accordance with His will. What He wants from speech is truth-telling and the abandonment of its opposite, lying. What He wants from sight is that it should be directed to what He loves and away from what displeases Him, such as the objects of lust. In the case of hearing, He wants us to listen only to the exhortations of religion and to what has been written by God, while we should not listen to anything that will invoke God's anger. As for the hands, He wants them not to keep what He has given them but to use it in ways that will please Him, as opposed either to clinging on to it or spending it on acts of disobedience. What He wants from feet is that they should be used to pursue what is good, such as in the search for education, and that they should only walk in His way. The other lusts followed by mankind proceed from the body at the command of the soul and these are of two kinds, procreative lust and the lust of the belly. Of the former, God only approves of what is licit, while what is unlawful angers Him. Lusts of the belly comprise eating and drinking. Here what wins His approval is that no one should take anything except what he has been allowed, whether this portion be small or great, and that God should be praised and thanked for this, while He is angered by whoever takes what is not his due. As you know, God, the universal Creator, only approves of what is good, and He has ordered every limb of the body to carry out His commands, for He is the All-Knowing, the All-Wise.'

The prince then said: 'Tell me, did God's foreknowledge extend to the fact that it would be thanks to Adam that the fruit of the tree which he had been forbidden to eat would be eaten, with the results that we know, which led him to turn from obedience to disobedience?' 'Yes indeed, wise prince,' Shimas answered, 'for He knew this before Adam was created, and the proof of that is that He warned him not to eat the fruit of the tree, telling him that this would be an act of disobedience. This was an act of justice and fairness on His part lest Adam might have any plea to advance against Him. When Adam fell into the abyss of error, exposing himself to great opprobrium and reproach, the effects of this extended to his descendants. God then sent out His prophets and apostles and gave them books, so that they could teach us His laws, giving us clear exhortations and regulations, detailing matters for us and making clear the path that leads to God. They showed us what it is that we should do and what we should leave undone, but we are endowed with the power to choose. Those who act according to these ordinances are in the right and are the gainers, but whoever flouts them and does what runs counter to them disobeys God and is a loser both in this world and the next. This is the path of good and evil. As you know, God has power over all things and He approved and willed the creation of lusts. He ordered us to satisfy these in a manner that is lawful in order that they may be good for us, but when we employ them unlawfully, they do us harm. Every good thing that comes to us comes from Almighty God, while we ourselves, His creation, are responsible for the evils and not our Creator, Who is far above such things.'

Night 916

Morning now dawned and Shahrazad broke off from what she had been allowed to say. Then, when it was the nine hundred and sixteenth night, SHE CONTINUED:

I have heard, O fortunate king, that the young son of King Jali'ad put these questions to the vizier Shimas, and he answered them. The prince then said: 'I understand what you have told me about the things that relate firstly to the Almighty and then to His creation, but tell me about something that perplexes and amazes me. I am astonished at how heedless Adam's children are of the afterlife and how they forget about it in their love for the present world, which they know that they will have to abandon as they take their humble leave of it.' Shimas replied: 'Yes, the way that the world changes and betrays its people shows that no fortunate man will continue to live in comfort, while the distress of the unfortunate will not last for ever. As the world changes, no earthly ruler, whatever his power and however much he is envied, can be secure, as his circumstances will alter and he will quickly be removed, for no one can securely profit from its vanities. When we realize that, we can see that those who are in the worst position are the ones who are deceived by the present world and neglect the world to come. The happiness that they enjoy cannot compensate for the fear, misery and terror that they will face when they leave this world. We know that were a man to realize what awaits him when death comes and he is removed from the pleasures and comforts that he enjoys here, he would renounce all worldly things, and we are certain that the world to come will be better and more advantageous for us.' 'Wise vizier,' replied the prince, 'the gleam of your lamp has cleared away the darkness that shrouded my heart, directing me to the paths that you yourself have trodden in the pursuit of truth and providing me with light by which to see.'

At this point, one of the sages who were present got up and said: 'When spring comes, the hare as well as the elephant must look for pasture. In your questions and explanations, I have heard what I don't think that I have ever heard before, and that prompts me to put a question to you both. What are the best worldly gifts?' The prince answered: 'Health, a livelihood obtained lawfully and a virtuous son.' The next question was: 'What is great and what is small?' to which the prince answered: 'The great is what is submitted to by the smaller, while the small is what endures the greater.' Then the sage asked: 'In what four things do all creatures have an equal share?' The prince said: 'Food and drink, the pleasure of sleep, desire for women and the pangs of death.' 'What are the three things whose ugliness cannot be set aside?' asked the questioner, and the prince answered: 'Stupidity, a mean nature and lying.' 'All lying is bad,' said the questioner, 'but which lie is better than others?' The prince said: 'The lie that protects the speaker from harm and brings advantage.' 'Although all truth is good,' the sage went on, 'which truth is ugly?' 'This is where the speaker boasts conceitedly about himself,' answered the prince. 'What is the foulest of the foul?' the man asked. 'This is where the speaker boasts of qualities that he does not possess,' replied the prince, and when he was then asked to define a fool, he said: 'A fool is someone whose only concern is for what he can stuff into his belly.'

'Your majesty,' Shimas now said, 'you are our king and we, your servants and subjects, would like you to nominate your son to succeed you as your heir.' At that, the king urged all who were present, both scholars and others, to remember what they had heard and to act on it, as well as to obey the orders of his son, whom he appointed as heir to his throne. All the citizens of his state, scholars, men of courage, the old and the young, as well as everyone else, were made to swear that they would not oppose him or disobey his orders.

When the prince was seventeen, a grave illness brought his father to the point of death, and being sure that he was about to die, he told his household: 'My illness is fatal, so summon my relatives and my son and order all my subjects to assemble here.' His servants went out and summoned those who were near at hand, while sending a proclamation to notify those who lived at a distance, with the result that everyone came into the king's presence. They asked him how he was and about the state of his disease, to which he replied: 'This is my last illness; by God's decree, the arrow of fate has pierced me. This is the last day of my earthly life and the first of my life in the next world.' He told his son to come near and when he did, the boy wept so bitterly that his tears almost soaked the bed. There were tears in the king's eyes, too, and everyone else who was present joined in the weeping. 'Do not shed tears, my son,' the king said. 'I am not the first to suffer this fate, which comes to all of God's creation. Fear God and perform good deeds which will go ahead of you to what is the goal of all creatures. Do not yield to passion, and occupy yourself with calling on God's Name, while you stand or sit, wake or sleep. Set truth before your eyes. This is my last word to you and so farewell.'

Night 917

Morning now dawned and Shahrazad broke off from what she had been allowed to say. Then, when it was the nine hundred and seventeenth night, SHE CONTINUED:

I have heard, O fortunate king, that King Jali'ad gave these instructions to his son and nominated him as his successor to the throne. 'Father,' said the prince, 'you know that I have always obeyed you and kept to your instructions, carrying out your orders and looking for your approval. You have been a good father to me, so how could I do anything of which you might disapprove after your death? You brought me up well, but now you are about to leave me and, although I cannot bring you back, to obey your instructions will make me happy and bring me the greatest share of good fortune.' The king, who was near to his end, said: 'My little son, there are ten things through which God will help you in this world and the next. When you are angry, suppress your anger; when misfortune visits you, endure; in speaking, tell the truth; keep your promises; be just in your judgements; be forgiving when you have the power; be generous to your officers; spare your enemies; be charitable to them and do them no harm. There are ten other things which will bring you God's help in relation to your subjects. Be just in what you share out; when you are right to punish do not act tyrannically; keep to your agreements; accept advice; do not be stubborn; make sure that your subjects abide by divine laws and praiseworthy traditions; arbitrate justly among the people so that high and low alike may love you, while the arrogant and the evil-doers fear you.' Then he turned to the learned men and the emirs who had been present when he had appointed his son as his heir, and said: 'Take care that you do not disobey the commands of your king and neglect what he says, for this would lead to the destruction of your country, the break-up of your society, physical harm and the loss of your wealth, leaving your enemies to exult over your misfortunes. You know that you swore allegiance to me and it is this allegiance that must now be transferred to the prince, so that the covenant that there was between us will now be between you and him. You must listen to him and obey him, as this will ensure that your affairs flourish. If you support him as firmly as you supported me, all will go well with you and you will be successful, for he is your king who is entrusted with your prosperity. Farewell.'

The death throes now intensified and, being unable to speak any longer, the king embraced his son and kissed him. Then, after he had given thanks to God, his soul left him and he died, mourned by all his subjects throughout the kingdom. They covered him in a shroud and buried him with full honours and respect. Then, when they returned from the funeral with the prince, they dressed him in royal robes, crowned him with his father's crown, put the signet ring on his finger and sat him on the royal throne.

For a short while, the young king followed his father's example by acting with clemency, justice and beneficence, but then the lure of the world and its lusts led him astray. He indulged in its pleasures, turning his attention to its vanities and abandoning the compact that his father had made with him. In disobedience to his father, he neglected his kingdom and pursued a path of self-destruction. So great was his love of women that whenever he heard of a beautiful girl, he would send for her and marry her, until he had collected more wives than Solomon, the king of the children of Israel. He would spend every day alone with a number of them, and would stay with this chosen group for a whole month at a time without leaving them or asking about the governance of his kingdom. He did not investigate the wrongs of which his subjects complained and if they sent written petitions he would not reply.

When his subjects saw this and were confronted by his persistent neglect of their affairs and his disregard for their interests as well as for affairs of state, they became sure that they would soon be faced with disaster. In their distress, they met and exchanged criticisms of the king. Then they suggested to each other that they had better go to the chief vizier, Shimas, and explain their position, telling him how the king was acting and asking him to advise them, as otherwise calamity would not be long in overtaking them. 'This king,' they said to him, 'has been dazzled by the delights of this world, which have ensnared him. He has turned his face to what is false, and is doing his best to ruin his kingdom, which will involve the ruin of the common people and the destruction of all that we have. This is because we can wait for days and months without seeing him and without his issuing any instructions either to you, the vizier, or to anyone else. No case can be taken to him; he will not look into the administration of justice; and he is so careless that he will not examine the condition of a single one of his subjects. We have come to you to tell you about the real state of affairs because you are our senior and better qualified than the rest of us. No catastrophe should affect a land in which you live, as no one has more power than you to reform the king. Go and talk to him, for it may be that he will listen to you and turn back to God.'

Shimas got up and went to look for whatever page he could find. 'My good boy,' he said to one of them, 'please get me permission to go to the king, as there is something that I have to tell him to his face and then listen to his answer.' 'By God, master,' replied the page, 'for a whole month he has not allowed anyone to see him, and for all this time I've not set eyes on his face. But I shall direct you to someone who can do this for you. Get hold of the servant who stands by his head and brings him food from the kitchen; when he comes out to the kitchen to fetch the food, ask him and he will do what you want.' Shimas went to the kitchen door and sat there for a short while before the servant arrived. As he was about to go in, Shimas said to him: 'My son, I want to meet the king to tell him something that particularly concerns him. When he is in a good mood after finishing his meal, would you be kind enough to ask him to allow me in, so as to talk to him about something he should find appropriate?' 'To hear is to obey,' the servant replied, and he then took the food to the king. When the latter had finished eating and was feeling cheerful, the servant told him that Shimas was standing at the door and wanted permission to enter in order to tell him about something of particular concern to him. The king, filled with alarm and disquiet, told the man to bring Shimas in.

Night 918

Morning now dawned and Shahrazad broke off from what she had been allowed to say. Then, when it was the nine hundred and eighteenth night, SHE CONTINUED:

I have heard, O fortunate king, that the king ordered his servant to bring Shimas in. The man did this, and when Shimas came into the king's presence, he prostrated himself before God and then kissed the king's hands, calling down blessings on him.

'What has made you ask to see me, Shimas?' said the king, and Shimas replied: 'It has been a long time since I last looked on your face and I have felt a great longing for you. Now that I am in your presence, I have something to tell you – you, whom God has aided with every favour.' 'Say whatever you think fit,' the king told him, and at that Shimas began:

'Your majesty knows that in spite of your youth the Almighty has provided you with more knowledge and wisdom than He has bestowed on any king before you. He completed His favours by giving you the kingship, and it is not His wish that you should abandon these gifts of His to look for something else out of disobedience to Him. You should not use the treasures that you have in order to fight against Him, but you should keep His injunctions in mind and obey His commands. In the last few days, I have seen that you have forgotten what your father told you, rejecting the pledge that you gave him and neglecting his words of advice. You have abandoned his just ordinances and, in your ingratitude, you have not remembered how gracious God has been to you.' 'How is that, and what makes you say this?' the king asked, and Shimas replied: 'You have given up taking an interest in matters of state as well as the affairs of your subjects which God has entrusted to you, and you have turned to the petty lusts of this world that your soul has made you think of as desirable. It has been said that what a monarch must preserve are the interests of his kingdom, of religion and of his subjects. My advice, then, your majesty, is that you should look carefully at what awaits you in the future, for in this way you will find the clear path that leads to salvation. Don't turn to the small pleasures that pass away and that lead to the abyss of destruction, for if you do, you will suffer the fate of the fisherman.' 'What was that?' asked the king. SHIMAS WENT ON:

I have heard that a fisherman was in the habit of going to a river to fish. When he got there one day and was walking along the dyke that bordered it, he caught sight of an enormous fish. 'There is no point in my staying here,' he told himself. 'I must go off and follow this fish wherever it goes until I catch it, for then I won't have to fish again for days.' He then stripped off his clothes and went down into the water behind the fish and the current carried him on until he managed to grasp it, but when he turned, he found that he was far away from the shore. Although he saw where the current had taken him, he didn't let go of the fish. Rather than going back, he put his life at risk by keeping hold of it with both hands and letting himself float with the current. This carried him on until it flung him into a whirlpool from which no one had ever come out alive. He began to shout: 'Save me; I'm drowning.' Some river wardens came up and asked what had come over him to make him put himself in such terrible danger. 'I abandoned the clear path of salvation,' he told them, 'and followed my destructive desires.' 'How did you come to do this and involve yourself in such danger?' they asked. 'You've known for a long time that no one who enters this whirlpool can ever be saved, so why didn't you let go of what you were holding and look for safety? In that way, you would have got out alive and avoided falling into this peril from which there is no escape. As it is, none of us can rescue you from it.' The fisherman despaired of his life and he lost the fish that he had been holding, which had led him on to total destruction.

'I have only told you this parable, your majesty,' said Shimas, 'so that you may abandon the contemptible distraction which diverts you from your real interests, and turn your attention to the responsibility that you have been given to look after the affairs of your subjects and the administration of your realm. In that way, no one will be able to find fault with you.' 'What are you telling me to do?' the king asked. Shimas said: 'Tomorrow, if you are in good health, allow the people to come to you; look into their affairs and make an apology to them. Then promise them good treatment and good conduct on your part.' 'What you have said is right, Shimas,' the king told him, 'and tomorrow, God willing, I shall act on your advice.'

Shimas left the king's presence and told the people everything that he had said. Then, next morning, the king came out from his private apartments and, after allowing the people to enter, he began to apologize to them and to promise that he would do what they wanted. They were satisfied with this and went off, each returning to his own home, but the king's favourite wife, to whom he showed particular honour, came in and found that he had lost colour and was brooding over his affairs because of what the chief vizier had told him. 'Your majesty,' she asked, 'why do I find you so disturbed? Is there something the matter with you?' 'No,' he told her, 'but I have been so sunk in my pleasures that I have neglected my duty, and I wonder how I came to be so neglectful of my affairs and those of my subjects. Were I to continue like this, very soon I would lose my kingdom.' His wife replied: 'I see that your governors and your viziers have misled you. They want to injure you and to scheme against you in order to see that you get no pleasure, enjoyment or rest from your kingship. Instead, they want you to pass your whole life in protecting them from hardship so that you will have to spend all your time in thankless labour. You will be like a man who killed himself for somebody else's benefit or like the boy and the thieves.' When the king asked her about this story, SHE SAID:

It is said that one day seven thieves went out to steal something, as was their habit. Their route took them past an orchard in which there was a walnut tree with ripe nuts. When they were going in, they found a young boy standing there and they asked him whether he would like to come with them and climb the tree. He could then eat his fill of the walnuts and throw down some to them. He agreed to this and accompanied them into the orchard.

Night 919

Morning now dawned and Shahrazad broke off from what she had been allowed to say. Then, when it was the nine hundred and nineteenth night, SHE CONTINUED:

I have heard, O fortunate king, that the boy agreed to do what the thieves wanted and went into the orchard with them. They then said to each another: 'We had better see which of us is the lightest and the smallest and make him climb the tree.' It was pointed out that none of them were as thin as the boy, and so they sent him up, but said: 'Don't touch any of the nuts on the tree in case somebody sees you and punishes you.' 'What am I to do, then?' the boy asked, and they told him: 'Sit in the middle of the tree and shake each of the branches so hard that the nuts fall off. We shall pick them up, and when you have finished and climbed down, you can take your share of what we have collected.' So the boy went up the tree and started to shake every branch that he could reach, so that the nuts fell off and the thieves gathered them up. While they were doing this, the owner came and stood beside them to ask what they were doing with his tree. 'We haven't taken anything from it,' they said, 'but as we were passing, we saw this boy up there and, since we thought that he must be the owner, we asked him for some nuts to eat. He shook some down from the branches, but as for us, we've done nothing wrong.' 'What have you to say for yourself?' the owner asked the boy, and he replied: 'They are lying. I shall tell you the truth, which is that we all came here and they told me to climb the tree in order to shake nuts down for them from the branches. I only did what they told me.' 'You have brought great trouble on yourself,' said the owner. 'Did you take the opportunity to eat any of the nuts yourself?' When the boy insisted that he had eaten nothing, the owner said: 'I know now that you are a stupid fool, as you have done your best to ruin yourself for the good of others.' He told the thieves that they could go off as he had no case against them, but he seized the boy and beat him.

'In just the same way,' continued the king's wife, 'your viziers and state officials want to ruin you for their own benefit, and they are treating you as the thieves treated the boy.' 'That is true,' said the king, 'and you are right in what you say. I shall not go out to meet them and I shall not give up my pleasures.'

He spent the most agreeable of nights with his wife and then, when morning came, the vizier collected the state officials and the other citizens who were with them and brought them to the king's door. They were happy and joyful, but the door stayed shut; the king did not come out to them, nor did he allow them to go in. When they had given up hope of seeing him, they said to Shimas: 'Excellent vizier, perfect in your wisdom, what do you think of this young boy who is not only unintelligent but who has added lying to his other faults? See how he has broken his promise to you, and the fact that he has not kept his word has to be added to the list of his defects. But we hope that you will go to him again to find out what kept him and why he didn't come out, although we know well enough that this kind of thing is an example of his depraved nature, for he is now hardened in his ways.'

Shimas set off and when he had come into the king's presence and greeted him, he said: 'Why is it that I see that you have abandoned the great task with which you should concern yourself, in order to turn to paltry pleasures? You are like a man who was so delighted with the milk of his milk-camel that he neglected to hold on to its reins. One day, when he had forgotten to do this as he went to milk it, the camel, realizing what was happening, pulled away and made off into the desert. So the man lost both the milk and the camel, and his loss exceeded his gain. Your majesty should consider what is best for you and for your subjects. It is not a good idea for someone to keep sitting by the kitchen door because he needs food, and similarly, however fond he is of women, he should not always sit with them. Just as a man only needs enough food and drink to ward off hunger and thirst, so, if he is intelligent, out of twenty-four hours in the day, it is enough for him to pass two with women, while he should devote the rest to his own affairs and those of his dependants. If he stays and spends longer than this with them, it will harm both his mind and his body, for women do not prompt men to do good or act as guides to this. Neither what they say nor what they do is to be accepted; indeed, I have heard that many men have been destroyed because of women. There was, for instance, a man who died because, in an encounter with his wife, he did what she told him.' 'How was that?' asked the king, AND SHIMAS SAID:

According to the story, a man had a wife whom he loved and honoured. He used to listen to what she had to say, and act on her advice. This man had a garden which he had newly planted with his own hands, and he would go to it every day in order to look after it and water it. One day his wife asked him what he had planted in it, and he said: 'Everything that you are fond of and want to have, and I am doing my best to cultivate it and water it.' 'Would you take me and show me round,' she asked, 'so that I can look at it and then call down a blessing on you? For my prayers are answered.' Her husband agreed, but told her to wait until he could come and take her on the following day. The next morning, he went there with her, but, as they were going in, two young men looked at them from a distance and one said to the other: 'The man is an adulterer and she is an adulteress. They must have gone into the garden in order to make love.' So they followed the pair to see what would happen, and stood at the side of the garden.

As for the man and his wife, when they had gone in and stopped, he said to her: 'Now pray for the blessing that you promised to call down on me.' 'No,' she said, 'not until you do for me what women want from men.' He was indignant, saying: 'Isn't what I do at home enough? Here I would be afraid of disgrace; you are distracting me from what I should be doing. Aren't you afraid that someone may see us?' 'There's no need to worry about that,' she told him, 'as we are not doing anything disgraceful or forbidden, and there's no need to hurry when it comes to the watering of the garden, as you can do that any time you want.' She would listen to no excuse or argument and kept on pressing him to lie with her, until he did. When the two young men saw this, they leapt out and seized hold of the pair, saying: 'You are adulterers and we shan't let you go. Unless we can have the woman, we shall report the matter to the authorities.' 'Damn you!' said the man. 'This is my wife and I own this garden.' But they paid no attention and jumped on the woman. She screamed and called to her husband to help her, saying: 'Don't let them rape me.' He moved towards them, shouting for help, but one of the pair turned back at him, struck him with his dagger and killed him, after which he and his companion raped the woman.

Night 920

Morning now dawned and Shahrazad broke off from what she had been allowed to say. Then, when it was the nine hundred and twentieth night, SHE CONTINUED:

I have heard, O fortunate king, that after the young man had killed the woman's husband, he and his companion went back to her and raped her. Shimas went on: 'I have told you this story, your majesty, so that you may realize that no man should listen to what a woman says, obey her in any matter or take her advice. Take care not to dress yourself in ignorance after you have worn the clothing of wisdom and learning, and not to follow evil counsel after you have learned how to reach what is both right and advantageous. You should not pursue trifling pleasures, which will lead you to ruin and to disasters piled on disasters.'

When the king heard this, he promised that, God willing, he would come out to meet his people on the following day. Shimas went to tell this to those of the leading men of the kingdom who were there, but when the king's wife heard what Shimas had said, she went to the king and told him: 'The king's subjects are his slaves, but now I see that, king though you are, you are the slave of your subjects in your awe of them and you are afraid lest they harm you. They want to test what you are really like, and if they find that you are weak, they will despise you, whereas if they see that you are courageous, they will respect you. This is what evil viziers, being masters of guile, do with their kings. I can show you clearly what they are really planning, and if you agree to do what they want, they will make you abandon your own interests and follow theirs. Then they will shift you from one position to another until they ruin you, and your position will be like that of the merchant and the thieves.' The king asked about this, AND HIS WIFE SAID:

I have heard that there was once a wealthy merchant who went off to sell his goods in a certain city, and when he got there he settled in a rented house. Thieves, who used to keep an eye on merchants in order to steal their goods, noticed him and went to his house, hoping to find a way in. They failed and so their leader promised to solve the problem for them. He went off and dressed as a doctor, carrying a bag containing medicines over his shoulder. He went around crying: 'Who needs a doctor?' until he got to the merchant's house. He found the man sitting down to his morning meal and asked him if he needed a doctor. 'No, I don't,' the man replied, 'but sit down and share my meal,' at which the thief took his place opposite him and started to eat with him.

The merchant was a hearty eater, and this led the thief to tell himself: 'This is my opportunity.' He then said: 'In return for your kindness to me, I must give you some advice which I cannot keep from you. I see that you are a man who eats a lot, and this will lead to a stomach complaint, which, unless you look for a cure quickly, will prove fatal.' The merchant said: 'I have a healthy body and my digestion works fast, so even though I eat a lot, I owe God grateful thanks that I suffer from no sicknesses.' 'That's what you think,' replied the thief, 'but I know that there is a disease lurking within you, and if you take my advice you will try to cure yourself.' 'Where am I going to find someone who knows how to cure me?' the merchant asked, and the thief replied: 'It is God Who cures, but a doctor like me treats patients as best he can.' 'Show me the medicine I should take,' said the merchant, 'and let me have some of it.' The thief gave him powder mixed with a large quantity of aloes, telling him to dose himself with it that night. The merchant took the powder and used some of it in the evening. He found the taste of the aloes unpleasant, but that did not put him off, and when he had taken the dose, he found that it relaxed him that night.

The following evening, the thief came back with medicine that contained an even larger dose of aloes, some of which he gave to the merchant. When the merchant took it, it gave him diarrhoea, but he put up with this and made no objection. The thief saw that he had taken his words seriously and was prepared to entrust him with his life. When he was sure that the man was not going to disregard his advice, he went off and came back with some deadly poison, which he gave to him. His victim took it and drank it down, and when he had done so, it cleared out everything that was in his stomach and lacerated his entrails, as a result of which he died. The thieves then came and took all his goods.

'I have told you this, your majesty,' continued the king's wife, 'simply in order to stop you from accepting anything that this trickster says, lest the consequences involve you in ruin.' 'You are right,' the king told her, 'and I shall not go out to meet my subjects.' The next morning, the people gathered together and came to the palace gate, where they sat for most of the day until, having despaired of his ever coming out, they went back to Shimas and said: 'Wise philosopher and skilful scholar, don't you see that this ignorant boy does nothing but lie to us again and again?

The right thing to do is for us to depose him and replace him with someone else. In that way, things will be set right and we shall prosper. Nevertheless, go back to him a third time and let him know that the only thing that keeps us from rising against him and removing him from the throne is the kindness that his father showed us and the oaths that he made us take. Tomorrow every last one of us will gather together with our weapons in our hands and we shall smash down the gate of his citadel. If he comes out to meet us and does what we want, then we shall do him no harm, but otherwise we shall break in, kill him and give the kingdom to someone else.'

Shimas went back to the king and said: 'Your majesty, what are you doing to yourself, engrossed as you are in your lusts and pleasures? I wonder who it is who is prompting you to behave like this, but if it is you who are wronging yourself, then the sound wisdom and eloquence that, in my experience, you used to possess has deserted you. I would like to know who it is of those around you who has led you from knowledge to ignorance, from good faith to harshness and from mildness to severity, stopping you from accepting my advice and making you shun me. Three times I have given you advice and you have refused to accept it, although what you rejected was good. What is this heedless pursuit of amusement and who is leading you astray? You should know that your subjects have agreed to break in, kill you and give your throne to someone else. Do you have the power to deal with them all and to escape from their hands or to bring yourself back to life after you have been killed? If you can do all this, then you are safe enough and you don't need me to tell you anything, but if you still need worldly things, including your kingdom, come back to your senses, set the kingdom in order, make a show of power and excuse yourself to your subjects. For they intend to take away your throne and hand it to another, as they are determined to disobey you and rebel. They are led to this by what they can see of your youth and your obsession with lusts and pleasures. Even if stones have been in the water for a long time, when they are taken out and struck against one another, sparks will be dashed from them. Similarly, your many subjects are acting in concert against you in order to transfer the monarchy and achieve their purpose, which is to destroy you. Your position will be like that of the jackals and the wolf.'

Night 921

Morning now dawned and Shahrazad broke off from what she had been allowed to say. Then, when it was the nine hundred and twenty-first night, SHE CONTINUED:

I have heard, O fortunate king, that Shimas told the king: 'They will achieve their purpose, which is to kill you and your position will be like that of the jackals and the wolf.' The king asked about this story, AND THE VIZIER TOLD HIM:

It is said that one day a pack of jackals went out to look for something to eat, and in the course of their wanderings they came upon a dead camel. 'We have found enough to serve us by way of food for a long time,' they said to themselves, 'but we can't be sure of fair play, as the strong ones among us may use their strength to destroy the weak. What we need is someone to arbitrate between us, in return for a share in our find, so as to stop the strong from dominating the weak.' While they were consulting about this, a wolf came up to them and they said to each other: 'If this idea is sound, then let us take the wolf as an arbitrator, as he is the strongest of creatures and his father before him used to be our ruler. We put our hope in God that he will act justly with us.' So they went to the wolf and told him what they had decided. 'Act as an arbitrator for us,' they said, 'and give each of us what we need every day in the way of food so as to keep the strong from oppressing the weak, which might lead us to kill one another.'

The wolf agreed to this and took their affairs in hand. That day, he gave them enough to satisfy them, but next day he said to himself: 'If I divide this camel among these weaklings, the only thing I get will be what they have assigned to me, whereas if I eat it all myself they won't be able to harm me, as they are my flock and the flock of my family. Who is going to stop me from taking this for myself, and in any case might it not be that it was God Who brought it to me rather than any kindness on their part? The best thing would be for me to take it all, and from now on I'm not going to let them have anything.' The next morning, the jackals came to him as usual to get their daily ration, but when they asked the wolf for this, he told them that he had nothing left to give them. They went away in the saddest of states, saying: 'God has caused us great grief through this vicious traitor who has no fear of Him, while we ourselves have no power and are helpless.'

Then they told each other: 'It may be that it was hunger that forced the wolf to act like this, so let him eat his fill today, and tomorrow we can go to him again.' So next day they went to him and said: 'Wolf, we appointed you as our leader so that you should give each of us our share of food, dividing it fairly between the weak and the strong. Then, when there is no more left, it will up to you to do your best to get some more. In that way, we shall always be under your protection as your subjects. We are hungry, not having eaten for two days, so give us our rations and you will be free to do what you want with whatever is left over.' The wolf made no reply, treating them with even greater harshness, and refusing to relent in spite of all their efforts. So they told themselves that the only thing they could do would be to go to the lion, throw themselves on his mercy, and make over the camel to him. 'If he gives us any of it, this will be thanks to his kindness and if not, then he has a better right to it than this foul wolf.' They then went to the lion and told him of their experience with the wolf. 'We are your servants,' they said, 'and we are here to ask for your help to free us from the wolf, after which we shall remain in your service.' When the lion heard this, in his zeal to serve Almighty God he became angry, and he went with the jackals to find the wolf. On seeing him approach, the wolf tried to escape, but the lion bounded after him, seized him and tore him to pieces, after which he allowed the jackals to take back their prey.

'This shows us that no king should despise his subjects,' continued the vizier. 'So accept my advice and keep the promise that you made earlier. Remember that, before his death, your father told you to listen to counsel. This is my last word to you.' 'I hear what you say,' the king told him, 'and tomorrow, God willing, I shall come out to meet them.'

Shimas left and told the citizens that the king had accepted his advice and promised to come out the following day. But when the king's wife heard what Shimas had said and was certain that the king was definitely proposing to go out to meet his subjects, she hurried up to him and said: 'I am amazed at how you submit to your slaves and obey them. Don't you realize that these viziers of yours are, in fact, your slaves? How is it that you have raised them to such a height, leading them to imagine that it is they who have conferred the kingdom on you and it is they who are responsible for your eminence? They think that they have given you gifts, while, in reality, they can't do you the slightest harm. It is your duty to show them no deference whatsoever, while, for their part, they have to defer to you and carry out your orders. How is it, then, that you can show yourself to be so afraid of them? There is a saying that no one whose heart is not hard as iron is fit to be a king. It is your clemency that has led these people astray so that they have had the impudence to cast off their allegiance to you, although they should be compelled to submit and forced into obedience. If you rush to accept what they say, overlooking what they have done, or do the slightest thing for them against your will, they will press you harder; you will arouse their greed and this will become a habit with them. If you take my advice, you will not promote any of them, accept anything that they say or encourage their impudence, for otherwise you will be like the shepherd with the thief.' The king asked about this story, AND HIS WIFE TOLD HIM:

It is said that one night, while a shepherd was watching over his flock, he was approached by a thief who wanted to steal some of his sheep. The thief found that the shepherd guarded them so well, neither sleeping at night nor neglecting them by day, that although he tried all night long, he could not take a single one of them. Finding himself baffled, he went out into the country where he killed a lion, which he then skinned, stuffing the skin with straw. He brought the stuffed beast and set it up on a high point in the open country where it was in the shepherd's sight and he could make out what it was. He then went up to the shepherd and said: 'The lion has sent me to demand some of these sheep for his supper.' 'Where is the lion?' the shepherd asked, to which the thief replied: 'Look up. There he is, standing over there.' When the shepherd looked, he thought that this was a real lion and became terrified.

Night 922

Morning now dawned and Shahrazad broke off from what she had been allowed to say. Then, when it was the nine hundred and twenty-second night, SHE CONTINUED:

I have heard, O fortunate king, that when the shepherd saw the shape of the lion, he thought that it was real and was terrified. 'Brother,' he said, 'I cannot disobey; take what you want.' The thief did so, but then the shepherd's terror made him greedier, and at short intervals he would come back to scare the man by saying: 'The lion needs this and intends to do that,' after which he would take as many sheep as he wanted. This went on until he had destroyed the greater part of the flock.

'I have only told you this, your majesty,' continued the king's wife, 'in case your clemency and mildness mislead the ministers of state and rouse their greed. The right course to follow is to kill them rather than let them go on with what they are doing to you.' The king agreed that he would accept her advice and would not listen to what his counsellors said, or go out to confront his subjects.

The next morning, the viziers, ministers and leading citizens assembled, each man carrying arms. They set off to storm the royal palace, kill the king and replace him with another. When they got there, they asked the gatekeeper to open the door for them and when he refused, they sent for fire to burn down the gates so as to force their way in. When the gatekeeper heard what they were saying, he hurried off and told the king that the people had collected at the door and that when he had refused to open it as they had demanded, they had sent for fire and were intending to set it alight. 'Then they will burst in and kill you,' he added, 'and so what are your orders?' The king said to himself: 'I have fallen into deadly danger.' He sent for his wife and when she came, he said: 'Every single thing that Shimas told me has turned out to be true, and my subjects, high and low, have come here to kill both me and you. When the gatekeeper refused to open the door for them, they sent for fire to burn it down, as they intend to burn the palace with us inside it. What do you advise me to do?' 'No harm will come to you,' the woman replied, 'and don't be afraid of them, for this is a time when fools rise against their kings.' 'But what do you tell me to do,' asked the king, 'and how am I to deal with this business?'

She said: 'My advice is that you tie a bandage around your head and pretend to be ill. Then send for Shimas, and when he sees the state you are in, say that you had wanted to go to meet the people today but your illness prevented you. Tell him to go out himself to let them know what has happened and to promise that you will come next day to settle their business for them and look into their affairs, in order to quieten them and to calm their anger. Tomorrow morning, summon ten of your father's slaves, strong men to whom you can trust your life and who will listen to your orders and obey them, and who will keep your secret out of love for you. Get them to stand by your head and only to allow people in to see you one by one. Whoever comes in is to be seized on your orders and killed. When the slaves have agreed to this, have your throne placed in your audience chamber and open the door. Your subjects will be glad to see that, and they will approach without any evil intent in order to ask permission to enter your presence. As I said, you are to let them come in one by one, and then you can do what you want with them. But first of all start by killing Shimas, their leader, for he is the chief vizier and the administrator of the kingdom. When you have disposed of him, then kill all the others, one after the other, until you have left none of those whom you know are about to break their oath of allegiance to you or who you fear might attack you. If you do this, they won't be able to oppose you any longer, and, as far as they are concerned, you can be totally at your ease. Your reign will be untroubled and you will be able to do what you want. You can be assured that nothing will be of more use to you than this plan.'

The king said: 'This is good advice and the right way to proceed, and I shall certainly do what you suggest.' So he called for a bandage, which he tied around his head, and then, pretending to be ill, he sent for Shimas. Shimas came, and the king said: 'You know my affection for you and that I follow your advice, for you, more than anyone else, are like a brother or a father to me, and you know that I am ready to do whatever you tell me. You told me to go out to meet my subjects and to sit in judgement on the cases that they bring to me. I was quite sure that this was good advice and I had intended to go to meet them yesterday, when this illness attacked me and I was unable to sit up. This has made them angry with me, I hear, so that they are thinking of harming me, little though I deserve it, as they don't know how ill I am. So do you go out and tell them; give them my excuses, for I shall agree to their demands and do what they want. Settle the affair for me and give them a guarantee on my behalf, for you have acted as an advisor both to me and to my father before me, and you are in the habit of settling disputes. God willing, I shall come out to meet them tomorrow, as it may be that by then I shall have got rid of my illness, and the sincerity of my intention and the good that I have in mind to do for them may bring with them a blessing.'

Shimas prostrated himself in reverence to God, called down blessings on the king and kissed his hands. He happily went out to the people and when he told them what he had heard, he forbade them to carry out their plan of attacking the palace, explaining to them the king's excuse for not having come out. The king, he said, had promised to come the next day and to do what they wanted. As a result, the crowd dispersed back to their homes.

Night 923

Morning now dawned and Shahrazad broke off from what she had been allowed to say. Then, when it was the nine hundred and twenty-third night, SHE CONTINUED:

I have heard, O fortunate king, that Shimas went out to the council and told the counsellors that the king would come out to them the next day and do what they wanted. As a result, they dispersed back to their own homes.

So much for them, but as for the king, from a group of enormous slaves who had been in his father's service he selected ten huge, resolute and powerful men. He sent for them and told them: 'You know how my father favoured and promoted you, treating you with kindness and generosity. I shall advance you even further and I shall tell you the reason for this since, as far as I am concerned, you are under the protection of God. First, though, I must ask you whether you will carry out the orders that I give you and keep my secret from everyone. If you obey me, I shall give you more than you can wish for.' Speaking with one voice, all ten said: 'We shall do whatever you tell us to do; we shall disregard no instruction you give us, for you are our commander.' 'May God be good to you,' he replied. 'I shall now tell you why it is that I have singled you out for additional favour. You know how well my father treated his subjects; how he got them to swear allegiance to me and how they assured him that they would never break their word or disobey me. You saw what they did yesterday when they surrounded me and wanted to kill me, and now I want to do something to them. When I observed their behaviour yesterday, I could see that only exemplary punishment would keep them from doing this kind of thing again. I must then charge you with the task of killing secretly those whom I point out to you so that, by executing their leaders, I may save my lands from harm and disaster. In order to do this, I shall sit here tomorrow in this room and allow them to enter one after the other, coming in by one door and going out by another. You ten are to stand before me, waiting for my signal. Everyone who comes in is to be seized, taken into that room there and killed, after which the bodies are to be concealed.' 'We hear and shall obey your command,' they said, after which he gave them generous gifts and dismissed them.

When the night had passed, he sent for these men in the morning and told them to set up his throne. He put on his royal robes and, with his legal textbook in his hand, he ordered the door to be opened. This was done, and as the ten slaves stood before him, the herald called out: 'Let all office holders come before the king.' At that, the viziers, commanders and chamberlains came forward and each took up his proper place. They were then told to enter the king's presence one by one, the first to go in being Shimas, as this was the customary privilege of the grand vizier. He entered and was standing in front of the king when, before he knew what was happening, he was surrounded by the ten slaves, who seized him, took him into the next room and killed him. They then dealt with the rest of the viziers, the scholars and men of piety, killing them one after the other until they had despatched them all. The king then ordered his executioners to kill all the remaining men of courage and strength, and every single man distinguished for his energy perished. Only the lowest of the rabble were left, and these were driven away, each returning to his family.

The king was now left alone to enjoy his pleasure. He devoted himself to his lusts, following the path of oppression, injustice and wrongdoing until he had outstripped all the evil tyrants of earlier times. In his lands, gold and silver were mined, together with sapphires and other gems, and this was a source of envy among the neighbouring rulers, who had been waiting for some disaster to strike him. One of these kings now told himself: 'This is what I have been looking for. Now I can take the kingdom from that foolish boy thanks to the fact that he has killed all his principal officers, together with every brave champion that was in his land. This is my chance to seize what he holds, as he is a juvenile with no experience in war. He has no good sense of his own and there is no one left who can guide him or help him. Today I shall open the way to attacking him by writing him a mocking letter, criticizing him for what he has done in order to see how he will reply.'

So he wrote: 'I have heard how you treated your viziers, scholars and men of valour, and how you have brought disaster on yourself by your tyranny and wrongdoing, as you no longer have any power to repel an attack. God has helped to give me the upper hand over you and so you must listen to what I have to say and do what I tell you, which is to build an impregnable castle for me in the middle of the sea. If you cannot do that, then save yourself and leave your kingdom, for I am going to send twelve regiments against you from the borders of India, each of twelve thousand fighting men, who will cross your frontiers, plunder your goods, kill your men and take your women as captives. I shall appoint my vizier, Badi', as commander and tell him to lay siege to your city until he captures it, and I have told my servant, whom I am sending to you with this letter, that he is to stay no more than three days with you. If you do what I tell you, you will escape with your life, but if not, I shall do what I have said and despatch my army against you.' He then sealed this letter and handed it to his messenger, who took it to the king's city and gave it to him as soon as he was admitted into his presence.

When he had read it, the king felt weak, dismayed and confused. Faced with the certainty of destruction, he could find no one from whom to ask advice or who could aid and support him. He went to his wife and when she saw how pale he was, she asked what was wrong. 'I am no longer a king,' he told her, 'but only a king's slave,' and, opening the letter, he read it out to her. Having heard what it said, she started to weep and wail, tearing her clothes, and when he asked her whether she had any advice or could think of some way out of his difficulties, she said: 'Women's wiles don't work when it comes to war. They have neither judgement nor strength, as these, together with stratagems, are the prerogatives of men in cases like these.' On hearing this, the king was filled with bitter regret and sorrow for the excesses that he had committed against his followers and his ministers of state.

Night 924

Morning now dawned and Shahrazad broke off from what she had been allowed to say. Then, when it was the nine hundred and twenty-fourth night, SHE CONTINUED:

I have heard, O fortunate king, that when the king heard what his wife had to say, he was filled with regret and sorrow for the excesses that he had committed, in that he had killed his viziers and the leading citizens. He wished that he had died before this terrible news had reached him and he told his wives: 'What has happened to me, thanks to you, is like what happened to the partridge with the tortoises.' When they asked what this was, HE SAID:

It is said that a number of tortoises lived on a wooded island where there were fruits and streams. One day, it happened that a partridge, which was flying past, landed there, being overcome by heat and weariness, and when it saw where the tortoises lived it went there for shelter. These tortoises were in the habit of feeding in various parts of the island before coming back home, and this time when they got back from their pastures, it was to discover the partridge. The sight of it filled them with wonder, as God had made it seem beautiful to them. They praised their Creator and were filled with love for the partridge, which delighted them. They told each other: 'There is no doubt that this is one of the most beautiful of birds,' and all of them began to treat it with kindness and attention. When it saw how fond they were of it, it came to be on friendly terms with them, and each evening, after it had flown to wherever it wanted, it would come back to spend the night with them before going off again in the morning. It went on doing this for some time until the tortoises found that when it was away they felt lonely. They realized that they would only see it at night-time and that, in spite of their great love for it, it would fly off every morning out of their ken. So they told each other: 'We love this partridge; it has become a friend of ours and we cannot bear to be parted from it. How can we arrange for it to stay with us all the time, because when it flies off it stays away for the whole day and we only see it at night?'

One of them said: 'Relax, sisters, for I shall see to it that the partridge does not leave you even for the blink of an eye.' 'If you manage to do that, we shall all be your slaves,' the other tortoises promised. So when the partridge came back from where it had been feeding and settled down among them, the wily tortoise went up to it, called down blessings on it and congratulated it on its safe return. 'Sir,' it said, 'you know that God has supplied you with our love, as well as installing love for us in your heart. You have become a friend for us in this wilderness, and the best times for lovers are when they are together, while it is a misfortune for them to be far apart. But you abandon us at dawn and only come back to us at sunset, leaving us very lonely. That causes us great distress and we feel very disturbed because of our great fondness for you.' 'Yes,' agreed the partridge, 'I am very fond of you, too, and feel a greater longing for you than you do for me, nor is it easy for me to part from you, but there is nothing that I can do about it. I am a bird with wings and I cannot stay here with you all the time, as this is not in my nature. Feathered creatures only settle down at night, and that is to sleep. Then in the morning they fly off and go wherever takes their fancy.' 'That is true,' agreed the tortoise, 'but for most of the time such creatures have no rest, and they only get a quarter as much of the good things of life as they get of its troubles. What everyone most desires is comfort and ease. God has set love and friendship between us and you; we are afraid lest one of your enemies may hunt you down, and if you are killed we shall be deprived of the sight of your face.' 'True enough,' replied the partridge, 'but what do you advise me to do about it?' 'I think that you should pull out the wing-feathers that allow you to fly so fast, and then you can sit at your ease with us, eating our food and drinking what we drink on this island with all its trees and ripe fruits. Then we can all stay together in this fertile place enjoying each other's company.'

The partridge was inclined to accept this as it wanted to rest, and so it plucked out its wing-feathers one by one, following what it thought was the good advice given by the tortoise. It then settled down to live with the tortoises, being content with scanty pleasures and transitory delights. Then in the middle of all this a weasel happened to pass by, and when it looked closely it could see that the partridge's wings were featherless so that it could not rise from the ground. This delighted it, and it said to itself: 'This is a fat bird with few feathers,' and so it approached the partridge and pounced on it. The partridge shrieked and called to the tortoises for help, but rather than helping they moved away, huddling against each other when they saw their friend in the weasel's grip. As the weasel tormented it, they were choked with tears and when the partridge asked if all they could do was to weep, they said: 'Brother, there is nothing that we can do about the weasel, as we don't have the strength.' The partridge was filled with sorrow when it heard this and despaired of its life. Despite this, it said: 'This is not your fault but mine, in that I did what you told me and plucked out the feathers that allowed me to fly. I deserve to be killed because I obeyed you; I cannot blame you at all.'

The king then told his women: 'I do not reproach you, for it is I myself who deserve blame and punishment for not remembering that it was you who were responsible for the sin of our father, Adam, because of which he was driven out of Paradise. I forgot that you are the root of all evil, and in my stupidity and faulty judgement I managed things so badly that I did what you told me and killed my viziers and the officers of state who used to advise me on everything. It was they who ensured my glory and gave me strength to face all my problems. Now I can find no substitutes to replace them and so here I am in deadly peril.'

Night 925

Morning now dawned and Shahrazad broke off from what she had been allowed to say. Then, when it was the nine hundred and twenty-fifth night, SHE CONTINUED:

I have heard, O fortunate king, that the king blamed himself and said: 'I obeyed you in my folly and killed my viziers, for whom I have found no replacements. If God does not provide me with a good counsellor who can show me some way of escape, I am lost.' He then got up and went to his bedroom, having lamented the fate of his viziers and his sages, exclaiming: 'How I wish that I could see those lions with me now for even a single hour, so that I might make my excuses to them and complain to them of the danger that I have to face now that they have gone!' All day long, he remained drowning in a sea of cares, tasting neither food nor drink. Then, when night fell, he got up, changed into tattered clothes in order not to be recognized, and went out to wander round the city in the hope of hearing some words of comfort.

While he was roaming the streets, he came across two twelve-year-old boys sitting by themselves alongside a wall. He noticed that they were talking to each other and so he went up close to hear what they were saying. One of them was telling the other that his father had been talking the night before about the fact that his crops had dried up prematurely from lack of rain, while the city had suffered many other misfortunes. 'Do you know why?' asked his companion. 'No,' said the other, 'but if you do, then tell me.' 'I do know,' replied his companion, 'and I shall tell you what the reason is. According to one of my father's friends, our king killed his viziers and ministers of state, although they had done nothing wrong, merely because he was besotted by the love of women. The viziers tried to stop him but he was not to be turned back and, thanks to what his women said, he ordered their execution and he even killed Shimas, my father, his chief advisor, who had served both him and his father before him as vizier. But you will soon see what God will do to him because of these crimes and how He will avenge the dead.' 'What can God do to him now that they have been killed?' asked the boy, and the other replied: 'I can tell you that the ruler of Outer India, who despises our king, has sent him a letter of reproach, telling him to build a palace for him in the middle of the sea and adding: "If you fail to do this, I shall send out against you twelve regiments, each of a

hundred thousand fighting men, under the command of my vizier, Badi',
who will seize your kingdom, kill your men and capture you and your
women." The messenger who brought this letter gave our king three
days in which to reply to it, and you must know, my brother, that this
king is a stubborn tyrant and a man of great strength and power with
huge numbers of subjects. Unless our king can think of some way of
restraining him, he will be destroyed and then the Outer Indians will
seize what we have, kill our men and take our women as captives.'

When the king heard what the boys were saying, he became even more
agitated and turned towards them, saying to himself: 'This boy has
got this knowledge from somewhere, as he has been telling his friend
something that he never learned from me. The letter that came from the
king of Outer India is with me and I have kept it secret, telling nobody
else about it. How did the boy come to know of it? I shall approach him,
speak to him and ask for help, praying to God that, through him, I may
find a way to save myself.'

He went up to the boy courteously and said: 'My dear boy, what is
this that you said about our king? He did very wrong when he killed his
viziers and ministers of state, but, in fact, the greatest injury that he did
was to himself and to his subjects. You were quite right in what you
said, but, tell me, from where did you learn that the king of Outer India
had written to him, reproaching him and threatening him, as you said?'
The boy replied: 'I know this because the ancients have told us that no
secret is hidden from God and that in the children of Adam there is a
spiritual power through which such secrets are revealed.' 'True, my boy,'
said the king, 'but is there anything that our king can do to ward off this
great disaster from himself and from his realm?' 'Yes, there is,' replied
the boy, 'and if he sends me a message and asks me how to defend
himself against his enemy and escape from these plots, I shall tell him
how to save himself through the power of Almighty God.' The king said:
'And who is going to tell that to the king, so as to get him to send for
you?' The boy said: 'I have heard that he is looking for people of
experience and sound judgement. If he sends for me, I shall go to him
with his messengers and tell him what will help him to ward off this
disaster. On the other hand, he may prefer to ignore the crisis and to
spend his time enjoying himself with his women. In that case, if I wanted
to tell him how to save himself and went to him on my own account, he
would order my execution, just as he killed his former viziers. So by
introducing myself to him I would bring about my own death. People

would despise me and think poorly of my intelligence, and I would become one of those to whom the saying applies: "If a man's knowledge exceeds his intelligence, then, learned as he is, his folly will destroy him."'

When the king heard what the boy had to say, he could clearly see and be assured of his wisdom and excellence, and he was certain that it was through him that he himself and his subjects would escape from danger. So he spoke again to the boy and said: 'Who are you and where do you live?' 'Our house is next to this wall,' replied the boy, and the king, after taking note of the place, said goodbye to him and went off happily to his palace. When he was back in his apartments, he changed into his own clothes, called for food and drink and gave orders that his women were to be kept away from him. He ate and drank, giving thanks to God and praying for deliverance and help, as well as for pardon and forgiveness for what he had done to the scholars and leaders of his realm. He offered his sincere repentance to God, and undertook to fast, as well as to multiply his prayers and to provide votive offerings. Then he summoned one of his personal attendants and, having described where it was that the boy lived, he told him to fetch him with all due courtesy.

This man went to the boy and told him: 'You are summoned by the king, who wants to confer a benefit on you. He has a question to put to you, and afterwards he will reward you and send you back home.' 'What does he need that has made him send for me?' the boy asked, and the servant said: 'He has done it because he wants you to answer a question.' 'I willingly hear and obey,' the boy said, after which he went with the servant to the king. When he came into his presence, he prostrated himself before God and called down blessings on the king, who returned his greeting and told him to sit, which he did.

Night 926

Morning now dawned and Shahrazad broke off from what she had been allowed to say. Then, when it was the nine hundred and twenty-sixth night, SHE CONTINUED:

I have heard, O fortunate king, that when they boy came to the king and greeted him, he was told to sit down, which he did. 'Do you know who talked with you yesterday?' asked the king. 'Yes,' the boy replied. 'Where is he?' the king asked. The boy answered: 'He is the man who is talking to me now,' and the king replied: 'You are right, my dear boy.'

He then ordered a chair to be placed beside his own, on which he seated the boy, and he then had food and drink brought in. When they had chatted for a while, he said: 'Vizier, when you were talking to me yesterday, you mentioned that you had a scheme which could save us from the stratagems of the king of Outer India. What is this scheme and how can we defend ourselves against the evil that he is planning? Tell me and I shall appoint you as the first counsellor of my kingdom. I shall choose you as my vizier, follow all your advice and give you a magnificent reward.' 'Keep your reward to yourself, your majesty,' replied the boy, 'and get your women, who told you to kill my father, Shimas, and the other viziers, to advise you what to do.' When he heard this, the king was ashamed and he said with a sigh: 'My dear boy, is it true that Shimas was your father, as you said?' 'He was indeed my father and I am certainly his son,' the boy answered, at which the king shed tears of humility and said, after asking pardon from God: 'Boy, I acted out of folly, but the plans and schemes of women are very evil. I ask you to forgive me, and I shall set you in your father's place with an even higher position. When this present threat is over, I shall present you with a golden torque, mount you on the finest of my horses and get a herald to walk before you, announcing: "This is the august youth whose throne is second only to that of the king." As for what you said about the women, I promise you that I shall punish them whenever Almighty God wishes it. So reassure me by telling me your plan.' 'Swear to me,' said the boy, 'that you will not do anything contrary to the advice that I give you, and guarantee that I shall have nothing to fear from you.' The king replied: 'I pledge myself to you before God that I shall do whatever you tell me; you shall be my counsellor and I shall carry out your instructions. I call on Almighty God to bear witness to what I say.'

At this, the boy relaxed and felt free to speak. 'Your majesty,' he said, 'the plan that I propose is that you should wait for the courier to return for his answer at the end of the three-day period that you have been allowed. When he comes to ask for it, put him off for another day. He will excuse himself to you, saying that his king has allowed him only a fixed number of days, and he will try to get you to change your mind. You must then dismiss him and tell him to wait for another day, without specifying what day that will be. He will go off angrily and head for the centre of the city, where he will say in public: "Citizens, I am the envoy of the king of Outer India, a man of might, before whose resolution iron is softened. He has sent me with a letter to your king and has given me a fixed timetable, threatening to punish me if I fail to return after the allotted period of days. I went to your king with my letter and when he saw it he asked me to wait for three days, after which he would give me his reply. As an act of courtesy and to please him, I agreed, but when the three days were up and I went to ask for his answer, he put me off for another day. I cannot wait any longer, and so I shall set off back to my master to tell him what has happened. I call on you to witness what has happened between your king and me." What the man says will be reported to you, and you must then send for him and speak to him gently, saying: "You are doing your best to destroy yourself. What has led you to criticize me in front of my subjects, making yourself liable to instant execution? But, as the ancients have said, forgiveness is part of a noble nature. The fact that I have delayed giving you a reply is not due to any weakness on my part but because I have been too busy to have time to spare for writing to your king." Then ask for the letter that the man brought and read it over again. When you have finished, burst out laughing and say: "Have you another letter with you, apart from this one, so that I may reply to it as well?" The envoy will say that he has not, and when you repeat the question a second and a third time, he will tell you: "I have nothing at all apart from this."

'At that point, you must say: "This king of yours must be a fool to phrase his letter in such a way as to prompt me to set out against him with my army to attack his lands and seize his kingdom. I shall not punish him this time for the discourtesy of his letter, as he must be both unintelligent and feeble-minded. Considering the power that I have, the fitting thing to do is to start by giving him due warning not to repeat this kind of nonsense, for if he wants to put himself at risk by doing the same kind of thing again, he will deserve the disaster that will quickly overtake him. In my opinion, this master of yours must be an ignorant fool who does not consider the consequences of his actions and who has no intelligent vizier capable of sound judgement to give him advice. Had he any sense, he would have consulted such a vizier before sending me this ludicrous note, but I shall send him a reply to match it, or to better it, by handing it to a schoolboy to answer." Do you then send for me, and when I have been brought before you, allow me to read the letter and to compose a reply.'

The king relaxed, approving of the boy's suggestion and being struck with admiration for his scheme. He made him generous gifts, appointed him to his father's position, and sent him happily away. Then, when the

three-day delay arranged with the envoy had ended, the latter returned
to the king to ask for his answer. The king told him to wait for another
day, at which the envoy retired to the far end of the audience chamber
and expressed himself in undiplomatic language, just as the boy had said
he would. He then went out to the market place, where he addressed the
citizens, telling them that he was an envoy sent by the king of Outer
India with a letter to their own king, who was putting off giving a reply.
He added that the waiting period specified by his master had come to an
end and that the king now had no excuse, a point to which he told his
audience to bear witness. The king, on hearing of this, sent to have the
man brought before him, and then said: 'You are doing your best to
destroy yourself. Are you not supposed to be carrying a message from
one king to another, between whom there are shared secrets? How is it
that you go out and tell these secrets to the common people? You deserve
to be punished by me, but I shall overlook the matter so that you can
take back an answer to your foolish master. In this case, what is most
appropriate is that the answer should be written by an insignificant
schoolboy.' He then sent for Shimas's son, and when the boy appeared
before him in the presence of the envoy, he prostrated himself before
God and prayed that the king might enjoy prolonged glory and a long
life. The king then threw the letter over to him, saying: 'Read this and
write a quick reply.'

The boy took it, and when he had read it he smiled and asked the king
laughingly: 'Was it because of this that you sent for me?' 'Yes,' replied
the king and the boy readily agreed to answer it. He produced an
inkstand and paper and wrote as follows ...

Night 927

Morning now dawned and Shahrazad broke off from what she had been
allowed to say. Then, when it was the nine hundred and twenty-seventh
night, SHE CONTINUED:

I have heard, O fortunate king, that when the boy took the letter and
read it, he immediately brought out an inkstand and paper. He then
wrote: 'In the Name of God, the Compassionate, the Merciful, greetings
to one who has been granted clemency and the protection of the Merciful.
Let me inform you, you who lay claim to the empty title of great king,
that your letter has reached me. I have read it and noted its strange,

nonsensical babblings, which have convinced me of your folly and the injustice that you wish to impose on me. You have reached out for more than you can grasp. Were it not that I feel pity for all God's creatures and for your subjects, I would not hold back from attacking you. Your envoy went into the market place here and broadcast the contents of your letter to high and low alike, so making himself liable to punishment. I have spared his life as an act of mercy towards him and because it is for you to excuse him, but this was not out of respect for you. What you said about my having executed my viziers, together with the scholars and leaders of my land, is true enough, but I had a reason for that, and for each of them that I killed I had a thousand more of the same kind, with more knowledge, understanding and intelligence. There is no child here who is not filled with learning, and I can replace every one who was killed with more experts in his field than I can count. Each one of my soldiers can stand against a regiment of yours, while, when it comes to money, my land produces gold and silver. For me, precious stones are as common as pebbles, and I cannot describe to you how handsome and wealthy my people are. How, then, can you have been so presumptuous as to tell me to build you a palace in the middle of the sea? This is an egregious piece of folly and comes from the weakness of your intellect, for if you had any sense you would have studied the force of the waves and the strength of the winds before asking me to do this for you. As for your claim that you will defeat me, this is absurd. How can someone like you commit an outrage against me and hope to take my kingdom? Almighty God will ensure my victory because you are an unjust aggressor, and you should realize that you deserve to be punished both by God and by me. For my part, however, I fear God both in regard to you and to your subjects, and because of this I shall not attack you until I have given you warning. If you yourself fear Him, then be quick to send me this year's tribute, as this is the only thing that will stop me attacking you with one million, one hundred thousand gigantic fighting men accompanied by elephants. In the centre of them will be my vizier, whose orders will be to lay siege to your city for three years, in place of the three-day delay that you allowed your envoy. I shall then take over your kingdom, but will kill no one apart from you, and the only women I shall take as captives will be yours.'

Shimas's son added a sketch of himself to the letter and wrote alongside it: 'This reply has been written by the youngest pupil at school.' He then sealed it and handed it to the king, who gave it to the envoy. The man took it, kissed the king's hands and left his presence, giving thanks to God as well as to the king for his clemency, and wondering, as he went, at the boy's intelligence. He got back to his master three days after the end of the stipulated waiting period, to find that, because he was late, a meeting of the royal council had been convened.

On entering the king's presence, he prostrated himself before him and handed him the letter. The king took this and then asked him why he had been delayed and how things were with Wird Khan. The envoy told his story, and the account that he gave of all that he had seen and heard astounded his master, who exclaimed: 'Damn you, what is all this that you tell me about a king like Wird Khan?' The envoy replied: 'Mighty king, here am I standing before you, but open the letter and, when you read it, you will see what is true and what is false.' At that, the king opened and read the letter and looked at the portrait of the boy who had written it. This convinced him that he was about to lose his throne and, as he was at a loss to know what to do, he turned to his viziers and his ministers of state and told them what had happened. He read the letter out to them, as a result of which they were filled with dismay and terror. Although they tried to calm the king's fears with empty words, their hearts were fluttering with fear.

Then Badi', the chief vizier, said: 'Your majesty should know that there is no profit to be got from what my fellow viziers may say. My own view is that you should write a letter of apology to Wird Khan, telling him: "I am as fond of you as I was of your father before you, and the letter that I sent with my envoy was only intended as a test of your resolution and courage, as well as of your knowledge of theoretical and practical affairs, together with your grasp of coded secrets and the whole range of virtues that you possess. We pray that the Almighty may bless you in your kingdom, build up the fortresses of your city and add to your authority, in that you can both protect yourself and bring to fruition the affairs of your subjects." Then send this letter to him with another envoy.'

The king exclaimed: 'I swear by God, the Omnipotent, that this is a great wonder. How can Wird Khan be a great king, ready to wage war, after having executed his sages, counsellors and army commanders? How can his kingdom flourish after that, so as to allow him to send out so huge a force? Even more remarkable than this is the fact that young schoolboys there can compose a reply like this for their king. My own greed led me to light this fire which threatens me and my subjects, and the only way that I can put it out, as far as I can see, is to do as my vizier advises.' He then prepared a splendid gift, together with large numbers of eunuchs and servants, and he wrote the following letter: 'In the Name of God, the Compassionate, the Merciful. Wird Khan, great king, son of my dear brother, Jali'ad, may God have mercy on him and preserve you, your reply to my letter has reached me. When I read it and noted its points, I was pleased with what I found there, this being all that I had prayed God to provide for you. I ask Him to exalt you, to build up the pillars of your state, and to grant you victory over your enemies who wish to harm you. You must know, your majesty, that your father was my friend and that throughout his life we were joined by solemn pledges. Neither of us ever experienced anything but good from the other. When he died, I was overjoyed that it was you who took his place on the throne, but when news came of what you had done to your viziers and the leaders of your land, I was afraid that some other king might hear of this and be tempted to attack you. I thought that you were neglecting your own interests, including the defences of your fortresses, and that you were not paying proper attention to the affairs of your kingdom, and so I sent you a letter to rouse you from this state. When I saw the reply that you sent me, my anxiety for you was put at rest, may God give you enjoyment of your kingdom and aid you in your affairs. Goodbye.' He then finished preparing his gift for Wird Shah and sent it off to him with an escort of a hundred riders.

Night 928

Morning now dawned and Shahrazad broke off from what she had been allowed to say. Then, when it was the nine hundred and twenty-eighth night, SHE CONTINUED:

I have heard, O fortunate king, that the king of Outer India prepared a present for King Wird Khan and sent it off with a hundred riders. They rode to Wird Shah, greeted him and handed him the letter. When he had read it, he assigned a suitable lodging to the leader of the escort, treating him with honour and accepting the gift. News of this spread among the citizens, and the delighted king sent for Shimas's young son, whom he honoured. He summoned the leader of the escort, and having called for the king's letter, he gave it to the boy, who opened it and read it. Wird Shah, full of joy, began to reproach the envoy, who was kissing the king's hands, making excuses and calling on God to grant him prolonged life and continued favour. Wird Shah thanked him for this, treating him with the greatest honour and presenting him with gifts, while giving the other members of the escort what was suitable for their rank. He prepared presents to send off with them, and he ordered Shimas's son to compose a reply to what the king had written. The boy did this with a courteous address to the king of Outer India, a brief mention of peace between their kingdoms, and a reference to the well-mannered behaviour of the envoy and his men. When he had finished, he presented the draft to the king, who said: 'Read it out, dear boy, so that we may know what you have written.' The boy did this in the presence of the hundred riders of the escort, and the king, together with all those present, were filled with admiration for what he had said and how he had expressed it. Wird Shah then sealed it and handed it to the envoy, before sending him off with a squadron of his own men to escort him to the frontier.

So much for the king and the boy, but as for the envoy, he was astonished by what he saw of the boy's learning and he gave thanks to God for the speedy conclusion of his mission and for the fact that Wird Shah had agreed to peace. He rode on and, having reached his master, he handed over the gifts and presents, after which he gave him the letter and told him what he had seen. The king was delighted and gave thanks to Almighty God. He showered honours on the envoy, thanking him for the zeal with which he had carried out his task, and giving him promotion. From that time on, he lived in peace and security, enjoying tranquillity and ever-increasing happiness.

So much for him, but as for Wird Khan, he no longer strayed from God's path, but turned from his evil ways in sincere repentance, abandoning his women altogether and devoting the whole of his attention to the good of his state and the welfare of his subjects, prompted by the fear of God. He appointed the son of Shimas to his father's post as vizier, taking him as his chief counsellor of state and the guardian of his secrets. On his instructions, his capital was adorned with decorations for seven days, as were all his other cities, and the delighted citizens were freed from fear, looking forward joyfully to being treated justly and fairly. They called down divine blessings on the king and on the vizier who had freed them from their anxiety.

The king then asked his new vizier to put his kingdom in order, to set to rights the affairs of his subjects and to restore things to their former state, appointing leaders and administrators. The vizier told him: 'In my opinion, before you do anything else you must start by cutting out from your heart all disobedience to God, abandoning the pleasures and the wrongdoing you indulged in before, as well as your obsession with women. If you return to what caused you to sin in the first place, then this second error will have a worse result than the first.' 'What was the prime cause of my sins which I should now root out?' asked the king. The vizier, young as he was in years but old in wisdom, told him: 'Your majesty, you should know that this was your love of women, your attraction to them and the fact that you accepted their advice and fell in with their plans. Love of women alters the pure light of reason and corrupts a sound nature. There are clear proofs that bear witness to what I say, and if you think about them and carefully examine the factual evidence, you will find that you yourself are your own best advisor and you will have no need at all for me to tell you anything. Do not become obsessed by thinking about women, and remove the images you have of them from your mind, since Almighty God through Moses, His prophet, told us not to have too much to do with them. A wise king once told his son: "My son, when you succeed me to the throne, do not frequent the society of women lest they lead your heart astray and corrupt your judgement." In short, if you are over-fond of their company this will lead to love, and love impairs judgement. An example of this is what happened to our lord Solomon, the son of David, on both of whom be peace, whom God had singled out to receive more knowledge, wisdom and dominion than had been granted to any other king before him. Women had been the cause of his father's sin, and although there are many other examples, I have only mentioned Solomon because as a ruler he had no match and all the kings of the earth obeyed him. The love of women is the root of all evil; none of them has good judgement, and so a man should restrict his dealings with them to matters of necessity and not give rein to his emotions where they are concerned, as this will lead him to ruin and destruction. If you follow my advice, all your affairs will prosper, but if you abandon it, then you will repent when repentance will serve no purpose.'

The king replied: 'I have given up my excessive passion for women . . .'

Night 929

Morning now dawned and Shahrazad broke off from what she had been allowed to say. Then, when it was the nine hundred and twenty-ninth night, SHE CONTINUED:

I have heard, O fortunate king, that King Wird Khan told his vizier: 'I have given up my passion for women, and I have entirely abandoned my preoccupation with them. But what should I do to them to repay them for their actions, in that it was their scheming that brought about the execution of your father, Shimas? This was not something that I wanted and I cannot understand how I came to agree with them to have it done.' Sighing, he exclaimed in regret for the loss of his vizier, for the soundness of his judgement and the excellence of his administration, and he lamented the deaths of Shimas's fellow viziers and ministers, men of unfailingly good counsel.

The young vizier replied: 'The fault was not that of the women alone, your majesty. They are like attractive sale goods which take the fancy of those who look at them. They are sold to whoever wants to buy them, but no one is forced to do this, and it is the buyer who is in the wrong, particularly if he knows that the goods can be harmful. I warned you, and before me my father used to warn you, but you would take no advice.' The king replied: 'As I told you, vizier, I accept that I was in the wrong, and my only excuse is that this was ordained by providence.' The vizier said: 'You must know, your majesty, that Almighty God, Who created us, granted us the ability and free will with which to make our choices, so that we only act if we wish to do so, and not if we do not. He never ordered us to do harm, lest we should sin and so deserve punishment in matters where we should have acted rightly. For in all things, He only orders us to do what is good and He forbids us to do evil. It is we by our own choice who do whatever it is that we choose, right or wrong.' 'That is true,' agreed the king, 'and the fault was mine because I followed my lusts. I warned myself about that many times, as did your father, Shimas, but my passions overcame my intellect. Can you think of any way to stop me from falling into the same fault again, so that this time my intellect can get the upper hand over my lusts?' The vizier replied that he could and explained: 'What you must do is to exchange the garment of folly for that of justice; you must disobey your desires in obedience to your Master, and return to the path followed by that just king, your father. You must do your duty in the service of God and of your subjects, preserving your religion, protecting your people and exercising self-control. You should not put your subjects to death, but look to the consequences of your actions, abandoning evil-doing, injustice, tyranny and wickedness, and in their place employing justice, fair dealing and humility. You must obey the commands of the Almighty and constantly show compassion to His creatures over whom He has set you as His deputy, taking pains to do what will earn you their blessings. As long as you act like this, your days will be free of trouble, and God, in His mercy, will forgive you and make you a figure of awe for all who see you. Your enemies will be crushed and Almighty God will rout their armies, while you yourself will be acceptable to Him and venerated and loved by His creation.'

Wird Shah replied: 'You have given me new life, illumining my heart with your sweet speech and clearing away the blindness from my eyes. I am determined that, with God's help, I shall do everything that you tell me, abandoning my former wicked lusts and allowing my cramped soul to expand, as I move from fear to security. You have a right to be happy because of this, since, although I am old, I have become a son to you, while you, young as you are, are to me a beloved father. It has become my duty to do all that I can to follow your orders. I give thanks to God for His grace and to you, because it is through you that He has shown me favour with guidance and good counsel, dispelling my cares and grief. It is to you, your superior knowledge and the excellence of your planning, that my subjects owe their safety. From now on, you are the administrator of my kingdom and it is only my position on the throne that gives me any superiority over you. I shall not oppose anything that you do and, in spite of your youth, no one can gainsay your words, as you are old in intelligence and full of knowledge. I thank God for granting you to me in order to lead me from the crooked path of destruction into the way of righteousness.'

'Fortunate king,' replied the vizier, 'you must know that it is not a merit on my part to have given you advice, for, as I am a product of your bounty, whatever I say or do is part of my duty; nor is this true of me alone, for my father before me was overwhelmed by your great generosity. We both share alike in your bounty and your favour, and how could we fail to acknowledge this? You, your majesty, are our shepherd and our ruler; it is you who fight for us against our enemies; you are our guardian and our defender, exerting yourself to defend us, so that even if we gave our lives in your service, this would not settle our debt of gratitude. We address our prayers to Almighty God, Who set you over us as our ruler, asking Him to grant you long life and success in all your undertakings. May He subject you to no trials in your lifetime, but bring you to your goal and make you an object of veneration until you die. May He help you to show generosity so that you act as a leader for all men of learning, while conquering all who oppose you. May learned and courageous men be found in your lands, while fools and cowards are removed from them. May your subjects be spared times of scarcity and distress and may the seeds of friendship and affection be sown among them. May you be granted success in this world, coupled with blessings in the world to come, through God's bounty and generosity, combined with the mysteries of His grace. For He is Omnipotent; nothing is too difficult for Him; to Him all things return and to Him they make their way.'

Wird Shah was filled with delight and, moved by deep affection for the boy, he said: 'Know, vizier, that you have become like a brother, a son and a father to me; nothing but death will divide us and all that I have is at your disposal. If I leave behind no heir, it is you who will take my place on the throne, for you have a greater right to it than any other of my subjects. I shall appoint you as my successor in the presence of all the leaders of the land, nominating you as my heir, if Almighty God wills it.'

Night 930

Morning now dawned and Shahrazad broke off from what she had been allowed to say. Then, when it was the nine hundred and thirtieth night, SHE CONTINUED:

I have heard, O fortunate king, that King Wird Khan said to the son of Shimas the vizier: 'I shall appoint you as my deputy and name you as my successor, calling all the leaders of the land to witness this, through the help of Almighty God.' He then called for his scribe and ordered him to send a summons to all the principal officers of state and he had a proclamation made in the city to all the citizens, high and low alike. A meeting of emirs, commanders, chamberlains and other officers of state, together with all the learned sages, was convened in his presence. He held court in magnificent style with a banquet of unparalleled splendour, to which he invited all ranks of society. Everyone came to enjoy themselves, eating and drinking for a whole month, and he then presented robes to all his retainers and to the poor, as well as giving generous presents to the men of learning. From these latter he chose a number, approved by Shimas's son, and when they had appeared, he told the boy to select six to serve under him as assistants, while he himself was to act as the seventh, and chief, vizier. The boy picked six of the oldest and the most intelligent, with the greatest knowledge and the quickest grasp of affairs. When they had been presented to him, the king gave them their robes of office and told them: 'You are to be my viziers under the authority of the son of Shimas, whose instructions and orders you must never disregard, since, even if he is the youngest of you, he is the most intelligent.' He then seated them on the ornate chairs normally used by viziers and assigned them revenues and allowances. They were instructed to pick out from the leading men of the kingdom, assembled at the banquet, soldiers suitable for state service, from whom he could appoint officers commanding units of men, thousands, hundreds and tens, who would then receive the salaries and allowances that were customary for men of rank. This was quickly done, and the king ordered lavish gifts to be given to everyone else who was there, and that they should be sent back to their homes with honour and distinction. The governors were instructed to be just in their treatment of the people and to show sympathy in their dealings with both rich and poor, giving them help from the treasury in accordance with their positions. The viziers prayed that God might extend his glory and prolong his life, after which the city was adorned with decorations for three days in gratitude for the favour that He had showed the king.

So much for the king and his vizier, the son of Shimas, as regards their organization of the kingdom, its emirs and its governors. There remained the matter of the favourite concubines and the other women whose trickery and deceit had led to the execution of the viziers and the near-destruction of the kingdom. When those who had come to the royal court from the city and the other towns had been sent back to their homes after their affairs had been set in order, the king ordered his young but intelligent vizier to summon his colleagues. When they had all come, the king held a meeting with them in private, at which he said: 'You must know, viziers, that I strayed from the right way, drowning in my own folly, turning from good counsel, breaking my solemn word and opposing my advisors. All this was caused by these women who tricked and deceived me with their specious and lying words. I accepted what they said because the sweetness and softness with which they spoke led me to think that this was good advice, whereas, in fact, it was deadly poison. I know for certain that what they did was to bring ruin and destruction on me, and they deserve a just punishment at my hands that will make an example of them for those who can learn from their lesson. So what advice can you give me about putting them to death?'

The vizier replied: 'Great king, I told you at the start that women are not the only ones at fault but that this is shared between them and the men who obey them. These women, however, do deserve to be punished, for two reasons. The first of these is to enable you, the great king, to keep your word, and the second is that they dared to deceive you, thrusting themselves into what was no concern of theirs, matters on which they were not qualified to speak. Although they deserve to die, what has happened to them is punishment enough. Reduce them to the position of servants, but here, as elsewhere, it is for you to command.' One of the other viziers sided with Shimas's son, but another advanced and prostrated himself before the king, after which he said: 'May God prolong the days of your majesty's life. If you are determined that these women should die, then do what I suggest.' The king asked what this was, and the man replied: 'The best thing would be for you to order one of your slave girls to take the women who deceived you into the room where the viziers and learned men were killed. They should be confined there with just enough food and drink to keep body and soul together, and never allowed to leave. The bodies of those of them who die are to be left there, just as they are, until every last one of them is dead. This is the least of punishments for them, as it was they who were responsible for this great upheaval, and they were the root cause of all the disasters and disturbances that have taken place in our time. They illustrate the truth of the saying: "Even though he may be safe for a long time, whoever digs a pit for his brother will fall into it."'

The king accepted his advice and did what he had suggested. He sent for four strong slave girls and handed over the women to them with orders that they were to be taken to the execution chamber and confined there with a small allowance of poor food and bad water. For their part, they were filled with sorrow and regret for what they had lost, grieving bitterly, while God punished them with disgrace in this world, while preparing punishment for them in the world to come. There they stayed in the dark, stinking room, where every day some of them would die, until none were left. News of this spread throughout the land.

This is the end of the story of the king, his viziers and his subjects. Praise be to God, Who destroys peoples and gives life to dry bones and Who is eternally worthy of supreme honour and veneration.

A story is also told that in the city of Alexandria there were two men, one a dyer by the name of Abu Qir and the other a barber named Abu Sir. They were neighbours in the market, where their shops stood side by side. The dyer was a very bad man, a cheat and a liar, the temples of his head looking as though they had been carved out of rock or cut from the lintel of a Jewish synagogue. However disgraceful his conduct, he was never ashamed. Whenever anyone gave him anything to be dyed, he would ask for payment in advance, pretending that he was going to buy the necessary dyes. The customer would hand over the money and Abu Qir would take it and spend it on food and drink, and when the customer had gone, he would sell whatever had been left with him, using what he got for it on food, drink and other such things. His own food was always of the best and he would only drink the finest of wines such that would rob men of their wits. When his customer returned, Abu Qir would say: 'If you come back to me tomorrow before sunrise, you'll find your stuff ready.' The man would go off, telling himself: 'From one day to the next is not a long wait,' but when he came back, as he was told, the following day, Abu Qir would put him off once more, saying: 'I couldn't do the job yesterday as I had guests whom I had to look after until they went.' He would then promise to have the job finished for the customer if he came back again the following day before sunrise. This time, when he did, the excuse would be that Abu Qir's wife had given birth that night and he had had to spend the whole day looking after things. He would promise faithfully that the stuff could be picked up the following day, and then, when the customer came, he would be met with yet another excuse, with Abu Qir swearing that the job would be finished.

Night 931

Morning now dawned and Shahrazad broke off from what she had been allowed to say. Then, when it was the nine hundred and thirty-first night, SHE CONTINUED:

I have heard, O fortunate king, that whenever the customer came back to the dyer, he would find yet another excuse, swearing that the job would be finished. He would keep on promising and swearing until eventually the restive customer would say: 'How many times are you going to tell me to come back tomorrow? Give me back my stuff; I don't want it dyed.' Abu Qir would then tell him: 'By God, brother, I am ashamed, but I shall have to tell you the truth, for God harms all those who damage other people's goods.' 'Tell me what has happened,' the man would say and Abu Qir would reply: 'I made a most splendid job of dyeing your material and hung it out on a rope to dry, but then it was stolen and I don't know who took it.' If the owner were a charitable man, he would say: 'God will give me something else in return,' but if he were ill disposed, he would keep on threatening Abu Qir with open shame and disgrace, yet even if he took his complaint to the judge, he would get nothing out of him.

Abu Qir kept on doing this kind of thing until he made himself notorious and people started warning each other against him. He became proverbial for his conduct, and the only customers who went to him were those who didn't know about him. Every day people would abuse and revile him, and because of this his business ground to a halt. He started going into the shop of his neighbour, Abu Sir the barber, where he would sit opposite his own premises with his eyes fixed on the door. If he saw anyone who didn't know his reputation standing there holding something to be dyed, he would leave the barber's shop and ask him what he wanted. If the man asked him to take the stuff and dye it for him, he would say: 'What colour do you want it?' For, in spite of his bad qualities, he had the skill to dye in all colours and it was only because he could never act honestly with anyone that he had been reduced to poverty. He would then take the material, demand payment in advance and tell the customer to come back for it next day. When the man had handed over the money and left, Abu Qir would take the material to the market, where he would sell it and use what he got for it to buy meat, vegetables, tobacco, fruit and whatever else he needed. Then, if he saw at his shop door someone who had given him something to be dyed, he would not put in an appearance but would keep out of sight. This went on for some years until it happened that one day he took something from an overbearing fellow, which he sold and then spent the money that he got for it. The owner of the material came back day after day, but never found Abu Qir there, as in these circumstances he would make his escape into the shop of Abu Sir. At last the owner, after failing to catch him in his shop, could think of nothing else to do but to go to the *qadi.* The *qadi* sent an officer with him and, in the presence of a number of Muslims, he nailed up the shop door and put a seal on it, as there was nothing to be seen in it except for broken pots, and he could find nothing to replace the lost material. The officer took the key and told the neighbours: 'Tell the dyer to produce this man's property and then he can have his key back.' He and the owner then went off on their way.

Abu Sir said to Abu Qir: 'What is wrong with you? Every time someone brings you something, you lose it for him. Where did this man's stuff go?' When Abu Qir told him that it had been stolen, Abu Sir said: 'It is surprising that whatever anyone gives you is stolen. Do all the thieves meet at your shop? I think you're lying, so tell me the true story.' Abu Qir admitted that nobody had stolen anything at all from him, and when Abu Sir asked him what he did with what he was given, he confessed: 'Whenever someone gives me something, I sell it and spend what I get for it.' 'Does God permit that?' asked Abu Sir, and Abu Qir replied: 'It is only poverty that makes me do this, because my business is not doing well and, as a poor man, I haven't any money.' He started to tell Abu Sir how bad business was and how little work he had, and Abu Sir replied with the same complaint, saying: 'I have no equal in my field in this city, but nobody wants to be shaved by me because I am poor. I hate this profession, brother.' Abu Qir echoed this, saying that he, too, hated his job because it was bringing in no work, and he went on: 'But why should we stay here? The two of us should travel to other countries, because the skills that we have will be in demand wherever we go. As we go, we can smell the breezes and rest from this burden of care.'

Abu Qir kept on telling Abu Sir what a good thing it would be to go travelling until he became enthusiastic about it and the two of them agreed to set off.

Night 932

Morning now dawned and Shahrazad broke off from what she had been allowed to say. Then, when it was the nine hundred and thirty-second night, SHE CONTINUED:

I have heard, O fortunate king, that Abu Qir kept on telling Abu Sir what a good thing it would be to go travelling until he became enthusiastic about it. Abu Qir was delighted that Abu Sir wanted to come and so he recited these lines:

Leave your land to look for advancement;
Travel, for travel has five advantages –
The dispelling of care, the gain of a living,
Knowledge, culture and a noble companion.
You may say that travel brings care and trouble,
Parting from friends and the facing of hardships,
But it is better for a young man to die than to live
Where he is despised and surrounded by envious slanderers.

When they had made up their minds to go, Abu Qir said to his companion: 'Neighbour, we are now brothers and there is no distinction between us. We must recite the Fatiha to solemnize our oath that if only one of us finds work, he should use his earnings to buy food for the other, and whatever money is left over should be put in a box. Then, when we get back to Alexandria, we can divide it justly and fairly between us.' Abu Sir agreed to this and the two of them confirmed the terms that Abu Qir had proposed by his recitation from the Quran. Abu Sir then locked up his shop and gave the keys to the man from whom he had hired it, while Abu Qir left his key with the *qadi*'s officer, leaving the shop locked and sealed. They then got their things together and set off in the morning, boarding a galleon that was putting out to sea that day. They were aided by the lucky chance that among the hundred and twenty men on board, not counting the captain and the crew, there was not a single barber. When the sails were set, Abu Sir said to his companion: 'Brother, we are at sea and we need something to eat and to drink, as we have only a small stock of provisions. It may be, however, that someone will call me to shave him and I shall do that in return for a loaf of bread, half a *fidda*, or a drink of water, which will benefit us both.' 'All right,' said Abu Qir, after which he put his head down and went to sleep.

For his part, Abu Sir got up and, taking his equipment and his bowl, with a rag over his shoulder in place of a towel, because he was so poor, he walked around among the passengers. One of them called to him to come and shave him, which he did, and when he had finished, the man offered him half a *fidda*. 'Brother,' said Abu Sir, 'I don't need this coin, for here at sea it would do me more good if you were to offer me a loaf of bread, as I have a companion and we have little in the way of provisions.' So the man gave him a loaf as well as a piece of cheese, and filled his bowl with fresh water. Abu Sir took this and went to Abu Qir, to whom he said: 'Take this loaf and eat it with the cheese and drink the water in the bowl.' When Abu Qir had done this, Abu Sir went off with his shaving gear, his rag over his shoulder and his bowl in his hand. He made his way through the passengers again, and this time he shaved one man for two loaves and another for a piece of cheese. There was a demand for his services and he started to ask for two loaves and half a *fidda* from everyone who wanted to be shaved, as there was no other barber on board. By sunset, he had acquired thirty loaves and thirty coins, as well as cheese, olives and fish roes, all of which he had collected because people would give him whatever he asked. He shaved the captain, and complained to him of his shortage of provisions for the journey. The captain cordially invited him to bring his friend and dine with him every evening, adding: 'There is no need for you to worry while you are sailing with us.'

Abu Sir then went back to Abu Qir, whom he found still asleep, but, on being roused, he discovered large quantities of bread, cheese, olives and fish roes placed by his head. 'Where did you get this?' he asked and Abu Sir replied: 'From the bounty of Almighty God.' Abu Qir was about to start eating, when Abu Sir told him: 'Don't touch any of this, brother, but leave it for another time. I have to tell you that I shaved the captain, and when I complained to him that we were short of supplies, he said: "You will be welcome to bring your friend every evening and dine with me." We shall have our first meal with him tonight.' 'I'm seasick,' complained Abu Qir, 'and I can't get up. So let me eat from what is here, while you go to the captain by yourself.' Abu Sir agreed to that and sat watching as the other ate. He saw him gouging out his food in hunks, like rocks being quarried from a mountain, and swallowing it down like an elephant that hasn't eaten for days. He would stuff in a second mouthful before he had swallowed the first, glowering like a *ghul* at what was in front of him, and snorting as a hungry bull snorts over its straw and beans.

Just then, a sailor came up and said: 'Master, the captain says: "Bring your friend and come to supper."' Abu Sir asked Abu Qir if he was going to come with him, but when the latter insisted that he could not walk, Abu Sir went off alone. He found the captain sitting with twenty or more different dishes set out in front of him, waiting with his companions for their two guests. When he saw Abu Sir, he asked where his friend was and, on being told that he was seasick, he said: 'That will do him no harm; it will wear off. But come yourself and eat with us, for I have been waiting for you.' He then put a kebab plate on one side and filled it with all kinds of foods until there was enough on it to satisfy ten, and when Abu Sir had finished his meal, he told him to take the plate with him to his friend. So he took it off and found Abu Qir gnawing away at his food like a camel, cramming in one mouthful after another as fast as he could. 'Didn't I tell you not to eat,' said Abu Sir, 'as the captain is a very generous man? I told him that you were seasick but look at what he has sent you.' 'Hand it over,' said Abu Qir, and he took the plate impatiently from Abu Sir, eager to wolf it down, as well as anything else he could get, like a dog showing its teeth, a ravenous lion, a *rukh* pouncing on a pigeon, or a half-starved man who sees food and starts to devour it.

Abu Sir left him while he went to drink coffee with the captain, and when he got back he discovered that Abu Qir had eaten everything on the plate and thrown it away empty.

Night 933

Morning now dawned and Shahrazad broke off from what she had been allowed to say. Then, when it was the nine hundred and thirty-third night, SHE CONTINUED:

I have heard, O fortunate king, that when Abu Sir came back to Abu Qir, he found that he had eaten everything on the plate and thrown it away empty. Abu Sir picked it up and gave it to one of the captain's servants, after which he went back to his companion and slept until morning. The next day, he set about shaving his clients, and whatever he was given, he would pass on to Abu Qir, who sat there eating and drinking, only getting up to relieve himself. Every night he would be given a loaded plate from the captain's table, and things went on like this for twenty days until the galleon anchored in the harbour of a city. The two companions disembarked and went into the city where they rented a room in a *khan*. Abu Sir furnished the place, buying all that they needed and fetching meat, which he cooked. Abu Qir had been asleep all the time since he entered the room, only waking up when Abu Sir roused him and placed the food in front of him. He would eat it, saying: 'Don't blame me, for I am a dyer,' and then go back to sleep again.

This went on for forty days, on each of which the barber would take the tools of his trade and go round the city, making what money fell to his lot and coming back to rouse Abu Qir from his sleep. Abu Qir would then fall on his food like a man who could never have enough, after which he would go back to sleep. Another forty days passed in this way, with Abu Sir telling Abu Qir: 'Sit up and when you have rested, go out and look around the city, for it is a splendid sight and there is no other place to match it.' Every time he said this, Abu Qir would say: 'Don't blame me, for I am a dyer,' and Abu Sir would be unwilling to annoy him or to say anything that might hurt him.

On the forty-first day, Abu Sir fell ill and was unable to go on his rounds, and so he hired the doorkeeper of the *khan* to look after them and to bring them food and drink. This continued for four days, during which Abu Qir ate and slept, but then Abu Sir's illness became so much worse that he lost consciousness. It was at this point that Abu Qir, consumed by hunger, got up and rummaged through his companion's clothes where he found a number of dirhams which he pocketed. Then he locked Abu Sir in the room and went off without a word to anyone, unseen by the doorkeeper, who happened to be in the market at the time. He went to the market, where he got himself expensive clothes to wear, and he then started on a sight-seeing tour of the city.

He found this to be unequalled in its splendour, although the only colours worn by its inhabitants were white and blue. He came to a dyer's shop where everything was blue, and so he produced a handkerchief and told the dyer that he would pay to have it dyed. When the man quoted him a price of twenty dirhams, he said: 'At home we would do that for two dirhams.' 'Go back home, then, and have it dyed there,' said the man. 'I will only do it for twenty and for nothing less.' 'What colour do you propose to dye it?' asked Abu Qir. 'Blue,' replied the man, at which Abu Qir said: 'But I want it red.' 'I don't know how to do that,' the man told him. 'Then green,' said Abu Qir, and the man replied: 'I don't know how to do that either.' 'What about yellow?' asked Abu Qir, and when the man made the same reply, he started to list the other colours, one by one. 'In our country there are forty expert dyers, no more and no less,' the man told him, 'and when one of us dies, the rest teach his son. If he has no son, we have to remain one short, while if someone has two sons we teach one of them and if he dies, then we teach his brother. This craft of ours is regulated and the only colour that we know how to produce is blue.' Abu Qir said: 'I am a dyer and I can produce all the colours. I want you to take me on as a salaried assistant and I'll teach you how to manage them all, so that you can boast that you are better than all your colleagues.' 'We will never accept a stranger in our guild,' the dyer told him. 'Then I shall open a shop on my own,' said Abu Qir, but the man told him: 'You won't able to do that.' Abu Qir left him and went to another dyer, who gave him the same answer, after which he went on from one to another until he had visited all forty of them. As none of them was willing to hire him as an assistant or to accept him as an expert, he went to the master of the guild and told him of his proposal, but the man merely repeated: 'We don't accept strangers in our guild.'

Abu Qir, in a fury, now took his complaint up to the king of the city, saying: 'King of the age, I am a stranger here and a dyer by trade.' After telling the king of his experiences with the local dyers, he went on: 'I can produce various shades of red, including that of a rose or a jujube, and green like that of a field, a pistachio, olive oil or a parrot's wing. My blacks can be charcoal or kohl black and my yellows can vary from orange to lemon.' He continued listing all the various colours and then said: 'King of the age, the dyers in your city cannot produce a single one of these, as all their knowledge goes no further than blue. In spite of that, they have refused to hire me or to accept me as a master craftsman.' 'This is true,' said the king, 'but I shall open a workshop for you and provide you with capital. You need fear no trouble from the others, for if any one of them interferes with you, I shall hang him over his own shop door.' He then ordered builders to go through the city with Abu Qir, and when they came to a place that took his fancy, whether it was a shop, a *khan* or anything else, they were to expel its owner and build him a workshop to his own specifications. They were to do exactly what he told them, without contradiction. The king then presented Abu Qir with a splendid robe and gave him a thousand dinars for his personal expenses until the building work was finished. He also gave him two mamluks to act as servants and a horse with ornamental trappings.

Abu Qir put on the robe and rode off on the horse as though he were an emir. On the king's orders, a house was cleared and furnished for him.

Night 934

Morning now dawned and Shahrazad broke off from what she had been allowed to say. Then, when it was the nine hundred and thirty-fourth night, SHE CONTINUED:

I have heard, O fortunate king, that, on the king's orders, a house was cleared and furnished for Abu Qir. The day after he had settled in, he rode out through the city, preceded by the architects, and he continued to look around until he found a place he liked. When he had given his approval, the owner was evicted and brought before the king, who paid him the price of his property with a satisfactory bonus. Building work started, and Abu Qir continued to give the workers detailed instructions until they had built him a dye house that had no equal. He then went to tell the king of its completion, adding that in order that work could start there, he needed money for the dyes. 'Take these four thousand dinars as capital,' the king said, 'and then show me what you can produce.' Abu Qir went off with the money to the market, where he found dye-stuffs is such quantities that they were virtually worthless. He bought everything that he needed, and when the king sent him five hundred pieces of. material, he dyed them in various colours and then spread them out to dry in front of the door of his workshop. The passers-by were amazed to see something that they had never in their lives seen before, and a crowd formed, looking and asking questions. 'Master,' people would say, 'what are these colours called?' and he would tell them: 'This is red, this is yellow and this is green,' going through the names of the full spectrum. People would then fetch him material and say in what colour they wanted it dyed, telling him to name his own price.

When he had finished the work commissioned by the king, he took it to the royal court. The king was delighted and gave him a lavish reward, after which all his guards started to bring stuff, specifying the dye colours of their choice, and when he did what they wanted, they would shower gold and silver on him. His reputation spread and his workshop became known as the King's Dye House. Riches poured in from all sides, and as none of the other dyers had anything to say, they would come to kiss his hands and to excuse themselves for their earlier behaviour. They would offer their services, asking to be taken on as his servants, but he refused to take any of them and, instead, bought slaves, male and female, and acquired great wealth.

So much for him, but as for Abu Sir, he had been locked in his room by Abu Qir, who had taken his money and gone off, leaving him unconscious. He lay shut in there for three days, after which the doorkeeper came to the door and found it locked. Having not seen either Abu Qir or Abu Sir throughout the day and having heard no word of them, he said to himself that they must have left without paying the rent for their room, or they might have died, or something else might have happened to them. While he was outside the locked door, he heard Abu Sir moaning in the room, and as he could see the key in the lock, he opened the door and went in. On finding Abu Sir groaning, he said: 'I mean you no harm. Where is your companion?' 'By God,' Abu Sir replied, 'I have only just recovered my senses today after having been ill, and I have been calling out without getting any reply. Please, brother, for God's sake, look for the purse beneath my head; take out five *nusfs* and buy me some food, for I am dying of hunger.' The man put out his hand and took the purse but, finding it empty, had to tell Abu Sir that there was nothing left in it. Abu Sir realized that Abu Qir must have gone off after taking what was in it, and he asked: 'Have you seen my companion?' 'Not for three days,' the doorkeeper said, adding, 'and I thought that the two of you must have gone.' 'We didn't go,' said Abu Sir, 'but I think that my companion coveted my money and took it off with him when he saw that I was ill.' He began to weep and wail, but the doorkeeper said: 'Don't worry; God will repay him for what he did.' He then cooked some soup, which he ladled into a dish and gave to Abu Sir, and he continued to look after him for two months, using his own money to pay the costs, until at last Abu Sir broke out in a sweat and was cured of his illness.

He then got to his feet and said to the doorkeeper: 'If God enables me, I shall reward you for the kindness you have shown, but it is only He, through His grace, who gives rewards.' The doorkeeper replied: 'Praise be to God that you are better. I only did what I did for you because I hoped to win His favour.' Abu Sir then left the *khan*, and as he made his way through the markets, fate led him to the one where Abu Qir had his workshop. By its door he saw materials dyed in various colours, which were the centre of attention for a large crowd. He asked one of the townsfolk what the place was and why people were crowding around it. 'This is the King's Dye House,' the man told him, 'which he set up for a foreigner called Abu Qir. Whenever he has dyed a piece of clothing, we gather round to look at it, because we don't have any dyers here who can produce these colours.' The man went on to tell of Abu Qir's experience with the local dyers and of how he had complained to the king, who had helped him by having this workshop built for him, as well as by making him gifts.

When Abu Sir had heard the full story, he was glad, saying to himself: 'Praise be to God Who has given Abu Qir this opportunity to set himself up as a master craftsman. It is excusable that, being so busy with his profession, he should completely forget about me, but when he was out of work, I was helpful and generous to him and, when he sees me, he will gladly repay my generosity.' He then moved forward towards the workshop door, where he saw Abu Qir seated on a thick mattress that had been spread on top of a stone bench. He was regally dressed, with four black slaves and four white mamluks, themselves splendidly clothed, standing in front of him. Ten slaves were busy working as craftsmen, for after Abu Qir had bought them he had taught them the craft of dyeing. He himself was lolling among cushions like a great vizier or a king in his magnificence, taking no hand in the work himself but contenting himself with saying: 'Do this' and 'Do that.'

Abu Sir stood in front of him, thinking that he would be glad to see him and would greet him hospitably and look after his interests. But when their eyes met, Abu Qir exclaimed: 'You disgusting fellow, how many times have I told you not to stand at my workshop door? Are you trying to bring public disgrace on me, you thief? Seize him!' At this, the slaves ran after him and when they had caught hold of him, they threw him down and gave him a hundred strokes on his back before turning him over and giving him another hundred on his front. 'You filthy traitor,' said Abu Qir. 'After today if I see you standing here again, I'll send you straight off to the king, who will hand you over to the *wali* to have your head cut off. Be off with you, and may God give you no blessing.' When Abu Sir had gone off dejectedly because of the beating and the contempt with which he had been received, the bystanders asked Abu Qir what he had done. He replied: 'He is a thief who steals materials from people . . .'

Night 935

Morning now dawned and Shahrazad broke off from what she had been allowed to say. Then, when it was the nine hundred and thirty-fifth night, SHE CONTINUED:

I have heard, O fortunate king, that Abu Qir had Abu Sir beaten and thrown out. He told the bystanders that this was a thief who stole materials from people, adding: 'How many times has he done this to me and I have said to myself: "May God forgive him; he is a poor man." Not wanting to cause trouble for him, I would pay the owners the price of the stolen goods and tell him, gently, to keep away, but he would never listen. If he comes back once more, I shall send him to the king, who will have him executed, so that he won't be able to go on harming people.' This led the bystanders to call out abuse after the retreating Abu Sir.

So much for Abu Qir, but as for Abu Sir, when he got back to the *khan* he sat there thinking about what his companion had done to him. He stayed until the pain of his beating had worn off and then he went out again and made his way through the markets. It occurred to him to visit the baths and he asked one of the townsfolk to tell him the way. 'What are baths?' the man asked. 'Places where people wash and remove dirt from themselves,' Abu Sir replied, adding, 'and they are one of the greatest delights in the world.' 'Go to the sea,' said the man, and when Abu Sir insisted that it was the baths that he wanted, the man said: 'We know nothing about them, whatever they may be. We all go to the sea and even the king, when he wants to wash, goes there.' Abu Sir now realized that there were no baths in the city and that the inhabitants knew nothing about them or how they functioned. So he went to the king's court and, after kissing the ground in front of the king and calling down blessings on him, he said: 'I am a stranger here and a bath keeper by trade. When I came to your city I wanted to go to a bath house, but I could not find a single one. How is it that this can be true of a beautiful city like this, when baths are a worldly paradise?' 'What are baths?' the king asked, and in reply Abu Sir started describing them for him, adding: 'Your city will not be perfect until you have a bath house here.'

The king now welcomed Abu Sir warmly, clothed him in a robe of unmatched beauty and presented him with a horse and two black slaves. He gave him four slave girls and two mamluks, and had a house furnished for him, treating him with even more generosity than he had shown to

Abu Qir. He sent builders with him, instructing them to build a bath house wherever he chose. Abu Sir set off with them through the centre of the city until he came across a spot that he liked and pointed it out to the builders. They worked under his direction until they had completed a bath house of unequalled splendour, and on his instructions they painted it so beautifully that it was a delight to the eyes. He then returned to the king and told him that the building work and the painting were finished and that all that was lacking were the furnishings. The king gave him ten thousand dinars, which he took and used for these furnishings, together with towels, which he hung on ropes. All the passers-by stood to stare at it, bewildered by the beauty of the painting, and crowds gathered to look at something they had never seen in their lives before. 'What is this?' they would ask as they looked, and Abu Sir would astonish them by saying: 'This is a bath house.' He then heated the water and got things working, with water spurting from the fountain, captivating the minds of the spectators. Abu Sir asked the king for ten mamluks who had not yet reached the age of puberty, and the king presented him with ten who were as beautiful as moons. He started to massage them and told them to do the same to the clients, and he then released perfumes before sending out a crier to invite the people to visit what was now called the King's Baths.

Customers flocked in and Abu Sir would tell the mamluks to wash them. They would go down into the bathing pool before coming out and sitting in the vestibule, where the mamluks would massage them as Abu Sir had taught them. For three days, people were given free access to use the baths as they wanted, and on the fourth day the king came to pay a visit. He and his grandees rode up, and when he had taken off his clothes and entered the baths, Abu Sir came in and massaged him, removing the dirt from his body in rolls like lamp wicks. The king was delighted to see this, and when he clapped his hands to his body, it felt soft and clean. Abu Sir then mixed rosewater into the king's bath, and when he came out his body was moist and he felt more vigorous than he had ever done in his life. He took his seat in the vestibule, where the mamluks massaged him, as the censers diffused the scent of aloes wood. 'Master,' said the king, 'is this a bath house?' 'Yes,' replied Abu Sir, and the king swore: 'It is only these baths that make this a proper city,' and he went on to ask: 'What charge are you going to make for each of your customers?' 'Whatever you tell me,' Abu Sir said. The king ordered him to be presented with a thousand dinars and told him to make this the fee for everyone using the baths. 'I beg your pardon, your majesty,' said Abu Sir, 'but these people are not all alike. Some of them are rich and some are poor, and if I charge a thousand dinars a head, the baths will go out of business, as the poor won't be able to afford so much.' 'What will you do about charging then?' the king asked, and Abu Sir said: 'I shall leave this to their sense of honour, so that everyone who can will give as much as he feels like, and the fee that I get will be in proportion to the customer's means. In that case, people will flock in; the rich will give what matches their status and the poor will give as much as they can. This will keep the baths busy and prosperous, while the thousand dinars that you suggested is a royal gift which not everyone can afford.'

The leading courtiers agreed with him, saying: 'That is so, your august majesty. Do you suppose that everyone is like you?' 'True enough,' replied the king, 'but this man is a poor stranger and it is our duty to treat him generously, as he has made for us in our city this bath house, the like of which I had never seen. It is he who has added this adornment to our city and has increased its importance, so that however greatly we reward him it will not be too much.' They said: 'If you want to be generous, use your own money, and your generosity to the poor will be reflected in the low charge for the baths, for which your subjects will bless you. As for the thousand dinars, we are the leaders of the state and we would not want to pay as much, so what about the poor?' The king told them: 'This time each one of you can give a hundred dinars together with a mamluk, a slave girl and a black slave.' 'Yes, we shall do that,' they said, 'but from now on everyone who goes in will only have to pay what he wants.' The king agreed, and each one of the courtiers gave Abu Sir a hundred dinars, together with a slave girl, a mamluk and a black slave.

There were four hundred courtiers who accompanied the king to the baths that day . . .

Night 936

Morning now dawned and Shahrazad broke off from what she had been allowed to say. Then, when it was the nine hundred and thirty-sixth night, SHE CONTINUED:

I have heard, O fortunate king, that four hundred of the leading courtiers accompanied the king to the baths that day, and so what they paid amounted in total to forty thousand dinars, in addition to four hundred mamluks, four hundred black slaves and four hundred slave girls – a large enough gift indeed. On top of this, the king presented Abu Sir with ten thousand dinars, ten mamluks, ten slave girls and ten black slaves. At that, he came forward and kissed the ground in front of the king, saying: 'O fortunate and well-directed king of sound counsel, where is there room for me to accommodate all these mamluks and slaves, male and female?' The king said: 'I only told my courtiers to give these to you in order to provide you with a large sum of money. It may be that you think with longing of your own country and your family, and you may want to go home. You will then have got enough wealth from here on which to live back there for the rest of your life.' 'Your majesty,' said Abu Sir, 'may God grant you glory. This number of mamluks and slaves is only suitable for a king, and if you were to order me to be presented with money, it would be better for me than to have such an army. They would eat, drink and have to be clothed, and however much money I made, it would not be enough to meet their expenses.' The king laughed and said: 'By God, you are right. They do form a huge host and you cannot possibly afford to maintain them. Would you then sell them to me for a hundred dinars each?' Abu Sir agreed to this and the king sent word to his treasurer to fetch him money. When this had been done, he paid Abu Sir the purchase price in full and then returned the mamluks and slaves to their owners, saying: 'Whoever recognizes his slave, slave girl or mamluk is to take them back, as this is a gift from me to you.' They did as they were told and each one of them reclaimed his own property. 'May God relieve you from care, your majesty, as you have relieved me from these *ghuls*, whose bellies God alone could fill.' The king laughed, recognizing the truth of his words, and then, with his courtiers in attendance, he left the baths and returned to his palace. Abu Sir still had in his service twenty black slaves, twenty mamluks and four slave girls, and he spent the night counting his gold, before putting it into bags and sealing them.

The next morning, he opened up the baths and sent out a crier to let the people know that whoever came to wash there would only have to give what he felt like and what his generosity suggested. He himself took his seat beside a chest, where he was besieged by customers, and as every one who came put into it as much as he could spare, by the evening, thanks to the grace of Almighty God, the chest was full. It was after that that the queen decided that she wanted to visit the baths, and when Abu

Sir heard of this, in order to accommodate her, he divided the day into two periods, with the time from sunrise to noon being restricted to men, while from noon until sunset was kept for women. When the queen arrived, he stationed a slave girl behind the chest with four others, whom he had taught and who were already skilled bath attendants. The queen admired all this and, as a mark of her pleasure, she put a thousand dinars in the chest.

Abu Sir's reputation spread through the city; as he treated all his customers, rich and poor alike, with respect, he was deluged with benefits from all sides. He made the acquaintance of the king's guards and acquired companions and friends. The king would visit him one day each week, presenting him with a thousand dinars. On other days, his visitors were both the men of importance and the poor, and he did his best to please everyone, showing them the greatest politeness. One day the king's captain came and Abu Sir stripped and went in with him to massage him, treating him with all courtesy. When he came out, Abu Sir produced sherbet and coffee for him, and when the captain wanted to give him something in return, Abu Sir swore that he would accept nothing at all. His courtesy and kindness left the captain with a feeling of obligation, and he was at a loss to know how to repay him for this.

So much for Abu Sir, but as for Abu Qir, he heard everyone talking about the baths. They all told him: 'There is no doubt that these are a worldly paradise and, God willing, you must come with us tomorrow to see their magnificence.' Abu Qir told himself that he had better go with the others to see what it was that had so taken their fancy. He put on his most splendid clothes and rode off on his mule towards the baths, accompanied by four black slaves and four mamluks, who walked behind and in front of him. He dismounted by the door of the baths and, on entering, he detected the scent of aloes and saw people going in and coming out, while the benches were crowded with rich and poor alike. When he got to the entrance hall, he was seen by Abu Sir, who got up and greeted him gladly. Abu Qir said: 'Is this how gentlemen keep their promises? I opened a dyer's shop, and am now the city's leading dyer, an acquaintance of the king's, well off and influential. You have never come to me or asked about me or said: "Where is my friend?" I have been looking for you in vain, sending out my slaves and mamluks to search through the *khans* and other such places, but they failed to get on your track and no one could tell them anything about you.' 'Didn't I come to you,' said Abu Sir, 'only to be called a thief, beaten and put to public disgrace?' Abu Qir showed distress and said: 'What are you saying? Was it you whom I beat?' 'Yes it was,' Abu Sir told him, at which Abu Qir swore with a thousand oaths that he had not recognized him. 'Someone who looked like you had been coming to me every day and stealing people's belongings and I thought that you were that man.' He went on to express his regrets, striking his hands together and exclaiming: 'There is no might and no power except with God, the Exalted, the Omnipotent. I wronged you, but I wish you had introduced yourself and told me who you were. Since you didn't do that, the fault is yours, particularly since I was overwhelmed by pressure of work.' 'God forgive you, my friend,' said Abu Sir. 'This was something that had been preordained and all authority belongs to God. Come in; take your clothes off; wash and enjoy yourself.' 'For God's sake, forgive me, brother,' said Abu Qir. 'May God clear you of responsibility and pardon you,' replied Abu Sir, 'as this was something ordained for me from past eternity.'

Abu Qir now asked him how he had reached so powerful a position, to which he replied: 'God, Who helped you, helped me. I went to the king and told him about baths, after which he told me to build a bath house.' 'You know the king and so do I,' said Abu Qir . . .

Night 937

Morning now dawned and Shahrazad broke off from what she had been allowed to say. Then, when it was the nine hundred and thirty-seventh night, SHE CONTINUED:

I have heard, O fortunate king, that, after Abu Qir and Abu Sir had blamed each other, Abu Qir said: 'You know the king and so do I. If God wills, I shall see to it that he loves and honours you for my sake even more than he has done when he didn't know that you were a friend of mine. I shall tell him that, and ask him to look after your interests.' 'There is no need for this,' Abu Sir told him, 'since there is a God Who moves men's hearts to tenderness and Who has inspired the king and all his court with affection for me.' He went on to tell of the gifts that he had been given and of all that had happened to him. Then he said: 'Take off your clothes behind the chest and when you go into the baths, I shall come in with you to give you a massage.' So Abu Qir stripped and Abu Sir accompanied him to the baths, where he massaged him, soaped him down and then gave him back his clothes, busying himself in attendance on him until he came out. When he did, Abu Sir brought him food and sherbet, and everyone was astonished at the distinction with which he treated him. When, at the end of this, Abu Qir wanted to give him something, Abu Sir exclaimed: 'No, no, you are my friend and there is no difference between us.' Then Abu Qir said: 'My friend, these baths are splendid, but there is something missing.' When Abu Sir asked what this was, he told him that he needed to stock a mixture of arsenic and lime to use as a simple depilatory. 'Make some of this,' he said, 'and then, when the king comes, produce it for him and tell him how easily it makes hair fall out, after which he will love and honour you even more.' 'That is true,' Abu Sir replied, 'and, God willing, I shall make some.'

On leaving, Abu Qir mounted his mule and rode off to the king. When he came into his presence, he said: 'I have some advice for you, king of the age.' The king asked what this was, and Abu Qir replied: 'It has come to my notice that you have built a bath house.' 'Yes,' said the king, 'I was approached by a foreigner and I had this built for him, just as I had the dye house built for you. It is a magnificent place which adds to the attractions of the city.' He went on to speak of its beauties, and when Abu Qir asked whether he had entered it himself, he said that he had. 'Praise be to God who saved you from the evil schemes of that wicked bath keeper, the enemy of religion!' exclaimed Abu Qir. 'What do you know of him?' the king asked. Abu Qir replied: 'I have to tell you, your majesty, that if you go there after today, you will die.' When the king asked why that was, he said: 'The bath keeper is an enemy of yours and an enemy of religion. The only reason that he got you to have the bath house built was in order to poison you there. He has prepared something for you, and if you go in, he will tell you: "This is a medicinal ointment and its application will cause hairs to fall out easily." Far from being a medicine, it is noxious in the extreme, as it is, in fact, a deadly poison. That evil man was promised by the Christian king that if he killed you, his wife and children, whom the king has been holding as captives, would be freed. I myself was a fellow captive of his in the lands of the Christians, but after I had opened a workshop and dyed stuff for them in various colours, they interceded for me with their king. He asked me what I wanted, and when I told him that I wished to be freed, he let me go. It was after that that I came here, and when I saw this fellow in the baths I asked him how he had got away, together with his wife and children. He said: "While my family and I were held there as captives, I went to a meeting of the king's court and I was standing there among the audience when I heard a list of kings being read out. When the king of this city was mentioned, the Christian king sighed and said: 'No one in this world has ever defeated me except for that man, and if anyone can produce a scheme to get him killed for me, I shall give him everything he asks for.' I went up to him and said: 'If I manage to find a way to kill him, will you free me, together with my wife and children?' 'Yes,' replied the king, 'and I shall grant all your wishes.' We agreed, and he sent me off here in a galleon. I made an approach to the king, who had this bath house built for me, and all that remains for me to do now is to kill him and then go back to the Christian king, win freedom for my wife and children, and ask him to grant me a wish." "How do you plan to kill him?" I asked, and he said: "It is simple – very simple indeed. I have made a mixture for him containing poison, and when he comes to the baths, I shall tell him to take it and use it to smear over his nether parts in order to make the hairs fall out. After he has done this, it will take the poison a day and a night to have its effect, but when it gets to the heart it will kill him and that will be the end." When I heard what he had to say, I was afraid for you, and as you have been so good to me, I had to tell you.'

The king, on hearing this, became furiously angry and, after telling Abu Qir to keep the affair secret, he decided to visit the baths in order to settle the matter beyond doubt. When he got there, Abu Sir stripped himself as usual and came to attend to his massage. He said: 'Your majesty, I have made something to clean away your lower hairs.' The king told him to bring it out for him, and when he did, the king detected an unpleasant smell and was certain that this must be poison. He called to his guards to seize Abu Sir and when they had done so, he came out full of rage. No one knew why he was so angry; he told nobody the reason and nobody dared ask him. When he had put his clothes back on, he went to his court and had Abu Sir brought before him with his hands tied behind his back. He summoned his captain, and when the man came he said: 'Take this foul fellow; put him in a sack with two *qintars* of unslaked lime and then tie up its mouth. Place him in a rowing boat and bring it underneath my palace. You will see me sitting by my window and you are then to ask me: "Shall I throw him into the water?" I will say yes and you will toss him in so that the lime will be slaked and he will die through being both drowned and burned.' 'To hear is to obey,' replied the captain.

He took Abu Sir from the king's presence and brought him to an island opposite the royal palace. Then he said to him: 'I once went to your baths where you treated me with extreme respect and looked after me to my great satisfaction. You swore that you would not accept any fee from me and I formed a great affection for you. So tell me what have you done to the king and what evil deed it is that has led him to become so angry with you that he has ordered me to put you to this cruel death?' Abu Sir replied: 'I haven't done anything and I know of no fault that I have committed to deserve this.'

Night 938

Morning now dawned and Shahrazad broke off from what she had been allowed to say. Then, when it was the nine hundred and thirty-eighth night, SHE CONTINUED:

I have heard, O fortunate king, that the captain asked Abu Sir why the king was angry with him. 'By God, brother,' he replied, 'I haven't earned this by anything wrong that I did to him.' The captain said: 'Your standing with the king was higher than anyone had ever achieved before, but every favourite is an object of jealousy. Perhaps someone grudged you this good fortune and said something to the king that made him angry with you. But you are welcome here and no harm will come to you. You were good to me, although we didn't know each other, and so I am going to set you free, but when I do that, you will have to stay with me on this island until I can send you off on a galleon leaving for your homeland.' Abu Sir kissed the captain's hand in gratitude, and the latter got lime, which he put into a sack, adding a large stone, as big as a man, and saying: 'I take refuge with God.' He then gave Abu Sir a net and told him to cast it into the sea in the hope of catching fish, as it was his responsibility to produce fish every day for the king's kitchens. He explained: 'This bad business of yours has kept me from fishing, and I'm afraid that the kitchen boys may come looking for fish and not find any, unless you manage to get hold of something. I shall go off myself beneath the palace and trick the king by pretending that I have thrown you into the water.' Abu Sir agreed to this and prayed that God might aid him.

The captain put the sack in a boat and brought it underneath the palace, where he saw the king sitting by a window. When he asked whether he should throw Abu Sir in, the king said yes, and gestured with his hand. As he did so, something gleaming fell from his hand into the water and this turned out to be the royal signet ring, which had talismanic powers. Whenever the king was angry with someone and wanted to kill him, he would gesture with his right hand with the ring on it and the ring would then produce a lightning flash which would strike the man at whom he was pointing, removing his head from his shoulders. This ring was the source of the king's power over his troops, as it allowed him to overcome the most powerful of men. When it fell off, he said nothing. He could not let it be known that it had fallen into the sea for fear that his troops would rise against him and kill him, and so he kept quiet.

So much for him, but as for Abu Sir, after the captain had left, he took the net and cast it into the sea. When he pulled it in, it was full of fish; the same thing happened when he made a second cast, and this went on and on until there was a great pile of fish in front of him. 'By God,' he said to himself, 'it is a long time since I ate any fish,' and so he picked out a large fat one, saying: 'When the captain comes, I will ask him to fry this for me so that I can have it for my meal.' He cut it open with a knife that he had with him, but the knife stuck in its gills and there he discovered the king's ring. The fish had swallowed it and destiny had then led it to the island where it had fallen into Abu Sir's net.

Abu Sir, not knowing the ring's powers, took it and put it on his little finger. Just then, two of the king's kitchen boys came for fish and they asked Abu Sir where the captain had gone. 'I don't know,' he said, making a gesture with his right hand, and as he did so their heads fell off their shoulders. Astonished by this unfortunate fate of theirs, he wondered who had killed them. As he was thinking the matter over, up came the captain, to find a large pile of fish, together with the two dead kitchen boys. He noticed the ring on Abu Sir's finger and called out: 'Brother, don't move your ring hand or else you will kill me.' Abu Sir was amazed, and when the captain came up to him and asked who had killed the two boys, he told him that he didn't know. 'That has to be true,' the captain said, 'but tell me about the ring. Where did you get it from?' 'I saw it in the gills of this fish,' Abu Sir told him, and again the captain said: 'This is true, for I saw it glittering as it fell into the sea from the royal palace just as the king gestured towards what he thought was you and told me to throw you into the water. When he did that, I tossed the sack overboard and the ring must have fallen from his finger into the water, where this fish swallowed it. God must have brought the fish here so that you might have the good luck to catch it. Do you have any idea of its properties?' When Abu Sir said no, the captain explained: 'Fear of this ring is the only reason why the king's troops obey him. It has talismanic powers and when he is angry with anyone and wants to kill him, he makes a gesture with it and the man's head falls from his shoulders. Lightning flashes from it, and when it reaches the victim, he dies on the spot.'

When Abu Sir heard what the captain had to say, he was delighted and asked to be taken back to the city. The captain agreed, adding: 'I'm no longer afraid that the king may harm you, because if you point at him with the intention of killing him, his head will fall off in front of you, and if you wanted to kill not only him but all his men as well, you could do that with no difficulty.' He then took Abu Sir on board his boat and set off with him for the city.

Night 939

Morning now dawned and Shahrazad broke off from what she had been allowed to say. Then, when it was the nine hundred and thirty-ninth night, SHE CONTINUED:

I have heard, O fortunate king, that the captain took Abu Sir on board his boat and set off with him to the city. When he got there, he went to the royal palace and, on entering the audience chamber, he found the king sitting there in a state of deep distress, with his guards, to whom he had not been able to reveal the ring's loss, ranged in front of him. On seeing Abu Sir, he exclaimed: 'Didn't I have you thrown into the sea? How did you manage to escape?' 'King of the age,' Abu Sir replied, 'when you gave your command, your captain took me to an island where he asked me what I had done to lead you to order my execution. I told him that I knew of no wrongdoing on my part, and he suggested that perhaps the great favour that you had shown me might have made someone envious enough to try to rouse your anger by slandering me to you. In return for the kindness that I had shown him when he came to my baths, he promised to arrange for my escape and to send me back home. He then put a stone in the boat instead of me and dropped it into the sea, but when you gestured towards me, your ring fell from your hand into the water, where a fish swallowed it. I was on the island fishing and that fish formed part of my catch. I took it out and was intending to grill it, but when I cut it open I saw the ring inside it and so I took it and put it on my finger. Then two kitchen boys came for fish and I gestured towards them, not knowing the ring's qualities, at which their heads fell off. The captain recognized what was on my finger and told me about its talismanic powers. I have brought it to you because you treated me well and were extremely generous to me, kindnesses for which I cannot show ingratitude. Here is your ring. Take it, and if I have done anything to deserve death, tell me what it was and then kill me, for you will be free of any blood guilt.' He then removed the ring from his finger and handed it to the king.

The king, on seeing the service that Abu Sir had done him, recovered his good spirits and, after taking the ring and putting it on his finger, he rose to his feet and embraced him. 'You are one of the best and most virtuous of men,' he said. 'Do not blame me but forgive me for what I did to you. If anyone else had got hold of this ring, he would never have given it back to me.' Abu Sir said: 'If you want my forgiveness, your majesty, tell me what it was that I did wrong which led you to order my execution.' The king replied: 'By God, I am now convinced that you are innocent of any wrongdoing since you have done me this favour. It was the dyer who accused you,' and he told him what Abu Qir had said. 'By God, your majesty,' Abu Sir protested, 'I don't know the king of the Christians and never in my life have I been to their lands, nor have I ever thought of killing you. This dyer was a friend and neighbour of mine in Alexandria, where we found it hard to make a living. Because of that, we left after having made an agreement, confirmed on the Quran, that whichever of us found work should provide food for the other while he was without a job.' He then told the king what had happened and how Abu Qir had taken his money, leaving him sick in his room in the *khan*. He explained how the doorkeeper had spent his own money on him during his illness until God had brought about his cure, and how he had then wandered around the city taking the tools of his trade with him as usual. On his way, he had seen a crowd outside a workshop and when he looked at the door, there he saw Abu Qir sitting on a bench. He had gone in to greet him, but had been maltreated and beaten as a thief, receiving a painful thrashing.

When he had told the king all the particulars of his story from beginning to end, he added that it was Abu Qir who had told him to manufacture the ointment for the king's use, saying that this was the one thing lacking in the baths. 'I must tell you, your majesty,' he went on, 'that there is no harm in it. We manufacture it in our own country where it is one of the accessories of a bath house. I had forgotten this, and when

Abu Qir came to me and I treated him well, he reminded me of it and told me to make some. Send for the doorkeeper of the *khan*, as well as for the workmen in the dye house, and ask all of them about what I have told you.' The king sent for them and when they had arrived, in answer to his questions they confirmed the facts. He then ordered Abu Qir to be brought to him barefooted, bare-headed and with his hands tied behind his back.

Abu Qir had been sitting at home, happy in the thought that Abu Sir was dead, when before he knew what was happening the king's guards had pounced on him and, after beating him on the nape of his neck, they tied his hands and brought him before the king. There he was met by the sight of Abu Sir sitting beside the king, with the doorkeeper of the *khan* and his own workmen standing in front of him. 'This is his companion,' said the doorkeeper, 'the man who stole his money and abandoned him with me when he lay sick in his room.' When he had completed his story, Abu Qir's workmen said: 'That is the man our master told us to seize and beat.'

Abu Qir's villainy became clear to the king, who realized that he deserved a worse punishment than any that could be inflicted by Munkar and Nakir. He ordered his guards to take him and parade him around the city and through the market ...

Night 940

Morning now dawned and Shahrazad broke off from what she had been allowed to say. Then, when it was the nine hundred and fortieth night, SHE CONTINUED:

I have heard, O fortunate king, that when the king heard the evidence of the doorkeeper and the craftsmen, he was convinced of Abu Qir's villainy. He told his guards to take him and parade him around the city, after which he was to be put in a sack and thrown into the sea. 'Your majesty,' said Abu Sir, 'please accept my intercession for him, as I forgive him for how he treated me.' 'Even if you do,' replied the king, 'I cannot forgive him for what he did to me,' and he then shouted: 'Seize him!' The guards took hold of Abu Qir and, after parading him around the town, they put him in a sack with lime and threw him into the sea, where he died, both drowned and burned.

The king now promised Abu Sir that he would grant him whatever he wanted. Abu Sir told him: 'I wish that you would send me back home, for I no longer want to stay here.' So the king, after having added greatly to the wealth that he already had and the gifts he had been given, presented him with a galleon laden with good things and crewed by mamluks, whom he assigned to him. This was after he had offered to appoint him as vizier, a post which he refused. He then took his leave of the king and set off in the galleon, in which everything, including the crew, was his own property. They came to land at Alexandria and anchored off shore, after which they disembarked. One of the mamluks called Abu Sir's attention to a large and heavy sack that was lying on the shore, saying: 'Its mouth is tied up and I don't know what is in it.' Abu Sir came across and, on opening the sack, he found inside the corpse of Abu Qir which had been carried there by the sea. He removed it and buried it near Alexandria, making the tomb a place of pilgrimage supported by religious endowments. Over its entrance he inscribed the following lines:

> It is by his actions that a man is known;
> What the noble and the freeborn do reflects their lineage.
> Slanderers will be slandered in their turn,
> And others will say of you the things you say.
> Avoid obscenities within your speech,
> Whether this be in earnest or in jest.
> Dogs that are well behaved are kept as pets,
> But ignorant lions must be held in chains.
> Corpses float on the surface of the open sea,
> While pearls are hidden beneath the sands.
> No sparrow will compete against a hawk
> Unless through recklessness and lack of sense.
> Written on the pages of the windy sky
> You find: 'Do good and so it will be done to you.'
> Look for no sugar in the colocynth,
> For every taste reverts to its own root.

Abu Sir lived on for some time and when he died he was buried beside his companion, Abu Qir. The place was known by both their names but is now commonly called Abu Qir. This is as much as I know of their story. Praise be to the Eternal God, Whose will governs the progress of the nights and days.

*

A story is also told of a fisherman named 'Abd Allah, whose large family comprised nine children and their mother. 'Abd Allah himself was a very poor man whose only possession was his fishing net, and every day he used to go to the seashore in order to fish. If he only caught a few, he would sell his catch and spend what God had given him on his children, while if he caught a lot he would cook a good dish of fish and buy fruit, continuing to spend money until it had all gone, while saying to himself: 'God will provide tomorrow's food tomorrow.' His wife then gave birth to their tenth child on a day when he had been left with no money at all. She asked him to go to find her something to eat and he told her: 'I shall go to the shore today to try my luck, relying on the blessing of the Almighty and on the good fortune brought by this newborn baby.' 'Put your trust in God,' she said, and so he took his net and went off.

He cast the net, invoking the baby's good fortune and saying: 'My God, make it easy and not difficult for me to find something on which to nourish him, and let it be a lot rather than a little.' He waited for a time but when he pulled the net in, it turned out to be full of rubbish, sand, pebbles and weeds, and there was not a single fish to be seen. He made a second cast, waited, but still found no fish when he brought the net in, and the same thing happened with the third, fourth and fifth casts. He moved to another spot and set about praying to God to grant him something, but although he carried on like this until evening, he caught not even the smallest sprat. At this, he asked himself wonderingly: 'Did God create this child without providing food for him? This can never be, as He Who has made mouths to open guarantees them food to eat, for He is the Generous Provider.'

With his net over his shoulder, he went back disheartened and preoccupied with worries about his family, whom he had left without food, and in particular about his wife, because she had just given birth. 'What shall I do,' he said to himself as he walked along, 'and what am I going to say to my children tonight?' He passed by a baker's oven where he saw a large crowd, as this was a time when prices were high and people had not much food. They were offering money to the baker, but the crowd was so thick that he could not pay attention to any of them. 'Abd Allah stood watching, and the scent of the warm bread filled him with hungry longing. The baker saw him and called out: 'Fisherman, come here.' When 'Abd Allah had gone up to him, he asked him if he wanted bread. 'Abd Allah kept silent, but the baker said: 'Speak and don't be embarrassed. God is generous and if you haven't any money with you,

I'll give you bread and wait for you to have better luck.' 'By God, master,' 'Abd Allah said, 'I've no money with me, but if you give me enough bread for my family, I'll leave my net with you until tomorrow as a pledge.' The baker replied: 'My poor fellow, this net is your shop and your means of livelihood. If you pledge it, what will you use to fish? Tell me how much will be enough for you.' 'Ten *nusfs*' worth,' 'Abd Allah answered, at which the baker gave him that amount of bread and presented him with ten *nusfs* in cash, saying: 'Take these coins and use them to buy a cooked meal. You will then owe me twenty and tomorrow you can bring me the equivalent of this in fish, or, if you don't catch anything, come for your bread and I'll give you another ten and wait until you find yourself better off . . .'

Night 941

Morning now dawned and Shahrazad broke off from what she had been allowed to say. Then, when it was the nine hundred and forty-first night, SHE CONTINUED:

I have heard, O fortunate king, that the baker told the fisherman: 'Take what you need and I shall wait until you find yourself better off. Then you can pay what you owe me in fish.' 'God reward you,' said 'Abd Allah, 'and shower benefits on you.' He then took the bread and the coins and went off happily. He bought what he could and returned to his wife, whom he found comforting the children, who were crying with hunger, and telling them: 'Your father is on his way with something to eat.' When he went in, he put down the bread for them and, as they ate, he told his wife what had happened. 'God is generous!' she exclaimed.

The next day, he took his net and left the house, saying: 'My Lord, I ask You to provide something for me today that may clear my debt to the baker,' but when he got to the sea and started to cast his net and draw it in, there were no fish. This went on all day, and when he had still caught nothing, he went back full of sorrow. His way home led past the baker's oven and as this was the only road that he could take, he decided to walk so fast that the baker would not see him. When he got to the oven and saw the crowd there, he quickened his pace as he was ashamed to be seen. But the baker looked up and shouted: 'Fisherman, you have forgotten to come for your bread and your spending money.'

'By God, I didn't forget,' said 'Abd Allah, 'but I was ashamed to meet you because I caught nothing today.' 'Don't be embarrassed,' the baker told him. 'Didn't I tell you to take your time until better days come?' He then gave him the bread and the money, after which 'Abd Allah went back home and told his wife what had happened. 'God is gracious,' she told him, 'and if it is His will, you will be lucky and will be able to pay off your debt.'

Things went on like this for forty days, on each of which he went off to the shore at sunrise and stayed until sunset, before going back without fish and getting bread and cash from the baker. Never once did the baker mention fish to him or stop helping him out, as other people would have done. He kept on handing over the bread and the ten coins, and whenever 'Abd Allah said: 'Brother, how much do I owe you?' he would tell him to go away, saying: 'This isn't the time to draw up an account. Wait until you are better off and then I'll settle with you.' 'Abd Allah would bless him and go off gratefully.

On the forty-first day, he said to his wife: 'I intend to cut up my net and take a rest from this way of life.' His wife asked why, and he told her: 'It seems as though I can get no living from the sea. How long can this go on? By God, I melt with shame in front of the baker and I shall never go to the sea again. Nothing else can stop me from having to pass his oven, as it is on the only way that I can take to get there. Every time I pass him, he calls to me and gives me bread and cash, but how long can I go on getting deeper and deeper into his debt?' His wife replied: 'Praise be to the Exalted God, Who has filled the man's heart with pity for you so that he gives you food. What is it that you don't like about this?' He told her: 'I owe him a great deal of money and the time must come when he will ask me for it.' 'Has he said anything hurtful to you?' she asked. 'No,' 'Abd Allah replied, 'and he won't give me a reckoning, telling me to wait until things get better for me.' 'Well then,' she said, 'if he does ask you for the money, tell him to wait for the good times to come that you both hope for.' 'And when will that be?' he asked, to which she replied: 'God is gracious.' 'That is true,' he said.

He then picked up his net and set off for the sea, saying: 'Lord, let me catch even a single fish which I can give the baker.' He cast his net into the sea and when he began to pull it in he found it so heavy that he became exhausted as he hauled at it. Then, when he got it out, he was disgusted to find that in it was the bloated and stinking carcass of a donkey. He freed it from the net and exclaimed: 'There is no might and no power except with God, the Exalted, the Omnipotent! I can't go on telling my wife that I can no longer get a living from the sea and asking her to let me give up fishing, only to have her tell me that God is gracious and that good will come. Is this dead donkey good?' In deep sorrow, he went off to another spot in order to get away from the smell of the carcass and there he took his net and made another cast. He waited for a time and when he began to pull, he again found it heavy and this time he had to go on straining at it until blood dripped from his hands. When it was clear of the water, he saw in it a human shape, which he took to be one of the *'ifrits* of our lord Solomon, who used to imprison them in brass bottles and throw them into the sea. He supposed that in the course of time one of these bottles must have broken, so letting out the *'ifrit*, who had then been caught in his net. So he started to run away, calling out: 'Mercy, mercy, *'ifrit* of Solomon!' But the creature called to him from inside the net: 'Don't run off, fisherman. I am a human being like you and if you set me free I shall reward you.'

When 'Abd Allah heard this, his panic subsided and he went back and said: 'Aren't you an *'ifrit* from among the *jinn*?' 'No,' said the creature, 'I am a human, who believes in God and His Prophet.' 'Who was it then who threw you into the sea?' 'Abd Allah asked, and the creature told him: 'I am one of the children of the sea, and I was wandering around when you threw your net over me. My people obey God's laws and we show pity to His creatures, for otherwise, if I were not afraid of being counted among the disobedient, I would cut your net. As it is, I am content with the fate that God has decreed for me, and if you set me free, you will be like my master and I your prisoner. Are you prepared to free me for God's sake and to come to an agreement with me to become my companion? I shall then come to meet you here every day and you shall bring me a present from the fruits of the land. For you have grapes, figs, melons, peaches, pomegranates and so on, and anything that you bring will be acceptable. For our part, we have coral, pearls, chrysolites, emeralds, sapphires and other gems, and I shall use these to fill the basket in which you bring me fruit. What do you say to this, brother?' 'We should recite the Fatiha to confirm this agreement between us,' said 'Abd Allah, and when they had both done this, he freed the merman from the net. He then asked him his name. 'I am 'Abd Allah of the sea,' the merman replied, and he went on: 'If you don't see me when you come here, call out: "Where are you, 'Abd Allah of the sea?" and I shall be with you immediately . . .'

Night 942

Morning now dawned and Shahrazad broke off from what she had been allowed to say. Then, when it was the nine hundred and forty-second night, SHE CONTINUED:

I have heard, O fortunate king, that 'Abd Allah of the sea said to him: 'If you don't see me when you come here, call out: "Where are you, 'Abd Allah of the sea?" and I shall be with you immediately. So tell me, what is your own name?' The fisherman replied: 'My name is 'Abd Allah,' and the merman told him: 'You can be 'Abd Allah of the land and I am 'Abd Allah of the sea. Stay here until I go and fetch you a present.' Land 'Abd Allah agreed and sea 'Abd Allah disappeared back under the waves.

At that, 'Abd Allah the fisherman regretted having freed the merman from his net, saying to himself: 'How am I to know that he will ever come back to me? He must be laughing at me for having let him go. Had I kept him, I could have shown him to the townspeople and got money from them all, and I could have taken him to the houses of the great folk.' In his regret, he kept telling himself: 'You have lost your catch,' but then, while he was feeling sorry for himself, back came the merman with his hands full of pearls, coral, emeralds, sapphires and other jewels. 'Take these, my brother,' he said, 'and don't blame me, for I have no basket which I could fill for you.' The delighted fisherman took the jewels from him and the merman told him to come there every day before sunrise. He then took his leave and went off back into the sea.

The fisherman went happily into the city, walking on until he came to the baker's oven. 'Brother,' he said, 'I have been lucky, so tell me how much I owe you.' 'There is no need for that,' the baker told him. 'If you have got something with you, you can give it to me, but if not, take away your bread and your cash and wait for the good times to come.' 'They have come, my friend,' 'Abd Allah said, 'thanks to God's grace. I must owe you a lot, but take this,' and he passed over a handful of pearls, coral, sapphires and other gems. This was half of what he had with him, and when he had given it to the baker, he asked him for cash which he could use as spending money that day until he was able to sell the other jewels. The baker handed over all the money that he had, together with all the bread that was in his basket. He was delighted with what he had been given and he told 'Abd Allah: 'I am your slave and your servant.' Then he picked up all the bread that he had and walked behind him to his house, carrying it on his head. He handed the bread over to 'Abd Allah's wife and children and then went off to the market, from where he brought back meat, vegetables and all sorts of fruit. He abandoned his oven and spent the whole day in attendance on 'Abd Allah, looking after his needs. 'You are putting yourself to too much trouble,' 'Abd Allah protested, but the baker said: 'This is my duty as I have become your servant and you have overwhelmed me with your kindness.' 'It was you who were kind to me,' 'Abd Allah corrected him, 'at a time when I was in need and prices were high.' The baker spent the night with him and enjoyed a good meal, after which the two became firm friends. When 'Abd Allah told his wife of his encounter with the merman, she was pleased but told him to keep the affair secret lest he fall into the hands of the authorities. He said: 'Even if I keep it from everyone else, I can't hide it from the baker.'

The next morning, he took a basket which he had filled the evening before with all kinds of fruit and carried it before sunrise to the seashore. He put it down and called to the merman, who immediately answered: 'Here I am,' and came out to meet him and to be presented with the fruit. This he carried into the sea, diving under the surface, and after a short delay he came out again, bringing back the basket filled with precious stones and gems of all kinds. 'Abd Allah carried it away on his head and when he got to the baker's oven, the baker said: 'Master, I've baked you forty cakes and sent them to your house. I'm now going to bake some special bread for you and when it is ready, I shall bring it to your house and then go and fetch you vegetables and meat.' 'Abd Allah took three handfuls of jewels from the basket and presented them to him before going home and putting down the basket. He picked out one precious gem from each of the various types that were there and took them to the jewellers' market. He stopped by the shop of the superintendent of the market and asked him to buy the gems. The man asked to be shown them, and when he had seen them he said: 'Do you have any more?' 'A basket full of them,' 'Abd Allah replied. 'Where do you live?' the man said and, when 'Abd Allah had told him, the superintendent ordered his servants to seize him, saying: 'This is the thief who stole the property of the queen.' On his instructions, his men then beat 'Abd Allah and tied his hands behind his back. The superintendent and everyone else in the jewellers' market got up, saying: 'We've caught the thief.' One of them was claiming: 'It was this ruffian who stole So-and-So's goods,' while another said: 'He stole everything that was in So-and-So's house.'

Everyone had a different story to tell, but all the while 'Abd Allah stayed silent, saying nothing and giving no replies until he was brought before the king. Then the superintendent said: 'King of the age, when the queen's necklace was stolen, you sent word to us and asked us to arrest the man responsible for the theft. I have distinguished myself by my hard work on this case and I have caught the thief for you. Here he is, standing in front of you, and these are the jewels that we have recovered from him.' The king said to the chief eunuch: 'Take these; show them to the queen and ask her whether they are what she lost.' When the eunuch took them to the queen, she was filled with admiration for what she saw, but she sent a message to the king saying: 'I found my necklace in my own apartments and these jewels don't belong to me. In fact, they are finer than mine. Don't mistreat the man who has them . . .'

Night 943

Morning now dawned and Shahrazad broke off from what she had been allowed to say. Then, when it was the nine hundred and forty-third night, SHE CONTINUED:

I have heard, O fortunate king, that the king's wife sent a message to him, saying: 'These are not mine, and the jewels are finer than those of my necklace. Don't mistreat the man who has them, but if he is willing to sell them, buy them from him to make into a necklace for your daughter, Umm al-Su'ud.' When the eunuch went back to tell the king what the queen had said, he cursed the superintendent of the market and all his associates with the damnation of 'Ad and Thamud, but they protested: 'King of the age, we knew this man to be a poor fisherman, and as we didn't see how he could have so many jewels, we thought that he must have stolen them.' 'You vile creatures,' said the king, 'did you think that a Muslim could have too much good fortune? Why didn't you ask him? God Almighty may have provided him with this from a source for which he is not accountable, so how is it that you make him out to be a thief and disgrace him in public? Be off, may God grant you no blessing.' They left in fear.

So much for them, but as for the king, he said: 'Fisherman, may God, Who has been generous to you, grant you His blessing. I guarantee your safety. Tell me the truth about where you got these gems, for, king as I am, I have none to match them.' 'I have a basket full of them,' 'Abd

Allah told him and he then went on to explain about his friend, the merman. 'I have an agreement with him,' he explained, 'by which I am to bring him a basket of fruit every day and in return he will fill it for me with these jewels.' 'Fisherman,' said the king, 'this is your good fortune, but wealth needs to be supported by rank. For the present, I can see to it that you don't fall into the power of others, but I may be deposed or die and my successor would then kill you out of greed and love for the things of this world. So I would like to marry you to my daughter, appointing you as my vizier and nominating you as my successor, in order that no one may hope to rob you after my death.'

The king then ordered his servants to take 'Abd Allah to the baths, and when they had done this and he had been washed, they clothed him in royal robes and brought him back to the king, who appointed him as a vizier. He then sent his couriers, his guards and the wives of his leading courtiers to 'Abd Allah's house, where they dressed his wife and his children regally and mounted the woman on a palanquin. All the wives, together with the soldiers, couriers and guards, walked ahead of her as they conducted her to the palace, with her youngest child in her lap. The elder children were brought to the king, who greeted them kindly and hugged them, before seating them by his side. There were nine of them, all boys, while the king's only child was his daughter, Umm al-Su'ud. The queen treated 'Abd Allah's wife with courtesy and kindness and appointed her as one of her advisors. Then the king ordered a marriage contract to be drawn up between 'Abd Allah and his daughter, stipulating as her dowry all the precious jewels in 'Abd Allah's possession. The wedding celebrations started and on the king's instructions a proclamation was made that the city be adorned with decorations in honour of the event.

The morning after the marriage had been consummated and the bride deflowered, the king looked out of his window and saw 'Abd Allah carrying a basket of fruit on his head. 'Son-in-law,' he called out, 'what is that you have with you and where are you off to?' 'I am going to my friend, the merman,' 'Abd Allah replied. 'This is no time to be going to your friend,' objected the king, but 'Abd Allah told him: 'I'm afraid that if I don't keep my appointment with him, he will think me a liar and say that worldly success has made me forget him.' 'True enough,' replied the king, 'so go off to him, and may God aid you.' As 'Abd Allah walked through the city on his way to meet the merman, some people recognized him and said: 'There is the king's son-in-law on his way to exchange fruit for gems.' Others, who didn't know him, would say: 'How much are you asking for a pound of fruit? Come over here and sell it to me.' He would tell them: 'Wait till I return,' so as not to offend anyone. Then he went off and met the merman, who exchanged the fruit that he was given for jewels.

Things went on like this, but every day, when 'Abd Allah passed the bakery, he found it closed. After ten days, during which he had not seen the baker and had found the bakery shut, he said to himself: 'This is surprising. Where can he have got to?' He asked his neighbour about this, saying: 'What has God done with the baker?' and the man told him: 'He's ill and can't leave his house.' 'Where does he live?' 'Abd Allah asked, after which he went where he had been directed, to ask after him. He knocked on the door of the house, and when the baker looked out of the window and saw his friend the fisherman with a filled basket on his head, he went down and opened the door for him. 'Abd Allah threw himself on him, embracing him tearfully and saying: 'How are you, my friend? I have been going past your bakery every day only to find it locked up. Then I asked your neighbour, who told me that you were sick, and so I got him to tell me where you lived in order that I could come to see you.' The baker replied: 'May God reward you well! I'm not sick, but I heard that you had been arrested by the king, as some people had falsely accused you of being a thief. I was afraid for myself and so I locked up the bakery and hid away.' 'I was arrested,' 'Abd Allah agreed, and he then told the baker the story of his encounter with the king and the superintendent of the market, explaining that the king had married his daughter to him and appointed him as his vizier. He went on: 'Take what is in the basket as your share and don't be afraid.'

When he had set the baker's fears at rest, he went back to the king with his empty basket and the king said: 'It seems as though you didn't meet your friend the merman today.' 'Abd Allah replied: 'I did go to him, but I passed on the jewels that he gave me to my friend the baker, who had done me a service.' 'Who is this baker?' the king asked, and 'Abd Allah told him that this was a kind man who had done him a service in the days of his poverty, never neglecting or disappointing him. 'What is his name?' the king asked, and 'Abd Allah told him: 'He is 'Abd Allah the baker, while I am 'Abd Allah of the land and my friend is 'Abd Allah of the sea.' 'I too am called 'Abd Allah,' the king told him, 'and all 'Abd Allahs are brothers. So summon your baker friend in order that I may appoint him as my left-hand vizier.' The message was sent and when the baker arrived, the king gave him a vizier's robe and confirmed his appointment and that of 'Abd Allah the fisherman as vizier of the left and of the right respectively.

Night 944

Morning now dawned and Shahrazad broke off from what she had been allowed to say. Then, when it was the nine hundred and forty-fourth night, SHE CONTINUED:

I have heard, O fortunate king, that the king appointed his son-in-law, 'Abd Allah of the land, the vizier of the right and the baker vizier of the left. Things went on like this for a whole year with 'Abd Allah the fisherman going every day with a basket of fruit and bringing it back filled with precious gems. When there were no longer any garden fruits, he started to take raisins, almonds, hazelnuts, walnuts, dried figs and so on, and the merman would accept whatever he brought, returning the basket filled with gems. One day, in the course of this exchange, 'Abd Allah sat on the shore while the merman was nearby in the water. They had been talking about a range of subjects when the conversation turned to tombs. The merman said: 'We are told, brother, that on land you have the grave of the Prophet, may God bless him and give him peace. Do you know where this is?' 'Yes,' said his friend and when the merman asked where it was, he told him: 'In a city called Tayyiba.'* 'And do the land folk make pilgrimages to it?' 'Yes,' said 'Abd Allah. 'I congratulate you people,' said the merman, 'on being able to visit this noble and merciful Prophet, for those who make this pilgrimage win the right to his intercession. Have you yourself visited it, my brother?' 'No,' said the fisherman, 'for I was a poor man and I could not pay the expenses of the journey. It is only since I met you and you have been so good to me that I have become rich, but now I must go there after I have made my pilgrimage to the sacred House of God. The only thing that stops me is my love for you, as I cannot bear to leave you for a single day.' 'Do you put your love for me before a visit to the grave of Muhammad – may God bless him and give him peace?' his friend asked, adding: 'He will intercede for you before God on Judgement Day, so saving you from hell-fire and allowing you to enter Paradise. Are you going to give up your

* Medina.

pilgrimage to his grave all because of worldly love?' 'No, by God,' the fisherman replied, 'for as far as I am concerned, this takes precedence over everything else, and I would like your permission to go there this year.'

The merman agreed to this and said: 'When you stand over his grave, greet him for me. There is also something that I want to entrust to you. If you come into the sea with me, I shall take you to my city, entertain you as a guest in my house and give you this to deposit at the Prophet's grave. You are then to say: "Apostle of God, 'Abd Allah of the sea greets you and has sent you this gift, hoping that you will intercede for him to save him from hellfire."' The fisherman said: 'Brother, you were born in the water; you live there and it does you no harm, but would you be injured if you left it and came out on land?' 'Yes,' said the other, 'my body would dry up and the land breezes would kill me.' 'I am in the same position,' the fisherman told him. 'I was born on land and live there. If I go into water, I swallow it and it will choke me to death.' 'Don't be afraid,' his friend told him. 'If you smear your body with an ointment that I shall bring you, the water won't harm you, even if you spend the rest of your life wandering round in the sea, lying down to sleep and getting up afterwards.' 'If that is the case, well and good,' said 'Abd Allah, 'so bring me the ointment and let me try it out.' 'Very well,' replied the merman.

He then took the basket and disappeared into the sea, returning soon afterwards with what looked like cow's grease, golden yellow in colour, with a clean smell. When the fisherman asked what it was, he said: 'It comes from the liver of a species of fish called the *dandan*, a huge creature and one of our worst enemies. It is larger than any animal you have on land, and were it to come across a camel or an elephant, it would swallow them up.' 'What does this sinister beast eat?' asked the fisherman, and his friend told him: 'It eats sea creatures, and you may have heard the proverb: "Like fish in the sea, the strong eat the weak."' 'True enough,' said the fisherman, 'but are there many of these *dandans* in the sea?' 'More than anyone can count, except Almighty God,' replied the other, at which the fisherman said: 'If I go down there with you, I would be afraid that one of them might meet me and gobble me up.' 'There is no need to be fearful,' his friend told him, 'for if one of them were to catch sight of you, it would recognize that you are a son of Adam and be frightened away. There is nothing in the sea that alarms it so much as humans, for to eat one of you means instant death for it, human fat being a deadly poison for its species. The only way that we manage to collect its liver grease is when one of you falls into the sea and is drowned, for the appearance of the corpse changes and its flesh may become torn. In that case, the *dandan* will eat it, thinking that it belongs to a sea creature, and as a result it will die. Then, when we come across its carcass, we take its liver grease and rub it over our bodies to help us travel around in the sea. If there are a hundred *dandans*, two hundred, a thousand or even more within range of a human voice, one sound from this will kill them all immediately . . .'

Night 945

Morning now dawned and Shahrazad broke off from what she had been allowed to say. Then, when it was the nine hundred and forty-fifth night, SHE CONTINUED:

I have heard, O fortunate king, that the merman told 'Abd Allah: 'If a thousand of these creatures or even more hear a single shout from a human, they will die instantly. Not a single one of them will be able to get away, wherever he happens to be.' 'I put my trust in God,' said 'Abd Allah, and he stripped off his clothes and buried them in a hole that he had dug on the shore. He rubbed the grease over his body from the crown of his head to the soles of his feet, and then went down and plunged into the water. When he opened his eyes, he found that he was unaffected by it and so he started to walk wherever he chose, coming up if he wanted or else going down into the depths. He saw the water stretched over him like a tent and found that it was doing him no harm. 'What do you see, brother?' asked his companion. 'Nothing but good,' 'Abd Allah replied, adding: 'What you said was true, for the water is not harming me.' On his instructions, he followed the merman and the two of them walked from place to place, with 'Abd Allah looking at water mountains rising in front of him and to the right and left. He gazed at them with pleasure, as well as at the various kinds of fish, large and small, that were disporting themselves there. Among them were things that looked like buffaloes; others looked like cows; some were like dogs; while a number had what resembled human forms. Whenever the two companions came near any of them, they would take flight at the sight of 'Abd Allah, and when he asked his friend why this was, he replied: 'It is because they are afraid of you, as all God's creatures fear man.'

'Abd Allah continued to inspect the wonders of the sea until he and his friend arrived at a lofty mountain. He was walking alongside it when all of a sudden there was a loud cry and he turned to see a black shape, as big or bigger than a camel, diving down on him from the mountain and bellowing as it did so. 'What is it?' he asked and the merman said: 'This is a *dandan* which is coming down to look for me, as it wants to eat me. Shout at it before it gets to us or else it will carry me off as its prey.' 'Abd Allah shouted and the beast fell dead. When he saw the carcass, 'Abd Allah exclaimed: 'Glory and praise be to God. I used neither sword nor knife, and how could so huge a creature die merely because it couldn't bear the sound of my voice?' 'There is no need to be surprised,' said his friend, 'for even if there had been a thousand or two thousand of them, they would all have succumbed to that.'

The two walked on to a city, whose inhabitants were women, with no men to be seen. 'What is this place and what are these women?' 'Abd Allah asked, and the merman replied: 'It is the City of Women, peopled only by mermaids.' When 'Abd Allah questioned him whether there were any males there at all, and was told that there were none, he went on to ask: 'If that is so, how do they manage to conceive and produce children?' His friend explained: 'These are women who have been exiled here by the king of the sea, and they can neither conceive nor give birth. He sends here those with whom he is angry. They cannot leave and if they try, any sea creature that saw them would eat them, but in the other cities you can find both men and women.' 'So there are other cities in the sea as well as this?' said 'Abd Allah. 'There are many of them,' the merman answered, and when 'Abd Allah asked whether there was a king of the sea, he confirmed that as well.

'Abd Allah said: 'Brother, I have seen many marvels in the sea.' 'What do you mean?' asked his friend. 'Have you not heard the proverb that there are more marvels on sea than on the land?' 'True enough,' said 'Abd Allah, looking at the women, whose faces, he saw, gleamed like moons; they had women's hair, with hands and legs in the middle of their bodies, and they had fish-like tails. After showing him the mermaids, 'Abd Allah's companion took him away and led him to another city, which he found full of people of both sexes, whose shapes, including their tails, were like those of the inhabitants of the first city. Unlike the custom among land folk, there was no trading done there, and the people were naked, not concealing their private parts. When 'Abd Allah asked about this, his friend explained that this was because the sea people had no materials to make clothing.

'What do they do when they marry?' 'Abd Allah asked. 'They don't marry,' replied the merman, 'but whenever a man is attracted to a woman, he does what he wants with her.' 'That's unlawful,' said 'Abd Allah. 'Why doesn't he ask for her hand, give her a dowry, celebrate the wedding and marry her in accordance with the will of God and of His Apostle?' His friend explained: 'We don't all share the same religion. Some of us are Muslims, who believe in the unity of God, while others are Christians, Jews and so on, although it is the Muslims in particular who get married.' 'Since you have no clothes and no trade, what is the dowry of your women? Do you give them precious gems?' 'Abd Allah asked. 'To us those are just valueless stones,' the other replied. 'Whoever wants to marry is asked for a certain type of fish and he has to catch a thousand of them or two thousand, or more or fewer, depending on the agreement that he has made with the bride's father. When he has produced them, both families get together to attend a banquet and it is after this that the bridegroom is brought to the bride. He feeds her on fish that he has caught himself, and if he cannot do this, she catches them and feeds him.' 'What happens if someone commits adultery?' 'Abd Allah asked, to which his friend replied: 'If any woman is proved to be guilty of this, she is banished to the City of Women. If her liaison has left her pregnant, they leave her until she has given birth. Then, if the child is a girl, mother and child are both exiled and the child is called adulteress, daughter of an adulteress, and she remains a virgin until she dies, while if it is a boy, they take him to the king of the sea, who has him killed.'

This astonished 'Abd Allah, who was then taken by his friend from one city to another, until he had been shown eighty in all, each of whose inhabitants differed from those of the others. He asked if there were any more cities in the sea, and his companion replied: 'What have you seen of the cities of the sea and its marvels? I swear by the noble Prophet, the compassionate and merciful, that if I were to guide you for a thousand years and show you a thousand cities each day, with a thousand marvels in every one of them, I would not have shown you one carat's weight of the twenty-four carats of the cities of the sea and their wonders. All that I have shown you is no more than the districts of my own country.' 'If that is so, my friend,' said 'Abd Allah, 'I have seen enough, for I am getting tired of eating fish. I have been with you for eighty days, and every morning and evening you have given me nothing but fish, neither grilled nor cooked in any other way, but raw.' 'What do you mean by cooking and grilling?' the merman asked, and 'Abd Allah explained: 'We grill fish over the fire and cook it in various ways, making a number of different dishes from it.' 'How could we get fire?' asked the other. 'We know nothing about grilling or cooking or anything else of the kind.' When 'Abd Allah went on to tell him about fish cooked in olive or sesame oil, he repeated: 'How could we get oil like that? We sea dwellers know nothing about the things you have talked about.' 'True enough,' said 'Abd Allah, 'but, brother, although you have taken me to many cities, you have not shown me your own.' 'We have gone a long way past it,' the other said, 'as it is close to the shore from where we started. Rather than taking you to it, I brought you here as I wanted to give you a tour of the cities of the sea.' 'I've seen enough of them now,' replied 'Abd Allah, 'and what I want to look at is your own city.' His companion agreed to this and took him back.

When 'Abd Allah got there, he found that the city was smaller than the others he had seen. After he had entered it, his companion took him to a cave. 'This is my house,' he said, 'for all the houses here are mountain caves, both large and small. This is true throughout the cities of the sea, for whoever wants to construct a house goes to the king and tells him where he wants to live. The king sends him off with a number of so-called "borer" fish, which are hired in exchange for a certain quantity of other fish. These "borers" have beaks that can penetrate solid rock, and when they come to the mountain which the man has chosen as the site for his house, they set to work excavating, while the would-be house owner catches fish on which to feed them until the cave has been dug out. The "borers" then leave and the cave is occupied by the owner. All sea people live like this and all their transactions and the services that they do for one another are done in exchange for fish, and it is only fish that they eat.'

'Abd Allah was now invited to go into the cave, and when he had done so, the merman called to his daughter. The naked girl that came to meet him had a rounded face like a moon, long hair, heavy buttocks, kohl-darkened eyes, a slender waist and a tail. When she caught sight of 'Abd Allah with her father, she said: 'Father, what is this tailless creature you have brought with you?' He told her: 'This is my land friend and it is from him that I have been bringing you fruits from the land. Come and greet him.' At that, the girl came up and when she had given him a well-expressed and eloquent greeting, her father said: 'Bring food for our guest, whose arrival has brought us blessing.' She produced two large fish, each the size of a lamb, and when his host told him to eat, he did so unwillingly, prompted only by hunger, as he was tired of a diet of fish, the only food that they had. Soon afterwards, the merman's wife came in, a beautiful woman accompanied by two small boys, each of whom was nibbling a small fish as a man would nibble a cucumber. Seeing 'Abd Allah with her husband, she said: 'What is this tailless creature?' Both boys and their sister, together with their mother, peered at 'Abd Allah from behind, laughing at him and exclaiming: 'Yes, by God, he hasn't got a tail.' 'Brother,' said 'Abd Allah, 'have you brought me here to be a laughing-stock for your children and your wife?'

Night 946

Morning now dawned and Shahrazad broke off from what she had been allowed to say. Then, when it was the nine hundred and forty-sixth night, SHE CONTINUED:

I have heard, O fortunate king, that 'Abd Allah said to the merman: 'Brother, have you brought me here to be a laughing-stock for your children and your wife?' 'Please excuse this,' said his host, 'but we don't have anyone here who is without a tail, and if any such person is found, the king summons him in order to laugh at him. But don't blame these young children and this woman, for they are short of intelligence.' He then shouted at his family to be quiet and they stayed silent out of fear.

While he was trying to reassure 'Abd Allah by talking to him, ten large, strong and thick-set mermen arrived and said to him: 'The king has heard that you have with you one of the tailless land creatures.' 'Yes,' he replied, 'he is this man, a friend of mine, who is here as my guest and whom I intend to take back to the land.' The newcomers told him: 'We cannot go off without him, but if you have anything to say, then take him with you and speak to the king rather than to us.' 'Brother,' his host said to 'Abd Allah, 'my excuse must be obvious to you, as we cannot disobey the king. Come to him with me and I'll do my best to see that you get off free, if God wills it. There is no need to be afraid, for when he sees you he'll recognize that you come from the land and then he is bound to treat you generously and put you back on shore.' 'I shall do whatever you say,' 'Abd Allah replied, 'and I shall go with you, relying on God.'

He went off with his host to the king, who laughed at the sight of him and said: 'Welcome, tailless.' There was general laughter among the courtiers, every one of whom exclaimed: 'Yes, by God, he hasn't got a tail!' At this, his host went up to the king and told him about 'Abd Allah, saying: 'He is from the land and is a friend of mine. He cannot live among us because he only likes to eat fish grilled or cooked, and so I would like your permission to return him to the shore.' 'If that is so,' replied the king, 'and if he cannot live with us, then you have my leave to take him home after I have entertained him.' He ordered a meal to be brought, and when fish of all kinds and descriptions had been produced, 'Abd Allah ate obediently, and at the end of the meal the king told him to make a wish. He asked for jewels and the king said: 'Take him to the jewel house and let him choose whatever he needs.' His host took him there and when he had made his choice, he brought him back to his city, where he produced a purse and said: 'I entrust this to you to take to the grave of the Prophet – may God bless him and give him peace.' 'Abd Allah took it without knowing what was in it, and then left with his host, who was intending to take him to the shore.

On his way, he heard singing and sounds of merriment; there were dishes of fish laid out and people were eating and singing cheerfully. He asked his host whether this was a wedding, but was told that, far from that, someone had died. 'Do you really sing joyfully and eat when someone dies?' he said. 'Yes,' replied his host, 'and what do you do on the land?' 'Abd Allah said: 'When one of us dies, we mourn for him and weep, while the women strike their faces and tear their clothes out of grief for him.' His host stared at him and said: 'Give me back what I entrusted to you.' When 'Abd Allah had done this, his host took him to the shore and said: 'This is the end of our friendship and from this day on you will not see me again and I shall not see you.' 'Why do you say that?' 'Abd Allah asked, and the other said: 'Do you land people not represent a deposit left by God?' 'Yes, we do,' agreed 'Abd Allah, and the merman then said: 'Why then should you find it a serious business, worthy of tears, if God takes back his deposit, and how can I entrust you with something for the Prophet? You are happy when a child is born to you, but the life within it is deposited there by God, and so why should you find it hard and grieve when He takes it back? There is nothing to be gained from association with you.' At that, he left 'Abd Allah and went back into the sea.

'Abd Allah put on his things, took the jewels and went off to the king, who greeted him eagerly and happily, saying: 'How are you, my son-in-law, and why have you stayed away from me for so long?' 'Abd Allah astonished him by telling him what had happened to him and what he had seen of the wonders of the sea. He went on to quote what the merman had said to him, and the king commented: 'It was you who made a mistake in telling him what you did.' For a time, 'Abd Allah kept on visiting the shore and calling to the merman, but as no answer ever came and the merman never appeared, he gave up hope of seeing him again. Meanwhile he, his father-in-law the king and their family continued to enjoy the happiest and most beneficent of lives until they were visited by the destroyer of delights and the parter of companions and they all died. Praise be to the Living God, Who does not die, the Lord of majesty and kingship, the Omnipotent, Who in His omniscience is kind towards His servants.

A story is also told that one night, when the caliph Harun al-Rashid could not sleep, he summoned Masrur and told him to bring Ja'far quickly. When Masrur had gone off and fetched the vizier, the caliph said to him: 'Ja'far, I find that I can't sleep tonight and I don't know what can cure this.' 'Commander of the Faithful,' Ja'far replied, 'wise men have said that to look at a mirror, to enter the baths and to listen to singing banishes cares and troubles.' 'I have done all that,' said the caliph, 'but it hasn't helped, and I swear by my pure ancestors that unless you find some way of curing me, I shall cut off your head.' 'Will you follow my advice?' Ja'far asked. 'What is it?' the caliph asked, and Ja'far said: 'Come down the Tigris with me in a boat as far as a place called Qarn al-Sarat. It may be that we shall hear or see something that we have never come across before, for it is said that cares can be dispelled by one of three things: to see what one has never seen before; to hear what one has never heard before; or to go to a place where one has never been before. Perhaps this will cure you of your restlessness.'

The caliph got up and took with him Ja'far and his brother, al-Fadl, his drinking companion Ishaq, Abu Nuwas, Abu Dulaf and Masrur the executioner.

Night 947

Morning now dawned and Shahrazad broke off from what she had been allowed to say. Then, when it was the nine hundred and forty-seventh night, SHE CONTINUED:

I have heard, O fortunate king, that when the caliph got up from his

place, together with Ja'far and the rest of his companions, they all went
to the dressing room where they put on merchants' clothes before going
to the Tigris and embarking on a gilded boat. The current took them
downstream until, when they had reached their destination, they heard
a girl singing these lines to her lute:

> The wine is here; the nightingale sings on the branch.
> How long, I ask, will you hold back from joy?
> Wake, for this life is merely ours on loan.
> Accept your wine from the hands of a dear friend,
> Whose eyelids languorously droop.
> I planted a fresh rose upon his cheek,
> And pomegranates blossomed in his hair.
> Were that cheek scratched, the scratch would seem to be
> Extinguished ashes, while the cheek is fire.
> The censurer says: 'Forget your love for him.'
> How could I do that, when the down betrays him?

'What a lovely voice!' exclaimed the caliph on hearing this, and Ja'far
said: 'Never in all my life, master, have I heard anything sweeter or
better than this singing, but to hear it from behind a wall is only to get
half the pleasure. How would it be if we listened from behind a curtain?'
'Come, Ja'far,' the caliph said, 'let us go in, uninvited as we are, to the
owner of the house in the hope of catching sight of the singer.' 'To hear
is to obey,' Ja'far replied. They disembarked from the boat, and when
they asked permission to enter the house, a handsome, well-spoken and
eloquent young man came out to greet them, saying: 'You are most
welcome to enter, and you are conferring a favour on me.' When he led
them in, they found a square-built house, with a gilded roof and walls
adorned with lapis lazuli. In it was a hall with a splendid couch where a
hundred girls were sitting, all of whom came down when their master
called to them. He then turned to Ja'far and said: 'Sir, I don't know
which of you is the senior but I invite him in God's Name to sit at the
top and to seat his companions each according to their rank.' The visitors
took their proper places, and Masrur stood in front to serve them. Their
host said: 'With your permission, my guests, I shall have something to
eat brought for you.' They agreed to this, and he told his slave girls to
fetch food. Four of them, with their clothes belted around their waists,
brought in a table on which was a remarkable mixture of creatures that
either moved on the land, flew, or swam in the sea, including sandgrouse,

quails, chickens and pigeons, while around the edges appropriate lines
of poetry were inscribed.

When the guests had eaten their fill and washed their hands, their host
asked if there was anything more they wanted so he might have the
honour of supplying it. 'Yes there is,' they told him. 'The reason why we
came here was that we heard a voice from the far side of the wall of your
house and we would like to hear it again and to make the acquaintance
of the singer. It would be extremely kind of you to grant us this favour
and we would then go back where we came from.' 'You will be welcome,'
the young man said, and he then turned to a black slave girl and told
her to fetch her mistress. The girl went off and brought back a chair
which she set down, and then she went out again, returning this time
with another girl, as beautiful as the full moon, who took her seat on
the chair. The black girl handed her a satin bag from which she took out
a lute set with sapphires and other gems, whose pegs were made of gold.

Night 948

Morning now dawned and Shahrazad broke off from what she had been
allowed to say. Then, when it was the nine hundred and forty-eighth
night, SHE CONTINUED:

I have heard, O fortunate king, that the girl came forward, sat down
on the chair and took from its bag the lute set with sapphires and other
gems, whose pegs were made of gold. She tightened its strings until it
was properly tuned; she and her lute were as the poet has described:

She cradled it in her lap, with its gleaming pegs,
Like a tender mother with her child.
As her right hand moved to touch the strings,
With her left hand she tuned its pegs.

She clasped it to her breast, bending over it like a mother with her baby,
and as she touched it, it moaned like a child calling for its mother. Then
she touched the strings and started to recite these lines:

Time has generously given me the one I love;
I say reproachfully: 'Pass round the cup and drink.'
When wine like this has mixed with a man's heart,
It always brings him to delight and joy.

Its bearer carried it like a gentle breeze;
Have you yet seen a moon carry a star?
How many nights I spent together with this moon
That gleams above the Tigris through the dark,
And as it sinks down the western sky
It stretches a golden sword above the stream!

When she had finished, she wept bitterly and everyone in the house wept so loudly that they almost died. There was not one who did not lose his senses, tear his clothes and slap his face because of the beauty of her singing. 'This shows that she must be a lover who has been parted from her beloved,' said Harun al-Rashid. Her master said: 'She has lost both her mother and her father,' but the caliph told him: 'These tears are not for the loss of parents; this is distress for a missing beloved.' He himself was entranced by the girl's singing and he said to Ishaq: 'By God, I have never seen her like.' Ishaq replied: 'She is so wonderful that I cannot contain my own delight.'

While this was going on, the caliph was watching the owner of the house, noting the elegance of his nature together with his handsomeness. His face, however, showed a trace of pallor. The caliph turned and called him. 'Here I am, master,' the young man replied, and the caliph then asked: 'Do you know who we are?' When he said no, Ja'far asked: 'Would you like me to tell you the names of all your visitors?' 'Yes,' replied the young man, and Ja'far explained: 'This is the Commander of the Faithful, the descendant of the Lord of the apostles,' and he went on to introduce the rest of the company by name. After that, the caliph said: 'I would like you to tell me about the paleness of your face. Is this newly acquired or has it been like that since birth?' The young man replied: 'Commander of the Faithful, mine is a strange and remarkable story which, were it engraved with needles on the inner corners of the eyes, would serve as a warning to those who take heed.' The caliph said: 'Tell it to me, young man, and it may be that I shall be able to put things right for you.' 'Lend me your ears, then, and pay attention,' the young man told him, and the caliph replied: 'Produce your story, for you have made me eager to hear it.' THE YOUNG MAN BEGAN:

You must know, then, Commander of the Faithful, that I am a seafaring merchant, originally from Oman. My father was a rich trader who had thirty ships, used for sea trade, which he hired out each year for thirty thousand dinars. He was a noble-hearted man and he taught me to write, as well as instructing me in every other necessary skill. When he was on the point of death, he called for me and gave me the usual injunctions before God gathered him to His mercy – may He prolong the life of the Commander of the Faithful. He had had a number of associates whose trading voyages he would finance. One day, I happened to be sitting at home with a number of merchants when a servant of mine came in to tell me that there was someone at the door who wanted permission to enter. When I had given permission, in came a man carrying a covered basket on his head. He put it down before me and, after he had removed the covering, I saw that it contained fruits that were out of season and elegant rarities such as were not to be found in my own country. I thanked him for this and gave him a hundred dinars, sending him off full of gratitude.

I distributed the contents of the basket among those of my friends who were there, asking them where such things came from. They told me that it must have been Basra, but, although they went on to sing Basra's praises and describe its beauties, they all agreed that there was no finer place or people than Baghdad and its inhabitants. They started to tell me about Baghdad, the good qualities of its citizens, its excellent climate and its fine layout, until I felt a longing to see it. Having set my hopes on this, I began to sell my property and my possessions; I disposed of the ships for a hundred thousand dinars and sold all my slaves, male and female alike, so that when I collected all the money I had, it came to a million dinars, not counting jewels and other precious stones. I hired a ship and loaded it with my money and my goods and then set sail. After a voyage of some days and nights, I reached Basra. Some time later, I hired another ship, to which I transferred my property, and sailed in a few days to Baghdad. There I asked where the merchants lived and which was the most pleasant quarter, and when I was told that this was Karkh, I went there and hired a house in what was called Saffron Street, to which I moved all my possessions.

One day, after I had been there for some time, I set off on a sight-seeing tour, taking some money with me. This was on a Friday and I got to the al-Mansur mosque where the Friday prayer was being held. When this was over, I went out with a group of people to a place called Qarn al-Sarat and there I caught sight of a fine, tall building with a large window overlooking the shore. I went towards it with my companions and I saw seated there an old man wearing fine clothes, who diffused a pleasant scent of perfume. His beard was combed so as to be split over his chest in two strands, like silver branches, and he was surrounded by four slave girls and five pages. 'What is the old man's name,' I asked somebody, 'and what is his profession?' 'This is Tahir ibn al-'Ala',' I was told: 'He keeps girls and whoever goes to his house gets food and drink, as well as a sight of lovely women.' 'By God,' I exclaimed, 'I have spent a long time looking for something like this.'

Night 949

Morning now dawned and Shahrazad broke off from what she had been allowed to say. Then, when it was the nine hundred and forty-ninth night, SHE CONTINUED:

I have heard, O fortunate king, that the young man said: 'By God, I have spent a long time looking for something like this.' HE CONTINUED:

I went up to greet Tahir, saying: 'Sir, there is something that I want from you.' He asked me what this was and I told him that I wanted to be his guest that night. 'You will be very welcome,' he said, adding: 'I have many girls, for some of whom the nightly fee is ten dinars, others for whom it is forty and others for whom I charge even more. Choose whichever of them you want.' I said that I would pick one of the ten-dinar girls, and I weighed him out three hundred dinars as a month's fee. He handed me over to a page who took me to the baths in the villa, where he attended efficiently to my needs, and when I came out he led me to a room and knocked on the door. Out came a girl to whom he said: 'Take in your guest.' She gave me a warm welcome and laughed happily as she led me into a wonderful room embellished with gold. When I looked at her more closely, I found that she looked like the full moon, while the two maids who attended on her were like stars. She sat me down and seated herself beside me and, at her signal, the maids brought in a table on which were various kinds of meat, including chicken, quail, sandgrouse and pigeon. We both ate until we had had enough, and never in my life have I had a more delicious meal. After we had finished, the table was removed and wine was brought in, together with scented flowers, sweetmeats and fruits.

At the end of a month spent with her, I went back to the baths before returning to the old man to tell him that I now wanted a girl whose charge was twenty dinars a night. 'Weigh out the gold,' he told me and so I went off for it and then weighed him out six hundred dinars to cover a month. He called a page and told him: 'Take your master,' at which he took me first to the baths and then to the door of a room on which he knocked, telling the girl who came out to look after her guest. She gave me the warmest of greetings and told her four maids to bring food. The table that they brought was set with all kinds of dishes, and when I had finished eating from them and the table had been removed, she took a lute and sang these lines:

Breath of musk from the land of Babel,
I conjure you by my love to carry my message.
I know this land well as the dwelling of my dear ones;
How noble it is, this dwelling of theirs!
In it is one whose love tortures her lovers
And for whom their love is of no avail.

I spent a month with her, and then went back to tell the old man that I wanted a forty-dinar girl. Again, he asked for the money and this time I weighed out twelve hundred dinars for him as a month's fee, a month which passed as quickly as a single day because of the girl's beauty and my pleasure in her company. When I went back to the old man, it was in the evening and I heard a hubbub with raised voices. When I asked him what this was, he told me: 'This night marks our greatest celebration when everyone comes out to enjoy looking at one another. Would you like to go on the roof to see them?' I said that I would, and when I got there I saw a splendid curtain veiling a large room in which there was a couch spread with beautiful furnishings. On this was a lovely girl, whose beauty, grace and symmetrical form were such as to bewilder everyone who looked at her. Beside her stood a young man who was resting his hand on her neck and who was kissing her as she was kissing him. When I looked, Commander of the Faithful, her beauty dazzled me to such an extent that I could no longer control myself and I did not even know where I was.

After I had come down from the roof, I described her to the girl with whom I was and asked about her. 'What is she to you?' my girl asked, and when I said: 'She has robbed me of my wits,' she smiled and asked me if I wanted her. 'Yes, by God,' I told her, 'for she has captured my heart and my mind.' 'This is our mistress, the daughter of Tahir ibn al-'Ala,' my companion told me, 'and we are all her servants. But do you know how much it costs to spend twenty-four hours with her?' When I said no, the girl told me: 'Her fee is five hundred dinars, for she is someone for whom the hearts of kings ache.' 'By God,' I said, 'I shall spend all that I have on her,' and I passed the whole night suffering from the pangs of love. Then in the morning I went to the baths and when I had put on my most splendid robes – robes that were fit for a king – I went to Tahir and told him that I wanted the girl whose fee was five hundred dinars. He asked me for the money and I weighed out fifteen thousand dinars to cover the monthly fee. He took it and told the page to bring me to his mistress. I was escorted to one of the most beautiful apartments that I had ever seen anywhere. I went in, and at the sight of the lovely girl who was sitting there I was lost in wonder, for she was like a moon on its fourteenth night.

Night 950

Morning now dawned and Shahrazad broke off from what she had been allowed to say. Then, when it was the nine hundred and fiftieth night, SHE CONTINUED:

I have heard, O fortunate king, that the young man described the girl to the caliph as a moon on its fourteenth night. HE WENT ON:

With her beauty and grace, her symmetrical figure and her speech, which put to shame the notes of the lute, it seemed that it was to her the poet was referring in his lines:

> Love played with her emotions, and she said
> In the dark night of shadows:
> 'Will someone keep me company tonight,
> And pleasure me as a true lover should?'
> She clapped her hand between her thighs and sighed,
> A sigh of sorrow, mixed with grief and tears.
> 'A tooth-pick shows the beauty of the mouth;
> The male is a tooth-pick for the female part.
> Muslims, cannot your organs stand erect
> And is there no one to help my distress?'
> Beneath its coverings my tool stood up
> Calling to her: 'He comes to you; he comes!'
> As I unloosed her drawers, she said in fright:
> 'Who are you?' I replied: 'An answer to your call.'
> I thrust with what was sturdy as her arm,

Gently, but deep enough to hurt her hips.
We made love thrice before I rose. She said:
'That's how it's done,' and I said: 'Yes, it is.'

How well another poet has expressed it:

If the idolaters could see this girl,
She would replace the idols as their god.
Were she to spit into the salt sea,
Because of that, its water would turn fresh,
And were a monk to see her in the east,
He'd change his eastern custom and use that of the west.

Another has written equally well:

I looked at her, and all my inner thoughts
Became bewildered by her loveliness.
This led her to suspect I was in love,
A thought which brought the colour to her cheeks.

I greeted her and, after she had welcomed me, she took me by the
hand and sat me down beside her. The excess of my longing led me to
weep for fear we might have to part and so I tearfully recited these lines:

I love the nights of parting, not through joy,
But in the hope that Time will join us once again,
While days of union are the ones I hate,
As I can see that nothing stays the same.

She tried to entertain me with kind words, but, as I was drowning in
the sea of love, my passionate longing made me fear that, close as we
were together, we would soon be separated and, thinking of the pain of
being far apart, I recited:

When we were together, I thought of our parting,
Since tears, dyed red with blood, flowed from my eyes,
And so I wiped my eyes against her neck,
As camphor serves to staunch the flow of blood.

When she ordered food to be brought in, four swelling-breasted virgins
came with different foods, fruits and sweetmeats, along with scented
flowers and wine fit for a king, which they put in front of us. We ate and
then sat drinking, surrounded by aromatic plants in a room suitable for

royalty. A maid fetched the girl a silk bag, from which she took out a lute and, after placing it in her lap, she touched the strings. The plaintive sound they made was like a child calling for its mother and she recited:

Drink wine, but only from the hands of a fawn
Whose delicacy matches it, as it matches her.
No one takes pleasure in a cup of wine,
Unless the cupbearer is pure of cheek.

I stayed enjoying myself with this girl for so long that I ran through all my money and then, as I sat with her, I thought of having to leave her and rivers of tears poured down my cheeks until I could no longer distinguish night from day. She asked me the reason for this, and I said: 'Lady, every night since I came to you, your father has been taking five hundred dinars from me, and now I have no money left. The poet was right who said:

Poverty is exile in one's native land,
While wealth will make a stranger feel at home.'

She told me: 'If my father finds a merchant here who becomes poor, he entertains him as a guest for three days but then sends him away, and the man can never come back to us again. So don't tell anyone about this, and, if you keep it secret, I shall find a way to stay with you as long as God wills it, because of the depth of my love for you. I must tell you that I'm in charge of all my father's money, and he doesn't know how much he has. So every day I shall let you have a purse containing five hundred dinars, and you can give him this and tell him that, in future, you will pay him by the day. Whatever you hand over to him, he will give to me and I shall pass it back to you again, so that we can go on doing this as long as God wills.' I thanked her for that, kissed her hand and stayed with her on those terms for a whole year.

Then one day she gave a painful beating to a maid of hers, and the girl said: 'By God, as you have hurt me, so I will hurt your heart.' She then went to Tahir and told him everything that we had been doing. On hearing this, he got up immediately and came to me as I was sitting with his daughter. He addressed me by name and, when I answered, he said: 'When a merchant who stays with us has lost his money, I entertain him as a guest for three days, but you have been here for a year, eating and drinking and doing what you want.' He turned to his servants and told them to strip me of my clothes, which they did, and they then gave me tattered replacements, worth five dirhams, together with ten dirhams in cash. Then he said: 'Get out. I'm not going to beat you or pour abuse on you, but you had better be off, for if you stay here in this town, your blood will be on your own head.'

I left unwillingly with no notion of where to go; all the cares in the world had settled on my heart, filling me with melancholy. I said to myself: 'How is it that I set to sea with a million dinars, including the price of thirty ships, only to lose everything in the house of this ill-omened old man? Now I am forced to leave, naked and heartbroken, and there is no might and no power except with God, the Exalted, the Omnipotent.' I stayed in Baghdad for three days, neither eating nor drinking, and on the fourth day I found a ship heading for Basra. I went on board and paid the captain for my passage and then, when I got to Basra, I came to the market, suffering from the pangs of hunger. There I was seen by a grocer, who came up and embraced me, turning out to have been a friend of mine and of my father before me. He asked me how I was, and when I had told him everything that had happened to me, he said: 'By God, that wasn't sensible behaviour, but now, after what has happened to you, what are you thinking of doing?' I told him that I didn't know and he said: 'Stay with me and keep a record of my income and expenditure in return for two dirhams a day, as well as your food and drink.' I agreed to that and I remained with him for a year, by which time I had made a hundred dinars by trading. I hired an upper room by the river bank in the hope that a ship might arrive with merchandise from which I could buy something and then set off back to Baghdad.

As it happened, one day a ship did put in, and I went with all the other merchants to buy goods. Two men emerged from its hold and sat down on chairs that had been placed for them. When the merchants arrived in order to start dealing, these two said to some of their servants: 'Fetch the carpet.' When this had been done, another man brought out a pair of saddlebags from which he pulled out a sack, opened it and poured its contents on to the carpet – a dazzling collection of gems of all sorts, pearls, coral, sapphires and carnelians.

Night 951

Morning now dawned and Shahrazad broke off from what she had been allowed to say. Then, when it was the nine hundred and fifty-first night, SHE CONTINUED:

I have heard, O fortunate king, that the young man told the caliph about the merchants and the sack with its contents of jewels. HE CONTINUED:

One of the two who were sitting on the chairs turned to the merchants and told them that, being tired, this was all that he was going to sell that day. They started bidding against each other until they were offering four hundred dinars for the contents of the sack. Its owner, an old acquaintance of mine, asked me why I was not taking part in the bidding and I had to tell him, shamefacedly, that I had only a hundred dinars left in the world. He saw me weeping and, finding my plight hard to bear, he said to the others: 'I call you to witness that I have sold all the jewels and precious stones in the sack to this man for a hundred dinars, although I know that they are worth many thousands. This is a gift from me to him.' He then gave me the saddlebags, the sack, the carpet and all the jewels that lay on it, at which I thanked him and all the merchants present praised him.

I took all this and went off to the jewellers' market, where I sat down to trade. Among the treasures I had been given was a circular amulet, made by magicians, weighing half a *ratl*. It was bright red and covered on both sides with lines that looked like ants' tracks, but what purpose it served I didn't know. After I had spent a whole year trading, I took this out, telling myself: 'For a long time now I have known nothing about this or what use it is.' I passed it to the auctioneer and he took it round the market, but then came back to say that none of the merchants would offer more than ten dirhams for it. I told him that I would not sell for that price and he threw it back to me and went away. I tried again another day and this time the offer was fifteen dirhams, but again I took it angrily from the auctioneer and put it back among my things. Then one day as I was sitting there, a man came up who greeted me and asked my permission to look through my wares, which I allowed him to do. I was still angry at the fact that I couldn't sell the amulet, but when the man had finished looking, this was the only thing he picked on. When he saw it, he kissed his hand and gave thanks to God before asking me if I would sell it. I was becoming even angrier, but I told him that I would, and when he asked me its price, I said: 'How much will you give me for it?' When he said twenty dinars, I thought that he was making fun of me and I told him to go away. He then offered fifty, and when I said nothing, he raised the offer to a thousand dinars. All the while I stayed silent and made no reply. He laughed at my silence and said: 'Why don't you answer me?' I told him again to go away and I was about to quarrel with him while he kept raising his offer by a thousand dinars each time, and when I still made no reply he asked: 'Will you sell it for twenty thousand dinars?'

I still thought that he was laughing at me, but a crowd gathered around, every one of them telling me to sell it and saying: 'If he doesn't buy it, we shall all join up against him and give him a beating, before driving him out of town.' 'Do you really want to buy or are you making fun of me?' I asked the man, and he replied by asking: 'Do you really want to sell or is it you who are making fun of me?' 'I am ready to sell,' I told him and he replied: 'I offer thirty thousand dinars. Take them and let the sale proceed.' I said to the bystanders: 'You are witnesses to this, but I am only selling on condition that the buyer tells me what purpose the amulet serves.' When he had agreed to this, I said: 'I have sold it to you,' and he replied: 'God is my guarantee for the truth of what I have to say.' He produced the money for me, took the amulet, put it in his pocket, and then asked me if I was satisfied. I told him that I was, and he asked all those who were present to bear witness that the sale had been concluded and that the thirty thousand dinars had been paid over.

He then turned to me and said: 'You poor fellow, I swear by God that if you had delayed the sale, I would have taken my offer up to a hundred thousand dinars, or even a million.' When I heard this, the blood left my face, and since then it has been replaced by the pallor that you have noticed, Commander of the Faithful. I asked him the reason behind this and what purpose the amulet served. He told me: 'The king of India has a daughter, the loveliest girl ever seen, but she suffers from epilepsy. The king collected the masters of talismans, the sages and the soothsayers to be present at a council meeting, but they could not cure her. I happened to be there and I said: 'Your majesty, I know a man called Sa'd Allah of Babel who knows more than anyone else on earth about these matters and, if you want, you can send me to him.' The king agreed to this and when I asked him for carnelian, he gave me a large piece, together with a hundred thousand dinars, as well as a gift. I took that and set off for the land of Babel, where I asked for the *shaikh* and was directed to him. After he had accepted the cash and the gift which I presented to him, he took the carnelian and brought in an engraver who turned it into this amulet. The *shaikh* had spent seven months watching the stars before picking a time for the engraving to be done, and it was then that the talismanic inscriptions that you can see were put on it. I then took it to the king . . .'

Night 952

Morning now dawned and Shahrazad broke off from what she had been allowed to say. Then, when it was the nine hundred and fifty-second night, SHE CONTINUED:

I have heard, O fortunate king, that the young man told the king what the man had said to him. HE CONTINUED:

The man went on: 'When he placed it on his daughter she was instantly cured, although before that she had had to be tied up with four chains, and every night a slave girl would sleep with her, only to be found in the morning with her throat cut. The king was overjoyed with the instant cure that the amulet had produced; he presented me with a robe of honour and gave me a large quantity of money, while the amulet itself was set in the princess's necklace. It happened, however, that one day she went out to sea for a boat trip with her maids. One of them reached out to her playfully and the necklace broke and fell into the sea. After that, to the great grief of the king, the princess's illness returned, and so he provided me with a great deal of money and told me to go back to the *shaikh* to get him to replace the amulet. I went, only to find that the *shaikh* had died, and when I returned and told the king, he sent me out with ten others to scour the lands in the hope of finding some cure for the princess. God has now led me to you.' The man then took the amulet from me and went off. This, then, is the cause of the pallor of my face.

After that, I took all the money that I had and returned to Baghdad, where I settled in my former lodgings. The next morning, I dressed and went off to the house of Tahir ibn al-'Ala' in order to see my beloved, for whom my love had never ceased to increase, but when I got there it was to find the windows broken. I asked a servant: 'What happened to the old man?' He told me: 'Some time ago, a merchant called Abu'l-Hasan of Oman came to him and spent some time with his daughter. When he had run out of money, the old man sent him away broken-hearted, but the girl was deeply in love with him and she fell so gravely ill that she was on the point of death. At this, her father had the lands searched for Abu'l-Hasan, promising a hundred thousand dinars to anyone who fetched him, but no one found him or could discover any trace of him, as a result of which the girl is now close to death.' 'What about her father?' I asked and the servant told me that, because of his misfortune, he had sold his girls. I then said: 'Shall I lead you to Abu'l-Hasan the Omani?' and the servant said: 'For God's sake, please do.' So I told him: 'Go to her father and claim the reward by telling him that Abu'l-Hasan is standing at the door.'

The servant rushed off like a mule freed from a mill, and after a while he came back with the old man who, on seeing me, went back home and handed over the hundred thousand dinars. The servant took the money and left, calling down blessings on me, and then the old man came up and embraced me tearfully, saying: 'Where have you been, sir? My daughter is dying because you left her, so come inside with me.' When I went into the house, he prostrated himself in thankfulness, exclaiming: 'Praise be to God Who has reunited us.' He then went to his daughter and told her: 'God has cured you of your illness.' 'Father,' she said, 'I shall never be cured until I see the face of Abu'l-Hasan.' 'If you take something to eat and then go to the baths, I will reunite you with him.' 'Are you telling the truth?' she asked, and when he swore by the majesty of God that he was, she said: 'If I can look at his face, I shan't need food.' So he told his servant to fetch me in, and at the sight of me she first fell down in a faint and then, when she had recovered, she recited the following:

God may unite parted lovers who were sure that they would never
    meet again.

Then she sat up and said to me: 'I never thought that I would see your face again except in a dream.' She embraced me, shed tears and told me: 'Now is the time for food and drink,' both of which were then fetched for her. After I had spent some time with her and her father, she regained her former loveliness, and her father summoned the *qadi* and the notaries. He drew up a marriage contract between us and provided a great banquet. She is still my wife to this day.

The young man then left the caliph and returned with a beautiful boy, graceful and well formed, to whom he said: 'Kiss the ground before the

Commander of the Faithful.' When he had done this, the caliph, astonished by his beauty, glorified his Creator, after which he left the house
with his party, telling Ja'far: 'This was a remarkable story, and never
have I seen or heard anything more curious.' When he had taken his seat
in the royal palace, he sent for Masrur, and when Masrur answered his
summons, he told him to place in the hall the tribute money from Basra,
Baghdad and Khurasan. Masrur gathered this all together, and the sum
was so huge that no one could count it but God. He then instructed
Ja'far to fetch Abu'l-Hasan. Ja'far obediently did this and Abu'l-Hasan
kissed the ground before him apprehensively, afraid lest he had been
summoned because of something that he had done wrong while the
caliph was in his house. 'Omani,' said the caliph, and Abu'l-Hasan
replied: 'Here I am, Commander of the Faithful, may God forever show
you His favour.' 'Lift this curtain,' the caliph told him. He had previously
instructed his servants to hang a curtain over the money collected from
the three provinces and when Abu'l-Hasan did as he was told, he was
bewildered by the sight of so much money. The caliph asked him whether
it was more than the profit he had failed to make on the sale of the
amulet and he said: 'It is certainly many times more.' 'I call everyone
here to bear witness to the fact that I have given this money to this young
man,' the caliph proclaimed. Abu'l-Hasan kissed the ground and shed
tears in front of the caliph in his embarrassment and delight. As the tears
poured down over his cheeks, the blood returned to them and his face
became like the moon at its full. The caliph exclaimed: 'There is no god
but God! Glory be to the One Who changes one state into another, while
He Himself is eternal and unchanging!' He then called for a mirror and
when Abu'l-Hasan had been shown his own face in it, he prostrated
himself in thankfulness to Almighty God. The caliph then gave orders
for the money to be taken to his house and asked that he himself should
keep on visiting him as a drinking companion. He continued to pay
frequent visits to the palace until the caliph was gathered into the mercy
of God – praise be to Him, Who is eternal in the majesty of His kingdom.

A story is also told that al-Khasib, the lord of Egypt, had a son so
handsome that, out of fear for him, the only time that he allowed him
out was to go to the Friday prayer. One day, when the prayer had
finished and the boy, Ibrahim, was leaving the mosque, he passed an old
man who had with him a large quantity of books. He dismounted, sat
down beside the man and had started to turn over the books and inspect

them, when he came across the speaking likeness of a girl unsurpassed for beauty anywhere on earth, which stole away his wits, leaving him stunned. He asked the old man to sell him this and, for his part, the man kissed the ground before him, saying: 'Sir, it is yours for nothing.' Ibrahim gave him a hundred dinars and took the book with the picture in it. He began to spend his time staring at it, weeping over it night and day, without eating, drinking or sleeping. He then said to himself: 'If I asked the bookseller who it was who had painted this portrait, he might be able to tell me, and if the subject is still alive, I could go to find her. On the other hand, if it is only a picture, I will give up my infatuation and not torture myself for something that has no substance.'

Night 953

Morning now dawned and Shahrazad broke off from what she had been allowed to say. Then, when it was the nine hundred and fifty-third night, SHE CONTINUED:

I have heard, O fortunate king, that the young man said to himself: 'If I asked the bookseller who it was who had painted this portrait, he might be able to tell me, and if the subject of the portrait is still alive, I could go to find her. On the other hand, if it is only a picture, I will give up my infatuation and not torture myself for something that has no substance.'

Next Friday, he again passed the bookseller, who stood up to greet him. 'Uncle,' Ibrahim said, 'tell me who painted this picture.' The bookseller replied: 'Sir, the painter is a Baghdadi named Abu'l-Qasim al-Sandalani, who lives in a district known as Karkh, but I don't know whose portrait it is.' Ibrahim left him and, without telling any of the courtiers of his feelings, he performed the Friday prayer and then returned to his apartments. There he took a bag and filled it with thirty thousand dinars' worth of jewels, together with gold. He waited until morning and then left, still without telling anyone, to join a caravan. He came across a Bedouin and asked how far it was to Baghdad. 'My son,' the Bedouin told him, 'you have a very long way indeed to go, as Baghdad is a two months' journey away.' Ibrahim said: 'Uncle, if you take me there, I shall give you a hundred dinars, together with this mare that I am riding, which is worth another thousand.' The Bedouin replied: 'God is the guarantor of our agreement. You must spend the night with me.'

Ibrahim agreed, and at daybreak the Bedouin started off with him, travelling fast by the shortest route, because he was eager to get the mare that he had been promised.

The two of them carried on with their journey until they reached the walls of Baghdad, where the Bedouin said: 'Praise be to God that we are safe. This is Baghdad, master.' Ibrahim was delighted and dismounted from his mare, which he handed over to his companion, together with the hundred dinars. He then took his bag and started to ask about the merchants' centre in the district of Karkh. Destiny led him to a street where there were ten small houses, five on each side, and at the far end was a door with two leaves and a silver ring. Here there were two marble benches, spread with the finest coverings, and on one of them a handsome man of dignified appearance was sitting, dressed in splendid clothes, with five mamluks like gleaming moons standing before him. When he saw this, Ibrahim recognized what the bookseller had told him to look for, and so he went up and greeted the man. He, for his part, returned the greeting, welcomed his visitor, invited him to sit down and asked him about himself. Ibrahim told him: 'I am a stranger here and, if you would be good enough, I would like you to find me a house in this street where I could lodge.' 'Ghazala,' the man called, and the call was answered by a slave girl, to whom he said: 'Take some servants with you; go to one of the houses; clean it out, furnish it and put in it all the necessary utensils and so on for this handsome youth.' When the girl had done what he told her, he took Ibrahim to show him the house, and when Ibrahim asked him about the rent, he said: 'My handsome fellow, while you stay here, I shall charge you no rent at all.'

When Ibrahim had thanked him, his host summoned another slave girl, as beautiful as the sun, and told her to fetch a chess set. She brought it and, after a mamluk had set out the board, he asked Ibrahim: 'Will you play with me?' Ibrahim agreed and they played a number of games, which Ibrahim won. 'Well done!' exclaimed his host. 'This adds the final touch to your qualities, as you have beaten me, a thing which nobody else in Baghdad has done.' When the house had been fully furnished and equipped, he handed over the keys and said: 'Will you do me the honour of entering my house and sharing my meal?' Ibrahim agreed and went off with him into the house, which he found to be a beautiful building, adorned with gold and filled with all kinds of pictures, while the splendour of the furnishings and furniture beggared description. After having welcomed him, his host called for food, at which a table of Yemeni workmanship made in San'a' was brought in, and on this were placed various splendid and delicious types of exotic food.

When Ibrahim had eaten his fill and washed his hands, he began to inspect the house and its furnishings, and then he turned to look for the bag that he had brought with him. It was nowhere to be seen, and he said to himself: 'There is no might and no power except with the Omnipotent God on high. I have eaten one or two dirhams' worth of food and have lost a bag containing thirty thousand dinars. God help me.' He stayed silent, being unable to speak.

Night 954

Morning now dawned and Shahrazad broke off from what she had been allowed to say. Then, when it was the nine hundred and fifty-fourth night, SHE CONTINUED:

I have heard, O fortunate king, that the young man was so distressed to find that the bag was missing that he could not speak. His host then set out the chessmen again and asked him if he wanted another game. Ibrahim agreed, but when they played, he lost. He congratulated his host, but stopped playing and got up. 'What is the matter with you, young man?' asked his host, to which he replied: 'I want my bag.' His host got up and brought it out for him, saying: 'Here it is. Now will you give me another game?' 'Yes,' said Ibrahim and this time he won. 'While you were thinking about your bag, I beat you but when I produced it for you, it was you who beat me,' said the man, and he then asked: 'Where have you come from?' On being told 'Egypt' he went on to ask why he had come to Baghdad. At that, Ibrahim produced the picture and said: 'I must tell you, uncle, that I am the son of al-Khasib, the lord of Egypt. I saw this picture at a bookseller's and it robbed me of my wits. When I asked about the painter, I was told that he was someone called Abu'l-Qasim al-Sandalani, who lived in Saffron Street in the district of Karkh, and so I took some money with me and came here on my own without telling anyone what I was doing. To complete your kindness to me, I would ask you to direct me to this Abu'l-Qasim, so that I may ask him why he painted the picture and who is its subject. In return, I am prepared to give him anything he wants.' 'By God, my son,' his host replied, 'I am Abu'l-Qasim, and it is marvellous how fate has led you to me.'

When he heard this, Ibrahim got up and embraced him, kissing his head and his hands and imploring him in God's Name to tell him whose portrait it was. 'To hear is to obey,' said Abu'l-Qasim, and he rose and opened a cupboard from which he took a number of books, in each of which he had painted the same picture. 'I must tell you, my son,' he said, 'that this is a portrait of my cousin, who lives in Basra where her father is governor. His name is Abu'l-Laith and hers Jamila. There is no more beautiful girl on the face of the earth, but she shuns men and will not allow any man to be mentioned in her salon. I approached her father to ask for her hand, offering him a large sum of money, but he wouldn't accept. When Jamila came to hear what I had done, she was furious and sent me a message saying, among other things: "If you have any sense you will not stay in this town, for otherwise your blood will be on your own head." Because of this haughty disposition of hers, I left Basra broken-hearted, and it was then that I painted this portrait in books, which I sent out to different lands. I hoped that the picture might fall into the hands of a handsome youth like you, who might find a way to reach her and with whom she might fall in love. I proposed to get him to agree that, were he to win her, he was to let me look at her, even if this had to be from a distance.'

When Ibrahim heard this, he bent his head for a time in thought, and Abu'l-Qasim told him: 'My son, I have never seen anyone more handsome than you in Baghdad, and I think that, when she sets eyes on you, she will fall in love with you. If you achieve union with her, will you allow me at least a distant view of her?' Ibrahim agreed to this and his host said: 'In that case, stay with me until you set out on your journey.' 'I cannot stay,' objected Ibrahim, 'because the fire of love in my heart burns ever more fiercely.' But his host told him to wait patiently for three days and so give him time to fit out a ship to take him to Basra. So Ibrahim stayed until the ship was ready and loaded with everything that he might need in the way of food and drink and other supplies. At the end of the three days, Abu'l-Qasim said: 'Get ready to leave, for I have got you a ship, and on board it is everything that you may need. The ship is my property; the sailors are in my service; it carries enough to last you until you return and I have told the sailors to look after you until you are safely back again.'

When Ibrahim had gone on board, after saying goodbye to Abu'l-Qasim, he sailed down to Basra, where he got out a hundred dinars to give to the crew. They pointed out that they had already been paid by their master, but he told them to take the money as a gift, promising not to tell Abu'l-Qasim anything about it. So they accepted it and said goodbye, after which Ibrahim entered Basra and asked where the merchants stayed. On being told that they lodged in a hostel known as Khan Hamdan, he went to the market where the *khan* stood and where, because of his beauty, he became the focus of attention. He entered it with one of his sailors and asked for the doorkeeper, whom, on being directed to him, he found to be a very old and dignified man. After they had exchanged greetings, Ibrahim asked him: 'Uncle, do you have a good room to let?' 'Yes,' replied the doorkeeper and he took Ibrahim and the sailor and opened up for them an elegant room embellished with gold. 'Young man,' he said, 'this room will suit you,' after which Ibrahim took out two dinars and said: 'Take these in return for having unlocked the door.' The doorkeeper took the money and called down blessings on Ibrahim, who then told the sailor to return to the ship. He himself went into the room and the doorkeeper stayed in attendance on him, saying: 'Master, we are glad to have you here.'

Ibrahim gave him a dinar and asked him to buy bread, meat, sweetmeats and wine, and the man later returned from the market having bought all this for ten dirhams. He handed the change to Ibrahim, but Ibrahim delighted him by telling him to spend it on himself. Out of the food that had been brought, Ibrahim put some on a round of bread and ate it. He then told the doorkeeper to take the rest to his family. The man took it and went off to tell them: 'I think that there is nobody on the face of the earth more generous or sweeter-tempered than this Ibrahim who has come to lodge with us today, and if he stays here we shall become rich.'

He then went to Ibrahim's room and found him weeping. He sat down, massaged his legs and kissed them before asking: 'Master, why are you weeping? May God keep you from this.' 'Uncle,' Ibrahim replied, 'I would like to drink with you tonight.' 'To hear is to obey,' the man replied, and at that Ibrahim gave him five dinars and told him to use it to buy fruit and wine. He then produced another five dinars and said: 'Buy us dried fruits, sweet-scented flowers and five fat chickens, and fetch me a lute.' The man went off and when he had bought all these things, he told his wife to cook the food and strain the wine, doing all this as well as she could in return for Ibrahim's goodwill. When his wife had done it as well as could be wished for, he took everything and brought it to Ibrahim.

Night 955

Morning now dawned and Shahrazad broke off from what she had been allowed to say. Then, when it was the nine hundred and fifty-fifth night, SHE CONTINUED:

I have heard, O fortunate king, that when the wife of the doorkeeper of the *khan* had prepared food and drink, he brought it to the prince. The two of them ate and drank with enjoyment, until Ibrahim burst into tears and recited these lines:

> Friend, if in my striving I could give my life,
> My money and this world, with all it has,
> Together with the eternal joys of Paradise,
> For one hour of her union, my heart would make the bargain.

With a deep groan he fell unconscious, leaving the doorkeeper to sigh over him. When he had recovered, the man asked him: 'Master, what makes you weep and who was it that you were referring to in those lines? She can be nothing but dust beneath your feet.' Ibrahim got up and produced a bundle containing the finest of women's clothes, telling the man to take it to his womenfolk. He took it and handed it to his wife, and she then came with him to visit Ibrahim, who was still in tears. The woman said: 'You are breaking our hearts, so tell us which beautiful girl it is that you want and she will be your slave.' Ibrahim said: 'Uncle, I must tell you that I am the son of al-Khasib, the lord of Egypt, and I am in love with Jamila, the daughter of Abu'l-Laith, the governor.' The doorkeeper's wife exclaimed: 'For God's sake, don't say that, my brother, in case anyone hears us and we die. Nowhere on the face of the earth is there anyone more haughty than that girl and no one can mention a man's name to her, as she shuns all men. Forget about her and find someone else.' When Ibrahim heard that, he wept bitterly, but the doorkeeper said: 'My life is the only thing I have, but I'm prepared to risk it for love of you and I shall think of some plan to get you what you want.' He and his wife then left the room.

The next morning, Ibrahim went to the baths and put on a robe fit for a king, after which he was approached by the doorkeeper and his wife. They told him that in the city there was a hunchbacked tailor who worked for Lady Jamila, and they suggested that he should go to this man and tell him his problem in the hope that he might be able to suggest a way of solving it. Accordingly, Ibrahim went off to the man's shop, and when he entered it he found ten mamluks, splendid as moons, who exchanged greetings with him, welcomed him and gave him a seat. They were taken aback by his beauty and grace, and the hunchbacked tailor himself was left in a state of confusion. Ibrahim had deliberately torn his pocket, and he now asked for it to be sewn up. When the tailor had done this, using silken thread, he produced five dinars and gave them to him, after which he returned to his lodgings. 'What did I do for him that he thought was worth five dinars?' the man wondered, and he spent the night thinking about his beauty and his generosity.

The next morning, Ibrahim went back to the tailor's shop, and when he entered, he exchanged greetings and was given a warm welcome. After taking his seat, he again asked the tailor to sew up his pocket, which he had torn for a second time. 'Willingly, my son,' said the tailor, and this time, when the work had been done, Ibrahim presented him with ten dinars, leaving him bewildered, as before, both by his beauty and his generosity. 'By God, young man,' he said, 'there must be something more behind this than just the sewing up of a pocket. Tell me the truth, for if you are in love with one of these young mamluks, I swear to God that none of them is more beautiful than you. They are all as dust beneath your feet and they are all here as your slaves. But if it is something else, then tell me.' Ibrahim said: 'Uncle, this is not the place to talk, for mine is a strange and wonderful story.' 'In that case, come with me to a private room,' the tailor told him and, taking him by the hand, he led him into a room in the interior of the shop where he said: 'Now tell me about it.' So Ibrahim repeated his story from beginning to end, and the startled tailor exclaimed: 'Fear for your life, young man! That lady is haughty and shuns men, so guard your tongue or else you will bring destruction on yourself.' When he heard that, Ibrahim shed bitter tears and, catching hold of the hem of the tailor's robe, he said: 'Help me, uncle, or else I am a dead man. I have left my kingdom and the kingdom of my father and my grandfather, becoming a solitary stranger in a foreign land because I cannot bear to be without her.' When the tailor saw how Ibrahim had been affected, he had pity on him and said: 'My son, all I have is my own life and I am prepared to risk it for love of you, as you have wounded my heart, and so tomorrow I shall come up with a plan to gratify you.' At that, Ibrahim called down blessings on him and went off to the *khan*, where he told the doorkeeper what the tailor had said. 'He has done you a service,' the doorkeeper commented.

The next morning, Ibrahim put on his most splendid robes and taking with him a purse filled with dinars he went to greet the tailor. When he had sat down, he said: 'Uncle, keep the promise you made me.' The tailor told him: 'Go off at once and get three fat chickens, three ounces of cane sugar, two small jugs, with wine to pour into them, together with a cup. Put these in a bundle and, after morning prayer, hire a boat with a boatman and tell him that you want to go downriver below Basra. If he says that he can't take you further than one *parasang*, tell him that this is up to him, but when he has got that far, use your money to persuade him to bring you to where you want to go. The first garden that you see as you go on downstream will be that of Lady Jamila. When you are in sight of it, go to its gate, where you will find two high steps covered with brocade, with a hunchback like me sitting there. Complain to him of your sufferings and ask for his help in the hope that, out of pity for you, he may bring you to where you can catch a glimpse of the lady, even if that is only from a distance. This is the only plan I can think of, and if the hunchback doesn't take pity on you, then both you and I are dead men. This is my suggestion, but the affair is in the hands of Almighty God.' 'It is from Him that I seek help,' said Ibrahim, 'for His will is done and there is no might and no power except with Him.'

Ibrahim now left the tailor, went to his lodgings and put what he had been told to take with him into a small bundle. Then, in the morning, he went to the bank of the Tigris, where he found a boatman lying asleep. He woke the man and, after giving him ten dinars, he told him to take him down below Basra. The man said: 'Yes, on condition that I don't go farther than one *parasang*, for a single span more would mean death for both of us.' 'It is for you to say,' Ibrahim told him, and so the boatman took him on board and started off downstream. When he had got near Jamila's garden, he said: 'My son, I can't take you any further, for if I pass this point we shall both be dead.' At that, Ibrahim took out another ten dinars and said: 'Take this spending money to help you out.' The man was ashamed and said: 'I entrust the affair to Almighty God.'

Night 956

Morning now dawned and Shahrazad broke off from what she had been allowed to say. Then, when it was the nine hundred and fifty-sixth night,
SHE CONTINUED:

I have heard, O fortunate king, that when the young man gave the additional ten dinars to the boatman, the man said: 'I entrust my affair to Almighty God.' He then went further down the river and when he reached the garden, Ibrahim got up joyfully and jumped from the boat with a leap as long as a spear cast. He threw himself ashore, while the boatman made off upstream as fast as he could.

As Ibrahim walked on, he saw everything that the tailor had described. There was the garden, with its gate open, and there in the entrance was an ivory couch on which was seated a good-looking hunchback, wearing robes of gold brocade and holding in his hand a silver mace overlaid with gold. Ibrahim hurried up to him, bent over his hand and kissed it. The man was dazzled at the sight of Ibrahim's beauty and said: 'Who are you, my son? Where have you come from and who brought you here?' Ibrahim answered: 'Uncle, I am an ignorant young stranger,' and started to weep, at which the man felt sorry for him, sat him up on the couch and wiped away his tears. 'No harm will come to you,' he said. 'If you are in debt, may God settle your debts, and if you are afraid, may He protect you from what you fear.' 'I'm not afraid, nor am I in debt,' Ibrahim told him, 'for, thanks to God's help, I have plenty of money with me.' 'What do you want, then,' asked the man, 'that has led you to risk your life and your beauty in so deadly a place?'

Ibrahim told his story and explained the matter, after which the man bent his head towards the ground for a time before asking: 'Was it the hunchbacked tailor who directed you to me?' When Ibrahim confirmed that it was, he said: 'He is my brother and he is a man whom God has blessed.' He then went on: 'My son, had I not become fond of you and felt pity for you, both you and my brother, together with the doorkeeper of the *khan* and his wife, would all have lost your lives. You have to know that this garden has no match on the face of the earth. It is called the Garden of the Pearl and in my lifetime the only people to have entered it, apart from me, have been the sultan and its owner, Jamila. I have been here for twenty years and I have never seen anyone else come here. Once every forty days, the lady arrives by boat, goes up and enters it, accompanied by her maids, wearing a satin robe whose train is carried by ten maids using golden hooks. I have never myself set eyes on her, but as I have only my own life to lose, I am willing to risk it for your sake.' At that, Ibrahim kissed his hand, and the man told him: 'Sit here with me until I think of a plan.'

He then took Ibrahim by the hand and led him into the garden. When he entered, he thought that this must be Paradise, for what he saw were intertwined trees, towering date palms, gushing waters and birds singing their various songs. The hunchback took him to a pavilion and said: 'This is where Lady Jamila sits,' and when Ibrahim looked, he found it to be one of the most wonderful of retreats, adorned as it was with paintings in gold and lapis lazuli. It had four doors approached by five steps, and in the centre was a pool, down to which led steps of gold, studded with precious stones. A golden fountain stood in the middle of the pool, with statues, both large and small, from whose mouths water gushed. The passage of this water caused different sounds to come from the statues, and those who heard them would think that they were in Paradise. Around the pavilion ran a channel that operated silver water-scoops covered with brocade, while on its left was a window of silver lattice work that overlooked a green pasture where there were wild animals of all kinds, including gazelles and hares, while another window on the right gave a view over a field where there were birds of various sorts, astonishing the listeners with their different songs.

Ibrahim was delighted by this sight, and as he took his seat by the gate with the hunchback at his side, the latter asked him: 'What do you think of my garden?' 'It is an earthly paradise,' Ibrahim replied, at which the other laughed. The hunchback then got up and left for a time before bringing back a tray containing chicken, quails and other tasty foods, together with sugared sweetmeats. Putting it down before Ibrahim, he said: 'Eat your fill.' Ibrahim ate until he had had enough, and the hunchback was glad to see this, exclaiming: 'By God, this is what kings and princes do!' He then asked Ibrahim what was in his bundle, and when it had been opened to let him see, he said: 'Take it with you, for it will come in handy when Lady Jamila arrives, as I shan't then be able to bring you any food.' He got up and took Ibrahim by the hand to a place opposite Jamila's pavilion, where he made an arbour for him among the trees, saying: 'Come up here, and when she arrives, you will be able to see her while she can't see you. This is the most that I can do for you, and you must rely on God. You can drink your wine while she sings, and when she leaves, go back where you came from in safety, if God so wills it.' Ibrahim thanked him and wanted to kiss his hand, but the hunchback would not allow it, and when the provisions had been placed in the arbour, he told Ibrahim to look around the garden and eat its fruits, adding that the lady was due to come the next day. Ibrahim amused himself by wandering around and eating fruit, after which he spent the night with the hunchback.

The next morning at daybreak, after Ibrahim had performed the morning prayer, his host came to him looking pale and saying: 'My son, go up to the arbour, for the slave girls have come to prepare the pavilion and Lady Jamila will be following them . . .'

Night 957

Morning now dawned and Shahrazad broke off from what she had been allowed to say. Then when it was the nine hundred and fifty-seventh night, SHE CONTINUED:

I have heard, O fortunate king, that the gardener went to Ibrahim in the garden and told him to go up to the arbour, for the slave girls had come to prepare the pavilion and their mistress would be coming after them. He continued: 'Take care not to spit, clear your nose or sneeze, for otherwise you and I will both be lost.' Ibrahim left for the arbour and the hunchback went off, saying: 'God grant you safety, my son.' While Ibrahim was sitting there, five of the loveliest maids he had ever seen went into the pavilion, took off their outer clothes and washed it down, sprinkling it with rosewater, perfuming it with aloes and ambergris, and spreading out brocades. They were followed by fifty more, carrying musical instruments, and in the middle of them was Jamila covered by a canopy of red brocade, whose fringes were held up by maids using golden hooks. Because of this, by the time that Jamila had entered the pavilion, Ibrahim had had no view either of her or of what she was wearing. 'By God,' he said to himself, 'all my efforts have been wasted, but I shall have to wait patiently to see how things turn out.'

The maids now fetched food and drink, and when they had eaten and washed their hands, they brought out a chair, on which Jamila took her seat. While all the girls played on their instruments and sang with voices of unrivalled beauty, an elderly duenna emerged. She clapped her hands and danced, being pulled to and fro by the girls, but then a curtain was lifted and out came Jamila herself, laughing. Ibrahim could see her ornaments and her robes, together with the crown on her head, set, as it was, with pearls and other gems. She was wearing a necklace of pearls and round her waist was a belt made out of chrysolites, with ropes of sapphires and pearls. The girls got up and kissed the ground before her as she laughed. According to Ibrahim's account: 'At the sight of her unmatched beauty, I lost my senses in bewilderment and confusion and I collapsed in a faint. When I had recovered, I tearfully recited the following lines:

> When I see you, I cannot close my eyes,
> Lest their lids veil you from my sight.
> However much I look at you,
> My eyes cannot compass your loveliness.'

The duenna told ten of the girls to dance and sing, and Ibrahim, watching them, said to himself: 'I wish that Lady Jamila would dance.' When the ten had finished, they gathered around her and said: 'Lady, now that we are assembled here, we would like you to complete our pleasure by dancing, as we have never seen a more pleasant day.' 'The doors of heaven must have opened,' said Ibrahim to himself, 'as God has answered my prayer.' The girls kissed Jamila's feet, telling her: 'By God, we have never seen you in such a cheerful mood as today.' They continued to encourage her until she took off her outer dress and stood clothed in a shift of cloth-of-gold, embellished with jewels of all kinds, revealing jutting breasts like pomegranates and a face like a full moon. Her movements were such as Ibrahim had never seen in his life, and the novel style of her dancing and her remarkable innovations were enough to drive from the mind the dancing of bubbles in the wine glass and to recall the way in which a turban slips down from the head. Jamila was as the poet has described:

> She was created as she wished, evenly proportioned
> In the mould of beauty, neither too tall nor too short.
> It was as though she was made from a bright pearl
> And in every part of her body a moon of beauty shone.

Another poet has written:

> There is many a dancer like the branch of a *ban* tree,
> Whose movements almost steal away my soul.
> In the dance her feet are never still,
> As though they felt beneath them my heart's fire.

IBRAHIM SAID:

As I was watching her, she happened to turn in my direction. She caught sight of me, and at that, her colour changed. Telling her maids to carry on singing until she came back, she went and fetched a knife half a cubit long and came towards me, exclaiming: 'There is no might and no power except with God, the Exalted, the Omnipotent.' As she approached, I lost control of my senses, but when we were face to face, the knife dropped from her hand and she said: 'Glory to God, Who causes hearts to change.' Then she said: 'Take courage, young man, for you are safe from what you fear.' I started to weep and she wiped away my tears, asking me who I was and what had brought me to her garden. I kissed the ground in front of her and clutched at the bottom of her robe, while she repeated that no harm would come to me, adding: 'I have never looked with pleasure on any male apart from you, so tell me who you are.'

I then told her my story from beginning to end, and she exclaimed in astonishment: 'For God's sake, tell me, are you really Ibrahim the son of al-Khasib?' When I replied that I was, she threw herself on me and said: 'Sir, it is because of you that I have shunned men. I was told that in Egypt there was a boy of unsurpassed beauty, and I fell in love with your description. You captivated my heart because of what I had heard of your dazzling charms, and such was my longing for you that it was as the poet described:

> In my love for you my ear outstripped my eye,
> For this is something that takes place at times.

Praise be to God, Who has let me see your face. I swear that had this been anyone else but you, I would have crucified the gardener, the doorkeeper of the *khan*, the tailor and anyone else who had sheltered you.' She went on: 'How can I manage to get you something to eat without my maids knowing?' 'I have food and drink with me,' I told her, opening my bundle in front of her. She took out a chicken and started feeding me with mouthfuls from it, while I did the same for her, thinking, as I saw what she was doing, that this was all a dream. I then brought out the wine and we drank. All the time that she was with me, her maids were singing, and this went on from morning until noon. Then she got up and told me to get a boat and wait for her in such-and-such a place, saying that she could not bear to be parted from me. 'I have a boat,' I told her. 'It is my property and the sailors, whom I've hired, will be waiting for me.' 'That's what is wanted,' she said, and she then went back to her maids . . .

Night 958

Morning now dawned and Shahrazad broke off from what she had been allowed to say. Then, when it was the nine hundred and fifty-eighth night, SHE CONTINUED:

I have heard, O fortunate king, that Lady Jamila went back to her maids and told them to get up, as they were all going back to the palace. They objected to leaving so soon, pointing out that they usually stayed there for three days, but Jamila explained: 'I feel a great heaviness as though I was ill, and I'm afraid it may get worse.' 'To hear is to obey,' they said, and, after putting on their outer clothes, they went down to the river bank and boarded their boat.

It was now that the hunchbacked gardener came up to Ibrahim, knowing nothing of what had happened. 'You didn't have the luck to enjoy a sight of her?' he suggested, adding: 'She usually stays here for three days and I was afraid that she might have seen you.' 'No,' said Ibrahim, 'she didn't see me and I didn't see her, for she never came out of the pavilion.' 'That must be true,' the gardener said, 'for had she caught sight of you, we would both be dead. So stay here with me until she comes back next week, when you will be able to look your fill on her.' Ibrahim told him: 'Sir, I've brought money with me which I'm afraid to lose, and I've left behind people who may try to profit by my absence.' 'It is hard for me to part from you,' said the gardener, embracing Ibrahim and saying goodbye. For his part, Ibrahim went back to the *khan* where he was lodging, met the doorkeeper and took back his money. 'I hope that the news is good,' the man said, but Ibrahim told him that he had found no way to get what he wanted and that he now intended to go back to his own people. The man took a tearful farewell of him, before carrying his possessions down and accompanying him to his boat.

Ibrahim now set off for the rendezvous given by Jamila and waited there for her. When the night had darkened, she arrived dressed as a swashbuckler with a bushy beard and a belt tied around her waist. In one hand she carried a bow and arrows and in the other a naked sword. 'Are you the son of al-Khasib, lord of Egypt?' she asked and when he said that he was, she said: 'And what kind of an evil-minded creature are you, coming here in order to seduce the daughters of kings? Get up and answer the summons of the sultan.' IBRAHIM SAID:

I fell down in a faint, and the boatmen shrivelled up in their skins, dying of fear. When she saw the effect that she had had on me, she pulled off her false beard, threw away the sword, unloosed her belt and showed herself to me as Lady Jamila. 'By God,' I told her, 'you almost stopped my heart.' I then told the sailors to get the boat under way as fast as they could, at which they unfurled the sails and set off in a hurry.

It took us no more than a few days to reach Baghdad, and when we got there we saw another boat by the river bank. Its crew shouted out to ours, calling them by name and congratulating them on their safe return. They brought their boat alongside and, when I looked, I could see that in it was Abu'l-Qasim al-Sandalani. On seeing me, he said: 'This is what I had hoped for. I commit you to God's protection wherever you go, but now I have to leave on an errand.' He was holding a candle and when, in answer to his question, I told him that I had got what I wanted, he brought this close to us. At the sight of him, Jamila's expression changed and she turned pale, while, when he saw her, he said: 'Go under the protection of God. I have to leave for Basra on the sultan's business, but it is those who are present who get the gift.' He then produced a box of sweetmeats which he threw into our boat, and which he had impregnated with the drug *banj*. I said: 'Eat some of these, my darling,' but Jamila wept and asked me if I knew who the man was. 'Yes,' I told her, 'it was al-Sandalani.' 'He is my cousin,' Jamila told me. 'He once asked my father for my hand but I wasn't prepared to accept him and now he is going to Basra and may tell my father about us.' I said: 'He won't get to Basra, lady, before we reach Mosul,' not knowing what the future held for us.

I then ate one of the sweetmeats, but as soon as it settled in my stomach my head struck the ground. Some time later, when it was nearly dawn, I sneezed; the drug cleared from my nostrils and I opened my eyes to find myself lying half-naked among some ruins. I slapped my face, realizing that this must have been a trick played on me by al-Sandalani. I had no idea where to go, dressed as I was only in my drawers, but I got up and walked a little way, when suddenly I saw the *wali* coming, accompanied by guards with swords and sticks. In my alarm, I took cover in a ruined bath house which I found there, but once inside I stumbled over something and the hand with which I had touched the obstacle turned out to be covered in blood. I wiped my hand on my drawers without knowing what it was that I had touched, but after stretching out my hand a second time, I found that it was a corpse. I lifted its head but then let it drop again, exclaiming: 'There is no might and no power except with God, the Exalted, the Omnipotent!' I hid myself away in one of the corners of the bath house, but as I did so, the *wali* came to the door and ordered his guards to go in and search the place. Ten of them entered, carrying torches, and in my panic I crouched behind a wall from where I could see the corpse.

This turned out to be that of a girl, wearing expensive clothes, her face as lovely as the moon, with her head lying in one place and her body in another. When I looked at her, I began to quake, and the *wali* entered and repeated his order to have the whole bath house searched. The guards came to where I was, and one of them, catching sight of me, advanced, carrying a knife half a cubit long. As he came near, he called out: 'Glory be to God, Who created this beautiful face.' Then, after asking me where I had come from, he took hold of my hand and said: 'Young man, why did you kill this girl?' 'By God,' I told him, 'I didn't kill her and I don't know who did, for I only came in here because I was afraid of you.' I then told him my story and added: 'For God's sake, don't wrong me, for I have enough trouble of my own.'

The man brought me before the *wali* who, seeing the bloodstains on my hand, said: 'There is no need for further proof; cut off his head!'

Night 959

Morning now dawned and Shahrazad broke off from what she had been allowed to say. Then, when it was the nine hundred and fifty-ninth night, SHE CONTINUED:

I have heard, O fortunate king, that IBRAHIM SAID:

The man brought me before the *wali* who, seeing the bloodstains on my hand, said: 'There is no need for further proof; cut off his head!' On hearing this, I wept bitterly and recited these lines:

We move on a predestined path,
And each must walk as fate decrees.
Where we must die has been ordained,
And in no other land will we meet death.

With a deep groan I fell down in a faint and the executioner, moved by pity, exclaimed: 'By God, this is not the face of a murderer!' The *wali*, however, repeated his order and his men made me kneel on the execution mat and blindfolded me. The executioner took his sword and asked the *wali*'s permission to proceed. He was about to strike while I cried out: 'Alas for a poor stranger!' when suddenly up came a band of horsemen and a voice shouted: 'Let him alone, executioner. Stay your hand!'

There was a remarkable reason for this. Al-Khasib, the lord of Egypt, had sent his chamberlain to bring precious gifts to the caliph, Harun al-Rashid, together with a letter in which he had written: 'My son has been missing for a year and I have heard that he is in Baghdad. I am writing this in the hope that the caliph may be good enough to seek news of him, do his best to find him and then send him back to me with my chamberlain.' When the caliph had read this, he instructed the *wali* to look into the matter and the two of them continued to make enquiries until the *wali* heard that Ibrahim was in Basra. He told this to the caliph, who supplied the Egyptian chamberlain with a letter and told him to make his way to Basra, escorted by a number of the vizier's servants. So eager was the chamberlain to track down his master's son that he left immediately, and it was at this point that he came across Ibrahim on the execution mat.

The *wali* recognized the chamberlain when he saw him and dismounted respectfully. 'Who is this young man and what is his crime?' the chamberlain asked. The *wali* explained the position and the chamberlain, who had not recognized Ibrahim, said: 'His is not the face of a murderer,' and ordered him to be released from his bonds. When this had been done, he said: 'Bring him up to me.' When the *wali* led Ibrahim forward, the hardship and terrors that he had experienced had robbed him of his beauty, and so it was that the chamberlain said: 'Tell me about this business, young man, and explain your connection with the dead girl.' Ibrahim recognized him when he saw him, and said: 'Damn you, don't you know me? I am Ibrahim, the son of your master, and it may be that you've come to search for me.' The chamberlain stared at him closely before he could be sure who he was, but he then fell down at Ibrahim's feet, and, seeing this, the *wali* turned pale. 'Damn you, you wicked fellow,' said the chamberlain, 'did you mean to kill the son of my master, al-Khasib, the lord of Egypt?' The *wali* kissed the skirt of his robe and said: 'Master, how was I to recognize him when I saw him looking like this, with the murdered girl by his side?' 'You're not fit to hold your office,' the chamberlain told him. 'This young man is fifteen years old and never in his life has he killed as much as a sparrow, so how can he now turn out to be a murderer? Why didn't you wait until you had asked him about himself?'

The chamberlain and the *wali* now began a search for the murderer, and when the baths were searched again, he was found and brought to the *wali*, who then set off for the palace and told the caliph all that had happened. The caliph, for his part, ordered the murderer to be executed and then summoned Ibrahim. When Ibrahim appeared before him, he said with a smile: 'Tell me everything that has happened to you.' So Ibrahim told him his story from beginning to end and the caliph, moved by his account, summoned Masrur, his executioner, and told him to go immediately to break into the house of Abu'l-Qasim al-Sandalani and to fetch Jamila. When Masrur did this, he found the girl half-dead and tied up by her own hair. He freed her and brought both her and al-Sandalani to the caliph, who was struck with wonder at the sight of her beauty. He then turned to al-Sandalani and said: 'Take him; cut off the hands with which he beat this girl and then crucify him, before handing over his wealth and his possessions to Ibrahim.'

These orders had been carried out when Jamila's father, Abu'l-Laith, the governor of Basra, arrived to ask for the caliph's help against Ibrahim, complaining that he had carried off his daughter. The caliph pointed out that it was Ibrahim who was responsible for saving her from torture and death and then, after having summoned Ibrahim, he asked Abu'l-Laith: 'Will you not accept this young man, the son of the lord of Egypt, as a husband for your daughter?' 'To hear you is to obey, Commander of the Faithful,' said Abu'l-Laith, at which the caliph called for the *qadi* and the notaries, married Jamila to Ibrahim and, after presenting him with all al-Sandalani's wealth, sent him back to his own country. The two lived together in the most perfect joy and pleasure until they were visited by the destroyer of delights and the parter of companions. Praise be to the Living God, Who never dies.

A story is also told that al-Mu'tadid bi'llah, a high-minded and noble ruler, had six hundred viziers in Baghdad and knew everything about the affairs of his subjects. At that time, when the caliph wanted to go round to observe his subjects, he would disguise himself as a merchant. One day he went out with Ibn Hamdun to inspect his subjects and listen to the latest news. In the noonday heat they turned off a street into a small lane, at the head of which was a fine, tall house that spoke well for its owner. As the two of them sat down by the door to rest, out came two eunuchs like moons on the fourteenth night, one of whom said to the other: 'I wish that some guest would ask leave to enter, because my master will only eat when there are other guests with him, and up till now I've not seen a soul.' The caliph was astonished by this and said: 'This shows that the owner of this house is a generous man. We shall have to go in to see his liberality for ourselves, and this may result in some favour coming to him from me.' He then told the eunuch to ask if his master would be prepared to receive strangers.

The eunuch went in to tell his master, who was pleased and came out to welcome the visitors himself. He turned out to be a man with a handsome face and a fine figure, wearing a shirt of Nisabur silk and a gilded cloak; he smelt of perfume and on his hand he wore a sapphire ring. At the sight of al-Mu'tadid and his companion, he exclaimed: 'Welcome to the guests who have done the very great honour of visiting me.' When they went in, they saw what looked like a corner of Paradise, such as would make a man forget his family and his homeland.

Night 960

Morning now dawned and Shahrazad broke off from what she had been allowed to say. Then, when it was the nine hundred and sixtieth night, SHE CONTINUED:

I have heard, O fortunate king, that when the caliph and his companion entered, they saw what looked like a corner of Paradise, such as would make a man forget his family and his homeland. Inside was a garden with a dazzling collection of trees of all sorts, and the rooms were magnificently furnished. They all took their seats and Ibn Hamdun noted that as al-Mu'tadid sat there looking at the house and the furnishings, his expression changed. IBN HAMDUN SAID:

I knew what he looked like when he was pleased and when he was angry, and I wondered what had happened to annoy him. Servants brought in a golden bowl, and when we had washed our hands they produced a silk cloth on which was set a cane table. When the covers were removed from the dishes, we saw foods as splendid as the flowers at the height of spring, presented singly or in pairs. 'In God's Name, gentlemen,' said our host, 'I am tormented by hunger, so do me the favour of eating some of this food like the noble men you are.' He started to put in front of us bits that he had torn from a chicken, all the while laughing, reciting poetry, telling stories and producing witticisms that were in keeping with a social gathering.

We ate and drank and then moved to another room of astonishing magnificence, from which exuded a fresh scent. Here we were served with fresh fruits and delicious sweetmeats, such as would add to joy and remove sorrow. In spite of this, the caliph continued to frown; he smiled at none of these delights, although in general he was fond of the kind of pleasurable entertainment that dispels care, and I knew that he was neither envious nor tyrannical. So I wondered to myself what it was that had caused him to glower and to refuse to be cheered. Then the servants brought in a tray of drinks, such as cement friendships, with wine that had been strained and goblets of gold, crystal and silver. Our host knocked with a bamboo staff on the door of a room and, when it opened, out of it came three swelling-breasted virgins with faces like the midday sun, one being a lute player, the second a harpist and the third a dancer. While we were being served fruit, both dried and fresh, a curtain of brocade with silken tassels and golden rings was lowered between us and the three girls.

The caliph paid no attention to any of this and said to the host, who had no idea who he was: 'Are you a descendant of the Prophet?' 'No, sir,' the man replied. 'I am of merchant stock and my name is Abu'l-Hasan 'Ali, the son of Ahmad of Khurasan.' 'My good man,' the caliph said, 'do you know who I am?' Abu'l-Hasan replied: 'No, I know nobody of your distinction.' I told him: 'This is the Commander of the Faithful, al-Mu'tadid bi'llah, the grandson of al-Mutawakkil.' At this, the man got up, trembling with fear, and kissed the ground before him, saying: 'Commander of the Faithful, I implore you by your pure ancestors to forgive me for any shortcomings that you have seen on my part or any lack of respect that I may have shown you.' The caliph replied: 'Your hospitality could not have been bettered, but there is something that I hold against you. If you tell me the truth and it makes sense to me, then you will be safe from me, but, failing that, I shall convict you on the grounds of clear proof and I shall punish you more severely than I have ever punished anyone.' 'God forbid that I should try to deceive you, Commander of the Faithful,' Abu'l-Hasan said, 'but what fault is it that you have found with me?' The caliph replied: 'Ever since I came into your house, I have been looking at its beauty, its utensils, its furnishings and decorations, as well as at your clothes. On all of them is the name of my grandfather, al-Mutawakkil.' 'Yes,' said Abu'l-Hasan, 'and, as you know – may God grant you His aid – truth is both your undergarment and your cloak, and in your presence no one can speak anything but what is true.' The caliph told him to sit down, and when he had done so, the caliph asked for his story.

Abu'l-Hasan said: 'May God assist you, Commander of the Faithful, and encompass you with His grace! You must know that in Baghdad nobody lived in easier circumstances than I and my father, but I would ask you now to pay attention, listen and look, so that I may tell you the reason for what you have found objectionable in me.' The caliph told him to go on, AND HE SAID:

I should tell you that my father functioned in the markets of the bankers, the apothecaries and the drapers, in each of which he had a shop, an agent and goods of various kinds. In his shop in the bankers' market he kept a private room, while the shop itself was used for business. The extent of his fortune passed all reckoning and I was his only son. He showed me great love and tenderness, and when he was on the point of death he summoned me and told me to look after my mother and to fear Almighty God. Then he died – may God have mercy on him and preserve the Commander of the Faithful – and I busied myself with my own pleasures, eating and drinking, and passing my time with friends and companions. My mother tried to stop me and criticized me, but I paid no attention to her until all my money had gone. I sold every one of my properties, until there was nothing left except for the fine house in which I lived. I told my mother that I was intending to sell this too, and she said: 'If you do that, my son, you will be disgraced and you won't know where to go for shelter.' I replied: 'The house is worth five thousand dinars and with this I shall buy another one for a thousand, and use the rest as trading capital.' She then asked me if I would sell her the house for that price, and when I told her that I would, she went to a chest, opened it and took out a porcelain jar with five thousand dinars in it. It seemed to me that the whole house was full of gold but she told me: 'Don't think that this is your father's money; rather, it came from my own father. I stored it up for emergencies, since in your father's time I had no need of it.'

I took the money from her, Commander of the Faithful, and returned to my old ways, eating, drinking and associating with friends until it had all gone, for I had not listened to my mother or taken her advice. Then I told her again that I wanted to sell the house and she said: 'I told you not to do that, my son, as I knew that you would need the house, so how is it that you are trying to sell it again?' 'Don't go on and on about this,' I told her, 'for I have to sell it.' 'Sell it to me for fifteen thousand dinars,' she told me, 'on condition that I take charge of your affairs myself.' So I sold it to her at that price and on that condition, and she then looked out my father's agents and handed each of them a thousand dinars. She took charge of the money and of the income and expenditure, giving me a certain amount of trading capital and telling me to take my place in my father's shop. I did as she said and went to his room in the bankers' market. My friends came to me and I did business with them, making a good profit and amassing a considerable sum of money. When my mother saw how well I was doing, she showed me what she had stored up in the way of gems, precious stones, pearls and gold. I recovered the properties that I had squandered and became as wealthy as I had been before.

Things stayed like this for some time. I supplied merchandise to my father's agents and had a second room built within the shop. Then one day, as I was sitting there as usual, I was approached by a girl unsurpassed in beauty, who asked: 'Is this the shop of Abu'l-Hasan 'Ali, the son of Ahmad of Khurasan?' I told her that it was, and when she asked: 'Where is he?' I said, bewildered as I was by her beauty: 'I am Ahmad.' She sat down and said: 'Tell your boy to weigh me out three hundred dinars.' When he had done this on my instructions, she took the money and went off, leaving me distracted. 'Do you know her?' my servant asked, and when I said no, he said: 'Then why did you tell me to weigh out the money?' 'By God,' I told him, 'I was so smitten by her beauty and grace that I didn't know what I was saying.' Unbeknown to me, he got up and followed her, only to return in tears with his face showing the mark of a blow. 'What happened to you?' I asked, and he said: 'I followed the girl to see where she was going, but when she realized what I was doing, she came back and struck me, almost knocking out my eye.'

A month passed during which I didn't see her and, as she never returned, the love that I felt for her left me in a state of confusion. Then, at the end of the month, back she came and greeted me, making me ecstatic with joy. She asked me about myself and said: 'Perhaps you said to yourself: "How did that cheat manage to go off with my money?"' 'By God, lady,' I told her, 'my money and my life are yours.' At that, she unveiled herself and sat down to rest, with the ornaments that she was wearing dancing over her face and her breast. 'Weigh me out three hundred dinars,' she told me. 'To hear is to obey,' I said, and when I had done it, she took the money and left. I told my servant to follow her, which he did, but to no avail.

For some time she did not come back, but then as I was sitting in the shop one day, she approached me and, after some conversation, she said: 'Weigh me out five hundred dinars, for I need it.' I wanted to say: 'Why should I give you my money?' but so great was my love that I could not speak and every time I looked at her I trembled and turned pale, forgetting what I was going to say. It was as the poet has described:

Whenever I see her suddenly,
I am taken aback with scarcely a word to say.

I weighed out the money and this time, when she had left with it, I followed her myself until she came to the gem market. She stopped there beside a shopkeeper and picked up a necklace, before turning and seeing me. 'Pay him five hundred dinars,' she said to me, and when the man came up and flattered me, I told him to give her the necklace as I would pay the cost. 'To hear is to obey,' he said and she then took it and left.

Night 961

Morning now dawned and Shahrazad broke off from what she had been allowed to say. Then, when it was the nine hundred and sixty-first night, SHE CONTINUED:

I have heard, O fortunate king, that ABU'L-HASAN AL-KHURASANI SAID:

I told him to give her the necklace as I would pay the cost. She then took it and left. I followed her as far as the Tigris, where she got into a boat, leaving me pointing at the ground to indicate that I wanted to kiss it in front of her. She sailed away laughing, and I stood there watching until she went into a palace, which, when I looked at it closely, turned out to be that of the caliph al-Mutawakkil. I went home with all the cares in the world weighing down my heart. She had got three thousand dinars from me and I said to myself: 'She has taken my money, robbed me of my wits and it may be that I shall die of love for her.' When I got home, I told my mother everything that had happened and she said: 'My son, take care to have nothing more to do with this girl lest you meet your death.'

When I went to my shop I was approached by my agent in the apothecaries' market, a very old man, who said: 'Master, why is it that you seem so changed and are looking distressed? Tell me what has happened.'

So I told him all about my encounters with the girl and he said: 'My son, she is one of the slave girls in the caliph's palace. She is his favourite, and so consider the money as a gift to charity and don't concern yourself about her. If she approaches you again, take care lest she try to get something more from you, and send word to me, so that I may think of something to save you from ruin.' He then went off, leaving me with a fire burning in my heart.

At the end of the month, to my great joy she came again. 'What led you to follow me?' she asked, and I told her that it was because of the great love that I felt for her. Then I broke down in tears in front of her and she wept out of pity for me. 'By God,' she said, 'if you feel passion, I feel it more, but what can I do? The best thing is for me to see you once a month.' Then she passed me a note and told me to take it to a man whom she named, saying: 'He is my agent, and you are to get from him the sum that I have written down here.' I told her that I'd no need for money, repeating that I would give up my wealth and my life for her, and she promised that, however difficult it might be, she would arrange some way of bringing me to her. Then she took her leave of me and went off.

I went to the old apothecary and told him what had happened. He went with me to al-Mutawakkil's palace, and I saw that it was the place into which the girl had gone. He was at a loss to know what to do next, but then he turned and saw a tailor with his workmen in a shop opposite a palace window overlooking the river bank. 'It is through this man that you will get what you want,' he said. 'First, tear your pocket and go to him to ask him to stitch it up, and when he has done that, give him ten dinars.' 'To hear is to obey,' I replied, and I went up to the tailor, taking with me two pieces of Rumi brocade which I told him to make into four jackets, two with long sleeves and two without. After he had cut them out and sewn them, I paid him well over the usual price, and when he was about to hand the clothes to me, I told him to keep them for himself and for anyone else who was there with him. After that, I spent a long time sitting with him, and I gave him other fabric to cut out, telling him to hang the finished clothes in the front of his shop where people might see them and buy them. He did as I suggested, and if anyone from the palace admired any of them, I would give it to him as a present, even if he was only the doorkeeper.

One day the tailor said to me: 'Tell me the truth about yourself, my son. You have had me make a hundred expensive robes, and, although each of them was worth a lot of money, you gave most of them away. That's not the behaviour of a merchant, for merchants keep track of their cash. How much money must you have to allow you to make gifts like these, and what must you earn in a year? Give me an honest answer, for I may be able to help you to get what you want.' Then he added: 'I conjure you in God's Name; are you not a lover?' 'Yes, I am,' I told him. 'Who is the girl?' he asked, and I said: 'She is one of the palace slave girls.' 'God curse them,' he said. 'How many men they seduce!' He asked if I knew her name and when I said that I did not, he told me to describe her. After I had done so, he exclaimed in horror: 'She is the caliph's lute player and his favourite. But she does have a mamluk, and if you befriend him it may be that through him you will be able to get to her.'

As we were talking, that same mamluk came out of the palace gate, looking like a moon on its fourteenth night. In front of me were the clothes that had been made for me by the tailor out of variously coloured brocades. The mamluk started to look at them and inspect them, before approaching me. I stood up and greeted him, and when he asked who I was, I told him that I was a merchant. 'Are you selling these clothes?' he asked, and when I told him that I was, he chose five robes and asked me how much they were. He was pleased when I said to him: 'They are a present from me to you, in order that we may become friends.' I then went back to my house where I picked out for him a robe studded with sapphires and other gems, worth three thousand dinars, which I brought to him. He accepted it and took me into a room inside the palace and asked me: 'What do the merchants call you?' 'One of their own,' I told him, but he said to me: 'I'm suspicious of you.' When I asked why, he said: 'Because you have given me a valuable present to win me over, and I'm sure that you must be Abu'l-Hasan al-Khurasani, the banker.' I burst into tears and he said: 'Why are you weeping? The one for whom you are shedding these tears is more deeply and passionately in love with you than you are with her, and the story of her dealings with you is common knowledge among all the palace girls.' Then he asked me what I wanted, and I said that I wanted his help in my sad situation. He told me to come back next day, and I then went home.

The next morning, I set off to meet him and went to his room. When he came, he told me: 'I have to tell you that yesterday, when she had finished attending on the caliph and was back in her room, I told her all about you. She is determined to meet you, so sit here with me until the end of the day.' I did that, and when it had grown dark, he brought me a shirt of cloth of gold together with one of the caliph's robes, which he made me put on, and these, together with the scent with which he had perfumed me, made me look like the caliph himself. He then led me to a hall with two series of rooms, one on each side, which he told me belonged to the caliph's favourites. 'As you pass them, place a bean by each door,' he instructed me, 'for this is what the caliph is in the habit of doing every night . . .'

Night 962

Morning now dawned and Shahrazad broke off from what she had been allowed to say. Then, when it was the nine hundred and sixty-second night, SHE CONTINUED:

I have heard, O fortunate king, that the mamluk told Abu'l-Hasan: 'As you pass them, place a bean by each door – for this is what the caliph is in the habit of doing – until you come to the second passage on the right, where you will see a door with a marble threshold. When you get there, feel it with your hand, or, if you prefer, count the doors' – and he explained how many there were – 'and then go through it' – and he described how it was marked. 'Your friend will see you and take you to her room; when it is time for you to leave, I will look to God to help me, even if I have to carry you out in a chest.' ABU'L-HASAN WENT ON:

When he had gone back and left me, I began to count the doors as I walked, placing a bean by each of them, but when I was halfway along, I heard a great noise and saw lighted candles. The lights were coming in my direction and when they were close, I stared and discovered that here was the caliph himself, surrounded by slave girls with tapers in their hands. I heard one girl say to another: 'Have we got two caliphs, sister? I could smell the scent of the caliph's perfume as he went past my room, and he put a bean by my door as usual, but now I can see the light of his candles and here he is coming towards us.' 'It is very strange,' the other answered, 'for no one would dare dress up as the caliph.' The lights were getting closer and I was trembling, when a eunuch called out to the girls to come. They turned into one of the rooms, but then left again and walked on until they came to the room of my beloved. I could hear the caliph ask whose it was and when they told him that it belonged to Shajarat al-Durr, he said: 'Call her.' She came out in answer to the summons and kissed the caliph's feet. 'Will you drink tonight?' he asked and she replied: 'Were it not for your presence and for the privilege of looking at you, I would not, as I don't feel like drinking this evening.' The caliph then told the eunuch to instruct the treasurer to give her a particular necklace, after which he told his entourage to go into her room.

The tapers preceded him as he entered, but then a girl at the head of the procession, the radiance of whose face outshone the light of the candle she was carrying, came up to me, saying: 'Who is this?' She took hold of me and brought me into one of the rooms before repeating: 'Who are you?' I kissed the ground before her and said: 'For God's sake, my lady, don't shed my blood, but have mercy on me and seek God's favour by saving my life.' I wept in fear for my life and she said: 'You must be a thief.' 'By God, I am not,' I told her. 'Do I look like one?' She replied: 'If you tell me the truth, I shall protect you.' So I said: 'I am a lover, foolish and stupid, who has been prompted by passion and folly to do what you can see has led me into this predicament.' 'Stay here until I come back,' she told me, and then she went out and brought back women's clothing, in which she dressed me then and there. She told me to follow her and brought me to her room. I went in as she had told me, and she took me to a couch covered with a magnificent rug, saying: 'Sit down, and no harm will come to you. Aren't you Abu'l-Hasan al-Khurasani, the banker?' 'Yes,' I replied, and she said: 'May God preserve you if you're telling the truth and are not a thief, for otherwise you are certain to be killed for dressing up as the caliph and wearing his perfume. But if you really are Abu'l-Hasan, you are safe from all harm, as you are the friend of my sister, Shajarat al-Durr. She is always talking about you and telling us how your expression never changed when she took your money and how you followed her to the river bank and pointed at the ground as a sign of your respect for her. She is more inflamed by love for you than you are for her, but how did you get here? Was it or was it not on her instructions that you risked your life, and what did you hope to get by meeting her?' 'By God, lady,' I said, 'it was I who risked my life, and all that I wanted from such a meeting was to look at her and to hear her voice.' 'Very well,' she said, and I repeated: 'God is my witness, lady, that I had no intention of committing any sin with her.' 'Because of that,' she told me, 'God has saved you and has made me pity you.'

Then she said to her maid: 'Go to Shajarat al-Durr and tell her: "Your sister sends you her greetings and asks if you will be good enough to visit her as usual this evening, since she is feeling depressed."' The maid went off to Shajarat al-Durr, but then came back with a message: 'She says: "May God do me the favour of granting you long life, and may He make me your ransom. Had you invited me at any other time, I would have had no hesitation, but I can't accept now as the caliph has a headache and you know the position I hold with him."' The girl said to her maid: 'Go back and tell her that she must come because of a secret that we share.' The maid went and after a time she returned with Shajarat al-Durr, whose face was gleaming like the moon. My rescuer welcomed her with an embrace and then said: 'Come out, Abu'l-Hasan, and kiss my sister's hands.' I was in a closet within the room and when I came out and she saw me, she threw herself on me and clasped me to her breast. Then she asked me how I came to be dressed as the caliph, with all his finery and his perfume, and she wanted to know everything that had happened to me. I gave her an account of my experiences and of the fear and apprehension that I had endured. 'I'm distressed by what you have suffered for my sake,' she told me, 'but I thank God, Who has brought you at last to safety, for in my room and the room of my sister you will be quite safe.'

She took me to her room and then said to her sister: 'I made a pact with him that I would have no unlawful intercourse with him, but now that he has risked his life and faced such terrors, I am earth for him to tread and dust beneath his feet.'

Night 963

Morning now dawned and Shahrazad broke off from what she had been allowed to say. Then, when it was the nine hundred and sixty-third night, SHE CONTINUED:

I have heard, O fortunate king, that the girl said to her sister: 'I made a pact with him that I would have no unlawful intercourse with him, but now that he has risked his life and faced such terrors, I am earth for him to tread and dust beneath his feet.' ABU'L-HASAN WENT ON:

'If this is what you intend,' her sister said, 'may God Almighty save you.' Shajarat al-Durr then told her: 'You will see what I shall do in order to be united with him lawfully, for I must try to achieve this, even at the risk of my own life.' While we were talking, we suddenly heard a great commotion, and when we turned we could see that the caliph was on his way to her room, such was his fondness for her. She took me and hid me in a cellar, closing the trapdoor over me, before going out to greet the caliph. He sat down and she stood in front of him, waiting on him, and then ordering wine to be brought in. The caliph was in love with a girl named al-Banja, the mother of al-Muʿtazz, but she had broken with him and he in turn had broken with her. In the pride of her beauty, she would not seek for a reconciliation, while for his part, his pride as caliph and ruler kept him from doing so himself, for he refused to humiliate himself before her. Although his heart was on fire with love for her, he tried to distract himself with other slave girls like her, whom he would visit in their rooms. He liked Shajarat al-Durr's singing and when he now told her to sing for him, she took her lute, tightened the strings and recited these lines:

> I wondered at how Time worked so hard to part us,
> And then, when our union was over, how it ceased to move.
> I left you and men said: 'He does not know what love is,'
> And then I visited you and they said: 'He has no patience.'
> Let love for her increase my passion every night;
> Forgetfulness brought by Time will not come till Judgement Day.
> Her skin is like silk and her speech is soft,
> With no foul words and no solemnity.
> To her eyes God said: 'Be,' and they were,
> While on the hearts of men they act like wine.

When the caliph heard this he was stirred with delight, and I, in my cellar, was so moved that, had it not been for the grace of Almighty God, I would have shouted out and we would have been exposed. Then she recited the following lines:

> I kiss my love, for whom my soul still longs,
> But can I get nearer him than an embrace?
> I kiss his mouth to cool my heat,
> But the violence of my passion grows still more.
> It is as though my ardour will not be cured
> Until I see that our two souls have merged.

The delighted caliph told her to make a wish for him to grant, at which she asked for her freedom, as this would bring him his own reward from God. He replied: 'I set you free for His sake.' She kissed the ground before him, and he said: 'Take your lute and sing for me something about the slave girl who holds me in the bonds of love and whose favour I seek, just as my subjects seek mine.' So she picked up the lute and recited:

Mistress of beauty, you have carried off my piety;
Whatever happens, I cannot do without you,
Whether I win you by humility, as befits a lover,
Or by that dignity which befits my rank.

The caliph was enchanted and said: 'Take up your lute again and sing something that tells of how I am placed as regards three girls who hold the reins of my heart and rob me of sleep: you, the girl who has abandoned me and another, whom I shall not name and who is beyond compare.' Shajarat al-Durr took her lute, struck up a tune and sang:

Three lovely ones are holding my heart's reins,
Where they have occupied the highest place.
Although I need obey no ruler among men,
I must obey these three rebellious girls.
The king of love has helped them conquer me,
And his authority outweighs my own.

The caliph was amazed at how well these lines reflected his position, and his delight led him to seek reconciliation with the girl who had broken with him. He went off to her room, where news of his coming had already been brought, and she received him, kissing the ground before him and then kissing his feet, after which the two were reconciled.

So much for the caliph, but as for Shajarat al-Durr, she went to me and said: 'The blessing of your coming has brought me freedom, and I hope that God will help me to arrange things so that I may lawfully be joined with you.' I for my part praised God, and while they were talking Shajarat al-Durr's mamluk arrived and was told what had happened. 'Praise be to God,' he said, 'Who has brought this to a happy ending, and I pray that He may complete this by bringing you out safely.'

At that point, Shajarat al-Durr's sister, whose name was Fatir, arrived and Shajarat al-Durr asked her how they were to get me safely out of the palace, adding: 'God has granted me the gift of freedom, thanks to the blessing brought by his arrival.' Fatir said: 'The only way to do this is to dress him as a woman,' and she fetched women's clothing and made me put it on. I left the room there and then, but when I had got to the middle of the palace, I came upon the caliph who was sitting with his eunuchs standing before him. The caliph looked at me and, thinking that there was something very wrong, he told his servants to hurry after the 'girl' who was just going out and fetch her to him. When they had brought me to him and raised my veil, the caliph looked at me and recognized me as a man. In answer to his questions, I told him the whole story, keeping nothing back. After listening to what I had to say, the caliph thought the matter over, before getting up quickly and going to the room of Shajarat al-Durr. 'How did you come to prefer a merchant to me?' he asked, and she kissed the ground before him and gave him a truthful account of her story from beginning to end. He sympathized with her and pitied her, as her excuse was based on the circumstances of love.

When he had left, Shajarat al-Durr's eunuch came in and said: 'You can be easy in your mind, for when your lover appeared before the caliph and was questioned, he told the same story as you, word for word.' The caliph, meanwhile, had gone back and now summoned me again and asked: 'How did you dare to approach my palace?' I said: 'I was carried away by my folly and my love, Commander of the Faithful, and I also relied on your generous clemency.' I then burst into tears and kissed the ground in front of the caliph, who said: 'I forgive you both.'

He told me to sit down, and when I had done so, he summoned the *qadi* Ahmad ibn Abi Duwad* and married me to Shajarat al-Durr, giving instructions that all her property was to be transferred to me. A wedding procession escorted her to join me in her room, and when I came out, three days later, everything there was taken to my house. So everything that you have been so shocked to see here is what came to me through my bride.

She told me one day: 'Al-Mutawakkil is a generous man but I'm afraid that he may change his mind or be reminded by someone who envies us, and so I want to do something to make sure that this will bring us no harm.' When I asked her what she had thought of doing, she said: 'I'm going to ask his permission to make the pilgrimage to Mecca and repent to God for having been a singer.' I said that this seemed to me to be an excellent plan and, while we were talking, a messenger from the caliph arrived to summon her, for he was very fond of her singing. She went to attend on him, and when he told her not to desert him she said: 'To hear is to obey.'

One day she went to him after having received the usual summons,

* 'Da'ud' in the Arabic text.

but before I knew it, she was back in tears, with her clothes torn. I was startled and repeated the formula: 'We belong to God and to Him do we return,' thinking that the caliph must have ordered our arrest. 'Is al-Mutawakkil angry with us?' I asked, but she said: 'Where is al-Mutawakkil? His reign is over and no trace remains.' 'Tell me what really happened,' I said, and she explained: 'He was sitting behind the curtain drinking with al-Fath ibn Khaqan and Sadaqa ibn Sadaqa, when his son, al-Muntasir, burst in on him with a band of Turks and killed him. Joy was changed to calamity and good fortune was replaced by tears and wailing. I ran away with my maid and God brought us to safety.' I got up immediately and went downriver to Basra, where I heard that war had broken out between al-Muntasir and al-Musta'in, and it was to Basra that, in my alarm, I moved my wife and all my wealth.

This is my story, Commander of the Faithful, and I have added nothing to it nor subtracted anything from it. Everything that you saw in my house that carries the name of your grandfather, al-Mutawakkil, is what he gave us in his generosity, and our fortune comes from your noble ancestors, for you are people of liberality and a mine of bounty.

'The caliph was delighted by this and astonished by the story,' continued Ibn Hamdun, 'after which Shajarat al-Durr and her children came out and kissed the ground before him, filling him with wonder at their beauty. He called for an inkstand and wrote out an order freeing our properties from tax for twenty years.'

The caliph was pleased with Abu'l-Hasan and took him as a drinking companion until they were parted by Time and exchanged palaces for tombs. Praise be to the Forgiving God.

A story is also told, O fortunate king, that in the old days there was a merchant named 'Abd al-Rahman, to whom God had given a daughter and a son. The daughter was called Kaukab al-Sabah because of her loveliness, and for the same reason the son was called Qamar al-Zaman.* When their father noted their beauty and grace, together with the splendour of their shapely forms, he feared that, when others saw them, there might be envious rumours, and that cunning and debauched persons might scheme against them. As a result, for fourteen years he kept them

---

* Kaukab al-Sabah means 'Morning Star' and Qamar al-Zaman 'Time's Moon'.

out of sight in a villa where the only people to see them were their parents and a slave girl who looked after them. Both their parents used to read them what God had revealed in the Quran, the mother reading to her daughter and the father to his son, until they had memorized all of it. They also learned calligraphy and arithmetic, as well as various other branches of art, from their parents, as they needed no other teacher.

When Qamar al-Zaman had reached maturity, 'Abd al-Rahman's wife asked him how long he proposed to go on hiding him away. 'Is he a boy or a girl?' she queried, and when her husband replied: 'A boy,' she said: 'In that case, why don't you take him to the market and let him sit in the shop so that he may get to know people and they may get to know him and recognize him as your son? You should teach him how to buy and sell so that, if anything happens to you, he can be acknowledged as your son and take over what you leave behind. If things go on as they are and he later claims to be the son of 'Abd al-Rahman the merchant, far from believing him, people will say: "We have never seen you and we never knew that 'Abd al-Rahman had a son." Then the authorities will seize your money and your son will be deprived of his inheritance. In the same way, I am going to introduce Kaukab al-Sabah to society in the hope that someone suitable will ask for her hand and we shall be able to marry her off and hold a wedding feast for her.' 'I am afraid of people looking at them,' said 'Abd al-Rahman . . .

Night 964

Morning now dawned and Shahrazad broke off from what she had been allowed to say. Then, when it was the nine hundred and sixty-fourth night, SHE CONTINUED:

I have heard, O fortunate king, that when his wife had spoken, the merchant told her that he was afraid of people looking at them: 'Because I love them and lovers are very jealous. How well the poet wrote:

You made me jealous of my eyes and of myself –
Of you, of where you are and of Time itself.
If I could set you in my eyes and keep you there
For ever, you would not tire me by being close,
And if I had you with me every day
Till the Last Judgement, that would not be enough.'

His wife said replied: 'Put your trust in God, for no harm will come
to those whom He protects. So take him with you to the shop today.'
She then dressed Qamar al-Zaman in splendid clothes, until he bewitched
all who saw him, leaving distress in lovers' hearts. His father went off
with him to the market, and he so captivated the onlookers that they
came up to kiss his hand and to greet him. His father started to scold
those who were following and who gawped at him, then one of them
said: 'The sun has risen from over there and has come to shine in the
market'; another remarked: 'It is the moon that has risen'; while a third
said: 'It is the new moon marking the feast that has appeared to God's
servants.' They went on talking about the boy and calling down blessings
on him, until his embarrassed father, unable to stop them, began to vilify
and curse his wife because it was she who had made him bring him out.
As he turned, he could see crowds following, while other people were
going on ahead. He walked to his shop, opened it and took his seat, with
the boy sitting in front of him. As he looked, he could see that the street
was blocked with people, and all the passers-by would stop in front of
the shop to look at that beautiful face, unable to tear themselves away.
The knots of men and women who formed there seemed to be acting in
accordance with the lines:

> You created beauty in order to tempt us,
> And said: 'Fear Me, My servants.'
> You are beautiful and love beauty,
> So how can Your servants not fall in love?

At the sight of the mixed crowd of men and women standing and
staring at his son, 'Abd al-Rahman was ashamed and too perplexed to
know what to do. At that moment, a wandering dervish, marked out
as one of God's pious servants, came towards him from the edge of
the market. When he saw Qamar al-Zaman sitting there like a branch
of a *ban* tree growing on a mound of saffron, he shed tears and recited
these lines:

> I saw a branch on a sand hill,
> That was like a shining moon.
> I asked his name. 'Lu'lu',' he said;
> 'For me? For me?' I asked. He said: 'No, no.'*

---

* A play on words – *li li* ('for me, for me') and *la la* ('no, no').

He then walked on slowly, smoothing his grey hairs with his right hand and making his way through the heart of the crowd, which parted in awe of him. When he looked again at the boy, he was astonished and dazzled, as though it was to him that the poet's lines applied:

> The handsome boy was seated there,
> With the new moon of 'Id al-Fitr* in his face,
> When there appeared a dignified old man
> Whose paces were deliberate and slow,
> Showing the marks of an ascetic life.
> He had experience of nights and days,
> Tasting both licit and illicit love,
> A lover both of women and of men,
> Thin as a tooth-pick, a bag of worn-out bones.
> He was a Persian in the art of love
> When he was in the company of boys.
> For love of women he was an 'Udhri,
> Wickedly skilful in both of these fields.
> Zainab to him was just the same as Zaid.
> Love for the beauties filled him, as he wept
> Over the traces of deserted camps.
> The force of longing made him seem a branch,
> Blown in this way and that by the east wind.
> Only what has no feelings can stay firm.
> He was a man experienced in love,
> Clear-sighted and awake in its affairs.
> He faced its hardships and its easiness,
> Embracing the gazelles of either sex,
> With age no barrier to the love he felt.

The dervish came up to the boy and gave him a twig of sweet basil. 'Abd al-Rahman put his hand in his pocket and passed him a handful of coins, saying: 'Take your share, dervish, and go on your way.' The man took the money and sat down on the shop bench in front of the boy, looking at him, weeping and sighing constantly. His tears flowed like springs of water and people looked critically at him, some of them saying: 'All dervishes are corrupt,' while others said: 'This man's heart is consumed by love for the boy.' When 'Abd al-Rahman saw what was

---

* The festival marking the end of Ramadan.

happening, he rose and told his son: 'Get up and let us close the shop and go home; there is no need for us to do any trading today. May God repay your mother for what she has done, as it is she who is responsible for all this.' He then said to the dervish: 'Stand up so that I may lock the shop.' The dervish did as he was told and 'Abd al-Rahman closed up the shop and walked away with his son. A crowd of people, including the dervish, followed them until they got home, and when the boy had gone in, his father turned to the dervish and asked what he wanted and why he was weeping. The man said: 'Sir, I would like to be your guest tonight, for all guests are guests of Almighty God.' 'Abd al-Rahman welcomed him and invited him in . . .

Night 965

Morning now dawned and Shahrazad broke off from what she had been allowed to say. Then, when it was the nine hundred and sixty-fifth night, SHE CONTINUED:

I have heard, O fortunate king, that the dervish said to Qamar al-Zaman's father, the merchant: 'I am the guest of God.' The merchant then welcomed him and invited him in, saying to himself: 'This dervish is in love with my son and wants to debauch him. I shall have to kill him tonight and bury him in secret, but if he intends no evil, he can have his guest provisions.'

He brought both the dervish and Qamar al-Zaman into a room and told Qamar al-Zaman privately: 'Sit beside this man and flirt with him after I have left you. If he tries to debauch you, I shall be watching from the window that overlooks the room and I'll come down and kill him.' When the boy was left alone with the dervish, he took a seat beside him and the man looked at him, sighing and weeping. When the boy spoke to him, he would give a friendly answer but would tremble and turn towards him with sighs and tears. When their supper was brought in, he ate with his eyes still fixed on the boy and his tears still flowing. By the end of the first quarter of the night, they had finished talking and time had come for sleep. 'Abd al-Rahman now said: 'Look after your uncle, the dervish, my boy, and do whatever he says.' He was then about to leave when the dervish said: 'Sir, take your son with you or sleep here with us.' 'Abd al-Rahman refused and said: 'My son will sleep in this room with you, and if you happen to want anything, he will see to it and be at your service.' He then went out and left them alone, going to the window overlooking the room.

So much for him, but as for Qamar al-Zaman, he went up to the dervish and teased him by flaunting himself before him. 'What are you saying, my boy?' the dervish asked angrily. 'I take refuge in God from Satan the damned. Oh my God, this is a sin which disgusts You. Keep away from me, boy.' He got up and sat down at a distance, but Qamar al-Zaman came after him, threw himself on him and said: 'Why do you deny yourself the delights of union, dervish? I love you with all my heart.' The dervish became even angrier and said: 'If you don't stay away from me, I shall call your father and tell him about you.' 'My father knows what I'm like,' Qamar al-Zaman told him, 'and he cannot stop me. Do what I want; why are you avoiding me? Don't I please you?' The dervish said: 'By God, my son, I wouldn't do that even if I were to be hacked to pieces with sharp swords.' Then he recited:

My heart is in love with the lovely ones,
Male and female alike, and I am not slack.
Morning and evening I spend watching them,
But I am neither sodomite nor adulterer.

He then burst into tears and said: 'Get up and open the door so that I can go off on my way, for I am not going to stay here to sleep.' He rose to his feet, but the boy clung to him, saying: 'Look at my bright face, my rosy cheeks, the tenderness of my body and my soft lips.' He showed the dervish a thigh that would put to shame both wine and cupbearer, turning on him a gaze that would overcome an enchanter's magic. In his beauty and soft coquetry, he was as a poet has described:

I have not forgotten him since he rose to show
A leg that was like a gleaming pearl.
No wonder my emotions were aroused.
The Resurrection comes when legs are bared.*

He then bared his breast and said: 'Look at my nipples which are lovelier than a maiden's breasts, while my saliva is sweeter than sugar from the sugar-cane. Forget about piety and asceticism; give up abstinence and devotion; take the opportunity to have union with me; enjoy my beauty and have no fear, for you are safe from harm. Give up this evil habit of dull stupidity.'

* c.f. Quran 68.42.

He started to show the dervish his hidden charms, trying to bend him to his will as he twisted his body, but all the while the dervish kept his face turned away, saying: 'I take refuge in God. Be ashamed of yourself, boy; this is a sin which I am not going to commit, even in a dream.' When the boy pressed him, he slipped away from him and, turning in the direction of Mecca, he started to pray. Seeing this, the boy left him until he had performed two *rak'as* and pronounced the *salam*, but just as he was about to go up to him again, the dervish embarked on another prayer, again performing two *rak'as*. He kept on doing this a third, fourth and fifth time until the boy said: 'What is all this praying? Are you thinking of being carried up on a cloud? You're wasting this opportunity by passing the whole night in the prayer niche.' He then threw himself on the man and started kissing him between his eyes. 'Shame the devil, boy,' exclaimed the dervish, 'and obey the Merciful God!' 'If you don't do what I want,' the boy told him, 'I'll call out to my father and tell him that you wanted to lie with me, and then he will come and beat you until he has broken your bones and stripped them of flesh.'

While all this was going on, 'Abd al-Rahman was watching and listening until he was sure that the dervish meant no harm. 'Had he been a libertine, he would never have been able to resist such pressure,' he told himself. For his part, the boy continued with his provocation and every time the dervish tried to pray, he interrupted until the man became very angry indeed and started to beat him, making him burst into tears. At that point, his father came in and, after wiping away his tears and consoling him, he said to the dervish: 'Brother, as you are so ascetic a man, why did you carry on weeping and sighing at the sight of my son? Was there some reason for that?' 'Yes, there was,' the dervish confirmed, and 'Abd al-Rahman continued: 'When I saw that, I was suspicious of you and I told my son to do what he did in order to test you, for if I had seen you trying to lie with him, I would have come in and killed you. Now, after watching what you did, I realize that you are a very pious man, but please tell me why it was that you shed those tears.' The dervish sighed and said: 'Sir, don't reopen a healed wound,' but 'Abd al-Rahman insisted.

'You have to know,' the man then said, 'that I am a wandering dervish, travelling through the lands in order to learn from the works of God, Who created night and day. One Friday morning, I happened to enter the city of Basra . . .'

Night 966

Morning now dawned and Shahrazad broke off from what she had been allowed to say. Then, when it was the nine hundred and sixty-sixth night, SHE CONTINUED:

I have heard, O fortunate king, that THE DERVISH SAID TO THE MERCHANT:

Know that I am a wandering dervish. I happened to enter the city of Basra one Friday morning, and there I found the shops open and filled with goods of all kinds, together with food and drink, but there was nobody in them, neither man, woman, boy or girl. There were not even any dogs or cats in the streets and the markets; there were no sounds and no people. I said to myself in my astonishment: 'Where can they have taken their dogs and cats, and what has God done with them all?' As I was hungry I took some warm bread from a baker's oven, and then went into the shop of an oil seller, where I spread the bread with butter and honey and ate it. Next, I drank what I wanted from a shop that sold drinks and after that I caught sight of an open coffee house. I went in and discovered pots full of coffee standing on the fire, although there was no one there. I drank as much as I wanted, saying to myself: 'This is very strange. It is as though all the inhabitants had just died or fled away in fear of some disaster that allowed them no time even to close their shops.'

While I was thinking the matter over, I was alarmed to hear the sound of a drum being beaten, and I hid myself away. Then, peering through a crack, I saw girls, lovely as moons, walking in pairs through the market with uncovered heads and unveiled faces. There were eighty of them, in forty pairs, and with them was a young girl mounted on a horse that could scarcely move because of the weight of gold, silver and jewels that both it and its rider were wearing. The rider herself was bare-faced and bare-headed; she wore the most splendid jewellery and the most magnificent clothes; there was a jewelled collar round her neck and hanging over her breast were necklaces of gold; she had bracelets that shone like stars on her wrists; while on her feet she wore golden anklets set with precious stones. She was surrounded on all sides by her maids, while in front of her was a girl with a huge emerald-hilted sword, suspended by fastenings of gold studded with precious stones.

When the rider reached a spot opposite me, she reined in her horse and said: 'Girls, I heard a sound from inside this shop. Search it, for

there may be someone hidden there trying to look at us while we are
unveiled.' The girls made a search of the shop that was in front of the
coffee house where I was hiding, cowering in fear, and then I saw them
coming out with a man. 'Lady,' they said, 'we found a man there, and
here he is before you.' At that, the rider said to the girl with the sword:
'Cut off his head,' and when this had been done, they left the corpse
sprawled on the ground and passed on. I was terrified by what I had
seen, but my heart was filled with love for the lady.

An hour later, the inhabitants appeared and all the shopkeepers went
back to their shops, while others made their way through the markets.
As they gathered around to inspect the corpse, I came out of my hiding
place. No one paid any attention to me and when I tried to make cautious
enquiries about the lady with whom I had fallen in love, no one would
tell me anything about her. So I left Basra filled with lovesickness, and
when I came across your son, I saw that he bore an astonishing likeness
to the lady. Because he reminded me of her, he stirred up in me the fires
of passionate love and this was why I wept.

Then, in floods of tears, the dervish said: 'For God's sake, sir, open the
door for me and let me go on my way.' 'Abd al-Rahman opened the door
and the man left.

So much for him, but as for Qamar al-Zaman, what he had heard the
dervish say had filled him with love and longing for the girl, and so next
morning he said to his father: 'All merchants' sons travel to get what
they want and in each case their fathers provide them with trade goods
that they can take with them in order to make a profit. Why haven't you
done this for me, letting me go off with trade goods to try my luck?'
'Abd al-Rahman said: 'My son, it is only poor merchants whose sons
travel in the hope of getting a profit and making their fortunes. I am
wealthy and I want no more, so why should I send you away? I cannot
bear to be parted from you for a single hour, especially as your unequalled loveliness and perfection give me such cause for fear.' But
Qamar al-Zaman said: 'You will have to give me merchandise to take
with me, father, or else, before you know it, I shall have run away – if
necessary, without money or goods. So if you want to keep me happy,
give me trade goods so that I may make my way to other lands.'

When his father saw that the boy had set his heart on going, he went
to his wife and told her: 'Your son wants me to provide him with goods
to take abroad with him, in spite of the fact that this kind of travel is a

wearisome business.' 'Why should you dislike the idea?' his wife asked, pointing out that all young merchants were in the habit of priding themselves on the journeys they had made and the profits they had won. 'Most merchants are poor,' he objected, 'and are looking for wealth, whereas I am a rich man.' 'It does no harm to get more of a good thing,' she pointed out, adding: 'And if you don't supply him with merchandise yourself, I shall use my own money to equip him.' 'Abd al-Rahman repeated his fear that foreign travel would lead the boy into trouble, but she countered that there was nothing wrong with it if it brought in a profit. 'Otherwise,' she said, 'he will go off and we shan't be able to find him when we look for him, which will bring us public disgrace.' Her husband accepted her advice and got ready ninety thousand dinars' worth of merchandise for his son, while his wife provided him with a purse containing forty precious gemstones, each worth at least five hundred dinars. 'Keep these, my son,' she told him, 'as they may come in handy for you.' Qamar al-Zaman took all this and set off for Basra . . .

Night 967

Morning now dawned and Shahrazad broke off from what she had been allowed to say. Then, when it was the nine hundred and sixty-seventh night, SHE CONTINUED:

I have heard, O fortunate king, that Qamar al-Zaman took all this and set off for Basra with the jewels in a belt that he fastened around his waist. He travelled on until, when he was within a single stage of Basra, he was attacked by Bedouin, who stripped him and killed his men and his servants. He smeared his face with blood and lay among the corpses so that the attackers, thinking that he was dead, left without approaching him. When they had gone off, taking all his goods, he got up from where he was lying and walked as far as Basra with the gemstones in his belt, these being the only things that he still had. As it happened, it was on a Friday that he entered the city and he found the place deserted, just as the dervish had described. Although the shops were still open and filled with goods, there was no one in the markets, and so he ate and drank and began to look around. As he was doing this, he heard the sound of a drum and hid himself in a shop, from which he watched the girls as they arrived. At the sight of the young rider, he fell so deeply in love that the force of his passion left him unable to stand.

After a time, the inhabitants appeared and filled the markets, at which Qamar al-Zaman made his way to a jeweller and sold him one of his forty gems for a thousand dinars. He then went back to where he had come from in order to pass the night, and in the morning, after changing out of what he had been wearing, he visited the baths and emerged looking like a full moon. He sold four more gems for a total of four thousand dinars and then, dressed in the most splendid clothes, he went to look around the city streets. In one of the markets he found a barber's shop, which he entered to have his head shaved. After striking up a friendship with the barber, he said: 'Father, I am a stranger here. When I came into the city yesterday, I found it deserted with no one here, either human or *jinn*. Then I caught sight of a procession of girls, among whom a young lady was riding,' and he went on to describe what he had seen. 'My son,' the barber said, 'have you told anyone else about this?' When Qamar al-Zaman said no, the barber warned him: 'Then take care not to talk about it in front of anyone besides me, for not everyone can hold their tongue and keep a secret. You are a young boy, and I'm afraid that word might spread, from mouth to mouth, until it reached the people involved in this business, and they would kill you. What you saw is something that has not been seen or heard of in any city but this, and the people here are close to death because of the distress it brings them. Every Friday morning, they have to shut away the dogs and cats to stop them from straying through the markets, while they themselves must all go inside the mosques and shut the doors. No one can walk through a market or look out of a window, but no one knows why. Tonight, however, I shall ask my wife what the reason is because, as a midwife, she enters the great houses and knows what goes on there. God willing, if you come to me tomorrow, I shall be able to pass on what she tells me.'

Qamar al-Zaman took out a handful of gold and told the barber to take it to his wife, saying: 'She has become a mother to me.' He then produced another handful and said: 'Take this for yourself.' 'My son,' said the barber, 'sit where you are while I go and ask my wife, and then I shall be able to tell you the true story.' He left Qamar al-Zaman in his shop and went to tell his wife about him, saying: 'I want you to tell me what is really happening in this city, so that I may let this young merchant know about it, for he is passionately concerned about why it is that the citizens and their animals have to stay away from the markets on Friday mornings. I think that he must be in love, and as he is a liberal and generous man, if we tell him about this, he will give us a great reward.' His wife said: 'Go and fetch him. Ask him to come to your wife, who is a mother to him, for she will greet him and tell him what he wants to know.'

The barber went back to his shop, where he found Qamar al-Zaman sitting waiting for him. He gave him his news and, after passing on his wife's message, he took him to her. She welcomed him and asked him to sit, after which he produced a hundred dinars and gave them to her, saying: 'Tell me, who is this girl?' She said: 'My son, you should know that the sultan of Basra was given a jewel by the king of India, which he wished to have pierced. He summoned all the jewellers and told them what he wanted done, saying that whoever did the job successfully could have whatever he wished for, but if he split the jewel, his head would be cut off. They were all afraid and they said: "King of the age, gems split easily and there are few who can bore a hole through them successfully, as in most cases they will crack. Don't make us attempt something that is not possible for us. We cannot pierce this jewel, but our master knows more about the matter than we do." The sultan asked who this was and they told him that he was Master 'Ubaid. "He is the expert in the field," they said, "a very wealthy man and a skilled craftsman. Summon him and tell him what you want done." The sultan did this, telling him his conditions for the task, and 'Ubaid took the jewel and pierced a hole through it just as the king wanted. The king then told him to make a wish but he asked for a delay until the following day, because he wanted to consult his wife. It was his wife whom you saw in the procession, and 'Ubaid is deeply in love with her. Because of this he will do nothing without consulting her and this was why he asked for a delay. When he approached her, he told her that the sultan had promised to grant him a wish because he had pierced a hole in his jewel and that he had asked him to wait until he had consulted her.

' "What do you want me to wish for?" he asked, and she replied: "We have so much money that no fire could burn it all, so ask the sultan to have it proclaimed in the streets of Basra that the people are to go to the mosques two hours before the Friday prayer, and that everyone, old and young alike, must stay either in a mosque or at home with the doors closed. I shall then ride through the city with my slave girls; no one is to look at me from a window or through a lattice, and anyone I come across I shall kill." Her husband went to the sultan and told him of this wish, which he granted. When the proclamation was made . . .'

Night 968

Morning now dawned and Shahrazad broke off from what she had been allowed to say. Then, when it was the nine hundred and sixty-eighth night, SHE CONTINUED:

I have heard, O fortunate king, that the barber's wife said: 'The king granted the jeweller his wish and when the proclamation was made, the people of Basra said that they were afraid lest the cats and dogs harm their goods, and as a result the sultan gave orders that these must be kept inside until the people returned from Friday prayer. Two hours before that, 'Ubaid's wife would ride out in a procession with her maids through the city streets, and no one was to pass through the market or to look from a window or through a lattice. This is the reason behind the custom and I have now told you about the lady. But do you only want to know about her or do you want to meet her?'

When he told her that he wanted to meet her, she asked: 'What do you have with you in the way of costly treasures?' 'Mother,' he said, 'I have four sorts of gems. Of the first class, each is worth five hundred dinars; each of the second is worth seven hundred; each of the third is worth eight hundred; while in the case of the fourth, each is worth a thousand.' 'Are you willing to sacrifice four of them?' she asked, and when he told her that he was ready to give them all, she said: 'Go at once and fetch one worth five hundred dinars and ask for the shop of 'Ubaid, the master of the jewellers. When you come to him, you will find him sitting in his shop splendidly dressed, with his craftsmen standing in front of him. Greet him; take your seat in his shop and then produce your gem. Tell him to set it in a gold ring which is not to be big but is to weigh exactly one *mithqal* and no more, and is to be of good workmanship. Then hand him twenty dinars and give one dinar to each of the workmen. Sit and talk to him for a while, and if any beggar approaches you, display your generosity by giving him a dinar so that the jeweller may grow fond of you. Then get up and go back to spend the night at home and next morning bring a hundred dinars for your father, the poor barber.' 'I shall do that,' said Qamar al-Zaman.

He then left the barber's wife and went to his lodgings to fetch a gem worth five hundred dinars, after which he made his way to the jewellers' market and asked for the shop of Master 'Ubaid. He was directed to it and when he got there he saw 'Ubaid, a dignified-looking man, splendidly dressed, sitting behind four craftsmen. The two of them exchanged greetings, after which 'Ubaid welcomed him and invited him to sit down. When he had done this, he produced his gem and told the jeweller that he wanted him to set it for him in a well-fashioned gold ring weighing exactly one *mithqal* and no more. He then presented him with twenty dinars and said: 'This is to pay for cutting the gem, and over and above that I shall pay later for the ring itself.' Having secured 'Ubaid's goodwill, he handed a dinar to each of the craftsmen, so winning theirs. While he sat talking with 'Ubaid, he gave a dinar to every beggar who approached him, leaving them marvelling at his generosity.

The equipment that 'Ubaid had in his shop was duplicated in his home, and when he had any unusual task, he was in the habit of working on it at home, to keep his craftsmen from learning how to do it. His young wife used to sit in front of him and, looking at her as she sat there, he would produce all kinds of remarkable pieces, fit only for kings. As he sat at home fashioning the gem for Qamar al-Zaman's ring with astonishing artistry, she saw it and asked: 'What are you going to do with this gem?' He said: 'I want to set it in a gold ring, for it is worth five hundred dinars.' 'Whose is it?' she asked and he told her: 'It belongs to a handsome young merchant with eyes that wound, cheeks like red anemones, from which fire is struck, and a mouth like the seal of Solomon. He has lips that are red as coral and a neck like that of a gazelle. His complexion is pink and white and he is elegant, graceful and generous.' He told her what Qamar al-Zaman had done and started to describe his beauty and grace before turning to his generosity and his perfection. He kept on talking about these qualities until, thanks to him, his wife had fallen in love with his description – and there is no greater pimp than a husband who describes a man to his wife as both handsome and generous with his money. Flooded with passion, she asked: 'Does he have any of my charms?' to which her husband answered: 'He has them all and he can be described in the same terms as you. He must be about your age and if I were not afraid of offending you, I would say that he was a thousand times more beautiful than you.'

His wife stayed silent, but the fire of love blazed in her heart. Her husband meanwhile kept on enumerating the charms of Qamar al-Zaman as he talked to her, until he had finished fashioning the ring. He handed it to her and when she put it on, it exactly fitted her finger. She told him that she had fallen in love with the ring and wanted to have it for herself. 'I shan't take it off,' she said but he replied: 'Have patience.

Its owner is a generous man and I shall ask him to sell it to me. If he does, I'll bring it to you, or if he has another gem, I'll buy that for you and make it into a ring like this one.'

Night 969

Morning now dawned and Shahrazad broke off from what she had been allowed to say. Then, when it was the nine hundred and sixty-ninth night, SHE CONTINUED:

I have heard, O fortunate king, that the jeweller said to his wife: 'Have patience. Its owner is a generous man and I shall ask him to sell it to me. If he does, I'll bring it to you, or if he has another gem, I'll buy that for you and make it into a ring like this one.'

So much for 'Ubaid and his wife, but as for Qamar al-Zaman, he spent the night in his lodgings and the next morning he took a hundred dinars and went to the barber's elderly wife. He offered her the money, but she told him to give it to her husband, which he did. Then she asked him whether he had done what she had told him, and when he said that he had, she instructed him to go back to the jeweller. 'When he gives you the ring,' she went on, 'put it on the tip of your finger and then take it off quickly and tell him that he has made a mistake, for it is too tight. He will say: "Do you want me to break it up and make it wider?" Say: "There is no need to remake it. Take it and give it to one of your slave girls." Then bring out another gem for him, worth seven hundred dinars, and tell him to take this one and set it in a ring for you, as it is better than the other. This time give him thirty dinars, and hand the craftsmen two dinars each, explaining that this is for the gem-cutting and that you will pay later for the ring. After that, go back to spend the night in your lodgings, and when you return the next morning, bring two hundred dinars and I shall then put the last touch to my scheme.'

So Qamar al-Zaman went off to the jeweller, who welcomed him and sat him down in his shop. As he was sitting there, he asked: 'Have you finished the job?' 'Yes,' said 'Ubaid, and he produced the ring. Qamar al-Zaman took it and put it on his finger tip, but then took it off quickly and said: 'You have made a mistake, master.' He threw it back to the man, saying: 'It is too narrow for me.' 'Shall I make it larger?' 'Ubaid asked, but Qamar al-Zaman said: 'No; keep it as a present and give it to one of your slave girls, for it is worth a mere five hundred dinars and there is no need to refashion it.' He then produced another gem, worth seven hundred, and told him to set it, handing him thirty dinars and giving each of his craftsmen two. 'Ubaid said that he would not ask for the money until the work had been done, but Qamar al-Zaman explained that this was for the gem-cutting and that he would pay the fee for the ring later. He then went off, leaving the jeweller and his assistants astonished by his generosity.

'Ubaid returned to tell his wife that he had never seen anyone more generous than this young man, adding: 'And you are in luck, as he gave me the ring for free, telling me to hand it to one of my slave girls.' He told her the whole story and said: 'I don't think that he can really be the son of a merchant. He is more likely to be a prince or the son of a sultan.' The more he praised Qamar al-Zaman the more his wife's passion for him increased. She put on the ring and her husband fashioned another that was a little bigger. When he had finished, she slipped it on over the first one. 'See how well the two of them look on my finger,' she said, adding, 'and I wish they were both mine.' 'Have patience,' he told her, 'and I may buy you the second one.'

The next morning, he took the ring and set off for his shop while, for his part, Qamar al-Zaman went to the barber's wife and gave her two hundred dinars. 'Go to the jeweller,' she told him, 'and when he hands you the ring, put it on your finger and then take it off quickly and say: "You have got it wrong, master. This one is too big, and when a master craftsman is approached by someone like me with a job to be done, he should take measurements. Had you measured my finger you would not have made a mistake." Then take out another gem, worth a thousand dinars, and tell him to take it and set it in a ring, and to pass the other one to one of his slave girls. Give him forty dinars and hand each of his craftsmen four dinars, repeating that this is for cutting the gem and that you will pay his fee later. Come here after that and hand my husband three hundred dinars to help him with his daily expenses, for he is a poor man.' 'To hear is to obey,' said Qamar al-Zaman.

He then set off to see 'Ubaid, who welcomed him and gave him a seat. He produced the ring, which Qamar al-Zaman put on his finger and then took off quickly, telling him that a craftsman, asked to make something for a customer like him, should take measurements and had he done so he would not have gone wrong. 'Take the ring,' he said, 'and give it to one of your slave girls.' Then he produced a gem worth a thousand dinars and told him to set it in a ring that would fit his finger.

'Ubaid agreed that he was right and measured his finger, after which Qamar al-Zaman brought out forty dinars and handed them to him for the gem-cutting, promising to pay the fee later. 'You have already paid more than enough by way of a fee,' objected 'Ubaid, 'and you have been exceedingly generous.' 'Never mind about that,' Qamar al-Zaman told him, and while they sat talking he handed a dinar to each beggar who passed. He then left the jeweller and went on his way.

So much for him, but as for 'Ubaid, he went home and told his wife: 'What a generous man that young merchant is! I have never seen anyone more open-handed, more handsome or more sweetly spoken.' He continued to heap praises on Qamar al-Zaman and to speak of his charms and his generosity until his wife said: 'You insensitive man, you know these qualities of his and he has given you two valuable rings, so you should invite him here and entertain him in a friendly fashion. If you do this and he comes to our house, he may do you a great deal of good, but even if you don't want to entertain him, still invite him here and I shall do the entertaining at my own expense.' 'Are you saying this because you know me for a miser?' her husband asked. 'No, you aren't a miser,' she said, 'but you have no sensitivity. Invite him here tonight and don't come back without him. If he refuses, press him by swearing that, otherwise, you will have to divorce me.' 'Ubaid gave her his promise, and then, after he had made the ring, he went to sleep. On the third day, he went off to his shop in the morning and took his seat.

So much for him, but as for Qamar al-Zaman, he took three hundred dinars to the barber's wife and handed them to her husband as before. She said: 'The jeweller may invite you home today. If he does and you spend the night with him, let me know in the morning what happened and bring four hundred dinars with you to give to my husband.' 'To hear is to obey,' he said, for whenever he ran out of money he would sell some of his gems. He then set off to visit 'Ubaid, who rose to greet him, receiving him with open arms as a friend. He produced the ring for him and when Qamar al-Zaman found that it fitted his finger, he exclaimed: 'God bless you, master craftsman, this is fine workmanship, but the stone is not to my liking . . .'

Night 970

I have heard, O fortunate king, that Qamar al-Zaman told the jeweller: 'This is fine workmanship, but the stone is not to my liking because I have a better one. So do you take this and give it to one of your slave girls.' Then he produced another one and gave the man a hundred dinars, saying: 'Take this as your fee and don't hold it against me that I have put you to such trouble.' 'Ubaid replied: 'Merchant, you gave me the rings on which I worked, and a lot more besides. I have become so fond of you that I cannot bear to be parted from you, so please console me by passing the night as my guest.' Qamar al-Zaman agreed, but said that he would have to go back to his lodgings to give instructions to his servants and to tell them not to expect him as he would not be spending the night there. 'In which *khan* are you lodging?' asked 'Ubaid and, on being told, he said: 'I shall come to fetch you.' Qamar al-Zaman agreed to this and the jeweller went to the *khan* before sunset, as he was afraid that his wife would be angry if he went home without him.

When he then took Qamar al-Zaman home, his wife was enthralled when she saw him enter. He and her husband took their seats in a room of unparalleled splendour where they talked until supper time, and after they had eaten and drunk, they were served with coffee and sherbet. Their conversation went on until it was time for the evening prayer and when they had performed this, a maid brought them something to drink in two cups. When they drank this they were both overcome by sleep. It was then that 'Ubaid's wife came in and, finding them asleep, she gazed with astonishment at the beauty of Qamar al-Zaman's face, exclaiming: 'How can a lover of beautiful girls fall asleep?' She rolled him over on to his back and straddled his chest; then, overwrought with passion for him, she showered kisses on his cheeks leaving traces that enhanced their reddish bloom. She went on sucking his lips until she drew blood from his mouth but none of this quenched her passion or satisfied her thirst and she continued kissing and embracing him, wrapping her legs round his, until the east lightened with the approach of dawn. At that, she put four knucklebones in his pocket before going off and leaving him. She then sent in her maid with something that looked like snuff which she put into the nostrils of the two sleepers, at which they sneezed and woke up. 'Prayer is a duty, my masters,' the maid told them, 'so get up to perform the morning prayer,' and she fetched them a bowl and a jug of water.

'Master,' said Qamar al-Zaman, 'it is getting late and we have overslept.' 'This is a room that lends itself to sound sleep,' 'Ubaid told him, 'and whenever I come to sleep here, this is what happens.' Qamar al-Zaman agreed with him and then started to perform the ritual ablution, but when he put water on his face, his cheeks and lips smarted. 'Strange!' he exclaimed. 'If we slept so soundly in the heavy air of this room, why are my cheeks and my lips burning?' When he pointed this out, 'Ubaid said: 'I think that this must be because of mosquito bites.' 'Strange!' repeated Qamar al-Zaman. 'Has the same thing happened to you?' 'No,' acknowledged his host, 'but when I have a guest like you, the following morning he does complain of mosquito bites. I suppose that that is because young men like you have no beards, whereas if you had one like mine, the mosquitoes wouldn't harm you. It must be this that keeps them off, for they aren't fond of bearded men.' 'That must be right,' agreed Qamar al-Zaman. The maid then brought them breakfast, and when they had eaten they both left.

Qamar al-Zaman went off to the barber's wife, who when she caught sight of him said: 'I can see from your face that you were lucky, so tell me what you saw.' 'I didn't see anything,' he answered. 'The jeweller and I had our supper in a room in his house and after we had performed the evening prayer we fell asleep and didn't wake up until morning.' The woman laughed and said: 'What are those marks I see on your cheeks and your lips?' 'It was the mosquitoes in the room that did this,' he told her. 'That may be so,' she said, 'but was your host bitten too?' 'No,' admitted Qamar al-Zaman, 'but he told me that his mosquitoes only bite the beardless and don't attack bearded men. Every beardless guest that he has complains next morning of their bites, while nothing of the kind happens to those who have beards.' 'All right,' she said, 'but did you come across anything apart from that?' He told her that he had discovered four knucklebones in his pocket. 'Show them to me,' she said, and when he offered them to her she took them and laughed. 'It was your lady love who put these in your pocket,' she told him, and when he asked: 'How is that?' she said: 'She is telling you by this sign that, were you in love, you would not have fallen asleep, for lovers don't sleep. She is saying that you are still young; you are clearly only good for playing with these bones and so what has prompted you to fall in love with a beautiful girl? She must have come to you at night and, finding you asleep, covered your cheeks with love-bites, leaving you this token. But that won't be enough for her and she is bound to send her husband to invite you again tonight. When you go with him, don't be in a hurry to fall asleep; take five hundred dinars with you and then come back and tell me what happened so that I can finalize my plans for you.' 'To hear is to obey,' said Qamar al-Zaman before setting off for his *khan*.

So much for him, but as for 'Ubaid's wife, she asked her husband whether his guest had gone. He told her that he had, adding: 'I was ashamed because the mosquitoes attacked him last night and bit his cheeks and lips.' 'That's what they normally do in that room,' she agreed, 'for they only like beardless boys. But invite him again tonight.' So 'Ubaid went off to Qamar al-Zaman's *khan*, gave him the invitation and brought him back to the same room. The two of them ate and drank, before performing the evening prayer, and the maid then brought in a cup for each of them.

Night 971

Morning now dawned and Shahrazad broke off from what she had been allowed to say. Then, when it was the nine hundred and seventy-first night, SHE CONTINUED:

I have heard, O fortunate king, that the maid brought in a cup for each of them. They both drank and promptly fell asleep. 'Ubaid's wife came in and exclaimed: 'Clod, how is it that you are asleep when you claim to be a lover and lovers never sleep?' She mounted on his breast and did not stop kissing, biting, sucking and grappling with him until morning. Then she put a knife in his pocket and, when dawn had broken, she sent in her maid, who woke them up. Qamar al-Zaman's cheeks were red as fire and his lips were like red corals thanks to how they had been sucked and kissed. 'It looks as though the mosquitoes attacked you again,' 'Ubaid said, but Qamar al-Zaman disagreed, because whoever knows the real reason for something no longer complains about it. Then he discovered the knife in his pocket and stayed silent.

After he had had his breakfast and drunk his coffee, he left the jeweller's house and set off for his *khan*. He took five hundred dinars and went to tell the barber's wife what had happened. 'I fell asleep in spite of myself,' he said, 'and in the morning the only thing that I found in my pocket was a knife.' 'God protect you from her this coming night,' she said, 'for she is telling you that if you go to sleep again, she will cut your throat. You will be invited back there tonight and that is what will happen to you if you fall asleep.' 'What can I do?' he asked, and she said: 'Tell me what you had to eat and drink before you went to sleep.' He said: 'After the usual kind of supper, a maid came in with a cup for each of us and it was after I had drunk mine that I fell asleep and didn't wake until morning.' 'The mischief is in the cup,' she said, 'so take it from the girl but don't drink it until her master has drunk his and has fallen asleep. When she gives it to you, ask her for some water and when she has gone to fetch it in a jug, empty out the cup behind your pillow and pretend to be sleeping. Then when she brings in the jug, she will suppose that you drank what was in the cup and are asleep. She will leave you and then after a while you will see what happens next. Take care not to disobey me.' 'To hear is to obey,' said Qamar al-Zaman and he left for his *khan*.

So much for him, but as for 'Ubaid's wife, she pointed out to her husband that, according to the laws of hospitality, a guest had to be entertained for three nights. He went to give Qamar al-Zaman the invitation and took him back to the same room in his house as before. When they had eaten and performed the evening prayer, in came the maid with a cup for each of them. 'Ubaid drank and so fell asleep, while Qamar al-Zaman did not. 'Aren't you going to drink, master?' asked the girl, and he told her to fetch him a jug of water because he was thirsty. When she went off to do that, he poured out what was in the cup behind his pillow and then lay down. The girl came back and, on finding him supposedly asleep, she told her mistress that he had drunk up the cup and was lying down. For her part, 'Ubaid's wife said to herself: 'He is better dead than alive,' and, taking a sharp knife, she went up to him saying: 'You fool, this is the third time that you have paid no attention to the signs I left you, and so now I shall slit your belly.' As she approached him, knife in hand, he opened his eyes and got up laughing. 'You didn't work out what the last sign meant by your own wits,' she said, 'but some shrewd person guided you. So tell me where it was that you learned about it.' He replied: 'It was from an old woman,' and he told her the story of his dealings with the barber's wife. She said: 'When you leave tomorrow, go and ask her whether she has any more schemes besides this. If she says that she has, tell her to do her best to arrange for you to approach me openly, but if she says no and that this is the most she can do, then let her be. Tomorrow night, my husband will come to invite you again, and you can then come back with him and tell me and I shall know what next to do.'

Qamar al-Zaman agreed, and he spent the rest of the night hugging and embracing her, joining the preposition with what it governed by agreement, adding the connection to what was to be connected and leaving the husband as the elided ending of a word. This went on until morning and 'Ubaid's wife then said: 'One night is not enough for me, and neither is a day, or a month or a year. I want to stay with you for the rest of my life, but you will have to wait for me to play a trick on my husband good enough to baffle the intelligent and get us what we want. I shall make him suspicious of me so that he will divorce me, and I can then marry you and go with you to your own country. I shall see to it that you take with you all his wealth and his treasures, and for your sake I shall bring ruin on his houses and see that no trace of him remains. You must listen to what I say and obey me without contradiction.' 'To hear is to obey,' Qamar al-Zaman replied. 'There will be no contradiction.'

She then told him to go back to the *khan* and when her husband came to invite him, he was to say: 'Brother, a man can become tiresome, and if someone pays too many visits, he disgusts both the generous and the miserly. How can I go to you every evening and sleep in the same room as you? Even if you don't become annoyed with me, your womenfolk may, because I am keeping you away from them. If you want my friendship, get me a house next to yours and then you can sometimes spend the evening with me until it is time to go to bed, and at other times I can do the same with you. I shall then go back to my own house and you will be able to go to your women; this is better than keeping you away from them every night.' She went on: 'After that he will come to ask for my advice, and I will tell him to evict our neighbour, for the house in which he lives is our property and he only rents it. When you have moved in, God will make it easy for us to plan the rest. Off you go now and do what I have told you.' 'To hear is to obey,' replied Qamar al-Zaman, and she then left him and went away, while he pretended to be asleep.

After a time, the maid came and woke both him and the jeweller, who asked him whether he had been troubled by the mosquitoes. When he said no, his host said: 'You must be getting used to them.' They then had breakfast and drank coffee before going off about their affairs. Qamar al-Zaman went to the barber's wife and told her what had happened . . .

Night 972

Morning now dawned and Shahrazad broke off from what she had been allowed to say. Then, when it was the nine hundred and seventy-second night, SHE CONTINUED:

I have heard, O fortunate king, that Qamar al-Zaman went to the old woman and told her what had happened, explaining: 'She said this and that to me and I said that and this to her.' He then asked whether she herself could think of any better scheme to allow him to meet his beloved openly. 'My son,' she said, 'this is the end of my planning; I have run out of schemes,' at which he left her and went back to the *khan*. The following evening, when 'Ubaid came to invite him home, Qamar al-Zaman said: 'I can't go with you.' 'Why is that?' asked the jeweller, adding: 'I am so fond of you that I can't bear to be parted from you, so please, for God's sake, come with me.' Qamar al-Zaman said: 'If what you really want is a long association with me and that we should remain friends, get me a house next to yours. We can then take it in turns to spend our evenings together before going off to sleep each in our own house.' 'I do own a house next to my own,' 'Ubaid said, 'so come with me tonight and tomorrow you can have vacant possession of it.'

Qamar al-Zaman went with him, and after they had had supper and performed the evening prayer, the jeweller drank from the drugged cup and fell asleep, while the visitor's cup had not been doctored and Qamar al-Zaman drank it and stayed awake. 'Ubaid's wife then came and sat talking to him until morning, while her husband lay stretched out like a dead man. When he woke up in the normal way, he sent for his tenant and asked him to quit the house as he needed it himself. The man did as he was told and Qamar al-Zaman settled in, moving all his belongings there. 'Ubaid spent the evening with him and then went home.

The next day, his wife sent for a skilled builder and offered him money to construct an underground passage for her, leading from her own quarters to the house of Qamar al-Zaman, with a subterranean trapdoor. Then, before Qamar al-Zaman knew what was happening, in she came with two purses filled with money. 'Where have you come from?' he asked, and she showed him the tunnel before telling him to take the purses that were full of her husband's money. She then stayed, indulging herself in love-play with him until morning. 'Wait for me,' she told him, 'and when I have gone to rouse him and seen him off to his shop, I'll come back to you.'

While he sat waiting for her, she returned to her husband and woke him up. When he had risen and performed the ritual ablution and the prayer, he left for his shop and she then took four purses and went through the tunnel to Qamar al-Zaman. 'Take these,' she said, and she stayed with him until each of them went their separate ways, she to her house and he to the market. When he got back at sunset, he found that he now had ten purses, together with jewels and other treasures. 'Ubaid then came and took him back to his room, where they spent the evening together before the maid came in as usual with the two cups, which she passed to them. Her master fell asleep, but Qamar al-Zaman was unaffected, as his cup had not been drugged. 'Ubaid's wife came in to toy with him, while the maid started to transfer her master's goods to his house by way of the tunnel. This went on until morning, when the maid woke her master and poured coffee for the two of them, after which they each left.

On the third day, 'Ubaid's wife brought out a knife which belonged to her husband, who had made it himself. He valued it at five hundred dinars, as it had no equal in the beauty of its workmanship, and in spite of the number of would-be buyers, he kept it in a chest and could not bring himself to sell it to anyone. 'Take this,' she said. 'Put it in your belt and go and sit with my husband. Then bring it out and say: "Master, take a look at this knife which I have bought today, and tell me if I have made a good bargain or a bad one." He will recognize it, but he will be too embarrassed to say: "That is mine." If he asks you where you got it from and how much you paid, tell him: "I saw two Levantines quarrelling, and one said to the other: 'Where have you been?' 'With my mistress,' replied the other, and he went on: 'Every time I met her she used to hand me cash, but today she said that she had none to give and, instead, she told me to take this knife that belonged to her husband. I took it from her and I want to sell it.' I admired the knife and when I heard what the man said, I asked him if he would sell it to me. He agreed and I bought it from him for three hundred dinars. Was that cheap, do you think, or dear?" See what my husband says and after you have talked to him for a while, leave him and hurry back to me. You will find me sitting in the tunnel waiting for you and you must then give me back the knife.' 'To hear is to obey,' said Qamar al-Zaman.

He took the knife, put it in his belt and went to 'Ubaid's shop. He greeted the jeweller, who welcomed him and invited him to take a seat. Then, noticing the knife in his belt, 'Ubaid said to himself in astonishment: 'That's my knife. Who gave it to this merchant?' Then he thought again and said: 'Is it really my knife or one that looks like it?' At that point, Qamar al-Zaman took the knife from his belt and said: 'Master, take a look at this.' On taking the knife, he recognized it beyond all doubt, but was too embarrassed to claim it as his.

Night 973

Morning now dawned and Shahrazad broke off from what she had been allowed to say. Then, when it was the nine hundred and seventy-third night, SHE CONTINUED:

I have heard, O fortunate king, that when the jeweller took the knife from Qamar al-Zaman he recognized it but was too embarrassed to claim it as his. Instead, he said: 'Where did you buy it from?' after which Qamar al-Zaman repeated what he had been told to say. 'Ubaid told him: 'It was cheap at the price, as it's worth five hundred dinars.' There was fire in his heart and his hands could no longer carry on with their work. While Qamar al-Zaman went on talking, he was drowning in a sea of care and for every fifty words spoken to him he would reply with no more than one. His heart was tortured, his body shaken and his thoughts filled with gloom. He was as the poet has described:

I did not know what to say when they wanted to talk to me;
They spoke, but saw that my thoughts were somewhere else,
For, drowning in a bottomless sea of care,
I could not tell the difference between male and female.

On seeing that he had changed colour, Qamar al-Zaman said: 'You must be busy,' and, getting up, he hurried home, where he saw 'Ubaid's wife standing at the entrance to the tunnel waiting for him. 'Have you done what I told you?' she said when she saw him. He told her that he had, and she asked: 'What did he say to you?' He replied: 'He told me that it was cheap at the price, as it was worth five hundred dinars, but his colour changed and so I left him. What happened after that I don't

know.' 'Give me the knife and don't concern yourself about it,' she told
him, and she then took it, put it back in its place and sat down.

So much for her, but as for her husband, Qamar al-Zaman had left
him with his heart on fire and filled with suspicion. 'I must get up and
look for the knife in order to replace doubt with certainty,' he told
himself, and at that he went back to his wife, puffed up like a snake. She
asked what was wrong and he said: 'Where is my knife?' 'In the box,'
she told him, and then she beat her breast and said: 'My sorrow! Have
you quarrelled with someone and are you here to fetch the knife to use
against him?' 'Fetch it for me and let me see it,' he told her. 'If you swear
that you are not going to use it against anyone,' she said, and when he
had done that, she opened the box and brought it out for him. He started
to turn it over and over, saying: 'That's strange,' and then he told her to
take it and put it back in its place. She asked him what lay behind this,
and he said: 'I saw that our friend had one like it.' He went on to tell
her the whole story, adding: 'When I saw it in its box, this dispelled my
suspicions.' 'You thought ill of me, did you?' she asked. 'You took me
for the mistress of the Levantine, who had given him the knife?' He
agreed, but added that when he had seen the knife, his suspicions had
been removed. 'There is no good left in you,' she told him, but he went
on making excuses until he had won her over, after which he went back
to his shop.

On the following day, his wife gave Qamar al-Zaman her husband's
watch, which he had made with his own hand and which had no equal,
telling him to return to his shop and sit with him. She said: 'Tell him: "I
came across the fellow whom I saw yesterday, this time with a watch in
his hand. He asked me if I wanted to buy it, and I asked him where he
had got it from. He told me that he had been with his mistress and it
was she who had given it to him. I paid him eighty-five dinars for it, so
look and tell me whether it was cheap at the price or too expensive."
Then see what he has to say, and when you leave, hurry back and hand
it to me.' Qamar al-Zaman went off and did what he was told and, after
looking at the watch, 'Ubaid, who was filled with renewed suspicion,
told him that it was worth seven hundred dinars. Qamar al-Zaman then
left and returned to the lady, handing over the watch before her husband
arrived, swollen with rage. 'Where's my watch?' he demanded. 'It's here,'
she said, and when he told her to produce it, she fetched it for him.
'There is no might and no power except with God, the Exalted, the
Omnipotent!' he exclaimed, at which she said: 'You obviously have

something to say, so tell me what it is.' He replied: 'What can I say? I am bewildered by what has happened.' Then he recited:

> I am at a loss but the Merciful God is doubtless here with me;
> Sorrows surround me, but I do not know their source.
> I shall show patience, until patience itself finds out
> That what I suffer is bitterer than aloes juice,
> Or rather that is nothing to the patience which I show.
> What I endure is hotter than burning coals.
> What I do here is not of my own wish,
> But God's command orders me to endure.

He went on: 'Woman, first of all I saw our friend the merchant with my knife. I recognized it because the way in which it was fashioned was an invention of my own and there is not another one like it. What he told me filled me with grief, but I then came back and saw that it was still here. The next thing was that I saw him with my watch, which again was fashioned after a pattern of my own, and there is nothing like it in all Basra. He told me another disturbing story, which has left me bemused, and I don't know what has happened to me.' She said: 'What you mean is that I am the friend and mistress of that merchant and gave him your things, and you came to question me because you wanted to test whether I was playing you false. If you had not seen the knife and the watch here with me, you would have been sure that I had betrayed you. As you have thought so ill of me, I shall never eat or drink water with you again, for I loathe you so much that you are now prohibited to me.' 'Ubaid started to try to win back her favour, and when he had done this, he went off to his shop, full of regret at what he had said to her. He sat there . . .

Night 974

Morning now dawned and Shahrazad broke off from what she had been allowed to say. Then, when it was the nine hundred and seventy-fourth night, SHE CONTINUED:

I have heard, O fortunate king, that when 'Ubaid left his wife, he started to regret what he had said to her. He then went to his shop and sat there in a state of great agitation and extreme concern, half believing and half disbelieving what she had said, and in the evening he went home alone without Qamar al-Zaman. 'What have you done with the merchant?' asked his wife, and when he said that he was in his own house, she said: 'Has your friendship cooled?' 'By God,' he told her, 'what has happened has made me dislike him.' But she insisted that he went to fetch him as a favour to her. So he got up and went to Qamar al-Zaman's house, where he recognized his own belongings scattered around the place. Fire was rekindled in his heart and he began to sigh, leading Qamar al-Zaman to ask him why he was so pensive. For his part, 'Ubaid was too embarrassed to tell him that these things were his property and to ask who had brought them to him. All that he said was: 'I have been disturbed, so come home with me where we can divert ourselves.' 'Leave me where I am,' said Qamar al-Zaman, 'for I am not going to go with you,' but the jeweller swore that he must, and eventually took him off. They ate their supper and spent the evening together, with Qamar al-Zaman talking and his host sunk in a sea of care, speaking only one word to the other's hundred.

When the maid had brought in the two cups as usual and when they had both drunk, the jeweller lay down to sleep, unlike Qamar al-Zaman, whose drink had not been drugged. 'Ubaid's wife came in and said to Qamar al-Zaman: 'What do you think of this drunken cuckold who is lying there unconscious with no knowledge of women's wiles? I have to trick him into divorcing me. Tomorrow I shall dress up as a slave girl and walk behind you to the shop. You are to tell him: "Master, today I went to the Yasiriya *khan* where I saw this slave girl. I bought her for a thousand dinars and I would like you to take a look at her and tell me whether she was cheap at the price or not.' Uncover my face and breasts and let him look at me, before you take me back to your house. I shall go home through the tunnel to see how we can bring the matter to a finish.' They then spent a night of unmixed pleasure in each other's company, relaxing and enjoying their love-play until it was morning, when she went to her own room. She sent her maid to wake the two men, who got up and performed the morning prayer. Then, when they had breakfasted and drunk their coffee, 'Ubaid left for his shop and Qamar al-Zaman went home.

At that point, the lady emerged from the tunnel dressed as a slave girl, which, in fact, she was by birth. As Qamar al-Zaman set out for the shop, she walked behind him, and they carried on in this way until they got there. Having greeted the jeweller and taken his seat, Qamar al-Zaman said: 'Master, today I went into the Yasiriya *khan* to look around and, as I liked this girl whom I saw with the auctioneer, I bought her for a thousand dinars. I'd like you to inspect her to see whether she was cheap at the price or not.' He then unveiled her and 'Ubaid saw that here was his wife, dressed in all her finery, with her most splendid ornaments, painted with kohl and dye, just as she used to present herself to him at home. He recognized her beyond a doubt through her face, her clothes and her jewellery, which he had fashioned himself, while on her fingers she was wearing the rings that he had recently made for Qamar al-Zaman. Everything led him to be certain that this was his wife, and when he asked her her name, she said 'Halima', which was the name of his wife. In his amazement, he asked Qamar al-Zaman how much he had paid for her and on being told 'a thousand dinars', he exclaimed: 'Then you got her for nothing, as a thousand dinars is less than the price of her rings, with her clothes and jewellery costing you nothing!' 'May God bring good news to you as well,' said Qamar al-Zaman, 'and as you admire her, I shall take her back home.' 'Do as you wish,' replied the jeweller. So Qamar al-Zaman went back with Halima, who passed through the tunnel and took her seat in her own room.

So much for her, but as for her husband, with fire burning in his heart he said to himself: 'I shall go to look for my wife. If she is at home, then this is only a matter of resemblance – glory be to Him, Who has no match! – but if not then this "slave girl" must be her without any doubt.' He got up and ran back home, only to find his wife sitting there wearing the same clothes and ornaments that he had seen on her in his shop. Clapping his hands together, he exclaimed: 'There is no might and no power except with God, the Exalted, the Omnipotent!' Halima said: 'Man, have you gone mad or what has happened? This isn't your normal behaviour; something must have befallen you.' 'If you want me to tell you, you mustn't feel hurt,' he answered, and when she told him to go on, he said: 'Our friend, the merchant, has bought a slave girl whose figure is just like yours. She is exactly the same height as you; her name is the same; not only is she like you in every single respect, but she wears rings like yours on her fingers, and her jewellery is the same as yours. When he showed her to me, I thought that she was you, and I am now completely bewildered. I wish that we had never set eyes on that merchant or taken him as a friend. I wish that he had never left his own country and that we had never got to know him, as he has made a misery of my life, which used to be so happy. Where there was once good faith, now there is estrangement, as I have begun to have doubts.' 'Look me in the face,' she said. 'Maybe it was I who was with him and he was my lover. I may have dressed up as a slave girl and concocted a plan for him to show me to you in order to trick you.' 'What are you saying?' he exclaimed. 'I would never suspect you of doing anything of the kind.'

'Ubaid did not know about the wiles of women and how they treat men, and he had never heard the lines:

> You were carried away by a heart that delighted in lovely girls,
> Soon after your youth had passed, when grey hairs were at hand.
> I am distressed by Laila, distant as she is;
> Impediments and hardships come between us.
> If you ask me about women, I am the expert
> On the diseases that they cause, and the doctor.
> When a man's hair is grey and he has little wealth,
> There is no share at all for him in their love.

Nor the lines:

> Disobey women, for that is how best to follow God;
> No man will prosper who allows women to lead him.
> They would keep him from perfecting his good qualities
> Were he to spend a thousand years in search of knowledge.

Nor these:

> Women have been created for us as devils;
> God is my refuge from all devilish wiles.
> Those who have been afflicted by their love
> Waste all their efforts in this world and the next.

Halima said to him: 'I shall stay sitting here. Do you go this moment to knock on his door and see that you get in quickly. If when you are there you see the slave girl, then she will be a girl who looks like me, and glory be to God, Who has no match. But if you don't, then I must be the one whom you saw with him and your suspicions will be confirmed.' 'That's right,' he said and he left her and went out. She for her part passed through the tunnel to sit with Qamar al-Zaman, whom she told what had happened. 'Open the door at once,' she said, 'and let him see me.' While they were speaking, a knock came on the door and when Qamar al-Zaman asked: 'Who is there?' 'Ubaid answered: 'It is I, your friend. I was delighted for you when you showed me your slave girl in the market, but this wasn't enough for me, so open the door and let me look at her again.' 'Certainly,' replied Qamar al-Zaman, and when he opened the door, the jeweller saw his wife sitting there. She got up and kissed his hand, as well as that of Qamar al-Zaman, and she talked to him for a time while he stared at her. He could find no difference at all between her and his wife, and he exclaimed: 'God creates what He wills!' He left for home, full of suspicion, only to discover his wife sitting there for, when he had opened the door, she had got there first through the tunnel.

Night 975

Morning now dawned and Shahrazad broke off from what she had been allowed to say. Then, when it was the nine hundred and seventy-fifth night, SHE CONTINUED:

I have heard, O fortunate king, that when 'Ubaid went out of the door, the girl got home first through the tunnel and sat down in her room. When he came in, she asked him what he had seen and he said: 'I saw the girl sitting with her master, and she looks just like you.' 'That is enough of suspicion,' she said. 'Go off to your shop and don't think ill of me again.' 'Yes indeed,' he replied, 'but don't blame me for my behaviour.' 'God forgive you,' she told him and, after kissing her on both cheeks, he went back to his shop.

Halima went through the tunnel to Qamar al-Zaman, taking with her four purses. 'Get ready to leave in a hurry,' she told him, 'and be prepared to carry off this money without delay, while I do what I have planned.' He went off and bought mules, prepared the baggage and got ready a litter. When he had bought mamluks and eunuchs and had taken everything out of the city so that there was nothing left to hinder him, he went to Halima and told her that everything was ready. She said: 'I for my part have brought the rest of his money and all his treasures to you, and I've left him nothing useful, great or small. All this is because of my love for you, heart's darling, and I would give my husband a thousand times over to ransom you. You must now go and say goodbye to him and tell him: "I intend to set off in three days' time and I have come to take my leave of you. Tell me what I owe you for the rent of this house so that I may pay you and settle my debt." See what he says to that and then come back and tell me, for I can do no more. I've been scheming in order to make him angry enough to divorce me, but I can see that he is still attached to me and I can't think of anything better for us to do than to leave for your own country.' 'How splendid it would be if dreams came true!' he exclaimed.

Qamar al-Zaman then went and sat down with the jeweller in his shop. 'Master,' he said, 'I intend to leave in three days' time and I have come to say goodbye to you. Please tell me how much I owe you as rent for the house so that I can settle it and pay you.' 'What are you talking about?' asked 'Ubaid. 'It is you who have done me a favour. I swear that I shall not take any rent from you, for you have brought down blessings on us. You are going to leave us lonely when you go, and were I not forbidden from doing so, I would stand in your way and stop you from returning to your family and your own country.' He then said goodbye to Qamar al-Zaman, and after they had both shed bitter tears, he locked up his shop, saying to himself: 'I must go with my friend.' Wherever Qamar al-Zaman went on some errand, 'Ubaid went with him, and whenever he went to Qamar al-Zaman's house, he would find Halima there. She would stand in front of them to serve them, and then, when he went to his own house, he would find her sitting there. So for three days, whichever of the two houses he visited, he would come across her.

Halima now told Qamar al-Zaman that she had now moved out all her husband's treasures, wealth and furnishings and the only thing she had left was the slave girl who had been in the habit of bringing in the cups for them to drink. She said: 'I cannot be parted from her. In addition to being a relative of mine, she is very dear to me and she keeps my secrets. I propose to give her a beating, pretending to be angry with her, and when my husband comes I shall say: "I can't abide this girl any longer and I refuse to stay in the same house as her, so take her and sell her." When he does this, do you buy her yourself so that we can take her with us.' Qamar al-Zaman agreed to this and Halima beat the girl. When the jeweller came home he found her in tears and when he asked why she was weeping, she said: 'My mistress beat me.' He went to his wife and asked: 'What did the damned girl do to make you beat her?' She replied: 'I have only one thing to say to you and that is that I cannot stand the sight of her any longer. Take her and sell her, or else divorce me.' He said: 'I shall sell her and not disobey you.' As he was taking her off to his shop, he passed by Qamar al-Zaman.

After he had left with the girl, Halima slipped quickly through the tunnel to Qamar al-Zaman, who put her into the litter before her husband arrived. When he came and Qamar al-Zaman saw the girl with him, he asked about her, and 'Ubaid said: 'This is my slave girl who used to bring us in drinks, but she offended her mistress, who became angry with her and has told me to sell her.' Qamar al-Zaman said: 'As her mistress dislikes her, she can't stay with her, but you can sell her to me so that she may remind me of you and she can act as a maid for Halima, my slave girl.' 'All right; take her,' said 'Ubaid and when Qamar al-Zaman asked how much he wanted, he said: 'I shan't take anything from you, as it is you who have done us a favour.' Qamar al-Zaman accepted the gift and then called to Halima, to tell her to kiss the jeweller's hand. She emerged from the litter, kissed his hand and then remounted, while he kept his eyes fixed on her. 'I commend you to God's protection, master,' Qamar al-Zaman said. 'Let me go with a clear conscience.' 'May God clear you of all debts and bring you safely back to your family,' replied 'Ubaid. He then took his leave of Qamar al-Zaman and went back to his shop in tears. He found it hard to part from a man who had been his friend, for friendship has its rights, but at the same time he was glad to be freed from the suspicion that he had formed about his wife, as now Qamar al-Zaman had left his suspicions seemed groundless.

So much for him, but as for Qamar al-Zaman, Halima told him: 'If you want to be safe, let us not go by any of the usual ways.'

Night 976

Morning now dawned and Shahrazad broke off from what she had been allowed to say. Then, when it was the nine hundred and seventy-sixth night, SHE CONTINUED:

I have heard, O fortunate king, that when Qamar al-Zaman set off, the girl told him: 'If you want to be safe, let us not go by any of the usual ways.' 'To hear is to obey,' he replied, and he then chose a track that was not normally used, going through one country after another until he reached the Egyptian border. He then wrote a letter and sent it by courier to his father, 'Abd al-Rahman the merchant. His father was sitting with his colleagues in the market, his heart consumed with sadness for his son, from whom he had not heard since the day he left. It was just then that the courier arrived and asked the merchants whether there was anyone called 'Abd al-Rahman among them. 'What do you want of him?' they asked, and he replied: 'I have a letter from his son, Qamar al-Zaman, whom I left at al-'Arish.' 'Abd al-Rahman was delighted and relieved, and the others shared in his joy and congratulated him on his son's safe return. He then took the letter and read it. It was headed: 'From Qamar al-Zaman to the merchant 'Abd al-Rahman,' and it went on: 'I send my greetings to you and to all the other merchants. If you ask how I am, the answer is that, with the help of God, to Whom be praise and thanks, I have made a profit by trading and I have come back safely and in good health.'

'Abd al-Rahman now began to celebrate, giving banquets, together with a series of entertainments and parties, where there were musical instruments and remarkable diversions of every kind. When his son reached al-Salihiya, he went out to meet him, together with all the other merchants, and when they met the two embraced and the father clasped his son to his breast and wept until he fainted. When he had recovered, he said: 'This is a blessed day, my son, as God, the Omnipotent Protector, has reunited us.' Then he recited these lines:

To be near the beloved is final happiness,
As the cups of delight circulate among us.
Welcome and welcome again
To Time's radiance and the moon of moons.

Shedding tears of sheer delight, he recited more lines:

The shining moon of Time* has now appeared,
Gleaming brightly, returning from his travels.
His hair is black as the night of his absence,
But the sun rises from the buttons of his shirt.

The other merchants now came up to greet him, and they saw that he had with him a quantity of loads as well as servants, together with a litter placed within a wide enclosure. When Qamar al-Zaman had been escorted into the house, Halima emerged from this litter, enchanting all who saw her, as 'Abd al-Rahman observed. They opened up for her a great villa that looked like a treasure chamber which had been freed from a talismanic spell. At the sight of her, Qamar al-Zaman's mother fell under her spell, thinking happily that she must be a queen and the wife of a king, but in answer to her question, Halima said: 'I am your son's wife.' 'Since he has married you,' she said, 'we must hold a great celebration in your honour and in his.'

---

* i.e. Qamar al-Zaman.

So much for her, but as for 'Abd al-Rahman, when the others had dispersed, each going off on his own way, he went to his son and said: 'My boy, what is this girl you have with you, and how much did you pay for her?' Qamar al-Zaman replied: 'She is no slave girl, father, and it was because of her that I went away.' 'How was that?' his father asked, and Qamar al-Zaman replied: 'She is the girl whom the dervish described to us the night he spent here. From that moment, all my hopes were set on her; it was for her sake that I wanted to go on my travels, and this led to my being stripped on the road by Bedouin, who took all my goods. I was the only one to get to Basra' – and he went on to tell his father the details of what had happened from start to finish. 'And after all that, did you marry her?' his father asked. 'No,' said Qamar al-Zaman, 'but I have promised to do so.' 'Do you intend to marry her?' his father pressed him, and he said: 'If you tell me to, I shall, and if not, I shall not.' 'If you do,' his father said, 'I shall wash my hands of you both in this world and the next and I shall be furiously angry with you. How can you marry her after what she did to her husband? After the way she treated him for your sake, she would do the same to you for the sake of someone else, for she is treacherous, and traitors should not be pardoned. If you disobey me, I shall be angry with you, but if you listen to what I have to say, I shall try to find you a lovelier girl, pure and chaste, and marry you to her, even if I have to spend all my money on it. I shall give you a feast of unparalleled splendour and I shall be able to boast of you both. It is better for people to say: "So-and-So married So-and-So's daughter," than to have them say: "He married a slave girl of no birth or breeding."'

He continued to press Qamar al-Zaman not to marry Halima, quoting examples, anecdotes, poems, proverbs and moral exhortations, until Qamar al-Zaman told him: 'If this is how things stand, I don't feel myself bound to marry her.' At that, his father kissed him between the eyes and exclaimed: 'You are truly my son and I swear by your life that I shall marry you to a girl who has no match!' He then locked Halima and her slave girl in a room, putting them in the charge of a black maid who was to take them their food and drink. He told her: 'You and your slave girl will remain shut up here until I sell you, if I can find someone who wants to buy you. If you disobey me, I shall kill you both, as you are treacherous and there is no good in you.' 'Do as you wish,' she replied, 'for I deserve everything that you do to me.' He then locked the door on them and told his womenfolk that no one was to visit them or speak to them, apart from the black maid, who was to pass their food and drink through the window. Halima sat with her slave girl, weeping and regretting what she had done to her husband.

So much for her, but as for 'Abd al-Rahman, he sent out marriage brokers to search for a well-born bride for his son. Whenever they came across a girl during the course of their search, they would hear of one who was even lovelier, and this went on until they came to the house of the *shaikh* al-Islam, whose daughter they found to have no equal in all Cairo. She was a beautiful and graceful girl with a perfect figure, a thousand times more lovely than Halima. When they told 'Abd al-Rahman about her, he went to her father, accompanied by the leading citizens, and asked him for her hand. A marriage contract was drawn up; wedding celebrations were held on a grand scale and banquets prepared. On the first day, the *faqihs* were invited to join a feast of great splendour, while on the following day invitations went out to all the merchants. Drums were beaten, pipes played and not only the immediate neighbourhood but the entire quarter was lit up with lamps. The celebrations lasted for forty days, on each of which a different group were invited, whether scholars, emirs, beys or governors. Every day 'Abd al-Rahman would sit to receive his guests with his son by his side, looking at them as they ate. Such feasting had never been seen before.

On the last day, the poor and the down-and-outs were invited from far and near, and they came in groups to eat. As 'Abd al-Rahman was sitting there with Qamar al-Zaman beside him, in came Halima's husband, 'Ubaid, the jeweller, with one of the groups. He was half-naked, weary and showing the strains of travel, but Qamar al-Zaman recognized him at once and told his father to look at the poor man who had just come through the door. 'Abd al-Rahman noted his tattered clothes and ragged gown, worth no more than two dirhams. With his pallor and covered with dust, he looked like a broken-down pilgrim and he was groaning like a man both sick and needy, stumbling and swaying to right and left as he walked. He could best be described by the lines of the poet:

Poverty always diminishes a man,
As the sun pales before it sets.
He tries to hide among the crowds;
When he is alone, his tears flood down.
If he is absent, no one cares;
When he is present, he receives no share.

Within his own family a man struck down
By poverty will find himself a stranger.

Another poet has said:

On his way the poor man finds all things against him;
The earth itself closes its doors to him.
You find him hated, though he does no wrong;
You see hostility but not its cause.
At the sight of a rich man you find the dogs
Turning towards him, as they wag their tails,
But if it is a poor wretch whom they see
They come to bark at him and bare their teeth.

How well yet another poet has expressed it:

When glory and good luck befriend a man,
Disasters and misfortunes keep away.
Friends come as uninvited parasites,
And even supposed guardians act as pimps.
He farts and people take it as a song,
While any smell he makes is sweet perfume.

Night 977

Morning now dawned and Shahrazad broke off from what she had been
allowed to say. Then, when it was the nine hundred and seventy-seventh
night, SHE CONTINUED:

I have heard, O fortunate king, that Qamar al-Zaman told his father
to look at the poor man. 'Who is he, my son?' asked 'Abd al-Rahman,
and Qamar al-Zaman told him: 'This is 'Ubaid, the master jeweller, the
husband of the woman who is locked up here.' 'Is he the man whom
you told me about?' his father asked, and Qamar al-Zaman said: 'Yes;
it is him beyond a doubt.'

The reason for 'Ubaid's arrival was that after he had said goodbye to
Qamar al-Zaman, he set off back to his shop, where he had in hand a
delicate job that occupied him for the rest of the day. In the evening, he
locked the place up and went home. The door opened as he put his hand
on it, and when he went in he could find neither his wife nor the slave
girl. The place was in the worst of states, fitting the lines:

The hive that had been full of bees
Was empty when they left.
It looked as though no one had ever lived in it,
Or its inhabitants had met some fate.

On finding the house empty, he turned to the right and left and went round the house like a madman, but could find no one there. Then, on opening the door of his strongroom, he discovered that neither money nor treasures were left in it. This roused him from his befuddled state and brought him back to his senses, as he realized that it must have been his wife who had plotted against him and betrayed him. He wept at what had happened, but kept it quiet, lest any of his enemies might gloat or his friends be distressed, for he knew that if he gave the secret away, he would be disgraced and people would censure him. He told himself: 'Say nothing about your troubles and, whatever you do, remember what the poet said:

If a man finds it hard to keep a secret,
The one with whom he shares it finds it harder.'

He locked up his house and went to his shop. There he put one of his craftsmen in charge, telling him: 'My friend the young merchant has invited me to go with him on a trip to Cairo, swearing that he will not leave until he can take me and my wife along with him. You, my boy, will look after the shop for me in my absence, and if the sultan asks about me, tell him that I have gone with my wife to Mecca.' He then sold some of his belongings and bought camels, mules and mamluks, as well as a slave girl, whom he put in a litter. Ten days later, he left Basra and his friends, who came to say goodbye, thought as he left that he was taking his wife with him on his pilgrimage. Everyone was glad that God had rescued them from having to shut themselves up in a mosque or at home every Friday. Some of them prayed that he might never come back to Basra so as to save them from ever having to do this again, since it had caused the people great inconvenience. 'I don't think that he will get back from his journey,' one of them said, 'because of the curses of the citizens'; while another added: 'If he does, things will not go so well for him.' There were great celebrations in the city in place of the great gloom that there had been, and even the cats and dogs found life easier. When Friday came, however, the usual proclamation was made that people were to go to the mosque two hours before the prayer or were to hide themselves away in their homes and not let the cats and dogs go out. In their annoyance, the citizens gathered together and made their way to the sultan's court where they stood before him and said: 'King of the age, the jeweller has taken his wife and left on pilgrimage. The reason why we had to shut ourselves away has gone, so why should we do it any longer?' 'How can that disloyal fellow have left without letting me know?' exclaimed the sultan. 'When he comes back things will be put right, but meanwhile go back to your shops and carry on your business, for this order no longer applies.'

So much for the sultan and the people of Basra, but as for 'Ubaid the jeweller, he had travelled for ten stages before what had happened to Qamar al-Zaman on his way to Basra happened to him and he was attacked near Baghdad by Bedouin, who stripped him and seized everything he had with him. It was only by pretending to be dead that he managed to escape, and when the Bedouin had left, he had to walk, naked as he was, to a town. There charitable people were moved by God to pity him, and they gave him some ragged clothes with which to cover himself. He then started to go from one place to the next, begging for his bread, until he reached Cairo, the guarded city, and there, consumed by hunger, he went around begging in the markets. Someone told him: 'Poor fellow, there is a wedding feast going on in a house nearby, and if you go there you'll get food and drink, for today they are providing a meal for the poor and the strangers.' 'I don't know the way,' he said and the man replied: 'Follow me and I'll show you.' So 'Ubaid followed him to the house, and his guide then said: 'This is it. Go inside and don't be afraid, for the door of a house of feasting is open to all.'

When 'Ubaid went in, Qamar al-Zaman saw him and recognized him, after which he told his father. 'Leave him for the moment, my son,' his father said. 'He may be hungry, so let him eat his fill and relax, and after that we shall send for him.' While the two of them waited, 'Ubaid had his meal and washed his hands, before drinking coffee and sugared drinks mixed with musk and ambergris. Then, when he was about to leave, a servant came to say: 'Stranger, 'Abd al-Rahman the merchant wants a word with you.' 'Ubaid asked who 'Abd al-Rahman was and when he was told that this was his host, he went back, thinking that he was going to be given a present. Then, when he went up to him, he caught sight of his friend, Qamar al-Zaman, and almost fainted with embarrassment. Qamar al-Zaman got to his feet and greeted him with a hug, after which they both shed floods of tears. Qamar al-Zaman sat him down by his side, but his father exclaimed: 'You have bad manners, boy; this is no way to greet a friend. Start by sending him to the baths and providing him with a suitable set of clothes, and after that you can sit with him and talk together.'

Qamar al-Zaman summoned servants, telling them to escort his guest to the baths, and he sent him splendid clothes worth a thousand dinars or more, so that when he had been washed and dressed, he had the appearance of a merchant prince. While he was in the baths, the other people at the feast asked Qamar al-Zaman who he was and where they had got to know each other. He said: 'This is my friend who lodged me in his house and did me innumerable favours, treating me with the greatest generosity. He is a prosperous man of high rank, a jeweller by profession, unrivalled in his craft. He stands high in the affection of the sultan of Basra and his authority is unquestioned.'

Qamar al-Zaman continued his eulogy, speaking of what 'Ubaid had done for him and saying: 'I am ashamed because I don't know how to reward him in return for the favours that he showed me.' He went on showering praises on him until all who were present had formed the highest opinion of him and he had become a figure of respect in their eyes. They all said that they would do their best for him and would receive him hospitably for Qamar al-Zaman's sake. They added: 'But we would like to know the reason for his coming to Cairo, why it was that he left his native land and what it was that brought him here in such a state.' 'There is nothing surprising about that,' Qamar al-Zaman replied, 'for all men are subject to the decrees of fate and in this world they can never be free from disasters. The lines of the poet are true:

Time hunts men down; be not deceived
By offices or ranks.
Beware of slips; avoid distressful things,
For it is in Time's nature to destroy.
The smallest things bring many fortunes low
And every change they suffer has its cause.

You must know that when I entered Basra I was in an even worse state than this. When he came to Cairo he had rags with which to cover himself, but in Basra I had to shield my nakedness with one hand in front and one behind, and it was only God and this noble man who helped me. The reason was that the Bedouin had stripped me, taking my camels, my mules and my baggage and killing my servants and my men.

I lay among the corpses until they went off and left me, thinking that I was dead. Then I got up and walked into Basra, naked as I was, and it was there that this man met me, gave me clothes, lodged me in his house and supported me with money. All that I have brought here with me comes from God's bounty and from his. When I left for home, he showered me with gifts and so it was that I came back here in this happy state. When I parted from him he was enjoying good fortune and a position of power, but it may be that, after that, he was overtaken by one of the misfortunes of Time, which has led him to forsake his family and his homeland, and it would not be surprising if on his way here he suffered the same kind of disaster that happened to me. It is for me now to repay him for his generosity, and to act in accordance with the poet's lines:

> You who have a rosy view of Time,
> Are you aware what it can do?
> In all you plan to do, act generously,
> And you will meet a generous return.'

While they were talking in this way, 'Ubaid appeared, looking like a merchant prince, and they all rose to greet him and to seat him in the place of honour. Qamar al-Zaman said: 'My friend, this is a blessed and a happy day. There is no need to tell me what happened, as it happened to me before you. If the Bedouin stripped you and robbed you, you have no need to worry, for it is with money that we ransom ourselves. I was naked when I entered Basra and you clothed me and were generous to me. Your many kindnesses have left me in your debt. I shall repay you . . .'

Night 978

Morning now dawned and Shahrazad broke off from what she had been allowed to say. Then, when it was the nine hundred and seventy-eighth night, SHE CONTINUED:

I have heard, O fortunate king, that Qamar al-Zaman told 'Ubaid the jeweller: 'When I entered your country, I was naked and you clothed me. You showed me many kindnesses and I shall repay you and do for you as much, or more, as you did for me. So be happy and console yourself.' He continued to reassure 'Ubaid, not allowing him to speak, lest he talk of his wife and what she had done to him. Instead, he kept producing moral admonitions, proverbs, poems, anecdotes, tales and stories in order to divert him, until he for his part realized that his host was wanting to keep him from talking about his private affairs. So he kept those hidden and comforted himself with the tales and anecdotes that he heard, quoting the lines:

Were you to read what fate inscribes
On Time's forehead, you would shed tears of blood.
If Time's right hand preserves a man from harm,
Its left will pour him out a cup of doom.

It was after this that Qamar al-Zaman and his father took 'Ubaid into the women's quarters. When they were alone there, 'Abd al-Rahman said: 'The only reason we stopped you speaking was that we were afraid lest this might bring disgrace on you and on us. Now that we are here in private, tell me what took place between you, your wife and my son.' 'Ubaid told him the whole story from beginning to end, and 'Abd al-Rahman asked: 'Was the fault that of your wife or of my son?' 'By God,' replied 'Ubaid, 'your son did nothing wrong. Men lust after women and it is for women to stop them. The blame rests with my wife, who betrayed me through her actions.' 'Abd al-Rahman took his son aside and said: 'My boy, we have put his wife to the test and found her to be a traitress. What I want to do now is to test her husband to see whether he is a man of honour or a pimp.' Qamar al-Zaman asked how he proposed to do that and he said: 'I am going to try to reconcile him with his wife. If he agrees and forgives her, I shall cut him down with my sword and then kill her and her maid, for it is not good that a pimp and an adulteress should live, but if he turns away from her, I shall marry him to your sister and give him more money than you took from him.'

'Abd al-Rahman then went back to 'Ubaid and said: 'Master, it takes patience to associate with women and whoever loves them needs to be magnanimous. They are always picking quarrels with men and annoying them, as they are proud of being the fair sex, boasting of their own superiority and despising men. This is particularly true when they see that their husbands love them, and they respond to this with pride, coquetry and misdeeds of every kind. If a man becomes angry when he sees his wife doing something that he dislikes, there can be no association between the two of them, for the only men whom women find acceptable are the magnanimous and long-suffering. If a man is not prepared to put

up with his wife and to overlook her evil deeds, then he cannot successfully associate with her. It is said that were women up in the sky, men
would crane their necks to look at them, and whoever forgives from a
position of strength finds his reward with God. You have lived for a long
time with this woman, who has been your wife and your companion,
and you should forgive her, as this is one of the signs of a successful
relationship. Women are lacking in sense and in religiosity, but if they
sin, they may repent and, God willing, not repeat their offence. My
advice to you is to make peace with your wife, and I shall then give you
back more than you had before. The two of you will be welcome to stay
with me and you will meet with the best of treatment, but if you want
to go home I shall present you with an acceptable gift. Here is the litter
waiting and you can put your wife and her maid in it and set off on your
journey. Many things happen between husbands and wives, and you
should be gentle rather than rough.'

'And where is my wife, sir?' 'Ubaid asked. 'Here in this house,' 'Abd
al-Rahman replied, 'so go up to her and be good to her, as I have
recommended. There is no need to cause trouble for her; when my son
brought her here, he wanted to marry her, but I stopped him and put
her in a room, locking the door on her. I told myself: "It may be that
her husband will come and I can then hand her over to him, for she is a
lovely woman and no husband could possibly abandon a wife like her."
Things have worked out as I thought they would, and I thank Almighty
God that you have been reunited with her. As for my son, I have found
him another wife. The banquets and entertainments that I have been
giving are to celebrate his marriage and tonight I shall bring him to his
bride. This is the key to your wife's room. Take it and when you have
opened the door, you can go in and enjoy yourself with her and her
maid. Food and drink will be brought to you and you can stay until you
have had enough.' 'God reward you well, master,' said 'Ubaid. He then
took the key and went off cheerfully, leaving 'Abd al-Rahman to think
that he approved of his advice and was prepared to accept it. He himself
took his sword and followed without being seen, after which he stood
there to see what happened between husband and wife.

So much for him, but as for 'Ubaid, when he went in, he found his
wife shedding bitter tears because Qamar al-Zaman had married someone else. Her maid was saying: 'How many times did I give you advice,
mistress, telling you that no good would come to you from the boy and
that you should break off your affair with him? But you didn't listen to

me and, instead, you robbed your husband of all his wealth and gave it to the boy. You then left home for love of him and came here, only to find that he forgot all about you and married someone else, so that at the end of your affair with him you find yourself locked up.' 'Damn you!' said Halima. 'Even if he has got another wife, one day he will think of me. I cannot forget the nights I spent with him, and at all events I console myself with the poet's lines:

> Will you not think of one
> Who thinks of none but you?
> Do not ignore his state,
> Who forgets himself through you.

He is bound to remember the intimacy we enjoyed together and ask about me, and even if I die as a prisoner I shall not stop loving him, for he is my darling and the doctor who can treat what ails me. My hope is that he will come back to me and we can be happy together.'

When 'Ubaid heard this, he burst in and said: 'Traitress, you have as much hope of getting him as the devil has of reaching Paradise. You had all these evil qualities and I didn't know about them, for had I realized that you had even one of them, I wouldn't have kept you with me for a single hour. Now that I know for certain what you are like, I am going to kill you for your faithlessness, even if I myself am killed in return.' Seizing her with both hands, he recited these lines:

> They drove away my faithful love, those lovely girls,
> By their misdeeds, and did not heed my rights.
> How many times I fell in love before, but now,
> After this wrong, my love has turned to hate.

He squeezed her windpipe with such violence that he broke her neck. The maid screamed out in distress for her mistress, but 'Ubaid said: 'You whore, this is all your fault. You knew how lascivious she was but you said nothing to me.' He then took hold of her and strangled her, while all the time 'Abd al-Rahman, sword in hand, was standing behind the door, watching and listening.

Having killed his wife and her maid, 'Ubaid was filled with foreboding, fearing what might happen because of what he had done and saying to himself: 'When this merchant learns that I have killed these two under his own roof, he is bound to kill me, but I pray to God that I may die as a true Muslim.' In his confusion, he had no idea what to do, but at that moment 'Abd al-Rahman entered. 'No harm will come to you,' he said, 'for you have earned your safety. You see this sword in my hand – if you had made your peace with your wife, I had intended to kill both of you, but after what you have done, I welcome you, and in return for it I shall marry you to my daughter, Qamar al-Zaman's sister.' When he had taken him off, he sent for a woman to wash the corpses, and word spread that Qamar al-Zaman had brought two girls with him from Basra but that both had died. People began to address the usual condolences to him, saying: 'May you yourself live,' and: 'May God give you a replacement.' Then when the dead had been washed and dressed in shrouds, they were buried without anyone knowing the real story.

So much for 'Ubaid, his wife and her maid, but as for 'Abd al-Rahman, he summoned Shaikh al-Islam, together with the leading citizens, and told him to draw up a marriage contract between his daughter, Kaukab al-Sabah, and Master 'Ubaid the jeweller, confirming that her dowry had been paid to him in full. The contract was written out, drinks were brought for the company and both weddings shared the same celebrations. Qamar al-Zaman's bride, the daughter of al-Islam, and his sister, Kaukab al-Sabah, were carried in procession in the same litter, and that same evening both bridegrooms, Qamar al-Zaman and 'Ubaid, were brought to their brides. 'Ubaid discovered that Kaukab al-Sabah was a thousand times more lovely than Halima. He took her virginity and next morning accompanied Qamar al-Zaman to the baths.

He spent a pleasant and enjoyable time with his wife's family, but then he went to tell 'Abd al-Rahman that he was feeling a longing for his homeland, where he had properties and revenues and where he had left one of his craftsmen as his agent. He said: 'I have it in mind to go home, sell what I own and then come back to you.' He asked for, and received, 'Abd al-Rahman's permission to do this, while the latter added: 'You are not to be blamed, as love for one's country is part of true faith and whoever is without value in his own land is valueless elsewhere. It may be, however, that if you go back home without your wife, you will enjoy it there and you will be torn between coming back to her and staying where you are. It would be best to take her with you and if after that you want to come back here, do so and you will both be welcome. In this country we don't recognize divorce; a woman never marries twice and she doesn't abandon a man wantonly.' 'Uncle,' 'Ubaid objected, 'I'm afraid that your daughter may not want to go with me to my own country.' 'My son,' replied 'Abd al-Rahman, 'we have no women here who disobey their husbands and we know of none who become angry with them.' 'God bless you and your women,' replied 'Ubaid, and he then went to his wife and said: 'I want to go to my own country, but what do you say?' She replied: 'While I was a virgin, my father told me what to do, but now that I am married, it is for my husband to say, and I shall not disobey him.' 'May God bless you and your father,' 'Ubaid exclaimed, 'and show His mercy to the womb that bore you and the loins that engendered you!'

After that, he repaired his equipment and made ready for the journey. 'Abd al-Rahman showered him with gifts, and when they had said goodbye to each other, 'Ubaid started off with his wife, travelling on until he got to Basra. There his friends and relatives came out to meet him, under the impression that he had been in the Hijaz. Some were glad to see him back but others were saddened, as people told one another that Fridays would become a misery again, with the inhabitants confined in the mosques and their own homes, and even the cats and dogs shut up.

So much for 'Ubaid, but as for the sultan of Basra, when he heard of his return he summoned him angrily and said in rough tones: 'How was it that you went off without telling me? Did you think that I was not in a position to give you any help in your pilgrimage to the Holy House of God?' 'Forgive me, master,' said 'Ubaid, 'but I did not, in fact, make the pilgrimage,' and he went on to tell him in detail what had happened between him, his wife and 'Abd al-Rahman, the Cairene merchant. He explained how 'Abd al-Rahman had married him to his daughter and how he had brought her with him to Basra. The king said: 'Were it not for my fear of Almighty God, I would have you killed and then marry this high-born lady myself, whatever huge sums it might cost me, as she should properly be the wife of a king. However, may God, Who has given her to you, bless your marriage. Treat her well.' He then made generous gifts to 'Ubaid before he left, and after that he stayed with his wife for five years before being gathered into the mercy of God.

At that point, the king asked for the widow's hand in marriage, but she did not accept, telling him: 'Your majesty, among my people no woman remarries after losing her husband, and now that my husband is dead I shall not marry you or any other man, even were you to kill me.' The king then sent to ask whether she wanted to go back home, to which she replied: 'God will repay you for whatever good you do.' The king had all 'Ubaid's wealth collected and added a gift appropriate to his own rank. He provided an escort of five hundred riders and put Kaukab al-Sabah in the charge of one of his viziers, a man noted for his goodness and virtue, who brought her back to her father. She remained single for the rest of her life until she and all her family were dead.

If this woman was unwilling to exchange her dead husband for a king, how does she compare with one who exchanged her living husband for a young man of whose background she knew nothing, especially when this was an illicit union, unsanctioned by marriage? Whoever thinks that all women are alike is suffering from a disease of madness for which there is no cure. Praise be to the Sovereign King, Who lives and never dies.

A story is also told that one day when the caliph Harun al-Rashid was inspecting the tax revenues of his lands, he discovered that while every other district had sent in what it owed to his treasury, in that year Basra had not. He summoned a council meeting to discuss this and sent for Ja'far, his vizier, whom he told of the situation, saying that nothing at all had come from Basra. 'Perhaps something distracted the governor's attention and made him forget to send it,' Ja'far said, but the caliph pointed out: 'The tax should have arrived twenty days ago. What excuse can he have for not forwarding it during all that time or at least giving the reason?' Ja'far suggested that, if the caliph wanted, an envoy could be sent. 'Send him my drinking companion Abu Ishaq al-Mausili,' ordered the caliph, to which Ja'far answered: 'To hear is to obey both God and you, Commander of the Faithful.'

When Ja'far had left the palace, he summoned Abu Ishaq and provided him with official accreditation, telling him: 'Go to 'Abd Allah ibn Fadil, the governor of Basra, and find out what has kept him from sending in his taxes. Then collect what is owed and bring it to me at once, for the caliph has inspected the revenues of every region in his empire and has discovered that only Basra has failed to send in its dues. If you find that it is not ready, and if the governor makes an excuse, bring him with you so that he can explain himself to the caliph.' 'To hear is to obey,' said Abu Ishaq, and he then took a force of five thousand riders and travelled to Basra. When 'Abd Allah learned of his arrival, he went out to meet him with his own troops and escorted him into the city. Abu Ishaq himself was taken to 'Abd Allah's palace, while his men camped outside the city, where they were supplied with everything that they needed.

When Abu Ishaq arrived in the council chamber, he took his seat with 'Abd Allah at his side, surrounded by the Basran dignitaries in order of rank. After greetings had been exchanged, 'Abd Allah asked him whether there was any particular reason for his visit. 'Yes, there is,' Abu Ishaq told him. 'I have come for the tax that is overdue, as the caliph has asked about it.' 'I wish you hadn't put yourself to this trouble and undergone the fatigues of the journey,' said 'Abd Allah, 'for the tax is ready in full and I had decided to send it off tomorrow. Now that you are here, I shall hand it over to you in four days' time, after you have been entertained for three days. Just now, however, I must present you with a gift as a partial return for your kindness and that of the caliph.' Abu Ishaq agreed to this, and when the formal meeting had broken up, 'Abd Allah took him to room of unparalleled splendour in his palace. A table was set for him and his guests; when they had eaten and drunk with enjoyment and pleasure, it was removed and after they had washed their hands, coffee and drinks were brought in and they sat drinking together until a third of the night had passed. Bedding was then spread for Abu Ishaq on an ivory couch inlaid with gleaming gold. He lay down and 'Abd Allah lay on another couch by his side.

Abu Ishaq found it difficult to sleep and so he began to brood about poetic metres and the rules of composition, because, as one of the caliph's most intimate companions, he was a master both of poetry and of witty anecdotes. He stayed awake composing poetry until midnight, but as he lay there 'Abd Allah got up, tightened his belt and opened a cupboard. From this he took a whip and, with a lighted candle in his hand, he went out through the door of the room, thinking that his guest was asleep.

Night 979

Morning now dawned and Shahrazad broke off from what she had been allowed to say. Then, when it was the nine hundred and seventy-ninth night, SHE CONTINUED:

I have heard, O fortunate king, that 'Abd Allah ibn Fadil went out through the door of the room, thinking that Abu Ishaq was asleep. For his part, Abu Ishaq asked himself in surprise: 'Where can he be going with this whip? Maybe he intends to punish someone, but at all events I must go after him to see what he's going to do now.' So he got up and followed him, inching forward so as not to be seen. He saw 'Abd Allah opening a closet from which he took a table with four plates of food, together with bread and a jug of water. As he carried the table and the jug, Abu Ishaq followed him until he entered a room. Abu Ishaq hid behind the door, through cracks in which he could see that this was a spacious room, splendidly furnished, in the centre of which was a couch inlaid with gleaming gold. To this two dogs were attached with golden chains.

Abu Ishaq saw his host set the table down at the side of the room and then tuck up his sleeves and release the first dog. It started to twist around in his arms and put its nose down as though it was kissing the ground in front of him, whimpering quietly all the while. He then tied its legs and threw it down, before taking out the whip and beating it painfully and mercilessly as it writhed around, unable to escape. He continued to strike it until it ceased to howl and became unconscious, after which he put it back where it had been and tied it up. When he had treated the second dog in the same way, he took out a handkerchief with which he wiped away the dogs' tears, consoling them, calling down blessings on them and saying: 'Don't hold this against me. By God, I am not doing it of my own free will and I don't find it easy. It may be that God will rescue you from this distress.'

While all this was going on, Abu Ishaq was standing there watching and listening, filled with amazement. 'Abd Allah then brought out the food and started to feed the dogs with his own hand until they had had enough, after which he wiped their muzzles. When he had fetched the jug and poured them water, he picked up the table, the jug and the candle and was about to leave the room when Abu Ishaq went ahead of him, returning to his couch, where he pretended to be sleeping. 'Abd Allah had not seen him and did not realize that he had been followed and observed. So he replaced the table and the jug in the closet, and went back to his room, where he opened the cupboard and put the whip back in its place. Then he took off his things and went to sleep.

So much for him, but as for Abu Ishaq, he spent the rest of the night thinking about what he had seen and he was too full of astonishment to be able to sleep. He kept asking himself what the reason for 'Abd Allah's behaviour could be and he went on like this until it was morning. He and his host then got up to perform the dawn prayer, and when they had eaten the breakfast that was brought them and drunk their coffee, they went to the council chamber. Abu Ishaq spent the whole day thinking about the strange events but he said nothing and asked 'Abd Allah no questions. On the following night, the same thing happened. 'Abd Allah beat the dogs and then fondled them, giving them food and water,

watched all the while by Abu Ishaq, who saw the pattern of the first
night repeated. On the third night, it happened again, and then on the
fourth day 'Abd Allah brought the tax money to Abu Ishaq, who took it
and went off, without making any reference to the dogs.

When he reached Baghdad and had handed over the money, the caliph
asked why it had not been sent in time. Abu Ishaq told him: 'Commander
of the Faithful, I found that 'Abd Allah had got it ready and was on the
point of sending it off, and had I been a day later, I would have met it
on the way. But I saw him doing something extraordinary, the like of
which I have never seen in my life.' The caliph asked what this was and
Abu Ishaq told him the whole story of the dogs, before adding: 'I saw
him doing this three nights in a row, beating the dogs and then fondling
and comforting them before giving them food and water. I was watching
him from my hiding place.' 'Did you ask him the reason?' the caliph
asked, and when Abu Ishaq said no, the caliph ordered him to go back
to Basra to fetch 'Abd Allah together with the dogs. 'Please don't ask me
to do that,' said Abu Ishaq, 'because 'Abd Allah was very hospitable to
me and it was only by accident and not through any intention of mine
that I found out about this and told you, so how can I go back to fetch
him? If I did, I would be too ashamed to say anything. The proper thing
to do would be for you to send someone else with a letter signed by you
to summon him here with the dogs.' 'Were I to send anyone else,'
objected the caliph, 'he might deny the whole thing and claim: "I have
no dogs," whereas if I send you, you can say that you saw them with
your own eyes and he will not be able to deny it. So either you go to
him and bring him back with the dogs, or else I shall have to put you
to death.'

Night 980

Morning now dawned and Shahrazad broke off from what she had been
allowed to say. Then, when it was the nine hundred and eightieth night,
SHE CONTINUED:

I have heard, O fortunate king, that the caliph Harun al-Rashid said
to Abu Ishaq: 'Either you go to him and bring him back with the dogs,
or else I shall have to put you to death.' 'To hear is to obey, Commander
of the Faithful,' replied Abu Ishaq, adding: 'God provides for us and it
is to Him that we properly entrust our affairs. It has been well said that

a man's misfortunes are brought on him by his own tongue, and it was I who injured myself by telling you about this. Write an official note for me and I shall go and fetch him.'

The caliph wrote the note and Abu Ishaq set off for Basra. When he presented himself, 'Abd Allah said: 'God protect me against any bad news that you are bringing back, Abu Ishaq. Why are you here again so soon? Did the caliph refuse to accept the tax money because it was not enough?' 'No,' replied Abu Ishaq, 'I've not come back because of the money, which he accepted as the full amount. I hope that you will not blame me, but I have done you a wrong – something that was decreed for me by Almighty God.' 'Tell me what you did, Abu Ishaq,' said 'Abd Allah, 'for you are my friend and I shan't hold it against you.' Abu Ishaq explained: 'I have to tell you that when I was staying with you, I followed you for three nights in a row as you got up at midnight and beat your dogs before coming back. This astonished me, but I was too embarrassed to ask you about it. I did, however, mention it to the caliph, by chance, without meaning to do so, and he made me come back to you. Here is his note. Had I known that it would lead to this, I wouldn't have said anything to him, but this was ordained by fate.' 'Abd Allah interrupted his excuses to say: 'Now that you have told him, I shall confirm what you said lest he think that you were lying, for you are a friend of mine. But had he been told this by anyone else, I should have denied it and claimed that the story was false. As it is, I shall go with you and take the dogs, even if this brings about my death.' 'May God protect you,' said Abu Ishaq, 'as you have protected my reputation with the caliph.' 'Abd Allah then fetched a suitable gift for the caliph and took the dogs, secured with golden chains, each carried on a camel, after which they all set off for Baghdad.

On entering the caliph's presence, he kissed the ground before him and was allowed to sit. When he had done this, the dogs were brought in and the caliph asked about them. For their part, they were kissing the ground in front of him, wagging their tails and shedding tears, as though they were addressing a complaint to him. This astonished him and he said to 'Abd Allah: 'Tell me about these dogs. Why do you beat them and then treat them with respect?' 'Caliph of God,' replied 'Abd Allah, 'these are not dogs. Rather, they are two handsome and shapely young men, my full brothers.' 'If they were once men, how did they come to be transformed into dogs?' the caliph asked. 'Abd Allah replied: 'If you will permit me, Commander of the Faithful, I shall tell you what happened.'

'Do so,' the caliph told him, 'and take care not to lie, for this is a characteristic of hypocrites. Keep to truth, which is the vessel of salvation and the mark of men of piety.' 'The dogs will be my witnesses in whatever I tell you,' said 'Abd Allah. 'If I lie, they will testify to that, and if I tell you the truth, they will confirm it.' 'As dogs, they can neither speak nor answer,' objected the caliph, 'and so how can they give evidence either for you or against you?' 'Brothers,' said 'Abd Allah to the dogs, 'if I tell a lie, raise your heads and glare, while if I am telling the truth, lower your heads and look down.' HE WENT ON:

You must know, caliph of God, that we are three full brothers. Our father's name was Fadil, because his mother gave birth to twins, one of whom died immediately, while he survived and so was called Fadil* by his father. He was brought up well, and when he reached manhood he was married to my mother and his father then died. Fadil's wife had given birth first of all to this brother of mine whom Fadil named Mansur, after which she conceived again and gave birth to my other brother, Nasir, and then finally she gave birth to me. I was given the name 'Abd Allah. Our father brought us up until we had reached manhood and, on his death, he left us a house and a shop filled with coloured materials of all kinds from India, Rum and Khurasan, as well as with other goods, together with sixty thousand dinars in cash. When he died, we washed his corpse, and made a magnificent tomb for him, where we buried him, entrusting him to the mercy of God. We arranged for prayers to be said for his soul and the Quran to be recited, and we gave alms in his name for a full forty days.

At the end of this period, I gathered together the merchants and leading citizens and provided a lavish meal. When they had finished eating, I said: 'Merchants, this world is transitory while the next is eternal; glory be to God, Who is everlasting, while His creatures pass away! Do you know why I have brought you together here on this blessed day?' 'Praise be to God,' they replied, 'for it is He Who knows what is hidden from us.' 'My father has died leaving behind a sum of money,' I told them, 'and I am afraid that some of you may have claims on his estate through debts or pledges or something else. If there is anything that he owed to anyone, I want to clear the debt, and so if any of you is owed money, please tell me how much this is so that I can settle it on my father's behalf.' ''Abd Allah,' they told me, 'this world's goods are no substitute

* 'The Leftover'.

for the world to come. We are not fraudsters; all of us know how to distinguish right from wrong; we fear Almighty God and we do not defraud orphans. We know that your father, may God have mercy on him, never pressed anyone to return his money, while, for his part, he never allowed his own debts to go unpaid. We always used to hear him say how afraid he was of misappropriating other people's goods and he used to pray constantly: "My God, in You is my trust and my hope. Do not allow me to die with unpaid debts." It came naturally to him to settle what he owed without being asked for it, while he would never press for the payment of what was owed him but would tell the debtor: "Take your time." If the man were poor, he would freely cancel the debt and if he died, but was not poor, your father would say: "God has forgiven him what he owed me." We can all testify to the fact that no one is owed anything.' 'God bless you,' I told them.

I then turned to these brothers of mine and said: 'Brothers, our father owed no one anything and he has left us money, materials, a house and a shop. Each of the three of us is entitled to a third of the total. Shall we agree not to divide it but to keep it all in common and eat and drink together, or shall we split up the materials and the money, each taking his share?' They wanted to follow this second course.

At this point, he turned to the dogs and said: 'Is that what happened?' and they lowered their heads and looked down, as though to say yes.

'ABD ALLAH WENT ON:

I got the *qadi* to send an official to oversee the division, and he divided up the money, the materials and everything that our father had left. I got the house and the shop in exchange for part of my share in the money, and as we had all agreed, my brothers took their own shares from the money and the materials. I then opened up the shop with a stock of materials, to which I had added by using the money that had come to me together with the house and the shop, until the shop was full. There I sat buying and selling, while my brothers bought materials, hired a ship and set out on a trading voyage to distant parts. 'May God aid them,' I said, 'but as for me, my livelihood comes to me here, and no price can be set on a quiet life.'

I stayed like this for a whole year, enjoying prosperity and making large profits until I had acquired as much as our father had left to all three of us. Then one day I happened to be sitting in the shop wearing two furs, one sable and one of grey squirrel, as it was winter and the weather was very cold. While I was there, in came my brothers, each wearing no more than a ragged shirt, shivering, their lips white with cold. I was so distressed to see them in such a state . . .

Night 981

Morning now dawned and Shahrazad broke off from what she had been allowed to say. Then, when it was the nine hundred and eighty-first night, SHE CONTINUED:

I have heard, O fortunate king, that 'ABD ALLAH IBN FADIL TOLD THE CALIPH:

I was so distressed to see them shivering that I was almost out of my mind in my grief for them. I got up and embraced them, shedding tears for their plight, and then I handed one of them the sable and the other the squirrel fur. I took them to the baths and provided each of them with clothes suitable for a wealthy merchant, which they put on after they had washed. I then took them back home and, seeing that they were hungry, I produced a meal for them and we all ate together, while I humoured them and consoled them.

'Abd Allah turned again to the dogs and they confirmed what he had said by lowering their heads and looking down. HE CONTINUED:

I then asked them what had happened and where their money had gone. They said: 'We sailed off to a city called Kufa, where we sold for ten dinars materials that had cost us half a dinar, and for twenty dinars what had cost us a single dinar. We made a huge profit and then bought Persian silks for ten dinars a piece, each of which was worth forty in Basra. Next we went to a city called al-Karkh, where we traded at a great profit and acquired a large sum of money.' They went on telling me about the places they had been to and the profits they had made until I said: 'Since you enjoyed such success, how is that I see you coming back here without a thing?' 'Brother,' they said, sighing, 'we were unlucky and travelling is a dangerous business. We collected our wealth and our goods and when we had loaded everything on board ship, we sailed off on a course for Basra. On the fourth day of our voyage, the sea became disturbed, foaming and frothing as the swollen waves dashed together with fiery sparks. We were at the mercy of the winds, which drove our ship against the projecting spur of a mountain. It broke up and we were plunged into the water, losing all our goods in the sea. We struggled in the water for a day and a night until God sent us another ship, whose crew took us on board, and on this we sailed from place to place, begging what food we could get and enduring great hardship. We started to strip off our clothes and to sell them for food until we came near Basra. When we got there, we were consumed by regret, for had we been able to save what we had, we would have fetched riches to rival those of the king – but this was what God had decreed.'

I told them: 'Don't be distressed, my brothers, for wealth is used to ransom lives and if a man is safe, he has made a profit. As God has decreed your safe return, this is all that could be wished for, since poverty and wealth are no more than shadowy fantasies. How well the poet has expressed it:

If you have managed to save your head,
Wealth is no more than the clipping of a fingernail.'

I went on: 'Let us suppose that our father has just died today and that all the money that I now have is what he left us. I am happy to divide it equally with you.' I then got the *qadi* to send an official, and after I had brought him all that I had, he divided it between the three of us, with each of us taking a third share. I then said: 'When a man stays at home, God blesses him by providing him with his daily bread. Each of you should open a shop and stay there to earn his living, for he is bound to get what is destined to come to him.'

I did my best to help them do this, filling their shops with merchandise, and I told them to buy and sell, while keeping their money to themselves and not spending any of it. All that they might need in the way of food, drink and so on, I promised to supply for them. I continued to treat them generously, not allowing them to use any of their own money, and they would conduct their business by day before returning in the evening to spend the night in my house. Whenever I sat talking with them, they would start to sing the praises of foreign travel, reciting its advantages and describing the profits they had made, in the hope of inducing me to join with them in an expedition abroad.

Here 'Abd Allah again asked the dogs whether this was true, and again they confirmed it by lowering their heads and looking down. HE WENT ON:

They kept on trying to persuade me, telling me of the great profits that could be made in foreign parts and insisting that I should accompany them, until eventually, in order to please them, I said that I would go. We agreed to a partnership and hired a ship, which we loaded with all kinds of precious materials and trade goods of various sorts, as well as everything else that we might need. We put out from Basra, heading for the open sea with its boisterous waves, which brings destruction on all who set out upon it, while those who emerge from it are naked as newborn children. Our voyage took us on to a city where we traded profitably and from where we sailed on to another. We continued to go from place to place and from city to city, doing good business, until we had accumulated a very large sum of money and made a handsome profit. Eventually our captain dropped anchor by a mountain and told his passengers that, by way of relaxation, they could go ashore for the day in the hope of finding water.

Everyone disembarked, including me, and each of us went off on his own to look for water. I started to climb the mountain, and on my way I caught sight of a white snake that was fleeing with a deformed and terrifying-looking black snake in pursuit. The black snake caught up with the white one and pressed against it, seizing it by the head and wrapping its tail around the tail of its victim. The white snake cried out and I realized that it was about to be raped. Feeling sorry for it, I took up a lump of flint, weighing five or more *ratls*, and with this I struck the black snake and crushed its head. Then, before I knew what was happening, the white snake turned into a perfectly formed and lovely young girl, like a gleaming moon. She came up to me, kissed my hand and said: 'May God grant you double shelter, shelter from shame in this world and from hellfire in the world to come on the Day of Judgement, "a day when a man who comes to God will not be helped by wealth or children, but only by an innocent heart".'* She went on: 'Mortal, you have saved my honour and done me a service for which I must reward you.' She then pointed at the ground, which opened for her and when she had gone down into it, it closed over her again, making me realize that she must be one of the *jinn*. As for the black snake, fire spread through its corpse and burned it to ashes.

I was astonished by this and when I got back to my companions, I told them what I had seen. We spent the night there and the next morning the captain weighed anchor, hoisted the sails and had the ropes coiled.

* Quran 26.88–9.

We sailed off until we were out of sight of land, and we went on for another twenty days, during which we saw no land and no birds. Then the captain told us that our stock of fresh water was exhausted and when we suggested landing to look for more, he said: 'By God, I have strayed from my course and I don't know how to get to land.' We shed tears of distress, praying that Almighty God might send us guidance and passing an unhappy night. How well was this expressed by the poet:

How many a night of misery did I spend,
Such as would whiten the hairs of a suckling child,
But as soon as morning dawned, there came
Help from on high and speedy victory.

When dawn broke and the light spread, we were overjoyed to see a lofty mountain, and when we reached it the captain told us to go ashore to look for water. We all landed and began to search but were saddened when we failed to find any water at all. At that point, I climbed to the top of the mountain and on its far side I caught sight of a rounded depression an hour's journey away or more. I called up my companions and, when they came, I pointed it out to them and said: 'I can see a lofty and strongly built city there, with walls and towers; there are hills and meadows and it is bound to have water and other good things. Let us go there to fetch water and to buy what we need by way of provisions, including meat and fruit, before we go back to the ship.' The others said that they were afraid, pointing out: 'The people there may be enemies of religion, infidel polytheists, who will seize us and hold us prisoner, or else kill us, in which case we shall be responsible for our own deaths, having brought this on ourselves. Whoever is deceived into risking disaster wins no thanks, as the poet has said:

As long as the earth and sky remain the same,
The risk taker is not praised, even if he escapes unhurt.

We are not going to endanger ourselves.' 'I've no authority over you,' I told them, 'but I shall take my brothers and go off there.' But even my brothers refused to come, saying that they were afraid, and so I said: 'For my own part, I have made up my mind to go, trusting in God and content with the fate He allots me.' I told my brothers to wait until I came back from the city . . .

Night 982

Morning now dawned and Shahrazad broke off from what she had been allowed to say. Then, when it was the nine hundred and eighty-second night, SHE CONTINUED:

I have heard, O fortunate king, that 'ABD ALLAH IBN FADIL SAID:

I told my brothers to wait for me to come back from the city, and then I left them and walked to the city gate. The city itself I discovered to be strangely designed and remarkably built, with high walls, strong towers and lofty palaces. Its gates were made of Chinese iron, embellished and engraved in an astonishing fashion. When I entered the gate, I saw a stone bench on which a man was sitting with a brass chain wrapped round his arm from which fourteen keys were dangling. I realized that this must be the gatekeeper and that the city must have fourteen gates. So I went up and greeted him, but he did not return my greeting, and although I repeated it a second and a third time, he still made no reply. I put my hand on his shoulder and said: 'Why don't you answer my greeting? Are you asleep or deaf, or is it because you are not a Muslim?' When he still gave no answer and stayed motionless, I looked more carefully and discovered that he was made of stone. 'How remarkable it is,' I told myself, 'that this stone has been carved into so perfect a likeness of a man that all that it lacks is the power of speech.'

I left him and went into the city where I saw someone standing on the road, but when I went up and looked more closely, I found that he too was made of stone. After that, as I walked through the streets, I would go up and stare at the people whom I came across, but they all turned out to be stone. There was an old woman carrying on her head a bundle of clothes to be washed, and here again, when I looked at her from close at hand, both she and her bundle of clothes turned out to be stone. In the market I came across an oil seller with his scales set up and various types of goods, such as cheeses and so forth, there in front of him, all of stone. The other tradesmen were seated in their shops, while elsewhere some people were standing and others sitting. I saw stone men, women and children, and when I got to the merchants' market, everything that I saw – including the seated merchants and the goods that filled their shops – was stone, while their fabrics were as insubstantial as spiders' webs. I started to look at them, but whenever I touched a robe, it crumbled to dust in my hands. I opened one of the chests that I found and in it were purses of gold. The purses themselves disintegrated at my touch but the gold was unchanged. I carried off as much as I could, saying to myself: 'If only my brothers had been with me, they could have taken as much of this as they needed and they could have helped themselves to these ownerless treasures.'

I went into another shop, where there was even more, but as I couldn't carry anything else, I left and went on from market to market, looking at all kinds of stone creatures, including dogs and cats. When I got to the goldsmiths' market, I saw men sitting in their shops holding some of their wares in their hands, while other pieces were in baskets. At the sight of this, I threw away my gold and took as many of these as I could carry. From there I passed on to the jewellers' market and saw the owners sitting in their shops, each with a basket of precious stones in front of him, containing sapphires, diamonds, emeralds, balas rubies and other gems. As the jewellers themselves were all of stone, I threw away the ornaments and picked up as many gems as I could carry, still regretting the fact that my brothers were not there to take what they could.

After leaving the jewellers' market, I passed a huge door, finely decorated and most elegantly ornamented. Inside it were benches on which sat eunuchs, soldiers, guards and officials, all splendidly dressed and all of stone. I touched one of them and the clothes that he was wearing melted from his body like a spider's web. I went through the door and discovered a palace unequalled in the splendour of its architecture and workmanship, its council chamber filled with men of rank, viziers, leading citizens and emirs, seated on chairs and all of stone. On a golden chair, studded with pearls and other gems, a man was sitting, splendidly dressed and wearing an imperial crown set with precious jewels that shone as brightly as the day, and when I went up to look at him, he too was of stone. From there I went to the door of the women's quarters and, on entering, I found their audience chamber, where there was another chair of red gold studded with pearls and other gems. On this sat a queen with a crown set with precious gems, and on chairs around her sat women beautiful as moons wearing the most magnificent of coloured robes, while to serve them eunuchs were standing with their arms crossed over their breasts. The room itself would dazzle all who looked at it with its decorations, its remarkable paintings and its sumptuous furnishings. Hanging there were magnificent lamps of pure crystal, and in every crystal globe was a unique jewel beyond all price.

I threw away what I was carrying and started to collect as many of these gems as I could carry. I didn't know what to take and what to leave, since what I saw there looked like a city's treasure house, but then I caught sight of a little door that was standing open, leading to a flight of stairs. I went through and climbed up forty steps and then I heard someone reciting the Quran in a melodious voice. I walked in the direction of the voice and found myself at the door of a room where there was a silk curtain worked with threads of gold, set with pearls, coral, sapphires and emeralds, all gleaming like stars. The voice was coming from the far side of the curtain and so I went up and lifted it, to discover a door ornamented with bewildering beauty. Then, when I went in, I found what looked like a version of a talismanic treasure chamber set on the surface of the earth. Inside was a girl like a sun shining in a clear sky, dressed in the most splendid of robes and decked out with the richest of jewels. She was lovely, shapely and perfect in her elegance, with a slender waist and heavy buttocks. Her saliva could cure the sick; her eyelids drooped languorously; and it was as though it was to her that the poet was referring in his lines:

> I greet the figure that the robes enclose,
> With roses in the gardens of her cheeks.
> The Pleiades seem fixed upon her brow,
> With other stars a necklace on her breast.
> Were she to wear a rosebud dress,
> The rose leaves would draw blood from her soft flesh,
> And were she to spit once in the salt sea,
> The sea would all be honey sweet.
> Were an old man, leaning on a staff, to lie with her,
> Old as he was, he would hunt lions down.

At the sight of this girl, I fell in love. I went up to her and found that she was seated on a high dais, reciting the book of God, the Great and Glorious, from memory, and her voice was like the sound made by the gates of Paradise when opened by Ridwan. The words came from her lips like scattered jewels, while her lovely face was radiant and blooming, fitting the poet's description:

> Your words and qualities bring me delight,
> Enhancing the desire and longing that I feel.
> You have two qualities that melt the lover's heart,
> David's melodiousness and Joseph's face.

When I heard her melodious recitation, my heart, slain by her glance, stammered the words of the Quran: 'Peace, a saying of the Compassionate Lord'.* I looked and was so bemused that I could not greet her properly, for I was as the poet has described:

Longing confused me so I could not speak;
The entrance to her sanctuary cost me my life.
I only listen to the critic's words
In order to call on her to be my witness.

I then nerved myself to endure love's terrors and said: 'Peace be upon you, guarded lady and sheltered pearl. May God preserve the foundations of your prosperity and exalt the pillars of your glory.' She replied: 'Peace, greeting and honour be on you, 'Abd Allah ibn Fadil. I welcome you, my darling and the delight of my eye.' 'How did you know my name?' I asked her. 'Who are you and how is that that all the people in this city have been turned to stone? Please tell me what has happened, for what amazes me about the city and its inhabitants is that you are the only living creature here. For God's sake, tell me the truth about this.' She said: 'Sit down, 'Abd Allah, and, God willing, I shall tell you all about myself and about this city and its people. There is no might and no power except with God, the Exalted, the Omnipotent.'

I took a seat by her side, AND SHE WENT ON:

You must know, 'Abd Allah, that I am the daughter of the king of this city and it was my father whom you saw sitting on the high throne in the council chamber, surrounded by his officers of state and the leading men of his kingdom. He was a man of great power, with an army of one million, one hundred and twenty thousand men. He had twenty-four thousand emirs, all of whom were governors holding state offices, and he ruled over a thousand cities, not to mention towns, estates, fortresses, citadels and villages. A thousand Bedouin emirs owed him allegiance, each of whom commanded twenty thousand riders, while his money, treasures, precious stones and jewels were such as had never been seen or heard of before . . .

---

* Quran 36.58.

Night 983

Morning now dawned and Shahrazad broke off from what she had been allowed to say. Then, when it was the nine hundred and eighty-third night, SHE CONTINUED:

I have heard, O fortunate king, that THE DAUGHTER OF THE KING OF THE CITY OF STONES TOLD 'ABD ALLAH:

My father had wealth and treasures such as had never been seen or heard of before. On the battlefield he was a conqueror of kings and a destroyer of brave heroes, feared by tyrants and a subduer of emperors. For all that, however, he was an unbeliever, a polytheist who worshipped idols rather than the True Lord, as did all his men.

One day he was seated on his throne surrounded by his grandees, when suddenly in came a man, the radiance of whose face illumined the council chamber. My father looked at him and saw that he was dressed in green, as well as being tall, with hands that hung down below his knees; his dignity was awe-inspiring and light shone from his face. 'False oppressor,' he said to my father, 'how long will you deceive yourself with your idolatry and neglect to worship the Omniscient King? Say: "I bear witness that there is no god but God and that Muhammad is His servant and His messenger." Accept Islam, both you and your people, and abandon your worship of idols, which can do you no good and cannot intercede for you. The only true object of worship is God, Who raised up the heavens without pillars and unfolded earth and sea out of pity for His servants.' My father said: 'Who are you, who rejects the worship of idols and dares to speak as you have done? Are you not afraid that the idols may be angry with you?' 'The idols are stones,' the man replied. 'Their anger cannot hurt me, nor would their approval help me. Bring out the idol whom you worship and tell every one of your people to bring his own. When you have them all here, call on them to show their anger against me, while I call on my Lord to show his anger against them, and then you will see the difference between the wrath of the Creator and that of the created. You yourselves fashioned your idols, and devils then entered into them. It is these devils who speak to you from within them; the idols themselves are things that have been made, while my God is the Maker, the Omnipotent. If you can see what is true, follow it, and abandon what you see to be false.' The people said: 'Bring a proof of your Lord for us to see,' to which he replied: 'Bring me proofs of your own gods.' My father gave orders that everyone who had an idol that he worshipped was to fetch it, and all these were then brought into the council chamber.

So much for them, but as for me, I was sitting behind a curtain, looking down at my father's council chamber. I had an idol of my own, man-sized and made of green emerald. When my father asked for it, I sent it to him and it was placed beside his own one, which was made of sapphire. The vizier's idol was made of diamonds, while as for those belonging to the army officers and others, some were of hyacinth gems, others of carnelian, coral, aloes wood or ebony. There were a number of silver idols, together with others of gold. Everyone had whatever he could afford, while as for the common soldiers and the citizens, some of theirs were of flint and others of wood, pottery or clay, and they were all of different colours – yellow, red, green, black or white.

The stranger said to my father: 'Call on your idol and on these others to show their anger against me.' So they arranged them in the form of a court, with my father's idol placed on a golden throne, with mine by its side at the head of the room, while all the others were set out in order of precedence, according to the rank of their worshipper. My father began to prostrate himself to his idol, saying: 'My god, you are the gracious lord and among the idols none is greater than you. You know that this man has come here to attack your divinity and to mock you, claiming that he has a god who is stronger than you and telling us to abandon your service for that of this god of his. My god, show your anger against him.' He kept on imploring his idol, but the idol made no reply and when it did not speak, he said: 'My god, this is not your custom. When I spoke to you, you used to reply to me, so why do you now stay silent and say nothing? Are you not paying attention or are you asleep? Rouse up! Help me and speak to me!' He shook it, but it still neither spoke nor moved from its place, and the stranger said to my father: 'Why is it saying nothing?' He replied: 'I think that it cannot be paying attention or else is sleeping.' 'Enemy of God,' said the stranger, 'how can you worship a god who cannot speak and who has no power, in place of One Who is at hand to answer prayer and is never absent, heedless or asleep? He cannot be grasped by the imagination of men; He sees but is not seen and He has power over all things. Your god cannot protect itself from harm and within it lurks an accursed devil who misleads and deceives you. This devil has now left, so worship the True God, and acknowledge that there is no other god who deserves your veneration and your service but Him, and He is the only good. As for this god of yours, as he cannot even protect himself, how can he protect you? Look and see with your own eyes how powerless he is.' He then went up to the idol and started to strike it on the neck until it fell to the ground. My father was angry and called out to those who were there: 'This unbeliever has struck my god. Kill him.' They wanted to rise and strike him, but not one of them was able to move from his place. The stranger then offered them conversion to Islam, but they refused, at which he said: 'I shall show you the anger of my God.' 'Show us, then,' they said, and he spread out his hands and called: 'My Lord and God, in You is my trust and my hope. Answer my prayer and curse these evil-doers who accept the good things You give them but worship other gods. I pray You, Who are the mighty Truth, the Creator of night and day, to turn them into stone, for nothing is beyond Your power as You are the Omnipotent.' At that, God turned everyone in the city to stone.

As for me, when I saw the proof that the stranger brought, I surrendered myself to God and was saved from the fate of the others. The stranger came up to me and said: 'Your good fortune has been preordained by God in accordance with His will.' He started to teach me, and I gave him my faithful pledge. I was seven years old at the time and I am now thirty. I said to the stranger: 'Sir, thanks to your pious prayer, everything in this city and all its people have been turned to stone, while I have been saved because I was converted to Islam at your hands and you are my *shaikh*. Tell me your name; help me and provide me with food to eat.' He told me that his name was Abu'l-'Abbas al-Khidr and with his own hand he planted a pomegranate tree for me. It grew large, sprouted leaves, flowered and instantly produced a single pomegranate. Al-Khidr said: 'Eat what God has provided for you and give Him the worship that is His due.'

He then taught me about Islam and instructed me how to pray and how to worship God, as well as teaching me how to recite the Quran. For twenty-three years after that I have been worshipping God here, and every day the tree produces a pomegranate for me, which I eat and which sustains me until the next one grows. Al-Khidr, on whom be peace, visits me every Friday and it is he who told me your name and gave me the good news that you would come to visit me here. He told me that, when you came, I was to treat you with honour and do whatever you told me, without disobedience. I was to be your wife and you were to be my husband and I was to go with you wherever you wanted. As soon as I saw you I recognized you, and this is the story of the city and of its people.

She then showed me the pomegranate tree on which there was a single fruit, half of which she ate. The other half she gave to me and never have I come across anything that tasted sweeter, purer or more delicious. I asked her whether she was content to follow the instructions of her *shaikh*, al-Khidr, on whom be peace, and to go home with me as my wife to live in the city of Basra. 'Yes,' she said, 'if this is the will of Almighty God. I shall listen to you and obey you without dispute.' We made a binding promise, and after that she took me to her father's treasury, from which we removed as much as we could carry. We then left the city and walked back to my brothers, whom I found searching for me. 'Where have you been?' they said. 'You have been so slow in coming back that we were worried about you.' For his part, the captain said: ''Abd Allah, you have stopped us putting to sea although the wind has been favourable for some time now.' 'That doesn't matter,' I said, 'and it may be that it was a good thing to delay. It was not for nothing that I stayed away, and I got all that I could have hoped for. How well the poet expressed it when he said:

> I do not know, when I make for a land
> In search of profit, what I shall find there.
> Will it be the good for which I look,
> Or will it be the harm which looks for me?'

I went on: 'See what I found when I was away,' and I showed them the treasures that I had with me, explaining to them what I had seen in the City of Stones. 'Had you done what I told you,' I added, 'and come with me, you could have got quantities of this treasure.'

Night 984

Morning now dawned and Shahrazad broke off from what she had been allowed to say. Then, when it was the nine hundred and eighty-fourth night, SHE CONTINUED:

I have heard, O fortunate king, that 'Abd Allah said to his brothers: 'Had you gone with me, you could have got quantities of this treasure.' 'By God,' they said, 'even if we had gone, we would never have dared to intrude on the king.' I told my brothers: 'It's all right. I've enough for all of us and everyone can have a share in it.' I then divided up what I had according to our numbers, giving a share to each of my brothers and another to the captain, while I took the same amount for myself and gave what I could spare to the servants and the sailors. Everyone was delighted and called down blessings on me except for my brothers, who were the only people not to be pleased with what they had been given. I could see by the way their faces fell and their eyes rolled that they were in the grip of greed. 'It looks as though what I gave you was not enough,' I told them, 'but after all, we are brothers; there is no difference between us, and my money is yours. When I die, you will be my only heirs.'

After I had tried to win them over, I brought the princess on board the ship and took her to the cabin, where I had food sent to her, and after that I sat talking with my brothers. They asked me what I intended to do with this lovely girl and I said: 'I propose to draw up a marriage contract for her when I get to Basra and then give a huge wedding feast before consummating the marriage.' One of them said: 'She is so remarkably beautiful that I have fallen in love with her, and I would like you to give her to me, so that I can marry her myself.' 'It's the same with me,' my second brother said. 'Let me have her so that I can marry her.' I told them that I had made a binding promise to marry her, adding: 'Were I to give her to one of you, I would be breaking the pledge that we made between us and it might well distress her, as she only came with me on condition that I would marry her. How could I then marry her off to someone else? You say that you love her, but I love her more, for I was the one who found her. As for my giving her to one of you, that is something that will never happen, but if we get to Basra safely I shall search out two of the most eligible girls in the city and ask them in marriage for you, paying the dowries from my own money. I shall arrange for a joint wedding feast and the three of us can consummate our marriages on the same night. But leave this girl alone, for she is mine.' They said nothing and I thought that they had accepted what I had said.

As we sailed on, heading for Basra, I continued to supply the princess with food and drink. She never came out of the cabin and I slept between my brothers on deck. Things went on like this for forty days until we were delighted to discover that we were in sight of Basra. I had no suspicions of my brothers and was at ease with them, for no one knows what the future holds except Almighty God. As I was lying fast asleep that night, suddenly, before I knew what was happening, I found myself lifted up by my brothers, one holding me by the legs and the other by the arms, and it turned out that, because of the princess, they had decided to throw me overboard. Finding myself in this position, I asked them why they were doing this to me, and they said: 'You boor, how could you sell our goodwill for the sake of a girl? Because of that we are going to throw you into the sea' – which they then did.

'Abd Allah turned to the dogs and said: 'Have I told the truth or not, brothers?' To the astonishment of the caliph, they lowered their heads and began to whine as though to confirm what he had said. 'ABD ALLAH CONTINUED:

When they threw me in, I sank to the bottom before rising back up to the surface. Then all of a sudden a huge bird, as large as a man, swooped down, snatched me up and flew off with me into the upper air. When I opened my eyes, I found myself in a lofty and strongly built palace adorned with splendid paintings and with strings of jewels of all shapes and colours. Maids were standing there, arms crossed over their breasts, and among them a lady was sitting on a throne of red gold adorned with pearls and other gems. So brightly did the jewels gleam on the robes she was wearing that no one could look at them without shielding their eyes. Around her waist was a jewelled girdle such as no money could buy, and on her head was a triple crown of bewildering magnificence, stealing hearts and delighting the eyes.

At this point, the bird that had carried me off shook itself and turned into a girl as lovely as the shining sun. As I stared at her, I recognized that it was she who had been on the mountain in the form of a snake when the other snake had attacked her and wound its tail around hers, before I killed it with a stone, seeing that it had got the better of her. The lady sitting on the throne asked the girl: 'Why have you brought this mortal here?' and she said: 'Mother, this is the man who saved my honour among the daughters of the *jinn*.' She then asked me whether I knew who she was and when I said no, she told me: 'I was on the mountain when the black snake attacked me, trying to rape me, and you killed it.' I said: 'I saw nothing but a white snake and a black one,' but she explained: 'The white snake was I, and I am Sa'ida, the daughter of the Red King of the *jinn*, while the lady seated here is my mother, Mubaraka, the Red King's wife. The black snake that was struggling with me and trying to rape me was Darfil, the misshapen vizier of the

Black King. It so happened that he had seen me and fallen in love with me, but when he asked my father for my hand, my father sent him a message to say: "Who are you, miserable vizier, to marry the daughter of a king?" This enraged Darfil, who swore that he would rape me in order to confound my father, and because of that he started to track me, following me wherever I went. He and my father fought great wars and faced enormous difficulties, but my father could not get the upper hand, as Darfil was a powerful and wily opponent. Every time my father was pressing him hard and was about to overcome him, he would escape and my father could not catch him. Every day I changed into a different shape and colour, but whatever shape I took, he would take its opposite, and wherever I fled, he would sniff me out and catch up with me. When I took the form of a snake, I had been suffering the greatest of hardships, but he turned himself into a huge serpent and came after me. He caught up with me and, after a struggle, he had mounted me and was about to have his way with me when you came up and struck him dead with your stone. I then showed myself to you as a girl and told you that you had done me a favour such as only the lowest of the low would allow to pass unrepaid. When I saw that your brothers had plotted against you and thrown you into the sea, I came as fast as I could to rescue you, since both my parents owe you a debt of gratitude.'

She then said to her mother: 'Mother, honour this man in return for his defence of my honour.' 'You are welcome, mortal,' the queen said, 'for you have done us a service that has earned you our favour.' She then ordered me to be given a precious robe from a talismanic hoard, together with a quantity of jewels and precious stones. Then she gave instructions that I was to be taken to the king in his council chamber, and there I saw him seated on a throne surrounded by various types of *jinn*. At the sight of him, I had to look away because of the splendour of his jewels, but when he saw me he rose to his feet, as did his guards, in a sign of respect. He welcomed me warmly, showing me the greatest honour and lavishing his favours on me. Then he told his servants to take me back to his daughter, who was to return me to the place where she had rescued me. When they had brought me to her, she flew off with me and with the gifts that I had been given.

So much for me and the *jinn* princess, Sa'ida, but as for the captain of the ship, he had been roused by the splash when my brothers threw me into the sea. 'What has fallen overboard?' he asked, and my brothers started beating their breasts with tears in their eyes, calling out: 'Alas for the loss of our brother! He went to relieve himself over the side of the ship, but he fell into the sea.' They then laid their hands on my money, but quarrelled over the girl, each saying: 'No one else shall have her.' As they continued to quarrel, they forgot about me and how I had been lost, and they stopped pretending to be sorry. It was at that point that Sa'ida landed with me in the middle of the ship.

Night 985

Morning now dawned and Shahrazad broke off from what she had been allowed to say. Then, when it was the nine hundred and eighty-fifth night, SHE CONTINUED:

I have heard, O fortunate king, that while they were engaged in quarrelling, Sa'ida landed with 'Abd Allah in the middle of the ship. HE WENT ON:

When my brothers saw me, they embraced me with a show of gladness and said: 'How did you survive your accident? We were concerned about you.' Sa'ida said: 'Had you really been fond of him, you would not have thrown him into the sea while he was asleep. Now choose how you want to die.' At that, she took hold of them and was about to kill them when they cried out, asking me to protect them. I tried to intervene, saying to Sa'ida: 'I appeal to your honour; don't kill my brothers,' but she continued to insist that she was going to do it because of their treachery. I went on trying to placate and conciliate her and eventually she said: 'For your sake, I shall not kill them but I am going to put a spell on them.' She took a bowl into which she poured some seawater and, after pronouncing some unintelligible words, she said: 'Leave your human shape and become dogs.' Then she sprinkled them with the water and they became the dogs that you can see now.

'Abd Allah turned to them and asked them if he had been telling the truth, and they lowered their heads as if to say that he had. HE THEN CONTINUED:

After Sa'ida had transformed my brothers, she announced to everyone on board: 'Know that 'Abd Allah ibn Fadil has become my brother and I shall visit him once or twice every day. If anyone thwarts him or disobeys his orders, injuring him with hand or tongue, I shall treat him as I have treated these two traitors and he will spend the rest of his life in the shape of a dog, from which he will find no escape.' Everyone said: 'Lady, we are all his servants and his slaves, and we shall never do anything to oppose him.' Then she told me: 'When you get to Basra, check everything you own and if anything is missing, let me know and I shall fetch it for you from whoever has it, wherever it is, and I shall turn whoever took it into a dog. Then, when you have stored away all your goods, put collars round the necks of these two traitors and place them in a prison by themselves, tied to the leg of a couch. You must visit them every night at midnight and beat each of them unconscious, and if on any night you fail to do this, I shall come myself and beat first you and then them.' 'To hear is to obey,' I replied. She told me to tie them up until I got to Basra, and so I put ropes around their necks and fastened them to the mast. She then left.

The next day, we entered Basra and while the merchants who came out to meet me greeted me, not one of them asked about my brothers. They looked at the dogs and asked me what I was doing with them, now that I had brought them there. I told them that I had looked after them on the voyage and was taking them with me, and they laughed, not realizing who the 'dogs' really were. I put them in a room and busied myself that night in sorting out the bundles containing fabrics and precious stones. I had been distracted by the merchants who had come to greet me and so I did not beat my brothers, tie them up, or do them any harm. Then I fell asleep, but before I knew what was happening Sa'ida had arrived. 'Didn't I tell you to put chains on their necks and to beat them?' she said. She took hold of me, brought out a whip and thrashed me until I lost consciousness and then she went to the room where the dogs were and whipped them both until they were almost dead. 'Beat them like this every night,' she said, 'and if you miss a single night, I shall beat you again.' 'My lady,' I told her, 'tomorrow I shall put chains around their necks and when night comes I shall beat them and I shall go on doing this every single night.' In spite of her insistence, next morning I did not find it easy to think of putting chains around their necks and so I went to a goldsmith and got him to make chains of gold for them. I used these as collars and tied up the dogs as I had been ordered and on the following night I beat them reluctantly.

This adventure took place during the caliphate of al-Mahdi, the fifth in the line* of al-'Abbas, whose favour I won by sending him gifts, so

* In reality, third in line.

that he appointed me governor of Basra. Some time later, I told myself that Sa'ida's anger might have cooled and so I let a night pass without beating my brothers, but she came and gave me a thrashing whose burning pain I shall remember for the rest of my life. From that time on, I have beaten them every night and this went on throughout the rest of al-Mahdi's caliphate, after which you became caliph and confirmed me in my position as governor of Basra. For twelve years, I have been doing this every night against my will, and when I have beaten them I console them, excusing myself and giving them food and water. They remain tied up and no one knew anything about them until you sent me Abu Ishaq to fetch the tax money. He discovered my secret and returned and told you. You then sent him back again to summon me and my brothers, and I brought them to you obediently. You asked me for the truth and I have now told you my story.

The caliph Harun al-Rashid was amazed by the plight of the dogs and he asked: 'Have you or have you not now forgiven your brothers for what they did to you?' 'Abd Allah replied: 'May God forgive them and clear them of all responsibility both in this world and the next, but I need them to forgive me for having beaten them every night for twelve years.' The caliph told him: 'If Almighty God wills it, I shall do my best to free them and to have them returned to their original human shapes. I shall reconcile you, so that for the rest of your lives you may live in loving brotherhood and as you have forgiven them, they must forgive you. So take them back to your lodgings and don't beat them tonight. Tomorrow everything will turn out for the best.' 'Abd Allah objected that if he stopped for a single night, Sa'ida would come and whip him, something that he no longer had the strength to endure. 'Have no fear,' the caliph said, 'as I shall sign a letter for you to give to her when she comes. If she excuses you when she has read it, the credit will be hers, but if she refuses to obey me, then entrust yourself to God and let her beat you. If she does this because you neglected to beat your brothers on one night and if she disobeys me by whipping you, as certainly as I am the Commander of the Faithful, I shall settle the matter with her.'

The caliph now wrote on a piece of paper two fingers' width in breadth and sealed the message with his seal. ''Abd Allah,' he said, 'when Sa'ida comes to you, tell her that the caliph, the ruler of mankind, has ordered you not to beat your brothers and has written this note for her, sending her his greeting. Then give her the letter and don't be afraid.' He got

'Abd Allah to swear that he would not beat the dogs, and 'Abd Allah, for his part, went back to his lodgings saying to himself: 'What can the caliph do about the daughter of the ruler of the *jinn* if she disobeys him and whips me tonight? But I shall put up with a beating and spare my brothers tonight, whatever pain this brings me.' He then reasoned to himself, saying: 'Were he not relying on some great support, the caliph would not have told me not to beat them.'

When he reached his lodgings, he removed the collars from his brothers' necks, exclaiming: 'My reliance is on God!' He started to comfort them, telling them: 'No harm is going to come to you, for the caliph, the sixth in the line of al-'Abbas,* has undertaken to rescue you. You have my forgiveness and, God willing, the time has come; you can be happy to hear that this blessed night will bring you freedom.' At that, they began to yelp as dogs do . . .

Night 986

Morning now dawned and Shahrazad broke off from what she had been allowed to say. Then, when it was the nine hundred and eighty-sixth night, SHE CONTINUED:

I have heard, O fortunate king, that 'Abd Allah said to his brothers: 'I bring you good news of happiness and delight.' When they heard this they began to yelp as dogs do, rubbing their cheeks against his feet as though they were calling down blessings on him and abasing themselves before him. He was sorry for them and started to stroke their backs until it was supper time. When the table was laid, he told them to sit and they sat there eating with him, to the amazement of the servants who were astonished to see how he was treating them. 'He must be mad or unhinged!' they exclaimed. 'How can the governor of Basra, a man greater than any vizier, be eating with dogs? Doesn't he know that the dog is an unclean beast?' As they watched the dogs eating decorously with 'Abd Allah, they had no idea that these were his brothers, and they kept staring until all three had finished their meal. 'Abd Allah then washed his hands and the dogs stretched out their paws and started to wash them, to the accompaniment of astonished laughter from all those who were standing there. 'In all our lives,' they told each other, 'we have

* In reality, fifth in line.

never seen dogs washing their paws after eating.' The dogs then sat down on seats beside 'Abd Allah, but none of the servants dared question this. Things went on like that until midnight when the servants were sent off to bed and the dogs fell asleep, each on his seat. The servants were saying to each other: ''Abd Allah has fallen asleep with the dogs.' 'If he ate with them at table, then they can sleep with him,' said another, adding: 'But this is the behaviour of a madman.' They themselves would not touch what had been left on the table, saying: 'How can we eat dogs' leftovers?' and they took everything there and threw it away, exclaiming: 'This is dirty!'

So much for them, but as for 'Abd Allah, he saw the earth suddenly split open and Sa'ida emerged. 'Why have you not beaten your brothers tonight,' she asked, 'and why have you removed their collars? Are you trying to thwart me because you don't take me seriously? I shall now beat you and then turn you into a dog yourself, just like them.' 'My lady,' he replied, 'I conjure you by the inscription on the seal of Solomon, son of David, on both of whom be peace, to bear with me until I have told you the reason for this, after which you can do what you want.' 'Tell me, then,' she said and he explained: 'I didn't beat them because I was ordered not to do that tonight by the Commander of the Faithful, Harun al-Rashid, the king of mankind, who made me swear to it. He sends you his greetings and he has given me a note written in his own hand which I am to pass to you. I did what he told me, as it is our duty to obey the caliph. Here is the note; take it and read it and then do what you want.' 'Give it to me, then,' she told him, and when he had passed it to her, she opened it and read: 'In the Name of God, the Compassionate, the Merciful; from the king of mankind, Harun al-Rashid to Sa'ida, daughter of the Red King – to continue: this man has forgiven his brothers and has abandoned his claim on them. I have ordered that they be reconciled and when this happens, their punishment should be lifted. If you oppose my decrees, I shall oppose yours and break your ordinances, while if you obey me and carry out my orders, I shall carry out yours. I tell you not to act against 'Abd Allah's two brothers, and if you believe in God and His Prophet, it is for you to obey my authority. If you forgive them, I shall repay you with all that God allows me, and as a sign of your acquiescence, remove their enchantment so that they may come to me tomorrow as free men. If you do not, I shall set them free in spite of you through the help of Almighty God.'

Sa'ida told 'Abd Allah that she would do nothing until she had gone to show the letter to her father, after which she would return quickly with his answer. She then pointed at the ground and went down into a crack which opened up in it, while 'Abd Allah in his delight invoked God's blessing on the caliph. Sa'ida for her part went to tell her father what had happened, and when she showed him the caliph's letter he kissed it and placed it on his head before reading it. When he found what was in it, he said: 'My daughter, we are bound to obey the authority of the king of men, whose writ extends over us. As we cannot disobey him, go back and set those two men free immediately, telling them that you are doing this because of the caliph's intercession. Were he to become angry with us, he could destroy every last one of us, so don't impose a burden on us that we cannot bear.' 'If he did become angry with us, what could he do?' Sa'ida asked and her father replied: 'There are many ways in which he has power over us. In the first place, he is superior in that he is a human; in the second, he is the caliph appointed by God; while thirdly, he constantly performs the dawn prayer with two *rak'as*. Were all the *jinn* from the seven lands to join together, they would not be able to do him any harm. If he lost his temper with us and gave a single cry as he was performing the *rak'as* of his dawn prayer, we would all have to gather obediently before him as sheep before a butcher. Were he to tell us to leave our lands and move to a desert, we could not stay where we were, and if he wanted to destroy us and told us to kill each other, that is what we would have to do, as we could not disobey him. If we disobey him, he will burn the lot of us and we shall have no way to escape him. In the same way, we have to obey any of God's servants who constantly perform two *rak'as* at dawn. You must not bring destruction on us for the sake of two men; go and free them before we become the victims of the caliph's just anger.'

Sa'ida went back to 'Abd Allah and let him know what her father had said, telling him to kiss the caliph's hands for her and to ask for his approval. She then brought out her bowl and, putting water into it, she recited a spell with unintelligible words. She sprinkled the dogs with the water and at her command they changed shape and reverted to their original human form, as the spell was lifted. Both of them recited the formula: 'I bear witness that there is no god but God and that Muhammad is the Prophet of God.' They then prostrated themselves before 'Abd Allah, kissing his hands and his feet and begging his forgiveness, while he for his part begged theirs. They sincerely repented, saying: 'We were tempted by Iblis and carried away by greed. We were justly punished by God, but forgiveness is one of the qualities of the noble.'

They continued to try to placate 'Abd Allah with tears and prot-
estations of regret for what had happened. It was then that he asked
them: 'What did you do with my wife, whom I brought from the City of
Stones?' They said: 'When Satan tempted us and we threw you into the
sea, we quarrelled and each of us insisted that it was he who was going
to marry her. She said: "Don't quarrel over me for I shall belong to
neither of you. My husband is lost in the sea and I shall follow him,"
after which she threw herself overboard and died.' 'Abd Allah exclaimed:
'This was a martyr's death! There is no might and no power except with
God, the Exalted, the Omnipotent!' He wept bitterly and said: 'You were
wrong in what you did, as you have deprived me of my wife.' 'We were
at fault,' they acknowledged, 'but God punished us and this is something
that He had decreed before we were born.' 'Abd Allah accepted their
excuse, but Sa'ida said: 'Are you really going to forgive them after what
they have done to you?' 'Whoever is in a position of power and forgives
is rewarded by God,' he told her. 'Be on guard against them,' she warned
him, 'for they are treacherous.' Then she said goodbye to him and left.

Night 987

Morning now dawned and Shahrazad broke off from what she had been
allowed to say. Then, when it was the nine hundred and eighty-seventh
night, SHE CONTINUED:

I have heard, O fortunate king, that after Sa'ida had warned 'Abd
Allah against his brothers, she took her leave of him and went on her
way. 'Abd Allah and his brothers spent a joyful and relaxed night, eating
and drinking, and in the morning 'Abd Allah took them to the baths.
When they came out, he presented each of them with costly robes; he
called for food and when this was brought, they all ate. The servants
recognized his brothers when they saw them and, after greeting them,
they congratulated 'Abd Allah on his reunion with his dear kinsmen and
asked where they had been all this time. He told them: 'It was they whom
you saw in the shape of dogs and, God be praised, they have been freed
from that imprisonment and painful torment.' Then he went off to
present them to the caliph in his audience chamber, kissing the ground
before him and praying that God would prolong his glory and fortune,
removing all hardship and misfortune. The caliph welcomed him and
asked what had happened to him. 'Commander of the Faithful,' 'Abd

Allah replied, 'may God enhance your dignity, when I took my brothers home I was easy in my mind because you had undertaken to rescue them, telling myself that when kings set their minds on something, they succeed because divine providence aids them. I unfastened their collars, putting my trust in God, and the three of us shared the same table. When my servants saw me eating with what seemed to be dogs, they thought that I was becoming feeble-witted, and they said to each other: "He must have gone mad! How can the governor of Basra, a man more important than the vizier himself, eat with dogs?" They threw away all the food that had been left, saying: "We're not going to eat dogs' leftovers," and they kept on talking about my lunacy. I could hear what they were saying, but I said nothing to them, for they didn't know that these were my brothers. At bedtime I dismissed them and lay down to sleep, but before I knew what was happening, the earth split open and out came Sa'ida, the Red King's daughter. She was angry with me and her eyes were like fire.'

He went on to tell the caliph all the details of his encounter with Sa'ida, what her father had said and how she had restored his brothers from their shape as dogs to human form. He ended by saying: 'And here they are in front of you, Commander of the Faithful.' The caliph turned and saw two young men, radiant as moons. 'May God reward you on my behalf, 'Abd Allah,' he said, 'for you have told me a useful thing that I did not know, and, God willing, never, as long as I live, will I neglect to perform these two *rak'as* before dawn.' He then reprimanded the brothers for what they had done to 'Abd Allah, but when they had made their excuses to him, he said: 'Shake hands and forgive each other, and may God excuse what has happened in the past.' He then turned to 'Abd Allah and said: 'Take your brothers as your assistants and give them your instructions.' For his part, he ordered them to obey their brother and, after giving them each a substantial gift, he told them to return to Basra. They left his court happily, while he himself was glad to have profited from the adventure by learning the importance of continuing to perform two *rak'as* before dawn. 'It is a true saying,' he exclaimed, 'that some people's misfortune can be to the benefit of others!'

So much for the caliph, but as for 'Abd Allah, he and his brothers left Baghdad in honour, dignity and pomp and travelled to Basra, from where the city dignitaries came out to meet him. The city was adorned with decorations and 'Abd Allah's party was escorted into it in a procession of unparalleled splendour. The people were there, calling down blessings on him, while he was scattering gold and silver for them, but in all this noisy welcome no one paid any attention to his brothers. The two became jealous and envious, and although he treated them as gently as a doctor would treat an inflamed eye, this only increased their hatred and envy, as the poet has said:

> I treat all men with courtesy, but the envious
> Are seldom to be won by this and such success is rare.
> How can you please a man who envies you good fortune?
> His only pleasure would be if it left.

He provided them with the loveliest of concubines and assigned them forty eunuchs, forty attendants, forty slave girls and forty slaves, both black and white. Each of them was given fifty fine horses, together with a retinue of followers; revenues and allowances were assigned them and 'Abd Allah appointed them as his assistants, telling them: 'We are all equal and there is no difference between you and me . . .'

Night 988

Morning now dawned and Shahrazad broke off from what she had been allowed to say. Then, when it was the nine hundred and eighty-eighth night, SHE CONTINUED:

I have heard, O fortunate king, that 'Abd Allah assigned allowances to his brothers and appointed them as his assistants, saying: 'We are all equal and there is no difference between you and me. Ours is the authority here after God and the caliph, and it is for you to exercise this authority in Basra whether I am absent or present. Your orders must be carried out, but these orders must accord with piety. Be careful to avoid injustice, which, if prolonged, brings destruction, while to act with justice over a long period ensures prosperity. Do not harm God's servants, for they will curse you; word of this will reach the caliph and you will bring disgrace on yourselves and on me. Injure no man and if you are tempted by people's goods, take what you need and more from mine. You know well enough how clearly injustice is condemned in the verses of the Quran. How admirably the poet has expressed it in these lines:

> Injustice lies hidden in a young man's soul,
> Concealed by nothing but his lack of power.

The man of wisdom takes nothing in hand,
Until he sees the time is suitable.
The wise man's tongue is in his heart,
But, as for fools, their hearts are in their mouths.
The man who is not greater than his mind
Is killed by what is smallest in himself.
His own essential self is well concealed,
But what is hidden shows in what he does,
And if his origins are bad,
Nothing his mouth reveals is good.
Whoever imitates a fool in what he does
Becomes his equal in his foolishness.
If you broadcast your secrets far and wide,
Your enemies will become aware of them.
Your own affairs should be enough for you,
So leave what is no business of your own.'

He continued to admonish his brothers, urging them to do good and to refrain from injustice until he thought that because of his advice he had won their love. He began to rely on them more and more, showing them ever-increasing respect, but this merely added to their envy and hatred of him. They conferred with each other, and Nasir said to Mansur: 'How long are we going to be under the authority of our brother, 'Abd Allah, while he enjoys the authority of the emirate? He was a merchant and has now become an emir, rising from obscurity to grandeur. Nothing of the kind has happened to us and we are powerless and valueless, while he laughs at us. He has made us his assistants, but what does that mean? Are we not his servants and subordinates? We shall never rise or achieve importance as long as he has his health. The only way to get what we want is to kill him and seize his wealth, for we can't get hold of this until he is dead. Then we shall have power and be able to take all the jewels, precious stones and treasures that he has stored up and divide them between ourselves. After that, we can prepare a gift for the caliph and ask him for the governorship of Kufa, so that you can be governor of Basra while I have Kufa, or the other way round, and both of us will be seen clearly as men of importance. That can't happen, however, unless we manage to kill him.'

'You're right,' said Mansur, 'but how are we going to do that?' 'We shall hold an entertainment at one of our houses to which we shall invite him,' Nasir told him, 'and then we shall do all we can to look after him. We shall spend the evening talking to him, telling him stories and rare anecdotes, until, when he is tired out, we shall make up a bed for him on which to lie down. When he does and has fallen asleep, we can kneel on him and strangle him, before throwing him into the river. The following morning, we can say: "His *jinn* sister came to him while he was sitting talking to us and said: 'You scum, how dare you complain of me to the Commander of the Faithful? Do you think that we are afraid of him? He may be a king, but we are kings too. If he doesn't treat us with proper respect, we shall kill him in the cruellest way, but before doing this I shall kill you and wait to see what the caliph will do.' Then she seized him and disappeared through a crack in the earth, leaving us fainting, and when we recovered we didn't know what had become of him." Afterwards we can send news of this to the caliph and he will appoint us in his place. Later we can provide him with a splendid gift and ask him for the governorship of Kufa. One of us can then stay here while the other takes over Kufa and we shall get what we want through the enjoyment of a successful and victorious rule.' 'This is good advice,' said Mansur and so they agreed to kill 'Abd Allah.

After preparing for the entertainment, Nasir said to 'Abd Allah: 'I am your brother and I would like you and Mansur to do me the favour of coming to eat at my house, so that I can boast about it and people will say: "The emir 'Abd Allah went as a guest to his brother's house as an act of kindness."' 'Abd Allah agreed, adding: 'There is no difference between you and me, and your house is my house. However, you have invited me and only the ignoble refuse to accept hospitality.' He turned to Mansur and asked: 'Will you come with me to eat at Nasir's house in order to please him?' Mansur insisted that he would only go on condition that 'Abd Allah swore that, after he left Nasir's house, he would go and eat with him, adding: 'Is Nasir your brother and not I? If you do something to please him, you should please me as well.' 'Abd Allah agreed willingly and said: 'From Nasir's house I shall go on to yours, for as he is my brother, so are you.'

Nasir kissed 'Abd Allah's hand, left the audience chamber and prepared to receive his guests. The next day, 'Abd Allah rode to his house, taking with him a number of his men as well as his brother Mansur. After entering, he took his seat, as did his escort, together with Mansur. Nasir welcomed them and when the food had been brought in, they ate, drank and enjoyed themselves. Tables and plates were then removed and the company washed their hands, after which they spent the rest of the day until nightfall eating, drinking and amusing themselves, and when they had had their supper they performed the sunset and the evening prayers. They sat drinking together in a room by themselves, while 'Abd Allah's men were elsewhere, and first Mansur would tell a story and then Nasir would follow it with another, while 'Abd Allah listened. They kept on telling anecdotes, stories and remarkable histories until at last 'Abd Allah could stay awake no longer.

Night 989

Morning now dawned and Shahrazad broke off from what she had been allowed to say. Then, when it was the nine hundred and eighty-ninth night, SHE CONTINUED:

I have heard, O fortunate king, that, after a long evening and seeing that 'Abd Allah wanted to sleep, his brothers made up a bed for him and he took off his clothes and lay down. They lay down beside him on another bed and waited until they could see that he was sound asleep, when they got up and kneeled on his chest. He woke and, seeing what they were doing, he said: 'What is this, brothers?' 'We are no brothers of yours,' they told him. 'We don't recognize you, you boor, and it is better that you should die.' Putting their hands around his throat, they started to throttle him until he lost consciousness, and then, as he no longer moved, they thought that he was dead.

The room in which they were overlooked the river, and it was into this they then threw him. When he fell, however, God sent a dolphin to his rescue. There was a window in the kitchen that overlooked the river and this dolphin was in the habit of swimming beneath it. Scraps from slaughtered carcasses used to be tossed down into the water and the dolphin would go there to collect them. That day, because of the entertainment a large quantity of leftovers had been thrown in and the dolphin had been strengthened by eating more than its usual daily rations. When it heard the splash of something falling in, it swam up fast and came upon a man. Through God's guidance, it supported 'Abd Allah on its back and swam out with him until it was mid-stream, carrying on until it reached the opposite shore, where it deposited its burden on the bank.

As it happened, the place where the dolphin left 'Abd Allah was by a high road, and a caravan, which happened to be passing, saw the body lying on the shore. The travellers said: 'Here is a drowned man thrown up by the river,' and a number of them gathered round to look. The caravan leader was a virtuous man, with a wide knowledge of all branches of learning, an expert doctor and a sound judge of men. He asked what had happened and they told him that they had found the corpse of a drowned man. He went to examine the body and then said: 'This young man is still alive. He belongs to the upper classes and must have been brought up in splendour and luxury. God willing, there is hope for him yet.' He took 'Abd Allah, put clothes on him to keep him warm, and for the next three stages of the caravan's journey he tended him gently until he had regained consciousness. 'Abd Allah was suffering from shock and was very weak, as a result of which the caravan leader had to keep tending him with herbs he knew about.

He continued this treatment until the caravan had covered the distance of a thirty-day journey from Basra, reaching a Persian city called Auj. Here they halted in a *khan*, where a bed was made up for 'Abd Allah, but as he lay there he disturbed everyone by moaning constantly all night long. The next morning, the doorkeeper went to the caravan leader and asked him about the sick man they had with them, complaining of the disturbance. The leader told him: 'This is someone whom I found half drowned on the river bank. I treated him but I have not been able to cure him fully.' 'Show him to the *shaikha*, Rajiha,' the doorkeeper said, and when the leader asked him who this was, he explained: 'We have this beautiful virgin *shaikha* named Rajiha, to whom all our sick are brought. They spend the night with her and in the morning they are cured and whatever malignant condition they had has vanished.' The leader asked to be guided to her and was told to pick up his patient, which he did, and the doorkeeper then walked ahead of them until they came to a hermitage. There people could be seen going in with votive offerings, while others were coming out looking happy. The doorkeeper went in and came up to the curtain behind which the *shaikha* sat, to ask her permission to bring in a sick man. 'Bring him in behind the curtain,' she said, and he then told 'Abd Allah to enter.

When he came in and looked at the woman, he discovered that she was none other than his promised wife whom he had brought from the City of Stones. He recognized her and she recognized him, and when they had exchanged greetings, he asked how she came to be there. She said: 'When I saw your brothers throw you into the sea and then start quarrelling over me, I jumped overboard. I was rescued by my *shaikh*,

Abu'l-'Abbas al-Khidr, who brought me to this hermitage and gave me permission to cure the sick. He then announced throughout the city that all who were sick should go to the *shaikha*, Rajiha, and he told me to stay here until, in the fullness of time, my husband would arrive. I lay my hands on everyone who comes to me and the following morning they are cured. My reputation has spread far and wide; people bring me votive offerings and not only do I have great wealth but I enjoy honour and respect, as everyone in these parts comes to ask for my blessing.' She then laid her hands on 'Abd Allah and he was cured through the power of Almighty God.

Al-Khidr, on whom be peace, would come to visit her every Friday. She and 'Abd Allah had met on a Thursday evening and, when it grew dark and they had had a splendid meal, she sat with him, awaiting al-Khidr's arrival. Suddenly he appeared and he then transported them both from the hermitage to 'Abd Allah's palace in Basra, where he left them. In the morning, as 'Abd Allah stared out and recognized where he was, he heard a commotion and, looking from the window, he saw his two brothers crucified each on a separate cross.

The reason for this was that the morning after they had thrown him into the river, they had told a tearful story of how 'Abd Allah had been snatched away by the *jinn* princess. They prepared a present which they sent to the caliph, together with the news, and they asked him to confirm them as governors of Basra. He summoned them to Baghdad, where, in answer to his questions, they told him the story they had prepared. He, for his part, was furiously angry and that night before dawn he performed two *rak'as* as usual and then summoned the tribes of the *jinn*. When they obediently answered his call, he asked them about 'Abd Allah and they swore to him that none of them had harmed him and that they knew nothing about him. Sa'ida then arrived and told him what had really happened, after which he dismissed the *jinn*. The following day, he had Nasir and Mansur beaten and, when each of them had accused the other, the angry caliph ordered them to be taken back to Basra and crucified in front of 'Abd Allah's palace.

So much for them, but as for 'Abd Allah himself, he ordered their bodies to be buried and then rode off to Baghdad where he astonished the caliph with the full story, from beginning to end, of what his brothers had done to him. The *qadi* and the notaries were summoned and they proceeded to draw up a marriage contract between 'Abd Allah and the princess whom he had fetched from the City of Stones. The marriage was consummated and he stayed with her in Basra until they were visited by the destroyer of delights and the parter of companions. Praise be to the Living God, Who never dies.

A story is also told, O fortunate king, that in Cairo, the guarded city, there was a cobbler named Ma'ruf, who used to patch old shoes. His wife, Fatima, was an evil-minded, vicious and shameless intriguer who was nicknamed 'Dung'. She dominated her husband, and every day she would hurl abuse and a thousand curses at him, while for his part he was afraid of her evil nature and of the harm that she might do him. He was a sensible man, anxious to protect his honour, but he was poor. When he did well in his work, he would spend his earnings on her, but when there was only a little, she would take her revenge on him in the evening, ruining his health and making his night as black as the book of her deeds.* She fitted the poet's lines:

How many a night have I passed with my wife,
Spending it in the greatest misery!
When I first lay with her,
I wish I had brought poison and poisoned her.

One of the things that she did to Ma'ruf was to ask him to bring her a *kunafa* pastry covered with honey in the evening. He said: 'If God helps me to get it, I'll fetch it for you, but today I made no money at all.' 'I don't know what you're talking about,' she said . . .

Night 990

Morning now dawned and Shahrazad broke off from what she had been allowed to say. Then, when it was the nine hundred and ninetieth night, SHE CONTINUED:

I have heard, O fortunate king, that Ma'ruf the cobbler said to his wife: 'If God helps me to find the money for it, I shall fetch it for you this evening. I made no money today, but God may supply it.' 'I don't know what you're talking about,' she said. 'Whether God helps you or not, don't come back to me without it, for if you do, I'll make your night as miserable as your luck was when you married me and fell into my

* The record kept for each individual and used on the Day of Judgement.

hands.' 'God is generous,' he replied, and he went out, grief oozing from every pore. After he had performed the morning prayer, he opened his shop and prayed: 'My Lord, I implore You to provide me with enough to buy this *kunafa* pastry, so as to save me from the wickedness of that evil woman this evening.' He then sat in his shop until noon without getting any business and becoming more and more afraid of what his wife would do. He got up, locked the shop and began to wonder about the pastry, as he didn't even have enough money to buy bread. He passed by the shop of a *kunafa* seller, and stood there distractedly with his eyes bathed in tears. The shopkeeper noticed him and asked why he was weeping. Ma'ruf told him his story and said: 'My wife is a domineering woman. She has asked me for a *kunafa* pastry, but although I sat in my shop until past noon, I didn't even get the price of a loaf of bread, and I am afraid of her.' The shopkeeper laughed and said: 'No matter! How many *ratls* of pastry do you want?' 'Five,' said Ma'ruf and the man told him: 'I have the butter, but I don't have proper honey but only treacle, although this is better, and what harm would it do if you have the pastry with this?' Ma'ruf was ashamed to object as he was not being pressed to pay, and so he agreed to the treacle. The man cooked him the pastry and drowned it with treacle, making it into a gift fit for a king. 'Would you like some bread and cheese?' he asked, and when Ma'ruf said that he would, he gave him four *nusfs*' worth of bread and half a *nusf*'s worth of cheese, to add to the ten which the *kunafa* cost. 'You owe me fifteen *nusfs*,' he told Ma'ruf, 'but go back to your wife and enjoy yourself. Take this other *nusf* to spend at the baths, and you don't have to repay me for two or three days until God helps you earn some money. There is no need for you to keep your wife short, and I'm prepared to wait until you have some spare cash, over and above what you need for your expenses.'

Ma'ruf took the pastry, the bread and the cheese, and went off contentedly, calling down blessings on his benefactor and praising God for His goodness. When he got back to his wife, she asked if he had brought the pastry and he said yes, and put it in front of her. When she looked at it and saw that it was coated with treacle, she said: 'Didn't I tell you to bring one made with honey? You haven't done what I wanted – you had it made with treacle.' He tried to excuse himself, pointing out that he had had to buy it on credit. 'Nonsense,' she said. 'I only eat this pastry if it's made with honey.' In her anger she threw it in his face, telling him: 'Go off, you pimp, and get me another!' Then she struck him on the temple and knocked out one of his teeth, so that the blood ran down over his chest. This infuriated him and in return he gave her a light blow on the head, at which she took hold of his beard and started to call for help. The neighbours came in and, after freeing Ma'ruf's beard from her grasp, they began to blame and accuse her, saying: 'All of us are happy to eat this pastry made with treacle. It is disgraceful for you to bully the poor man like this.' Then they tried to humour her until eventually they managed to reconcile husband and wife, but when they had gone she swore that she wouldn't touch any of the pastry. For his part, Ma'ruf was suffering from the pangs of hunger and he said to himself: 'She may have sworn not to eat it, but I shall,' and he fell to. When she saw him eating, she said: 'God willing, this food will turn to poison and destroy your unmentionable body.' 'Not if you say so,' he told her, and he laughed and went on eating, saying: 'You swore that you wouldn't eat any of it, but God is generous and if He wills it, tomorrow night I shall fetch you a honey pastry and you can eat it all yourself.'

He then tried to console her, but she kept on cursing him with foul and abusive language all night long. When morning came, she rolled up her sleeve in order to strike him, but he said: 'Wait and I'll bring you another pastry.' He then left for the mosque and, having performed his prayer, he went on to open his shop, where he took his seat. He had scarcely settled down before two of the qadi's officers came to summon him before their master, telling him that his wife had laid a complaint against him. Recognizing her from the description that they gave, he exclaimed: 'May God bring her misery!' He then got up and went with the two to the qadi, where he found his wife with her arm bandaged and her veil stained with blood. She was weeping and wiping away her tears. 'Man,' said the qadi, 'are you not afraid of Almighty God? How can you strike this woman, breaking her arm, knocking out her tooth and behaving like this?' 'If I did hit her or knock out her tooth,' said Ma'ruf, 'then pass whatever sentence you want on me, but the truth is as follows, and the neighbours had to make peace between us' – and he told what had happened from beginning to end. The qadi was a kindly man and he produced a quarter of a dinar and told Ma'ruf to have a honey pastry made for his wife, after which they could be reconciled. 'Give it to her,' said Ma'ruf, and when she had taken it, the qadi tried to make peace between them. 'Woman,' he said, 'obey your husband and do you, in turn, be gentle with her.'

After the qadi had reconciled them, they left, each going off in a different direction. Ma'ruf went and sat in his shop, and while he was there the *qadi*'s men came back and demanded to be paid for their services. 'The *qadi* took nothing from me,' Ma'ruf objected, 'and, in fact, he gave me a quarter of a dinar.' 'It's no concern of ours whether he gave or took, but unless you give us our due, we'll take it from you by force.' They started to drag him around the market, and in order to give them half a dinar, he had to sell his tools. When they had left him, he sat sadly, hand on cheek, for without tools he could do no work. While he was sitting there, two ugly-looking fellows came in and told him to report to the *qadi* because his wife had laid a complaint against him. He told them that the *qadi* had reconciled them, but they said: 'We are from another *qadi*, and she has made her complaint to him.' He went off with them, calling on God to settle his account with her, and when he saw her, he said: 'My good woman, were we not reconciled?' 'There can be no reconciliation between the two of us,' she replied, at which he went to the *qadi* and told his story, pointing out that his colleague had just settled their differences. 'You harlot,' the *qadi* said, 'in that case, why have you come to complain to me?' She said: 'He hit me after that,' and the *qadi* said: 'Make it up; don't hit her again and she will not disobey you again.' When they had been reconciled once more, the *qadi* told Ma'ruf to pay his officers their fee. He did this and then went back and opened his shop, where he sat like a drunken man, reeling from the worries from which he was suffering.

As he was sitting there, a man came up to him and said: 'Hide yourself, Ma'ruf. Your wife has laid a complaint against you to the High Court and the bailiff is after you.' He got up, locked the shop and fled away in the direction of Bab al-Nasr. All that he had left from the sale of his lasts and his tools were five silver *nusfs*, four of which he used for bread and one for cheese. He was determined to keep out of the way of his wife, but this was on a winter's afternoon and when he came out from among the rubbish heaps, rain poured down on him as though from the mouths of water skins, soaking his clothes. He went to the 'Adiliya mosque and from there he spotted some ruins, among which there was a deserted building that no longer had any door. Wet through as he was, he went in to shelter from the rain, tears pouring from his eyes. He started to complain of his plight, saying: 'Where can I go to escape from that harlot? I pray to God to provide me with someone who might take me to a far-off land where she wouldn't know how to get to me.'

Suddenly, as he was sitting there weeping, the wall split open and out

came a tall and ghastly shape. 'Man,' it said, 'why have you disturbed
me tonight? I have been living here for two hundred years and I have
never seen anyone who came in and acted as you have done. I feel pity
for you, so tell me what you want and I shall do it for you.' 'Who and
what are you?' Ma'ruf asked him, and he replied: 'I am a *jinni*, the
familiar spirit of this place.' Ma'ruf then told him the whole story of his
dealings with his wife, and the *jinni* asked: 'Do you want me to take you
somewhere where your wife will not know how to follow you?' 'Yes,'
said Ma'ruf, at which the *jinni* told him to climb on to his back. When
he had done so, the *jinni* carried him off, flying with him from evening until dawn, before setting him down on the summit of a lofty
mountain . . .

Night 991

Morning now dawned and Shahrazad broke off from what she had been
allowed to say. Then, when it was the nine hundred and ninety-first
night, SHE CONTINUED:

'I have heard, O fortunate king, that the *jinni* flew off with Ma'ruf
and set him down on a lofty mountain, saying: 'When you go down, you
will see the gate of a city. Enter it, and your wife will never know how
to find her way to you or be able to reach you.' He then went off, leaving
the bewildered Ma'ruf there in a state of perplexity until the sun rose.
He then told himself: 'I have to get up and climb down to the city, as it
won't do me any good to stay sitting here.' When he reached the foot of
the mountain, he saw before him a city with high walls, imposing palaces
and finely decorated buildings, a delight to the eye. He went in through
the city gate, confronting a sight that would gladden the heart of the
sorrowful, and as he started to walk through the market, the inhabitants
gathered around to look at him, staring in surprise at his clothes, which
were not like theirs. One of them asked if he was a stranger, and when
he said that he was, the man asked where he had come from. 'From
Cairo, the fortunate city,' he said. 'Did you leave it some time ago?' the
man said, and when Ma'ruf replied: 'Yesterday afternoon,' he laughed
and called to the bystanders: 'Come and look at this fellow and hear
what he has to say.' 'What does he say?' they asked. 'He claims to be
from Cairo and to have left it only yesterday afternoon,' the man told
them, at which they all burst out laughing and said: 'Man, are you mad?

Do you claim to have left Cairo yesterday afternoon, arriving here this morning, when there is a distance of a full year's journey between the two cities?' 'You are the madmen here, not me,' Ma'ruf replied. 'I'm telling the truth; this is Cairene bread which I have with me and it's still fresh.' They started to look at the bread with astonishment, for it was not the same as theirs. The crowd grew bigger and, as people began to tell each other to come and see the bread from Cairo, Ma'ruf became a focus for gossip, with some people believing him and others making fun of him as a liar.

It was at that point that a merchant came up, riding on a mule and followed by two black slaves. He dispersed the bystanders, saying: 'Aren't you ashamed to be crowding around this stranger, making fun of him and laughing at him? What business is it of yours?' He went on heaping blame on them until he had driven them away, as none of them could answer him back. He then said to Ma'ruf: 'Come here, brother. None of this shameless crowd will do you any harm.' He took him off and brought him to a large and ornate house, where he seated him in a parlour fit for a king. On his instructions, his servants opened a chest and brought out a set of clothes such as a rich trader would wear. Ma'ruf was a handsome man, and when he had put these on he looked like a merchant prince. His host called for food and a table laden with all types of splendid dishes was set before them.

When they had eaten and drunk, his host asked his name, to which he replied that he was Ma'ruf, a cobbler by trade, who patched old shoes. 'Where do you come from?' asked his host. 'Cairo,' replied Ma'ruf, to which his host said: 'From what quarter?' 'Do you know Cairo, then?' Ma'ruf asked, to which the man replied that he was himself a Cairene. 'I come from Red Street,' Ma'ruf told him, and when he was asked who he knew there, he gave a long list of names. 'Do you know Shaikh Ahmad, the apothecary?' asked the man. 'He is my next-door neighbour,' Ma'ruf told him. 'Is he well?' was the next question, and when Ma'ruf said that he was, the man went on: 'How many children does he have?' 'Three,' said Ma'ruf. 'Mustafa, Muhammad and 'Ali.' 'And how are they getting on?' the man asked, and Ma'ruf replied: 'Mustafa is doing well as a learned schoolteacher; Muhammad is an apothecary who opened a shop next door to his father after getting married, and his wife has given birth to a son, Hasan.' 'May God bring news as good as this to you yourself,' said the man, and Ma'ruf went on: 'As for 'Ali, he was my boyhood friend. We were always playing together and we used to go to

the church disguised as Christians, steal their books and use the money
we got from selling them to buy ourselves treats. Once the Christians
saw us and caught us with one of their books. They complained to our
families, telling 'Ali's father that if he didn't stop his son robbing them,
they would lay a complaint before the sultan. To placate them, his father
gave him a beating and as a result 'Ali immediately ran away. No one
knows where he went and for twenty years he has not been heard of.' 'I
am 'Ali, son of Shaikh Ahmad, the apothecary,' said his host, 'and you
are my friend Ma'ruf.'

After they had exchanged greetings, 'Ali asked Ma'ruf why he had
come there from Cairo, and Ma'ruf told him about his wife, Dung
Fatima, and how she had treated him. 'When she had harmed me too
many times,' he explained, 'I ran away from her in the direction of Bab
al-Nasr. Rain was falling on me and so I went into a deserted building
in the 'Adiliya, where I sat weeping. It was then that an *'ifrit*, the familiar
spirit of the place, came out and asked me about myself. I told him about
the state I was in and he took me on his back and flew with me all night
long between the heavens and the earth, before setting me down on the
mountain here and telling me about the city. I walked down and when
I came into the city, people crowded around me, asking me questions. I
told them that I had left Cairo yesterday but they didn't believe me. Then
you came and drove them away, after which you brought me here. This
is why I left Cairo, but why was it that you came here?' 'Ali said:
'Youthful folly got the better of me at the age of seven, and from then
on I wandered from country to country and city to city until I arrived
here at Ikhtiyan al-Khutan. I found the people generous and sympathetic,
willing to trust a poor man and let him have credit, believing whatever
he said. I told them that I was a merchant, that my goods were following
on behind me and that I wanted a place where they could be stored.
They believed me and cleared out a place for me. I then asked whether
anyone would lend me a thousand dinars, to be repaid when my goods
came, explaining to them that there were some things that I needed
before they arrived. They gave me what I asked for and I set off to the
merchants' market, where I found and bought goods. The next day, I
sold them at a fifty-dinar profit and bought some more. I began to make
friends with the people, winning their affection by generous treatment,
and through my trading I made a lot of money. Remember the proverb:
"This world is all swagger and deceit." In a place where no one knows
you, you can do whatever you want. If you tell everyone who asks that

you're a poor cobbler running away from his wife and that you left Cairo yesterday, no one will believe you and you'll be a laughing-stock, however long you stay here. If you say that you were carried by an *'ifrit*, they will shy away; no one will approach you because they'll say that you are possessed and that harm will come to anyone who goes near you. That rumour would be bad for us both, because people know that I myself come from Cairo.'

'What am I going to do, then?' asked Ma'ruf. 'God willing, I'll teach you,' 'Ali replied, and he went on: 'Tomorrow I shall give you a thousand dinars, a mule to ride and a black slave to walk in front of you. He will bring you to the gate of the merchants' market, and when you go in you will find me sitting with them. As soon as I see you, I'll get up to greet you, kissing your hand and making you seem to be a person of the greatest importance. I'll ask you about various types of materials and when I say: "Have you brought such-and-such a fabric with you?" you must tell me that you have a lot of it. If they ask me about you, I shall sing your praises and tell them what a great man you are. I shall say that they should get you a storeroom and a shop, and as I'll stress that you are both rich and generous, whenever a beggar comes up to you, give him what you can. That will lead them to believe me, and as they will be convinced both of your importance and of your generosity, you will win their affection. After that, I shall invite them to a party in your honour and when you meet them, I shall introduce you to them all . . .'

Night 992

Morning now dawned and Shahrazad broke off from what she had been allowed to say. Then, when it was the nine hundred and ninety-second night, SHE CONTINUED:

I have heard, O fortunate king, that 'Ali the merchant said to Ma'ruf: 'I shall invite them to a party in your honour and when you meet them, I shall introduce you to them all. You can then start to trade with them and carry on normal business, and in no time you will be rich.'

The next morning, 'Ali handed Ma'ruf the thousand dinars, together with a suit of clothes, mounted him on a mule and gave him a black slave, telling him that there was no need for him to worry about repaying him. 'You are my friend,' he said, 'and it is up to me to treat you generously. Don't trouble yourself; forget how your wife treated you and don't tell anyone about her.' 'May God reward you,' replied Ma'ruf, and mounted the mule. The slave walked in front of him to the merchants' market, where 'Ali was already sitting, surrounded by all the others. Catching sight of the newcomer, he got up and threw himself on him, exclaiming: 'Here is Ma'ruf, the benevolent and generous. What a blessed day this is!' He kissed Ma'ruf's hand in front of all his colleagues, saying: 'Brothers, Ma'ruf the merchant has done you the favour of joining you.'

They greeted him and the respect that 'Ali showed him led them to believe that here was a man of importance. They greeted him after 'Ali had helped him dismount from his mule, and he took them aside privately, one after the other, and started to sing Ma'ruf's praises. 'Is he a merchant?' they asked. 'Yes, indeed,' 'Ali told them. 'He is the greatest of them all, and as far as wealth is concerned, no one has more. His fortune, together with those of his father and his forefathers, is famous among the merchants of Cairo and he has associates in Hind, Sind and Yemen. He is also an extremely generous man and so, bearing in mind the position he holds, you should show him respect and do what you can for him. I can tell you that it is not trade that has brought him here, but an urge to see foreign parts. He doesn't need to leave home to look for profit, as he has so much money that no fire could burn all of it. As for me, I am one of his servants.'

He went on eulogizing Ma'ruf and setting him on a pinnacle above the heads of all others, and his audience started talking to one another about his qualities. They flocked around him, bringing him food and drink for his breakfast, and even the senior merchant came and greeted him. In the presence of the others, 'Ali started to ask: 'Master, have you by any chance brought any material of such-and-such a kind with you?' To which Ma'ruf would answer: 'Yes, I have a lot of it.' Earlier that day 'Ali had showed him various types of costly fabrics and had taught him the names of both what was expensive and what was cheap. So when someone asked whether he had any yellow broadcloth, he said that he had it in quantity, and he gave the same answer when he was asked about cloth that was as red as gazelle's blood, replying in the same way about everything they asked. ' 'Ali,' said one of the merchants, 'I can see that if this fellow countryman of yours wanted to transport a thousand loads of precious fabrics, he would be able to do it.' 'Ali told him: 'If he took all that from a single one of his warehouses, it would still look full.'

As they were sitting there, a beggar came round the merchants, some of whom gave him various small coins, while most gave nothing at all. When the man came to Ma'ruf, he pulled out a handful of gold coins and passed them over. The beggar called down blessings on him and went off, leaving the admiring merchants to exclaim: 'That was a kingly gift, for he gave the beggar gold without even counting it, something that he would only have done if he were very prosperous and wealthy.' He then gave another handful of gold to a poor woman who approached him, and she too blessed him and went off to tell other poor beggars. These came up one after the other, and 'Ali kept on handing gold to every one of them, until he had given away a thousand dinars.

At that point, he clapped his hands together and recited the formula: 'God suffices for us and to Him we entrust our affairs.' The senior merchant asked him if anything was the matter, to which he replied: 'Most of the people here seem miserably poor and had I known that, I would have brought some money with me in my saddlebags to give to them. I'm afraid that I may be away from home for a long time and although it is not in my nature to turn away a beggar, I haven't any more gold. If one of them comes up to me, what am I to say to him?' 'Say: "May God sustain you," ' the man told him, but Ma'ruf objected: 'I am not in the habit of doing that, and this is something that worries me. What I want is a thousand dinars to give as alms until my baggage comes.' 'That's no problem,' the man said, and he sent off one of his servants to fetch a thousand dinars, which he then presented to Ma'ruf. Ma'ruf started to give these away to any poor person who passed by, until the time came for the noon prayer, and when they all went into the mosque to pray, he scattered the coins that were left over the heads of the congregation. When they realized what he was doing, they blessed him, while the merchants were astonished by his liberality. He then turned to another one of them, borrowed a thousand dinars and gave these away as well, while 'Ali looked on, unable to say a word. This went on until the afternoon prayer, when Ma'ruf entered the mosque, performed the prayer and distributed the rest of his money. By the time that the market was closed, he had received and given away five thousand dinars. All those from whom he had borrowed money were told to wait until his goods arrived, when they could be paid in gold if they wanted or else in fabrics, if they preferred, from his huge stock.

That evening, 'Ali invited Ma'ruf to a reception with all the other merchants, seating him in the place of honour. His talk was all about fabrics and jewels, and whatever anyone mentioned to him, he claimed to have it in bulk. The next day, he set out for the market and started turning to the merchants, borrowing money from them and distributing it to the poor. At the end of twenty days of this he had borrowed sixty thousand dinars, and no goods had arrived for him nor had anything happened to protect him from his creditors. They, for their part, were getting restive and saying: 'Nothing has come for him,' and asking: 'How long is he going to go on taking people's money and giving it away to the poor?' 'I think that we should talk to 'Ali, his compatriot,' said one of them, and so they went to 'Ali and pointed out that Ma'ruf's goods had not arrived. 'Wait,' 'Ali told them, 'for they are bound to come soon,' but when he was alone with Ma'ruf, he said: 'Did I tell you to toast the bread or to burn it? The merchants are clamouring for their money and they tell me that you owe them sixty thousand dinars, which you have taken and given away to the poor. How can you pay them back, as you are doing no trading?' 'What's all this about?' Ma'ruf asked. 'What is sixty thousand dinars to me? When my goods come, I shall pay them back in fabrics or in gold and silver, whichever they prefer.' 'In God's Name,' 'Ali said, 'do you have any goods?' 'Plenty,' Ma'ruf told him. 'May God and the saints repay you for this disgusting behaviour!' exclaimed 'Ali. 'Wasn't it I who taught you to say this? I'll tell everyone about you.' 'Go away and don't talk so much,' said Ma'ruf. 'Am I a poor man? I have a huge supply of goods and when they come, my creditors will be repaid twice over. I don't need them.' 'Ali grew angry and said: 'You mannerless lout, I'll teach you to tell me shameless lies.' 'Do what you want,' Ma'ruf told him. 'They'll have to wait until my goods arrive, and then they can have what they are owed and more.'

'Ali left him and went off, saying to himself: 'I started by praising him and if I now criticize him, I'll be seen as a liar and fit the proverb: "Whoever follows praise with criticism is a liar twice over."' While he was in this state of perplexity, the merchants came up and asked if he had spoken to Ma'ruf. He said: 'I'm too embarrassed to approach him. He owes me a thousand dinars, but I can't talk to him about this. You didn't ask my advice when you gave him money and so you can't blame me for this. Go and ask him yourselves, and if he doesn't repay you, then bring a complaint to the king, for if you tell him that you've fallen victim to a fraudster, he will come to your rescue.' So they went to the king and explained what had happened, saying: 'King of the age, we don't know what to do about this over-generous merchant.' They described Ma'ruf's behaviour and went on: 'Everything that he gets he distributes in handfuls to the poor. Were he short of money, he could never bring himself to give away such amounts of gold to them, whereas if he is wealthy, then it is the arrival of his baggage that will show whether he has been speaking the truth. He claims that he has come on ahead of it, but we ourselves have seen no trace of it. Whenever we talk of a certain type of fabric, he claims to have it in quantity, but although time has passed, we have heard nothing of the arrival of these goods of his. He owes us sixty thousand dinars, all of which he has given away to the poor, who are full of praise for him and extol his generosity.'

The king was a greedy man, more covetous than Ash'ab,* and when he heard how generous Ma'ruf was, his greed was aroused and he said to his vizier: 'If this merchant were not a very wealthy man, he would not be so generous. His baggage is bound to come, and when these merchants gather around him he will distribute large sums of money to them, money to which I have a better right. I want to make a friend of him and take him as a companion until his baggage arrives, and then I shall get whatever he gives them. He can have my daughter as his wife and I shall add his wealth to my own.' 'King of the age,' said the vizier, 'in my opinion this man is a fraudster and fraudsters bring ruin on the houses of the covetous.'

Night 993

Morning now dawned and Shahrazad broke off from what she had been allowed to say. Then, when it was the nine hundred and ninety-third night, SHE CONTINUED:

I have heard, O fortunate king, that the vizier told the king: 'In my opinion this man is a fraudster and fraudsters bring ruin on the houses of the covetous.' The king said: 'I shall test him, vizier, and find out whether he is a cheat or a truthful man and whether he has been brought up in luxury or not.' When the vizier asked how he proposed to do that, the king said: 'I shall send him a summons and when he comes here and sits down, I shall treat him politely and give him a jewel that I have. If he recognizes what it is and how much it is worth, then he must be a rich and prosperous man, but if he doesn't, then he is a fraud and a parvenu and I shall have him put to the worst of deaths.'

* Ash'ab was a notoriously greedy servant of the caliph 'Uthman (644–55).

Ma'ruf went to the palace in answer to a summons from the king. They exchanged greetings and the king seated him at his side. 'Are you Ma'ruf the merchant?' he asked, and when Ma'ruf said yes, the king went on: 'The merchants claim that you owe them sixty thousand dinars. Is what they say true?' Ma'ruf confirmed that it was, and when the king asked why he did not repay them, he replied: 'When my baggage train arrives, I'll give them twice as much as I owe, and they can have it in gold, silver or goods, whichever they prefer. Whoever is owed a thousand dinars will get two thousand back, because he saved me from disgracing myself among the poor, for I am a man of substance.' The king then said: 'Take this jewel and tell me what kind it is and what is its worth.' He handed over a gem the size of a hazelnut which he had bought for a thousand dinars; he was very proud of it, having no other like it. Ma'ruf took it in his hand and squeezed it between his thumb and forefinger. It proved too delicate to bear the pressure and so it shattered. 'Why have you destroyed my jewel?' asked the king, but Ma'ruf only laughed and said: 'King of the age, that was no jewel, but only a bit of mineral worth a thousand dinars. How can you call it a jewel, when a real jewel is something worth seventy thousand? This can only be described as a piece of mineral, and I myself am not concerned with any gem that is not the size of a walnut, since for me such a thing is valueless. How can you be a king and call a thousand-dinar piece like this a gem? You have an excuse, however, as your people are poor and have no valuable treasures.' The king said: 'Do you have any of the kind of jewels you have talked about?' and when Ma'ruf claimed to have many of them, greed got the better of him and he asked: 'Will you give me some of these real ones?' 'When my baggage comes, you can have them in plenty,' promised Ma'ruf, 'and as I have quantities of whatever you can ask for, I shall not charge you anything for them.' The delighted king dismissed the other merchants, telling them to wait until Ma'ruf's baggage arrived, when they could come back and he personally would pay them off. They then left.

So much for Ma'ruf and the merchants, but as for the king, he went to the vizier and told him to talk with Ma'ruf in a friendly way and to tell him about the king's daughter, so that a marriage might be arranged that would allow them to share in his fortune. The vizier said: 'King of the age, there is something about this man that I don't like. I think that he is a fraud and a liar, and I advise you not to talk like this, lest you lose your daughter for no return.' The vizier himself had earlier asked for the hand of this princess and her father had been ready to give her to him, but when she heard of it she had refused to accept him. So now the king said: 'Traitor, you don't want any good to come to me because my daughter turned you down when you asked for her hand. You hope to stand in the way of her marriage and would like her to stay unwed in order to give you a chance of winning her yourself. Listen to what I have to say. This is nothing to do with you and how can this man be a cheat and a liar? He could tell the price that I paid for my jewel and he broke it because he didn't like it. He has quantities of jewels and when he marries my daughter and sees how lovely she is, she will charm him, and in his love for her he will shower her with gems and treasures. What you want to do is to deprive both her and me of all these good things.'

The vizier stayed silent, as he was afraid of the king's anger, but he said to himself: 'Set the dogs on the cows.' He then made a friendly approach to Ma'ruf and said: 'His majesty is fond of you and he has a beautiful daughter whom he would like to marry to you. What have you to say?' Ma'ruf agreed to the offer, but added: 'Wait until my baggage arrives, for royal princesses need large dowries suitable for their rank and condition. At the moment I have no money, and the king had better wait until my goods come. As I am a rich man, for the princess's dowry I shall give five thousand purses of gold, and then I shall have to have a thousand purses to distribute to the poor and needy on the wedding night, with a thousand more for those who walk in the wedding procession. With another thousand I shall give a banquet to the troops and others, and I must have a hundred gems to present to the bride on the morning after the wedding, as well as a hundred for the slave girls and the eunuchs, since each of them must have one as a token of the bride's high rank. Then I shall need to clothe a thousand of the naked poor, as well as giving alms. This cannot be done until my baggage comes, but when it does, as I have plenty, I shan't have to worry about these expenses.'

The vizier went off and told all this to the king, who said: 'If this is what he proposes to do, how can you say that he is a fraudster and a liar?' 'I still do,' insisted the vizier, but the king reprimanded him harshly, swearing to kill him if he did not stop. He then told him to go back and bring Ma'ruf to him, saying that he would arrange things himself.

The vizier went to tell Ma'ruf of the king's summons. 'To hear is to obey,' Ma'ruf replied, and when he came, the king told him: 'Don't make this excuse. My own treasury is full, so take the keys: spend all you need; give away what you want; clothe the poor and do as you please. There is no need for you to concern yourself about the princess and the slave girls, for when your baggage comes you can be as generous as you like with your wife, and, until this happens, I am prepared to wait for her dowry, and I shall never be separated from you.' On his instructions, Shaikh al-Islam drew up a marriage contract between his daughter and Ma'ruf and began to prepare the wedding celebrations. The town was adorned with decorations, drums were beaten and tables laid with foods of all kinds. Performers arrived and Ma'ruf sat on a chair in a room with players, dancers, gymnasts, jugglers and mountebanks exhibiting their skills before him. On his instructions, the treasurer would fetch gold and silver and he would go round the audience with handfuls of money for the performers, gifts for the poor and needy and clothes for the naked.

This was a noisy celebration and the treasurer could scarcely keep up with the demands on the treasury's reserves, while as for the vizier, his heart was almost bursting with rage yet he could not speak. 'Ali was astounded at the amount of money being given away and he exclaimed: 'May God and the saints split your head, Ma'ruf! Wasn't it enough for you to waste the money of the merchants, that you had to do the same to the wealth of the king?' 'That is nothing to do with you,' replied Ma'ruf, 'and when my baggage comes, I'll give him back twice as much.' So he went on throwing away the money, telling himself: 'Something will happen to protect me; what will be will be, for no one can escape fate.'

The celebrations carried on for forty days and on the forty-first a magnificent bridal procession was organized, with all the emirs and the soldiers walking in front of the bride. When they brought her to Ma'ruf, he began to scatter gold over the heads of the people and huge amounts of money were spent. He was then escorted to the princess and took his seat on a high couch; the curtains were lowered, the door shut and everyone there went out, leaving Ma'ruf with his bride. For a time he sat there sadly, striking one hand against the other and reciting the formula: 'There is no might and no power except with God, the Exalted, the Omnipotent.' 'God preserve you, my master,' said the princess. 'Why are you so sad?' 'How can I fail to be sad,' he told her, 'when your father has thrown me into confusion? What he has done to me is like burning crops while they are still green.' 'What is it that he has done?' she asked. 'Tell me.' Ma'ruf explained: 'He has brought me to you before my baggage has arrived. I had wanted to have at least a hundred jewels to distribute to your maids, one for each of them, so that they might have the pleasure of saying: "My master gave this to me on the night he went in to my lady." This is something that would have added to your status and increased your reputation. Because I have so many jewels, I have not been accustomed to limiting myself when it comes to giving them away.' The princess replied: 'There is no need to vex yourself or to worry about that. I'm not going to hold it against you and I can wait until your baggage arrives, while as for my maids, you needn't bother about them. Take off your clothes now and relax, for when your goods are here, we can get the jewels and whatever else there is.'

So Ma'ruf got up and undressed before sitting back on the couch in order to dally and play with his bride. He put his hand on her knee and she sat down on his lap, thrusting her lip into his mouth. This was an hour to make a man forget his father and his mother. He put his arms around her, squeezing her tightly and drawing her close to his breast. He sucked her lip until honey dripped from her mouth, and when he put his hand beneath her left armpit, both their bodies felt the urge for union. He touched her between her breasts before moving his hand down between her thighs. He got between her legs and set about the two tasks, exclaiming: 'Father of the two veils!' before priming the charge, lighting the fuse, adjusting the compass and then applying the fire. All four corners of the tower were demolished as the strange adventure, which none can question, took place and the bride gave the shriek that is unavoidable.

Night 994

Morning now dawned and Shahrazad broke off from what she had been allowed to say. Then, when it was the nine hundred and ninety-fourth night, SHE CONTINUED:

I have heard, O fortunate king, that as the bride gave the shriek that is unavoidable, Ma'ruf took her virginity. This was a night standing outside the ordinary span of life, comprising, as it did, the union of beauties, with embraces, love-play, sucking and copulation lasting until morning. Ma'ruf then got up and went to the baths, from which he emerged wearing a royal robe. When he entered the audience chamber, all present rose to their feet to greet him respectfully and courteously, congratulating him and calling down blessings on him. He took his seat beside the king and asked: 'Where is the treasurer?' 'Here he is in front of you,' they told him, and he then gave instructions that robes of honour were to be given to all the viziers, emirs and officers of state. When all that he asked for had been fetched, he sat distributing gifts to everyone who came to him, in accordance with the man's status.

Things went on like this for twenty days, during which neither his baggage nor anything else made its appearance. The treasurer became extremely disgruntled and approached the king when Ma'ruf was absent and the king and the vizier were sitting by themselves. He kissed the ground before the king and said: 'King of the age, there is something that I must tell you, as you might blame me if I failed to bring it to your notice. You should know that the treasury is empty, or rather there is so little money left there that in ten days' time there will be none at all and we shall have to close it.' The king turned to the vizier and said: 'My son-in-law's baggage has been delayed and there has been no news of it,' but the vizier laughed and said: 'May God deal kindly with you, king of the age. You don't realize what this lying trickster is doing. I take my oath that he has no baggage at all, and there is no plague to get rid of him for us. He has gone on and on playing his tricks on you until he has managed to squander your riches and marry your daughter, all for nothing. How long will you let him get away with it?' The king asked how he could find out the real truth and the vizier told him: 'The only person who can discover a man's secrets is his wife. Send for your daughter to come and sit behind the curtain here so that I can ask her how things really stand and get her to test him and find out how he is placed.' 'There can be no harm in that,' the king said, adding, 'and I swear that if I discover him to be an impostor, I shall put him to the foulest of deaths.'

He brought the vizier to his sitting room and then sent for his daughter. She came and sat behind the curtain, all this being while Ma'ruf was away. She asked her father what he wanted and he told her to speak to the vizier. When she put the question to him, he said: 'My lady, you must know that your husband has squandered your father's wealth and has married you without paying a dowry. He keeps on making promises to us, but he never keeps them; there is no news of his baggage and, in short, we want you to tell us about him.' 'He talks a lot,' she replied, 'but although he is forever coming to me and promising me jewels, treasures and precious stuffs, I've not seen anything.' The vizier asked:

'My lady, could you discuss things with him this evening in the give-and-take of conversation? Get round to saying: "Don't be afraid to tell me the truth. You are my husband and I would do nothing to hurt you. If you tell me how things really stand, I'll think of some way to get you out of your difficulties." Then say whatever you think best as you talk to him and let him see that you love him. If you get him to confess, come and tell us the truth.' 'I know how to test him, father,' she agreed.

She went off and after supper when her husband came to her, as usual, she got up and, putting her arms under his, she did her best to ensnare him – and how well women are able to do this when there is something they want from a man! She went on flattering him with words that were sweeter than honey until she had stolen away his wits. When she saw that he was entirely taken up with her, she said: 'My darling, the delight of my eyes and the fruit of my heart, may God never deprive me of you and may Time never separate us. Love for you has lodged in my heart; the fire of my passion for you has consumed my entrails and I can never fail in my duty towards you. I want you to tell me the truth, for lying is unhelpful as it can never succeed all the time. How long are you going to go on trying to trick my father? I'm afraid that you will be exposed before I can think of some plan, and he will use violence against you. If you tell me the truth, all will be well for you and you need fear no harm. How many times are you going to claim to be a wealthy merchant with a baggage train on the way? You have been talking again and again about this baggage for a very long time, but no word has come of it and your face shows how worried you are. If it's not true, tell me and, God willing, I shall think of some way to get you out of the difficulty.' 'My lady,' he replied, 'I shall tell you the truth and you can then do whatever you want.' 'Tell me, then,' she said, 'but be sure to stick to the truth, for this is the vessel of salvation, and beware of lying, which disgraces the liar. How well the poet has put it:

Stick to truth even though it threatens you with death by fire;
Seek God's approval, for the foolish man
Angers his Lord and seeks to please His slaves.'

Ma'ruf then said: 'I must tell you, my lady, that I am not a merchant; I have no baggage and there is nothing to protect me from my creditors. In my own land I was a cobbler and I had a wife known as Dung Fatima' – and he went on to tell her the story of his dealings with Fatima from beginning to end. She laughed and said: 'What a good liar and trickster you are!' 'My lady,' he said, 'God preserve you as a keeper of shameful secrets and a remover of anxieties.' She said: 'You duped my father and deceived him with all your bragging so that, thanks to his greed, he married me off to you and you then squandered his money. The vizier holds this against you and on innumerable occasions he has told my father that you are a fraud and a liar. My father has refused to accept this because the vizier once asked for my hand and I wouldn't take him as a husband. Now, as time goes on, my father finds himself in difficulties, and he has asked me to get you to confess. I've done that and your cover has been removed. That would make my father determined to do you a mischief, but I have become your wife and I'm not going to neglect you. If I tell him, he will know for certain that you are a swindler and a liar who has tricked princesses and wasted the wealth of kings. For him that would be an unforgivable sin and he would be certain to have you put to death. Then everybody would know that I married a fraud and I would be disgraced. Also, if my father has you killed, he may try to marry me to another man and that is something that I shall never accept, even if it costs me my life. So now get up, dress yourself as a mamluk, take fifty thousand dinars of my money and ride off on a good horse to some place where my father's writ does not run. You can set up as a merchant there and you must then write me a letter, giving it to a courier who is to deliver it to me secretly, so that I may know where you are. Then I'll send you whatever I can lay my hands on and you will be a wealthy man. When my father dies, I'll send for you and you can return with all honour and respect, while if either you or I die and are gathered into God's mercy, we shall be reunited on the Day of Resurrection. This is the right course to follow and as long as we both remain well, I shall not stop sending you letters and money. Now get up before day breaks and you find yourself at a loss, with destruction facing you on every side.' He said: 'My lady, I am under your protection. Let me lie with you before we say goodbye.' 'There's no harm in that,' she replied and so he lay with her, and then washed, before dressing as a mamluk. He told the grooms to saddle him a good horse and, having taken leave of his wife, he left the city as night was ending and rode away. Everyone who saw him thought that he must be one of the king's mamluks going out on an errand.

In the morning, the king, with his vizier, came to the sitting room and sent for the princess, who arrived behind the curtain. 'What have you to say?' her father asked her and she replied: 'I say: "May God blacken the face of the vizier in the same way that he would have disgraced me in the eyes of my husband."' When her father asked about this, she said: 'He came to me yesterday, but before I could say anything to him about this, in came Faraj the eunuch with a letter in his hand. He said: "There are ten mamluks standing beneath the palace window. They gave me this letter and told me: 'Kiss the hands of our master, Ma'ruf the merchant, for us and give him this letter. We are some of the mamluks who were with his baggage train, and when we heard that he had married the king's daughter, we came to tell him what happened to us on the way.'"' I took the letter and when I read it I found that it ran: "From the five hundred mamluks to our master, Ma'ruf the merchant. To continue: we have to tell you that after you left us, we were attacked by mounted Bedouin. There were some two thousand of them against our five hundred. There was a fierce battle, as they blocked our way and we had to go on fighting them for thirty days. This is why we have been so slow in reaching you . . ."'

Night 995

Morning now dawned and Shahrazad broke off from what she had been allowed to say. Then, when it was the nine hundred and ninety-fifth night, SHE CONTINUED:

I have heard, O fortunate king, that the princess told her father: 'A letter came to my husband from his servants with news that Bedouin had blocked their way and that this was why they had been slow in coming. It went on: "The Bedouin took two hundred loads of fabrics from the baggage and killed fifty of us." When my husband heard the news, he cursed the mamluks and exclaimed: "How could they fight with Bedouin for the sake of a mere two hundred loads of goods? How much is that? They shouldn't have wasted time on that, as the two hundred loads would be worth no more than seven thousand dinars. I shall have to go to them and hurry them on. What was taken will not diminish the baggage train or have any effect on me, and I can count it as alms given by me to the Bedouin." He was laughing as he left me and showing no signs of distress either at the goods that he had lost or at the death of his mamluks. When he went down, I looked out of the palace window and saw the ten mamluks that had brought the letter. They were splendid as moons, and my father has none like them; while the robes that they were wearing were each worth two thousand dinars. My husband went off with them in order to fetch his goods, and thank God I never had the chance to say anything about what you told me to ask him, as he would have laughed at me and at you and he might have begun to disparage me and to hate me. All the fault lies with your vizier who keeps on speaking improperly about my husband.' The king said: 'Daughter, your husband is a very wealthy man who never thinks about money. From the first day that he came here, he has been giving it away to the poor and, God willing, his baggage will arrive soon to our great benefit.' He was completely taken in by her scheme and kept on trying to reassure her, while heaping abuse on the vizier.

So much for the king, but as for Ma'ruf, he rode off through the desert in a state of confusion, not knowing where to go. He was distressed by the pain of parting from his wife and, faced with the torments of passion, he recited:

Time has betrayed our union and separated us;
The harshness of parting burns my heart until it melts.
My eyes shed tears at the loss of the beloved:
Now we are parted; when shall we meet again?
You moon that gleams so brightly, I am he
Whose heart is torn in pieces by your love.
I wish I had not met you even once,
For, after union, I have tasted pain.
My love for Dunya stays with me always,
And if I die of love, may she survive.
O brilliance of the radiant sun, shine on
A heart love's former favours have consumed.
Will days to come unite us once again,
Allowing us to meet in joyfulness,
Happy within her palace, where I may
Clasp in my arms the sand hill's branch?
You are the moon, whose sun sheds light,
And may your lovely face not cease to shine.
I am content with the distress of love,
For love's good fortune is pure misery.

When he had finished these lines, he shed bitter tears and, as he could see nowhere to go, he decided that death was preferable to life. In his perplexity, he wandered on like a drunkard and did not stop until noon, when he had come to a small village. Near it he could see a peasant ploughing with two oxen, and as he was very hungry, he went up to him. After the two of them had exchanged greetings, the peasant said: 'Welcome, master. Are you one of the king's mamluks?' Ma'ruf said that he was and he was then invited to eat at the peasant's house. Recognizing that here was a generous man, he said: 'Brother, I don't see that you have anything to give me to eat, so how can you invite me?' 'There are good things here, master,' the peasant told him, 'for the village is close at hand. Do you dismount and I'll go and fetch food for you and fodder for your horse.' 'If it's as near as that, I can get there as fast as you,' said Ma'ruf, 'and I can then buy what I want from the market and eat it.' The peasant said: 'It's only a little place with no market, and nobody trades there, so I beg you to do me the favour of staying with me. I'll go there and be back soon.'

Ma'ruf dismounted and the peasant started off to the village to fetch food, leaving him sitting there to wait for him. Ma'ruf said to himself: 'I've turned this poor fellow away from his work and so I'd better get up and do his ploughing for him, to let him have something in return for the time that I've made him waste.' He took hold of the plough and drove the oxen forward, but before he had gone far, the plough snagged on an obstacle. The oxen stopped, and for all his efforts Ma'ruf couldn't get them to move on. He looked at the plough and found that it had caught on a golden ring. Then, when he cleared the earth from round it, he found that the ring was set in the middle of a marble slab as big as a millwheel. He worked at it until he succeeded in shifting it and under it he discovered steps leading down into an underground chamber. Down he went and there he found a place as big as a bath house, with four side chambers. Of these, the first was filled from top to bottom with gold; the second held emeralds, pearls and corals; the third was full of sapphires, hyacinth gems and turquoises; while in the fourth were diamonds and various other types of precious stones. At the head of the room was a chest made of clear crystal, filled with unique gems, each as big as a walnut, and sitting on top of this was a small golden casket, the size of a lemon.

Ma'ruf was astonished and delighted by what he saw, but he wondered what could be in the casket. He opened it and discovered a golden ring inscribed with names and talismans in a script that looked like ants' tracks. He rubbed the ring and immediately a voice was heard saying: 'Here I am, master. Whatever you ask will be given you. Do you want to build a town, destroy a city, have a canal dug or anything else of the kind? Whatever you ask will be done, through the permission of the Omnipotent God, the Creator of night and day.' 'Who and what are you, creature of my Lord?' asked Ma'ruf, and the voice replied: 'I am the servant of the ring, bound to the service of its owner. I shall perform whatever task you ask of me, and I can have no excuse for not doing so as I am lord of the races of the *jinn*, with seventy-two tribes under my command, each numbering seventy-two thousand. Each of these controls a thousand *marids*; each *marid* controls a thousand *'auns*; each *'aun* controls a thousand devils and each devil controls a thousand *jinn*. All of these owe me allegiance and none of them can disobey me. I am bound to this ring by a spell and I cannot disobey its owner. You are now the owner and I am your servant, so ask whatever you want and I shall hear and obey you. Whenever you need me, whether on land or in the sea, rub the ring and you will find me with you, but take care not to rub it twice in a row or else fire from the names engraved on it will burn me up and you will have cause to regret my loss when I am dead. This is all that I have to tell you about myself.'

Night 996

Morning now dawned and Shahrazad broke off from what she had been allowed to say. Then, when it was the nine hundred and ninety-sixth night, SHE CONTINUED:

I have heard, O fortunate king, that the servant of the ring told Ma'ruf about himself. 'What is your name?' asked Ma'ruf, and the servant of the ring told him that it was Abu'l-Sa'adat. 'What is this place, Abu'l-Sa'adat,' he asked, 'and whose spell imprisoned you in this casket?' Abu'l-Sa'adat replied: 'This is known as the treasure chamber of Shaddad ibn 'Ad, who built Iram, City of the Columns, whose like has never been created in any land.* While Shaddad lived, I was his servant and this was his ring; he put it among his treasures but it has now fallen to your lot.' Ma'ruf asked him: 'Can you bring all this treasure to the surface?' 'With the greatest of ease,' replied Abu'l-Sa'adat, at which Ma'ruf instructed him to bring out the whole of it, leaving nothing behind. Abu'l-Sa'adat pointed at the ground, which split open and into which

---

* Quran 89.6–7.

he disappeared briefly, before two graceful and handsome young boys emerged, carrying golden baskets, themselves filled with gold. They emptied out the baskets, went off and fetched more, and kept on bringing up gold and jewels until, before an hour was up, they announced that there was nothing left in the treasure chamber. Abu'l-Sa'adat himself then came out and said: 'Master, I have checked and we have removed everything that was there.' 'Who are these handsome boys?' Ma'ruf asked him, to which he replied: 'They are my sons. I didn't think it worthwhile summoning the *'auns* for a task like this, so my sons have done what you wanted and they are honoured to have been of service to you. Now ask for something else.'

Ma'ruf said: 'Are you able to fetch me mules and chests, and can you then put the money in the chests and load them on the mules?' 'Nothing could be easier,' replied Abu'l-Sa'adat, who then gave a great cry, at which all eight hundred of his sons appeared in front of him. He told them: 'Some of you are to change into mules, while others are to become handsome mamluks, even the least of whom is to be better than any found with kings; some are to be muleteers and others servants.' They did what he told them, with seven hundred becoming mules and the rest servants. Abu'l-Sa'adat then summoned the *'auns* and, when they came, he ordered some of them to turn into horses, with golden saddles encrusted with gems. When Ma'ruf saw this, he asked: 'Where are the chests?' and after these had been fetched, he gave instructions that the gold and the various precious stones were to be packed separately. The chests were then loaded on to three hundred mules.

Then Ma'ruf asked Abu'l-Sa'adat whether he could fetch him bales of costly fabrics, and when he was asked whether he wanted these from Egypt, Syria, Persia, India or Rum, he said: 'Bring me a hundred bales from each of them, carried on a hundred mules.' 'Allow me time to arrange for my *'auns* to do that, master,' Abu'l-Sa'adat replied, and he then instructed each group of *'auns* to go to a different land to get what was needed; they were then to take the shape of mules and come back carrying the goods. 'How long do you need?' Ma'ruf asked, and Abu'l-Sa'adat said: 'However long it remains dark, for by dawn you shall have all that you want.' After Ma'ruf had agreed to this, Abu'l-Sa'adat gave orders that a tent was to be pitched for him and, when this had been done, a table laden with food was brought in. 'Take your seat in this tent, master,' said Abu'l-Sa'adat. 'You need have no fear, as my sons here will guard you, while I go to collect my 'auns and send them off to carry out your wishes.'

When Abu'l-Sa'adat had left, Ma'ruf sat in the tent with the food in front of him and Abu'l-Sa'adat's sons standing there in the shape of mamluks, servants and retainers. While he was seated in state, up came the peasant carrying a big bowl of lentils and a horse's nosebag full of barley. When he saw the tent with the mamluks standing with their hands on their breasts, he thought that the king must have come and halted there. He stopped, flabbergasted, saying to himself: 'I wish I had killed a couple of chickens and roasted them with cow's butter for the king.' He was intending to go back to do this in order to entertain the king, when Ma'ruf caught sight of him and called out to him. He then told the mamluks to fetch him and they brought him up, together with his bowl of lentils. 'What's this?' Ma'ruf asked him, and he said: 'This is your meal and here is fodder for your horse. Don't blame me, for I never thought that the king would come here. Had I known, I would have killed a couple of chickens and prepared a better meal.' 'The king has not come,' Ma'ruf told him. 'I am his son-in-law, but I quarrelled with him and he sent his mamluks to reconcile me with him. I'm on my way back to the city, but as you have produced this food for me without knowing who I was, I shall accept what you have brought. Even if it is lentils, this and only this is what I intend to eat.' He told the peasant to put the bowl in the middle of the table and ate from it until he had had enough, while the peasant ate his fill from the other splendid dishes that were there. Ma'ruf then washed his hands, and after he had allowed the mamluks to eat, they finished off what was left on the table. Then, when the peasant's bowl had been emptied, Ma'ruf filled it with gold and said: 'Take this home with you and then come to visit me in the city, where I shall treat you generously.' The man took the bowl, brimming with gold, and drove his two oxen back to the village as proudly as if he were cousin to the king.

Ma'ruf passed a night of unalloyed pleasure. The brides of the treasure* were brought for him and beat their tambourines, all the while dancing in front of him, making it a night that stood outside the ordinary span of life. The next morning, before he knew what was happening, a cloud of dust could be seen rising and when it cleared away there were seven hundred mules carrying bales of fabric, surrounded by servants –

* Beautiful guardians of magical treasure troves.

muleteers, baggage handlers and torch-bearers – with Abu'l-Sa'adat mounted on a mule playing the part of baggage master. In front of him was a palanquin with four ornamented poles of glistening red gold, set with gems. When he reached Ma'ruf's tent, he dismounted, kissed the ground and said: 'All that you asked for has been done. In this palanquin is a set of robes from the treasure chamber, the like of which no king possesses. Put it on; then get into the palanquin and tell us what you want done.'

Ma'ruf said: 'I want to write a letter for you to take to the city of Ikhtiyan al-Khutan, where you must go to my relative, the king, whose presence you are to enter in the shape of a human courier.' 'To hear is to obey,' said Abu'l-Sa'adat, and when Ma'ruf had written his letter and sealed it, he took it away and brought it to the king. He found the king telling the vizier: 'I'm worried about my son-in-law, as I'm afraid that the Bedouin may kill him. I wish that I knew where he has gone so that I could follow him with my troops, and I wish that he had told me before going.' 'May God be gentle to you in your foolishness,' the vizier replied. 'I swear that the man realized that we had woken up to what he was doing and then ran away, fearing disgrace, for he is nothing but a trickster and a liar.' It was at this point that the 'courier' came in, kissing the ground before the king and praying that he be granted long life as well as continued glory and prosperity. 'Who are you and what do you want?' asked the king, and the 'courier' said: 'I am a messenger sent to you by your son-in-law, who is on his way with his baggage train. He sent me with a letter for you, and here it is.'

The king took the letter and, on reading it, he found the following: 'The best of greetings to my uncle, the great king. I have arrived with my baggage, so come out and meet me with your troops.' 'May God blacken your face, vizier!' exclaimed the king. 'How many times have you cast aspersions on my son-in-law's honour, calling him a fraud and a liar? Here he is with his baggage, while you are nothing but a traitor.' The vizier looked down at the ground in shame and embarrassment and replied: 'King of the age, I only said that because his baggage took so long in coming, and I was afraid that all the money he spent would be lost and gone.' 'Traitor!' said the king. 'What is my wealth in comparison with what he is bringing? He is going to repay me many times over.' He ordered the city to be adorned with decorations and then went to his daughter and said: 'I have good news for you. Your husband is close at hand, bringing his baggage with him. He has sent me a letter to tell me about it, and I'm just on my way off to meet him.' The princess was astonished by this and said to herself: 'How amazing! I wonder if he was making a fool of me and laughing at me, or else testing me by telling me that he was a poor man. God be praised that I didn't fail in my duty to him.'

So much for Ma'ruf, but as for 'Ali the merchant, he saw the decorations in the city and asked about them. When people told him: 'The baggage of Ma'ruf the merchant, the king's son-in-law, has arrived,' he exclaimed: 'Great God, what calamity is this? He came as a poor man, running away from his wife, so how can he have got hold of baggage? It may be that the princess did something for him so as to avoid disgrace, as there is nothing that kings cannot do. At any rate, may God shelter him and preserve him from shame.'

Night 997

Morning now dawned and Shahrazad broke off from what she had been allowed to say. Then, when it was the nine hundred and ninety-seventh night, SHE CONTINUED:

I have heard, O fortunate king, that 'Ali the merchant asked why the city had been adorned with decorations, and when they told him what had happened, he prayed for Ma'ruf and said: 'May God shelter him and not expose him to disgrace.' All the other merchants were pleased and delighted at the prospect of getting their money back, and the king, for his part, collected his troops and went out. Abu'l-Sa'adat had come back to say that he had delivered the letter, and Ma'ruf gave the command: 'Load up.' Wearing his treasure-hoard robes he mounted his palanquin and moved off in a thousand times greater and more imposing state than the king. When he had got halfway, the king met him with his men, and on his arrival, when he saw the robes Ma'ruf was wearing and the palanquin in which he was riding, he threw himself on him, greeting him and praying that God preserve him. Every notable in the state joined in the greeting, as it was clear that Ma'ruf had been telling the truth and that there was nothing false about him.

He entered the city in a procession splendid enough to cause the gall bladder of the envious to burst, and the merchants hurried up to kiss the ground before him. 'Ali the merchant said to him: 'You have made a success of this, master of impostors, but you deserve your success, and may God in His grace grant you even more.' Ma'ruf laughed, and when he had entered the palace and taken his seat on the throne, he said: 'Place the gold in the treasury of my uncle, the king, and bring me the fabrics.' These were fetched by the servants, who started to open the bales, one after another, until, when seven hundred had been unpacked, he picked out the best and ordered that they be taken to the princess to distribute among her maids. He also sent her the contents of a chest, full of jewels, which were to be brought to her to be given to the eunuchs as well as the maids. The merchants to whom he owed money were repaid in fabrics and whoever had lent him a thousand dinars was given what was worth two thousand dinars or more. After that, he began to distribute money to the poor and needy, while the king, who was watching, was unable to object. This went on until he had given away everything that had been in the seven hundred bales and it was then that he turned to the soldiers, to whom he gave precious stones – emeralds, sapphires, pearls, corals, and so on. He gave away gems in handfuls, without counting, and the king said: 'That's enough, my son; there is nothing much left of your goods.' 'I have plenty,' said Ma'ruf, and as it was clear that he had been speaking the truth, no one could think that he was now lying and he, for his part, didn't care what he gave away, as Abu'l-Sa'adat could fetch him whatever he wanted.

The treasurer now came to the king and said: 'King of the age, the treasury is full and there is no room for the remaining gold and jewels. Where are we going to put the rest of them?' The king suggested where they could be stored. Meanwhile the princess, on seeing all this, was overjoyed as well as amazed, asking herself where it had all come from. The merchants, for their part, were delighted with what they had been given and they called down blessings on Ma'ruf. As for 'Ali, he started saying to himself in his astonishment: 'How did he manage to lie and cheat his way into the possession of all these treasures? If he had got them from the princess, he wouldn't have been giving them away to the poor. How fine are the lines of the poet:

When the King of kings gives gifts,
Do not ask the reason.
God gives to whom He wants,
So mind your manners.'

So much for him, but as for the king, he was astonished when he saw the lavish generosity with which Ma'ruf gave away money. Ma'ruf himself now went to his wife, who greeted him joyfully with smiles and laughter. She kissed his hand and said: 'Were you making fun of me or

testing me when you told me that you were a poor man, running away
from your wife? I praise God that I did not fall short in my duty to you,
my darling, for whether you are rich or poor, there is no one dearer to
me than you. Please tell me now what you meant by what you said.'
Ma'ruf answered: 'I did mean to test you to see whether your love was
genuine or merely a matter of money and a wish for worldly goods. You
showed me your sincerity, and since you really love me, I welcome you
and appreciate your integrity.'

He then went off by himself and rubbed the ring. Abu'l-Sa'adat
appeared and said: 'Here I am; ask for what you want.' Ma'ruf told him:
'I want robes for my wife from a treasure chamber, as well as jewellery,
including a necklace, set with forty incomparable gems.' 'To hear is to
obey,' Abu'l-Sa'adat replied, and when he had fetched what he had been
told to bring, Ma'ruf dismissed him and went with it to his wife. He
placed the treasures in front of her, telling her that he wanted her to put
them on, and when she saw them she was ecstatic with joy. Among them
she found a pair of golden anklets studded with gems – the work of
magicians – bracelets, earrings and a girdle, past all price. When she had
put on the robe and this jewellery, she said: 'Master, I would like to keep
these for feasts and holidays,' but he told her to wear them all the time,
adding: 'I have plenty more besides them.' Her maids were delighted to
see her wearing the robe and they kissed Ma'ruf's hands. He then left
them and went off by himself to rub the ring. This time, when Abu'l-
Sa'adat appeared, Ma'ruf told him to fetch a hundred robes together
with jewellery. 'To hear is to obey,' said Abu'l-Sa'adat and he fetched
the robes, each with its accompanying jewellery wrapped inside it.
Ma'ruf took these and called for the maids. When they came, he presented one robe to each of them, with the result that, when they had put
them on, they looked like the houris of Paradise, with the princess among
them resplendent as the moon among stars. One of them told the king
about this, and when he came to see his daughter, he found her and her
maids dazzling all who looked at them.

Full of amazement, he went off, summoned the vizier and, after telling
him what had happened, he added: 'What have you got to say to this?'
'King of the age,' replied the vizier, 'this is not the behaviour of a
merchant. A merchant will sit for years hoarding pieces of linen and only
sell them at a profit. How can men like that possibly acquire generous
habits like these, and how can they get hold of wealth and jewels such
as few kings possess or find themselves with quantities of goods like

these? There must be something behind this, and if you will listen to me, I shall show you how to discover the truth.' The king agreed and the vizier went on: 'Arrange to meet him and then talk to him in a friendly way. Say: "I am thinking of going out to a garden to enjoy myself with nobody but you and the vizier." When we are there, we can set out wine and get him to drink. When he does that, he will lose his wits and his good sense, and if we then ask him how things really stand with him, he is bound to tell us what he is hiding, for wine gives secrets away. How well the poet has expressed it:

> When the effects of the wine I drank crept on their way
> To where my secrets were, I cried out: "Stop! Enough!"
> For fear of being overpowered, lest those
> Who drank with me might learn what I kept hid.

When he tells us the truth, we shall discover what he really is, and then we can do exactly what we want, for as things stand at the moment I am afraid of what could happen to you. He might want the throne for himself and win over the army by his lavish generosity, before deposing you and seizing power himself.' 'True enough,' said the king.

Night 998

Morning now dawned and Shahrazad broke off from what she had been allowed to say. Then, when it was the nine hundred and ninety-eighth night, SHE CONTINUED:

I have heard, O fortunate king, that when the vizier devised this scheme for the king, the king could see he was right. The two of them agreed on this plan that night. The next morning, the king went to his audience chamber and, as he was sitting there, his servants and grooms came to him in a state of distress. He asked them what was wrong and they told him: 'King of the age, the grooms rubbed down the horses and gave both them and the baggage mules their feed, but this morning we found that they had all been stolen by the mamluks. We searched the stables but saw no trace of them, and when we went to the mamluks' quarters we found them empty and we have no idea how they can have got away.' The king was astonished as he had thought that these had all been real – horses, mules, mamluks and all – not knowing that they were, in fact, 'auns in the service of Abu'l-Sa'adat. 'Damn you,' he exclaimed, 'how can a thousand beasts, five hundred mamluks, and other servants as well, escape without anyone noticing?' They said that they could think of no answer, and so he told them: 'Go off and wait for your master to come out from the harem and then give him the news.' They went away and sat in dismay, until Ma'ruf came out and noticed how distressed they were. He asked the reason for this and they told him the news. He said: 'Were they so valuable that you should worry yourselves about them? Go off about your business.' He sat there laughing, showing no signs of anger or grief, while the king looked at the vizier and said: 'What kind of a man is this, for whom money has no value? There must be something behind this.'

They had talked for a while with him, when the king said: 'I want to go to a garden with the two of you to enjoy myself. What do you say?' Ma'ruf raised no objection and they went off to a garden which contained two kinds of every fruit, with flowing streams, tall trees and singing birds. They entered a pavilion that was calculated to clear sorrow from the heart, and there they sat talking, as the vizier told remarkable stories and amusing jokes, using language that stirred them to delight. Ma'ruf sat listening until the time came for their evening meal, when food was brought in as well as a pitcher of wine. After they had eaten and washed their hands, the vizier filled a wine cup, which he gave to the king, who drank it, and then, filling another, he said to Ma'ruf: 'Take wine, before whose dignity the heads of men's intelligence bow low.' 'What is it, vizier?' Ma'ruf asked, and the vizier said: 'This is the grey-haired virgin, long kept at home, who brings delight to men's hearts. It is of this that the poet has said:

> The barbarians' feet trampled it down,
> And it took its revenge on the heads of Arabs.
> It is poured by an infidel like the moon in darkness,
> Whose glances are the surest cause of sin.

How well another has expressed it in the lines:

> It is as though both wine and cupbearer,
> Who stands unveiling it among the guests,
> Danced like the morning sun, upon whose face
> The moon had placed twin stars as beauty spots.
> It is so delicate and subtly mixed
> That it flows through men's limbs like their souls.

Another good description is as follows:

> The moon of beauty clasped me all night long,
> Although the sun had not set in the glass.
> The fire to which the Magians bow down,
> Bowed down to me all night from the wine-jar.

Another poet has said:

> The wine moved through the drinkers' limbs
> Like signs of health returning to the sick.

Another has written:

> I wonder how those who press it ever die,
> When what they leave us is the water of life.

Better than these are the lines of Abu Nuwas:

> Don't blame me, for this just prompts me to sin,
> But cure my sickness with what was its cause –
> Wine in whose courts sorrows cannot dismount;
> Its touch would fill a rock with happiness.
> In the dark night, as it rests in its jug,
> The house gleams with its glittering light.
> It passes among young men before whom Time is humbled,
> But what it does to them is only what they want,
> Carried by a girl dressed as a boy,
> The darling of sodomites and fornicators alike.
> Tell the man who lays claim to knowledge:
> "You may know something, but there is much you do not know."

Best of all are the lines of Ibn al-Mu'tazz:

> May al-Jazira, shaded by its trees,
> And Dair 'Abdun be blessed with showers of rain.
> How often was I woken there to take a morning draught,
> At early dawn, before the birds had stirred,
> By the monks' voices as they went to prayer
> In their black habits, chanting mournfully.
> How many a handsome shape was there, adorned
> With coquetry, and eyelids that shrouded dark eyes.
> One of them came to visit me, concealed by night,

Cautious and fearful, hurrying along.
I rose to spread my cheek upon his path
Humbly, wiping away my footsteps with my robe.
The new moon's light almost brought us disgrace,
Small as it was, like the clipping of a fingernail.
What happened then, I am not going to tell;
Believe the best and ask for no more news.

How well another poet expressed it when he said:

This morning I'm the wealthiest of men,
Because I know that happiness will come.
For I have here a store of liquid gold,
Which I shall measure in a drinking cup.

Other excellent lines run as follows:

This is the only alchemy there is,
And what is said of other types is false.
One carat's weight of wine in one *qintar* of grief
Is instantly transmuted into joy.

Another has said:

The empty glass feels heavy when it comes,
But when we fill it up with unmixed wine,
It is so light it almost flies away,
For so our spirits make our bodies light.

Yet another has said:

The wine cup and the wine require respect;
It is their right that their rights be preserved.
When I am dead, bury me beside a vine;
It may be that its roots will wet my bones.
Don't lay me in the desert, for I fear
When I am dead that I may not taste wine.'

The vizier went on trying to persuade Ma'ruf to drink, telling him of all the attractive qualities of wine, reciting the lines written in its honour and telling him witty stories until he was prompted to sip from the edge of the wine cup. Soon this was all that he wanted, and as the vizier kept on filling up the cup, he kept on drinking with such pleasure that he became so fuddled that he could no longer tell right from wrong. When the vizier saw that he had passed the limits of sobriety and was completely drunk, he said: 'Ma'ruf, I wonder where you got these unique jewels. Not even the most powerful kings have any that can match them and never in my life have I seen a merchant who has acquired as much wealth as you have, nor one as generous as you. Your actions are those of a king and not of a merchant, and so, for God's sake, please tell me about yourself so that I may fully grasp the high position you occupy.' He went on wheedling away as best he could until Ma'ruf, who had now lost his wits completely, said: 'I am not a merchant or a king,' and went on to tell him his whole story from beginning to end. The vizier then pressed him to let him see the ring so that he could examine how it had been made, and in his drunken state Ma'ruf took it from his finger and handed it to him to inspect. The vizier took it and turned it over before saying: 'If I rub it, will the servant of the ring appear?' 'Yes, indeed,' Ma'ruf told him. 'He will come to you when you do that and you can then look at him.'

The vizier rubbed the ring and a voice spoke: 'Here am I, master. Ask, and your wish will be granted. Do you want a city destroyed or another built up, or do you want to have a king killed? I shall obediently do whatever you ask.' At that, the vizier pointed to Ma'ruf and said: 'Carry off this worthless creature and throw him down in the most desolate of deserts where he can find neither food nor drink, so that he may starve miserably to death without anyone knowing of his fate.' Abu'l-Sa'adat snatched up Ma'ruf and flew away with him between heaven and earth. Ma'ruf was sure that things were going to end badly, with his death, and he asked Abu'l-Sa'adat tearfully where he was taking him. He said: 'I am going to throw you down in the Empty Quarter, you ill-educated fellow. What man with a talisman like this can hand it to other people to look at? You deserve what has happened to you, and were it not for the fact that I fear God, I would drop you from a thousand fathoms up, so that before you reached the ground the winds would blow you to pieces.' He said nothing more until he reached the Empty Quarter, where he dropped Ma'ruf and went off, leaving him alone in the desert.

Night 999

Morning now dawned and Shahrazad broke off from what she had been allowed to say. Then, when it was the nine hundred and ninety-ninth night, SHE CONTINUED:

I have heard, O fortunate king, that the servant of the ring took Ma'ruf and dropped him in the Empty Quarter, before leaving him there by himself and going back.

So much for Ma'ruf, but as for the vizier, when he got hold of the ring, he said to the king: 'What do you think? Didn't I tell you that he was a liar and a trickster and you didn't believe me?' 'You were right, vizier, may God grant you health,' the king said, 'and now let me look at the ring.' The vizier turned to him angrily and spat in his face, saying: 'You fool, why should I give it to you and remain your servant, now that I have become your master. I am not going to let you stay alive.' At that, he rubbed the ring and when Abu'l-Sa'adat came, he said: 'Take this boor and throw him down where you put his son-in-law, the trickster.' So Abu'l-Sa'adat picked him up and as he was carrying him off, the king said: 'Creature of God, what crime have I committed?' 'I don't know,' answered Abu'l-Sa'adat, 'but this is what my master ordered me to do and I cannot disobey whoever holds the ring.' He flew on and deposited the king in the same place as Ma'ruf, before leaving him there and going back. Hearing the sound of Ma'ruf weeping, the king went up to him and told him what had happened. The two of them then sat shedding tears for their misfortune, having nothing either to eat or to drink.

This is what happened to them, but as for the vizier, after getting rid of them both, he left the garden, assembled all the troops and convened an assembly at which he told them the story of the ring and let them know what he had done with Ma'ruf and the king. He went on: 'Unless you appoint me as your ruler, I shall order the servant of the ring to carry all of you off and throw you into the Empty Quarter to die of hunger and thirst.' 'Don't harm us,' they said, 'for we are content to accept you as our ruler and we shall not disobey your commands.' In this way, they were forced to submit to him and he gave them robes of honour. He then began to ask Abu'l-Sa'adat for whatever he wanted and this was immediately fetched for him.

He took his seat on the throne and, when he had received the homage of the troops, he sent a message to the princess, telling her: 'Prepare yourself, for I want you and am coming to lie with you tonight.' The princess burst into tears, finding it hard to bear the loss of her father and her husband, and she sent back to say: 'Wait for the end of the period that according to law I must delay after losing my husband, and you can then send me a note and sleep with me legally.' He wrote back: 'I don't recognize any waiting period, short or long. There is no need for letters; I see no difference between what is legal and what is not, and I shall certainly have you tonight.' At that, she sent back a letter welcoming him, but this was a trick on her part. The vizier was pleased and delighted to get her reply, as he was deeply in love with her. He ordered food to be provided for everyone, telling them: 'Eat; this is a wedding feast, for I propose to lie with the princess tonight.' The *shaikh* al-Islam objected that it was unlawful for him do this until the legal waiting period was over, after which a marriage contract could be drawn up, but he repeated: 'I don't recognize any waiting period; so say no more about it.' The *shaikh* was afraid that he might do him a mischief and so said nothing to him, but to the soldiers he said: 'This man is an irreligious unbeliever, a follower of no creed.'

That evening, the vizier went to the princess, whom he found wearing her most splendid robes with her most magnificent ornaments. When she saw him, she went up to him laughingly and said: 'This is a blessed night, although I would have thought it even better had you killed my father and my husband.' 'I shall certainly do that,' he promised, after which she made him sit down and started to flirt with him, making a show of affection. When she talked caressingly to him, smiling at him, he became carried away, while for her part she was using this flattery to deceive him so that she could get hold of the ring and, in place of his joy, bring down misfortune on his head. In what she did she was following the advice of the poet who said:

By my cunning I achieved what I could not get by the sword,
And I came back with spoils which had been sweet to win.

When the vizier saw how tenderly she smiled at him, he felt a surge of passion and wanted to take her, but when he came near, she drew away and burst into tears. 'Sir,' she said, 'don't you see that man looking at us? For God's sake, keep me from his sight, for how can you make love to me while he is watching?' The vizier said angrily: 'Where is he?' and she said: 'In the ring; he is raising his head and looking at us.' The vizier laughed, thinking that the servant of the ring was watching them, and he said: 'Don't be afraid; this is the servant of the ring, who is under my control.' 'But I am afraid of '*ifrits*,' she objected, 'so take it off and throw it away from me.' He removed the ring and put it on the pillow before approaching her again, but when he did, she used her foot to give him such a kick that he fell over backwards, unconscious. She then called out to her servants, and when they hurried in she told them to seize him. While forty of her maids took hold of him, she quickly snatched the ring from the pillow and rubbed it. 'Here I am, mistress,' said Abu'l-Sa'adat, coming forward, and she told him: 'Take this infidel, put him in prison and load him with fetters.' So Abu'l-Sa'adat took the vizier to the Dungeon of Anger before returning to tell the princess.

She now asked him where he had taken her father and her husband, and when he told her that he had left them in the Empty Quarter, she ordered him to fetch them back immediately. 'To hear is to obey,' he said, and he then flew away and did not stop until he had reached the Empty Quarter, where he found the two of them sitting in tears and complaining to one another. 'Don't be afraid,' he called to them. 'Relief has come.' He told them what the vizier had done and explained: 'I imprisoned him with my own hands in obedience to my mistress's commands and she then instructed me to bring you back.' They were delighted by this news and Abu'l-Sa'adat flew off with them, bringing them back to the princess after no more than an hour. She rose to greet them, before seating them and producing food and sweetmeats. They spent the rest of the night with her and next morning she clothed them both in splendid robes and said to her father: 'Father, resume your seat on the royal throne but make my husband your chief vizier and tell the army what has happened. Then bring the old vizier out of prison, put him to death and burn the body. He was an infidel who wanted to lie with me by way of fornication rather than legal marriage, and he said himself that he was an unbeliever and a man of no religion. You must also be sure to treat your son-in-law well, now that you have appointed him as your chief vizier.' 'To hear is to obey, daughter,' her father replied, 'but give me the ring or else give it to your husband.' 'It is not for you or for him,' she said, adding: 'I will keep it and I shall probably guard it better than either of you. Whatever you want you can ask for from me and I shall then get the servant of the ring to fetch it for you. As long as I am alive and well you need fear no harm, and after my death you can do what you want with it.' 'That is the proper thing to do, my daughter,' her father agreed.

The king then went with Ma'ruf to the council chamber. His men had spent a troubled night because of the princess, as they were distressed to think that the vizier might have slept with her by way of fornication rather than after a legal marriage, and they were also concerned about the harm that he had done to the king and his son-in-law. They were afraid of a breach in Islamic law, as it was clear to them that the vizier was an unbeliever. When they assembled in the audience hall they began to criticize Shaikh al-Islam and ask him why he had not kept the vizier from fornication. His answer was: 'The man is an unbeliever, but he holds the ring and there is nothing that either you or I can do about him. God will repay him for what he has done, but meanwhile say nothing lest he kill you.' As they were gathered there talking, in came the king together with Ma'ruf . . .

Night 1000

Morning now dawned and Shahrazad broke off from what she had been allowed to say. Then, when it was the thousandth night, SHE CONTINUED:

I have heard, O fortunate king, that the soldiers sat in the audience hall talking angrily about the vizier and what he had done to the king, his son-in-law and his daughter. At that point, in came the king together with Ma'ruf, his son-in-law, and at this sight they rose to their feet delightedly and kissed the ground in front of him. After the king had taken his seat on his throne, he relieved their distress by telling them what had happened. On his orders the city was adorned with decorations and the vizier was taken from prison. As he passed the soldiers he was met with curses, insults and abuse and when he was brought before the king he was sentenced to be put to the most hideous of deaths. He was killed and his body burned, while his wretched soul was consigned to hell. How well the poet's words describe him:

> May God's mercy shun the place where his bones lie,
> And may Munkar and Nakir never quit that spot.

Ma'ruf was now appointed as chief vizier, and things went well and happily for them all.

This went on for five years and then in the sixth the king died and the princess appointed Ma'ruf as his successor, but did not give him the

ring. During this period she had conceived and given birth to a remarkably beautiful and perfectly formed son. The child stayed in the care of
nurses until he was five years old, and it was then that his mother
succumbed to a fatal illness. She sent for her husband and told him: 'I
am ill.' 'May God make you well again, heart's darling,' he exclaimed,
but she said: 'I may die,' adding: 'You don't need to be told to look after
your son, but I must tell you to guard the ring, as I am afraid both for
you and for him.' 'No harm can come to one whom God guards,' he
told her, and at that she took off the ring and gave it to him. On the
following day, she died and was gathered into the mercy of God, while
Ma'ruf remained as king, administering the kingdom.

It happened that one day, after he had waved his handkerchief, sending
his guards back to their own quarters, he went to his sitting room and
sat there for the rest of the day until it grew dark. Then his leading
courtiers arrived as usual to drink with him, and they stayed there,
happily relaxing, until midnight. They then asked him for permission to
leave and when this had been granted, they went off to their own homes.
The maid whose duty it was to make up Ma'ruf's bed came in and spread
out the mattress for him, before removing his robes and giving him his
nightshirt. When he lay down, she began to massage his feet until he had
fallen fast asleep, when she left him and went off to her own bed.

So much for her, but as for the sleeping Ma'ruf, he suddenly woke up
in alarm to discover something beside him in the bed. He repeated the
formula: 'I take refuge in God from the accursed devil,' before opening
his eyes to discover an ugly woman lying beside him. 'Who are you?' he
asked, and she said: 'Don't be alarmed; I am your wife, Dung Fatima.'
When he looked at her he recognized her by her misshapen features and
her long teeth. 'How did you get in here and who brought you to this
country?' he asked. 'Where are you now?' she asked in return, and he
said: 'In the city of Ikhtiyan al-Khutan,' adding: 'When did you leave
Cairo?' 'This very hour,' she said, and when he asked her how that was,
she explained: 'When I quarrelled with you and Satan tempted me to do
you an injury, I lodged a complaint against you with the magistrates.
They looked for you but could not find you and the *qadis* asked questions
but failed to discover you. Two days passed and then I began to regret
what I'd done, realizing that the fault was mine, but repentance did me
no good. I sat for some days weeping over your departure. Then, as I
was running short of money, I had to start begging for my bread from
rich and poor alike, and, since you left me, it has only been in this

humiliating way that I got any food at all. I was reduced to the worst of straits and every night I'd sit there weeping over your loss and lamenting the shame and humiliation, together with the misery and distress, that I had had to endure in your absence.'

Ma'ruf stared at her in astonishment as she started to tell him everything that had happened to her. She went on to say: 'I spent all yesterday going round begging, but nobody would give me anything and everyone whom I approached, asking for a crust, would hurl abuse at me and refuse to give me anything. When it grew dark I had to spend the night with nothing to eat, racked by hunger pangs, and, finding my sufferings hard to bear, I sat there in tears. At that moment, a figure appeared in front of me and said: "Woman, why are you weeping?" I replied: "I had a husband who used to spend money on me and do what I wanted, but I have lost him. I don't know where he went but since he left I have been unable to cope." He asked me the name of my husband and when I told him that it was Ma'ruf, he said: "I know the man. You must know that he is now the king of a city, and if you want, I shall take you to him." I said: "I appeal to you to do that," and he picked me up and flew off with me between heaven and earth until he brought me to this palace. "Go into this room," he said, "and there you will see your husband asleep on the bed." So I went in and found you in this lordly state. I hope that you will not abandon me, for I am your companion, and I praise God for reuniting me with you.' 'Did I leave you or did you leave me?' Ma'ruf asked. 'It was you who went from one *qadi* to another to complain about me, and to crown it all, you lodged a complaint against me with the High Court, setting the bailiff on me from the citadel so that I was forced to flee.'

He went on to tell her everything that had happened to him, explaining how he had become king after having married the princess, who had died, leaving a son who was now seven years old. Fatima said: 'What happened was predestined by the Almighty God. I have repented and I appeal to your honour not to abandon me. Let me eat my bread here with you as an act of charity.' She kept on abasing herself before him until he began to pity her. 'If you repent of your misdeeds,' he said, 'you can stay with me and you will be treated well, but if you try to harm me in any way, I shall have you killed. There is no one whom I fear, so you needn't think of complaining to the High Court and having the bailiff sent down from the citadel. I am now a king and people fear me, while I myself fear no one except Almighty God. I have a ring with which

I can command the *jinn*, and when I rub it, its servant, Abu'l-Sa'adat, appears and brings me whatever I ask for. If you want to go back to Cairo, I shall provide you with money enough to last you for the rest of your life and send you back there quickly, but if you prefer to stay with me, I shall give you a palace of your own, furnish it with choice silks and assign twenty slave girls to wait on you. I shall provide you with regular supplies of delicious foods, as well as splendid robes, and you will live in the lap of luxury as a queen until either of us dies. What do you say to this?' 'I want to stay with you,' she told him and then she kissed his hand, repenting of her evil deeds.

He then gave her the sole use of a palace and presented her with slave girls and eunuchs, as she assumed the status of a queen. Ma'ruf's son used to visit both her and his father, but she disliked him as he was no son of hers, and when the boy saw her dislike and anger, he avoided her and returned her detestation. As for Ma'ruf, he was too busy paying court to beautiful slave girls to think about his wife, who had become an old woman whose scanty hair was grey and whose misshapen appearance was uglier than that of a spotted snake. Added to this was the fact that she had treated him as badly as possible, and, as the proverb has it: 'Ill treatment cuts off the roots of desire and sows hatred in the heart.' How well the poet has expressed it in the lines:

Take care not to alienate the heart by wrongs;
To win it back, when it has shied away, is hard.
When love is lost, this heart becomes
A broken glass that cannot be repaired.

It had not been for any laudable quality of hers that Ma'ruf had given her shelter, but rather this was a generous action through which he hoped to win the approval of Almighty God.

At this point, Dunyazad said to Shahrazad, her sister: 'How pleasant are these words, which have a greater effect on the heart than bewitching glances, and how splendid are these remarkable books and strange anecdotes.' Shahrazad replied: 'They cannot compare to what I shall tell you on this coming night, if the king allows me to live.' The next morning, when day broke the king was happily looking forward to the rest of the story and he said to himself: 'By God, I am not going to kill this woman until I have heard it.' He went to his throne room and the vizier came in as usual with the shroud under his arm. The king spent the whole day there delivering judgements among his subjects and then he came to his harem and went as usual to his wife, Shahrazad, the daughter of the vizier.

Night 1001

Morning now dawned and Shahrazad broke off from what she had been allowed to say. Then, when it was the thousand and first night, she continued: 'I have heard, O fortunate king, that the king came to his harem and went to his wife, Shahrazad, the daughter of the vizier.' Dunyazad now said to her sister: 'Finish the story of Ma'ruf for us.' 'With pleasure,' replied Shahrazad, 'if the king allows me to tell it.' 'I do,' said the king, 'because I want to hear the rest of it.' 'I have heard, your majesty,' said Shahrazad, 'that Ma'ruf did not bother to sleep with his wife, Fatima, and it was only in order to win favour with Almighty God that he provided her with food.' SHE WENT ON:

When she saw that he was not going to approach her and was busying himself with other women, she hated him and was overcome by jealousy. At this point, Iblis inspired her with the idea of taking the ring from him, killing him and reigning as queen in his place. So one night she left her own apartments and went to those of her husband. As providence had decreed, Ma'ruf was sleeping with a lovely, graceful and shapely concubine. Before Fatima had left her quarters she had found out that when he wanted to make love, his piety would lead him to remove the ring from his finger, out of respect for the great names inscribed on it, and place it on the pillow, only putting it back on after he had ritually purified himself. He was also in the habit of dismissing his concubines out of fear for the ring, and when he went to the baths he would keep the door of his room locked until he returned and put it back on again, after which those who wanted could enter without hindrance.

All this Fatima knew, and so she came out at night in order to go in while he was fast asleep and to steal the ring without being seen. It so happened, however, that just as she did so, Ma'ruf's son had come out to relieve himself, carrying no light with him. As he was sitting in the dark on the seat of the privy, having left the door open, he caught sight of Fatima hurrying in the direction of his father's quarters. He said to himself: 'I wonder why this witch has left her own apartments in the dark of night and is heading for those of my father. There must be a reason behind this,' and he followed her, keeping out of sight. He had a short sword of polished steel, of which he was so proud that he never went to his father's court without it. His father used to laugh when he saw it and exclaim: 'Good God, that is a big sword you have, my son, but you have never yet taken it into battle or used it to cut off a head!' 'I shall certainly cut off one that deserves it,' he would say, and his father would laugh at his words. Now, while he was trailing Fatima, he drew this sword from its sheath and followed her until she went into his father's room. He stood by the door and watched as she searched, saying to herself: 'Where has he put the ring?' The prince realized what she was looking for, and waited for her to find it, until, with an exclamation – 'Here it is!' – she picked it up. As he hid behind the door she turned to leave the room and when she had come out, she looked at the ring and turned it over in her hand. She was just about to rub it when he raised his arm and struck her a blow on the neck with his sword. With a single shriek she fell down dead.

Ma'ruf was roused, seeing his wife lying in a pool of blood and his son with a drawn sword in his hand. 'What is this, my son?' he asked and the boy replied: 'How many times have you said: "That is a big sword you have, but you have never gone into battle with it or used it to cut off a head"? I would promise to cut off one that deserved it, and now that is just what I have done for you.' He told his father what had happened, and Ma'ruf started to look for the ring. At first he couldn't find it but he kept on searching Fatima's body until he noticed that her hand was closed over it. He removed it and said: 'There is no doubt at all that you are my true son. May God gladden you, both in this world and the next, as you have saved me from this vile woman whose only purpose was to destroy me. How well the poet has expressed it:

When God grants to a man His aid,
Everything that he wants will come to him,
But where this aid has been withheld,
However hard he tries, he only harms himself.'

Ma'ruf now called out for his servants and when they hurried in, he told them what his wife had done and instructed them to take away her body and put it somewhere until morning. They did this, and then, on his orders, a number of eunuchs took charge of it, washed it and clothed it in a shroud. They made a tomb for her and there she was buried, her journey from Cairo having ended in her grave. How well the poet has put it:

> We travel on the path destined for us,
> And no man can avoid his destiny.
> Fate has decreed where we are going to die,
> And no one dies in any other land.

Another poet expressed it well when he wrote:

> When I go out to look for fortune in a land,
> I do not know which of two fates will follow;
> It may be that the good I seek will come,
> Or else misfortune may be seeking me.

After this, Ma'ruf sent for the peasant who had entertained him when he was a fugitive, and when the man came he appointed him as his chief vizier and counsellor. As Ma'ruf discovered, he turned out to have a beautiful daughter, who combined good qualities with a distinguished lineage and an excellent reputation. He married her and after a time he found a wife for his son. They continued to enjoy the most prosperous, pleasant and enjoyable of lives until they were visited by the destroyer of delights, the parter of companions, the ravager of prosperous lands, who orphans children. Praise be to the Living God, Who never dies and in Whose hand are the keys of dominion and power.

During this period Shahrazad had had three sons by the king and when she finished the story of Ma'ruf, she got to her feet before kissing the ground in front of the king. 'King of the age and unique ruler of this time,' she said, 'I am your servant and for a thousand and one nights I have been telling you stories of past generations and moral tales of our predecessors. May I hope to ask you to grant me a request?' 'Ask and your wish will be granted,' he told her, and she then called to the nurses and the eunuchs, telling them to fetch her children, which they quickly did. Of the three boys one could walk, another was at the stage of crawling and the third was still a suckling. When they were brought in, she took them and placed them before the king. Then she kissed the ground again and said: 'King of the age, these are your children and my wish is that as an act of generosity towards them you free me from sentence of death, for if you kill me, these babies will have no mother and you will find no other woman to bring them up so well.' At that, the king shed tears and, gathering his sons to his breast, he said to her: 'Even before the arrival of these children, I had intended to pardon you, as I have seen that you are a chaste and pure woman, freeborn and God-fearing. May God bless you, your father and mother, and your whole family, root and branch. I call God to witness that I have decided that no harm is to come to you.' At this, she kissed his hands and feet in her delight, exclaiming: 'May God prolong your life and increase your dignity and the awe that you inspire!'

Joy spread through the palace and from it to the city and this was a night that stood outside the ordinary span of life, whiter than the face of day. In the morning, the happy king, overwhelmed by his good fortune, summoned his troops and when they came he presented a splendid and magnificent robe of honour to his vizier, Shahrazad's father. 'May God shelter you,' he said, 'because you gave me your noble daughter as a wife, and it is thanks to her that I have turned in repentance from killing the daughters of my subjects. I have found her noble, pure, chaste and without sin; God has provided me with three sons by her and I give thanks to Him for this great good fortune.'

He then presented robes of honour to all the viziers, emirs and ministers of state. On his instructions, the city was adorned with decorations for thirty days and he did not ask any of the citizens for contributions from their own funds, as all the costs and expenses were borne by the royal treasury. The splendour of the decorations had never been matched before; drums were beaten; pipes sounded and every entertainer displayed his skill. The king showered gifts and presents on them; he gave alms to the poor and needy and his bounty extended to all his subjects living in his realm. He and his kingdom continued to enjoy prosperity, happiness, pleasure and joy, until they were visited by the destroyer of delights and the parter of companions. Praise be to God, Whom the passage of time does not wear away. He is not subject to change; His attention cannot be distracted and He alone possesses the qualities of perfection. Blessing and peace be on the leader of His choice, the best of His creation, Muhammad, the lord of mankind. It is through him that we pray to God to bring us to a good end.

The story of Aladdin, or
The magic lamp

In the capital city of a rich and vast kingdom in China whose name I cannot at the moment recall, there lived a tailor called Mustafa, whose only distinguishing feature was his profession. This Mustafa was very poor, his work hardly producing enough to live on for him, his wife and a son whom God had given him.

The son, who was called Aladdin, had received a very neglected upbringing, which had led him to acquire many depraved tendencies. He was wicked, stubborn and disobedient towards his father and mother, who, once he became little older, could no longer keep him in the house. He would set out first thing in the morning and spend the day playing in the streets and public places with small vagabonds even younger than himself.

As soon as he was of an age to learn a trade, his father, who was not in a position to make him learn any trade other than his own, took him into his shop and began to show him how to handle a needle. But he remained unable to hold his son's fickle attention, neither by fear of punishment nor by gentle means, and could not get him to sit down and apply himself to his work, as he had hoped. No sooner was Mustafa's back turned than Aladdin would escape and not return for the rest of the day. His father would punish him, but Aladdin was incorrigible, and so, much to his regret, Mustafa was forced to leave him to his dissolute ways. All this caused Mustafa much distress, and his grief at not being able to make his son mend his ways resulted in a persistent illness of which, a few months later, he died.

Aladdin's mother, seeing how her son was not going to follow in his father's footsteps and learn tailoring, closed the shop so that the proceeds from the sale of all the tools of its trade, together with the little she could earn by spinning cotton, would help provide for herself and her son.

Aladdin, however, no longer restrained by the fear of a father, paid so little attention to his mother that he had the effrontery to threaten her when she so much as remonstrated with him, and now abandoned himself completely to his dissolute ways. He associated increasingly with children of his own age, playing with them with even greater enthusiasm. He continued this way of life until he was fifteen, with his mind totally closed to anything else and with no thought of what he might one day become. Such was his situation when one day, while he was playing in the middle of a square with a band of vagabonds, as was his wont, a stranger who was passing by stopped to look at him.

This stranger was a famous magician who, so the authors of this story tell us, was an African, and this is what we will call him, as he was indeed from Africa, having arrived from that country only two days before.

Now it may be that it was because this African magician, who was an expert in the art of reading faces, had looked at Aladdin and had seen all that was essential for the execution of his journey's purpose, or there might have been some other reason. Whatever the case, he artfully made enquiries about Aladdin's family and about what sort of fellow he was. When he had learned all that he wanted to know, he went up to the young man and, drawing him a little aside from his companions, asked him: 'My son, isn't your father called Mustafa, the tailor?' 'Yes, sir,' replied Aladdin, 'but he has been dead a long time.'

At these words, the magician's eyes filled with tears and, uttering deep sighs, he threw his arms round Aladdin's neck, embracing and kissing him several times. Aladdin, seeing his tears, asked him why he was weeping. 'Ah, my son,' exclaimed the magician, 'how could I stop myself? I am your uncle and your father was my dear brother. I have been travelling for several years and now, just when I arrive here in the hope of seeing him again and having him rejoice at my return, you say that he is dead! I tell you it's very painful for me to find I am not going to receive the comfort and consolation I was expecting. But what consoles me a little in my grief is that, as far as I can remember them, I can recognize his features in your face, and that I was not wrong in speaking to you.' Putting his hand on his purse, he asked Aladdin where his mother lived. Aladdin answered him straight away, at which the magician gave him a handful of small change, saying: 'My son, go and find your mother, give her my greetings and tell her that, if I have time, I will go and see her tomorrow, so that I may have the consolation of seeing where my brother lived and where he ended his days.'

As soon as the magician had left, his newly invented nephew, delighted with the money his uncle had just given him, ran to his mother. 'Mother,' he said to her, 'tell me, please, have I got an uncle?' 'No, my son,' she replied, 'you have no uncle, neither on your late father's side nor on mine.' 'But I have just seen a man who says he is my uncle on my father's side,' insisted Aladdin. 'He was his brother, he assured me. He even began to weep and embrace me when I told him my father was dead. And to prove I am telling the truth,' he added, showing her the money he had been given, 'here is what he gave me. He also charged me to give you his greetings and to tell you that tomorrow, if he has the time, he will come and greet you himself and at the same time see the house where my father lived and where he died.' 'My son,' said his mother, 'it's true your father once had a brother, but he's been dead a long time and I never heard him say he had another brother.' They spoke no more about the African magician.

The next day, the magician approached Aladdin a second time as he was playing with some other children in another part of the city, embraced him as he had done on the previous day and, placing two gold coins in his hand, said to him: 'My son, take this to your mother; tell her I am coming to see her this evening and say she should buy some food so we can dine together. But first, tell me where I can find your house.' Aladdin told him where it was and the magician then let him go.

Aladdin took the two gold coins to his mother who, as soon as she heard of his uncle's plans, went out to put the money to use, returning with abundant provisions; but, finding herself with not enough dishes, she went to borrow some from her neighbours. She spent all day preparing the meal, and towards evening, when everything was ready, she said to Aladdin: 'My son, perhaps your uncle doesn't know where our house is. Go and find him and, when you see him, bring him here.'

Although Aladdin had told the magician where to find the house, he was nonetheless prepared to go out to meet him, when there was a knock on the door. Opening it, Aladdin discovered the magician, who entered, laden with bottles of wine and all kinds of fruit which he had brought for supper and which he handed over to Aladdin. He then greeted his mother and asked her to show him the place on the sofa where his brother used to sit. She showed him and immediately he bent down and kissed the spot several times, exclaiming with tears in his eyes: 'My poor brother! How sad I am not to have arrived in time to embrace you once more before your death!' And although Aladdin's mother begged him to sit in the same place, he firmly refused. 'Never will I sit there,' he said, 'but allow me to sit facing it, so that though I may be deprived of the satisfaction of seeing him there in person as the head of a family which is so dear to me, I can at least look at where he sat as though he were present.' Aladdin's mother pressed him no further, leaving him to sit where he pleased.

Once the magician had sat down in the place he had chosen, he began to talk to Aladdin's mother. 'My dear sister,' he began, 'don't be surprised that you never saw me all the time you were married to my brother Mustafa, of happy memory; forty years ago I left this country, which is mine as well as that of my late brother. Since then, I have travelled in India, Arabia, Persia, Syria and Egypt, and have stayed in the finest cities, and then I went to Africa, where I stayed much longer. Eventually, as is natural – for a man, however far he is from the place of his birth, never forgets it any more than he forgets his parents and those with whom he was brought up – I was overcome by a strong desire to see my own family again and to come and embrace my brother. I felt I still had enough strength and courage to undertake such a long journey and so I delayed no longer and made my preparations to set out. I won't tell you how long it has taken me, nor how many obstacles I have met with and the discomfort I suffered to get here. I will only tell you that in all my travels nothing has caused me more sorrow and suffering than hearing of the death of one whom I have always loved with a true brotherly love. I observed some of his features in the face of my nephew, your son, which is what made me single him out from among all the children with whom he was playing. He will have told you how I received the sad news that my brother was no longer alive; but one must praise God for all things and I find comfort in seeing him again in a son who retains his most distinctive features.'

When he saw how the memory of her husband affected Aladdin's mother, bringing tears to her eyes, the magician changed the subject and, turning to Aladdin, asked him his name. 'I am called Aladdin,' he replied. 'Well, then, Aladdin,' the magician continued, 'what do you do? Do you have a trade?'

At this question, Aladdin lowered his eyes, embarrassed. His mother, however, answered in his place. 'Aladdin is an idle fellow,' she said. 'While he was alive, his father did his best to make him learn his trade but never succeeded. Since his death, despite everything I have tried to tell him, again and again, day after day, the only trade he knows is acting the vagabond and spending all his time playing with children, as you saw for yourself, mindless of the fact that he is no longer a child. And if you can't make him feel ashamed and realize how pointless his behaviour is, I despair of him ever amounting to anything. He knows his father left nothing, and he can himself see that despite spinning cotton all day as I do, I have great difficulty in earning enough to buy us bread. In fact, I have decided that one of these days I am going to shut the door on him and send him off to fend for himself.'

After she had spoken, Aladdin's mother burst into tears, whereupon the magician said to Aladdin: 'This is no good, my nephew. You must think now about helping yourself and earning your own living. There are all sorts of trades; see if there isn't one for which you have a particular inclination. Perhaps that of your father doesn't appeal to you and you would be more suited to another: be quite open about this, I am just trying to help you.' Seeing Aladdin remain silent, he went on: 'If you want to be an honest man yet dislike the idea of learning a trade, I will provide you with a shop filled with rich cloths and fine fabrics. You can set about selling them, purchasing more goods with the money that you make, and in this manner you will live honourably. Think about it and then tell me frankly your opinion. You will find that I always keep my word.'

This offer greatly flattered Aladdin, who did not like manual work, all the more so since he had enough sense to know that shops with these kinds of goods were esteemed and frequented and that the merchants were well dressed and well regarded. So he told the magician, whom he thought of as his uncle, that his inclination was more in that direction than any other and that he would be indebted to him for the rest of his life for the help he was offering. 'Since this occupation pleases you,' the magician continued, 'I will take you with me tomorrow and will have you dressed in rich garments appropriate for one of the wealthiest merchants of this city. The following day we will consider setting up a shop, as I think it should be done.'

Aladdin's mother, who up until then had not believed the magician was her husband's brother, now no longer doubted it after hearing all the favours he promised her son. She thanked him for his good intentions and, after exhorting Aladdin to make himself worthy of all the wealth his uncle had promised him, served supper. Throughout the meal, the talk ran upon the same subject until the magician, seeing the night was well advanced, took leave of the mother and the son and retired.

The next morning, he returned as he had promised to the widow of

Mustafa the tailor and took Aladdin off with him to a wealthy merchant who sold only ready-made garments in all sorts of fine materials and for all ages and ranks. He made the merchant bring out clothes that would fit Aladdin and, after putting to one side those which pleased him best and rejecting the others that did not seem to him handsome enough, said to Aladdin: 'My nephew, choose from among all these garments the one you like best.' Aladdin, delighted with his new uncle's generosity, picked one out which the magician then bought, together with all the necessary accessories, and paid for everything without bargaining.

When Aladdin saw himself so magnificently clothed from top to toe, he thanked his uncle profusely with all the thanks imaginable, and the magician repeated his promise never to abandon him and to keep him always with him. Indeed, he then took him to the most frequented parts of the city and in particular to those where the shops of the rich merchants were to be found. When he reached the street which had the shops with the richest cloths and finest fabrics, he said to Aladdin: 'As you will soon be a merchant like these, it is a good idea for you to seek out their company so that they get to know you.' The magician also showed him the largest and most beautiful mosques and took him to the *khans* where the foreign merchants lodged and to all the places in the sultan's palace which he was free to enter. Finally, after they had wandered together through all the fairest places in the city, they came to the *khan* where the magician had taken lodgings. There they found several merchants whom the magician had got to know since his arrival and whom he had gathered together for the express purpose of entertaining them and at the same time introducing them to his so-called nephew.

The party did not finish until towards evening. Aladdin wanted to take leave of his uncle to return home, but the magician would not let him go back alone and himself accompanied him back to his mother. When his mother saw Aladdin in his fine new clothes, she was carried away in her delight and kept pouring a thousand blessings on the magician who had spent so much money on her child. 'My dear relative,' she exclaimed, 'I don't know how to thank you for your generosity. I know my son does not deserve all you have done for him and he would be quite unworthy of it if he was not grateful to you or failed to respond to your kind intention of giving him such a fine establishment. As for myself, once again I thank you with all my heart; I hope that you will live long enough to witness his gratitude, which he can best show by conducting himself in accordance with your good advice.'

'Aladdin is a good boy,' the magician replied. 'He listens to me well enough and I believe he will turn out well. But one thing worries me – that I can't carry out what I promised him tomorrow. Tomorrow is Friday, when the shops are closed, and there is no way we can think of renting one and stocking it at a time when the merchants are only thinking of entertaining themselves. So we will have to postpone our business until Saturday, but I will come and fetch him tomorrow and I will take him for a walk in the gardens where all the best people are usually to be found. Perhaps he has never seen the amusements that are to be had there. Up until now he has only been with children, but now he must see men.' The magician took his leave of mother and son and departed. Aladdin, however, was so delighted at being so smartly turned out that he already began to anticipate the pleasure of walking in the gardens that lay around the city. In fact, he had never been outside the city gates and had never seen the surroundings of the city, which he knew to be pleasant and beautiful.

The next day, Aladdin got up and dressed himself very early so as to be ready to leave when his uncle came to fetch him. After waiting for what seemed to him a very long time, in his impatience he opened the door and stood on the doorstep to see if he could see the magician. As soon as he spotted him, Aladdin told his mother and said goodbye to her, before shutting the door and running to meet him.

The magician embraced Aladdin warmly when he saw him. 'Come, my child,' he said to him, smiling, 'today I want to show you some wonderful things.' He took him through a gate which led to some fine, large houses, or rather, magnificent palaces, which all had very beautiful gardens that people were free to enter. At each palace that they came to, he asked Aladdin whether he thought it beautiful, but Aladdin would forestall him as soon as another palace presented itself, saying: 'Uncle, here's another even more beautiful than those we have just seen.' All the while, they were advancing ever deeper into the countryside and the wily magician, who wanted to go further still in order to carry out the plan he had in mind, took the opportunity of entering one of these gardens. Seating himself near a large pool into which a beautiful jet of water poured from the nostrils of a bronze lion, he pretended to be tired in order to get Aladdin to take a rest. 'Dear nephew,' he said to him, 'you, too, must be tired. Let's sit here and recover ourselves. We shall then have more strength to continue our walk.'

When they had sat down, the magician took out from a cloth attached to his belt some cakes and several kinds of fruit which he had brought with him as provisions, and spread them out on the edge of the pool. He shared a cake with Aladdin but let him choose for himself what fruits he fancied. As they partook of this light meal, he talked to his so-called nephew, giving him numerous pieces of advice, the gist of which was to exhort Aladdin to give up associating with children, telling him rather to approach men of prudence and wisdom, to listen to them and to profit from their conversation. 'Soon you will be a man like them,' he said, 'and you can't get into the habit too soon of following their example and speaking with good sense.' When they had finished eating, they got up and resumed their walk through the gardens, which were separated from each other only by small ditches which defined their limits without impeding access – such was the mutual trust the inhabitants of the city enjoyed that there was no need for any other boundaries to guard against them harming each other's interests. Gradually and without Aladdin being aware of it, the magician led him far beyond the gardens, making him pass through open country which took them very close to the mountains.

Aladdin had never before travelled so far and felt very weary from such a long walk. 'Uncle,' he asked the magician, 'where are we going? We have left the gardens far behind and I can see nothing but mountains. If we go any further, I don't know if I'll have enough strength to return to the city.' 'Take heart, my nephew,' replied the bogus uncle. 'I want to show you another garden which beats all those you have just seen. It's not far from here, just a step away, and when we get there you yourself will tell me how cross you would have been not to have seen it after having got so close to it.' Aladdin let himself be persuaded and the magician led him even further on, all the while entertaining him with many amusing stories in order to make the journey less tedious for him and his fatigue more bearable.

At last they came to two mountains of a moderate height and size, separated by a narrow valley. This was the very spot to which the magician had wanted to take Aladdin so that he could carry out the grand plan which had brought him all the way from the furthest part of Africa to China. 'We are not going any further,' he told Aladdin. 'I want to show you some extraordinary things, unknown to any other man, and when you have seen them, you will thank me for having witnessed so many marvels that no one else in all the world will have seen but you. While I am making a fire, you go and gather the driest bushes you can find for kindling.'

There was such a quantity of brushwood that Aladdin had soon amassed more than enough in the time that the magician was still starting up the fire. He set light to the pile and the moment the twigs caught fire, the magician threw on to them some incense that he had ready at hand. A dense smoke arose, which he made to disperse right and left by pronouncing some words of magic, none of which Aladdin could understand.

At the same moment, the earth gave a slight tremor and opened up in front of Aladdin and the magician, revealing a stone about one and a half feet square and about one foot deep, lying horizontally on the ground; fixed in the middle was a ring of bronze with which to lift it up. Aladdin, terrified at what was happening before his very eyes, would have fled if the magician had not held him back, for he was necessary for this mysterious business. He scolded him soundly and gave him such a blow that he was flung to the ground with such force that his front teeth were very nearly pushed back into his mouth, judging from the blood which poured out. Poor Aladdin, trembling all over and in tears, asked his uncle: 'What have I done for you to hit me so roughly?' 'I have my reasons for doing this,' replied the magician. 'I am your uncle and at present take the place of your father. You shouldn't answer me back.' Softening his tone a little, he went on: 'But, my child, don't be afraid. All I ask is that you obey me exactly if you want to benefit from and be worthy of the great advantages I propose to give you.' These fine promises somewhat calmed Aladdin's fear and resentment, and when the magician saw he was completely reassured, he went on: 'You have seen what I have done by virtue of my incense and by the words that I pronounced. Know now that beneath the stone that you see is hidden a treasure which is destined for you and which will one day make you richer than the greatest kings in all the world. It's true, you are the only person in the world who is allowed to touch this stone and to lift it to go inside. Even I am not allowed to touch it and to set foot in the treasure house when it is opened. Consequently, you must carry out step by step everything I am going to tell you, not omitting anything. The matter is of the utmost importance, both for you and for me.'

Aladdin, still in a state of astonishment at all he saw and at what he had just heard the magician say about this treasure, which was to make him happy for evermore, got up, forgetting what had just happened to him, and asked: 'Tell me then, uncle, what do I have to do? Command me, I am ready to obey you.' 'I am delighted, my child, that you have made this decision,' replied the magician, embracing him. 'Come here, take hold of this ring and lift up the stone.' 'But uncle, I am not strong enough – you must help me,' Aladdin cried, to which his uncle replied: 'No, you don't need my help and we would achieve nothing, you and I, if I were to help you. You must lift it up all by yourself. Just say the names of your father and your grandfather as you hold the ring, and lift. You will find that it will come without any difficulty.' Aladdin did as the magician told him. He lifted the stone with ease and laid it aside.

When the stone was removed, there appeared a cavity about three to four feet deep, with a small door and steps for descending further. 'My son,' said the magician to Aladdin, 'follow carefully what I am going to tell you to do. Go down into this cave and when you get to the foot of the steps which you see, you will find an open door that will lead you into a vast vaulted chamber divided into three large rooms adjacent to each other. In each room, you will see, on the right and the left, four very large bronze jars, full of gold and silver – but take care not to touch them. Before you go into the first room, pull up your gown and wrap it tightly around you. Then when you have entered, go straight to the second room and the third room, without stopping. Above all, take great care not to go near the walls, let alone touch them with your gown, for if you do, you will immediately die; that's why I told you to keep it tightly wrapped around you. At the end of the third room there is a gate which leads into a garden planted with beautiful trees laden with fruit. Walk straight ahead and cross this garden by a path which will take you to a staircase with fifty steps leading up to a terrace. When you are on the terrace, you will see in front of you a niche in which there is a lighted lamp. Take the lamp and put it out and when you have thrown away the wick and poured off the liquid, hold it close to your chest and bring it to me. Don't worry about spoiling your clothes – the liquid is not oil and the lamp will be dry as soon as there is no more liquid in it. If you fancy any of the fruits in the garden, pick as many as you want – you are allowed to do so.'

When he had finished speaking, the magician pulled a ring from his finger and put it on one of Aladdin's fingers, telling him it would protect him from any harm that might come to him if he followed all his instructions. 'Be bold, my child,' he then said. 'Go down; you and I are both going to be rich for the rest of our lives.'

Lightly jumping into the cave, Aladdin went right down to the bottom of the steps. He found the three rooms which the magician had described to him, passing through them with the greatest of care for fear he would die if he failed scrupulously to carry out all he had been told. He crossed the garden without stopping, climbed up to the terrace, took the lamp alight in its niche, threw away the wick and the liquid, and as soon as this had dried up as the magician had told him, he held it to his chest. He went down from the terrace and stopped in the garden to look more closely at the fruits which he had seen only in passing. The trees were all laden with the most extraordinary fruit: each tree bore fruits of different colours – some were white; some shining and transparent like crystals; some pale or dark red; some green; some blue or violet; some light yellow; and there were many other colours. The white fruits were pearls; the shining, transparent ones diamonds; the dark red were rubies, while the lighter red were spinel rubies; the green were emeralds; the blue turquoises; the violet amethysts; the light yellow were pale sapphires; and there were many others, too. All of them were of a size and a perfection the like of which had never before been seen in the world. Aladdin, however, not recognizing either their quality or their worth, was unmoved by the sight of these fruits, which were not to his taste – he would have preferred real figs or grapes, or any of the other excellent fruit common in China. Besides, he was not yet of an age to appreciate their worth, believing them to be but coloured glass and therefore of little value. But the many wonderful shades and the extraordinary size and beauty of each fruit made him want to pick one of every colour. In fact, he picked several of each, filling both pockets as well as two new purses which the magician had bought him at the same time as the new clothes he had given him so that everything he had should be new. And as the two purses would not fit in his pockets, which were already full, he attached them to either side of his belt. Some fruits he even wrapped in the folds of his belt, which was made of a wide strip of silk wound several times around his waist, arranging them so that they could not fall out. Nor did he forget to cram some around his chest, between his gown and his shirt.

Thus weighed down with such, to him, unknown wealth, Aladdin hurriedly retraced his steps through the three rooms so as not to keep the magician waiting too long. After crossing them as cautiously as he had before, he ascended the stairs he had come down and arrived at the entrance of the cave, where the magician was impatiently awaiting him. As soon as he saw him, Aladdin cried out: 'Uncle, give me your hand, I beg of you, to help me climb out.' 'Son,' the magician replied, 'first, give me the lamp, as it could get in your way.' 'Forgive me, uncle,' Aladdin rejoined, 'but it's not in my way; I will give it you as soon as I get out.' But the magician persisted in wanting Aladdin to hand him the lamp before pulling him out of the cave, while Aladdin, weighed down by this lamp and by the fruits he had stowed about his person, stubbornly refused to give it to him until he was out of the cave. Then the magician, in despair at the young man's resistance, fell into a terrible fury: throwing a little of the incense over the fire, which he had carefully kept alight, he uttered two magic words and immediately the stone which served to block the entrance to the cave moved back in its place, with the earth above it, just as it had been when the magician and Aladdin had first arrived there.

Now this magician was certainly not the brother of Mustafa the tailor, as he had proudly claimed, nor, consequently, was he Aladdin's uncle. But he did indeed come from Africa, where he was born, and as Africa is a country where more than anywhere else the influence of magic persists, he had applied himself to it from his youth, and after forty years or so of practising magic and geomancy and burning incense and of reading books on the subject, he had finally discovered that there was somewhere in the world a magic lamp, the possession of which, could he lay hands on it, would make him more powerful than any king in the world. In a recent geomantic experiment, he had discovered that this lamp was in an underground cave in the middle of China, in the spot and with all the circumstances we have just seen. Convinced of the truth of his discovery, he set out from the furthest part of Africa, as we have related. After a long and painful journey, he had come to the city that was closest to the treasure, but although the lamp was certainly in the spot which he had read about, he was not allowed to remove it himself, he had ascertained, nor could he himself enter the underground cave where it was to be found. Someone else would have to go down into it, take the lamp and then deliver it into his hands. That is why he had turned to Aladdin, who seemed to him to be a young boy of no consequence, just right to carry out for him the task which he wanted him to do. He had resolved, once he had the lamp in his hands, to perform the final burning of incense that we have mentioned and to utter the two magic words that would produce the effect which we have seen, sacrificing poor Aladdin to his avarice and wickedness so as to have no witness. The blow he gave Aladdin and the authority he had assumed over him were only meant to accustom him to fear him and to obey him precisely so that, when he asked him for the famed lamp, Aladdin would immediately give it to him, but what happened was the exact opposite of what he had intended. In his haste, the magician had resorted to such wickedness in order to get rid of poor Aladdin because he was afraid that if he argued any longer with him, someone would hear them and would make public what he wanted to keep secret.

When he saw his wonderful hopes and plans forever wrecked, the magician had no other choice but to return to Africa, which is what he did the very same day, taking a roundabout route so as to avoid going back into the city he had left with Aladdin. For what he feared was being seen by people who might have noticed him walking out with this boy and now returning without him.

To all appearances, that should be the end of the story and we should hear no more about Aladdin, but the very person who had thought he had got rid of Aladdin for ever had forgotten that he had placed on his finger a ring which could help to save him. In fact it was this ring, of whose properties Aladdin was totally unaware, that was the cause of his salvation, and it is astonishing that the loss of it together with that of the lamp did not throw the magician into a state of complete despair. But magicians are so used to disasters and to events turning out contrary to their desires that all their lives they forever feed their minds on smoke, fancies and phantoms.

After all the endearments and the favours his false uncle had shown him, Aladdin little expected such wickedness and was left in a state of bewilderment that can be more easily imagined than described in words. Finding himself buried alive, he called upon his uncle a thousand times, crying out that he was ready to give him the lamp, but his cries were in vain and could not possibly be heard by anyone. And so he remained in the darkness and gloom. At last, when his tears had abated somewhat, he descended to the bottom of the stairs in the cave to look for light in the garden through which he had passed earlier; but the wall which had been opened by a spell had closed and sealed up by another spell. Aladdin groped around several times, to the left and to the right, but could find no door. With renewed cries and tears, he sat down on the steps in the cave, all hope gone of ever seeing light again and, moreover, in the sad certainty that he would pass from the darkness where he was into the darkness of approaching death.

For two days, Aladdin remained in this state, eating and drinking nothing. At last, on the third day, believing death to be inevitable, he raised his hands in prayer and, resigning himself completely to God's will, he cried out: 'There is no strength nor power save in Great and Almighty God!'

However, just as he joined his hands in prayer, Aladdin unknowingly rubbed the ring which the magician had placed on his finger and of whose power he was as yet unaware. Immediately, from the ground beneath him, there rose up before him a *jinni* of enormous size and with a terrifying expression, who continued to grow until his head touched the roof of the chamber and who addressed these words to Aladdin: 'What do you want? Here am I, ready to obey you, your slave and the slave of all those who wear the ring on their finger, a slave like all the other slaves of the ring.'

At any other time and on any other occasion, Aladdin, who was not used to such visions, would perhaps have been overcome with terror and struck dumb at the sight of such an extraordinary apparition, but now, preoccupied solely with the danger of the present situation, he replied without hesitation: 'Whoever you are, get me out of this place, if you have the power to do so.' No sooner had he uttered these words than the earth opened up and he found himself outside the cave at the very spot to which the magician had led him.

Not surprisingly, Aladdin, after so long spent in pitch darkness, had difficulty at first in adjusting to broad daylight, but his eyes gradually became accustomed to it. When he looked around, he was very surprised not to find any opening in the ground; he could not understand how all of a sudden he should find himself transported from the depths of the earth. Only the spot where the kindling had been lit allowed him to tell roughly where the cave had been. Then, turning in the direction of the city, he spotted it in the middle of the gardens which surrounded it. He also recognized the path along which the magician had brought him and which he proceeded to follow, giving thanks to God at finding himself once again back in the world to which he had so despaired of ever returning.

When he reached the city, it was with some difficulty that he dragged himself home. He went in to his mother, but the joy of seeing her again, together with the weak state he was in from not having eaten for nearly three days, caused him to fall into a faint that lasted for some time. Seeing him in this state, his mother, who had already mourned him as lost, if not dead, did all she could to revive him. At last Aladdin recovered consciousness and the first words he addressed to her were to ask her to bring him something to eat, for it was three days since he had had anything at all. His mother brought him what she had, and, placing it before him, said: 'Don't hurry, now, because that's dangerous. Take it easy and eat a little at a time; eke it out, however much you need it. I don't want you even to speak to me; you will have enough time to tell me everything that happened to you when you have quite recovered. I am so comforted at seeing you again after the terrible state I have been in since Friday and after all the trouble I went to to discover what had happened to you as soon as I saw it was night and you hadn't come home.'

Aladdin followed his mother's advice and ate and drank slowly, a little at a time. When he had finished, he said to his mother: 'I would have been very cross with you for so readily abandoning me to the mercy of a man who planned to kill me and who, at this very moment, is quite certain either that I am no longer alive or that I will die at first light. But you believed him to be my uncle and so did I. How could we have thought otherwise of a man who overwhelmed me with both affection and gifts and who made me so many other fair promises? Now, mother, you must see he is nothing but a traitor, a wretch and a cheat. In all the gifts he gave me and the promises he made he had but one single aim – to kill me, as I said, without either of us guessing the reason why. For my part, I can assure you that I didn't do anything to deserve the slightest ill treatment. You will understand this yourself when you hear my faithful account of all that happened since I left you, right up to the time he came to execute his deadly plan.'

Aladdin then began to tell his mother all that had happened to him since the previous Friday, when the magician had come to take him with him to see the palaces and gardens outside the city, and what had happened along the way until they came to the spot by the two mountains where the magician's great miracle was to take place. He told her how, with some incense cast into the fire and a few words of magic, the earth had opened up, straight away, revealing the entrance to a cave which led to a priceless treasure. He did not leave out the blow he had received from the magician, nor how, once the magician had calmed down a little, he had placed his ring on Aladdin's finger and, making him countless promises, had got him to go down into the cave. He left out nothing of all that he had seen as he passed through the three rooms, in the garden and on the terrace from where he had taken the magic lamp. At this, he pulled the lamp from his clothes to show to his mother, together with the transparent fruits and those of different colours which he had gathered in the garden on his return and with which he had filled the two purses that he now gave her, though she did not make much of them. For these fruits were really precious stones; in the light of the lamp which lit up the room they shone like the sun and glittered and sparkled in such a way as to testify to their great worth, but Aladdin's mother was no more aware of this than he was. She had been brought up in very humble circumstances and her husband had never been wealthy enough to give her jewels and stones of this kind. Nor had she ever seen such things worn by any of her female relatives or neighbours. Consequently, it is not surprising that she should regard them as things of little value – a pleasure to the eye, at the very most, due to all their different colours – and so Aladdin put them behind one of the cushions of the sofa on which he was seated. He finished the account of his adventures by telling her how, when he returned to the entrance to the cave, ready to come out, he had refused to hand over to the magician the lamp that he wanted to have, at which the cave's entrance had immediately closed up, thanks to the incense which the magician had scattered over the fire that he had kept lit and to the words he had pronounced. Aladdin could not go on without tears coming to his eyes as he described to her the wretched state in which he found himself after being buried alive in that fatal cave, right up to when he emerged and returned to the world, so to speak, as the result of having touched the ring (of whose powers he was still unaware). When he had come to the end of his story, he said to his mother: 'I don't need to tell you any more; you know the rest. That was my adventure and the danger I was in since you last saw me.'

Aladdin's mother listened patiently and without interrupting to this wonderful and amazing story which at the same time was so painful for a mother who loved her son so tenderly despite all his faults. However, at the most disturbing points when the magician's treachery was further revealed, she could not prevent herself from showing, with signs of indignation, how much she hated him. As soon as Aladdin had finished, she broke out into a thousand reproaches against the impostor, calling him a traitor, trickster, murderer, barbarian – a magician, an enemy and a destroyer of mankind. 'Yes, my son,' she added, 'he's a magician and magicians are public menaces; they have dealings with demons through their spells and their sorcery. Praise the Lord, Who wished to preserve you from everything that his great wickedness might have done to you! You should indeed give thanks to Him for having so favoured you. You would have surely died had you not remembered Him and implored Him for His help.' She said much more besides, all the while execrating the magician's treachery towards her son. But as she spoke, she noticed that Aladdin, who had not slept for three days, needed some rest. She made him go to bed and went to bed herself shortly afterwards.

That night Aladdin, having had no rest in the underground cave where he had been buried and left to die, fell into a deep sleep from which he did not awake until late the following day. He arose and the first thing he said to his mother was that he needed to eat and that she could not give him a greater pleasure than to offer him breakfast. 'Alas, my son,' she sighed, 'I haven't got so much as a piece of bread to give you – yesterday evening you ate the few provisions there were in the house. But be patient for a little longer and I will soon bring you some food. I have some cotton yarn I have spun. I will sell it to buy you some bread and something else for our dinner.' 'Mother,' said Aladdin, 'leave your cotton yarn for some other occasion and give me the lamp I brought yesterday. I will go and sell it and the money I get will help provide us with enough for both breakfast and lunch, and perhaps also for our supper.'

Taking the lamp from where she had put it, Aladdin's mother said to her son: 'Here it is, but it's very dirty. With a little cleaning I think it would be worth a little more.' So she took some water and some fine sand in order to clean it, but no sooner had she begun to rub it than all of a sudden there rose up in front of them a hideous *jinni* of enormous size who, in a ringing voice, addressed her thus: 'What do you want? Here am I, ready to obey you, your slave and the slave of all those who hold the lamp in their hands, I and the other slaves of the lamp.'

But Aladdin's mother was in no state to reply; so great was her terror at the sight of the *jinni*'s hideous and frightening countenance that at the first words he uttered she fell down in a faint. Aladdin, on the other hand, had already witnessed a similar apparition while in the cave, and so, wasting no time and not stopping to think, he promptly seized the lamp. Replying in place of his mother, in a firm voice he said to the *jinni*: 'I am hungry, bring me something to eat.' The *jinni* disappeared and a moment later returned, bearing on his head a large silver bowl, together with twelve dishes also of silver, piled high with delicious foods and six large loaves as white as snow, and in his hands were two bottles of exquisite wine and two silver cups. He set everything down on the sofa and then disappeared.

This all happened so quickly that Aladdin's mother had not yet recovered from her swoon when the *jinni* disappeared for the second time. Aladdin, who had already begun to throw water on her face, without effect, renewed his efforts to revive her, and whether it was that her wits which had left her had already been restored or that the smell of the dishes which the *jinni* had brought had contributed in some measure, she immediately recovered consciousness. 'Mother,' said Aladdin, 'don't worry. Get up and come and eat, for here is something to give you heart again and which at the same time will satisfy my great hunger. We mustn't let such good food grow cold, so come and eat.'

Aladdin's mother was extremely surprised when she saw the large bowl, the twelve dishes, the six loaves, the two bottles and the two cups, and when she smelt the delicious aromas which came from all these dishes. 'My son,' she asked Aladdin, 'where does all this abundance come from and to whom do we owe thanks for such great generosity? Can the sultan have learned of our poverty and had compassion on us?' 'Mother,' Aladdin replied, 'let us sit down and eat; you need it as much as I do. When we have eaten, I will tell you.' They sat down and ate with all the more appetite in that neither had ever sat down before to such a well-laden table.

During the meal, Aladdin's mother never tired of looking at and admiring the large bowl and the dishes, although she did not know for sure whether they were of silver or some other metal, so unaccustomed was she to seeing things of that kind, and, to tell the truth, as she could not appreciate their value, which was unknown to her, it was the novelty of it all that held her admiration. Nor did her son Aladdin know any more about them than she did.

Aladdin and his mother, thinking to have but a simple breakfast, were still at table at dinner time; such excellent dishes had given them an appetite and while the food was still warm, they thought they might just as well put the two meals together so as not to have to eat twice. When this double meal was over, there remained enough not only for supper but for two equally large meals the next day.

After she had cleared away and had put aside those dishes they had not touched, Aladdin's mother came and seated herself beside her son on the sofa. 'Aladdin,' she said to him, 'I am expecting you to satisfy my impatience to hear the account you promised me.' Aladdin then proceeded to tell her exactly what had happened between the *jinni* and himself while she was in a swoon, right up to the moment she regained consciousness.

Aladdin's mother was greatly astonished by what her son told her and by the appearance of the *jinni*. 'But, Aladdin, what do you mean by these *jinn* of yours?' she said. 'Never in all my life have I heard of anyone I know ever having seen one. By what chance did that evil *jinni* come and show itself to me? Why did it come to me and not to you, when it had already appeared to you in the treasure cave?'

'Mother,' replied Aladdin, 'the *jinni* who has just appeared to you is not the same as the one that appeared to me; they look like each other to a certain extent, being both as large as giants, but they are completely different in appearance and dress. Also, they have different masters. If you remember, the one I saw called himself the slave of the ring which I have on my finger, while the one you have just seen called himself the slave of the lamp which you had in your hands. But I don't believe you can have heard him; in fact, I think you fainted as soon as he began to speak.'

'What?' cried his mother. 'It's your lamp, then, that made this evil *jinni* speak to me rather than to you? Take it out of my sight and put it wherever you like; I don't want ever to touch it again. I would rather have it thrown out or sold than run the risk of dying of fright touching it. If you were to listen to me, you would also get rid of the ring. One should not have anything to do with *jinn*; they are demons and our Prophet has said so.'

Aladdin, however, replied: 'Mother, with your permission, for the moment I am not going to sell – as I was ready to do – a lamp which is going to be so useful to both you and me. Don't you see what it has just brought us? We must let it go on bringing us things to eat and to support us. You should see, as I have seen, that it was not for nothing that my wicked and bogus uncle went to such lengths and undertook such a long and painful journey, since it was to gain possession of this magic lamp, preferring it above all the gold and silver which he knew to be in the rooms as he told me and which I myself saw. For he knew only too well the worth and value of this lamp than to ask for anything other than such a rich treasure. Since chance has revealed to us its merits, let's use it to our advantage, but quietly and in a way that will not draw attention to ourselves nor attract the envy and jealousy of our neighbours. I will take it away, since the *jinn* terrify you so much, and put it somewhere where I can find it when we need it. As for the ring, I can't bring myself to throw it away either; without the ring, you would never have seen me again. I may be alive now but without it I might not have lasted for very long. So please let me keep it carefully, always wearing it on my finger. Who knows whether some other danger may happen to me that neither of us can foresee and from which it will rescue me?' Aladdin's reasoning seemed sound enough to his mother, who could find nothing to add. 'My son,' she said, 'you can do as you like. As for myself, I wouldn't have anything to do with *jinn*. I tell you, I wash my hands of them and won't speak to you about them again.'

The next evening, there was nothing left after supper of the splendid provisions brought by the *jinni*. So, early the following day, Aladdin, who did not want to be overtaken by hunger, slipped one of the silver dishes under his clothes and went out to try to sell it. As he went on his way, he met a Jew whom he drew aside and, showing him the dish, asked him if he wanted to buy it. The Jew, a shrewd and cunning man, took the dish, examined it and, discovering it to be good silver, asked Aladdin how much he thought it was worth. Aladdin, who did not know its value, never having dealt in this kind of merchandise, happily told him that he was well aware what it was worth and that he trusted in his good faith. The Jew found himself confused by Aladdin's ingeniousness. Uncertain as to whether Aladdin knew what the dish was made of and its value, he took out of his purse a piece of gold, which at the very most was equal to no more than a seventy-second of the dish's true value, and gave it to him. Aladdin seized the coin with such eagerness and, as soon as he had it in his grasp, took himself off so swiftly that the Jew, not content with the exorbitant profit he had made with this purchase, was very cross at not having realized that Aladdin was unaware of the value of what he had sold him and that he could have given him far less for it. He was about to go after the young man to try to recover some change from his gold, but Aladdin had run off and was already so far away that he would have had difficulty in catching up with him.

On his way home, Aladdin stopped off at a baker's shop where he bought some bread for his mother and himself, paying for it with the gold coin, for which the baker gave him some change. When he came to his mother, he gave it her and she then went off to the market to buy the necessary provisions for the two of them to live on for the next few days.

They continued to live thriftily in this way; that is, whenever money ran out in the house, Aladdin sold off all the dishes to the Jew – just as he had sold the first one to him – one after the other, up to the twelfth and last dish. The Jew, having offered a piece of gold for the first dish, did not dare give him any less for the rest, for fear of losing such a good windfall, and so he paid the same for them all. When the money for the remaining dish was completely spent, Aladdin finally had recourse to the large bowl, which alone weighed ten times as much as each dish. He would have taken it to his usual merchant but was prevented from doing so by its enormous weight. So he was obliged to seek out the Jew, whom he brought to his mother. The Jew, after examining the weight of the bowl, there and then counted out for him ten gold pieces, with which Aladdin was satisfied.

As long as they lasted, these ten gold coins were used for the daily expenses of the household. Aladdin, who had been accustomed to an idle life, had stopped playing with his young friends ever since his adventure with the magician and spent his days walking around or chatting with the people with whom he had become acquainted. Sometimes he would call in at the shops of the great merchants, where he would listen to the conversation of the important people who stopped there or who used the shops as a kind of rendezvous, and these conversations gradually gave him a smattering of worldly knowledge.

When all ten coins had been spent, Aladdin had recourse to the lamp once again. Taking it in his hand, he looked for the same spot his mother had touched and, recognizing it by the mark left on it by the sand, he rubbed it as she had done. Immediately the selfsame *jinni* appeared in front of him, but as he had rubbed it more lightly than his mother had done, the *jinni* consequently spoke to him more softly. 'What do you want?' he asked in the same words as before. 'Here am I, ready to obey you, your slave and the slave of all those who hold the lamp in their hands, I and the other slaves of the lamp.'

'I'm hungry,' answered Aladdin. 'Bring me something to eat.' The *jinni* disappeared and a little later he reappeared, laden with the same bowls and dishes as before, which he placed on the sofa and promptly disappeared again.

Aladdin's mother, warned of her son's plan, had deliberately gone out on some errand in order not to be in the house when the *jinni* put in his appearance. When she returned a little later and saw the table and the many dishes on it, she was almost as surprised by the miraculous effect of the lamp as she had been on the first occasion. They both sat down to eat and after the meal there was still plenty of food for them to live on for the next two days.

When Aladdin saw there was no longer any bread or other provisions in the house to live on nor money with which to buy any, he took a silver dish and went to look for the Jew he knew in order to sell it to him. On his way there, he passed in front of the shop of a goldsmith, a man respected for his age, an honest man of great probity. Noticing him, the goldsmith called out to him and made him come in. 'My son,' he said, 'I have frequently seen you pass by, laden, just like now, on your way to a certain Jew, and then shortly after coming back, empty-handed. I imagine that you sell him something that you are carrying. But perhaps you don't know that this Jew is a cheat, even more of a cheat than other Jews, and that no one who knows him wants anything to do with him. I only tell you this as a favour; if you would like to show me what you are carrying now and if it is something I can sell, I will faithfully pay you its true price. Otherwise, I will direct you to other merchants who will not cheat you.'

The hope of getting more money for the dish made Aladdin draw it out from among his clothes and show it to the goldsmith. The old man, who at once recognized the dish to be of fine silver, asked him whether he had sold similar dishes to the Jew and how much the latter had paid him for them. Aladdin naively told him he had sold the Jew twelve dishes, for each of which he had received only one gold coin from him. 'The robber!' exclaimed the goldsmith, before adding: 'My son, what is done is done. Forget it. But when I show you the true value of your dish, which is made of the finest silver we use in our shops, you will realize how much the Jew has cheated you.'

The goldsmith took his scales, weighed the dish and, after explaining to Aladdin how much an ounce of silver was worth and how many parts there were in an ounce, he remarked that, according to the weight of the dish, it was worth seventy-two pieces of gold, which he promptly counted out to him in cash. 'There, here is the true value of your dish,' he told Aladdin. 'If you don't believe it, you can go to any of our goldsmiths you please and if he tells you it is worth more, I promise to pay you double that. Our only profit comes from the workmanship of the silver we buy, and that's something even the most fair-minded Jews don't do.'

Aladdin thanked the goldsmith profusely for the friendly advice he had just given him which was so much to his advantage. From then on, he only went to him to sell the other dishes and the bowl, and the true price was always paid him according to the weight of each dish. However, although Aladdin and his mother had an inexhaustible source of money from their lamp from which to obtain as much as they wanted as soon as supplies began to run out, nonetheless they continued to live as frugally as before, except that Aladdin would put something aside in order to maintain himself in an honest manner and to provide himself with all that was needed for their small household. His mother, for her part, spent on her clothes only what she earned from spinning cotton. Consequently, with them both living so modestly, it is easy to work out how long the money from the twelve dishes and the bowl would have lasted, according to the price Aladdin sold them for to the goldsmith. And so they lived in this manner for several years, aided, from time to time, by the good use Aladdin made of the lamp.

During this time, Aladdin assiduously sought out people of importance who met in the shops of the biggest merchants of gold and silver cloth, of silks, of the finest linens and of jewellery, and sometimes joined in their discussions. In this way, he completed his education and insensibly adopted the manners of high society. It was at the jewellers', in particular, that he discovered his error in thinking that the transparent fruits he had gathered in the garden where he had found the lamp were only coloured glass, learning that they were stones of great price. By observing the buying and selling of all kinds of gems in their shops, he got to know about them and about their value. But he did not see any there similar to his in size and beauty, and so he realized that instead of pieces of glass which he had considered as mere trifles, he was in possession of a treasure of inestimable value. He was prudent enough not to speak about this to anyone, not even to his mother; and there is no doubt that it was by keeping silent that he rose to the heights of good fortune, as we shall see in due course.

One day, when he was walking around in a part of the city, Aladdin heard a proclamation from the sultan ordering people to shut all their shops and houses and stay indoors until Princess Badr al-Budur, the daughter of the sultan, had passed on her way to the baths and had returned from them.

This public announcement stirred Aladdin's curiosity; he wanted to see the princess's face but he could only do so by placing himself in the house of some acquaintance and looking through a lattice screen, which would not suffice, because the princess, according to custom, would be wearing a veil over her face when going to the baths. So he thought up a successful ruse: he went and hid himself behind the door to the baths, which was so placed that he could not help seeing her pass straight in front of him.

Aladdin did not have to wait long: the princess appeared and he watched her through a crack that was large enough for him to see without being seen. She was accompanied by a large crowd of her attendants, women and eunuchs, who walked on both sides of her and in her train. When she was three or four steps from the door to the baths, she lifted the veil which covered her face and which greatly inconvenienced her, and in this way she allowed Aladdin to see her all the more easily as she came towards him.

Until that moment, the only other woman Aladdin had seen with her face uncovered was his mother, who was aged and who never had such beautiful features as to make him believe that other women existed who were beautiful. He may well have heard that there were women of surpassing beauty, but for all the words one uses to extol the merits of a beautiful woman, they never make the same impression as a beautiful woman herself.

When Aladdin set eyes on Badr al-Budur, any idea that all women more or less resembled his mother flew from his mind; he found his feelings were now quite different and his heart could not resist the inclinations aroused in him by such an enchanting vision. Indeed, the princess was the most captivating dark-haired beauty to be found in all the world; her large, sparkling eyes were set on a level and full of life; her look was gentle and modest, her faultless nose perfectly proportioned, her mouth small, with its ruby lips charming in their pleasing symmetry; in a word, the regularity of all her facial features was nothing short of perfection. Consequently, one should not be surprised that Aladdin was so dazzled and almost beside himself at the sight of so many wonders hitherto unknown to him united in one face. Added to all these perfections, the princess also had a magnificent figure and bore herself with a regal air which, at the mere sight of her, would draw to her the respect that was her due.

After the princess had entered the baths, Aladdin remained for a while confused and in a kind of trance, recalling and imprinting deeply on his mind the image of the vision which had so captivated him and which had penetrated the very depths of his heart. He eventually came to and, after reflecting that the princess had now gone past and that it would be pointless for him to stay there in order to see her when she came out of the baths, for she would be veiled and have her back to him, he decided to abandon his post and go away.

When he returned home, Aladdin could not conceal his worry and confusion from his mother, who, noticing his state and surprised to see him so unusually sad and dazed, asked him whether something had happened to him or whether he felt ill. Aladdin made no reply but slumped down on the sofa, where he remained in the same position, still occupied in conjuring up the charming vision of the princess. His mother, who was preparing the supper, did not press him further. When it was ready, she served it up near to him on the sofa, and sat down to eat. However, noticing he was not paying any attention, she told him to come to the table and eat and it was only with great difficulty that he agreed. He ate much less than usual, keeping his eyes lowered and in such profound silence that his mother was unable to draw a single word out of him in reply to all the questions she asked him in an attempt to discover the reason for such an extraordinary change in his behaviour. After supper, she tried to ask him once again the reason for his great gloom but was unable to learn a thing and Aladdin decided to go to bed rather than give his mother the slightest satisfaction in the matter.

We will not go into how Aladdin, smitten with the beauty and charms of Princess Badr, spent the night, but will only observe that the following day, as he was seated on the sofa facing his mother – who was spinning cotton, as was her custom – he spoke to her as follows: 'Mother,' he said, 'I am breaking the silence I have kept since my return from the city yesterday because I realize it has been worrying you. I wasn't ill, as you seemed to think, and I am not ill now, but I can't tell you what I was feeling then, and what I am still feeling is something worse than any illness. I don't really know what this is, but I'm sure that what you are going to hear will tell you what it is.' He went on: 'No one in this quarter knew, and so you, too, cannot have known, that yesterday evening the daughter of the sultan, Princess Badr, was to go to the baths. I learned this bit of news while walking around the city. An order was proclaimed to shut up the shops and everyone was to stay indoors, so as to pay due respect to the princess and to allow her free passage in the streets through which she was to pass. As I was not far from the baths, I was curious to see her with her face uncovered, and so the idea came to me to go and stand behind the door to the baths, thinking that she might remove her veil when she was ready to go in. You know how the door is placed, so you can guess how I could see her quite easily if what I imagined were to happen. And indeed, as she entered she lifted her veil and I had the good fortune and the greatest satisfaction in the world to see this lovely princess. That, then, mother, is the real reason for the state you saw me in yesterday when I came home and the cause for my silence up till now. I love the princess with a passion I can't describe to you; and as this burning passion grows all the time, I feel it cannot be assuaged by anything other than the possession of the lovely Badr; which is why I have decided to ask the sultan for her hand in marriage.'

Aladdin's mother listened fairly carefully to what her son told her, up to the last few words. When she heard his plan to ask for the princess's hand, she could not help interrupting him by bursting out laughing. Aladdin was about to go on but, interrupting him again, she exclaimed: 'What are you thinking of, my son? You must have gone out of your mind to talk to me about such a thing!'

'Mother,' replied Aladdin, 'I can assure you I have not lost my senses but am quite in my right mind. I expected you would reproach me with madness and extravagance – and you did – but that will not stop me telling you once again that I have made up my mind to ask the sultan for the princess's hand in marriage.'

'My son,' his mother continued, addressing him very seriously, 'I can't indeed help telling you that you quite forget yourself; and even if you are still resolved to carry out this plan, I don't see through whom you would dare to make this request to the sultan.' 'Through you yourself,' Aladdin replied without hesitating. 'Through me!' exclaimed his mother, in surprise and astonishment. 'I go to the sultan? Ah, I would take very great care to avoid such an undertaking! And who are you, my son,' she continued, 'to be so bold as to think of the daughter of your sultan? Have you forgotten that you are the son of a tailor, among the least of his capital's citizens, and of a mother whose forebears were no more exalted? Don't you know that sultans don't deign to give away their daughters in marriage even to the sons of sultans, unless they are expected to reign one day themselves?' 'Mother,' replied Aladdin, 'I have already told you that I had foreseen all that you have said or would say, so despite all your remonstrances, nothing will make me change my mind. I have told you that through your mediation I would ask for Princess Badr's hand in marriage: this is a favour I ask of you, with all the respect I owe you, and I beg you not to refuse, unless you prefer to see me die rather than give me life a second time.'

Aladdin's mother felt very embarrassed when she saw how stubbornly he persisted in such a foolhardy plan. 'My son,' she said, 'I am your mother and, as a good mother who brought you into the world, there is nothing right and proper and in keeping with our circumstances that I would not be prepared to do out of my love for you. If it's a matter of speaking about marriage to the daughter of one of our neighbours, whose circumstances are equal or similar to ours, then I would gladly do everything in my power; but again, to succeed, you would need to have some assets or income, or you should know some trade. When poor people like us want to get married, the first thing they need to think about is their livelihood. But you, not reflecting on your humble status, the little you have to commend you and your lack of money, you aspire to the highest degree of fortune and are so presumptuous as to demand no less than the hand in marriage of the daughter of your sovereign – who with a single word can crush you and bring about your downfall. I won't speak of what concerns you; it is you who should think what you should do, if you have any sense. I come to what concerns me. How could such an extraordinary idea as that of wanting me to go to the sultan and propose that he give you the princess's, his daughter's, hand in marriage ever have come into your head? Supposing I had the – I won't say courage – effrontery to present myself to his majesty to put such an extravagant request to him, to whom would I go to for an introduction? Don't you think that the first person to whom I spoke about it would treat me as a mad woman and throw me out indignantly, as I deserved? And what about seeking an audience with the sultan? I know there is no difficulty when one goes to him to seek justice and that he readily grants it to his subjects when they ask him for it. I also know that when one goes to ask him a favour, he grants it gladly, when he sees that one has deserved it and is worthy of it. But is that the position you are in and do you think you merit the favour that you want me to ask for you? Are you worthy of it? What have you done for your sultan or for your country? How have you distinguished yourself? If you haven't done anything to deserve so great a favour – of which, anyhow, you are not worthy – how could I have the audacity to ask him for it? How could I so much as open my mouth to propose it to the sultan? His majestic presence alone and the brilliance of his court would make me dry up immediately – I, who used to tremble before my late husband, your father, when I had to ask him for the slightest thing. There is something else you haven't thought about, my son, and that is that one does not go to ask a favour of the sultan without bearing a present. A present has at least this advantage that, if, for whatever reason, he refuses the favour, he at least listens to the request and to whoever makes it. But what present do you have to offer? And if you had something worthy of the slightest attention from so great a ruler, would your gift adequately represent the scale of the favour you want to ask him? Think about this and reflect that you are aspiring to something which you cannot possibly obtain.'

Aladdin listened quietly to everything his mother had to say in her attempt to make him give up his plan. Finally, after reflecting on all the points she had made in remonstrating with him, he replied to her, saying: 'Mother, I admit it's great rashness on my part to carry my pretensions as far as I am doing, and that it's very inconsiderate of me to insist with such heat and urgency on your going and putting my proposal of marriage to the sultan without first taking the appropriate measures for you to obtain a favourable and successful audience with him. Please forgive me, but don't be surprised if, in the strength of the passion which possesses me, I did not at first envisage all that could help me procure the happiness I seek. I love Princess Badr beyond anything you can imagine, or rather, I adore her and will continue to persevere in my plan to marry her – my mind is quite made up and fixed in this matter. I am grateful to you for the opening you have just given me; I see it as the first step which will help me obtain the happy outcome I promise myself. You tell me that it is not customary to go before the sultan without bearing him a present, and that I have nothing which is worthy of him. I agree with you about the present, and I admit I hadn't thought about it. As for your telling me that I have nothing I can possibly offer him, don't you think, mother, that what I brought back with me the day I was saved from almost inevitable death could not make a very nice gift for the sultan? I am talking about what I brought back in the two purses and in my belt, which you and I both took to be pieces of coloured glass. I have since learned better and I can tell you, mother, that these are jewels of inestimable value, fit only for great kings. I discovered their worth by frequenting jewellers' shops, and you can take my word for it. None of all those I have seen in the shops of our jewellers can compare in size or in beauty to those we possess, and yet they sell them for exorbitant prices. The fact is that neither you nor I know what ours are worth, but however much that is, as far as I can judge from the little experience I have gained, I am convinced that the present will please the sultan very much. You have a porcelain dish large enough and of the right shape to contain the jewels; fetch it and let's see the effect they make when we arrange them according to their different colours.'

Aladdin's mother fetched the porcelain dish and Aladdin took out the stones from the two purses and arranged them in it. The effect they made in full daylight, by the variety of their colours, their brilliance and sparkle, was such as to almost dazzle them both and they were greatly astonished, for neither of them had seen the stones except in the light of a lamp. It is true that Aladdin had seen them hanging on the trees like fruit, which must have made an enchanting sight; but as he was still a boy, he had only thought of these stones as trinkets to be played with, and that is the only way he had thought of them, knowing no better.

After admiring for some time the beauty of the jewels, Aladdin spoke once more. 'Mother,' he said, 'you can no longer get out of going and presenting yourself to the sultan on the pretext of not having a present to offer him; here is one, it seems to me, which will ensure you are received with the most favourable of welcomes.'

For all the beauty and splendour of the present, Aladdin's mother did not think it was worth as much as Aladdin believed it to be. Nonetheless she thought it would be acceptable and she knew she had nothing to say to the contrary; but she kept thinking of the request Aladdin wanted her to make to the sultan with the help of this gift and this worried her greatly. 'My son,' she said to him, 'I don't find it difficult to imagine that the present will have its effect and that the sultan will look upon me favourably; but when it comes to my putting the request to him that you want me to make, I feel I won't have the strength and I will remain silent. My journey will have been wasted as I will have lost what you claim is a gift of extraordinary value. I will come home completely embarrassed at having to tell you that you are disappointed in your hopes. I have already explained this to you and you should realize that this is what will happen. However,' she added, 'even if it hurts me, I will give in to your wish and I will force myself to have the strength and courage to dare to make the request you want me to make. The sultan will most probably either laugh at me and send me away as a madwoman or he will quite rightly fly into a great rage of which you and I will inevitably be the victims.'

Aladdin's mother gave her son several other reasons in an attempt to make him change his mind; but the charms of Princess Badr had made too deep an impression on his heart for anyone to be able to dissuade him from carrying out his plan. Aladdin continued to insist his mother go through with it; and so, as much out of her love for him as out of fear that he might resort to some extreme measure, she overcame her aversion and bowed to her son's will.

As it was too late and the time to go the palace for an audience with the sultan that day had passed, the matter was put off until the following day. For the rest of the day, mother and son spoke of nothing else, Aladdin taking great care to tell his mother everything he could think of to strengthen her in the decision which she had finally made, to go and present herself to the sultan. Yet, despite all his arguments, his mother could not be persuaded that she would ever succeed in the matter, and, indeed, one must admit she had good reason to doubt. 'My son,' she said to Aladdin, 'assuming the sultan receives me as favourably as I wish for your sake, and assuming he listens calmly to the proposal you want me to put to him, what if, after this friendly reception, he should then ask about your possessions, your riches and your estates? For that's what he will ask about before anything else, rather than about you yourself. If he asks me about that, what do you want me to reply?'

'Mother,' said Aladdin, 'let's not worry in advance about something which may never happen. Let's first see what sort of reception the sultan gives you and what reply he gives you. If he happens to want to know all you have just suggested, I will think of an answer to give him, for I am confident that the lamp, which has been the means of our subsistence for the past few years, will not fail me in time of need.'

Aladdin's mother could think of nothing to say to this. She agreed that the lamp might well be capable of greater miracles than simply providing them with enough to live on. This thought satisfied her and at the same time removed all the difficulties which could have stopped her carrying out the mission she had promised her son. Aladdin, who guessed what she was thinking, said to her: 'Mother, above all remember to keep the secret; on it depends all the success you and I expect from this affair.' They then left each other to have some rest; but Aladdin's mind was so filled with his violent passion and his grand plans for an immense fortune that he was unable to pass the night as peacefully as he would have wished. Before daybreak, he rose and immediately went to wake his mother. He urged her to get dressed as quickly as possible in order to go to the palace gate and to pass through it as soon as it was opened, when the grand vizier, the other viziers and all the court officials entered the council chamber where the sultan always presided in person.

Aladdin's mother did everything her son wanted. She took the porcelain dish containing the jewels, wrapped it in two layers of cloth, one finer and cleaner than the other, which she tied by all four corners in order to carry it more easily. She then set out, to Aladdin's great satisfaction, and took the street which led to the sultan's palace. When she arrived at the gate, the grand vizier, accompanied by the other viziers and the highest-ranking court officials, had already entered. There was an enormous crowd of all those who had business at the council. The gate opened and she walked with them right up into the council chamber, which was a very handsome room, wide and spacious, with a grand and magnificent entrance. She stopped and placed herself in such a way as to be opposite the sultan, with the grand vizier and the nobles who had a seat at the council to the right and left of him. One after the other, people were called according to the order of the requests that had been presented, and their affairs were produced, pleaded and judged until the time the session usually adjourned, when the sultan rose, dismissed the council and withdrew to his apartments where he was followed by the grand vizier. The other viziers and the court officials withdrew, as did all who were there on some particular business, some happy to have won their case, others less satisfied as judgement had been made against them, and still others left in the hope of their case being heard at the next session.

Aladdin's mother, seeing that the sultan had risen and withdrawn and that everyone was leaving, concluded rightly that he would not reappear that day and so she decided to return home. When Aladdin saw her coming in with the present destined for the sultan, he did not know at first what to think. Afraid that she had some bad news for him, he did not have the strength to ask her about her trip. The good woman, who had never before set foot in the sultan's palace and who had not the slightest acquaintance with what normally happened there, helped him out of his difficulty by saying to him with great naivety: 'My son, I saw the sultan and I am quite sure he, too, saw me. I was right in front of him and nobody could prevent him seeing me, but he was so occupied with all those talking to the right and left of him, that I was filled with pity to see the trouble he took to listen patiently to them. That went on for such a long time that I think he finally became weary; for he arose all of a sudden and withdrew quite brusquely, without wishing to listen to the many other people who were lined up to speak to him. I was, in fact, very pleased because I was beginning to lose patience and was very tired from standing up for so long. However, all is not lost and I intend to return there tomorrow; perhaps the sultan will be less busy.'

However great his passion, Aladdin had to be content with this excuse and remain patient. But he at least had the satisfaction of seeing that his mother had taken the most difficult step, which was to stand before the sultan; he hoped that she would follow the example of those whom she saw speaking to him, and not hesitate to carry out the task with which she was charged when she found an opportunity to speak to him.

The next day, arriving early, as she had done the previous day, Aladdin's mother again went to the sultan's palace with the present of gems; but her journey once again proved futile. She found the door of the council chamber closed, council sessions being held only every other day, and realized that she would have to return the following day. This news she reported back to Aladdin, who had to remain patient. She returned six more times to the council chamber, on the appropriate days, always placing herself in front of the sultan, but with the same lack of success as on the first occasion. She would perhaps have returned a hundred more times, all to no avail, had not the sultan, who had seen her standing in front of him at each session, finally paid attention to her. Her lack of success is hardly surprising in that only those who had petitions to present approached the sultan, one by one, to plead their cause, whereas Aladdin's mother was not among those lined up before him.

At last, one day, after the council had risen and he had returned to his apartments, the sultan said to his vizier: 'For some time now I have noticed a certain woman who comes regularly every day that I hold my council session. She carries something wrapped up in a cloth and remains standing from the beginning of the audience to the end, always deliberately placed in front of me. Do you know what she wants?'

The grand vizier, who knew no more about her than the sultan but did not wish to appear to be stuck for an answer, replied: 'Sire, your majesty knows well how women often raise complaints about matters of no importance: this one, apparently, has come to complain to you about having been sold bad flour, or about some other, equally trivial, wrong.' But the sultan was not satisfied with this reply and said: 'On the next council day, if this woman comes again, be sure to have her summoned so that I can hear what she has to say.' To this the grand vizier replied by kissing the sultan's hand and raising it above his head to indicate that he was prepared to die if he failed to carry out the sultan's command.

Aladdin's mother had by now become so accustomed to going to the council and standing before the sultan that she did not think it any trouble, as long as she made her son understand that she was doing everything she could to comply with his wishes. So she returned to the palace on the day of the next session and took up her customary position at the entrance of the chamber, opposite the sultan.

The grand vizier had not yet begun to bring up any case when the sultan noticed Aladdin's mother. Feeling compassion for her, having seen her wait so long and so patiently, the sultan said to him: 'First of all, in case you forget, here is the woman I was telling you about; make her come up and let us begin by hearing her and getting her business out of the way.' Immediately, the grand vizier pointed the woman out to the chief usher, who was standing ready to receive his orders, and commanded him to fetch her and bring her forward. The chief usher went up to her and made a sign to follow him to the foot of the sultan's throne, where he left her, before taking his place next to the grand vizier.

Aladdin's mother, having learned from the example of the many others she had seen approach the sultan, prostrated herself, with her forehead touching the carpet that covered the steps to the throne, and remained thus until the sultan ordered her to rise. When she rose, the sultan asked her: 'My good woman, for some time now I have seen you come to my council chamber and remain at the entrance from the beginning to the very end of the session – so what brings you here?'

Hearing these words, Aladdin's mother prostrated herself a second time; standing up again, she said: 'King of all kings, before I reveal to your majesty the extraordinary and almost unbelievable business which brings me before your exalted throne, I beg you to pardon me for the audacity, not to say the impudence, of the request I am going to make to you – a request so unusual that I tremble and am ashamed to put it to my sultan.'

The sultan, to allow her to explain herself in complete freedom, ordered everyone to go out of the council chamber, except the grand vizier. He then told her she could speak and explain herself without fear. But Aladdin's mother, not content with the sultan's kindness in sparing her the distress she would have endured in speaking in front of so many people, wished to protect herself from what she feared would be his indignation at the unexpected proposal which she was going to put to him, and continued: 'Sire, I dare to entreat you that if you find the request I am going to put to your majesty in any way offensive or insulting, you will first assure me of your forgiveness and grant me your pardon.' 'Whatever it is,' replied the sultan, 'I now forgive you and assure you that no harm will come to you. So speak out.'

Having taken all these precautions because of her fear of arousing the sultan's anger at receiving a proposal of so delicate a nature, Aladdin's mother then went on to relate faithfully how Aladdin had first seen Princess Badr, the violent passion which the sight of her had inspired in him, what he had said to her; and how she had done everything she could to talk him out of a passion so harmful not only to his majesty but also to the princess, his daughter, herself. 'But my son,' she continued, 'far from profiting from my advice and admitting his audacity, has obstinately persisted in his purpose. He even threatened that he would be driven to do something desperate if I refused to come and ask your majesty for the hand of the princess in marriage. And it was only with extreme reluctance that I finally found myself forced to do him this favour, for which I beseech your majesty once more to pardon not only me but also my son, Aladdin, for having deigned to aspire to so elevated a union.'

The sultan listened to this speech very gently and kindly, showing no sign of anger or indignation, nor making fun of her request. But before giving her an answer, he asked her what it was she had brought wrapped in a cloth, whereupon she immediately took the porcelain dish, which she had set down at the foot of the throne before prostrating herself, unwrapped it and presented it to him.

One can hardly describe the sultan's surprise and astonishment when he saw such a quantity of precious gems, so perfect, so brilliant and of a size the like of which he had never seen before, crammed into this dish. For a while he remained quite motionless, lost in admiration. When he had recovered, he received the present from the hands of Aladdin's mother, exclaiming ecstatically: 'Ah! How beautiful! What a splendid present!' When he had admired and handled virtually all the jewels, one by one, examining each gem to assess its distinctive quality, he turned towards his grand vizier and, showing him the dish, said to him: 'Look, don't you agree you won't find anything more splendid or more perfect in the whole world?' The grand vizier was dazzled. 'So, what do you think of such a present?' the sultan asked him. 'Isn't it worthy of the princess, my daughter, and can't I then give her, at a price like that, to the man who asks me for her hand in marriage?'

These words roused the grand vizier into a state of strange agitation. Some time ago, the sultan had given him to understand that it was his intention to bestow the princess in marriage to one of his sons, and so he feared, and with some justification, that the sultan, dazzled by such a sumptuous and extraordinary gift, would now change his mind. He went up to the sultan and whispered into his ear: 'Sire, one can't disagree that the present is worthy of the princess; but I beg your majesty to grant me three months before you come to a decision. Before that time, I hope that my son, on whom you have been so kind as to indicate you look favourably, will be able to present her with a much more valuable gift than that offered by Aladdin, who is a stranger to your majesty.'

The sultan, although he was quite sure that his grand vizier could not possibly come up with enough for his son to produce a gift of similar value to offer the princess, nonetheless listened to him and granted him this favour. Turning, then, to Aladdin's mother, he said: 'Go home, good woman, and tell your son that I agree to the proposal you have made on his behalf; but I can't marry the princess, my daughter, to him before I have furnishings provided for her, and these won't be ready for three months. At the end of that time, come back.'

Aladdin's mother returned home, her joy being all the greater because she had first thought that, in view of her lowly state, access to the sultan would be impossible, whereas she had in fact obtained a very favourable reply instead of the rebuffs and resulting confusion she had expected. When Aladdin saw his mother come in, two things made him think that she was bringing good news: one was that she was returning earlier than usual, and the other was that her face was all lit up and she was smiling. 'So, Mother,' he said to her, 'is there any cause for hope, or must I die of despair?' Having removed her veil and sat down beside him on the sofa, she replied: 'My son, I'm not going to keep you in a state of uncertainty and so will begin at once by telling you that far from thinking of dying you have every reason to be happy.' She went on to tell him how she had received an audience, before everyone else, and that was the reason she had returned so early. She also told him what precautions she had taken not to offend the sultan in putting the proposal of marriage to Princess Badr, and of the very favourable response she had received from the sultan's own mouth. She added that, as far as she could judge from indications given by the sultan, it was above all the powerful effect of the present which had determined that favourable reply. 'I least expected this,' she said, 'because the grand vizier had whispered in his ear before he gave his reply and I was afraid he would deflect any goodwill the sultan might have towards you.'

When he heard this, Aladdin thought himself the happiest of men. He thanked his mother for all the trouble she had gone to in pursuit of this affair, whose happy outcome was so important for his peace of mind. And although three months seemed an extremely long time such was his impatience to enjoy the object of his passion, he nonetheless prepared himself to wait patiently, trusting in the sultan's word, which he considered irrevocable.

One evening, when two months or so had passed, with him counting not only the hours, days and the weeks, but even every moment as he waited for the period to come to an end, his mother, wanting to light the lamp, noticed that there was no more oil in the house. So she went out to buy some. As she approached the centre of the city, everywhere she saw signs of festivity: the shops, instead of being shut, were all open and were being decorated with greenery, and illuminations were being prepared – in their enthusiasm, every shop owner was vying with each other in their efforts to display the most pomp and magnificence. Everywhere were demonstrations of happiness and rejoicing. The streets themselves were blocked by officials in ceremonial dress, mounted on richly harnessed horses, and surrounded by a milling throng of attendants on foot. Aladdin's mother asked the merchant from whom she was buying her oil what this all meant. 'My good woman, where are you from?' he replied. 'Don't you know that the son of the grand vizier is to marry Princess Badr, daughter of the sultan, this evening? She is about to come out of the baths and the officials you see here are gathering to accompany her procession to the palace, where the ceremony is to take place.'

Aladdin's mother did not wish to hear any more. She returned home in such haste that she arrived almost breathless. She found Aladdin, who little expected the grievous news she was bringing, and exclaimed: 'My son, you have lost everything! You were counting on the sultan's fine promises – nothing will come of them now.' Alarmed at these words, Aladdin said to her: 'But, mother, in what way will the sultan not keep his promise to me? And how do you know?' 'This evening,' she replied, 'the son of the grand vizier is to marry Princess Badr, in the palace.' She went on to explain how she had learned this, telling him all the circumstances so as to leave him in no doubt.

At this news, Aladdin remained motionless, as though he had been struck by a bolt of lightning. Anyone else would have been quite overcome, but a deep jealousy prevented him from staying like this for long. He instantly remembered the lamp which had until then been so useful to him: without breaking out in a pointless outburst against the sultan, the grand vizier or his son, he merely said to his mother: 'Maybe the son of the grand vizier will not be as happy tonight as he thinks he will be. While I go to my room for a moment, prepare us some supper.'

Aladdin's mother guessed her son was going to make use of the lamp to prevent, if possible, the consummation of the marriage, and she was not deceived. Indeed, when Aladdin entered his room, he took the magic lamp – which he had removed from his mother's sight and taken there after the appearance of the *jinni* had given her such a fright – and rubbed it in the same spot as before. Immediately, the *jinni* appeared before him and asked: 'What is your wish? Here am I, ready to obey you, your slave and the slave of all those who hold the lamp, I and the other slaves of the lamp.'

'Listen,' Aladdin said to him, 'up until now, you have brought me food when I was in need of it, but now I have business of the utmost importance. I have asked the sultan for the hand of the princess, his daughter; he promised her to me but asked for a delay of three months. However, instead of keeping his promise, he is marrying her tonight to the son of the grand vizier, before the time is up: I have just learned of this and it's a fact. What I demand of you is that, as soon as the bride and bridegroom are in bed, you carry them off and bring them both here, in their bed.' 'Master,' replied the *jinni*, 'I will obey you. Do you have any other command?' 'Nothing more at present,' said Aladdin, and the *jinni* immediately disappeared.

Aladdin returned to his mother and had supper with her, calmly and peacefully as usual. After supper, he talked to her for a while about the marriage of the princess as if it were something which no longer worried him. Then he returned to his room, leaving his mother to go to bed. He himself did not go to sleep, however, but waited for the *jinni*'s return and for the order he had given him to be carried out.

All this while, everything had been prepared with much splendour in the sultan's palace to celebrate the marriage of the princess, and the evening passed in ceremonies and entertainments which went on well into the night. When it was all over, the son of the grand vizier, after a signal given him by the princess's chief eunuch, slipped out and was then brought in by him to the princess's apartments, right to the room where the marriage bed had been prepared. He went to bed first. A little while after, the sultana, accompanied by her ladies and by those of the princess, her daughter, led in the bride, who, as is the custom of brides, put up a great resistance. The sultana helped to undress her and put her into bed as though by force; and, after having embraced her and saying goodnight, she withdrew, together with all the women, the last to leave shutting the door behind her.

No sooner had the door been shut than the *jinni* – as faithful servant of the lamp and punctual in carrying out the commands of those who had it in their hands – without giving the bridegroom time to so much as caress his wife, to the great astonishment of them both, lifted up the bed, complete with bride and groom, and transported them in an instant to Aladdin's room, where he set it down.

Aladdin, who had been waiting impatiently for this moment, did not allow the son of the grand vizier to remain lying with the princess but said to the *jinni*: 'Take this bridegroom, lock him up in the privy and come back tomorrow morning, a little after daybreak.' The *jinni* immediately carried off the son of the grand vizier from the bed, in his nightshirt, and transported him to the place Aladdin had told him to take him, where he left the bridegroom, after breathing over him a breath which he felt from head to toe and which prevented him from stirring from where he was.

However great the passion Aladdin felt for Princess Badr, once he found himself alone with her, he did not address her at length, but declared passionately: 'Don't be afraid, adorable princess, you are quite safe here, and however violent the love I feel for your beauty and your charms, it will never go beyond the bounds of the profound respect I have for you. If I have been forced to adopt such extreme measures, this was not to offend you but to prevent an unjust rival from possessing you, contrary to the word in my favour given me by your father, the sultan.'

The princess, who knew nothing of the circumstances surrounding all this, paid little attention to what Aladdin had to say and was in no state to reply to him. Her terror and astonishment at so surprising and unexpected an adventure had put her into such a state that Aladdin could not get a word out of her. He did not leave it at that but decided to undress and then lie down in the place of the son of the grand vizier, his back turned to the princess, after having taken the precaution of putting a sword between them, to show that he deserved to be punished if he made an attempt on her honour.

Happy at having thus deprived his rival of the pleasure which he had flattered himself he would enjoy that night, Aladdin slept quite peacefully. This was not true of the princess, however: never in all her life had she spent so trying and disagreeable a night; and as for the son of the vizier, if one considers the place and the state in which the *jinni* had left him, one can guess that her new husband spent it in a much more distressing manner.

The next morning, Aladdin did not need to rub the lamp to summon the *jinni*, who came by himself at the appointed hour, just when Aladdin had finished dressing. 'Here am I,' he said to Aladdin. 'What is your command?' 'Go and bring back the son of the grand vizier from the place where you put him,' said Aladdin. 'Place him in this bed again and carry it back to the sultan's palace, from where you took it.' The *jinni* went to fetch the son of the grand vizier, and when he reappeared, Aladdin took up his sword from the bed. The *jinni* placed the bridegroom next to the princess and, in an instant, he returned the marriage bed to the same room in the sultan's palace from where he had taken it.

It should be pointed out that, all the while, the *jinni* could not be seen by either the princess or the son of the grand vizier – his hideous shape would have been enough to make them die of fright. Nor did they hear any of the conversation between Aladdin and him. All they noticed was how their bed shook and how they were transported from one place to another; which was quite enough, as one can easily imagine, to give them a considerable fright.

The *jinni* had just restored the nuptial bed to its place when the sultan, curious to discover how his daughter, the princess, had spent the first night of her marriage, entered her room to wish her good morning. No sooner did he hear the door open than the son of the grand vizier, chilled to the bone from the cold he had endured all night long and not yet having had time to warm up again, got up and went to the closet where he had undressed the previous evening.

The sultan approached the princess's bed, kissed her between the eyes, as was the custom, and asked her, as he greeted her with a smile, what sort of night she had had; but raising his head again and looking at her more closely, he was extremely surprised to see that she was in a state of great dejection and neither by a blush spreading over her face nor by any other sign could she satisfy his curiosity. She only gave him a most sorrowful look, which indicated either great sadness or great discontent. He said a few more words to her but, seeing that he could get nothing more from her, he decided she was keeping silent out of modesty and so retired. Nevertheless, still suspicious that there was something unusual about her silence, he went straight away to the apartments of the sultana and told her in what a state he had found the princess and how she had received him. 'Sire,' the sultana said to him, 'this should not surprise your majesty; there's no bride who does not display the same reserve the morning after her wedding night. It won't be the same in two or three days: she will then receive her father, the sultan, as she ought. I am going to see her myself,' she added, 'and I will be very surprised if she receives me in the same way.'

When the sultana had dressed, she went to the princess's room. Badr had not yet risen, and when the sultana approached her bed, greeting and embracing her, great was her surprise not only to receive no reply but also to see the princess in a state of deep dejection, which made her conclude that something she could not understand had happened to her daughter. 'My daughter,' she said to her, 'how is it that you don't respond to my caresses? How can you behave like this to your mother? Don't you think I don't know what can happen in circumstances like yours? I would really like to think that that's not what's in your mind and something else must have happened. Tell me quite frankly; don't leave me weighed down by anxiety for a moment longer.'

At last, the princess broke her silence and gave a deep sigh. 'Ah! My dear and esteemed mother,' she exclaimed, 'forgive me if I have failed to show you the respect I owe you. My mind is so preoccupied with the extraordinary things that happened to me last night that I have not yet recovered from my astonishment and terror and I hardly know myself.' She then proceeded to tell her, in the most colourful detail, how shortly after she and her husband had gone to bed, the bed had been lifted up and transported in a moment to a dark and squalid room where she found herself all alone and separated from her husband, without knowing what had happened to him; how she had seen a young man who had addressed a few words to her which her terror had prevented her understanding, who had lain beside her in her husband's place, after placing a sword between them; and how her husband had been restored to her and the bed returned to its place, all in a very short space of time. 'All this,' she added, 'had just taken place when the sultan, my father, came into the room; I was so overcome by grief that I had not the strength to reply even with a single word, and so I have no doubt he was angry at the manner in which I received the honour he did me by coming to see me. But I hope he will forgive me when he knows of my sad adventure and sees the pitiful state I'm still in.'

The sultana listened calmly to everything the princess had to say, but she did not believe it. 'My daughter,' she said, 'you were quite right not to talk about this to the sultan, your father. Take care not to talk about it to anyone – they will think you mad if they hear you talk like this.' 'Mother,' she rejoined, 'I can assure you that I am in my right mind. Ask my husband and he will tell you the same thing.' 'I will ask him,' replied the sultana, 'but even if his account is the same as yours, I won't be any more convinced than I am now. Now get up and clear your mind of such fantasies; a fine thing it would be if you were to let such a dream upset the celebrations arranged for your wedding, which are set to last several days, not only in this palace but throughout the kingdom! Can't you already hear the fanfares and the sounds of trumpets, drums and tambourines? All this should fill you with pleasure and joy and make you forget the fantastic stories you've been telling me.' The sultana then summoned the princess's maids and, after she had made her get up and seen her set about getting dressed, she went to the sultan's apartments and told him that some fancy had, indeed, entered the head of his daughter, but that it was nothing. She sent for the son of the vizier to discover from him a little about what the princess had told her; but he, knowing himself to be greatly honoured by his alliance with the sultan, decided it would be best to conceal the adventure. 'Tell me, son-in-law,' the sultana said to him, 'are you being as stubborn as your wife?' 'My lady,' he replied, 'may I enquire why you ask me this?' 'That will do,' retorted the sultana. 'I don't need to hear anything more. You are wiser than she is.'

The rejoicings continued in the palace all day, and the sultana, who never left the princess, did all she could to cheer her up and make her take part in the entertainments and amusements prepared for her. But the princess was so struck down by the visions of what had happened to her the previous night that it was easy to see she was totally preoccupied by them. The son of the vizier was just as shattered by the bad night he had spent but, fired by ambition, he concealed it and, seeing him, no one would have thought he was anything else but the happiest of bridegrooms.

Aladdin, knowing all about what had happened in the palace and never doubting that the newly-weds would sleep together, despite the misadventure of the previous night, had no desire to leave them in peace. So, after nightfall, he had recourse once again to the lamp. Immediately, the *jinni* appeared and greeted him in the same way as on the other occasions, offering him his services. 'The son of the grand vizier and Princess Badr are going to sleep together again tonight,' explained Aladdin. 'Go, and as soon as they are in bed, bring them here, as you did yesterday.'

The *jinni* served Aladdin as faithfully and as punctually as on the previous day; the son of the grand vizier spent as disagreeable a night as the one he had already endured and the princess was as mortified as before to have Aladdin as her bedfellow, with the sword placed between them. The next day, the *jinni*, following Aladdin's orders, returned and restored the husband to his wife's side; he then lifted up the bed with the newly-weds and transported it back to the room in the palace from where he had taken it.

Early the next morning, the sultan, anxious to discover how the princess had spent the second night, and wondering if she would receive him in the same way as on the previous day, went to her room to find out. But no sooner did the son of the grand vizier, more ashamed and mortified by his bad luck on the second night, hear the sultan come in than he hastily arose and hurled himself into the closet.

The sultan approached the princess's bed and greeted her, and after embracing her in the same way as he had the day before, asked her: 'Well, my dear, are you in as bad a mood this morning as you were yesterday? Tell me what sort of night you had.' But the princess again remained silent, and the sultan saw that her mind was even more disturbed and she was more dejected than the first time. He had no doubt now that something extraordinary had happened to her. So, irritated by the mystery she was making of it and clutching his sword, he angrily said to her: 'My daughter, either you tell me what you are hiding from me or I will cut off your head this very instant.'

At last, the princess, more frightened by the tone of her aggrieved father and his threat than by the sight of the unsheathed sword, broke her silence, and, with tears in her eyes, burst out: 'My dear father and sultan, I beg pardon of your majesty if I have offended you and I hope that in your goodness and mercy anger will give way to compassion when I give you a faithful account of the sad and pitiful state in which I spent all last night and the night before.' After this preamble, which somewhat calmed and softened the sultan, she faithfully recounted to him all that had happened to her during those two unfortunate nights. Her account was so moving that, in the love and tenderness he felt for her, he was filled with deep sorrow. When she had finished her account, she said to him: 'If your majesty has the slightest doubt about the account I have just given, you can ask the husband you have given me. I am convinced your majesty will be persuaded of the truth when he bears the same witness to it as I have done.'

The sultan now truly felt the extreme distress that such an astonishing adventure must have caused the princess and said to her: 'My daughter, you were very wrong not to have told me yesterday about such a strange affair, which concerns me as much as yourself. I did not marry you with the intention of making you miserable but rather with a view to making you happy and content, and to let you enjoy the happiness you deserve and can expect with a husband who seemed suited to you. Forget now all the worrying images you have just told me about. I will see to it that you endure no more nights as disagreeable and as unbearable as those you have just spent.'

As soon as the sultan had returned to his own apartments, he called for his grand vizier and asked him: 'Vizier, have you seen your son and has he not said anything to you?' When the vizier replied that he had not seen him, the sultan related to him everything Princess Badr had just told him, adding: 'I do not doubt my daughter was telling the truth, but I would be very glad to have it confirmed by what your son says. Go and ask him about it.'

The grand vizier made haste to join his son and to tell him what the sultan had said. He charged him to not conceal the truth but to tell him whether all this was true, to which his son replied: 'Father, I will conceal nothing from you. All that the princess told the sultan is true, but she couldn't tell him about the ill treatment I myself received, which is this: since my wedding I have spent the two most cruel nights imaginable and I do not have the words to describe to you exactly and in every detail the ills I have suffered. I won't tell you what I felt when I found myself lifted up four times in my bed and transported from one place to another, unable to see who was lifting the bed or to imagine how that could have been done. You can judge for yourself the wretched state I found myself in when I tell you that I spent two nights standing, naked but for my nightshirt, in a kind of narrow privy, not free to move from where I stood nor able to make any movement, although I could see no obstacle to prevent me from moving. I don't need to go into further detail about all my sufferings. I will not conceal from you that all this has not stopped me from feeling towards the princess, my wife, all the love, respect and gratitude that she deserves; but I confess in all sincerity that despite all the honour and glory that comes to me from having married the daughter of the sultan, I would rather die than live any longer in such an elevated alliance if I have to endure any further such disagreeable treatment as I have done. I am sure the princess feels the same as I do and will readily agree that our separation is as necessary for her peace of mind as it is for mine. And so, father, I beseech you, by the same love which led you to procure for me such a great honour, to make the sultan agree to our marriage being declared null and void.'

However great the grand vizier's ambition was for his son to become the son-in-law of the sultan, seeing how firmly resolved he was to separate from the princess, he did not think it right to suggest he be patient and wait a few more days to see if this problem might not be solved. He left his son and went to give his reply to the sultan, to whom he admitted frankly that it was only too true after what he had just learned from his son. Without waiting even for the sultan to speak to him about ending the marriage, which he could see he was all too much in favour of doing, he begged him to allow his son to leave the palace and to return home to him, using as a pretext that it was not right for the princess to be exposed a moment longer to such terrible persecution for the sake of his son.

The grand vizier had no difficulty in obtaining what he asked for. Immediately, the sultan, who had already made up his mind, gave orders to stop the festivities in his palace, the city and throughout the length and breadth of his kingdom, countering those originally given. In a very short while, all signs of joy and public rejoicing in the city and in the kingdom had ceased.

This sudden and unexpected change gave rise to many different interpretations: people asked each other what had caused this upset, but all that they could say was that the grand vizier had been seen leaving the palace and going home, accompanied by his son, both of them looking very dejected. Only Aladdin knew the secret and inwardly rejoiced at the good fortune which the lamp had procured him. Once he had learned for certain that his rival had abandoned the palace and that the marriage between him and the princess was over, he needed no longer to rub the lamp nor to summon the *jinni* to stop it being consummated. What is strange is that neither the sultan nor the grand vizier, who had forgotten Aladdin and his request, had the slightest idea that he had any part in the enchantment which had just caused the break-up of the princess's marriage.

Meanwhile, Aladdin let the three months go by that the sultan had stipulated before the marriage between him and Princess Badr could take place. He counted the days very carefully, and when they were up, the very next morning he hastened to send his mother to the palace to remind the sultan of his word.

Aladdin's mother went to the palace as her son had asked her and stood at the entrance to the council chamber, in the same spot as before. As soon as the sultan caught sight of her, he recognized her and immediately remembered the request she had made him and the date to which he had put off fulfilling it. The vizier was at that moment reporting to him on some matter, but the sultan interrupted him, saying: 'Vizier, I see the good woman who gave us such a fine gift a few months ago; bring her up – you can resume your report when I have heard what she has to say.' The grand vizier turned towards the entrance of the council chamber, saw Aladdin's mother and immediately summoned the chief usher, to whom he pointed her out, ordering him to bring her forward.

Aladdin's mother advanced right to the foot of the throne, where she prostrated herself as was customary. When she rose up again, the sultan asked her what her request was, to which she replied: 'Sire, I come before your majesty once more to inform you, in the name of my son Aladdin, that the three months' postponement of the request I had the honour to put to your majesty has come to an end and I entreat you to be so good as to remember your word.'

When he had first seen her, so meanly dressed, standing before him in all her poverty and lowliness, the sultan had thought that by making a delay of three months to reply to her request he would hear no more talk of a marriage which he regarded as not at all suitable for his daughter, the princess. He was, however, embarrassed at being called upon to keep his word to her but he did not think it advisable to give her an immediate reply, so he consulted his grand vizier, expressing to him his repugnance at the idea of marrying the princess to a stranger whose fortune he presumed was less than the most modest.

The grand vizier lost no time in telling the sultan what he thought about this. 'Sire,' he said, 'it seems to me there is a sure way of avoiding such an unequal marriage which would not give Aladdin, even were he better known to your majesty, grounds for complaint: this is to put such a high price on the princess that, however great his riches, he could not meet this. This would be a way of making him abandon such a bold, not to say foolhardy, pursuit, about which no doubt he did not think carefully before embarking on it.'

The sultan approved of the advice of the grand vizier and, turning towards Aladdin's mother, he said to her, after a moment's reflection: 'My good woman, sultans should keep their word; I am ready to keep mine and to make your son happy by marrying my daughter, the princess, to him. However, as I can't marry her before I know what advantage there is in it for her, tell your son that I will carry out my word as soon as he sends me forty large bowls of solid gold, full to the brim with the same things you have already presented to me on his behalf, and carried by a similar number of black slaves who, in their turn, are to be led by forty more white slaves – young, well built, handsome and all magnificently clothed. These are the conditions on which I am prepared to give him my daughter. Go, good woman, and I will wait for you to bring me his reply.'

Aladdin's mother prostrated herself in front of the sultan's throne and withdrew. As she went on her way, she laughed at the thought of her son's foolish ambition. 'Really,' she said to herself, 'where is he going to find so many golden bowls and such a large quantity of those coloured bits of glass to fill them? Will he go back to that underground cave with the entry blocked and pick them off the trees there? And all those slaves turned out as the sultan demanded, where is he going to get them from? He hasn't the remotest chance and I don't think he's going to be happy with the outcome of my mission.' When she got home, her mind was filled with all these thoughts, which made her believe Aladdin had nothing more to hope for, so she said to him: 'My son, I advise you to give up any thought of marrying the princess. The sultan did, indeed, receive me very kindly and I believe he was full of goodwill towards you; but the grand vizier, I am almost sure, made him change his mind, and I think you will think the same after you have heard what I have to say. After I reminded his majesty that the three months had expired and had begged him, on your behalf, to remember his promise, I noticed that he only gave the reply I am about to relate after a whispered conversation with his grand vizier.' Aladdin's mother then proceeded to give her son a faithful account of all that the sultan had said to her and the conditions on which he said he would consent to the marriage between him and the princess, his daughter. 'My son,' she said in conclusion, 'he is waiting for your reply, but, between ourselves,' she added with a smile, 'I believe he will have to wait for a long time.'

'Not so long as you would like to think, mother,' said Aladdin, 'and the sultan is mistaken if he thinks that by such exorbitant demands he is going to prevent me from desiring his daughter. I was expecting other insurmountable difficulties or that he would set a far higher price on my incomparable princess. But for the moment, I am quite content and what he is demanding is a mere trifle in comparison with what I would be in a position to offer him to obtain possession of her. You go and buy some food for dinner while I go and think about satisfying his demands – just leave it to me.'

As soon as Aladdin's mother had gone out to do the shopping, Aladdin took the lamp and rubbed it; immediately the *jinni* rose up before him and, in the same terms as before, asked Aladdin what was his command, saying that he was ready to serve him. Aladdin said to him: 'The sultan is giving me the hand of the princess his daughter in marriage, but first he demands of me forty large, heavy bowls of solid gold, filled to the brim with the fruits from the garden from where I took the lamp whose slave you are. He is also demanding from me that these forty bowls be carried by a similar number of black slaves, preceded by forty white slaves – young, well built, handsome and magnificently clothed. Go and bring me this present as fast as possible so that I can send it to the sultan before he gets up from his session at the council.' The *jinni* told him his command would be carried out without delay, and disappeared.

Shortly afterwards, the *jinni* reappeared, accompanied by the forty black slaves, each one bearing on his head a heavy bowl of solid gold, filled with pearls, diamonds, rubies and emeralds, all chosen for their beauty and their size so as to be better than those which had already been given to the sultan. Each bowl was covered with a silver cloth embroidered with flowers of gold. All these slaves, both black and white, together with all the golden dishes, occupied almost the whole of the very modest house, together with its small courtyard in front and the little garden at the back. The *jinni* asked Aladdin if he was satisfied and whether he had any other command to put to him, and when Aladdin said he had nothing more to ask him, he immediately disappeared.

When Aladdin's mother returned from the market and entered the house, she was very astonished to see so many people and so many riches. She put down the provisions she had bought and was about to remove the veil covering her face when she was prevented by Aladdin, who said to her: 'Mother, we have no time to lose; before the sultan finishes his session, it is very important you return to the palace and immediately bring him this present, Princess Badr's dowry, which he asked me for, so that he can judge, by my diligence and punctuality, the sincerity of my ardent desire to procure the honour of entering into an alliance with him.'

Without waiting for his mother to reply, Aladdin opened the door to the street and made all the slaves file out in succession, a white slave always followed by a black slave, bearing a golden bowl on his head, and so on, to the last one. After his mother had come out, following the last black slave, he closed the door and sat calmly in his room, in the hope that the sultan, after receiving the present he had demanded, would at last consent to receive him as his son-in-law.

The first white slave who came out of Aladdin's house made all the passers-by who saw him stop, and by the time eighty black and white slaves had finished emerging, the street was crowded with people rushing up from all parts of the city to see this magnificent and extraordinary sight. Each slave was dressed in such rich fabrics and wore such splendid jewels that those who knew anything about such matters would have reckoned each costume must have cost more than a million dinars: the neatness and perfect fit of each dress; the proud and graceful bearing of each slave; their uniform and symmetrical build; the solemn way they processed – all this, together with the glittering jewels of exorbitant size, encrusted and beautifully arranged in their belts of solid gold, and the insignias of jewels set in their headdresses, which were of a quite special type, roused the admiration of this crowd of spectators to such a state that they could not leave off staring at them and following them with their eyes as far as they could. The streets were so crowded with people that no one could move but each had to stay where he happened to be.

As the procession had to pass through several streets to get to the palace, a good number of the city's inhabitants, of all kinds and classes, were able to witness this marvellous display of pomp. When the first of the eighty slaves arrived at the gate of the first courtyard of the palace, the doorkeepers, who had drawn up in a line as soon as they spotted this wonderful procession approaching, took him for a king, thanks to the richness and splendour of his dress and they went up to him to kiss the hem of his garment. But the slave, as instructed by the *jinni*, stopped them and solemnly told them: 'We are but slaves; our master will appear in due course.'

Then this first slave, followed by the rest, advanced to the second courtyard, which was very spacious and was where the sultan's household stood during the sessions of the council. The palace officials who headed each rank looked very magnificent, but they were eclipsed in splendour by the appearance of the eighty slaves who bore Aladdin's present. There was nothing more beautiful, more brilliant in the whole of the sultan's court; however splendid his courtiers who surrounded him, none of them could compare with what now presented itself to his sight.

The sultan, who had been informed of the procession and arrival of the slaves, had given orders to let them in, and so, as soon as they appeared, they found the entrance to the council chamber open. They entered in orderly fashion, one half filing to the right, one half to the left. After they had all entered and had formed a large semicircle around the sultan's throne, each of the black slaves placed the bowl he was carrying on to the carpet in front of the sultan. All then prostrated themselves, touching the carpet with their foreheads. At the same time, the white slaves did the same. Then they all got up and the black slaves, as they rose, skilfully uncovered the bowls in front of them and stood with their hands crossed on their chests in great reverence.

Aladdin's mother, who had, meanwhile, advanced to the foot of the throne, prostrated herself before the sultan and addressed him, saying: 'Sire, my son, Aladdin, knows well that this gift he sends to your majesty is far less than Princess Badr deserves, but he hopes nonetheless that your majesty will be pleased to accept it and consider it acceptable for the princess; he offers it all the more confidently because he has endeavoured to conform to the condition which your majesty was pleased to impose on him.'

The sultan was in no state to pay attention to her compliments: one look at the forty golden bowls, filled to the brim with the most brilliant, dazzling and most precious jewels ever to be seen in the world, and at the eighty slaves who, as much by their handsome appearance as by the richness and amazing magnificence of their dress, looked like so many kings, and he was so overwhelmed that he could not get over his astonishment. Instead of replying to Aladdin's mother, he addressed the grand vizier, who likewise could not understand where such a great profusion of riches could have come from. 'Well now, vizier,' he publicly addressed him, 'what do you think about a person, whoever he may be, who sends me such a valuable and extraordinary present, someone whom neither of us knows? Don't you think he is fit to marry my daughter, Princess Badr?'

For all his jealousy and pain at seeing a stranger preferred before his son to become the son-in-law of the sultan, the vizier nonetheless managed to conceal his feelings. It was quite obvious that Aladdin's present was more than enough for him to be admitted to such a high alliance. So the vizier agreed with the sultan, saying: 'Sire, far from believing that someone who gives you a present so worthy of your majesty should be unworthy of the honour you wish to do him, I would be so bold as to say that he deserves it all the more, were I not persuaded that there is no treasure in the world precious enough to be put in balance with your majesty's daughter, the princess.' At this, all the courtiers present at the session applauded, showing that they were of the same opinion as the grand vizier.

The sultan did not delay; he did not even think to enquire whether Aladdin had the other qualities appropriate for one who aspired to become his son-in-law. The mere sight of such immense riches and the diligence with which Aladdin had fulfilled his demand without making the slightest difficulty over conditions as exorbitant as those he had imposed on him, easily persuaded the sultan that Aladdin lacked nothing to render him as accomplished as the sultan wished. So, to send Aladdin's mother back with all the satisfaction she could desire, he said to her: 'Go, my good woman, and tell your son that I am waiting to receive him with open arms and to embrace him, and that the quicker he comes to receive from me the gift I have bestowed on him of the princess, my daughter, the greater the pleasure he will give me.'

Aladdin's mother left with all the delight a woman of her status is capable of on seeing her son, contrary to all expectations, attain such a high position. The sultan then immediately concluded the day's audience and, rising from his throne, ordered the eunuchs attached to the princess's service to come and remove the bowls and carry them off to their mistress's apartment, where he himself went to examine them with her at his leisure. This order was carried out at once, under supervision of the head eunuch.

The eighty black and white slaves were not forgotten; they were taken inside the palace and, a little later, the sultan, who had been telling the princess about their magnificence, ordered them to be brought to the entrance of her apartment so that she could look at them through the screens and realize that, far from exaggerating anything in his account, he had not told her even half the story.

Meanwhile, Aladdin's mother arrived home with an expression which told in advance of the good news she was bringing. 'My son,' she said to him, 'you have every reason to be happy: contrary to my expectations – and you will recall what I told you – you have attained the accomplishment of your desires. In order not to keep you in suspense any longer, the sultan, with the approval of his entire court, has declared that you are worthy to possess Princess Badr. He is waiting to embrace you and to bring about your marriage. You must now think about how to prepare for this meeting so that you may come up to the high opinion the sultan has formed of you. After all the miracles I have seen you perform, I am sure nothing will be lacking. I must not forget to tell you also that the sultan is waiting impatiently for you, and so waste no time in going to him.'

Aladdin was delighted at this news and, his mind full of the enchanting creature who had so bewitched him, after saying a few words to his mother, withdrew to his room. Once there, he took the lamp which had hitherto been so useful to him in fulfilling all his needs and wishes, and no sooner had he rubbed it than the *jinni* appeared before him and immediately proceeded to offer him his services as before. 'O *jinni*,' said Aladdin, 'I have summoned you to help me take a bath and when I have finished, I want you to have ready for me the most sumptuous and magnificent costume ever worn by a king.' No sooner had he finished speaking than the *jinni*, making them both invisible, lifted him up and transported him to a bath made of the finest marble of every shade of the most beautiful colours. Without seeing who was waiting on him, he was undressed in a spacious and very well-arranged room. From this room he was made to go into the bath, which was moderately hot, and there he was rubbed and washed with several kinds of perfumed waters. After he had been taken into various rooms of different degrees of heat, he came out again transformed, his complexion fresh, all pink and white, and feeling lighter and more refreshed. He returned to the first room, but the clothing he had left there had gone; in its place the *jinni* had carefully set out the costume he had asked for. When he saw the magnificence of the garments which had been substituted for his own, Aladdin was astonished. With the help of the *jinni*, he got dressed, admiring as he did so each item of clothing as he put it on, for everything was beyond anything he could have imagined.

When he had finished, the *jinni* took him back to his house, to the same room from where he had transported him. He then asked Aladdin whether he had any other demands. 'Yes,' replied Aladdin, 'I want you to bring me as quickly as possible a horse which is finer and more beautiful than the most highly valued horse in the sultan's stables; its trappings, its harness, its saddle, its bridle – all must be worth more than a million dinars. I also ask you to bring me at the same time twenty slaves as richly and smartly attired as those who delivered the sultan's present, who are to walk beside me and behind me in a group, and twenty more like them to precede me in two files. Bring my mother, too, with six slave girls to wait on her, each dressed at least as richly as the princess's slave girls, and each bearing a complete set of women's clothes as magnificent and sumptuous as those of a sultana. Finally, I need ten thousand pieces of gold in ten purses. There,' he ended, 'that's what I command you to do. Go, and make haste.'

As soon as Aladdin had finished giving him his orders, the *jinni* disappeared; shortly afterwards, he reappeared with the horse, the forty slaves – ten of whom were each carrying a purse containing a thousand pieces of gold, and the six slave girls – each one bearing on her head a different costume for Aladdin's mother, wrapped up in a silver cloth, and all this he presented to Aladdin. Of the ten purses Aladdin took four, which he gave to his mother, telling her she should use them for her needs. The remaining six he left in the hands of the slaves who were carrying them, charging them to keep them and throw out handfuls of gold from them to the people as they passed through the streets on their way to the sultan's palace. He also ordered these six slaves to walk in front of him with the others, three on the right and three on the left. Finally, he presented the six slave girls to his mother, telling her that they were hers to use as her slaves and that the clothes they brought were for her.

When Aladdin had settled all these matters, he told the *jinni* as he dismissed him that he would call him when he needed his services and the *jinni* instantly disappeared. Aladdin's one thought now was to reply as quickly as possible to the desire the sultan had expressed to see him. So he despatched to the palace one of the forty slaves – I will not say the most handsome, for they were all equally handsome – with the order to address himself to the chief usher and ask him when Aladdin might have the honour of prostrating himself at the feet of the sultan. The slave was not long in carrying out his task, returning with the reply that the sultan was awaiting him with impatience.

Aladdin made haste to set off on horseback and process in the order already described. Although this was the first time he had ever mounted a horse, he appeared to ride with such ease that not even the most experienced horseman would have taken him for a novice. In less than a moment, the streets he passed through filled with an innumerable crowd of people, whose cheers and blessings and cries of admiration rang out, particularly when the six slaves with the purses threw handfuls of gold coins into the air to the left and right. These cheers of approval did not, however, come from the rabble, who were busy picking up the gold, but from a higher rank of people who could not refrain from publicly praising Aladdin for his generosity. Anyone who could remember seeing him playing in the street, the perpetual vagabond, no longer recognized Aladdin, and even those who had seen him not long ago had difficulty making him out, so different were his features. This is because one of the properties of the lamp was that it could gradually procure for those who possessed it the perfections which went with the status they attained by making good use of it. Consequently, people paid more attention to Aladdin himself than to the pomp which accompanied him and which most of them had already seen that same day when the eighty slaves marched in procession, bearing the present. The horse was also much admired for its beauty alone by the experts, who did not let themselves be dazzled by the wealth or brilliance of the diamonds and other jewels with which it was covered. As the news spread that the sultan was giving the hand of his daughter, Princess Badr, in marriage to Aladdin, without regard to his humble birth, no one envied him his good fortune nor his rise in status, as they seemed well deserved.

Aladdin arrived at the palace, where all was set to receive him. When he reached the second gate, he was about to dismount, following the custom observed by the grand vizier, the generals of the armies and the governors of the provinces of the first rank; but the chief usher, who was waiting for him by order of the sultan, prevented him and accompanied him to the council chamber, where he helped him to dismount, despite Aladdin's strong opposition, but whose protests were in vain for he had no say in the matter. The ushers then formed two lines at the entrance to the chamber and their chief, placing Aladdin on his right, led him through the middle right up to the sultan's throne.

As soon as the sultan set eyes on Aladdin, he was no less astonished to see him clothed more richly and magnificently than he himself had ever been, than surprised at his fine appearance, his handsome figure and a certain air of grandeur, which were in complete contrast to the lowly state in which his mother had appeared before him. His astonishment and surprise did not, however, prevent him from rising from his throne and descending two or three steps in time to stop Aladdin from prostrating himself at his feet and to embrace him in a warm show of friendship. After such a greeting Aladdin still wanted to throw himself at the sultan's feet, but the sultan held him back with his hand and forced him to mount the steps and sit between the vizier and himself.

Aladdin now addressed the sultan. 'Sire,' he said, 'I accept the honours your majesty is so gracious as to bestow on me; but permit me to tell you that I have not forgotten I was born your slave, that I know the greatness of your power and I am well aware how much my birth and upbringing are below the splendour and the brilliance of the exalted rank to which I am being raised. If there is any way I can have deserved so favourable a reception, it is maybe due to the boldness that pure chance inspired in me to raise my eyes, my thoughts and my aspirations to the divine princess who is the object of my desires. I beg pardon of your majesty for my rashness but I cannot hide from you that I would die of grief if I were to lose hope of seeing these desires accomplished.'

'My son,' replied the sultan, embracing him a second time, 'you do me wrong to doubt for a single instant the sincerity of my word. From now on, your life is too dear to me for me not to preserve it, by presenting you with the remedy which is at my disposal. I prefer the pleasure of seeing you and hearing you to all my treasures and yours together.'

When he had finished speaking, the sultan gave a signal and immediately the air echoed with the sound of trumpets, oboes and drums. At the same time, the sultan led Aladdin into a magnificent room where a splendid feast was prepared. The sultan ate alone with Aladdin, while the grand vizier and the court dignitaries stood by during the meal, each according to their dignity and rank. The sultan, who took such great pleasure in looking at Aladdin that he never took his eyes off him, led the conversation on several different topics and throughout the meal, in the conversation they held together and on whatever matter the sultan brought up, Aladdin spoke with such knowledge and wisdom that he ended by confirming the sultan in the good opinion he had formed of him from the beginning.

Once the meal was over, the sultan summoned the grand *qadi* and ordered him immediately to draw up a contract of marriage between Princess Badr, his daughter, and Aladdin. While this was happening, the sultan talked to Aladdin about several different things in the presence of the grand vizier and his courtiers, who all admired Aladdin's soundness and the great ease with which he spoke and expressed himself and the refined and subtle comments with which he enlivened his conversation.

When the *qadi* had completed drawing up the contract in all the required forms, the sultan asked Aladdin if he wished to stay in the palace to complete the marriage ceremonies that same day, but Aladdin replied: 'Sire, however impatient I am fully to enjoy your majesty's kindnesses, I beg you will be so good as to allow me to put them off until I have had a palace built to receive the princess in, according to her dignity and merit. For this purpose, I ask you to grant me a suitable spot in the palace grounds so that I may be closer at hand to pay you my respects. I will do everything to see that it is accomplished with all possible speed.' 'My son,' said the sultan, 'take all the land you think you need; there is a large space in front of my palace and I myself had already thought of filling it. But remember, I can't see you united to my daughter soon enough to complete my happiness.' After he had said this, the sultan embraced Aladdin, who took his leave of the sultan with the same courtesy as if he had been brought up and always lived at court.

Aladdin remounted his horse and returned home the same way he had come, passing through the same applauding crowds who wished him happiness and prosperity. As soon as he got back and had dismounted, he went off to his own room, took the lamp and summoned the *jinni* in the usual way. The *jinni* immediately appeared and offered him his services. 'O *jinni*,' said Aladdin, 'I have every reason to congratulate myself on how precisely and promptly you have carried out everything I have asked of you so far, through the power of this lamp, your mistress. But now, for the sake of the lamp, you must, if possible, show even more zeal and more diligence than before. I am now asking you to build me, as quickly as you can, at an appropriate distance opposite the sultan's residence, a palace worthy of receiving Princess Badr, my wife-to-be. I leave you free to choose the materials – porphyry, jasper, agate, lapis lazuli and the finest marble of every colour – and the rest of the building. But at the very top of this palace, I want you to build a great room, surmounted by a dome and with four equal sides, made up of alternating layers of solid gold and silver. There should be twenty-four windows, six on each side, with the latticed screens of all but one – which I want left unfinished – embellished, skilfully and symmetrically, with diamonds, rubies and emeralds, so that nothing like this will have ever been seen in the world. I also want the palace to have a forecourt, a main court and a garden. But above all, there must be, in a spot you will decide, a treasure house, full of gold and silver coins. And I also want this palace to have kitchens, pantries, storehouses, furniture stores for precious furniture for all seasons and in keeping with the magnificence of the palace, and stables filled with the most beautiful horses complete with their riders and grooms, not to forget hunting equipment. There must also be kitchen staff and officials and female slaves for the service of the princess. You understand what I mean? Go and come back when it's done.'

It was sunset when Aladdin finished instructing the *jinni* in the construction of his imagined palace. The next day, at daybreak, Aladdin, who could not sleep peacefully because of his love for the princess, had barely risen when the *jinni* appeared before him. 'Master,' he said, 'your palace is finished. Come and see if you like it.' No sooner had Aladdin said he wanted to see it than in an instant the *jinni* had transported him there. Aladdin found it so beyond all his expectations that he could not admire it enough. The *jinni* led him through every part; everywhere Aladdin found nothing but riches, splendour and perfection, with the officials and slaves all dressed according to their rank and the services they had to perform. Nor did he forget to show him, as one of the main features, the treasure house, the door to which was opened by the treasurer. There Aladdin saw purses of different sizes, depending on the sums they contained, piled up in a pleasing arrangement which reached up to the vault. As they left, the *jinni* assured him of the treasurer's trustworthiness. He then led him to the stables where he showed him the most beautiful horses in the world and the grooms who were grooming them. Finally, he took him through storerooms filled with all the supplies necessary for both the horses' adornment and their food.

When Aladdin had examined the whole palace from top to bottom, floor by floor, room by room, and in particular the chamber with the twenty-four windows, and had found it so rich and magnificent and well furnished, beyond anything he had promised himself, he said to the *jinni*: 'O *jinni*, nobody could be happier than I am and it would be wrong for me to complain. But there's one thing which I didn't tell you because I hadn't thought about it, which is to spread, from the gate of the sultan's palace to the door of the apartment intended for the princess, a carpet of the finest velvet for her to walk on when she comes from the sultan's palace.' 'I will be back in a moment,' said the *jinni*. A little after his disappearance, Aladdin was astonished to see that what he wanted had been carried out without knowing how it had been done. The *jinni* reappeared and carried Aladdin back home, just as the gate of the sultan's palace was being opened.

The palace doorkeepers, who had just opened the gate and who had always had an unimpeded view in the direction where Aladdin's palace now stood, were astounded to find it obstructed and to see a velvet carpet stretching from that direction right up to the gate of the sultan's palace. At first they could not make out what it was, but their astonishment increased when they saw clearly Aladdin's superb palace. News of the marvel quickly spread throughout the whole palace. The grand vizier, who had arrived almost the moment the gate was opened, was as astonished as the rest at the extraordinary sight and was the first to tell the sultan. He wanted to put it down to magic but the sultan rebuffed him, saying: 'Why do you want it to be magic? You know as well as I do that it's the palace Aladdin has had built for my daughter, the princess; I gave him permission to do so in your presence. After the sample of his wealth which we saw, is it so strange that he has built this palace in such a short time? He wanted to surprise us and to show us what miracles one can perform from one day to the next. Be honest, don't you agree that when you talk of magic you are perhaps being a little jealous?' He was prevented from saying anything more, as the hour to enter the council chamber had arrived.

After Aladdin had been carried home and had dismissed the *jinni*, he found his mother had got up and was beginning to put on the clothes that had been brought to her. At about the time that the sultan had just left the council, Aladdin made his mother go to the palace, together with the slave girls who had been brought her by the *jinni*'s services. He asked her that if she saw the sultan she was to tell him that she had come in order to have the honour of accompanying the princess when she was ready to go to her palace towards evening. She left and although she and the slave girls who followed her were dressed like sultanas, the crowds watching them pass were not so large, as the women were veiled and wore appropriate overgarments to cover the richness and magnificence of their clothing. As for Aladdin, he mounted his horse and, leaving his home for the last time, without forgetting the magic lamp whose help had been so helpful to him in attaining the height of happiness, he publicly left for his palace, with the same pomp as on the previous day when he had gone to present himself to the sultan.

As soon as the doorkeepers of the sultan's palace saw Aladdin's mother, they told the sultan. Immediately the order was given to the bands of trumpets, cymbals, drums, fifes and oboes who had been stationed in different spots on the palace terraces, and all at once the air resounded to fanfares and music which announced the rejoicings to the whole city. The merchants began to deck out their shops with fine carpets, cushions and green boughs, and to prepare illuminations for the night. The artisans left their work, and the people hastened to the great square between the sultan's palace and that of Aladdin. But it was Aladdin's palace that first attracted their admiration, not so much because they were accustomed to see that of the sultan but because it could not enter into comparison with Aladdin's. The sight of such a magnificent palace in a place that, the previous day, had neither materials nor foundations, astonished them most and they could not understand by what unheard-of miracle this had come about.

Aladdin's mother was received in the sultan's palace with honour and admitted to the princess's apartments by the chief eunuch. When the princess saw her, she immediately went to embrace her and made her be seated on her sofa; and while her maidservants were finishing dressing her and adorning her with the most precious jewels, which Aladdin had given her, the princess entertained her to a delicious supper. The sultan, who came to spend as much time as he could with his daughter before she left him to go to Aladdin's palace, also paid great honour to Aladdin's mother. He had never seen her before without a veil, although she had spoken several times to him in public, but now without her veil, though she was no longer young, one could still see from her features that she must have been reckoned among the beautiful women of her day when she was young. The sultan, who had always seen her dressed very simply, not to say shabbily, was filled with admiration at seeing her clothed as richly and as magnificently as the princess, his daughter, which made him reflect that Aladdin was equally capable, prudent and wise in everything he did.

When night fell, the princess took leave of her father, the sultan. Their parting was tender and tearful; they embraced several times in silence, and finally the princess left her apartments and set out, with Aladdin's mother on her left, followed by a hundred slave girls, all wonderfully and magnificently dressed. All the bands of musicians, which had never stopped playing since the arrival of Aladdin's mother, now joined up and began the procession; they were followed by a hundred sergeants and a similar number of black eunuchs, in two columns, led by their officers. Four hundred of the sultan's young pages walked on each side of the procession, holding torches in their hands which, together with the light from illuminations coming from both the sultan's palace and Aladdin's, wonderfully took the place of daylight.

Accompanied in this fashion, the princess stepped on to the carpet which stretched from the sultan's palace to Aladdin's; as she advanced, the bands of musicians who led the procession approached and joined with those which could be heard on the terraces of Aladdin's palace. This extraordinary confusion of sounds nonetheless formed a concert which increased the rejoicing not only of the great crowd in the main square but also of those who were in the two palaces and indeed in the whole city and far beyond.

At last the princess arrived at the new palace and Aladdin rushed with all imaginable joy to receive her at the entrance to the apartments destined for him. Aladdin's mother had taken care to point out her son to the princess amid the officials who surrounded him, and the princess, when she saw him, found him so handsome that she was quite charmed. 'Lovely princess,' Aladdin addressed her, going up to her and greeting her very respectfully, 'if I have been so unlucky as to have displeased you by my rashness in aspiring to possess so fair a lady, the daughter of the sultan, then, if I may say so, you must blame your beautiful eyes and your charms, not me.' 'O prince – I can rightly call you this now – I bow to my father, the sultan's wishes; but it is enough for me to have seen you to tell you that I am happy to obey him.'

Overjoyed by such a satisfying reply, Aladdin did not keep the princess standing any longer after this unaccustomed walk. In his delight, he took her hand and kissed it, and then led her into a large room lit by innumerable candles where, thanks to the *jinni*, a magnificent banquet was laid out on a table. The plates were of solid gold and filled with the most delicious food. The vases, bowls and goblets, with which the side tables were well provided, were also of gold and of exquisite craftsmanship. All the other ornaments and decorations of the room were in perfect keeping with all this sumptuousness. Delighted to see so many riches gathered together in one place, the princess said to Aladdin: 'O prince, I thought that there was nothing in the world more beautiful than my father's palace; but seeing this room alone I realize how wrong I was.' 'Princess,' said Aladdin, seating her at the place specially set for her at the table, 'I accept such a great compliment as I ought; but I know what I should think.'

The princess, Aladdin and his mother sat down at table, and immediately a band of the most melodious instruments, played and accompanied by women, all of whom were very beautiful, started up and the music was accompanied by their equally beautiful voices; this continued uninterrupted until the end of the meal. The princess was so charmed that she said she had never heard anything like it in the palace of the sultan, her father. She did not know, of course, that these musicians were creatures chosen by the *jinni* of the lamp.

When the supper was over and the dishes had been swiftly cleared away, the musicians gave way to a troupe of male and female dancers who danced several kinds of dances, according to the custom of the land. They ended with two solo dances by a male and female dancer who, each in their turn, danced with surprising lightness and showed all the grace and skill they were capable of. It was nearly midnight when, according to the custom in China at that time, Aladdin rose and offered his hand to Princess Badr to dance with her and so conclude the wedding ceremony. They danced so well together that the whole company was lost in admiration. When the dance was over, Aladdin, still holding her by the hand, took the princess and together they passed through to the nuptial chamber where the marriage bed had been prepared. The princess's maidservants helped undress her and put her to bed, and Aladdin's servants did the same, then all withdrew. And so ended the ceremonies and festivities of the wedding of Aladdin and Princess Badr al-Budur.

The following morning, when Aladdin awoke, his servants came to dress him. They put on him a different costume from the one he had worn the day of the wedding, but one that was equally rich and magnificent. Next, he had brought to him one of the horses specially selected for him, which he mounted and then went to the sultan's palace, surrounded by a large troupe of slaves walking in front of him and behind him and on either side. The sultan received him with the same honours as on the first occasion, embraced him and, after seating him near him on the throne, ordered breakfast to be served. 'Sire,' said Aladdin, 'I ask your majesty to excuse me this honour today. I came to ask you to do me the honour of partaking of a meal in the princess's palace, together with your grand vizier and your courtiers.' The sultan was pleased to grant him this favour. He rose forthwith and, as it was not very far, wished to go there on foot, and so he set out, with Aladdin on his right, the grand vizier on his left and the courtiers following, preceded by the sergeants and principal court officials.

The nearer he drew to Aladdin's palace, the more the sultan was struck by its beauty. But he was much more amazed when he entered it and never stopped praising each and every room he saw. But when, at Aladdin's invitation, they went up to the chamber with the twenty-four windows, and the sultan saw the decorations, and above all when he caught sight of the screens studded with diamonds, rubies and emeralds – jewels all so large and so perfectly proportioned – and when Aladdin remarked that it was just as opulent on the outside, he was so astonished that he remained rooted to the spot.

After remaining motionless for a while, the sultan turned to the grand vizier standing near him and said: 'Vizier, can there possibly be such a superb palace in my kingdom, so near to my own palace, without my having been aware of it till now?' 'Your majesty may recall,' replied the grand vizier, 'that the day before yesterday you granted Aladdin, whom you accepted as your son-in-law, permission to build a palace opposite your own; that same day, at sunset, there was as yet no palace on that spot. Yesterday I had the honour to be the first to announce to you that the palace had been built and was finished.' 'Yes, I remember,' said the sultan, 'but I would never have thought this palace would be one of the wonders of the age. Where in the whole wide world can one find a palace built of layers of solid gold and silver rather than of stone or marble, where the windows have screens set with diamonds, rubies and emeralds? Never has anything like this been heard of before!'

The sultan wished to see and admire the beauty of the twenty-four screens. When he counted them, he found that only twenty-three of the twenty-four were each equally richly decorated but that the twenty-fourth, he was very surprised to discover, had been left unfinished. Turning to the vizier, who had made it his duty always to stay at his side, he said: 'I am surprised that a room of such magnificence should have been left unfinished.' 'Sire,' the grand vizier replied, 'Aladdin apparently was in a hurry and didn't have time to make this window like the rest; but I imagine he has the necessary jewels and that he will have the work done at the first opportunity.'

While this was going on, Aladdin had left the sultan to give some orders and when he came back to join him again, the sultan said to him: 'My son, of all the rooms in the world this one is the most worthy to be admired. But one thing surprises me – that is, to see this one screen left unfinished. Was it forgotten through carelessness or because the workmen didn't have time to put the finishing touches to so fine a piece of architecture?' 'Sire,' replied Aladdin, 'it's for neither of these reasons that the screen has remained in the state in which your majesty sees it. It was done deliberately and it was at my order that the workmen left it untouched: I wanted your majesty to have the glory of finishing this room and the palace at the same time. I beg that you will accept my good intentions so that I will be able to remember your kindness and your favours.' 'If that's how you intended it, I am grateful to you and shall immediately give the orders for it to be done.' And indeed he then summoned the jewellers with the greatest stock of jewels, together with the most skilled goldsmiths in the capital.

The sultan then went down from this chamber and Aladdin led him into the room where he had entertained the princess on the day of the wedding. The princess arrived a moment later and received her father in a manner which showed him how happy she was with her marriage. Two tables were laid with the most delicious dishes, all served in golden vessels. The sultan sat down at the first table and ate with his daughter, Aladdin and the grand vizier, while all the courtiers were served at the second table, which was very long. The sultan found the dishes very tasty and declared he had never eaten anything more exquisite. He said the same of the wine, which was, indeed, delicious. What he admired still more were four large side tables filled and laden with an abundance of flagons, bowls and goblets of solid gold, all encrusted with jewels. He was also delighted with the bands of musicians scattered around the room, while the fanfares of trumpets accompanied by drums and tambourines resounding at a suitable distance outside the room gave a most pleasing effect.

When the meal was over, the sultan was told that the jewellers and goldsmiths who had been summoned by his order had arrived. He went up again to the room with the twenty-four screens and, once there, showed the jewellers and goldsmiths who had followed him the window which had been left unfinished. 'I have brought you here,' he said, 'so that you will bring this window to the same state of perfection as the others; examine them well and waste no time in making it exactly the same as the rest.'

The jewellers and goldsmiths examined the twenty-three other screens very closely, and after they had consulted together and were agreed on what each of them would for his part contribute, they returned to the sultan. The palace jeweller, acting as spokesman, then said to the sultan: 'Sire, we are ready to employ all our skills and industry to obey your majesty, but many as we are, not one of our profession has such precious jewels and in sufficient quantities for such a great project.' 'But I have enough and more than enough,' said the sultan. 'Come to my palace; I can supply you with them and you can choose those you want.'

After the sultan had returned to the palace, he had all his jewels brought in to him and the jewellers took a great quantity of them, particularly from among those which Aladdin had given him as a present. They used all these jewels but the work did not seem to progress very much. They returned several times to fetch still more, but after a whole month they had not finished half of the work. They used all the sultan's jewels as well as some borrowed from the grand vizier; yet, despite all these, all they managed to do was at most to complete half of the window.

Aladdin, who knew that all the sultan's efforts to make this screen like the rest were in vain and that he would never come out of it with any credit, summoned the goldsmiths and told them not only to cease their work but even to undo everything they had done and to return to the sultan all his jewels together with those he had borrowed from the grand vizier.

In a matter of hours, the work that the jewellers and goldsmiths had taken more than six weeks to do was destroyed. They then departed and left Aladdin alone in the room. Taking out the lamp, which he had with him, he rubbed it and immediately the *jinni* stood before him. 'O *jinni*,' said Aladdin, 'I ordered you to leave one of the twenty-four screens in this room unfinished and you carried out my order; I have now summoned you here to tell you that I want you to make this window just like the rest.' The *jinni* disappeared and Aladdin went out of the room. When he went back in again a few moments later, he found the screen exactly as he wanted it and just like the others.

Meanwhile, the jewellers and goldsmiths had arrived at the palace and had been introduced and presented to the sultan in his apartments. The first jeweller, on behalf of all of them, presented the sultan with the jewels which they were returning and said: 'Sire, your majesty knows how long and how hard we have been working in order to finish the commission that you charged us with. The work was far advanced when Aladdin forced us not only to stop but even to undo all we had done and to return to your majesty all these jewels of yours and those of the grand vizier.' The sultan asked them whether Aladdin had told them why they were to do this, to which they replied no. Immediately, the sultan gave the order for a horse to be brought and when it came, he mounted and left with only a few of his men, who accompanied him on foot. On arriving at Aladdin's palace, he dismounted at the bottom of the staircase that led to the room with the twenty-four windows. Without giving Aladdin any advance notice, he climbed the stairs where Aladdin had arrived in the nick of time to receive him at the door.

The sultan, giving Aladdin no time to complain politely that his majesty had not forewarned him of his arrival and that he had thus obliged him to fail in his duty, said to him: 'My son, I have come myself to ask the reason why you wish to leave unfinished so magnificent and remarkable a room in your palace as this one.'

The real reason Aladdin concealed from him, which was that the sultan was not rich enough in jewels to afford such an enormous expense. However, in order to let the sultan know how far his palace surpassed not only his own, such as it was, but any other palace in the world, since the sultan had been unable to complete the smallest part of it, he said to him: 'Sire, it is true your majesty has seen this room, unfinished; but I beg you will come now and see if there is anything lacking.'

The sultan went straight to the window where he had seen the unfinished screen and when he saw that it was like the rest, he thought he must have made a mistake. He next examined not only one or two windows but all the windows, one by one. When he was convinced that the screen on which so much time had been spent and which had cost so many days' work had been finished in what he knew to be a very short time, he embraced Aladdin and kissed him between the eyes, exclaiming in astonishment: 'My son, what a man you are to do such amazing things and all, almost, in the twinkling of an eye! There is no one like you in the whole wide world! The more I know you, the more I admire you.' Aladdin received the sultan's praises with great modesty, replying: 'Sire, I am very honoured to merit your majesty's kindness and approval, and I can assure you I will do all I can to deserve them both more and more.'

The sultan returned to his palace in the same manner as he had come, not allowing Aladdin to accompany him. There he found the grand vizier waiting for him and the sultan, still filled with admiration at the wonders he had just seen, proceeded to tell him all about them in terms that left the vizier in no doubt that everything was indeed as he had described it. But it also confirmed him in his belief that Aladdin's palace was the effect of an enchantment – a belief that he had already conveyed to the sultan almost as soon as the palace appeared. The vizier started to repeat what he thought, but the sultan interrupted him: 'Vizier, you have already told me that, but I can see that you are still thinking of my daughter's marriage to your son.'

The grand vizier could see that the sultan was prejudiced against him and so, not wishing to enter into a dispute with him, made no attempt to disabuse him. Meanwhile, the sultan every day, as soon as he had arisen, regularly went to a small room from which he could see the whole of Aladdin's palace and he would come here several times a day to contemplate and admire it.

As for Aladdin, he did not remain shut up in his palace but took care to let himself be seen in the city several times a week, whether to go and pray at one mosque or another or, at regular intervals, to visit the grand vizier, who affected to pay court to him on certain days, or to do honour to the leading courtiers, whom he often entertained in his palace, by going to see them in their own houses. Every time he went out, he would instruct two of his slaves, who surrounded him as he rode, to throw handfuls of gold coins into the streets and squares through which he passed and to which a great crowd of people always flocked. Furthermore, no pauper came to the gate of his palace without going away pleased with the liberality dispensed there on his orders.

Aladdin passed his time in such a way that not a week went by without him going out hunting, whether just outside the city or further afield, when he dispensed the same liberality as he rode around or passed through villages. This generous tendency caused everyone to shower blessings on him and it became the custom to swear by his head. In short, without it giving any offence to the sultan to whom he regularly paid court, one may say that Aladdin, thanks to his affable manner and generosity, won the affection of all the people and that, in general, he was more beloved than the sultan himself. Added to all these fine qualities, he showed such valour and zeal for the good of the kingdom that he could not be too highly praised. He showed this on the occasion of a revolt which took place on the kingdom's borders: no sooner had he heard that the sultan was raising an army to put the revolt down than he begged the sultan to put him in command of it, a request which he had no difficulty in obtaining. Once at the head of the army, he marched against the rebels, and throughout this expedition he conducted himself so industriously that the sultan learned that the rebels had been defeated, punished and dispersed before he learned of Aladdin's arrival in the army. This action, which made his name famous throughout the kingdom, did not change his good nature, for he remained as amiable after as before his victory.

Several years went by in this manner for Aladdin, when the magician, who unwittingly had given him the means of rising to such heights of fortune, was reminded of him in Africa to where he had returned. Although he was till then convinced that Aladdin had died a wretched death in the underground cave where he had left him, the thought nonetheless came to him to find out exactly how he had died. Being a great geomancer, he took out of a cupboard a covered square box which he used to make his observations. Sitting down on his sofa, he placed the box in front of him and uncovered it. After he had prepared and levelled the sand with the intention of discovering if Aladdin had died in the cave, he made his throw, interpreted the figures and drew up the horoscope. When he examined it to ascertain its meaning, instead of discovering that Aladdin had died in the cave, he found that he had got out of it and that he was living in great splendour, being immensely rich; he had married a princess and was generally honoured and respected.

As soon as the magician had learned through the means of his diabolic art that Aladdin lived in such a state of elevation, he became red with rage. In his fury, he exclaimed to himself: 'That wretched son of a tailor has discovered the secret and power of the lamp! I took his death for a certainty and here he is enjoying the fruit of my labours and vigils. But I will stop him enjoying them much longer, or die in the attempt.' He did not take long in deciding what to do. The next morning, he mounted a barbary horse which he had in his stable and set off, travelling from city to city and from province to province, stopping no longer than was necessary so as not to tire his horse, until he reached China and was soon in the capital of the sultan whose daughter Aladdin had married. There he dismounted in a *khan* or public hostelry, where he rented a room and where he remained for the rest of the day and the night to recover from his tiring journey.

The next day, the first thing the magician wanted to find out was what people said about Aladdin. Walking around the city, he entered the best-known and most frequented place, where the most distinguished people met to drink a certain hot drink* which was known to him from his first journey. As soon as he sat down, a cup of this drink was poured and presented to him. As he took it, he listened to the conversation going on around him and heard people talking about Aladdin's palace. When he had finished his drink, he approached one of them, singling him out to ask him: 'What's this palace everybody speaks so well of?' To which the man replied: 'Where are you from? You must be a newcomer not to have seen or heard talk of the palace of Prince Aladdin' – that is how he was now called since he had married Princess Badr – 'I am not saying it is one of the wonders of the world, I say it is the only wonder of the world, for nothing so grand, so rich, so magnificent has ever been seen before or since. You must have come from very far away not to have heard talk of it. Indeed, the whole world must have been talking about

---

* i.e. tea.

it ever since it was built. Go and look at it and see if I'm not speaking the truth.' 'Excuse my ignorance,' said the magician. 'I only arrived yesterday and I have indeed come from very far away – in fact the furthest part of Africa, which its fame had not yet reached when I left. For in view of the urgent business which brings me here, my sole concern in travelling was to get here as soon as possible, without stopping and making any acquaintances. I knew nothing about it until you told me. But I will indeed go and see it; and so great is my impatience that I am ready to satisfy my curiosity this very instant, if you would be so kind as to show me the way.'

The man the magician had spoken to was only too happy to tell him the way he must take to have a view of Aladdin's palace, and the magician rose and immediately set off. When he reached the palace and had examined it closely and from all sides, he was left in no doubt that Aladdin had made use of the lamp to build it. Without dwelling on Aladdin's powerlessness as the son of a simple tailor, he was well aware that only *jinn*, the slaves of the lamp which he had failed to get hold of, were capable of performing miracles of this kind. Stung to the quick by Aladdin's good fortune and importance, which seemed to him little different from the sultan's own, the magician returned to the *khan* where he had taken up lodging.

He needed to find out where the lamp was and whether Aladdin carried it around with him, or whether he kept in some secret spot, and this he could only discover through an act of geomancy. As soon as he reached his lodgings, he took his square box and his sand which he carried with him on all his travels. When he had completed the operation, he found that the lamp was in Aladdin's palace and he was so delighted at this discovery that he was beside himself. 'I am going to have this lamp,' he said, 'and I defy Aladdin to stop me from taking it from him and from making him sink to the depths from which he has risen to such heights!'

Unfortunately for Aladdin, it so happened that he had set off on a hunting expedition for eight days and was still away, having been gone for only three days. This is how the magician learned about it. Having performed the act of geomancy which had given him so much joy, he went to see the doorkeeper of the *khan* under the pretext of having a chat with him. The latter, who was of a garrulous nature, needed little encouragement, telling him that he had himself just been to see Aladdin's palace. After listening to him describe with great exaggeration all the things he had seen which had most amazed and struck him and everyone in general, the magician said: 'My curiosity does not stop there and I won't be satisfied until I have seen the master to whom such a wonderful building belongs.' 'That will not be difficult,' replied the doorkeeper. 'When he is in town hardly a day goes by on which there isn't an opportunity to see him; but three days ago, he went out on a great hunting expedition which was to last for eight days.'

The magician needed to hear no more. He took leave of the doorkeeper, saying to himself as he went back to his room: 'Now is the time to act; I must not let the opportunity escape me.' He went to a lamp maker's shop which also sold lamps. 'Master,' he said, 'I need a dozen copper lamps – can you supply me with them?' The lamp maker told him he did not have a dozen but, if he would be patient and wait until the following day, he could let him have the whole lot whenever he wanted. The magician agreed to this and asked that the lamps be clean and well polished, and after promising him to pay him well, he returned to the *khan*.

The next day, the twelve lamps were delivered to the magician who gave the lamp maker the price he had asked for, without bargaining. He put them in a basket which he had specially acquired and, with this on his arm, he went to Aladdin's palace. When he drew near, he began to cry out: 'Old lamps for new!'

As he approached, the children playing in the square heard him from a distance and rushed up and gathered around him, loudly jeering at him, for they took him for a madman. The passers-by, too, laughed at what they thought was his stupidity. 'He must have lost his mind,' they said, 'to offer to exchange old lamps for new ones.' But the magician was not surprised by the children's jeers nor by what people were saying about him, and he continued to cry out to sell his wares: 'Old lamps for new!'

He repeated this cry so often as he went to and fro in front of and around the palace that Princess Badr, who was at that point in the room with the twenty-four windows, hearing a man's voice crying out something but unable to make out what he was saying because of the jeers of the children who followed him and who kept increasing in number, sent down one of her slave girls to go up to him and see what he was shouting.

The slave girl was not long in returning and entered the room in fits of laughter. Her mirth was so infectious that the princess, looking at her, could not stop herself from laughing too. 'Well, you crazy girl,' she said, 'tell me why you are laughing.' Still laughing, the slave replied: 'O princess, who couldn't stop himself laughing at the sight of a madman with a basket on his arm full of brand new lamps wanting not to sell them but to change them for old ones? It's the children, crowding around him so that he can hardly move and jeering at him, who are making all the noise.'

Hearing this, another slave girl interrupted: 'Speaking of old lamps, I don't know if the princess has observed that there is an old lamp on the cornice. Whoever owns it won't be cross to find a new one in its place. If the princess would like, she can have the pleasure of finding out whether this madman is really mad enough to exchange a new lamp for an old one without asking anything for it in return.'

The lamp the slave girl was talking about was the magic lamp Aladdin had used to raise himself to his present high state; he himself had put it on the cornice before going out to hunt, for fear of losing it, a precaution he had taken on all previous occasions. Up until now, neither the slave girls, nor the eunuchs, nor even the princess herself had paid any attention to it during his absence, for apart from when he went out hunting, Aladdin always carried it on him. One may say that Aladdin was right to take this precaution, but he ought at least to have locked up the lamp. Mistakes like this, it is true, are always being made and always will be.

The princess, unaware how precious the lamp was and that it was in Aladdin's great interest, not to mention her own, that no one should ever touch it and that it should be kept safe, entered into the joke. She ordered a eunuch to take the lamp and go and exchange it. The eunuch obeyed and went down from the room, and no sooner had he emerged from the palace gate when he saw the magician. He called out to him and, when he came up, he showed him the old lamp, saying: 'Give me a new lamp for this one here.'

The magician was in no doubt that this was the lamp he was looking for – there could be no other lamp like it in Aladdin's palace, where all the plates and dishes were either of gold or silver. He promptly took it from the eunuch's hand and, after he had stowed it safely away in his cloak, he showed him his basket and told him to choose whichever lamp he fancied. The eunuch picked one, left the magician and took the new lamp to the princess. As soon as the exchange had taken place, the square rang out again with the shouts and jeers of the children, who laughed even more loudly than before at what they took to be the magician's stupidity.

The magician let the children jeer at him, but not wanting to stay any longer in the vicinity of Aladdin's palace, he gradually and quietly moved away. He stopped crying out about changing new lamps for old, for the only lamp he wanted was the one now in his possession. Seeing his silence, the children lost interest and left him to go on his way.

As soon as he was out of the square between the two palaces, the magician escaped through the less frequented streets and, when he saw there was nobody about, he set the basket down in the middle of one, since he no longer had a use for either the lamps or the basket. He then slipped down another street and hastened on until he came to one of the city gates. As he made his way through the suburbs, which were extensive, he bought some provisions before leaving them. Once in the countryside, he left the road and went to a spot out of sight of passers-by, where he stayed a while until he judged the moment was right for him to carry out the plan which had brought him there. He did not regret the barbary horse he had left behind at the *khan* where he had taken lodgings, for he reckoned that the treasure he had acquired was fair compensation for its loss.

The magician spent the rest of the day in this spot until night was at its darkest. He then pulled out the lamp from under his cloak and rubbed it. Thus summoned, the *jinni* appeared. 'What is your wish?' it asked him. 'Here am I, ready to obey you, your slave and the slave of all those who hold the lamp, I and the other slaves.' 'I command you,' replied the magician, 'this very instant to remove the palace that the other slaves of the lamp have built in this city, just as it is, with all the people in it, and transport it and at the same time myself to such-and-such a place in Africa.' Without answering him, the *jinni*, with the assistance of other *jinn*, like him slaves of the lamp, transported the magician and the entire palace in a very short time to the place in Africa he had designated, where we will leave him, the palace and Princess Badr, and describe, instead, the sultan's surprise.

As soon as the sultan had arisen, as was his custom he went to his closet window in order to have the pleasure of gazing on and admiring Aladdin's palace. But when he looked in the direction of where he had been accustomed to see this palace, all he could see was an empty space, such as had been before the palace had been built. Believing himself mistaken, he rubbed his eyes, but still saw nothing, although the weather was fine, the sky clear and the dawn, which was just breaking, had made everything sharp and distinct. He looked through the two windows on the right and on the left but could only see what he had been used to seeing out of them. So great was his astonishment that he remained for a long time in the same spot, his eyes turned towards where the palace had stood but was now no longer to be seen. He could not understand how so large and striking a palace as Aladdin's, which he had seen as recently as the previous day and almost every day since he had given permission to build it, had vanished so completely that no trace was left behind. 'I am not wrong,' he said to himself. 'It was there. If it had tumbled down, the materials would be there in heaps, and if the earth had swallowed it up, then there would be some trace to show that had happened.' Although he was convinced that the palace was no more, he nonetheless waited a little longer to see if, in fact, he was mistaken. At last he withdrew and, after taking one final look before leaving, he returned to his room. There he commanded the grand vizier to be summoned in all haste and sat down, his mind so disturbed with conflicting thoughts that he did not know what he should do.

The vizier did not keep the sultan waiting long; in fact, he came in such great haste that neither he nor his officials noticed as they came that Aladdin's palace was no longer there, nor had the doorkeepers, when they opened the palace gates, noticed its disappearance. When he came up to the sultan, the vizier addressed him: 'Sire, the urgency with which your majesty has summoned me makes me think that something most extraordinary has happened, since you are well aware that today is the day the council meets and that I must shortly go and carry out my duties.' 'What has happened is indeed truly extraordinary, as you will agree. Tell me, where is Aladdin's palace?' asked the sultan. 'Aladdin's palace?' replied the vizier in astonishment. 'I have just passed in front of it – I thought it was there. Buildings as solid as that don't disappear so easily.' 'Go and look through my closet window,' said the sultan, 'and then come and tell me if you can see it.'

The grand vizier went to the closet, and the same thing happened to him as had happened to the sultan. When he had quite convinced himself that Aladdin's palace no longer stood where it had been and that there did not appear to be any trace of it, he returned to the sultan. 'Well, did you see Aladdin's palace?' the sultan asked him. 'Sire,' he replied, 'your majesty may remember that I had the honour to tell you that this palace, which was the subject of your admiration, with all its immense riches, was the result of magic, the work of a magician, but your majesty would not listen to this.'

The sultan, unable to disagree with what the vizier had said, flew into a great rage which was all the greater because he could not deny his incredulity. 'Where is this wretch, this impostor?' he cried. 'Bring him at once so that I can have his head chopped off.' 'Sire,' replied the vizier, 'he took leave of your majesty a few days ago; we must send for him and ask him about his palace – he must know where it is.' 'That would be to treat him too leniently; go and order thirty of my horsemen to bring him to me bound in chains,' commanded the sultan. The vizier went off to give the sultan's order to the horsemen, instructing their officer in what manner to take Aladdin so that he did not escape. They set out and met Aladdin five or six miles outside the city, hunting on his return. The officer went up to him and told him that the sultan, in his impatience to see him, had sent them to inform him and to accompany him back.

Aladdin, who had not the slightest suspicion of the real reason which brought this detachment of the sultan's guard, continued to hunt but, when he was only half a league from the city, this detachment surrounded him and the officer addressed him: 'Prince Aladdin, it is with the greatest regret that we have to inform you of the order of the sultan to arrest you and bring you to him as a criminal of the state. We beg you not to think ill of us for carrying out our duty and we hope you will forgive us.'

Aladdin, who believed himself innocent, was very much surprised at this announcement and asked the officer if he knew of what crime he was accused, to which the officer answered that neither he nor his men knew anything about it. When he saw how few his own men were compared to the horsemen in the detachment and how they were now moving away from him, he dismounted. 'Here I am,' he said. 'Carry out your order. I have to say, though, that I don't believe I am guilty of any crime, either against the sultan himself or against the state.' A very long, thick chain was immediately passed around his neck and tied around his body in such a manner as to bind his arms. Then the officer went ahead to lead the detachment, while a horseman took the end of the chain and, following the officer, led Aladdin, who was forced to follow him on foot. In this manner, Aladdin was led towards the city.

When the horsemen entered the outskirts, the first people who saw Aladdin being led as a criminal were convinced he was going to have his head chopped off. As he was held in general affection, some took hold of their swords or other weapons, while those who had no weapons armed themselves with stones, and they followed the detachment. Some horsemen in the rear turned round to face the people as though to disperse them; but the crowd quickly grew to such an extent that the horsemen decided on a stratagem, being concerned to get as far as the sultan's palace without Aladdin being snatched from them. To succeed in this, they took great care to take up the entire street as they passed, now spreading out, now closing up again, according to whether the street was broad or narrow. In this way, they reached the palace square, where they all drew themselves up in a line facing the armed populace, until their officer and the horseman who led Aladdin had entered the palace and the doorkeepers had shut the gate to stop the people entering.

Aladdin was led before the sultan who was waiting for him on the balcony, accompanied by the grand vizier. As soon as he saw him, the sultan immediately commanded the executioner, whom he had ordered to be present, to chop off his head, without wanting to listen to Aladdin or receive an explanation from him. The executioner seized Aladdin and removed the chain which he had around his neck and body. On the ground he spread a leather mat stained with the blood of the countless criminals he had executed and made him kneel on it before tying a bandage over his eyes. He then drew his sword, sized him up before administering the blow and, after flourishing the sword in the air three times, sat down, waiting for the sultan to give the signal to cut off Aladdin's head.

At that moment, the grand vizier noticed that the crowd, who had broken through the horsemen and filled the square, were scaling the palace walls in several places and were beginning to demolish them in an attempt to breach them. Before the sultan could give the signal to the executioner, the vizier said to him: 'Sire, I beseech your majesty to reflect carefully on what you are about to do. You will run the risk of seeing your palace stormed, and should such a disaster occur, the outcome could be fatal.' 'My palace stormed!' exclaimed the sultan. 'Who would be so bold?' 'Sire,' replied the vizier, 'if your majesty were to cast a glance towards the walls of your palace and towards the square, you would discover the truth of what I say.'

On seeing the excited and animated mob, the sultan was so terror-stricken that he instantly commanded the executioner to put away his sword in its sheath and to remove the bandage from Aladdin's eyes and let him go free. He also ordered the guards to proclaim that the sultan was pardoning him and that everyone should go away. As a result, all the men who had already climbed on top of the palace walls, seeing what had happened, now abandoned their plan. They very quickly climbed down and, filled with joy at having saved the life of a man they truly loved, they spread the news to everyone around them and from there it soon spread to all the crowd assembled in the palace square. And when the guards proclaimed the same thing from the top of the terraces to which they had climbed, it became known to all. The justice the sultan had done Aladdin by pardoning him pacified the mob; the tumult died down and gradually everyone went home.

Finding himself free, Aladdin looked up at the balcony and, seeing the sultan, cried out in an affecting manner to him: 'Sire, I beseech your majesty to add one more favour to the one you have already granted me and to let me know what crime I have committed.' 'Crime! You don't know your crime?' exclaimed the sultan. 'Come up here and I'll show you.'

Aladdin went up on to the balcony, where the sultan told him to follow him, and without looking back, led him to his closet. When he reached the door, the sultan turned to him, saying: 'Enter. You ought to know where your palace stood; look all around and then tell me what has happened to it.' Aladdin looked and saw nothing. He could see the whole area which his palace had occupied but, having no idea how the palace could have disappeared, this extraordinary event put him into such a state of confusion that in his astonishment he could not utter a single word in reply.

'Go on, tell me where your palace is and where is my daughter,' the sultan repeated impatiently. Aladdin broke his silence, saying: 'Sire, I see very well and have to admit that the palace I built is no longer where it was. I see that it has disappeared but I cannot tell your majesty where it can be. I can assure you, however, that I had no part in this.' 'I am not so concerned about what happened to your palace,' the sultan continued. 'My daughter is a million times more valuable to me and I want you to find her for me, otherwise I will cut off your head and nothing will stop me.'

'Sire,' replied Aladdin, 'I beg your majesty to grant me forty days' grace to do all I can, and if in that time I don't succeed in finding her, I give you my word that I will offer my head at the foot of your throne so you can dispose of it as you please.' 'I grant you the forty days you ask for,' answered the sultan, 'but don't think to abuse this favour by believing you can escape my anger, for I will know how to find you, in whatever corner of the earth you may be.'

Aladdin left the sultan, deeply humiliated and in a truly pitiful state:

with head bowed, he passed through the palace courtyards without daring to raise his eyes in his confusion. Of the chief court officials, whom he had treated graciously and who had been his friends, not one for all their friendship went up to him to console him or to offer to take him in, but they turned their backs on him as much to avoid seeing him as to avoid being recognized by him. But even had they gone up to Aladdin to say something consoling to him or to offer to help him, they would not have known him for he no longer knew himself, being no longer in his right mind. This was evident when he came out of the palace as, without thinking what he was doing, he went from door to door and asked passers-by, enquiring of them whether they had seen his palace or could give him any news of it. Consequently, everyone became convinced that Aladdin had gone out of his mind. Some only laughed, but the more reasonable, and in particular those linked to him either by friendship or business, were filled with compassion. He stayed three days in the city – walking hither and thither and only eating what people offered him out of charity – unable to decide what to do.

Finally, in the wretched state he was in and feeling he could no longer stay in a city where he had once cut such a fine figure, he left and went out into the countryside. Avoiding the main roads and after crossing several fields in a state of great uncertainty, eventually, at nightfall, he came to the bank of a river. There, greatly despondent, he said to himself: 'Where shall I go to look for my palace? In which province, which country or part of the world shall I find it and recover my dear princess, as the sultan demands of me? I will never succeed. It's best if I don't go to all of this wearisome effort, which will in any case come to nothing, but free myself of all this bitter grief that torments me.' Having made this resolution, he was on the point of throwing himself into the river but, being a good and faithful Muslim, he thought he ought not to do this before first performing his prayers. Wishing to prepare himself, he approached the river in order to wash his hands and face according to custom, but as the bank sloped at that point and was damp from the water lapping against it, he slipped and would have fallen into the river had he not been stopped by a small rock which protruded about two feet above the ground. Fortunately for him, he was still wearing the ring which the magician had put on his finger before he had gone down into the cave to remove the precious lamp that had now been taken away from him. As he caught hold of the rock, he rubbed the ring quite hard against it and immediately the same *jinni* who had first appeared to him in the cave in which the magician had shut him up appeared once again, saying: 'What is your wish? Here am I, ready to obey you, your slave and the slave of all those who wear the ring on their finger, I and the other slaves of the ring.'

Delighted at this apparition, which he had so little expected in his despair, Aladdin replied: 'Save me a second time, *jinni*, and either tell me where the palace I built is or bring it back immediately to where it was.' 'What you ask of me,' replied the *jinni*, 'is not within my power to bring back; I am only the slave of the ring. You must address yourself to the slave of the lamp.' 'If that's the case,' said Aladdin, 'I command you, by the power of the ring, to transport me to the place where my palace is, wherever it is in the world, and set me down underneath the windows of Princess Badr.' Hardly had he finished speaking than the *jinni* transported him to Africa, to the middle of a meadow where the palace stood, not far from a large town, and set him down right underneath the windows of the princess's apartments, where he left him. All this happened in a moment. Despite the darkness of the night, Aladdin easily recognized his palace and the princess's apartments; but as the night was already advanced and all was quiet in the palace, he moved off a little way and sat down at the foot of a tree. There, filled with hope as he reflected on the pure chance to which he owed his good fortune, he found himself in a much more peaceful state than he had been in since the time when he had been arrested and brought before the sultan and had been delivered from the recent danger of losing his life. For a while he entertained himself with these agreeable thoughts but eventually, not having slept at all for five or six days, he could not stop himself being overwhelmed by sleep and fell asleep at the foot of the tree where he was sitting.

The next morning, as dawn was breaking, Aladdin was pleasantly awoken by the singing of the birds, not only those that roosted on the tree beneath which he had spent the night but all those on the luxuriant trees in the very garden of his palace. When he cast his eyes on that wonderful building, he felt a joy beyond words at the thought that he would soon be its master once more and possess once again his dear princess. He got up and, approaching the princess's apartments, walked for a while underneath her windows until it was light and he could see her. As he waited, he searched his mind for a possible cause for his misfortune, and after much thought he became convinced that it all came from his having left the lamp out of his sight. He blamed himself for his negligence and his carelessness in letting it leave his possession for a single moment. What worried him still more was that he could not imagine who could be so jealous as to envy him his good fortune. He would have soon guessed had he known that such a man and his palace were both in Africa, but the slave of the ring had not mentioned this, while he himself had not even asked about it. The name of Africa alone should have reminded him of the magician, his avowed enemy.

That morning, Princess Badr arose earlier than she had done ever since the wily magician – now the master of the palace, the sight of whom she had been forced to endure once a day but whom she treated so harshly that he had not yet been so bold as to take up residence there – had by his cunning kidnapped her and carried her off to Africa. When she was dressed, one of her slave girls, looking through the lattice screen, spotted Aladdin and ran to tell her mistress. The princess, who could not believe the news, rushed to the window and saw her husband. She opened the screen and, hearing the sound, Aladdin raised his head. Recognizing her, he greeted her with great delight. 'So as to waste no time, someone has gone to open the secret door for you; enter and come up,' said the princess and closed the screen.

The secret door was beneath the princess's apartments. Finding it open, Aladdin entered and went up. It is impossible to describe the intensity of their joy at their reunion after believing themselves parted for ever. They embraced several times and showed all the signs of love and affection one can imagine after such a sad and unexpected separation. After many an embrace mingled with tears of joy, they sat down. Aladdin was the first to speak: 'Before we talk about anything else, dear princess, I beg you, in the Name of God, in your own interest and that of your worthy father, the sultan, and no less of mine, to tell me what has happened to the old lamp that I put on the cornice in the room of the twenty-four windows before I went off to hunt.' 'Ah, my dear husband!' sighed the princess. 'I did indeed suspect that all our troubles came from the loss of that lamp and what distresses me is that I myself am the cause of it.' 'Dear princess,' said Aladdin, 'don't blame yourself; it's all my fault and I should have taken greater care in looking after it. Let's now think only about how to repair the damage, and so please be so good as to tell me how it happened and into whose hands the lamp fell.'

The princess then proceeded to tell Aladdin how the old lamp had been exchanged for a new one, which she ordered to be brought in for him to see; and how the following night she had found the palace had been transported and, the next morning, she woke to find herself in an unknown country, where she was now talking to him, and that this was Africa, a fact she had learned from the very mouth of the traitor who had transported her there by his magic arts.

'Dear princess,' Aladdin interrupted her, 'you have already told me who the traitor is by explaining that I am with you in Africa. He is the most perfidious of all men. But now is not the time nor the place to give you a fuller picture of his evil deeds. I only ask you to tell me what he has done with the lamp and where he has put it.' 'He carries it with him carefully wrapped up close to his chest,' the princess answered, 'and I can bear witness to that since he pulled it out and unwrapped it in my presence in order to show it off.'

'Princess, please don't be annoyed with me for wearying you with all these questions,' said Aladdin, 'for they are as important to you as they are to me. But to come to what most particularly concerns me, tell me, I beg you, how you yourself have been treated by this wicked and treacherous man.' 'Since I have been here,' replied the princess, 'he only visits me once a day, and I am sure that he does not bother me more often because these visits offer him so little satisfaction. Every time he comes, the aim of all his conversation is to persuade me to break the vows I gave you and to make me take him for my husband, by trying to make me believe that I should not hope ever to see you again; that you are no longer alive and that the sultan, my father, has had your head chopped off. He adds, to justify himself, that you are ungrateful and that you owed your good fortune only to him, and there are a thousand other things I will leave him to tell you. And as all he gets from me in reply are tears and moans, he is forced to depart as little satisfied as when he came. However, I have no doubt that his intention is to let the worst of my grief and pain pass, in the hope that I will change my mind, and, finally, if I persist in resisting him, to use violence. But, dear husband, your presence has already dispelled my worries.'

'Princess,' said Aladdin, 'I am confident that it is not in vain and your worries are over, for I believe I have found a way to deliver you from your enemy and mine. But to do that, I have to go into the town. I will return towards noon and will then tell you what my plan is and what you will need to do to help make it succeed. I must warn you not to be astonished if you see me return dressed in different clothes, but give the order not to have me kept waiting at the secret door after my first knock.' The princess promised him that someone would be waiting for him at the door, which would be opened promptly.

When Aladdin had gone down from the princess's apartments, leaving by the same door, he looked around and saw a peasant who was setting off on the road which led into the country. The peasant had already gone past the palace and was a little way off, so Aladdin hastened his steps; when he had caught up with him, he offered to change clothes with him, pressing him until the peasant finally agreed. The exchange was done behind a nearby bush; and when they had parted company, Aladdin took the road back to the town. Once there, he went along the road leading from the city gate and then passed into the most frequented streets. On coming to the place where all the merchants and artisans had their own particular street, he entered that of the apothecaries where he sought out the largest and best-supplied store and asked the merchant if he had a certain powder which he named.

The merchant, who, from his clothes, imagined Aladdin to be poor and that he had not enough money to pay him, said he had but that it was expensive. Aladdin, guessing what was on the merchant's mind, pulled out his purse and, showing him gold coins, asked for half a drachm of the powder. The merchant weighed it out, wrapped it up and, as he gave it to Aladdin, asked him for a gold coin. Aladdin handed it to him and, stopping only long enough in the town to eat something, he returned to his palace. He did not have to wait at the secret door which was immediately opened to him, and he went up to the princess's apartments.

'Princess,' he said, 'the aversion you have shown me you feel for your kidnapper will perhaps make it difficult for you to follow the advice I'm going to give you. But allow me to tell you that it is advisable that you should conceal this and even go against your own feelings if you wish to deliver yourself from his persecution and give the sultan, your father and my lord, the satisfaction of seeing you again. If you want to follow my advice,' continued Aladdin, 'you must begin right now by putting on one of your most beautiful dresses, and when the magician comes, you must be prepared to welcome him as warmly as possible, without affectation or strain, and with a happy smile, in such a way that, should there still remain a hint of sadness, he will think it will go away with time. In your conversation, give him to understand that you are doing your best to forget me; and so that he should be all the more persuaded of your sincerity, invite him to have supper with you and indicate to him that you would be very pleased to taste some of the best wine his country has to offer. He will then have to leave you to go and find some. Then,

while waiting for him to return, when the food is laid out, pour this
powder into one of the goblets out of which you usually drink. Put it
aside and tell the slave girl who serves you your drink to bring it to you
filled with wine after you have given her a pre-arranged signal. Warn her
to take care and not to make a mistake. When the magician returns and
you are seated at table and when you have eaten and drunk what you
think is sufficient, ask for the goblet containing the powder to be brought
to you and exchange it for his. He will think you will be doing him such
a favour that he won't be able to refuse you, but no sooner will he have
emptied it than you will see him fall down backwards. If you don't like
drinking from his goblet, just pretend to drink; you can do so without
fear, for the effect of the powder will be so swift that he won't have time
to notice whether you are drinking or not.'

When Aladdin had finished speaking, the princess said: 'I must admit
I find it very distasteful to have to agree to make advances towards the
magician, even though I know I must; but what can one not resolve to
do when faced with a cruel enemy! I will do as you advise, for on it
depends my peace of mind no less than yours.' Having made these
arrangements with the princess, Aladdin took his leave and went to
spend the rest of the day near the palace, waiting for night to fall before
returning to the secret door.

From the moment of her painful separation, Princess Badr – inconsolable not only at seeing herself separated from her beloved husband,
Aladdin, whom she had loved and whom she continued to love more
out of inclination than out of duty, but also from the sultan, her father,
whom she cherished and who loved her tenderly – had remained very
neglectful of her person. She had even, one may say, forgotten the
neatness which so becomes persons of her sex, particularly after the first
time the magician had come to her and she discovered through her slave
girls, who recognized him, that it was he who had taken the old lamp in
exchange for a new one. Following her discovery of this outrageous
swindle, he had become an object of horror to her, and the opportunity
to take the revenge on him that he deserved, and sooner than she had
dared hope for, made her content to fall in with Aladdin's plans. Thus,
as soon as he had gone, she sat down at her dressing table; her slave girls
dressed her hair in the most becoming fashion and she took out her most
glamorous dress, searching for the one which would best serve her
purpose. She then put on a belt of gold mounted with the largest of
diamonds, all beautifully matched; to go with it she chose a necklace all

of pearls, of which the six on both sides of the central pearl – which was the largest and the most precious – were so proportioned that the greatest queens and the wives of the grandest sultans would have thought themselves happy to have a string of pearls the size of the two smallest pearls in the princess's necklace. The bracelets of diamonds interspersed with rubies wonderfully complemented the richness of the belt and the necklace.

When the princess was fully dressed, she looked in her mirror and consulted her maids for their opinion on her dress. Then, having checked that she lacked none of the charms which might arouse the magician's mad passion, she sat down on the sofa to await his arrival.

The magician came at his usual hour. As soon as she saw him come into the room of the twenty-four windows where she awaited him, she arose apparelled in all her beauty and charm, and showed him to the place of honour where she wanted him to take his seat so as to sit down at the same time as he did, a mark of courtesy she had not shown him before. The magician, more dazzled by the beauty of the princess's sparkling eyes than by the brilliance of the jewellery with which she was adorned, was very surprised. Her stately air and a certain grace with which she welcomed him were so different from the rebuffs he had up till then received from her. In his confusion, he would have sat down on the edge of the sofa but, seeing the princess did not want to take her seat until he had sat down where she wished him to, he obeyed.

Once the magician was seated where she had indicated, the princess, to help him out of the embarrassment she could see he was in, was the first to speak. Looking at him in such a manner as to make him believe that he was no longer odious to her, as she had previously made him out to be, she said: 'No doubt you are astonished to see me appear so different today compared to what you have seen of me up till now. But you will no longer be surprised when I tell you that sadness and melancholy, griefs and worries, are not part of my nature, and that I try to dispel such things as quickly as possible when I discover there is no longer a reason for them. Now, I have been thinking about what you told me of Aladdin's fate and, knowing my father's temper, I am convinced as you are that Aladdin cannot have avoided the terrible effects of his wrath. And I can see that if I persist in weeping for him all my life, all my tears would not bring him back. That is why, after I have performed for him the final rites and duties that my love dictates, now that he is in his grave it seems to me that I should look for ways of consoling myself.

This is the reason for the change you see in me. As a start, to remove any cause for sadness, I have resolved to banish it completely; and, believing you very much wish to keep me company, I have given orders for supper to be prepared for us. However, as I only have wine from China and I am now in Africa, I fancied trying some that is produced here and I thought that, if there is any, you would be able to procure some of the best.'

The magician, who had thought it would be impossible for him to be so fortunate as to find favour so quickly and so easily with the princess, told her he could not find words sufficient to express how much he appreciated her kindness. But to finish a conversation which he would otherwise have found difficult to bring to an end, once he had embarked on it, he seized upon the subject of African wine that she had mentioned. He told her that one of the main advantages of which Africa could boast was that it produced excellent wines, particularly in the region where she now found herself; he had a seven-year-old cask which had not yet been opened, whose excellence, not to set too high a value on it, surpassed the most exquisite wines in the whole of the world. 'If my princess will give me leave, I will go and fetch two bottles and I will be back immediately,' he added. 'I would be sorry to put you to such trouble,' the princess replied. 'It might be better if you sent someone else.' 'But I shall have to go myself,' said the magician, 'as only I know where the key of the storeroom is, and only I know the secret of how to open it.' 'If that is the case,' said the princess, 'then go, but come back quickly. The longer you take, the more impatient I will be to see you again, and bear in mind that we will sit down to eat as soon as you return.'

Filled with hope at his imagined good fortune, the magician did not so much run as fly to fetch his seven-year-old wine, and returned very quickly. The princess, well aware that he would make haste, had herself put the powder that Aladdin had brought her into a goblet which she had set aside and had then started to serve the dishes. They sat down opposite each other to eat, the magician so placed that his back was turned to the refreshments. The princess presented him with all the best dishes, saying to him: 'If you wish, I will entertain you with singing and music, but as there are only the two of us, it seems to me that conversation would give us more pleasure.' The magician regarded this as one more favour granted him by the princess.

After they had eaten a few mouthfuls, the princess asked for wine to be brought. She then drank to the health of the magician, after which she said to him: 'You were right to sing the praises of your wine. I have never drunk anything so delicious.' 'Charming princess,' replied the magician, holding the goblet he had just been given, 'my wine acquires an extra virtue by your approval.' 'Then drink to my health,' said the princess, 'and you will see for yourself what an expert I am.' The magician drank to the princess's health, saying to her as he handed back the goblet: 'Princess, I consider myself fortunate for having kept this wine for such a happy occasion, and I, too, must admit that I have never before drunk any so excellent in so many ways.'

At last, when they had finished eating and had drunk three more times of the wine, the princess, who had succeeded in charming the magician with her gracious and attentive ways, beckoned to the slave girl who served the wine and told her to fill her goblet with wine and at the same time fill that of the magician and give it to him. When they both had their goblets in their hands, the princess said to the magician: 'I don't know what one does here when one is in love and drinks together, as we are doing. Back home, in China, lovers exchange goblets and drink each other's health.' As she was speaking, she gave him the goblet she had, in one hand, while holding out the other to receive his. The magician hastened to make this exchange, which he did all the more gladly as he regarded this favour the surest sign that he had completely won over the heart of the princess. His happiness was complete. Before he drank to her, holding his goblet in his hand, he said: 'Princess, we Africans are by no means as skilled in the refinements of the art and pleasures of love as the Chinese. I learned from you something I did not know and at the same time I have learned how much I should appreciate the favour you grant me. Dear princess, I will never forget this: by drinking out of your goblet, I rediscovered a life I would have despaired of, had your cruelty towards me continued.'

Princess Badr, bored by the magician's endless ramblings, interrupted him, saying: 'Let us drink first; you can then say what you wish later.' At the same time, she raised the goblet to her mouth but only touched it with her lips, while the magician, who was in a hurry to drink first, drained his goblet without leaving a drop behind. In his haste to empty the goblet, when he finished, he leaned backwards a little and remained a while in this pose until the princess, her goblet still only touching her lips, saw his eyes begin to roll as he fell, lifeless, on his back.

She had no need to order the secret door to be opened for Aladdin: as soon as the word was given that the magician had fallen upon his back, her slave girls, who were standing several paces from each other outside the room and all the way down to the foot of the staircase, immediately opened the door. Aladdin went up and entered the room. When he saw the magician stretched out on the sofa, the princess got up and was coming towards him to express her joy and embrace him, but he stopped her, saying: 'Princess, now is not yet the time. Please would you go back to your apartments and see that I am left alone while I try to arrange to have you transported back to China as quickly as you were brought from there.'

As soon as the princess and her slave girls and eunuchs were out of the room, Aladdin closed the door and, going up to the magician's lifeless corpse, opened up his shirt and drew out the lamp which was wrapped up as the princess had described to him. He unwrapped it and as soon as he rubbed it, the *jinni* appeared with his usual greeting. '*Jinni*,' said Aladdin, 'I have summoned you to command you, on behalf of the lamp in whose service you are, to have this palace transported immediately back to China, to the same part and to the same spot from where it was brought here.' The *jinni* nodded to show he was willing to obey and then disappeared. Immediately, the palace was transported to China, with only two slight shocks to indicate the removal had taken place – one when the palace was lifted up from where it was in Africa and the other when it was set down again in China, opposite the sultan's palace. All this happened in a very space of short time.

Aladdin went down to the princess's apartments. He embraced her, saying: 'Princess, I can assure you that tomorrow morning, your joy and mine will be complete.' Then, as the princess had not yet finished eating and Aladdin was hungry, she had the dishes – which had hardly been touched – brought from the room of the twenty-four windows. She and Aladdin ate together and drank of the magician's fine old wine, after which, having no doubt enjoyed conversation which must have been very satisfying, they withdrew to her apartments.

Meanwhile, the sultan, since the disappearance of Aladdin's palace and of Princess Badr, had been inconsolable at having lost her, or so he thought. Unable to sleep by night or day, instead of avoiding everything that could keep him in his sorrow, he, on the contrary, sought it out all the more. Whereas previously he would only go in the morning to his closet to enjoy gazing at the palace – of which he could never have his fill – now he would go there several times a day to renew his tears and plunge himself into ever deeper suffering by the thought that he would never again see what had given him so much pleasure and that he had lost what he held dearest in the world. Dawn was just breaking when the sultan came to this room the morning that Aladdin's palace had just been restored to its place. He was lost in thought as he entered it and filled with grief as he glanced sadly at the spot, not noticing the palace at first as he was expecting to see only an empty space. When he saw that the space was no longer empty, he thought at first that this must be the effect of the mist. But when he looked more closely, he realized that it must be, without doubt, Aladdin's palace. Sadness and sorrow immediately gave way to joy and delight. He hastened to return to his apartments where he gave orders for a horse to be saddled and brought to him, and as soon as it was brought, he mounted and set off, thinking he could not arrive fast enough at Aladdin's palace.

Aladdin, expecting this to happen, had got up at first light and had taken out of his wardrobe one of his most magnificent costumes, put it on and gone up to the room of the twenty-four windows, from where he could see the sultan approaching. He went down and was just in time to welcome him at the foot of the staircase and to help him dismount. 'Aladdin,' the sultan said to him, 'I can't speak to you before I have seen and embraced my daughter.' Aladdin then led the sultan to the princess's apartments, where she had just finished dressing. He had already told her to remember that she was no longer in Africa but in China, in the capital of her father, the sultan, and next to his palace once again. The sultan, his face bathed in tears of joy, embraced her several times, while the princess, for her part, showed him how overjoyed she was at seeing him again.

For a while the sultan was unable to speak, so moved was he at having found his beloved daughter again after having so bitterly wept for her loss, sincerely believing she must be dead. The princess, too, was in tears, in her joy at seeing her father again. Finally, the sultan said to her: 'My daughter, I would like to think that it's the joy of seeing me again which makes you seem so little changed, as though no misfortune had happened to you. But I am convinced you have suffered a great deal, for one is not carried off with an entire palace as suddenly as you were without great alarm and terrible anguish. I want you to tell me all about it and to hide nothing from me.'

The princess was only too happy to tell him what he wanted to know. 'Sire,' she said, 'if I appear to be so little changed, I beg your majesty to bear in mind that I received a new life early yesterday morning thanks to Aladdin, my beloved husband and deliverer whom I had looked on and mourned as lost to me and whom the joy of seeing and embracing again has all but restored me. Yet my greatest distress was to see myself snatched both from your majesty and from my dear husband, not only because of my love for my husband but also because of my worry that he, innocent though he was, should feel the painful consequences of your majesty's anger, to which I had no doubt he would be exposed. I suffered only a little from the insolence of my kidnapper – whose conversation I found disagreeable, but which I could put an end to, because I knew how to gain the upper hand. Besides, I was as little constrained as I am now. As for my abduction, Aladdin had no part in it: I alone – though totally innocent – am to blame for it.'

In order to persuade the sultan of the truth of what she said, she told him in detail all about the African magician, how he had disguised himself as a seller of lamps who exchanged new lamps for old ones, and how she had amused herself by exchanging Aladdin's lamp, not knowing its secret and importance; how, after this exchange, she and the palace had been lifted up and both transported to Africa together with the magician; how the latter had been recognized by two of her slave girls and by the eunuch who had exchanged the lamp for her, when the magician first had the effrontery to come and present himself to her after the success of his audacious enterprise, and to propose marriage to her; how she had suffered at his hands until the arrival of Aladdin; and what measures the two of them had taken to remove the lamp which the magician carried on him and how they had succeeded, particularly by her dissimulation in inviting him to have supper with her; and, finally, she told him of the poisoned goblet she had offered to the magician. 'As for the rest,' she concluded, 'I leave it to Aladdin to tell you about it.'

Aladdin had little more to tell the sultan. 'When the secret door was opened and I went up to the room of the twenty-four windows,' he said, 'I saw the traitor stretched out dead on the sofa, thanks to the virulence of the poison powder. As it was not proper for the princess to remain there any longer, I begged her to go down to her apartments with her slave girls and eunuchs. As soon as I was there alone, I extracted the lamp from the magician's clothing and made use of the same secret password he used to remove the palace and kidnap the princess. By that means the palace was restored to where it had formerly stood and I had the happiness of bringing the princess back to your majesty, as you had commanded me. I don't want to impose upon your majesty but if you would take the trouble to go up to the room, you would see the magician punished as he deserves.'

The sultan, to convince himself that this was really true, got up and went to the room and when he saw the magician lying dead, his face already turned livid thanks to the virulent effect of the poison, he embraced Aladdin very warmly, saying: 'My son, don't think ill of me for my conduct towards you – I was forced to it out of paternal love and you must forgive me for being overzealous.' 'Sire,' replied Aladdin, 'I have not the slightest cause for complaint against your majesty, since you did only what you had to do. This magician, this wretch, this vilest of men, he is the sole cause of my fall from favour. When your majesty has the time, I will tell you about another wicked deed he did me, no less foul than this, from which it is only by a particular favour of God that I was saved.' 'I will indeed make time for this and soon, but let us think only of rejoicing and have this odious object removed.'

Aladdin had the magician's corpse taken away and gave orders that it be thrown on to a dunghill for the birds and beasts to feed on. The sultan, meanwhile, after having commanded that tambourines, drums, trumpets and other musical instruments be played to announce the public rejoicing, proclaimed a festival of ten days to celebrate the return of Princess Badr and Aladdin with his palace. Thus was Aladdin faced for a second time with almost inevitable death, yet managed to escape with his life. But it was not the last time – there was to be a third occasion, the circumstances of which we will now tell.

The magician had a younger brother who was no less skilled in the magic arts; one may even say that he surpassed him in wickedness and in the perniciousness of his schemes. They did not always live together nor even stay in the same city, and often one was to be found in the east and the other in the west. But every year they did not fail to inform each other, by geomancy, in what part of the world and in what condition they were, and whether one of them needed the assistance of the other.

Some time after the magician had failed in his attempt to destroy Aladdin's good fortune, his younger brother, who had not heard from him for a year and who was not in Africa but in some far-off land, wanted to know in what part of the world his brother resided, how he was and what he was doing. Wherever he went this brother always carried with him his geomancy box, as had his elder brother. Taking the box, he arranged the sand, made his throw, interpreted the figures and finally made his divination. On examining each figure, he found that his brother was no longer alive, that he had been poisoned and had died a sudden death, and that this had happened in the capital city of a kingdom in China, situated in such and such a place. He also learned that the man who had poisoned him was someone of good descent who had married a princess, a sultan's daughter.

Having learned in this way of his brother's sad fate, the magician wasted no time in useless regret, which could not restore his brother to life, but immediately resolving to avenge his death, he mounted a horse and set off for China. He crossed plains, rivers, mountains and deserts, and after a long and arduous journey, without stopping, he finally reached China and shortly afterwards the capital city whose location he had discovered by geomancy. Certain that this was the place and that he had not mistaken one kingdom for another, he stopped and took up lodgings there.

The day after his arrival, this magician went out into the city, not so much to see its fine sights – to which he was quite indifferent – as to begin to take the necessary steps to carry out his evil plan, and so he entered the most frequented districts and listened to what people were saying. There, in a place where people went to spend the time playing different kinds of games, some playing while others stood around chatting, exchanging news and discussing the affairs of the day or their own, he heard people talking of a woman recluse called Fatima, about her virtue and piety and of the miracles she performed. Believing this woman could be of some use to him for what he had in mind, he took one of the men aside and asked him to tell him particularly who this holy woman was and what sort of miracles she performed.

'What!' the man exclaimed. 'Have you never seen or even heard of her? She is the admiration of the whole city for her fasting, her austerity and her exemplary conduct. Except for Mondays and Fridays, she never leaves her little cell, and on the days she shows herself in the city she does countless good deeds, and there is not a person with a headache who is not cured by a touch of her hands.'

The magician wished to know no more on the subject, but only asked the man where in the city the cell of this holy woman was to be found. The man told him, whereupon – after having conceived and drawn up the detestable plan which we will shortly reveal and after having made this enquiry – so as to make quite sure, he observed this woman's every step as she went about the city, never leaving her out of his sight until evening when he saw her return to her cell. When he had made a careful note of the spot, he went back to one of the places we have mentioned where a certain hot drink is drunk and where one can spend the whole night should one so wish, especially during the days of great heat when the people in such countries prefer to sleep on a mat rather than in a bed.

Towards midnight, the magician, after he had settled his small bill with the owner of the place, left and went straight to the cell of this holy woman, Fatima – the name by which she was known throughout the city. He had no difficulty in opening the door, which was fastened only with a latch, and entered, without making a sound, and closed it again. Spotting Fatima in the moonlight, lying asleep on a sofa with only a squalid mat on it, her head leaning against the wall of her cell, he went up to her and, drawing out a dagger he wore at his side, woke her up. When poor Fatima opened her eyes, she was very astonished to see a man about to stab her. Pressing the dagger to her heart, ready to plunge it in, he said to her: 'If you cry out or make the slightest sound, I will kill you. Get up and do as I say.'

Fatima, who had been sleeping fully dressed, got up trembling with fear. 'Don't be afraid,' said the magician. 'All I want is your clothes. Give them to me and take mine instead.' They exchanged clothes and after the magician had put hers on, he said to her: 'Paint my face like yours so that I look like you and so that the colour doesn't come off.' Seeing that she was still trembling, he said to her, in order to reassure her and so that she might be readier to do what he wanted: 'Don't be afraid, I say. I swear by God that I will spare your life.' Fatima let him into her cell and lit her lamp. Dipping a brush into a liquid in a certain jar, she brushed his face with it, assuring him that the colour would not change and that his face was now the same colour as hers. Then she put her own headdress on his head and a veil, showing him how to conceal his face with it when he went through the city. Finally, after she put around his neck a large string of beads which hung down to the waist, she placed in his hand the same stick she used to walk with. 'Look,' she said to him, handing him a mirror, 'you will see you couldn't look more like me.' The magician looked just as he wanted to look, but he did not keep the oath he had so solemnly sworn to the saintly woman. In order to leave no trace of blood, he did not stab her but strangled her and when he saw that she had given up the ghost, he dragged her corpse by the feet to a cistern outside her cell and threw her into it.

Having committed this foul murder, the magician, disguised as Fatima, spent the rest of the night in her cell. The next day, an hour or two after sunrise, he left the cell, even though it was not a day when the holy woman would go out, quite sure that no one would stop and question him about it but ready with an answer if they did. One of the first things he had done on his arrival in the city was to go and look for Aladdin's palace, and as it was there he intended to put his plan into action, he went directly to it.

As soon as people saw what they thought to be the holy woman, a large crowd gathered around the magician, some asking for his prayers, some kissing his hands – the more reserved among them kissing the edge of his garment – and others, whether they had a headache or merely wanted to be protected from one, bowing their heads for him to lay his hands on them, all of which he did, mumbling a few words in the guise of a prayer. In fact, he imitated the holy woman so well that everyone believed it was really her. After frequent stops to satisfy such requests – for while this sort of laying-on of hands did them no harm, nor did it do them any good – he finally arrived in the square before Aladdin's palace where, the crowd being even greater, people were ever more eager to get close to him. The strongest and most zealous forced their way through to get to him and this caused such quarrels that they could be heard from the palace, right from the room with the twenty-four windows where Princess Badr was sitting.

The princess asked what all the noise was about and as no one could tell her anything about it, she gave orders for someone to go and see and report back to her. Without leaving the room, one of her slave girls looked out through a screen and came back to tell her that the noise came from the crowd of people who gathered around the saintly lady, to be cured of headaches by the laying-on of her hands. Now the princess, who had heard a lot about the holy woman and the good she did but had never yet seen her, was curious to talk to her. When she expressed something of her desire to the chief eunuch, who was present, he told her that if she wished, he could easily have the woman brought in – she had only to give the command. The princess agreed, and he immediately chose four eunuchs and ordered them to fetch the so-called holy woman.

As soon as the crowd saw the eunuchs come out of the gates of Aladdin's palace and make for the disguised magician, they dispersed and the magician, finding himself once more alone and seeing the eunuchs coming for him, stepped towards them, delighted to see his deceit was working so well. One of the eunuchs then said to him: 'Holy lady, the princess wants to see you; come, follow us,' to which the pseudo

Fatima replied: 'The princess does me a great honour; I am ready to obey her,' and followed the eunuchs, who had already set out back to the palace.

The magician, whose saintly dress concealed a wicked heart, was then led into the room of the twenty-four windows. When he saw the princess, he said to her: 'May all your hopes and desires be fulfilled,' and he began to launch into a long string of wishes and prayers for her health and prosperity. Under the cloak of great piety, he used all the rhetorical skills of the impostor and hypocrite he was to ingratiate himself into the princess's favour, which was all the more easy to achieve because the princess, in her natural goodness of heart, believed everyone was as good as she was, especially those who retreated from the world in order to serve God.

When 'Fatima' had finished her long harangue, the princess thanked her, saying: 'Lady Fatima, I thank you for your prayers and good wishes; I have great confidence in them and hope that God will fulfil them. Come, sit yourself beside me.' 'Fatima' took her seat with affected modesty. 'Holy lady,' the princess went on, 'there is something I ask you to grant me – please don't refuse it me – which is that you stay with me and tell me about your life, so that I can learn by your good example how I should serve God.' 'Princess,' replied 'Fatima', 'I beg you not to ask me something I can't consent to, without being distracted from and neglecting my prayers and devotions.' 'Don't worry about that,' the princess reassured her. 'I have several rooms which are not occupied. Choose the one you like and you shall perform all your devotions there as freely as if you were in your cell.'

Now the magician's only aim had been to enter Aladdin's palace, for he could more easily carry out there his pernicious plan under the auspices and protection of the princess than if he had been forced to go back and forth between the palace and the holy woman's cell. Consequently, he did not put up much resistance in accepting the princess's kind offer. 'Princess,' he said to her, 'however much a poor wretched woman like myself has resolved to renounce the pomp and grandeur of this world, I dare not presume to resist the wishes and commands of so pious and charitable a princess.' In reply, Badr rose from her seat and said to the magician: 'Get up and come with me, and I will show you the empty rooms I have, so that you may choose.' The magician followed the princess, and from among all the neat and well-furnished apartments she showed him he chose the one which he thought looked the humblest, saying hypocritically that it was too good for him and that he only chose it to please her.

The princess wanted to take the villain back to the room with the twenty-four windows to have him dine with her. The magician realized, however, that to eat he would have to uncover his face, which he had kept veiled until then, and he was afraid that the princess would then recognize that he was not the holy woman Fatima she believed him to be, and so he begged her earnestly to excuse him, telling her he only ate bread and some dried fruit, and to allow him to eat his modest meal in his room. She granted him his request, replying: 'Holy lady, you are free to do as you would do in your own cell. I will have some food brought you, but remember I expect you as soon as you have finished your meal.'

After the princess had dined, 'Fatima' was informed of it by one of her eunuchs and she went to rejoin her. 'Holy lady,' the princess said, 'I am delighted to have with me a holy lady like you, who will bring blessings to this place. Incidentally, how do like this palace? But before I show you round it, room by room, tell me first what you think of this room in particular.'

At this request, 'Fatima', who, in order better to perform her part, had affected to keep her head lowered, looking to neither right nor left, at last raised it and surveyed the room, from one end to the other; and when she had reflected for a while, she said: 'Princess, this room is truly wonderful and so beautiful. Yet, as far as a recluse such as myself can judge who does not know what the world thinks is beautiful, it seems to me that there is something lacking.' 'What is that, holy lady?' asked the princess. 'Tell me, I beseech you. I myself thought, and I have heard other people say the same, that it lacked nothing, but if there is anything it does lack, I shall have that put right.'

'Princess,' the magician replied, with great guile, 'forgive me for taking the liberty but my advice, if it is of any importance, would be that if a *rukh*'s egg were to be suspended from the middle of the dome, there would be no other room like this in the four quarters of the world and your palace would be the wonder of the universe.' 'Holy lady, what sort of bird is this *rukh* and where can one find a *rukh*'s egg?' Badr asked. 'Princess,' replied 'Fatima', 'this is a bird of prodigious size which lives on the summit of Mount Qaf. The architect of your palace will be able to find you one.'

After she had thanked the so-called holy woman for what she believed to be her good advice, the princess conversed with her on other things, but she did not forget the *rukh*'s egg, which she intended to mention to Aladdin as soon as he returned from hunting. He had been gone for six days and the magician, who was well aware of this, had wanted to take advantage of his absence, but Aladdin returned that same day, towards evening, just after 'Fatima' had taken her leave of the princess to retire to her room. As soon as he arrived, he went up to the princess's apartments, which she had just entered, and greeted and embraced her, but she seemed a little cold in her welcome, so he said to her: 'Dear princess, you don't seem to be as cheerful as usual. Has something happened during my absence to displease you and cause you worry and dissatisfaction? For God's sake, don't hide it from me; there's nothing that were it in my power I would not do to make it go away.' 'It's nothing, really,' replied Badr, 'and I am so little bothered by it that I didn't think it would show on my face enough for you to notice. However, since, contrary to my intentions, you have noticed a change in me, I won't hide from you the cause, which is of very little importance. Like you,' she continued, 'I thought that our palace was the most superb, the most magnificent, the most perfect in all the world. But I will tell you now about something that occurred to me when I was looking carefully around the room of the twenty-four windows. Don't you agree that it would leave nothing to be desired if a *rukh*'s egg were to be suspended from the middle of the dome?' 'Princess,' replied Aladdin, 'it is enough that you should find it lacks a *rukh*'s egg for me to agree with you. You shall see by the speed with which I put this right how there is nothing I would not do out of my love for you.'

Aladdin left the princess at once and went up to the room of the twenty-four windows; there he pulled out the lamp, which he always carried with him wherever he went, ever since the danger he had run into through neglecting to take this precaution, and rubbed it. Immediately the *jinni* stood before him and Aladdin addressed him, saying: '*Jinni*, what this dome lacks is a *rukh*'s egg suspended from the middle of its dome; so I command you, in the name of the lamp I am holding, to repair this deficiency.'

No sooner had Aladdin uttered these words than the *jinni* uttered such a terrible cry that the room shook and Aladdin staggered and nearly fell down the stairs. 'What, you miserable wretch!' cried the *jinni* in a voice which would have made the most confident of men tremble. 'Isn't it enough that I and my companions have done everything for you, but you ask me, with an ingratitude that beggars belief, to bring you my master and hang him from the middle of this dome? For this outrage you, your wife and your palace, deserve to be reduced to cinders on the spot. But it's lucky you are not the author of the request and that it does not come directly from you. The man really behind it all, let me tell you, is the brother of your enemy, the African magician, whom you destroyed as he deserved. This man is in your palace, disguised in the clothes of the holy woman Fatima, whom he has killed. It's he who suggested to your wife to make the pernicious demand you have made of me. His plan is to kill you – you must be on your guard.' And with these words, he disappeared.

Aladdin did not miss a single of the *jinni*'s final words; he had heard about the holy woman Fatima and he knew all about how she supposedly cured headaches. He returned to the princess's apartments, saying nothing about what had just happened to him, and sat down, telling her that he had been seized all of a sudden with a severe headache, upon which he put his hand up to his forehead. The princess immediately gave orders for the holy woman to be summoned and, while she was being fetched, she told Aladdin how she had come to be in the palace where she had given her a room.

'Fatima' arrived, and as soon as she appeared, Aladdin said to her: 'Come in, holy lady, I am very glad to see you and very fortunate to find you here. I've got a terrible headache which has just seized me and I ask for your help, as I have faith in your prayers. I do hope you will not refuse me the favour you grant to so many who suffer from this affliction.' On saying this, he stood up and bowed his head, and 'Fatima' went up to him, but with her hand clasping the dagger she had on her belt underneath her dress. Aladdin, observing her, seized her hand before she could draw it out and, stabbing her in the heart with his own dagger, he threw her down on the floor, dead.

'My dear husband, what have you done?' shrieked the astonished princess. 'You have killed the holy woman!' 'No, my dear,' replied Aladdin calmly, 'I have not killed Fatima but a scoundrel who would have killed me if I hadn't forestalled him. This evil fellow you see,' he said as he removed his veil, 'is the one who strangled the real Fatima – this is the person whom you thought you were mourning when you accused me of killing her and who disguised himself in her clothes in order to murder me. And for your further information, he was the brother of the African magician, your kidnapper.' Aladdin went on to tell her how he had discovered all this, before having the corpse removed.

Thus was Aladdin delivered from the persecution of the two brothers who were both magicians. A few years later, the sultan died of old age. As he had left no male children, Princess Badr al-Budur, as the legitimate heir, succeeded him and transferred to Aladdin the supreme power. They reigned together for many years and were succeeded by their illustrious progeny.

'Sire,' said Shahrazad when she had finished the story of the adventures which had happened through the medium of the wonderful lamp, 'your majesty will no doubt have seen in the person of the African magician a man abandoned to an immoderate passion, desirous to possess great treasures by wicked means, a man who discovered vast quantities of them which he could not enjoy because he made himself unworthy of them. In Aladdin, by contrast, your majesty sees a man of humble birth rising to royalty itself by making use of those same treasures, which came to him without him seeking them, but who used them only in so far as he needed them for some purpose he had in mind. In the sultan he will have learned how a good, just and fair-minded monarch faces many a danger and runs the risk even of losing his throne when, by a gross injustice and against all the laws of fairness, he dares, with unreasonable haste, to condemn an innocent man without wanting to hear his pleas. And finally, your majesty will hold in horror the abominations of those two scoundrel magicians, one of whom sacrifices his life to gain treasure and the other his life and his religion in order to avenge a scoundrel like himself, both of whom receive due punishment for their wickedness.'

Glossary

Many of the Arabic terms used in the translation are to be found in *The Oxford English Dictionary*, including 'dinar', 'ghazi' and 'jinn'. Of these the commonest – 'emir' and 'vizier', for instance – are not entered in italics in the text and, in general, are not glossed here. Equivalents are not given for coins or units of measure as these have varied throughout the Muslim world in accordance with time and place. The prefix 'al-' (equivalent to 'the') is discounted in the alphabetical listing; hence 'al-Mansur' is entered under 'M'. Please note that only the most significant terms and figures, or ones mentioned repeatedly, are covered here.

al-'Abbas *see* 'Abbasids.

'Abbasids the dynasty of Sunni Muslim caliphs who reigned in Baghdad, and for a while in Samarra, over the heartlands of Islam, from 750 until 1258. They took their name from al-'Abbas (d. 653), uncle of the Prophet. From the late ninth century onwards, 'Abbasid rule was nominal as the caliphs were dominated by military protectors.

'Abd Allah ibn Abi Qilaba the discoverer of the legendary city of Iram.

'Abd al-Malik ibn Marwan the fifth of the Umaiyad caliphs (r. 685–705).

'Abd al-Qadir al-Jilani (*c.*1077–1166) a Sufi writer and saint.

Abu Bakr al-Siddiq after the death of the Prophet, Abu Bakr was the first to become caliph (r. 632–4). He was famed for his austere piety.

Abu Hanifa (699–767) a theologian and jurist; founder of the Hanafi school of Sunni religious law.

Abu Hazim an eighth-century preacher and ascetic.

Abu Ja'far al-Mansur *see* al-Mansur.

Abu Muhammad al-Battal a legendary hero of popular tales, in which he plays the part of a master of wiles.

Abu Murra literally, 'the father of bitterness', meaning the devil.

Abu Nuwas Abu Nuwas al-Hasan ibn Hani (*c.*755–*c.*813), a famous, or notorious, poet of the 'Abbasid period, best known for his poems devoted to love, wine and hunting.

Abu Tammam (*c.*805–45) a poet and anthologist of the 'Abbasid period.

'Ad the race of 'Ad were a pre-Islamic tribe who rejected the prophet Hud and who consequently were punished by God for their impiety and arrogance.

'Adi ibn Zaid (d. *c.*600) a Christian poet in Hira.

Ahmad ibn Hanbal (780–855) a *hadith* scholar (student of traditions concerning the Prophet) and a legal authority; founder of the Hanbali school of Sunni religious law.

al-Ahnaf al-Ahnaf Abu Bakr ibn Qais, a *shaikh* of the tribe of Tamim. A leading general in the Arab conquests of Iran and Central Asia in the seventh century, he also had many wise sayings attributed to him.

'A'isha (d. 687) the third and favourite wife of the Prophet.

'Ali 'Ali ibn Abi Talib, cousin of the Prophet and his son-in-law by virtue of his marriage to Fatima. In 656, he became the fourth caliph and in 661 he was assassinated.

'Ali Zain al-'Abidin Zain al-'Abidin meaning 'Ornament of the Believers' (d. 712), the son of Husain and grandson of the caliph 'Ali, he was recognized as one of the Shi'i imams.

alif the first letter of the Arabic alphabet. It takes the shape of a slender vertical line.

Allahu akbar! 'God is the greatest!' A frequently used exclamation of astonishment or pleasure.

aloe aloe was imported from the Orient and the juice of its leaves was used for making a bitter purgative drug.

aloes wood the heartwood of a South-east Asian tree, it is one of the most precious woods, being chiefly prized for its pleasant scent.

al-Amin Muhammad al-Amin ibn Zubaida (d. 813), the son of Harun al-Rashid, succeeding him as caliph and reigning 809–13. He had a reputation as an indolent pleasure lover.

al-Anbari Abu Bakr ibn Muhammad al-Anbari (855–940), *hadith* scholar and philologist.

'Antar 'Antar ibn Shaddad, legendary warrior and poet of the pre-Islamic period who became the hero of a medieval heroic saga bearing his name.

ardabb a dry measure.

Ardashir the name of several pre-Islamic Sasanian kings of Persia. A great deal of early Persian wisdom literature was attributed to Ardashir I (d. 241) and there were many legends about his early years and his reign.

al-Asma'i (740–828?) an expert on the Arabic language and compiler of a famous anthology of Arabic poetry. Harun al-Rashid brought him from Basra to Baghdad in order to tutor his two sons, al-Amin and al-Ma'mun.

Atiya *see* Jarir ibn 'Atiya.

'aun a powerful *jinni*.

Avicenna the Western version of the Arab name Ibn Sina (980–1037), a Persian physician and philosopher, the most eminent of his time, whose most famous works include *The Book of Healing* and *The Canon of Medicine*.

balila stewed maize or wheat.

ban tree Oriental willow.

banj frequently used as a generic term referring to a narcotic or knock-out drug, but sometimes the word specifically refers to henbane.

banu literally, 'sons of', a term used to identify tribes or clans, e.g. the Banu Quraish.

Barmecides *see* Harun al-Rashid, Ja'far.

Bilal an Ethiopian contemporary of the Prophet and early convert to Islam. The Prophet appointed him to be the first muezzin.

Bishr al-Hafi al-Hafi meaning 'the man who walks barefoot' (767–841), a famous Sufi.

bulbul Eastern song thrush.

Chosroe in Persian 'Khusraw', in Arabic 'Kisra' – the name of several pre-Islamic Sasanian kings of Persia, including Chosroe Anurshirwan – 'the blessed' (r. 531–79).

Dailamis Dailam is a mountainous region to the south of the Caspian Sea whose men were celebrated as warriors.

daniq a medieval Islamic coin equivalent to a sixth of a dirham.

dhikr a religious recitation, particularly a Sufi practice.

dhimmi a non-Muslim subject, usually a Christian or a Jew, living under Muslim rule.

Di'bil al-Khuza'i (765–860) a poet and philologist who lived in Iraq and who was famous for his satirical and invective poetry.

dinar a gold coin. It can also be a measure of weight.

dirham a silver coin, approximately a twentieth of a dinar.

diwan council of state, council hall or reception room.

fals plural *flus*, a low-value copper coin.

faqih a jurisprudent, an expert in Islamic law.

faqir literally, 'a poor man', the term also is used to refer to a Sufi or Muslim ascetic.

al-Fath ibn Khaqan (d. 861) the caliph al-Mutawwakil's adoptive brother, chief scribe and general.

Fatiha literally, the 'opening'; the first *sura* (chapter) of the Quran.

Fatima (d. 633) daughter of the Prophet. She married 'Ali ibn Abi Talib. The Fatimid caliphs of Egypt, whose dynasty lasted from 909 to 1171, claimed descent from her.

fidda silver, a small silver coin.

flus see *fals*.

ghazi a holy warrior, a slayer of infidels or participant on a raiding expedition.

ghul a cannibalistic monster. A *ghula* is a female *ghul*.

Gog and Magog evil tribes dwelling in a distant region. According to legend, Alexander the Great built a wall to keep them from invading the civilized parts of the earth, but in the Last Days they will break through that wall.

hadith a saying concerning the words or deeds of the Prophet or his companions.

Hafsa daughter of 'Umar ibn al-Khattab, she married the Prophet in 623 and died in 665.

hajj the annual pilgrimage to Mecca.

al-Hajjaj ibn Yusuf al-Thaqafi (*c.*661–714) a governor of Iraq for the Umaiyad caliph 'Abd al-Malik ibn Marwan, he was notorious for his harshness, but famous for his oratory.

al-Hakim bi-amri-'llah Fatimid caliph in Egypt (r. 996–1021), he was notorious for his eccentricities and capricious cruelty. After his murder, he became a focus of Druze devotion.

al-Hariri (1054–1122) a poet, prose writer and government official. He is chiefly famous for his prose masterpiece, the *Maqamat*, a series of sketches involving an eloquently plausible rogue.

Harun al-Rashid (766–809) the fifth of the 'Abbasid caliphs, reigning from 786. In Baghdad, he presided over an efflorescence of literature and science and his court became a magnet for poets, musicians and scholars. Until 803, the administration was largely in the hands of a Persian clan, the Barmecides, but in that year, for reasons that are mysterious, he had them purged. After his death, civil war broke out between his two sons, al-Amin and al-Ma'mun. In retrospect, Harun's caliphate came to be looked upon as a golden age and in the centuries that followed numerous stories were attached to his name.

Harut a fallen angel who, together with another fallen angel, Marut, instructed men in the occult sciences (Quran 2.102).

Hasan of Basra Hasan ibn Abi'l-Hasan of Basra (642–728), a preacher and early Sufi ascetic to whom many moralizing sayings were attributed.

Hatim of Tayy a pre-Islamic poet of the sixth century, famed for his chivalry and his generosity. Many anecdotes and proverbs have been attributed to him.

hijri calendar the Muslim calendar, dating from the Hijra, or year of Muhammad's emigration from Mecca to Medina, each year being designated AH – *anno Hegirae* or 'in the year of the Hijra'.

Himyar a pre-Islamic kingdom in southern Arabia.

Hind India.

Hisham ibn 'Abd al-Malik the tenth of the Umaiyad caliphs (r. 724–43).

houri a nymph of the Muslim Paradise. Also a great beauty.

Iblis the devil.

Ibn 'Abbas 'Abd Allah ibn al-'Abbas (625–86 or 688), a cousin of the Prophet and transmitter of many traditions concerning him.

Ibn Zubair 'Abd Allah ibn Zubair (624–92), a grandson of the Prophet and a leading opponent of the Umaiyads. He was besieged by the Umaiyad caliph 'Abd al-Malik in Mecca (where the Ka'ba is situated) and he was eventually killed.

Ibrahim Abu Ishaq al-Mausili (742–804) a famous musician and father of the no less famous musician Ishaq al-Mausili. Like his son, he features in a number of *Nights* stories.

Ibrahim ibn Adham (730–77) a famous Sufi ascetic.

Ibrahim ibn al-Mahdi (779–839) the son of the caliph al-Mahdi and brother of Harun al-Rashid. From 817 to 819, Ibrahim set himself up as the rival of his nephew al-Ma'mun for the caliphate. He was famous as a singer, musician and a poet and as such he features in several *Nights* tales.

'Id al-Adha the Feast of Immolation, also known as Greater Bairam, is celebrated on the 10th of Dhu'l-Hijja (the month of *hajj* or pilgrimage). During this festival those Muslims who can afford it are obliged to sacrifice sheep, cattle or camels.

'Id al-Fitr the Feast of the Fast Breaking, marking the end of Ramadan.

Ifranja Europe; literally, 'the land of the Franks'.

'ifrit a kind of *jinni*, usually evil; an *'ifrita* is a female *jinni*.

imam the person who leads the prayers in a mosque.

Iram 'Iram, City of the Columns' is referred to in the Quran. Shaddad, king of the Arab tribe of 'Ad, intended Iram to rival Paradise, but God punished him for his pride and ruined his city.

Ishaq ibn Ibrahim al-Mausili (757–850) was the most famous composer and musical performer in the time of Harun al-Rashid. Like his father, Ibrahim al-Mausili, he features in a number of *Nights* stories.

Ja'far the Barmecide a member of a great Iranian clan which served the 'Abbasid caliphs as viziers and other functionaries. In the stories, he features as Harun's vizier, though in reality it was his father, Yahya, who held this post. For reasons that are mysterious, Ja'far and other members of his clan were executed in 803.

Jamil Buthaina Jamil ibn Ma'mar al-'Udhri (d. 701), a Hijazi poet who specialized in elegiac love poetry, famous for his chastely unhappy passion for Buthaina.

Jarir ibn 'Atiya (d. 729) a leading poet of the Umaiyad period, famous for his panegyric and invective verse.

Jawarna Zara, a port on the east coast of the Adriatic.

jinni a (male) spirit in Muslim folklore and theology; *jinniya* is a female spirit. *Jinn* (the collective term) assumed various forms: some were servants of Satan, while others were good Muslims and therefore benign.

Joseph features in the Quran as well as the Bible. In the Quran, he is celebrated for his beauty.

jubba a long outer garment, open at the front, with wide sleeves.

Ka'h al-Ahbar (d. c.653) a Jew who converted to Islam and a leading transmitter of religious traditions and an expert on biblical lore.

Ka'ba the cube-shaped holy building in Mecca to which Muslims turn when they pray.

kaffiyeh a headdress of cloth folded and held by a cord around the head.

khan an inn, caravanserai or market.

khalanj wood tree heath (*Erica arborea*), a hard kind of wood.

Khalid ibn Safwan (d. 752) a transmitter of traditions, poems and speeches, famous for his eloquence.

al-Khidr 'the Green Man', features in the Quran as a mysterious guide to Moses as well as appearing in many legends and stories. In some tales, this immortal servant of God is guardian of the Spring of Life, which gives eternal life to those who drink from it.

Khurasan in the medieval period, this designated a large territory that included eastern Persia and Afghanistan.

Kuthaiyir (660–723) a Hijazi poet who specialized in the theme of unfulfilled love, since the object of his passion, 'Azza, was married to another man.

Luqman a pre-Islamic sage and hero famed for his longevity. Many fables and proverbs were attributed to him.

Magian a Zoroastrian, a fire worshipper. In the *Nights*, the Magians invariably feature as sinister figures.

al-Mahdi (b. *c.*743) the 'Abbasid caliph who reigned from 775 to 785.

mahmal the richly decorated empty litter sent by a Muslim ruler to Mecca during the *hajj* (pilgrimage).

maidan an exercise yard or parade ground; an open space near or in a town.

maisir a pre-Islamic game of chance involving arrows and in which the stakes were designated parts of slaughtered camels.

Majnun Qais ibn Mulawwah al-Majnun ('the mad'), a (probably) legendary Arabian poet of the seventh century, famous for his doomed love for Laila. After she was married to another man, Majnun retired into the wilderness to live among wild beasts.

Malik the angel who is the guardian of hell.

Malik ibn Dinar an eighth-century Basran preacher and moralist.

mamluk slave soldier. Most mamluks were of Turkish origin.

al-Ma'mun (786–833) son of Harun al-Rashid and the 'Abbasid caliph from 813 until his death. He was famous for his patronage of learning and his sponsorship of the translation of Greek and Syriac texts into Arabic.

Ma'n ibn Za'ida (d. 769) a soldier, administrator and patron of poets under the late Umaiyads and early 'Abbasids.

mann a measure of weight.

al-Mansur (r. 754–75) 'Abbasid caliph.

marid a type of *jinni*.

Maslama ibn 'Abd al-Malik (d. 738) son of the Umaiyad caliph 'Abd al-Malik ibn Marwan and a leading general who headed a series of campaigns against the Byzantines.

Masrur the eunuch who was sword-bearer and executioner to Harun al-Rashid.

al-Mausili *see* Ibrahim Abu Ishaq al-Mausili *and* Ishaq ibn Ibrahim al-Mausili.

mithqal a measure of weight.

months of the Muslim year from the first to the twelfth month, these are: (1) al-Muharram, (2) Safar, (3) Rabi' al-awwal, (4) Rabi' al-akhir, (5) Jumada al-ula, (6) Jumada al-akhira, (7) Rajab, (8) Sha'ban, (9) Ramadan, (10) Shawwal, (11) Dhu'l-Qa'da, (12) Dhu'l-Hijja.

Mu'awiya Mu'awiya ibn Abi Sufyan, first of the Umaiyad caliphs (r. 661–80). He came to power after the assassination of 'Ali.

al-Mubarrad Abu al-'Abbas al-Mubarrad (c.815–98), a famous Basran grammarian and philologist.

muezzin the man who gives the call to prayer, usually from the minaret or roof of the mosque.

muhtasib market inspector with duties to enforce trading standards and public morals.

Munkar and Nakir two angels who examine the dead in their tombs and, if necessary, punish them.

al-Muntasir 'Abbasid caliph (r. 861–2).

al-Musta'in 'Abbasid caliph (r. 862–6).

al-Mustansir 'Abbasid caliph (r. 1226–42).

al-Mutalammis sixth-century pre-Islamic poet and sage.

al-Mu'tatid bi'llah 'Abbasid caliph (r. 892–902).

al-Mutawakkil (822–61) 'Abbasid caliph, and great cultural patron, who reigned from 847 until he was assassinated by murderers probably hired by his son, who became the caliph al-Muntasir.

muwashshahat strophic poetry, usually recited to a musical accompaniment. This form of verse originated in Spain, but spread throughout the Islamic world.

nadd a type of incense consisting of a mixture of aloes wood with ambergris, musk and frankincense.

Nakir *see* Munkar and Nakir.

naqib an official whose duties varied according to time and place. The term was often used to refer to the chief representative of the *ashraf*, i.e. the descendants of 'Ali.

Al-Nu'man ibn al-Mundhir a fifth-century Arab ruler of the pre-Islamic Christian kingdom of Hira in Iraq.

nusf literally, 'a half'; a small coin.

parasang an old Persian measure of length, somewhere between three and four miles.

qadi a Muslim judge.

Qaf Mount Qaf was a legendary mountain located at the end of the world, or in some versions one that encircles the earth.

qintar a measure of weight, variable from region to region, equivalent to 100 *ratls*.

qirat a dry measure, but the term could also be used of a certain weight; also a coin, equivalent to a twenty-fourth of a dinar.

Quraish the dominant Arab clan in Mecca at the time of the Prophet.

rafidi literally, 'a refuser', a term applied to members of various Shi'i sects.

rak'a in the Muslim prayer ritual, a bowing of the body followed by two prostrations.

Ramadan the ninth month of the Muslim year, in which fasting is observed from sunrise to sunset. *See also* months of the Muslim year.

ratl a measure of weight, varying from region to region.

Ridwan the angel who is the guardian of the gates of Paradise.

Rudaini spear *see* Samhari spear.

rukh a legendary bird of enormous size, strong enough to carry an elephant (in English 'roc').

Rum/Ruman theoretically designates Constantinople and the Byzantine lands more generally, but in some stories the name is merely intended to designate a strange and usually Christian foreign land.

Rumi of Byzantine Greek origin.

Safar *see* months of the Muslim year.

Said ibn Jubair a pious Muslim and Quran reader of the Umaiyad period.

Sakhr an evil *jinni* whose story is related by commentators on the Quran.

Saladin (1138–93), Muslim political and military leader, famed for his chivalry and piety and for opposing the Crusaders. He took over Egypt and abolished the Fatimid caliphate in 1171; in 1174 he also became sultan of most of Syria. In 1187 he invaded the kingdom of Jerusalem, occupying the city and many other places. Thereafter he had to defend his gains from the armies of the Third Crusade.

salam meaning 'peace', the final word at the end of a prayer, similar to the Christian 'amen'.

Samhari spear opinions varied as to whether Samhar was the name of a manufacturer of spears, or whether it was the place where they used to be made. A 'Samhari spear' was a common metaphor for slenderness; likewise 'Rudaini spear', said to be related to Rudaina, the supposed wife of Samhar.

Sasanian the Sasanians were the Persian dynasty who ruled in Persia and Iraq from 224 until 637, when Muslim armies overran their empire.

Serendib the old Arab name for Ceylon or Sri Lanka.

Sha'ban *see* months of the Muslim year.

Shaddad ibn 'Ad legendary king of the tribe of 'Ad who attempted to build the city of Iram as a rival to Paradise and was punished by God for his presumption.

al-Shafi'i Muhammad ibn Idris al-Shafi'i (767–820), jurist and founder of the Shafi'i school of Sunni religious law, whose adherents are know as Shafi'ites.

shaikh a tribal leader, the term also commonly used to refer to an old man or a master of one of the traditional religious sciences or a leader of a dervish order. Similarly, a *shaikha* is an old woman or a woman in authority.

Shaikhs of the Fire Zoroastrian priests or elders.

shari'a shari'a law is the body of Islamic religious law.

sharif meaning 'noble', often used with specific reference to a descendant of the Prophet.

Shi'i an adherent of that branch of Islam that recognizes 'Ali and his descendants as the leaders of the Muslim community after the Prophet.

Sufi a Muslim mystic or ascetic.

Sufyan al-Thauri (716–78) born in Kufa, theologian, ascetic and transmitter of *hadiths* (sayings of the Prophet). He wrote on law and was a leading spokesman of strict Sunnism.

sunna the corpus of practices and teachings of the Prophet as collected and transmitted by later generations of Muslims, the *sunna* served as the guide to the practice of the Sunni Muslims and as one of the pillars of their religious law, supplementing the prescriptions of the Quran.

sura a chapter of the Quran.

sycamore a type of fig; also known as the Egyptian fig.

taghut a term designating pagan idols or idolatry. By extension, the word was used to refer to soothsayers, sorcerers and infidels.

tailasan a shawl-like garment worn over head and shoulders. It was commonly worn by judges and religious high functionaries.

Thamud an impious tribe in pre-Islamic Arabia whom Allah destroyed when they refused to pay heed to his prophet Salih.

'Udhri love this refers to the Banu 'Udhra. Several famous 'Udhri poets were supposed to have died from unconsummated love.

Umaiyads a dynasty of Sunni Muslim caliphs who ruled the Islamic lands from 661 until 750. The Umaiyads descended from the powerful Meccan tribe of the Quraish. In 750, they were overthrown by a revolution in favour of the 'Abbasids. One member of the family succeeded in escaping to Spain, where he set up an Umaiyad emirate.

'Umar 'Umar ibn 'Abd al-'Aziz, eighth Umaiyad caliph (r. 717–20), famed for his piety.

'Umar ibn al-Khattab (581–644) the second of the caliphs to succeed the Prophet (r. 634–44).

'umra the minor pilgrimage to Mecca, which, unlike the *hajj*, can be performed at any time of the year.

al-'Utbi (d. 1022) famous author of prose and poetry, worked in the service of the Ghaznavid court. (The Ghaznavids were a Turkish dynasty who ruled in Afghanistan, Khurasan and north-western India from the late tenth till the late twelfth century.)

waiba a dry measure.

wali a local governor.

witr a prayer, performed between the evening and the dawn prayers, which is recommended but not compulsory.

Yahya ibn Khalid the Barmecide a Persian who was a senior government official under the 'Abbasid caliphs al-Mansur and Harun. He was disgraced and executed in 805 for reasons that remain mysterious.

Zaid ibn Aslam a freed slave of 'Umar ibn al-Khattab.

Ziyad ibn Abihi ibn Abihi meaning 'Son of his Father' – the identity of his father being unknown (d. 676) – governor of Iraq under Mu'awiya.

Zubaida (762–831) the granddaughter of the 'Abbasid caliph al-Mansur and famous for her wealth. She became chief wife of the caliph Harun al-Rashid and was mother to al-Amin and al-Ma'mun, both later caliphs.

al-Zuhri Muhammad ibn Muslim al-Zuhri (d. 742), the transmitter of traditions concerning the Prophet and legal authority. He frequented the Umaiyad courts, where, among other things, he was a tutor.

Maps

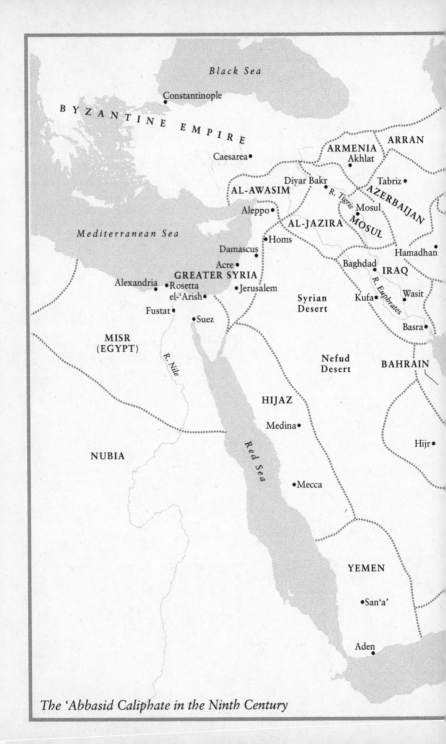

The 'Abbasid Caliphate in the Ninth Century

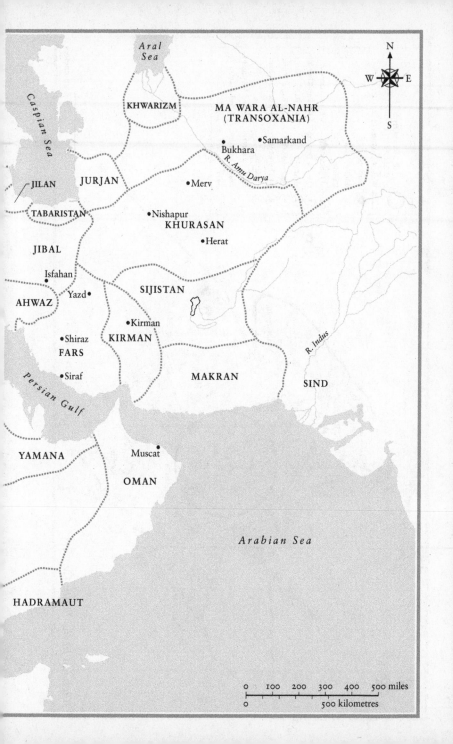

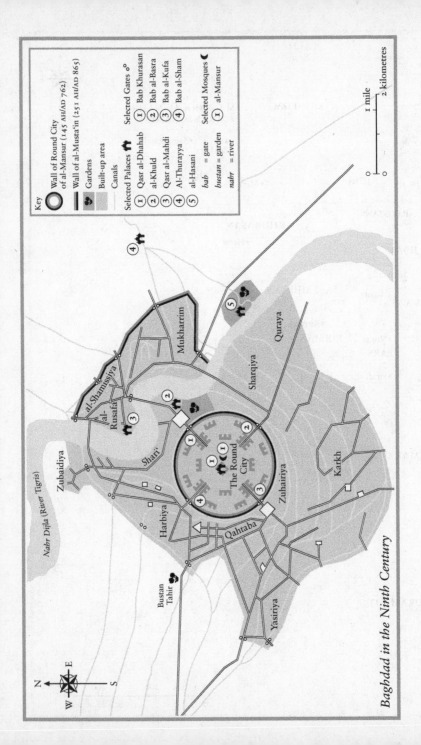

Key

⊚ Wall of Round City of al-Mansur (145 AH/AD 762)

▮ Wall of al-Musta'in (251 AH/AD 865)

🌳 Gardens

▨ Built-up area

  Canals

Selected Palaces 🏯
① Qasr al-Dhahab
② al-Khuld
③ Qasr al-Mahdi
④ Al-Thurayya
⑤ al-Hasani

bab = gate
bustan = garden
nahr = river

Selected Gates ∘°
① Bab Khurasan
② Bab al-Basra
③ Bab al-Kufa
④ Bab al-Sham

Selected Mosques ☾
① al-Mansur

Baghdad in the Ninth Century

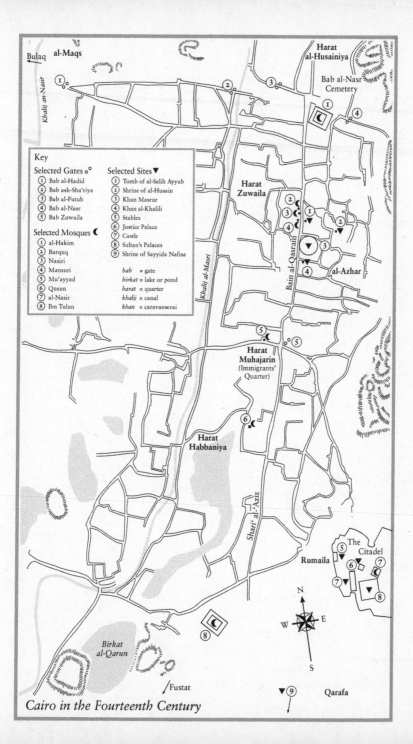

Cairo in the Fourteenth Century

Index of Nights and Stories

Bold *numbers indicate the Night, or series of Nights, over which a story is told.*
Stories told within a story are presented in brackets.

Volume One

Volume Two

Volume Three